American Seashells

American Seashells

The Marine Mollusca of the Atlantic
and Pacific Coasts of North America

R. Tucker Abbott, Ph.D

du Pont Chair of Malacology
Delaware Museum of Natural History

VNR VAN NOSTRAND REINHOLD COMPANY
New York / Cincinnati / Toronto / London / Melbourne

Van Nostrand Reinhold Company Regional Offices:
New York Cincinnati Chicago Millbrae Dallas

Van Nostrand Reinhold Company International Offices:
London Toronto Melbourne

Copyright © 1974 by Litton Educational Publishing, Inc.

Library of Congress Catalog Card Number: 74-7267
ISBN: 0-442-20228-8

Manufactured in the United States of America

Published by Van Nostrand Reinhold Company
450 West 33rd Street, New York, N.Y. 10001

Published simultaneously in Canada by Van Nostrand Reinhold Ltd.

15 14 13 12 11 10 9 8 7 6 5 4 3 2 1

Library of Congress Cataloging in Publication Data
Abbott, Robert Tucker, 1919-
  American seashells.
  Bibliography: p.
  1. Mollusks—North American. I. Title.
QL411.A19 1974 594′.04′7 74-7267
ISBN 0-442-20228-8

Contents

American Seashells

Introduction

This book was born during a long-overdue revision of the fifteen-year-old *American Seashells*. It soon became evident that there was a need for a more comprehensive treatment of our American marine mollusks. Since 1954 a new generation of serious amateurs and a growing number of graduate students in marine biology and oceanography have wanted a more scientific and more complete compendium.

Ideally, this manual should be a complete monograph of every marine species of mollusks of the Americas, but this would require the combined efforts of many malacologists over quite a number of years. Short of this, I have attempted to describe in detail about 2,000 species and to list the other 4,500 recorded from our shores. This has resulted in many unavoidable shortcomings. In leaning heavily upon the old checklists assembled by William H. Dall, Charles W. Johnson and others, I have doubtlessly absorbed their many errors and misunderstandings.

In some fortunate cases, I have been able to take advantage of the excellent research work being published in such journals as *Johnsonia, The Veliger, The Nautilus* and the *Bulletin of Marine Sciences of Miami*. Under certain families I have acknowledged the assistance of several experts such as Ruth D. Turner, Donald R. Moore, Richard E. Petit, Myra Keen, Hal Lewis, Kenneth Boss, Arthur Clarke, Rae Baxter, Ellen Crovo, Corinne Edwards, as well as members of the Miami Malacological Society and numerous friends who have forwarded much new information to me.

I am also indebted to Daniel Steger of St. Petersburg, Florida, for many new west Florida records and to Thomas L. McGinty of Boynton Beach, Florida, for southeast Florida information. Russell Jensen of the Delaware Museum of Natural History has kindly made available his records of 620 species found in Bermuda. Mr. and Mrs. Allen D. Russell of Plymouth, Massachusetts, gave many hours in suggesting useful changes. I would like to thank Neil M. Hepler of Florida for making the drawings of the shells showing the various conchological terms, and M. G. Jerry Harasewych for taking the color photographs.

Admittedly, this is an advanced and an expensive work. Alone, it would not serve the purpose of the original *American Seashells*, which was intended to be a bridge between the beginner and the professional or advanced amateur. There is still a place for an intermediate book, and I refer the interested beginner to my *Seashells of North America* (Golden Press, New York). In that book the biology and natural history of our mollusks and a general introduction to malacology are emphasized.

This manual includes many species found in the Gulf of California where active research is now being conducted by Arizona and California institutions. A much more competent and complete treatment of the Panamanian fauna is found in Myra Keen's *Sea Shells of Tropical West America*. The neighboring fauna of the Bahamas and West Indies is more completely treated in *Caribbean Seashells* by Germaine L. Warmke and R. T. Abbott. Students of the Canadian fauna will find Aurèle La Rocque's *Catalogue of the Recent Mollusca of Canada* a useful checklist. Eastern South America is adequately covered by E. C. Rios' 1970 *Coastal Brazilian Seashells*, Museu Oceanográfico de Rio Grande, Brasil.

GEOGRAPHICAL SCOPE OF THIS BOOK

This book is basically a taxonomic survey of the marine species found along the shores and continental shelf of North America. This vast area includes several quite different faunas, some of which blend in with those of neighboring parts of the world. The mollusks of the arctic seas are much the same in northern Alaska, Siberia, Norway, Greenland and Labrador. Many mollusks of the keys of southern Florida, Bermuda, the West Indies and northern Brazil are identical.

Certain large stretches of our coasts have peculiar oceanographic characteristics and unique faunas, thus suggesting recognition of such marine provinces as the Oregonian and Virginian, the Caribbean and the Panamanian. Mixtures of neighboring provinces are common. Many very adaptable species exist in several provinces.

The geographic ranges given in this manual are only approximations. A distribution map for each species would be better, but even then, bathymetric and ecologic conditions often determine the presence or absence of a species. Many of the Atlantic coast ranges read "Off North Carolina to Florida, Texas and the West Indies." These usually apply to Caribbean species that are abundant in the West Indies and perhaps common in southeast Florida, but are rare in North Carolina or southern Texas where offshore waters are semitropical. This sporadic distribution also occurs where the Californian and Panamanian provinces meet in Baja California. The term "south Florida" usually means both coasts from St. Petersburg and Jupiter Inlet south to Key West. The term "southeast Florida" refers to the warm-water region from Palm Beach County south to Key West. The "Lower Florida Keys" is a confusing term, for Floridians divide these keys into the Upper, Middle and Lower Keys. We mean the keys from Miami to Key West.

The Gulf of Mexico is a huge, diverse body of water, and its shell fauna greatly differs from one part to another. The species washed ashore along the western shores of Florida, from Clearwater south to Marco, are quite different from those of the Dry Tortugas or those of South Padre Island, Texas. The sea mounts miles off the coast of Galveston, Texas, support species more in common with Bermuda than with those of the muddy depths south of Mississippi. Several "tongues" of West Indian assemblages jut well into the Gulf, one example being the offshore area between Clearwater north to off Aucilla River in northwest Florida.

Collectors should be aware that long-range, as well as seasonal short-range, changes in the oceanic weather can affect the ranges of most species. Droughts along the Texas coast, a swerving of the Gulf Stream near the North Carolina coast, or a series of unusually cold winters in the Lower Florida Keys can influence the presence or absence of many species.

REFERENCES

The original references to most of the species in this manual have been repeatedly published in standard works, and for anyone wishing to do serious taxonomic research, a good library will contain them: Sherborn's "Index Animalium" for species published prior to 1851, and the "Zoological Record," section on "Mollusca," for the later ones. For convenience, I have attempted to give original references for most of the species described within the last 30 years. Special works, such as those listing all of the original references to the species of Dall, Pilsbry, Carpenter and others, are included in the bibliography. When known to me, I have added abbreviated references to biological studies under the appropriate species.

BASIC LITERATURE ON THE SYSTEMATICS OF THE MARINE MOLLUSKS OF NORTH AMERICA

A complete bibliography of all the genera and species mentioned in this book would probably amount to over 10,000 titles. Many extensive bibliographies have been assembled and published in the last 30 years. The following journals, books and pamphlets give bibliographies or contain original material which will lead the student to nearly all of the original descriptions.

SOURCE BOOKS

Addicott, Warren O. 1973: "Neogene Marine Mollusks of the Pacific Coast of North America: An Annotated Bibliography, 1797–1969." Index and abstracts of about 1200 articles on Tertiary fossils. *Geol. Survey Bull.* 1362.

Moore, Raymond C. (editor). 1960–1970: "Treatise on Invertebrate Paleontology." Part 9, "Mollusca." Geological Society of America. Several volumes on the classification and arrangement of mollusks by leading malacologists.

Neave, Sheffield A. 1939–1950: "Nomenclator Zoologicus." 5 vols. Zool. Soc. London. Lists genera of all animals and gives original citations.

Thiele, Johannes. 1929–1935: "Handbuch der Systematischen Weichtierkunde." 4 vols. Text on classification and arrangement of mollusks, becoming out-of-date.

Wenz, W. 1938–1944: "Handbuch der Paläozoologie (editor: O. Schindewolf). "Gastropoda," vol. 6, pts 1–7. Gebrüder Borntraeger, Berlin. Standard text on classification and arrangement.

AMERICAN JOURNALS

Johnsonia. Monographs of the Marine Mollusca of the Western Atlantic. W. J. Clench and Kenneth Boss, editors. Museum of Comparative Zoölogy, Cambridge, Mass. 02138.

The Nautilus, a quarterly devoted to the interests of conchologists. R. T. Abbott, editor. Delaware Museum of Natural History, Greenville, Del. 19807.

Occasional Papers on Mollusks. Department of Mollusks, Museum of Comparative Zoölogy, Cambridge, Mass. 02138.

Malacologia, an international journal of malacology. Museum of Zoology, University of Michigan, Ann Arbor, Mich. 48104.

The Veliger, a quarterly devoted to all aspects of mollusks from the Pacific region. Northern California Malacozoological Club, Berkeley, California (% Dept. Zoology, Univ. Calif., Berkeley, Ca. 94720).

LEADING REFERENCES

Abbott, R. Tucker. 1955: "American Seashells." 541 pp., 40 pls. Van Nostrand Reinhold, New York. Contains a large bibliography arranged by regions and subjects.

Abbott, R. Tucker. 1958: "The Marine Mollusks of Grand Cayman Island, West Indies." 138 pp., 5 pls., Monograph 11, Academy of Natural Sciences of Philadelphia.

Adams, Charles B. 1839–1852: See "The Western Atlantic marine mollusks described by C. B. Adams," by W. J. Clench and R. D. Turner. 1950, *Occasional Papers on*

Mollusks, vol. 1, no. 15, pp. 233–403, 20 pls.; and "The Eastern Pacific marine species described by C. B. Adams," *ibid.,* vol. 2, no. 20, pp. 21–136, 16 pls.

Burch, John Q. (editor). **1944–1946:** "Distributional list of the West American marine Mollusca from San Diego, California, to the Polar Seas." Part I, nos. 33–45; pt. II (Gastropoda), nos. 46–63. Mimeographed.

Bush, Katherine, J. 1885: "Additions to the shallow-water Mollusca of Cape Hatteras, N.C., dredged by the U.S. Fish Commission steamer "Albatross" in 1883 and 1884." *Connecticut Acad. Arts and Sci.,* vol. 6, pt. 11, pp. 453–480, 45 pls. (see also 1897, *ibid.,* vol. 10, pp. 97–144).

Carpenter, Philip P. [1958]: "Type specimens of marine mollusca described by P. P. Carpenter from the West Coast (San Diego to British Columbia)." By K. van Winkle Palmer. Geol. Soc. America, Memoir 76, 376 pp., 35 pls.

Carpenter, Philip P. 1864: "Supplementary report on the present state of our knowledge with regard to the Mollusca of the west coast of North America." Report British Assoc. Adv. Sci. for 1863, pp. 517–686.

Dall, William H. [1968]: "The Zoological Taxa of William Healey Dall." By K. J. Boss, J. Rosewater and F. A. Ruhoff. Bull. 287, U.S. National Museum, Washington. An indispensable alphabetical list of 5,302 molluscan names proposed by Dall, with their original citations.

Dall, William H. 1886 and 1889: "Reports on the results of dredging . . . Steamer 'Blake.'" *Bull. Museum Comparative Zoölogy,* vol. 12, pp. 171–318, pls. 1–9; vol. 18, pp. 1–492, pls. 10–40.

Dall, William H. 1890–1903: "Contributions to the Tertiary fauna of Florida." *Trans. Wagner Free Inst. Sci.,* vol. 3, pts. 1–6, 1,654 pp. 60 pls.

Dall, William H. and Charles T. Simpson. 1901: "The Mollusca of Puerto Rico." *U.S. Fish Commission Bull. for 1900,* vol. 20, pp. 351–524. pls. 53–58.

Dall, William H. 1921: "Summary of marine shellbearing mollusks of the northwest coast of America." Bulletin 112, U.S. Nat. Mus., 217 pp., 22 pls.

Gould, Augustus A. and W. G. Binney. 1870: "Report on the Invertebrata of Massachusetts." 524 pp., 27 pls. Boston.

Grant, Ulysses S., IV, and Hoyt R. Gale. 1931: "Catalogue of the marine Pliocene and Pleistocene Mollusca of California and adjacent regions." *Memoirs San Diego Soc. Nat. Hist.,* vol. 1, 1,036 pp., 32 pls.

Johnson, Charles W. 1934: "List of Marine Mollusca of the Atlantic Coast from Labrador to Texas." *Proc. Boston Soc. Nat. Hist.,* vol. 40, pp. 1–204. Large bibliography.

Keen, A. Myra. 1971: "Marine Shells of Tropical West America." 2nd edition. 750 pp. Stanford Univ. Press., Calif. Excellent bibliography.

La Rocque, Aurèle. 1953: "Catalogue of the Recent Mollusca of Canada." 406 pp. Bull. no. 129, National Museum of Canada, Ottawa. Large bibliography and numerous original citations.

Oldroyd, Ida S. 1925–1927: "The marine shells of the West Coast of North America." *Stanford Univ. Publ., Univ. Ser., Geol. Sci.* vol. 1, pt. 1, vol. 2, pts. 1–3.

Olsson, Axel A. and Anne Harbison. 1953: "Pliocene Mollusca of Southern Florida." Monograph 8, Academy of Natural Sciences of Philadelphia, 457 pp., 65 pls. [out of print].

Pilsbry, Henry A. [1962]: "New names introduced by H. A. Pilsbry in the Mollusca and Crustacea." By William J. Clench and Ruth D. Turner. Special publ. no. 4, Academy Natural Sciences of Philadelphia, 218 pp.

Smith, Allyn G. and M. Gordon. 1948: "The marine mollusks and brachiopods of Monterey Bay, California, and vicinity." *Proc. Calif. Acad. Sci.,* 4th ser., vol. 26, no. 8, pp. 147–245.

Verrill, Addison E. 1882: "Catalogue of the marine Mollusca added to the fauna of the New England region during the past ten years." *Trans. Conn. Acad.,* vol. 5, pp. 447–587, pls. 42–44, 57–58; Second Catalogue, *ibid.,* vol. 6, pp. 139–294, pls. 28–32; Third Catalogue, *ibid.,* vol. 6, pp. 395–452, pls. 42–44.

Verrill, Addison E. and K. J. Bush. 1898: "Revision of the deep-water Mollusca of the Atlantic Coast of North America, with descriptions of new genera and species." Bivalvia. *Proc. U.S. Nat. Mus.,* vol. 20, no. 1139, pp. 775–901, pls. 71–97.

Warmke, Germaine L. and R. T. Abbott. 1961: "Caribbean Seashells." 348 pp., 44 pls. Livingston Press, Wynnewood, Pa.

Weisbord, Norman. 1962: "Late Cenozoic Gastropods from northern Venezuela." *Bull. Amer. Paleontology,* Ithaca, N.Y., vol. 42, no. 193, 672 pp., 48 pls.; 1964, *Pelecypods,* vol. 45, no. 204, 564 pp., 59 pls. Synonymies and large bibliographies.

I

Systematics of the Mollusca

Systematics is the study of the various forms and kinds of organisms, including their identification, classification and evolution. The actual sorting of specimens into various species or subspecies and identifying them is known as taxonomy. To go beyond this and arrange the individual species into groups and the groups into an evolutionary system is the practice of classification. The professional field of systematics is complex and requires highly skilled biologists trained in many aspects of their particular fields, including ecology, morphology, evolution, data analyses, library and museum techniques, field methods, nomenclatural rules, as well as a basic knowledge of modern biology. The most comprehensive textbook on this subject is *Taxonomy—A Text and Reference Book*, by Richard E. Blackwelder, 1967, xiv + 698 pp., John Wiley & Sons, Inc., New York.

DIVERSITY OF INDIVIDUALS

While it is evident to most people who have looked over a box of shells that there are different kinds showing different forms, colors and sizes, it is not so often recognized that there can be tremendous variation from one individual to another, even within the same species. Furthermore, differences in the mollusks are by no means limited to the shell, but may also exist in the soft parts. The shell, produced by the soft mantle, is a "frozen" record of the physiological history of the mollusk. It not only gives the genetic imprint of the individual, but very often can tell us whether or not food was plentiful, if the salinity and temperature of the water was ideal, or if conditions were overcrowded or not. There are dozens of ways in which two individuals may differ although belonging to the same species:

1. Different *age* (a young *Strombus* lacks the large, flaring outer lip of the adult. A very old specimen may have a greatly thickened lip covered with an aluminumlike glaze).
2. Different *sex* (male *Cassis* are smaller and have fewer and larger knobs; male *Argonauta* are one-tenth the size of the female and do not make a "shell").
3. Different phases in a *life cycle* (a developing oyster goes from egg to trochophore, to veliger, to crawling spat, to attached young bivalve, to large, distorted, senile shell).
4. Were in differing positions in a *colony* (crowded mussels in the middle of a clump are dwarfed).
5. Were living in different physical *habitats* (ecophenotypes, such as high-domed *Patella* in protected waters, and low-domed individuals on wave-dashed rocks; or stenomorphs, such as dwarfed *Teredo* living in small wooden sticks).
6. Had responded in color to differing *backgrounds* (some color changes in squid and *Octopus*).
7. Were feeding or living on different *hosts* (yellow or purple gorgonians; nudibranchs obtaining different colors by feeding on colored coral polyps).
8. Were differently *parasitized* (distortion of tentacles of *Succinea* by parasitic larval form of trematode).
9. Deformed, diseased, or suffered an accident, resulting in abnormal color or shape of shell (tilted apex, double siphon, shell-less adult snail, colorless shell, etc.).

WHAT IS A SPECIES?

Volumes have been written in answer to this question, and the subject is one of continuous investigation by many biologists working with all forms of animals and plants. The concept of a species differs between the geneticist, evolutionist and taxonomist. There is no satisfactory or accurate definition that encompasses all concepts. A species in this book is a "taxonomic species" and is defined by R. E. Blackwelder (1967) as one which consists of "all the specimens which are, or would be, considered by a particular taxonomist to be members of a single kind as shown by the evidence or the assumption that they are as alike as their offspring or their hereditary relatives within a few generations."

The "biological species," as defined by Ernst Mayr (1942, "Systematics and the Origin of Species," Columbia University Press, New York), are groups of actually or potentially interbreeding individuals of natural populations, which are reproductively isolated from other such groups by geographical, physiological or ecological barriers. Unfortunately, this biological concept of species cannot as yet be used extensively in the field of mollusks, for malacology is largely in the purely descriptive and cataloguing stages, and the majority of species being described today are still based on the old-fashioned morphological species concept. However, many of our "taxonomic species" coincide with the definition of a "biological species."

Every population of mollusks is inherently different, and these differences, however minute, are morphological, physiological or genetic. One need only collect some members of a common species in several localities along our coast and carefully examine them in order to reach this conclusion. It is this factor of geographical variation, together with timely isolation and selection, which has been largely responsible for the evolutionary production of species. The development of species is a continuous and very gradual process, and when we settle upon a reasonably homogeneous series of populations and label them as, say, *Melongena corona*, we are merely "snapping a candid camera shot" of a species living today, one whose picture looked quite different several million years ago during the Pliocene period. Within the geographical range of this species we find a series of populations on the coast of Yucatan, Mexico, which seem to be attempting a "break-away" from the typical form, and to this geographical race the name *Melonga corona* subspecies *bispinosa* Philippi has been given. Perhaps in another million years, through fortuitous isolation (geographical or reproductive) and selection, it will merit recognition as a full species. Elsewhere throughout the range of *corona*, we find minor groups of variants, some that are individuals stunted by ecological conditions, others that are minor genetic variations which seem to crop up at random in all parts of Florida. These ecotypes, aberrations and varieties, although actors in the evolution game, do not warrant subspecific names.

Naturally, there are differences between individuals of the same species. These may not be just differences due to age, diet or ecologic conditions. They may be genetic differences of minor importance, such as the color pattern, number of spines, size of the ribs or curvature of the whorls. It is necessary to distinguish these differences from the more significant ones that are important in separating species. This is possible to determine only by a competent taxonomist using as many comparative characters and other types of data that he can assemble.

Polymorphism is the existence of individuals of more than one form within a species. These abrupt differences are genetic, and in most forms of life, particularly the insects and jellyfish, they may be due to different stages of development, alternating generations, castes (as in insects) or sex. Sexual dimorphism is often expressed in a structural difference between male and female. In prosobranch gastropods the male has an external penis, and the shell may or may not show prominent features differing from those of the female's shell. Some *Nerita, Columbella* and many banded land snails exhibit distinct and radically different color patterns from one individual to another.

Genetic variation may be due to rare instances of mutation (of genes, which produce new features), hybridization (which introduces new genes into the "species") or recombination of the genes. It is quite probable that many more species of mollusks regularly hybridize than has been generally admitted or recognized by today's malacologists. In all likelihood, several hundred so-called "good" species are nothing but hybrids.

Among the taxonomic characters, or individual features used by the taxonomist to separate groups, there are some that are convenient or useful in quickly distinguishing species. These are the diagnostic or key characters, sometimes major, sometimes minor, in the eyes of the observer. On the other hand, some features or forms are very obvious, but evidently not of sufficient weight or genetic significance to be used in characterizing or separating species. An albino or completely black color phase may be readily recognized, and, indeed, have been given a species name, such as *alba* or *nigra*. These and other phases of color and form are not on the level of species or subspecies, but in order to talk about them, taxonomists have retained the names as *forma*. Thus, a *Murex* shell found living at a great depth and producing long spines may bear the *forma* name *spinosa*, while its brothers, born and living on the shallow reef, may be spineless and bear the *forma* name *aspinosa*. In general, however, *forma* names are not popular, nor always useful, and they have no official place among the binomial names. There is an old adage (Hubbell, 1954) that "nothing should ever be named for the sake of naming it, but only in order that something may be said about it."

SUBSPECIES

The term subspecies is sometimes misunderstood and not infrequently misused by both amateur and professional malacologists. A concise and satisfactory definition is given by Ernst Mayr: "a subspecies is an aggregate of local populations of a species, inhabiting a geographical subdivision of the range of the species, and differing taxonomically from other populations of the species." While distinct morphological, as well as interbreeding, gaps exist between species, the differences between subspecies are overlapping, and sometimes cannot be expressed except in some percentage of cases, a figure often placed between 50% and 90%. However, a widely ranging species may exhibit a clinal variation in one or more characters, usually correlated with some environmental condition such as climate or altitude of habitat. Clines in themselves do not necessarily measure the existence or limits of subspecies. In fact, a species may exhibit two or three clines, one going from north to south (climactic, perhaps), and another going from east to west (altitudinal or soil-correlated, perhaps). Infra-subspecific names, such as *maxima, alba* and *elongata,* are sometimes used, when they serve a purpose, but they are not accepted on the level of subspecies or species names.

TAXON-TAXA

A relatively new term in taxonomy is taxon (plural: taxa). It is a group of animals having a recognized entity such as a subspecies, species, genus or family. Latin names are applied to taxa—the zoological names. Some taxa are at the species level, while others, like families and classes, are higher-category taxa.

TYPE SPECIMEN

In taxonomic practice, a single specimen is selected from among those originally used in the description of a new species or subspecies. It is an arbitrary selection which, nonetheless, serves several important functions. It serves as a check on the accuracy or completeness of the published description. It serves as a peg upon which the proposed name can be hung permanently. It demonstrates some of the characters of the species, but, of course, not all of the variations existing in nature. Although many terms for various kinds of types have been proposed and used, the following are the most important and probably the only really useful ones. These terms apply to the nominal or name-species, not to the real species in nature.

1. Primary types—the single nomenclatural kind.
   Holotype—the single specimen designated or indicated as the type-specimen of a species-group taxon. If a new species is based upon a single specimen, that specimen is the holotype.
   Lectotype—if a species has no holotype, one of the original specimens (syntypes or co-types) may be designated as the lectotype, and preferably a syntype that has been previously illustrated.
   Neotype—when all original specimens have been lost, a new specimen may be designated as the type-specimen, but this is done only in connection with revisionary work.
2. Secondary types—the other specimens of the original material from which the primary type must be selected.
   Syntype (formerly called co-type)—every specimen in a type-series in which no holotype has, as yet, been designated. When a syntype is selected to be the type-specimen, the remaining ones become known as paralectotypes.
3. Tertiary types—other specimens of the original material set aside as of special taxonomic interest to supplement the primary type.
   Paratype—every specimen in the original type series, other than the holotype.
   Allotype—another paratype, usually of the opposite sex to that of the holotype (used extensively in the Cephalopoda).
4. Other types.
   Topotype—a specimen coming from the type locality, but not part of the original material used in describing the species.
   Hypotype—a specimen, other than the holotype or paratypes, subsequently illustrated or described in order to give more formal information about the species.
   Type locality—the place where the type was collected.

CLASSIFICATION

The 6,500 species of mollusks in North American marine waters are classified under a generally accepted universal system in which groups of species are placed in larger groups, and these in turn, into higher categories. There is an attempt to make this arrangement more or less natural—that is, one in which the groups are recognized by having a maximum number of attributes in common. This may or may not coincide with an evolutionary or phylogenetic relationship. In any event, it is impossible to arrange the sequence of families and orders in any real phylogenetic order in a book that is not three-dimensional, as is the theoretical "family tree." Needless to say, classifications, being man-made, are subject to continuous changes as taxonomists gain more knowledge about the various groups. This sometimes leads to necessary name changes.

Among the various levels or categories most commonly used in the taxonomic classification of the Kingdom Animalia are the following:

Category	Ending used	Molluscan example	Author, date
Phylum	varies	Mollusca	Cuvier, 1797
Class	varies	Gastropoda	Cuvier, 1797
Subclass	-ia	Prosobranchia	Milne Edwards, 1848
Order	-a	Archaeogastropoda	Thiele, 1925
Suborder	-ina	Pleurotomariina	Cox and Knight, 1960
Superfamily	-acea	Fissurellacea	Fleming, 1822
Family	-idae	Fissurellidae	Fleming, 1822
Subfamily	-inae	Emarginulinae	Gray, 1834
Genus	varies	*Puncturella*	Lowe, 1827
Subgenus	varies	*Cranopsis*	A. Adams, 1860
species	varies	*cucullata*	(Gould, 1846)

NOMENCLATURE

The rules of nomenclature and a full explanation of them are too complex and lengthy to be adequately treated here. The best treatment for the serious student is found in R. E. Blackwelder's "Taxonomy," 1967, John Wiley & Sons, Inc., New York. The code of rules which zoological taxonomists attempt to abide by is found in a booklet entitled "International Code of Zoological Nomenclature adopted by the XV International Congress of Zoology" (1964). Copies of the latter may be obtained for about $5.00 by remitting an international money order to the International Trust for Zoological Nomenclature, % British Museum (Natural History), Cromwell Road, London, S.W. 7, England.

THE NAMES OF SPECIES

The names of species consist of two parts—the genus and the specific part of the binomial name, the trivial name. For example: *Littorina littorea.* The genus is *Littorina,* a group of marine periwinkles containing many species, including *littorea, scabra* and *ziczac.* Customarily, generic, specific and subspecific names are italicized, while the names of families and other higher categories are not.

Following the name is the author of the trivial name and the date of publication. If the trivial name is currently being used in combination with a genus other than the one originally used, the author's name and date are put in parentheses, as for example: *Littorina littorea* (Linné, 1758). In 1758 Linné described the species *littoreus*, in combination with the genus *Turbo*. In 1822, the species was placed in the genus *Littorina*, where it still stands. Parentheses are used in this case. Were *littorea* still kept in the original genus, the parentheses would be omitted: *Turbo littoreus* Linné, 1758. The author's name and date serve to distinguish different species that might have the same name; and they give a lead to the original publication. The second edition of *American Seashells* has parentheses in the appropriate places. There should never be an intervening punctuation mark between the name of the species and the author's name.

It is incorrect to place a comma between the trivial name and the author of the species (e.g., *Littorina littorea*, Linné), except in very special cases when the synonymic history of a species is being given in a formal, monographic study. For instance, in 1828, Wood described *Turbo tuberculatus*. A correct synonymic citation would appear as follows:

1828 *Turbo tuberculatus* Wood, Index Testaceologicus, p. 20, pl. 6, fig. 3.

If we were to add an additional and subsequent citation indicating that Orbigny had placed Wood's species in the genus *Littorina*, we would make the following citation:

1852 *Littorina tuberculata* Wood, Orbigny, in Sagra's Hist. Nat. de Cuba, vol. 2, p. 522.

Some workers abbreviate the above citation by eliminating "Wood" but note that a comma is added. This is done so that it does not appear that Orbigny is the author of the name *tuberculata*:

1852 *Littorina tuberculata*, Orbigny, in Sagra's. . . .

In modern usage, however, the first 1852 example above is preferred, since it clearly states that Orbigny used the name "*tuberculata* Wood." In general, it is best to give the full names in taxonomic reports, that is, genus, species, author, and date of publication: *Littorina littorea* (Linné, 1758). In popular accounts or ecological or biological reports the author and date may be omitted, if there is no doubt about the identification.

In the case of synonyms existing for the same taxon, the Law of Priority holds that the oldest available name shall be used. Homonymy exists when there are two trivial names of identical spelling under one genus name. The second or later name cannot be used even though it applies to another species. It must be re-named as long as it remains in the same genus. The reader is referred to Blackwelder (1967) for extensive details of the rules covering these subjects.

The gender of trivial names which are adjectives should agree with the gender of the genus. The gender of the subgenus does not affect the trivial name, unless it is later raised to generic level. Simple examples are *Conus albus*, *Conus niger*, *Vasum album*, *Vasum nigrum* and *Vasum globosum*. The last means the globular vase. However, *Vasum globulus* means the globe vase. Globe is a substantive masculine noun and is not declined like the adjective *globosum*. Trivial

names of females end in *-ae*, males in *-i*, and of more than one person in *-orum* (*Conus juliae*, *williami*, *clenchorum*).

Among the special terms frequently encountered in taxonomic literature are:

Nomen dubium, a name not applicable with any certainty to any known taxon.

Nomen nudum, an invalid name not accompanied by a description and/or illustration, or otherwise failing to satisfy the minimum requirements of the rules.

Nomen oblitum, a forgotten name not used in the last 50 years in the primary literature.

Nominate subspecies, the typical subspecies which contains the type specimen of the species and bears the same name as the species. (e.g., *Fasciolaria lilium* subspecies *lilium* G. Fischer, 1807).

When a zoological name is written and it is desirable to indicate the subgenus, the latter is included in parentheses: *Littorina* (*Littorinopsis*) *angulifera* Lamarck, 1822. Synonyms are indicated by a plus sign: *Littorina angulifera* Lamarck, 1822 (+ *aurantia* Philippi, 1846). Equal signs should be avoided since they can be interpreted both ways, thus the senior valid synonym is not always recognizable.

Sensu stricto (abbreviation is *s.s.*) means "in the strict sense" and is used in explaining that a subgenus bears the same name as the genus.

IDENTIFICATION FEATURES

These are the many morphological features exhibited in mollusks which are used for identifying species and in understanding the evolutionary relationships existing between members of the higher categories, such as genera, families or orders. It must be realized that in some groups of shells certain types of characters, such as number of spines, shapes of aperture or color markings, are used to distinguish species, while in other groups these will prove useless and reliance may have to be put on the number of folds in the columella, the number of teeth in the aperture or the sculpturing on the operculum. These key features are pointed out in their appropriate places throughout this book.

The verbal tools which are used in the study of mollusks are especially designed to assure a method as accurate as possible for telling apart the tens of thousands of living, and many more fossil, species of mollusks. It is impossible to avoid using technical names for various parts of the shell and its animal, such as apex, spire, whorls, operculum, for most of these words have no counterpart in everyday language. Familiarization with these few terms is gained easily and rapidly as trial identifications and references to the illustrated glossaries are made. Many of the technical terms explained below are not employed in this book, but they are presented for the sake of those readers who intend to use more advanced works.

GASTROPOD FEATURES

Shape of shell   It is this character that is instinctively used at first when identifying a snail shell, and little would be gained in discussing at length what our photographs so clearly demonstrate. However, the shape of the adult shell in some species may differ radically from its young stages as may be seen in the illustrations of the cowries or the American pelican foot (*Aporrhais*). Monstrosities caused by embryological defaults or by injury in early life have always been a source of error in identification, and in certain extreme cases many species have been erroneously described as new.

Parts of the shell   As the typical gastropod mollusk grows, it adds to the spiral shell and produces turns or **whorls.** The first few whorls, or nuclear whorls, are generally formed in the egg of the mollusk and usually differ in texture, color and sculpturing from the postnuclear whorls which are formed after the animal has hatched. When the nuclear whorls are marked off from the remainder of the whorls they are often referred to as the **protoconch.** The part of the shell produced after the protoconch is called the **teleoconch.** The last and largest whorl which terminates at the aperture of the shell is known as the **body whorl.** The **periphery** is an imaginary spiral area on the outside of the whorl, usually halfway between the suture and the base or at a point where the whorl has its greatest width. The giant Atlantic pyram shows a narrow color band on the periphery of the last whorl. The whorl just before the last whorl often has distinctive characters and has been differentiated by the name **penultimate whorl.** Above this the succeedingly smaller earlier whorls in the pointed apex of the shell are known as **apical whorls.** The rate of expansion of the growing whorls and the degree to which the succeeding whorls "drop" determine the shape of the shell. The sides of the whorls may be flat, globose, concave, channeled or ribbed. The juncture of each whorl against the other forms a **suture** at the top or above the shoulder of each whorl. The suture may be very fine—a mere tiny, spiral line—or it may be deeply channeled (see *Busycon canaliculata,* the channeled whelk). Sutures may be wavy, irregular, slightly or deeply indented or impressed.

The **anterior end** of the shell is that end which is in front when the animal is crawling. The aperture, the siphonal canal (when present), the head and the tentacles of the mollusk are at this end. The **posterior end** is the opposite, where the apex and nuclear whorls are located, hence it is sometimes referred to as the apical end. When we speak of a rib or a bar of color we mean the side nearest to the apex or away from the anterior end of the mollusk. The total distance between the two ends of the shells is known as the length, although this measurement is often called the height. The **width** or **diameter** is the measurement 90 degrees to the length or axis.

The gastropod shell is generally an expanding tube or **helicocone,** usually coiled, and closed at its smaller, original end.

The **aperture** of the shell is the hole or space at the end of the body whorl into which the mollusk can withdraw itself. The edge of the body whorl which borders the aperture is known as the **lip** (sometimes called the **peristome** or **peritreme** in technical works). Sometimes the lip is thickened greatly or flaring like an old-fashioned blunderbuss rifle. Any startling development of the lip is generally a sign of adulthood. If the lip thickens into an unusually large,

rounded, sharp rib it is known as a **varix.** Varices may be produced at various stages in the growth of a shell, and their number and position are used as identifying characters (see *Bursa,* the frog shells).

For the sake of convenience, the part of the lip which is away from the center of the shell or is not next to the axis of the shell is known as the **outer lip** or **labrum.** Opposite this on the other side of the aperture is the **inner lip, labium** or **parietal wall,** which may be thickened, armed with teeth (see *Nerita*) or have a parietal shield (see the helmet shells, *Cassis*). The inner lip is continuous with the thickened axis or **columella** of the shell about which the whorls are developed. In many kinds of marine gastropods, especially the murexes, the columella extends forward and forms the tubelike anterior **siphonal canal.** In a few genera there is a small **posterior canal** formed at the upper or posterior end of the aperture (see *Bursa*).

The outer lip in a few genera has a very characteristic notch or slit. It is longest in the very rare, large *Pleurotomaria* shell. The scar band formed behind the slit is called the **selenizone.** The scar or track formed behind the apertural notch or sinus in the Turridae is called an **anal fasciole.** The "stromboid notch" in the conchs is weak but distinct. In the abalones, *Haliotis,* the slit is replaced by a series of small, round anal holes. Nearly all the turrids are recognized by their "turrid notch" on the upper portion of the outer lip. The keyhole limpets, *Fissurella,* have reduced the slit to a single small hole which is located at the apex of their cap-shaped shells, although in their young stages the slit is well-developed at the edge of the shell.

The sculpturing on the exterior of the shell—ribs, nodules, cords, threads, indented lines, pits, spines, etc.—are grouped into two basic types: (1) The axial sculpture (or transverse sculpture), that is, any markings, ribs or lines

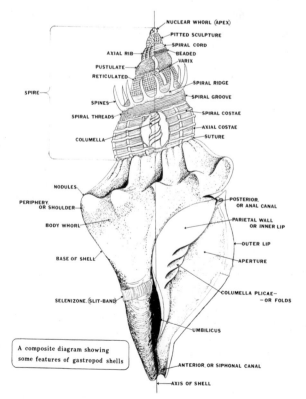

A composite diagram showing some features of gastropod shells

which run across the whorl in line with the axis of the shell or from suture to suture. Sometimes it is called longitudinal sculpture. Varices, growth lines and the outer lip are axial features. (2) The **spiral sculpture,** which is spirally arranged in the direction of the suture or in line with the direction of the growth of the whorls. Thus we often speak of spirally arranged color bands (as in the tulip shell, *Fasciolaria hunteria*), or axially arranged color streaks (as in the lightning whelk, *Busycon contrarium*).

When axial and spiral sculpturing are equally prominent and cross each other at right angles, a **cancellate** or decussate sculpture is produced. **Reticulate** sculpture is similar, but the lines do not cross at right angles.

Growth lines, mentioned in many of our descriptions, are the axial lines which run parallel to the edge of the apertural lip. They are irregularities in the shell, usually very small but sometimes coarse, which mark places where growth of the shell was stopped for a relatively long time. Sometimes the lip of the aperture becomes stained or slightly thickened during these brief rest periods (probably a few days apart), and, when additional growth takes place, these blemishes are left as growth lines.

The **umbilicus** is a hole or chink in the shell next to the base of the columella, which is formed because the whorls are not closely wound against each other at their anterior or basal end. The umbilicus may be quite large and deep as in the sundial shells, *Architectonica*. Commonly, there is a spiral cord in the umbilicus which may terminate in a button-like callus. Some species are differentiated by the size, position or color of this **umbilical callus** (see *Polinices duplicatus*). About a fourth of our marine species are umbilicated to some degree. A second or more anterior umbilicus may be formed at the base of siphonate shells and usually affects only the last whorl. This is called the **false umbilicus** or pseudumbilicus.

Teeth (not to be confused with the radular teeth in the animal's proboscis or mouth cavity) are often present in the aperture. The distorted shell, *Distorsio,* is an extreme example, but some shells have teeth on the parietal wall only (*Nerita*) or on the inside of the outer lip (*Cassis*).

The **periostracum** is a horny covering which overlays the exterior of the shell in many species and, like the shell, is secreted and shaped by the fleshy mantle of the animal. The periostracum (erroneously called the epidermis) may be very thin and transparent or only slightly tinted (as in some volutes, moon shells and the smaller conchs); or it may be like a thick coating of shellac which flakes off when dry (as in the pink conch, *Strombus gigas*). In a few buccinids, some frog shells (*Cymatium*) and the vase shells (*Vasum*), the periostracum may be very thick and often have clumps which simulate hairs and bristles. It is wholly absent in many groups, including the cowries, olives and marginellas. It is primarily a protective coating and prevents damage from boring sponges and water acids.

The **operculum** is a horny or calcareous plate firmly attached to the dorsal side of the posterior end of the foot. When the head and foot are withdrawn into the shell, this "trapdoor" is the last part to be pulled in, and it thus serves as a protection against enemies and, in many species, seals the shell from either noxious fluids or the drying effects of the sun and air. When the foot is extended and used in crawling, the operculum serves as a foot-pad on which the heavy shell may rest and rub without injury to the soft foot. The operculum is present in many families of marine mollusks, and it often serves as a useful identification character. It is absent in adults in the following families: Marginellidae Cancellariidae, Cypraeidae, Tonnidae, Haliotidae, Acmaeidae, Fissurellidae, Janthinidae and nearly all of the sea slugs (opisthobranchiates, nudibranchs, bullas, etc.). Some genera lack this organ, such as *Oliva* and *Cypraecassis,* although their close relatives, *Olivella, Ancilla, Phalium* and *Cassis* possess well-developed opercula. Nearly all *Voluta* are without the operculum, except our West Indian music volute. This is also true of the genera *Conus* and *Mitra,* whose various species may or may not possess one. In the Alaskan volute, *Volutharpa ampullacea* Dall, 15% have an operculum, 10% have only traces of the operculigenous area and 75% are without a trace of either. The presence or absence of this part of the animal is not always a good classificational character.

Many families, genera and species possess a characteristic type of operculum. Calcareous or hard, shelly opercula are found in the turban shells ("cat's eyes" of *Turbo*), the nerites and the natica moon shells. The color and sculpturing of these opercula are used for identification purposes. The liotias (Cyclostrematidae) possess a horny operculum which is overlaid by rows of calcareous beads.

THE RADULA

The minute teeth or radulae (also called the odontophore or lingual ribbon), located in the mouths of all classes of mollusks, except the clams, are so very distinctive in the various families, genera and species that they have been used as a fairly reliable identification criterion. Our present arrangement of the gastropod families is based largely upon the radula, although many other anatomical characters of the animal and shell are equally important. The Greek naturalist, Aristotle, mentioned the radula of snails as early as 350 B.C., but a fuller account was given by the Dutch naturalist, Swammerdam, in the seventeenth century. The Italian malacologist, Poli, was the first to figure the radulae of gastropods, cephalopods and chitons.

The radula is attached to the floor of the buccal cavity or inner mouth and consists of a ribbon-shaped membrane to which are attached many small, fairly hard teeth. The radula ribbon is maneuvered back and forth in somewhat

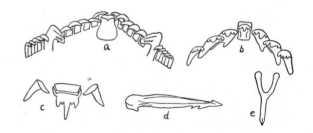

Types of radular teeth found in the prosobranch gastropods. **a,** rhipidoglossate (*Calliostoma*); **b,** taenioglossate (*Littorina*); **c,** rachiglossate (*Purpura*); **d,** toxoglossate (*Conus*); **e,** reduced rachiglossate (*Scaphella*). All greatly magnified and representing only a single transverse row of teeth.

licking fashion as the animal rasps its food. The teeth are arranged in transverse rows on the ribbon. The number of rows may vary from a dozen (in some nudibranchs) to several hundred. Each transverse row contains a specific number of teeth, depending on the family or group to which the snail belongs. In the taenioglossate snails (many families, including Cypraeidae, Strombidae, Cerithiidae and Littorinidae) there are generally only seven teeth in each row, but each of these teeth has a distinctive shape and a specific number of tiny cusps on its edges. The tooth in the center is called the **rachidian** or central. Flanking this tooth on each side is a **lateral.** Beyond each lateral there is first an **inner marginal** and finally an **outer marginal.** This makes seven teeth in all. In the rachiglossate snails (Muricidae, Buccinidae, Olividae, etc.) there are only three teeth per row—the rachidian and a strongly cusped lateral on each side. The four toxoglossate families (Conidae, Turridae, Terebridae and Cancellariidae) have lost their rachidians and laterals and have retained only the marginals.

The docoglossate snails (Acmaeidae and Patellidae) have less than twelve teeth per row but are peculiar in that there are two to four identical rachidians or centrals. In rhipidoglossate families (Trochidae, Fissurellidae, Neritidae) the radula is very complicated, and the very numerous laterals at the end of each row are called **uncini.** Among the gastropods which do not have a radula are the Pyramidellidae, Eulimidae, the genus *Coralliophila*, all *Magilus* and a few genera of nudibranchs.

We have figured several main types of gastropod radulae, but other examples have been included in the systematic section when they are of special use in identification. It is not expected that many amateurs will want to prepare and examine radulae, but, because so many serious private collectors and many biology students will find this identification too indispensable, we have included brief instructions on the preparation of radula slides.

Preparation of the radula In large specimens, such as the whelks or conchs, the proboscis may be slit open from above and the round buccal mass removed. Occasionally, the proboscis is withdrawn far inside the animal, but it is easily located below the thin skin on the dorsum just posterior to the tentacles. The flesh may be torn away with the aid of small dissecting needles until the glistening, wormlike radula pops out. In order to remove the last traces of flesh, the radula may be soaked in a saturated solution of potassium hydroxide (KOH) for a few minutes. A solution of common lye will do as well. Animals whose flesh has been hardened by a preservative will have to be carefully boiled for a few minutes or soaked overnight in KOH or lye. Small specimens may be dropped whole into this alkaline solution if only the radula is desired. Transfer the radula successively to several watchglasses of clean water in order to rid it of all traces of KOH. The radula may then be placed in one or two drops of water on a clean glass microscope slide and, by observation under the dissecting microscope, a few teeth may be teased apart with fine needles. Leave some of the ribbon intact to show the relative position of the teeth. Add a square cover slip for study under the compound microscope. In water mounts such as these, stains are usually unnecessary. This temporary preparation may be permitted to dry for a day, the cover slip gently lifted, a few drops of

euporol or mounting medium added, and the cover slip replaced to make a permanent slide. Some workers prefer to go from water to eosin stain to 96% alcohol and then to euporol, but this is an unnecessary elaboration. There are also excellent, permanent, plastic mounting mediums on the market. Canada balsam and glycerine jelly eventually deteriorate. Keep in mind that KOH or lye will burn flesh and eat holes in clothing.

BIVALVIA FEATURES

Shape of shell In most families of bivalves or pelecypods, the shape of the shell is extremely important as a species character, and only in a few groups, such as the oysters and mussels, is shape so variable within a species as to be of little taxonomic value. Shape of shell, as a whole, is of little value in determining families or genera, except in a few instances such as *Pecten*, *Spondylus* and *Pinna*.

Parts of the shell The two valves of a clam are bound together by a brown, chitinous **ligament,** and usually hooked together by a **hinge** which is furnished with interlocking **teeth.** The valves are kept closed by powerful, internal adductor **muscles** but kept spread open by the action of the ligament when the animal relaxes or after it is dead. Each valve is a shallow, hollow cone, with the apex, from which point growth of ahe valves commences, turned to one side. This apex is termed the **umbo** (plural: umbos or umbones) or **beak.** The interior cavity lying within the umbo is the **umbonal cavity.** The hinge and its teeth are usually just below the beak on the inside of the valve. The **prodissoconch** is the embryonic shell of the bivalve, and corresponds to the protoconch or nucleus of the gastropods. It is generally eroded away in adults, but when preserved it serves as a useful identification character, especially in such groups as the oysters. If the umbones curve in an anterior direction, they are described as being **prosogyrate;** if posteriorly, they are **opisthogyrate;** if each umbo is pointing towards the umbo of the other valve, they are **orthogyrate.**

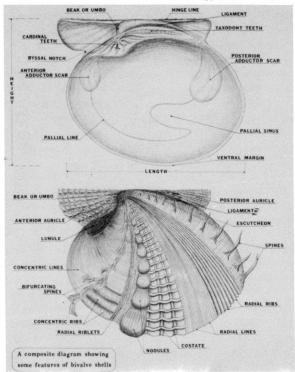

A composite diagram showing some features of bivalve shells

Right and left valves   It is important to distinguish one valve from the other and to determine which is the anterior or posterior end, for many identification features are used in relation to these orientations. The upper, or **dorsal margin** is located on the beak or hinge side; the **ventral margin** is the opposite side. The beaks usually are pointed or curved towards the anterior end, which is generally the less pointed end of the shell. The ligament in the great majority of cases is posterior to the beaks. When present, the heart-shaped impression called the **lunule** is anterior to the beaks. When a clam is placed on its ventral margins on the table with the dorsal hinge margin up, and with the anterior end away from the observer, the right valve is on the right, the left valve to the left. Another quick way is to observe the concave interior of a valve with the hinge margin away from the observer and to locate the U-shaped pallial sinus impression. If the sinus opens towards the left, it is a left valve, and vice versa for the right valve.

In most bivalves, the two valves are of the same size (**equivalve**) but in some genera one valve is larger and slightly overlaps the other (**inequivalve**). In *Ostrea, Pandora* and *Lyonsia,* the left valve is the larger; in *Corbula,* the right valve is the larger. A bivalve is said to be **equilateral** when the beak is midway between the anterior and posterior ends of the valve. Most bivalves, however, are **inequilateral,** with the beak placed nearer one end.

In many forms, the margins of the valves do not fit closely together, but have an opening called the **gape** somewhere along the margin. In the soft-shelled clam, *Mya,* the gape is posterior and through it protrudes the siphon (siphonal gape); in *Gastrochaena* it is anterior and large and serves for the passage of the foot. Some clams, such as *Solen* and *Ensis,* gape at both ends. In *Arca* there is a small notch or opening on the ventral margin for the passage of the anchoring organ of chitinous threads, the **byssus.** This is called the **byssal notch.**

The **ligament** is a brown, horny band located above the hinge, and is generally posterior to the beaks. As a rule, the greater part of the ligament is externally placed on the shell, but in some genera it may be partially or entirely internal. The ligament consists of two distinct parts, which may occur together in the same species or separately in others—the ligament proper and the internal cartilage or **resilium.** In most cases, the two portions are intimately connected with one another, but in some clams, such as *Mya* and *Mactra,* the cartilage is entirely separate (the resilium) and is lodged within the hinge in a spoon-shaped **chondrophore.** The external ligament is inelastic and insoluble in strong alkali (KOH). The cartilage is very elastic, slightly iridescent and soluble in KOH.

Muscle scars or impressions   The interior, concave surface of the valve possesses a number of useful identification features. The large muscles which serve to close the valves leave round impressions on the surface. When two muscles are present, as in the venus, lucine, tellin and other clams, they are known as the **anterior** and **posterior muscle scars** respectively. The fine, single-line impression produced by the muscular edge of the mantle is known as the **pallial line.** The pallial line may have a U-shaped notch at the posterior end of the valve indicating the presence of a siphon and its siphonal muscles. This is known as the **pallial sinus.** It is entirely absent in genera possessing no retractile siphons.

The hinge   This is one of the most important identification features in the bivalves, and often many hours of fruitless search can be avoided when the major types of hinges and their various parts are understood. There are many types of hinges, from those without teeth (**edentulous**) to those with a complex pattern. We have figured below some of the major types of hinges. The teeth are distinguished as **cardinals,** or those immediately below the umbo, and **laterals,** or those on either side of the cardinals. In many inequilateral bivalves, the teeth have become so distorted or set out of place that it is often difficult to distinguish the cardinals from the laterals or to determine which ones are absent. We have labeled the teeth in several groups in the systematic section of this book to overcome this difficulty. In *Chama,* for instance, the cardinals have been pushed up into the umbo and have become a mere ridge, while the strong anterior lateral has become nearly central and simulates a cardinal.

Sculpture   In many groups, such as the scallops (*Pecten*), sculpture is of paramount importance in determining species. In most other groups it is used in conjunction with other characters. These are two major types of sculpture—**concentric** and **radial**—and both of these may be present in many forms, such as ridges, ribs, nodules, spines, foliaceous processes (leaf-like), threads, beads and indented striae (fine lines). Concentric growth lines of varying degree of development are seen on most bivalves. They are always parallel to the margins of the valves, may be exceedingly fine or very coarse, and they generally indicate former growth and resting stages. **Radial sculpture,** running from the umbones to the lower or end margins of the valves, is exemplified in the ribs of *Cardium, Pecten* and others. Concentric and radial sculpture may occur together to form a **cancellate sculpture** as in *Chione cancellata.* In a few genera, such as *Poromya,* the valve's surface may be granulose, as if finely sugar-coated.

The **periostracum** or protective chitinous sheath overlaying the exterior of the valves is present in most bivalves. It may be extremely thin and transparent so that it imparts a high gloss to the shell, or it may be thick and matted or even very coarse and stringy so that the valves appear to be bearded, as in *Modiolus* and *Arca.*

AMPHINEURAN FEATURES

The chitons, or amphineurans, are strictly marine, bilaterally symmetric animals with an anterior mouth and a posterior anus. The foot is ventral and usually almost as long as the body. Radulae are present, but there are no cephalic eyes, tentacles or otocysts. The class was divided into the subclass Polyplacophora, in which there are eight shelly valves, and the subclass Aplacophora, or solenogasters, which lack shelly valves, and have a spiculose integument and a furrowlike foot. Today they are treated as two classes.

The eight shelly valves of a chiton are held in place by a band of muscular tissue, the **girdle.** The girdle may be smooth or ornamented by **scales,** sometimes overlapping and

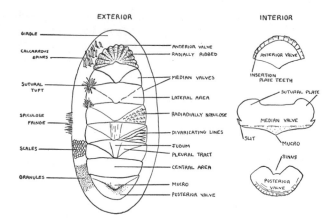

EXTERIOR / INTERIOR

resembling "split peas," or by **spicules** of various sizes, shapes and frequency. Some may resemble miniature bunches of glass fibers. The **valves** are calcareous, and for identification purposes are numbered from i at the anterior end, to viii at posterior end. The end valves are referred to as the **head valve** and the **tail valve**, respectively. Between them are six **intermediate** valves. The valves sometimes barely touch each other, but in many cases the entire rear edge of one may overlap the edge of the next valve, thus forming a **complete coverage.**

The valves consist of four layers, which are (1) the **periostracum,** a thin chitinous external, varnishlike layer; (2) the thick, not so hard, spongy calcareous **tegmentum;** (3) the harder **articulamentum** layer, including the insertion plates and sutural laminae; and (4) the lowest, or innermost, **hypostracum,** having a thin, different crystalline structure. In some highly developed chitons there may be an additional crystalline layer, the **mesostracum,** lying between the articulamentum and the tegmentum.

The dorsal or outer surface of the valves is divided sometimes into more or less recognizable areas which may bear distinguishing sculptural features. The **apex** or **beak** is the central point of the posterior edge of an intermediate valve. A similar projection at the anterior edge is termed a **false beak.** Beaked valves are termed **mucronate.** The **jugum** is a dorsal longitudinal ridge or band, the latter termed the **jugal area.** Shelly projections or narrow marginal extensions of the valve which project into the flesh of the girdle are called **insertion plates.** Teeth and slits may occur along these edges. Lobelike or platelike anterior projections on either side of the intermediate and tail valves are called **sutural laminae** or apophysis plates. Between the plates is a **sinus.**

SCAPHOPOD FEATURES

The tusklike hollow, tubular shell of the Scaphopoda is open at both ends. From the larger **anterior orifice** the foot and feeding tentacles or fleshy **captacula** (singular: captaculum) project. Water is drawn in and expelled through the smaller **posterior orifice.** The latter may have a shelly, solid or hollow tube or **terminal pipe** projecting beyond the edge. The posterior orifice may or may not have slots, slits or notches, some of which are characteristic of genera and species. The **dorsal face** of the tusk shell is the concave side; the **ventral face** is the convex side. The shell has three dif-

ferent layers, in contrast to one or two layers found in similarly shaped worm tubes. A radula is present, but there are no eyes, tentacles or gills.

NAMES AND NOMENCLATURE

In order to discuss the various kinds of mollusks, we must use standardized names which are understood or recognized by students in every part of the world. For this reason, Latin names, or latinized forms, are employed as the official medium for nomenclature. It is not at all necessary to have a knowledge of Latin or Greek in order to label a seashell. Nor is it supposed that one should attempt to remember the names, although it adds to the enjoyment of the study to absorb those of a few commoner species. In fact, it is not difficult to remember such scientific names as *Venus, Mitra, Oliva* and *Conus.* It may be of interest to beginners to know that few professional malacologists can remember more than a hundredth part of the total number of names. They, too, consult books to refresh their memories.

Popular names Popular or vernacular names in seashells are in great need of standardization, and while their use sometimes has its drawbacks, there is no reason they cannot become as acceptable to the amateur as have the popular names of birds, fishes and wild flowers. It is true that one species may be known by one name in New England and another in Florida, but these are generally names which are in use by local fishermen and not necessarily accepted by amateur shell collectors. In the face of so much name-changing in the scientific literature because of legalistic technicalities, the existence of a few provincial popular names seems little enough excuse for not attempting to standardize the common names of seashells. Throughout this book we have presented both scientific and popular names. The latter have been derived from several sources and listed only after careful consideration of the evidence. Private collectors, shell dealers, professionals and, in some cases, many popular books, both recent and old, have contributed to the final choice. In a few instances, alternate popular names which are well-entrenched along wide regions of our coast have been listed. Popularization of patronymic names, such as Clark's cone for *Conus clarki,* has been simple. Direct translations of the Latin have in many, but not all, cases been advisable. Many obvious direct translations have been avoided in order to avoid confusion with names already used for shells in other regions of the world. It is interesting to note that many popular names in use today were recorded by early eighteenth century writers, and that a few popular generic names are to be found in the writings of Aristotle and Pliny. We have not, of course, employed the rule of using the name first employed as is done in scientific nomenclature (rule of priority). It is hoped that this listing of popular names of American seashells will bring fuller enjoyment to the many amateurs who do not desire to "wrestle" with scientific names.

Name-changing There is nothing more annoying than having a well-known and frequently used scientific name changed; and the field of mollusks seems to be having its lion's share of tossing out of old friends for utter strangers. There are two basic kinds of changes—zoological and nomenclatorial. Everyone will condone the former, for it is obvious, as our knowledge increases, that certain genera or even species will be found to be mixtures, and this necessitates separating and applying new names. In this book, for example, *Fasciolaria gigantea* is changed to *Pleuroploca gigantea*. The horse conch, *P. gigantea*, does not have characters like those of the tulip shells, and it cannot be put in the genus *Fasciolaria* with such species as *F. tulipa* (Linné) and *F. hunteria* (Perry). For the same reason, what has been called by many workers *Ostrea virginica* is now *Crassostrea virginica*. *Venus mercenaria* is now *Mercenaria mercenaria*.

Nomenclatorial name-changing is hardest for everyone to accept. As not infrequently happens, a species may be given several different names inadvertently by various authors. The International Commission for Zoological Nomenclature has set up an extensive set of rules; among these is the rule of priority by which the earliest valid name is chosen if several names are available. Unfortunately, the earliest name may have been overlooked for many years, and its subsequent discovery will "knock out" one which has been in use for a long time. Thus, about 30 years ago the whelk genus *Fulgur* Lamarck, 1799, was abandoned for *Busycon* Röding, 1798. The same fate may be met by well-known species. Thus, *Busycon pyrum* (Dillwyn, 1817) now becomes *B. spirata* (Lamarck, 1816). It is believed that "rock bottom" will be reached some day, so that few, if any, further changes will occur. Nevertheless, it is with considerable regret that I change a number of familiar names in this book.

Occasionally, certain names are conserved or "frozen" by the International Commission if they are well-established and are in danger of being replaced by an earlier but obscurely known name. The following marine genera of mollusks are on the conserved list. *Aplysia, Arca, Argonauta, Buccinum, Bulla, Calyptraea, Columbella, Dentalium, Mactra, Modiolus, Mya, Mytilus, Neritina, Ostrea, Sepia, Spirula, Teredo* and *Turbinella*. Many others, including very familiar species names, need to be added to this list.

Pronunciation of scientific names   There is no official pronunciation established for names, and for certain words it may vary from one county to another. Many pronunciations not based on classical rules have become established and passed on from one generation of malacologists to another. A long-playing record giving the more or less accepted pronunciations of 1,000 common North American marine mollusks may be purchased from the Delaware Museum of Natural History, Greenville, Delaware 19807. A few examples, classical or not, are given below:

Oliva (**all**-eeva), *Eulima* (you-**lee**-mah), *Chiton* (**kite**-on), *Chama* (**kam**-ah), *Chione* (kigh-**own**-ee), *Cypraea* (**sip**-ree-ah), *Cyphoma* (sigh-**fo**-mah), *versicolor* (ver-**sik**-o-lor, said quickly), *Busycon* (**bue**-see-kon), *Janthina* (yan-**theena**), *Xenophora* (zen-**off**-fora), *gigas* (rhymes with "jibe gas": **ji**-gas), *conch* (konk), *radula* (**rad**-you-lah), *operculum* (oh-**perk**-you-lum),

smithi (**smith**-eye), *ruthae* (**rooth**-ee). The pronunciations of some of the authors are: Linné (lin-**ay**) or sometimes Linnaeus (lin-**ay**-us), Gould (goold), Deshayes (desh-**ayz**), Orbigny (or-**bee**-nee), Gmelin (**mel**-lan), Bruguière (broo-gui-**air**), Kiener (**keen**-er), Mighels (**my**-els), Couthouy (kooth-**wee**).

Common abbreviations of names of well-known authors
Although most popular and scientific books spell in full the names of authors of scientific designations, a large number of articles and most museum labels bear only abbreviations. For this reason, a short list of frequently seen examples is included:

A. Ads.—Arthur Adams
H. Ads.—Henry Adams
A. and H.—Alder and Hancock
Ag.—Aguayo
B. and S.—Broderip and Sowerby
Brod.—Broderip
Brug.—Bruguière
C. B. Ads.—C. B. Adams
Cl.—Clench
Con.—Conrad
Coop.—Cooper
Couth.—Couthouy
Cpr.—Carpenter
Dautz.—Dautzenberg
Desh.—Deshayes
Dkr.—Dunker
d'Orb.—Orbigny (d'Orbigny)
Esch.—Eschscholtz
Fer.—Férussac
Dill.—Dillwyn
G. and G.—Grant and Gale
Gld.—Gould
Gmel.—Gmelin
Hemp.—Hemphill
Hert.—Hertlein
L. or Linn.—Linné; Linnaeus
Lam. or Lk.—Lamarck
Midff.—Middendorff
Migh.—Mighels
Mts.—E. von Martens
Nutt.—Nuttall
Old.—Oldroyd
Orb.—Orbigny (d'Orbigny)
Pfr.—Pfeiffer
Phil.—Philippi
Pils.—Pilsbry
Q. and G.—Quoy and Gaimard
Raf.—Rafinesque
Röd.—Röding (or Roeding)
Rve.—Reeve
Sby. or Sow.—Sowerby
Swain.—Swainson
Val.—Valenciennes
Verr.—A. E. Verrill

II

The Univalves
(limpets, conchs, whelks and other snails)

Class GASTROPODA Cuvier, 1797

Head fused to a flattish, muscular ventral foot. Usually with a single, coiled or cup-shaped calcareous shell. A ribbon of radular teeth present in the majority of forms. The cerebral, pleural and pedal ganglia of the "brain" are distinct. The various organs of the pallial region, such as the gills, anus and kidney exit-tubes, face forward, instead of backward, because of a torsion or twisting that always takes place in the early embryonic stage. The class came into being during the Lower Cambrian. There are probably about 50,-000 living species, 60% being marine, the remainder being either land or fresh-water dwellers.

Subclass *Prosobranchia* Milne-Edwards, 1848

Order *Archaeogastropoda* Thiele, 1925

Suborder *Pleurotomariinae* Cox and Knight, 1960

Superfamily Pleurotomariacea Swainson, 1840

Family PLEUROTOMARIIDAE Swainson, 1840

Genus *Perotrochus* P. Fischer, 1885

For a recent review of the living pleurotomariids, see F. N. Bayer, 1965, *Bull. Marine Sci. Miami,* vol. 15, no. 4, pp. 737–796. Type: *quoyanus* (F. and B., 1856). No umbilicus; anal slit short.

Perotrochus quoyanus (Fischer and Bernardi, 1856) 1
Quoy's Pleurotomaria Color Plate 2

Gulf of Mexico and the West Indies; Bermuda.

1½ to 2 inches in length and width. Umbilicus sealed over. Sculpture of finely beaded, small spiral threads. Characterized by

a relatively short but wide slit at the periphery of the body whorl just behind the outer lip. Color dull orange-yellow with darker maculations. Base white. Interior slightly pearly. Dredged from 70 to 300 fathoms. One of our rarest seashells.

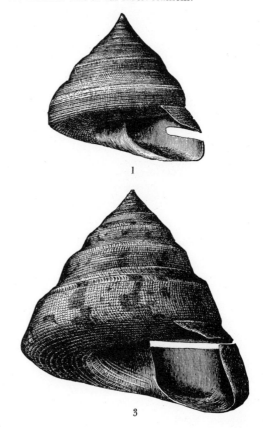

1

3

2   6   7

Perotrochus amabilis (F. M. Bayer, 1963)   **2**
Lovely Pleurotomaria

Gulf of Mexico and off Key West, Florida

2 to 3 inches, brightly colored with axial streaks and flames of rusty-red, some of which extend over onto the base where there are about 26 weak threads. Top of whorls with about 10 to 12 coarse threads above the slit-scar. Similar to *P. quoyanus,* but with coarser sculpture, slightly concave slope to the spire, and with a more indented umbilical region. Rare; deep water. (See F. M. Bayer, 1963, *Bull. Marine Sci. Gulf and Caribbean,* vol. 13)

Subgenus *Entemnotrochus* P. Fischer, 1885

Umbilicus very deep; anal slit about ½ whorl long. Operculum multispiral and small. Type: *adansonianus* Crosse and Fischer, 1861.

Perotrochus adansonianus (Crosse and Fischer, 1861)   **3**
Adanson's Pleurotomaria   **Color Plate 1**

Off Key West, Florida, to the Lesser Antilles; Bermuda.

3 to 7 inches in length, and slightly more in width. Umbilicus round, very deep. Sculpture of coarsely beaded, moderately small spiral threads. Slit on periphery of whorl narrow and very long (½ of a whorl). Color cream with a salmon blush and irregular, small patches of red. Base similarly colored. Occurs from 50 to 110 fathoms, and sometimes brought up in fish traps. This is a moderately common species in nature, but rare in collections. Scuba divers have been obtaining them on the drop-off wall at 350 to 450 feet on the east side of Andros Island, Bahamas.

Other Atlantic species:

4 *Perotrochus atlanticus* Rios and Matthews, 1968. 133 meters, off Sao Paulo, Brazil. Atlantic Pleurotomaria. Rare. (*Arq. Est. Biol. Mar. Univ. Fed. Caesar,* vol. 8, p. 65)

5 *Perotrochus gemma* F. M. Bayer, 1965. 100 fms., off Barbados. Jewel Pleurotomaria. Rare. (*Bull. Marine Sci.,* vol. 15)

6 *Perotrochus lucaya* F. M. Bayer, 1965. 366 meters, off Lucaya, Grand Bahama Island. Lucayan Pleurotomaria. Rare. (*Bull. Marine Sci.,* vol. 15)

7 *Perotrochus midas* F. M. Bayer, 1965. 600 to 700 meters, 15 miles northwest of Great Stirrup Cay, Berry Islands, Bahamas. King Midas Pleurotomaria. Rare. (*Bull. Marine Sci.,* vol. 15)

8 *Perotrochus (Mikadotrochus) notialis* Leme, 1969. 150 meters,

off Rio Grande do Sul, Brazil. Notable Pleurotomaria. Rare. (*Papéis Avulso de Zool.,* vol. 22, p. 225)

9 *Perotrochus pyramus* F. M. Bayer, 1967. Off Guadeloupe, Lesser Antilles. Pyramid Pleurotomaria. Rare. (*Bull. Marine Sci.,* vol. 17, p. 389)

Family Scissurellidae Gray, 1847

Shells minute, fragile, white, with an anal slit or orifice at the periphery of the whorl. Umbilicus usually well-developed. Sculpture of microscopic spiral or axial threads. Worldwide, but generally in temperate or warm seas. Occur in shallow to very deep waters.

Genus *Scissurella* Orbigny, 1824

Shells less than ⅓ inch, white, porcelaneous except for a thin nacreous layer inside; top-shaped; outer lip with a slit or hole. Operculum multispiral, with a central nucleus. Type: *laevigata* Orbigny, 1824. Mediterranean. *Schismope* Jeffreys, 1856, and *Woodwardia* Crosse and Fischer, 1861, are synonyms. A review of the Pacific coast species was given by J. H. McLean, 1967, *The Veliger,* vol. 9, p. 404.

Subgenus *Anatoma* Woodward, 1859

Spire elevated; slit on the middle of the whorl. Type: *crispata* Fleming, 1828. *Schizotrochus* Monterosato, 1877, is a synonym.

Scissurella crispata (Fleming, 1828)   **10**
Crispate Scissurelle

Arctic Seas to Florida and the West Indies. Europe.
Alaska to Baja California. Japan.

3.5 mm. (1.8 inch) in width and 3.0 mm. in length. 4 to 5 whorls. Fragile, frosty-white in color and sculptured by very delicate reticulations. Umbilicus small, round and very deep.

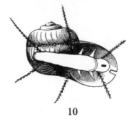

10

Periphery of whorls angulate and with two thin, sharp spiral lamellae. Between these there is an open slit running from the edge of the thin apertural lip back about ⅕ of a whorl. Uncommon from 60 to 1,215 fathoms. *S. striatula* Philippi, 1844; *aspera* Philippi, 1844; *angulata* Lovén, 1846; *japonica* A. Adams, 1862; *paucicostata* Jeffreys, 1865; *kelseyi* Dall, 1905; *chiricova* Dall, 1919, are synonyms. *S. keenae* J. H. McLean, 1970 (*Malacol. Rev.*, vol. 2, p. 117) is possibly a synonym.

Scissurella proxima Dall, 1927   **11**
Florida Scissurelle

South Carolina to off the Lower Florida Keys.

Shell 2 mm. or less, white, 4 whorls, the slit ¼ as long as the last whorl, rounded periphery; fairly high spire, so that the length of the entire shell is about equal to the width. Ridges bordering slit-scar are very weak. Minute umbilicus. Base weakly reticulated. Uncommon from 20 to 434 fathoms.

Scissurella alta Watson, 1886   **12**
Tall Scissurelle

Off south Florida to West Indies; Azores.

Extremely small (3 to 4 mm.), globose, transparent, base well-rounded, columella ½ as long as entire shell. Umbilicus chink-like. Top shoulder of whorl bears old slit-scar. Base reticulated. White. A rare species found from 390 to 450 fathoms.

Scissurella lamellata (A. Adams, 1862)   **13**
Lamellate Scissurelle

Prince William Sound, Alaska, to Japan.

3 to 4 mm., about as wide, with a spire angle of about 110 degrees, translucent white, 4 whorls. Slit-scar at the periphery has sharp edges and is open ⅕ of the last whorl. Axial slanting ribs sharp, broadly spaced. Lower columella slightly reflected. Umbilicus deep. Uncommon; fine sand bottoms, 25 to 75 fathoms.

Scissurella lyra S. S. Berry, 1947   **14**
Lyre Scissurelle

Farallon Islands to Carmel, California.

Very small, 1.0 mm. high, 1.4 mm. in diameter, depressed, fragile, 3½ whorls. Slit-scar, slightly above the periphery; edges sharp. Axial sculpture of retractively slanting, minute, raised lines above the slit band. Base with axial and spiral sculpture. Deep umbilicus bordered by a weak keel. Rare; on gravel on slopes, 7 to 16 fathoms. Present in Pleistocene of central California.

Scissurella soyoae (Habe, 1951)   **15**
Soyo Scissurelle

Prince William Sound, Alaska, to Japan.

1 to 2.5 mm. in length (2.3 × 2.8 mm.), yellowish, 4 whorls, rapidly expanding, rounded at the periphery. Slit-scar at the periphery, its edges not sharply raised, open ⅕ of the last whorl. Sculpture of fine axial and spiral lines. Columella reflected partially over the umbilicus. Uncommon; on fine sandy mud, 20 to 25 fathoms.

Other species:

16 *Scissurella* (*Incisura* Hedley, 1904) *lacuniformis* Watson, 1886. 390 fms., West Indies. Rare.

17 *Scissurella* (*Incisura*) *tabulata* Watson, 1886. 390 fms., West Indies. Rare.

18 *Scissurella* (*Scissurella*) *cingulata* O. G. Costa, 1861. Belt Scissurelle. West Indies. Bermuda. 1 mm. Common in algae on mangrove roots.

19 *Scissurella caliana* Dall, 1919. San Diego, California [not in the Scissurellidae, fide McLean, 1967].

Genus *Sinezona* Finlay, 1927

Similar to *Scissurella*, but the slit is closed at the margin of the outer lip, leaving a slotlike hole some distance back from the edge of the lip. Type: *brevis* (Hedley, 1904).

Sinezona rimuloides (Carpenter, 1865)   **20**
Rimmed Scissurelle

Farallon Islands, California, to Chile.

1 mm., white, as high as wide, with 3 rapidly increasing whorls. Shoulder carinate, bearing a spiral gutter and having a tear-drop hole back from the edge of the fragile outer lip. Above and below are numerous slanting, rounded riblets. Base with spiral lirae. Umbilicus chink broad. Common; in sand and gravel in shallow water near kelp. *S. simonsae* (Bartsch, 1946) is a synonym. This is *S. coronata* of authors, not Watson, 1866.

Family HALIOTIDAE Rafinesque, 1815

Genus *Haliotis* Linné, 1758

Shell ear-shaped, flat, with a nonprotruding spire of 2 to 5 whorls. Last whorl with a few open holes. Interior nacreous and iridescent. Type: *asinina* Linné, 1758. A popular genus both as food and as shell ornaments. In California, a sport fishing license is required of any person over 16 years of age. Legal minimum size for sportsmen is 7 inches (red abalone) or 6 inches (pink abalone), and only 5 specimens per person per day. There are other restrictions concerning the collecting of Pacific Coast abalones. (See A. G. Smith, 1960, *The Veliger*, vol. 3, no. 3, p. 83.)

Hybrids are not uncommon, thus sometimes making identification difficult, and suggesting that some of the so-called species may be ecologic forms or subspecies. See "Hybridization in the eastern Pacific abalones" by Owen, McLean and Meyer, 1971, *Bull. Los Angeles Co. Mus. Nat. Hist. Sci.*, no. 9, 37 pp. Twelve types of hybrid crosses have been recognized. About 2 individuals per thousand of commercial catches are hybrids. For effects of diet, see *The Veliger*, vol. 4, pp. 29 and 129 (1961).

Haliotis cracherodii Leach, 1814   **21**
Black Abalone   **Color Plate 1**

Coos Bay, Oregon, to Baja California.

6 inches in length, oval, and fairly deep. Outer surface smoothish, except for coarse growth lines. Usually 5 to 8 holes are open. External color bluish to greenish black. Interior pearly-white. A fairly abundant, edible species, although not fished commercially to any great extent. A littoral species which clings to rocks between tide marks. Some shells may lack the holes (unnecessarily named *H. c. holzneri* Hemphill, 1907, *H. c. imperforata* Dall, 1919 and *H. c. lusus* Finlay). A subspecies, *H. c. californiensis* Swainson, 1822, **(22)** occurs on Guadalupe Island and is characterized by 12 to 16 very small holes. *H. c. bonita* Orcutt, 1900, is the same as this subspecies. *H. c. splendidula* Williamson, 1892, is the typical *cracherodii*.

Haliotis rufescens Swainson, 1822   23
Red Abalone   Color Plate 1

Oregon to Baja California.

10 to 12 inches in length, oval, rather flattened. Outer surface rather rough, dull brick-red with a narrow red border around the edge of the shell. Interior iridescent blues and greens, with a large central muscle scar. 3 or 4 holes are open. The tentacles and body are black. Intertidal to 540 feet. Fished commercially from 20 to 50 feet, especially between Monterey and Point Conception. The legal minimum size for sportsmen is 7 inches, and the catch is limited to 5 specimens per person per day. This is a popular food, and when polished on the outside, makes an attractive mantel piece. For spawning, see J. G. Carlisle, Jr., 1962, *The Nautilus,* vol. 76, p. 44.

Haliotis corrugata Wood, 1828   24
Pink Abalone   Color Plate 1

Monterey, California, to Baja California.

5 to 7 inches in length, almost round, fairly deep, with a scalloped edge and strong corrugations on the outer surface. 3 or 4 large tubular holes are open. Exterior dull-green to reddish brown. Interior brilliant iridescent. The variety *diegoensis* Orcutt, 1919, is the same. Abundant in its southern range. Occurs intertidally to 180 feet, but mostly between 20 and 80 feet where it feeds on the giant kelp, *Macrocystis.* The legal minimum collecting size is 6 inches. Gray is not the author of this species.

Haliotis fulgens Philippi, 1845   25
Green Abalone   Color Plate 1

Point Conception, California, to Baja California.

7 to 8 inches in length, almost round, moderately deep, and sculptured with 30 to 40 raised, coarse spiral threads. Exterior dull reddish brown; interior iridescent blues and greens. 5 or 6 holes are open. Fished commercially in southern California, usually from 10 to 20 feet, rarely down to 25. The legal minimum size is 6¼ inches. *H. splendens* Reeve, 1846; *H. revea* Bartsch, 1942; and *H. turveri* Bartsch, 1942, are the same species in all likelihood.

Haliotis walallensis Stearns, 1899   26
Northern Green Abalone   Color Plate 1

British Columbia to La Jolla, California.

4 to 5 inches (rarely 7) in length, elongate, flattened, with numerous spiral threads. Exterior dark brick-red, mottled with pale bluish green. 5 or 6 holes are open, and their edges are not elevated. This is a small, relatively scarce species, found from 1 to 70 feet. Also called the flat abalone.

Haliotis sorenseni Bartsch, 1940   27
White Abalone   Color Plate 1

Point Conception, California, to Baja California.

Shell large, reaching 10 inches, deep, oval, relatively thin-shelled. Exterior reddish brown, usually covered with marine growths. 3 to 5 open holes with elevated rims. Interior bright, pearly white with iridescent tints of pink. Outer edge of lip with a narrow border of red. Muscle scar hard to see. Mantle has purple border anteriorly. Soft parts yellow or orange. A deep-water form, most abundant from 80 to 100 feet, but rarely 150; prevalent off southern California.

Haliotis kamtschatkana Jonas, 1845   28
Japanese Abalone   Color Plate 1

Japan, southern Alaska to Point Conception, California.

4 to 6 inches in length, elongate, with a fairly high spire. 4 or 5 holes open which have raised edges. Outside of shell rudely corrugated, but a few specimens may have weak, spiral cords. The body is mottled tan and greenish, some specimens with tinges of orange. Tentacles green and slender. This is a small species, uncommon in California, but increasingly abundant northward where it may be found intertidally. The southern subspecies of *kamtschatkana* is *assimilis* Dall, 1878.

Haliotis kamtschatkana subspecies *assimilis* Dall, 1878   29
Threaded Abalone   Color Plate 1

Point Conception, California, to Baja California.

4 to 5 inches in length, oval, fairly deep, with weak corrugations and weak to strong spiral threads. 4 or 5 holes open, tubular. Outer color mottled with brick-red, greenish blue and gray. *H. aulaea* Bartsch, 1940, is a little more corrugated than usual, and it may be this species. *H. smithsoni* Bartsch, 1940, appears to be a giant specimen of *assimilis* Dall.

Haliotis pourtalesii Dall, 1881   30
Pourtales' Abalone

Off North Carolina to Florida; off Texas to Brazil.

½ to 1 inch in length, elongate, with 22 to 27 wavy, spiral cords. Outside waxy yellow to light-brown with a few irregular patches of reddish orange. A light-orange band runs from each hole to the edge of the shell. Inside pearly-white. An uncommon species, and the only one recorded from our eastern coast. It has been dredged from 65 to 200 fathoms. Beware of young specimens from other oceans labeled as this species.

30

Other species:

31 *Haliotis barbouri* Foster, 1946. Praia de Copacabana, Brazil. See *Johnsonia,* vol. 2, no. 41. Rare. (Possibly a Polynesian specimen.)

32 *Haliotis (Padollus) dalli* Henderson, 1915. Galapagos Islands, Ecuador; offshore. Gorgona and Colombia, 10 to 20 fms. Uncommon.

32

33 *Haliotis (Padollus) roberti* J. H. McLean, 1970. Chatham Bay, Cocos Island, Costa Rica, 40 to 47 fms. *Malacol. Rev.,* vol. 2, p. 115. 18 mm.

33

Superfamily Fissurellacea Fleming, 1822

Family FISSURELLIDAE Fleming, 1822
(Keyhole Limpets)

Subfamily EMARGINULINAE Gray, 1834

Key to the Genera of Emarginulinae

a. Apex at the same level as the base of the shell:
  b. With an internal septum . . . . . *Zeidora*
  bb. Without internal septum . . . . . *Nesta*
aa. Apex above the base of the shell:
  c. Slit at anterior edge . . . . . *Emarginula*
  cc. Slit at anterior middle:
    d. Funnel around slit on inside . . *Puncturella*
    dd. No funnel around slit . . . . . *Rimula*

Genus *Zeidora* A. Adams, 1860

Small, fragile, deepsea, limpetlike snails with the apex hooked over at the posterior end. The anterior fissure is narrow and open for a short distance, but the closed scar extends over the curved back of the shell. Viewed from the underside, there is a *Crepidula*-like septum at the posterior end. Worldwide; 2 species in the Caribbean. No operculum present. Type: *calceolina* A. Adams, 1860. *Crepiemarginula* Seguenza, 1880; *Legrandia* Beddome, 1883 and *Zidora* Fischer, 1885, are synonyms. See I. P. Farfante, 1947, *Johnsonia*, vol. 2, no. 24.

Zeidora bigelowi Farfante, 1947   34
Bigelow's Zeidora

South coast of Cuba.

Shell only 2.5 mm. in length, delicate, semi-transparent, glossy. Height 40% of the length. Anal fissure ⅓ the length of the shell. Exterior cancellated by fine threads. Margin finely denticulated. Septum narrow and slightly arched, being 10% of the shell length. A rare species; 175 to 225 fathoms.

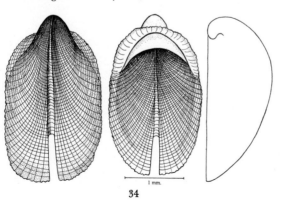

34

Other species:

35 *Zeidora naufraga* Watson, 1883. Off Puerto Rico; Azores; 390 fms. See *Johnsonia*, vol. 2, no. 24, pl. 42.

36 *Zeidora flabellum* (Dall, 1896). Clarion Island, off West Mexico, in 840 meters.

Genus *Nesta* H. Adams, 1870

Small, fragile, deepsea, limpetlike snails, similar to *Zeidora,* but lacking an internal shelf or septum. Anterior slit broad. One Red Sea, one Atlantic, and one Galapagos species known. Type: *candida* (H. Adams, 1879).

Subgenus *Laevinesta* Pilsbry and McGinty, 1952

A *Nesta* characterized by the wholly internal shell which is without concentric sculpture or marginal crenulations; marginal nodular tooth single, long and denticulated. Type: *atlantica* Farfante, 1947.

Nesta atlantica Farfante, 1947   37
Atlantic Nesta

Off south Florida to Barbados.

Shell completely enclosed in the yellow animal which is 3 times as large as the 10 mm.-long, smooth, thin shell. Anteriorly the mantle is notched in the middle, and digitated on the sides where it extends up over the shell. Shell cap-shaped, fragile, translucent white, with the hooked apex at the posterior end. Shell is shorter on the right anterior end, giving the whole shell a skewed effect. Slit U-shaped, short, and at the anterior end of a wide dorsally running, concave fasciole. Uncommon; lives on yellow sponge; 30 to 130 fathoms. For radula and animal details see *The Nautilus,* vol. 66, p. 1.

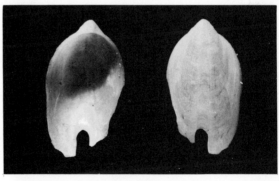

37

Other species:

38 *Nesta galapagensis* J. McLean, 1970. Galapagos Islands. 145 meters. *The Veliger,* vol. 2, p. 363, pl. 54.

Genus *Emarginula* Lamarck, 1801

Shell opaque-white, small, but strong; apex curved posteriorly but not down to the margin of the shell. Slit narrow and the same size along its length. No internal septum. In the interior of the shell, the anal fasciole has a long callus with a weak groove. Worldwide; most species deepsea. Seven Atlantic species. Type: *conica* Lamarck, 1801.

41a

Emarginula phrixodes Dall, 1927   39
Ruffled Emarginula

Off North Carolina to both sides of Florida to Brazil.

7 mm. in length, thin but strong, and with a small, narrow slit on the anterior slope of the shell near the margin. Base oval. Color translucent-white. Interior glossy. Concentric cords and 20 to 24 radial ribs across each other to form a knobby, cancellate pattern. Uncommon. Ranges from 6 to 1,075 fathoms.

Emarginula tuberculosa Libassi, 1859   40
Tuberculate Emarginula

Off Georgia to Florida to Brazil; Portugal; Azores. West Colombia; Galapagos.

18 mm. in length, white or buff, elevated; with a strongly curved anterior slope. Anal fasciole, which extends from the hooked, 1.5-whorled apex to the short slit, has numerous, arched lamellae. Surface has beaded cancellations between which are small depressed pits. Moderately common from 33 to 450 fathoms.

Emarginula pumila (A. Adams, 1851)   41
Pygmy Emarginula

Southeast Florida to Brazil; Bermuda.

12 mm. in length, thin to thick and strong. Variable in shape, particularly in height. White to buff in color. Slit at the anterior end is very short. Radial ribs 11 to 13, with 2 or 3 smaller ones in between. Crossed by small concentric ribs, thus sometimes forming small irregular pits. Margin strongly crenulate. Moderately common on rocks from 1 to 16 fathoms. The synonyms are: *rollandii* Fischer, 1856; *tumida* Dall, 1889.
   Emarginula dentigera Heilprin, 1889 **(41a)** from Bermuda and the Lower Florida Keys is whitish to greenish, high-spired, with 18 to 26 radial riblets, and is about 7 mm. in length. A fairly common, littoral species. *E. pileum* Heilprin, 1889, is a synonym. See M. C. Teskey, 1973, *The Nautilus*, vol. 87, p. 60.

Other species:

42 *Emarginula sicula* Gray, 1825. Both sides of Florida (100 to 130 fms.) to Barbados; Western Europe to Mediterranean. *E. cancellata* Philippi, 1836, is a synonym. See *Johnsonia*, vol. 2, no. 24, pl. 45.

43 *Emarginula crassa* Sowerby, 1812. Virgin Islands and Western Europe. See *Johnsonia*, vol. 2, no. 24, pl. 46. *E. magnifica* Pilsbry, 1891, is a synonym.

44 *Emarginula nordica* Farfante, 1947. 70 fms., South of Block Island, Rhode Island. See *Johnsonia*, vol. 2, no. 24, pl. 48.

45 *Emarginula velascoensis* Shasky, 1961. Gulf of California. 73 to 550 meters. *The Veliger*, vol. 4, no. 1, p. 18.

Genus *Rimula* Defrance, 1827

Shell with the fissure about at the middle of the anterior slope. Without an internal septum at the fissure. Shell white, fragile and with the apex hooked over at the posterior end. Radial riblets and concentric threads prominent. Type: *blainvillii* "Defrance" Blainville, 1824. *Rimularia* G. Fischer von Waldheim, 1834, is a synonym.

Rimula frenulata Dall, 1889   46
Bridle Rimula

Off North Carolina to both sides of Florida, and the West
Indies.

7 mm. in length, thin, very delicate. Anal slit in the middle
of the anterior slope of the shell and arrow-shaped. Base elongate-oval. Shell ⅓ high as long. Sculpture of fine cancellations.
Margin finely crenulate. Color translucent-white to cream or
rust, generally a deeper shade at the apex. The commonest
species of American *Rimula*, but rare in collections. Under
rocks; 1 to 150 fathoms, especially off the Miami area. *R. longa*
Pilsbry, 1943, is a synonym.

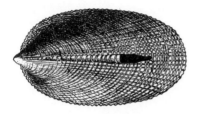

46

Other species:

47   *Rimula pycnonema* Pilsbry, 1943. 33 fms., off Palm Beach,
and west Florida to Barbados. *The Nautilus*, vol. 57, pl. 7.

47

48   *Rimula aequisculpta* Dall, 1927. Southeast Florida to Barbados. 1 to 100 fms.

49   *Rimula dorriae* Farfante, 1947. Off southeast Florida, 80 to
118 fms. *Johnsonia*, vol. 2, no. 24, pl. 51.

50   *Rimula californiana* S. S. Berry, 1964. Catalina Island, California, and Baja California. *Leaflets in Malacology*, vol. 1, no. 24,
p. 147.

51   *Rimula mexicana* S. S. Berry, 1969. Off Tule, east Baja California, 21 to 27 fms. *Leaflets in Malacology*, no. 26, p. 159. *R.
astricta* J. H. McLean, 1970, is a synonym.

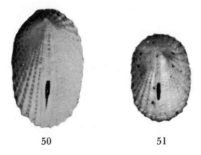

50   51

Genus *Hemitoma* Swainson, 1840

Small, thick, flattish, limpetlike snails, very similar to
Emarginula, but in *Hemitoma* the anal notch is reduced or
absent, but there is an internal anal groove. The apex is
central. *Subemarginula* Gray, 1847, is a synonym. There are
two Western Atlantic species, both shallow water; three in
California. Type: *octoradiata* (Gmelin, 1791).

Hemitoma octoradiata (Gmelin, 1791)   52
Eight-rayed Emarginula

Southeast Florida to Brazil; Bermuda (fossil).

1¼ inches in length, with 8 main, unbranched radial ribs
which are crudely nodulated. Secondary and tertiary ribs appear in older specimens. Interior of shell glossy olive-green to
purplish brown with a white margin. Interior shows a narrow,
deep groove at the anterior third. Animal of vivid turquoise,
magenta and greens. Common on rocks just below low tide
mark. The synonyms are: *tricostata* Sowerby, 1824; *listeri* Anton, 1839; *clausa* Orbigny, 1842; *depressa* Sowerby, 1863; *guadaloupensis* Sowerby, 1863; and *rubida* A. H. Verrill, 1950.

52

Other species of *Hemitoma*:

53   *Hemitoma bella* Gabb, 1865. Alaska to Santa Cruz to San
Pedro, California, 5 to 20 fms. (Synonym: *yatesii* Dall, 1901.)

—   *Hemitoma golischae* Dall, 1916, is a young *Fissurella volcano*
Reeve.

54   *Hemitoma natlandi* Durham, 1950. Jalisco, West Mexico, to
Colombia. 8 to 45 meters. (Synonyms: *scrippsae* Durham, 1950;
chiquita Hertlein and Strong, 1951.)

Subgenus *Montfortia* Récluz, 1843

White to greenish shells, with cancellate sculpturing and
with a tiny, subcentral apex that points towards the back.
Embryonic shell covered with a light-brown periostracum.
Type: *australis* Quoy and Gaimard, 1832.

Hemitoma emarginata (Blainville, 1825)   55
Emarginate Emarginula

Southeast Florida to the Lesser Antilles.

1 inch; white in color; strongly cancelled by 8 main ribs
and not-so-strong concentric cords, thus forming squarish pits
on the surface. Interior anal groove moderately strong and ending as a small notch near the anterior margin. Synonyms appear
to be *retiporosa* (Dall, 1903) and *ostheimerae* (Abbott, 1958).
An uncommon rock-dweller found from low tide to 83 fathoms.

Other tropical Pacific species:

56   *Hemitoma* (*M.*) *hermosa* Lowe, 1935. Off Carmen Island, Gulf
of California, 20 fms.

55

Genus *Puncturella* R. T. Lowe, 1827

Shell with a cuplike septum inside, just in back of the slotlike fissure. Members of the typical subgenus, *Puncturella*, do not show an anal fasciole scar on the outer shell. In the subgenus *Cranopsis* A. Adams, 1860, the anal fasciole is always present. The open fissure is elongate. In the subgenus *Fissurisepta* Seguenza, 1863, the apical whorls are lost in the adult; the fissure at the summit is circular, and there is no groove from the slot to the anterior margin. About 60 worldwide, deepsea species. Type: *noachina* (Linné, 1771).

Puncturella noachina (Linné, 1771)   57
Linné's Puncturella

Circumpolar; south to Cape Cod.
South to the Aleutians.

½ inch in length, conical, laterally compressed, with an elliptical base. 21 to 26 primary radial ribs between each of which are added a smaller, secondary rib farther down. Margin crenulate. Tiny slit just anterior to the apex, and internally it is bordered by a funnel-shaped cup on each side of which is a minute, triangular pit. Color uniformly white, internally glossy. May be collected under rocks at lowest tides in its northern range but also occurs in waters over a mile deep. Common.

57

Puncturella galeata (Gould, 1846)   58
Helmet Puncturella

Aleutian Islands to Redondo Beach, California.

½ to ¾ inch in length, similar to *cucullata*, but with an almost smooth basal edge; with numerous, much finer radial ribs, and with the internal shelf behind the slit reinforced by a second, straight shelf. Commonly dredged in mud from 10 to 75 fathoms.

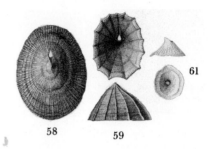

58   59

61

Subgenus *Cranopsis* A. Adams, 1860

The type is *pelex* A. Adams, 1860, from Japan.

Puncturella cucullata (Gould, 1846)   59
Hooded Puncturella

Alaska to La Paz, Mexico.

¾ to 1 inch in length, moderately strong. Apex small, elevated and hooked over toward the anterior end. Behind it is a small, elongate slit penetrating through the shell. Internally the slit is separated from the apex by a calcareous, convex shelf. Exterior with 14 to 23 major ribs and with 1 to 5 smaller radial ribs between the main ones. The fewer the ribs, the stronger they are. Shell dull-gray externally, glossy-white inside. The border is crenulated. Found at low tide in Alaska and dredged from 20 to 75 fathoms off southern California. Common.

Other species:

Subgenus *Puncturella:*

60 *Puncturella profundi* Jeffreys, 1877. Greenland to the Virgin Islands; France to the Azores. 294 to 1,450 fms. See *Johnsonia,* vol. 2, no. 24, pl. 56.

61 *Puncturella circularis* Dall, 1881. East Florida to Tobago Island. 380 to 580 fms. See *Johnsonia,* vol. 2, no. 24, pl. 57.

62 *Puncturella borroi* Farfante, 1947. Cuba to Argentina. 11 to 1,020 fms. *Johnsonia,* vol. 2, no. 24, pl. 57.

63 *Puncturella sportella* Watson, 1883. Georgia to the Lesser Antilles. 294 to 390 fms. See *Johnsonia,* vol. 2, no. 24, pl. 58.

64 *Puncturella oxia* Watson, 1883. Georgia to the Lesser Antilles. 294 to 440 fms. See *Johnsonia,* vol. 2, no. 24, pl. 58.

65 *Puncturella plecta* Watson, 1883. Georgia to the Virgin Islands; off Portugal. 294 to 390 fms. See *Johnsonia,* vol. 2, no. 24, pl. 59.

66 *Puncturella brychia* Watson, 1883. 1,340 fms., off Nova Scotia. See *Johnsonia,* vol. 2, no. 24, pl. 63.

67 *Puncturella pauper* Dall, 1927. 250 fms., off Guantanamo, Cuba. See *Johnsonia,* vol. 2, no. 24, pl. 59.

68 *Puncturella abyssicola* Verrill, 1885. 1,537 fms., south of Martha's Vineyard, Massachusetts. See *Johnsonia,* vol. 2, no. 24, pl. 62.

69 *Puncturella cooperi* Carpenter, 1864. Alaska to California. (Synonym: *eyerdami* Dall, 1924.)

70 *Puncturella caryophylla* Dall, 1914. 67 to 81 fms., off San Diego, California.

71 *Puncturella acuminata* Watson, 1883. South Carolina to Mexico and the Lesser Antilles, 159 to 390 fms. *P. triangulata* Dall, 1889, and *microphyma* (Dautzenberg and Fischer, 1927) are synonyms. See *Johnsonia,* vol. 2, no. 24, pl. 64.

72 *Puncturella tenuicola* Dall, 1927. 294 fms., off Cumberland Island, Georgia. See *Johnsonia,* vol. 2, no. 24, pl. 64.

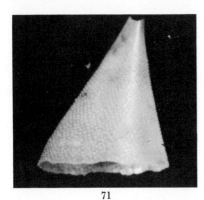

71

73 *Puncturella longifissa* Dall, 1914. Off Bering Islands, Bering Sea.

74 *Puncturella punctocostata* S. S. Berry, 1947 Monterey Bay, California, to the Gulf of California. 4 mm. (Synonym: *ralphi* S. S. Berry, 1947.)

75 *Puncturella asturiana* (Fischer, 1882). North Carolina to the Lesser Antilles, 100 to 1,103 fms.; off Spain and Portugal. See *Johnsonia,* vol. 2, no. 24, pl. 52.

75

76 *Puncturella antillana* Farfante, 1947. Cuba to Martinique. 210 to 400 fms. See *Johnsonia,* vol. 2, no. 24, pl. 52.

77 *Puncturella erecta* Dall, 1889. North Carolina to south Florida. 107 to 440 fms. *P. hendersoni* Dall, 1927, is a synonym.

77

78 *Puncturella billsae* Farfante, 1947. Both sides of Florida to Cuba. 60 to 225 fms. See *Johnsonia,* vol. 2, no. 24, pl. 54.

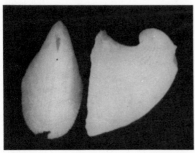

78

79 *Puncturella granulata* Seguenza, 1863. Florida Keys to Lesser Antilles; Azores. 80 to 1,075 fms. See *Johnsonia,* vol. 2, no. 24, pl. 24. *P. tuberculata* Watson, 1883, and *P. watsoni* Dall, 1889, are synonyms.

80 *Puncturella larva* Dall, 1927. 294 fms., off Fernandina, Florida. See *Johnsonia,* vol. 2, no. 24, pl. 54.

81 *Puncturella agger* Watson, 1883. Florida to Lesser Antilles. 136 to 1,075 fms. See *Johnsonia,* vol. 2, no. 24, pl. 55.

82 *Puncturella decorata* Cowan and McLean, 1968. Sitka, Alaska, 110 to 117 fms., to Cortez Bank, California, 60 fms. See *The Veliger,* vol. 11, p. 106.

83 *Puncturella multistriata* Dall, 1914. Aleutians to Prince William Sound (abundant), Alaska, to off San Diego, California.

84 *Puncturella major* Dall, 1891. Alaska.

85 *Puncturella expansa* (Dall, 1896). Deep water off Baja California to Panama and the Galapagos.

Subgenus *Fissurisepta:*

86 *Puncturella trifolium* Dall, 1881. 640 fms., Yucatan Straits, Mexico. See *Johnsonia,* vol. 2, no. 24, pl. 63.

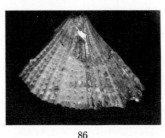

86

87 *Puncturella pacifica* Cowan, 1969. Off Triangle Island, British Columbia, 470 fms. See *The Veliger,* vol. 12, p. 24.

Subfamily DIODORINAE Odhner, 1932

Genus *Diodora* Gray, 1821

Keyhole limpets with the internal callus of the hole truncated and frequently minutely excavated behind; shell with its basal margin never raised at the ends. Central tooth of the radula wide. Compare with *Fissurella. Diadora* is a misspelling. Type: *D. graeca* (Linné, 1758), from Europe.

Diodora cayenensis **(Lamarck, 1822)**   **88**
Cayenne Keyhole Limpet

Maryland to south half of Florida and to Brazil; Bermuda.

1 to 2 inches in maximum diameter. Orifice just in front of and slightly lower than the apex. Many radial ribs with each fourth one larger. Color variable from whitish, pinkish to dark-gray. Interior white or bluish gray. Just behind the callus of the orifice on the inside there is a deep pit. *D. listeri* is much more coarsely sculptured. A common intertidal to moderately deep water species. *D. alternata* Say, 1822; *fumata* Reeve, 1850; *larva* Reeve, 1850; and *viminea* Reeve, 1850, are synonyms.

Diodora listeri **(Orbigny, 1842)**   **89**
Lister's Keyhole Limpet

South half of Florida to Brazil; Bermuda.

1 to 2 inches in maximum diameter. Similar to *D. cayenensis* but differs in that: (1) every second radial rib is larger; (2) concentric threads are more distinct, and by crossing the ribs, form little squares; (3) radial ribs often have nodules or scales. Color usually white, cream or gray, sometimes with obscure radial bands. Intertidal. Common in the West Indies.

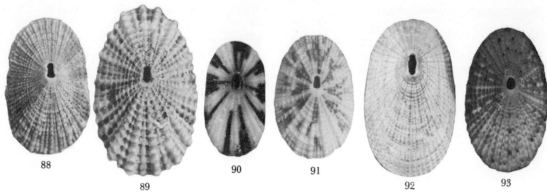

88   89   90   91   92   93

Diodora minuta (Lamarck, 1822)   90
Dwarf Keyhole Limpet

Southeast Florida and the West Indies to Brazil.

½ inch in maximum diameter, rather thin, depressed. Apex at anterior third of shell. Base elliptical, raised slightly at the center, so that the shell rests on its ends. Short front slope slightly concave, back slope convex. Orifice narrow and trilobated. Exterior shiny, with numerous, finely beaded radial ribs. Color white, with many of the ribs entirely or partly blackened. Margin very finely crenulate. Internal callus around hole frequently bounded by a black line. Not very common. Occurs in rocky areas, 1 to 72 fathoms, but has been picked up on beaches. Do not confuse with *D. dysoni* which has a black orifice. *D. elongata* C. B. Adams, 1845; *gemmulata* Reeve, 1850; and *variegata* Sowerby, 1862, are synonyms.

Diodora dysoni (Reeve, 1850)   91
Dysons' Keyhole Limpet

Florida, the Bahamas and West Indies to Brazil; Bermuda.

½ to ¾ inch in maximum diameter, depressed and with straight sides. Base ovate. Apex slightly in front of the middle and characterized by a blunt knob situated behind the posterior wall of the small, almost triangular, black orifice. Sculpture of 18 strong ribs with three smaller ones between, and with numerous concentric lamellae. Color milky-white or cream with 8 solid, broken or dotted black rays. Margin sharply crenulated with the denticles arranged in groups of four. Distinguished from *cayenensis* by the shape of the orifice. Moderately common on reefs under rocks; sometimes washed ashore. *D. microsticta* Dall, 1927, is a synonym.

Diodora sayi (Dall, 1899)   92
Say's Keyhole Limpet

Southeast Florida to Brazil.

½ to 1 inch, similar to *cayenensis*, but the apex is nearer the front end, the posterior slope is long and convex, the ribs are all about equal in size, and it lives at depths from 12 to 220 fathoms. Color uniformly white, cream or faintly olive. In the latter case there are 7 slightly darker rays, 3 on each side and 1 in front. Common.

Diodora jaumei Aguayo and Rehder, 1936   93
Jaumé's Keyhole Limpet

South Florida to the Lesser Antilles.

¾ inch, somewhat low, prettily sculptured with radial ribs of alternating size and beaded with squamose nodules. Apex a little in front of the middle. Slopes very slightly convex. Color white to buff with brown freckles. Resembles young *listeri*, but the "keyhole" is ovate in *jaumei* and there is no black line around the hole on the inner side. Uncommon, on rocks from low tide to 220 fathoms.

Diodora meta (von Ihering, 1927)   94
Meta Keyhole Limpet

South half of Florida; Brazil.

½ inch, oval-elongate, low-spired, pure-white throughout; "keyhole" 1 mm., almost circular and nearer the front end. Sculpture of 37 to 39 radial beaded ribs, between which is a single fine riblet. Concentric threads form nodules or scales on ribs. Uncommon; intertidal to 60 fathoms.

Diodora viridula (Lamarck, 1822)   95
Green Keyhole Limpet

Florida Keys to the lower Caribbean.

1 inch, narrow at the front end, alternating white and green rays on 18 to 20 main ribs. "Keyhole" long, narrow, trilobate and stained black. Interior polished bluish or greenish gray with white radial rays. Attached to intertidal rocks; common in Greater Antilles. *D. bicolor* Pilsbry, 1891, is a synonym.

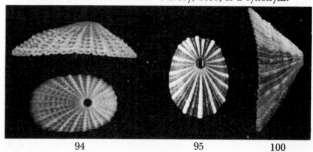

94   95   100

Other Atlantic species:

96 *Diodora fragilis* Farfante and Henriquez, 1947 (synonym: *D. delicata* F. and H., 1946, not E. A. Smith, 1889). Dredged near Havana, Cuba. See *Johnsonia*, vol. 3, no. 39, pl. 179.

97 *Diodora patagonica* Orbigny, 1847 (synonym: *D. metcalfii* Reeve, 1850). Trinidad to Argentina.

98 *Diodora harrassowitzi* von Ihering, 1927. Santa Catharina, Brazil. See *Johnsonia*, vol. 1, no. 11, pl. 3.

99 *Diodora bermudensis* (Dall and Bartsch, 1911). Bermuda. Off West Florida, 25 fms. See *Johnsonia*, vol. 1, no. 11, pl. 3.

100 *Diodora aguayoi* Farfante, 1943. Bermuda; Cuba to Barbados, 80 to 450 fms. *Johnsonia*, vol. 1, no. 11.

101 *Diodora arcuata* Sowerby, 1862. Off Palm Beach, 12 to 14 fms., to Lesser Antilles. See *Johnsonia*, vol. 1, no. 11, pl. 5.

102 *Diodora fluviana* Dall, 1889. Off Palm Beach, 100 fms., to Lesser Antilles, 170 fms. West Florida, 26 fms. See *Johnsonia*, vol. 1, no. 11, pl. 6.

103 *Diodora tanneri* Verrill, 1883. Delaware Bay to Barbados, 104 to 399 fms. West Florida, 100 fms. 2 inches, finely ribbed. See *Johnsonia*, vol. 1, no. 11, pl. 6. 35 mm.

104 *Diodora wetmorei* Farfante, 1945. 50 to 116 fathoms, off Lower Florida Keys to Curaçao. *Johnsonia*, vol. 1, no. 18, pl. 4.

Diodora aspera (Rathke, in Eschscholtz, 1833)   **105**
Rough Keyhole Limpet   **Color Plate 2**

Cook's Inlet, Alaska, to Magdalena Bay, Mexico.

1½ to 2½ inches in maximum diameter, slightly less than ⅓ as high. The roundish to slightly oval, flat-sided apical hole is 1/11 the length of the shell and about ⅓ back from the narrow, anterior end of the shell. Sculpture of coarse radial and weaker concentric threads. Color externally is grayish white with about 12 to 18 irregularly sized, purplish blue, radial color bands. Commonly found clinging to rocks at low tide. In the south, dredged no deeper than 20 fathoms, and often found on the stems of kelp. Feeds on bryozoans. *D. murina* Arnold, 1903, is a synonym.

Diodora arnoldi J. McLean, 1966   **106**

Neat-ribbed Keyhole Limpet

Crescent City, California, to San Martin Island, Mexico.

¾ inch in length, similar to *aspera,* but smaller, with a lower, more rounded apex, with convex sides, a narrower shell, and with finer, much neater cancellate sculpturing. Color white or with few, or many, broken radial rays of gray-black. The apical hole is nearer the anterior end. Moderately common on sublittoral rocks. This is *D. densiclathrata* of authors, not of Reeve. McLean (1966) has shown that *murina* Arnold, 1903, is actually a synonym of *aspera* Rathke, in Eschscholtz.

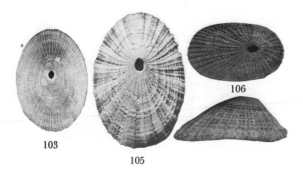

103
106
105

Other Pacific species:

107 *Diodora alta* (C. B. Adams, 1852). Baja California to Panama; Galapagos. 13 × 10 mm.

108 *Diodora crenifera* (Sowerby, 1835). Pacific coast of Central America. Now considered to be a *Lucapinella* by Keen, 1971.

109 *Diodora inaequalis* (Sowerby, 1835). (Synonym: *Rimula mazatlanica* Carpenter, 1857; *pluridenta* Mabille, 1895). Baja California to Panama; Galapagos.

110 *Diodora panamensis* (Sowerby, 1835). Panama, 6 to 10 fms.

111 *Diodora saturnalis* (Carpenter, 1864). Baja California to Ecuador. See Myra Keen for figs. of above 4 species.

112 *Diodora fusilla* S. S. Berry, 1959. Head of the Gulf of California, on gravel bottoms, 3 to 12 fathoms. 5 mm.

113 *Diodora* (*Stromboli* Berry, 1954) *beebei* (Hertlein and Strong, 1951). 40 to 100 fms., Gorda Banks, Gulf of California. Rare.

Genus *Megathura* Pilsbry, 1890

Shell large, perforation large. Type: *crenulata* Sowerby, 1825. Synonym: *Macrochasma* Dall, 1915.

Megathura crenulata (Sowerby, 1825)   **114**
Great Keyhole Limpet   **Color Plate 2**

Monterey, California, to Baja California.

2½ to 4 inches in length, ⅕ as high. Apical hole large, with rounded sides, ⅙ the length of the shell, and bordered externally by a white margin. Interior glossy-white. Basal edge finely crenulate. Exterior finely beaded and light mauve-brown. Animal much larger than the shell, with a massive, yellow foot and a black or brown mantle that nearly covers the entire shell. Common in many low-tide, rocky areas, such as breakwaters.

Genus *Lucapina* Sowerby, 1835

Shell thin, low-conic, with the apex in front of the middle. Orifice rather large, roundish. Margin finely crenulated. Fleshy mantle covers most of the shell; foot larger than shell. Type: *sowerbii* (Sowerby, 1835).

Lucapina sowerbii (Sowerby, 1835)   **115**
Sowerby's Fleshy Limpet

Southeast Florida and the West Indies to Brazil.

¾ inch in length, oblong in outline. With about 60 alternating large and small radiating ribs. Also with 9 to 13 raised, concentric threads. Color white to buff, with 7 to 9 small, splotched rays of pale-brown. Inside whitish; callus sometimes bounded by an olive-green streak. Outside of orifice not stained. Fairly common under rocks at low-tide zone to offshore. Usually occurs in pairs. Synonyms are *adspersa* Philippi, 1845, and *lentiginosa* Reeve, 1850.

Lucapina suffusa (Reeve, 1850)   **116**
Cancellate Fleshy Limpet

South half of Florida and the West Indies to Brazil. Bermuda.

1 to 1½ inches in length, oblong in outline. Much like *L. sowerbii,* but larger, a delicate mauve to pinkish, and with a bluish black orifice. Inside grayish to dirty-white. Cream to orange mantle covers entire shell. Not uncommon under rocks. Formerly called *L. cancellata* Sowerby, 1862, not his 1835. Farfante (1945) described a subspecies, *tobagoensis,* from Tobago, Lesser Antilles, in which the radial rays are 9- or 10-dotted.

Lucapina aegis (Reeve, 1850)   **117**
Aegis Fleshy Limpet

Florida, Bahamas and Cuba to Brazil.

1 to 1⅓ inches, base oblong, anterior slope nearly straight, posterior slope convex. The "keyhole" is narrow, oblong, contracted in the central portion, its sides produced upward in little points on either margin, and its length ⅙ to ⅐ that of the shell. Between each of the 40 radial ribs is an intermediate tiny riblet. Concentric lamellae numerous. Color olive-green with 8 darker broken rays. Mantle wine-red. Uncommon, intertidal to 10 fathoms.

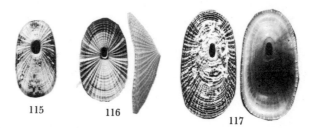

115   116   117

Lucapina philippiana (Finlay, 1930)   **118**
Philippi's Fleshy Limpet

Off southeast Florida to the Virgin Islands.

15 mm., elongate, narrow, with a large, oblong "keyhole" located in the anterior ⅓ of the shell. Finely sculptured with

about 38 beaded riblets of uneven sizes. Color cream or white and freckled with rusty-brown. Internal callus white, truncate behind. Uncommon; shore to 60 fathoms.

Lucapina eolis Farfante, 1945   119
Eolis Fleshy Limpet

Off Lower Florida Keys and Gulf of Mexico.

12 to 20 mm.; thin-shelled, depressed, well-sculptured. Similar to *aegis*, but in *eolis* the "keyhole" is smaller, narrower, widened in the middle, and placed nearer the front of the shell. Color always all-white. The internal callus of the "keyhole" is sharply, rather than roundly truncated behind. Anteriorly it is rounded, quite unlike the triangular effect of *aegis*. Rare; 27 to 90 fathoms.

118   119

Genus *Lucapinella* Pilsbry, 1890

Shell depressed, conical, less than ¾ inch, with a large orifice and thickened margins. Type: *callomarginata* (Dall, 1871).

Lucapinella limatula (Reeve, 1850)   120
File Fleshy Limpet

North Carolina to Florida and the West Indies. Brazil.

⅓ inch in length, resembling *Lucapina sowerbii*, but smaller with a proportionately larger apical hole which is sharp at its top edge and which is nearer the center of the shell. The ends of the shell are slightly turned up and the sides are slightly concave. Sculpture of about 2 dozen heavily scaled, radial ribs and numerous, fine, threadlike concentric ridges. Color whitish with weak mauve or brown discoloring. Commonly found off south Florida, from intertidal to 80 fathoms. *L. aculeata* (Reeve, 1850), is a synonym.

The subspecies (121) *hassleri* Farfante, 1943, from 17 to 30 fathoms off Argentina, is thicker-shelled, more elevated, and the "keyhole" is elongate-oval instead of somewhat triangular. Uncommon.

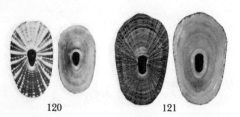

120   121

Lucapinella callomarginata (Dall, 1871)   122
Hard-edged Fleshy Limpet

Morro Bay, California, to Nicaragua.

¾ to 1 inch in length, narrower at the anterior end, quite flat. Base flat and usually with strong crenulations on the under edge. Sides slightly concave. Apical hole narrowly elongate, its length 2 times its width, slightly nearer the anterior end, about ⅕ the length of the shell and with flat inner sides. Sculpture coarsely cancellate with the radial ribs stronger and often scaled. Color dark-gray with irregular, darker, radial color-rays. Rather rare, under rocks in the intertidal zone. Feed upon the sponge, *Tetilla mutabilis* (R. L. Miller, 1968, *The Veliger*, vol. 11, p. 130).

Lucapinella eleanorae J. H. McLean, 1967   123
Eleanor's Fleshy Limpet

Guaymas, Mexico, to Ecuador.

14 to 18 mm. (¾ inch), reddish buff with radial bands of gray. 20 primary, 20 secondary and 40 tertiary imbricated ribs. Keyhole relatively small and oval. Uncommon; on rocks, 10 to 20 fathoms. *The Veliger*, vol. 9, p. 350

Lucapinella milleri S. S. Berry, 1959   124
Miller's Fleshy Limpet

Gulf of California to Mazatlan, Mexico.

6 to 9 mm., characterized by large keyhole (¼ the length of the shell); nearly parallel sides to the shell; posterior slope slightly concave; radial sculpture of wide primary and secondary imbricated ribs. Color white, gray to reddish with darker rays. *Leaflets in Malacology*, vol. 1, no. 8.

Other species:

125 *Lucapinella aequalis* (Sowerby, 1835). Mexico to Ecuador. See *The Veliger*, vol. 10, pl. 49. Ribs smoothish. 1 inch.

126 *Lucapinella henseli* (von Martens, 1900). Brazil to Argentina.

Genus *Megatebennus* Pilsbry, 1890

Type of the genus is *bimaculatus* (Dall, 1871).

Megatebennus bimaculatus (Dall, 1871)   127
Two-Spotted Keyhole Limpet

Alaska to Tres Marias Islands, Mexico.

¾ to ⅝ inch in length, low, with ends turned slightly up. Apical hole elongate-oval, located at the center of the shell and about ⅓ the length of the shell. Numerous radial and con-

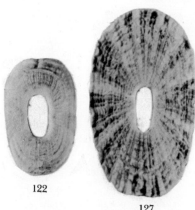

122

127

centric threads give a fine cancellate sculpturing. Color dark-gray to light-brown with a wide, darker ray on each side of the hole, and occasionally at each end. Interior white to gray-ish. Animal several times as large as the shell, variable in color—red, yellow or white. Common under stones at low tide.

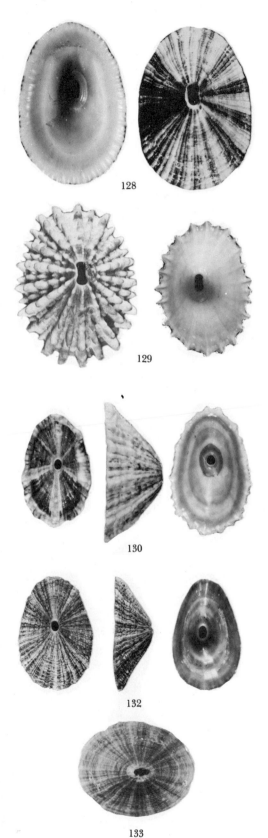

128

129

130

132

133

Subfamily FISSURELLINAE Fleming, 1822

Genus *Fissurella* Bruguière, 1789

Keyhole limpets with a swollen ridge or callus bordering the "keyhole" on the inner side of the shell. This callus is about the same width all around, whereas in the genus *Diodora* the callus is squared off or truncated in back. The central tooth of the radula is narrow. No operculum in this family. Type: *nimbosa* Linné, 1758.

Fissurella nimbosa (Linné, 1758)   **128**
Rayed Keyhole Limpet

Puerto Rico to Brazil.

1 to 2 inches, "keyhole" relatively large, oblong with its sides produced upward in little points on either side. Sculpture of numerous, irregular radiating grooves separating low ribs and crossed by fine growth lines. Color buff with 11 or 12 rays of purplish brown or reddish. Moderately common; intertidal rocks of the lower Caribbean. *F. balanoides* Reeve, 1850, is a synonym. Do not confuse with *Diodora viridula* whose "key-hole" is stained black.

Subgenus *Cremides* H. and A. Adams, 1854

Type of the subgenus is *alabastites* Reeve, 1849.

Fissurella nodosa (Born, 1778)   **129**
Knobby Keyhole Limpet

Lower Florida Keys (?) and the West Indies.

1 to 1½ inches in length. 20 to 22 strongly nodulated, radial ribs. Margin sharply crenulated. Interior pure-white. Orifice oblong. An intertidal rock-dweller. Rare in Florida (dead only) and the Bahamas; abundant in the West Indies. *Patella rudis* Röding, 1798 (and possibly *crusoe* Farfante, 1943), is a synonym.

Fissurella barbadensis (Gmelin, 1791)   **130**
Barbados Keyhole Limpet   **Color Plate 2**

Southeast Florida, Bermuda and the West Indies to Brazil.

1 to 1½ inches in length. With irregular radiating ribs. Orifice almost round. Inside with green and whitish concentric bands. Border of orifice deep-green, rarely with a reddish brown line. Outside grayish white to pinkish buff, generally with purplish lines between the small ribs. Commonly blotched with purple-brown. Lives on wave-dashed rocks. Rare in Florida. Common elsewhere. The synonyms are *porphyrozonias* Gmelin, 1791; *edititia* Reeve, 1849; *antillarum* Orbigny, 1842; *intensa* and *bermudensis* Pilsbry, 1890. For breeding habits see J. Ward, 1966, *Bull. Marine Sci.*, vol. 16, p. 685.

A similar species, (131) *F. angusta* (Gmelin, 1791), also inter-tidal and frequently covered with calcareous algae, occurs on the Florida Keys. The shell is flattish, pointed in front, and its internal callus is light-brown to reddish brown, but not bounded by a reddish line as in *barbadensis*. *F. schrammii* Fischer, 1857, is a synonym.

Fissurella rosea (Gmelin, 1791)   **132**
Rosy Keyhole Limpet

Southeast Florida and the West Indies to Brazil.

1 inch in length, thin, flattish, narrower at the anterior end. Orifice slightly oblong. Many radiating, small, rounded riblets. Color of alternating whitish to pale-straw and pinkish rays. Interior pale-green at the margins, blending to white in the center. Green orifice callus bordered by a pinkish line. **Rare**

in Florida. Common in beach drift elsewhere. Do not confuse with the larger, more elevated *F. barbadensis.* The synonyms are *radiata* Lamarck, 1822, var. *sculpta* Pilsbry, 1890, and *Lucapina itapema* von Ihering, 1927.

Fissurella volcano Reeve, 1849   133
Volcano Limpet

Crescent City, California, to Baja California.

¾ to 1 inch in length, ⅓ to almost ½ as high. Orifice at the very top, very slightly nearer the somewhat narrower anterior end, and elongate with deep, flat inner sides. Sculpture of numerous rather large, but low and rounded, radial ribs of varying sizes. Base of shell slightly crenulate and with color blotches. Exterior grayish white to dark-slate with numerous radial rays of mauve-pink. Interior glossy-white, often with a fine pink line around the callus at the apex. Foot of animal yellow; mantle with red stripes. Very common on rocky rubble at low tide. The variety *crucifera* Dall, 1908, is merely a color form with white radial bands. *Hemitoma golischae* Dall, 1916, is a synonym.

Fissurella rubropicta Pilsbry, 1890   134
Red-painted Keyhole Limpet

Baja California to Oaxaca, Mexico.

20 to 30 mm., oval-oblong, moderately high, sculpture and shape variable. Radial riblets uneven. Characterized by a dozen or so black to reddish radial short bars on a greenish gray exterior. Interior whitish green with a splotch of bright-pink. Callus around hole is greenish. Common; on intertidal rocks.

Other tropical Pacific species:

135 *Fissurella microtrema* Sowerby, 1835. Baja California to Ecuador. 20 to 30 mm. Exterior red to white; interior pale-green, darker around callus. (Synonyms: *chlorotrema* and *humilis* Menke, 1847; *rugosa* of authors, not of Sowerby, 1835.)

Subgenus *Clypidella* Swainson, 1840

Orifice widened in middle; shell ends raised; internal callus oval and white. Type: *punctata* Fischer, 1857.

Fissurella fascicularis Lamarck, 1822   136
Wobbly Keyhole Limpet

Southeast Florida and the West Indies.

¾ to 1½ inches in length. Both ends turned up (can be rocked back and forth on a flat table). The orifice towards the anterior end is keyhole or crosslike in shape. Color a faded magenta. Interior whitish, tinged with pale-green or pink. Inner callus of orifice white with a narrow red line. Uncommon in Florida. Moderately common elsewhere in small pot-holes in the intertidal zone.

Fissurella punctata Fischer, 1857   137
Punctate Keyhole Limpet

Off North Carolina to West Indies.

About 1 inch, similar to *fascicularis* but squared off anteriorly, with the "keyhole" much nearer the front end and usually with more (50 to 57) radial, rounded ribs. Uncommon.

Other species:

138 *Fissurella (Cremides) barbouri* Farfante, 1943. Bahamas to the Lesser Antilles. *Johnsonia,* vol. 1, no. 10, pl. 2.

139 *Fissurella (Cremides) clenchi* Farfante, 1943. Brazil to French Guiana. *Johnsonia,* vol. 1, no. 10, pl. 3.

136   137   138

140 *Fissurella (Clypidella) subrostrata* Sowerby, 1835. Lesser Antilles. Rare. See *Johnsonia,* vol. 1, no. 11, pl. 6.

Genus *Leurolepas* J. H. McLean, 1970

Shell oblong in outline, lacking radial sculpture, having a large oblong foramen. Margin of shell rests on one plane. Animal not capable of retraction within the confines of the shell. Rachidian tooth small and similar to the lateral teeth. Do not confuse with *Megatebennus,* which has a broad rachidian radular tooth. Type: *roseola* J. H. McLean, 1970.

Leurolepas roseola J. H. McLean, 1970   141
Smooth Keyhole Limpet

Tres Marias Islands, west Mexico.

10 to 12 mm., thin but sturdy. Foramen ¼ the length of the shell and forward of center. Radial sculpture absent, concentric growth lines thin and raised. Pink with fine white flecking and several broad lateral bands of brown. Internal callus around foramen is rounded and distinct. Not uncommon on rocky bottoms in 5 to 40 fathoms. (*The Veliger,* vol. 12, no. 3, p. 366.)

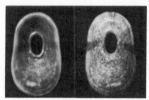

141

Suborder *Patellina* von Ihering, 1876
Superfamily Patellacea Rafinesque, 1815
Family PATELLIDAE Rafinesque, 1815

Limpetlike, flat snails living in intertidal rock area. Without feather-shaped gills, having instead a series of pronglike structures along the edge of the mantle. The genus *Patina* Gray, 1840, is absent from North America, although the common blue-rayed limpet of Europe (*Patina pellucida* Linné, 1758) has been found in ballast sand dumped at Newport, Rhode Island.

Genus *Patella* Linné, 1758

Members of the typical group of *Patella,* in the strict sense, are absent from North American waters. The gill projections competely encircle the body along the inner mantle edge. The interior of the shell is weakly iridescent. Type: *vulgata* Linné, 1758.

Subgenus *Ancistromesus* Dall, 1871

Shells are large, porcelaneous and without iridescence. Only one species is present in the Eastern Pacific. Type: *mexicana* B. and S., 1829.

Patella mexicana Broderip and Sowerby, 1829   **142**
Giant Mexican Limpet

Gulf of California to Peru.

Shell reaching 14 inches in length, heavy, porcelaneous white with numerous, irregular, small axial ribs. Soft parts black with white mottlings. Moderately common, but now seldom found over 6 inches in length. This is the largest living limpet in the world. *P. maxima* Orbigny, 1841, is a synonym.

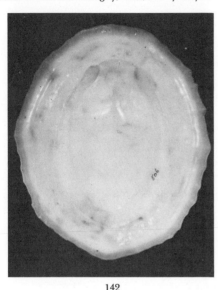

142

Family ACMAEIDAE Carpenter, 1857

Shells limpetlike, not nacreous; one finger-shaped **gill** plume present at the anterior end. Muscle scar horseshoe-shaped, open in front. Lottiidae Gray, 1840, is a *nomen oblitum.*

Genus *Lottia* Gray, 1833

Large, low limpetlike shells with the apex at one end; muscle impressions joined by a curved line. Type: *gigantea* (Sowerby, 1834).

Lottia gigantea (Sowerby, 1834)   **143**
Giant Owl Limpet

Crescent City, California, to Baja California.

3 to 4 inches in maximum diameter, oval in outline, low, with the apex close to the front end. Exterior dirty-brown, rough, commonly stained with algal green. Interior glossy, with a wide, dark-brown border. Center bluish with an "owl-shaped" whitish to brownish scar in the very center. Very common at or above high-tide line where the sea spray may reach them. In the south they grow to a large size. Frequently polished and used as souvenirs. A white-maculated color form has the synonymic name *albomaculata* Dall, 1910.

143

Genus *Scurria* Gray, 1847

Similar to *Lottia* in having a single anterior ctenidium, but differs in having the branchial, leaflike cordon continuous around the inner margin of the mantle. In *Lottia* the cordon is absent in the head area. The group is absent on the Atlantic coast, and only one species is present in the southwest. Type: *scurra* (Lesson, 1831).

Scurria mesoleuca (Menke, 1851)   **144**
Half-white Limpet
Southern Baja California and Mazatlan, Mexico, south to Ecuador.
20 to 35 mm., finely radially ribbed, interior blue-green. Intertidal; common. (Synonyms: *diaphana* Reeve, 1854; *floccata, striata* [Reeve's sp. 99, not 58], and *vespertina* Reeve, 1855).

Genus *Acmaea* Rathke, in Eschscholtz, 1833

Shells limpetlike, oval, apex subcentral, exterior smooth or radially ribbed; muscle scars joined by a thin anterior line. Type: *mitra* Rathke, in Eschscholtz, 1833. In *Acmaea sensu stricto,* the shell is white, the apex central, the proboscis produced at the lower anterior corners into two lappets. Radula without uncini teeth. For the biology of Californian *Acmaea* of the subgenus *Colsella,* see *The Veliger,* vol. 11, supplement for 1968, and vol. 4, p. 41 (1961).

Subgenus *Acmaea* Rathke, in Eschscholtz, 1833

Acmaea mitra Rathke, in Eschscholtz, 1833   **145**
White-cap Limpet

Alaska to Baja California in cold water.

1 inch in maximum diameter, thick, pure-white, conic in shape and with an almost round base. Apex pointed and near

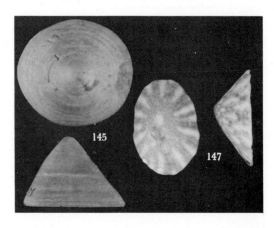

148

149

150

151

152

153

the center. Often covered with small, knobby nullipore growths. Commonly washed ashore. It lives in cold water below the low-tide level.

Acmaea funiculata Carpenter, 1864   146
Corded White Limpet

Shumagin Islands, Alaska, to Baja California.

¾ inch, half as high, white, similar to *mitra,* but usually smaller, with numerous, small, coarse, uneven, radial riblets. Common; found on rocky bottoms well below low-tide mark to 11 fathoms.

Acmaea rosacea Carpenter, 1864   147
Rosy Pacific Limpet

Ketchikan, Alaska, to Baja California.

⅓ inch, almost half as high as long, conical, moderately thin-shelled, apex a little more than ⅓ the way from the front margin. Posterior face convex. Surface smoothish or with fine radial riblets. The young are pinkish with white and brown spots; adults rosy-brown and dotted with white and rose. Interior white, blushed with rose. Common; sublittoral to 20 fathoms on hard bottoms. *A. semirubida* Dall, 1914, found south to Panama, may be a synonym.

Subgenus *Collisella* Dall, 1871

Members of this subgenus have a darkly colored exterior and a colored margin on the inner side. Type: *pelta* Rathke, in Eschscholtz, 1833. Proboscis frill without side lappets. Radula bounded below by a very small uncinid tooth. This is considered a genus by some workers.

Pacific Coast species:

Acmaea pelta Rathke, in Eschscholtz, 1833   148
Shield Limpet

Alaska to Baja California.

1 to 1½ inches in maximum diameter, elliptical in outline, with a moderately high apex which is placed ⅓ to almost ½ way back from the anterior end. With about 25 axial, weakly developed, radial ribs. Edge of shell slightly wavy. External color of strong black radial, often intertwining, stripes on a whitish cream background. Interior usually faint bluish white, with or without a dark-brown spot. Inner border edged with alternating black and cream bars. A common rock-dweller.

Acmaea digitalis Rathke, in Eschscholtz, 1833   149
Fingered Limpet

Aleutian Islands to Socorro Island, Mexico.

1¼ inch in maximum diameter, elliptical in outline; generally with a moderately high apex which is minutely hooked forward and which is placed ⅓ back from the anterior end of the shell. The 15 to 25 moderately developed, coarse, radiating ribs give the edge of the shell a slightly wavy border. Color grayish with tiny, distinct mottlings of white dots and blackish streaks and lines. Inside white with faint bluish tint and with a large, usually even, patch of dark-brown in the center. Edge of shell with a solid or broken, narrow band of black-brown. Common. Do not confuse this species with *A. scabra* which does not have the "hooked-forward" apex and is not glossy on its internal brown patch. Compare also with *persona.* Non *digitale* Röding, 1798, a *nomen oblitum.* The next available name would be *textilis* Gould, 1846, or *umbonata* Reeve, 1855. Prefers vertical rock surfaces where wave action is severe. A nonhoming species.

Acmaea strigatella (Carpenter, 1864)   150
Strigate Limpet

Vancouver, British Columbia, to northwest Baja California.

13 to 16 mm., smoothish, subovate in outline, 3 times as long as high; anterior end slightly narrower. Apex about ⅓ back from the anterior end. Surface tessellated with brown and white, with brown and white lines radiating from the apex which is sometimes eroded. Internal color bluish white to brownish, with the brown radial lines showing through near the margins. Animal white. Differs from its rock companion, *digitalis,* in having no ribbing and lacking the brown owl-shaped apical spot inside. Common; upper tide line. For radular details, see Fritchman, 1960, *The Veliger,* vol. 2, p. 53. *A. paradigitalis* Fritchman, 1960, is a synonym.

Acmaea scabra Gould, 1846   151
Rough Limpet

Southern Oregon to Baja California.

1¼ inch in maximum diameter, elliptical in outline, generally with a low apex which is placed ⅓ back from the front end. The 15 to 25 strong, coarse radiating ribs give the edge of the shell a strong crenulation. Color dirty gray-green. Underside of shell whitish, irregularly stained in the center with blackish brown. Edge of shell between the serrations is stained blackish to purplish brown. A common species found clinging to gently sloping rock surfaces high above the water line but within reach of the ocean spray. *A. spectrum* Nuttall is the same species. Do not confuse with the smaller *A. conus* which is evenly glossed, instead of coarse and dull, on its interior center. British Columbia and Washington records not confirmed. For biology, see S. B. Haven, 1971, *The Veliger,* vol. 13, p. 231. A homing species.

Acmaea conus Test, 1945   152
Test's Limpet

Point Conception, California, to Baja California.

¾ inch in maximum diameter. Shell low, and like *A. scabra,* is with distinct but widely spaced, radial ribs. Distinguished from *scabra* by its glossy, smooth interior which often has an evenly colored brown center. *A. scabra* has a rough interior center and the brown stain looks smeared. However, this species may be a form of *scabra.* It is very abundant south of La Jolla and is found with *A. scabra* and *A. digitalis.*

Acmaea limatula Carpenter, 1864   153
File Limpet

Puget Sound to Baja California.

1 to 1¾ inches in maximum diameter, elliptical to almost round in outline, low to quite flat. Characterized by radial rows of small beads which sometimes may be crowded together to form tiny, rough riblets. Exterior greenish black. Interior glossy-white, younger specimens having a blue tint. Patch of brown on inside generally weak or absent. Edge of shell usually with solid, black-brown, narrow band. Occasional albinos are cream-brown or tan on the outside. Compare with *A. scutum* which is smooth and has a barred band of color on its under edge.

Acmaea asmi Middendorff, 1849   154
Black Limpet

Alaska to Mexico, clinging to the gastropod, *Tegula.*

¼ inch in maximum diameter, high-conic, elliptical in outline, and solid black inside and out. In the northern part of its range, the black limpet is found living attached to the common snails, *Tegula funebralis* and *gallina.*

154

Acmaea triangularis (Carpenter, 1864)   155
Triangular Limpet

Southern California to Gulf of California.

¼ inch in maximum diameter, oblong in outline, side view distinctly triangular. Color whitish with 3 or 4 vertical, rather broad, brown stripes on each side. Found among coralline algae from the shore line down to several fathoms. Uncommon.

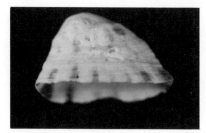

155

Acmaea fenestrata Reeve, 1855   156
Fenestrate Limpet

Alaska to Baja California.

1 to 1½ inches in maximum diameter, almost round in outline, rather high, and smoothish. The northern subspecies, *cribraria* Carpenter, 1866, found from Alaska to northern California, has interior with various shades of glossy, chocolate-brown, and with a narrow, solid black border. The exterior is plain dark-gray. The typical southern *fenestrata* Reeve, found from Point Conception south, has an external color pattern of regular dottings of cream on a gray-green background. Its interior has a small, brown apical spot surrounded by a bluish area and bordered at the margin of the shell with brown. Integrades occur near Point Conception. This species is the only Pacific *Acmaea* which lives among loose boulders that are set in sand. It only feeds when submerged. Common.

This species has been placed in a doubtfully valid subgenus *Notoacmea* Iredale, 1915, whose type is *pileopsis* Quoy and Gaimard, 1834, a New Zealand species.

Acmaea persona Rathke, in Eschscholtz, 1833   157
Mask Limpet

Aleutian Islands to Monterey, California.

1 to 1¾ inches in maximum diameter, with characters much the same as those of *digitalis,* but differing in being smoothish, larger, often slightly higher, and in having a strong tint of blue or blue-black inside. I am inclined to believe that Pilsbry is correct in considering *digitalis* a smaller, ribbed form of *persona,* despite the fact that recent workers place these two species in different subgenera. It is possible that colder waters allow the smooth *persona* form to express itself. The mask limpet is very common from Monterey north. It is an intertidal dweller where strong waves flush the shaded rock crevices. It feeds mostly during the ebb tide and is more active during dark hours.

Compare with the more southerly species (no. 150), *A. strigatella* (Carpenter).

Acmaea testudinalis scutum Rathke, in Eschscholtz, 1833   158
Pacific Plate Limpet

Alaska to Oregon (common) to Baja California (rare).

1 to 2 inches in maximum diameter, almost round in outline, quite flat, with the apex toward the center of the shell. Smoothish, except for very fine radial riblets in young specimens. External color greenish gray with slate-gray radial bands or mottlings. Interior bluish white with faint or darkish brown spot. Inner edge with band of alternating bars of black or brown and bluish white. This species was also known as *tessulata* Müller. The typical *testudinalis* from the Arctic seas and New England rarely, if ever, exceeds a size of 1½ inches, is not so round, and has a darker, more concentrated brown patch on the inside. Intergrades exist in Alaskan waters. The Pacific race was also named *patina* Rathke. *A. ochracea* (Dall, 1871) **(159)** is a solid, yellowish brown form, but McLean claims it as a good species.

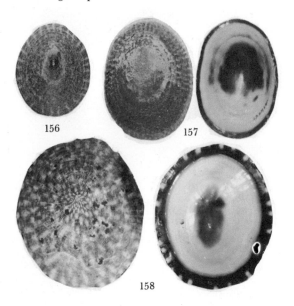

156   157

158

Acmaea insessa (Hinds, 1842)   160
Seaweed Limpet

Alaska to Baja California.

½ to ¾ inch in maximum diameter, narrowly elliptical, with a high apex, and colored a uniform, greasy light-brown. Abundant on the stalks or holdfasts of the large seaweeds, such as *Egregia*.

Acmaea paleacea Gould, 1853   161
Chaffy Limpet

Vancouver, B.C. to Baja California.

¼ inch in maximum diameter, very fragile, translucent-brown. 3 or 4 times as long as wide. Sides straight with fine, raised radial threads. Abundant on the narrow-leaved eel-grass of the open coast.

Acmaea depicta (Hinds, 1842)   162
Painted Limpet

Santa Barbara, California, to Baja California.

160   163

½ inch in maximum diameter, very narrow, 3 times as long as wide. Sides straight with brown vertical stripes on a whitish background. Smoothish. This species is found on the broad-leaved eel-grass of the estuaries. Abundant in certain localities, such as Mission Bay. *Acmaea gabatella* S. S. Berry, 1960, 10 fathoms off Reef Point, Orange County, California, may be an aberrant specimen of this species.

Acmaea instabilis (Gould, 1846)   163
Unstable Limpet

Alaska to San Diego, California.

1 to 1¼ inches in maximum diameter, oblong with a rather high apex. Sides compressed. Lower edge curved so that the shell rocks back and forth if put on a flat surface. Exterior dull, light-brown. Interior whitish with faint brown stain in the center and with a narrow, solid border of brown. Inhabits the stems or holdfasts of large seaweeds. Moderately common.

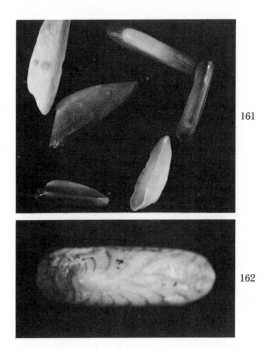

161

162

Atlantic Coast species (subgenus *Collisella*):

Acmaea testudinalis testudinalis (Müller, 1776)   164
Atlantic Plate Limpet

Arctic Seas to Long Island Sound, New York.

1 to 1½ inches in maximum diameter, oval in outline, moderately high with the apex nearly at the center of the shell. Smoothish except for a few coarse growth lines and numerous, very fine axial threads. Interior bluish white with a dark- to light-brown center and with short, radial brown bars at the edge. Exterior dull cream-gray with irregular

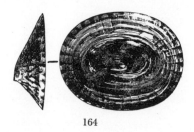

164

axial bars and streaks of brown. A common littoral species in New England. Formerly referred to as *A. tessulata* Müller, 1776. The form *alveus* Conrad, 1831 (**165**) is a thin, elongate, heavily mottled ecological variant which lives on eel-grass. *A. fergusoni* Wheat, 1901, is a synonym of this species.

Acmaea antillarum (Sowerby, 1831)   **166**
Antillean Limpet

Lower Florida Keys (rare) and the West Indies.

⅓ to 1 inch in maximum diameter, usually very flat, rather thin, oval in outline but narrower at the anterior end. Neatly sculptured with numerous radial threads. Color variable: exterior whitish with a few or many narrow or wide radial rays of brownish green. Interior glossy whitish with dark- or light-brown callus. Borders or sometimes the entire inside marked by numerous radial lines of purple-brown. These are often divided near the edge of the shell. Rarely recorded from Florida, but abundant in the West Indies on shore rocks. *A. candeana* Orbigny, 1845, and *A. tenera* C. B. Adams, 1845, are the same.

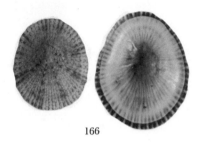

166

Acmaea pustulata (Helbling, 1779)   **167**
Spotted Limpet

Southeast Florida, the West Indies and Bermuda.

½ to 1 inch in maximum diameter, oval in outline, moderately flat with a pointed apex and rounded sides. Shell thick, with coarse axial ribs which are crossed by fine concentric threads. Interior glossy-white, with the central callus yellowish. Exterior chalk-white, dull. Sometimes flecked with red-brown dots and bars. Common. Formerly known as *punctulata* Gmelin, 1791. The small form living on turtle grass (**168**) is very thin, light-rose in color (or opaque-white with two brown side rays), with a tiny, sharp apex and is occasionally flecked with red (forma *pulcherrima* "Guilding" Petit, 1856).

This species may well belong to the subgenus or genus *Patelloida* Quoy and Gaimard, 1834. Its Gulf of California to Panama counterpart is (**169**) *Acmaea (Patelloida) semirubida* Dall, 1914.

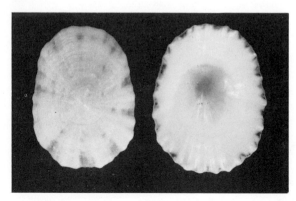

167

Acmaea leucopleura (Gmelin, 1791)   **170**
Black-ribbed Limpet

Florida Keys and the West Indies.

½ inch in maximum diameter, moderately high, with roundish sides, thick, with about 15 to 20 rather large, rounded, white radial ribs on a black-brown background. Sometimes completely white. Interior white, occasionally with a black-spotted edge and with a thickened central callus which is light-brown to black. *A. albicosta* C. B. Adams, 1845, *A. fungoides*, Röding, 1798 and *jamaicensis* (Gmelin, 1791) are the same. Moderately common in the West Indies, occasionally found on the Florida Keys.

There is a small ⅓-inch form (**171**) living attached to the underside of large gastropods, such as *Cittarium pica*, that has a higher apex and with numerous, alternating black and white radial riblets. It is common in the Bahamas and West Indies, but not found in Florida. It has been previously called the "dwarf suck-on limpet." Synonyms are *digitale* and *papillaris* Röding, 1798, *cubensis* Reeve, 1855, and possibly *simplex* Pilsbry, 1891.

170

Other Atlantic *Collisella*:

172 *Acmaea subrugosa* Orbigny, 1846. Brazil to Uruguay. (Synonym: *Lottia onychina* Gould, 1852.) See G. Righi, 1966, *Malacologia*, vol. 4, p. 269.

173 *Acmaea marcusi* Righi, 1966. Trinidade Island, south Brazil.

174 *Acmaea noronhensis* E. A. Smith, 1890. Fernando de Noronha Island, Brazil.

Other tropical Pacific *Collisella*:

175 *Acmaea atrata* Carpenter, 1857. Baja California to Acapulco, Mexico. 30 mm.

176 *Acmaea discors* (Philippi, 1849). Cape San Lucas, Baja California, to Panama. 40 mm. (Synonym: *aenigmatica* Mabille, 1895).

177 *Acmaea fascicularis* (Menke, 1851). Cape San Lucas, Baja California, to west Mexico. 20 to 30 mm.

178 *Acmaea mitella* Menke, 1847. Mazatlan, Mexico, to Colombia. 11 mm. (Synonym: *navicula* Reeve, 1854)

179 *Acmaea pediculus* (Philippi, 1846). Gulf of California to Gulf of Tehuantepec, Mexico. 20 to 30 mm. (Synonym: *corrugata* Reeve, 1854.)

180 *Acmaea strongiana* Hertlein, 1958. Gulf of California. *Bull. So. Calif. Acad. Sci.*, vol. 56, pt. 3.

181 *Acmaea turveri* Hertlein and Strong, 1951. Northern part of the Gulf of California. 18 mm. (Synonym: *fayae* Hertlein, 1958)

182 *Acmaea acutapex* S. S. Berry, 1960. Intertidal, head of the Gulf of California to Guaymas, Mexico.

Subgenus *Nomaeopelta* S. S. Berry, 1958

Limpets with thin shells, which have very fine radial threads. The soft parts like *Acmaea* and *Collisella* in lacking a branchial cordon. Cervical gill present. Mantle edge with numerous, tentaclelike, sensory papillae in small pigmented pockets. Type: *dalliana* (Pilsbry, 1891). The Eastern Pacific species are:

183 *Acmaea dalliana* (Pilsbry, 1891). Gulf of California. Uncommon; rocks at mid-tide zone. 20 to 30 mm.

184 *Acmaea stanfordiana* (S. S. Berry, 1957). Sonoran coast of west Mexico. Common; on rocks at mid-tide level. 20 to 30 mm. (Synonyms: *goodmani* S. S. Berry, 1960, and *concreta* S. S. Berry, 1963).

Genus *Pectinodonta* Dall, 1882

Deepsea, small limpets with low, somewhat arched shells. Apex blunt, subcentral. Animal blind. Type: *P. arcuata* Dall, 1882.

185 *Pectinodonta arcuata* Dall, 1882. Off Santa Lucia, Lesser Antilles, deep water. 15 mm.

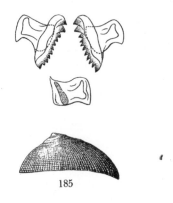

185

Family LEPETIDAE Gray, 1850

Genus *Lepeta* Gray, 1842

Small, flattish, uncoiled shells which are "hat-shaped," similar to *Acmaea*, but the embryonic nucleus is spiral; the animal has no external gills and the proboscis is produced into a labial process on each side. The radula has à median tooth, which in *Acmaea* is absent.

Lepeta caeca (Müller, 1776)   **186**
Northern Blind Limpet

> Arctic Seas to Cape Cod, Massachusetts.
> Alaska to Vancouver Island, British Columbia. Europe.

¼ to ½ inch in maximum diameter, moderately conic, with straight sides, oval-elongate in outline. Rather fragile, dull-white to brownish externally and with fine, granulose, crowded, radial threads. Interior white or tinged with pink.

186

Apex usually eroded. An uncommon cold-water species dredged in shallow water off New England from 2 to 600 fathoms.

Other species:

Subgenus *Cryptobranchia* Middendorff, 1851 (*Cryptoctenidia* Dall, 1918 is a synonym):

187 *Lepeta* (*C.*) *concentrica* Middendorff, 1851. Arctic Ocean to Forrester Island, Alaska.

188 *Lepeta* (*C.*) *caecoides* Carpenter, 1865. Arctic; Aleutians to Farallone Islands, California.

189 *Lepeta alba* (Dall, 1869) (and variety *instabilis* (Dall, 1869)). Bering Sea to Washington.

Genus *Propilidium* Forbes and Hanley, 1849

Small, white, cap-shaped, deepsea shells with a cancellate sculpture, a subcentral, coiled, forward-pointing apex, and with a tiny, triangular septum inside under the apex. Tentacles long, 2 short gills, filamented mantle edge and a long radular ribbon. Type: *ancyloides* Forbes, 1840, from Europe. Two deepsea Western Atlantic species:

190 *Propilidium elegans* Verrill, 1884. 1,395 fms., off Chesapeake Bay, Maryland.

191 *Propilidium pertenue* Jeffreys, 1882. 640 fms., off Rhode Island.

Superfamily Cocculinacea Thiele, 1909

Family COCCULINIDAE Dall, 1889

Genus *Cocculina* Dall, 1882

Deepsea, limpetlike, small, white shells with the apex pointing backward. Spiral nucleus usually worn off. Animal blind, with a single gill plume, with the anal opening above and behind the head. Penis extends from the inner side of the right tentacle. Type: *rathbuni* Dall, 1882. The rather rare deepsea species from American waters are:

192 *Cocculina rathbuni* Dall, 1882. Off Martha's Vineyard, Massachusetts, to the West Indies, 100 to 616 fms.

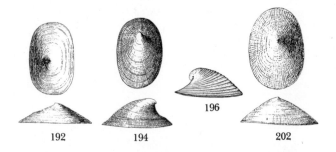

192   194   196   202

193 *Cocculina dalli* Verrill, 1884. Off Delaware Bay, 317 fms.

194 *Cocculina beanii* Dall, 1882. Off Rhode Island to the West Indies, 100 to 583 fms.

195 *Cocculina reticulata* Verrill, 1885. Off Virginia and North Carolina, 70 fms. Off Fernandina, Florida, 294 fms.

196 *Cocculina leptalea* Verrill, 1884. Off New Jersey to Florida, 292 to 2,033 fms.

197 *Cocculina conica* Verrill, 1884. Off Nova Scotia, 499 fms.

198 *Cocculina spinigera* Jeffreys, 1883. Off New Jersey to North Carolina, 335 to 843 fms.

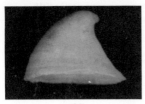

198

199 *Cocculina georgiana* Dall, 1927. Off Georgia, 440 fms.

200 *Cocculina rotunda* Dall, 1927. Off Georgia, 440 fms. Off Fernandina, Florida, 294 fms.

201 *Cocculina lissocona* Dall, 1927. Off Fernandina, Florida, 294 fms. Florida Keys, 68 to 135 fms.

202 *Cocculina portoricensis* Dall and Simpson, 1901. 310 fms., off San Juan Harbor, Puerto Rico.

203 *Cocculina agassizii* Dall, 1908. 140 fms., off Queen Charlotte Islands, British Columbia, to 556 fms., off Panama.

204 *Cocculina casanica* Dall, 1919. Alaska, 95 to 160 fms.

Family LEPETELLIDAE Dall, 1881

Genus *Lepetella* Verrill, 1880

Deepsea, small, smooth, oval, non-spiral shells with an elliptical aperture and a central apex. Soft parts like *Lepeta,* but with eyes. May live in old worm tubes. Type and only American species is **(205)** *tubicola* Verrill and Smith, 1880, from 130 to 388 fathoms off Cape Cod to 324 fathoms in the Gulf of Mexico.

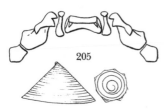

205

Genus *Addisonia* Dall, 1882

Deepsea, small, limpetlike, subconical shells with a strongly asymmetrical, curved apex. No epidermis or operculum. No eyes, 2 tentacles, gill-prongs along inner margin of mantle. Only 2 American species. Type: *paradoxa* Dall, 1882.

206 *Addisonia paradoxa* Dall, 1882. Off Rhode Island to Virginia, 50 to 640 fms.

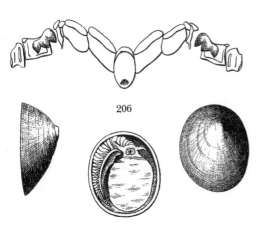

206

207 *Addisonia lateralis* (Requien, 1848). North Atlantic and Mediterranean, 50 to 640 fms. (Synonym: *Tylodina excentrica* Tiberi, 1857.)

Suborder *Trochina* Cox and Knight, 1960

Superfamily Trochacea Rafinesque, 1815

Family TROCHIDAE Rafinesque, 1815

Subfamily STOMATELLINAE Gray, 1840

Genus *Margarites* Gray, 1847

Shells less than 1/3 inch, trochoid in shape, round aperture; umbilicate; smooth (subgenus *Margarites s.s.*). Type: *Clio helicinus* Phipps, 1774. *Valvatella* E. A. Smith, 1889, and *Eumargarita* Fischer, 1885, are synonyms.

For a recent treatment of northern forms, see I. U. Galkin, 1955, "Monograph of the Mollusks of the Family Trochidae of the Far Eastern and Northern Seas of the U.S.S.R." (Opredeliteli po faune S.S.S.R., no. 57). However, the gender of the genus is masculine.

Subgenus *Margarites* Gray, 1847

***Margarites helicinus* (Phipps, 1774)**　　　　**208**
Helicina Margarite

Arctic Seas to Alaska and to Massachusetts.

1/4 to 1/3 inch, fragile, rounded whorls, soft translucent-brown; deep umbilicus. Exterior smooth. Aperture round, pearly within. Common; from 1 to 75 fathoms. Fabricius redescribed this species in 1780. Synonyms are; *excavatus, elevatus* Dall, 1919; *beringensis* and *albolineatus* both (E. A. Smith, 1899); *helicinus* (Fabricius, 1780); *campanulata* Packard, 1867; *arctica* Leach, 1819.

208

***Margarites olivaceus* (Brown, 1827)**　　　　**209**
Eastern Olive Margarite

Arctic Seas to Massachusetts.

2 to 4 mm., globose, dull, smoothish with numerous fine spiral striae. Suture well-impressed. Umbilicus deep. Translucent tan. *M. glauca* (Möller, 1842); *M. argentatus* (Gould, 1841); *harrisoni* Hancock, 1846; *grosvenori* Dall, 1926, are synonyms. Uncommon 1 to 80 fathoms.

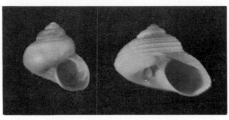

209　　　　　　211

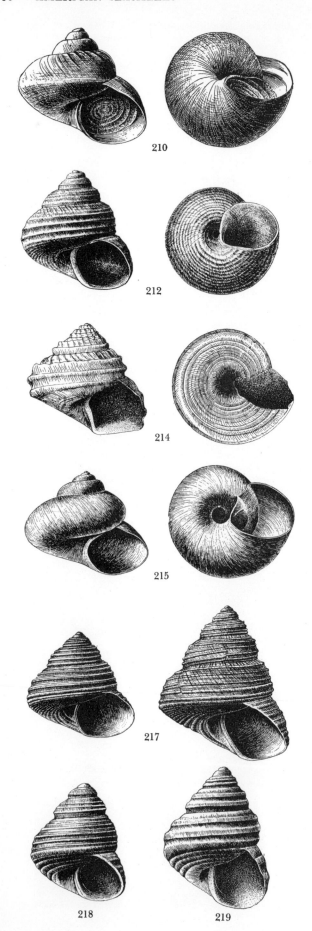

210

212

214

215

217

218   219

Margarites olivaceus marginatus Dall, 1919   **210**
Western Olive Margarite

North Pacific to Washington; to Japan.

Similar to typical *olivaceus*, but the spire is depressed, giving the shoulders a more acute angle and the shell a wider dimension. Common; 1 to 60 fathoms.

Subgenus *Omphalomargarites* Habe and Ito, 1965

Type of the subgenus is *vorticiferus* (Dall, 1873).

Margarites vorticiferus (Dall, 1873)   **211**
Vortex Margarite

Pacific Arctic Seas to off San Pedro, California; to Japan.

9 to 12 mm., depressed; 3 flattened, rapidly expanding whorls, which have a tendency in old specimens to overhang the suture. Periphery subcarinate. Umbilicus funnel-shaped. Surface with microscopic spiral threads crossed by faint growth lines. Color pinkish white to rosy-brown. Nacreous salmon within. Common; offshore 1 to 60 fathoms. Synonyms are: *sharpii* Pilsbry, 1898, and *escarinatus* Dall, 1919.

Subgenus *Pupillaria* Dall, 1909

Shells trochiform, dull, unicolor, strongly spirally striated. Type: *pupillus* Gould, 1849. Synonym: *Lirularia* Dall, 1909.

Margarites costalis (Gould, 1841)   **212**
Northern Rosy Margarite

Greenland to Cape Cod, Massachusetts.
Bering Strait to Port Etches, Alaska.

¼ to ⅜ inch in length, a little wider, with 5 evenly and well-rounded whorls. Narrowly and deeply umbilicate. Angle of spire about 90 degrees. Next to last whorl with 10 to 12 smoothish, raised, spiral threads. Columella and outer lip thin, sharp, the latter finely crenulate. Color rosy to grayish cream. White within the smoothish umbilicus. Aperture pearly-rose. Commonly dredged from 10 to 62 fathoms. *M. groenlandicus* Möller, 1842, is the same. Formerly known as *M. cinereus* (Couthouy, 1838), non Born, 1778, and *striatus* Broderip and Sowerby, 1829, non Leach, 1819.
Galkin (1955) has shown that the arctic form is *sordidus* Hancock, 1846, **(213)** and that there is a subarctic subspecies, the earliest name probably being *rudis* Dall, 1919 **(214)**.

Margarites groenlandicus (Gmelin, 1791)   **215**
Greenland Margarite

Arctic Seas to Massachusetts Bay.

½ inch in length, ¾ inch in width. Angle of spire 110 degrees. Whorls strongly rounded, aperture round, umbilicus wide and deep. Outer lip and columella very thin. Base smooth; top of whorls with about a dozen smooth spiral lirations or almost entirely smooth **(216)** (form *umbilicalis* Broderip and Sowerby, 1829). Nucleus glassy smooth. Suture finely impressed. Color glossy-cream to tan. Aperture pearly. Commonly dredged from 5 to 150 fathoms. Synonyms are: *striatus* Leach, 1819; *carneus* Lowe, 1826; *undulata* and *sulcata* Sowerby, 1838; *incarnatus* Couthouy, 1838.

Margarites pupillus (Gould, 1849)   **217**
Puppet Margarite

Bering Sea to San Diego, California.

⅓ to ½ inch in length, whorls 5 to 6, upper whorls with 5 or 6 smoothish, small, spiral threads, between or over which are microscopic, axial, slanting threads. Umbilicus a minute chink. Exterior dull, chalky whitish to yellowish gray. Aperture rosy to greenish pearl. Apex usually eroded. A common littoral species in the northern half of its range. Also dredged in 50 fathoms.

Margarites lirulatus (Carpenter, 1864)   **218**
Lirulate Margarite

Prince William Sound, Alaska, to the Coronado Islands.

¼ inch in length, 4 to 5 whorls, strong, semiglossy. Very variable in color (solid purple, whitish with dark-brown variegations and sometimes with a spiral row of dark squares on the periphery), and variable in the number and strength of the small, smooth, spiral cords. Base rounded. Umbilicus narrow but deep. Suture well-impressed. Interior iridescent. Common in shallow water. *M. parcipictus, obsoletus, subelevatus, funiculatus* and *conicus* Carpenter, 1864 **(219),** are forms of this species.

Margarites optabilis (Carpenter, 1864)   **220**
Choice Margarite

Santa Barbara to Coronado Islands, California.

5 mm., subconical, grayish with purplish red maculations, suture distinct, 2 large spiral cords in the spire with smaller ones in between. Decussate between the ribs. Periphery carinate. Umbilicus large, deep and with two threads running into it. An uncommon, attractive species; 3 to 30 fathoms.

Margarites salmoneus (Carpenter, 1864)   **221**
Salmon Margarite

Washington to southern California.

6 to 8 mm., of a deep salmon hue, with 3 purplish nuclear whorls; 8 spiral threads on the spire. No axial growth lines. Uncommon; 6 to 40 fathoms.

Margarites succinctus (Carpenter, 1864)   **222**
Tucked Margarite

Alaska to Baja California.

⅛ inch in length, 4 whorls, slightly wider than long, smoothish except for microscopic, weak threads or incised lines. Umbilicus small, round, deep. Exterior grayish brown, commonly with microscopic, brown, spiral lines. Aperture dark-greenish iridescent. Littoral on algae; common.

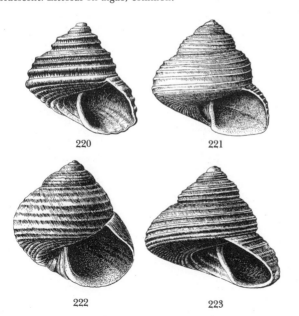

220   221

222   223

Other species:

223   *Margarites rhodia* Dall, 1920. Prince William Sound, Alaska, to Crescent City, California. (New name for *inflata* Carpenter, 1864.) 10 mm.

224   *Margarites healyi* Dall, 1919. North of Bering Strait.

225   *Margarites simbla* Dall, 1913. Off Santa Barbara, California. 14 mm.

226   *Margarites (Pupillaria) acuticostatus* Carpenter, 1864. Southern California. 5 mm.

227   *Margarites (P.) althorpensis* Dall, 1919. 14 fms., Port Althorp, Alaska.

228   *Margarites (P.) smithi* Bartsch, 1927. Monterey, California. Shore.

229   *Margarites (P.) keepi* A. G. Smith and Gordon, 1948. *Proc. Calif. Acad. Sci.,* vol. 26. 25 fms., in Monterey Bay, California.

230   *Margarites giganteus* (Leche, 1878). (Synonym: *probilofensis* Dall, 1919; *frielei* Krause, 1885.) Arctic seas to Japan and Aleutians. 2 to 90 fms. Type of subgenus *Margaritopsis* Thiele, 1906.

231   *Margarites (Cantharidoscops* Galkin, 1955) *frigidus* Dall, 1919. Arctic seas to Aleutians. 16 to 30 fms. Type of the subgenus.

232   *Margarites (M.) vahlii* (Möller, 1842). (Synonyms: *hypolispus* Dall, 1919, and variety *angulata* Galkin, 1955; *mighelsi* Rehder, 1937; *johnsoni* Dall, 1921; *acuminata* Mighels and Adams, 1842 (non Sowerby, 1838).) Circumpolar Arctic; Bering Sea; Japan; Greenland to the Gulf of Maine. 2 to 200 fms.

233   *Margarites (M.) multilineatus* De Kay, 1843. New York and Connecticut. From codfish stomachs.

234   *Margarites (M.) minutissimus* Mighels, 1843. Casco Bay, Maine, from haddock stomach.

235   *Margarites (Lirularia) bicostatus* J. H. McLean, 1964. South Coronado Island; Guadalupe Island, Baja California.

236   *Margarites (Lirularia) mionus* Dall, 1927. Off Georgia and east Florida, 294 to 440 fms.

237   *Margarites (Bathymophila* Dall, 1881) *euspirus* Dall, 1881. Florida Keys and West Indies, 390 to 805 fms. Gulf of Mexico, 220 fms. Type of the subgenus.

238   *Margarites scintillans* Watson, 1879. Bermuda, 1,075 fms.

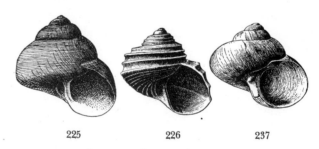

225   226   237

Genus *Basilissa* Watson, 1879

Deepsea, small (5 mm.), trochiform, umbilicate, nacreous. Outer lip with two sinuses. Columella with a strong tooth at the base. Aperture sub-rhomboidal. Operculum multispiral, corneous. Periphery with a row of beads or spiral channel. Type: *superba* Watson, 1879. The species are:

239   *Basilissa alta* Watson, 1879. (Synonym: *delicatula* Dall, 1881 and 1889.) Gulf of Mexico to the Lesser Antilles, 60 to 769 fms.

240   *Basilissa watsoni* Dall, 1927. Off North Carolina to the Gulf of Mexico, 400 fms.

241   *Basilissa (Ancistrobasis* Dall, 1889) *costulata* Watson, 1879. (Synonym or subspecies: *depressa* Dall, 1889.) Southeast Florida and Gulf of Mexico, 15 to 640 fms. Common. Type of the subgenus.

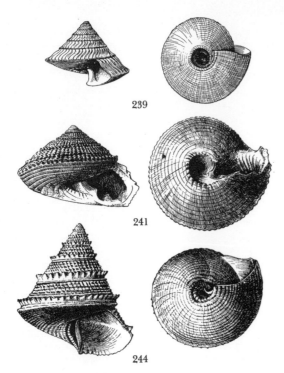

239

241

244

242 *Basilissa rhyssa* Dall, 1927. Off Georgia to Cuba, 220 to 440 fms.

243 *Basilissa* (*Thelyssa* F. M. Bayer, 1971) *callisto* F. M. Bayer, 1971. 2,700 meters, off Great Inagua Island, Bahamas.

244 *Basilissa cinctellum* (Dall, 1889). 174 fms., off Havana, Cuba. 8 mm.

Genus *Seguenzia* (Jeffreys) Seguenza, 1876

Deepsea, very small, fragile, trochoid shells with a strong anal notch at the upper end of the aperture. Spiral cords or carinae quite strong. Some of the species included in this genus may be larval, free-swimming forms of other prosobranch mollusks. Verrill reported one species to have a taenioglossate radula (see Dall, 1889, p. 274). Type: *monocingulata* Seguenza, 1876 (synonym: *formosa* Jeffreys, 1876). The American species are:

Atlantic species:

245 *Seguenzia carinata* Jeffreys, 1876. Off Florida and the West Indies, 675 to 1,125 fms.

246 *Seguenzia ionica* Watson, 1879. Gulf of Mexico to the West Indies, 390 to 1,568 fms.

247 *Seguenzia monocingulata monocingulata* Seguenza, 1876 (*formosa* Jeffreys, 1876). Gulf of Maine to the Gulf of Mexico, 100 to 2,033 fms.

248 *Seguenzia monocingulata nitida* Verrill, 1884. South of Martha's Vineyard, Massachusetts, 2,033 fms.

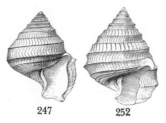

247   252

249 *Seguenzia trispinosa* Watson, 1888. Off North Carolina to the Gulf of Mexico, 294 to 675 fms.

250 *Seguenzia floridana* Dall, 1927. Off Georgia, Fernandina and Cape Florida, 193 to 440 fms.

251 *Seguenzia rushi* Dall, 1927. Off Georgia, 440 fms. Off Fernandina, Florida, 294 fms.

252 *Seguenzia eritima* Verrill, 1884. South of Martha's Vineyard, 1,290 to 2,033 fms.

253 *Seguenzia elegans* Jeffreys, 1876. Bermuda.

Pacific species:

254 *Seguenzia certoma* Dall, 1919. Off San Diego, California, 600 fms.

255 *Seguenzia giovia* Dall, 1919. Off Catalina Island, California, 118 fms.

256 *Seguenzia cervola* Dall, 1919. Off North Coronado Island, California, 692 fms.

257 *Seguenzia caliana* Dall, 1919. Off La Jolla and San Diego, California, 243 to 822 fms.

Genus *Euchelus* Philippi, 1847

Solid, thick, ½-inch or less, round aperture; thick outer lip; lirate within; inner lip with a tooth. Adults with no umbilicus; operculum few-whorled. Typical *Euchelus*, with its type *quadricarinatus* (Holten, 1802) is from the Indo-Pacific.

Subgenus *Mirachelus* Woodring, 1928

Small (5 to 6 mm.), solid, coarse sculpturing, without an umbilicus. Type: *corbis* Dall, 1889.

Euchelus guttarosea Dall, 1889   **258**
Red-spotted Euchelus

Southeast Florida to the Lesser Antilles. Gulf of Mexico, Bermuda.

4 to 6 mm., noduled, white with the upper surface of the whorls with small, distinct rose-red dots on the nodules. Nucleus smooth. Umbilicus absent. Columella nearly straight, with a strong tooth near its base. Inside of lip 6 to 8 toothed lirae. From just offshore to 450 fathoms. Moderately common in the Greater Antilles.

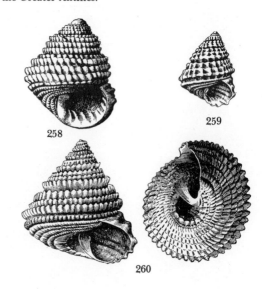

258   259

260

Other species:

259 *Euchelus (Mirachelus) corbis* (Dall, 1889). Gulf of Mexico, Florida and Cuba, 220 to 450 fms.

260 *Euchelus (Antillachelus* Woodring, 1928) *dentiferus* (Dall, 1889). Deepwater, West Indies. Type of the subgenus.

261 *Euchelus eucastus* (Dall, 1889). 440 fms., off Georgia.

Genus *Lischkeia* Fischer, 1879

Shells trochoid in shape, moderately thin-shelled; base flattened or slightly rounded; suture distinct; sculpture of irregular nodose spiral ribs and beaded threads. Mostly deepsea. Type: *Trochus moniliferus* Lamarck, 1816, non 1804. Is *alwinae* (Lischke, 1871).

Subgenus *Turcicula* Dall, 1881

Resembling *Lischkeia,* but with microscopic vermiculate sculpturing. Suture very wide and deep. One species, the type, known: *imperialis* Dall, 1881.

Lischkeia imperialis (Dall, 1881)　　　　**262**
Giant Imperial Margarite

Florida Straits to Lesser Antilles; Jamaica.

2 inches, 5 whorls, deep suture; row of blunt nodules on periphery of whorl; below it is another row of smaller, more numerous nodules. Base flattish; no umbilicus. Color ash-gray; nacreous within. Microscopic axial sculpture vermiculate. Very rare; 30 to 182 fathoms. (See Rehder, 1955, *Proc. Mal. Soc. London,* vol. 31, pl. 12.) *L. deichmannae* F. M. Bayer, 1971 (*Bull. Marine Sci.,* vol. 21, p. 121) is a synonym.

Subgenus *Bathybembix* Crosse, 1893

The subgenotype is *aeola* Watson, 1874, a Japanese species. *Bembix* Watson, 1879 (not Fabricius, 1775) is a synonym.

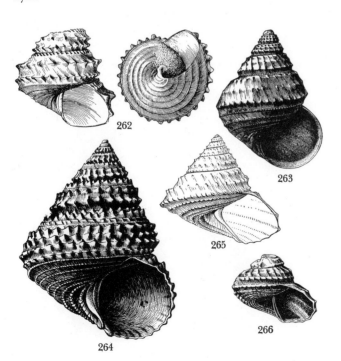

262　263　265　264　266

Lischkeia bairdii (Dall, 1889). Pl 3　　**263**
Baird's Spiny Margarite　　　　　Color Plate 2

Bering Sea to Coronado Islands, Mexico, and off Chile.

2 inches in length, moderately fragile, sculptured with varying number of spiral rows of fairly large beads. No umbilicus. Shell white with a thin, glossy, yellowish green periostracum. Interior of aperture pearly-white. This is a choice deep-water species much sought after by collectors. Moderately common in 10 to 600 fathoms in cold water.

Subgenus *Cidarina* Dall, 1909

The type of the subgenus is *cidaris* Carpenter, 1864.

Lischkeia cidaris (Carpenter, 1864)　　**264**
Adams' Spiny Margarite

Alaska to Baja California.

1 to 1½ inches in length, moderately solid, similar to *L. bairdii* but with a higher, flat-sided spire. The suture is usually more impressed. Color gray to grayish white. Moderately common from 20 to 350 fathoms. This is the type of the subgenus *Cidarina.*

Subgenus *Calliotropis* Seguenza, 1903

Type of the subgenus is *ottoi* (Philippi, 1844). Synonyms or subgenera: *Solaricida* Dall, 1919; *Solariellopsis* Schepman, 1908 (non Gregorio, 1886).

Lischkeia ottoi (Philippi, 1844)　　**265**
Otto's Spiny Margarite

Nova Scotia to North Carolina.

⅝ to ¾ inch in length, equally wide. Moderately thin with sharp lip. The round, narrow umbilicus is partially covered by the top of the columella. Color pearly-white. Sculpture of whorls in spire with 3 evenly spaced spiral rows of prickly beads. Suture wavy. Base of shell with 4 or 5 spiral threads which bear smaller, often obscure, beads. Nuclear whorls with axial lamellae. *Solariella regalis* Verrill and Smith, 1880, is the same species. Common from 50 to 100 fathoms.

Other species:

266 *Lischkeia (C.) carlotta* (Dall, 1902). Queen Charlotte Islands, British Columbia. 13 mm.

267 *Lischkeia (Solaricida* Dall, 1919) *ceratophora* (Dall, 1896). Off San Diego, California, to Mazatlan, Mexico. Deep water.

268 *Lischkeia (S.) equatorialis* (Dall, 1896). Off San Diego, California, to Ecuador. Deep water.

Subfamily PLANITROCHINAE Knight, 1956

Genus *Planitrochus* Perner, 1903

Depressed trochiform, with a sharp carinate periphery and wide umbilicus. Type: *P. amicus* Perner, 1903. This is an Upper Silurian genus, and the only known living species, provisionally placed here, is *Planitrochus disculus* (Dall, 1889) **(269)** from the Straits of Florida, 500 meters, and the Lesser Antilles. The shell is 6 to 10 mm. in diameter and it has a nacreous iridescence within its white, glossy shell.

Subfamily UMBONIINAE H. and A. Adams, 1854

Small, low-spired, glossy shells with a large, buttonlike callus wholly or partially covering the umbilicus. Operculum multispiral and corneous. This subfamily and its genus *Umbonium* Link, 1807, represents a large number of species in the Indo-Pacific area. It is doubtful if there is truly an American representative. Dall, in 1889, described *Umbonium bairdii* from off Yucatan in 200 to 640 fathoms, **(270)**. I suspect it may be a member of the Solariellinae. Some workers place *Microgaza* Dall in this subfamily, but that does not seem reasonable.

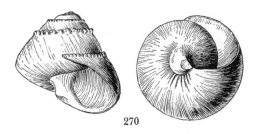

270

Subfamily SOLARIELLINAE Powell, 1951

The subfamily name Minoliinae Kuroda, Habe and Oyama, 1971, appears to be the same.

Genus *Solariella*, S. V. Wood, 1842

Small shells, less than ⅓ inch, trochoid in shape; rounded whorls; with strong spiral cords; wide, deep umbilicus; round aperture. Radula with an exceptionally small number of marginal teeth. Type: *maculata* S. V. Wood, 1842.

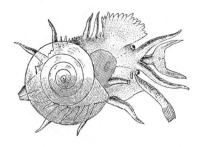

Subgenus *Machaeroplax* Friele, 1877

Rather high spire; strong spiral cords and with prominent axial growth striae. Numerous species. Type: *M. affinis* Jeffreys, 1870.

Solariella obscura (Couthouy, 1838)   271
Obscure Solarelle

Arctic Seas to off Chesapeake Bay, Virginia.
Alaska to Washington.

¼ inch in length, similar to *Margarites costalis*, but with whorls made more angular by one large, feebly beaded, spiral cord above the periphery. Base smoothish except for microscopic, spiral scratches. Umbilicus narrower and bordered by an angular rim. Color grayish to pinkish tan, often worn to reveal a pearly-golden color. Aperture pearly-white. Some specimens may have weak axial riblets below the strongest spiral cord on the periphery of the whorl. Commonly dredged

271

from 3 to 400 fathoms, especially on the Grand Banks. *S. bella* Verkrüzen, 1875, and *carinata* Verrill, 1882, are probably synonyms.

Galkin (1955) recognizes a smoothish form from the Bering Sea and northern Siberia **(272)** called *forma intermedius* (Leche, 1878). *S. micraulax* McLean, 1964, from Alaska may be an even more rounded, smoothish form of this very variable species **(273)**.

Solariella lacunella (Dall, 1881)   274
Channeled Solarelle

Virginia to south Florida.

⅜ inch in length, equally wide, thick, pure-white. Whorls convex. Aperture circular, internally pearly. Suture channeled. Whorls with 6 spiral cords, bottom 3 smooth, the upper ones axially beaded. Nuclear whorls glassy, with microscopic axial ribs. Umbilicus round, narrow, deep, lined with spiral rows of coarse beads. Very commonly dredged from a few feet to 200 fathoms. *S. maculata* Dall, 1881 (not Wood, 1842) is a synonym, as is *depressa* Dall, 1889 (Gulf of Mexico in 805 fms.).

274

Solariella lamellosa Verrill and Smith, 1880   275
Lamellose Solarelle

Massachusetts to Key West, Yucatan and the West Indies.

⅛ inch in length, similar to *S. lacunella*, but with a much deeper channel at the suture, below which are numerous, small axial, short lamellarlike ribs. Middle of whorl with a strong, sharp, smooth or beaded, spiral thread. Base of shell smoothish except for one smooth spiral thread near the periphery and one heavily beaded cord bordering the deep, round umbilicus. Entire shell with numerous microscopic incised lines. Very commonly dredged from 15 to 150 fathoms, but also recorded from 683 fathoms.

275

Solariella varicosa (Mighels and Adams, 1842)   **276**
Varicose Solarelle

Labrador to Gulf of Maine. Europe.
Alaska to San Diego.

4 to 6 mm.; as high as wide; grayish white; periphery with 3 to 5 spiral cords, above which finer, weaker ones cross 14 to 16 strong, oblique, distantly spaced axial ribs. Concave base with 12 to 16 arching ribs ending at the deep umbilicus as one or two rows of beads. Moderately common from 7 to 60 fathoms. A very distinctive little species. Synonyms are: *Margarita acuminata* Sowerby, 1838; *M. elegantissima* S. Wood, 1848; *M. paupercula* Dall, 1919.

Other Atlantic species:

277 *Solariella* (*Dentistyla* Dall, 1889) *asperrima* (Dall, 1881). 36 fms., off Dry Tortugas, Florida. Gulf of Mexico, 130 to 220 fms.

— *Solariella dentifera* (Dall, 1889) is a *Euchelus*.

278 *Solariella* (*Dentistyla*) *sericifila* (Dall, 1889). 92 fms., off Grenada, West Indies.

279 *Solariella pourtalesi* Clench and Aguayo, 1939. Off Cape Florida and West Indies. 193 to 888 fms. This is *amabilis* of Dall, not Jeffreys, 1865.

280 *Solariella scabriuscula* (Dall, 1881). Gulf of Mexico, 539 fms.

281 *Solariella crossata* (Dall, 1927). 294 fms., off Fernandina, Florida.

282 *Solariella periscopia* (Dall, 1927). North Carolina to Bahamas and Yucatan. 52 fms.

283 *Solariella* (*Micropiliscus* Dall, 1927) *constricta* Dall, 1927. 440 fms., off Georgia. Type of the subgenus.

284 *Solariella tiara* (Watson, 1879). 39 fms., off Culebra Island, Puerto Rico; 294 fms., off Fernandina, Florida; Bermuda.

285 *Solariella laevis* Friele, 1886. Europe; Greenland, 40 to 350 fms.

286 *Solariella aegleis aegleis* Watson, 1879. Florida Strait and Gulf of Mexico. (Synonyms: *lata* Dall, 1889; *rhina* Watson, 1879; *clavata* Watson, 1879.)

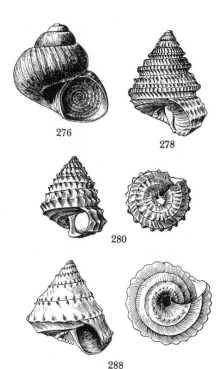

276

278

280

288

287 *Solariella infundibulum* (Watson, 1879). Massachusetts to North Carolina, 679 to 1,782 fms. Bermuda. Brazil.

287

288 *Solariella lissocona* (Dall, 1881). Gulf of Mexico, 331 fms.

289 *Solariella iris* (Dall, 1881). Sand Key, Florida, 119 fms. Gulf of Mexico, 200 fms.

290 *Solariella* (*Suavotrochus* Dall, 1924) *lubrica lubrica* Dall, 1881. Off Cedar Keys, Florida, 200 fms.; West Indies, 805 fms. Type of the subgenus. (Var. *iridea* Dall, 1889, (**290a**)). Off Cape Florida, 193 fms.).

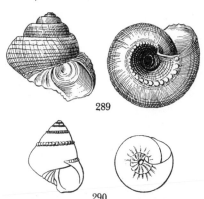

289

290

290 a

291 *Solariella calatha, anoxia* (**292**) and *tubula* (**293**) all Dall, 1927. All 294 to 440 fms., off Georgia and northern Florida.

294 *Solariella carvalhoi* Lopes and Cardoso, 1958. East and south Brazil. *Rev. Brasil. Biol.*, vol. 18, p. 59.

295 *Solariella actinophora* Dall, 1890. Off the Lesser Antilles, 1,019 fms. 7 mm.

Solariella permabilis Carpenter, 1864   **296**
Lovely Pacific Solarelle

Alaska to the Gulf of California; Japan.

½ to ¾ inch in length, equally wide, solid, semi-glossy. Aperture circular. Umbilicus fairly wide, round, very deep.

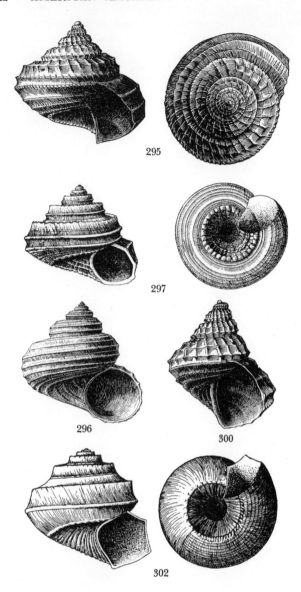

295

297

296

300

302

7 whorls, shouldered just below the suture by a flat shelf. Lower ⅔ of whorl with numerous weak spiral cords that are smoothish in the last whorl but crossed by numerous axial riblets in the early whorls. Color tan with light-mauve stains and mottlings. Interior iridescent. Moderately common from 20 to 339 fathoms.

Other Pacific coast species:

297 *Solariella triplostephanus* Dall, 1910. Off San Diego, California, to Panama. 10 to 12 fms. 6.5 mm.

298 *Solariella rhyssa* Dall, 1919. Catalina Island, California.

299 *Solariella nuda* Dall, 1896. British Columbia to Baja California. 50 to 455 fms.

— *Solariella micraulax* J. H. McLean, 1964, see under *obscurus* Couthouy.

300 *Solariella elegantula* Dall, 1925. Off La Paz, Baja California, 26 fms. 5.5 mm.

301 *Solariella lewisae* Willett, 1946. Kodiak Island, Alaska, 15 fms. *Bull. So. Calif. Acad. Sci.,* vol. 45, pt. 1.

302 *Solariella oxybasis* Dall, 1890. Off Santa Barbara, California. 13.5 mm.

Genus *Microgaza* Dall, 1881

Shell small, spire flat and low, with beads just below the suture, the remaining whorls smooth. Umbilicus wide and deep. Type: *rotella* Dall, 1881. Considered a subgenus of *Solariella* by some workers.

Microgaza rotella Dall, 1881   **303**
Dall's Dwarf Gaza

North Carolina to Florida, the West Indies and Brazil.

¼ inch in diameter, spire flat, surface smooth except for a spiral row of low pimples just below the suture. About 5 whorls. Umbilicus fairly wide, very deep, its squarish edge bearing numerous, neat, rounded creases. Columella straight. Color whitish gray with a beautiful opalescent sheen, especially inside the aperture. Top of whorls colored with chestnut, zebralike axial stripes. Very commonly brought up in dredging hauls off Miami from 50 to 100 fathoms. The form *inornata* Dall, 1881 (**304**) lacks the pimples just below the suture.

303

305

Subfamily CALLIOSTOMATINAE Thiele, 1921

Genus *Calliostoma* Swainson, 1840

Shells usually 2 inches or less in size, top-shaped, and have spiral rows of beads. Columella usually arched, sometimes truncate at the base. Umbilicus present or absent. Interior of aperture iridescent. Operculum circular, thin, horny and multispiral. Eggs are laid in gelatinous ribbons. The veliger stage of the young is passed within the egg membrane before hatching as crawling snails. There are about 40 Western Atlantic species, and possibly 30 in the Eastern Pacific. Type: *zizyphinum* Linné, 1758, of western Europe. Synonyms include: *Conulus* Nardo, 1841; *Ziziphinus* Gray, 1843; *Manotrochus* Fischer, 1885; *Jacinthinus* Monterosato, 1889; *Ampullotrochus* Monterosato, 1890; *Dymares* Schwengel, 1942; and *Eucasta* Dall, 1889.

Subgenus *Calliostoma* Swainson, 1840

Calliostoma occidentale (Mighels and Adams, 1842)   **305**
North Atlantic Top-shell

Nova Scotia to off New Jersey; Europe.

½ inch in length, equally wide. Whorls convex and with 3 or 4 strong spiral cords, the 2 lower ones smooth, the upper one beaded. Color pearly-white. No umbilicus. Outer lip fragile. Moderately common from 10 to 365 fathoms. Synonyms are: *quadricinctus* (Wood, 1842); *alabastrum* Lovén, 1846; *formosus* McAndrew and Forbes, 1847.

Calliostoma pulchrum (C. B. Adams, 1850)   **306**
Beautiful Top-shell

North Carolina to Florida, the Gulf of Mexico and the West Indies.

⅜ inch in length, ¾ as wide. Angle of spire about 50 degrees. Sides of whorls straight. Characterized by a pair of strong, spiral cords just above the suture which are white with distantly spaced red-brown dots. Rest of whorl pearly-

306

green with 6 or 7 very weak (or sometimes strong) beaded spiral threads. Columella almost upright, its inner side rounded, pearly. No umbilicus. Moderately common from 1 to 200 fathoms. *C. veliei* Pilsbry, 1900, is a synonym.

Calliostoma roseolum Dall, 1880   307
Dall's Rosy Top-shell

North Carolina to Mexico and Barbados.

½ inch in length, ¾ as wide. Angle of spire about 50 degrees. Sides of whorls well-rounded, and with 8 or 9 crowded spiral rows of numerous neat beads. Columella upright, strong, with a slight twist. Color of shell light orange-tan to cream, often with arched splotches of darker color running axially across the whorl. No umbilicus. Aperture pearly-rose. Moderately common; from 7 to 175 fathoms. *C. apicinum* Dall, 1881, is a synonym.

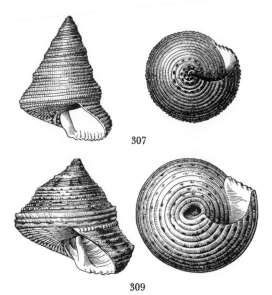

307

309

Calliostoma fascinans Schwengel and McGinty, 1942   308
Enchanting Top-shell

Off Florida to Mexico.

½ inch, imperforate, with 4 to 6 strongly beaded cords above the periphery, and 7 or 8 finely beaded threads on the flat base. Axial threads give shell a reticulate appearance. Angle of spire 50 degrees. Color white with reddish brown mottlings. Moderately common; from 37 to 69 fathoms.

308

Calliostoma yucatecanum Dall, 1881   309
Yucatan Top-shell

Off North Carolina to Texas; Mexico.

⅔ inch; spire low with an angle of about 85 degrees; umbilicus very deep, white and smooth; 3 or 4 cords on early whorls are beaded, becoming smoothish on last whorl and have between them microscopic spiral threads. Below periphery and over the base are 9 to 13 spiral threads. Moderately common; in 5 to 35 fathoms. *C. agalma* Schwengel, 1942, is a synonym.

Subgenus *Elmerlinia* Clench and Turner, 1960

Chitinous jaws pointed in front; outer two lateral radulae without denticulations. Umbilicus deep. Type: *jujubinum* Gmelin, 1791.

Calliostoma javanicum (Gmelin, 1791)   310
Chocolate-lined Top-shell   Color Plate 2

South Florida Keys and the West Indies.

¾ to 1 inch in length, slightly wider. Angle of spire about 70 degrees. Sides of whorls flat; periphery sharp; base flat. Umbilicus deep, smooth-sided, white. Whorls characterized by 10 spiral, beaded threads between each of which there is a dark-chocolate line. Base olive with about 5 or 6 fine brown, spiral lines. A very beautiful and moderately rare species much sought after by collectors. Occurs in shallow reef waters to 7 fathoms. *C. zonamestum* A. Adams, 1851, is a synonym.

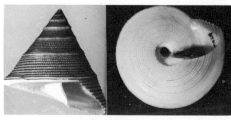

310

The similar (311) *Calliostoma barbouri* Clench and Aguayo, 1946, (north coast of Cuba) has a narrower, deep umbilicus, has less distinct, more numerous beaded threads on the spire and lacks distinct brown lines on the beaded base. The umbilicus is brownish red, while in *javanicum* it is white. Uncommon.

311

Calliostoma jujubinum (Gmelin, 1791)   312
Jujube Top-shell   **Color Plate 2**

North Carolina to Texas, the Bahamas and the West Indies to Brazil.

½ to 1¼ inches in length. Characterized by the deep, narrow, smooth-sided umbilicus which is bordered by a spiral, beaded thread, and by the swollen, rounded periphery of each whorl, which in the spire is located just above the suture. Color ranges from brownish cream to reddish purple and is often maculated with white splotches near the periphery. Typical *jujubinum* has a spire angle of about 50 degrees; the spiral threads on the whorls are weakly beaded, and the umbilicus is almost closed. Synonyms are: *lunatus* Röding, 1798; *perspectivus* Philippi, 1843; *alternatus* Sowerby, 1873, and *rawsoni* Dall, 1889. Common under rocks on offshore reefs.

C. *jujubinum tampaense* (Conrad 1846) **(313)** (North Carolina to both sides of Florida to Yucatan) varies in spire angle from 50 to 65 degrees, is not always so swollen at its periphery, and has 9 or 10 well-beaded spiral threads between each suture. It is possibly only a minor variant of *jujubinum*, but is more common on west Florida beaches.

312

Calliostoma adelae Schwengel, 1951   314
Adele's Top-shell

Southeast Florida.

¾ inch, very similar to *jujubinum*, but golden-tan, usually smaller, has all of the spiral cords beaded; white-blotched, peripheral cords on the keel are much larger than the others; has only 6 to 8 spiral cords of equal strength on the base, while in *jujubinum* there are 10 to 15, with the strongest ones near the umbilicus. Moderately common locally in shallow water in grass beds in 6 to 20 feet of water. Named after Adele Koto, an ardent shell collector.

314

Subgenus *Kombologion* Clench and Turner, 1960

Without a perforated umbilicus. Jaws rounded and fringed in front. Outermost marginal teeth not serrated. Type: C. *bairdii* Verrill and Smith, 1880.

Calliostoma bairdii Verrill and Smith, 1880   315
Baird's Top-shell

Massachusetts to North Carolina and to Florida.

1 to 1¼ inches in length, about as wide. Angle of spire about 70 degrees. Sides of spire straight to slightly convex. Base rather flat. Periphery angular. Sculpture of 6 or 7 spiral rows of small, neat beads, with those on the topmost row being the largest. Suture difficult to find. No umbilicus. Color brownish cream with faint maculations of light reddish. Not uncommon; from 43 to 250 fathoms.

315

C. *bairdii psyche* Dall, 1889 **(316)** (North Carolina to Key West, 30 to 130 fathoms) is usually ¾ inch in length, slightly wider with a spire angle of about 75 to 80 degrees, and the color is lighter and more pearly. Base with 3 or 4 spiral brown lines. It has a chinklike depression beside the umbilicus. Uncommon. C. *"subumbilicatum* Dall" **(317)** is a form of this species whose umbilicus is half open. C. *hendersoni* Dall, 1927, may be only a form of this subspecies.

C. *bairdii* subspecies *rosewateri* Clench and Turner, 1960 **(318)** (off Lesser Antilles to Colombia 150 to 200 fathoms) is 1¼ inches, an iridescent golden-yellow with green highlights and with patches of brown. Umbilical region slightly indented. Base almost smooth and with about 6 spiral brown lines. Uncommon.

316

C. *bairdii* subspecies *oregon* Clench and Turner, 1960 **(319)** (off west Florida to off Texas, 116 to 190 fathoms), similar to *rosewateri*, but only ¾ inch, golden-red, umbilicus closed or tiny and deep; base weakly beaded. 7 or 8 beaded cords between sutures.

319

Calliostoma sayanum Dall, 1889   320
Say's Top-shell

Off North Carolina to Key West, Florida.

1½ inches; deeply umbilicate, golden-brown with a red band at the periphery. Upper part of whorls slightly rounded and bearing 8 beaded cords of alternating sizes. Periphery rounded. Angle of spire about 80 degrees. Umbilicus white, deep, smooth and bordered by a beaded cord. Slightly convex base has about 15 beaded cords. Uncommon; deep water.

320

Calliostoma benedicti Dall, 1889   321
Benedict's Top-shell

Off North Carolina to off Mississippi.

½ to 1¼ inches, wider than high, umbilicus deep, shelved, smooth; spire concave in outline; about 10 spiral white beaded cords between sutures with brownish red lines in between. Periphery rounded. 19 to 21 threads on base, becoming weak near the umbilicus and with faint brown lines between. Originally described from a young specimen. Adults were named *C. springeri* Clench and Turner, 1960, 200 to 260 fathoms; rare.

321

Calliostoma euglyptum (A. Adams, 1854)   322
Sculptured Top-shell

North Carolina to Florida and Texas; Mexico.

¾ inch in length, equally wide. Angle of spire about 70 degrees. Sides of whorls slightly concave. Periphery well-rounded. No umbilicus. Whorls with 6 major, well-beaded spiral cords between each of which is a much smaller, weakly beaded thread. Color dull-rose, sometimes with axial flammules of cream. Nucleus pink or, when worn, dark-purple. Moderately common in some localities from low-tide mark to 32 fathoms. Rare in southeast Florida.

322

Calliostoma marionae Dall, 1906   323
Marion's Top-shell

Off Florida to Campeche, Mexico.

1 inch, with a beaded cord bordering the deep umbilicus; lower whorls oily smooth with 4 to 9 finely cut grooves. Base flat, with 12 to 15 cut grooves. Color pinkish brown with small patches of white. Uncommon offshore; from 23 to 90 fathoms. *C. faustum* Schwengel and McGinty, 1942, is a synonym.

Calliostoma sarcodum Dall, 1927   324
Flesh-colored Top-shell

Bahamas to Caribbean.

½ inch, similar to and possibly a subspecies of *euglyptum* (A. Adams), but the whorls are very rounded; umbilicate; very dark brownish red, with red spots on peripheral beaded cord. *C. jaumei* Clench and Aguayo, 1946, is a synonym. Uncommon on shallow reefs among live and dead *Porites* coral.

324   323

Other Atlantic species:

325 *Calliostoma* (*Calliostoma*) *jucundum* (Gould, 1849). Brazil to Argentina. See *Johnsonia*, vol. 4, no. 40, pl. 17. Uncommon.

326 *Calliostoma* (*Calliostoma*) *coppingeri* (E. A. Smith, 1880). (Synonym: *cymatium* Dall, 1889.) Uruguay to Argentina, 7 to 48 fms. See *Johnsonia*, vol. 4, no. 40, pl. 18. Common. 10 mm.

327 *Calliostoma* (*Elmerlinia*) *bullisi* Clench and Turner, 1960. 38 fms., off Cabo Orange, Brazil. Rare.

328 *Calliostoma* (*Kombologion*) *schroederi* Clench and Aguayo, 1938. Bahamas to Cuba, 145 to 230 fms. Rare. See *Johnsonia*, vol. 4, no. 40, pl. 18.

329 *Calliostoma* (*Kombologion*) *cubanum* Clench and Aguayo, 1940. 490 fms., off Cardenas, Cuba. See *Johnsonia*, vol. 4, no. 40, pl. 43. Very rare.

330 *Calliostoma* (*Kombologion*) *atlantis* Clench and Aguayo, 1940. 330 fms., off Pinar del Rio, Cuba. See *Johnsonia*, vol. 4, no. 40, pl. 44. Very rare and choice.

329

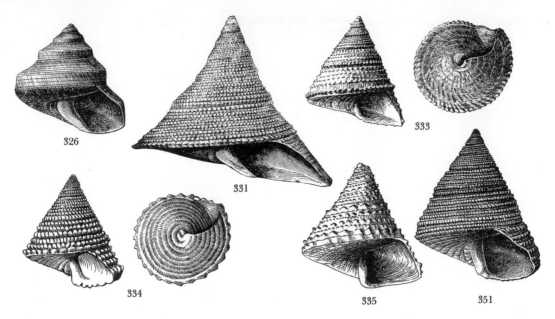

326

331

333

334

335   351

341

331 *Calliostoma (Kombologion) aurora* Dall, 1888. Off the Lesser Antilles, 140 to 576 fms. Rare and choice.

332 *Calliostoma (Kombologion) adspersum* (Philippi, 1851). Brazil; intertidal. (Synonym: *depictum* Dall, 1927.)

333 *Calliostoma (Eucasta* Dall, 1889) *indiana* Dall, 1889. 170 fms. off Grenada, West Indies. Type of the subgenus.

334 *Calliostoma sapidum* Dall, 1881. Off west Florida to Lesser Antilles, 66 to 805 fms. Rare. See *Johnsonia*, vol. 4 no. 40, pl. 34.

335 *Calliostoma orion* Dall, 1889. Off north Cuba in 80 fms. Rare. See *Johnsonia*, vol. 4, no. 40, pl. 35.

336 *Calliostoma echinatum* Dall, 1881. Off Havana, Cuba, in 80 fms. Choice and rare. See *Johnsonia,* vol. 4, no. 40, pl. 36.

341 *Calliostoma bigelowi* Clench and Aguayo, 1938. 205 to 230 fms., off the coast of Cuba. See *Johnsonia*, vol. 4, no. 40, pl. 53. Rare and very choice.

336

337 *Calliostoma carcellesi* Clench and Aguayo, 1940. Off Rio Negro Province, Argentina, 30 fms. See *Johnsonia*, vol. 4, no. 40, pl. 38.

338 *Calliostoma torrei* Clench and Aguayo, 1940. 385 fms. off Matanzas, Cuba. Very rare. See *Johnsonia*, vol. 4, no. 40, pl. 40.

339 *Calliostoma jeanneae* Clench and Turner, 1960. Dredged off Havana, Cuba. Very rare. See *Johnsonia*, vol. 4, no. 40, pl. 47.

340 *Calliostoma hassler* Clench and Aguayo, 1939. 35 fms., off Cabo Frio, Brazil. See *Johnsonia*, vol. 4, no. 40, pl. 48.

342 *Calliostoma militare* von Ihering, 1907. (Synonyms: *dalli* von Ihering, 1907; *amazonicum* Finlay, 1930; *iheringi* Dall, 1927; *quequensis* Carcelles, 1944.) Argentina.

343 *Calliostoma arestum* Dall, 1927. 440 fms., off Fernandina, Florida. Probably not a *Calliostoma*.

344 *Calliostoma blakei* Clench and Aguayo, 1938. 17 fms., off Cape Bermeja, Argentina. Brazil and Uruguay. Probably a *Photinula*.

345 *Calliostoma circumcinctum* Dall, 1880. 805 fms., off Havana, Cuba. Probably not a *Calliostoma*.

346 *Calliostoma halibrectum* Dall, 1927. 294 fms., off Brunswick, Georgia. Probably not a *Calliostoma*.

347 *Calliostoma kampsa* Dall, 1927. 440 fms., off Fernandina, Florida. Probably not a *Calliostoma*.

348 *Calliostoma tittarium* Dall, 1927. 440 fms., off Fernandina, Florida. Probably not a *Calliostoma*.

349 *Calliostoma trachystum* Dall, 1927. 440 fms., off Fernandina, Florida. Probably not a *Calliostoma*.

350 *Calliostoma brunneum* (Dall, 1881). Off Havana, Cuba, 3 to 15 fms.; Barbados. Has reticulated nuclear whorls. Subgenus *Fluxina* Dall, 1881 is a synonym of *Calliostoma*. (Synonym: *tejedori* Aguayo, 1949.) See A. Merrill, 1970, *The Nautilus*, vol. 83.

351 *Calliostoma rioensis* Dall, 1890. Off Brazil and Uruguay, 12 to 59 fms.

Pacific *Calliostoma*

***Calliostoma tricolor* Gabb, 1865**   352
Three-colored Top-shell

San Francisco, California, to Cape San Lucas, Mexico.

¾ to 1 inch in length, heavy for its size; whorls angular, with the upper third slightly concave to flat and the somewhat angular periphery flattish. Early whorls with minutely beaded threads, later whorls with fine, smoothish cords of various sizes. Nucleus tan to whitish. Color yellowish brown with a few spiral lines of alternating brown and white bars. Sometimes axially variegated. Dredged just offshore from 8 to 35 fathoms. Moderately common.

***Calliostoma gemmulatum* Carpenter, 1864**   353
Gem Top-shell

Cayucos, California, to the Gulf of California.

¾ inch in length, not as wide; characterized by its dark gray-green color and 2 extra-strong, beaded spiral cords. There are also 3 or 4 minor cords that are not so heavily beaded. Nucleus dark-tan. Moderately common in the littoral zone on rocks and wharf pilings. *C. formosum* Carpenter, 1864, is a synonym.

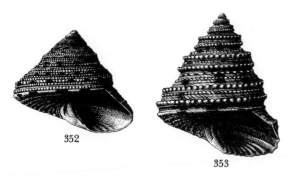

352

353

***Calliostoma supragranosum* Carpenter, 1864**   354
Granulose Top-shell

Monterey, California, to Baja California.

½ inch in length, solid, glossy; characterized by numerous, fine, spiral cords which are sometimes weakly beaded, and by a wide, rather flattish periphery. Nucleus tan. Color light yellowish brown, commonly with a spiral row of subdued white spots at the lower periphery. Interior brightly nacreous. Moderately common on rocks at low tide.

354

***Calliostoma annulatum* (Lightfoot, 1786)**   355
Ringed Top-shell   **Color Plate 2**

Alaska to San Diego, California.

1 to 1¼ inch in length, not quite so wide; characterized by its light weight, golden-yellow color with a mauve band at the periphery, and by the numerous, spiral rows of tiny, dis-

tinct beads (5 to 9 rows in the spire whorls). Nucleus pink. Dredged offshore and occasionally washed ashore. Formerly *C. annulatum* Martyn (non-binomial).

***Calliostoma canaliculatum* (Lightfoot, 1786)**   356
Channeled Top-shell   **Color Plate 2**

Alaska to San Diego, California.

1 to 1½ inches in length, not heavy, sides of whorl flat. Periphery of the last whorl sharp. Base of shell almost flat. Characterized by sharp, prominent, slightly beaded, spiral cords. Color yellowish tan. Nuclear whorls white. Moderately common offshore. Found on floating kelp weed. Formerly known as *C. canaliculatum* Martyn, and *doliarium* Holten, 1802. Humphrey was erroneously attributed the authorship of this species.

***Calliostoma variegatum* Carpenter, 1864**   357
Variable Top-shell

Alaska to southern California.

1 inch in length, similar to *canaliculatum*, but with smaller cords which are strongly beaded; nucleus pink; the sides of the spire slightly concave, and the periphery of the last whorl rounded. Uncommonly dredged in 15 to 400 fathoms.

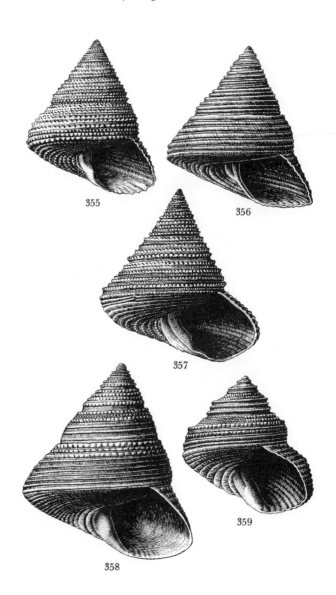

355

356

357

358

359

Calliostoma gloriosum Dall, 1871   358
Glorious Top-shell

San Francisco to San Diego, California.

1 inch in length, not quite so wide, rather light, with about 10 fine, spiral threads between sutures. The upper 5 are inclined to be minutely beaded. Periphery of last whorl moderately sharp. Columella white, fairly thick and with a swelling at the lower ⅔. Nuclear whorls white. Color of shell yellowish brown with darker, purplish brown, slanting, rather elongate spots arranged in 2 spiral series. Moderately common in shallow water to 25 fathoms, in kelp weed.

Calliostoma splendens Carpenter, 1864   359
Splendid Top-shell

Monterey to Baja California.

¼ to ⅓ inch in length, equally wide, with about 5 or 6 whorls which bear between sutures 5 strong spiral cords. The upper 2 or 3 are finely beaded, the lower 2 or 3 are smooth and cordlike. Between the cords, the shell is brilliant, iridescent-orange. General color a yellowish orange with large white maculations on the upper half of the whorls. Moderately common offshore, uncommonly washed ashore.

Calliostoma ligatum (Gould, 1849)   360
Western Ribbed Top-shell   **Color Plate 2**

Alaska to San Diego, California.

¾ to 1 inch in length, equally wide, rather heavy; whorls quite well-rounded; characterized by smooth, spiral, light-tan cords (6 to 8 on the spire whorls) on a background of chocolate-brown. Sometimes flushed with mauve. No umbilicus. Aperture usually pearly-white. A very common littoral species from northern California north. Formerly *C. costatum* Martyn (non-binomial). Dall named the color forms, *caeruleum* and *pictum* in 1919.

360

Calliostoma keenae J. H. McLean, 1970   361
Myra Keen's Top-shell

Off Laguna Beach, California, to Baja California.

15 mm., base imperforate, whorls well-rounded. Color drab-green or yellow-brown with brown flammules. Base delimited by a larger, white-and-brown spotted, spiral rib. First 3 post-nuclear whorls with 3 nonbeaded spirals. Penultimate whorl with about 15 minutely beaded spirals. In *C. supragranosum*

the early spirals are beaded. Uncommon; 24 to 68 fathoms. (*The Veliger*, vol. 12, p. 424.)

Other Pacific species:

362 *Calliostoma platinum* Dall, 1889. Queen Charlotte Strait, British Columbia, to San Diego, California. 50 to 414 fms.

362

363 *Calliostoma antonii* Koch (in Philippi, 1843). Baja California to Panama. Intertidal on rocks.

364 *Calliostoma eximium* (Reeve, 1843). Baja California to Ecuador. (Synonym: *T. versicolor* Menke, 1851.)

365 *Calliostoma marshalli* Lowe, 1935. Gulf of California to Sonora coast of Mexico. (Synonyms: *gemmuloides* Lowe, 1935; *angelenum* Lowe, 1935.)

366 *Calliostoma nepheloide* Dall, 1913. Gulf of California to Panama. 35 to 92 fms. Not uncommon. 25 mm.

367 *Calliostoma turbinum* Dall, 1895. Point Conception to San Diego, California. 35 to 75 fms.

368 *Calliostoma mcleani* Shasky and Campbell, 1964. Baja California to Ecuador. *The Veliger*, vol. 7, p. 117.

369 *Calliostoma gordanum* J. H. McLean, 1970. Gorda Banks, Baja California. 10 fms. *The Veliger*, vol. 12 no. 4, p. 422. 20 mm.

370 *Calliostoma sanjaimense* J. H. McLean, 1970. San Jaime Bank, west of Cape San Lucas, Baja California. 75 fms. *The Veliger*, vol. 12, no. 4, p. 423. 20 mm.

371 *Calliostoma leanum* (C. B. Adams, 1852). Gulf of California to Ecuador. (Synonym: *Trochus macandreae* Carpenter, 1857.)

372 *Calliostoma palmeri* Dall, 1871. Gulf of California.

Subfamily GIBBULINAE Stoliczka, 1868

Genus *Cittarium* Philippi, 1847

There is only one species in this genus, namely *C. pica* from the West Indies. Although fairly good specimens are found without their soft parts in southern Florida and Bermuda, this species has been extinct in those areas for several hundred years. Living individuals may be found

361   366   369   370

abundantly in the West Indies where they are used in chowders by some people. *Livona* Gray, 1847, published nine months after Philippi's name, is a synonym. *Livonae* (and *Livona*) Gray, 1842, is a *nomen dubium* and cannot be used, despite the opinion of Clench, Rehder and others. Type of the genus: *Turbo pica* Linné, 1758.

Cittarium pica (Linné, 1758)   **373**
West Indian Top-shell

Southeast Florida (dead) and the West Indies (alive).

2 to 4 inches in length, heavy, rather rough and with splotches of purplish black on dirty-white. Umbilicus round, narrow and very deep. Inner edge of lip with rich cobalt-blue mottlings. Operculum horny, large, round, multispiral and opalescent blue-green in life. An abundant rock-dwelling intertidal species feeding on algae. Fossil in Bermuda. Occasionally introduced to the Florida Keys, but usually does not survive. An *Acmaea* species lives on the underside of the shell. For biology, see H. A. Randall, 1964, *Bull. Marine Sci.* Gulf Caribbean, vol. 14, pp. 424–433.

373

Genus *Norrisia* Bayle, 1880

Large, solid, roundly depressed shells; nacreous within. Operculum round, multispiral, chitinous and ornamented with spiral rows of bristles. Type: *norrisi* Sowerby, 1838. One species only.

Norrisia norrisi (Sowerby, 1838)   **374**
Norris Shell

Monterey, California, to Baja California.

1½ inches in length, slightly wider, heavy, smoothish with a glossy finish, especially on the underside. Lip sharp. Aperture thickened within and pearly. Umbilicus ovate, very deep, colored a greenish blue on the columellar side, bordered on the other side by glossy black-brown which fades into rich chestnut over the remainder of the shell. Operculum, multispiral, externally ornamented with spiral rows of dense bristles. Animal tinged with red. Moderately common among the kelp weed beds.

374

Genus *Gaza* Watson, 1879

Shell thin, highly opalescent, deeply umbilicate. Operculum multispiral, amber. Type: *daedala* Watson, 1879.

Gaza superba (Dall, 1881)   **375**
Superb Gaza   **Color Plate 2**

Northern Gulf of Mexico to the West Indies.

1 to 1½ inches in width. Spire somewhat elevated. Color old-ivory with a golden sheen. Umbilicus very deep. Early whorls faintly wine-colored. Although formerly thought to be one of our rarest shells, it is now known to be relatively common in the Gulf of Mexico in 50 or more fathoms. It is indeed a beautiful species and difficult to obtain. *G. cubana* Clench and Aguayo, 1940, (**375a**) appears to be a dwarf, deep-water form.

Other species:

376 *Gaza* (*G.*) *fischeri* Dall, 1889. 475 fms., off Cuba to 423 fms. off St. Lucia, Lesser Antilles. *Johnsonia*, vol. 1, no. 12, pl. 3. Rare.

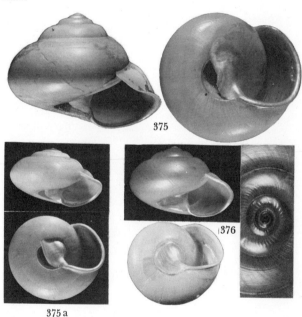

375

375a

377 *Gaza* (*Callogaza* Dall, 1881) *watsoni* (Dall, 1881). 117 to 500 fms., Cuba to Brazil. *Johnsonia*, vol. 1, no. 12, pl. 2. Rare.

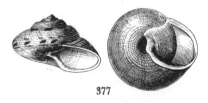

377

Subfamily MONODONTINAE Cossmann, 1916

Genus *Tegula* Lesson, 1835

Turbinate, ½ to 2 inches, solid, whorls flattish, with or without an umbilicus. Thickened or toothed columella. Operculum horny, multispiral. Type: *pellisserpentis* Wood, 1828. For radulae of California *Tegula*, see H. K. Fritch-

man, 1965, *The Veliger*, vol. 8, p. 11. For spawning, see
F. P. Bilcik, *The Veliger*, vol. 7, p. 233.

Subgenus *Agathistoma* Olsson and Harbison, 1953

Smooth or finely beaded spirally, with a narrow umbilicus.
Subgenotype: *viridulus* (Gmelin, 1791) (not *fasciata* Born,
1778).

Tegula fasciata (Born, 1778)　　378
Smooth Atlantic Tegula

South Florida and the West Indies to Brazil.

½ to ¾ inch in width. Surface smooth; color yellowish to
brown, with a fine mottling of reds, browns and blacks; often
with a narrow, pale, spiral band of color on the periphery.
Under the lens, spiral rows of alternating red and white, short
lines or dots may be seen. Some specimens may have zigzag
white bands. Interior of deep, round, smooth umbilicus and
the callus are white. Two teeth at the base of the columella.
Thick adults may have small teeth just inside the lower
margin of the aperture. Whorl may be slightly concave just
below the suture. In the young only, the umbilicus has two
deep spiral grooves. Moderately common under rocks at low
tide. Uncommon on West Florida coast. Synonyms are: *sub-
striatum* Pilsbry, 1889; *picta* Tenison-Woods, 1877; *maculo-
striata* C. B. Adams, 1845.

T. hotessieriana (Orbigny, 1842) (**379**) from the Lower Florida Keys to Brazil is similar, but rarely over ⅓ inch, with a
more rounded periphery, with smaller, neater, equal-sized,
smooth spiral threads, which are dark bluish black in color,
except for a whitish area around the narrow umbilicus. Uncommon. This is considered a variant of *fasciata* Born, by some
workers, and may well be immatures or dwarfs.

378

Tegula lividomaculata (C. B. Adams, 1845)　　380
West Indian Tegula

Florida Keys and the West Indies.

¾ inch in width and about ½ inch in length; not glossy.
Tops of whorls sculptured with about a dozen fairly regular,
small, spiral cords. The angular periphery of the whorls bears
the largest cord. Umbilicus round, deep and furrowed on its
sides by two spiral cords, the upper one ending at the columella in a fairly sizable bead. Columella set back quite far
at its upper half; the lower section bears the bead, and below

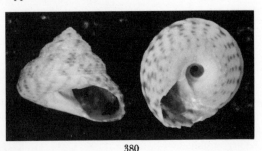

380

that there are several smaller, indistinct beads. Color of
shell grayish to brownish white with small mottlings of
reddish or blackish brown. Formerly *scalaris* Philippi, 1844;
canaliculatus Orbigny, 1842 (non Brocchi, 1814); *gundlachi*
Philippi, 1848; *turbinata* Tenison-Woods, 1877 (not Brocchi,
1814) and *indusi* "Gmelin." Common under rocks in the West
Indies, but uncommon in the Lower Florida Keys.

Tegula excavata (Lamarck, 1822)　　381
Green-based Tegula

Lower Florida Keys and Caribbean area.

½ inch in length and width. Characterized by its bluish
gray color, corrugated sculpture (weak spiral cords and oblique
lines of growth), concave base, thin outer lip and especially
by the blue-green to iridescent-green circle of color around
the very deep, round, narrow umbilicus. A variant exists in
some areas which lacks the green umbilical color and in which
the spiral cords are stronger and the shell with axial, slanting
bars of black-brown. Very common in the West Indies, along
the rocky shores. Believed to be rare in Florida.

381

Tegula gallina Forbes, 1850　　382
Speckled Tegula

San Francisco to the Gulf of California.

1 to 1½ inches in length, very similar to *funebralis*, but a
lighter, grayish green color with dense, zigzag, axial stripes of
purplish. The shell surface is also coarser. A common, southern species found among littoral rocks. The Guadalupe Island
forma multifilosa Stearns, 1893, lacks axial markings.

Tegula brunnea Philippi, 1848　　383
Brown Tegula

Oregon to Santa Barbara Islands, California.

1 to 1½ inches in length, similar to *funebralis*, but light
chestnut-brown in color with the base often glossy, brownish
white. The umbilicus is closed, but usually with a dimplelike
impression. Columella usually with only one small tooth near
the base. Common at dead low tide on rocks. Usually heavily
encrusted with algal growths. Synonyms are *fluctuatum* Dall,
1871, and *fluctuosum* Dall, 1919.

Tegula aureotincta Forbes, 1850　　384
Gilded Tegula

Southern third of California to Mexico.

¾ to 1 inch in length, heavy; dark grayish to gray-green;
characterized by a golden-yellow stain within the deep, round

382　　　384　　386

umbilicus, by the sky-blue band around the umbilicus, and by the 4 or 5 strong, smoothish, spiral cords on the periphery and the base. Top of whorls with weak, crude, slanting, axial wrinkles. A moderately common, littoral, rock-loving species.

Tegula funebralis (A. Adams, 1855)   385
Black Tegula

Vancouver, British Columbia, to Baja California.

1 to 1½ inches in length, heavy, dark purple-black in color; smoothish, but with a narrow, puckered band just below the suture. Weak spiral cords rarely evident; coarse growth lines present in large, more elongate specimens. Base rounded. Umbilicus closed or merely a slight dimple. Columella pearly, with two small nodules at the base. The head and tentacles are entirely black. The sole of the foot of males is usually light-cream in color, that of females usually brownish (see P. W. Frank, 1969, *The Veliger*, vol. 11, p. 440). A very common littoral, rock-loving species. Do not confuse with *T. gallina*. For biology, see *The Veliger*, 1964, vol. 6, supplement.

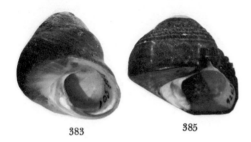

383   385

Tegula eiseni Jordan, 1936   386
Western Banded Tegula

Monterey, California, to Baja California.

¾ inch in length, heavy; whorls and spire convex. Umbilicus very deep, round and fairly narrow. Whorls with numerous, beaded, spiral cords. Outer lip sharp, but thickened and pearly within. Lower part of lip with about 8 small nodules opposite the spiral threads which run back into the aperture. Color rusty-brown with black flecks. Compare with *aureotincta* whose umbilical area is stained with greenish blue and golden-yellow. A moderately common littoral rock-dweller, living down to the subtidal region. Formerly known as *ligulata* Menke, 1850 (of authors). *T. mendella* J. H. McLean, 1964, is a synonym. True *ligulata* comes from Mazatlan, Mexico.

Tegula mariana Dall, 1919   387
Maria Tegula

Baja California to Peru.

8 to 10 mm. in length, somewhat wider, variable in shape, umbilicate, with a fine striate surface and with several noded keels on the periphery of the last whorl. Especially common along the west Mexican coast. Synonyms: *turbinata* Pease, 1869 (not A. Adams, 1859). *T. mariamadre* Pilsbry and Lowe, 1932, is a southern subspecies of *ligulata* (Menke, 1850).

Subgenus Stearnsium S. S. Berry, 1958

The type of this subgenus is *regina* (Stearns, 1892). See S. S. Berry, *Leaflets in Malacology*, vol. 1, no. 16, p. 92.

Tegula regina Stearns, 1892   388
Queen Tegula   Color Plate 2

Catalina Island to the Gulf of California.

1½ inches in length, slightly wider; 6 or 7 whorls, spire flat-sided; base slightly concave. With numerous slanting,

small, axial cords. The crenulated periphery slightly overhangs the suture of the whorls below. Base with strong, arched lamellae. Color dark purplish gray. Umbilical region stained with bright golden-yellow. An uncommon collector's item secured by diving. It has also been washed ashore on Catalina Island.

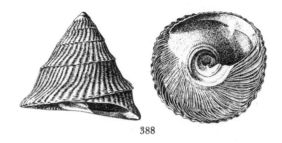

388

Subgenus Promartynia Dall, 1909

The type of this subgenus is *pulligo* (Gmelin, 1791).

Tegula pulligo (Gmelin, 1791)   389
Dusky Tegula

Alaska to Santa Barbara, California.

1 to 1½ inches in length, slightly wider. Resembles *brunnea*, but has a deep, round umbilicus and a thin, rather sharp columella. It is also very similar to *montereyi*, but its whorls are more rounded and its umbilicus is more smoothly rounded and without the white color and faint spiral ridges found in *montereyi*. This species is doubtfully placed here and perhaps should be considered a typical *Tegula*. *T. marcida* (Gould, 1853) is a synonym. Moderately common, especially in the north. Very large, heavy specimens from Oregon are *forma taylori* Oldroyd, 1922.

Tegula montereyi (Kiener, 1850)   390
Monterey Tegula

Bolinas Bay, California, to Santa Barbara Islands.

1 to 1½ inches in length, about as wide. Conical in shape, with very flat-sided whorls and spire. Base almost flat. Surface smoothish, except for almost obsolete spiral threads. Umbilicus very deep, lined with 1 or 2 weak spiral cords. Columella arched, and with 1 prominent, pointed tooth. This rather rare species resembles a large *Calliostoma*. It is found on kelp in moderately deep waters.

389   390

Other species:

391 *Tegula (Agathistoma) rubroflammulata* (Koch in Philippi, 1843). Gulf of California to Panama. (Synonym: *Trochus coronulatus* C. B. Adams, 1852)

392 *Tegula (Agathistoma) verdispira* J. H. McLean, 1970. Tres Marias Islands, Mexico, and Baja California. See *Malacol. Rev. for 1969*, vol. 2, p. 122.

393 *Tegula viridula* (Gmelin, 1791). Northern South America to eastern Brazil.

392

394 *Tegula (Omphalius) patagonica* (Orbigny, 1840). Rio de Janeiro, Brazil, to Argentina. (Synonym: *Minolia amblia* Dall, 1927)

395 *Tegula (Agathistoma) corteziana* McLean, 1970. Gulf of California to Guaymas, Mexico.

396 *Tegula (Agathistoma) felipensis* McLean, 1970. North end of the Gulf of California.

Genus *Turcica* A. Adams, 1854

Shell strong, top-shaped; suture deep. Columella with one or two strong teeth. Type: *T. monilifera* (A. Adams, 1854).

Turcica caffea (Gabb, 1865)   397
Gabb's Two-toothed Top-shell

Monterey to San Diego, California.

1 inch; heavy, beaded, deep suture; resembling a *Calliostoma*, but has two strong white folds on the middle of the columella. Common; 15 to 25 fathoms, and occasionally found in kelp washed ashore.

397

Other species:

398 *Turcica admirabilis* S. S. Berry, 1969. Off Morro Colorado, Sonora, Mexico, 20 to 30 fms., to Panama Bay. *Leaflets in Malacology*, no. 26, p. 160. 40 mm.

Subfamily HALISTYLINAE Keen, 1958

Genus *Halistylus* Dall, 1890

Very small, brownish, elongate, cyclindrical shells, either smooth or spirally striate. Aperture round, lip thickened. Operculum multispiral. Type: *columna* Dall, 1890. A strange trochid genus more resembling a pupoid land snail in shape. Radula similar to that of Umboniinae.

Halistylus pupoides (Carpenter, 1864)   399
Pupoid Halistyle

Forrester Island, Alaska, to Panama.

4 to 5 mm., thick-shelled, elongate, brownish, with 5 or 6 whorls with 8 or 9 spiral, sharp threads. Axial growth lines microscopic. Apertural rim thickened. Common; dredged on sandy bottoms from 8 to 40 fathoms. *H. subpupoides* (Tryon, 1887) is a synonym.

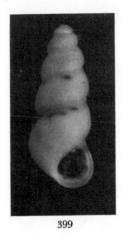   

400

399

Other species:

400 *Halistylus columna* (Dall, 1890). Off Brazil and Uruguay, 10 to 36 fms. 6 mm. Argentina.

Genus *Pseudostomatella* Thiele, 1924

Small, trochoid, wide-mouthed, spire short. Nacreous within. Operculum multispiral. Type: *Turbo papyraceus* Gmelin, 1791.

Pseudostomatella erythrocoma (Dall, 1889)   401
Dall's False Stomatella

Southeast Florida; Bahamas; Greater Antilles.

4 to 6 mm., resembles a miniature *Solariella*, trochoid in shape, yellowish, variegated with pink and opaque-white. 4 or 5 whorls, carinated by two rows of beads or nodules. Spiral threads present. Umbilicus present. Base round; aperture round. Moderately common; from offshore to 54 fathoms. *Margarita samanae* Dall, 1889, may be a synonym.

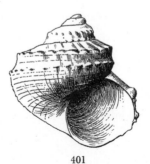

401

Other species:

402 *Pseudostomatella coccinea* (A. Adams, 1850). Scarlet false stomatella. Greater Antilles; Rare.

Genus *Synaptocochlea* Pilsbry, 1890

Shell ear-shaped, without holes, spire very short and submarginal; aperture very large; surface spirally striate or decussate. Operculum chitinous. Type: *montrouzieri* Pilsbry, 1890.

Synaptocochlea picta (Orbigny, 1842)   **403**
Painted False Stomatella

Bermuda, Florida Keys to West Indies; off Brazil.

2 to 4 mm., ear-shaped. Color variegated with white, reds and purples. With an oily glistening surface and semitranslucent. 3 whorls, nucleus white. With 15 to 22 weakly beaded spiral threads. Interior shiny, but not nacreous. Moderately common; intertidal to offshore among coral rubble. The form or subspecies *nigrita* Rehder, 1939, has stronger and fewer riblets, and is sometimes colored black. Albino shells rare.

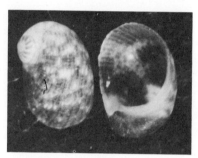

403

Family CYCLOSTREMATIDAE Fischer, 1885

The operculum in members of this family is round, multispiral, and with a horny base on top of which are numerous rows of tiny calcareous beads. The shells are porcelaneous. For various West American species, see A. M. Strong, 1934, *Trans. San Diego Soc. Nat. Hist.*, vol. 7, pp. 429–452. The family Liotiidae is synonymous.

Genus *Cyclostrema* Marryat, 1818

Shells white, small, discoidal and umbilicate. Type: *cancellatum* Marryat, 1818. Synonyms are: *Pseudoliotina* Cossman, 1925, and *Munditia* Finlay, 1926. *Cyclostrema* is a neuter, not feminine, word. Many species originally described in this genus by Dall, Verrill and Bush belong to the Vitrinellidae.

Cyclostrema cancellatum Marryat, 1818   **404**
Cancellate Cyclostreme

South Florida, the Bahamas to the Caribbean. Brazil.

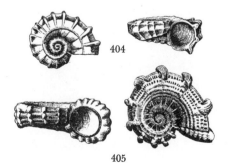

404
405

½ inch in diameter, flat-topped, 4 whorls, opaque-white. Widely and deeply umbilicate. Axial sculpture of 15 to 17 rounded, low ribs which encircle the entire whorl and are made nodulose in crossing the 12 smaller spiral cords. Periphery squarish, with a cord above, below and at the center. Rare; from 1 to 37 fathoms. Synonym: *acrilla* (Dall, 1889).

C. amabile (Dall, 1889), **(405)** from Cuba to Barbados, is much rarer and differs in being smaller, in having a thicker, more rounded lip, and in lacking axial cords on top of the whorls. 25 to 80 fathoms. Uncommon.

Cyclostrema tortuganum (Dall, 1927)   **406**
Tortugas Cyclostreme

Southeast Florida.

4 to 6 mm. in diameter, white, discoidal; spire flat. Periphery of whorl squarish, with numerous, fairly sharp axial riblets. Top edge of last whorl with fairly large nodes which become larger and more distant near the aperture. Base of whorl with similar but smaller nodes. Entire external surface has a microscopic "frosted" appearance. Uncommon; 16 to 35 fathoms.

Other species:

407   *Cyclostrema cookeanum* (Dall, 1918). 7 to 10 fms., off South Coronado Island, Baja California. See *Johnsonia*, vol. 2, no. 27, pl. 88.

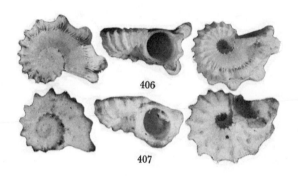

406
407

—   *Cyclostrema planorbis* (Dall, 1927). Is an *Omalogyra*.

408   *? Cyclostrema schrammii* Fischer, 1857. Guadeloupe Island, Lesser Antilles. Rare. May be an *Arene*.

409   *? Cyclostrema huesonicum* Dall, 1927. 90 to 95 fms., off Florida Keys. May be young in some other genus.

410   *Cyclostrema miranda* Bartsch, 1910, is a synonym of the British *Tornus subcarinatus* (Montagu, 1803) fide D. R. Moore, 1969. (*The Veliger*, vol. 12, p. 169)

Genus *Coronadoa* Bartsch, 1946

Shell minute, planorboid, with axial scalariform ribs. Doubtfully belongs in this subfamily. Type: *simonsae* Bartsch, 1946 **(411)** is the only known species. California.

Genus *Liotia* Gray, 1847

Shell small, turbinate, with a round, thickened, continuous outer lip; umbilicus usually deep. Cancellate, pitted sculpturing. Operculum as in *Cyclostrema*. Type: *cancellata* (Gray, 1828).

Liotia fenestrata Carpenter, 1864   **412**
Californian Liotia

Monterey, California, to San Martin Island, Mexico.

412

⅛ inch in diameter; spire low, shell solid; deeply and nar-rowly umbilicate. Aperture circular, pearly within. Ash-white in color. Characterized by heavy cancellate sculpturing which makes the shell appear pitted by rows of deep, squarish holes. Uncommonly dredged from 10 to 25 fathoms. *L. cookeana* Dall is not this species, as is commonly thought, but is a *Cyclostrema*.

Other species:

413 *Liotia microgrammata* Dall, 1927. 191 fms., Florida Straits off Havana, Cuba.

414 *Liotia lurida* Dall, 1913. San Joseph Island, Gulf of Cali-fornia. 5 mm.

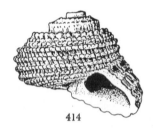

414

Genus *Sansonia* Jousseaume, 1892

Shell minute, turbinate, with a high, flat spire. Aperture round; outer lip thick and with a strong varix. Type: *Iphitus tuberculatus* Watson, 1886. *Mecoliotia* Hedley, 1899, and *Pickworthia* Iredale, 1917, are synonyms. Thiele (1928) un-wisely placed this genus in Aedorbidae.

Sansonia tuberculata (Watson, 1886)   414a
Tuberculate Sansonia

Off Miami, Florida, to Cuba and Puerto Rico.

Shell 1.5 to 1.7 mm. in length, solid, white, with 6 flat-sided whorls each bearing 3 rows of large, rounded beads. Base of shell with a row of large beads bordering a narrow umbilicus. Aperture round, the outer lip being very thick and varixed. Rare on rubble bottom from a depth of 100 to 713 meters. *Mecoliotia bermudezi* Clench and Aguayo, 1936 (*The Nautilus*, vol. 49, p. 91) is a Pleistocene Cuban synonym. Illustration and details are from Moore, D. R., 1963, *Bull. Marine Sci.*, vol. 13, p. 73.

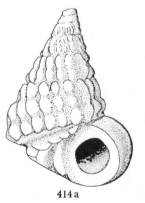

414a

Genus *Arene* H. and A. Adams, 1854

Shell small, depressed turbinate, thick-shelled with a rimmed, thickened round aperture. Sculpture of spinose spiral cords. Operculum multispiral with rows of tiny cal-careous beads. Type: *cruentata* (Mühlfeld, 1829).

Arene cruentata (Mühlfeld, 1829)   415
Star Arene

Southeast Florida and the West Indies. Gulf of Mexico.

¼ inch in length, ⅓ again as wide. 4 or 5 whorls angular, with the periphery bearing a series of strong, triangular spines which are hollow on their anterior edges. Color white to cream with small, bright-red patches on top of the whorls. Be-low the main row of spines there is a minor spiral row of smaller spines. Suture channeled. Aperture circular, pearly within. Umbilicus round, deep and bordered by 3 spiral, beaded cords. Uncommon under rocks.

The form *vanhyningi* Rehder, 1943 (**416**), from Sand Key, Key West, is pale gray-white with most red patches absent. It lacks fine, axial ridges on top of the whorl which are usually present in the typical form. Uncommon.

Arene venustula Aguayo and Rehder, 1936   417
Venuste Arene

Miami, Florida, to Puerto Rico.

Similar to *cruentata*, but smaller, much more squat, chalky-white and with two peripheral rows of blunt spines. The rows are very close to each other. Rare; 20 fathoms.

Subgenus *Marevalvata* Olsson and Harbison, 1953

Shells small, 3 mm., lacking the spiral cord bordering the umbilicus. Outer lip not thickened. Type: *tricarinata* (Stearns, 1872).

Arene tricarinata (Stearns, 1872)   418
Gem Arene

North Carolina to south half of Florida to Brazil.

⅛ inch or less, turbinate in shape; 3 spiral rows of neat, tiny beads on the squarish periphery. Suture minutely chan-neled and bounded below by a spiral row of whitish beads. Top slope of whorls and base of shell flattish. Axial threads on entire shell microscopic and crowded. Umbilicus round, deep, bordered by 7 to 9 distinct beads. Color of shell white to tan with minute speckling of red and/or brown. Commonly dredged from 3 to 100 fathoms. The name *gemma* Tuomey and Holmes, 1856, evidently refers to a different Miocene species.

415

417

418

Arene variabilis (Dall, 1889)   **419**
Variable Arene

North Carolina to Florida and the West Indies. Brazil.

3/16 inch in length, turbinate, similar to *A. tricarinata,* but pure-white in color, with scalelike beads, suture more deeply channeled, and with a more rounded periphery. 12 very weak beads bordering the more open umbilicus. The 3 spiral rows of beads on the whorl may be almost smooth in some specimens. Very commonly dredged from 20 to 270 fathoms.

419   421

Arene briareus (Dall, 1881)   **420**
Briar Arene

North Carolina to Florida and Yucatan.

6 to 8 mm. in length, not quite so wide; thick, rose in color. Whorls globose, the last with about 10 spiral cords of tiny, prickly beads. Suture deeply channeled. Umbilicus very narrow and deep. Moderately common from 18 to 85 fathoms. The shell illustrated in the first edition of "American Seashells" (pl. 17 u) was actually *A. bairdii* Dall. *A. perforata* (Dall, 1889) and *aspina* (Dall, 1889) are minor varieties of *briareus.*

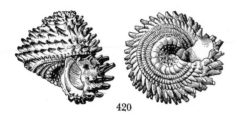

420

Arene bairdii (Dall, 1889)   **421**
Baird's Arene

North Carolina to Key West, Florida.

6 mm., similar to *briareus* but having much weaker sculpture, having noduled instead of spinose ridges. Last whorl has 6 basal, 2 peripheral and 3 subsutural fine revolving lines. Suture channeled. Outer lip thin. Umbilicus small, with a single, rounded spiral ridge within. Color whitish, weakly maculated or a solid reddish. A form with the minute nodules produced into squarish flutes was called *trullata* (Dall, 1889) **(422)**. Moderately common from 45 to 805 fathoms.

Subgenus *Macrarene* Hertlein and Strong, 1951

Shell depressed turbinate, with strong peripheral projections. The surface is sculptured with both spiral cords and axial ribs or threads. Six living Pacific species. Type: *californicus* (Dall, 1908).

Other species:

423 *Arene (Macrarene) lepidotera* (J. H. McLean, 1970). Revillagigedo Islands, Mexico. 10 mm. *Malacol. Rev. for 1969,* vol. 2, p. 125.

424 *Arene (Macrarene) californica* (Dall, 1908). Off Baja California, 113 fms. Cerros Island. (Synonym: *pacis* Dall, 1908)

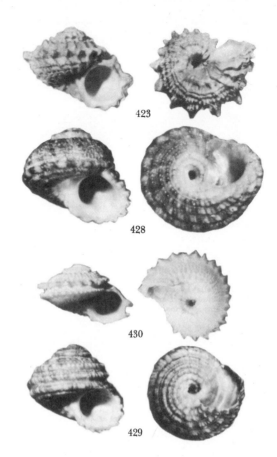

423

428

430

429

425 *Arene (Macrarene) diegensis* J. H. McLean, 1964. A Pliocene species from San Diego.

426 *Arene (Macrarene) coronadensis* (Stohler, 1959). North Island of the Coronado Islands, Baja California, 150 feet. *Proc. Calif. Acad. Sci.,* ser. 4, vol. 29, p. 439. Is *californicus* (Dall, 1908)?

427 *Arene (Macrarene) farallonensis* (A. G. Smith, 1952). S. E. Farallon Islands, California; Gulf of California. *Proc. Calif. Acad. Sci.,* ser. 4, vol. 27, p. 385.

428 *Arene adusta* J. H. McLean, 1970. Espiritu Santo Islands, Baja California, Mexico. 6 mm. in width. *Malacol. Rev. for 1969,* vol. 2, p. 123.

429 *Arene ferruginosa* J. H. McLean, 1970. South off White Friars, Mexico, 25 fms. *(loc. cit.,* p. 123)

430 *Arene stellata* J. H. McLean, 1970. Espiritu Santo Islands, Baja California, 3 to 25 fms. *(loc. cit.,* p. 125)

431 *Arene (Marevalvata) balboai* (Strong and Hertlein, 1939). Head of the Gulf of California to Ecuador, 18 to 37 meters.

432 *Arene miniata* (Dall, 1889). Bahamas to Barbados. Beach to 15 fms.

432

433 *Arene (Otollonia* Woodring, 1928) *fricki* (Crosse, 1865). Abundant, 3 to 20 meters, head of the Gulf of California to Ecuador. (Synonym: *Liotia rammata* Dall, 1918)

434 *Arene lucasensis* Strong, 1933. Cape San Lucas, Baja California.

435 *Arene carinata* Carpenter, 1857. Gulf of California.

436 *Arene (Marevalvata) riisei* "Dunker" Rehder, 1943. Panama to the Virgin Islands. *Proc. U.S. Nat. Mus.*, vol. 93, p. 192, pl. 19. Uncommon.

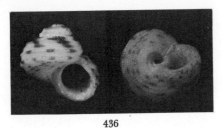

436

437 *Arene brasiliana* (Dall, 1927). Dredged off Cape Roque, Brazil. Rare.

Subfamily SKENEINAE Thiele, 1929

Shells very small, sculpture weak or absent; radula with only 4 or 5 marginal teeth. Operculum corneous, multispiral, without minute beads. This subfamily is a "dumping ground" for very small, little understood, deep-water snails, and over the years will doubtlessly lose "enlightened" genera and gain new strangers.

Genus *Skenea* Fleming, 1825

Minute, planorboid, smoothish shells with a low spire and round aperture. Type: *serpuloides* Montagu, 1808. *Delphionoidea* Brown, 1844, is a synonym. Consult *Skeneopsis* in the Rissoidae for the well-known *"Skenea" planorbis.* No Atlantic coast species, but the following Pacific coast species may belong here:

438 *Skenea californica* (Bartsch, 1907). Monterey to San Diego, California.

439 *Skenea concordia* (Bartsch, 1920). Olga, Washington.

440 *Skenea coronadoensis* (Arnold, 1903). San Diego to Baja California.

441 *Skenea carmelensis* A. G. Smith and M. Gordon, 1948. Carmel Bay, California. 1.7 mm. in diam. *Proc. Calif. Acad. Sci.*, series 4, vol. 26, p. 230.

Genus *Ganesa* Jeffreys, 1883

Small, turbinate, smooth or striate, whitish, deepsea shells. Type: *pruinosa* Jeffreys, 1883, of the North Atlantic. There are four species known from the Eastern Tropical Pacific and the Galapagos Islands. The Western Atlantic deepsea species are:

Subgenus *Lissospira* Bush, 1897

442 *Ganesa proxima* Tryon, 1888. Type of the subgenus. (Synonym: *affinis* Verrill, 1884.) Off North Carolina to Fernandina, Florida, 294 to 843 fms.

443 *Ganesa bushae* (Dall, 1927). Off Fernandina, Florida, 294 fms.

444 *Ganesa conica* (Dall, 1927). Off Fernandina, Florida, 294 fms.

445 *Ganesa depressa* (Dall, 1927). Off Fernandina, Florida, 294 fms.

446 *Ganesa valvata* (Dall, 1927). Off Georgia, 440 fms. Off Fernandina, Florida, 294 fms.

447 *Ganesa diaphana* Verrill, 1884. Off Martha's Vineyard, Massachusetts, 1,290 fms.

448 *Ganesa striata* Bush, 1897. Off Martha's Vineyard, Mass., 384 fms.

449 *Ganesa convexa* Bush, 1897. Off Delaware Bay, 630 fms.

450 *Ganesa verrilli* Tryon, 1888. (Synonym: *Cyclostrema cingulatum* Verrill, 1884.) Southeast of Nantucket, 547 fms.

451 *Ganesa ornata* Verrill, 1884. Off Cape Hatteras, 843 fms.

452 *Ganesa dalli* Verrill, 1882. (Synonym: *Cyclostrema trochoides* Verrill, not Sars; *Cyclostrema fulgidus* Dall, 1889.) South of Martha's Vineyard, 487 fms.

452

453 *Ganesa abyssicola* Bush, 1897. East of Georges Bank, 980 fms.

Subgenus *Granigyra* Dall, 1889

454 *Ganesa limata* (Dall, 1889). Type of the subgenus. Off Georgia to Florida Straits, 294 to 440 fms. Cuba, 310 fms.

455 *Ganesa radiata* Dall, 1927. Off Fernandina, Florida, 294 fms.

456 *Ganesa spinulosa* Bush, 1897. Off east Florida, 338 fms.

Genus *Haplocochlias* Carpenter, 1864

Shells 3 to 5 mm., very solid, turbinate, white to tan, resembling the land snails *Helicina,* with a thick, round outer lip; no umbilicus; 2 nuclear whorls, smooth. Postnuclear whorls finely, spirally striate. Outer lip flaring. Type: *cyclophoreus* Carpenter, 1864. Three species reported:

457 *Haplocochlias swifti* Vanatta, 1913. St. Thomas, Virgin Islands; Havana, Cuba; Bocos, Panama. *Proc. Acad. Nat. Sci. Phila.*, vol. for 1913, p. 23, fig. 3. Rare.

457

458 *Haplocochlias cyclophoreus* Carpenter, 1864. Gulf of California; Magdalena Bay, Baja California: Uncommon. 5 mm.

459 *Haplocochlias lucasensis* (Strong, 1934). Gulf of California. Intertidal to 10 meters. 1.7 mm.

Genus *Parviturbo* Pilsbry and McGinty, 1945

Shells minute, 2 mm., solid, turbinate, narrowly umbilicate, 1 or 2 smooth nuclear whorls, following gradually by spiral threads and axial striae. Operculum multispiral, chitinous. Radula rhipidoglossate. Foot with 3 pairs of long, ciliated epipodial cirri. Posterior end of foot not bifurcate as in the very similar *Parviturboides* Pilsbry and McGinty, 1950, a taenioglossate Vitrinellidae. Type: *rehderi* Pilsbry and McGinty, 1945.

Parviturbo rehderi Pilsbry and McGinty, 1945   **460**
Rehder's Parviturbo

South half of Florida to Panama.

1.7 mm. wide, 1.6 mm. high, 4 convex whorls with a deep suture. Whorls strongly angular, the last with 7 strong ones, the upper one with rather wide intervals above and below it, the rest with intervals about equal in width to the spiral ridges. Axial threads on the slopes of the ridges, but not on their tops. Moderately common; under rocks, intertidally.

Parviturbo acuticostatus (Carpenter, 1864)   **461**
Sharp-ribbed Parviturbo

Monterey, California, to the Gulf of California.

3 mm., whitish, 6 spiral ribs on the body whorl, with the interspaces smooth. A variable species with several varietal names. Those that have minute axial threads between the spiral cords are form *bristolae* Baker, 1929. Synonyms or minor varieties are: *stearnsii* (Dall, 1918) (Gulf of California); *radiata* (Dall, 1918); *Fossarus angiolus* Dall, 1919; *supranodosa* Strong, 1933 (Gulf of California). Formerly placed in *Arene*.

Other species:

462 *Parviturbo weberi* Pilsbry and McGinty, 1945. South Florida, 10 fms. *The Nautilus*, vol. 59, p. 56.

463 *Parviturbo francesae* Pilsbry and McGinty, 1945. Off Palm Beach, Florida, 50 fms. Southwest of Johns Pass, west Florida, 35 fms. *The Nautilus*, vol. 59, p. 56. 3 mm.

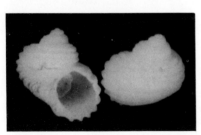

463

464 *Parviturbo calidimaris* Pilsbry and McGinty, 1945. South half of Florida, 11 fms. *The Nautilus*, vol. 59, p. 56.

465 *Parviturbo concepcionensis* (Lowe, 1935). Head of the Gulf of California to Panama, in 15 to 35 meters. 5.6 mm.

466 *Parviturbo stearnsii* (Dall, 1918). Baja California to Colombia. (Synonym: *Liotia heimi* Strong and Hertlein, 1939.) 2.2 mm.

Genus *Molleriopsis* Bush, 1897

Type of the genus is *abyssicola* Bush, 1897.

467 *Molleriopsis abyssicola* Bush, 1897. Off Georges Banks, Massachusetts, 1,769 fms.

468 *Molleriopsis sincera* (Dall, 1890). Off Florida, 294 fms. to off Para River, Brazil, 391 fms. 3 mm.

468

Genus *Dillwynella* Dall, 1889

Minute (4 mm.), solid shell (resembling a miniature *Norrisia* of California), with a white color, covered by a thin epidermis. Umbilical region pushed in slightly and bounded by a spiral cord. Operculum corneous, round, multispiral, with 5 whorls. Type and only known species (469): *Dillwynella modesta* Dall, 1889. Off Georgia, 440 fathoms; off Fernandina, Florida, 294 fathoms; off St. Lucia, Lesser Antilles, in 226 fathoms.

469

Genus *Leptogyra* Bush, 1897

Minute, deepsea, planorboid, shells with rounded whorls, round aperture and deep rounded umbilicus. Operculum horny. Type: *verrilli* Bush, 1897. The known species are:

470 *Leptogyra verrilli* Bush, 1897. Off Delaware Bay, 1,594 fms.

471 *Leptogyra inconspicua* Bush, 1897. Off Delaware Bay, 1,594 fms.

472 *Leptogyra eritmeta* Bush, 1897. Off Delaware Bay 1,594 fms.

473 *Leptogyra alaskana* Bartsch, 1910. Port Graham, Cook's Inlet, Alaska. 0.8 mm. *The Nautilus*, vol. 23, p. 136.

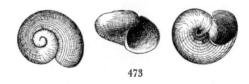

473

Family TURBINIDAE Rafinesque, 1815

Subfamily TURBININAE Rafinesque, 1815

Genus *Turbo* Linné. 1758

Operculum shelly. Type of the genus is *petholatus* Linné, 1758. *Laeviturbo* Cossmann, 1918, is a synonyn.

Subgenus *Marmarostoma* Swainson, 1829

Shells nodulose or spiny; operculum smooth or granular. Type: *chrysostomus* Linné, 1758. *Senectus* Swainson, 1840, is a synonym.

Turbo castanea Gmelin, 1791   **474**
Chestnut Turban   **Color Plate 2**

North Carolina to Florida, Texas and the West Indies. Brazil.

1 to 1½ inches in length. Color orangish, greenish, brown or grayish, commonly banded with flamelike white spots. Aperture white. Callus on columella heavy. Lower lip projects downward. Operculum calcareous. The form named *crenulatus* Gmelin, 1791, is merely less tuberculate. The name *castanea* is a noun in apposition and should not be declined to *castaneus*. A common, shallow-water species.

Subgenus *Taeniaturbo* Woodring, 1928

Spirally ribbed; operculum calcareous and smoothish. *Taenioturbo* is a misspelling. Type: *canaliculata* Hermann, 1781.

Turbo canaliculatus Hermann, 1781   **475**
Channeled Turban   **Color Plate 2**

Southeast Florida and the West Indies. Brazil.

2 to 3 inches in length. A deep smooth channel runs just below the suture. Surface glossy. 16 to 18 strong, spiral, smooth cords on body whorl. Aperture white. Umbilicus narrow. Operculum pale-brown inside with 3 or 4 whorls, and white, smoothish and convex on the outside. This is the handsomest *Turbo* in the Western Atlantic, and is considered uncommon in American waters. Formerly *T. spenglerianus* Gmelin, 1791. Shore to 60 fathoms.

Turbo cailletii Fischer and Bernardi, 1856   **476**
Filose Turban   **Color Plate 2**

Southeast Florida, Bahamas, and the Caribbean.

½ to ¾ inch, similar to *castanea*, but has about a dozen, smooth, spiral cords on the body whorl, is slightly umbilicate, has no beading, and has a broader lower half of the columella. Color maculated red, rarely golden-yellow. Formerly called *T. filosus* Wood, 1828, and Kiener, 1873. The rarest of the Caribbean turbans. 12 to 120 feet, near corals.

476

Subgenus *Callopoma* Gray, 1850

Operculum thick, shelly, granular, with a strong spiral central ridge, a deep pit and a marginal band of pustular ribs and deep grooves. Type: *fluctuosus* Wood, 1828.

477 *Turbo fluctuosus* Wood, 1828. Cedros Island, Baja California, Gulf of California; Ecuador and Peru. 2 inches.

Subgenus *Halopsephus* Rehder, 1943

Adult shell smoothish, polished, imperforate. Operculum smooth and polished at the center, granulose at the edges. See R. Robertson, 1958, *Jour. Wash. Acad. Sci.*, vol. 47, p. 316, for details. Type: *haraldi* Robertson, 1958.

Turbo haraldi Robertson, 1957   **478**
Harald's Dwarf Turban

Bahamas to Barbados and Panama.

Adult shell 11 to 25 mm., smoothish, imperforate, mouth round. Color reddish brown to apricot-orange with minute white flecks and a white columella and umbilical region. Nuclear whorls planate and white or pink; following early whorls are red with 2 carinae at the periphery, the upper one bearing short, flattened spines. Young (2 mm.) has row of strong beads bordering a deep umbilicus. Operculum calcareous, granular around the edges, smoothish at the center. Rare; on gravel and sponge bottom, 37 to 50 fathoms. *Turbo pulcher* Rehder, 1943, non Dillwyn, 1817, is this species. (*Jour. Wash. Acad. Sci.*, vol. 47, p. 316, and "Studies in Tropical American Mollusks," 1971, p. 129)

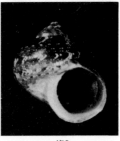

478

Subfamily ASTRAEINAE Davies, 1933

Genus *Astraea* Röding, 1798

Shells solid, flattened, coarse sculpturing, with or without an umbilicus; periphery usually spined; operculum calcareous. Type: *heliotropium* Martyn, 1784, from New Zealand.

Subgenus *Astralium* Link, 1807

Resembling a true *Astraea*, but without an umbilicus. Type: *calcar* Linné, 1758. The name *Calcar* Montfort, 1810, is a synonym.

Astraea phoebia Röding, 1798   **479**
Long-spined Star-shell   **Color Plate 2**

Southeast and off northwest Florida, and the West Indies. Bermuda. Brazil.

2 to 2½ inches in width; shell low, almost flat on its underside. Periphery of whorls with strong, flattened, triangular spines. Either with or without an umbilicus. Aperture silvery inside. A form which has an elevated spire and is less spinose was known as *A. spinulosa* Lamarck, 1822. Synonyms for this common, variable species are: *longispina* Lamarck, 1822; *titaina* Röding, 1798; *deplanatum* Link, 1807; *costulatus* Lamarck, 1822; *latispina* Philippi, 1844; *armatus* Philippi, 1849; *orichalceus* Philippi, 1849; *aster* and *heliacus* Philippi, 1850. Short-spined specimens of this species are often erroneously called *A. brevispina* Lamarck, 1822. The latter (Color Plate 2),

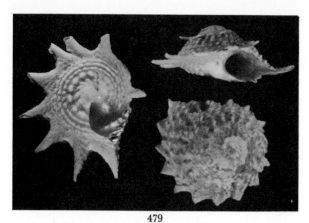

479

however, is a distinct species from the West Indies which is characterized by a splotch of bright orange-red around the umbilical region **(480).** Florida *phoebia*, pl. 2, may have a weak ring of pinkish orange, however. A common, shallow-water species found in turtle grass beds.

Subgenus *Lithopoma* Gray, 1850

Spire high, periphery nodulose or weakly scaled; no umbilicus. Type: *tuber* Linné, 1767. *Pachypoma* Gray, 1850, is a synonym.

***Astraea tecta americana* (Gmelin, 1791)**   **481**
American Star-shell   **Color Plate 2**

Southeast Florida.

1 to 1½ inches in length, ¾ as wide. Characterized by its sharp-angled spire, flat sides, white to cream color and the numerous, long, wavy, weak, axial ribs. Base of shell with 5 to 8 small, finely fimbriated, spiral cords, and a small ridge at the base of the columella which has about a dozen small axial ridges. Commonly found under rocks or in grassy areas at low tide. Operculum variable, but usually thick, convex and with a small or large dimple.

The typical subspecies, *tecta* (Lightfoot, 1786) **(482;** Color Plate 2), from the West Indies has stronger, longer and fewer axial ribs which extend to the flat base of the shell and are hollow at their ends. Synonyms are: *imbricata* Gmelin, 1791, and *corolla* Reeve, 1861. The subspecies *cubana* Philippi, 1849, from the Greater Antilles is intermediate between these two. Both moderately common at low water.

481

***Astraea caelata* (Gmelin, 1791)**   **483**
Carved Star-shell   **Color Plate 2**

Southeast Florida and the West Indies.

2 to 3 inches in length and width. Similar to *A. tuber,* but with 9 or 10 spiral rows of numerous, hollow, scalelike spines

on the lower ⅔ of the last whorl, 5 of which are on the base of the shell. Operculum thick, convex and finely pustulose. Body and foot cream with brownish red streaks. Moderately common in the West Indies.

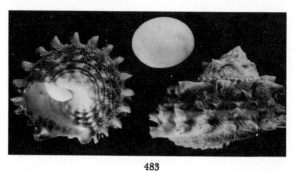

483

***Astraea tuber* (Linné, 1767)**   **484**
Green Star-shell   **Color Plate 2**

Southeast Florida and the West Indies.

1 to 2 inches in length, equally wide. Characterized by the peculiar green-and-white, cross-hatched color scheme, by the low, blunt, smooth axial ridges and by the smoothish base of the shell. Sometimes mottled in soft browns. Common below low water to 5 fathoms. Operculum with a thick, arched, tapering ridge on the exterior (like a large comma).

484

Subgenus *Pomaulax* Gray, 1850

Type of the subgenus is *Trochus japonicus* Dunker, 1844, from Japan.

***Astraea gibberosa* (Dillwyn, 1817)**   **485**
Red Turban

British Columbia to Magdalena Bay, Baja California.

1½ to 3 inches in length, heavy, brick-red to reddish brown in color. Characterized by 5 or 6 strong, spiral cords on the

485

flattish base. Operculum chitinous, green on inner side; outer side swollen, smooth, enamel-white. Formerly *A. inaequalis* Martyn, non-binomial. Among the many synonyms are *guadalupeana* S. S. Berry, 1957; *montereyensis* Oldroyd, 1927; *lithophorum* Dall, 1910; *magdalena* Dall, 1910; var. *depressum* Dall, 1909; *Turbo rutilus* C. B. Adams, 1852; *ochraceus* Philippi, 1846; and *diadematus* Valenciennes, 1846.

Subgenus *Megastraea* J. H. McLean, 1970

Shells large, periphery angular, sculpture rugose, periostracum brown and fuzzy. Operculum tri-ridged, the nucleus near the terminal. Type: *undosus* (Wood, 1828).

Astraea undosa (Wood, 1828)   486
Wavy Turban

Ventura, California, to Baja California.

2 to 3 inches in length; characterized by a strong, wavy, overhanging periphery, and by the dark-brown, fuzzy periostracum. Base concave, with 3 small, indistinct spiral cords. Outside of operculum with 3 strong, prickly ridges. Common in shallow water, especially around Todos Santos Bay, Baja California.

486

Astraea turbanica (Dall, 1910)   487
Rockpile Turban

Off South Coronado Island, Baja California.

5 to 6 inches high, similar to *undosa,* but much larger; the spire is more acute, the whorls descending more rapidly, so that the wavy or noduled periphery and one of the base spiral cords shows just above the suture. The periphery of the last whorl is rounded and slightly bicarinate. Operculum similar to that of *undosa,* with 3 strong curved ridges. *A. petrothauma* S. S. Berry, 1940 (*Bull. Amer. Paleont.,* vol. 25) is a synonym, as is *rupicollina* Stohler, 1959.

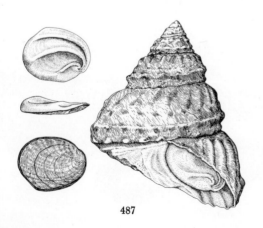

487

Other species:

488 *Astraea olivacea* (Wood, 1828). Western Mexico. 2 inches. (*Trochus melchersi* Menke, 1851, and *T. erythropthalmus* Philippi, 1848 are synonyms.) Color Plate 2.

Subfamily HOMALOPOMATINAE Keen, 1960

Genus *Homalopoma* Carpenter, 1864

Shells small, turbinate in shape. Operculum calcareous, oval, thick; its exterior with a thick, paucispiral whorl. Underside of operculum convex with multispiral, chitinous whorling. *Leptothyra* Pease, 1869, and *Leptonyx* Carpenter, 1864, are this genus. The gender of the name *Homalopoma* is neuter. Type: *sanguineus* (Linné, 1758).

Homalopoma albidum (Dall, 1881)   489
White Dwarf Turban

Off North Carolina to Cuba and Yucatan.

¼ inch in length, equally wide, very thick-shelled, resembling in shape a *Margarites*. Pure-white in color. Whorls rounded, 5 to 6 in number, each bearing 5 or 6 strong, rounded, spiral cords, the lower 2 being below the periphery of the whorl. Aperture and parietal wall glossy, slightly opalescent. Columella arched, with a small tooth in the middle and a smaller one usually at the base. No umbilicus. Commonly dredged from 35 to 450 fathoms.

489   499

H. linnei (Dall, 1889) **(490)**, from southeast Florida to Barbados (116 to 805 fathoms), has 8 smaller, beaded spiral cords on the upper part of the whorls and 10 on the base, otherwise it is very similar to *albidum*. It is quite rare. Rough specimens were given the varietal name, *limatum* (Dall, 1889).

Homalopoma carpenteri (Pilsbry, 1888)   491
Carpenter's Dwarf Turban

Alaska to Baja California.

¼ to ⅜ inch (5 to 9 mm.) in length, solid, globose. Pinkish red to brownish red in color. Last whorl and base with 15 to 20, evenly sized, smooth, spiral cords separated from each other by a space about half as wide as the cords. Base of pearly columella with 2 or 3 exceedingly weak nodules. A very common species frequently washed ashore and inhabited by small hermit crabs, from Monterey to Mexico. Do not confuse with *luridum*.

Homalopoma luridum (Dall, 1885)   492
Dark Dwarf Turban

Puget Sound to Baja California.

¼ inch (5 to 7 mm.) in length, similar to *carpenteri,* but half as large, black-brown in color, although occasionally whitish with red axial streaks. The spiral cords are usually fewer in number and more rounded. Moderately common in shallow water under rocks.

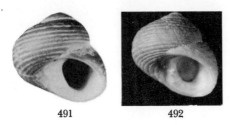

491 492

Homalopoma baculum Carpenter, 1864
Berry Dwarf Turban

493

Puget Sound to Baja California.

¼ inch or less in length, similar to *carpenteri* but with a flatter spire, and smoothish, except for numerous incised, spiral lines producing very weak threads. Color dark, rosy-brown. A moderately common shallow-water species, sometimes found with *carpenteri*. A thorough anatomical and life history study of this genus is needed to ascertain the validity of these species.

Other species:

494 *Homalopoma indutum* (Watson, 1879). Off North Carolina to Texas and the West Indies. 15 to 2,805 fms.

495 *Homalopoma paucicostatum* (Dall, 1871). (Synonym: *fene-stratum* Bartsch, 1919.) Monterey to Coronado Islands, California.

496 *Homalopoma juanensis* (Dall, 1919). San Diego to off Tia Juana, Baja California.

497 *Homalopoma* (*Panocochlea* Dall, 1908) *grippi* (Dall, 1911). 50 to 150 fms., Catalina Island and San Diego, California, to Baja California.

498 *Homalopoma engbergi* (Willett, 1929). Port Dick, Alaska, to Olga, Washington.

499 *Homalopoma* (*Leptothyropsis* Woodring, 1928) *philipiana* (Dall, 1889). 138 fms., off Dominica, Lesser Antilles. Type of the subgenus.

500 *Homalopoma* (*Cantrainea* Jeffreys, 1883) *panamense* (Dall, 1908). Off Cape San Lucas, Baja California, to Panama, 380 to 1,150 meters.

Genus *Moelleria* Jeffreys, 1865

Small (1 to 2 mm.), depressed, widely and deeply umbilicate, axial plicae fine, stronger below. Operculum calcareous, multispiral. Type: *costulata* Möller, 1842.

The American species are:

501 *Moelleria costulata* (Möller, 1842). Arctic European seas to New England, 4 to 30 fms. Off Fernandina, Florida, 294 fms.

501

502 *Moelleria drusiana* Dall, 1919. Amchitka Island to Glacier Bay, Alaska. Intertidal to offshore.

503 *Moelleria quadrae* Dall, 1897. Alaska to Queen Charlotte Islands, British Columbia. Intertidal to offshore.

Family PHASIANELLIDAE Swainson, 1840

Shell entirely porcelaneous, bulimoid in shape, usually smoothish. Operculum calcareous, usually paucispiral. Radula of the rhipidoglossate type. The subfamily Phasianellinae is not living in American waters. The subfamily Tricoliinae Robertson, 1958, has American species in the genera *Tricolia* and *Gabrielona*. Shells usually very small, less than ⅓ inch. For an excellent monograph, see R. Robertson, 1958, *Johnsonia*, vol. 3, no. 37.

Genus *Gabrielona* Iredale, 1917

Adult extremely small, 2.5 mm. or less, globose, wider than long; minutely umbilicate. Operculum calcareous, "naticoid" in shape, spirally ridged externally. Type: *nepeanensis* Gatliff and Gabriel, 1908. Only one living species in America. For a monograph, see R. Robertson, 1973, *Indo-Pacific Mollusca*, vol. 3, no. 14.

Gabrielona sulcifera Robertson, 1973
Caribbean Micro Pheasant

504

Cuba to the Lesser Antilles.

2.4 mm., whitish with red zigzag axial stripes and maculations. Whorls 4 and globose. Umbilicus chinklike. Early whorls microscopically spirally striate, including about 14 on the last whorl. Operculum "naticoid," calcareous and white. There is a minute denticle on the base of the columella. A seldom-found species. Erroneously identified as *brevis* Orbigny, 1842, by Dall and Robertson (1958).

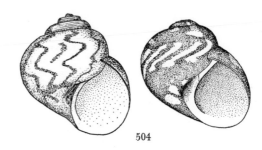

504

Genus *Tricolia* Risso, 1826

Shells about ⅓ inch or less, usually oval-elongate, usually smooth, but some species are spirally corded. Usually without an umbilicus. Red and orange dots and oblique spiral lines predominate. Operculum calcareous, smoothish and white. Type: *pullus* Linné, 1758, from Europe. 5 Western Atlantic, 1 Eastern Pacific species. Synonyms are: *Eudora* Gray, 1852; *Chromotis* H. and A. Adams, 1863; *Eucosmia* Carpenter, 1864; *Tricoliella* Monterosato, 1884; *Steganomphalus* Harris and Burrows, 1891; *Eulithidium* Pilsbry, 1898; *Usatricolia* Habe, 1956. Collected by sweeping *Zostera* turtle grass with nets.

Tricolia affinis (C. B. Adams, 1850)
Checkered Pheasant

505

Lower Florida Keys and the West Indies.

5 to 8 mm. in length, moderately elongate, smoothish except for microscopic spiral grooves in some specimens. Color rose to brownish, sometimes whitish. Always with numerous small

dots of pink, orange or a brownish color. Frequently with zigzag, axial bars of rose or brownish yellow. Often with irregular small spots or blotches of opaque-white. Umbilicus slitlike. Common; to 5 fathoms in turtle grass and coral rubble. *T. concinna* C. B. Adams, 1850, is the same. The following subspecies are recognized:

506 Subspecies *affinis* (C. B. Adams, 1850). Florida Keys, Bahamas, Greater Antilles to northern Lesser Antilles; Brazil. Abundant; has red dots which are paired up with a larger white dot. Operculum white. Umbilicus fairly wide.

507 Subspecies *pterocladica* Robertson, 1958. Northwest Florida; and southeast Florida. Brown and white spiral lines descend from suture at angle of 45 degrees. Operculum tinged with brown at the margin; umbilicus chinklike. Lives on red alga, *Pterocladia*. Uncommon.

508 Subspecies *beaui* Robertson, 1958. Lesser Antilles. Red oblique spiral lines, each bordered with a white line. Ground color pink to pale-orange. Uncommon to rare.

509 Subspecies *cruenta* Robertson, 1958. South Texas to Lower Caribbean to Brazil. Spire relatively low; covered with regularly spaced large red spots. Uncommon.

Tricolia adamsi (Philippi, 1853)   510
Adams' Pheasant

Bahamas and Caribbean.

2.5 to 3.8 mm., thin, subglobose, pink with irregularly placed, tiny red dots; white below the suture. Minute umbilicus. Upper whorls minutely striate. Operculum white, polished, convex on outer surface. Uncommon.

Tricolia thalassicola Robertson, 1958   511
Turtle Grass Pheasant

Off North Carolina to Florida and to Brazil.

2.5 to 7 mm. in length, oval, inflated, color of regularly spaced, brownish, orange or greenish spots and 7 pairs of axial, oblique flames of darker color just below the suture and also at the periphery. White between flames. Early whorls evenly spirally striate (smooth in *affinis*). Umbilicus fairly narrow. Operculum white, convex and nearly smooth. Usually lives on turtle grass (*Thalassia*); common; 1 to 35 fathoms.

Tricolia bella (M. Smith, 1937)   512
Shouldered Pheasant

Southeast Florida to the Caribbean and Brazil.

1/8 inch in length, spiral sculpture of numerous very small spiral cords, the largest being at the periphery of the whorl, thus giving the shell a slightly carinate shape. This carina is more pronounced in the early whorls and commonly bears a spiral row of tiny white dots. Rarely smooth with slightly angular whorls. Color variable, usually whitish gray with pink or brown axial mottlings and irregularly placed tiny dots of rose, yellow-brown or purplish brown. Umbilicus a mere chink. Operculum calcareous, convex, half smooth, the other half with fine, arched riblets. Common in shallow water among dead corals, down to a few fathoms. *T. pulchella* (C. B. Adams, 1845), non Récluz, 1843, and *adamsii* Reeve, 1857, are synonyms.

Tricolia tessellata (Potiez and Michaud, 1838)   513
Checkered Pheasant

Greater Antilles and lower Caribbean.

3 to 5 mm., oval, inflated, smooth; umbilicus chinklike; characterized by the oblique spiral lines arising from the suture at an angle of 15 degrees. Color variable—red, brown or blackish. Operculum white with greenish brown margin and

radially striate. An abundant, shallow-water West Indian species. Do not confuse with banded *thalassicola* or *affinis*.

Tricolia compta (Gould, 1855)   514
Californian Banded Pheasant

Crescent City, California, to the Gulf of California.

1/4 to 1/3 inch in length, resembling a moderately high-spired *Littorina*, but distinguished from that genus by its calcareous

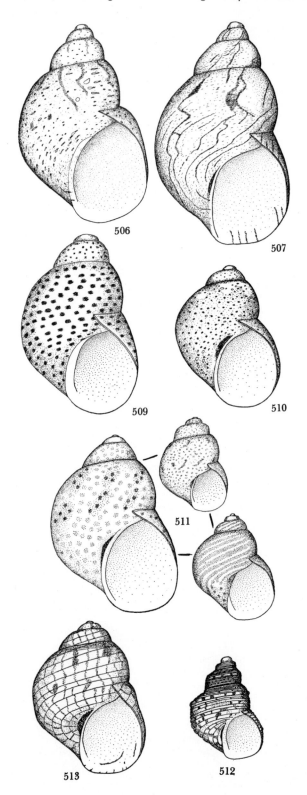

506   507

509   510

511

513   512

514 515

508

operculum. Shell smooth, in life covered by a thin, gray-green, translucent periostracum. Characterized by the numerous, spiral lines of blackish green, red, brown or purplish which slant slightly downward, so that they are not parallel to the suture. Axial zigzag, wider bands are also present. Very abundant on eel-grass in shallow bays. Frequently washed ashore. *T. producta* (Dall, 1908) is a synonym.

Tricolia variegata (Carpenter, 1864)   515
Pacific Micro Pheasant

Monterey to Baja California.

2 mm. in length, depressed turbinate in shape, with 4 to 5 whorls. Characterized by its very small size, and by about a dozen obliquely set, spiral bright-red lines. The top of the whorls may be solid red and with large opaque-white spots. Umbilicus a mere chinklike depression. Operculum calcareous and white. Common among weeds in tidepools and among kelps offshore. Synonyms are: *typica* Dall, 1908; *Eulithidium rubrilineatum* Strong, 1928; *punctata* Carpenter, 1864; *carpenteri* Dall, 1908.

Other species:

516 *Tricolia cyclostoma* (Carpenter, 1864). Gulf of California, 18 to 37 meters.

517 *Tricolia pulloides* (Carpenter, 1865). California to Cape San Lucas, Baja California. (Synonym: *Phasianella compta* var. *elatior* Carpenter, 1865)

518 *Tricolia substriata* (Carpenter, 1864). Catalina Island, California, to Cape San Lucas, Baja California.

Family NERITIDAE Rafinesque, 1815

Genus *Nerita* Linné, 1758

Shells solid, with a flat columella bearing teeth, folds or pimples. Operculum calcareous and with a peglike projection which inserts into the muscle of the snail. Egg capsules are 1 mm. or less, dome-shaped and laid on rocks. Type: *peloronta* Linné, 1758. Synonyms are *Neritarius* Duméril; 1806; *Dontostoma* Herrmannsen, 1847; *Tenare* Gray, 1858.

Nerita peloronta Linné, 1758.   519
Bleeding Tooth   Color Plate 3

Southeast Florida, Bermuda and the West Indies.

¾ to 1½ inches in length; grayish yellow with zigzags of black and red. Characterized by the blood-red parietal area which bears 1 or 2 whitish teeth. Operculum: underside coral-pink; one half of outer side smooth and dark-orange, other half smoothish or papillose and brownish green. Very abun-

dant along the rocky shores facing the open ocean. It is a popular souvenir.

Nerita versicolor Gmelin, 1791   520
Four-toothed Nerite   Color Plate 3

South ¾ of Florida and the West Indies. Bermuda.

¾ to 1 inch in length; dirty-white with irregular spots of black and red arranged in spiral rows; spirally grooved; outer lip spotted with red, white and black on margin. Parietal area slightly convex, white to yellowish and with 4 (rarely 5) strong teeth. Operculum: exterior brownish gray, finely papillose and slightly concave. Commonly associated with *N. peloronta*. *Nerita variegata* Karsten (1789) is invalid, since it appears in a non-binomial work.

Nerita tessellata Gmelin, 1791   521
Tessellate Nerite   Color Plate 3

Florida, the West Indies and Bermuda. Brazil.

¾ inch in length, irregularly spotted with black and white, sometimes heavily mottled; coarsely sculptured with spiral cords of varying sizes. Parietal area concave, bluish white and bearing 2 weak teeth in the middle. Operculum: exterior slightly convex, black in color. Commonly congregate in large numbers under rocks at low tide. Rare in northern Florida. Do not confuse with *N. fulgurans* Gmelin, 1791, whose operculum is bluish white to yellowish gray, not black.

Nerita fulgurans Gmelin, 1791   522
Antillean Nerite   Color Plate 3

Southeast Florida, Texas and the West Indies and Bermuda. Brazil.

¾ to 1 inch in length, very similar to *N. tessellata*, but with a lighter-colored, yellowish gray operculum. The spiral ridges on the shell are more numerous, the color patterns blurred, the aperture relatively wider, and the teeth more prominent. This is a salt to brackish-water inhabitant of protected shores, and is abundant only in certain restricted localities. It is seldom represented or properly labeled in private collections.

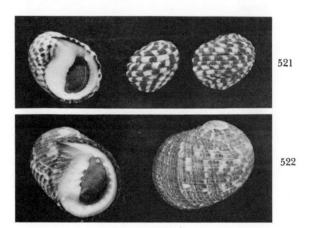

521

522

Nerita scabricosta Lamarck, 1822   523
Rough-ribbed Nerite   Color Plate 3

Baja California to Ecuador.

1 to 1½ inches, rotund, dark-gray, with rough-surfaced, strong spiral cords. Parietal wall with 4 or 5 irregular, large, glossy-white teeth. Common intertidally on rocks in the Gulf of California. Synonyms are *fuscata* Menke, 1829; *deshayesii* Récluz, 1841; *papilionacea* Valerciennes, 1832; *multijugis* Menke, 1847.

The subspecies *ornata* Sowerby, 1833 (524), found from Costa Rica south, is rounder and with smoothish spiral ribs.

Subgenus *Theliostyla* Mörch, 1852

Spire low, last whorl large, parietal area flattened and bearing pustules. Type: *albicilla* Linné, 1758, from the Indo-Pacific. Synonyms are *Natere* Gray, 1858, and *Ilynerita* von Martens, 1887.

Nerita funiculata Menke, 1851   525
Funiculate Nerite   Color Plate 3

Baja California to Peru.

¾ inch, with fine irregular spiral threads on the last large whorl. Spire low. Color dark-gray. Parietal region white, with numerous raised pustules. Columella with 3 or 4 very fine teeth at the center. Common; on intertidal rocks. *N. bernhardi* Récluz, 1855, is a synonym.

525

Genus *Puperita* Gray, 1857

Small, brackish water, globose shells with a non-glossy surface. Type: *pupa* (Linné, 1767).

Puperita pupa (Linné, 1767)   526
Zebra Nerite   Color Plate 3

Southeast Florida, Bermuda and the West Indies.

⅓ to ½ inch in length, thin, smooth, chalky-white with black, axial, zebralike stripes. Aperture and smooth operculum light-yellow. Lives in small, placid pools above the high-water mark. Common in the West Indies, rare in Florida.

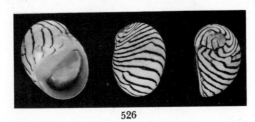

526

Puperita tristis (Orbigny, 1842)   526a
Melancholy Nerite

West Indies

About ⅓ inch, very similar to *pupa* but usually smaller; color pattern black with small white spots; aperture tends

526a

towards bluish yellow to bluish white, while in *pupa* it is light- to dark-orangish yellow. Some workers have considered this common littoral species to be only a dark variety of *pupa*, but H. D. Russell (1941) on the basis of radula differences, keeps it as a distinct species.

Genus *Neritina* Lamarck, 1816

Smaller and thinner-shelled than *Nerita*. Usually brackish, but some are fresh-water and oceanic. Inner lip with weak teeth. Type: *pulligera* (Linné, 1758). American forms belong to the subgenus *Vitta* Mörch, 1852, with its type: *virginea* (Linné, 1758). Synonyms are: *Laphrostoma* Rafinesque, 1815; *Neritella* Gray, 1848; *Chernites* Gistel, 1848; *Labialia* and *Onychina* Scudder, 1882.

Neritina virginea (Linné, 1758)   527
Virgin Nerite   Color Plate 3

Florida to Texas, the West Indies and Bermuda. Brazil.

½ inch in length, smooth, glossy, very variable in color pattern and shades—blacks, browns, purples, red, whites, olive —crooked lines, dots, mottlings, zebralike stripes and sometimes spirally banded. Parietal area smooth, convex, white to yellow, and with a variable number of small, irregular teeth. Operculum usually black. A very common, widespread inhabitant of intertidal, brackish-water flats. The dwarf form, *minor* Metcalf, 1904, is a synonym.

527

Neritina reclivata (Say, 1822)   528
Olive Nerite   Color Plate 3

Florida to Texas (rare) and the West Indies.

½ inch in length, glossy, often with the spire eroded away. Ground color brownish green, olive or brownish yellow with numerous axial lines of black-brown or lavender. Operculum black to slightly brownish. Common in brackish water and also found in fresh-water springs near the seashore in Florida.

A globose form or subspecies (529) with a short spire and more convex whorls replaces the higher-spired, typical form from Texas to Panama, but may also appear in eastern Florida. It has been named *floridana* Reeve, 1855, *rotundata* von Martens, 1865, and *sphaera* Pilsbry, 1931. Another synonym is *palmae* Dall, 1885.

Neritina clenchi Russell, 1940   530
Clench's Nerite

Florida, Central America and the Antilles.

¾ to 1 inch, smooth, glossy and very variable in coloration. Do not confuse with *virginea* which is not as globose or as

528   530

large. The edge of the parietal area opposite parietal teeth is outlined with dark orange-yellow, while in *virginea* it is outlined in black. Operculum black to pink. Prefers fresh to brackish water. Locally common.

Genus *Theodoxus* Montfort, 1810

Shells and opercula smooth, the latter without a peg. Type: *fluviatilis* (Linné, 1758).

Subgenus *Vittoclithon* H. B. Baker, 1923

Operculum smooth, with a small blunt peg. Type: *meleagris* (Lamarck, 1822).

Theodoxus luteofasciatus Miller, 1879   **531**
Painted Nerite   **Color Plate 3**

Gulf of California to Panama.

8 to 12 mm., glossy-smooth, round, with very variable, oblique stripes of gray, black, yellow and white. Parietal wall swollen, smooth and glossy-brown. Common; on mudflats, usually near brackish water and mangroves. Synonyms are *picta* Sowerby, 1832 (not Eichwald, 1830), and *usurpatrix* Crosse and Rischer, 1892, and several color forms by Miller, 1879.

Subfamily SMARAGDIINAE H. B. Baker, 1923

Genus *Smaragdia* Issel, 1869

Shells small, glossy, smooth, ovately oblique, usually green with red flecks or stripes. Type: *viridis* (Linné, 1758). *Gaillardotia* Bourguignat, 1877, is a synonym.

Smaragdia viridis (Linné, 1758)   **532**
Emerald Nerite

Southeast Florida, the West Indies and Bermuda.

¼ to ⅓ inch in length, glossy, smooth, pea-green, often with tiny chalk-white bars and rarely with purplish brown, narrow, zigzag bars. Locally common; in turtle grass from 1 to 10 fathoms. True *viridis* comes from the Mediterranean. Some workers separate our form as the subspecies *viridemaris* Maury, 1917. *N. weyssei* Russell, 1940, is a synonym.

532

Family PHENACOLEPADIDAE Pilsbry, 1900

Genus *Phenacolepas* Pilsbry, 1891

Small, hooked-over, whitish, limpetlike, porcelaneous shells. Muscle scar horseshoe-shaped. Type: *crenulata* (Broderip, 1834). Synonyms are *Scutella* Broderip, 1834; *Scutellina* Gray, 1847, and *Scutulina* Cossmann, 1912.

Phenacolepas hamillei (Fischer, 1857)   **533**
Hamille's Limpet

Florida to Texas to the Lesser Antilles. Bermuda. Brazil.

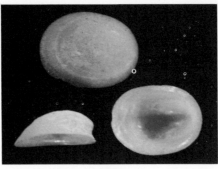

533

6 to 8 mm., limpet-shaped with the apex hooked over at the back. Nucleus minute, spiral and half-immersed, with a slight contraction where the conical shell begins. Exterior white or yellowish with numerous minutely spined radial riblets. Interior brilliantly polished. Moderately common just offshore. *P. antillarum* "Shuttleworth" Dall, 1889, is a synonym. The date of original description is sometimes erroneously listed as 1856.

Other species:

534 *Phenacolepas malonei* Vanatta, 1912. (Synonym: *magdalena* Dall, 1918.) Baja California and Gulf of California. Rare. 15 mm.

535 *Phenacolepas osculans* (C. B. Adams, 1852). (Synonym: *Scutellina navicelloides* Carpenter, 1856.) Gulf of California to Panama. Rare. 12 mm.

Suborder *Littorinina* Gray, 1840

Superfamily Littorinacea Gray, 1840

Family LACUNIDAE Gill, 1871

Genus *Lacuna* Turton, 1827

Rather fragile, smooth periwinkles characterized by a shelflike columella and a chinklike umbilicus. Periostracum smooth, fairly thin and light-brown. Operculum paucispiral and corneous. Cold-water inhabitants, usually dredged in areas of kelp weed upon which the females lay their jelly eggmasses. Veligers of some species are common in the plankton; others do not swim. Type: *Helix lacuna* (Montagu, 1803).

Lacuna vincta (Montagu, 1803)   **536**
Common Northern Lacuna

Arctic Ocean to Rhode Island.
Alaska to California.

¼ to ⅜ inch in length, 4 to 5 whorls, resembling a *Littorina*, but characterized by its fairly thin, but strong, translucent shell, its shelflike columella alongside of which is a long, narrow, deep umbilical chink. Outer lip fragile. Shell smooth except for microscopic, spiral scratches. Color light-tan to brown with the spire tinted with purplish rose. Com-

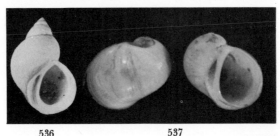

536   537

monly with 3 or 4 narrow brown bands. Often confused with *Litiopa* which has a bladelike ridge on the columella just inside the aperture. Common from low water to 25 fathoms. Alias *Trochus divaricata* Fabricius, 1780 (not Linné, 1758) and *solidula* Lovén, 1846; *pretusa* Conrad, 1829; *fusca* Gould, 1841. This is the type of the subgenus *Epheria* Leach in Gray, 1847.

Lacuna pallidula (da Costa, 1778)   537
Pale Lacuna

Greenland to Connecticut; Europe.

⅓ inch, with the last whorl greatly expanded; spire short. Umbilicus very deep, showing most of the interior of the spire. Yellowish green, not banded. Moderately common in New England in shallow water. *L. neritoidea* Gould, 1840, is probably a synonym.

Lacuna unifasciata Carpenter, 1856   538
One-banded Lacuna

Monterey, California, to Baja California.

¼ inch in length, moderately fragile, similar to the other lacunas, but characterized by its very narrow, long, chinklike umbilicus and by the carinate periphery of the whorl which bears a fine, dark-brown spiral line. Early whorls usually pinkish, remainder yellowish tan. Umbilicus and columella white. The peripheral carina may be weak or obsolete, and the color line may consist of a series of faint, slanting streaks of light reddish brown. Very common in littoral seaweed and kelp in southern California. *L. aurantiaca* Carpenter, 1856, is a synonym.

Lacuna carinata Gould, 1848   539
Carinate Lacuna

Alaska to Monterey, California.

⅜ to ½ inch in length, 3 to 4 whorls, moderately fragile. Aperture semilunar, large. Outer lip thin. Columellar chink large, long and white. Shell smooth, chalky-white, but always covered by a thin, yellowish brown, smooth periostracum. Common on kelp weed and eel-grass. *L. porrecta* Carpenter, 1864; *effusa* Carpenter, 1864; *exaequata* Carpenter, 1864; *puteoloides* Dall, 1919, and *striata* Gabb, 1861, a fossil, are the same. Do not confuse with *vincta*, which has a higher spire and much narrower, brownish tan umbilical chink.

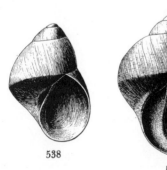

538   539

Lacuna variegata Carpenter, 1864   540
Variegated Lacuna

Drier Bay, Alaska, to Santa Monica, California.

¼ inch in length, similar to *unifasciata*, but having a very deep umbilical chink which is bordered by a sharp ridge. The spiral carina at the level of the suture is very small, but quite sharp. The yellowish tan shell has mottlings or oblique bands of darker color. Moderately common in eel-grass along the shore.

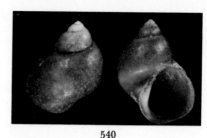

540

Lacuna succinea S. S. Berry, 1953   541
Amber Lacuna

San Pedro to San Diego, California.

8 to 10 mm., thin-shelled, with a tall, narrow apex. Periphery bears a strong, thin cord. Umbilicus long, deep and sharply bordered. Color light-brown, rarely with dim oblique lines of darker brown. Nuclear whorls whitish. Later whorls with microscopic spiral striae. Moderately common; shallow water. (*Trans. San Diego Soc. Nat. Hist.*, vol. 11, p. 411.) The name is a homonym of *L. succinea* Mörch, 1860.

Other species:

542 *Lacuna parva* (da Costa, 1778). Southern Cape Cod and Nantucket, Massachusetts.

543 *Lacuna crassior* (Montagu, 1803). (Synonym: *glacialis* Möller, 1842.) Greenland to Quebec. 96 fms. Alaska. Type of the subgenus *Temanella* Rovereto, 1899.

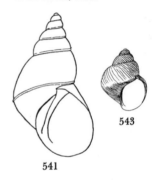

543

541

544 *Lacuna marmorata* Dall, 1919. Saginaw Bay, Alaska, to San Diego, California (and var. *olla* Dall, 1919, from Olga Washington).

545 *Lacuna* (*Boetica* Dall, 1918) *vaginata* Dall, 1918. 53 to 199 fms., off southern California.

Genus *Aquilonaria* Dall, 1886

A rather thin, littorinidlike, 5-mm., snail with a thin periostracum, globose whorls and weak columellar pillar. Type and only species: *Aquilonaria turneri* Dall, 1886 (**546**). Arctic Ocean: near Bering Strait and Labrador, deep water.

Genus *Haloconcha* Dall, 1886

Shells rather thin, turbinate, broader than high, with a large last whorl. Type: *reflexa* Dall, 1884. Synonyms: *Lacunella* Dall, 1884, non Deshayes, 1864; *Lacunaria* Dall, 1884, non Conrad, 1866.

Haloconcha reflexa (Dall, 1884)   547
Reflexed Haloconch

Alaska and the Bering Sea.

½ inch in length, fragile, 3 whorls; body whorl large. Resembles a *Velutina*. Early whorls purplish brown, last whorl translucent, light chestnut-brown. Shell covered by a thin varnish of periostracum of the same color. At the edge of the thin, smooth outer lip, the periostracum is neatly curled back to form a minute ridge. Umbilicus slitlike, with the periostracum puckered along its length. Operculum littorinid, and is withdrawn well within the glossy-brown aperture. Not uncommon in shallow water.

Other species:

548 *Haloconcha minor* Dall, 1919. Aleutian Islands eastward to Chirikoff Island, Alsaka.

Family LITTORINIDAE Gray, 1840

Subfamily LITTORININAE Gray, 1840

Genus *Littorina* Férussac, 1822

Small, ovately conical shells, usually without an umbilicus. Operculum horny, thin, paucispiral and tan to brown. Penis simple or multipronged and attached to the right side of the body behind the right tentacle. 2 tentacles, slender and tapering. Musculature of the foot divided longitudinally into 2 lateral halves. Radula ribbon long. Eggs shed externally or kept within the oviduct where they hatch, brood and grow. Popularly known as periwinkles (arising from the early English "penny winkle"), they serve as food for man, birds and crabs. Type: *littorea* (Linné, 1758). For a recent taxonomic treatment, see Joseph Rosewater, 1970, *Indo-Pacific Mollusca*, vol. 2, no. 11, and Joseph Bequaert, 1943, *Johnsonia*, vol. 1, no. 7. Synonyms include *Litorina* Menke, 1828; *Neritrema* Récluz, 1869; *Littorivaga* Dall, 1918; *Algaroda* Dall, 1918; *Neritotrema* Wenz, 1939; *Ezolittorina* Habe, 1958.

Littorina littorea (Linné, 1758)   **549**
Common Periwinkle

   Labrador to Maryland. Western Europe.

¾ to 1 inch in length, thick, smoothish. Gray to brownish gray in color. Inside of aperture chocolate-brown. Columella and inner edge of aperture whitish. In young or perfect specimens there are fine, irregularly spaced, spiral threads with microscopic, wavy wrinkles in between. A favorite food in

Europe. Very common along the rocky shores of New England. It has spread southward from Nova Scotia to Maryland in 130 years. Its lenticular, microscopic egg capsules are free-floating in the sea. Rarely and accidentally introduced (but not established as yet) in Puget Sound and San Francisco. (*The Veliger*, vol. 11, p. 283). For biology see F. R. Hayes, 1929, *Contrib. Canadian Biol. Fish.*, n. s., vol. 4, pp. 413–430; J. N. Gowanloch and F. R. Hayes, 1926, *ibid.*, vol. 1, pp. 133–162.

Littorina obtusata (Linné, 1758)   **550**
Northern Yellow Periwinkle   **Color Plate 3**

   Labrador to Cape May, New Jersey. Northwest Europe.

⅓ to ½ inch in length, equally wide, with a low spire; smoothish. Color variable but usually a uniform, bright, brownish yellow or orange-yellow. Sometimes with a white or brown spiral band. Columella whitish. Operculum bright yellow to orange-brown. This is *L. palliata* Say, 1822; *peconica* S. Smith, 1860; *littoralis* (Linné, 1758); and *arctica* Möller, 1842. A common coastal species associated with rockweeds, upon which it deposits its jelly eggmasses. It migrates to deeper water in early winter, so as to escape the grinding action of shore ice, returning to higher levels in the spring.

Littorina saxatilis (Olivi, 1792)   **551**
Northern Rough Periwinkle

   Arctic Seas to Cape May, New Jersey.
   Arctic Seas to Puget Sound.

¼ to ½ inch in length, resembling a "distorted, small *L. littorea*." Adults characterized by poorly developed, smoothish, fine spiral cords. Color drab-gray to dark-brown. Interior of aperture chocolate-brown. Females give birth to live, shelled young. Often found with *L. obtusata*, but not so common. This is *L. rudis* (Maton, 1797); *groenlandica* Menke, 1830; *tenebrosus* (Montagu, 1803); *obligatus* (Say, 1822); *vestita* (Say, 1822); *castanea* Deshayes, 1843; *davidus* (Röding, 1798); *jugosa* (Montagu, 1803); and *nigrolineata* Gray, 1839.

Littorina sitkana Philippi, 1846   **552**
Sitka Periwinkle

   Bering Sea to Puget Sound, Washington.

¾ inch in length, solid, sharp lip, characterized by about a dozen strong spiral threads on the body whorl. Columella whitish. Shell dark grayish to rusty-brown; some with 2 or 3 wide spiral bands of whitish. A common littoral species of the north. Rarely found in northern California. The name is sometimes misspelled *sitchana*.

Littorina planaxis Philippi, 1847   **553**
Eroded Periwinkle

   Puget Sound to Baja California.

½ to ¾ inch in length, usually badly eroded; grayish brown with bluish white spots and flecks. Characterized by the eroded, flattened area on the body whorl just beside the columella. Interior of aperture chocolate-brown with a white spiral band at the bottom. A common littoral, rock-loving species. Do not confuse with the smoother, higher-spired *L. scutulata*. For articles on biology, see *The Veliger*, 1964, vol. 7.

Littorina scutulata Gould, 1849   **554**
Checkered Periwinkle

   Alaska to Baja California.

½ inch in length, moderately slender, semigloss finish and smooth. Color light to dark reddish brown with small, irregular spots of bluish white. Columella white; interior of aperture whitish brown. A common littoral species. Compare with *L. planaxis*. *L. plena* Gould, 1849, is a synonym.

In southern California there is a color form or subspecies, *pullata* Carpenter, 1864, (555) in which the background color is darker brown with numerous, fine, spiral, incised white lines.

554

Littorina ziczac (Gmelin, 1791)   556
Zebra Periwinkle   **Color Plate 3**

Southeast Florida and the West Indies; Bermuda. Introduced to Pacific Panama.

½ to ¾ inch, fairly thick and strong. Base of shell weakly angulate. Columella brown. Color whitish to bluish gray with irregular, slanting zigzag stripes of brown. Upper whorls have 20 to 26 fine spiral lines. Do not confuse with *L. angustior* Mörch or *lineolata* Orbigny, 1840. Common; found in the intertidal zone in the crevices of rocks.

Subgenus *Melarhaphe* Menke, 1828

The type of this subgenus is *neritoides* (Linné, 1758) from the eastern Atlantic.

Littorina mespillum (Mühlfeld, 1824)   557
Dwarf Brown Periwinkle

Florida Keys and the Caribbean area; Bermuda.

¼ inch in length, somewhat shaped like *obtusata*. Characterized by its dark-brown periostracum, glossy-brown columella and aperture, by its tiny, chinklike umbilicus, and by the presence, in some specimens, of rows of small, round blackish spots. A whitish form with prominent purple-brown dots occurs in some colonies (form *minima* (Wood, 1828)). Other synonyms are: *fusca* Pfeiffer, 1840; *naticoides* Orbigny, 1842 (resembling a miniature *Natica*); and *gundlachi* Philippi, 1849. Common in "splash-pools" from high-tide line to 6 or 7 feet above.

Littorina meleagris (Potiez and Michaud, 1838)   558
White-spotted Periwinkle

Southeast Florida, Texas to the lower Caribbean. Bermuda.

¼ to ⅓ inch, with a high pointed spire, thin periostracum, inner lip forming a slight callus on the body whorl. Umbilicus chinklike. Aperture reddish brown. Exterior brown with large, irregular white spots, sometimes arranged in spiral rows. Moderately common on intertidal rocks. Synonyms are *guttata* Philippi, 1847; *punctata* Pfeiffer, 1840; and *hidalgoi* Arango, 1880.

Subgenus *Austrolittorina* Rosewater, 1970

Type of the subgenus is *unifasciata* Gray, 1826, of Australia.

Littorina lineolata Orbigny, 1840   559
Lineolate Periwinkle

Texas to Florida; Caribbean.

⅓ inch, similar to *ziczac*, but broader, heavier, its aperture about ½ or more of the entire length of the shell. Upper whorls with only 8 to 11 spiral lines. Zebra, brown stripes present. Operculum elongate, its nucleus small. Sides of foot light yellowish gray. Sole of foot gray. Common; lives in intertidal area. Erroneously called *floccosa* Mörch by Abbott in 1968. ("Seashells of North America," Golden Press, New York)

Littorina angustior (Mörch, 1876)   560
Angust Periwinkle

Texas to south Florida; Bermuda to Brazil.

¼ to ⅓ inch, similar to *ziczac* and *lineolata* but the shell is more carinate at the base, more elongate, with its aperture

557

558

556

559

560

562

563

564

less than ½ the total length of the entire shell. Has 6 to 9 spiral lines on the upper whorls. Sides of the foot mottled black on gray. Operculum paucispiral, but almost round. Common; on rocks near the high-tide line. *L. lineata* Orbigny, (1841) (not Gmelin, 1791), and *carinata* Orbigny, 1841, are synonyms. Erroneously called *lineolata* Orbigny by Abbott in 1968.

Littorina aspera Philippi, 1846   **561**
Aspera Periwinkle   **Color Plate 3**

Baja California to Panama and to Ecuador.

½ to ¾ inch, whitish with axial, slanting flames of brown. Aperture and columella chocolate-brown. Sculpture of numerous, sharp, spiral ribs. Moderately common; on intertidal rocks. Synonyms: *parvula* Philippi, 1849; *apicina* Menke, 1850; *dubiosa* C. B. Adams, 1852; *philippii* Carpenter, 1857; *penicillata* Carpenter, 1864; *alba, latistrigata* and *subsuturalis* von Martens, 1901.

Subgenus *Littoraria* Gray, 1834

Type of the subgenus is *zebra* (Donovan, 1825) of the tropical eastern Pacific.

Littorina nebulosa (Lamarck, 1822)   **562**
Cloudy Periwinkle

Texas to Florida; Caribbean to Brazil. Bermuda.

½ to ¾ inch, dull surface, spire usually eroded, following whorls spirally striate. Columellar area long, wide, smooth, without an umbilicus and characteristically tinged with purplish mauve, as is the aperture sometimes. A color form, *tessellata* Philippi, 1847, has a checkered brown pattern on a grayish white background. Moderately common.

Subgenus *Littorinopsis* Mörch, 1876

Characterized by strong but lightweight shells with spiral sculpture. Usually ovoviviparous. Mangrove dwellers. Type of the subgenus is *angulifera* (Lamarck, 1822).

Littorina angulifera (Lamarck, 1822)   **563**
Angulate Periwinkle

South half of Florida to Texas and to Brazil; Bermuda. Introduced to Pacific Panama.

About 1 inch in length; thin-shelled but strong. First 2 or 3 whorls smooth, remainder with many fine, spiral grooves. Last whorl sometimes carinate. Color variable—whitish, yellowish or orange- to red-brown—with darker, wavy, vertical, oblique stripes. Columella pale purplish with whitish edges. Operculum pale-brown. Common in mangrove areas where the waters are calm and brackish. It is found high above the high-tide mark clinging to wharf pilings, and is often seen on the trunks and branches of mangrove trees. Possibly introduced to the Pacific side of Panama, although Rosewater (1970) recognizes an eastern Pacific subspecies, *aberrans* Philippi, 1846. *L. scabra* (Linné, 1758) is from the Indo-Pacific. Synonyms are *L. lineata* Gmelin, 1791; *flavescens* Philippi, 1847; *strigata* Philippi, 1847; and *striata* Schumacher, 1817 (not Vallot, 1801). For biology, see R. E. Lenderking, 1954, *Bull. Marine Sci. Gulf and Carib.*, vol. 3, no. 4, pp. 273–296.

Littorina irrorata (Say, 1822)   **564**
Marsh Periwinkle

New York to central Florida to Texas.

About 1 inch in length, thick-shelled, with numerous, regularly formed spiral grooves. Outer lip strong, sharp, slightly flaring, and with tiny grooves on the inside. Color usually grayish white with tiny, short streaks of reddish brown on the spiral ridges. Aperture yellowish white. Callus of inner lip and the columella pale reddish brown. Commonly found in large numbers among the sedges of brackish water marshes. not recorded alive south of Indian River (east Florida) or Punta Rassa (west Florida). The synonyms are: *sulcata* Lamarck, 1822; *lunata* H. C. Lea, 1845; *sayi* Reeve, 1858; and *carolinensis* Conrad, 1863.

Littorina modesta Philippi, 1846   **565**
Conspersa Periwinkle

Baja California to Ecuador.

½ to ¾ inch, cream-white with small, purplish brown dots. Spiral sculpture of close, flat-topped spiral threads or cords. Columella whitish with a brown stain. Interior of aperture light-brown. Common. Synonyms: *conspersa* Philippi, 1847; *puncticulata* Philippi, 1847; *albida* Philippi, 1848.

Littorina albicarinata McLean, 1970   **566**
White-keeled Periwinkle

Gulf of California.

5 mm., turbinate, grayish with brown maculations, upper whorls with 2 thin, whitish, spiral carinations between which are about 6 fine incised lines. Base with 4 strong cords. Columella whitish; aperture dark-brown. Common; in crevices near barnacles in the mid-intertidal zone from the head of the Gulf to Guaymas, Mexico. (*Malacol. Rev. for 1969*, vol. 2, p. 127.)

Other Tropical Pacific species:

567 *Littorina (Littorinopsis) fasciata* Gray, 1839. Baja California to Ecuador. 30 mm. Color Plate 3.

568 *Littorina pullata* Carpenter, 1864. Baja California to Panama.

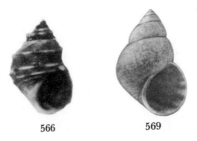

566   569

Genus *Algamorda* Dall, 1918

Shell thin, whorls well-rounded; outer lip thin. Thin brown periostracum present. Type: *newcombiana* (Hemphill, 1876).

Algamorda newcombiana (Hemphill, 1876)   **569**
Newcomb's Periwinkle

Humboldt Bay, California.

5 to 7 mm., with 4 or 5 well-rounded whorls. Light tan with 2 spiral bands of darker brown. Periostracum thin, brown. Lives in intertidal mud areas near salt marsh sedges. Locally uncommon, and perhaps now extinct due to pollution.

Genus *Nodilittorina* von Martens, 1897

Small, solid littorinids having spiral rows of low, rounded nodules. Columella usually flattened. Operculum paucispiral. Type: *pyramidalis* (Quoy and Gaimard, 1833).

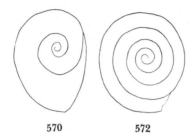

570   572

Subgenus *Echinolittorina* Habe, 1956

Type of the subgenus is *tuberculata* (Menke, 1828).

Nodilittorina tuberculata (Menke, 1828)   570
Common Prickly-winkle

South Florida, the West Indies and Bermuda.

½ to ¾ inch in length. Shell rounded at the base. Several spiral rows of small, fairly sharp nodules on the whorls. Columella flattened, forming a slightly dished-out shelf. Color brownish gray. Operculum paucispiral. A common rock-dwelling species found near the high-tide line. Do not confuse with the extremely similar *Echininus nodulosus* (Pfeiffer 1839) which has a multisprial operculum, and whose columella is not shelved. Erroneously listed in Johnsonia and other books as *Tectarius tuberculatus* Wood, 1828. *N. dilatata* (Orbigny, 1841); *trochiformis* (Dillwyn, 1817); and *thiarella* (Anton, 1839), are synonyms.

Subfamily TECTARIINAE Rosewater, 1970

Genus *Tectarius* Valenciennes, 1833

True *Tectarius* is limited to the Indo-Pacific. *Hamus* "Klein" H. and A. Adams, 1858, appears to be a synonym. Type: *coronatus* Valenciennes, 1833.

Subgenus *Cenchritis* von Martens, 1900

Type of this subgenus is *muricatus* (Linné, 1758).

Tectarius muricatus (Linné, 1758)   571
Beaded Periwinkle

South Florida, the West Indies and Bermuda.

½ to 1 inch in length. Shell thick, with 11 rows of neat, rounded, whitish, evenly spaced beads on the last whorl. Columella grooved; umbilicus a narrow, oblique slit. Color of outer shell ash-gray. Interior dark-tan. Operculum paucispiral. One of the commonest West Indian littoral species, usually found well out of water on the rock cliffs.

570   572

571

Subfamily ECHININAE Rosewater, 1970

Genus *Echininus* Clench and Abbott, 1942

The operculum is chitinous and multispiral. Type: *Trochus cumingii* Philippi, 1846. Synonym is: *Nina* Gray, 1850, non Horsfield, 1829.

Subgenus *Tectininus* Clench and Abbott, 1942

Shell without an umbilicus. Type: *nodulosus* (Pfeiffer, 1839).

Echininus nodulosus (Pfeiffer, 1839)   572
False Prickly-winkle

Southeast Florida and the West Indies. Bermuda.

½ to 1 inch in length. Base of shell squarish. Whorls with 2 spiral, carinate rows of sharp nodules in addition to 2 or 3 rows of smaller, blunt nodules. Columella not shelved. Color grayish brown. Operculum multispiral. Lives well above high-tide mark on rocky shores. Be sure not to confuse with *Nodilittorina tuberculata* whose beads are lined up axially one under the other. Synonyms are *scabra* Anton, 1839; *antoni* Philippi, 1847; *pfeifferianus* Weinkauf, 1882.

Superfamily Rissoacea Gray, 1847

Family RISSOIDAE Gray, 1847

This is a difficult family to classify. A proposed revision was attempted by Eugene Coan in *The Veliger*, vol. 6, no. 3, pp. 164–171 in 1964, but much work needs yet to be done in this complicated group.

Genus *Rissoa* Fréminville, 1814

Small (2 to 4 mm.), conic, whitish shells with axial riblets and spiral threads, rarely smooth. Tentacles with microscopic cilia at the ends. Operculum corneous, paucispiral, with the nucleus at one end. Outer lip usually thickened. Type: *ventricosa* Desmarest, 1814. Synonyms are: *Anatasia* and *Apanthausa* Gistel, 1848. The genus does not appear to be present in American waters, although many species have been assigned to *Rissoa* by former workers. Most of these American species are now in the genera *Cingula*, *Alvania*, *Onoba* or *Rissoina*. True *Rissoa* is common in northern Europe. The Atlantic coast species, all deep-water and doubtfully belonging to this genus are:

Subgenus *Mirarissoina* Woodring, 1928

Type of this subgenus is *lepida* Woodring, 1928.

Rissoa bermudezi Aguayo and Rehder, 1936.   573
Bermudez's Risso

Southeast Florida and Cuba.

3 to 5 mm., 8 to 10 whorls; outer lip thick; with a deep anal notch, which is almost closed by a strong node on the parietal callus. With 20 to 26 spiral striations between the 18 to 21 axial ribs per whorl. Uncommon; 16 to 40 fathoms. (See *The Nautilus*, vol. 51, p. 35.)

Other species:

574 *Rissoa syngenes* Verrill, 1884. Off Cape Hatteras, North Carolina, 142 fms., to off Fernandina, Florida, 294 fms.

575 *Rissoa xanthias* Watson, 1885 (an *Alvania?*). Off Georgia and northeast Florida, 294 to 440 fms.

576 *Rissoa pompholyx* Dall, 1927. Off Georgia, 440 fms.

577 *Rissoa listera* Dall, 1927. Off Fernandina, Florida, 294 fms.

578 *Rissoa pyrrhais* Watson, 1885. Florida Strait to the Lesser Antilles, 390 to 780 fms.

579 *Rissoa* (*Setia* H. and A. Adams, 1852) *curta* Dall, 1927. Off Fernandina, Florida, 294 fms.

580 *Rissoa toroensis* Olsson and McGinty, 1958. Off west Florida, 30 to 130 fms., to Panama.

Genus *Alvania* Risso, 1826

Shells 5 mm. or less in length; ovate to elongate-conic; thick. Spire short; apex sharp; 5 postnuclear whorls sculptured spirally and axially. Apertural rim continuous and buttressed on the right by a thick varix with its interior bearing minute denticles or crenulations. Operculum corneous, brown and paucispiral. Free-swimming veligers are common in the plankton. Type: *Turbo cimex* Linné, 1758. There are about 21 Atlantic coast species and about 20 Pacific species from Alaska to the Gulf of California. The subgenus *Willettia* Gordon, 1939, was erected for the Pacific coast species with its type, *Alvania montereyensis* Bartsch, 1911. The entire group is in need of revision before assigning all of them to subgenera.

Several of the species listed here may be synonyms or homonyms. Coan (1964) places the Pacific coast species in the genus *Alvinia* Monterosato, 1884 (not *Alvania*) and in the subgenus *Willettia*. The type of *Alvinia* is *weinkauffi* Monterosato, 1877, of the Mediterranean.

Alvania acuticostata (Dall, 1889)   **581**
Sharp-ribbed Alvania

Off North Carolina to Mississippi to Barbados.

2 to 4 mm., conical, thin, whitish, with neat, strong axial ribs (14 on the last whorl) which bear a tubercle just below the impressed suture. Aperture roundish, with a thickened outer lip. Spiral sculpture of 4 or 5 threads on the base. No umbilicus. Nucleus yellow, conical, with its third whorl spirally grooved and crossed by microscopic axial raised growth lines. Moderately common; 32 to 683 fathoms.

Alvania precipitata (Dall, 1889)   **582**
Precipitate Alvania

Gulf of Mexico to Florida Straits.

2 to 4 mm., similar to *acuticostata*, but lower-spired, with a thin outer lip with 22 ribs on the last whorl. Base with

numerous spiral grooves. Columella moderately thickened. Rare in 30 to 670 fathoms.

Alvania pelagica (Stimpson, 1851)   **583**
Carinate Alvania

Gulf of St. Lawrence to off Florida.

3 mm., 4 whorls, brownish, conic with 4 or 5 spiral cords on the body whorl giving a carinate outline. Upper cord located near the periphery is beaded by short axial riblets that extend from suture to suture. This is *carinata* Mighels and Adams, 1842 (non da Costa, 1777, non Philippi, 1847). Moderately common; shallow water in the north; down to 355 fathoms in the south.

Alvania brychia (Verrill, 1884)   **584**
Jan-mayen's Alvania

Arctic Canada to North Carolina. Europe.

4 or 5 mm., ovate, translucent glossy-brown with last ¼ of the last whorl and the aperture white. One nuclear whorl which is large, smooth, tan. Postnuclear whorls with a carinated shoulder, rarely noduled and above which on the spire whorls are a dozen, smooth, rounded axial plicae or ribs. 2 or 3 spiral cords below the periphery. Operculum fills the aperture and is paucispiral, thin and chitinous. Common; 10 to 600 fathoms. *A. sibirica* Leche, 1878, is a synonym, as is the well-known name *janmayeni* (Friele, 1877). *Rissoa americana* Friele, 1886, and *Cingula bryanti* Johnson, 1926, are also synonyms.

Alvania gradata (Orbigny, 1842)   **585**
Shouldered Alvania

West Indies.

3 mm., short, stout. Color white, or tinged with orange. Suture deeply channeled. 2 nuclear whorls smooth, the remaining ones with very strong axial ribs (12 to 14 on the last whorl) between which are microscopic spiral, wavy striae. Aperture oval-round. Operculum brown, paucispiral. Moderately common in shallow, offshore waters.

Alvania auberiana (Orbigny, 1842)   **586**
West Indian Alvania

Florida, Texas and the Caribbean; Bermuda.

2 mm., globose, slightly umbilicate. Color yellowish white. 2 nuclear whorls smooth; 3 postnuclear whorls with axial and spiral sculpture of about the same intensity, giving a reticulate appearance. Whorls slightly carinate. Base with spiral cords only. Common in 6 to 20 feet of water. *Rissoa minuscula* Verrill and Bush, 1900, from Bermuda may be this species.

586

Other Atlantic species:

587 *Alvania areolata* Stimpson, 1851. Gulf of St. Lawrence to off Martha's Vineyard, Massachusetts, 10 to 130 fms. Yellowish; 3.5 mm.

581   582   584

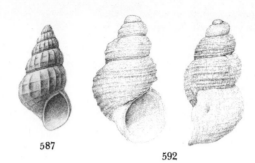

587

592

588 *Alvania apicina* Verrill, 1884. Southeast of Nantucket, 1,608 fms.

590 *Alvania campta* (Dall, 1927). Off Georgia, 440 fms. (not *compta*).

591 *Alvania canonica* (Dall, 1927). Off Fernandina, Florida, 294 fms.

592 *Alvania castanea* Möller, 1842. Labrador to North Carolina, 2 to 102 fms.

593 *Alvania exarata* Stimpson, 1851. (Synonym: *arenaria* Mighels and Adams, 1842, non Montagu, 1808)

594 *Alvania globula* Möller, 1842. Gulf of St. Lawrence, 60 fms.

595 *Alvania harpa* Verrill, 1882. Off Massachusetts, 160 to 487 fms. 3 mm., umbilicate, numerous weak axial plicae; yellowish.

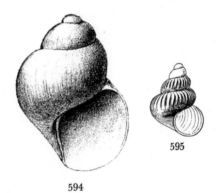

595

594

596 *Alvania karlini* A. H. Clarke, 1963, *Bull. Nat. Mus. Canada*, no. 185, p. 94. Chukchi Rise, North Canadian Basin, 290 fms.

597 *Alvania latior* Mighels and Adams, 1842. Casco Bay, Maine.

598 *Alvania leptalea* Verrill, 1884. South of George's Bank, 858 fms.

599 *Alvania lampra* (Dall, 1927). Off Fernandina, Florida, 294 fms.

600 *Alvania multilineata* Stimpson, 1851. Nova Scotia to Cape Cod, Massachusetts, 5 fms.

601 *Alvania sandersoni* Verrill, 1884. Off Cape Hatteras, North Carolina, 142 fms.

602 *Alvania turgida* "Jeffreys" Verrill, 1880. Off Martha's Vineyard, Massachusetts, 487 fms.

603 *Alvania wyvillethomsoni* (Friele, 1877). Chukchi Rise, North Canadian Basin to Norway, 290 to 389 fms.

604 *Alvania chiriquiensis* Olsson and McGinty, 1958. Western Caribbean. *Bull. Amer. Paleontol.*, vol. 39, no. 177.

605 *Alvania nigrescens* Bartsch and Rehder, 1939. Old Providence Island, Caribbean. 3.1 mm. Smithsonian Miscellaneous Collections.

606 *Alvania aberrans* (C. B. Adams, 1850). West Indies to Brazil.

Subgenus *Alvinia* Monterosato, 1884

The type of this subgenus to which most of the Pacific coast species belong is *Rissoa weinkauffi* Monterosato, 1877, of the Mediterranean. The subgenus *Willettia* Gordon, 1939, with *Alvania montereyensis* Bartsch, 1911, as type may be insignificant. *Willetia* Haas, 1943, is a misspelling.

Alvania compacta Carpenter, 1864   607
Compact Alvania

Port Etches, Alaska, to Baja California.

2 to 3 mm., brownish, ovate-conic, with slightly rounded whorls, and a well-impressed suture. 2½ nuclear whorls, which are strongly rounded and smooth. Postnuclear whorls with 20 to 30 fine protractively slanting axial riblets, crossed by about 6 to 8 spiral threads, and forming slight beads where they cross. Rounded base with about 8 to 10 weak spiral threads. Inner lip slightly swollen and purple-brown. Common; low-tide mark to 100 fathoms on rock bottoms and on the shells of abalones. Synonyms: *acutelirata* Carpenter, 1866; *iliulinkensis* Bartsch, 1911, and possibly *californica* Bartsch, 1911.

607

Alvania purpurea Dall, 1871   608
Purpurea Alvania

Monterey, California, to Baja California.

1.5 to 2.2 mm., light-brown, rather pointed, its whorls carinated by 2 spiral cords crossed by about 20 axial riblets giving an openly cancellate appearance. Suture deep. Base with about 5 spiral threads. Chinklike umbilicus. 2 nuclear whorls, smooth. Common. Among algae from intertidal zone to 30 fathoms.

Alvania rosana Bartsch, 1911   609
Santa Rosa Alvania

Monterey to Catalina Island, California.

2 to 2.6 mm., broadly ovate, yellowish white. 2½ nuclear whorls, smooth, rounded. Postnuclear whorls rounded, bearing 22 to 24 fairly strong, nearly vertical, axial riblets which override 6 or 7 spiral threads. Base rounded with about 8 spiral threads. Umbilicus chinklike. Entire lip slightly thickened. Suture moderately impressed. Common with *compacta*; dredged on gravel bottom from 25 to 75 fathoms.

Alvania oldroydae Bartsch, 1911   610
Oldroyd's Alvania

Monterey, California, to Baja California.

1.4 to 1.6 mm., yellowish white, broadly ovate. 1½ nuclear whorls, smooth, rounded. Postnuclear whorls well-rounded, with 3 spiral cords between the impressed sutures, crossed by 24 to 30 slightly slanting riblets, giving a closely cancellate, or even sub-beaded, appearance with squarish pits between. Inner and outer lips thickened, making the peritreme complete. Suture strongly impressed. Umbilicus narrow, deep. Common on sublittoral gravel bottoms.

Alvania aequisculpta Keep, 1887   611
Evenly-sculptured Alvania

Monterey, California, to Baja California.

3 to 3.5 mm., elongate-conic, light-yellow to whitish. 2 nuclear whorls, marked with 6 microscopic spiral threads and numerous axial striae. Postnuclear whorls slightly carinate by 3 spiral cords crossed by 14 to 18 axial riblets, giving the surface a cancellate appearance. The intersections result in distinct tubercles. Base with 3 or 4 spiral threads, the upper 2 or 3 being crossed by axial threads. Outer and inner lip thickened. Common; low-tide mark to 70 fathoms on gravel bottom. *Rissoa grippiana* Dall, 1908, is a synonym.

Alvania cosmia Bartsch, 1911   612
Cosmic Alvania

Los Angeles County, California, to Baja California.

2 to 2.2 mm., similar to *aequisculpta,* white, cancellate sculpturing, reticulated nuclear whorls, but a smaller shell, with only 2 spiral cords between sutures, and the 3 spiral cords on the base are not crossed by axial riblets. Common; sublittoral to several fathoms. Could be variant or males of *aequisculpta.*

Alvania microglypta Haas, 1943   613
Microglypta Alvania

Monterey, California.

1 mm., creamy-white, solid, 4½ whorls. Nuclear whorl globular, partially immersed. Whorls with 3 or 4 strong, irregularly noduled keels. Aperture oval, doubly lipped. Base of shell concave to flat, smooth, shiny. Found in tidepools. Rare. (*Zool. Series Field Mus. Nat. Hist.,* vol. 29, p. 2)

Other Pacific species:

614 *Alvania carpenteri* (Weinkauff, 1885). (Synonym: *reticulata* Carpenter, 1866.) Forrester Island, Alaska, to Baja California. 2 mm.

615 *Alvania almo* Bartsch, 1911. Santa Barbara Island to Reef Point, California. 1.5 mm.

616 *Alvania bartolomensis* Bartsch, 1917. San Bartolome Bay, Baja California.

617 *Alvania* (*Lapsigyrus* S. S. Berry, 1958) *contrerasi* Jordan, 1909. Baja California. Recent and Pleistocene. Coan (1964) raises Berry's taxon to generic level and adds it to a new subfamily, Phosinellinae.

618 *Alvania keenae* Gordon, 1939. Moss Beach, San Mateo County, California.

619 *Alvania montereyensis* Bartsch, 1911. Sitka, Alaska, to Cayucos, California. Intertidal. 2.3 mm.

620 *Alvania sanjuanensis* Bartsch, 1920. Prince William Sound, Alaska, to Seattle, Washington.

621 *Alvania* (*Alvinia*) *albolirata* (Carpenter, 1864). Gulf of California.

622 *Alvania* (*Alvinia*) *herrerae,* (**622a**) *gallegosi* and (**622b**) *lucasana* (Baker, Hanna and Strong, 1930). Baja California and Tres Marias Islands. (*Proc. Calif. Acad. Sci.,* ser. 4, vol. 19)

623 *Alvania* (*Alvinia*) *electrina* (Carpenter, 1864). Cape San Lucas.

624 *Alvania* (*Alvinia*) *monserratensis* (Baker, Hanna and Strong, 1930). Monserrate Island, Gulf of California.

625 *Alvania kyskaensis* Bartsch, 1917. Kyska Harbor, Aleutians.

626 *Alvania castanella* Dall, 1886. Kyska to Atka, Aleutians. 2.4 mm.

627 *Alvania dalli* Bartsch, 1927. Afognak Island, Alaska, on nullipores.

628 *Alvania aurivillii* Dall, 1886. Kyska to Adakh Island, Aleutians. 4.2 mm.

628

629 *Alvania bakeri* Bartsch, 1910. Port Graham, Cook's Inlet, Alaska. 1.4 mm. *The Nautilus,* vol. 23, p. 137.

630 *Alvania pedroana* Bartsch, 1911. Redondo Beach to Mission Bay, California. 2.2 mm.

631 *Alvania filosa* Carpenter, 1865. Hoonah, Alaska, to Vancouver Island, British Columbia.

616 619 625 626

629 630 632 633

632 *Alvania alaskana* Dall, 1886. Nunivak Island, to Windfall Harbor, Alaska. 2.9 mm.

633 *Alvania dinora* Bartsch, 1917. Forrester Island, Alaska.

634 *Alvania burrardensis* Bartsch, 1921. Burrard Inlet, British Columbia.

Genus *Cingula* Fleming, 1828

Extremely small shells, conic-ovate; aperture round, peristome complete; whorls moderately rounded. Smooth or with weak spiral sculpture. Nuclear whorls smooth. Umbilicus slitlike. In some species, there is no free-swimming larval stage. There are about 15 confusing species on the west coast of America, most of which are found in Alaskan waters. Type: *Turbo cingillus* Montagu, 1803. Synonyms are: *Cingilla* Monterosato, 1884; *Crisilla* Cossmann, 1921. Coan, 1964, erected the subfamily *Cingulinae* to contain this genus, but gave no definitions or reasons.

Cingula montereyensis Bartsch, 1912   635
Monterey Cingula

Moss Beach to Monterey, California.

4 mm. in length, light-brown, smooth, elongate-ovate. Suture slightly indented. Uncommon from shore to 15 fathoms.

Cingula eyerdami Willett, 1934   636
Eyerdam's Cingula

Afognak to Evans Island, Alaska.

2 to 3 mm., elongate-ovate, grayish, with white nuclear whorls. Whorls rounded; suture impressed. Base well-rounded, narrowly umbilicate. Smooth, except for microscopic spiral striations. Found on nullipores on rocks at low tide. *The Nautilus*, vol. 47, p. 103.

Other species:

637 *Cingula (Cingula) forresterensis* Willett, 1934. Forrester Island, Alaska. *The Nautilus*, vol. 47, p. 104.

638 *Cingula (C.) alaskana* Bartsch, 1912. Aleutians to Afognak Island, Alaska.

639 *Cingula (C.) katherinae* Bartsch, 1912. Windfall Harbor, Admiralty Island, Alaska, to Queen Charlotte Island, British Columbia.

640 *Cingula (C.) aleutica* Dall, 1886. Pribilof Islands to Alaska.

641 *Cingula (Nodulus) megalomastomus* Olsson and McGinty, 1958. Bocas Island, Atlantic Panama. *Bull. 177, Amer. Paleont.*, p. 27.

642 *Cingula (Nodulus) fernandinae* Dall, 1927. Off Fernandina, Florida. 294 fms.

643 *Cingula castanea* (Möller, 1842). Arctic Seas to Labrador, 1 to 80 fms. Northern Europe.

644 *Cingula moerchi* Collin, 1887. Arctic Canada, Greenland to Siberia, 1 to 130 fms.

645 *Cingula (N.) cerinella* (Dall, 1887). Atka to Afognak Island, Alaska. Intertidal on nullipores.

646 *Cingula (N.) asser* (Bartsch, 1910). Atka to Cook's Inlet, Alaska.

647 *Cingula (N.) kyskensis* (Bartsch, 1911). Kyska Island, Aleutians, to Afognak, Alaska. Intertidal on nullipores.

648 *Cingula (N.) palmeri* (Dall, 1919). Pribilof Islands, Bering Sea.

649 *Cingula (Falsincingula* Habe, 1958) *martyni* Dall, 1887. Aleutians to Korovia Bay, Alaska.

650 *Cingula (F.) robusta scipio* Dall, 1887. Pribilof and Aleutian Islands, to Middleton Island, Alaska.

651 *Cingula (N.) stewardsoni* (Vanatta, 1909). Fairyland, Bermuda. *The Nautilus*, vol. 23, p. 65.

Subgenus *Onoba* H. and A. Adams, 1854

Cingula aculeus Gould, 1841   652
Pointed Cingula

Nova Scotia to New Jersey.

Extremely small, 2.5 mm. in length, elongate, about 5 whorls, no umbilicus. Whorls rounded. Suture well-impressed. Aperture ovate with a slightly flaring lip. Color light- to rusty-brown. Spiral sculpture of numerous, microscopic incised lines. Below the suture there are numerous, short, axial riblets. Common in shallow water. The type of the subgenus *Onoba* is the common British *semicostata* Montagu, 1803.

Cingula jacksoni Bartsch, 1953   653
Jackson's Cingula

Chesapeake Bay, Maryland.

2.4 mm. long, 1.1 mm. wide, with 5 whorls. Nuclear whorl rounded, smooth. Last whorl with 26 microscopic spiral threads. Umbilicus a narrow chink. Similar to *aculeus* but narrower and with a more pronounced umbilical chink. *The Nautilus*, vol. 67, p. 40.

Subgenus *Microdochus* Rehder, 1943

Seems closest to *Onoba* Adams, 1852, differing markedly from it in being ovate-conic and not cylindrical, in having a shallower suture and a more open umbilicus, and in never having axial riblets. Type: *floridanus* Rehder, 1943. Coan (1964) considers this a subgenus of *Amphirissoa* Dautzenberg and Fischer, 1897.

Cingula floridana (Rehder, 1943)   654
Florida Cingula

Florida Keys and Gulf of Mexico to Puerto Rico.

1 to 2.4 mm., ovate-conic, thin, translucent-brown, 4 whorls (nuclear whorls smooth), convex, finely and evenly spirally lirate, the sculpture commencing imperceptibly; whorls slightly appressed to the preceding whorl at the sutures. Narrow umbilicus bordered by a sharp, fine keel. Columella narrowly reflected. Operculum paucispiral, thin, transparent. Uncommon just offshore to 30 fathoms.

Genus *Amphithalamus* Carpenter, 1865

Extremely small shells, less than 2 mm. in length, smooth, except for a faint cord or spiral thread on the periphery.

Nucleus large, of 1½ whorls which are finely pitted like a thimble. The most striking character is a thin bridge separating the inner lip from the open umbilicus. Type: *inclusus* Carpenter, 1864. There are three species in southern California. A fourth, *A. stephensae* Bartsch, 1927, comes from Baja California.

Periphery without spiral line . . .
> **(655)** *lacunatus* Carpenter, 1864 (San Pedro south).

Periphery with thread or cord . . .
> Periphery angulate . . .
>> **(656)** *inclusus* Carpenter, 1864 (San Pedro south).
> Periphery rounded . . .
>> **(657)** *tenuis* Bartsch, 1911 (Monterey south).

Amphithalamus vallei Aguayo and Jaume, 1947   658
del Valle's Tiny Snail

Florida Keys, Bahamas, to the Lesser Antilles.

1 mm. in length, almost as wide, ovate, with 1½ large, pitted nuclear whorls, and 2 well-rounded postnuclear smooth whorls. Chestnut-brown in color, with the base whitish and bearing a fine spiral line and a larger spiral ridge in the umbilical region. Periphery of whorl with a suggestion of a carination. The last part of the last whorl is detached from the previous whorl. Aperture ovate, lip entire. Common; in the red algae, *Bostrychia,* found on intertidal rocks, mangrove roots and wharf pilings.

Genus *Nannoteretispira* Habe, 1961

Shells minute, translucent-white, cylindrical, whorls slowly increasing in size. Aperture round. Type: *japonica* Habe, 1961. *Venus,* vol. 21, p. 273. One species known from southern California.

Nannoteretispira kelseyi (Bartsch, 1911)   659
Kelsey's Alvania

Southern California to Baja California.

1.5 to 2.0 mm., very slender, cylindro-conic, translucent-white. 2 nuclear whorls, round, smooth. Postnuclear whorls slightly rounded, with fine lines of growth and microscopic spiral striations. Aperture round, somewhat trumpet-shaped. Moderately common; in gravel from low-tide mark.

Genus *Floridiscrobs* Pilsbry and McGinty, 1949

Shells small, 3 mm., solid, ovate-conic, imperforate, of few whorls (the apex lost by erosion), smooth, and with a dark periostracum. Last whorl contracting slightly at the aperture, the suture descending steeply there. Peristome continuous, thick, the parietal margin not appressed but separated from the adjoining whorl by a groove. Operculum thin, oblong, with a long, narrow scar of muscular attach-

660

ment. Type: *Amphithalamus dysbatus* Pilsbry and McGinty, 1949.

660 *Floridiscrobs dysbatus* (Pilsbry and McGinty, 1949). Snake Creek, Plantation Key, Upper Florida Keys. Salt water inlet. 3.4 × 2 mm. *The Nautilus,* vol. 63, p. 14.

Genus *Benthonella* Dall, 1889

Deepsea, small (5 to 9 mm.), elongate, glistening-white, extremely thin. Nuclear whorls brown, polished, with a single carina above the periphery. Umbilicate. Three Atlantic species known. Type: *gaza* Dall, 1889.

Benthonella nisonis Dall, 1889   661
Nison Benthonella

Off both sides of Florida.

5 to 9 mm., elongate, resembling a *Niso,* waxen-white, with 3 dark-brown, polished nuclear whorls and 8 slightly rounded adult whorls. Suture distinct. Occasional axial growth-stoppage line present. Base rounded. Umbilicus small. Aperture rounded in front, peristome thin, discontinuous, slightly reflected, especially at the inner lip, which is arcuated. Last whorl ½ the length of the shell, its width almost as great. Uncommon; 40 to 948 fathoms.

661   662

Other species:

662 *Benthonella gaza* Dall, 1889. Off Georgia to Cuba and the Gulf of Mexico, 6 to 463 fms.

663 *Benthonella fischeri* Dall, 1889. Gulf of Mexico and West Indies, 940 to 1,060 fms.

Genus *Anabathron* Frauenfeld, 1867

Shells very minute (less than 2 mm.), elongate, scalariform, angulate shoulder, round aperture. Operculum corneous. Type: *contabulata* Fruenfeld, 1867. Only one American species, doubtfully belonging to this genus.

Anabathron muriel Bartsch and Rehder, 1939   664
Muriel's Anabathron

Aleutian Islands.

0.8 mm. in length, elongate, thin, semitransparent, white. One nuclear whorl, inflated, slightly tilted. Upper whorls with one strong keel; last whorl with 3, the lower one bordering a funnel-shaped umbilicus. Surface with microscopic reticulations. Peristome slightly thickened. Rare; offshore. (*The Nautilus,* vol. 57)

Family RISSOINIDAE Stimpson, 1865

Genus *Rissoina* Orbigny, 1840

Shells small, usually less than ⅛ inch in length, generally white in color, with strong or weak axial ribs, occasionally with fine spiral, incised lines. Aperture semilunar and somewhat flaring. Operculum corneous, thick, paucispiral, with a claviform process on the inner surface. We have presented nearly all of the species known to both sides of the United States in the form of a key. Type: *inca* Orbigny, 1840. Over 20 subgenera, some probably not justified, have been recognized in this variable genus.

The following names are treated as subgenera: *Phosinella*, Mörch, 1876 (type: *pulchra* C. B. Adams, 1850); *Schwartziella* Nevill, 1881 (type: *orientalis* Nevill, 1881); *Zebinella* Mörch, 1876 (type: *decussata* Montagu, 1803); *Moerchiella* Nevill, 1884 (type: *gigantea* Deshayes, 1850); *Costalynia* Laseron, 1956 (type: *cardinalis* Brazier, 1877).

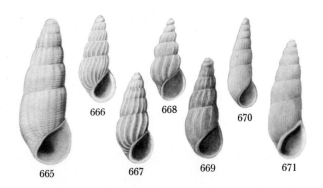

665   666   667   668   669   670   671

Key to the Pacific Coast *Rissoina*

A. Color pure-white or bluish white . . . . . B
   Color yellow to light-red; 6 mm., off Redondo Beach
   to Baja California
             (665) *kelseyi* (Dall and Bartsch, 1902)
B. Axial ribs strong, less than 20 on the last whorl . . C
   Axial ribs weak, numerous . . . . D
C. Interspaces with silky, wavy crinkles; 3 mm.; Coronado
   Islands . . . . . **(666)** *cleo* Bartsch, 1915
   Interspaces smooth, 3 mm.; Catalina Islands, south . .
            (667) *californica* Bartsch, 1915
D. Whorls decidedly inflated; 3 mm.; Monterey south . .
          (668) *bakeri* Bartsch, 1902
   Whorls not inflated . . . . E
E. With very fine, numerous axial threads (48 to 55 on last
   whorl) . . . . F
   With coarse riblets (36 on last whorl; 14 on next to last);
   3 mm. Alaska to Monterery . . . . .
         (669) *newcombei* Dall, 1897
F. Shell slender; 2 mm.; San Pedro south . . .
         (670) *dalli* Bartsch, 1915
   Shell not as slender; 3.5 mm.; Redondo Beach south .
       (671) *coronadoensis* Bartsch, 1915

673   674   682

Key to the Atlantic Coast *Rissoina* and *Zebina*

A. Shell sculptured with riblets or spiral lines . . . B
   Shell smooth, glossy-white; 4 mm.; Carolinas to Florida,
   Texas and to Brazil. (syn.: *laevigata* C. B. Adams,
   1850). . **(672)** *Zebina browniana* (Orbigny, 1842)
B. With axial ribs more prominent than spiral threads   C
   With axial ribs not more prominent than spiral
   threads . . . . F
C. Axial ribs only . . . . D
   Axial ribs and spiral threads both present . . . E
D. 4.5 to 6.0 mm.; white or stained yellow; 16 to 22 ribs;
   South Florida and the West Indies; Bermuda; Brazil
    (673) *R. (Schwartziella)* *bryerea* (Montagu, 1803)
   3.0 to 5.0 mm.; white; 11 to 14 ribs; suture sometimes
   deep; North Carolina to Florida to Brazil (brackish
   waters) . . **(674)** *R. (S.)* *catesbyana* Orbigny, 1842
E. 4 to 5 mm.; ribs strong but disappearing on base; spiral
   threads strongest on base; white to rusty; southeast
   Florida, Texas and the West Indies . . . .
    (675) *R. (Zebinella)* *multicostata* (C. B. Adams, 1850)
   6 to 7 mm.; 25 to 28 low, weak ribs, spirally striated or
   pitted between; glossy; white to yellowish; North
   Carolina to Texas and the West Indies . . . .
    (676) *R. (Zebinella)* *decussata* (Montagu, 1803)
F. Sculpture strongly cancellate . . . . . . G
   Not strongly cancellate; low, spiral threads dominant;
   axial ribs faint; weakly cancellate; 5 to 10 mm.;
   southeast Florida and the West Indies
    (677) *R. (Morchiella)* *striosa* (C. B. Adams, 1850)
G. 5 to 7 mm.; white; strongly cancellate; depressed interspaces large and square; southeast Florida, south
   Texas and the West Indies; Brazil; Bermuda
    (678) *R. (Phosinella)* *cancellata* Philippi, 1847
   4 to 4.5 mm.; glassy-white; depressed interspaces small,
   rounded; Texas to the West Indies . . . .
    (679) *R. (Phosinella)* *sagraiana* Orbigny, 1842

Rissoina catesbyana Orbigny, 1842   680
Catesby's Risso

North Carolina to Texas to Brazil; Bermuda.

3 to 4 mm., white, elongate, with a slender, styliform protoconch (and pelagic larval stage). Prominent tooth inside the outer lip. Suture usually deep. 11 to 14 strong axial ribs. No spiral striae present. Thrives in *Thalassia* weed beds in sheltered bays and lagoons. Not found on outer reefs. Synonyms: *scalarella* C. B. Adams, 1845; *bermudensis* Peile, 1926. (See D. R. Moore, 1969, *Trans. Gulf Coast Assoc. Geol. Soc.*, vol. 19, p. 425)

Other species:

681 *Rissoina mayori* Dall, 1927. Off Georgia, 440 fms.; off Miami, Florida, 58 fms.

682 *Rissoina (Atlantorissoina* Kosuge, 1965) *chesnelii* (Michaud, 1830). Appears to be limited to Jamaica. This name has been misapplied to what is now known as the common *catesbyana* Orbigny, 1842.

683 *Rissoina (Phosinella) shaeferi* McGinty, 1962. Barbados; Panama; Virgin Islands. *The Nautilus*, vol. 76, p. 42.

684 *Rissoina (Schwartziella) fischeri* Desjardin, 1949. Cuba and Bermuda. *Jour. de Conchyl.*, vol. 89, p. 199. 5 mm. Rare.

685 *Rissoina hannai* A. G. Smith and Gordon, 1948. 25 fms., Carmel Bay, California.

686 *Rissoina keenae* A. G. Smith and Gordon, 1948. 5 to 15 fms., Point Pinos, Monterey Bay, California. Both *Proc. Calif. Acad. Sci.*, series 4, vol. 26, p. 226 and 227.

About 40 species live from Baja California to Peru. Consult: Paul Bartsch, 1915, "Recent and fossil mollusks of the genus *Rissoina* from the West Coast of America." *Proc. U.S. Nat. Mus.*, vol. 49, pp. 33–63, pls. 28–33.

Genus *Zebina* H. and A. Adams, 1854

Small rissoinids without axial ribs or spiral threads. Two or three toothlike nodes may be formed on the inside of the outer lip. Type: *coronata* (Mohrenstern, 1860). *Cibdezebina* Woodring, 1928; and *Iopsis* Gabb, 1873, are synonyms.

Zebina browniana (Orbigny, 1842)   **687**
Smooth Risso

North Carolina to the West Indies. Bermuda.

4 to 5 mm., smooth, with nearly flat whorls. Color white, sometimes with 2 or 3 pale-brown bands; highly polished. Apex acute, aperture rather small; outer lip thickened, and with 2 small nodes, one near the base, the other near the posterior canal. Common; in sand in shallow water to 3 fathoms. Synonyms: *sloaniana* (Orbigny, 1842); *laevigata* (C. B. Adams, 1850); *laevissima* (C. B. Adams, 1850).

687

Gulf of California *Rissoina*:

688 *Rissoina firmata* (C. B. Adams, 1852). Gulf of California to Panama. (Synonyms: *scalariformis* (C. B. Adams, 1852) and *excolpa* Bartsch, 1902.) 5 mm. 14 axial ribs; 40 spiral threads.

689 *Rissoina stricta* Menke, 1850. Gulf of California to Panama; Galapagos. Common. 7 to 9 mm. 18 to 28 axial ribs. 15 spiral threads on base. (Synonyms: *favilla* Bartsch, 1915; *io* Bartsch, 1915; *gisna* Bartsch, 1915; *fortis* C. B. Adams, 1852; *mazatlanica* Bartsch, 1915; *dina* Bartsch, 1915)

690 *Rissoina expansa* Carpenter, 1865. Guaymas to Mazatlan, Mexico. 9 mm.

691 *Rissoina peninsularis* Bartsch, 1915. Cape San Lucas and Gulf of California. 5 to 6 mm. (Synonym: *townsendi* Bartsch, 1902.) 32 to 52 axial ribs on last whorls.

692 *Rissoina barthelowi* Bartsch, 1915. Concepcion Bay, Baja California, 2 to 4 fms. 7 mm. 38 axial ribs on last whorls.

693 *Rissoina* (*Sulcorissoina* Kosuge, 1965) *lapazana* Bartsch, 1915. La Paz, Baja California. 6 mm.

694 *Rissoina*(?) *histia* Bartsch, 1915. Off La Paz, Baja California, 9 to 10 fms. 2.8 mm.

695 *Rissoina burragei* Bartsch, 1915. Gulf of California. 4 mm.

696 *Rissoina nereina* Bartsch, 1915. West Baja California. 4 mm.

697 *Rissoina mexicana* Bartsch, 1915. Gulf of California. 2.7 mm.

698 *Rissoina cerrosensis* Bartsch, 1915. Off Cerros Island, Baja California. 2.4 mm.

699 *Rissoina basilirata;* (**700**) *melanelloides;* (**701**) *porteri* all Baker, Hanna and Strong, 1930. Baja California, Mexico.

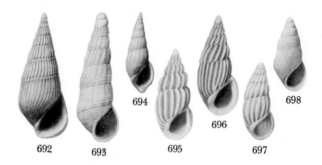

694   698
692   693   695   696   697

Genus *Crepitacella* Guppy, 1867

Shells small (9 to 10 mm.), whitish, with axial riblets on the early whorls and gradually fading out on the last whorl. Whorls usually shouldered. 2 nuclear whorls, inflated, and turned a little to one side. Anal fasciole broad; umbilicus chinklike. Most are deep water and rare. *Dolophanes* Gabb, 1872, is a synonym. Type: *cepula* Guppy, 1867, from the Miocene of Jamaica.

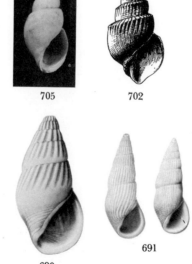

705   702

688   689   690   691

702 *Crepitacella gabbi* (Dall, 1889). 785 fms., off St. Vincent, West Indies.

703 *Crepitacella columbella* (Dall, 1881). 158 to 220 fms., off northern Cuba.

704 *Crepitacella leucophlegma* (Dall, 1881). Deep water, off Havana, Cuba.

705 *Crepitacella vestalis* Rehder, 1943. 10 mm. Off southeast Florida, 18 to 35 fms. *Proc. U.S. Nat. Mus.*, vol. 93, p. 194, pl. 20.

Subfamily BARLEEIINAE Thiele, 1925

Genus *Barleeia* Clark, 1853

Shells small, conic-ovate, whorls rounded, usually smooth or faintly striated; aperture oval, entire. Outer lip thin. Nuclear whorls microscopically "thimble-pitted." Operculum calcareous, with annular growth lines and with an internal riblike process. Tentacles fairly short, broad, with large eyes at the outer bases of the tentacles. Foot emarginate behind. Type: *unifasciata* Montagu, 1803. Small single eggs are attached to algae.

Although well-represented in the Pacific coast and in northwestern Europe, the genus has not been recorded from the Atlantic coast of the United States.

Do not confuse some species with *Assiminea* which has a strong but microscopic spiral thread below the suture.

Barleeia californica Bartsch, 1920   **706**
California Barley Shell

San Pedro, California, to Baja California.

2 to 2.5 mm., strong, glossy, subovate, light-brown with a broad spiral whitish band at the periphery. Much more ovate than *haliotiphila*. Common; low-tide mark to just offshore among kelp beds.

Barleeia haliotiphila Carpenter, 1864   **707**
Abalone Barley Shell

British Columbia to Baja California.

2 to 3.5 mm., strong, glossy, elongate-ovate, light-brown. 2 nuclear whorls, rounded, marked by curved, axial rows of pinpoints arranged in spiral series. Base of shell slightly angular at the whorl's periphery. Surface with microscopic, closely spaced, spiral striations. Apertural peritreme thickened, circu-

lar and complete. *B. oldroydae* (Bartsch, 1911) is a synonym. Common; shore to 15 fathoms, on rocks, abalone shells and kelp holdfasts.

Barleeia subtenuis Carpenter, 1864   **708**
Carpenter's Barley Shell

Alaska to central Baja California.

2.0 to 2.7 mm., ovate, dark-brown, thin-shelled; whorls flatly convex, suture minutely impressed. Outer lip slightly thickened. Nuclear whorls swollen. Umbilicus chinklike. This species lacks the white band of *californica* and has a broader shell. Moderately common; shallow, sublittoral waters. Synonyms: *rimata* Carpenter, 1864; *coronadoensis* Bartsch, 1920; and *sanjuanensis* Bartsch, 1920.

Subgenus *Pseudodiala* Ponder, 1967

Type of this subgenus is *acuta* (Carpenter, 1864).

Barleeia acuta (Carpenter, 1864)   **709**
Acute Barley Shell

British Columbia to Baja California.

3 to 4 mm., elongate-conic, with a flat-sided spire, a rather strong basal carination, shiny, and with numerous, broad, dark-brown axial color blotches. Uncommon; shallow water. *B. marmorea* (Carpenter, 1864), *dalli* Bartsch, 1920, and possibly *bentleyi* Bartsch, 1920, are synonyms.

Other species:

710 *Barleeia bentleyi* Bartsch, 1920. Venice, California, to Cape San Lucas, Baja California.

711 *Barleeia carpenteri* Bartsch, 1920. Newport Beach, California, to Cape San Lucas, Baja California. 4.8 whorls. 1.6 mm.

712 *Barleeia orcutti* Bartsch, 1920. Baja California and the Gulf of California. 5.5 whorls. 2.2 mm.

713 *Barleeia alderi* (Carpenter, 1864). Orange County, California, to Mexico.

Family ASSIMINEIDAE H. and A. Adams, 1856

· Details of the anatomy, ecology and life history of this family are found in Abbott, 1958, *Proc. Acad. Nat. Sci. Philadelphia*, vol. 110.

Genus *Assiminea* Fleming, 1828

Small, ovate, smoothish shells with rounded whorls, simple outer lip and with or without an umbilicus. Operculum paucispiral, chitinous. Eyes on small, stubby peduncles. Most are brackish water and intertidal. Type: *grayana* Fleming, 1828. *Syncera* Gray is a *nomen dubium*.

Subgenus *Angustassiminea* Habe, 1943

The type of this subgenus is *castanea* Westerlund, 1883.

Assiminea succinea (Pfeiffer, 1840) ·   **714**
Atlantic Assiminea

Boston, Massachusetts, to Texas and to Brazil. Bermuda.

1 to 2.5 mm., ovate, smooth, light translucent-brown. About 5 whorls, rounded. There is a microscopic, single, spiral, raised thread running just below the neatly impressed suture. Columella flattens out over the parietal wall to form a narrow, slightly raised callus. Umbilicus very small, chinklike. Outer

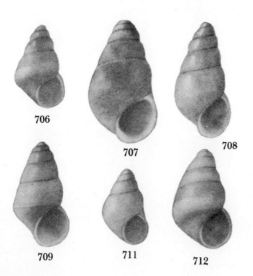

706   707   708

709   711   712

714

lip thin, sharp and smooth. Rarely banded with lighter brown. Common; intertidal, mudflats. Synonyms: *modesta* H. C. Lea, 1845; *concolor* (C. B. Adams, 1850); *concinna* (C. B. Adams, 1850) and *auberiana* of Dall, not Orbigny, 1842.

Assiminea californica (Tryon, 1865)   **715**
California Assiminea

Vancouver Island, British Columbia, to Baja California.

2 to 3.5 mm., glossy-brown, ovate-conic, parietal wall glazed. Similar to *succinea*. The only Pacific coast species. *A. translucens* Carpenter, 1864, and *magdalensis* (Bartsch, 1920) are synonyms. Found on mud in marsh areas among the grass, *Monanthocloe littoralis*.

Family HYDROBIIDAE Stimpson, 1865

Minute aquatic, elongate-conic, horn-colored shells with a thin, paucispiral, chitinous operculum. Tentacles long and with microscopic cilia. Penis in males large and arising on the back, behind the right tentacle.

Genus *Hydrobia* Hartmann, 1821

Shells minute, 2 to 4 mm. in length, translucent-brown, with a blunt apex and rounded whorls. Most species are brackish water, intertidal in habit. Type: *stagnalis* Baster, 1765, from Europe.

Hydrobia totteni Morrison, 1954   **716**
Minute Hydrobia

Labrador to New Jersey.

2 to 4 mm. in length, with 4 or 5 rounded whorls. The apical whorls are usually eroded. Color translucent-brown. Axial growth lines microscopic and closely spaced. Outer lip thin. Head dusky-brown to blackish. White granules are embedded in the tentacles and neck. Black spot just anterior to each eye. Penis simple, sickle-shaped, with microscopic cilia on the distal end and penial lobes on the inner base. Abundant in shallow pools near shore or in marshes. Associated with green algae, *Littorina littorea* and *Elysia*. For details, see George M. Davis, 1966, *Venus*, vol. 25, p. 27. *Turbo minutus* Totten, 1834, is a synonym and homonym.

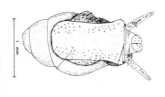

716

Genus *Littoridina* Eydoux and Souleyet, 1852

Shells small, about 3 to 5 mm., ovate-conic, smoothish, brown or whitish. Whorls moderately rounded. Operculum chitinous, thin, paucispiral. Periostracum very thin and translucent-gray. Suture moderately impressed. Type: *gaudichaudii* Souleyet, 1852.

Subgenus *Texadina* Abbott and Ladd, 1951

Peristome constricted and more rounded than in typical *Littoridina*. The last third of the whorl more rapidly descending. Type: *sphinctostoma* Abbott and Ladd, 1951. *Jour. Wash. Acad. Sci.*, vol. 41, p. 335.

Littoridina sphinctostoma Abbott and Ladd, 1951   **717**
Small-mouthed Hydrobid

Mississippi to Mexico.

2 to 3.5 mm., ovate-conic to fusiform, very narrowly umbilicate, 5 to 7 whorls, translucent-gray (live) to opaque-white (dead). Glossy nuclear whorls not distinguishable from postnuclear whorls. Axial sculpture of widely spaced, weak growth lines. Last whorl sometimes rapidly descending. Aperture ovate to round and sometimes slightly constricted. Common; in sheltered, brackish to fresh water lagoons in muddy sand in water 1 to 3 feet.

Family ANAPLOCAMIDAE Dall, 1895

Genus *Anaplocamus* Dall, 1895

Both the genus and family were proposed for *A. borealis* Dall, 1895 (*Proc. U.S. Nat. Mus.*, vol. 24, p. 550, pl. 38, fig. 4), a fresh-water snail of the pleurocerid genus *Anculosa* which had accidentally fallen into a dredging sample from Unimak Island, Alaska, while in the process of being curated at the U.S. National Museum. As pointed out by H. A. Rehder (*The Nautilus*, vol. 56, p. 49) the name *Anaplocamus* should be expunged from the marine mollusk lists.

Family TRUNCATELLIDAE Gray, 1840

Small, ovate to elongate shells, mostly land-dwelling or associated with brackish water, having a paucispiral chitinous operculum, a reduced set of gills on the inner mantle wall, 2 moderately long tentacles, with the eyes at the base.

Genus *Truncatella* Risso, 1826

Shells less than 10 mm., elongate-cylindrical and, in the adult form, with the top whorls knocked off. Usually a translucent, glossy-tan, smooth or with fine axial riblets. Aperture surrounded by a complete, thickened peritreme. Outer lip thickened into a varix. Type: *subcylindrica* Linné, 1758, of Europe. They live high up on the beach under rotted leaves where there is shade. Our treatment of the Caribbean species follows A. de la Torre, 1960 (*The Nautilus*, vol. 73, p. 79), rather than Clench and Turner, 1948 (*Johnsonia*, vol. 2, no. 25). The accepted species are very variable, and I suspect fewer should be recognized.

Truncatella pulchella Pfeiffer, 1839   **718**
Beautiful Truncatella

South half of Florida, West Indies; Bermuda.

3 to 6 mm., horn-colored, outer lip greatly thickened, with 17 to 40 axial riblets. No microscopic spiral lines between ribs. Ribs are either very weak, obsolete (typical form) or more often quite strong (forma *bilabiata* Pfeiffer, 1840), the latter being the dominant form in Bermuda. Abundant, in rotted leaves and shady humid crevices above the high-tide mark. Synonyms: *bairdiana* C. B. Adams, 1852; *capillacea* Pfeiffer, 1859.

Truncatella caribaeensis Reeve, 1842   **719**
Caribbean Truncatella

South half of Florida; to Texas; West Indies; Bermuda.

6 to 9 mm., differing from *pulchella* in being larger and having a thin outer lip. I find it difficult to believe this a different species, but I follow de la Torre, Clench and Turner, and Pilsbry. Abundant. This is what Clench and Turner (1948) erroneously called "*pulchella* Pfeiffer." *T. succinea* C. B. Adams, 1845, is a synonym.

Truncatella californica Pfeiffer, 1857   **720**
California Truncatella

Southern California and Mexico.

4 to 5 mm., cylindrical, translucent-tan; nuclear whorl involuted. 3 or 4 postnuclear whorls, smoothish except for numerous fine axial riblets. Columella thick. Common; along the shore under damp detritus and rocks.

Other species:

721 *Truncatella stimpsoni* Stearns, 1872. Catalina Island, California, to Baja California.

Subgenus *Tomlinitella* Clench and Turner, 1948

With less than 16 axial riblets which are very strong. Type: *scalaris* (Michaud, 1830). Synonym: *Tomlinella* Clench and Turner, 1948, non Viader, 1938.

Truncatella scalaris (Michaud, 1830)   **722**
Ladder Truncatella

Florida Keys, West Indies; Bermuda.

4 to 5 mm., with microscopic spiral threads between the riblets which may be 8 to 11 per whorl (typical form) or 12 to 16 (forma *clathrus* Lowe, 1832, and *piratica* Clench and Turner, 1948, from Bermuda). The outer lip has 2 adjoining varices in all forms. Locally uncommon.

718   722   723

Truncatella bahamensis Clench and Turner, 1948   **723**
Bahama Truncatella

Bahama Islands.

4 to 5 mm., rather narrow, with about 20 very strong, rather widely spaced, bladelike riblets. Outer lip thick, with a double varix. Locally common.

Other species:

724 *Truncatella barbadensis* Pfeiffer, 1856. Lesser Antilles. 40 riblets on last whorl.

Family RISSOELLIDAE Gray, 1850

Genus *Rissoella* J. E. Gray, 1847

Shells usually less than 2 mm., thin, semitranslucent, whitish, broadly ovate; whorls inflated; smooth or with slanting riblets and fine spiral threads. Umbilicate or not. Peritreme (mouth rim) completely circular. Operculum corneous, thin, with the nucleus in the middle and a short rib on the underside, which proceeds from the nucleus in the direction of the outer margin. Type: *glabra* Alder, 1844, non Brown, and now *diaphana* Alder, 1848. *Jeffreysia* Forbes and Hanley, 1850, is a synonym. Egg capsules are attached to algae. No free-swimming larval stage. Several species are known in northern Europe. The Tropical Eastern Pacific has 8 species assigned to this genus. For a world list see Robertson, 1961, *The Nautilus*, vol. 74 and 75. Undetermined subgenera include *Jeffreysilla* Thiele, 1925 (?synonym: *Phycodrosus* Rehder, 1943); *Jeffreysiella* Thiele, 1912 (?synonym: *Heterorissoa* Iredale, 1912); *Jeffreysiopsis* Thiele, 1912; and *Jeffreysina* Thiele, 1925.

Rissoella caribaea Rehder, 1943   **725**
Caribbean Risso

Florida Keys to Puerto Rico.

Extremely small, 1 to 1.7 mm., broadly ovate, glassy, transparent (alive) to white (dead). 4 whorls, smooth; umbilicus narrow and bordered by a weak keel. Operculum corneous, transparent, ovate and with a peglike extension on the straight columella side. Color of animal's body is black. Head has one set of "false tentacles" and one pair of true tentacles. Abundant in algae on mangrove prop-roots and dredged alive to 6 fathoms.

Rissoella galba Robertson, 1961   **726**
Yellow-bodied Risso

Bahama Islands.

Extremely small, 0.5 to 0.7 mm., similar to *caribaea*, but half as large, with a wider umbilicus that is bordered by a

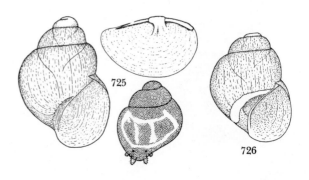

725   726

white band, and having a yellow, rather than black, body. Lives in algae on mangrove roots with *caribaea,* but is uncommon. (Illust. from Robertson, 1961, *The Nautilus,* vol. 74, pl. 9)

Other species:

727 *Rissoella californica* Bartsch, 1927. San Clemente to San Martin Islands, Baja California.

728 *Rissoella tumens* (Carpenter, 1856). Gulf of California to Mazatlan, Mexico.

729 *Rissoella johnstoni* Baker, Strong and Hanna, 1930. Cape San Lucas, Baja California.

730 *Rissoella excolpa* Bartsch, 1920. Baja California.

731 *Rissoella bifasciata* (Carpenter, 1856). Mexico.

732 *Rissoella bakeri* Strong, 1938. Guadalupe Island, Baja California.

733 *Rissoella anguliferens* (de Folin, 1870). Panama Bay.

734 *Rissoella hertleini* A. G. Smith and Gordon, 1948. 10 fms., off Cabrillo Point, Monterey Bay, California.

Family SKENEOPSIDAE Iredale, 1915

Genus *Skeneopsis* Iredale, 1915

Shells minute, planorboid, red-brown, with a multispiral operculum. Sexes separate. Lack pallial and metapodial tentacles. Gills greatly reduced. Type: *planorbis* (Fabricius, 1780).

Skeneopsis planorbis (Fabricius, 1780)　　　　**735**
Planorb Skenea

Greenland to Florida; northern Europe.

Extremely small, 1 mm. in diameter, planorboid, red-brown in color; apex almost flat, whorls round and rapidly increasing in size. Aperture round. Surface smooth. Umbilicus very wide and deep. Operculum corneous, multispiral. Common in shallow water from lower tideline to 16 fathoms. Often found among sponges and old oyster shells. Albuminous egg capsules (0.4 mm., ovoid) attached to filamentous algae. No free-swimming larvae. Formerly put in *Skenea.* Drawing courtesy of Hubendick and Warén, 1971.

Other species:

736 *Skeneopsis alaskana* Dall, 1919. Pribilof and Unalaska Islands, Alaska.

735

737

Family OMALOGYRIDAE P. Fischer, 1885

Genus *Omalogyra* Jeffreys, 1867

Shell extremely small, less than 2 mm., planorboid; apex submerged; whorls expand on the same plane; aperture clasping both sides of the periphery. Operculum few-whorled, nucleus central. Animals hermaphrodite and have large eggs. Gills and osphradium lost. Penis is a retractile tubular structure lying within a sac. *Homalogyra* Jeffreys, 1867, is the same. Type: *atomus* (Philippi, 1841).

Omalogyra atomus (Philippi, 1841)　　　　**737**
Atom Snail

Maine to Rhode Island; northern Europe.

The smallest New England marine snail known, reaching a diameter of 0.5 mm. Shell red-brown, flat, spire sunken; whorls round and encompassing the former ones on the same plane. Umbilicus very wide. Suture deep. Aperture round; operculum few-whorled, nucleus central. Intertidal in lower rock pools. Rarely found because of its small size. It probably has a wide range.

Other species:

738 *Omalogyra* (*Ammonicera* Vayssière, 1893) *planorbis* (Dall, 1927). Labrador to off northwest Florida, 294 to 1,622 fms. 1.3 mm. Formerly placed in *Lippistes* and *Cyclostrema.* See D. R. Moore, 1971, *The Nautilus,* vol. 84, p. 113, fig. 1.

738

739 *Omalogyra* (*Ammonicera*) *densicostata* (Jeffreys, 1884). Off Portugal; off Labrador, 994 and 1,095 fms.; Azores.

Family CYCLOSTREMELLIDAE D. R. Moore, 1966

Heterostrophic protoconch, planispiral shell, paucispiral operculum. See Moore, 1966, *Bull. Marine Sci.,* vol. 16, p. 480. Now placed in the Pyramidellidae (see R. Robertson, 1973, *The Nautilus,* vol. 87, p. 88).

Genus *Cyclostremella* Bush, 1897

Shell minute, thin, planorbiform, semitransparent, whorls convex, nearly symmetrically coiled, forming a concavely depressed spire and large umbilical cavity. Suture deep. Mouth round. Type: *humilis* Bush, 1897. Radulae absent. Formerly placed in the Rissoacea (see R. Robertson, 1973, *The Nautilus,* vol. 87, p. 88).

740 *Cyclostremella humilis* Bush, 1897. Off Cape Hatteras, North Carolina, to Texas and to the Lesser Antilles, 15 to 16 fms. Common. 2 mm. Illus. from D. Moore, 1966.

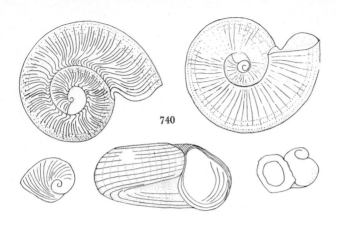

740

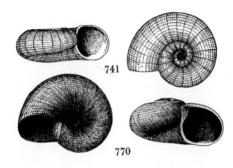

741

770

741 *Cyclostremella californica* Bartsch, 1907. San Pedro to San Diego, California. Pleistocene. 2.3 mm. in diameter.

742 *Cyclostremella concordia* Bartsch, 1920. Olga, Orcus Island, Washington. *Jour. Wash. Acad. Sci.,* vol. 10.

—*Cyclostremella dalli* Bartsch, 1911, see *Vitrinella.*

Family VITRINELLIDAE Bush, 1897

Subfamily VITRINELLINAE Bush, 1897

Shells very small, usually less than 5 mm., porcelaneous, white or translucent, wider than high. Operculum chitinous, circular, multispiral. Penis present in the male. Gills set deep within the mantle cavity. Radula taenioglossate. Do not confuse with the family Tornidae (Aedorbidae).

Genus *Vitrinella* C. B. Adams, 1850

Shell minute, thin, depressed, umbilicate, and with 3 to 4 subtubular whorls. The umbilicus has rather flattened walls and is usually bounded by a spiral cord. The rounded aperture is oblique, with a thin lip, its upper margin arching forward. Columella only moderately thickened. Operculum corneous, thin, multispiral. The tentacles are microscopically ciliated. There are many species in American waters with quite a number of genera and subgenera. Type: *helicoidea* C. B. Adams, 1850, *Tomura* Pilsbry and McGinty, 1945, is a synonym.

Subgenus *Vitrinella* C. B. Adams, 1850

Vitrinella helicoidea C. B. Adams, 1850   **743**
Helix Vitrinella

North Carolina to Florida, Texas, West Indies. Bermuda.

2 mm. in diameter, planorboid, 4 whorls, spire moderately raised. Translucent-white, glossy, smooth. Umbilicus round, very deep, moderately wide, bounded by a small, spiral, smoothish thread. Wall of umbilicus flattish. Columella strong, braced on the whorl above by a small, spreading callus. Outer lip thin, sharp. Living animal translucent-white, pinkish around the head. Not uncommon in shallow water, living under rocks.

Vitrinella hemphilli Vanatta, 1913   **744**
Hemphill's Vitrinella

Cedar Keys, Northwest Florida.

2.5 mm. wide, 1.6 mm. high, depressed, hyaline, white, polished, with faint growth striae; spire acute, composed of 4 convex whorls which are slightly concave below the suture. Parietal wall with a thin callus; columella narrow. Umbilicus deep, round, bordered by a weak angle. Uncommon; shallow water.

Vitrinella praecox Pilsbry and McGinty, 1946   **745**
Premature Vitrinella

Southeast Florida to Panama.

1.6 mm. in width, 0.8 mm. in height, grayish white (but sometimes stained reddish brown), depressed, with 3 weak spiral threads just above the periphery, and sometimes with several on the middle of the rounded base. Umbilicus ⅓ the size of the entire shell, with a strong cord overhanging it. Columella thickened below. Adults may be smoothish. Found under stones in the intertidal zone.

Vitrinella thomasi (Pilsbry, 1945)   **746**
Tom McGinty's Vitrinella

Florida to Texas.

1 mm. wide, 0.6 mm. high, cinnamon-buff, with about a dozen widely separated, arching, low, short riblets extending from a spiral subsutural ridge to the whorl's periphery. Microscopic spirals on the periphery. Strong spiral cord bounds the deep, fairly wide umbilicus. Outer lip thin. Uncommon; shallow water.

Vitrinella filifera Pilsbry and McGinty, 1946   **747**
Threaded Vitrinella

Southeast Florida.

1.3 mm. in width, 0.7 mm. in height, depressed, umbilicus ¼ the size of the shell. Last whorl with a spiral cord just below the suture. Deep umbilicus bounded by a cord which becomes weaker near the aperture. Rare; shallow water.

Vitrinella bicaudata Pilsbry and McGinty, 1946   **748**
Two-tailed Vitrinella

Lower Florida Keys.

1.2 mm. in width, 0.8 mm. in height, similar to *helicoidea,* but with a higher spire, relatively larger aperture and smaller umbilicus. Type of the subgenus *Tomura* Pilsbry and McGinty, 1945. Rare; living under stones in the intertidal area.

Vitrinella floridana Pilsbry and McGinty, 1946   **749**
Florida Vitrinella

    Southeast Florida to Texas.

    2 mm. in width, 1 mm. in height; umbilicus 0.6 mm. White, depressed, smooth throughout. Without a cord bordering the umbilicus. Not uncommon; under stones; intertidal.

Vitrinella terminalis Pilsbry and McGinty, 1946   **750**
Terminal Vitrinella

    Northwest Florida.

    2.4 mm. in width, 1.7 mm. in height. Resembles a *Turbo* in shape; spire fairly high and conic. Whorls convex. Suture deep. Columellar margin extremely thick, reflected over part of the umbilicus, passing into a rather thick but thin-edged parietal callus. A few coarse spiral threads are on the periphery of the last whorl. Rare; dredged in 19 fathoms. *The Nautilus,* vol. 60, no. 1 (1946).

Vitrinella texana D. Moore, 1965   **751**
Texas Vitrinella

    Texas coast.

    1.7 mm. wide, 0.8 mm. high, white, fragile, depressed, umbilicus narrow but deep. Periphery forms an angle with the base of the shell. Glassy nuclear whorls 1¾. Last whorl with fine spiral grooves above and with numerous (30 to 36) short, radial riblets below. Spiral grooves in the umbilicus. Uncommon; shallow water. *The Nautilus,* vol. 78, p. 76.

Vitrinella oldroydi Bartsch, 1907   **752**
Oldroyd's Vitrinella

    Cayucos, California, to Baja California.

    2.1 mm. in diameter, 0.8 mm. high, discoidal, whitish, semi-transparent. Nuclear whorls not differentiated. Surface of shell smooth, shiny, with occasional growth lines. Whorls rounded. Suture distinct. Openly umbilicate to the very apex. Aperture oblique, broadly oval; outer lip thin. Parietal wall covered by a rather strong callus. Usually the apex is eroded. Not uncommon; in sand just offshore.

Subgenus *Docomphala* Bartsch, 1907

    The type of this subgenus is *stearnsi* Bartsch, 1907.

Vitrinella stearnsi Bartsch, 1907   **753**
Stearns' Vitrinella

    Monterey Bay, California.

    3 to 4 mm. in diameter, 1 to 1.5 mm. high, discoidal. 1½ nuclear whorls smooth, followed by 1½ whorls with strong, oblique axial riblets, the remaining whorl being smoothish. Umbilicus wide, deep, extending to the apex and bounded by a slight angulation. Columella angulated in the middle. Uncommon.

Other Atlantic species:

754 *Vitrinella blakei* Rehder, 1944. Pleistocene, Wailes Bluff, St. Mary's County, Maryland. Recent: Bocas, Atlantic Panama. *The Nautilus,* vol. 57, p. 97.

755 *Vitrinella georgiana* Dall, 1927. Off Georgia, 440 fms.

756 *Vitrinella massarita* Dall, 1927. Off Fernandina, Florida, 294 fms.

757 *Vitrinella cerion* Dall, 1927. Off Georgia, 440 fms.

758 *Vitrinella rhyssa* Dall, 1927. Off Fernandina, Florida, 294 fms.

759 *Vitrinella(?) carinifex* Dall, 1927. Off Georgia, 440 fms. Last 5 described in *Proc. U.S. Nat. Mus.,* vol. 70.

760 *Vitrinella tryoni* Bush, 1897. Off Cape Hatteras, North Carolina, 16 fms.

761 *Vitrinella diaphana* (Orbigny, 1842). Off Cape Hatteras, North Carolina, 15 fms. Cuba.

762 *Vitrinella carinata* (Orbigny, 1842). Off Cape Hatteras, North Carolina, 16 fms.; Cuba.

763 *Vitrinella elegans* Olsson and McGinty, 1958. Costa Rica to Bocas, Atlantic Panama. *Bull. Amer. Paleontol.,* vol. 39, no. 177.

Other Pacific species:

764 *Vitrinella williamsoni* Dall, 1892. San Pedro to San Diego, California.

765 *Vitrinella eshnaurae* Bartsch, 1907. Central to southern California, 10 to 35 fms. (as *eshnauri*). 2.3 mm. diam.

766 *Vitrinella alaskensis* Bartsch, 1907. Unalaska, Alaska. 1.6 mm. diam.

767 *Vitrinella (Docomphala) berryi* (Bartsch, 1907). Monterey to South Coronado Island, 12 to 20 fms. 2.2 mm. diam.

768 *Vitrinella (Docomphala) columbiana* Bartsch, 1921. Departure Bay, British Columbia. *Proc. Biol. Soc. Wash.,* vol. 34, p. 39.

769 *Vitrinella (Docomphala) smithi* Bartsch, 1927. White's Point, Los Angeles County, to South Coronado Island.

770 *V. (Vitrinellops* Pilsbry and Olsson, 1952) *dalli* (Bartsch, 1911). Head of the Gulf of California. 1.3 mm.

771 *V. (Vitrinellops) lucasana* (Baker, Hanna and Strong, 1938). Gulf of California.

772 *V. (Vitrinellops) tiburonensis* Durham, 1942. Gulf of California.

Genus *Scissilabra* Bartsch, 1907

    Vitrinella-like, minute shells with the middle of the outer lip deeply and broadly notched, the center of the notch coinciding with the periphery of the shell. Type: *dalli* Bartsch, 1907.

Scissilabra dalli Bartsch, 1907   **773**
Dall's Scissilabra

    Monterey, California, to the Gulf of California.

    2 mm. in diameter, 0.8 in height, depressed, lenticular, acutely angled at periphery, having 3½ transparent, shiny whorls. Suture distinct. Umbilicus not large, but very deep and bounded below by a carina. Outer lip broadly and strongly notched. Uncommon; shallow water.

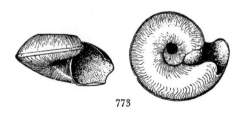

773

Genus *Circulus* Jeffreys, 1865

    Shell small, thin, circular, depressed, with a wide and deep umbilicus. Nucleus transparent, smooth, 2 whorls.

Outer lip not stepped forward at the top. Upper edge of outer lip serrate. Sculpture of strong spiral threads. Type: *duminyi* Requien, 1848, which may be the European living *striatus* (Philippi, 1836). See Vera Fretter, 1956, "The anatomy of the prosobranch, *Circulus striatus* (Philippi) and a review of its systematic position." *Proc. Zool. Soc. London*, vol. 126, p. 369.

The genus *Lydiphnis* Melvill, 1906, from the Gulf of Oman (type: *euchilopteron* Melvill and Standen) has 3 spiral carinae of which the upper one projects forward at the lip. Some east American *Circulus* were put in this genus by various workers.

Circulus multistriatus (Verrill, 1884)   774
Threaded Vitrinella

North Carolina to Florida; Caribbean.

5 mm. in diameter, planorboid, well-compressed, 4 whorls, opaque-white, with a glossy sheen. Outer surface covered with numerous, crowded, spiral, incised lines. Umbilicus with rounded sides, deep, rather narrow. 50 to 100 fathoms. Uncommon.

774

Circulus suppressus (Dall, 1889)   775
Suppressed Vitrinella

South half of Florida to Texas.

1.7 mm. wide, 0.75 mm. high, a little excavated toward the periphery where there are 3 sharp strong keels with deep gutters between them. Umbilicus small but open, bordered by a carinal thread. Aperture subcircular, prolonged into a little channel at its upper junction with the body whorl. Surface marked with fine lines of growth. Abundant; shallow water under rocks. Placed in *Cyclostremiscus* by some authors, perhaps justifiably.

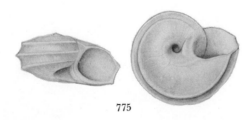

775

Other Atlantic species:

776 *Circulus margaritiformis* Dall, 1927. Off Fernandina, Florida, 294 fms.

777 *Circulus translucens* Dall, 1927. Off Georgia, 440 fms.

778 *Circulus hendersoni* Dall, 1927. Off Georgia, 440 fms.

779 *Circulus dalli* Bush, 1897. Off Cape Hatteras, North Carolina, 43 fms.

780 *Circulus semisculptus* (Olsson and McGinty, 1958). Southern Florida and the western Caribbean.

781 *Circulus cubanus* Pilsbry and Aguayo, 1933. Varadero, Matanzas, Cuba. *The Nautilus*, vol. 46, p. 120.

Other Pacific species:

782 *Circulus rossellinus* Dall, 1919. San Diego to Baja California.

783 *Circulus cosmius* Bartsch, 1907. Catalina Island, California, to Ecuador. (Synonym: *cerrosensis* Bartsch, 1907.) 2.5 mm.

783

Genus *Cyclostremiscus* Pilsbry and Olsson, 1945

Shells about 2 to 3 mm., solid, depressed, umbilicate, with 1 to 2 smooth nuclear whorls, the remaining with several spiral angles or carinae, with axial riblets. Outer lip not thickened externally. Type: *panamensis* (C. B. Adams, 1852). This genus could well be considered a subgenus of *Circulus*, differing only in having axial ribs.

Cyclostremiscus schrammii (Fischer, 1857)   784
Schramm's Vitrinella

Caribbean.

1 to 1.5 mm., discoidal, white, translucent; 2½ whorls, the last one being elevated above the spire. Each whorl has 5 keeled rows of minute nodules. Umbilicus wide and deep. Axial riblets between the nodes. (See J. Houbrick, 1967, *The Nautilus*, vol. 80, p. 131)

Cyclostremiscus pentagonus (Gabb, 1873)   785
Trilix Vitrinella

Off North Carolina to Florida, Texas and the West Indies.

3 mm. wide, 1 mm. high, yellow-white, lustrous, disc-shaped, with the small, prominent, glassy nucleus rising a little above the adult whorls. 4 whorls, periphery tricarinate, nearly smooth between them. Umbilicus very deep and funnel-shaped, and spirally grooved within. Aperture nearly circular. Operculum circular, multispiral, light-yellow. Abundant; 7 to 17 fathoms. *C. trilix* (Bush, 1885) and probably *liratus* (Verrill, 1882) are synonyms. Erroneously called *supranitidus* Wood.

785

Subgenus *Ponocyclus* Pilsbry, 1953

The type of this subgenus, lacking axial sculpture, is *beauii* (Fischer, 1857). Monograph 8, Acad. Nat. Sci. Phila., p. 426.

Cyclostremiscus beauii (Fischer, 1857)   **786**
Beau's Vitrinella

North Carolina to Florida and the West Indies. Brazil.

⅓ inch in diameter, strong, opaque-white, depressed, 4 whorls. Top of whorls rounded, slightly concave just below the fine suture; bearing 5 or 6 major, smooth, spiral threads on top with numerous, much finer threads between. Periphery bordered above and below by a major cord. Umbilicus widely funnel-shaped, deep. Outer lip crenulate above. Not uncommon in shallow water. One of our largest American vitrinellid species.

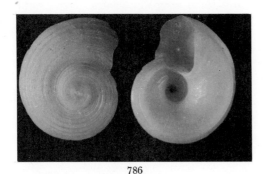

786

Atlantic species:

787 *Cyclostremiscus jeannae* (Pilsbry and McGinty, 1945). South half of Florida to Texas, Costa Rica and Panama, shore to 25 fms. *The Nautilus*, vol. 59, p. 82.

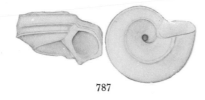

787

788 *Cyclostremiscus ornatus* Olsson and McGinty, 1958. Florida and the Caribbean. *Bull. Amer. Paleontol.*, no. 177.

Other Pacific species:

789 *Cyclostremiscus xantusi* (Bartsch, 1907). Cape San Lucas, Baja California. 1.4 mm. diam.

790 *Cyclostremiscus baldridgei* (Bartsch, 1911). Gulf of California. 4.5 mm.

791 *Cyclostremiscus gordanus* (Hertlein and Strong, 1951). Gulf of California.

792 *Cyclostremiscus lowei;* (**792a**) *spiceri;* (**792b**) *spiritualis* (all Baker, Hanna and Strong, 1938). Baja California and Gulf of California.

Genus *Discopsis* De Folin and Périer, 1870

Shell very small and depressed; with a single prominent peripheral keel. Type: *omalos* De Folin and Périer, 1870.

793 *Discopsis omalos* De Folin and Périer, 1870. Off Fernandina, Florida, 294 fms., to Guadaloupe, West Indies.

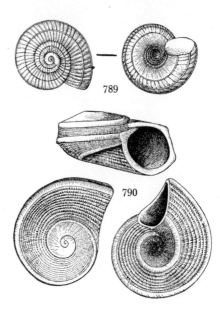

789

790

Subgenus *Alleorus* Strong, 1938

Suture beaded; peripheral carina smooth. Type: *deprellus* (Strong, 1938).

794 *Discopsis (Alleorus) deprellus* (Strong, 1938). Gulf of California.

Genus *Vitrinorbis* Pilsbry and Olsson, 1952

Type of this genus is *callistus* Pilsbry and Olsson, 1952, from Ecuador.

Vitrinorbis diegensis (Bartsch, 1907)   **795**
San Diego Vitrinella

Monterey Bay, California, to Baja California.

1.0 mm. in diameter, 0.3 mm. in height, discoidal, strongly carinate by a spiral cord on the center of the periphery, above which on the angular top of the whorl are numerous, slanting, tiny axial riblets. The latter also extend from the peripheral keel to a spiral cord on the middle of the angulate base. Umbilicus wide and very deep. Common; in coarse sand near kelp weed.

795

Other species:

796 *Vitrinorbis elegans* Olsson and McGinty, 1958. Western Caribbean. *Bull. Amer. Paleontol.*, no. 177.

Genus *Aorotrema* Schwengel and McGinty, 1942

Shells 3 mm., almost as high as wide; few whorls, their tops flat, the last whorl strongly carinate at the top and periphery of the whorl. Umbilicus broadly funnel-shaped. Do not confuse with young *Turbo* or *Astraea* which have a calcareous operculum. Type: *pontogenes* Schwengel and McGinty, 1942.

797 *Aorotrema pontogenes* (Schwengel and McGinty, 1942). 18 to 20 fms., off Destin, northwest Florida. Rare. 1.3 mm. × 1.9 mm. *The Nautilus,* vol. 56, p. 17.

798 *Aorotrema erraticum* Pilsbry and McGinty, 1945. 1.5 miles off Cape Florida, 12 fms. *The Nautilus,* vol. 59, p. 11. 1.5 mm. × 0.9 mm.

799 *Aorotrema cistronium* (Dall, 1889). Off North and South Carolina, 22 to 63 fms. Also Pliocene. 2.3 mm. × 2 mm.

799

800 *Aorotrema humboldti* (Hertlein and Strong, 1951). Gulf of California to Costa Rica.

Genus *Anticlimax* Pilsbry and McGinty, 1946

Shells 2 to 4 mm., wider than high, with a dome-shaped spire and carinate periphery. Radial oblique wavy ribs on base. Aperture triangular. Umbilical spiral cord thick, sometimes buttonlike. For a review of the living and fossil species see Pilsbry and Olsson, 1950, *Bull. Amer. Paleontol.,* vol. 33, no. 135. Synonyms: *Climacia* Dall, 1903, not McLachlan, 1869; *Climacina* Aguayo and Borro, 1946, not Gemmellaro, 1878. Type: *calliglyptum* Dall, 1903.

Anticlimax athleenae Pilsbry and McGinty, 1946   **801**
Athleen's Dome Vitrinella

Lower Florida Keys.

2.6 mm. wide, 1.7 mm. high, thin-shelled, spire dome-shaped; last whorl with peripheral keel and numerous fine spiral striae. Base with 8 weak, oblique, radial waves. Umbilicus bordered by a spiral cord. Suture well-impressed. Rare; shallow water.

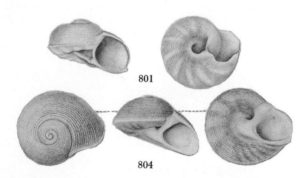

801

804

Other species:

802 *Anticlimax (Anticlimax) schumoi* (Vanatta, 1913). British Honduras. *Proc. Acad. Nat. Sci. Phila.,* for 1913, p. 24.

803 *Anticlimax occidens* Pilsbry and Olsson, 1952. Punta Penasco, Sonora, west Mexico; Baja California, 21 fms. *Proc. Acad. Nat. Sci. Phila.,* vol. 104, p. 59.

Subgenus *Subclimax* Pilsbry and Olsson, 1950

Like *Anticlimax,* but there is a buttonlike callus partially or wholly covering the umbilicus. Type: *hispaniolensis* Pilsbry and Olsson, 1950.

Anticlimax pilsbryi McGinty, 1945   **804**
Cupola Dome Vitrinella

Southeast Florida to Texas.

2.6 mm. wide, 1.7 mm. high, solid, white; base flat; whorls spirally striate. Outer lip thick. Umbilicus almost entirely filled with a buttonlike extension of the columella pillar. Last whorl with a smooth, weak keel at the periphery. Base with about 10 short, oblique, very low radial waves. Rare; 80 fathoms. (*The Nautilus,* vol. 59, p. 79.) *A. tholus* Pilsbry and McGinty, 1946, is a synonym.

Genus *Episcynia* Mörch, 1875

Shells 3 or 4 mm., depressed, with a low conoidal spire of about 5 convex whorls with a minutely serrated peripheral keel; umbilicus deep, scalar, bounded by an angle or keel. Surface nearly smooth under a thin periostracum which bears spiral fringes of filaments above and below the periphery. Type: *inornata* (Orbigny, 1842).

Episcynia multicarinata (Dall, 1889)   **805**
Fringed Vitrinella

North Carolina to Florida and the Caribbean.

3.4 mm. wide, 2 mm. high, 5 whorls, with very minute serrations on the narrow peripheral carina. 3 spiral rows of flattened epidermal filaments on the periphery. Base with strong radial riblets. Uncommon; 15 to 60 fathoms. May be *inornata.*

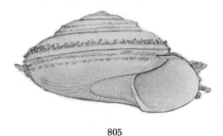

805

Episcynia inornata (Orbigny, 1842)   **806**
Hairy Vitrinella

Texas and the Greater Antilles.

2 mm., depressed, vitreous-white, whorls 5, convex, with a minutely serrate peripheral keel. Umbilicus deep, bounded by an angle or keel. Surface nearly smooth under the periostracum, which bears spiral fringes of filaments above and below the periphery. Not uncommon along the Texas coast, and offshore 6 to 12 fathoms.

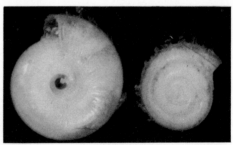

806

Episcynia devexa Keen, 1946   **807**
Devex Vitrinella

Central California.

4 mm. in diameter, 3 mm. high, whitish, 5 whorls, with a microscopic carina which has chitinous fringes above and below it. Last ⅓ of body whorl rapidly descending. Base convex, wrinkled near the deep, scalar umbilicus, the latter bounded by a rough keel. Rare; 2 to 3 fathoms. *The Nautilus,* vol. 60, p. 9.

Other species:

808 *Episcynia medialis* Keen, 1971. Guaymas and Banderas Bay, Mexico, 18 meters. Keen species 352.

Genus *Parviturboides* Pilsbry and McGinty, 1950

Shells minute, resembling *Turbo* in shape, with 2 smooth nuclear whorls abruptly giving place to spiral cords and fine axial threads. Umbilicus small, narrow, bounded by a spiral cord. Columellar margin thickened. Operculum chitinous, thin, multispiral; radula taenioglossate. Tentacles with stiff cilia at the tips and a gutter of fine cilia along their inner sides. Posterior end of foot bifurcate and with stiff cilia. Eyes not on pedicles. Do not confuse with *Parviturbo* Pilsbry and McGinty, 1945, which has a rhipidoglossate radula, epipodial cirri, nonbifurcate posterior foot, and belongs in the Trochidae. Type: *interruptus* (C. B. Adams, 1850).

Parviturboides interruptus (C. B. Adams, 1850)   **809**
Interrupted Vitrinella

South half of Florida, Texas and the West Indies.

1 mm. wide, 0.8 mm. high; white, with 3 or 4 larger spiral cords on the upper whorl, 5 or 6 smaller ones on the base. Axial striae between the cords are microscopic. 2 nuclear whorls, smooth. Animal illustrated by D. Moore, 1962, *Bull Marine Sci.,* vol. 12, p. 697. Moderately common; on sponges, 4 to 50 fathoms. Synonyms: *sanibelense* (Pilsbry, 1945); *zacalles* Mazyck, 1913. (Drawing from D. R. Moore, 1972, *Bull Marine Sci.,* vol. 22)

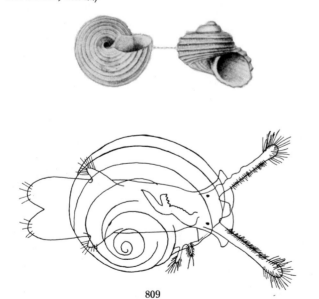

809

Other species:

810 *Parviturboides copiosus* (Pilsbry and Olsson, 1945). Guaymas, Mexico to Ecuador.

Genus *Didianema* Woodring, 1928

Shells minute, less than 2 mm., white, globose, umbilicate. Within the aperture and visible on the inner lip and parietal wall lies a narrow shelf against which the operculum probably fits. Type: *tytha* Woodring, 1928. Fossils only.

Subgenus *Diagonaulus* Pilsbry and McGinty, 1945

Less than 2 mm. subglobose, whorl rounded; a small funnel-shaped umbilicus bounded by a blunt ridge. Nucleus 2 smooth whorls; postnuclear whorls strongly, retractively corrugated. Small ledge within the inner lip. Operculum corneus, thin, slightly concave, multispiral. Type: *pauli* Pilsbry and McGinty, 1945.

811 *Didianema (Diagonaulus) pauli* Pilsbry and McGinty, 1945. 18 to 20 fms., off Destin, northwest Florida. 1.4 mm. × 1.2 mm.

Genus *Solariorbis* Conrad, 1865

Shells 2 to 4 mm., white, strong, depressed, with 3 or 4 whorls, either rounded or angular, and usually with some spiral striation, the grooves typically punctate; apical whorls level and smooth. The umbilicus has a spiral ridge or a thickening of the wall, ending in a callus lobe or ledge at the columellar margin. Aperture rounded, with a small groove at the top. Outer lip rather thin and evenly curved. Type: *depressus* Lea, 1833.

Solariorbis infracarinata Gabb, 1881   **812**
Gabb's Vitrinella

South half of Florida and Texas; Caribbean.

1.8 mm. wide, 1 mm. high, biconvex, carinate; white; umbilicus widening to ⅕ the shell's diameter. Last whorl excavated above the peripheral keel with a group of fine spiral striae above that. There are 2 strong cords and several fine spiral striae below the keel. Base smooth. Uncommon; shallow water. Also found in the Florida Pliocene. *S. euzonus* Pilsbry and McGinty, 1950, is a synonym.

Other species:

813 *Solariorbis corylus* Olsson and McGinty, 1958. Bocas Island; Colon; Atlantic Panama. *Bull. Amer. Paleontol.,* vol. 39, p. 28.

814 *Solariorbis decipiens* Olsson and McGinty, 1958. Bocas Island, Atlantic Panama. *Bull. Amer. Paleontol.,* vol. 39, p. 28.

815 *Solariorbis contracta* (Vanatta, 1913). Monkey River, Honduras. *Proc. Acad. Nat. Sci. Phila.,* p. 25.

816 *Solariorbis bartschi* (Vanatta, 1913). Porto Barrios and Livingston, Guatemala. *Proc. Acad. Nat. Sci. Phila.,* p. 26.

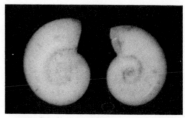

816

817 *Solariorbis hondurasensis* (Vanatta, 1913). Belize and Monkey River, British Honduras. *Proc. Acad. Nat. Sci. Phila.,* p. 26.

818 *Solariorbis semipunctus* D. Moore, 1965. Campeche Bank, Mexico, 18 meters; Haiti. *The Nautilus,* vol. 78, p. 78.

819 *Solariorbis blakei* Rehder, 1944. South Carolina to Texas and the Caribbean.

820 *Solariorbis mooreana* Vanatta, 1904. Texas.

821 *Solariorbis arnoldi* Bartsch, 1927. San Pedro, California. *Proc. U.S. Nat. Mus.* vol. 70, p. 32.

822 *Solariorbis (Hapalorbis* Woodring, 1957) *liriope* (Bartsch, 1911). Gulf of California. Type of the subgenus.

823 *Soloriorbis (Systellomphalus* Pilsbry and Olsson, 1941) *elegans* Pilsbry and Olsson, 1952. Guaymas, Mexico, to Peru.

Genus *Pleuromalaxis* Pilsbry and McGinty, 1945

With fine spiral striae and strong, widely spaced radial riblets. Type: *balesi* Pilsbry and McGinty, 1945.

Pleuromalaxis balesi Pilsbry and McGinty, 1945   **824**
Bales' False Dial

Southeast Florida, Texas and the Caribbean.

1.8 mm. in diameter, 0.6 mm. in height, 3 to 4 whorls, semi-translucent-white to burnt-sienna. Sculpture of fine, spiral striae and strong, widely spaced, radial ribs. Peripheral zone flattened or concave between 2 projecting nodulose keels. Under rocks. Moderately common to rare. This genus was formerly placed in the family Architectonicidae.

Other species:

825 *Pleuromalaxis (Paurodiscus* Rehder, 1935) *lamellifera* (Rehder, 1934). Florida Straits, 205 fms. 3 mm. × 1 mm. *The Nautilus,* vol. 48, p. 128. Type of the subgenus.

825

826 *Pleuromalaxis (Pleuromalaxis) pauli* (Olsson and McGinty, 1958). Bocas, Atlantic Panama. *Bull. Amer. Paleontol.,* vol. 39, no. 177, p. 30.

827 *Pleuromalaxis (Calodisculus* Rehder, 1934) *retifera* (Dall, 1892). Pliocene of the Caloosahatchee, Florida. Type of the subgenus.

Subfamily TEINOSTOMINAE Cossmann, 1917

Genus *Teinostoma* H. and A. Adams, 1854

Shells usually about 2 to 3 mm. in diameter, depressed, glossy, white, usually smooth, and with an umbilical callus. They are very distinctive little shells, but require a high-powered lens for their inspection. Operculum circular, multispiral, chitinous. Type: *politum* A. Adams, 1853. The key to the Florida species is from Pilsbry and McGinty, 1945, *The Nautilus,* vol. 59, p. 2.

Key to the Florida *Teinostoma*

1. Umbilical callus encircled by a keel. 1.7 mm.; Palm Beach to Cape Florida. 12 to 50 fms. . . . . .
    . . . . **(828)**
    T. (Annulicallus) lituspalmarum Pils. and McG., 1945
2. Umbilical callus and columellar lobe not closing the umbilicus completely; 3 mm.; southeast Florida. 80 fms.
    (829) *T. (Ellipetylus) cocolitoris* Pils. and McG., 1945

3. Umbilicus closed by the callus, which passes smoothly into the base . . . (subgenus *Idioraphe* Pilsbry, 1922)
    A. Periphery strongly carinate; 2 mm.; Destin, Florida. 20 fms.; Panama . . . . . .
        . . . . **(830)** *goniogyrus* Pils. and McG., 1945
    AA. Periphery rounded or indistinctly rounded.
        B. Surface spirally striate:
            C. Umbilical callus extremely convex and thick; 2 mm. Palm Beach to Cape Florida . . . .
                . . . . **(831)** *pilsbryi* McGinty, 1945
            CC. Umbilical callus strong, slightly convex:
                D. Strongly spirally striate throughout; 2.3 mm.; Key Largo; Bahamas . . . .
                    . . . . **(832)**
                    clavium Pils. and McG., 1945
                DD. Weakly striate above only; 1.5 mm.; southeast Florida and the Caribbean . . . . . .
                    . . . . **(833)**
                    nesaeum Pils. and McG., 1945
        BB. No spiral striations.
            C. Diameter 1.8 to 2.2 mm.:
                D. Rather globose, h/d ratio 75; shore to 50 fms.; southeast Florida, Texas . . . . . .
                    . . . . **(834)**
                    parvicallum Pils. and McG., 1945
                DD. Depressed, h/d ratio about 50:
                    E. Callus large. Lake Worth . .
                        . . . . . **(835)** *obtectum* Pils. and McG., 1945
                    EE. Callus small. Florida to Texas
                        . . . . . **(836)** *biscaynense* Pils. and McG., 1945
            CC. Diameter 0.7 to 1.0 mm.; callus thick; Lower Keys to Texas. . . . . .
                . **(837)** *lerema* Pils. and McG., 1945

Teinostoma supravallatum (Carpenter, 1864)   **838**
Upright Teinostome

Monterey, California, to Baja California.

1.8 mm. in diameter, 1.1 in height, lenticular in side view, glassy, white, whorls either smoothly rounded or with a carination on the shoulder. Base has a weak carination. Columellar pad large, moderately swollen. Abundant; in coarse sand near eel-grass in the sublittoral zone to 30 fathoms. *T. invallatum* (Carpenter, 1864) is the smooth form.

Other Atlantic species:

839 *Teinostoma (Annulicallus) carinicallus* Pilsbry and McGinty, 1946. South half of Florida, shore to 70 feet; Bocas, Panama. 2.7 mm. × 1.8 mm. *The Nautilus,* vol. 60, p. 17.

840 *Teinostoma (Pseudorotella* Fischer, 1857) *minuscula* (Bush, 1897). Off North Carolina, 14 fms.

841 *Teinostoma (Pseudorotella) semistriata* (Orbigny, 1842). Florida Keys and the West Indies.

842 *Teinostoma (Pseudorotella) solida* (Dall, 1889). Off Georgia, 440 fms.; off Fernandina, Florida, 294 fms.; Florida Straits, 310 fms.

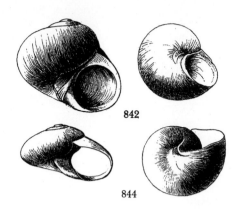

842

844

843 *Teinostoma (Pseudorotella) floridensis* (Dall, 1927). Off Fernandina, Florida, 294 fms.

844 *Teinostoma (Pseudorotella) reclusa* (Dall, 1889). Off North Carolina to Florida Keys and Yucatan, 12 to 640 fms.

845 *Teinostoma cryptospira* Verrill, 1884. Off North Carolina to the West Indies, 30 to 150 fms.

845

846 *Teinostoma multistriata* Verrill, 1884. Off North Carolina to the West Indies, 3 to 142 fms.

847 *Teinostoma megastoma* (C. B. Adams, 1850). North Carolina to the western Caribbean. Common. 2 mm. in diameter.

847

Other Pacific species:

848 *Teinostoma politum* A. Adams, 1853. Baja California to Ecuador. Type of the genus.

849 *Teinostoma (Pseudorotella* Fischer, 1857) *amplectans* Carpenter, 1857. Baja California, the Gulf, to Ecuador. (Synonym or southern subspecies: *americanum* Pilsbry and Olsson, 1945; *cecinella* Dall, 1919, is also probably a synonym.)

850 *Teinostoma salvania* Dall, 1919. Off South Coronado Island, California.

851 *Teinostoma bibbiana* Dall, 1919. San Diego, California.

852 *Teinostoma sapiella* Dall, 1919. San Pedro to San Diego, California. Above 3 in *Proc. U.S. Nat. Mus.,* vol. 56, p. 369.

853 *Teinostoma gallegosi* Jordan, 1936. Magdalena Bay, Baja California.

Superfamily Tornacea Kuroda, Habe and Oyama, 1971

Family TORNIDAE Sacco, 1896

Shells resembling vitrinellids in being small, white, generally depressed and umbilicate. However, the operculum is oval and paucispiral; there is no penis; the gill extends well

out of the right side of the aperture and may even curve around the margin of the shell. Alias Adeorbiidae Monterosato, 1884.

Genus *Tornus* Turton and Kingston, 1830

Shells 2 to 3 mm., solid, depressed, 3 or 4 whorls. Widely umbilicate, with strong beaded spiral cords crossed by strong axial riblets. Protoconch smooth, 1½ whorls, not elevated. Outer lip crenulated. Aperture subtrigonal. Operculum ovate, paucispiral, chitinous. Type: *subcarinatus* (Montagu, 1803). *Adeorbis* Wood, 1842, is a synonym. D. R. Moore, 1969 (*The Veliger,* vol. 12, p. 169) reports that *Cyclostrema miranda* Bartsch, 1910, is a synonym of *subcarinatus* and has untrustworthy locality data. One Pacific species:

854 *Tornus calianus* (Dall, 1919). San Diego, California.

Genus *Macromphalina* Cossmann, 1888

Shells 3 or 4 mm., depressed, white, auriform, with a wide umbilicus. Axially fimbriated. Protoconch of 2 or 3 smooth whorls projecting upward at a slant. Aperture ovate. Operculum paucispiral. Type: *problematicus* (Deshayes) of the Paris Eocene. Synonyms: *Gyrodisca* Dall, 1896, whose type is *depressa* (Jeffreys); *Chonebasis* Pilsbry and Olsson, 1945. *Macromphalina pilsbryi* Olsson and McGinty is a *Vanikoro.* Keen, 1971, p. 454, places this genus in the family Fossaridae.

Macromphalina palmalitoris **Pilsbry and McGinty, 1950**   855
Palm Beach Vitrinella

Off Palm Beach, Florida, to Texas.

2 mm. wide, 1 mm. high, white, thin-shelled, fragile, rounded periphery. 2 nuclear whorls, smooth, convex and pimplelike. Last adult whorl rounded, with retractively curved, well-spaced, axial riblets and fine spiral striae. Rare, 88 fathoms.

855

Macromphalina floridana **D. Moore, 1965**   856
Floridian Vitrinella

South half of Florida.

3 mm. wide, 1.5 mm. high, depressed, periphery strongly carinate, white, fragile, with a tilted 2-whorl protoconch. Axial sculpture of numerous, thin, slanting axial lamellarlike riblets between which are microscopic spiral threads. About 46 riblets on dorsal half of whorl, about 34 on base of last whorl. Umbilicus widely open. Uncommon; shallow water. *The Nautilus,* vol. 78, p. 75.

Macromphalina adamsii **(Fischer, 1857)**   857
Adams' Vitrinella

Florida Keys and the West Indies.

8 mm. wide, 7 mm. high, well-rounded whorls, translucent tan-white, shining, widely umbilicate, 3½ whorls; upper part

axially striate, smoothish on the base. Aperture subovate. Outer lip smooth and fragile. Uncommon.

Macromphalina californica (Dall, 1903)   858
Californian Vitrinella

Santa Barbara Channel, California.

5.5 mm. high, 5.5 mm. wide, whitish. Whorls 2½, the last much the largest, rounded above, with a prominent suture, below with a wide funicular umbilicus bordered externally by an obtuse carina. With coarse, slanting, axial threads crossed by finer spiral striae. Outer lip thin, smooth. Uncommon. Formerly and erroneously considered a *Megalomphalus* (Vanikoroidae).

858   862

Other Atlantic species:

859 *Macromphalina caro* (Dall, 1927). Off Fernandina, Florida. Deep water.

860 *Macromphalina pilsbryi* Olsson and McGinty, 1958. Boca, Atlantic Panama. *Bull. Amer. Paleontol.*, vol. 39, no. 177, p. 34 is a *Vanikoro.*

861 *Macromphalina pierrot* Gardner, 1948. Port Aransas, Texas, living. Pliocene of North Carolina. U.S.G.S. Prof. Paper 199-B, p. 195.

Other Pacific species:

862 *Macromphalina occidentalis* Bartsch, 1907. Point Abreojos, Baja California. 1.7 mm. in diameter.

Genus *Cochliolepis* Stimpson, 1858

Shells about 3 mm., thin-shelled, depressed, widely and openly umbilicate, with a few rapidly enlarging, more or less enveloping, smoothish whorls. Aperture oblique. Live associated with marine polychaete worms. Type: *parasitica* Stimpson, 1858. The following subgenera have been proposed, and may be synonyms: *Naricava* Hedley, 1913; and *Tylaxis* Pilsbry, 1953.

Cochliolepis parasitica Stimpson, 1858   863
Parasitic Scale Snail

South Carolina to Texas and the West Indies.

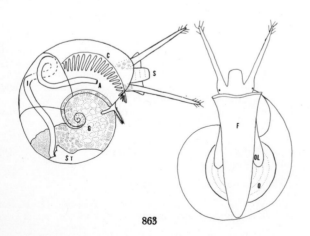

863

2 to 3 mm. in diameter, thin, discoidal, flattened on top, with 3 rapidly increasing whorls. Color translucent-white. Umbilicus wide and deep. Peristome thin, the upper margin arched forward. Sculpture of microscopic growth lines. Soft parts brick-red. Moderately common; can be found under the scales of the annelid worm, *Acoetes lupina.* It is an herbivore (see D. R. Moore, 1972, *Bull. Marine Sci.*, vol. 22, p. 100). *C. nautiliformis* (Holmes, 1860) is a synonym.

Cochliolepis striata Dall, 1889   864
Striate Scale Snail

Florida to Texas.

5 to 7 mm., in diameter, 1.5 mm. high, 3 adult whorls and a globular nucleus which is almost enveloped by the last whorl. Sculpture of numerous strong, minute, spiral striae. Umbilicus very wide and deep. Shallow water; not uncommon near Texas inlets.

Family CHORISTIDAE Verrill, 1882

Placement of this family is provisional. It may be a tectibranch. Kuroda, Habe and Oyama, 1971, created a superfamily for this, Choristiacea.

Genus *Choristes* Carpenter, 1872

Shells 5 to 10 mm., heliciform, low spire; large ventricose, rounded last whorl. Periostracum remains on shell as new whorls are added. Operculum chitinous, paucispiral. Radula taenioglossate. Frontal tentacles united by a fold. Simple posterior tentacles present. Jaw well-developed; pharynx large and retractile. Eyes absent. Penis small, papilliform, swollen at base, and located just below the right posterior tentacle. Two small flat cirri behind and beneath the operculum. Type: *elegans* Carpenter, 1872, a fossil. Above description based on the only known living American species. There are two Japanese species, one of which is parasitic on the egg capsules of the shark.

865 *Choristes tenera* Verrill, 1882. Off Martha's Vineyard, Massachusetts, 255 fms. Found in empty skate egg case. *Trans. Conn. Acad.*, vol. 5, pl. 58, fig. 27.

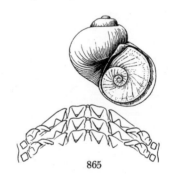

865

Family CAECIDAE Gray, 1850

These tiny, tusk-shaped gastropods are very abundant in shallow-water grass beds or around coral reefs. Some live on the open sand bottom of the shallow continental shelf. The caecums begin life in a normal snail-like manner as freeswimming veligers with a tiny spiral shell, but when they take up life on the bottom, the shell grows as a simple, slightly curved tube. As the animal grows, it forms an internal septum, and the spiral apex is dropped off. As growth continues, another septum is formed at the rear of the snail, and the second stage is dropped off. The septum is usually armed with a spike, or mucro. The operculum is thin, circular, horny and multispiral. Descriptions are for the third-stage adults only.

D. R. Moore (1962) has shown that this family belongs in the superfamily Rissoacea (*Bull. Marine Sci. Gulf and Caribbean,* vol. 12, no. 4, p. 695). Shrimp feed upon *Caecum* (*The Nautilus,* vol. 71, p. 152). I am indebted to Dr. Moore for supplying the taxonomy and descriptions of the Atlantic species.

Genus *Caecum* Fleming, 1813

The type of the genus is *imperforatum* Kanmacher, 1798, of Europe (*trachea* Montagu, 1803, is a synonym). *Micranellum* Bartsch, 1920, may be considered a synonym.

Subgenus *Caecum* Fleming, 1813

Caecum pulchellum Stimpson, 1851   866
Beautiful Little Caecum

New Hampshire to Brazil.

About 2 mm. in length, light-tan to white in color, 20 to 30 axial rings about as wide as the interspaces between. The septum is slightly convex, the mucro weak and projecting only slightly. There is no terminal varix. A very abundant species in grass beds in sheltered bays and lagoons. Some synonyms are: *regulare* Carpenter, 1858; *capitanum* Folin, 1874; *conjunctum* Folin, 1867; *curtatum* Folin, 1867; *dux* Folin, 1871.

Caecum bipartitum Folin, 1870 (**867**), found offshore in Texas and the western Gulf of Mexico, may be only a form of *pulchellum*. *Caecum bipartitum* is about the same size and shape, but the axial rings disappear on the anterior part of the shell. Moderately common. Its synonyms include *contractum, instructum* and *triornatum* all Folin, 1870.

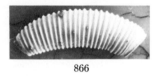

866

Caecum textile Folin, 1867   868
Textile Caecum

Florida Keys and West Indies.

Very similar in shape and size to *C. pulchellum,* but the axial rings are very low and close-set and number around 35 to 45. There are fine longitudinal striations that show up best between the axial rings. The septum is almost flat, and the mucro low and slightly pointed. Synonym: *leptoglyphos* Folin, 1881. Moderately common; in sand in shallow water.

Caecum tornatum Verrill and Bush, 1900   869
Bermuda Caecum

Bermuda only.

About 2.5 mm. in length; slender and with around 20 widely spaced axial rings, the last 2 or 3 enlarged and forming a varix. The mucro is large and projecting. The second stage is very variable in length, diameter and sculpture, and is probably responsible for several specific names in the Bermuda area. Moderately common in shallow water.

Caecum condylum D. R. Moore, 1969   870
Condylum Caecum

Off the Texas coast; Virgin Islands.

About 3 mm.; characterized by about 100 very fine axial rings and a swollen anterior end. Rare.

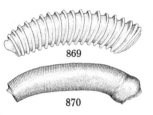

869

870

Caecum californicum Dall, 1885   871
California Caecum

Monterey, California, to Baja California.

2 to 3 mm. in length. With 30 to 40 moderately developed, evenly spaced, rounded or squarish axial rings. Lip of aperture slightly thickened. Color a glossy, olive-brown. Common.

Caecum dalli Bartsch, 1920   872
Dall's Caecum

San Diego to Baja California.

About 3 mm. in length, usually with 18 to 24 moderately developed, evenly spaced, round or squarish axial rings. Lip of aperture usually heavily developed in adults. Color tan. The number of raised rings varies from specimen to specimen, often in the same locality, and diligent search will usually bring to light any number desired. Extremes have been unwisely named (15 rings—*grippi* Bartsch, 1920; 17 to 19 rings—*licalum* Bartsch, 1920; and 19 to 22 rings—*diegense* Bartsch, 1920).

Caecum crebricinctum Carpenter, 1864   873
Many-named Caecum

Monterey to Baja California.

6 mm. in length. Color pinkish brown to chalky-white with occasional darker brown mottlings. With about 100 fine, squarish, closely set axial rings. Plug with a rather long, oblique spur. Spur sometimes eroded down to a small sharp pimple (form *oregonense* Bartsch, 1920). Irregularities occur in the expansion of the tube; sometimes there is a more rapid expansion toward the anterior end (forms named as species: *pedroense* Bartsch, 1920, and *barkleyense* Bartsch, 1920). *C. catalinense* Bartsch, 1920, is probably this species, since many of the paratypes do not have the anterior end supposedly "bulbously expanded," and many specimens have about 100 axial rings, and not 75 as claimed. *C. rosanum* Bartsch, 1920, appears to be a very long specimen (7 mm.) with sharply defined rings. Common.

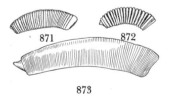

871   872

873

Subgenus *Elephantulum* Carpenter, 1857

Medium to large species with longitudinal or axial sculpture, or a combination of both in varying degrees of strength. Mucro usually strong and pointed. Type: *ab-*

normale Carpenter, 1857. (Designated by Cossmann, 1912, p. 151). *Elephantanellum* Bartsch, 1920, and *Quadrulata* Folin, 1867 are synonyms.

Caecum floridanum Stimpson, 1851   874
Florida Caecum

North Carolina to Brazil.

3 to 4 mm. in length, opaque-white, with about 15 to 30 strong axial rings, the last 3 or 4 being quite large. The septum is recessed, and the mucro is a slender sharp spike. Common; 2 to 18 fathoms. Also known as: *irregulare* Folin, 1867; *phronimum* Folin, 1867; *cayosense* Rehder, 1943; and *puntagordanum* Weisbord, 1962.

874

Caecum imbricatum Carpenter, 1858   875
Imbricate Caecum

Florida to Texas; Bahamas and West Indies.

Similar to *C. floridanum,* but the axial rings are flattened and the terminal ones are enlarged only a little more than the others; fairly strong longitudinal ridges. The septum is even with the end of the shell, and the mucro is prominent and triangular. Common; shallow water. Synonyms are: *formulosum* Folin, 1868; *coronatum* Folin, 1867; *insigne* Folin, 1867; and *sculptum* Folin, 1881.

Caecum insularum D. R. Moore, 1970   876
Island Caecum

Puerto Rico and Virgin Islands.

2.5 to 3.4 mm., similar to *imbricatum,* but only slightly curved, with weaker longitudinal ribs, and without a varix around the aperture. With 35 to 50 low, flattened, closely spaced annular ribs. The mucro is strong and forms a right-angled triangle when seen from the side. Color white to brownish. Common; shallow, warm, oceanic waters surrounding small islands. (*Bull. Marine Sci.,* vol. 20, no. 2, p. 369.)

Caecum cooperi S. Smith, 1860   877
Cooper's Atlantic Caecum

South of Cape Cod to west Florida; and Texas.

4 to 5 mm. in length, slender, glossy, opaque-white; with about 15 strong, longitudinal ribs. Axial, raised rings are prominent near the aperture and sometimes give the shell a cancellate appearance at the anterior end. Apical plug with a fairly long, pointed prong. Common, offshore on sand. *Caecum smithi* J. G. Cooper, 1872, is a synonym.

Caecum tortile Dall, 1892   878
Carolina Caecum

North and South Carolina.

4 to 5 mm., slender, very similar to *cooperi* but without the strong longitudinal ribs. Uncommon.

Caecum clava Folin, 1867   879
Blade Caecum

Texas and the West Indies.

2 to 3 mm., similar to *cooperi,* but smaller, has strong longitudinal ribs and has a swollen anterior end. Uncommon; offshore.

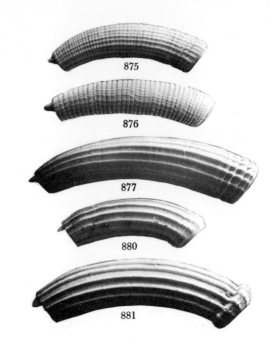

875

876

877

880

881

Caecum plicatum Carpenter, 1858   880
Plicate Caecum

Florida and West Indies; Bermuda.

About 2.5 to 3.5 mm. in length, color pattern a mottled white and tan; about 15 thick longitudinal ridges, and weak axial rings near the aperture, the last 2 or 3 forming a moderately strong varix. The mucro is low, but strong and somewhat pointed. Some specimens, however, are white and without any trace of color pattern. Also known as: *decussatum* Folin, 1868; *termes* Heilprin, 1889; *obesum* Verrill and Bush, 1900; and *biminicola* Pilsbry, 1951 (the last in *The Nautilus,* vol. 65, p. 69)

Caecum cycloferum Folin, 1867   881
Swollen-mouthed Caecum

Southeast United States and the West Indies.

Large, 4 to 5.5 mm., with 13 to 24 longitudinal ridges, and with a very abrupt, strong varix around the aperture. Septum flattened; mucro low but strong. Synonyms include *gracilis* Folin, 1870; *coronellum* Dall, 1892; and *clenchi* Olsson and McGinty, 1958. Moderately common.

Caecum carpenteri Bartsch, 1920   882
Carpenter's Caecum

San Pedro to Gulf of California.

3.5 to 4.8 mm. in length. First ½ to first ¾ of shell smooth, but at the apertural end developing about a dozen small, sharply defined axial rings. Longitudinal sculpture microscopic or absent. Color translucent-white to gray. This species is doubtfully placed in this subgenus.

Caecum heptagonum Carpenter, 1857   883
Heptagonal Caecum

Gulf of California to Panama.

2.0 to 2.5 mm. in length. Opaque-white. 7-sided in cross-section. The 7 longitudinal ribs are strong, raised, and the spaces between them are flat or slightly concave. There are about 30 deeply cut circular lines around the shell, cutting across the ribs. Lip of aperture with 1 or 2 swollen axial rings. *C. hexagonum* Bartsch, 1920, is a misspelling.

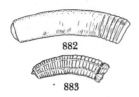

882

883

Subgenus *Brochina* Gray, 1857

Slender, smooth, cylindrical, thin-shelled species with a convex septum, mucro small and projecting from right side of septum. Type: *glabrum* Montagu, 1803. According to D. R. Moore, 1970, Journ. de Conchyl., vol. 108, no. 1, *Armata* Folin, 1875, is a synonym.

Caecum vestitum Folin, 1870   **884**
Vera Cruz Caecum

Florida Keys, lower Gulf of Mexico; West Indies.

About 2 to 2.5 mm. in length, slender and cylindrical; the shell is smooth or with incipient annulations at the anterior end; a slight varix is formed around the aperture. The septum is a little convex, the mucro is only a small projection on the right side. Shell almost transparent when alive, but covered by a brownish periostracum. Synonyms are *buccina* Folin, 1870, *carmenensis* Folin, 1870, and *veracruzanum* Folin, 1870. Common.

(885) *Caecum antillarum* Carpenter, 1858, from the Bahamas and Florida Keys, is similar, but is smaller, 1.3 to 1.7 mm., mottled in appearance, with no trace of a varix, and with a somewhat weaker mucro. Uncommon.

(886) *Caecum heladum* Olsson and Harbison, 1953, from the west coast of Florida, has weak longitudinal and axial sculpture, 2 to 3.5 mm. Uncommon.

Caecum johnsoni Winkley, 1908   **887**
Johnson's Caecum

Massachusetts to North Carolina.

Similar to *C. vestitum* Folin, but with a more convex septum which has a roughened surface. There are many fine axial rings which are obscure in some specimens. Operculum flat,

spiral turns not visible. This species is sometimes misidentified as *C. glabrum* Montagu, 1803, a European species.

Caecum carolinianum Dall, 1892   **888**
Carolina Caecum

North Carolina to southern Florida.

About 4.5 mm. in length, glossy, cream-white. Smooth except for microscopic growth lines. Apical plug sunk in at the posterior end of the shell and with a sharp, hornlike projection. Aperture minutely constricted. Moderately common; shallow water.

Subgenus *Fartulum* Carpenter, 1857

Shells very small, about 2 mm. in length, fragile, smooth, except for microscopic growth lines; not swollen in the middle; and with a nonconstricted aperture facing to one side (oblique). Periostracum thick. Type: *laeve* C. B. Adams, 1852. *Defolinia* Weisbord, 1962, and *Levia* Folin, 1875, are synonyms.

Caecum ryssotitum Folin, 1867   **889**
Minute Caecum

Texas and the West Indies to Brazil.

About 1.3 to 2.0 mm. in length; glassy-white but covered with tan periostracum in life. Cylindrical, except for a tapered section at the posterior end and a thickening around the aperture. The aperture is very oblique, compressing the thickened area on the ventral side behind the aperture. The mucro is flattened, giving a low triangular outline when viewed from the side. Common; shallow water. *Caecum tomaculum* Weisbord, 1962, is this species. Similar in shape to *C. (Meioceras) cornucopiae*, but without color pattern and with a thickened, wrinkled area behind the ventral side of the aperture.

Caecum orcutti Dall, 1885   **890**
Orcutt's Caecum

San Pedro to Baja California.

2.0 to 2.5 mm. in length. Smooth, except for fine, circular scratches. Shell stubby, slightly compressed laterally; aperture oblique; apical plug dome-shaped. Color translucent-tan to yellow-brown. Moderately common.

Caecum occidentale Bartsch, 1920   **891**
Western Caecum

Alaska to Baja California.

2.2 to 3.5 mm. in length. Smooth, except for fine, circular scratches. Shell elongate, round in cross-section. Aperture moderately oblique; apical plug dome-shaped with a tiny pimple on one side. Color translucent-tan to light-brown. Old specimens are whitish, often with a purplish stain. The shell has a white band behind the aperture. *C. hemphilli* Bartsch, 1920, and *C. bakeri* Bartsch, 1920, are probably diminutive forms of this species. The development of the small pimple on top of the dome-shaped plug is variable. Common.

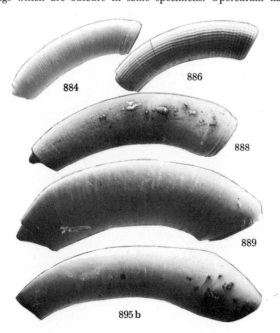

884

886

888

889

895 b

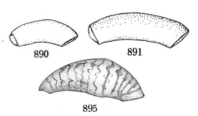

890

891

895

Other species:

891a *Caecum (Caecum) gurgulio* Carpenter, 1858. Bahamas and West Indies. 2 mm., with about 36 low, flat, close-set ridges; periostracum not noticeable. Shallow water.

892 *Caecum (Micranellum) profundicolum* Bartsch, 1920. Off San Diego, California. 55 to 199 fms.

893 *Caecum (Elephantulum) liratocinctum* Carpenter, 1857. Gulf of California to Panama.

894 *Caecum (Fartulum) bakeri* (Bartsch, 1920). Magdalena Bay to Gulf of California.

Subgenus *Meioceras* Carpenter, 1858

Shells 2 to 4 mm. in length, very bulbous in the middle, smooth, and with an oblique, constricted aperture. Resembles a miniature *Cadulus* (Scaphopoda). Type: *cornucopiae* (Carpenter, 1858).

Caecum nitidum Stimpson, 1851   895
Little Horn Caecum

South Florida and the Gulf coast to the West Indies and Brazil.

2 to 3 mm. in length, glassy, translucent-white with irregular specks or mottlings of chalk-white; bulbous in the center; apex with a lopsided, rounded plug which has a tiny projection on the highest side. Some forms have a single, moderately well-raised, circular hump around the middle of the shell. Uncommonly dredged just offshore where they live in mats of green algae.

Synonyms: *rotundum, marmoratum, subinflexum, tumidissimum, carpenteri, bitumidum, moreleti, deshayesi, crossei, undulosum, coxi*, all Folin, 1869; *fischeri* and *imiklis* Folin, 1870; *cingulatum* Gabb, 1892; *constrictum* Gabb, 1873; *contractum* Folin, 1874; *lermondi* Dall, 1924; *apanium* (Woodring, 1928); and *amblyoceras* (Woodring, 1959). The second stage of growth of this species is an open spiral, resembling a cow's horn.

(895a) *Caecum cornucopiae* Carpenter, 1858, is small, cylindrical and with a short straight second stage. Found abundantly in the Bahamas, southeast Florida and the West Indies around coral reefs. Some specimens very similar to some *nitidum* on the Florida reefs. Synonyms: *nebulosum* (Rehder, 1943); *mariae* (Folin, 1881); *cornubovis* Carpenter, 1858; *constrictum* (Pilsbry and Aguayo, 1933); *leoni* Folin, 1874; *bermu-*

(895b) *Caecum cubitatum* Folin, 1868, from North Carolina to Texas to Brazil, is an offshore species. It is small, slender and tapering with an enlarged area near the aperture. Mucro is small, slender and pointed. Second stage is slightly twisted. Fairly common in depths from 15 to 100 fathoms. *C. tenerum* Folin, 1869, is a synonym.

Superfamily Cerithiacea Fleming, 1822

Family TURRITELLIDAE Clarke, 1851

Genus *Tachyrhynchus* Mörch, 1868

Shells elongate, many whorls, spirally grooved, resembling a *Turritella*, but the operculum is multispiral with a central nucleus. Type: *reticulatus* (Mighels and Adams, 1842). The genus name is now declared masculine by Declaration 39, 1958, of the Inter. Comm. Zool. Nom. The original spelling left out the third "h," which Mörch added in 1875.

Tachyrhynchus erosus (Couthouy, 1838)   896
Eroded Turret-shell

Arctic Canada to Cape Cod, Massachusetts.
Alaska to British Columbia.

¾ to 1 inch in length, elongate, ¼ as wide; 8 to 10 rounded whorls. No umbilicus. Aperture round; columella smooth, slightly arched. Whorls with 5 or 6 smooth, flat-topped, spiral cords between sutures. Color cream to chalky-white, with a thin, polished, gray-brown periostracum. Operculum round, multispiral, chitinous, dark-brown. Common from 7 to 75 fathoms. *T. major* Dall, 1919, is a synonym.

(897) *T. reticulatus* (Mighels and Adams, 1842) (Arctic Ocean to Maine; and Alaska) is similar, but usually has spiral cords only on the base and one just above the suture, and has about 18 to 10 axial, rounded ribs per whorl. Some specimens show fine, spiral, incised lines. Synonym: *lactea* (Möller, 1842). Common from 2 to 1,245 fathoms.

Tachyrhynchus lacteolus (Carpenter, 1865)   898
Milky Turret-shell

Alaska to Baja California.

½ inch in length, similar to *erosus*, but ⅓ as wide as long, and the cords between sutures are finely beaded. The beads are arranged more or less in axial rows. The last ⅓ of the body whorl bears weak, nonbeaded spiral cords. Uncommon; 15 to 18 fathoms. This species differs from *reticulatus* in its smaller size, less slender shape, less convex whorls, and much finer sculpturing. *T. subplanatus* (Carpenter, 1865) is a synonym.

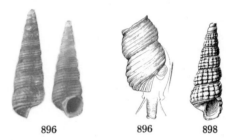

896   896   898

Other species:

899 *Tachyrhynchus pratomus* Dall, 1919. Alaska to Baja California, 20 fms.

900 *Tachyrhynchus stearnsii* Dall, 1919. San Pedro, California.

Genus *Turritellopsis* G. O. Sars, 1878

Shells very similar to those of *Turritella*, usually spirally grooved. Radula without marginal teeth. Cold-water, boreal seas. Type: *acicula* Stimpson, 1851.

Turritellopsis acicula (Stimpson, 1851)   901
Needle Turret-shell

Arctic Seas to Massachusetts.
Arctic Seas to San Diego, California.

8 to 10 mm., whitish, elongate-turreted, 10 whorls rounded and bearing 3 or 4 large, strong spiral cords between which are minute axial costae. Slightly umbilicate. Aperture round, outer lip sharp. Moderately common; 5 to 40 fathoms on sand bottoms. *T. stimpsoni* Dall, 1919, is a synonym.

Other species:

902 *Turritellopsis floridana* Dall, 1927. Off Fernandina, Florida, 294 fms.; Gulf of Mexico, 169 fms.

Genus *Turritella* Lamarck, 1799

Shells elongate, narrow, of many whorls, solid, whorls flat, rounded or carinate. Aperture round; outer lip thin. Columella narrow and strongly curved. No siphonal canal. Operculum circular, multispiral and with bristles along the edge. Type: *terebra* (Linné, 1758).

Turritella variegata (Linné, 1758)   903
Variegated Turret-shell

West Indies and southern Gulf of Mexico.

3 to 4 inches, pale-brown and variegated with purplish to reddish brown. Whorls nearly flat, but having the suture well-marked by a slight carina above and below. Common; in shallow bays in the West Indies. Rare in Gulf of Mexico dredgings.

Subgenus *Torcula* Gray, 1847

Type of this subgenus is *exoleta* (Linné, 1758).

Turritella acropora Dall, 1889   904
Boring Turret-shell

North Carolina to Florida, Texas and the West Indies.

1 inch in length, resembling *exoleta,* but with convex whorls, and with numerous, fine, spiral threads a few of which, at the periphery, are slightly larger than the others. There is a very weak series of riblets just below the suture. Color purplish to brownish orange. Common just offshore.

Turritella exoleta (Linné, 1758)   905
Eastern Turret-shell

Off North Carolina to the West Indies. Brazil.

2 to 2½ inches in length, long, slender and with a sharp apex. Each whorl with a large, coarse cord above and below, with the part between the cords concave and occasionally crossed by microscopic, arched, brown, scalelike lamellae. Base of shell concave. Color glossy-white to cream with sparse, axial flammules of light yellow-brown. Moderately common from 1 to 100 fathoms.

Turritella cooperi Carpenter, 1864   906
Cooper's Turret-shell

Monterey, California, to Baja California.

1 to 2 inches in length, 17 to 20 slightly convex whorls. Base concave. Columella and outer lip fairly fragile. Whorls with 2 or 3 small, spiral cords and usually with a number of much smaller, variously sized threads. Color orangish to yellowish white with darker, axial flammules. Moderately common just offshore. *T. jewetti* Carpenter, 1864, is of dubious status and may be limited to the Pleistocene. The fossil was also given the name *T. hemphilli* Merriam, 1941.

Turritella orthosymmetrica S. S. Berry, 1953   907
Symmetrical Turret-shell

Southern half of California.

1 inch, compactly coiled, with straight, smooth slopes. Suture narrowly channeled. Color grayish, clouded with brown. Differing from *cooperi* in being flat-sided, having a tightly coiled spire, acute basal keel, quadrate-angled aperture and 12 to 20 microscopic spiral threads over the whorls and base. Moderately common; 10 to 50 fathoms.

Turritella mariana Dall, 1908   908
Maria's Turret-shell

Catalina Island to Panama Bay, Panama.

1½ to 2½ inches in length, similar to *cooperi,* but with the whorls slightly concave due to the more prominent, irregularly beaded spiral cords. The aperture is not circular as in *cooperi.* Its color is usually much lighter. Uncommon; 20 to 40 fathoms.

Other species:

909 *Turritella yucatecanum* Dall, 1881. Yucatan Straits, Mexico. 16 mm.

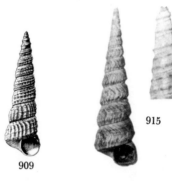

910 *Turritella anactor* S. S. Berry, 1957. Gulf of California.

911 *Turritella clarionensis* Hertlein and Strong, 1951. Gulf of California to Panama, 40 to 55 fms.

912 *Turritella lentiginosa* Reeve, 1849. Gulf of California. Common.

913 *Turritella nodulosa* King and Broderip, 1832. Baja California to Ecuador.

914 *Turritella willetti* J. H. McLean, 1970. Baja California. 15 to 40 fms. *The Veliger,* vol. 12, no. 3, p. 312.

915 *Turritella parkeri* J. H. McLean, 1970. Bahia de La Paz, Baja California, 40 to 90 fms. 48 mm. *Malacol. Rev. for 1969,* vol. 2, p. 127.

916 *Turritella hookeri* Reeve, 1849. Eastern Brazil, 5 to 33 fms.

917 *Turritella leucostoma* Valenciennes, 1832. Baja California, the Gulf, south to Panama. (Synonyms: *tigrina* Kiener, 1843; *cumingii* Reeve, 1849; *dura* Mörch, 1860)

903   904   905   906   908

Genus *Vermicularia* Lamarck, 1799

Shell beginning as *Turritella*-like snails, but soon becoming uncoiled as it grows. May grow in sponges, or among other specimens of the same species, or move about more or less freely. Type: *lumbricalis* (Linné, 1758). There is no zoological justification for accepting the family or subfamily Vermiculariinae. Consult also the Vermetidae worm shells.

Vermicularia spirata (Philippi, 1836)   918
West Indian Worm-shell

Southeast Florida and the West Indies. Bermuda.

Evenly and closely spiraled for about ¼ inch, then becoming random and drawn out in its wormlike, dextral coiling. Shell rather thin, colored a translucent- to opaque-amber, orange-brown or yellowish. Early whorls dark, smooth, except for 1 (rarely 2) smooth, spiral cord on the middle of the whorl. Subsequent whorls with 2 major cords which soon lose their prominence. Smaller threads present, especially on the base of the shell. Common in shallow water, the adults partially embedded in sponges and colonial ascidians. Also attaches to the tree coral, *Oculina*. This is not the common west Florida species usually called "*spirata*" in other books. See *knorrii* and also *fargoi*.

For anatomy and relationships in the worm-shells, see the excellent works by J. E. Morton (1951) in the Transactions of the Royal Society of New Zealand. For ecology of *spirata*, see S. J. Gould, 1969, *Bull. Marine Sci.*, vol. 19, p. 432.

Vermicularia knorrii (Deshayes, 1843)   919
Florida Worm-shell

North Carolina to Florida and the Gulf of Mexico. Bermuda.

Differing from *spirata* in having the early, evenly coiled part pure-white in color. The later whorls are very similar to *spirata*. Common in sponge masses, and frequently washed ashore. In Bermuda they occur in small clumps on dead coral rock in about 8 feet of water.

Vermicularia fargoi Olsson, 1951   920
Fargo's Worm-shell

West Coast to Florida to Texas.

Similar to *spirata* and *knorrii*, but the "turritella" or wound stage is ¾ to 1 inch in length; the shell is thicker and sturdy,

its color a drab grayish to yellowish brown. Early whorls tan to brown, with 2 (sometimes 3) spiral cords. Subsequent whorls with 3 major, brown-spotted, thick cords. Aperture with a squarish columella corner. Minute minor threads are between the main cords. Commonly found crawling on mud flats. A race occurs in Texas in which the "turritella" stage is much more slender.

Vermicularia fewkesi Yates, 1890   921
Fewkes' Worm-shell

Santa Barbara, California, to Cedros Island, Baja California.

Shell rather small, usually less than 1 inch. Early whorls 5 mm., closely coiled, whitish, with 2 spiral cords. Later whorls loosely coiled and with a strong cord on the base. Diameter of adult aperture about 4 mm. Not uncommon; in sublittoral areas under kelp.

Other species:

922 *Vermicularia pellucida* (Broderip and Sowerby, 1829). White form called *eburnea* (Reeve, 1842). Gulf of California to Panama, 2 inches.

923 *Vermicularia frisbeyae* J. H. McLean, 1970. Jalisco, Mexico, to Panama. 18 to 60 fms. 60 mm. *The Veliger*, vol. 12, no. 3, p. 311.

924 *Vermicularia radicula* Stimpson, 1851. Southern Cape Cod, Massachusetts, to Connecticut. May be the northern form of *V. spirata* (no. 918).

Family SILIQUARIIDAE Anton, 1839

Genus *Siliquaria* Bruguière, 1789

Shell a long, coiled, wormlike tube with a series of tiny holes arranged in a single long, longitudinal row. Type: *anguina* Linné, 1758. The genus *Tenagodus* Guettard, 1770, is rejected as nonbinomial.

Siliquaria squamata Blainville, 1827   925
Slit Worm-shell

Off North Carolina to the West Indies; Bermuda. Brazil.

A small wormlike shell with detached whorls throughout. Characterized by the long row of small holes or elongate slits on the middle of the whorl. Early whorls smooth, white; later whorls becoming very spinose and stained with brown. The coiling is very irregular and loose. Grows to about 5 or 6 inches in length. *S. angullae* Mörch, 1860, is a synonym. Lives embedded in yellow sponge, down to 400 fathoms. For shell morphology, see S. J. Gould, 1966, *Amer. Mus. Novitates*, no. 2263.

Other species:

926 *Siliquaria modesta* Dall, 1881. Challenger Bank, off Bermuda, 20 fms. West Indies.

919   920

918

925   926

Family MATHILDIDAE Dall, 1889

This family and the Architectonicidae are placed superfamily Architectoniacea by Kuroda, Habe and Oyama, 1971.

Genus *Mathilda* Semper, 1865

Shell small, brown, resembling miniature *Bittium* or *Cerithium,* but with the nuclear whorl tilted on its axis. Sculture of spiral cords, often beaded. Type: *quadricarinata* (Brocchi, 1814) from the Pliocene of Europe.

Subgenus *Fimbriatella* Sacco, 1895

Spiral cords prominent and beaded. Type: *fimbriatum* Michelotti, 1847.

Mathilda barbadensis Dall, 1889 927
Barbados Mathilda

Southeast Florida and the West Indies.

4 to 6 mm., light-brown, with about 7 adult whorls having 2 to 4 strongly beaded, raised, spiral cords. Umbilicus present. Anterior edge of aperture slightly produced and flaring. Uncommon; 50 to 100 fathoms.

Mathilda yucatecana (Dall, 1881) 928
Yucatan Mathilda

Georgia to west coast of Florida to Yucatan.

7 to 8 mm., light-brown, similar to *barbadensis,* but with fine spiral and axial threads forming a reticulate pattern on the angulate whorls. Moderately common; 13 to 640 fathoms.

928

927

Other species:

929 *Mathilda rushii* Dall, 1889. Off Fernandina, 294 fms., and Florida Strait, 465 fms.

930 *Mathilda scitula* Dall, 1889. Off Cape Hatteras, North Carolina, 49 to 63 fms.

931 *Mathilda georgiana* Dall, 1927. Off Georgia, 440 fms.

932 *Mathilda lacteosa* Dall, 1927. Off Georgia, 440 fms.

933 *Mathilda globulifera* Dall, 1927. Off Fernandina, Florida, 294 fms.

934 *Mathilda granifera* Dall, 1927. Off Georgia, 440 fms.

935 *Mathilda amaea* Dall, 1927. Off Georgia, 440 fms., and Fernandina, Florida, 294 fms.

936 *Mathilda hendersoni* Dall, 1927. Off Fowey Light and Turtle Harbor, Florida, 25 to 50 fms.

937 *Mathilda (Tuba* H. C. Lea, 1838 = *Gegania* Jeffreys, 1884) *jeffreysii* Dall, 1889. Off Fernandina, Florida, 294 fms.

Family ARCHITECTONICIDAE Gray, 1850

Shells coiled dextrally almost in a flat plane. Umbilicus wide. Operculum chitinous. Larval shell coils to the left, a condition called hyperstrophy.

Subfamily ARCHITECTONICINAE Gray, 1850

Genus *Architectonica* Röding, 1798

Shells low turbinate, with a large, deep umbilicus. Sculpture of spiral beaded threads. Aperture relatively large. Type: *perspectiva* (Linné, 1758) from the Indo-Pacific. *Solarium* Lamarck, 1799, is a synonym.

Architectonica nobilis Röding, 1798 938
Common Sundial   Color Plate 3

North Carolina to Florida, Texas and the West Indies. Brazil. Baja California to Peru.

1 to 2 inches in diameter, heavy, cream with reddish brown spots which are especially prominent just below the suture. Sculpture of 4 or 5 spiral cords which are usually beaded. Umbilicus round, deep and bordered by a heavily beaded, spiral cord. Operculum corneous, thin, paucispiral, brown and with lamellate growth lines. Moderately common in sand below low-water line to a depth of 20 fathoms. Usually associated with sea pansies. Known for years as *A. granulata* Lamarck, 1816, which, however, is a later name. Other synonyms are: *granosum* Valenciennes, 1832; *quadriceps* Hinds, 1844; *verrucosum* Philippi, 1849; *valenciennesi* Mörch, 1860.

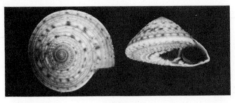

938

Subgenus *Discotectonica* Marwick, 1931

Top of whorls and spire flattish. Periphery very sharp and thin. Colors subdued. Type: *acutissima* Sowerby, 1914, of Japan. *Acutitectonica* Habe, 1961, is a synonym.

Architectonica peracuta (Dall, 1889) 939
Keeled Sundial

North Carolina to both sides of Florida and the West Indies.

¾ inch in diameter, similar to *nobilis,* but smaller, with the spire much flatter, whorls almost smooth, periphery very sharp, and without color spots. Rare; in 45 to 73 fathoms.

(939a) The Pacific counterpart is *A. placentalis* (Hinds, 1844). It ranges from Magdalena Bay, Baja California, to

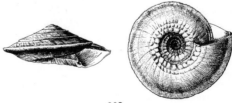

939

Panama, and is scarcely discernible from *peracuta*. Some workers consider them the same (see Marche-Marchad, 1969, *Bull. Inst. Fondamental d'Afrique Noire*, vol. 31, p. 461).

Subfamily HELIACINAE Troschel, 1867

Genus *Heliacus* Orbigny, 1842

Shells planorboid, with a fairly high spire, strongly sculptured, with a deep, wide umbilicus. The thick, corneous, conical operculum is multispiral, with a central nucleus, and it coils counterclockwise. Members of this genus are closely associated with the colonial "sea anemone," *Zoanthus*. Type: *heberti* Deshayes, 1830. *Torinia* Gray, 1842, is a nude name. *Torinia* Gray, 1847, is a synonym.

Heliacus cylindricus (Gmelin, 1791)   **940**
Atlantic Cylinder Sundial

Southeast Florida and the West Indies to Brazil. Bermuda.

⅜ inch in length, equally wide; spire high; umbilicus narrow, round, very deep, bordered inside with 3 spiral, beaded cords. Columella with 4 small, depressed, spiral lines. Tops of whorls with 4 spiral cords of closely packed, small beads. Color dark-gray to reddish brown with a cream base and with white spots on the periphery. Very rarely one finds an adult specimen that coils "sinistrally" and with its whorls slightly detached, resembling a *Vermicularia*. For details see Robertson and Merrill, 1963, *The Veliger*, vol. 6, no. 2, pp. 76 to 79. Synonyms are: *heberti* (Deshayes, 1830); *cyclostomum* (Menke, 1830); *aethiops* (Menke, 1830); and *nubilum* (Philippi, 1849). Uncommon at low tide to 25 fathoms. Feeds and lives on the colonial zoanthid coelenterate, *Palythoa*.

940

Heliacus bicanaliculatus (Valenciennes, 1832)   **941**
Pacific Cylinder Sundial

Puertecitos, Gulf of California, to Acapulco, Mexico.

¾ inch, turban-shaped, with 4 or 5 spiral maculated cords on each whorl in the spire. Umbilicus deep and containing 2 spiral ribs. Operculum multispiral with fringed whorls. Uncommon; intertidal in sand where there are zoanthid sea anemones. *H. radiatus* Menke, 1851, is a synonym, as is *chiquita* Pilsbry and Lowe, 1932.

Heliacus perrieri (Rochebrune, 1881)   **942**
Channelled Sundial

Southeast Florida and the West Indies. Brazil.
Mexico to Panama.

15 to 20 mm., spire low and slightly convex. Color uniformly light- to dark-brown. Suture deep. Umbilicus wide, very deep and funnel-shaped. Aperture round. Whorls round and with numerous irregularly sized spiral cords, usually well-beaded on the spire and at the periphery of the whorl. Operculum chitinous, very thick, top surface with flaky brown spiral flaps. Uncommon. Western Atlantic specimens are larger and with a more rounded periphery than the Indo-Pacific *infundibuliformis* (Gmelin, 1791). Rarely, abnormal specimens are loosely

942   943

coiled with rapidly descending whorls. *H. fallaciosa* Marche-Marchad, 1956, is a synonym.

Heliacus bisulcatus (Orbigny, 1842)   **943**
Orbigny's Sundial

North Carolina to Texas and to Brazil. Bermuda.

¼ to ½ inch in diameter, spire flattened, each whorl with 5 crowded rows of neat, tiny, squarish beads. Periphery with a major, and below it a minor, beaded cord. Base rounded and with about 7 wide cords bearing beads. Umbilicus quite wide and very deep. Nuclear whorl glassy-white. Color of shell brown to dull-cream. Operculum conic, chitinous. Uncommon from intertidal to 200 fathoms on rock bottom, associated with soft corals. Also found living in West Africa.

(944)   The subspecies *mazatlanicus* Pilsbry and Lowe, 1932, scarcely discernible from *bisulcatus*, is the Pacific analog or "twin" and has a more rounded periphery. It occurs in the Gulf of California down to Mazatlan, Mexico. It is brown in color and reaches 6 mm. in diameter. Rare.

Subgenus *Solatisonax* Iredale, 1931

Shell with a strong, single-keeled periphery and basal axial plications radiating from the umbilicus. Operculum concave and spirally compressed. Type: *injussa* Iredale, 1931. Information courtesy of Arthur Merrill.

Heliacus borealis (Verrill and Smith, 1880)   **945**
Boreal Sundial

Massachusetts to the lower Caribbean. Off Portugal.

10 to 13 mm., strong, yellowish brown, with a strong, rounded, nodulous cord at the periphery. Umbilicus deep and with radiating axial plications. Deep water down to 440 fathoms; uncommon.

Other species:

946 *Heliacus planispira* Pilsbry and Lowe, 1932. Puertecitos, Gulf of California to Panama. Low-tide line; rare.

945

947

947 *Heliacus sigsbeei* (Dall, 1889). 5 mm. Florida Strait, 310 fms. Off Barbardos.

948 *Heliacus architae* (Costa, 1844). Southern California, Gulf of California to Ecuador. (Synonym: *panamensis* Bartsch, 1918.)

Genus *Pseudomalaxis* P. Fischer, 1885

Shell discoidal, with a squarish last whorl bounded by 2 subequal keels. Fine spiral striae present. Operculum thin, multispiral. Type: *zanclaea* (Philippi, 1844). *Discosolis* Dall, 1892, is a synonym.

Pseudomalaxis nobilis Verrill, 1885   **949**
Noble False Dial

Virginia to southeast Florida and the West Indies.

10 mm. in diameter, dull-white to chestnut mottled, planorboid, with a very flat spire and a wide concave, nonumbilicate base. Periphery of shell flat, bordered above and below by 1 or 2 spiral cords of small beads. The nuclear whorl is sinistral. Aperture squarish. Whorls touching. Operculum round, multispiral with a chitinous pimple on the inside. A rare and choice collector's item. Deep water. 70 fathoms. Formerly placed in the Vitrinellidae.

Subgenus *Spirolaxis* Monterosato, 1913

Whorls coiling on the same plane but not touching each other. Type: *centrifuga* (Monterosato, 1890). *Aguayodiscus* Jaume and Borro, 1946, is a synonym.

Pseudomalaxis centrifuga Monterosato, 1890   **950**
Exquisite False Dial

Texas; West Indies and the Eastern Atlantic.

3 to 4 mm. in diameter, dull-white, planorboid, with square whorls, bearing granular cords on the periphery. Similar to *nobilis*, but the whorls are entirely disjunct. Rare, 25 fathoms. Synonyms: *zanclaea* Gray, 1853 (non Philippi, 1844); *macandrewi* Iredale, 1911; *exquisita* Dall and Simpson, 1901.

950

949

Other species:

— *Pseudomalaxis lamellifera* Rehder, 1934, see *Pleuromalaxis* in the Vitrinellidae.

951 *Pseudomalaxis clenchi* Jaume and Borro, 1946. 3 mm. off Havana, Cuba, 230 fms.

Genus *Philippia* Gray, 1847

Shell shiny, turbinid in shape, with a narrow umbilicus bordered by a row of beads. Operculum with a subcentral nucleus and having a calcareous process underneath. Type: *lutea* (Lamarck, 1822) from the Indo-Pacific. For biology see R. Robertson, 1970, *Pacific Sci.*, vol. 24, pp. 66–83; and R. Robertson, 1964, *Proc. Acad. Nat. Sci.*, vol. 116, pp. 1–27.

Subgenus *Psilaxis* Woodring, 1928

Type of this subgenus is *krebsii* (Mörch, 1875).

Philippia krebsii (Mörch, 1875). Pl. 4   **952**
Krebs' Sundial

North Carolina to southeast Florida to Brazil. Bermuda.

½ inch in diameter, similar to *A. nobilis*, but glossy-smooth on top except for 2 microscopic spiral threads above the suture, with a smooth rounded base and with its deep umbilicus bordered by 2 beaded spiral rows, the innermost having about 30 barlike beads (in contrast to 12 in *nobilis*). Operculum chitinous, brown and multispiral. Uncommon from 16 to 63 fathoms.

952

Family VERMETIDAE Rafinesque, 1815

Long, coiling shells, resembling "worm tubes," usually attached to hard surfaces. The axis of the globose nuclear whorl is at right angles to that of the following whorls. Operculum, when present, multispiral, corneous and the edges of the whorls free. An excellent review of the classifications was given by A. M. Keen, 1961, *Bull. British Mus. Nat. Hist.*, vol. 7, no. 3, pp. 183–213. There are no true *Vermetus* living in the United States. The name *Spiroglyphus* Daudin, 1800 should not have been rejected by Keen. Consult also the genus *Vermicularia* in the family Turritellidae.

Genus *Petaloconchus* H. C. Lea, 1843

Colonial wormlike tubes; sometimes characterized by spiral cords inside on the columella of the middle whorls. There may be fairly regular scars of broken tube ends along the course of a coil, where a vertical feeding tube has been abandoned and replaced by one growing off at a different angle. Type: *P. sculpturatus* Lea, 1843, a Miocene species. Recent species all belong to the subgenus *Macrophragma* Carpenter, 1857. Operculum chitinous, concave, multispiral; 2 to 4 nuclear whorls, yellowish to white.

Subgenus *Macrophragma* Carpenter, 1857

Type of this subgenus is *macrophragma* Carpenter, 1857. Cross-section of tube shows 2 large and 1 small longitudinal ridges.

Petaloconchus varians (Orbigny, 1841)   **953**
Variable Worm-shell

Florida to Brazil; Bermuda.

This is a very variable species, showing many individual as well as colonial differences depending upon the conditions under which they grow. In sheltered areas, colonies grow loosely and upwards, much like a clump of staghorn or organ-pipe coral. These massive, darkly colored colonies, with elongate, pillarlike, crowded tubes take the form name of *nigricans* (Dall, 1884). They form large reefs or banks offshore on the west coast of Florida. Individual forms which have grown flat to the surface of shells, and look like "squashed *Turritella*," have the form name *floridana* Olsson and Harbison, 1953. The usual West Indian form consists of conglomerate, moderately compacted masses of individuals. The surface of the tube is coarsely reticulate. Color varies from brown to orange. Common; attached to intertidal rocks, sometimes in tidepools, where they feed during high tide.

ish; pure-white in color. Uncommon; offshore to 805 fathoms. The form *mcgintyi* Olsson and Harbison, 1953, merely has the early whorls "squashed" instead of well-rounded in forms that were not crowded when they grew. The latter is fairly common in southeast Florida.

Other species:

956 *Petaloconchus (M.) montereyensis* Dall, 1919. Monterey to San Diego, California.

957 *Petaloconchus (M.) contortus* (Carpenter, 1856). Gulf of California to Panama; not common.

958 *Petaloconchus (M.) macrophragma* (Carpenter, 1856). Baja California to Panama.

959 *Petaloconchus (M.) flavescens* (Carpenter, 1857). Gulf of California to Mazatlan, Mexico.

953   956

Genus *Spiroglyphus* Daudin, 1800

Solitary or colonial snails, corroding a trench in the substrate, in which the lower part of each coil is embedded; coiling planorboid nuclear whorls. Adults usually white with brown stainings. Growth rings usually prominent and rising to a ridge along the outer edge of the coils. 2 nuclear whorls dark-brown and may have a wavy sinusigerous outer lip. Operculum concave with fringed, chitinous multispiral whorls and with a buttonlike core attached to the underside. Type: *annulatus* Daudin, 1800. *Siphonium* Mörch, 1859 (non Link, 1807); *Vermitoma* Kuroda, 1928; *Bivonia* Gray, 1847; *Veristoa* Iredale, 1937; and *Dendropoma* Mörch, 1861, are synonyms.

Petaloconchus compactus (Carpenter, 1864)   954
Compact Worm-shell

Washington to California.

Small, whitish worm-shell, usually attached to shells and rocks, the entire mass about 1 inch. Tubes ⅛ inch in diameter, with coarse, axial, ringlike growth lines and 2 to 4 week longitudinal riblets. No internal lamellae noticed. Operculum thin, reddish. Moderately common from just below low-tide mark to 50 fathoms. *P. complicatus* Dall, 1908, may be a "free" form of this species which occurs in self-attached clumps, rather than being coiled against a hard substrate. *P. montereyensis* Dall, 1919, seems close to this species, and may be only an ecologic variant of it. Its tube diameter is 2 mm., and has diagonal wrinkles on the early part of tube.

Spiroglyphus annulatus Daudin, 1800   960
Corroding Worm-shell

Southeast Florida to the Lesser Antilles. Bermuda.

Shell minute, up to ⅓ inch, regularly coiled for the first 3 or 4 whorls and sunk into its shell or limestone substrate. Rarely on wood. Tube only 2 mm. in diameter, with strong, crowded, circular fimbriations. Color dark-brown to gray with brown longitudinal bands or rarely white. Operculum circular, multispiral, underside reddish and with a swollen pimple in the middle. Very abundant on intertidal surfaces, including other shells, throughout the West Indies. They feed by entrapping plankton on retractable mucus threads. *Dendropoma corrodens* (Orbigny, 1842) is a synonym.

Petaloconchus erectus (Dall, 1888)   955
Erect Worm-shell

Florida to Brazil; Bermuda.

1 to 2 inches; tube ⅛ inch in diameter; attached to shells or coral, coiled for several whorls, then straightening out and rising upward for about 1 inch. Surface malleated or smooth-

Spiroglyphus lituellus (Mörch, 1861)   961
Flat Worm-shell

Forrester Island, Alaska, to San Diego, California.

A small worm-tube mollusk found adhering to rocks and the shells of abalones in a tightly wound, flat spiral. The last whorl may grow up on top of the previous whorls and be erect for ¼ of an inch. Aperture circular, about ⅛ inch in diameter. Shell solid, with 2 large, scaled cords which give a somewhat squarish cross-section to the whole shell. Hollow scales and fimbriations present elsewhere. Color cream to purplish gray. Operculum horny, multispiral and brown. Moderately common.

955

Subgenus *Novastoa* Finlay, 1927

Like *Spiroglyphus* but living in large, intertwined masses. Probably not a valid zoological division. Type: *lamellosa* (Hutton) from New Zealand.

Spiroglyphus irregularis (Orbigny, 1842)   **962**
Irregular Worm-shell

Southeast Florida to Brazil. Bermuda.

Similar to *P. varians,* but greatly coiled, and with heavier, larger shells which are strongly rugose. A strong, scaled rib runs along the dorsal surface of the tube. Diameter of tube ¼ to ⅓ inch, brownish within. Color grayish. Operculum reddish brown, multispiral, fringed. Solitary or colonial. Has been called *nebulosum* by some workers. Occurs in compact masses attached to rocks and other shells.

961

962

Other species:

963 *Spiroglyphus (N.) nebulosus* (Dillwyn, 1817). West Indies. An undetermined species of dubious identity. See *irregularis.*

964 *Spiroglyphus rastrus* (Mörch, 1861). California.

Genus *Stephopoma* Mörch, 1860

Shell small, corkscrew whorling; without internal septa or ledges. Operculum chitinous, concave and with chitinous bristles. Type: *Vermetus roseum* Quoy and Gaimard, 1832.

965 *Stephopoma myrakeenae* Olsson and McGinty, 1958. Western Caribbean.

Genus *Serpulorbis* Sassi, 1827

Shells among the largest of the family; coiling of nearly concentric loopings, especially in the early whorls; later whorls disjunct. 2 to 4 globose, smooth, nuclear whorls. Operculum not persent. Type: *arenaria* Linné, 1758. Synonyms: *Serpuloides* Gray, 1850; *Aletes* Carpenter, 1857; *Anguinella* Conrad, 1846; *Thylacodes* Mörch, 1862; *Tetranemia* Mörch, 1859.

Serpulorbis decussatus (Gmelin, 1791)   **966**
Decussate Worm-shell

North Carolina to the West Indies. Brazil.

1 to 3 inches, heavy, worm-tube in appearance; coarsely sculptured with irregular longitudinal cords. Yellowish to brown. Attached to stones and other shells. Moderately common. Down to 17 fathoms. *S. riisei* (Mörch, 1862) may be a synonym in which the sculpture is more pebbly.

966   967

Serpulorbis squamigerus (Carpenter, 1856)   **967**
Scaled Worm-shell

California to Peru.

Grows in large, twisted masses. The shelly tubes are circular, ¼ to ½ inch in diameter. Scultpure of numerous, minutely scaled or rough, longitudinal cords. Color gray to pinkish gray. The last part of the shell which usually stands erect for ½ inch is smoothish. No operculum present. A very common, colonial species found in masses on wharf pilings or attached to rocks below the low-water line.

Other species:

968 *Serpulorbis cruciformis* (Mörch, 1862). Gulf of California.

969 *Serpulorbis (Cladopoda* Gray, 1850) *margaritaceus* (Chenu, 1844). West Mexico.

970 *Serpulorbis oryzata* (Mörch, 1862). Guaymas to Acapulco, Mexico.

Genus *Tripsycha* Keen, 1961

White shells of moderate size, with the early whorls firmly attached and coiling as in *Serpulorbis.* Last whorl lax and uncoiled. Operculum slightly concave, with an appressed spiral lamina of several volutions. Type: *tripsycha* (Pilsbry and Lowe, 1932) of West Mexico **(971)**. *Bull. Brit. Mus. Nat. Hist.,* vol. 7, p. 196.

971

Subgenus *Eualetes* Keen, 1971

Type of the subgenus is *Vermetus centiquadrus* Valenciennes, 1846. *The Veliger,* vol. 13, p. 296.

972 *Tripsycha centiquadra* (Valenciennes, 1846). **Gulf of California to Panama.** Diameter of aperture 5 mm. Operculum half as small. (Synonyms: *?peronii* Valenciennes, 1846; *sutilis* Mörch, 1862.)

973 *Tripsycha tulipa* (Chenu, 1843). Panama.

Family PLANAXIDAE Gray, 1850

Genus *Planaxis* Lamarck, 1822

Shells small, thick-shelled. Sculpture and color patterns spirally arranged. Columella truncated below. Type: *sulcatus* Born, 1778, from the Indo-Pacific.

Planaxis lineatus (da Costa, 1778)   974
Dwarf Atlantic Planaxis

Southeast Florida to Brazil. Bermuda.

¼ inch in length, thick and strong; glossy-smooth when the thin, smoothish, translucent periostracum is worn away. Color whitish cream with neat, spiral bands of brown (10 in last whorl, 5 showing in whorls above); rarely all black-brown. Whorls in top of spire with 4 or 5 small spiral cords, later becoming obsolete. Aperture slightly flaring, enamel-white with 10 brown dots on the edge of the outer lip. Nuclear whorls very small, glossy, translucent-brown and sharply pointed. Common from the mid-tide line to a few feet in quiet areas in sand and among small rocks and broken shell.

974   975

Subgenus *Supplanaxis* Thiele, 1929

Type of this subgenus is *nucleus* (Bruguière, 1789).

Planaxis nucleus (Bruguière, 1789)   975
Black Atlantic Planaxis

Southeast Florida and the West Indies. Bermuda.

½ inch in length, resembling a thick, polished, dark-brown *Littorina* periwinkle. Characterized by 5 strong spiral cords which are developed on the outside of the body whorl only in the region behind the slightly flaring lip. 3 other cords are present just below the suture. Columella area dished; reinforced by the round, pillarlike columella. A small pimple is present near the posterior canal in the aperture. Outer lip with strong crenulations on the inside. Periostracum a soft gray-black felt. A common littoral species along the open ocean front in the West Indies, which bears its young in a brood pouch. Rare in Florida.

(975a) *Planaxis obsoletus* Menke, 1851, is similar, and occurs from the Gulf of California to Salina Cruz, Mexico. (Synonyms are *acutus* Menke, 1851, non Krauss, 1848, and *nigritella* Forbes, 1852.)

Family MODULIDAE Fischer, 1884

Genus *Modulus* Gray, 1842

Shell trochoid in shape, with a stout tooth at the base of the columella. Operculum multispiral, chitinous. Type: *modulus* (Linné, 1758).

Modulus modulus (Linné, 1758)   976
Atlantic Modulus

North Carolina to Texas and to Brazil. Bermuda.

About ½ inch in length. Characterized by the small, projecting-toothlike, frequently brownish spine located on the lower end of the columella. Base of shell with about 5 strong, spiral cords. Top of whorls with low, slanting, axial ribs. Color grayish white with beach-worn specimens often exhibiting flecks of purple-brown. Found abundantly among weeds in shallow, warm waters. Synonyms include *floridanus* Conrad, 1869; *filosus* (Helbling, 1779); *perlatus* Gmelin, 1791, *pisum* Mörch, 1876; and *tasmanica* Tenison-Woods, 1877.

(977) A very similar species, *Modulus disculus* (Philippi, 1846), with a violet aperture, occurs from the Gulf of California to Panama. *M. dorsuosus* Gould, 1853, and possibly *cerodes* (A. Adams, 1851) are synonyms.

976   978

Modulus carchedonius (Lamarck, 1822)   978
Angled Modulus

Caribbean.

½ inch, similar to *modulus*, but the periphery of the shell well-angulated, the spiral cords smaller and neater, in lacking the strong, axial ribs, and in never having the columella tooth colored. Uncommon; shallow water. *M. angulata* C. B. Adams, 1845, is a synonym.

(979) A similar species, *Modulus catenulatus* (Philippi, 1849), occurs on mud flats from the Gulf of California to Peru.

Family POTAMIDIDAE H. and A. Adams, 1854

Subfamily POTAMIDINAE H. and A. Adams, 1854

Genus *Cerithidea* Swainson, 1840

The horn shells are intertidal mud-lovers. The shells are elongate and with 10 to 15 convex whorls. Axial ribs are more prominent on the early whorls. Outer lip flares. Operculum horny, thin, paucispiral and with its nucleus at the center. Type: *decollata* Linné, 1758.

Subgenus *Cerithideopsis* Thiele, 1929

The type of this subgenus is *pliculosa* (Menke, 1829). It was founded on radular characters.

Cerithidea costata (da Costa, 1778)   980
Costate Horn Shell

South Florida and the West Indies.

½ inch in length, translucent, pale yellowish brown. With 9 to 12 very convex whorls. Axial, curved ribs are round and distinct on the early whorls, fading out on the last two whorls. No old varices present. A common shallow-water, mud-loving species. Synonyms are: *lafondii* Michaud, 1829; *ambiguum* C. B. Adams, 1845; *salmacidum* Morelet, 1849; *petitii* Schramm, 1869; and *pupoidea* Mörch, 1876.

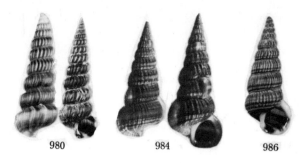

980   984   986

(981) The subspecies *C. costata turrita* Stearns, 1873, the Turret Horn Shell from the Tampa-Sanibel region, has 15 to 20 (instead of 25 to 30) axial ribs on the next to the last whorl.

(982) The subspecies or forma *beattyi* Bequaert, 1942, occurs in the Bahamas to Trinidad and is characterized by beaded axial ribs caused by 2 or 3 spiral grooves on the whorls.

(983) The Eastern Pacific equivalent of this species is the handsome, 1-inch, bright *Cerithidea montagnei* (Orbigny, 1837) which is abundant in mangrove swamps from Baja California to Ecuador. *C. reevianum* C. B. Adams, 1852, is a synonym.

Cerithidea pliculosa (Menke, 1829)   984
Plicate Horn Shell

Texas, Louisiana and the West Indies. Not Florida.

1 inch in length, brownish black in color. 11 to 13 slightly convex whorls. Several yellowish, former varices are present. Numerous spiral threads make the axial ribs slightly nodulose. Locally common on mud flats. Synonyms are: *iostoma* Pfeiffer, 1839; *lavalleanum* Orbigny, 1842; and *varicosa* Mörch, 1876.

(985) The subspecies *veracruzensis* Bequaert, 1942, found from Tampico, Mexico, to Nicaragua, is characterized by well-beaded axial riblets which give the surface of the whorls a reticulate appearance. Common.

Cerithidea scalariformis (Say, 1825)   986
Ladder Horn Shell

South Carolina to the Florida Keys and the West Indies. Bermuda (rare).

¾ to 1¼ inches in length. Pale russet-brown to slightly violaceous, usually with many conspicuous, dirty-white, spiral bands. 10 to 13 moderately convex whorls. Many coarse, axial ribs present which stop abruptly below the periphery of the whorl at a sharply marked, rounded spiral ridge. Base of shell with 6 to 8 spiral ridges. No former varices. Common on mud flats, usually with *Melampus*. Synonyms are: *tenuis* Pfeiffer, 1839, and *hanleyana* Reeve, 1866.

Cerithidea californica Haldeman, 1840   987
California Horn Shell

Bolinas Bay, California, to the Gulf of California.

987

1 to 1½ inches in length, resembling our photo of *C. pliculosa* from the Atlantic. 11 whorls, spirally and weakly threaded, and axially strongly ribbed (12 to 18 ribs per whorl). Dark-brown in color with 1 or 2 yellowish white, swollen varices on the spire. A very common species found in large colonies on mud flats. In Mexico, the typical form is replaced by the subspecies *mazatlanica* Carpenter, 1856 **(938)** (? or *hegewischii* Philippi, 1848). *C. albonodosa* Gould and Carpenter, 1857, is an ecologic form.

Other Pacific species:

988 *Cerithidea valida* (C. B. Adams, 1852). Gulf of California to Ecuador. (Synonyms: *varicosum* Sowerby, 1834, non Valenciennes, 1832; *fortiusculum* Bayle, 1880; *aguayoi* Clench, 1934; *meta* Li, 1930.)

Genus *Rhinocoryne* von Martens, 1900

Shell fusiform, turreted, with a spiny periphery and a strong, short siphonal canal. Operculum horny, circular, almost multispiral. Type: *humboldti* Valenciennes, 1832.

Rhinocoryne humboldti (Valenciennes, 1832)   989
Humboldt's Potamid

Sonora, Mexico, to Chile.

1 to 1½ inches, solid, with a sharp apex. Whorls spirally striate; periphery of whorls angular and bearing a single row of sharp nodules. Color grayish to brown. Rarely with two rows of nodules (forma *lamarckii* Valenciennes, 1832). Common in estuaries and on sandy bottoms offshore. *Cerithium pacificum* Sowerby, 1833, is a synonym.

Subfamily BATILLARIINAE Thiele, 1929

Genus *Batillaria* Benson, 1842

Cerithium-like in appearance. Siphonal canal very short and twisted to the left. Outer lip smooth inside. Operculum round, multispiral and horny, while in *Cerithidea* and *Cerithium* it is paucispiral. Type: *zonalis* (Bruguière, 1792).

Batillaria   *Cerithium* RTA

Batillaria minima (Gmelin, 1791)   990
False Cerith

South half of Florida to Brazil. Bermuda.

½ to ¾ inch in length, resembling the dwarf cerith, *C. lutosum*. Color varies from black, gray to whitish, and often has black or white spiral lines. Finely nodulose with coarse axial swellings and uneven spiral threads. The siphonal canal is very short and twisted slightly to the left. Operculum multispiral. A very common intertidal species. Dwarf, feebly sculptured forms were given the names *rawsoni* Mörch, 1876, *degenerata* Dall, 1894, and *albocoopertum* Davis, 1904. Other synonyms are *clathratum* Menke, 1828, *nigrescens* Menke, 1828, and *C. septemstriatum* Say, 1832.

990

991

Batillaria zonalis (Bruguière, 1792)   991
Cuming's False Cerith

Washington to Monterey, California.
Southwest Pacific; Japan.

10 to 34 mm., slender, turreted, spirally threaded, with traces of numerous varices on the early whorls. Color light-brown to dirty-black with about 5% of the specimens having a narrow white band on the 2 most posterior spiral threads. Abundant; found at high-water mark in saline to brackish mud flats amid the green alga *Enteromorpha*. Introduced from Japan prior to 1931. See J. H. McLean, 1960, *The Veliger*, vol. 2, p. 61. Synonyms are: *cumingi* Crosse, 1862; *multiformis* Lischke, 1869; *aterrima* (Dunker, 1877); *atramentaria* (Sowerby, 1855). For anatomy, see A. L. Driscoll, 1972, *The Veliger*, vol. 14, no. 4, p. 375.

Family CERITHIIDAE Fleming, 1822

Subfamily CERITHIINAE Fleming, 1822

Genus *Cerithium* Bruguière, 1789

The operculum is horny, thin, brown and paucispiral. Most species in the genus are shallow-water dwellers. Type: *nodulosum* Bruguière, 1789. *Thericium* Monterosato, 1890, is a synonym, but is used as a subgenus by some workers.

Cerithium atratum (Born, 1778)   992
Florida Cerith

North Carolina to the south half of Florida; Texas. Brazil.

1 to 1½ inches in length, elongate. Spire pointed, with 2 or 3 white, former varices on each whorl. Siphonal canal well-developed. With several spiral rows of 18 to 20 neat beads per whorl between which are fine, granulated spiral threads. Color whitish with mottlings and specklings of reddish brown. Distinguished from *C. litteratum* by its more elongate shape and neater, smaller, more numerous beads. Common in shallow water. *C. floridanum* Mörch, 1876, is a synonym.

Cerithium muscarum Say, 1832   993
Fly-specked Cerith

South half of Florida and the West Indies.

1 inch in length, moderately elongate. Siphonal canal rather long and twisted to the left. 9 to 11 nodulated axial ribs on each whorl. Base of shell with a very strong spiral cord, often nodulated. Former varices rarely present. Apertural side of body whorl convex. Color slate- to brown-gray, usually with brown to reddish specks in spiral rows. Common in shallow, warm waters which are brackish.

Cerithium litteratum (Born, 1778)   994
Stocky Cerith

Southeast Florida, Bermuda and the West Indies to Brazil.

1 inch in length, half as wide; siphonal canal short. Aperture side of body whorl slightly flattened. Usually 1 weak, former varix present. With numerous coarse spiral threads, and with a spiral row of 9 to 12 sharp, prominent nodules just below the suture. Sometimes a second, smaller row of spines is on the periphery. Color whitish with spiral rows of many black or reddish squares. Common in shallow water in sand and weedy areas. *C. literatum* is a misspelling.

The heavier, smoothish form with orange-brown to rust-brown mottlings is found in deeper offshore reef waters in Bermuda and bears the form name *semiferrugineum* Lamarck, 1822.

Cerithium eburneum Bruguière, 1792   995
Ivory Cerith

Southeast Florida, the Bahamas and Greater Antilles.

¾ to 1 (rarely 1½) inches in length, variable in shape, but usually moderately elongate. Each whorl has 4 to 6 spiral rows of from 18 to 22 small rounded beads. The beads are slightly larger in the middle row. There are usually a number of fairly large, former varices. Color variable: all white or cream, or with reddish brown blotches. Very common in shallow water. *C. versicolor* C. B. Adams 1850, is this species.

The genetic *forma*, *algicola* C. B. Adams, 1845 (996), is characterized by each whorl having the middle spiral row of 9 to 12 beads fairly large and pointed. These large beads may be axially drawn out to form low ribs. Former varices are not often present. Common in the West Indies to Brazil. *C. litteratum* has its strongest row of spine-like beads just below the suture.

Cerithium guinaicum Philippi, 1849   997
Schwengel's Cerith

Florida, Bahamas and Caribbean.
West Africa.

1 to 1½ inches, somewhat stubby, with 9 or 10 large, rounded, elongate ribs or nodes per whorl which on the last whorl are limited to the upper ⅓. Color white to cream with

992   993   994   995   996

999

998

a few irregularly placed, burnt smudges of black or brown. Smooth, small rounded spiral threads (15 to 20 between the wavy sutures) cover the exterior of the shell. Minute white beads may be present. Aperture white, with a black-lined smudge on the inside of the outer wall. Apex usually eroded. Locally moderately common; on reef shallows clinging to rocks. *C. auricoma* Schwengel, 1940, is a synonym. *C. moenensis*

997   1000   1001

Gabb, 1881, is a different Pliocene species from Costa Rica. (see Rehder, 1940, *The Nautilus*, vol. 54, p. 72). Also a synonym is *C. stantoni* Dall, 1907.

Cerithium lutosum Menke, 1828   998
Dwarf Cerith

South half of Florida to Texas and the West Indies. Bermuda.

⅓ to ½ inch in length, not elongate. Apertural side of body whorl sometimes flat. 1 or 2 former varices on last whorl. 3 or 4 spiral rows of even-sized fine beads on the whorls of the spire. Color dark brown-black, but sometimes whitish with heavy speckling and mottlings and bands of reddish brown. Very common under rocks in warm water. Do not confuse with *Batillaria minima*. This was formerly known as *variabile* C. B. Adams, 1845 (not Deshayes, 1833), *ferrugineum* Say, 1832 (not Bruguière, 1792), *mutabile* C. B. Adams, 1845 (not Lamarck, 1804), *versicolor* C. B. Adams, 1850 (not Philippi, 1848). Other synonyms are: *sagrae* Orbigny, 1842; *bermudae*, *rissoide* and *thomasiae* all Sowerby in Reeve, 1865.

A curious stubby, smoothish, axially flamed form found in tide pools in east Florida and Bimini was given the name **(999)** *biminiense* Pilsbry and McGinty, 1949 (*The Nautilus*, vol. 63, p. 13).

Other Atlantic species:

1000 *Cerithium chara* Pilsbry, 1949. 7 miles off Hudson, northwest Florida, 3 fms. *The Nautilus*, vol. 63, p. 66.

1001 *Cerithium lymani* Pilsbry, 1949. Off Hudson, northwest Florida. *The Nautilus*, vol. 63, p. 66.

1002 *Cerithium caribbaeum* M. Smith, 1946. Colon, Atlantic side of the Canal Zone. *The Nautilus*, vol. 60,, p. 60.

Cerithium stercusmuscarum Valenciennes, 1833   1003
Pacific Fly-specked Cerith

Baja California to Peru.

¾ to 1 inch, broadly fusiform, heavy; bluish slate-gray with fine specklings of white. There is a spiral row of strong, dis-

tantly spaced, pointed tubercles on the center of the whorl midway between the sutures. Fine spiral threads are also present. One of the commonest snails in the Gulf of California. Found in sand near rock ledges near lagoons. *C. irroratum* Gould, 1851, and *exaggeratum* Pilsbry and Lowe, 1932, are synonyms.

Other Pacific species:

1004 *Cerithium* (*Ochetoclava* Woodring, 1928) *gemmatum* Hinds, 1844. Magdalena Bay, Baja California, the Gulf of California to Ecuador. Common offshore.

1005 *Cerithium maculosum* Kiener, 1841. Magdalena Bay, Baja California, and the Gulf of California. (Synonym: *alboliratum* Carpenter, 1857)

1006 *Cerithium menkei* Carpenter, 1857. Gulf of California to Ecuador, under rocks in the intertidal area. (Synonym: *interruptum* Menke, 1851)

1007 *Cerithium uncinatum* (Gmelin, 1791). Gulf of California to Ecuador, mostly offshore. (Synonym: *famelicum* C. B. Adams, 1852)

Genus *Liocerithium* Tryon, 1887

About ½ inch, slender, early whorls cancellate, the later ones having only well-incised spiral grooves. Interior of aperture with spiral lirae. Type: *judithae* Keen, 1971.

Liocerithium judithae Keen, 1971   1008
Judith's Cerith

Magdalena Bay to Mazatlan, Mexico.

½ to ¾ inch, with flat, spiral cords, the upper 2 or 3 being whitish with squarish black dots. Shell olive-green to gray. Inside of outer lip with strong spiral lirations. Common under intertidal rocks. Synonyms are *incisum* Sowerby, 1855 (not Hombron and Jacquinot, 1854) and *sculptum* of authors, not Sowerby, 1855.

Genus *Fastigiella* Reeve, 1848

Shell cerithid in shape, white, spirally lirate, with a funnel-shaped umbilicus. Although placed in the Cerithiidae, it may prove to be in a totally different family, possibly the Pyramidellidae. Type: *carinata* Reeve, 1848.

Fastigiella carinata Reeve, 1848   1009
Carinate False Cerith

Bahamas and northern Cuba.

20 to 40 mm., alabaster-white, chalky, with 11 whorls. 3 narrow, strong, raised, squarish spiral cords on the spire

1009

whorls, 7 on the last, with 2 or 3 intermediate threads. Siphonal fasciole conspicuous; umbilicus fairly deep. Upper angle of the aperture narrowly channeled. Rare; shallow water. Soft parts unknown. *F. poulsenii* Mörch, 1877, is probably a synonym. Illustrated specimen collected by Sue Abbott on Andros Island.

Subfamily BITTIINAE Cossmann, 1906

Genus *Bittium* Gray, 1847

Shell small, very slender, spire high and body whorl small. Whorls varicose. Nucleus of about 3 glassy, smoothish whorls. Aperture ovate, the anterior canal broad and stout. Sculpture of axial ribs, overridden by spiral threads that are swollen into beads on the ribs. Type: *reticulatum* (da Costa, 1778). In the subgenus *Bittium,* the nuclear whorls may have 2 spiral lirations. For a review of the Pacific coast species, see P. Bartsch, 1911, *Proc. U.S. Nat. Mus.,* vol. 40, no. 1826.

Subgenus *Stylidium* Dall, 1907

Nuclear whorls smooth. Spiral sculpture predominating over the axial. Type of this subgenus is *eschrichtii* (Middendorff, 1849).

Bittium eschrichtii (Middendorff, 1849)   1010
Giant Pacific Coast Bittium

Alaska to Crescent City, California.

½ to ¾ inch in length, dirty whitish gray in color with an undertone of reddish brown. About a dozen whorls. With wide, flat-topped, raised spiral cords between which are depressed, squarish, spiral furrows ½ as wide as the cords. 4 to 5 cords between sutures. Common below low water. *B. icelum* Bartsch, 1911, is a synonym. The subspecies *montereyense* Bartsch, 1911 (**1011**), (Crescent City south to Baja California) is glossy, whitish with brown maculations and is proportionately shorter.

Subgenus *Semibittium* Cossmann, 1896

Nuclear whorls smooth. Spiral and axial sculpture equally strong. Type of this subgenus is *cancellatum* Lamarck, 1804.

Bittium quadrifilatum Carpenter, 1864   1012
Four-threaded Bittium

Monterey, California, to Baja California.

⅜ inch in length, similar to *attenuatum,* but earliest whorls with about a dozen smooth axial ribs which, however, in sub-

sequent whorls become beaded as 4 or 5 small spiral threads cross them. The sculpturing may become faint at the very last ⅓ of the last whorl. Color reddish brown to gray. A very common littoral species.

Subgenus *Lirobittium* Bartsch, 1911

Nuclear whorls with two spiral lirations. Postnuclear whorls without varices. Type of this subgenus is (*catalinense* Bartsch, 1907) *attenuatum* Carpenter, 1864.

Bittium attenuatum Carpenter, 1864   1013
Slender Bittium

Forrester Island, Alaska, to Baja California.

⅓ inch in length, slender, yellowish brown to dark-brown. Sculpture variable. Nuclear whorls with 2 smooth spiral cords. Early whorls have 4 or 5 spiral rows of small beads, sometimes arranged axially. In the last whorl, the cords gradually become smooth and flat-topped and resemble those of *eschrichtii.* Common just offshore to 35 fathoms. *B. esuriens* Carpenter, 1864; *boreale* Bartsch, 1911; *multifilosum* Bartsch, 1907; *latifilosum* Bartsch, 1911; and possibly *subplanatum* Bartsch, 1911; *catalinense* Bartsch, 1907; and *inornatum* Bartsch, 1911, are synonyms.

Bittium interfossum (Carpenter, 1864)   1014
White Cancellate Bittium

Monterey, California, to Baja California.

¼ inch in length, pure-white; whorls in spire with 2 rows of sharp beads connected by small axial and spiral threads or small cords. Base of shell with 3 very strong, rounded, smooth spiral cords. Moderately common under rocks at low tide. *B. fortior* (Carpenter, 1864) is a synonym.

Other species:

1015   *Bittium (Semibittium) armillatum* (Carpenter, 1864). Santa Barbara, California, to Baja California. Also Pleistocene. (Synonyms: *ornatissimum* Bartsch, 1911; *purpureum* (Carpenter, 1864).) 9 mm.

1016   *Bittium (Semibittium) rugatum* (Carpenter, 1864). San Pedro to Baja California. (Synonym: *asperum* Carpenter, 1864, not Gabb, 1861.) 12 mm.

1017   *Bittium (Lirobittium) munitum* (Carpenter, 1864). Alaska to California. (Synonym: *munitoide* Bartsch, 1911.) 8 mm.

1018   *Bittium (Stylidium) paganicum* (Dall, 1919). Off Monterey, California, in 356 fms.

1019   *Bittium vancouverense* Dall and Bartsch, 1910. Drier Bay, Alaska, to Vancouver Island, British Columbia. 7.8 mm.

1020   *Bittium challisae* Bartsch, 1917. North Queen Charlotte Island to San Juan Island, British Columbia.

1010   1011   1012   1013   1014   1015   1016   1017

1019

1035   1037   1036

1021 *Bittium sanjuanense* Bartsch, 1917. San Juan Island, Strait of Georgia, British Columbia.

1022 *Bittium (Semibittium) serra* Bartsch, 1917. San Diego to Baja California, 75 fms.

1023 *Bittium asperum* Gabb, 1861, forma *lomaense* Bartsch, 1911. Catalina Island to San Diego, California, 50 to 75 fms.

1024 *Bittium (Lirobittium) larum* Bartsch, 1911. Common; 25 to 50 fms., Santa Monica, California, to Baja California. 10 mm.

1025 *Bittium oldroydae* Bartsch, 1911. Destruction Island, Washington, to Baja California. 13.3 mm.

1026 *Bittium fetellum* Bartsch, 1911. Off Catalina Island, California. 16 fms.

1027 *Bittium tumidum* Bartsch, 1907. Monterey to San Pedro, California.

1028 *Bittium (Bittium) johnstonae* Bartsch, 1911. Laguna Beach, California, to Baja California. 8 mm.

1029 *Bittium mexicanum* Bartsch, 1911. Gulf of California. 6 mm.

1030 *Bittium bartolomensis* Bartsch, 1917. San Bartolome Bay, Baja California. 6.6 mm.

1031 *Bittium santamariensis* Bartsch, 1917. Santa Maria Bay, Baja California. 4.9 mm.

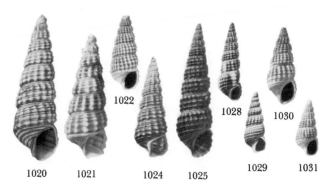

1020   1021   1024   1025   1022   1028   1030   1029   1031

1032 *Bittium nicholsi* Bartsch, 1911. Gulf of California. 6 mm.

1033 *Bittium cerralvoense* Bartsch, 1911. Gulf of California. 8 mm.

1034 *Bittium arenaense* Hertlein and Strong, 1951. Arena Bank, Gulf of California.

Subfamily DIASTOMINAE Cossmann, 1895

Genus *Diastoma* Deshayes, 1850

Shell small, turreted, whorls with numerous axial riblets and with occasional intermediate varices. Aperture narrow above. Type: *costellatum* (Lamarck, 1804).

Diastoma alternatum (Say, 1822)   1035
Alternate Bittium

Gulf of St. Lawrence to Virginia.

Adults very small, ⅛ to ¼ inch in length, light- to dark-brown in color, sometimes translucent or with specklings. Suture impressed, whorls rounded. Sculpture on top whorls either cancellate or with 4 or 5 spiral rows of beads, or occasionally with axial, nodulated ribs. Base with small spiral cords. Outer lip flaring, thin and sharp. Columella short, twisted at the base and stained brown. Very abundant from tidal flats to 20 fathoms. *B. nigrum* Totten is a synonym.

The giant, ecologic form, *virginicum* Henderson and Bartsch, 1914 (**1036**), from Chincoteague, Virginia, is similar, but very elongate, with more whorls, a much more flaring and basally projecting lip, and with a large, whitish, former varix on the body whorl.

Diastoma varium (Pfeiffer, 1840)   1037
Variable Bittium

Maryland to Florida, Texas and to Brazil.

Adults similar to *alternatum*, but smaller (⅛ inch), nearly always with a former, thickened varix. The aperture is proportionately smaller and the base of the apertural lip is squarish instead of rounded. The last ⅓ of the body whorl is generally destitute of sculpturing. Common in eel-grass just below low tide.

Diastoma fastigiatum (Carpenter, 1864)   1038
Californian Bittium

Southern California to the Gulf of California.

5 to 9 mm., elongate-conic, brownish to pinkish white, the smooth 3 or 4 nuclear whorls usually eroded. Suture well-impressed. 4 to 7 Spiral threads, crossing 16 to 24 axial riblets per whorl, sometimes slightly beaded at their intersections. Base of last whorl with 6 to 8 spiral threads and no axial riblets. Columella slightly angled or truncated at the base. Uncommon; shallow water. Synonyms: *chrysalloideum* Bartsch, 1911; *oldroydae* Bartsch, 1911; *stearnsi* Bartsch, 1911.

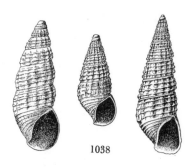

1038

Genus *Finella* A. Adams, 1869

Shell small, slender. Nucleus slender, of 3 to 4 glassy, smooth whorls. Aperture subcircular. Lower part of the outer lip extended and flaring. Umbilicus very narrow and very small. With obscure, narrow, curved axial ribs and prominent spiral threads. *Alabina* Dall, 1902, is a synonym. This genus is put in the family Diastomidae by some workers. The type is *pupoides* A. Adams, 1860, from Korea. *Obtortio* Hedley, 1899, and *Eufenella* Kuroda and Habe, 1952 (replacement for *Fenella* A. Adams, 1864, non Westwood, 1840), are synonyms.

Finella dubia (Orbigny, 1842)   **1039**
Dubious Alabine

North Carolina to Florida and to Brazil. Bermuda.

3 mm., white (rarely brown-stained), elongate, 9 or 10 whorls, the first 3 smooth, the next 3 with 1 or 2 carinations, the remaining rounded and with numerous delicate, curved axial riblets on the upper half of the whorls. Spiral sculpture of numerous, irregular threads. Base with spiral threads. Smooth-humped varices rarely present in spire. *Bittium cerithioides* Dall, 1889; and *yucatecanum* Dall, 1881, appear to be the same. Common; dredged offshore to 40 fathoms.

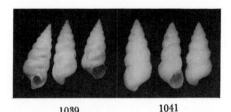

1039   1041

Finella tenuisculpta (Carpenter, 1864)   **1040**
Sculptured Alabine

San Pedro, California, to Baja California.

¼ inch in length, slender, 8 or 9 whorls, ashen-gray with a light-brown undertone. Outer lip thin; umbilicus small. Spiral sculpture of 4 to 6 weak cords or threads. Axial sculpture of weak, obsolete or sometimes strong, very tiny, rounded riblets. *F. t. diegensis* Bartsch, 1911, is a strongly sculptured form of this species. *A. phalacra* Bartsch, 1911, is also a synonym.

Subgenus *Caloosalaba* Olsson and Harbison, 1953

The type of this subgenus is *adamsi* (Dall, 1889).

Finella adamsi (Dall, 1889)   **1041**
Adams' Alabine

Florida and the Caribbean. Bermuda.

4 mm., white, elongate, with coarse, somewhat beaded sculpture caused by axial riblets and spiral cords. 10 whorls. Protoconch small, 3 smooth whorls. Axial sculpture becomes weak on the last whorls. Moderately common; 8 to 25 feet. This is the type of *Caloosalaba,* and was formerly placed in *Alabina.*

Other Atlantic species:

1042 *Finella longinqua* (Haas, 1949). 1,700 fms., off Bermuda. *Bull. Inst. Catalana Hist. Nat.,* vol. 37, p. 71.

1043 *Finella nigricans* (Bartsch and Rehder, 1939). Old Providence Island, Caribbean. *Smithsonian Misc. Coll.,* vol. 98, p. 8.

Other Pacific species:

1044 *Finella californica* (Dall and Bartsch, 1901). San Pedro, California. Pleistocene.

1045 *Finella barbarensis, hamlini, phanea,* and *io,* all southern California, all Bartsch, 1911, *Proc. U. S. Nat. Mus.,* vol. 39. Some Pleistocene.

1046 *Finella(?) calena* Dall, 1919. Off San Luis Obispo Bay, California, in 252 fms.

Subfamily LITIOPINAE H. and A. Adams, 1854

Genus *Litiopa* Rang, 1829

Shells small, fusiform, smooth, thin-shelled, brown, with a truncated lower end to the columella. Type: *melanostoma* Rang, 1829. Synonym: *Bombyxinus* Lesson, 1834.

Litiopa melanostoma Rang, 1829   **1047**
Brown Sargassum Snail

Pelagic in floating sargassum weed.
Both southern coasts of United States. Bermuda. Brazil.

³⁄₁₆ to ¼ inch in length, fragile, light-brown; moderately elongate, with 7 whorls, the last being quite large. Nuclear whorls extremely small. Surface glossy, smooth, except for numerous, microscopic, incised spiral lines. Characterized by the strong ridge just inside the aperture on the columella. Often washed ashore with floating sargassum weed, upon which it lives and lays its eggs, and frequently dredged in a dead condition at any depth. This is *L. bombix* Kiener, 1833, *L. bombyx* "Rang," and *maculata* Rang, 1829.

1047

Subfamily CERITHIOPSINAE H. and A. Adams, 1854

Genus *Cerithiopsis* Forbes and Hanley, 1849

Shells small, brown, elongate, with spiral rows of small, rounded beads. Siphonal canal short, slightly twisted. Nuclear whorls smooth. Operculum chitinous, thin. Type: *tubercularis* Montagu, 1803. The genus name is neuter.

Cerithiopsis greeni (C. B. Adams, 1839)   **1048**
Green's Miniature Cerith

Cape Cod to both sides of Florida. Bermuda. Brazil.

1048

1050

⅛ inch in length, elongate, slightly fusiform in shape, glossy-brown in color. 9 whorls, the first 3 embryonic, translucent-brown and smooth, the remainder with 2 or 3 spiral rows of large, glassy beads connected by weak spiral and axial threads. Columella arched in young specimens, but straight and continuous with the short siphonal canal in adults. Lip in adults smoothish, slightly flaring. *C. virginicum* Henderson and Bartsch, 1914, and *C. vanhyningi* Bartsch are possibly variations of this species. Common in shallow water.

Cerithiopsis crystallinum Dall, 1881　　　**1049**
Crystal Miniature Cerith

Both sides of Florida and the West Indies.

10 to 16 mm., elongate, snow-white, with about 20 whorls (with the nuclear one usually lost). Suture well-indented. Whorls slightly convex and bearing 3 or 4 spiral rows of distinct, raised, somewhat pointed beads. Base slightly concave and with spiral threads. Operculum concave, red-brown, circular, paucispiral, and chitinous. Common. 6 to 450 fathoms.

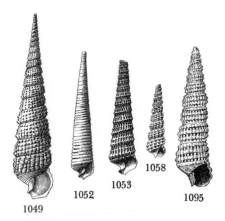

1049　1052　1053　1058　1095

Subgenus *Laskeya* Iredale, 1918

Type of this subgenus is *costulatum* (Möller, 1842).

Cerithiopsis emersoni (C. B. Adams, 1838)　　　**1050**
Awl Miniature Cerith

Massachusetts to the West Indies, and to Brazil.

½ to ¾ inch in length, rather strong, slender and with about 14 whorls. Sides of whorls flattish, with 3 rows of distinct, raised, roundish beads (about 28 per row on the last whorl). There may be faint axial riblets connecting the beads. The middle row of beads may be reduced to a mere thread in specimens from southern localities. Base concave and with fine axial growth lines. Color chocolate-brown, with the beads a lighter shade. Some shells become eroded and colored an ash-gray or chalky-brown. Formerly and erroneously thought to be *subulatum* (Montagu, 1808). Common; from 1 to 33 fathoms.

Cerithiopsis carpenteri Bartsch, 1911　　　**1051**
Carpenter's Miniature Cerith

Crescent City, California, to Ensenada, Mexico.

¼ to ⅓ inch in length, dark chocolate-brown with whitish beads. Whorls in spire with 3 spiral rows of evenly sized, glassy, rounded beads. Base of shell with 2 large, smoothish, spiral cords. *C. grippi* Bartsch, 1917, and *C. pedroanum* Bartsch, 1907, are possibly dwarf forms of this species whose beaded sculpture is more variable than is generally suspected. Common; shallow water.

1096　1051　1087　1097　1102

Other Atlantic species:

1052 *Cerithiopsis sigsbeanum* Dall, 1881. Florida Strait, 220 fms. Gulf of Mexico, 200 to 220 fms. Common.

1053 *Cerithiopsis martensii* Dall, 1881. Gulf of Mexico, 229 fms.

1054 *Cerithiopsis tubercularis floridana* Dall, 1892. Key West, Florida.

1055 *Cerithiopsis pulchellum* Jeffreys, 1858. Off Georgia (non C. B. Adams, 1850).

1056 *Cerithiopsis georgianum* Dall, 1927. Off Georgia, 440 fms.

1057 *Cerithiopsis (Metaxia* Monterosato, 1884) *metaxae* Della Chiaje, 1829. North Carolina to the West Indies, 2 to 200 fms.; variety *taeniolata* Dall, 1889, off North Carolina, 15 to 52 fms.

1058 *Cerithiopsis (Metaxia) abruptum* Watson, 1880. Off North Carolina to the West Indies, 15 to 100 fms.

1059 *Cerithiopsis (Laskeya) costulatum* Möller, 1842. Greenland to Eastport, Maine.

1060 *Cerithiopsis (Onchodia* Dall, 1924) *benthicum, argenteum* (**1061**), *althea* (**1062**), *apicinum* (**1063**), *docata* (**1064**), *decorum* (**1065**), *elima* (**1066**), *eliza* (**1067**), *elsa* (**1068**), *honora* (**1069**), *leipha* (**1070**), *merida* (**1071**), *petala* (**1072**), and *serina* (**1073**) all (Dall, 1927), all off Fernandina, Florida, 294 and 440 fms.

1074 *Cerithiopsis (Stilus* Jeffreys, 1885) *vitreum* (Dall, 1927). Off Georgia, 440 fms; off Fernandina, Florida, 294 fms.

1075 *Cerithiopsis rugulosum* (C. B. Adams, 1850). Caribbean. Gulf of Mexico, 20 to 70 fms. Common. Bermuda. (Synonym: *bermudensis* Verrill and Bush, 1900.)

1076 *Cerithiopsis latum* (C. B. Adams, 1850). Greater Antilles and Central America.

1077 *Cerithiopsis pupa* Dall and Stimpson, 1901. Greater Antilles and western Caribbean. Bermuda.

1078 *Cerithiopsis bicolor* (C. B. Adams, 1845). Caribbean.

1079 *Cerithiopsis fusiforme* (C. B. Adams, 1845). Gulf of Mexico; west Florida, shore to 90 fms. Common.

1080 *Cerithiopsis movilla, ara* (**1081**), *pesa* (**1082**), *vicola* (**1083**) and *io* (**1084**) all Dall and Bartsch, 1911. Bermuda. Proc. U.S.N.M., vol. 40.

1085 *Cerithiopsis iontha, cynthia* and *hero* all Bartsch, 1911. Bermuda. *Proc. U.S. Nat. Mus.*, vol. 41, pp. 303–304.

1086 *Cerithiopsis gemmulosa* (C. B. Adams, 1847). West Indies to Brazil.

Other Pacific species:

1087 *Cerithiopsis charlottensis* Bartsch, 1917. Queen Charlotte Sound, British Columbia, to Puget Sound, Washington.

1088 *Cerithiopsis paramoea* Bartsch, 1911. Neah Bay, Washington.

1089 *Cerithiopsis stejnegeri* Dall, 1884. Aleutian Islands, Alaska. Variety *dina* Bartsch, 1911. Sitka, Alaska.

1090 *Cerithiopsis* (*Cerithiopsidella* Bartsch, 1911) *columnum* Carpenter, 1864. Vancouver, British Columbia, to Monterey Bay, California.

1091 *Cerithiopsis* (*C.*) *fraseri* Bartsch, 1921. Vancouver Island, British Columbia.

1092 *Cerithiopsis* (*C.*) *onealense* Bartsch, 1921. Off O'Neal Island, Puget Sound, Washington.

1093 *Cerithiopsis* (*C.*) *signa* Bartsch, 1921. Off O'Neal Island, Puget Sound, Washington.

1094 *Cerithiopsis* (*C.*) *willetti* Bartsch, 1921. Forrester Island, Alaska, to Puget Sound, Washington. Above 4 species in *Proc. Biol. Soc. Wash.*, vol. 34, pp. 34–36.

1095 *Cerithiopsis* (*C.*) *stephensae* Bartsch, 1909. Port Frederick, Alaska, to Puget Sound, Washington (as *stephensi*). 9 mm.

1096 *Cerithiopsis* (*C.*) *truncatum* Dall, 1886. Unalaska, Alaska, to Puget Sound, Washington. 3.1 mm.

1097 *Cerithiopsis berryi* (3.4 mm.), *cesta* (**1098**), *arnoldi* (**1099**), *diomedeae* (**1100**), *montereyensis* (**1101**) all California, all Bartsch, 1911, Proc. U.S. Nat. Mus., vol. 40.

1102 *Cerithiopsis bakeri* Bartsch, 1917. South Coronado Island, Baja California.

1103 *Cerithiopsis* (*Cerithiopsida* Bartsch, 1911) *diegensis*, *rowelli* (**1104**), *gloriosum* (**1105**), *antemundum* (**1106**) all southern California, all Bartsch, 1911, *Proc. U.S. Nat. Mus.*, vol. 40.

1107 *Cerithiopsis* (*Cerithiopsidella*) *cosmia* Bartsch, 1907. Monterey, California, to Baja California.

1108 *Cerithiopsis* (*C.*) *antefilosum* and *alcima* (**1109**) both southern California, both Bartsch, 1911.

1110 *Cerithiopsis ingens* and *tumidum* (**1111**) both Monterey, California, both Bartsch, 1907, *Proc. U.S. Nat. Mus.*, vol. 33.

1112 *Cerithiopsis oxys*, *halia* (**1113**), *aurea* (**1114**) all Baja California, all Bartsch, 1911.

1115 *Cerithiopsis bristolae*; *kinoi* (**1116**); *subgloriosa* (**1117**), *porteri* (**1118**) all Baker, Hanna and Strong, 1938. All Gulf of California.

Genus *Cerithiella* Verrill, 1882

Shells small, elongate-turreted, with cancellate sculpturing, a concave smoothish base, and a twisted, elongate, narrow siphonal canal. Type: *metula* (Lovén, 1846). Synonyms: *Lovenella* G. O. Sars, 1878, non Hicks, 1869; *Newtonia* Cossmann, 1892; *Newtoniella* Cossmann, 1893; *Cerithiolinum* Locard, 1903.

1119 *Cerithiella whiteavesii* Verrill, 1880. Gulf of St. Lawrence, 110 to 200 fms. (Synonym: *costulatus* Whiteaves, 1901, non Möller, 1842.)

1119

1120 *Cerithiella producta* Dall, 1927. Off Fernandina, Florida, 294 fms.

Genus *Alaba* H. and A. Adams, 1853

Shells small, thin but strong, elongate-turreted, with somewhat convex whorls. Spiral sculpture of weak incised lines. Type: *incerta* (Orbigny, 1842).

Alaba incerta (Orbigny, 1842)   **1121**
Varicose Alaba

Southeast Florida and the West Indies; Bermuda. Brazil.

5 to 9 mm., elongate, thin-shelled, translucent, with 10 to 12 rounded whorls. After the 5th whorl there may or may not be 2 or 3 large, swollen, rounded varices per whorl. 2 nuclear whorls tan or white, next 2 whorls black-purple and with 2 dozen axial riblets per whorl. Later whorls flattish, smoothish, except for weak, incised lines just above the suture. Color tan with white and/or brown dots. Operculum corneous, thin, paucispiral. *Rissoa tervaricosa* C. B. Adams, 1845, and *melanura* C. B. Adams, 1850, are synonyms. Common; in sand, *Thalassia* weeds and rubble, 1 to 2 fathoms in oceanic water.

Other species:

1122 *Alaba jeanettae* Bartsch, 1910. San Pedro, California, to the Gulf of California. 5 mm.

1123 *Alaba catalinensis* Bartsch, 1920. Catalina Island, California.

1124 *Alaba supralirata* Carpenter, 1856. Gulf of California to Panama. 7 mm.

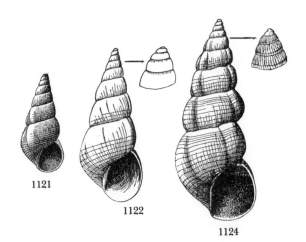

1121

1122

1124

1125 *Alaba serrana* A. G. Smith and M. Gordon, 1948. Carmel Bay, California, 25 fms. 5.2 mm. *Proc. Calif. Acad. Sci.*, series 4, vol. 26, p. 225.

Genus *Seila* A. Adams, 1861

Shell small, very slender, whorls flat-sided, nucleus glassy-smooth and of about 3 whorls. Small, short siphonal canal. Sculpture of strong spiral cords between which lie microscopic axial threads. The type is *dextroversus* A. Adams and Reeve, from Japan.

Seila adamsi (H. C. Lea, 1845)   **1126**
Adams' Miniature Cerith

Massachusetts to Florida, Texas and to Brazil. Bermuda.

¼ to ½ inch in length, resembling a miniature *Terebra*, with about a dozen whorls. Long, slender, flat-sided, dark-brown to light orange-brown in color, and characterized by 3

1126

strong, squarish, spiral cords on each whorl (4 on the last whorl). Occasionally with minute axial threads showing between the spiral cords. Base of shell smoothish, concave. Outer lip fragile, wavy and sharp. Suture indistinct. This is *S. terebralis* (C. B. Adams, 1840) (non Lamarck, 1804) and *terebellum* C. B. Adams, 1847 (non Brown, 1831). Common from shore to 40 fathoms.

Seila montereyensis Bartsch, 1907   1127
Monterey Miniature Cerith

Monterey, California, to the Gulf of California.

⅜ to ½ inch in length, yellowish to reddish brown. Whorls and spire flat-sided. Whorls in spire with 3 raised, flat-topped, evenly spaced, smooth cords between which are numerous, microscopic, axial threads. Last whorl with 5 cords. Base smoothish, concave. Common from low tide to 35 fathoms.

Other Atlantic species:

1128 *Seila subalbida* Dall, 1927. Off Fernandina, Florida, 294 fms.

Other Pacific species:

1129 *Seila diadema* Bartsch, 1907. Montery to San Diego, California.

1130 *Seila assimilata* (C. B. Adams, 1852). Gulf of California to Panama. (Synonyms: *Cerithiopsis kanoni* and *moreleti* De Folin, 1867.)

Family TRIPHORIDAE Gray, 1847

Genus *Triphora* Blainville, 1828

Shell left-handed (sinistral), very small, and slender. Aperture subcircular. Siphonal canal short, curved backward, slightly emarginate, upper part almost or completely closed. Posterior canal very slightly developed. Sculpture of spiral rows of neat beads, often joined by axial threads. Type: *gemmata* Blainville, 1828. Synonyms: *Tristoma* Menke, 1830; *Biforina* Bucquoy, Dautzenberg and Dollfus, 1884. The valid genus *Triforis* Deshayes, 1834, is limited to the Eocene of Europe and represented by the type and only species, *plicatus* Deshayes, 1834.

Triphora nigrocincta (C. B. Adams, 1839)   1131
Black-lined Trifora

Massachusetts to Florida, Texas to Brazil; Bermuda.

⅛ to ¼ inch in length, left-handed, with 10 to 12 slightly convex whorls; dark chestnut-brown with 3 spiral rows of prominent, grayish, glossy beads. Darker band of black-brown

1131   1132   1133   1135

is just below the suture. Aperture and columella brown. A common species found on seaweed at low tide. Sometimes considered a subspecies of *perversa* from Europe.

Triphora turristhomae (Holten, 1802)   1132
Thomas' Trifora

North Carolina to Brazil; Bermuda.

6 to 7 mm., sinistral, elongate-fusiform, with 15 or 16 whorls, the last enwrapping 2 former siphonal canals. Aperture almost round. Nuclear whorls brown, 3 in number, with numerous axial microscopic fimbriations and 2 fine spiral lines. Postnuclear whorls with 2 spiral rows of large, glossy, bulbous, closely packed, rounded beads. The lower row crossed by a prominent spiral band of yellow-brown. Moderately common; 1 to 5 fathoms in sand. Synonyms: *mirabile* (C. B. Adams, 1850) and *bermudensis* Bartsch, 1911.

Triphora decorata (C. B. Adams, 1850)   1133
Mottled Trifora

Southeast Florida, the West Indies to Brazil and Bermuda.

7 to 9 mm., in length, left-handed, with about 20 flat-sided whorls which bear 3 spiral rows of large beads (28 per row per whorl). Color of shell cream to gray with large, irregular maculations of reddish brown. Moderately common from 1 to 57 fathoms. *T. olivacea* Dall, 1889, is a synonym.

T. ornata (Deshayes, 1832) from the same area is very similar, but half as large, the spire slightly concave instead of being flat, and the beads more numerous and smaller.

Triphora melanura (C. B. Adams, 1850)   1134
White Atlantic Trifora

North Carolina to both sides of Florida and to Brazil.

5 to 6 mm., sinistral, pure-white, except for the 4 brownish nuclear whorls which bear 2 spiral rows of microscopic beads and fine axial threads. 11 postnuclear whorls white, with 3

1134

crowded rows of small, roundish beads which are joined by weak spiral and axial cords. Sides of spire and whorls flattish to slightly convex. Aperture round. Moderately common; 1 to 50 fathoms.

Triphora pulchella (C. B. Adams, 1850)   **1135**
Beautiful Trifora

South Florida and the West Indies to Brazil.

6 mm. in length, left-handed, spire slightly convex; 15 whorls slightly convex, and with 3 spiral rows of beads which are joined axially and spirally by small, low, smooth threads. Suture well-indented. Upper third of whorl, including beads, colored light-brown, lower two thirds white. Uncommon, in shallow water down to 56 fathoms.

Triphora pedroana (Bartsch, 1907)   **1136**
San Pedro Trifora

Redondo Beach, California, to Baja California.

5 mm. or less in length, slightly fusiform with very slightly convex sides to the spire. Suture almost impossible to see. Color glossy yellow-brown with 2 rows of glassy, whitish, rounded beads. A third much weaker row of beads, or an additional spiral thread, may appear in the last 2 or 3 whorls. Axial threads connecting the beads are weak and form small pits. Fairly common under stones along the low-tide zone.

Other Atlantic species:

1137 *Triphora lilacina* (Dall, 1889). Florida Keys, 6 fms. West Florida, 30 to 100 fms.

1138 *Triphora longissima* (Dall, 1881). North Carolina to the West Indies, 175 to 450 fms. West Florida, 40 to 220 fms. 26 mm.

1139 *Triphora triserialis* (Dall, 1881). North Carolina to the West Indies, 125 to 154 fms. Variety *asper* Jeffreys, 1885. Off Georgia, 440 fms.; Florida Strait, 125 to 731 fms.; variety *intermedia* (**1139a**) Dall, 1881. Florida Strait. Off West Florida, 30 to 50 fms. 8 to 11 mm.

1137   1139

1140 *Triphora colon* (Dall, 1881). Florida Strait, 450 fms. 12 mm.

1141 *Triphora bigemma* (Watson, 1880). Off Georgia, 294 fms. West Indies, 390 fms. Variety *hircus* (Dall, 1881) (**1142**). Gulf of Mexico, 640 fms. 12 mm.

1143 *Triphora abrupta* (Dall, 1881). Gulf of Mexico, 640 fms.

1144 *Triphora torticula* (Dall, 1881). Gulf of Mexico, 640 fms. 10.5 mm.

1145 *Triphora pyrrha* Henderson and Bartsch, 1914. Chincoteague Island, Virginia.

1146 *Triphora cylindrella* (Dall, 1881). Gulf of Mexico, 640 fms. 6.5 mm.

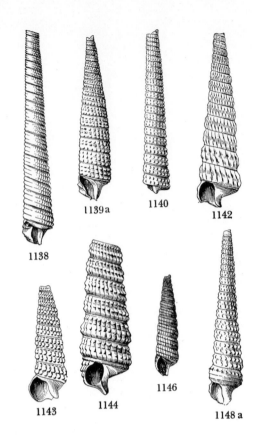

1138   1139a   1140   1142
1143   1144   1146   1148 a

1147 *Triphora rushii* (Dall, 1889). Florida Strait, 200 fms.

1148 *Triphora* (*Strobiligera* Dall, 1924) *inflata* (Watson, 1880). Off Fernandina, Florida, 294 fms; Gulf of Mexico, 640 fms. Variety *ibex* (**1148a**) (Dall, 1881). Cuba.

1149 *Triphora* (*Strobiligera*) *pompona, dinea* (**1150**), *gaesona* (**1151**), *enopla* (**1152**), *meteora* (**1153**), *compsa* (**1154**), and *sentoma* (**1155**), all (Dall, 1927). Off Georgia, 440 fms., and/or Fernandina, Florida, 294 fms.

1154

1156 *Triphora* (*Biforina* Bucquoy, Dautzenberg and Dollfus, 1889) *indigena, caracca* (**1157**), *georgiana* (**1158**), all Dall, 1927. Off Georgia, 440 fms.

Other Pacific species:

1159 *Triphora carpenteri* (Bartsch, 1907). Neah Bay, Washington. 7 mm.

1160 *Triphora montereyensis* (Monterey); *callipyrga* (**1161**) (southern California) 5.2 mm.; *catalinensis* (**1162**) (Catalina and Laguna Beach) 5 mm.; *stearnsi* (**1163**) (San Diego to the Gulf of

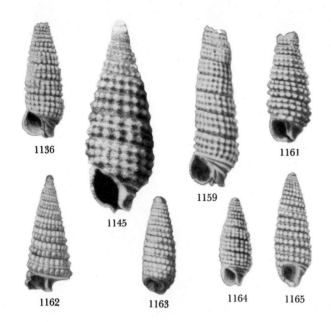

1136 1145 1159 1161 1162 1163 1164 1165

California) 4 mm.; *peninsularis* (**1164**) (San Diego to Baja California) 4 mm. (all Bartsch, 1907), *Proc. U.S. Nat. Mus.*, vol. 33, no. 1569.

1165 *Triphora hemphilli* (Bartsch, 1907). Point Abreojos, Baja California. 5 mm.

1166 *Triphora chamberlini;* (**1167**) *contrerasi;* (**1168**) *escondidensis;* (**1169**) *evermanni;* (**1170**) *hannai;* (**1171**) *johnstoni;* (**1172**) *oweni;* (**1173**) *pazensis;* (**1174**) *slevini;* (**1175**) *vanduzeei* all Fred Baker, 1926, *Proc. Calif. Acad. Sci.*, ser. 4, vol. 15. All Gulf of California.

1176 *Triphora cookeana* Baker and Spicer, 1935. Gulf of California.

1177 *Triphora stephensi* Baker and Spicer, 1935. Gulf of California.

Suborder *Ptenoglossa* Gray, 1853

Family JANTHINIDAE Leach, 1823

Genus *Janthina* Röding, 1798

Pelagic snails with fragile, purple, trochoid-shaped shells. Produce a bubble float. Type: *janthina* (Linné, 1758). *Ianthina* is an emendation.

Janthina janthina (Linné, 1758)   **1178**
Common Purple Sea-snail   **Color Plate 3**

Pelagic in warm waters; both coasts of the United States. Brazil. Bermuda.

1 to 1½ inches in diameter. Whorls slightly angular. Two-toned, with purplish white above and deep purplish violet below. Outer lip very slightly sinuate. Common after certain easterly blows along the southeastern United States, especially from April to May. This is *J. fragilis* Lamarck, 1801, *striulata* and *contorta* Carpenter, 1857, and *carpenteri* Mörch, 1860.

Subgenus *Violetta* Iredale, 1929

Type of this subgenus is *globosa* Swainson, 1822.

Janthina globosa Swainson, 1822   **1179**
Elongate Janthina   **Color Plate 3**

Cast ashore along both southern coasts of United States. Bermuda.

¾ to 1 inch, glossy violet to purplish; whorls moderately rounded. Aperture elongate, with the outer lip slightly sinuate, and the columella extending straight down to form a point with the base of the outer lip (not rounded as in *pallida*). Tentacles black with pale tips. Moderately common. *J. prolongata* Blainville, 1822, is a synonym (by a few months only). Early books erroneously used the name *globosa* for the pale, globose *pallida*.

Janthina pallida (Thompson, 1840)   **1180**
Pallid Janthina   **Color Plate 3**

Palagic; worldwide, warm seas.

¾ to 1 inch, globose; base of aperture rounded and without the slight projection seen in *globosa*. Color light whitish violet and not very glossy. Sinus keel, or scar, can be seen in whorls of the spire. Tentacles pale throughout. Up to 400 elongate egg capsules are attached to the foot. Moderately common; appears in the spring months in eastern Florida. Formerly, and erroneously, identified by American workers as *globosa* Swainson.

Subgenus *Jodina* Mörch, 1860

Type of this subgenus is *exigua* Lamarck, 1816.

Janthina exigua Lamarck, 1816   **1181**
Dwarf Purple Sea-snail   **Color Plate 3**

Cast ashore in most warm seas.

¼ inch in length. Whorls slightly flattened from above. Outer lip with a prominent notch. Light-violet, banded at the suture. Fairly common. *J. bifida* Nuttall is probably this species.

Genus *Recluzia* Petit, 1853

Like *Janthina*, but the shell is brown, bulimoid and the animal yellow. Type: *rollandiana* Petit, 1853.

Recluzia rollandiana Petit, 1853   **1182**
Recluzia Snail   **Color Plate 3**

Florida; Texas; Brazil.
Tropical Eastern Pacific.

½ to 1 inch (resembling a fresh-water *Lymnaea*), high-spired, thin-shelled but strong. Whorls globose, chocolate-brown. Minutely umbilicate. Body and tentacles sulfur-yellow. Float of brown bubbles is 2 or 3 inches long. Brown egg cap-

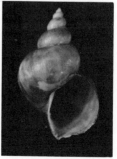

1182

sules numerous on the underside of the float. Feeds on the floating sea anemone, *Minyas*. Occasionally washed ashore in April and May on Texas and Florida beaches. *Limnaea palmeri* Dall, 1871, may be a synonym.

Other species:

1183 *Recluzia insignis* Pilsbry and Lowe, 1932. Panama.

1184 *Recluzia palmeri* (Dall, 1871). Gulf of California. Originally described as a fresh-water snail, *Lymnaea*.

Superfamily Epitoniacea S. S. Berry, 1910

Family EPITONIIDAE S. S. Berry, 1910

Genus *Sthenorytis* Conrad, 1862

Shell heavy, turbinate, no umbilicus; face of the round aperture is offset 40 degrees from the axis of the shell; axial blades strong and numerous. Type: *expansa* (Conrad, 1842), a fossil. Synonyms are: *Pseudosthenorytis* Sacco, 1891, and *Stenohyscala* Boury, 1912.

Sthenorytis pernobilis (Fischer and Bernardi, 1857)   **1185**
Noble Wentletrap

North Carolina to southeast Florida and to Barbados.

1 to 1¾ inches in length, solid, pure-white to grayish; angle of spire about 50 degrees. The 10 whorls are globose and each bears about 14 very large, thin, bladelike ribs. Apertural rim round, solid. Operculum circular, black, with 5 or 6 whorls. A very choice collector's item, occasionally dredged from 50 to 805 fathoms. It is the only member of the genus in Western Atlantic waters. *S. belaurita* (Dall, 1889), *cubana* Bartsch,, 1940, *hendersoni* Bartsch, 1940 and *epae* Bartsch, 1940, are minor forms of this rare species.

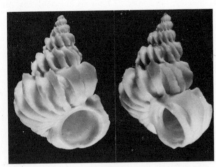

1185

(1186) The Gulf of California comparable species, occurring in 48 to 80 fathoms, is *S. dianae* (Hinds, 1844). 35 mm., and 10 varices on the last whorl. *Epitonium paradisi* Hertlein and Strong, 1951, is a synonym.

Other species:

1187 *Sthenorytis turbinus* (Dall, 1908). Off Cape San Lucas, Baja California, 75 fms.; north of San Pedro Nolasco Island, Gulf of California, 60 fms.; Galapagos. *The Veliger*, 1966, vol. 8, p. 311. (Synonym: *toroense* Dall, 1912.)

Genus *Cirsotrema* Mörch, 1852

Shells elongate, without an umbilicus, having a spiral cord on the base; sculpture of fine spiral striae crossed by

axial riblets, thus giving a curious "pitted" appearance to the surface. Operculum paucispiral. Type: *varicosa* Lamarck, 1822. *Cirsotremopsis* Thiele, 1928, is a synonym.

Cirsotrema dalli Rehder, 1945   **1188**
Dall's Wentletrap

North Carolina to Florida and to Brazil.

1 to 1½ inches in length, rather slender, with a quite deep suture, thus giving the whorls a shouldered appearance. No umbilicus. Color a uniform, chalky grayish white. Outer lip with a thickened varix. Whorls with numerous groups of

1188

foliated costae. Surface pitted with small holes when the costae or ribs are closely crowded. Uncommon from 18 to 75 fathoms. *C. arcella* Rehder, 1945, is believed to be the young of this species.

Other species:

1189 *Cirsotrema pilsbryi* (McGinty, 1940). 40 to 100 fms., off southeast Florida. See *Johnsonia*, vol. 2, no. 29, pl. 98.

1190 *Cirsotrema togatum* Hertlein and Strong, 1951. Gulf of California to Costa Rica; Galapagos Islands.

1191 *Cirsotrema vulpinum* (Hinds, 1844). Gulf of California to Panama. (Synonym: *pentedesmium* S. S. Berry, 1963.)

Genus *Acirsa* Mörch, 1857

Shells elongate, white, smoothish with weak axial undulations and weak spiral cut lines. Periphery weakly carinate. Resembles a white *Turritella*. Type: *costulata* Mighels and Adams, 1842.

Acirsa borealis (Lyell, 1842)   **1192**
Northern White Wentletrap

Greenland to Massachusetts; Aleutian Islands.

1¼ inches, elongate, chalky-white to yellowish, no umbilicus, 8 to 10 whorls convex. Spiral cut lines numerous, sometimes filled with brown periostracum. Axial ribs very weak and inconspicuous. 2 nuclear whorls, smooth. Operculum paucispiral. Uncommon; offshore down to 50 fathoms. *A. ochotensis* Middendorff, 1849 and *costulata* Mighels and Adams, 1842 (non Borson, 1825) are synonyms.

1192

Other species:

1193 *Acirsa menesthoides* (Carpenter, 1864). Baja California.

1194 *Acirsa exopleura* Dall, 1917. Baja California.

Genus *Opalia* H. and A. Adams, 1853

Shells solid, without an umbilicus; axial ribs usually strong, but, if weak, they form crenulations along the lower edge of the suture. Spiral sculpture of fine threads usually marked with "pin-prick" pittings. Operculum paucispiral. Type: *australis* Lamarck, 1822. Synonym: *Psychrosoma* Tapparone-Canefri, 1876.

Opalia wroblewskii (Mörch, 1876)   1195
Wroblewski's Wentletrap

Forrester Island, Alaska, to off San Diego, California.

1 to 1¼ inches in length, slender, heavy; looks beachworn; grayish white in color, often stained purple from the animal's dye gland. With 6 to 8 low, pronounced, axial, wide ribs. Base of shell bounded by a strong, smooth, low, spiral cord. In well-preserved specimens, particularly those from California, the outer layer of the shell is minutely punctate. Fairly common. *O. chacei* Strong, 1937, is probably a form of this species. It is smaller, wider and proportionately heavier than typical specimens. Moderately common from low tide to 50 fathoms.

Subgenus *Dentiscala* de Boury, 1886

Opalia hotessieriana (Orbigny, 1842)   1196
Hotessier's Wentletrap

South Florida to Brazil. Bermuda.

⅓ to ½ inch in length, moderately slender. Characterized by 10 to 14 large, square notches along the suture of each whorl. Ribs are rather weak. Surface, in fresh specimens, microscopically pitted. Color grayish white. Not uncommon from low water to 90 fathoms. The synonyms are: *crassicostata* Sowerby, 1844; *grossicostata* Nyst, 1871; *crassicosta* C. B. Adams, 1845, and *scaeva* Mörch, 1874.

(1197) *O. crenata* (Linné, 1758) (same range, but also the Eastern Atlantic) is larger, its whorls more strongly shouldered, and the notches at the suture are much weaker and more numerous.

Opalia funiculata (Carpenter, 1857)   1198
Scallop-edged Wentletrap

Southern California to west Mexico.

½ to ¾ inch in length, dull whitish, 7 to 8 whorls, moderately slender. Characterized by the smoothish sides of the whorls and by the spiral ramp below the suture which bears 12 to 14 short, horizontal ribs per whorl. Early whorls may have weak axial ribs running from suture to suture. Spinal sculpture of microscopic, numerous scratches. *O. crenimarginata* (Dall, 1917), *insculpta* Carpenter, 1864, and *nesiotica* Dall, 1917, are this species. Very common among rocks at the bases of the large green sea anemone upon which it feeds.

1195   1196   1197   1198

Subgenus *Nodiscala* de Boury, 1889

Opalia with strong axial ribs. Basal ridge absent. Type: *bicarinata* Sowerby, 1844. Synonym: *Punctiscala* de Boury, 1889.

Opalia pumilio (Mörch, 1874)   1199
Pumilio Wentletrap

Off North Carolina to Florida and to Brazil. Bermuda.

5 to 10 mm., elongate, with 14 to 16 rounded costae on the body whorl. Outer lip very thick and rounded. No basal ridge present. There may be 2 or 3 former varices in the spire. Microscopically pitted. 2½ nuclear whorls, smooth, amber in color. The form *morchiana* (Dall, 1889) **(1200)**, has 9 to 15 axial costae and is strongly angulate at the periphery of the whorl. Fairly common from just offshore to 100 fathoms. The synonyms are: *subvaricosa* Mörch, 1874; *nodosocarinata* (Dall, 1889); *dunkeri* de Boury, 1889; *semivaricosa* de Boury, 1889; *linteatum* Schwengel, 1943; and *barbadensis* de Boury, 1913.

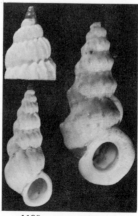

1199   1200

Opalia spongiosa (Carpenter, 1864)   **1201**
Spongy Wentletrap

Oregon to the Gulf of California. Galapagos Islands.

7 to 9 mm., elongate; whitish; surface microscopically pin-pointed; axial costae very low, rounded and have the appearance of being tucked in at the deep suture. Operculum pauci-spiral. *O. retiporosa* (Carpenter, 1864); and *crosseana* (Tapparone-Canefri, 1876), are synonyms. Common; offshore 10 to 50 fathoms on sand and shale bottoms.

Other species:

1202   *Opalia* (*Cylindriscala*) *watsoni* (de Boury, 1911) (synonym: *funiculata* Watson, 1883). Off Key West to off Brazil. 120 to 350 fms.

1203   *Opalia* (*Opalia*) *abbotti* Clench and Turner, 1952. 70 to 385 fms. Off southeast Florida and Cuba.

1201   1202   1203

1204   *Opalia* (*Dentiscala*) *burryi* Clench and Turner, 1950. Off southeast Florida. 50 to 100 fms.

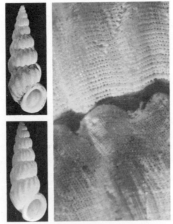

1204

1205   *Opalia* (*Nodiscala*) *aurifilia* (Dall, 1889). Off Cape Hatteras, North Carolina, to Martinique. 30 to 170 fms.

1206   *Opalia* (*Nodiscala*) *eolis* Clench and Turner, 1950. Off southeast Florida to Lesser Antilles. 70 to 250 fms.

1207   *Opalia* (*Opalia*) *evicta* de Boury, 1919. Alaska to Baja California. May be synonym of *montereyensis* (Dall, 1907).

1208   *Opalia* (*Nodiscala*) *retiporosa* (Carpenter, 1864). Catalina Island, California, to Baja California.

1209   *Opalia* (*Dentiscala*) *crenatoides* (Carpenter, 1864). Southern California (rare) to Nicaragua. Galapagos Islands. (Synonym: *golischi* Baker, Hanna and Strong, 1930.)

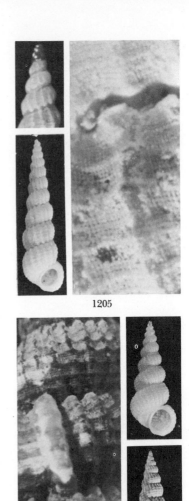

1205

1206

1210   *Opalia* (*Nodiscala*) *espirita* (Baker, Hanna and Strong, 1930). Gulf of California to Chiapas, Mexico.

1211   *Opalia* (*Opalia*) *leeana* (Verrill, 1883). 146 fms., 185 miles east of Barnegat Bay, New Jersey.

— *Opalia discobalaria* (Dall, 1889), is a worn *Cerithidea.*

1211   1214

1212 *Opalia (Cylindriscala) acus* (Watson, 1883). 390 fms., off Puerto Rico; Azores to Gibraltar, down to 1,000 fms.

1213 *Opalia (Cylindriscala) tortilis* (Watson, 1883). 390 fms., off Puerto Rico.

1214 *Opalia (Cylindriscala) andrewsii* (Verrill, 1882). 100 fms., off New Jersey to off Cuba, 500 fms. 5 mm.

1215 *Opalia (Opalia) montereyensis* (Dall, 1907). Monterey to San Pedro, California, 25 fms.

1216 *Opalia (Nodiscala) bullata* (Carpenter, 1864). Southern California to Nicaragua. (Synonyms: *tremperi* Bartsch, 1927, and *ordenanum* Lowe, 1932.)

1217 *Opalia (Nodiscala) sanjuanensis* (Lowe, 1932). Gulf of California to Nicaragua. (Synonym: *Dentiscala clarkei* Olsson and M. Smith, 1951.)

Genus *Amaea* H. and A. Adams, 1854

Shells without an umbilicus; whorls joined at the suture. Basal ridge present. Low axial costae and low spiral threads. Type: *magnifica* Sowerby, 1853.

Amaea mitchelli (Dall, 1896)   **1218**
Mitchell's Wentletrap

Texas coast to Panama.

1½ to 2½ inches in length, thin but strong; without an umbilicus. With about 15 rather strongly convex, pale-ivory whorls which have a dark brownish band at the periphery and a solid brown area below the basal ridge. About 22 low, irregular costae per whorl. Numerous spiral threads are fine, and produce a weak, reticulated pattern. Not very common, but occasionally washed up on Texas beaches. It is associated with sea anemones.

Subgenus *Scalina* Conrad, 1865

Shells deepsea, beautifully reticulated by costae and spiral ridges. Basal ridge present. Type: *staminea* Conrad, 1865. Synonyms: *Elegantiscala* de Boury, 1911; and *Ferminoscala* Dall, 1908.

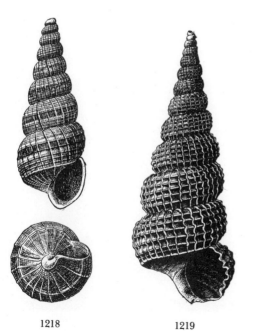

1218   1219

Amaea retifera (Dall, 1889)   **1219**
Reticulate Wentletrap

North Carolina to both sides of Florida to Brazil.

1 inch in length, elongate, thin but strong; with about 16 whorls which are beautifully reticulated by strong, sharp threads. Color straw to pale-brown with 2 light and narrow brownish bands, one above and one below the periphery. Commonly dredged off Florida from 13 to 120 fathoms.

(1220) The Baja California to Colombia comparable species (1½ inches) is *Amaea (Scalina) ferminiana* Dall, 1908. *Scala weigandi* Böse, 1910, is a synonym. Rare.

Other species:

1221 *Amaea (Amaea) contexta* DuShane, 1970. West Mexico, 5 to 10 fms. 15 mm. (*Contrib. in Sci., Los Angeles,* no. 185, p. 5.)

1221

1222 *Amaea (Scalina) brunneopicta* (Dall, 1908). Cedros Island, Baja California, the Gulf of California to Galapagos Islands. (Synonym: *Englisia nebulosa* Dall, 1919.)

1223 *Amaea (Scalina) deroyae* DuShane, 1970. Gulf of California and Galapagos Islands.

1224 *Amaea (Scalina) tehuanarum* DuShane and McLean, 1968. Gulf of California to the Gulf of Tehuantepec, Mexico.

Genus *Epitonium* Röding, 1798

Shells white, with strong, bladelike costae, whorls sometimes not touching; usually umbilicate. Aperture edge at about the same axis as that of the shell. Type: *scalare* Linné, 1758. Numerous species. The typical subgenus lacks spiral sculpturing. Synonyms are: *Nitidiscala* de Boury, 1909 (sometimes misspelled *Nitidoscala*); *Anguliscala, Viciniscala, Lamelliscala* and *Eburniscala* all de Boury, 1909.

Subgenus *Epitonium* Röding, 1798

Epitonium krebsii (Mörch, 1874)   **1225**
Krebs' Wentletrap

Off South Carolina to Brazil. Bermuda.

½ to ¾ inch in length, stout. With umbilicus fairly narrow to wide, and very deep. 7 to 8 whorls attached by the costae (10 to 12 per whorl). China-white, rarely with a trace of brown to pinkish brown undertones. No spiral sculpturing. Operculum brown, paucispiral. Moderately common from a few feet to 160 fathoms. *E. swifti* (Mörch, 1874), *E. contorquatum* (Dall, 1889), and *electa* (Verrill and Bush, 1900) are this species.

1225   1226

Do not confuse with *E. occidentale* Nyst from the same areas. It is not so stout, has 12 to 15 costae per whorl, a very small umbilicus or none, and the shoulder of the whorls is somewhat flattened.

Epitonium occidentale (Nyst, 1871)   1226
Western Atlantic Wentletrap

Off southeast Florida to Barbados; Bermuda. Brazil.

½ to 1 inch; umbilicus absent or very small. 12 to 15 costae on the body whorl. Top of the shoulder is flattened, giving the whorl an inset appearance. Shoulders of costae pointed, sometimes forming a spine. Fairly common; just offshore to 61 fathoms. Synonyms are: *tenuis* Sowerby, 1844 (non Gray); *micromphala* Mörch, 1874; *aurita* Mörch, 1874.

Epitonium albidum (Orbigny, 1842)   1227
Bladed Wentletrap

Bermuda; south Florida to Argentina; West Africa.

½ to ¾ inch, rather light in structure, white; whorls not touching each other; with 12 to 14 low, bladelike costae on the last whorl, but they are not angled at the shoulder and generally fuse with those of the whorl above. Rarely there is a yellow-brown band just below the suture. Fairly common; just offshore and often cast on the beaches. Synonyms are: *ligatum* C. B. Adams, 1850; *quindecimcostatum* Mörch, 1874; *gradatella* Mörch, 1874; *undecimcostata* Mörch, 1874.

1227

Epitonium tollini Bartsch, 1938   1228
Tollin's Wentletrap

West Coast of Florida.

½ inch in length, slender, no umbilicus. 9 to 10 whorls strongly convex; suture deep. Each whorl has from 11 to 16 costae which are not shouldered, but are rounded, on top. They often line up one below the other. Outer lip thick and reflected. Inner lip much smaller. Color china-white, with the first few whorls a very faint amber-brown. Fairly common just off the outer beaches. Do not confuse with *E. humphreysi* whose costae are angular at the top. It is possible that this is only a narrow variant of *humphreysi*.

1228

Epitonium humphreysi (Kiener, 1838)   1229
Humphrey's Wentletrap

Cape Cod, Massachusetts, to Florida and to Texas.

½ to ¾ inch in length, fairly slender, thick-shelled, and without an umbilicus. Color dull-white. Suture deep. The 9 to 10 convex whorls each have about 8 or 9 costae that are somewhat angled at the shoulder. Costae usually thick and strong. Outer and lower part of the apertural lip thickened and slightly flaring. Common from shore to 52 fathoms. *Scala sayana* Dall, 1889, is a synonym. Do not confuse with *E. angulatum* which is not so slender, is glossier, has thinner costae that are usually more angular at the shoulders.

Epitonium angulatum (Say, 1830)   1230
Angulate Wentletrap

New York to Florida and to Texas. Bermuda.

1229   1230

¾ to 1 inch in length, moderately stout to somewhat slender, strong and without an umbilicus. 8 whorls with about 9 or 10 strong but thin costae which are very slightly reflected backwards and which are usually angulated at the shoulder, especially in the early whorls. The costae are usually formed in line with those on the whorl above and are fused at their points of contact. Outer lip thickened and reflected. Color china-white. One of the commonest Atlantic wentletraps found in shallow water to 25 fathoms. *Scalaria turbinata* Conrad, 1837, is a synonym. Do not confuse with *E. humphreysi*. Reported from Brazil by E. C. Rios, 1970.

Epitonium foliaceicostum (Orbigny, 1842)   **1231**
Wrinkled-ribbed Wentletrap

Southeast Florida to Brazil.

½ to ¾ inch in length, moderately stout, without an umbilicus, and similar to *E. angulatum*, except that the 7 or 8 costae per whorl are thinner, more highly developed and usually quite angular. Moderately common from low water to 120 fathoms. Alias *muricata* Sowerby, 1844, *spina-rosae* Mörch, 1874, *pretiosula* Mörch, 1874, and *novemcostata* Mörch, 1874. This may well be only a subspecies of *angulatum* (Say, 1830).

1231   1232

Epitonium unifasciatum (Sowerby, 1844)   **1232**
One-banded Wentletrap

Southeast Florida to Barbados.

½ inch; elongate, white with a subsutural band of brown. 7 to 9 costae on the body whorl. No basal ridge present (as in the similar *lamellosum* (Lamarck)). Costae are low, narrow and rounded. Locally common in the West Indies just offshore.

Epitonium indianorum (Carpenter, 1864)   **1233**
Money Wentletrap

Forrester Island, Alaska, to Baja California.

1 inch in length, slender, pure-white, of 11 whorls, each of which has 13 or 14 sharp costae which are slightly bent back-

1234

wards. The tops of the costae are slightly pointed. Fairly common offshore on gravel bottom in association with sea anemones. From Monterey south this species is found from 25 to 71 fathoms. It is found just below low-tide mark from British Columbia north to Alaska.

Epitonium tinctum (Carpenter, 1864)   **1234**
Tinted Wentletrap

Vancouver, British Columbia, to Baja California.

½ inch, somewhat stout (its width is about ⅖ of its length), with 10 to 12 strong varicose axial ribs on the last whorl. White with a brown band in the sutural area. Readily distinguished from the elongate *indianorum* by its stout form, color band and spined or angulated costae. *E. subcoronatum* (Carpenter, 1866) and *fallaciosum* Dall, 1921, are synonyms. The doubtful *crebricostatum* (Carpenter, 1864) is believed to be this species. The forma *bormanni* Strong, 1941, **(1235)** lacks the strong coronation on the shoulder of the costae, has 1 or 2 extra costae per whorl and has a very faint tan subsutural color band. This is an abundant intertidal species associated with small sea anemones that may irritate the skin of the collector.

Epitonium cooperi Strong, 1930   **1236**
Cooper's Wentletrap

Vancouver, British Columbia, to Baja California.

½ to ¾ inch; fairly thin-shelled and delicate; length 2½ times the width. Pure-white. Top of costae (11 or 12 on the last whorl) pointed. Characterized by loosely coiled, convex whorls and its deep suture. Moderately common; dredged from 25 to 75 fathoms.

Epitonium sawinae (Dall, 1903)   **1237**
Sawin's Wentletrap

British Columbia, to Catalina Island, California.

⅓ to ½ inch, elongate, deeply sutured, pure-white, with numerous costae (19 to 32 per whorl), usually angled to spinose at the top, although sometimes rounded and smoothish. A variable species with *catalinensis* Dall, 1917, probably a synonym. Common; on gravel, shale or mud from 10 to 50 fathoms.

Subgenus *Cycloscala* Dall, 1889

Shell small (9 mm.) and with its sharp costae fluted. Type: *echinaticostum* Orbigny, 1842. One Caribbean species.

Epitonium echinaticostum (Orbigny, 1842)   **1238**
Widely-coiled Wentletrap

Bermuda; Florida to Barbados. Brazil.

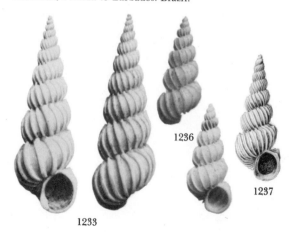

1236   1237   1233

6 to 10 mm., fragile; widely umbilicate; whorls sometimes detached except for the touching of the fluted or scalloped axial costae. Nuclear whorls 2½, glassy and sometimes set at

1238

an angle. Moderately common; 1 to 200 fathoms. Synonyms of the variable species are: *soluta* Mörch, 1874; *volubilis* Mörch, 1874; *blandii* Mörch, 1874; *dunkeriana* Dall, 1889; and *inconstans* de Boury, 1913. Dan Steger reports it common in Florida bays and inlets.

Subgenus *Gyroscala* de Boury, 1887

Shells without an umbilicus; with a distinct basal ridge; strongly costate; color white with suffusions of brown. Type: *lamellosum* (Lamarck, 1822). *Pictoscala* Dall, 1917, and *Depressiscala* de Boury, 1909, are synonyms.

Epitonium lamellosum (Lamarck, 1822)   1239
Lamellose Wentletrap

South half of Florida and the Caribbean. Bermuda.

¾ to 1¼ inches in length, with an umbilicus. 11 whorls whitish with irregular, brownish markings. Costae thin, high, always white. Characterized by a fairly strong, raised, spiral thread on the base of the shell. Moderately common from low

1239   1240

water to 33 fathoms. Alias *E. clathrum* of authors, not Linné, 1758. Synonyms are: *pseudoscalaris* Philippi, 1836; *monocycla* Kiener, 1839; *commutata* Monterosato, 1877.

Epitonium rupicola (Kurtz, 1860)   1240
Brown-banded Wentletrap

Cape Cod, Massachusetts, to Florida and to Texas.

½ to 1 inch in length, moderately stout to slender, and without an umbilicus. Color whitish or yellowish with 2 brownish, spiral bands on each side of the suture. Color often diffused. About 11 globose whorls, each of which has from 12 to 18 weak or strong costae. Former, thickened varices sometimes present. Base of shell with a single, fine, spiral thread. Formerly known as *lineatum* Say, 1822, *reynoldsi* Sowerby, 1916, *fischeriana* Tapparone-Canefri, 1876, and *sublineata* de Boury, 1918. Common from low water to about 20 fathoms.

Subgenus *Asperiscala* de Boury, 1909

Possessing spiral sculpture ranging from low cords to very fine cut lines; axial ribs rounded or bladelike. Basal ridge absent. 2 or 3 nuclear whorls, white or tan, smooth. Type: *bellastriata* Carpenter, 1864. Synonyms are *Cinctiscala, Decussiscala* and *Sodaliscala* all de Boury, 1909.

Epitonium apiculatum (Dall, 1889)   1241
Apiculate Wentletrap

Off North Carolina; Puerto Rico.

4.5 mm. in length, 9 whorls attached by costae only. 11 on the body whorl. Curiously, the first 3 postnuclear whorls have numerous axial riblets crossed by spiral threads. No spiral sculpture on the last whorls. Uncommon; 8 to 49 fathoms. This is not *apiculatum* Dall, 1917, from the tropical Eastern Pacific (now *bakhanstranum* Keen, 1962, *The Veliger*, vol. 4, p. 179).

Epitonium multistriatum (Say, 1826)   1242
Many-ribbed Wentletrap

Off Massachusetts to Texas and Florida. Bermuda.

½ inch, light-weight, elongate, no umbilicus; pure-white; axial blades numerous, crowded (16 to 19 on the last whorl). No basal ridge. Spiral sculpture of numerous cut lines, not crossing the blades. 3 or 4 nuclear whorls, tiny. Offshore to 120 fathoms.

A west coast of Florida subspecies, *matthewsae* Clench and Turner, 1952 **(1243)**, differs very little except in being narrower, and in having the whorls coiled more tightly together (in the typical form the whorls do not touch each other except by the costae). Both uncommon offshore from 8 to 120 fathoms. The synonyms are: *leptalea* (Bush, 1885), *elliotti* (Mazyck, 1913), and *virginicum* Henderson and Bartsch, 1914. Reported from Bermuda by R. Jensen (*in litt.*).

Epitonium novangliae (Couthouy, 1838)   1244
Couthouy's Wentletrap

Virginia to Texas to Brazil; Bermuda.

½ inch, umbilicate, somewhat elongate. Later whorls attached only by the costae (9 to 16 on the body whorl). Color white, sometimes with light-brown banding. The shoulder of the body whorl costae may have a fine, backwardly slanting hook. Spiral threads cross axial threads to give a fine, reticulated surface between the costae. No basal ridge. Operculum paucispiral and dark-brown. A widely spread, fairly common species found from just offshore to Brazil. "New England" is a misnomer for this species, as it has never been found alive

1241   1242   1244   1245   1246

north of Virginia. Synonyms are: *uncinaticosta* Orbigny, 1842; *aeospila* Mörch, 1874; *muscapedia* Dall, 1889; and *bahamensis* Piele, 1926.

Epitonium candeanum (Orbigny, 1842)   1245
Candé's Wentletrap

East Florida to Brazil; Bermuda.

8 to 10 mm., very similar to *novangliae*, but lacks the angles or hooks on the shoulders of the costae, and has more costae (18 to 25 on the body whorl). Spiral sculpture quite coarse. Uncommon; low water line to 300 fathoms. Synonyms are *turrita* Nyst, 1871; *antillarum* de Boury, 1909; *marcoense* Dall, 1927.

Epitonium bellastriatum (Carpenter, 1864)   1246
Beautifully Threaded Wentletrap

Monterey, California, to Baja California.

⅓ to ½ inch, squat, with sharp, clear spiral sculpture over the entire surface. Axial costae 15 to 17 on the last whorl. The shoulders of the costae are weakly spined. White, sometimes stained with tan blotches. Very common offshore; 10 to 36 fathoms in sand and shale.

Subgenus *Boreoscala* Kobelt, 1902

Shells without umbilicus, strongly sculptured with low, thick costae, and with coarse, flattened spiral ridges. Basal ridge or cord present. Type: *greenlandicum* Perry, 1811. Synonyms are: *Arctoscala* Dall, 1909; *Liriscala* de Boury, 1909; *Pyramiscala* de Boury, 1909.

Epitonium greenlandicum (Perry, 1811)   1247
Greenland Wentletrap

Alaska to Graham Island, British Columbia.
Greenland to Long Island, New York.

1 to 2½ inches; solid, elongate, chalky-gray; ridgelike costae (9 to 12 on body whorl) sometimes very broad. Spiral sculpture usually prominent. 9 spiral cords on base. Operculum dark-brown, paucispiral. Common; 10 to 130 fathoms. Synonyms or forms are: *loveni* A. Adams, 1856; *norvegicum* Clench and Turner, 1952; *similis* Sowerby, 1813; *subulata* Couthouy, 1838; *planicosta* Kiener, 1839; *crebricostata* Sars, 1878.

Epitonium blainei Clench and Turner, 1953   1248
Blaine's Wentletrap

West and southeast Florida.

1 to 1⅓ inches, 9 whorls, suture deep; all-white; basal ridge strong; without an umbilicus. Similar to *greenlandicum*, but has nodulose axial costae. Rare; offshore to 22 fathoms.

Other Atlantic species:

1249 *Epitonium (E.) venosum* (Sowerby, 1844). Cuba to Venezuela. Rare.

1250 *Epitonium (E.) georgettina* (Kiener, 1839). Brazil to Argentina; 17 to 55 fms.

1251 *Epitonium (E.) fractum* Dall, 1927. Off Fernandina, Florida, to Key West. 30 to 325 fms. See *Johnsonia*, vol. 2, no. 30, pl. 125.

1252 *Epitonium (E.) dallianum* (Verrill and Smith, 1880). 192 fms., off New Jersey to 168 fms.; off North Carolina and to 120 fms. off Miami, Florida.

1253 *Epitonium (E.) eulita* (Dall and Simpson, 1901). 25 to 30 fms., Mayaguez Harbor, Puerto Rico. 4 mm.

1254 *Epitonium acutum* (Pfeiffer, 1840). Cuba. An unrecognized species.

1255 *Epitonium (E.) filare* Mörch, 1874, is an unrecognized species.

1256 *Epitonium permodesta* (Dall, 1889), is an unrecognized species.

1257 *Epitonium (Asperiscala) rushii* (Dall, 1889). Off North Carolina to Key West, Florida. 38 to 100 fms.

1258 *Epitonium (A.) turritellulum* Mörch, 1874. (Synonyms: *riisei* Mörch, 1874, *stylina* Dall, 1889.) Greater Antilles. Brazil.

1259 *Epitonium (A.) tenuistriatum* Orbigny, 1840. Uruguay to Argentina. Brazil.

1247   1248   1251   1252

1260 *Epitonium (A.) frielei* (Dall, 1889). 63 fms., off North Carolina to Puerto Rico.

1260

1261 *Epitonium (A.) tiburonense* Clench and Turner, 1952. Southwest Haiti and Puerto Rico.

1262 *Epitonium (A.) denticulatum* (Sowerby, 1844). (Synonyms: *centiqadra* and *octocostata* Mörch, 1874.) Southeast Florida to the Virgin Islands. 3 to 805 fms. Brazil, 40 to 65 fms.

1262

1263 *Epitonium (A.) pourtalesii* (Verrill and Bush, 1880). 80 fms., off New Jersey to 350 fms. off Barbados. Bermuda.

1264 *Epitonium (A.) babylonium* (Dall, 1889). 87 fms., off North Carolina to 400 fms., off Cuba. 30 mm.

1265 *Epitonium (A.) polacium* (Dall, 1889). 115 to 229 fms., off Key West, Florida.

1266 *Epitonium (A.) sericifiilum* (Dall, 1889). Honduras and off Port Aransas, Texas. See *Johnsonia*, vol. 2, no. 31, pl. 152.

1267 *Epitonium (A.) championi* Clench and Turner, 1952. Massachusetts to South Carolina. A form of *E. greenlandicum* Perry?

1263   1264   1267

1268 *Epitonium (Boreoscala) magellanicum* (Philippi, 1845). (Synonym: *douvillei* Fenaux, 1937.) Argentina to Chile. 30 to 55 fms.

1269 *Epitonium (B.) pandion* Clench and Turner, 1952. (Synonym: *gracilis* Verrill, 1880.) 547 fms., off New Jersey to off Cape Hatteras, North Carolina, in 843 fms.

1270 *Epitonium (Depressiscala) nitidella* (Dall, 1889). Off North Carolina to Alabama; Florida to Barbados. 32 to 117 fms. Bermuda (R. Jensen).

1271 *Epitonium (D.) nautlae* Mörch, 1874. (Synonyms: *scipio* Dall, 1889; *teres* Bush, 1885.) 15 to 66 fms. Off North Carolina to both sides of Florida; Bahamas; Cuba and Vera Cruz, Mexico.

1269   1270   1271

1272 *Epitonium (E.) arnaldoi* Tursch and Pierret, 1964. Off Punta de Juatinga, Brazil, 50 meters. *The Veliger*, vol. 7, p. 36.

1273 *Epitonium (E.) mauryi* Tursch and Pierret, 1964. Off Punta de Juatinga, Brazil, 50 meters. *The Veliger*, vol. 7, p. 36.

Other Pacific species:

1274 *Epitonium (E.) persuturum* Dall, 1917. San Diego, California.

1275 *Epitonium (E.) densiclathratum* Dall, 1917. Washington to San Diego, California. 1 to 50 fms.

1276 *Epitonium (E.) columbianum* Dall, 1917. Alaska to Baja California. 10 to 40 fms.

1277 *Epitonium (E) californicum* Dall, 1917. Los Angeles County, California, to the Gulf of California. 15 to 25 fms.

1278 *Epitonium (E.) rectilaminatum* (Dall, 1907). Monterey, California, to the Gulf of California. 10 to 25 fms.

1279 *Epitonium (E.) caamanoi* Dall and Bartsch, 1910. Ketchikan, Alaska, to San Pedro, California. 10 fms. 10 mm. Memoir 14 N, Canada Dept. Mines, p. 13.

1280 *Epitonium (E.) regiomontanum* (Dall, 1917). 30 fms., Monterey Bay, California.

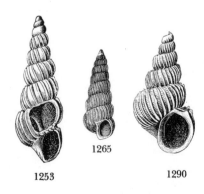

1253   1290

1265

1281 *Epitonium (E.) berryi* (Dall, 1907). 12 fms., Monterey Bay, California. May be a young form of *rectilaminatum* Dall.

1282 *Epitonium (E.) diegense* Dall, 1917. San Diego, California. Rare.

1283 *Epitonium (E.) catalinae* Dall, 1908. Alaska to Vancouver Island to San Diego, California. 18 to 75 fms.

1284 *Epitonium (E.) regum* Dall, 1927. Point Reyes to San Diego, California. 15 to 61 fms.

1285 *Epitonium (E.) acrostephanus* (Dall, 1908). Barkley Sound and Puget Sound to Coronado Islands, California. 15 to 203 fms.

1286 *Epitonium (E.) tabulatum* Dall, 1917. Monterey to Coronado Islands, California. 10 to 44 fms.

1287 *Epitonium (E.) orcuttianum* Dall, 1917. San Diego, California.

1288 *Epitonium (E.) barbarinum* Dall, 1919. San Diego to Panama.

1289 *Epitonium (E.) zephyrium* Dall, 1917. San Diego, California. (Synonym: *basicum* Dall, 1917.)

1290 *Epitonium (Asperiscala) lowei* (Dall, 1906). Southern California, 30 to 40 fms., to Punta Abreojos, Baja California.

1291 *Epitonium (A.) cookeanum* Dall, 1917. San Diego to the Gulf of California.

1292 *Epitonium (A.) arnoldi* Dall, 1917. San Pedro, California.

1293 *Epitonium (A.) lagunarum* Dall, 1917. Laguna Beach, California.

1294 *Epitonium (A.) tinctorum* Dall, 1917. Magdalena Bay, Baja California.

1295 *Epitonium (A.) canna* Dall, 1919. Gulf of California. (Synonym: *reedi* Bartsch, 1928.)

1296 *Epitonium (A.) longinosanum* DuShane, 1970. Rancho Inocentes, Baja California del Sur, Mexico. 16.5 mm. (*Contrib. in Sci., Los Angeles*, no. 185, p. 2).

1297 *Epitonium (A.) macleani* DuShane, 1970. Off Punta Ventana, Baja California del Sur, Mexico, 15 to 25 fms. 7 mm. (*loc. cit.*, p. 3).

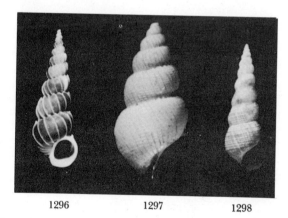

1296   1297   1298

1298 *Epitonium (A.) cerralvoensis* DuShane, 1970. Cerralvo Island, Baja California del Sur, Mexico, 4 to 20 fms. 12 mm. (*loc. cit.*, p. 3). An *Acirsa?*

1299 *Epitonium (A.) acapulcanum* Dall, 1917. Magdalena Bay and the Gulf of California to the Galapagos Islands. (Synonyms: *xantusi* Dall, 1917; *keratium* Dall, 1919; *strongi* Bartsch, 1928; *slevini* Strong and Hertlein, 1939.)

1300 *Epitonium (A.) centronium* (Dall, 1917). Gulf of California.

1301 *Epitonium (A.) emydonesus* Dall, 1917. Gulf of California to the Galapagos Islands. (Synonyms: *imperforatum* Dall, 1917; *manzanillense* Hertlein and Strong, 1951.)

1302 *Epitonium (A.) eutaenium* (Dall, 1917). Gulf of California to the Galapagos Islands. (Synonym: *vivesi* Hertlein and Strong, 1951.)

1303 *Epitonium (?A.) gaylordianum* Lowe, 1932. Gulf of California to Panama.

1304 *Epitonium (A.) habeli* Dall, 1917. Gulf of California to the Galapagos Islands. (Synonym: *kelseyi* Baker, Hanna and Strong, 1930.)

1305 *Epitonium (A.) huffmani* DuShane and McLean, 1968. Gulf of California to Ecuador.

1306 *Epitonium (A.) thylax* Dall, 1917. Gulf of California to Panama.

1307 *Epitonium (A.) tinctorium* Dall, 1919. Gulf of California to Panama.

1308 *Epitonium (A.) walkerianum* Hertlein and Strong, 1951. San Felipe, Gulf of California, to Nicaragua.

1309 *Epitonium (Asperiscala) billeeana* (DuShane and Bratcher, 1965). Gulf of California to Manzanillo, Mexico; Galapagos Islands. 7 × 4 mm. On orange coral, *Tubastrea. The Veliger*, vol. 8, p. 160; vol. 10, p. 87.

1310 *Epitonium (Nidiscala) obtusum* (Sowerby, 1844). Southern Baja California, Gulf of California to Ecuador. (Synonyms: *suprastriata* and *tiara* Carpenter, 1857.)

1311 *Epitonium (Nidiscala) bakhanstranum* Keen, 1962. Gulf of California to Panama. (Synonym: *apiculatum* Dall, 1917 (not Dall, 1889), from the Caribbean.)

1312 *Epitonium (N.) callipeplum* Dall, 1919. Magdalena Bay, Baja California, and Mazatlan, Mexico.

1313 *Epitonium (N.) colpoicum* Dall, 1917. Gulf of California to Mazatlan, Mexico.

1314 *Epitonium (N.) compradora* Dall, 1917. Gulf of California.

1315 *Epitonium (N.) cumingii* (Carpenter, 1856). Gulf of California to Panama. (Synonym: *gissleri* Strong and Hertlein, 1939.)

1316 *Epitonium (N.) durhamianum* Hertlein and Strong, 1951. Gulf of California to Nicaragua.

1317 *Epitonium (N.) elenense* (Sowerby, 1844). Magdalena Bay and the Gulf of California to Ecuador. (Synonyms: *raricostata* Carpenter, 1857 (not Wood, 1828); *carpenteri* Tapparone-Canefri, 1876; *phanium* Dall, 1919.)

1318 *Epitonium (N.) gradatum* (Sowerby, 1844). Gulf of California to Ecuador.

1319 *Epitonium (N.) hexagonum* (Sowerby, 1844). Magdalena Bay; Gulf of California to Panama.

1320 *Epitonium (N.) pazianum* Dall, 1917. Southern California to Peru. (Synonyms: *cylindricum* and *musidora* Dall, 1917.)

1321 *Epitonium (N.) politum* (Sowerby, 1844). San Pedro, California, to Ecuador and the Galapagos Islands. (Synonyms: *appressicostatum* Dall, 1917; *implicatum* Dall and Ochsner, 1928; *pedroanum* Willett, 1932.)

1322 *Epitonium (N.) roberti* Dall, 1917. Southern Gulf of California to Panama.

1323 *Epitonium (N.) syorum* DuShane and McLean, 1968. Gulf of California to La Libertad, Ecuador.

1324 *Epitonium (Hirtoscala* Monterosato, 1890) *reflexum* (Carpenter, 1856). Gulf of California to Manzanillo, Mexico.

1325 *Epitonium (H.) replicatum* (Sowerby, 1844). Gulf of California to the Galapagos Islands. (Synonyms: *bialatum* Dall, 1917; *wurtzbaughi* Strong and Hertlein, 1939; *oerstedianum* Hertlein and Strong, 1951.)

Genus *Alexania* Strand, 1928

A smooth epitoniid shell with rounded whorls. The side of the foot extends up over part of the shell on each side. Radula ptenoglossate. Semiparasitic on sea anemones of the genus *Aiptasiomorpha.* They lay egg capsules which become sand-covered. Larvae pelagic. Type: *natalensis* Tomlin, 1926. Synonyms: *Alexandria* Tomlin, 1926, non Pfeffer, 1881; *Tomlinula* Strand, 1932; *Stenacme* Pilsbry, 1945; and *Habea* Kuroda, 1943.

Alexania floridana (Pilsbry, 1945)   1326
Smooth Florida Wentletrap

Florida (North Carolina to Texas?)

6 mm., ovate, fragile, buff with a paler band below the suture and a white basal area. Smoothish. Spire acuminate with the first 3 being small and pupiform. Suture almost channeled. Umbilicus chinklike or closed. Operculum chitinous, thin, paucispiral. (*The Nautilus,* vol. 58, p. 112.) Lives intertidally on beaches of pebbles and rocks in association with the introduced Japanese sea anemone, *A. luciae* (Verrill), which occurs from North Carolina to Texas. (*The Nautilus,* vol. 72, p. 68, and vol. 78, p. 140).

1326

Subfamily NYSTIELLINAE Clench and Turner, 1952

Shells similar to those in Epitoniinae, but usually elongate and sometimes with whorls completely unattached. The first nuclear whorl, only, is smooth and followed by 3 brown, axially costate nuclear whorls. Radular teeth few in number with large heavy bases and a pronglike extension (in Epitoniinae the numerous teeth are thin, and very long and narrow).

Genus *Nystiella* Clench and Turner, 1952

Shells elongate, resembling *Turritella,* usually costate and with spiral threads. Nuclear whorls brown and axially costate. Basal cord present. All species are deepsea, and we list them here (all are illustrated in *Johnsonia,* vol. 2, no. 31):

1327 *Nystiella opalina* (Dall, 1927). (Synonyms: *lavaratum* Dall, 1927; *dromio* Dall, 1927.) Off Fernandina, Florida, to St. Kitts, Lesser Antilles, from 294 to 440 fms.

1328 *Nystiella concava* (Dall, 1889). 15 fms., off Key West, Florida. Rare.

1329 *Nystiella cania* (Dall, 1927). East coast of Florida from 440 fms., off Fernandina to 150 fms. off the Lower Florida Keys.

1330 *Nystiella azelotes* (Dall, 1927). 440 fms., off Fernandina, Florida.

1331 *Nystiella atlantis* Clench and Turner, 1952. 63 to 265 fms., southeast Florida to Cuba.

Subgenus *Eccliseogyra* Dall, 1892

Shell freely coiled, 5 to 10 mm., nuclear whorl brown, the first smooth, the next 2 to 5 axially costate. Type: *Delphinula nitida* Verrill and Smith, 1885. Synonym: *Solutiscala* de Boury, 1909.

1332

1332 *Nystiella nitida* (Verrill and Smith, 1885.) Off eastern U.S.; Brazil; West Africa; Portugal; 1,000 to 4,600 meters. (Synonyms: *Scalaria vermetiformis* Watson, 1886; *dissoluta* Locard, 1897.) Formerly listed as *Liotia* (*Laxispira*) *nitida.* See Rex and Boss, 1973, *The Nautilus,* vol. 87, p. 93, who erroneously placed it in *Epitonium.*

1327   1328   1329   1330   1331

Subgenus *Foratiscala* de Boury, 1909

Shells 5 to 10 mm., whorls angled and attached. Basal ridge weak or absent. Axial riblets and spiral threads weak. The 2 rare Western Atlantic species (illustrated in *Johnsonia,* vol. 2, no. 331) in this subgenus are:

1333 *Nystiella (F.) formosissima* (Jeffreys, 1884). 120 fms., off Lower Florida Keys; 730 to 940 fms., south of Alabama; 994 fms., off Portugal, Europe.

1334 *Nystiella (F.) pyrrhias* (Watson, 1886). 390 fms., off Culebra Island, Puerto Rico.

Suborder Gymnoglossa Gray, 1853
(Aglossa Thiele, 1929)

Family MELANELLIDAE Bartsch, 1917

Genus *Melanella* Bowdich, 1822

Shell elongate, white, with an oily, glossy surface. Whorls numerous, slightly convex. Apex sometimes bent to one side. Not umbilicated. *Eulima* Risso, 1826, is a synonym. Type: *dufresnii* Bowdich, 1822 (is *arcuata* Sowerby?). There are many named forms in this group, and their speciation is in need of revision. Some are parasites of holothurians, starfish and sea urchins.

Melanella jamaicensis (C. B. Adams, 1845)   **1335**
Jamaica Melanella

Virginia to Florida and the West Indies.

6 to 9 mm., slender, glossy, milk-white, nearly straight, except for the apex which is slightly bent. 12 to 14 whorls, the last sometimes faintly subangulate. Aperture ovate, narrow above and well advanced in the middle. Common; 1 to 200 fathoms.

Subgenus *Balcis* Leach, 1847 (in Gray)

Type: *alba* Da Costa, 1778 (*montagui* Leach, 1847, is a synonym). These snails attach themselves to various genera of holothurian "sea cucumbers," generally at the anal or oral end and suck the blood of their hosts. Synonym: *Vitreolina* Monterosato, 1884.

Melanella intermedia (Cantraine, 1835)   **1336**
Cucumber Melanella

New Jersey to Brazil. Bermuda.
Europe.

6 to 12 mm., elongate, with 10 to 13 whorls which taper very gradually to a sharp apex. Glossy white. Aperture narrow, outer lip thin and sharp. Common; found living as a parasite on the sea cucumber, *Holothuria impatiens.*

Melanella conoidea Kurtz and Stimpson, 1851   **1337**
Conoidal Melanella

Florida and the West Indies.

4 to 6 mm., slender. Last whorl faintly subangulate at the base. Highly polished, all-white or slightly touched with brown. Moderately common. *Eulima alba* Calkins, 1878, is a synonym.

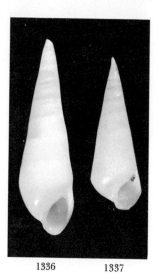

1336   1337

Melanella micans (Carpenter, 1864)   **1338**
Carpenter's Melanella

Alaska to Baja California.

9 to 12 mm., rather straight, elongate, with about 15 flattened whorls. Parietal wall covered with a moderately thick glaze. Common; 1 to 30 fathoms.

(1339) The subspecies *borealis* Bartsch, 1917, occurring from Kodiak Island, Alaska, to Vancouver Island, British Columbia, is uniformly more slender. 12 whorls; length 11.3 mm.; width 3.3 mm.

Melanella rutila (Carpenter, 1864)   **1340**
Rutila Melanella

Vancouver Island to Baja California.

6 to 7 mm., straight, slender, polished. Periphery of the last whorl rounded, the base sloping in such a way as to lend the left outline a somewhat flattened appearance. Parietal wall with weak callus. Common; on starfish from 1 to 360 fathoms.

1338   1340

1339

Other Atlantic species:

1341 *Melanella arcuata* C. B. Adams, 1850. North Carolina to the West Indies.

1342 *Melanella carolii* Dall, 1889. North Carolina to the West Indies. (Synonym: *affinis* C. B. Adams, 1850, non Philippi, 1844.)

1341

1343 *Melanella elongata* Bucquoy, Dollfuss and Dautzenberg, 1883. North Carolina to the Florida Keys, 22 to 100 fms. (Synonyms: *distorta* Verrill, 1880, non Philippi, 1844; *Eulima perversa* Bush, 1909.)

1344 *Melanella gibba* De Folin, 1867. North Carolina and the Gulf of Mexico.

1345 *Melanella gracilis* (C. B. Adams, 1850). North Carolina to the Gulf of Mexico and the West Indies. (See *hypsela*.)

1344   1345   1359

1346 *Melanella subcarinata* Orbigny, 1842. North Carolina to Florida and the West Indies.

1347 *Melanella albida, anachorea* **(1348)**, *callistemma* **(1349)**, *cinca* **(1350)**, *corrida* **(1351)**, *fernandinae* **(1352)**, *ira* **(1353)**, *ophiodon* **(1354)**, *parallella* **(1355)**, *penna* **(1356)**, *stamina* **(1357)** and *versus* **(1358)** all Dall, 1927, all off Georgia and Fernandina, Florida, 294 to 440 fms.

1359 *Melanella hypsela* (Verrill and Bush, 1900). Bermuda; Havana, Cuba. *Trans. Conn. Acad. Art. Sci.*, vol. 10, p. 526, fig. 8 (not 9). (Is this *gracilis* C. B. Adams, 1850?)

1360 *Melanella bermudezi* Pilsbry and Aguayo, 1933. Varadero, Matanzas, Cuba. *The Nautilus*, vol. 46, p. 117.

1361 *Melanella amblytera, atypha* **(1362)**, *compsa* **(1363)** and *engonia* **(1364)** (Verrill and Bush, 1900). Bermuda. *Trans. Conn. Acad. Sci.*, vol. 10, p. 526.

Other Pacific species:

1365 *Melanella* (*Balcis*) *montereyensis* (Trinidad to Monterey Bay), 5 mm.; *peninsularis* **(1366)** (San Diego to Magdalena Bay, Baja California) 5 mm.; *lastra* **(1367)** (San Pedro to Magdalena Bay); *columbiana* **(1368)** (Baranoff Island, Alaska, to Departure Bay, British Columbia) 9.5 mm.; *comoxensis* **(1369)** (Comox, Vancouver Island, British Columbia) 7 mm.; *macra* **(1370)** (Departure Bay to Seattle, Wash.) 7.5 mm.; *berryi* **(1371)** (Monterey Bay to Catalina Island, Calif.); *grippi* **(1372)** (San Pedro, Calif., to Point Abreojos, Baja Calif.) 8 mm.; *catalinensis* **(1373)** (San Rosa Island, Calif., to San Hipolito Point, Baja California) all Bartsch, 1917, *Proc. U.S. Nat. Mus.*, vol. 53.

1374 *Melanella* (*Balcis*) *thersites* Carpenter, 1864. Monterey, California, to San Geronimo Island, Baja California. *M. bistorta* (Vanatta, 1899) and *M. lowei* (Vanatta, 1899) are synonyms.

1375 *Melanella* (*Melanella*) *randolphi* Vanatta, 1899. Aleutian Islands to Puget Sound. 7 mm.

1376 *Melanella compacta* Carpenter, 1864. San Pedro, California, to Point Abreojos, Baja California. 7 mm.

1377 *Melanella* (*Melanella*) *mexicana* Bartsch, 1917. Gulf of California to Acapulco, Mexico. 6.4 mm.

1378 *Melanella* (*Melanella*) *oldroydi* (San Pedro to Point Abreojos, Baja California) 9 mm.; *californica* **(1379)** (Catalina Island and San Martin, California); *hemphilli* **(1380)** (San Diego, California, to Point Abreojos, Baja California) 8.3 mm.; *tacomaensis* **(1381)** (Tacoma, Wash.) 5 mm., all Bartsch, 1917, *Proc. U.S. Nat. Mus.*, vol. 53.

1382 *Melanella* (*Sabinella* Monterosata, 1890) *bakeri* Bartsch, 1917. San Diego, California. 2.7 mm.

1383 *Melanella ptilocrinicola* (Bartsch, 1907). Off British Columbia, 1,588 fms. 9.5 mm. Parasitic on the crinoid, *Ptilocrinus pinnatus*.

1384 *Melanella rosa* Willett, 1944. Off Redondo Beach, California, 125 fms. *Bull So. Calif. Acad. Sci.*, vol. 43, p. 72.

1385 *Melanella* (*Balcis*) *titubans* (S. S. Berry, 1956). Anacapa Island, California, 46 to 58 fms. *Jour. Wash. Acad. Sci.*, vol. 46, p. 155.

1386 *Melanella* (*Balcis*) *delmontensis* (A. G. Smith and M. Gordon, 1948). Off Del Monte, California, 10 fms. *Proc. Calif. Acad. Sci.*, series 4, vol. 26, p. 219. 4.5 mm.

Genus *Strombiformis* Da Costa, 1778

Shell small, transparent, elongate, glossy, with an umbilical depression. Type: *glabra* (Da Costa, 1778). *Leiostraca*

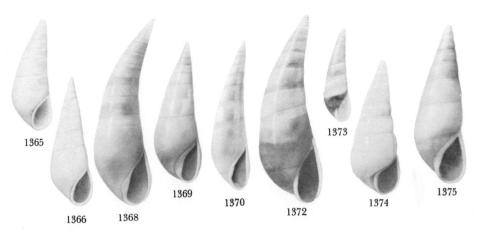

1365   1369   1373   1375
1366   1368   1370   1372   1374

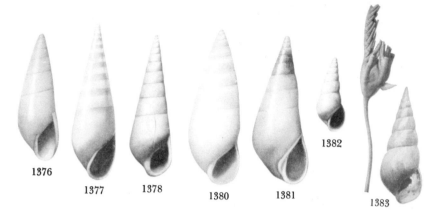

H. and A. Adams, 1853, is a synonym. The genus name is masculine.

Strombiformis auricinctus Abbott, 1958   1387
Gold-striped Melanella

North Carolina to Florida and the West Indies.

3 to 5 mm., slender, polished, 7 to 10 whorls. Spire somewhat constricted near the summit. Suture inside the whorl shows through the transparent shell. Umbilicus broad and shallow, bounded on the left by a raised keel. Outer lip advanced in the middle. A single, narrow, faint orange band occurs just above the suture. Base of columella may be tipped with orange. Uncommon; 1 to 100 fathoms. Formerly called *acuta* Sowerby, 1834, but that species comes from the Eastern Pacific.

Strombiformis bilineatus Alder, 1848   1388
Two-lined Melanella

North Carolina to the West Indies.

6 to 8 mm., polished, white, brownish at the base; slightly flattened in antero-posterior diameter; imperforate. 10 flattened whorls with 2 faint spiral yellow lines, one just below the suture, the other between. Columella concave. Moderately common.

Strombiformis hemphilli (Dall, 1884)   1389
Hemphill's Melanella

West coast of Florida; Brazil.

5 mm., slender, conic, chestnut-brown, translucent and polished. About 8 flat-sided whorls. Suture distinct, with a thin, narrow spiral thread below the suture. Columella slightly concave. Moderately common; shallow water. Host unknown.

Strombiformis patula (Dall and Simpson, 1901)   1390
Large-mouthed Melanella

Off Georgia to the West Indies.

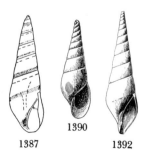

4 to 6 mm., translucent-white. About 9 whorls, with a rapidly diminishing spire, blunt at the extreme tip. Characterized by its large, flaring aperture. Moderately common; 2 to 440 fathoms.

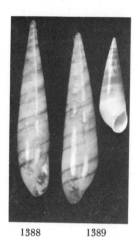

1388   1389

Other Atlantic species:

1391 *Strombiformis bifasciatus* Orbigny, 1842. Florida Keys and the West Indies.

1392 *Strombiformis fusus* (Dall, 1889). Off Fernandina, Florida, Gulf of Mexico and West Indies, 294 to 640 fms.

1393 *Strombiformis elatus* Dall, 1927. Off Georgia, 440 fms.

1394 *Strombiformis rectiuscula* (Dall, 1890). Off Fernandina, Florida, 294 fms. (Synonym: *stenostoma* Dall, 1889, non Jeffreys, 1858.)

Strombiformis californicus Bartsch, 1917   1395
Californian Melanella

Catalina Island to San Diego, California.

11 mm., with 13 flat-sided whorls. Elongate, narrow, polished. Early whorls yellowish white, succeeding ones light-brown, marked with a dark-brown band at the periphery. A second band occurs a little below the middle of the whorl. Outer lip edged with brown. Pale-brown growth streaks present on whorls. Parietal wall callused. Uncommon; 14 to 60 fathoms.

Strombiformis almo Bartsch, 1917   1396
Almo's Melanella

Santa Rosa Island to San Diego, California.

7 mm., broadly elongate-conic, polished, whitish with a broad chestnut-brown band around the middle of the whorls. 10 whorls slightly convex. Uncommon; 53 to 113 fathoms on sandy mud bottom.

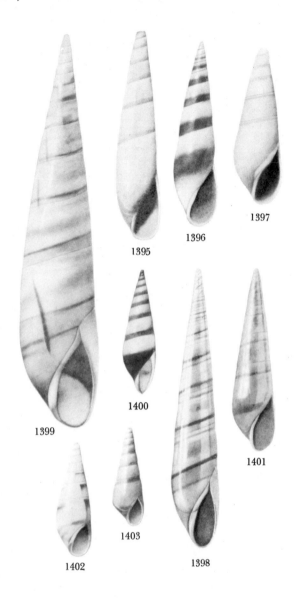

1395
1396
1397
1399
1400
1401
1402
1403
1398

Other Pacific species:

1397 *Strombiformis alaskensis* Bartsch, 1917. Unalaska, Aleutians. 4.2 mm.

1398 *Strombiformis lapazana* Bartsch, 1917. Off La Paz and in the Gulf of California. 9 to 26 fms. 7.8 mm.

1399 *Strombiformis townsendi* Bartsch, 1917. Gulf of California. 1 to 2 fms. 11 mm.

1400 *Strombiformis fuscostrigata* (Carpenter, 1864). Cape San Lucas, Baja California. 4.7 × 1.3 mm.

1401 *Strombiformis barthelowi* Bartsch, 1917. Santa Maria Bay, Baja California. 5 × 1.3 mm.

1402 *Strombiformis hemphilli* Bartsch, 1917. Point Abreojos, Baja California. 3.1 × 1.1 mm. This name is a homonym, see no. 1389.

1403 *Strombiformis burragei* Bartsch, 1917. Gulf of California. 3 fms. 2.7 × 0.9 mm.

Genus *Turveria* S. S. Berry, 1956

Shell small, glossy, whitish, ovoid, smooth, with rounded whorls, impressed suture and an ovate, narrow aperture. Parasitic on the underside of sea biscuits. Type: *encopendema* S. S. Berry, 1956.

Turveria encopendema S. S. Berry, 1956   **1404**
Turver's Snail

Cholla Bay, Gulf of California.

4 to 5 mm., ovoid, smooth, with slightly rounded whorls. Color whitish with a brown band just below the slightly impressed suture. Found attached to the giant sea biscuit, *Encope grandis. Amer. Midland Naturalist*, vol. 56, p. 355.

Genus *Couthouyella* Bartsch, 1909

Shell elongate-conic, about ½ inch, solid, white. Nuclear whorls dextral, columella simple. Surface spirally grooved. Animal without radula; eyes sessile at the outer bases of the tentacles. Operculum thin, horny, paucispiral. Type: *striatula* (Couthouy, 1839).

Couthouyella striatula (Couthouy, 1839)   **1405**
Couthouy's Snail

Nova Scotia to south Massachusetts.

12 to 15 mm., elongate-conic, strong, milk-white. Nuclear whorls dextral, 1½, smooth. Postnuclear whorls flat-sided, with numerous, low, flat-topped, irregularly sized spiral cords. Base rounded, with about 8 spiral cords. Suture well-impressed. Aperture ovate, its posterior angle acute. Outer lip thick within, curving to a sharp edge. Common offshore from 7 to 204 fathoms.

1405

Genus *Haliella* Monterosato, 1873

Shells small, very elongate, with about 5 to 8 whorls. Columella with obsolete fold. Aperture narrow and elongate. Type: *stenostoma* (Jeffreys, 1858). Two deep-water American species:

1406 *Haliella abyssicola* Bartsch, 1917. Off San Diego, California, 822 fms., to off Point Pablo, Baja California, 284 fms. 10 mm.

1407 *Haliella lomana* Dall, 1908. Off Point Loma, San Diego County, California, 650 fms. 20 mm.

Genus *Scalenostoma* Deshayes, 1863

Shell small, elongate with a high, straight-sided apex. Periphery of whorl with a strong raised keel or cord. Type: *carinatum* Deshayes, 1863, from the Indian Ocean. Syno-

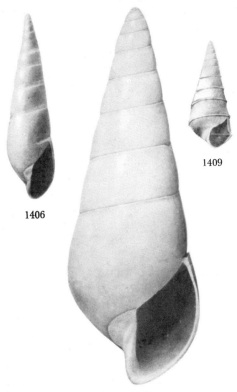

1406

1409

1407

nyms: *Amblyspira* Dall (in Guppy and Dall) 1896; *Subeulima* Souverbie, 1875. Two American species:

1408 *Scalenostoma immaculata* Dall, 1927. Off Fernandina, Florida, 294 fms.

1409 *Scalenostoma babylonia* (Bartsch, 1912). San Hipolito Point, Baja California. 3 × 1.2 mm.

Genus *Niso* Risso, 1826

Shell flat-sided, acutely conic, with a glossy surface. Umbilicus deep. Outer lip simple. Operculum corneous, thin, transparent-tan. Type: *eburnea* (Risso, 1826), Pliocene of Italy. For a review of the Eastern Pacific species, see W. K. Emerson, 1965, *Amer. Mus. Novitates*, no. 2218.

Niso hendersoni Bartsch, 1953   **1410**
Henderson's Niso   **Color Plate 3**

North Carolina to both sides of Florida.

1 inch, regularly elongate-conic, with 15 polished, shiny whorls bearing squarish brown blotches just above and below

1410   1411   1413

the brown-lined suture. Umbilicus open, funnel-shaped and bounded by a raised sharp keel and a slender brown line. A very attractive species occasionally dredged from 15 to 111 fathoms. Operculum chitinous, translucent-yellow, with a terminal apex, and filling the entire aperture. *The Nautilus,* vol. 67, p. 38. Formerly miscalled *splendidula* (Sowerby, 1834) but might be considered a subspecies of it.

Niso aeglees Bush, 1885   **1411**
Brown-lined Niso

North Carolina to Texas and the West Indies to Brazil.

10 to 14 mm., broadly conic, with about 11 whorls, flat-sided, and with an acute spire. Periphery of whorls angulate at which point there is a thin, reddish dark-brown spiral line. Color cocoa-brown with occasional flamelike patches of darker color. Former outer lips marked with an axial reddish brown line. Umbilicus deep. Common; 7 to 107 fathoms. Formerly miscalled *interrupta* (Sowerby, 1834), an Eastern Pacific species. *N. tricolor* Dall, 1889, is a synonym.

Niso portoricensis Dall and Simpson, 1901   **1412**
Puerto Rico Niso

Caribbean to Brazil.

8 to 12 mm., acutely conical, flat-sided, with about 12 flat-sided, brilliantly glossy whorls. Color chestnut-brown with a distinct light-tan narrow band at the suture and one on the middle of the base. Thin brown line at the periphery. Umbilicus deep and funnel-shaped. Columella dark-brown with a whitish base. Operculum thin, light-brown, transparent. Uncommon; dredged in Mayaguez Harbor, Puerto Rico.

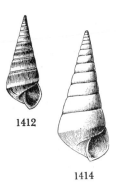

1412

1414

Niso hipolitensis Bartsch, 1917   **1413**
Hipolito Niso

San Diego, California, to the Gulf of California.

3 mm., with 10 flat-sided whorls. Narrowly umbilicate. Apex yellowish white; base white with a broad median brown band. Anterior half of aperture white. Suture feebly impressed. Periphery of the last whorl angulated. Uncommon. See *Amer. Mus. Novitates*, no. 2218, figs. 9 and 10, by W. K. Emerson, 1965.

Other Atlantic species:

1414 *Niso interrupta albida* Dall, 1889. Santa Lucia, West Indies. 8 mm.

1415 *Niso microforis* Dall, 1927. Off Georgia, 440 fms.

Other Pacific species:

1416 *Niso lomana* Bartsch, 1917. Santa Rosa Island to Point Loma, California.

1417 *Niso baueri* Emerson, 1965. Gulf of California.

1418 *Niso splendidula* (Sowerby, 1834). Angel de la Guarda Island, Gulf of California, to Ecuador, 6 to 45 fms. Very similar to *hendersoni* Bartsch of the Atlantic.

1419 *Niso interrupta* (Sowerby, 1834). Gulf of California to Panama, 11 to 57 fms.

1420 *Niso* (*Neovolusia*) *excolpa* Bartsch, 1917. Magdalena Bay, Baja California, to the Gulf of California.

1421 *Niso* (*Neovolusia* Emerson, 1965) *imbricata* (Sowerby, 1834). Santa Elena, Ecuador, 6 to 8 fms. See *Amer. Mus. Novitates,* no. 2218, p. 7, fig. 11 (1965).

Family STILIFERIDAE H. and A. Adams, 1853

Genus *Stilifer* Broderip and Sowerby, 1832

Small, globose, parasitic shells living embedded in the arms of starfish or the spines of sea urchins. Early whorls small and much smaller than the last, rather bulbous whorls. Type: *astericola* Broderip, 1832.

1422 *Stilifer stimpsoni* Verrill, 1872. Off Georges Bank, Massachusetts, 60 fms., to off east Florida. Among spines of sea urchin.

1422   1425   1429

1423 *Stilifer curtus* Verrill, 1882. Off Martha's Vineyard, Massachusetts, 410 fms.

1424 *Stilifer verrilli, minima* and *minuta* all Dall, 1927. All off Fernandina, Florida, 294 fms.

1425 *Stilifer subulatus* Broderip and Sowerby, 1832. Caribbean. (Synonyms? *bibsae* Usticke, 1959; *thomasiae* Sowerby, 1878. See "Caribbean Seashells," p. 59, fig. 13d.)

Genus *Hypermastus* Pilsbry, 1899

Small melanellids with a mucronate apex, pupiform outline, and with the inner lip appressed to the attenuated base of the preceding whorl. Type: *coxi* Pilsbry, 1899. Synonym: *Lambertia* Souverbie, 1869, non Robineau-Desvoidy, 1863. One species from the Americas:

1426

1426 *Hypermastus cookeanus* (Bartsch, 1917). San Hipolito Point, Baja California. 4 whorls; 3.7 × 2 mm. *Proc. U.S. Nat. Mus.,* vol. 53, no. 2207, p. 354.

Genus *Cythnia* Carpenter, 1864

Embedded in starfish. Similar to *Stilifer,* but the nuclear whorls are normal, not pupiform, and the operculum is multispiral. Type: *asteriaphila* Carpenter, 1857. *Cythna* is a misspelling. One United States species.

1427 *Cythnia albida* Carpenter, 1864. San Diego, southern California. Parasitic on starfishes.

1428 *Cythnia asteriaphila* Carpenter, 1864. Cape San Lucas, Baja California.

Genus *Mucronalia* A. Adams, 1862

Shells small, white, smooth, globose last whorl, attenuated apex. Aperture oval-elongate. Parasitic on spines of sea urchins. Type: *bicincta* A. Adams, 1862.

Subgenus *Pelseneeria* Kohler and Vaney, 1908

Mucronalia nidorum Pilsbry, 1956   **1429**
Pilsbry's Sea Urchin Snail

Southeast Florida.
Baja California.

4 mm., white, without an umbilicus, smooth, glossy. Upper whorls somewhat attenuated and tilting. Last whorl quite large. Aperture ⅔ times the length. Peristome entire. Inner margin concave. Columella thickened. Operculum thin, long-ovate. These snails live in the spines of the club-spined sea urchin, *Eucidaris tribuloides* (Lamarck) and cause the spine to be shortened and swollen. A pitlike nest is created and may contain 1 to 3 snails. Rare to uncommon; 25 fathoms. *The Nautilus,* vol. 69, p. 110, and *Amer. Malacol. Union Annual Rep. for 1967,* p. 74.

Other species:

1430 *Mucronalia mammillata* Dall, 1927. Off Fernandina, Florida, 294 fms.

1431 *Mucronalia suava* Dall, 1927. Off Georgia, 440 fms.

1432 *Mucronalia bulimaloides* Dall, 1927. Off Georgia, 440 fms.

Genus *Athleenia* Bartsch, 1946

Shell minute (2 mm.), elongate-ovate, thin. 3 nuclear whorls, smooth. Postnuclear whorls tabulatedly shouldered at the summit. Aperture ovate. Columella simple and short. Type: *burryi* Bartsch, 1946. Originally placed in the Stiliferidae, but it may be in the Rissoidae. *Jour. Wash. Acad. Sci.,* vol. 36, no. 1.

1433

Athleenia burryi Bartsch, 1946   1433
Burry's Athleenia

Southeast Florida.

2.1 mm. long, 0.8 mm. wide, with 8 whorls. 3½ nuclear whorls, rounded and smooth, forming a styliform apex. Adult whorls strongly tabulated above. Peristome slightly thickened at the edge and slightly angulated at the posterior shoulder. Parietal wall covered by a thin callus. Surface smooth, except for very fine growth lines. Uncommon; 66 fathoms.

Family ACLIDIDAE G. O. Sars, 1878

P. Bartsch, 1947, gave a review of the east American species of these minute, elongate-slender, white shells in the *Smithsonian Misc. Coll.*, vol. 106, no. 20. I do not follow him in his raising of *Graphis* and *Hemiaclis* to generic status.

Genus Aclis Lovén, 1846

Minute (2 to 8 mm.), elongate, slender, many-whorled, slightly umbilicate, white shells with convex whorls; bulbous, dextral nuclear whorl, and with a thin, chitinous operculum. Usually spirally striate. Type: *supranitida* (S. V. Wood, 1842).

Aclis eolis Bartsch, 1947   1434
Eolis Aclis

Off the Lower Florida Keys.

3 to 4 mm., elongate, glossy white. 2 nuclear whorls, rounded, smooth. Upper part of adult whorls smooth. Lower half with 3 equal-sized spiral cords. Base well-rounded, smooth, narrowly umbilicate. Uncommon; 65 to 95 fathoms.

Other species:

1435 *Aclis (Aclis) hypergonia* Schwengel and McGinty, 1942. Off Lantana, Florida, 550 ft. *The Nautilus*, vol. 56, p. 17.

1436 *Aclis (?Aclis) striata* Verrill, 1880. Bay of Fundy. Off Newport, Rhode Island, 100 fms.

Subgenus Hemiaclis G. O. Sars, 1878

Shells polished, marked only by faint growth lines and a hint of spiral striations. Umbilicus open or closed. No columella fold present. Type: *ventrosa* Sars, 1878. Synonym: *Stilbe* Jeffreys, 1884. Many of the east American species were originally described as *Aclis*.

Aclis verrilli Bartsch, 1911   1437
Verrill's Aclis

Labrador to off Massachusetts.

5 mm., elongate-ovate, with globose lower whorls, and the apex coming rapidly to a narrow point. Yellow-white. 4 nuclear whorls. Postnuclear whorls with 6 feeble spiral threads and numerous growth lines; old varical lines sometimes prominent. Narrowly umbilicate. Base with 7 spiral lines. Columella edge flaring. Rare; 349 to 984 fathoms. Formerly called *walleri* Jeffreys. a European species.

Aclis lata Dall, 1889   1438
Lata Aclis

Off Georgia to West Florida and the West Indies.

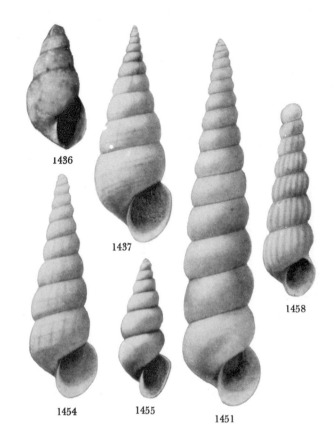

3 to 5 mm., greenish white; 13 whorls, the last being somewhat globose. Umbilicus deep. Suture bounded below by a fine line. Uncommon; 20 to 440 fathoms.

Other species:

1439 *Aclis (Hemiaclis) pyramida* Dall, 1927. Off Georgia, 440 fms. Rare.

1440 *Aclis (Hemiaclis) conula* Dall, 1927. Off Fernandina, Florida, 294 fms. Rare.

1441 *Aclis (Hemiaclis) stilifer* and *pendata* (**1442**) Dall, 1927. Both off Georgia, 440 fms., and Fernandina, Florida, 294 fms.

1443 *Aclis (Hemiaclis) marguerita* (Bartsch, 1947). Off Fernandina, Florida, 294 fms. *Smithsonian Misc. Coll.*, vol. 106, p. 20.

1444 *Aclis (Hemiaclis) hyalina* Watson, 1880. Off Pernambuco, Brazil, 350 fms.

1445 *Aclis (Hemiaclis) benedicti* (Bartsch, 1947). Off Fernandina, Florida, 294 fms.; Georgia, 440 fms.

1446 *Aclis (Hemiaclis) tenuis* Verrill, 1882. Off Georges Bank, 1,769 fms., and Martha's Vineyard, Massachusetts, 100 fms.

1447 *Aclis (Hemiaclis) tanneri* (Bartsch, 1947). Off Long Island, New York, 810 fms.

1448 *Aclis (Hemiaclis) fernandinae, georgiana* (**1449**) and *limata* (**1450**), all Dall, 1927. All off Fernandina, Florida, 294 fms., and Georgia, 440 fms.

1451 *Aclis (Hemiaclis) dalli* Bartsch, 1911. Off Cuba, 780 fms. 7.8 mm.

1452 *Aclis (Hemiaclis) sarissa* Watson, 1880. Off Pernambuco, Brazil, 350 fms.

1453 *Aclis (Hemiaclis) acuta* Jeffreys, 1884. Off Labrador, 1,622 fms.

1454 *Aclis (Hemiaclis) carolinensis* Bartsch, 1911. Off Cape Hatteras, North Carolina, 63 fms. 4.7 mm.

1455 *Aclis (Hemiaclis) rushi* Bartsch, 1911. Off Fowey Rocks, Florida, 150 to 200 fms. 2.7 mm.

Subgenus *Costaclis* Bartsch, 1947

Shells 6 to 17 mm. Axial ribs usually strong and confined to the early postnuclear whorls. White. Nucleus of 2 or 3 small, smooth whorls. Type: *nucleata* (Dall, 1889) from the West Indies. Mostly deep-water species.

1456 *Aclis (Costoclis) egregia* Dall, 1889. Off Fernandina, Florida, 294 fms.; Lesser Antilles, 796 fms. 12 mm.

1457 *Aclis (Costaclis) nucleata* Dall, 1889. Lesser Antilles, 464 to 496 fms. Off Fernandina, Florida, 294 fms. 9 to 17 mm.

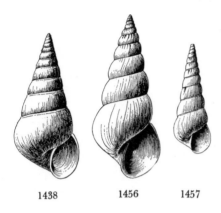

1438   1456   1457

1458 *Aclis (Costaclis) cubana* (Bartsch, 1911). Off Cuba, 780 fms. 4 mm.

1459 *Aclis (Costaclis) rhyssa* Dall, 1927. Off Georgia, 440 fms., and Fernandina, Florida, 294 fms. 6 mm.

Subgenus *Graphis* Jeffreys, 1867

Minute (2 to 3 mm.), slender, elongate, without an umbilicus and with reticulate sculpturing. Type: *Turbo unicus* Montagu, 1803. Synonyms: *Cioniscus* Jeffreys, 1869; *Pseudochemnitzia* O. Boettger, 1901; *Miraclis* O. Boettger, 1906.

Aclis shepardiana Dall, 1919   **1460**
Shepard's Aclis

San Pedro to San Diego, California.

3 to 4 mm., slender, yellowish, about 9 whorls, including the blunt, smooth nucleus. Coils of the spire rather lax. Axial sculpture of minute, closely set plications which fade out on the base of the shell. Spiral striae form an inconspicuous reticulation. No umbilicus. Uncommon; sublittoral.

Aclis underwoodae (Bartsch, 1947)   **1461**
Underwood's Aclis

Tampa Bay, Florida.

2 to 3 mm., elongate, yellowish white. One blunt nuclear whorl, smooth. Rounded 8 postnuclear whorls with strong, wavy axial ribs (30 on last whorl). Spiral sculpture of about 9 threads on the lower $\frac{2}{3}$ of each whorl. No umbilicus. Uncommon; in shallow brackish waters near mangroves.

Other Pacific species:

1462 *Aclis turrita* Carpenter, 1864. San Pedro, California.

1463 *Aclis occidentalis* Hemphill, 1894. San Diego, California.

1464 *Aclis californica* Bartsch, 1927. San Clemente Island, Baja California. 5 mm.

Genus *Schwengelia* Bartsch, 1947

Small aclids with elongate-turreted shells, having a strong shoulder extending over the posterior portion of the whorls limited anteriorly by a spiral cord. Anterior portion of the whorl bears additional spiral threads. Base narrowly umbilicate. Outer lip flaringly expanded in the middle. Type: *hendersoni* (Dall, 1927).

Schwengelia hendersoni (Dall, 1927)   **1465**
Henderson's Aclis

Off both coasts of Florida.

2 to 5 mm., vitreous or milk-white. Nucleus of 1.5 rounded, smooth whorls. Early and middle adult whorls with 4 strong spiral cords which weaken on the last 2 whorls. Suture minutely indented. Top of shoulder carinate. Base with faint spiral threads. Columella with a slightly reflected left edge. Umbilicus narrow, deep. *Aclis floridana* Bartsch, 1911, is possibly a synonym. Uncommon; 5 to 40 fathoms.

Schwengelia floridana (Bartsch, 1911)   **1466**
Florida Aclis

Southeast Florida.

3 mm., white, narrowly conic; 1½ nuclear whorls, smooth. Postnuclear whorls with a sloping, strong shoulder bounded below by a fairly strong carina. Suture well-impressed. Sculpture of feeble spiral lines and axial growth lines. Rare; 150 to 200 fathoms.

1468

1466

Genus *Bermudaclis* Bartsch, 1947

Shells minute (2 mm.), elongate-pupoid, thin, translucent with 6 to 8 rounded whorls. Nucleus with one blunt whorl. Spiral threads crossed by growth lines form a reticulate sculpture. No umbilicus. Type: *bermudensis* (Dall and Bartsch, 1911). Two Atlantic species:

1467 *Bermudaclis tampaensis* Bartsch, 1947. Tampa Bay, Florida; shallow water. 1.7 to 2 mm.

1468 *Bermudaclis bermudensis* (Dall and Bartsch, 1911). Bermuda. 2.1 mm. *Proc. U.S. Nat. Mus.*, vol. 40, p. 278.

Genus *Henrya* Bartsch, 1947

Shell minute, thin, elongate-pupoid. Nucleus of 1 rounded whorl. Growth lines very strong on the early adult whorls. Periphery strongly rounded. Umbilicus chinklike.

Columella slightly twisted. Type: *henryi* Bartsch, 1947. Live near mangroves and saline ponds.

Henrya morrisoni Bartsch, 1947 1469
Morrison's Aclis

Tampa to Marco, west Florida.

1.5 mm., minute, pupoid, slender, milk-white. Nucleus of 1 blunt whorl. Whorls well-rounded, only slightly increasing in size. Fine, slanting, growth lines conspicuous. Suture constricted. Umbilicus absent. Uncommon; shallow water near mangrove swamps.

Other species:

1470 *Henrya goldmani* Bartsch, 1947. Saline lagoon, near Progresso, Yucatan, Mexico.

1471 *Henrya henryi* Bartsch, 1947. Saline lakes in the Bahamas. All *Smithsonian Misc. Coll.*, vol. 106, no. 20.

[Order Entoconchidida]

The order Parasita Fischer, 1888, containing the degenerate, parasitic families of snails, Entoconchidae and Enteroxenidae, is now placed in the Opisthobranchia (see G. Mandahl-Barth, 1941, *Vidensk, Meddel. Dansk Naturhist. Foren.*, vol. 104; and E. S. Tikasingh and I. Pratt, 1961, *Systematic Zool.*, vol. 10). This group was formerly placed among the Aglossa of the prosobranchs by J. Thiele, 1929.

Superfamily Carinariacea Blainville, 1818
(Heteropoda Lamarck, 1801)

Dioecious, pelagic gastropods, most bearing simple shells. A review was given by J. J. Tesch, 1949, *Dana Report* no. 34.

Family ATLANTIDAE Wiegmann and Ruthe, 1832

The family name Atalantidae Gray 1840, is synonymous. These are small, fragile, discoidal shells belonging to pelagic snails.

Genus *Atlanta* Lesueur, 1817

Shell small, fragile, discoidal, transparent-white, compressed. Protoconch dextral. Operculum corneous, subtrigonal, with a small apical nucleus. Type: *peronii* Lesueur, 1817. Synonyms: *Atlanta* Orbigny, 1834; *Atlantidea* Pilsbry, 1922; *Steira* Eschscholtz, 1825.

Atlanta peronii Lesueur, 1817 1472
Peron's Atlanta

Atlantic and Pacific warm waters; pelagic.

½ inch in diameter, planorboid, compressed from above, fragile, transparent and glassy. Later whorls openly coiled but connected by a sharp peripheral keel. Outer lip notched

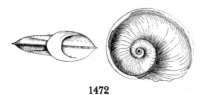

1472

in the region of the shell. Often washed ashore after storms, and frequently brought up in dredge hauls. 5 other pelagic species have been reported from American offshore oceanic waters. Synonyms: *rosea* Souleyet, 1852; *steindachneri* Oberwimmer, 1898. *A. gaudichaudi* Souleyet, 1852, may only be a form.

Atlanta brunnea Gray, 1850 1473
Brown Atlanta

Gulf of Mexico to mid-Atlantic; Eastern Pacific Mediterranean and Indo-Pacific.

2 mm., intense buff color, inflated shell. Spire elevated and with microscopic spiral lines. Aperture roundish. Fissure in lip very short. First ½ of body whorl with 5 to 7 wavy spiral lines above and below the peripheral carina. *A. fusca* Souleyet, 1852, is a synonym. Moderately common; pelagic.

Other species:

1474 *Atlanta lesueuri* Souleyet, 1852. Worldwide; pelagic. (Synonym: *oligogyra* Tesch, 1906.)

1475 *Atlanta quoyii* Gray, 1850. Atlantic and Indo-Pacific. (Synonyms: *quoyana* Souleyet, 1852; *inflata* Souleyet, 1852.)

1476 *Atlanta helicinoides* Souleyet, 1852. Atlantic and Indo-Pacific (Synonym?: *depressa* Souleyet, 1852.)

1477 *Atlanta inclinata* Gray, 1850. Atlantic and Pacific. (Synonyms: *inclinata* Souleyet, 1852; *gibbosa* Souleyet 1852; *macrocarinata* Bonnevie, 1920.)

1478 *Atlanta pulchella* Verrill, 1884. Off Massachusetts and Delaware.

Genus *Protatlanta* Tesch, 1908

Shell right-handed, chalky-translucent with a few very indistinct spiral lines on both sides. Large umbilicus on the under side. Peripheral keel thin, cartilaginous, but extending only ½ the circumference of the last whorl. No slit on outer lip. Operculum roundish, thin, with a spiral at the narrower end. Type and only species: *souleyeti* (E. A. Smith, 1888). *Protoatlanta* Wenz, 1941, is a misspelling.

1479 *Protatlanta souleyeti* (E. A. Smith, 1888). Caribbean; central Atlantic. Off Cape Hatteras. Off Bermuda. Indo-Pacific. 1 to 1.5 mm. (not including keel). Common. (Synonym: *Atlanta lamanonii* Souleyet, 1852, non Eschscholtz, 1825.)

Genus *Oxygyrus* Benson, 1835

Shell fragile, compressed, narrowly umbilicate on both sides; nucleus not visible; back rounded, keeled only near the aperture. Type: *keraudrenii* (Lesueur, 1817). Synonyms: *Helicophlegma* Orbigny, 1385; *Ladas* Cantraine, 1841.

Oxygyrus keraudrenii (Lesueur, 1817) 1480
Keraudren's Atlanta

Atlantic warm waters; pelagic. Bermuda.
Off California; pelagic. Panama.

½ inch in diameter, planorboid, nuclear whorls not visible; narrowly umbilicate on both sides. Whorls keeled only near the aperture. Body whorl near the aperture and the keel are corneous. No apertural slit. Operculum small, trigonal and lamellar. A common pelagic species, and the only one reported from our Atlantic waters. Synonyms: *rangii* Orbigny, 1836; *inflatus* Benson, 1835.

1480

Family CARINARIIDAE Blainville, 1818

Genus *Carinaria* Lamarck, 1801

Type of the genus is *vitrea* (Gmelin, 1791), which is *cristata* Linné, 1766. *Tithyonia* Tiberi, 1880, is a synonym. A pelagic group. Swims rapidly forward by undulating the foot and caudal appendage. Of the 4 recognized species, only one occurs in the Atlantic, but 2 are found west of California. See Tesch, 1949, *Dana Report* no. 34.

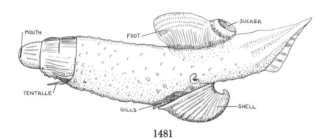

1481

Carinaria lamarcki Péron and Lesueur, 1810   1481
Lamarck's Carinaria

Gulf of Mexico; Western Atlantic. Bermuda.
Eastern Pacific and Indo-Pacific.

Body up to 10 inches in length, tissues transparent; proboscis large and purple. Shell ⅕ the size of the animal, cap-shaped, very thin, fragile and transparent. Its apex is hooked. The shell is borne below the animal. This is a valuable collector's item, and in former years it brought fancy prices. Synonyms are *C. atlantica* Adams and Reeve, 1848; *mediterranea* Blainville, 1825; *oceanica* Vayssière, 1904; *challengeri* Bonnevie, 1920; *cymbium* Guérin, 1829; *australis* Quoy and Gaimard, 1832; *punctata* Orbigny, 1836; *grimaldi* Vayssière, 1904.

1481

Carinaria galea Benson, 1835   1482
Helmet Carinaria

West of California (130° W)
Indo-Pacific; Galapagos.

Body only 1 to 2½ inches; right tentacle vestigial. No distinct tubercles on the body. Shell has a distinctly higher elevation than in *lamarcki* and has a very broad keel, which measures about ½ the length of the shell mouth, and runs up, gradually diminishing in height, to the summit. Uncommon. *C. gaudichaudi* Souleyet, 1852, is a synonym. Not recorded from the Atlantic.

Carinaria cithara Benson, 1835   1483
Harp Carinaria

West of California (130° W)
Indo-Pacific.

Body 2 inches; right tentacle absent. A few tubercles present on the body. Shell in the form of a high rectangle; plications nearly straight. Nuclear whorls with 2 fine, spiral lines. Common. *C. latidens* Dall, 1919, from the North Pacific is probably this species. *C. macrorhynchus* Tesch, 1906, is a synonym.

Genus *Cardiapoda* Orbigny, 1835

Animal resembling that of *Carinaria;* tail simple and pointed. Shell minute, cartilaginous, with a greatly expanded lip which is bilobed in front. Spire enveloped behind. Type: *placenta* (Lesson, 1830). *Carinairoida* Souleyet, 1852, is a synonym.

Cardiapoda placenta (Lesson, 1830)   1484
Flat Cardiapod

North Pacific; pelagic.
Gulf of Mexico, Caribbean, West Africa.
Tropical eastern Pacific and Indo-Pacific.

Body 1 inch. Gills numerous, more than 20, arranged in a row around the visceral nucleus. Swim-fin large. Tail ending in a star-shaped expansion. Shell about 2 mm., very fragile, of 3 planorboid coils, and with a flaring, winged outer lip. Abundant. Synonyms: *pedunculata* Orbigny, 1836; *trachydermon, sublaevis* and *acuta* all Tesch, 1906.

Family PTEROTRACHEIDAE Gray, 1840

No shell. Animal elongate, cylindrical, translucent, with a ventral fin. Gills exposed on the posterior part of the back. Pelagic.

Genus *Pterotrachea* Forskål, 1775

Head with a large proboscis. Ventral fin narrowed at the base, usually with a small sucker. Tail keeled, sometimes pinnate. Branchial processes numerous, conical and slender. Tentacles very short. Type: *coronata* Forskål, 1775. Synonyms: *Firola* Bruguière, 1791; *Anops* Orbigny, 1835; *Eupterotrachea* Bonnevie, 1920; *Euryops* Tesch, 1906; *Hypterus* Rafinesque, 1814; and *Heterodens* Bonnevie, 1920.

1485 *Pterotrachea keraudrenii* Eydoux and Souleyet, 1832. Pelagic. Atlantic. Rare. Questionable record.

1486 *Pterotrachea scutata* Gegenbaur, 1855. Has a large, gelatinous hyaline sheath over the anterior part of the trunk. 3 to 4 inches. Caribbean; east coast U.S.; Mediterranean and Indo-Pacific. *Firola gegenbauri* Vayssière, 1904, is a synonym.

1487 *Pterotrachea coronata* Forskål, 1775. Gulf of California; Indo-Pacific.

Genus *Firoloida* Lesueur, 1817

Resembling *Pterotrachea,* head tapering, furnished in the males with 2 slender tentacles. With or without small branchial filaments. Egg-tube regularly annulated. Tail fin small and with a minute sucker in the males. Pelagic. Type: *desmarestia* Lesueur, 1817. Synonyms: *Cerophora* Orbigny, 1835; *Demarestia* Griffith and Pidgeon, 1834; *Firolella* Troschel, 1855. One species:

1488 *Firoloida demarestia* Lesueur, 1817. North Atlantic (below 40° N); Caribbean and Sargasso Sea (abundant); Bermuda. Indo-Pacific. Gulf of California. (Synonyms: *lesueuri* and *gaimardi* Orbigny, 1835; *blainvilleana* and *aculeata* Lesueur, 1817; *gracilis* and *vigilans* Troschel, 1855; *liguriae* Issel, 1910; *kowalewskyi* Vayssière.) *F. demarestia* was an original misspelling for *desmarestia.*

Superfamily Hipponicacea Troschel, 1861

Family HIPPONICIDAE Troschel, 1861

Genus *Hipponix* Defrance, 1819

Shell cap-shaped, apex (originally of minute coiled nuclear whorls) eroded away, sculpture of radial ribs. Muscular impression on the inside is horseshoe-shaped, opening anteriorly. Most species form a calcareous base upon which the animal sits. Periostracum usually fur-like. *Hipponyx* is an incorrect spelling. Type: *cornucopiae* Lamarck, 1802, an Eocene species. *Amalthea* Schumacher, 1817 (not Rafinesque, 1815) is a synonym.

Hipponix antiquatus (Linné, 1767)   **1489**
White Hoof-shell

Florida to Brazil.
Graham Island, British Columbia, to Peru.

½ inch in size, white, heavy for its size, cap-shaped, and usually with a poorly developed spire which may be located either at one end of the shell or near the center. The nuclear whorls are spiral and glassy-white. There is a horseshoe-shaped muscle scar inside the shell. Axial sculpture of prominent, rugose ribs which are crossed by microscopic, incised lines. Periostracum absent or very thin and light-yellowish. Moderately common. Found clinging under rocks and other shells in the intertidal zone.

Some Pacific northwest specimens are limpetlike in shape, flattish, circular, gray-white, with the apex near the center of the shell, and with smoothish, strong, circular cords **(1490)** (form *cranoides* Carpenter, 1864). Another form bears foliaceous concentric lamellae which are finely striate axially **(1491)** (*serratus* Carpenter, 1856, from Monterey to Panama). Keen uses the name *panamensis* C. B. Adams, 1852, for the Eastern Pacific forms.

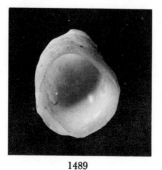

1489

Hipponix subrufus subrufus (Lamarck, 1819)   **1492**
Orange Hoof-shell

Florida and the West Indies; west Panama. Brazil.

½ inch in size, similar to *antiquatus,* but usually stained with light orange-brown, and with numerous, small spiral cords crossing concentric ridges of about the same size. This frequently gives a beaded surface. Periostracum fairly heavy, tufted and light-brown. Moderately common; on rocks in shallow water.

Hipponix subrufus tumens (Carpenter, 1865)   **1493**
Pacific Orange Hoof-shell

Crescent City, California, to Baja California.

Very close in characters to the Atlantic *subrufus subrufus,* but the shell is white in color (although the periostracum is yellow-brown), with more prominent spiral threads, and with coarser spiral threads in the young. Found littoral in crevices, and dredged commonly down to 40 fathoms.

Hipponix pilosus (Deshayes, 1832)   **1494**
Bearded Hoof-shell

Baja California to Ecudaor.

1 inch, apex nearer the posterior end, somewhat pointed anteriorly where, on the dorsal slope, the yellowish brown periostracum is in shaggy, straight, raised rows. The edge of the shell is strongly serrated with cut lines. Radial riblets nodulated. A common shoreline species found among rocks. *H. barbatus* (Sowerby, 1835) is a synonym.

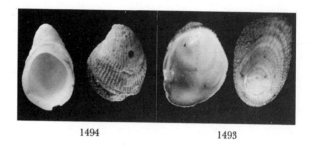

1494   1493

Genus *Malluvium* Melvill, 1906

Shell small, cap-shaped, with a hooked-over, small apex at the posterior end. A thick shelly platform is secreted upon which the animal sits. Some species live on the spines of sea urchins. Type: *lissus* (E. A. Smith). One American species:

Malluvium benthophilum (Dall, 1889)   **1495**
Dall's Hoof-shell

Southeast Florida to the West Indies.

5 to 8 mm., narrow, apex hooked-over, white to yellowish, and, except for crude growth lines, is smoothish. The shelly platform is oval, thick and whitish. A rare deepsea species found attached to the spines of large sea urchins.

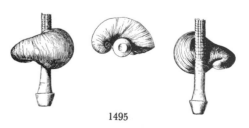

1495

Family FOSSARIDAE Troschel, 1861

Genus *Fossarus* Philippi, 1841

Shells 2 to 10 mm., white, umbilicate, subglobose, spirally corded; spire rather short, last whorl large; aperture semi-circular; columellar margin nearly straight; outer lip rounded, undulated. Operculum subconcentric; chitinous. Type: *ambiguus* (Linné, 1758), from the Eastern Atlantic.

Fossarus elegans Verrill and Smith, 1882   1496
Elegant Fossarus

Massachusetts to Florida.

2 to 3 mm. in length, turbinate in shape, with 4 whorls, chalky-white to gray in color and characterized by its delicate sculpturing which consists of 2 strong carinae on the periphery and 3 smaller ones below and a large one bordering the chink-like umbilicus. Between the cords are numerous distinct, arched riblets. Outer lip thickened by a large varix. 2 or 3 smaller, former varices commonly present on the last whorl. The base of the arched columella is projecting. Uncommon; 70 to 140 fathoms.

1496   1500   1501

Fossarus orbignyi Fischer, 1864   1497
Orbigny's Fossarus

Bahamas to Brazil. Bermuda.

2 mm. in length, equally wide, pure-white except for the pimplelike, reticulated, translucent-brown nuclear whorls. 3 postnuclear whorls strongly shouldered, the last bearing 6 to 7 strong, raised, squarish, smoothish, spiral cords, between which are 3 or 4 microscopic threads. Umbilicus narrow, slit-like and fairly deep. Not uncommon from low tide to 2 fathoms, in sand. *Narica sulcata* Orbigny, 1842 (in part) is a synonym. Reported from Bermuda by R. Jensen (*in litt.*). *F. tridentata* Nowell-Usticke, 1959, is the same.

1497

Other species:

1498 *Fossarus parcipictus* Carpenter, 1864. Baja California to Panama.

1499 *Fossarus lucanus* Dall, 1919. Cape San Lucas, Baja California.
— *Fossarus angiolus* Dall, 1919, is a *Parviturbo*.

1500 *Fossarus* (*Gottoina* A. Adams, 1863) *bellus* (Dall, 1889). Off North Carolina, 15 to 107 fms. Florida Keys. 3.5 mm.

1501 *Fossarus* (*G.*) *compactus* (Dall, 1889). Off North Carolina and the Florida Keys, 49 to 107 fms. 2.3 mm.

Genus *Iselica* Dall, 1918

Shells less than 8 mm., umbilicate, elevated spire, with spiral cords or cancellate; columella with a small median tooth or swelling. Type: *Isapis anomala* C. B. Adams, 1850. Keen, 1971, p. 770, places this genus in the Pyramidellidae. The soft parts are unknown.

Iselica anomala (C. B. Adams, 1850)   1502
Anomalous Fossarus

East Florida to the West Indies. Brazil.

3 to 5 mm., ovate, white, sculptured with prominent spiral cords (usually 7 or 8 on the last whorl) and with numerous fine, axial lines between the cords. Aperture large, ovate. Columella with a weak, transverse fold. Uncommon from 3 to 294 fathoms.

Iselica fenestrata (Carpenter, 1864)   1503
Fenestrate Fossarus

Puget Sound, Washington, to Nicaragua.

6 mm. in length, especially wide, columella with a small median tooth; sculpture of 9 sharp, spiral cords on the body whorl. Color light yellowish brown. An abundant snail among beds of *Mytilus* mussels, and dredged on rocky bottoms down to 25 fathoms.

1503

Iselica obtusa (Carpenter, 1864)   1504
Obtuse Fossarus

Vancouver, British Columbia, to San Liego, California.

6 mm. in length, not as wide; similar to *fenestrata* but with 7 spiral cords on the body whorl. Color a glossy, pale-rose. Uncommon from 5 to 25 fathoms, usually on a gravel bottom. *I. obtusa laxa* Dall, 1919, from Vancouver Island, is a synonym.

Other species:

1505 *Iselica maculosa* (Carpenter, 1857). Gulf of California.

1506 *Iselica ovoidea* (Gould, 1856). Gulf of California.

1502   1507   1504

Family VANIKOROIDAE Gray, 1845

Genus *Vanikoro* Quoy and Gaimard, 1832

Shell rather thin, usually white, with a smooth columella, usually umbilicate; sculptured spirally and axially. Operculum paucispiral. Type: *Sigaretus cancellatus* Lamarck, 1822. *Merria* Gray, 1839, is a synonym. *Vanicoro* is a misspelling.

Vanikoro oxychone Mörch, 1877   **1507**
West Indian Vanikoro

    Southeast Florida and the West Indies; Bermuda.

    ⅓ inch in length, solid, strong and pure white. With 3 whorls. Characterized by its large aperture, by its deep, narrow, arched umbilicus and straight, rounded, pillarlike columella and by the 10 to 12 beaded, spiral cords on the last whorl. Small axial threads tend to give a slightly cancellate sculpture. Apex glassy and smooth. Suture well-indented. Uncommon in shallow water to several fathoms. *Macromphalina pilsbryi* Olsson and McGinty, 1958, is a synonym (*Bull. Amer. Paleontol.*, no. 177).

Other species:

1508   *Vanikoro sulcata* (Orbigny, 1842). West Indies. Common in shallow water.

1509   *Vanikoro aperta* (Carpenter, 1864). Baja California to Guaymas, Mexico. Rare.

Family CAPULIDAE Fleming, 1822

Genus *Capulus* Montfort, 1810

Shell cap-shaped, thin-shelled, with a small spiral, hooked apex and a very large, open aperture. Type: *ungaricus* (Linné, 1767). Synonym: *Pileopsis* Lamarck, 1812.

Capulus ungaricus (Linné, 1767)   **1510**
Fool's Cap Limpet

    Greenland to off Florida. Bermuda.
    Arctic Seas to Mediterranean.

    2 inches, cap-shaped, strong, pale-flesh to white, with an ashy-brown, shaggy periostracum. Irregularly sculptured with coarse striae. Apex spiral, inclined to one side. Animal whitish yellow, the mantle reddish and bordered by a bright yellow or orange fringe. Tentacles long, yellowish and eyes near the outside of the base. Adheres to rocks and shells; uncommon from 15 to 458 fathoms. See C. M. Yonge, 1938, "Evolution of ciliary feeding in the Prosobranchia, with an account of feeding in *Capulus ungaricus.*" *Jour. Mar. Biol. Assoc. United Kingdom*, vol. 22, p. 453. Synonym: *thorsoni* Nordsieck, 1969, p. 225. The Gulf of Mexico species are less than 10 mm.

1510

Subgenus *Krebsia* Mörch, 1877

Type of this subgenus is *incurvatus* (Gmelin, 1791).

Capulus incurvatus (Gmelin, 1791)   **1511**
Incurved Cap-shell

    North Carolina to Florida and to Brazil.

½ inch in size, cap-shaped, white to cream, and with a very large, circular to slightly oval aperture. 1½ to 2 whorls. Spire small, usually tightly coiled, but sometimes partially free. Early whorls usually with small spiral cords, but these are frequently worn smooth. Sculpture of small, irregular, rounded growth lines which are crossed by numerous spiral cords which may be rounded or sharp. Periostracum thick, light-brown, with spirally arranged rows of small tufts. Muscle scar within the aperture is horseshoe-shaped with the swollen end just inside the columella. Uncommon on rocks just below low water. I believe that *C. intortus* (Lamarck, 1822), is merely a variant of this species. Compare with *Hipponix antiquatus* which is much heavier, lacks the spiral cords and is more coarsely sculptured.

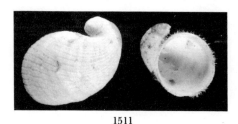

1511

Capulus californicus Dall, 1900   **1512**
Californian Cap-shell

    Redondo Beach, California, to Baja California.

    1½ inches in diameter, ⅓ as high, obliquely ovate, fairly thin and with a small, hooked-over apex. Shell white, covered by a soft, fuzzy, light-brown periostracum. Interior glossy-white. A rather rare species found in 20 to 30 fathoms attached to *Pecten diegensis*.

Other species:

1513   *Capulus sericeus* Burch and Burch, 1961. Off Guaymas, Mexico, 100 fms. *The Nautilus*, vol. 75, p. 19.

1514   *Capulus* (*Hyalorisia* Dall, 1889) *galeus* Dall, 1889. Off Barbados, Lesser Antilles. Type of the subgenus. 18 mm.

1512   **1514**

Genus *Thyca* H. and A. Adams, 1854

Shell small, cap-shaped, white, without periostracum; septum reduced; spire hooked; exterior ridged. Parasitic on starfish and crinoids. Mainly an Indo-Pacific group, usually in deep water. Type: *astericola* A. Adams and Reeve, 1850.

Subgenus *Bessomia* S. S. Berry, 1959

Apex submarginal, protoconch of 3 whorls; septal shelf sculptured. Type: *callista* S. S. Berry, 1959.

1515   *Thyca* (*Bessomia*) *callista* S. S. Berry, 1959. On starfish; Bahia San Carlos, Gulf of California, Mexico. 7.4 mm.

Superfamily Crepidulacea Fleming, 1822

Family TRICHOTROPIDAE Gray, 1850

Genus *Trichotropis* Broderip and Sowerby, 1829

Shells about 1 inch, turbinate, thin-shelled but strong, spirally carinate. Aperture large. Periostracum horny, usually with bristlelike hairs on the carinae. Operculum chitinous, paucispiral. Type: *bicarinata* (Sowerby, 1825).

Trichotropis bicarinata (Sowerby, 1825)   **1516**
Two-keeled Hairy-shell

Arctic Ocean to Queen Charlotte Islands, British Columbia. Arctic Ocean to Newfoundland.

1½ inches in length, equally wide, with about 4 whorls. Characterized by 2 strong, spiral carinae at the periphery, by the wide, flattened columella and by the flaky, brown periostracum which is grossly spinose on the carinae. Moderately common just offshore in cold water. Synonyms are *sowerbiensis* Lesson, 1832; *tenuis* E. A. Smith, 1877; *hjorti* Friele, 1878.

Trichotropis insignis Middendorff, 1849   **1517**
Gray Hairy-shell

Alaska to northern Japan.

1 inch in length, similar to *T. bicarinata* but smaller, with a much heavier shell, weakly carinate with other numerous, uneven, spiral threads, and with a thin, grayish periostracum. Both this species and *bicarinata* are easily distinguished from the more common *cancellata* by their much shorter spires and large, flaring apertures. Uncommon just offshore.

Subgenus *Ariadnaria* Habe, 1961

Type of this subgenus is *borealis* Broderip and Sowerby. *Ariadna* Fischer, 1864 (now Audouin in Savigny, 1826) is a synonym.

Trichotropis borealis Broderip and Soweby, 1829   **1518**
Boreal Hairy-shell

Arctic Seas to Maine.
Arctic Seas to British Columbia.

½ to ¾ inch in length, with 4 or 5 carinate whorls. Shell not very strong, chalky-white and covered with thick, brownish periostracum which has hairy spicules on the region of the shell's 3 major spiral cords. Umbilicus chinklike, bordered by a large, spiral cord. Spire usually eroded badly. Numerous, crowded axial threads present. A common cold-water species found from below low water to 90 fathoms. *T. costellata* Couthouy, 1838; *conica* Möller, 1842; *atlantica* Möller, 1842; *inermis* Hinds, 1843; *saintjohnensis* Verkrüzen, 1877, are synonyms.

Subgenus *Turritopis* Habe, 1961

Type of this subgenus is *cedonulli* A. Adams of Japan.

1516   1517   1518   1519

Trichotropis cancellata (Hinds, 1843)   **1519**
Cancellate Hariy-shell

Bering Sea to Oregon.

¾ to 1 inch in length, 5 to 6 rounded whorls bearing between sutures 4 to 5 strong spiral cords, between which there may be small axial ribs which produce a cancellate sculpturing. Spire high, rather pointed. Aperture a little more than ⅓ the length of the shell. Periostracum thick, brown and with long spicules over the region of the cords. Commonly dredged in cold, shallow water. One year old snails, 15 to 24 mm. long, are males, but turn into females in the second year and die after laying gelatinous egg capsules. For biology, see C. M. Yonge, *Bio. Bull. for 1962*, pp. 158–181.

Other species:

1520 *Trichotropis* (*Iphinoe* H. and A. Adams, 1854) *coronata* Gould, 1860. Arctic Ocean and Bering Sea, 20 to 30 fms. 24 mm.

1521 *Trichotropis* (*Iphinoe*) *kroyeri* Philippi, 1848. Arctic Ocean; Bering Sea to the Shumagin Islands. Northern Europe. Japan. 30 mm.

1522 *Trichotropis* (*Iphinoe*) *kelseyi* Dall, 1908. Off San Diego, California, 359 fms., to off Pt. San Quentin, Baja California.

1523 *Trichotropis* (*Iphinoe*) *nuda* Dall, 1927. Off Fernandina, Florida, 294 fms.

1524 *Trichotropis turrita* Dall, 1927. Off Fernandina, Florida, 294 fms.

1525 *Trichotropis* (*Cerithioderma* Gabb, 1860) *migrans* Dall, 1881. Florida Strait, 80 fms. West Florida, 70 to 200 fms.

1520   1521   1525

1526 *Trichotropis* (*Provana* Dall, 1918) *lomana* Dall, 1918. Off Point Loma, San Diego County, California, 650 fms. Type of the subgenus.

Genus *Torellia* Lovén (in Jeffreys) 1867

Resembling a small *Littorina*, 10 to 15 mm., globose whorls, pointed spire, narrowly umbilicate, aperture rim entire; thin-shelled, chalky-white, with a pale-brown velvety periostracum. 6 whorls spirally striate. Type: *vestita* Jeffreys, 1867.

1527 *Torellia vestita* Jeffreys, 1867. Gulf of Maine, 150 fms., to off Martha's Vineyard, Massachusetts, 86 to 146 fms. (Synonym: *Recluzia aperta* Jeffreys.)

1528 *Torellia fimbriata* Verrill and Smith, 1882. Gulf of Maine, 52 to 90 fms.; south of Martha's Vineyard, Massachusetts, 142 to 258 fms.

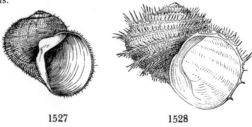

1527   1528

1529 *Torellia ammonia* Dall, 1919. Southwest of Sannakh Island, Alaska, 41 fms.

1530 *Torellia vallonia* Dall, 1919. Nazan Bay, Atka Island, Aleutians. Both *Proc. U.S. Nat. Mus.,* vol. 56, p. 355.

1531 *Torellia (Haloceras* Dall, 1889) *cingulata* Verrill. West Indies. Deep water.

Genus *Dolophanes* Gabb, 1872

The genus *Dolophanes* is now considered a synonym of *Crepitacella* Guppy, 1867, and is placed in the family Rissoinidae. See pages 77 and 78 for a listing of the known species.

Family CREPIDULIDAE Fleming, 1822

Subfamily CALYPTRAEINAE Blainville, 1824

Genus *Calyptraea* Lamarck, 1799

Shell conical with straight sides; apex usually central and may be slightly spiral. Aperture basal with an internal spiral diaphragm which has a twisted columellar border. Type: *Patella chinensis* Linné, 1758.

Calyptraea centralis (Conrad, 1841)   1534
Circular Cup-and-saucer

North Carolina to Texas and the West Indies. Brazil.

¼ to ½ inch in diameter, cap-shaped, with a circular base, and pure-white in color Apex central, small, minutely coiled and glossy-white. The shelly cup is attached to the inside of the shell and is flattish, arises near the center of the shell and flares out to the edge. Its free side is thickened into a columellalike, rounded edge. Commonly dredged in shallow water, especially off southeast Florida. Formerly known as *C. candeana* Orbigny, 1842.

Calyptraea fastigiata Gould, 1856   **1535**
Pacific Chinese Hat

Alaska to southern California.

⅓ to 1 inch in diameter, about ½ to ⅓ as high; the outline of the base of the shell is perfectly circular, and the apex is at the center of the shell. Interior glossy-white with the sinuate edge of the internal cup arising at the apex of the shell as a thickened, twisted columella and ending in fragile attachment near the edge of the shell. Young forms or those coming from shallower, warmer water (*C. contorta* Carpenter, 1865) are relatively higher-spired. Exterior chalky-white with a thin, brownish periostracum. Dredged moderately commonly from 10 to 75 fathoms.

(1536) A similar southern species, *Calyptraea mamillaris* Broderip, 1834 (synonym: *regularis* C. B. Adams, 1852) is common from Baja California, the Gulf, and south to Peru from shore to 20 fathoms. The spire is purplish brown, the size up to 1½ inches.

Calyptraea burchi A. G. Smith and Gordon, 1948   **1537**
Burch's Chinese Hat

Monterey Bay, California.

⅔ inch, very close to (if not the same species) *fastigiata,* but the interior is light yellow-brown with flecks and flames of darker brown. Nuclear whorls yellow-brown. 15 to 40 fathoms on shale bottom.

Other Pacific species:

1538 *Calyptraea conica* Broderip, 1834. Magdalena Bay and Gulf of California to Ecuador, offshore to 37 meters. (Synonyms: *sordida* Broderip, 1834; *aspersa* C. B. Adams, 1852.)

1539 *Calyptraea subreflexa* (Carpenter, 1856). Gulf of California to Acapulco, Mexico.

Genus *Cheilea* Modeer, 1793

Shell cap-shaped; with an internal cup support that is attached only at its base and has ⅓ of it neatly cut out. Type: *equestris* (Linné, 1758). *Mitrularia* Schumacher, 1817; *Lithedaphus* Owen, 1842; and *Calyptra* H. and A. Adams, 1854, are synonyms.

Cheilea equestris (Linné, 1758)   **1540**
False Cup-and-saucer

Both sides of Florida, the West Indies to Brazil.
Gulf of California to Chile.

½ to 1 inch in size, cap-shaped, dull-white, and with an internal, delicate, deep cup which has its anterior ⅓ neatly sliced away. The base of the cup is attached near the center of the inside of the shell but slightly off in the direction in which the apex of the shell points. Exterior has small, axial corrugations or tiny cords, rarely spinose. Nucleus minute,

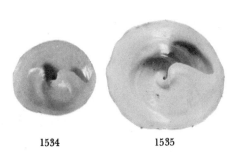

1534   1535

1540

spiral and glassy-white. Uncommon, except in the West Indies, where it may be locally common. Found on rocks below the low-tide line to 11 fathoms. Synonyms are *C. cepacea* (Broderip, 1834); *planulata* (C. B. Adams, 1852); and *tortilis* Reeve, 1858.

Cheilea corrugata (Broderip, 1834)   1540a
Corrugate Cup-and-saucer

Gulf of California to Peru.

1½ inches in diameter, cap-shaped, apex near the middle. Numerous radial ribs are rugose and becoming scally near the margin of the shell. Internal cup is deep and has ⅓ sliced away. Color yellow-brown. Uncommon; offshore to 14 fathoms.

Genus *Crucibulum* Schumacher, 1817

Shell cap-shaped with a complete cuplike support, which is attached by its base or along one side. Only 2 species in the Western Atlantic, but at least 8 in the Tropical Eastern Pacific. Type: *auricula* (Gmelin, 1791). *Trelania* Gray, 1867; and *Calypeopsis* Lesson, 1850, are synonyms.

Crucibulum auricula (Gmelin, 1791)   1541
West Indian Cup-and-saucer

Off South Carolina to Texas; West Indies to Brazil.

1 inch in diameter, similar to *C. striatum,* but the edges of the inner cup are entirely free. The edges of the main shell are crenulated, the external ribs are coarser, and the interior is sometimes pinkish. The outer surface may show coarse diagonal ribs, if the specimen has lived attached to a scallop or other ribbed mollusk. Uncommonly dredged in shallow water to 20 fathoms, and occasionally washed ashore.

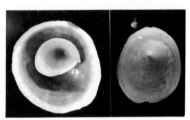

1541

Crucibulum spinosum (Sowerby, 1824)   1542
Spiny Cup-and-saucer

Southern California to Chile.

¾ to 1 inch in diameter, variable in height (⅓ to ¾ as high), and usually with an almost circular base. Exterior with a smoothish apical area, the remainder of the shell with radial rows of small prickles or sometimes erect, tubular spines. Interior glossy, chestnut-brown, sometimes with light radial rays, and with a delicate white cup attached by one side. A very common species from low water to 30 fathoms, attached to stones and dead shells. Albino shells are sometimes found. Synonyms are: *peziza* Wood, 1828; *tubifera* Lesson, 1830; *maculata* Broderip, 1834; and *hispida* Broderip, 1834.

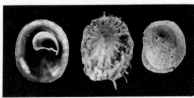

1542

Other species:

1543 *Crucibulum lignarium* (Broderip, 1834). (Synonym: *trigonalis* Adams and Reeve, 1850.) Gulf of California to Ecuador.

1544 *Crucibulum scutellatum* (Wood, 1828). (Synonyms: *imbricata* Sowerby, 1824 (not G. Fischer, 1807); *rugosa* Lesson, 1830; *corrugatum* Gould and Carpenter, 1857; *broderipii* Carpenter, 1857.) Cedros Island, Baja California, the Gulf and south to Ecuador. Shore to 15 fms.

1545 *Crucibulum umbrella* (Deshayes, 1830). (Synonym: *rude* Broderip, 1834.) Gulf of California to Panama. Uncommon.

1546 *Crucibulum cyclopium* S. S. Berry, 1969. Playa Miramar, Santiago Bay, Colima, Mexico, to Costa Rica. 65 mm. in diameter.

1547 *Crucibulum monticulus* S. S. Berry, 1969. Mazatlan, Mexico. 29 mm. in diameter. Both in *Leaflets in Malacology*, no. 26, p. 161.

Subgenus *Dispotaea* Say, 1824

A *Crucibulum* in which the interior cup is attached to the main shell by at least ⅓ of its side. Type: *striatum* Say, 1824.

Crucibulum striatum Say, 1824   1548
Striate Cup-and-saucer

Nova Scotia to both sides of Florida.

1 inch in diameter, cap-shaped, base round, edge smoothish and the slightly twisted apex near the center of the shell. Interior of shell with a small, shelly cup, of which only ⅔ is free from attachment to the main shell. Apex wax color and smooth; remainder of exterior with small, wavy, radial cords. Interior glossy, yellow-white or tinted with light orange-brown. Commonly dredged in shallow water from 3 to 189 fathoms.

1548

Other Pacific *Dispotaea:*

1549 *Crucibulum personatum* Keen, 1958. Acapulco, Mexico, to Panama. *Bull. Amer. Paleontol.*, vol. 38, no. 172.

1550 *Crucibulum serratum* (Broderip, 1834). Nicaragua to Ecuador.

1552 *Crucibulum subactum* S. S. Berry, 1963. Off Teacapan, Sinaloa, Mexico, 25 to 35 fms. *Leaflets in Malacology*, no. 23, p. 144.

1553 *Crucibulum concameratum* Reeve, 1859. Gulf of California to Acapulco, Mexico; common in 7 to 90 meters. (Synonym: *castellum* S. S. Berry, 1963.)

Genus *Crepipatella* Lesson, 1830

Slipper shells resembling true *Crepidula,* but the deck is not attached on one side. Type: *dilatata* (Lamarck, 1822).

Crepipatella lingulata (Gould, 1846) 1554
Pacific Half-slipper Shell

Bering Sea to Panama.

½ to ¾ inch in diameter, thin, almost circular, low and with the apex near the edge of the shell. Characterized by its tannish to mauve-white, glossy interior which has a shallow deck which is attached to the main part of the shell only along one side. The middle of the deck often has a weakly raised ridge. Exterior wrinkled and brownish. A very common species found on rocks and on the shells of living gastropods.

1554

Other species:

1555 *Crepipatella orbiculata* (Dall, 1919). 52 to 55 fms., Bering Sea to off San Diego, California. *Proc. U.S. Nat. Mus.,* vol. 56, p. 351.

1556 *Crepipatella* (*Verticumbo* S. S. Berry, 1940) *charybdis* (S. S. Berry, 1940). 50 fms., off Redondo Beach, California; also Pleistocene, San Pedro, California. May be synonym of *orbiculata*.

Subfamily CREPIDULINAE Fleming, 1822

Genus *Crepidula* Lamarck, 1799

Shells limpetlike, coiled apex at posterior end; interior with a shelly platform covering the posterior portion of the soft body. No operculum. Type: *fornicata* (Linné, 1758).

Crepidula fornicata (Linné, 1758) 1557
Common Atlantic Slipper-shell

Canada to Florida and to Texas.
Introduced to the State of Washington.

¾ to 2 inches in size. Shelly deck extending over the posterior ½ on the inside of the shell. The deck is usually concave and white to buff. Its edge is strongly sinuate or waved in 2 places. Exterior dirty-white to tan, sometimes with brownish blotches and rarely with long color lines. Variable in shape, rarely quite flat, sometimes high and arched. They may be corrugated if the individual has lived attached to a scallop or ribbed mussel. A common littoral species. Individuals usually stack up on top of one another. When collecting on the west coast of Florida, do not confuse with *C. maculosa*. *C. fornicata* has been introduced to the west coast of the United States and Europe.

1557

Crepidula maculosa Conrad, 1846 1558
Spotted Slipper-shell

Both sides of Florida to Mexico; Bahamas.

Resembling *C. fornicata*, but often spotted with small, mauve-brown blotches and sometimes streaked. The edge of the deck is straight or only very slightly convex. There is an oval muscle scar on the inside of the shell just below and in front of the right anterior edge of the deck and the main shell. The young are very much like southern forms of *C. convexa* Say. Rediscovered in 1952 by Dale Stingley (see *The Nautilus,* vol. 65, p. 84).

1558

Crepidula convexa Say, 1822 1559
Convex Slipper-shell

Massachusetts to Florida, Texas and the West Indies. Bermuda.
California (introduced).

¼ to ½ inch in size, usually highly arched and colored a dark reddish to purplish brown. Interior, including the deck, chestnut to bluish brown. Some specimens may be spotted. The edge of the deck is almost straight. There is a small muscle scar inside the main shell on the right side just under the outer corner of the deck (see also *maculosa*). Some specimens are thick and heavy, others quite fragile, the latter type found attached to other shell. Common just offshore down to 116 fathoms. Introduced to the Pacific coast prior to 1899 (see H. E. Vokes, 1935, *The Nautilus,* vol. 49, p. 37). Some forms are ⅓ inch long, thin-shelled, usually dark-brown or translucent-tan, and with a white deck. It is found in overcrowded colonies on eel-grass where specimens become long and narrow. *C. acuta* Lea, 1842, and *glauca* Say, 1822, are synonyms. For eggs and growth, see *Biol. Bull.,* vol. 141, pp. 514–526, 1971.

Crepidula onyx Sowerby, 1824 1560
Onyx Slipper-shell

Southern California to Peru.

1 to 2 inches in length, fairly thick-shelled, characterized by its glossy, dark-chocolate to whitish brown interior, and by the large, slightly concave, pure-white deck which has a sinuate free edge. Very common from shallow estuaries to 50 fathoms on rocks, on other shells, or stacked up on top of each other. Synonyms are *cerithicola* C. B. Adams, 1852, and *lirata* Reeve, 1859.

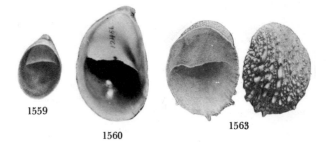

1559   1560   1563

Crepidula excavata Broderip, 1834   1561
Excavated Slipper-shell

Monterey, California, to Panama.

Females 1 inch in size, males ½ as large, rather thin; back strongly arched with the apex distinct and hooked under itself near the posterior margin of the shell. Characterized by its light brownish white color, by the straight or slightly curved edge of the interior deck and by a weak muscle scar on each side just under the deck. Found commonly attached to rocks and other shells in shallow water. Synonyms are: *naticarum,* Williamson, 1905, and *norrisiarum* Williamson, 1905, the latter form being found on the snail, *Norrisia.* See MacGinitie and MacGinitie, 1964, *The Veliger,* vol. 7, p. 34.

1561

Crepidula adunca Sowerby, 1825   1562
Hooked Slipper-shell

Graham Island, British Columbia, to southern California.

½ to 1 inch; highly arched with a sharp, hooked apex. Laterally compressed, giving a triangular appearance from the sides. Color chestnut-brown, sometimes with indistinct rays or rows of spots. Interior brown. Periostracum light-brown. A common species in the northern half of its range, sometimes on *Calliostoma* and *Tegula.* Also dredged from 20 fathoms. The eggs are relatively large and the young are brooded.

Subgenus *Bostrycapulus* Olsson and Harbison, 1953

Type of this subgenus is *aculeata* (Gmelin, 1791). *Sandalium* Schumacher, 1817, non Retzius, 1788, is a synonym.

Crepidula aculeata (Gmelin, 1791)   1563
Spiny Slipper-shell

North Carolina to Florida, Texas and to Brazil. Bermuda. Central California to Chile.

½ to 1 inch in size, similar to *fornicata,* but characterized by its rough, spinose exterior, thinner and flatter shell and by its irregular edges. Color whitish, although often heavily mottled with reddish brown. The exterior is sometimes stained green by algal growths. A common species found attached to stones, mangroves and other shells. Occasionally dredged. On the Pacific coast it is uncommon; sometimes found on kelp holdfasts. Synonyms include *Calyptraea echinus* and *hystrix* Broderip, 1834.

Other species:

1564 *Crepidula (Crepidula) arenata* (Broderip, 1834). La Jolla, California, to Chile; beaches to offshore in 55 fms.

1565 *Crepidula (C.) incurva* (Broderip, 1834). Baja California, the Gulf, to Peru. Offshore down to 10 fms. (Synonym: *coei* S. S. Berry, 1950.)

1566 *Crepidula (C.) lessonii* (Broderip, 1834). Head of the Gulf of California to Paita, Peru.

1567 *Crepidula striolata* Menke, 1851. Gulf of California to Panama. (Synonyms: *squama* Broderip, 1834; *nivea* C. B. Adams, 1852; *nebulata* Mabille, 1895.)

1568 *Crepidula uncata* Menke, 1847. Gulf of California.

1569 *Crepidula grandis* Middendorff, 1849. Aleutian Islands to Sitka, Alaska.

Subgenus *Ianacus* Mörch, 1852

Pure-white slipper shells. Type: *plana* Say, 1822. *Jenacus* Mörch, 1852, is an emendation. *Janacus* is a misspelling.

Crepidula plana Say, 1822   1570
Eastern White Slipper-shell

Canada to Texas and to Brazil. Bermuda.

½ to 1½ inches in size, very flat, either convex or concave, and always a pure milky-white. The apex is very rarely turned to one side. It commonly attaches itself to the inside of large, dead shells, and rarely, if ever, "piles up" like *fornicata.* A common shallow-water species.

Crepidula nummaria Gould, 1846   1571
Western White Slipper-shell

Alaska to Panama.

¾ to 1½ inches in length, characterized by its glossy-white underside, flattened shell, large deck which usually has a weak, raised ridge (or sometimes a hint of an indentation) running from the apical end forward to the leading edge. Exterior with or without a yellowish periostracum. Found in rock crevices and apertures of dead shells.

1570   1571

Family XENOPHORIDAE Philippi, 1853

Genus *Xenophora* G. Fischer, 1807

This group of gastropods is noted for its peculiar habit of cementing to its own shell fragments of other shells, stones, bits of coral and coal. The animals resemble those of the *Strombus* conchs, but the operculum is much wider and not sickle-shaped. Shell strong and without an umbilicus. B. R. Bales once humorously observed: "It is generally admitted that the camouflage of *Xenophora* is for protection rather than ornamentation, for it would be inconceivable that a female *Xenophora* would call over the back fence to her girl friend with 'Come and see the perfect dream of a shell I picked up today and tell me if I have it on straight.' " Type: *conchyliophora* Born, 1780. There are 3 species in the Western Atlantic, one in the Eastern Pacific. This family was formerly placed in the superfamily Strombacea, but is now in the Crepidulacea. See Morton, J. E., 1958, *Proc. Malacol. Soc. London,* vol. 33, p. 89.

Xenophora conchyliophora (Born, 1780)   **1572**
Atlantic Carrier-shell   **Color Plate 4**

North Carolina to the West Indies; Bermuda. Brazil.

2 inches in diameter, not including foreign attachments. No umbilicus. From above, the shell with its attached rubble and shells looks like a small heap of marine trash. It will attach any kind of shell to itself, but in some areas has access to only one kind, say *Chione cancellata*. Animal deep maroon-red. Seasonally not uncommon; in 6 to 30 feet of water. Erroneously called *trochiformis* Born, 1778, which belongs to the genus *Trochita* (*Crepidulidae*). Moderately common at the northern end of Harrington Sound, Bermuda.

Xenophora robusta Verrill, 1870   **1573**
Robust Carrier-shell

Pacific coast of Mexico.

2 to 3 inches in diameter, not including attachments. Shell solid, pyramidal in shape, without an umbilicus and colored a rich-brown on the base and within the glossy aperture. Usually very heavily bedecked with worn bivalve shells and a few gastropod shells. Moderately common offshore; 10 to 28 fathoms.

Genus *Tugurium* P. Fischer, 1876

Shells very thin, whitish, umbilicate and with an overhanging ledge accompanying each whorl. They gather only a relatively few shells and foreign objects. Outer surface shows oblique, ripplelike marks. Deepwater. Type: *exutus* Reeve, 1843, from the Indo-Pacific.

Tugurium caribaeum (Petit, 1856)   **1574**
Caribbean Carrier-shell

Off South Carolina to the West Indies and Brazil.

2 to 3 inches in diameter, ½ as high, rather fragile, white, with a small umbilicus and bearing few shells. Characterized on the underside by a milk-white, spiral swelling which covers the sutural area. Moderately common; from 75 to 300 fathoms.

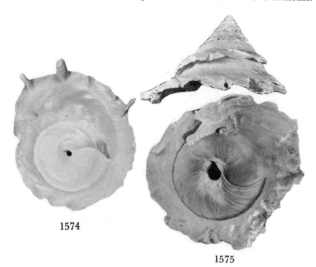

1574

1575

Subgenus *Trochotugurium* Sacco, 1896

Similar to *Tugurium* but with strong, curved growth ridges on the base, thus causing the suture to be finely crenulate. Umbilicus usually large, and inside which sometimes is found a living parchment tube-worm. Type: *borsoni* Sismonda, 1847, a European fossil.

Tugurium longleyi (Bartsch, 1931)   **1575**
Longley's Carrier-shell

North Carolina to Barbados.

4 to 6 inches in diameter, grayish white, with strongly overhanging or shingled edges to the whorls. Base with strong, fine, curved growth lines, causing the suture to be finely crenulate. Uncommon; 70 to 450 fathoms.

Family APORRHAIDAE Mörch, 1852

Genus *Aporrhais* da Costa, 1778

Type of the genus is *pespelicanis* Linné, 1758.

Aporrhais occidentalis Beck, 1836   **1576**
American Pelican's Foot

Arctic Canada to off North Carolina.

2 to 2½ inches in length, spire high, whorls well-rounded and with about 15 to 25 curved axial ribs per whorl. Many minute spiral threads present. Outer lip greatly expanded and its edge heavily thickened. Color ashen-gray to yellowish white. Operculum small, corneous, brown, clawlike, but with smooth edges. Commonly dredged off New England from a few to 350 fathoms. This species is the type of the subgenus *Arrhoges* Gabb, 1868.

The form *mainensis* C. W. Johnson, 1926 **(1577)**, (Nova Scotia to Mt. Desert) differs in having 14 axial ribs, instead of about 22 to 25 as in the typical form, but specimens intergrade. The form *labradorensis* C. W. Johnson, 1930 **(1578)**, is smaller, more slender and with up to 29 ribs per whorl. It is found off Labrador from 7 to 60 fathoms.

1576

Family STROMBIDAE Rafinesque, 1815

Genus *Strombus* Linné, 1758

Shells large, with a flaring outer lip which has a small U-shaped indentation or "stromboid notch" at the base. Operculum sickle-shaped. They feed on marine algae. Eggs are laid in long, intertwining, thin tubes of gelatinous material. For a more complete treatment, see Abbott, 1960, *Indo-Pacific Mollusca,* vol. 1, no. 2. Type: *pugilis* Linné, 1758.

Strombus pugilis Linné, 1758   **1579**
West Indian Fighting Conch   **Color Plate 4**

Southeast Florida to the West Indies. Brazil.

3 to 4 inches in length. Always with spines on the last whorl, but those on the next to the last whorl are nearly always the largest. Shoulder of outer lip nearly always turns slightly upwards. Color a rich cream-orange to salmon-pink throughout, except for a cobalt-blue splotch of color on the end of the canal. Periostracum very thin and velvety. This is

primarily a West Indian species, and apparently will interbreed with the mainland species, *S. alatus.* An aberrant form which has clublike spines was unnecessarily named *sloani* Leach, 1814, and *peculiaris* M. Smith, 1940. Dwarf forms, under 2 inches, were given the name *nicaraguensis* Fluck, 1905. Common; sand and grass bottoms, 6 to 30 feet.

Strombus alatus Gmelin, 1791   **1580**
Florida Fighting Conch   **Color Plate 4**

North Carolina to both sides of Florida and to Texas.

3 to 4 inches in length. With or without short spines on the shoulder of the last whorl. Shoulder of outer lip slopes slightly downward. Color a dark reddish brown, often mottled with orange-brown or having zigzag bars of color on the shiny parietal wall. Periostracum very thin and velvety. A very common shallow-water species, especially on the west coast of Florida. Not found in the West Indies. *S. undulatus* Küster, 1845, is a synonym. Malformed specimens may have a deeply channeled suture. Do not confuse with *S. pugilis.*

Strombus gracilior Sowerby, 1825   **1581**
Eastern Pacific Fighting Conch   **Color Plate 4**

Gulf of California to Peru.

2½ to 3½ inches, similar to the Atlantic *pugilis,* but the spines on the periphery of the whorls are small and blunt. Color brownish yellow, with a light periostracum and a whitish central band of color. Aperture white, edged with orange-brown. Outer lip turns slightly upward at the top, or posterior end. Common on soft bottoms from intertidal to 45 meters.

Subgenus *Lentigo* Jousseaume, 1886

Body whorl with several spiral rows of low, irregular nodules. Type: *lentiginosus* Linné, 1758, from the Indo-Pacific.

Strombus granulatus Swainson, 1822   **1582**
Granulated Conch   **Color Plate 4**

Gulf of California to Ecuador.

2 to 3½ inches, spire slender and tuberculate. Whorls with 5 spiral cords, the uppermost bearing the largest blunt spines. Inside of outer lip with numerous granulations. Common in shallow water down to 75 meters.

Subgenus *Tricornis* Jousseaume, 1886

Shell large, with a heavy outer lip and bearing strong knobs and having a smooth columella. Type: *tricornis* Lightfoot, 1786.

Strombus gigas Linné, 1758   **1583**
Pink Conch

Southeast Florida and the West Indies; Bermuda.

6 to 12 inches in length. Characterized by its large size, large and flaring outer lip and the rich shades of pink, yellow and orange in the aperture. Periostracum fairly thick and horny. It flakes off in dried specimens. A malform with flattened spines was named *horridus* M. Smith, 1940. A form with a deep channel at the suture occasionally turns up in the Bahamas and Florida. It was named *canaliculatus* L. Burry, 1949. See R. Robertson, 1962, *The Nautilus,* vol. 75, p. 128. *S. samba* Clench, 1937, is merely an old specimen with a thickened, gray lip. *S. gigas verrilli* McGinty, 1946, is a synonym, from Lake Worth, Florida. Very common in the West Indies, becoming uncommon in the Florida Keys from over-fishing. 6 to 40 feet of water, usually near *Thalassia* eel-grass. Also called

1583

the queen conch. See J. E. Randall, 1964, "Contributions to the biology of the Queen conch, *Strombus gigas,*" *Bull. Marine Sci.,* vol. 14, pp. 246–295.

Strombus costatus Gmelin, 1791   **1584**
Milk Conch   **Color Plate 4**

South Florida and the West Indies; Bermuda. Brazil.

4 to 7 inches in length. Shell usually very heavy and with low, blunt spines. Spire without knobs. Parietal wall and thick outer lip highly glazed with cream-white enamel. Outer shell a yellowish white, rarely orange or mauve. The periostracum in dried specimens flakes off. Common in the West Indies, in shallow water to 20 feet. *S. spectabilis* A. H. Verrill, 1950, *inermis* Swainson, 1822; *accipitrinus* Lamarck, 1822; and *leidyi* Heilprin, 1887 (a Bermuda form with a high spire), are synonyms. In Bermuda, it is called the harbour conch. Rare in Florida, generally offshore in the northwest and southeast.

Strombus raninus Gmelin, 1791   **1585**
Hawk-wing Conch   **Color Plate 4**

Southeast Florida and the West Indies; Bermuda. Brazil.

2 to 4 inches in length. Shell bluntly spinose with the last 2 spines on the last whorl by far the largest. Outer lip points upward at the top. Color of outer shell a brownish gray with chocolate-brown mottlings. Aperture cream-colored with a salmon-pink interior. Common in the West Indies. *S. bituberculatus* Lamarck, 1822, *lobatus* Swainson, 1822; *rarimus* Bosc, 1801 and *nanus* Bales, 1938 (dwarfs) are all this species.

Strombus gallus Linné, 1758. Pl. 5   **1586**
Rooster-tail Conch   **Color Plate 4**

Southeast Florida (rare) and the West Indies; Bermuda. Brazil.

4 to 6 inches in length, characterized by the long extension of the posterior end of the outer lip and the rather high spire. Aperture and inside of outer lip usually salmon-brown to yellowish. Outer shell usually maculated with brown and yellow, rarely purplish. This species is not at all common, although it

may be obtained in fair numbers in certain parts of the West Indies in 6 to 30 feet of water.

Strombus peruvianus Swainson, 1823   **1587**
Peruvian Conch   **Color Plate 4**

Tres Marias Islands, Mexico, to northern Peru.

4 to 5 inches in length, heavy, with a strong wing at the top of the outer lip. Aperture orange-brown. Exterior brownish and with brown flaky periostracum. Spire low and variegated with brown and white. Moderately common in shallow water.

Strombus galeatus Swainson, 1823   **1588**
Giant Eastern Pacific Conch

Gulf of California to Ecuador.

6 to 8 inches, heavy, smooth-shouldered, with a lip that flares upward toward the apex. Last whorl with numerous spiral, low cords. Periostracum brown and flaky when dry. Formerly common, but becoming rarer as over-collecting continues. Shallow water.

Superfamily Cypraeacea Rafinesque, 1815

Family VELUTINIDAE Gray, 1840

The family name Lamellariidae Orbigny is a synonym.

Subfamily LAMELLARIINAE Orbigny, 1841

Genus *Lamellaria* Montagu, 1815

Shell with a large last whorl, buried within the flesh, and very thin; surrounded by a white or red body. Most species feed upon colonial tunicates. Type: *perspicua* (Linné, 1758).

Lamellaria perspicua (Linné, 1758)   **1589**
Transparent Lamellaria

Southeast Florida to Brazil; Eastern Atlantic.
Sonora coast of west Mexico; Chile.

Shell 4 to 5 mm., fragile, smooth, transparent, almost without growth lines, and with 2 whorls. Nuclear whorl is slightly elevated. Suture impressed. Columella open, so that the apex can be seen from below. Parietal wall bears a crescent-shaped thickening. Foot bilabiate in front. Radula with 41 rows; central tooth inverted-V-shaped, with 5 denticles; inner lateral with 5 large denticles; outer lateral with 9 to 11 fine ones. Animal sometimes red to reddish brown. Moderately common; tidepools to 30 fathoms. A lower Caribbean to Brazil and to Chile subspecies was described: *mopsicolor* Eveline Marcus, 1958 **(1590)**. *Marsenia rangi* Bergh, 1853, is Linné's species in all likelihood, and *Lamellaria cochinella* Louise Perry, 1939, is probably a synonym.

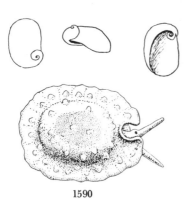

1590

Lamellaria leucosphaera Schwengel, 1942   **1591**
White-ball Lamellaria

West Florida to the Virgin Islands.

Shell 16 mm. high, 15 mm. wide, naticoid in shape, thin, transparent, with a milky cloudiness, faintly iridescent. Nucleus small, flat, glassy. Columella spirally gyrate. Last whorl rapidly descending, making a produced spire. Animal grayish, gelatinous, with the mantle covering the body and shell and notched anteriorly where the siphonal canal protrudes. Mantle peppered with small black dots and appears granular. Radula with 58 rows. *The Nautilus*, vol. 56, p. 62.

Lamellaria diegoensis Dall, 1885   **1592**
San Diego Lamellaria

Monterey, California, to Sonora, Mexico.

15 to 20 mm. in length, equally wide, quite fragile and transparent-white in color. 3 whorls moderately globose, the last large. Aperture very large. Columella very thin. Surface smoothish, except for fine, irregular growth lines. Periostracum thin, clear and glossy. The huge soft parts are bright-red to orange with black papillae. Uncommon sublittoral to offshore, where it lives on ascidians.

1592

Lamellaria rhombica Dall, 1871   **1593**
Rhombic Lamellaria

Washington to Baja California.

15 mm., similar to *diegoensis*, but it is much fatter, thicker-shelled and is opaque-white in color. Columella thick and ridgelike. Moderately common; lives offshore, but is commonly washed onto the beach. Placed in the subgenus *Marsenina* Gray, 1850. *Lamellaria sharonae* Willett, 1939, whose animal is reported as reddish with dark spots, and whose shell is only 7 mm., may be dwarfs or young of this species.

Other species:

1594 *Lamellaria pellucida pellucida* Verrill, 1880. 86 to 156 fms., off Martha's Vineyard, Massachusetts; off New Jersey to Panama.

1595 *Lamellaria pellucida gouldi* Verrill, 1882. 224 to 458 fms., off Martha's Vineyard, Massachusetts, to off Maryland.

1596 *Lamellaria fernandiniae* Dall, 1927. 294 fms., off Fernandina, Florida.

1597 *Lamellaria inflata* (C. B. Adams, 1852). Gulf of California to Panama. Rare.

1598 *Lamellaria koto* Schwengel, 1944. West Florida. *The Nautilus*, vol. 58, p. 17. 16 × 18 mm.

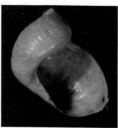

1598

1599 *Lamellaria stearnsii* Dall, 1871. Prince William Sound, Alaska; Puget Sound to Monterey, California. For its biology, see M. T. Ghiselin, 1963, *The Veliger*, vol. 6, p. 123, pl. 16.

1600 *Lamellaria stearnsii orbiculata* Dall, 1871. Alaska to the Gulf of California.

1601 *Lamellaria digueti* Rochebrune, 1895. San Pedro, California, to the Gulf of California.

Subfamily VELUTININAE Gray, 1840

Genus *Velutina* Fleming, 1821

Shell small, thin, subglobose, with 2 or 3 rapidly enlarging whorls; aperture large; outer lip thin; periostracum velvety or powdery. No operculum. The animal is small and fits within the shell. It lives and feeds upon tunicates and solitary ascidians. The egg capsules are sunk into the ascidians. There is a free-swimming echinospira larva. Type: *velutina* (Müller, 1776).

Velutina velutina (Müller, 1776)   **1602**
Smooth Velutina

Arctic Canada to Cape Cod, Massachusetts.
Alaska to Monterey, California. Europe.

½ to ¾ inch in length, very thin and fragile, translucent-amber, and covered with a thick, brownish periostracum which is spirally ridged. Columella arched and narrow. Common offshore from 3 to 50 fathoms. Also commonly found in fish stomachs. *V. capuloidea* Blainville, 1825; *schneideri* Friele, 1886; and *rupicola* Conrad, 1866, are synonyms. *Helix laevigata* Linné has been abandoned.

1602   1604

Subgenus *Velutella* Gray, 1847

Resembling *Velutina*, but the aperture is proportionately larger, the whole shell is obliquely set. Shell flexible, translucent, membraneous and smooth. Spire small. Type: *plicatilis* (Müller, 1776).

Velutina plicatilis (Müller, 1776)   **1603**
Oblique Velutina

Northern Europe to Halifax, Nova Scotia.
Arctic Ocean to Pribilofs and Aleutians; Alaska.

½ to 1 inch, yellowish, obliquely oblong, with 2½ membraneous whorls. Has a calcareous incrustation in the innermost whorls which may be spirally striate. Shape variable. Spire may be covered by the last whorl (form *cryptospira* Middendorff, 1849). Other synonyms: *coriacea* Pallas, 1788; *sitkensis* (A. Adams, 1851). Lives on hydroids. From shore to 180 fathoms; common.

Subgenus *Limmeria* H. and A. Adams, 1853

Shell elongate-globose; aperture large and expanded. Columella flattened and shelflike. Type: *undata* (Brown, 1839).

Velutina undata (Brown, 1839)   **1604**
Undate Velutina

Arctic Ocean to Massachusetts.
Alaska; Europe.

7 to 8 mm., similar to *velutina*, but more elongate, minutely reticulated, with a flattened shelflike columella, and usually with fine spiral bands of brown. Uncommon; intertidal to 616 fathoms, but sometimes washed ashore. Also commonly found in fish stomachs. *V. zonata* Gould, 1841, is a synonym.

Other species:

1605 *Velutina prolongata* Carpenter, 1865. Bering Sea to Monterey, California.

1606 *Velutina conica* Dall, 1887. Aleutian Islands to Kodiak and to Forrester Island, Alaska.

1607 *Velutina granulata* Dall, 1919. Monterey, California, in 55 fms.

1608 *Velutina lanigera* Möller, 1842. Northern Europe; Greenland; Point Barrow, Alaska, to Petrel Bank, Bering Sea. Baffin Island, Canada.

1609 *Velutina rubra* Willett, 1919. Forrester Island and the north coast of Prince William Sound, Alaska.

Genus *Marsenina* J. E. Gray, 1850

Shell moderately large, but quite thin. Aperture very large. Columella concave. Spire small with 2 or 3 whorls. Periostracum thin. Mantle split down the center and covering the outer shell. Type: *glabra* (Couthouy, 1832).

Species:

1610 *Marsenina ampla* Verrill, 1880. Eastport, Maine.

1611 *Marsenina glabra* (Couthouy, 1832). Arctic Canada to off Cape Cod, Massachusetts, 15 to 34 fms. (Synonyms: *prodita* Loven, 1846; and *micromphala* Bergh, 1853.)

1612 *Marsenina globosa* L. Perry, 1939. West Florida.

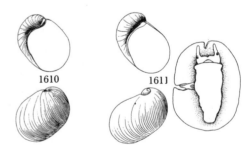

1610   1611

Genus *Onchidiopsis* Bergh, 1853

Animal about 1 inch, sluglike in appearance, having a warty back under which is a thin, slipperlike, vestigial, membraneous shell. Tentacles long with an eye at the outer base. Type: *groenlandica* Bergh, 1853. Four species reported from North America:

1613 *Onchidiopsis glacialis* (M. Sars, 1851). Northern Europe; Greenland; Arctic Canada; Point Barrow, Alaska (MacGinitie, 1959, *Proc. U.S. Nat. Mus.*, vol. 109, no. 3412). (Synonyms: *groenlandica* Bergh, 1853; *pacifica* Bergh, 1887.)

1614 *Onchidiopsis corys* Balch, 1910. Off Fish Island, Labrador, 75 fms.

1615 *Onchidiopsis hannai* Dall, 1916. St. Paul Island, Pribilof Islands, Bering Sea.

1616 *Onchidiopsis kingmaruensis* H. D. Russell, 1942. Baffin Island, Canada. *Can. Jour. Res.*, vol. 20, p. 50.

Genus *Capulacmaea* M. Sars, 1859

Shell ½ to 1 inch, limpet-shaped, oval in outline, low spire near the anterior end. Shell white, calcareous, thin, covered with a brown periostracum. Interior of shell sometimes rayed with purplish. Type: *Pilidium commodum* (Middendorff, 1851). *Piliscus* Lovén, 1859, is a synonym. One species reported from North America:

1617 *Capulacmaea commodum* (Middendorff, 1851). Point Barrow, Alaska, to the Sea of Okhotsk, U.S.S.R. Off Nova Scotia, 150 fms. Northern Europe. (Synonym: *Capulus radiatus* M. Sars, 1851.) Erroneously placed in *Pilidium* Middendorff, 1849, a true limpet.

Family ERATOIDAE Gill, 1871

Subfamily ERATOINAE Gill, 1871

Genus *Erato* Risso, 1826

Small, 2 to 8 mm., glossy shells, resembling miniature marginellids. Long aperture bordered on both sides by minute teeth. They feed upon colonial tunicates. They have a free-swimming echinospira larva. Type: *cypraeola* (Brocchi, 1814).

Subgenus *Hespererato* Schilder, 1932

Type of this subgenus is *vitellina* (Hinds, 1844).

Erato maugeriae Gray, 1832   **1618**
Mauger's Erato

North Carolina to Florida and to Brazil.

1/16 inch in length, resembling a small *Marginella,* but the curled-in, thickened outer lip has a row of about 15 small even-sized teeth. Upper end of the outer lip is well-shouldered. Shell glossy, tan with a pinkish or yellowish undertone. Apex bulbous. Commonly dredged on either side of Florida from 2 to 63 fathoms. Lives in algae.

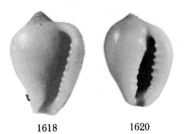

1618   1620

Erato columbella Menke, 1847   **1619**
Columbelle Erato

Monterey, California, to Panama.

¼ inch in length, glossy-smooth, slate-gray in color with a whitish, thickened outer lip. Spire elevated, nucleus brown. Outer lip markedly shouldered above, and bearing about a

dozen extremely small teeth. Siphonal canal stained inside with purple-brown. Not uncommon from shore to 50 fathoms and among eel-grass. Occasionally washed ashore with kelp weed. *Erato leucophaea* Gould, 1853, is a synonym.

Erato vitellina (Hinds, 1844)   **1620**
Apple Seed Erato

Bodega Bay, California, to Baja California.

½ inch in length, resembling a "beach-worn *Columbella*," and glossy-smooth. Body whorl with a large purple area bounded by a faint whitish line; remainder of shell, including the spire which is often glazed over, is dark brownish cream. Columella arched, bearing 5 to 8 small, whitish teeth. Lower ¾ of slightly incurled outer lip is with 7 to 10 small, whitish teeth. Moderately common in fairly shallow water. Occasionally washed ashore with kelp weed.

Erato albescens Dall, 1905   **1621**
Whitish Erato

Santa Rosa Isand, California, to Baja California.

½ inch, whitish, thin and polished, with 4 whorls. Spire low and covered with a glaze. Columella smooth, short, twisted. Outer lip thickened, obscurely margined, and with about 9 denticles within. Rare; offshore.

Other species:

1622 *Erato scabriuscula* Sowerby, 1832. Gulf of California to Ecuador. (Synonyms: *cypraeola* Sowerby, 1832; *granum* Kiener, 1843.)

Subfamily TRIVIINAE Troschel, 1863

Genus *Trivia* Broderip, 1837

Resembling miniature cowries (*Cypraea*), but characterized by strong wrinkles or riblets running around the shell from the slitlike aperture to the center of the back of the shell. The highly decorated red to yellow mantle extends up over the outside of the shell. Found in rocky areas where they are associated with compound ascidians in which they lay their eggs. There is a free-swimming echinospira larva in some species. We have carefully reviewed and included most, if not all, of the Western Atlantic species, but have not followed the Schilderian use of numerous genera, such as *Pusula* Jousseaume, 1884. Type: *monacha* (da Costa, 1778).

Subgenus *Pusula* Jousseaume, 1884

The type of this subgenus is *radians* (Lamarck, 1810). The ends are blunt, dorsal furrow present in most species and the rib interspaces granular.

Trivia pediculus (Linné, 1758)   **1623**
Coffee Bean Trivia   **Color Plate 3**

North Carolina to Florida and to Brazil. Bermuda.

½ inch in length, characterized by its tan to brownish pink color with 3 pairs of large, irregular, dark-brown spots on the back, and in having 16 to 19 (usually 17) ribs crossing the outer lip. The center pair of spots on the back are the largest. Some specimens may be quite pink. Common; intertidal to 25 fathoms.

In some areas, a dwarf forma of this species occurs **(1624)** (named *pullata* Sowerby, 1870) which is ¼ inch in length,

with a pink base, 13 to 17 riblets on the outer lip, and often with the brown mottlings spread over most of the back. Do not confuse this form with the species *suffusa* which is light-pink, with a white (not pink) outer lip crossed by 19 to 24 riblets, and with a pink blotch on each side of the anterior canal.

Trivia suffusa (Gray, 1832)   1625
Suffuse Trivia

Southeast Florida and the West Indies to Brazil.

¼ to ⅓ inch in length, elongate-globular, bright-pink with suffused brownish splotches and fine specklings. Anterior canal with a weak pinkish blotch on each side. Riblets on back somewhat beaded. Dorsal groove fairly well-impressed. Outer lip white and crossed by 18 to 23 (usually 20) riblets. It is not as globose as *quadripunctata*. Quite common in the Bahamas and Lesser Antilles; 12 to 50 fathoms. *T. armandina* Kiener, 1843 is the same.

Trivia maltbiana Schwengel and McGinty, 1942   1626
Maltbie's Trivia

North Carolina to Florida and the Caribbean.

¼ to ½ inch in length, globose, slightly flattened above, and characterized by its pale tannish pink, translucent color, by its fine riblets and by having 24 to 28 ribs crossing the outer lip. Areas between the ribs are microscopically granular. Nuclear whorls visible through the last whorl. The dorsal groove is slight and the riblets nearly cross it. Moderately common just offshore to 50 fathoms. *The Nautilus*, vol. 56, p. 16.

Trivia quadripunctata (Gray, 1827)   1627
Four-spotted, Trivia

Southeast Florida, Yucatan and the West Indies. Bermuda.

⅛ to ¼ inch in length, very similar to *suffusa*, but smaller, brighter pink, and with 2 to 4 very small, dark red-brown dots on the center line of the back. Riblets very fine, 19 to 24 crossing the outer lip. A very common species frequently found on beaches with the color dots worn away and the pink background rather faded. The riblets on the back are never pustulose as they tend to be in *suffusa*, nor is there any fine color speckling. The mantle is black in many specimens.

Trivia nix (Schilder, 1922)   1628
White Globe Trivia

Florida and the West Indies to Brazil.

⅜ inch in length, globular, pure-white in color. Characterized by about 22 to 26 riblets. Back with a strong groove inter-

rupting the riblets. Alias *T. nivea* Gray. This is the largest and most globular of the white species found in the Western Atlantic. It is moderately uncommon in the West Indies. Rare in Florida and Brazil, 31 to 50 fathoms.

Trivia californiana (Gray, 1827)   1629
Californian Trivia

Crescent City, California, to Acapulco, Mexico.

9 to 11 mm. in length, rotund, and characterized by its mauve color, white, slightly depressed crease on the midline of the back and by the fairly coarse riblets crossing over the entire shell (outer lip with about 15). A common littoral species, often washed ashore with seaweed. Also lives as deep as 40 fathoms. Probable synonyms are *californica* Sowerby, 1832; and *elsiae* Howard and Sphon, 1960.

Trivia sanguinea (Sowerby, 1832)   1630
Sanguin Trivia

Gulf of California.

9 to 15 mm., similar to *californiana*, but usually larger, with more and finer riblets (outer lip with about 20), and without the prominent white streak on the back. There is a stain of bright red along the dorsal furrow. Common; intertidal rocks to offshore.

1630

Trivia solandri (Sowerby, 1832)   1631
Solander's Trivia

Palos Verdes, California, to Peru.

13 to 16 mm. in length, rotund, and characterized by the strong, raised, smooth riblets running over the lip and up onto the back. Dorsal groove deep, cream-colored and flanked by 8 to 10 cream nodules on each side. Ground color of shell dark purplish brown. Moderately common in the littoral zone under rocks.

Trivia radians (Lamarck, 1810)   1632
Radiating Trivia

Baja California to Ecuador.

15 to 21 mm. in length, similar to *solandri*, but flatter, larger and with a brownish spot which discolors the central groove on the back. Moderately common; intertidal under rocks. Synonyms are *rota* Weinkauff, 1881; and *circumdata* Schilder, 1931.

Trivia pacifica (Sowerby, 1832)   1633
Eastern Pacific Trivia

Gulf of California to Ecuador.

9 to 12 mm., somewhat pyriformly ovate, pinkish to rose, with a large faint brown blotch on the back. Dorsal line narrow. Ribs fine and numerous. Subtidal; uncommon. This species is placed in the subgenus *Niveria* Jousseaume, 1884, by some workers.

Other species:

1634 *Trivia (Pusula) myrae* Campbell, 1961. Northern Gulf of California to Mazatlan, Mexico. *The Veliger*, vol. 4, p. 25; vol. 9, p. 35.

Subgenus *Dolichupis* Iredale, 1930

The type of this subgenus is *producta* Gaskoin, 1836, from the Indo-Pacific.

***Trivia antillarum* Schilder, 1922**	**1635**

Antillean Trivia

Southeast Florida and the Antilles. Brazil.

⅛ to ¼ inch in length, characterized by its deep reddish or brownish purple color. Elongate-globular in shape. Riblets smooth. With or without a faint dorsal groove over which the riblets usually cross. Outer lip with 18 to 22 teeth. Dredged from 30 to 100 fathoms and rarely cast upon the beach. Formerly *T. subrostrata* Gray, 1824.

1635

***Trivia candidula* Gaskoin, 1835**	**1636**

Little White Trivia

North Carolina to Florida to Barbados.

⅛ to ¼ inch in length, characterized by its fairly globular shape, pure-white color, somewhat rostrate ends and by the smooth riblets that pass over the back. There is no dorsal furrow. Many specimens have only a few rather strong riblets of which 17 cross the inside of the outer lip. Another common form has more riblets (20 to 24 over the outer lip). It has been named *leucosphaera* Schilder, 1931 (*globosa* of authors, not Gray). The forms intergrade. *Trivia nix* is also white, but is larger, more globose and with a strong dorsal groove interrupting the ribs.

***Trivia ritteri* Raymond, 1903**	**1637**

Ritter's Trivia

Monterey, California, to Baja California.

⅜ inch in length, globular, pure-white in color. Characterized by about 15 fine riblets that run over the bottom, sides and back of the shell without being interrupted by a dorsal groove. Uncommonly dredged on gravel bottom from 25 to 60 fathoms.

1637

Family CYPRAEIDAE Rafinesque, 1815

Genus *Cypraea* Linné 1758

Glossy, colorful, ovate shells with a long narrow aperture bounded on both sides by inrolled, toothed lips. Mantle extends over the dorsum of the shell and is retractable into the aperture. Some authors have used the many proposed subgenera as genera. Type: *tigris* Linné, 1758, from the Indo-Pacific.

Subgenus *Macrocypraea* Schilder, 1930

Shell large, elongate, rounded at the sides and with white spots on a tan or brown background. Type: *zebra* (Linné, 1758).

***Cypraea zebra* Linné, 1758**	**1638**
Measled Cowrie	**Color Plate 5**

Southeast Florida to Brazil.

2 to 3½ inches in length, oblong, light-faun to light-brown, with large, round, white dots over the back. Toward the base of the shell these white dots have a brown center. The shell is darker brown, narrower and less inflated than *cervus*. Moderately common in intertidal waters. Formerly called *C. exanthema* Linné, 1767. A light orangish form, probably due to being buried in sand for some time, was described from Cuba (form *vallei* Jaume and Borro, 1946). A southeast Brazilian form was named *dissimilis* Schilder, 1924.

***Cypraea cervus* Linné, 1771**	**1639**
Atlantic Deer Cowrie	**Color Plate 5**

Off North Carolina to Florida and Cuba. Bermuda.

3 to 7 inches in length, similar to *zebra,* but usually with smaller and more numerous white spots, with a more inflated and larger shell, and seldom has ocellated spots on the base of the shell. Moderately common from low tide to several fathoms. *C. peilei* (Schilder, 1932), a Pleistocene form from Bermuda, is a synonym. Collectors in the Lower Florida Keys are urged to restrict their take to 1 or 2 specimens, since this species is being seriously depleted. It may possibly hybridize with *zebra.* For biology, see M. E. Crovo, 1971, *The Veliger*, vol. 13, p. 292.

***Cypraea cervinetta* Kiener, 1843**	**1640**

Panamanian Deer Cowrie

Lower Gulf of California to Peru.

1½ to 4 inches, similar to *zebra,* but the base is darker, with a more bluish brown hue and with a dark-brown blotch on the columella of some specimens. Rare in its northern range; fairly common around Panama; shallow water.

Subgenus *Luria* Jousseaume, 1884

Type of this subgenus is *lurida* (Linné, 1758) from the Mediterranean.

***Cypraea cinerea* Gmelin, 1791**	**1641**
Atlantic Gray Cowrie	**Color Plate 5**

North Carolina to Florida and the West Indies. Bermuda. Brazil.

¾ to 1½ inches in length, rotund, with its back brownish mauve to light orange-brown which may be flecked with tiny, black-brown specks. Base cream to old-ivory with light mauve-brown between some of the teeth, or sometimes with tiny flyspecks of brown. A moderately common species found under rocks on reefs.

Subgenus *Zonaria* Jousseaume, 1884

Type of this subgenus is *zonaria* (Gmelin, 1791) from West Africa.

Cypraea annettae Dall, 1909	**1642**
Annette's Cowrie	**Color Plate 5**

Baja California, the Gulf, to Peru.

1 to 1½ inches, elongate-pear-shaped. Dorsal surface pale bluish white with heavy, overlapping spots of dark and light chestnut-brown. Base tan to light-brown, the teeth (about 22 on the outer lip) being light-cream. Fairly common among rocks in shallow water. Rare or absent from southern Mexico to Panama; common in Ecuador. Synonyms: *sowerbyi* Kiener, 1845; *ferruginosa* Kiener, 1843.

1642

Subgenus *Erosaria* Troschel, 1863

Type of this subgenus is *erosa* Linné, 1758, from the Indo-Pacific.

Cypraea spurca acicularis Gmelin, 1791	**1643**
Atlantic Yellow Cowrie	**Color Plate 5**

North Carolina to Yucatan and the West Indies. Bermuda. Brazil.

½ to 1¼ inches in length; back irregularly flecked and spotted with orange-brown and whitish. Base and teeth ivory-white. Lateral extremities often with small pie-crust indentations. Distinguished from *cinerea* in being flatter and without color on the base. A moderately common species found under rocks at low tide. True *spurca* Linné, 1758, is from the Mediterranean.

Subgenus *Siphocypraea* Heilprin, 1886

The type of this subgenus is *problematica* Heilprin, 1886. Gardner, 1948, created the genus *Akleistostoma,* and Woodring, 1957, created *Muracypraea.*

Cypraea mus Linné, 1758	**1644**
Mouse Cowrie	**Color Plate 5**

North coast of Colombia and Venezuela.

2 inches in length, mouse-gray, and has a pair of irregular black-brown stripes on the back. It is frequently deformed

with 1 or 2 small bumps on the back (varieties *tuberculata* Gray, 1828, and *bicornis* Sowerby, 1870). For a full treatment, see Coomans, 1963, *Studies Fauna Curaçao and Other Caribbean Islands,* vol. 15, no. 68, pp. 51–71.

Subgenus *Propustularia* Schilder, 1927

The type of this subgenus is *surinamensis* Perry, 1811.

Cypraea surinamensis Perry, 1811	**1645**
Surinam Cowrie	**Color Plate 5**

Southeast Florida to Brazil.

1 to 1½ inches, somewhat resembling *spurca,* but larger, with the sides and base an orange-buff to apricot. The top is whitish with many small blotches and speckles of orange-brown. About 20 to 24 apricot-colored teeth on each lip, slightly whitish between. Formerly, this was thought to be a very rare species. Deep water; rare in Florida (off Boynton Beach, 60 feet; off Matecumbe Key, 150 feet); uncommon in the Caribbean and Brazil. Found in fish stomachs. Synonyms are: *bicallosa* Gray, 1832; *aubryana* Jousseaume, 1869; *ingloria* Crosse, 1878; *callosa* Weinkauff, 1881; *barbadensis* (fossil) Schilder, 1941. See W. K. Emerson and W. E. Old, Jr., *The Nautilus,* vol. 79, p. 26.

Subgenus *Neobernaya* Schilder, 1927

Type of this subgenus is *spadicea* Swainson, 1823. Schilder and Wenz consider this a subgenus of *Zonaria* Jousseaume.

Cypraea spadicea Swainson, 1823	**1646**
Chestnut Cowrie	**Color Plate 5**

Monterey, California, to Cerros Island, Baja California.

1 to 2 inches in length, half as high. Base white, with about 20 to 23 teeth on each side of the aperture. Sides bluish to mauve-gray, above which there is dark-chocolate fading on top to light chestnut-brown with a bluish undertone. Moderately common at certain seasons at low tide among seaweed, under rock ledges and also down to 25 fathoms. Also called the California brown cowrie (*The Veliger,* vol. 4, p. 215).

Other tropical Pacific species:

1647 *Cypraea (Erosaria) albuginosa* Gray, 1825. Southern part of the Gulf of California; Panama to Ecuador.

1648 *Cypraea (Luria) isabellamexicana* Stearns, 1893. Cape San Lucas, Baja California; Tres Marias Islands to the Galapagos. See *The Veliger,* vol. 3, p. 111.

1649 *Cypraea (Pseudozonaria) arabicula* Lamarck, 1811. San Hipolito Point, Baja California; southern Gulf of California to Peru; Galapagos. Color Plate 5.

1650 *Cypraea (Pseudozonaria) nigropunctata* Gray, 1828. Galapagos; Ecuador and Peru. (Synonyms: *irina* Kiener, 1843; *gemmula* Weinkauff, 1881; *massauensis* Schilder, 1922.)

1651 *Cypraea (Pseudozonaria) robertsi* Hidalgo, 1906. Central part of the Gulf of California to Paita, Peru; Galapagos. For a review of the above species, see C. N. Cate, 1969, *The Veliger,* vol. 12, p. 103.

Family OVULIDAE Fleming, 1822

Mostly tropical species associated with gorgonians, madrepore corals and alcyonarians. For a revision of the Eastern Pacific Ovulidae, see C. N. Cate, 1969, *The Veliger,* vol. 12, p. 95. *Amphiperatidae* is a synonym. A thorough and well-illustrated monograph on this family appeared after this book went to press, and time did not permit incorporation of all necessary changes. Consult C. N. Cate, 1973, Supplement to *The Veliger*, vol. 15.

Subfamily EOCYPRAEINAE Schilder, 1927

Genus *Jenneria* Jousseaume, 1884

Cowrielike shell, about one inch, with strong pustules on the dorsum and with strong teeth on the base on both sides of the aperture. Type: *Cypraea pustulata* Lightfoot, 1786. Only one species in the eastern Pacific. The subfamily Sulcocypraeinae Schilder, 1936, is a synonym of Eocypraeinae.

Jenneria pustulata (Lightfoot, 1786)   **1652**
Pustulate Cowrie   **Color Plate 3**

West coast of Mexico to Ecuador.

½ to ¾ inch, characterized by a gray dorsum bearing about 100 small orange pustules each of which are encircled with a brown line. Base with about 15 to 18 fine raised, white teeth. Moderately common in shallow water, particular in the region of Panama. Usually associated with stony corals.

Subfamily PEDICULARIINAE Gray, 1853

Genus *Pedicularia* Swainson, 1840

Shell with a large aperture. No operculum. Shell strong; reticulate sculpturing. Type: *sicula* Swainson, 1840.

Subgenus *Pediculariella* Thiele, 1925

Type of the subgenus is *californica* Newcomb, 1864.

Pedicularia decussata (Gould, 1855)   **1653**
Decussate Pedicularia

Georgia to southeast Florida and the West Indies.

¼ to ½ inch in length, moderately thick-shelled, with a long and flaring aperture, and pure-white in color. Sculpture of fine reticulations with the spiral threads the strongest. Columella a straight ridge with the parietal wall concavely dished. The entire shell has a distorted, "squeezed" appearance. Nuclear whorls obese, translucent-brown, reticulated, and with a sinuate lip when in its free-swimming, larval stage. An uncommon species found clinging to coral stems in moderately deep water. This is the only Eastern American species in this genus. *P. albida* Dall, 1881, is a synonym.

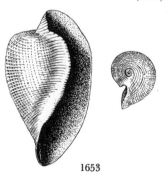

1653

Pedicularia californica Newcomb, 1864   **1654**
Californian Pedicularia

Farallon Islands to San Diego, California.

7 to 11 mm. in length, solid, aperture greatly enlarged and flaring. Apex hidden by the expanded lip. Early whorls showing minute decussations, the rest of the shell with small spiral threads. Interior uneven and glossy. Color rose with the outer lip whitish. Uncommon. Found attached to red hydrocoralline, *Allopora californica* Verrill. The form *ovuliformis* Berry, 1946, may represent the male of the species.

Subfamily OVULINAE Fleming, 1822

Genus *Primovula* Thiele, 1925

Shell small, elongate, spindle-shaped, with a rostrate anterior and posterior end. Outer lip marginate and finely toothed within. Type: *beckeri* (Sowerby, 1900) from South Africa.

Subgenus *Pseudosimnia* Schilder, 1927

Type of this subgenus is *carnea* (Poiret, 1789).

Primovula carnea (Poiret, 1789)   **1655**
Dwarf Red Ovula

Southeast Florida, the West Indies and the Mediterranean.

⅓ to ½ inch in length. This species resembles a miniature cowrie. The body whorl is rotund, pink to yellow in color and with numerous, fine spiral, incised lines. Aperture narrow, arched and with a canal at each end. Outer lip curled in like that of a cowrie, and with about 20 small, rounded, whitish teeth. Upper parietal wall with a large, rounded, short ridge or tooth. Apex not showing. Rare from 25 to 100 fathoms. *P. vanhyningi* M. Smith, 1940, is probably a synonym.

1655

Subfamily SIMNIINAE Schilder, 1927

Genus *Simnia* Risso, 1826

Shells spindle-shaped, about 1 inch or less, elongate, ends short. Outer lip narrow and slightly thickened. Type: *nicaeensis* Risso, 1826. *Neosimnia* Fischer, 1884, is a synonym. Do not confuse with immature *Cyphoma* whose lip has not thickened as yet.

Simnia acicularis (Lamarck, 1810)   **1656**
Common West Indian Simnia

North Carolina to southeast Florida and the West Indies. Bermuda.

½ inch in length, narrow, glossy, thin-shelled but strong, and with a long, toothless aperture. Color deep-lavender or

1656

yellowish. Columella area flattened or sometimes slightly dished and, in adults, always bordered by a long, whitish ridge, one inside the aperture, the other on the body whorl. Posterior end of columella sometimes slightly swollen. A common species which attaches itself and its tiny egg-capsules to purple or yellow seafans.

Simnia uniplicata (Sowerby, 1848)   1657
Single-toothed Simnia

Virginia to both sides of Florida and to Brazil.

½ to ¾ inch in length, similar to *acicularis,* but with only the innermost, longitudinal ridge on the columella, and with a twisted, spiral plication at the posterior end of the columella. Moderately common on seafans, varying in color according to the *Gorgonia* on which it lives.

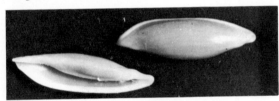

1657

Simnia piragua (Dall, 1889)   1658
Dall's Treasured Simnia

Between Jamaica and Haiti.

1 inch in length, extremely narrow, with the ends greatly produced. Columella area bordered by 2 longitudinal ridges, the inner one tinted with rose. Remainder of shell yellowish white. One of the rarest of the Western Atlantic mollusks. 23 fathoms.

Simnia avena (Sowerby, 1832)   1659
Western Chubby Simnia

West coast of Mexico to Peru.

½ inch in length, oblong, with short extremities. Lower end of the aperture wide, upper end narrow where the columella has a spiral swelling. Inner and lower part of the columella with a long, light-colored ridge. Exterior of whorls with numerous, microscopic, wavy, incised scratches. Color mauve to deep-rose with the varix and extremities a lighter pink. Uncommon.

1659

Simnia loebbeckeana (Weinkauff, 1881)   1660
Loebbeck's Simnia

Monterey, California, to west Baja California.

19 to 24 mm. in length, translucent yellowish, rarely pinkish, rather fragile, and fusiform in shape with the extremities narrow and the middle gently swollen. Columella rounded, usually smoothish, but sometimes with a hint of flattening and subsequent thickening of the lower part of the columella. Upper end of the columella with a weak, spiral fold. Dorsum smooth. Not uncommonly dredged in association with seafans from 20 to 50 fathoms. *Ovula barbarensis* Dall, 1892, and *catalinensis* S. S. Berry, 1916, may be synonyms.

Simnia rufa (Sowerby, 1832)   1661
Inflexed Simnia

Gulf of California to Panama.

15 to 24 mm. in length, a very vivid and dark lavender-rose or reddish purple. Rarely white with orange margins. Shell elongate; columella flattened, bordered within and also somewhat on the body whorl by a long, axial, lighter-colored ridge. Abundant; on gorgonians and seafans. Synonyms: *inflexum* Sowerby, 1832; *californica* Reeve, 1865; *neglecta* Reeve, 1865.

Simnia aequalis vidleri (Sowerby, 1881)   1662
Vidler's Simnia   **Color Plate 3**

Monterey Bay, California, to Panama.

15 to 22 mm. in length, narrowly elongate, pink to rose with orange extremities and a thickened, smooth, whitish outer lip. Sculpture of fine, incised, spiral lines at the ends. Columella ridge moderately strong and whitish. Moderately common; lives on red gorgonians at depths of 5 to 14 fathoms. *S. bellamaris* S. S. Berry, 1946, may be a synonym.

(1663) The typical *aequalis* Sowerby, 1832, is limited to the upper half of the Gulf of California. Has yellow terminal tips on white shell, and has a weak carinal ridge on the forward half of the columella. Common. *Neosimnia quaylei* Lowe, 1935; *variabilis* Reeve, 1865; and *tyrianthina* S. S. Berry, 1960 (dwarf male?), are synonyms.

Genus *Cyphoma* Röding, 1798

Shells elongate, with a thick, inrolled, smooth outer lip. Dorsum with a central cross-ridge. Live on seafans and sea rods. Lay eggs in single, dome-shaped, clear capsules. Type: *gibbosum* (Linné, 1758). The genus name is neuter.

Cyphoma gibbosum (Linné, 1758)   1664
Flamingo Tongue   **Color Plate 3**

North Carolina to southeast Florida and to Brazil. Bermuda.

¾ to 1¾ inch in length, glossy-smooth, chubby, and colored a rich cream-orange to apricot-buff except for a small whitish rectangle on the back. Callus on sides of shell indistinct and extending high up on the back with poorly defined edges. Mantle of animal pale-flesh with numerous squarish, black rings. Fairly common on searod gorgonians below low water from 3 to 30 feet. Please do not over-collect this species. It is becoming seriously depopulated everywhere by collectors.

Cyphoma macgintyi Pilsbry, 1939   1665
McGinty's Cyphoma

North Carolina to the both sides of Florida; Bahamas; Texas; Bermuda.

Similar to *gibbosum*, but more elongate, whitish with tints of lilac or pink on the back. The side callus on the right is thick and narrow. Aperture cameo-pink. Mantle with numerous solid spots which are roughly round or in the shape of short bars. Not uncommon. Feeds on searods, but will sometimes eat sea pansies. T. McGinty (*in litt.*) records it in 12 fathoms off Cameron, Louisiana. A thick variety with a stronger dorsal ridge was named *robustior* F. M. Bayer, 1941 (*The Nautilus*, vol. 55, p. 45).

Cyphoma signatum Pilsbry and McGinty, 1939   1666
Fingerprint Cyphoma

Lower Florida Keys; Bermuda (rare); Bahamas to Brazil.

Similar to *macgintyi*, but with the transverse ridge on the back much weaker, and the anterior end of the aperture more dilated than in the 2 preceding species. Color light-buff with a cream-buff tint deep inside the aperture. Mantle pale-yellow with numerous, crowded, long, black transverse lines. Uncommon in Florida on searod gorgonians from 15 to 30 feet.

1666

Cyphoma alleneae (C. N. Cate, 1973)   1667
Allene's Cyphoma

Lower Florida Keys.

20 to 30 mm. in length; shell similar to that of *macgintyi*, but with a pinkish buff color on either side of the narrow white ridge on the back. The mantle is very distinct, having curious, contorted graffiti marks resembling histological cross-sections. Oddly, this species is associated with corals, not sea whips. Uncommon. (Supplement to vol. 15, *The Veliger*, p. 67.) Named after Mrs. A. L. Snow of Marathon, Florida.

Other species:

1668 *Cyphoma* (*Cyphoma*) *emarginatum* (Sowerby, 1830). Baja California to Manzanillo, Mexico. ¾ inch. Color Plate 3.

1669 *Cyphoma* (*Pseudocyphoma* Cate, 1973) *intermedium* (Sowerby, 1828). Florida, West Indies to Brazil; Bermuda. 30 mm. Color Plate 3.

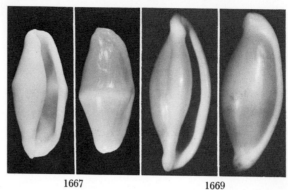

1667   1669

1670 *Cyphoma* (*Pseudocyphoma*) *aureocinctum* (Dall, 1899). Off Cuba to the Lesser Antilles, 123 meters on white gorgonium. 18.5 mm.

[Note: The **heteropods,** or superfamily **Atlantacea,** normally appearing here are now placed after the family Aclididae. See species **1472.**]

Superfamily Naticacea Gray, 1840

Family NATICIDAE Gray, 1840

Subfamily POLINICINAE Gray, 1847

Genus *Polinices* Montfort, 1810

Shells solid, glossy, ovate, with a buttonlike callus partially or completely filling the aperture. Operculum chitinous, translucent, thin, filling the aperture. Bore into clams and other snails. Lay a circular sand-collar egg capsule. Type: *mammilla* Linné, 1758. The genus name is masculine.

Polinices lacteus (Guilding, 1834)   1671
Milk Moon-shell

North Carolina to both sides of Florida to Brazil. Bermuda.

¾ to 1½ inches in length, glossy, milk-white, umbilicus deep with its upper portion covered over by the heavy callus of the parietal wall. Periostracum thin, smooth, yellowish. Operculum corneous, thin, transparent, either wine-red or amber. Common in sandy, intertidal areas.

Polinices uber (Valenciennes, 1832)   1672
Uber Milk Moon-shell

Baja California, the Gulf, to Chile.

½ to 2 inches, pure-white, glossy, umbilicus usually mostly covered with a thick parietal callus. Height of spire variable. Common from 1 to 50 fathoms. Synonyms include *Natica virginea* Récluz, 1850, and *ovum* Menke, 1850.

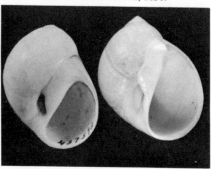

1671   1672

Polinices hepaticus (Röding, 1798)   **1673**
Brown Moon-shell   **Color Plate 4**

Southeast Florida (rare), Texas, West Indies to Brazil.

1 to 2 inches in length, heavy, glossy-smooth, with a deep, white umbilicus and small, low spiral callus. Exterior tan to orange-brown. Operculum corneous, thin, amber-brown. Formerly *P. brunneus* Link, 1807. Found in shallow water.

Polinices uberinus (Orbigny, 1842)   **1674**
Dwarf White Moon-shell

North Carolina to southeast Florida and the Caribbean. Brazil.

½ inch in length, very similar to *lacteus,* but the umbilical opening is larger, the callus is button-shaped and located against the columella near the center, and there is a large, rounded ridge running back from the callus into the umbilicus. Commonly dredged from 15 to 100 fathoms. Rarely in beach drift.

Polinices immaculatus (Totten, 1835)   **1675**
Immaculate Moon-shell

Gulf of St. Lawrence to North Carolina.

⅜ inch in length, subovate, smooth, milk-white and glossy when deprived of its thin greenish yellow periostracum. The ivory-white, thickened callus does not encroach upon the small, round, deep umbilicus. Operculum corneous, thin, light-brown. Commonly dredged off New England, and often found in fish stomachs.

Polinices bifasciatus (Griffith and Pidgeon, 1834)   **1676**
Two-striped Moon-snail

Gulf of California to Panama.

1 to 1½ inches, ovate, solid, heavy, brown to tan with 2 narrow white bands around the periphery. Umbilicus deep, but partially covered with a brown parietal callus. Locally common; on mud flats intertidally.

Subgenus *Neverita* Risso, 1826

Umbilicus partially blocked by a large brown button. Type: *josephinia* Risso, 1826, from the Mediterranean.

Polinices duplicatus (Say, 1822)   **1677**
Shark Eye   **Color Plate 4**

Cape Ann, Massachusetts, to Florida and the Gulf States.

1 to 2½ inches in length, glossy-smooth; umbilicus deep but almost covered over by a large, buttonlike, brown callus. Color slate-gray to tan; base of shell often whitish. Columella white. The shell is generally flattened and much wider than high, but

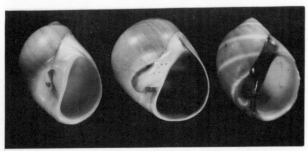

1673   1679   1676

some specimens are as wide as high and globose in shape. Operculum corneous, brown and thin. This is a very common sand-lover found along our eastern coasts. Compare young specimens with *Natica livida.* Diet can change the color of the parietal callus (H. J. Turner, Jr., 1958, *The Nautilus,* vol. 72, p. 1.)

Subgenus *Hypterita* Woodring, 1957

The type of this subgenus is *helicoides* (Gray, 1825).

Polinices helicoides (Gray, 1825)   **1678**
Helicoid Moon-shell

Baja California, the Gulf, to Peru.

2 inches across, very flattened (1 inch high), spire low. Color grayish brown to tan. Microscopic spiral striae present. Umbilicus partially covered with a brownish tonguelike callus. Moderately common from just offshore to 20 fathoms. *P. glaucus* Lesson, 1830, is a synonym, as are *patula* Sowerby, 1824, and *bonplandi* Valenciennes, 1832.

1678

Subgenus *Glossaulax* Pilsbry, 1929

Type of this subgenus is *reclusianus* (Deshayes, 1839).

Polinices reclusianus (Deshayes, 1839)   **1679**
Recluz's Moon-shell

Crescent City, California, to the Gulf of California.

1½ to 2½ inches in length, very heavy for its size. Spire moderately to quite well elevated. Exterior semiglossy, grayish with rusty-brown or greenish stains. Characterized by a large, tonguelike callus, brownish or white in color, which may or may not cover the entire umbilicus. There is a strong white, reinforcing callus at the top of the inside of the aperture. Operculum translucent, reddish brown. The shape of shell and degree of development of the umbilical callus is variable, and has received various names—*altus* Pilsbry, 1929 and *imperforatus* Dall, 1909. A common shallow-water species also found as deep as 25 fathoms.

Polinices draconis (Dall, 1903)   **1680**
Drake's Moon-shell

Trinidad, California, to Baja California

2 to 2½ inches in length, very similar to *Lunatia lewisii,* but with a wider, more elongate umbilicus, and with a very small, almost obsolete callus above the umbilicus. The periostracum is usually thick within the umbilicus. Uncommon in waters from 10 to 25 fathoms.

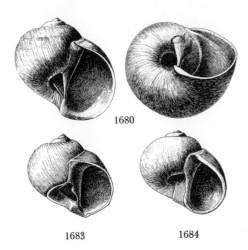

1680

1683   1684

Other species:

1681 *Polinices leptaleus* Watson, 1881. Off Fernandina, Florida, to the West Indies, 450 to 640 fms.

1682 *Polinices nanus* Möller, 1842. Greenland to off Block Island, Rhode Island, 22 to 115 fms.; off Fernandina, Florida. 294 fms. Arctic seas to off San Diego, California, 640 fms.

1683 *Polinices (Neverita) nubilus* Dall, 1889. Gulf of Mexico and the West Indies, 140 to 200 fms.

1684 *Polinices bahamiensis* (Dall, 1925). Grand Bahama Bank, 33 fms. 7 mm.

1686 *Polinices otis* (Broderip and Sowerby, 1829). Gulf of California to Ecuador. (Synonyms: *salongonensis* Récluz, 1844; *fusca* Sowerby, 1833; *Ruma subfusca* Dall, 1919 and *unimaculatus* Reeve, 1855.)

Genus *Calinaticina* Burch and Campbell, 1963

Shell thin, roundly barrel-shaped and oblique; light-brown. Umbilicus deep. Sculpture of fine spiral striae. Periostracum thin, silky-brown. Operculum small, crescent-shaped. For radula see Burch and Campbell, 1963, *Proc. Malacol. Soc. London*, vol. 35, p. 221. Type of the genus is *Sigaretus oldroydii* Dall, 1897.

Calinaticina oldroydii (Dall, 1897)   1687
Oldroyd's Fragile Moon-shell

Oregon to San Diego, California.

1½ to 3 inches in length, resembling *Lunatia lewisii*, but much lighter in weight, with a more pointed spire, without the

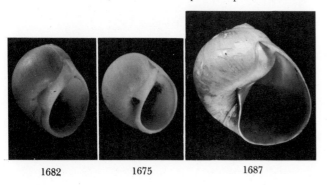

1682   1675   1687

heavy, brownish callus and having, instead, the upper part of the columella expanded into a white, thin area which partially obscures the umbilicus. Microsculpturing on shell exterior is prominent. Moderately common; dredged offshore 30 to 70 fathoms. For biology, see W. Ingram, 1941, *The Nautilus*, vol. 54, p. 136.

Genus *Sigatica* Meyer and Aldrich, 1886

Umbilicate naticids without an umbilical button and having some strong spiral grooves below the suture. Type: *boettgeri* (Meyer and Aldrich, 1886) of the Missouri Eocene.

Sigatica carolinensis (Dall, 1889)   1688
Carolina Moon-shell

North Carolina to southeast Florida and the West Indies.

¼ inch in length, white, glossy, ovate, fairly thin-shelled; umbilicus deep, round, without a callus. Characterized by 2 smooth nuclear whorls, followed by 3 whorls which are finely grooved by about 20 spiral lines. Suture well-channeled. Operculum paucispiral, corneous, its early whorls thickened and raised somewhat. *S. holograpta* McGinty, 1940 (*The Nautilus*, vol. 53, p. 110) is so similar that it may well be this species. Dredged 20 to 95 fathoms; not uncommon.

Sigatica semisulcata (Gray, 1839)   1689
Semisulcate Moon-shell

South Carolina to the West Indies.

8 to 10 mm., pure-white, resembling the common *Polinices lacteus*, but is broader, thinner-shelled, without an umbilical callus, with a well-impressed suture, below which are 4 to 6 microscopic incised spiral lines. Uncommon; 10 to 60 feet. *Natica fordiana* Simpson, 1889, is a synonym.

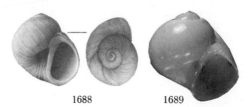

1688   1689

Genus *Lunatia* Gray, 1847

Shell rather large but very strong. Whorls regularly enlarging and rounded. Umbilicus small. Operculum corneous, paucispiral and filling the aperture. Type: *heros* (Say, 1822).

Lunatia heros (Say, 1822)   1690
Common Northern Moon-shell

Gulf of St. Lawrence to off North Carolina.

2 to 4½ inches in length, not so wide; globular in shape; umbilicus deep, round, not very large, and only slightly covered over by a thickening of the columellar wall. Color dirty-white to brownish gray. Aperture glossy, whitish or with tan or purplish brown stains. Periostracum thin, light yellow-brown. Operculum corneous, light-brown and thin. A very common intertidal species in the New England area. The egg case is a wide, circular ribbon of sand, about the thickness of an orange peel and easily bent when damp. The tiny eggs are embedded in the ribbon.

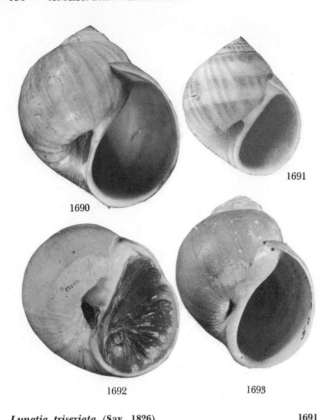

1690

1691

1692

1693

Lunatia triseriata (Say, 1826)   **1691**
Spotted Northern Moon-shell

Gulf of St. Lawrence to North Carolina.

½ inch in length, similar to young *heros* but the last whorl usually has 3 spiral rows of 12 to 14 bluish or reddish brown, squarish spots. The borders of the egg case are crenulated in contrast to the smooth borders of that in *heros*. This is a moderately deep-water species. Not uncommon from 1 to 63 fathoms.

Lunatia lewisii (Gould, 1847)   **1692**
Lewis' Moon-shell

British Columbia to Baja Colifornia.

3 to 5 inches in length, moderately heavy. Whorls globose, slightly shouldered a little distance below the suture. Umbilicus deep, round and narrow. Characterized by the brown-stained, rather small, buttonlike callus partially obscuring the top edge of the umbilicus. A very common species found in shallow water to 25 fathoms. They are more commonly found in the summer months. Do not confuse with *Polinices draconis*.

Lunatia pallida (Broderip and Sowerby, 1829)   **1693**
Pale Northern Moon-shell

Arctic Seas to off North Carolina.
Arctic Seas to California.

1½ to 1¾ inches in length, not quite so wide, smooth, pure-white in color and covered with a thin, yellowish white periostracum. Parietal wall moderately thickened with a white glaze. Umbilicus almost closed to slightly open. Commonly dredged offshore in cold northern waters. Deepest record: 1,245 fathoms. In the Atlantic, this species rarely exceeds 1 inch in length. *P. groenlandica* Möller, 1842; *borealis* Gray, 1839; *Natica pusilla* Gould, 1841; *monterona* Dall, 1919, are synonyms, as are *canonica* Dall, 1919; and *caurina* Gould, 1847.

Other species:

1694   *Lunatia levicula* (Verrill, 1880). Gulf of Maine, 26 to 100 fms. Light-weight; 2 inches. Uncommon.

1694

1695   *Lunatia tenuis* (Récluz, 1850). Off Cape Fear, North Carolina, to the West Indies, 84 to 640 fms.

1696   *Lunatia fringilla* (Dall, 1881). Gulf of Mexico, 382 to 840 fathoms. (Variety *perla* (Dall, 1889) **(1697),** Florida to the West Indies, 294 to 424 fms).

1696   1697

1698   *Lunatia politiana* (Dall, 1919). Petrel Bank, Bering Sea.

Genus *Bulbus* Brown, 1839

Shell rather thin, round to egg-shaped. Early whorls small. Umbilicus very small or absent. Type: *smithii* (Brown, 1839). *Acrybia* H. and A. Adams, 1853, is a synonym.

Bulbus smithii (Brown, 1839)   **1699**
Smith's Moon-snail.

Gulf of St. Lawrence to Georges Bank, Massachusetts.

1 inch, globular, inflated whorls, light-weight, fragile, and with a bright straw-colored or golden periostracum covering the shell whose surface is minutely reticulated by spiral threads and growth lines. Aperture very large. Umbilical region about the middle of the left margin greatly indented. Columella is a narrow, thick, rounded, ivory, twisted pillar. *Natica flava* Gould, 1840, is a synonym. A high-spired form from Japan was called *elongatus* Habe and Ito, 1965.

1699

Other species:

1700   *Bulbus fragilis* (Leach, 1819). Arctic Ocean to the Shumagin Islands and Aleutians. (Synonym: *aperta* Middendorff, 1851.)

Genus *Amauropsis* Mörch, 1857

This is considered a subgenus of *Bulbus* by Wenz and others, but we have kept it as a genus. It has a channeled suture. Type: *islandica* (Gmelin, 1791).

Amauropsis islandica (Gmelin, 1791)   1701
Iceland Moon-shell

Arctic Seas to off Virginia.

1 to 1½ inches in length, ¾ as wide, rather thin, but strong. Suture smooth and narrowly channeled. Shell smooth, yellowish white and covered with a thin, yellowish brown periostracum which flakes off when dry. Umbilicus absent or a very slight slit. Operculum paucispiral, horny, translucent-brown and with microscopic, spiral lines. A moderately common, cold-water species found just offshore down to 80 fathoms. *N. heliocoides* Johnston, 1835, is a synonym.

Its counterpart, *A. purpurea* Dall, 1871 (1702), common in Alaska, is very similar, but ¾ inch in length and with a greenish and darker periostracum. This may be *islandica*.

1701

Genus *Elachisina* Dall, 1918

Small, elevated, rotund naticoid shells with spiral sculpture and a small umbilicus. Type: *grippi* Dall, 1918. Only 1 species known:

1703   *Elachisina grippi* Dall, 1918. Off San Diego, California, 16 to 20 fms.

Genus *Gyrodes* Conrad, 1860

Shell naticoid, deeply and widely umbilicate, with a fairly deep suture bounded by an axially ribbed cord below. Type: *crenata* Conrad, 1860, from the Cretaceous. One living species doubtfully placed in this genus:

1704   *Gyrodes depressa* Seguenza, 1874. Off Cape Lookout, North Carolina, 15 to 52 fms.

Subfamily SININAE Woodring, 1928

Genus *Sinum* Röding, 1798

Shells strong but thin; aperture very large. Apical whorls small. Operculum corneous. Type: *haliotideum* (Linné, 1758) from West Africa. Synonyms: *Sigaretus* Lamarck, 1799; *Cryptostoma* Blainville, 1818.

Sinum perspectivum (Say, 1831)   1705
Common Baby's Ear

Maryland to Florida to Texas and West Indies. Bermuda. Brazil.

1 to 2 inches in maximum diameter, but very flat, with very large white aperture and strongly curved columella. Numerous fine spiral lines on top of whorls. Color dull-white with a light-brown, thin periostracum. Animal fully envelopes the shell. Commonly found in shallow, sandy areas, especially in the Carolinas and the west coast of Florida. It makes a wide track in the sand at low tide.

1705

Sinum maculatum (Say, 1831)   1706
Maculated Baby's Ear

Carolinas to Florida; West Indies.

Similar to *perspectivum*, but shell not so flat, with weaker spiral sculpture, and colored dull-brown or with yellowish brown maculations. The soft parts are whitish with livid purple maculations, while those of *perspectivum* are yellowish cream. Uncommon. Shallow water.

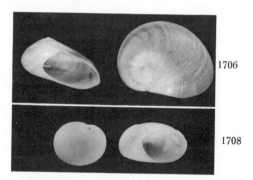

1706

1708

Sinum scopulosum (Conrad, 1849)   1707
Western Baby's Ear

Monterey to Todos Santos Bay, Baja California.

1 to 1¼ inches in length, 4 whorls, the early ones being very smooth, the last whorl very large. Numerous spiral grooves can be seen with the naked eye. Shell chalky-white, but usually covered with a thin, yellowish, translucent periostracum. The spire is more elevated and the whorls more inflated than those in *S. debile* Gould, 1853, from Catalina Island to the Gulf of California. *S. scopulosum* is moderately common. *S. californicum* Oldroyd, 1917, is a synonym.

1707

Other species:

1708 *Sinum minor* (Dall, 1889). Florida Keys and the West Indies, 54 to 84 fms. Rare.

1709 *Sinum keratium* Dall, 1919. Catalina Island, California.

1710 *Sinum debile* Gould, 1853. Catalina Island, California, to the Gulf of California. (Synonym: *pazianum* Dall, 1919.)

1711 *Sinum grayi* (Deshayes, 1843). West Mexico to Panama. (Synonym: *cortezi* J. and R. Burch, 1964.)

Genus *Haliotinella* Souverbie, 1875

Body like *Sinum,* elongate, whitish, with a large long propodium. Shell auriculate, thin; huge aperture, small nucleus. Columella thickened and arching inward in the region below the apex of the shell. Type: *montrouzieri* Souverbie, 1875.

Haliotinella patinaria (Guppy, 1876)   1712
Fingernail Sinum

Southeast Florida and the Caribbean.

Shell 12 to 14 mm. long, fragile, narrow (⅓ wide as long), with an aperture as large as the shell itself. 2 nuclear whorls, the first one brown, the next glossy opaque-white, smooth and rounded and sitting up on the apex of the shell. Columella long, a thickened glossy ridge which ends just under the apex; an umbilical chink is present at its top left. Periostracum thin and yellow. Interior glistening white. Operculum chitinous, oval, paucispiral. Animal 2 inches. An amazing evolutionary development in the *Natica-Sinum* line. Uncommon; shallow water in sandy mud among turtle grass. (Details in Marcus, 1965, *Bull. Marine Sci.*, vol. 15, p. 211.)

1712

Subfamily NATICINAE Gray, 1840

Genus *Natica* Scopoli, 1777

Shells globose, porcelaneous, glossy, strong, with a small umbilicus, and with a shelly operculum. Type. *vitellus* Linné, 1758, of the Indo-Pacific.

Subgenus *Natica* Scopoli, 1777

Natica livida Pfeiffer, 1840   1713
Livid Natica

Southeast Florida, Caribbean to Brazil. Bermuda.

7 to 17 mm., glossy-smooth, exterior lead-gray with vague, spiral, darker-gray bands. A subsutural white color band is crossed by short oblique wrinkles. Aperture and columella

brown; callus which almost fills the umbilicus characteristically dark to light chocolate-brown. Operculum smooth, white and calcareous. Moderately common on intertidal sand flats to 50 feet. Do not confuse with *Polinces duplicatus,* which is much flatter and has a corneous operculum, but which also has a brown to purplish brown callus. *N. jamaicensis* C. B. Adams, 1850, is a synonym.

1713   1719

Natica marochiensis (Gmelin, 1791)   1714
Morocco Natica

Southeast Florida to Brazil.

15 to 27 mm., higher than broad, early whorls blue-black, body whorl blurred yellowish gray with 4 or 5 spiral rows of chevron-shaped, blue-gray color marks. Narrow white band just below suture. Base and umbilical callus white. Operculum smooth, with a raised sharp rim, white with yellow stains. Locally common on intertidal sand flats. *N. maroceana* Dillwyn, 1817, is a synonym.

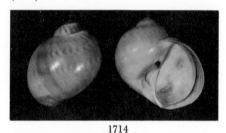

1714

Subgenus *Naticarius* Duméril, 1806

The type of this subgenus is *canrena* (Linné, 1758). The operculum has spiral cords.

Natica canrena (Linné, 1758)   1715
Colorful Atlantic Natica   **Color Plate 4**

North Carolina to Key West and the West Indies. Bermuda. Brazil.

1 to 2 inches in length, glossy-smooth, except for weak wrinkles near the suture. Color pattern variable; sometimes

1715

with axial, wavy, brown lines and with 4 spiral rows of arrow-shaped or squarish brown spots. Umbilicus and its large round, internal callus white. Exterior of hard operculum with about 10 spiral grooves. Moderately common; lives in sand in shallow water. A voracious carnivore most active at night. *Naticarius verae* Rehder, 1947, from west Florida (*The Nautilus,* vol. 61, p. 19) is probably only a banded color form of this species.

Subgenus *Glyphepithema* Rehder, 1943

The type of this subgenus is *idiopoma* Pilsbry and Lowe, 1932, of the Eastern Tropical Pacific. Operculum calcareous, with 1 or 2 broad ribs.

Natica floridana (Rehder, 1943)   **1716**
Florida Natica

Southeast Florida to Panama. Brazil.

15 to 22 mm., superficially resembling the common *Natica canrena,* but with a thicker rough, axially fimbriated periostracum; without axial streaks of brown; possessing 2 broad, irregular, spiral bands of light-brown, each bounded above and below by oblong (not arrow-shaped) small brown spots. Umbilical callus smaller and centrally located within the umbilicus. Operculum with 2 broad and 2 small ribs. Shore to 48 feet; in sand; uncommon. *Proc. U.S. Nat. Mus.,* vol. 93, p. 196.

Natica cayennensis Récluz, 1850   **1717**
Cayenne Natica   **Color Plate 4**

West Indies to Brazil.

20 to 27 mm., similar to *canrena,* but smaller, with strong wrinkles near the suture, pattern more blurred on the whorls, base white. Aperture violet within. The operculum is quite different: irregularly pustulose and with a large, swollen, arching ridge in the center. *N. haysae* Usticke, 1959, is the same. Moderately common in the Greater Antilles.

1716   1717

Subgenus *Tectonatica* Sacco, 1890

Type of this subgenus is *tectula* (Bonelli, 1847) of the Miocene. *Cryptonatica* Dall, 1892, is a synonym.

Natica clausa Broderip and Sowerby, 1829   **1718**
Arctic Natica

Arctic Canada to North Carolina.
Arctic Ocean to off California.

1 to 1¼ inches in length, fairly thin, smooth, yellow-white, with a smooth, gray to yellowish brown periostracum. Umbilicus sealed over by a small, flat callus. Operculum calcareous, thin, slightly concave, smooth, white and paucispiral. Commonly dredged in moderately deep water, and occasionally

found intertidal north of Massachusetts. Deepest record is 1,240 fathoms. The sand-collar egg-case has smooth edges, and has a pimpled surface caused by the small compartments of the young. Synonyms are *consolidata* Couthouy, 1838; *septentrionalis* Möller, 1842; *algida* Gould, 1848; *acosmita* Dall, 1919; *aleutica* Dall, 1919; *russa* Gould, 1859; and *salimba* Dall, 1919.

Natica pusilla Say, 1822   **1719**
Southern Miniature Natica

Maine to Florida, the Gulf States, and to Brazil.

¼ to ⅓ inch in length, glossy-smooth, similar to *clausa,* but more ovate, often with a small, open chink next to the umbilical callus, and is a much smaller shell. Nucleus of calcareous operculum often stained with brown. Color white, but often with weak, light-brown markings. Commonly dredged in shallow water to 18 fathoms. Do not confuse with *livida.*

Other species:

1720 *Natica (Natica) castrensis* Dall, 1889. Florida Keys to the West Indies, 27 to 100 fms.

1721 *Natica sanctivincentii* Brooks, 1933. St. Vincent, Lesser Antilles. Annals Carnegie Mus., vol. 21, p. 413.

1722 *Natica (Natica) perlineata* Dall, 1889. Florida Strait to the West Indies, 70 to 229 fms.

1723 *Natica (Tectonatica) micra* Haas, 1953. Ilha Grande, Brazil. Fieldiana, vol. 34, p. 206.

1724 *Natica (Tectonatica) affinis* Gmelin, 1791. Arctic Seas; Siberia; Greenland.

1725 *Natica (Tectonatica) janthostoma* Deshayes, 1841. Bering Sea to Japan; Kamchatka. 50 mm.

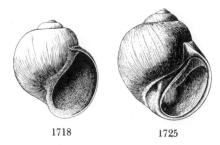

1718   1725

1726 *Natica (Natica) chemnitzii* Pfeiffer, 1840. Baja California, throughout the Gulf to northern Peru. Common. For feeding habits, see J. J. Gonor, 1965, *The Veliger,* vol. 7, p. 228.

Genus *Stigmaulax* Mörch, 1852

Whorls inflated, axial ribs strong, crowded and numerous. Umbilicus very wide, and with a large, round cord or funicle emerging near the base. Operculum calcareous and with a broad, thick, rough central rib. Type: *sulcatus* (Born, 1778).

Stigmaulax sulcatus (Born, 1778)   **1727**
Sulcate Natica

Southeast Florida and the West Indies. Brazil.

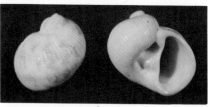

1727

½ to ¾ inch, strong, rounded whorls, with numerous, crowded axial ribs which may or may not be crossed by numerous incised lines, often giving a beaded surface to the ribs. Umbilicus with a large, white spiral, rounded cord in the lower section. Exterior white or suffused with tan, or marbled with spotches of light- to dark-brown. Aperture brownish within. Base white. Sculpture and coloration variable, the whitish form with practically no spiral lines being typical, *sulcata*, of which *rugosus* (Gmelin, 1791) is a synonym. The more colorful form, usually with a reticulate sculpture, is *cancellatus* Hermann, 1781 (synonym: *cancellatus* Gmelin, 1791). Uncommon, shallow water. *Natica (Stigmaulax) cubana* Dall, 1927, is probably a young specimen.

Other species:

1728 *Stigmaulax broderipiana* Récluz, 1844. Baja California to Peru. Uncommon. (Synonym: *iostoma* Menke, 1847.)

1729 *Stigmaulax elenae* Récluz, 1844. Baja California to Ecuador. Uncommon. Color Plate 4.

Superfamily Tonnacea Peile, 1926

Family CASSIDAE Swainson, 1832

A worldwide group of large shells usually having strong, glossy varices and a broad, well-developed parietal shield. Operculum corneous, fan-shaped. Many members of this family feed on sea-urchins and sand dollars. The family name was formerly spelled *Cassididae*. For a detailed treatment of the family, see Abbott, 1968, *Indo-Pacific Mollusca*, vol. 2, no. 9.

Genus *Sconsia* Gray, 1847

Shell fusiform, solid, with fine, spiral, incised or cut lines. With one or more former low varices. Columella with faint denticles; no parietal shield; no umbilicus; no gutter behind the short siphonal canal. Type: *striata* (Lamarck, 1816). Only living in the Western Atlantic.

Sconsia striata (Lamarck, 1816)   **1730**
Royal Bonnet   **Color Plate 6**

Southeast Florida to off Texas and to Brazil.

1½ to 2½ inches in height. Shell hard, polished, often with numerous fine, spiral incised lines. Usually 2 old varices are present. With 4 or 5 spiral rows of brownish spots. Locally moderately common; from 50 to 255 fathoms. Synonyms are *grayi* A. Adams, 1855; *barbudensis* Higgins and Marrat, 1877.

Other species:

1731 *Sconsia nephele* F. M. Bayer, 1971. Off Grenada, Lesser Antilles, in 18 meters. *Studies in Trop. Amer. Moll.*, p. 139.

Genus *Morum* Röding, 1798

Shells small, somewhat cylindrical, knobbed, with a strong, pimpled parietal shield and a nipple-like apex. Operculum small; radula absent. Type: *oniscus* Linné, 1767.

Morum oniscus (Linné, 1767)   **1732**
Atlantic Morum

Southeast Florida to Brazil. Bermuda.

¾ to 1 inch in height. Whorls with 3 spiral rows of rather prominent, bulbous, low tubercles. Parietal wall glazed over and ingrained with numerous white dots which are developed

1732

1733

into minutely raised pustules. Color (with thin, velvety, gray periostracum removed) whitish with specklings or mottlings of brown or black-brown. Nucleus papilliform, white or pink. Operculum very small, corneous, and with its nucleus on the side. Just below low-tide mark under coral slabs. Synonyms are *conoidea* Scopoli, 1786; *purpureum* Röding, 1798; and *triseriata* Menke, 1830. Egg capsules resembling dried corn seeds set in a row (R. C. Work, 1969, *Bull. Marine Sci.*, vol. 19, p. 657).

Subgenus *Cancellomorum* Emerson and Old, 1963

Sculpture is cancellate. Type of the genus is *grande* A. Adams, 1855, of Japanese waters.

Morum dennisoni (Reeve, 1842)   **1733**
Dennison's Morum

Off North Carolina to off Texas to Brazil.

2 inches, whorls shouldered and bearing fine nodules. Parietal shield orange to reddish brown with numerous fine, raised white pimples. Outer lip reflected and bears numerous weak lirae on the inner edge. Rare; 33 to 130 fathoms.

Other species:

1734 *Morum (Cancellomorum) matthewsi* Emerson, 1967. 12 to 40 fms., off northeast Brazil. 1 inch. Parietal wall purple-brown with white lirae. (*The Veliger*, vol. 9, no. 3, p. 289, pl. 39). Color Plate 6.

1735 *Morum (Cancellomorum) veleroae* Emerson, 1968. Costa Rica and the Galapagos Islands, 55 to 90 meters.

1736 *Morum (Morum) tuberculosum* (Reeve, 1842). Gulf of California to Peru, intertidal. 12–20 mm. Moderately common in shallow water. (Synonym: *M. xanthostoma* A. Adams, 1854). Color Plate 6.

Genus *Phalium* Link, 1807

These are miniature helmet shells which rarely exceed a length of 5 inches. The Scotch Bonnet of the Atlantic coast (*Phalium granulatum*) is well-known to most collectors. It is now the official shell of the state of North Carolina. This genus differs from *Cassis* in having much smaller shells which do not have an extended, upturned siphonal canal and do not develop a massive parietal shield. Typical *Phalium* which has 4 or 5 tiny spines on the base of the outer lip (as for example in the Indo-Pacific type, *P. glaucum* Linné) is not represented in American waters. Our 2 species belong to the subgenus *Tylocassis* which lacks these tiny spines. Operculum as in *Cassis*, but more fan-shaped.

Subgenus *Tylocassis* Woodring, 1928

Parietal wall covered with numerous, small, raised pimples. Type: *granulatum* Born, 1778.

Phalium granulatum (Born, 1778)   **1737**
Scotch Bonnet   **Color Plate 6**

North Carolina to Texas and Brazil; Bermuda.

1½ to 3 inches in length, with about 20 spiral grooves on the body whorl, or entirely smooth. Weak axial ribs sometimes present which make the shell coarsely beaded. Lower parietal area pustulose. Outer lip may be greatly thickened occasionally. Not uncommonly washed ashore. It is also present on the west coast of Central America as the subspecies *centiquadratum* (Valenciennes, 1832). *Semicassis abbreviata* Lamarck, 1822; *inflatum* Shaw, 1811; *cicatricosum* Gmelin, 1791 (**1738),** Color Plate 6 (smooth form); and *peristephes* Pilsbry and McGinty, 1939, are among the synonyms. This species is the official state shell of North Carolina.

The subspecies *centiquadratum* (Valenciennes, 1832) (**1739;** Color Plate 6) is very similar, but usually has sharper nodules on the shoulder. It is common and occurs from the Gulf of California to Peru. Synonyms are: *doliata* Valenciennes, 1832; and *lactea* Kiener, 1835.

Subgenus *Echinophoria* Sacco, 1890

Like *Phalium* but without spines on the outer lip, and with a thin, raised strong thread running along the dorsal surface of the outside of the siphonal canal. Type: *intermedium* Brocchi, 1814, a European fossil.

Phalium coronadoi (Crosse, 1867)   **1740**
Coronado Bonnet

North Carolina to Cuba and Venezuela.

3 to 4 inches, globose, fairly thin-shelled, with small knobs on the shouldered whorls; umbilicate; grayish white with a tan periostracum. Body whorl with 9 blunt knobs and 3 low

1740

---

spiral swellings below with smaller knobs. Umbilicus narrow and deep. Operculum tan, chitinous and fan-shaped. Rare; 18 to 124 fathoms.

Genus *Casmaria* H. and A. Adams, 1853

Shells small, glossy, whitish, with a smooth columella shield and having minute prickles on the base of the outer lip. Umbilicus absent. Type: *erinaceus* (Linné, 1758).

Casmaria ponderosa atlantica Clench, 1944   **1741**
Atlantic Casmaria   **Color Plate 6**

Off Palm Beach, Florida, and the Caribbean. Bermuda.

1 inch, smooth, whitish, glossy, with 4 rows of weak, squarish red-brown spots on the last whorl, and with 5 to 9 minute prickles on the lower half of the outer lip. Nuclear and early whorls smooth, not reticulated as in *Phalium*. Rare: 1 to 83 fathoms. The nominate subspecies, *ponderosa*, comes from the Indo-Pacific.

1742 *Casmaria erinaceus* (Linné) subspecies *vibexmexicana* (Stearns, 1894). West coast of Mexico. Uncommon.

Genus *Cassis* Scopoli, 1777

The helmet shells are large, handsome mollusks which have been used by man for centuries. Large numbers of cameos are still cut from them, the meat is often used in chowders and the uncut shells serve as attractive doorstops or mantelpieces. In the Pacific, they are sliced in half and the body whorl used either as a cooking container or boat-bailer. The half dozen known species are found only in the West Indies and Indo-Pacific area. They live in moderately deep water and although sometimes are obtained in knee-deep waters, they usually must be dived for in 10 to 20 feet of water. The helmet shells are carnivorous and include the sea urchins in their diet. For a detailed account of cassid feeding habits, see Don Moore, 1956, *The Nautilus,* vol. 69, pp. 73–76. Operculum elongate-oval to semicircular, corneous and concentric. Type: *cornuta* Linné, 1758, from the Indo-Pacific.

Cassis tuberosa (Linné, 1758)   **1743**
King Helmet

North Carolina to Brazil; Bermuda.

Adults 4 to 9 inches in length, massive, with a finely reticulated sculpture. Color brownish cream with black-brown patches on the lip and a large patch of brown at the center of the parietal shield. This species may be easily confused with the flame helmet (*Cassis flammea* Linné) which occurs in the Bahamas and Antilles. The latter lacks the reticulated sculpture, lacks brown color between the teeth on the outer lip, has a rounded (not triangular) parietal shield and is from 3 to 5 inches in length. Rare in Florida, common to the south in shallow water to 30 feet. Also found live in the Cape Verde Islands, off West Africa.

Cassis madagascariensis Lamarck, 1822   **1744**
Queen or Emperor Helmet

North Carolina to the Greater Antilles; Bermuda.

Adults 4 to 9 inches in length, massive. 3 spiral rows of large blunt spines; the topmost spine of the first row generally the largest. Color pale-cream on the outer surface. Parietal

1743

1744

1745

1746

shield and outer lip pale- to deep-salmon. Teeth white, brown sometimes between them. Sole and edge of foot dirty-yellow; tentacles black-brown with cream-orange tips. Egg case in the shape of a round bun and consists of 300 or so tall, tubular, squarish capsules, each 1 inch long and containing about 2,500 embryos (see D'Asaro, 1970, *Bull. Marine Sci.*, vol. 19, p. 905). Moderately common from 3 to 10 fathoms in the Bahamas. Very rare in Florida where it is replaced by the Clench's helmet, a form (1745) called *spinella* Clench, 1944. The latter turns up in Bermuda, North Carolina and in parts of the West Indies. Feeds on heart urchins and *Diadema* sea urchins.

Cassis flammea (Linné, 1758)   **1746**
Princess or Flame Helmet

Lower Florida Keys to Brazil; Bermuda.

3 to 5 inches, lacks reticulated sculpturing, has an ovate parietal shield, and lacks brown between the teeth of the outer lip. It is rare in Florida, common in the West Indies. Lives in shallow water.

Genus *Cypraecassis* Stutchbury, 1837

Shell oblong, heavy, predominantly orange-brown, with numerous, fine teeth on the columella, and with 4 or 5 large, orangish nuclear whorls. Operculum small, round, absent in some adults. Type: *rufa* Linné, 1758. Only 1 species in the Caribbean. Feeds on sea urchins.

Cypraecassis testiculus (Linné, 1758)   **1747**
Reticulated Cowrie-helmet   **Color Plate 6**

North Carolina to Brazil; Bermuda.

1 to 3 inches in length. Body whorl closely sculptured by small, distinct, longitudinal ridges which are crossed by a dozen or so spiral grooves, thus producing a reticulated surface. The shoulder of the body whorl in a very few specimens may have pinched-up, low tubercles or ribs. It is only a form. Entire animal light brownish orange, with underside of foot smeared with darker shades of orange. No periostracum. No operculum in the adult. Eggs laid under small rocks in tan capsules resembling miniature stovepipes. Reef inhabitant, below low-water level to 20 feet. Feeds on echinoderms, especially *Echinometra*.

Other species:

1748 *Cypraecassis tenuis* (Wood, 1828). Pl. 9. Baja California to Clipperton Island; Galapagos; Ecuador. 5 inches. Moderately common. Color Plate 6.

Subgenus *Levenia* Gray, 1847

Upper ⅓ of the outer lip reflected inwardly. One known species, the type, *coarctata* Sowerby, 1825.

Cypraecassis coarctata (Sowerby, 1825)   **1749**
Contracted Cowrie-helmet   **Color Plate 6**

Gulf of California to Peru.

2 to 3 inches; cylindrical, with the upper ⅓ of the outer lip pinched inwardly and the lower ½ reflected and bearing about 15 denticles on the inner edge. Color of brown and gray maculations. Parietal glaze orangish. Periostracum present in fresh specimens. Moderately common offshore from 2 to 6 fathoms; rarely intertidally.

Family CYMATIIDAE Iredale, 1913

We have largely followed the excellent monographic series of this family by Clench and Turner, 1957, in *Johnsonia*, vol. 3, no. 36. I wish to thank Mr. Hal Lewis for many improvements in this section. The family name is a conserved one.

Genus *Fusitriton* Cossmann, 1903

Shell rather large, fusiform, solid, thin, with nodulated axial ribs on the early whorls of the spire. Siphonal canal moderately long and open. Upper parietal wall with a strong internal ridge. Periostracum heavy and spiculose. Suture slightly indented. Outer lip weakly thickened. Type: *cancellatus* (Lamarck, 1816).

Fusitriton oregonensis (Redfield, 1848)   1750
Oregon Triton

Bering Sea to San Diego, California.

4 to 5 inches in length, about 6 whorls. Characterized by its fusiform shape, convex whorls, which each bear 16 to 18 axial ribs nodulated by the crossing of smaller spiral pairs of threads. The epidermis is heavy, spiculose, bristlelike and gray-brown. Aperture and siphonal canal interiors are enamel-white. Enamel, single tooth on parietal wall near the top of the aperture. Operculum chitinous, thick, brown. A common offshore species in its northern range. For egg-laying habits, see Faye Howard, 1962, *The Veliger*, vol. 4, p. 106, color plate 39.

1750

Genus *Cymatium* Röding, 1798

Subgenus *Cymatium* Röding, 1798

The type of the genus is *femorale* Linné, 1758. The shell bears 3 thick winged varices per whorl, giving the heavy shell a triangular shape.

Cymatium femorale (Linné, 1758)   1751
Angular Triton   Color Plate 7

Southeast Florida and the West Indies. Bermuda. Brazil.

3 to 8 inches in length, with 2 or 3 former varices; outer lip flaring, thickened into a noduled varix which is drawn up to a point posteriorly. Columella with 1 small fold and above it sometimes several much smaller ones. Color varies from brownish to reddish orange. Not uncommon in the West Indies in shallow water eel-grass. Rare in Florida and Bermuda. 2 to 65 fathoms in Brazil.

A very similar species, *Cymatium tigrinum* (Broderip, 1833) (1752) has a very heavy periostracum, reaches 7 inches, and is rare from La Paz, Baja California, and more common from Acapulco south to Ecuador.

Subgenus *Septa* Perry, 1810

The type of this subgenus is *rubeculum* (Linné, 1758).

Cymatium pileare (Linné, 1758)   1753
Atlantic Hairy Triton   Color Plate 7

South Carolina to Texas and to Brazil. Bermuda. Gulf of California to Panama.

1½ to 4 inches in length; old varices strong, beaded, and spaced ⅔ of a whorl apart. Spiral sculpture of a dozen or so squarish, irregularly sized, weakly beaded cords. Aperture orange-brown with the parietal area dark-brown between the white teeth. Periostracum flat with axial blades of fine hairs, light-brown. The embryonic shell is about 4 mm. in length, glossy-brown, with a flaring lip which has a small stromboid notch. *Dissentoma prima* Pilsbry, 1945, is this species. *C. martinianum* Orbigny, 1845, and *C. velei* Calkins, 1878, are also synonyms. Also common in the Indo-Pacific. Common in shallow water.

Do not confuse with *C. nicobaricum* which has just inside its outer lip a series of single, rather large, whitish teeth, instead of smaller, paired, yellowish brown teeth.

Cymatium vestitum Hinds, 1844   1754
Panamanian Hairy Triton

Gulf of California to Ecuador.

2 to 2½ inches, solid, ovate-elongate, resembling *pileare*. Spiral cords are often beaded and occur in pairs. Shoulder bears 4 or 5 low knobs between the varices. Inner lip blotched with black-brown and light-tan, over which are about 20 white or cream spiral, raised lirae. Outer lip with 7 pairs of short white teeth on brown staining. Periostracum thick, hairy, occurring in axial lamellae. Moderately common.

Cymatium vespaceum (Lamarck, 1822)   1755
Dwarf Hairy Triton

North Carolina to the West Indies. Bermuda. Brazil.

1754   1756

1 inch in length, with only 1 or no former varix. Whorls squarish at the shoulder where there are 2 spiral rows of prominent beads. The last whorl has only 1 row of about 6 to 8 rather large tubercles in addition to spiral and axial threads. Siphonal canal moderately long, slender. Color whitish with 1 or 2 orange-brown bars on the varix. Periostracum flat, with thin axial blades of bristles. Uncommon below low-water line. *C. gracile* Reeve, 1844, is believed to be a juvenile *pfeifferianum* (Reeve) from the Indo-Pacific. *C. pharcidum* Dall, 1889, appears to be a high-spired, finely beaded form of this species, or may be a valid species **(1756)**.

Cymatium krebsii Mörch, 1877   **1757**
Krebs' Hairy Triton   **Color Plate 7**

Off North Carolina to Florida and the Caribbean.

2 to 3 inches, solid, fusiform, with a small, ovate aperture surrounded by a white peristome. The columella bears 2 large and several smaller plicae. Outer lip with 6 or 7 large teeth. Varices 5 to 7, strong and nodulose. Between them are 5 or 6 axial ridges, crossed by 6 or 7 spiral cords. Nuclear whorls 3 or 4, straw-yellow and with microscopic axial striae. Periostracum thick, rough and light-brown. Uncommon 14 to 80 fathoms.

Cymatium amictum tremperi Dall, 1907   **1758**
Tremper's Hairy Triton.

San Diego, California, to the Gulf of California.

2½ inches, similar to *krebsii*. Periostracum black. 1 or 2 coarse threads between the large spiral cords. Rare. *C. amictoideum* Keen, 1971, is a synonym.

Cymatium rubeculum occidentale
Clench and Turner, 1947   **1759**
Atlantic Ruby Triton

Southeast Florida and the West Indies. Brazil.

1 inch, stout, solid, with a short siphonal canal, brownish red to light yellow-brown with 1 or 2 spiral bands of color. Outer lip with 8 denticles. Inner lip with 14 or 15 strong, long lamellae. With 3 to 5 ribbed varices crossed by 7 or 8 beaded cords. Between the varices are 3 to 5 weak axial ridges. Do not confuse with young specimens of *pileare*. Rare; low tide to 40 fathoms. Typical *rubeculum* (Linné, 1758), comes from the Indo-Pacific and is brilliant-red or orange-brown.

Subgenus *Cymatriton* Clench and Turner, 1957

The type of this subgenus is *nicobaricum* (Röding, 1798).

Cymatium nicobaricum (Röding, 1798)   **1760**
Gold-mouthed Triton   **Color Plate 7**

Southeast Florida to Brazil. Bermuda. Indo-Pacific.

¾ to 2½ inches in length; coarsely corrugated by spiral, noduled cords; varices spaced ⅔ of a whorl apart. Shell ash-gray with brown flecks and characterized by an orange mouth with white teeth. 6 nuclear whorls, strongly convex, amber and microscopically axially striate. A common West Indian shallow water species which is also abundant in the Indo-Pacific. Synonyms are: *chlorostomum* Lamarck, 1822; *pulchellus* C. B. Adams, 1850; and *pumilio* Mörch, 1877.

Subgenus *Linatella* Gray, 1857

Type of this subgenus is *cingulatum* (Lamarck, 1822).

1759   1763

1766   1767

Cymatium cingulatum (Lamarck, 1822)   **1761**
Poulsen's Triton   **Color Plate 7**

Off North Carolina to Texas to Brazil. Bermuda.

2 to 3 inches, lightweight, with 18 to 20 flattened spiral cords, the one at the angular shoulder being knobbed in some specimens. Former varices rare. Color yellowish to brown, rarely axially streaked. Aperture large; columella smooth. Outer lip crenulate. Periostracum thin, tan, flaking off easily when dried. Rare in Florida; common just offshore and on the beaches of northern Mexico near Texas. *C. poulsenii* Mörch, 1877 and *peninsulum* M. Smith, 1937, are synonyms.

A very similar species, *Cymatium wiegmanni* Anton, 1839 **(1762)**, occurs from Baja California to Peru. I follow Keen in rejecting the trivial name *cynocephalum* Lamarck, 1816, for this species. Common.

Subgenus *Tritoniscus* Dall, 1904

The type of this subgenus is *cutaceus* (Linné, 1758). Synnoyms are *Aquillus* Montfort, 1810; *Dolarium* Schlueter, 1838; *Neptunella* Gray, 1853; *Tritoniscus* Dall, 1904; *Turritriton* Dall, 1904; and *Particymatium* Iredale, 1936.

Cymatium labiosum (Wood, 1828)   **1763**
Lip Triton

Off North Carolina to both sides of Florida; and, the West Indies, and Brazil; also Indo-Pacific.

Shell ¾ inch in length, much like *Cymatium vespaceum,* but heavier, with a much shorter siphonal canal, slightly umbilicated and with strong spiral cords on the base of the shell. Uncommon in intertidal reef areas. Synonyms are: *rutilum* Menke, 1843; *loroisi* Petit, 1852; *strangei* Adams and Angas, 1864; and *orientalis* G. and H. Nevill, 1874.

Subgenus *Gutturnium* Mörch, 1852

Type of this subgenus is *muricinum* (Röding, 1798).

Cymatium muricinum (Röding, 1798)   **1764**
Knobbed Triton   **Color Plate 7**

Southeast Florida and the West Indies to Brazil. Bermuda. Also Indo-Pacific.

1 to 2 inches in length; characterized by a thickened cream parietal shield and a long, bent-back siphonal canal. Color ash-gray, sometimes dark-brown with a narrow cream, spiral band. Not uncommon in intertidal sand and seaweed areas. Interior of aperture brownish red to yellowish white. *C. tuberosum* Lamarck, 1822; *pyriformis* Conrad, 1849; *productum* Gould, 1852; *antillarum* Orbigny, 1842; and *Litiopa obesa* C. B. Adams, 1850 (an embryonic shell), are synonyms.

Subgenus *Ranularia* Schumacher, 1817

Type of this subgenus is *longirostrum* Schumacher, 1817. *Tritonocauda* Dall, 1904, is a synonym.

Cymatium moritinctum caribbaeum
Clench and Turner, 1957   **1765**
Dog-head Triton   **Color Plate 7**

South Carolina to Florida and the West Indies. Bermuda. Brazil.

1½ to 2½ inches in length; with globular whorls which are squarish at the shoulder. Siphonal canal long and slender. Usually with only 1 former varix. Apical whorls cancellate; last whorl with slightly noduled, spiral cords. Outer lip and parietal wall peach to flesh color. Parietal wall with an oval splotch of dark-brown, over which run light-orange spiral cords. Uncommon in Florida and Bermuda. Shallow water in weeds and sand. *Triton cynocephalum* Lamarck, 1816, is a species in the subgenus *Linatella.* Typical *C. moritinctum* (Reeve, 1844) is from the Indo-Pacific. True *sarcostomum* (Reeve, 1844) comes only from the Indo-Pacific.

Cymatium testudinarium rehderi Verrill, 1950   **1766**
Rehder's Triton

Cuba to the Lesser Antilles.

2½ to 3½ inches, light reddish brown, the varices having alternating bands of brown and white. Inner lip glossy, thickened, reddish brown with long cream plicae which extend across the entire parietal area. In *sarcostomum* and *caribbaeum* these plicae are well within the aperture. Last whorl with 5 or 6 corrugated varices, between which are 5 or 6 knobbed axial ridges crossed by about 7 spiral, beaded cords. Siphonal canal fairly long. Rare; offshore to 200 fathoms. Typical *testudinarium* (Adams and Reeve, 1848) is an Indo-Pacific species. Formerly, this subspecies was erroneously associated with *pyrum* Linné from the Indo-Pacific.

Subgenus *Monoplex* Perry, 1811

Type of this subgenus is *parthenopeum* (von Salis, 1793).

Cymatium parthenopeum (von Salis, 1793)   **1767**
Giant Hairy Triton

Off North Carolina to Texas to Brazil; Bermuda. Gulf of California to Galapagos Islands; Japan.

3 to 6 inches, with a very heavy, fuzzy, brown periostracum. Shell light-brown, rarely with light spiral bands. Columella dark reddish brown with strong, white, raised plicae. Outer lip with 5 to 7 brown spots which have 2 to 4 tiny white denticles. Not uncommon, usually occurring singly, low-tide mark to 35 fathoms. *C. costatum* Born, 1778 (not Pennant, 1777); *australasiae* Perry, 1811; *americana* Orbigny, 1842; *brasiliana* Gould, 1849; *echo* Kuroda and Habe, 1939; and *keenae* (Beu, 1970) are synonyms.

Other species:

1768 *Cymatium lignarium* (Broderip, 1833). Gulf of California to Peru.

1769 *Cymatium gibbosum* (Broderip, 1833). Gulf of California to Peru. (Synonym: *adairense* Dall, 1910.)

Genus *Distorsio* Röding, 1798

Cymatiid shells with gnarled, distorted apertures. Periostracum usually quite hairy. Operculum chitinous, irregular, with the nucleus submarginal. Type is *anus* (Linné, 1758). Synonyms are *Distortrix* Link, 1807; *Persona* Montfort, 1810; and *Rhysema* Clench and Turner, 1957. An excellent recent review by Hal Lewis, 1972, appeared in *The Nautilus,* vol. 86, nos. 2–4. There are 3 Caribbean species.

Distorsio clathrata (Lamarck, 1816)   **1770**
Atlantic Distorsio   **Color Plate 7**

North Carolina to Texas and the Caribbean. Brazil.

¾ to 3½ inches in length; whorls distorted, aperture with grotesque arrangement of the teeth; siphonal canal twisted. Whorls with coarse reticulate pattern. Parietal shield glossy, reticulated with raised threads, colored white to brownish white. Differs from *constricta macgintyi* in having a less distorted body whorl which is more evenly rounded and more evenly knobbed or reticulated. The parietal wall is generally reticulated instead of pustuled. Moderately common; 5 to 65 fathoms. Frequently brought in by shrimp fishermen. For breeding and embryology, see D'Asaro, 1969, *Malacologia,* vol. 9, no. 2, pp. 368–382.

1770   1771   1772

Distorsio constricta macgintyi Emerson and Puffer, 1953　　1771
McGinty's Distorsio

North Carolina to Florida and to Brazil. Bermuda.

1 to 2 inches in length, very close to *clathrata*, but the body whorl is very distorted, bulging and with cruder nodules. The upper and inner corner of the aperture usually has only 1 small, short, white tooth, while in *clathrata* there are usually 2 fairly large, obliquely set teeth. The lower parietal wall has a deep, smooth, wide groove separating the 2 axial rows of teeth. Commonly dredged from 25 to 125 fathoms. Formerly called *D. floridana* Olsson and McGinty, 1951 (not *floridana* Gardner, 1947, a Miocene fossil). Typical *constricta* Broderip is from the Eastern Pacific, ranging from the Gulf of California to Ecuador. The name *mcgintyi* must be modified to *macgintyi* according to the rules.

Distorsio constricta constricta (Broderip, 1833)　　1772
Constricted Distorsio

Gulf of California to Ecuador.

1½ inches, very similar to the Atlantic *macgintyi*, but the strong plica is absent in the parietal embayment. Uncommon; offshore to 40 fathoms.

Distorsio perdistorta Fulton, 1938　　1774
Very Distorted Distorsio

Off west Florida to off Barbados.
Japan and Madagascar.

2½ to 3 inches, similar to *clathrata*, but the shell is whitish and the cords and ribs orange-brown; the shield is oblong and

1774

has weak sculpturing, while in *clathrata* it is oval-round and is strongly sculptured. The outer lip is smoothish, while in *clathrata* it is strongly denticulated. It is uncommon from 155 to 274 meters in depth. *D. horrida* Kuroda and Habe, 1964, is a synonym.

1775

Other species:

1775 *Distorsio decussata* (Valenciennes, 1832). Southern part of the Gulf of California to Ecuador; offshore to 45 fathoms. Uncommon.

Subfamily CHARONIINAE Powell, 1933

Genus *Charonia* Gistel, 1848

This is the large and well-known Triton's trumpet. Type: *tritonis* Linné, 1758, of the Indo-Pacific. Synonyms are *Triton* Montfort, 1810; *Tritonium* Röding, 1798; *Buccinatorium* Mörch, 1877; and *Eutritonium* Cossmann, 1904.

Charonia variegata (Lamarck, 1816)　　1776
Trumpet Triton　　**Color Plate 7**

Southeast Florida to Brazil; Bermuda.

Adults 6 to 13 inches in length. The early whorls are purplish pink. In old specimens these are usually lost. The plicae on the dark columella are thin and strongly raised while in the closely resembling Pacific *tritonis* Linné, 1758, the plicae are very broad and scarcely raised. Adults usually have a swollen, angular shoulder on the last whorl, a feature which distinguishes our Atlantic subspecies from the typical *tritonis*. *C. atlantica* Bowdich, 1822, is a synonym of the Pacific subspecies, despite the name. Rare in Florida; moderately common in the West Indies below low water near reefs. Synonyms are *nobilis* Conrad, 1848; *commutatus* Kobelt, 1876; and *seguenzae* Aradas and Benoit, 1871. This species also occurs in the eastern Mediterranean, Cape Verde Islands, and St. Helena.

Family BURSIDAE Thiele, 1925

Genus *Bursa* Röding, 1798

Shells similar to the Cymatiidae, but there is a short or long posterior siphonal canal extending from the upper corner of the aperture. Type: *bufonia* (Gmelin, 1791).

Subgenus *Lampasopsis* Jousseaume, 1881

Type of this subgenus is *rhodostoma* (Sowerby, 1841). *Annaperenna* Iredale, 1936, is a synonym.

Bursa thomae (Orbigny, 1842)　　1777
St. Thomas Frog-shell　　**Color Plate 7**

Off South Carolina to Brazil.

½ to 1 inch in length. Characterized by the varices being placed axially one below the other and by the delicate lavénder aperture. The posterior siphonal canal is prominent and not attached to the body whorl. The outer lip bears 8 or 9 white, elongate teeth. Operculum thin, chitinous, concentric with the nucleus near the center. Moderately common under rocks on offshore reefs.

Subgenus *Colubrellina* P. Fischer, 1884

Members of this subgenus are not laterally compressed. Type: *condita* Gmelin, 1791 (*candisata* (Lamarck, 1816)) from the Philippines.

Bursa tenuisculpta (Dautzenberg and Fischer, 1906)　　1778
Fine-sculptured Frog-shell

Southeast Florida and the West Indies.

1778

2 to 3 inches in length; with 5 to 7 spiral rows of numerous, evenly sized beads. Old varices spaced ⅔ of a whorl apart so that the varices do not line up under each other. Color dull ash-gray. Dredged on occasions, 115 fathoms. A taller-spired subspecies was described from Natal Brazil, *B. natalensis* Coelho and Matthews, 1970 (*Bol. Mus. Nacional, n.s., Zoologia,* no. 279).

Bursa finlayi McGinty, 1962   **1779**
Finlay's Frog-shell

Southeast Florida and Cuba.

2 to 3 inches, thin-shelled but strong, with heavy nodules at the shoulder. Nucleus of 4 whorls, with fine axial riblets and 3 spiral threads. Final ⅓ whorl of nucleus smooth. Spire elongate and pointed. Spirally sculptured with small beaded lines forming a fine cancellate surface. Color straw, with brown suffusions. Aperture orchid within. Parietal wall has some brown between the many irregular spiral white lirae. Uncommon; 70 to 125 fathoms. *The Nautilus,* vol. 76, p. 39, pl. 3.

Bursa corrugata (Perry, 1811)   **1780**
Gaudy Frog-shell

Southeast Florida to Brazil. Bermuda.
Baja California to Ecuador.

2 to 3 inches in length; flattened laterally; with 2 prominent, knobbed varices on each whorl. Just in front of each varix is a sharp frill. There are generally 1 or 2 rows of blunt nodules on the whorls. Spire usually decollated. This is a rare species in the Atlantic, but is more frequently encountered on the west coast of Central America. Alias *caelata* Broderip, 1833; *ponderosa* Reeve; and *louisa* M. Smith, 1948; and *semigranosa* Lamarck, 1822. *B. calcipicta* Dall, 1908, from 66 fathoms, off Pacific Panama may also be a synonym. For breeding and embryology, see D'Asaro, 1969, *Malacologia,* vol. 9, no. 2, pp. 351–368.

Bursa granularis cubaniana (Orbigny, 1842)   **1781**
Granular Frog-shell   **Color Plate 7**

Southeast Florida and the West Indies to Brazil.

¾ to 2 inches in length, flattened laterally. Varices axially placed one below the other. Color orange-brown with 3 narrow, white bands which appear as prominent white squares on the varices. Spiral sculpture of several rows of small beads, those on the periphery of the whorl having the largest beads. Teeth in aperture white. Uncommon. I consider this merely a subspecies of the common Indo-Pacific *granularis* Röding, 1798 (*granifera* Lamarck, 1816; *elegans* Perry, 1811; *rubicola* Perry, 1811; and *livida* Reeve, 1844).

Subgenus *Marsupina* Dall, 1904

Type of this subgenus is *bufo* (Bruguière, 1792). Synonym: *Buffo* Montfort, 1810, non Lacépède, 1787.

Bursa bufo (Bruguière, 1792)   **1782**
Chestnut Frog-shell   **Color Plate 7**

Southeast Florida to Surinam and Brazil.

1 to 2 inches in length, flattened laterally; with strong, rounded varices, 2 on each whorl and lined up axially one under the other. Surface covered with spiral rows of numerous, small beads. Posterior siphon has one wall next to the body whorl. Color yellowish with diffused markings of orange-brown. Rare in Florida; abundant offshore in the lower Caribbean. Dredged in moderately deep water, 28 to 50 fathoms. Alias *B. crassa* Dillwyn, 1817, and *spadicea* (Montfort, 1810).

Subgenus *Crossata* Jousseaume, 1881

Type of this subgenus is *ventricosa* (Broderip, 1833).

Bursa californica (Hinds, 1843)   **1783**
Californian Frog-shell   **Color Plate 7**

Monterey, California, to the Gulf of California.

3 to 5 inches in length, moderately heavy, tan-cream in color and with about 6 whorls, each of which has 2 varices, one opposite the other. The last varix has 4 to 5 large nodules; in the spire only 1 nodule shows. Between the varices are 2 stout spines. White aperture with a posterior canal almost the size of the anterior (siphonal) canal. Lip crenulate. Common offshore, occasionally washed ashore. A scavenger.

An immature specimen from Guaymas, Mexico, has been given the subspecies name, *sonorana* S. S. Berry, 1960 (*Leaflets in Malacology,* vol. 1, no. 19).

Family TONNIDAE Peile, 1926

A full treatment of this interesting family was given by Ruth Turner, 1948, *Johnsonia,* vol. 2, no. 26.

Genus *Tonna* Brünnich, 1772

Shells large, rather thin but strong, with spiral ridges, and with 3 to 4 smooth, golden-brown embryonic whorls. Operculum lost in the adult stage. Type: *galea* Linné, 1758. *Dolium* Lamarck, 1801, and *Cadus* Röding, 1798, are synonyms. *Macgillivrayia* Forbes, 1852, is based on a pelagic larval specimen of a *Tonna.*

Tonna maculosa (Dillwyn, 1817)   **1784**
Atlantic Partridge Tun   **Color Plate 6**

Southeast Florida and the West Indies to Brazil. Bermuda.

2 to 5 inches in length, thin but strong. Nuclear whorls golden-brown and glassy-smooth. Periostracum thin and usually flakes off in dried specimens. *Dolium album* Conrad is only an albino form. *Tonna perdix* Linné is not this species, but an Indo-Pacific shell which has a more pointed spire, clearer squares of color and fewer spiral ribs. Our species is fairly common, especially in the West Indies. Adults do not have an operculum in this genus. Synonyms are: *sulphurea* C. B. Adams, 1849; *pennatum* Mörch, 1852; *album* Conrad, 1854.

Tonna galea (Linné, 1758)   **1785**
Giant Tun

North Carolina to Texas and the West Indies. Brazil.
Mediterranean and Indo-Pacific.

5 to 7 inches in length, thin but rather strong, although
the lip is easily broken. Ground color whitish to light coffee-
brown, sometimes slightly mottled. With 19 to 21 broad,
flattish ribs.

The subspecies, *brasiliana* Mörch, 1877 **(1786),** only from
Brazil, has a pushed-down, flattish spire. Uncommon; below
low water to 18 fathoms. Also known from the Mediterranean
and the Indian Ocean. *Dolium antillarum* Mörch, 1877, and
tenue Menke, 1830, are synonyms.

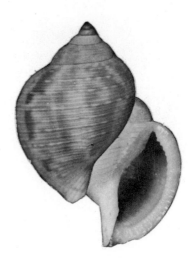

1788

Outer lip thickened, light purple, and its inner edge crenulated. Periostracum very thin, light-straw color. Uncommon;
50 to 80 fathoms.

Subfamily OOCORYTHINAE Fischer, 1885

Retaining the operculum in the adult; nuclear whorls
small. For details, see R. D. Turner, 1948, *Johnsonia,* vol.
2, no. 26.

Genus *Oocorys* Fischer, 1883

The type of this genus is *sulcata* Fischer, 1883.

1785

Genus *Eudolium* Dall, 1889

Shell without an umbilicus, with strong spiral ridges, reflected outer lip, weak parietal shield, with large, dark,
horny nuclear whorls. Operculum lacking. Type: *crosseanum* (Monterosato, 1869).

Eudolium crosseanum (Monterosato, 1869)   **1787**
Crosse's Tun   **Color Plate 6**

Off New Jersey to the Lesser Antilles.
Mediterranean to South Africa.

2 to 3½ inches in length, moderately thin-shelled, but
strong. Each of the 6 whorls bears numerous, spiral ridges and
fine threads. Nuclear whorls smooth, dark-brown. Outer lip
turned back, slightly thickened and with its inner edge crenulated. Color white to light-cream, with the ridges straw-yellow.
Periostracum thin and light yellowish brown. No operculum
in adults. Uncommon from 96 to 300 fathoms. Very rare in
private collections.

Eudolium thompsoni McGinty, 1955   **1788**
Thompson's Tun

Off the Florida Keys and in the Gulf of Mexico.

1½ to 2 inches, fragile, 6 whorls, the 3 nuclear ones being
yellowish brown. Color white, streaked and blotched with light
shades of brown. Sculpture of numerous alternating fine spiral
ridges of 3 sizes, which are crossed by very fine axial threads.

Oocorys bartschi Rehder, 1943   **1789**
Bartsch's False Tun

Off southeast Florida.

4½ inches, thin-shelled but strong, with about 40 even-
sized, squarish, narrow spiral cords on the last whorl. Color
peach or pale-flesh. Nuclear whorls white and smooth. Operculum chitinous, ovate. Uncommon; 79 to 140 fathoms.

1789

Oocorys barbouri Clench and Aguayo, 1939   **1790**
Barbour's False Tun

Off northern Cuba.

2 inches, 7 whorls, color white with a broad pink-brown
band on the upper ⅓ of the whorl. Spire elevated; whorls
shouldered and bearing numerous fine beads. Body whorl with
about 28 flattened spiral cords. Rare; 260 to 1,000 fathoms.

1790

Oocorys caribbaea Clench and Aguayo, 1939   1791
Caribbean False Tun

North coast of Cuba.

1½ inches, color ivory-yellow, ovate, suture not greatly indented, last whorl globose and with about 20 thin spiral cords which are finely beaded. Axial riblets are rather distinct on the shoulder. Outer lip smooth and white. Periostracum straw-yellow. Rare; from 615 to 1,800 fathoms.

1791

Oocorys verrillii (Dall, 1889)   1792
Verrill's False Tun

Off Grenada, Lesser Antilles.

1 inch, very heavy, globose, with about 20 very strong spiral cords which are weakly beaded. Suture appears to be channeled because 1 cord runs just below it. Parietal wall thick and with spiral ridges. Outer lip thick, flaring and with spiral lirae within. Very rare; 73 fathoms.

1792

Oocorys sulcata Fischer, 1883   1793
Sulcate False Tun

North Carolina to the Lesser Antilles. West Africa.

2 inches, thin-shelled, but strong. White with an ivory-colored periostracum. Outer lip smooth, reflected. Body whorl with 26 to 30 fine spiral ridges crossed by numerous, fine growth lines. Operculum chitinous, paucispiral. Uncommon; 88 to 2,512 fathoms.

1793

Subgenus *Benthodolium* Verrill and S. Smith, 1884

Type of this subgenus is *abyssorum* Verrill and S. Smith, 1884.

Oocorys abyssorum (Verrill and S. Smith, 1884)   1794
Deepsea False Tun

Off New Jersey to northwest Florida.

1¾ inches, thin, rather chalky, with a brownish yellow periostracum. Parietal wall forms a slight shield and forms an umbilicus. Last whorl with 30 to 35 very fine, raised threads. Axial growth lines give shell a fine reticulated appearance. Operculum chitinous, paucispiral. Rare; 169 to 2,221 fathoms.

1794

Genus *Dalium* Dall, 1889

Shell small and very solid, resembling *Sconsia*, but with a higher spire and distinct parietal callus and a subsutural concavity. Operculum chitinous, weakly spiral. Radula similar to that of *Oocorys*. Type: *solidum* Dall, 1889. Do not confuse the genus name with *Dolium* Lamarck, which is a synonym of *Tonna* Röding. For anatomy, see F. M. Bayer, 1971, "Studies Tropical American Mollusks," p. 145. Now placed in Oocorythinae.

Dalium solidum Dall, 1889   1795
Solid Dalium

Lesser Antilles to Colombia.

1795

1½ inches, solid, fusiform, without an umbilicus, 6 whorls. Sculpture of strong incised spiral grooves. Color dull-white to brownish. Parietal wall with a strong, raised callus. Columella without plicae. Operculum subspiral, chitinous. Rare. Deep water, usually over 1,000 meters.

Family FICIDAE Conrad, 1867

Genus *Ficus* Röding

Shells turnip-shaped, thin-shelled but strong, elongate, and without an operculum. *Pyrula* Lamarck, 1799, is a synonym. Type: *ficus* Linné, 1758, of the Indo-Pacific. The genus name is feminine.

Ficus communis Röding, 1798   **1796**
Common Fig Shell   **Color Plate 6**

North Carolina to the Gulf of Mexico.

3 to 4 inches in length, thin, rather fragile, and with spiral threads which are sometimes made reticulate by axial threads. Uncommon, except on the west coast of Florida where it is washed ashore in great numbers. No operculum present. Formerly known as *Pyrula* and *Ficus papyratia* Say, but the latter name is preceded by two earlier names, *communis* Röding, 1798, and *reticulata* Lamarck, 1816 (as well as 1822).

Carol's fig shell **(1797)** (named after Mrs. Richard W. Foster), 2½ inches, *Ficus carolae* Clench, 1945, is rare, but may be only a deep-water form of *communis*. It is irregularly spotted with reddish brown on the outside and the inside of the shell, which is more elongate than the shallow-water *communis*. It was first discovered by the late Leo L. Burry of Pompano off Key Largo, Florida, in 100 fathoms. Its range extends to the Campeche Banks, Mexico. *F. floridensis* Olsson and Harbison, 1953, from the Pliocene, may be the same.

Other species:

1798 *Ficus howelli* Clench and Farfante, 1940. 175 to 225 fms., off Bahia de Cochinos, Cuba, to Trinidad in 115 meters. See *Johnsonia*, vol. 1, no. 18, pl. 1. Very rare. See F. M. Bayer, 1971, p. 149.

1799 *Ficus atlantica* Clench and Aguayo, 1940. 450 fms., off Sao Salvador, Bahia, Brazil. See *Johnsonia*, vol. 1, no. 18, pl. 1. Very rare.

1800 *Ficus ventricosa* (Sowerby, 1825). Magdalena Bay and the Gulf of California to Peru. 3 to 4 inches. The only *Ficus* in the Eastern Pacific. Color Plate 6.

Superfamily Muricacea da Costa, 1776

Family COLUMBARIIDAE Tomlin, 1928

Deep water, spindle-shaped shells, called pagoda shells. For a review, see T. A. Darragh, 1969, *Proc. Royal Soc. Victoria*, vol. 83, pp. 63–119. His genera are usually reduced to subgenera. Do not confuse the name *Columbarium* with *Colubraria* (in Buccinidae), nor with *Columbella* (Dove Shells).

Genus *Columbarium* von Martens, 1881

Shells 1 to 2 inches, fusiform, with a very long siphonal canal. Whorls spined or carinate at the shoulder. 2 nuclear whorls bulbous and smooth. Operculum chitinous, leaf-shaped, with the nucleus at the narrow end. Radulae with a tricuspid central flanked on each side by a simple hooklike marginal tooth (see F. M. Bayer, 1971, figs. 39 and 43). Type of the genus: *Pleurotoma spinicincta* von Martens, 1881. Atlantic species were dealt with by W. J. Clench, 1944, *Johnsonia*, vol. 1, no. 15, vol. 3, no. 39, and by F. M. Bayer, 1971, "Studies Tropical American Mollusks," pp. 169–183.

1801 *Columbarium* (*Histricosceptrum* Darragh, 1969) *bartletti* Clench and Aguayo, 1940. Off Jamaica, in 528 to 562 meters. 40 mm.

1802 *Columbarium* (*Histricosceptrum*) *atlantis* Clench and Aguayo, 1938. Off Matanzas, northern Cuba, in 230 fms. 35 mm.

1803 *Columbarium* (*Fulgurofusus* Grabau, 1904) *bermudezi* Clench and Aguayo, 1938. Off the north coast of Cuba and between Florida and the Bahamas, 525 to 897 meters. See F. M. Bayer, 1971, pl. 38 figs. B–D.

1804 *Columbarium* (*Fulgurofusus*) *brayi* Clench, 1959. Southern Caribbean from Panama to Venezuela, in 300 to 641 meters. Common.

1805 *Columbarium* (*Peristarium* F. M. Bayer, 1971) *electra* F. M. Bayer, 1971. South-southeast of Key West, Florida, in 604 meters. Type of the subgenus.

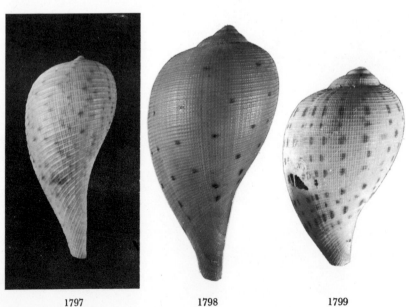

1797   1798   1799

1801   1802   1803

1806   *Columbarium (Peristarium) merope* F. M. Bayer, 1971. Southwest of the Marquesas Keys, in 512 to 584 meters.

1807   *Columbarium (Peristarium) aurora* F. M. Bayer, 1971. Off southeast Florida, in 247 to 403 meters.

Family MURICIDAE da Costa, 1776

A review of the higher classification of this family was made by Emily Vokes, 1963, *Malacologia*, vol. 2, no. 1, pp. 1–41.

Subfamily RAPANINAE Gray, 1853

Genus *Forreria* Jousseaume, 1880

Shell large, solid, broadly fusiform and with numerous long axial blades which form open spines on the shoulder of the whorls. Type: *belcheri* (Hinds, 1843).

Forreria belcheri (Hinds, 1843)   **1808**
Giant Forreria   **Color Plate 9**

    Morro Bay, California, to Baja California.
    3 to 6 inches in length, solid, smoothish, cream-brown; surface with 10 prominent, pointed, scalelike spines on the shoulder of each whorl. These are the tops of the varices which flatten out and are welded closely to the lower part of the whorl. Former siphonal canals prominent to the left of a narrow, not deep umbilicus. Interior enamel-white. Common in intertidal areas near oyster bars. Also down to 15 fathoms.

Subfamily MURICINAE da Costa, 1776

Genus *Murex* Linné, 1758

With a rather long, usually straight, slim siphonal canal, and with 3 rounded, usually spinose varices per whorl. There are about a dozen living American species in the typical subgenus whose type is the Indo-Pacific *tribulus* Linné, 1758. Fossil and some recent species are reviewed by Emily Vokes in the "Tulane Studies in Geology," vol. 1–8

(1963–1969). A catalog of species is found in the Bulletin 61, no. 268, of American Paleontology, Ithaca, N.Y., 1971.

Subgenus *Murex* Linné, 1758

Murex cabritii Bernardi, 1859   **1809**
Cabrit's Murex   **Color Plate 8**

    Off South Carolina to the Caribbean; Gulf of Mexico.

    1 to 2½ inches in length. The long, slender siphonal canal bears 3 rows of 3 to 7 long slender spines. Each rounded varix bears 3 or 4 sharp, slightly curved spines, between which are 1 to 3 small spiral cords. Between the varices are 3 axial rows of 6 to 8 even-sized, elongate, rounded whitish beads. Color yellow, white or blushed with pink-rose, rarely with a darker spiral color band. Fairly common in 18 to 76 fathoms. *M. maculatus* Verrill, 1950, not Reeve, 1845, is a synonym.
    A form having only 2 or 3 spines on the siphonal canal and having minute spiral brown lines was named *M. donmoorei* Bullis, 1964 (**1809a**). It is 1 to 2 inches long, brownish gray, and found 11 to 40 fathoms off British Guiana. A curious form off west Florida resembles young *cabritii* but is less than 1½ inches in length, lacks the brown spiral lines and has 3 long straight spines on each varix and only 1 or 2 on the canal. Erroneously called *tryoni* by many dredging fans.
    The Pacific counterpart of *cabritii* is *Murex elenensis* Dall, 1909 (**1810**) (*plicatus* Sowerby, 1834, not Gmelin, 1791) which ranges from Baja California to Ecuador, reaches 3 inches and is creamy-yellow in color. Uncommon. Color Plate 8.

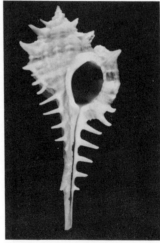

1810

Murex tryoni Hidalgo in Tryon, 1880   **1811**
Tryon's Murex

    Gulf of Mexico and the Caribbean.

    1 inch, rather fragile, all-white, with its siphonal canal less than ½ the entire length of the shell and bearing 1 or 2 spines. Varices with 3 short straight spines. Between the varices are very minute beads arranged in 6 or 7 axial rows. A rare species found from 70 to 200 fathoms.

Murex recurvirostris Broderip, 1833   **1812**
Bent-beak Murex

    Eastern Pacific and Western Atlantic.

    This is a widely distributed, common species that is broken up into several subspecies and has numerous sculptural and color forms. No two experts seem to agree on how to treat this interesting complex, and our interpretation may not be acceptable to some workers.

subspecies *rubidus* F. C. Baker, 1897   **1813**
Rose Murex

Off North Carolina to Florida and the Bahamas.

1 to 2 inches, with a short siphonal canal, spineless varices, and with 2, rarely 3, axial beaded ridges between the varices. The posterior ridge is the largest. Siphonal canal bears 3 rows of 1 or 2 spines only. Color usually orange, pink or rarely brownish with lighter spiral bands. Common on coral rubble bottom from 2 to 39 fathoms. Rarely washed ashore.

The Yucatan form (or possibly also a subspecies) *sallasi* Rehder and Abbott, 1951 **(1814)**, has 4 to 5 axial rows of fine beads between the varices, has a short canal and is brightly colored with whites, rose and yellows. Rarely with one spine on the varix, and rarely with spiral brown lines (approaching *cabritii* form *donmoorei*). Moderately common offshore.

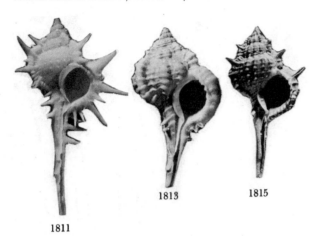

1813   1815

1811

subspecies *woodringi* Clench and Farfante, 1945   **1815**
Woodring's Murex

Caribbean. Brazil.

1 to 3 inches, dirty-brown to gray with long siphonal canal bearing 1, rarely 2, hooked spines. Usually with 3 axial rows of beads between the weakly spined varices. Common in muddy, shallow water areas. *M. thompsoni* Bullis, 1964, from French Guiana, may be this species in which the siphonal canal has been broken off and then repaired. The name *messorius* Sowerby, 1844, applies to a Pacific coast shell (see E. Vokes, 1965, p. 197).

1814

subspecies *recurvirostris* Broderip, 1833   **1816**
Bent-beak Murex   **Color Plate 8**

Baja California to Ecuador.

2 to 3 inches, brown to gray in color, siphonal canal fairly long and with 3 rows of 1 or 2 short spines. 2 to 4 (usually 3) axial ridges bearing nodules are between the varices which

bear 3 short, somewhat triangular spines. Spiral red-brown lines rarely present on the whorls. Moderately common offshore to 30 fathoms. Synonyms are *nigrescens* Sowerby, 1841, and *lividus* Carpenter, 1857. *M. tricoronis* S. S. Berry, 1960 (*funiculatus* Reeve, 1845, not Defrance, 1827) is an extra spiny form.

Subgenus *Siratus* Jousseaume, 1880

Similar to *Murex* but with a broader, shorter siphonal canal (including 1 or 2 former ones). Type: *senegalensis* Gmelin, 1791.

Murex beauii Fischer and Bernardi, 1857   **1817**
Beau's Murex   **Color Plate 8**

South Florida, the Gulf of Mexico, to Brazil.

3 to 5 inches in length. The spiny varices may have prominent, thin, wavy webs. Between the varices there are 5 or 6 rows of low, evenly sized and evenly spaced knobs. Color cream to pale brownish, rarely with spiral rows of yellow-brown flecks. Siphonal canal may be straight or bent. Common offshore from 100 to 200 fathoms. *M. branchi* Clench, 1953, is a synonym.

Murex formosus Sowerby, 1841   **1818**
Antillean Murex   **Color Plate 8**

Off Tortugas, Florida, to Barbados. Brazil.

2 to 4 inches, rather solid, spinose, and generally cream to rusty or purplish brown in color. Aperture sometimes tinted with purple. 3 varices per whorl have 3 to 6 spines, the top one usually being the longest. Between the varices are 3 to 4 axial rows of small (or in some forms large) beads or knobs. Moderately common from 27 to 190 fathoms. A very variable species. Formerly known as *antillarum* Hinds, 1834, a synonym. It is not unlikely that such species as *aguayoi*, *beauii*, *articulatus* (*finlayi*), *ciboney* and *springeri* will eventually be shown to be minor ecologic and/or genetic differences of *antillarum*.

Other species:

1819 *Murex aguayoi* Clench and Farfante, 1945. 185 to 235 fms., off Bahamas and Cuba. *Johnsonia*, vol. 1, no. 17, pl. 8. Rare.

1819

1820 *Murex ciboney* Clench and Farfante, 1945. 60 to 248 fms. Caribbean. *Murex springeri* Bullis, 1964, from off Surinam and Venezula may be this species. (*Tulane Studies in Zoology*, vol. 11, no. 4.) In fact both the above may be deep-water minor gene pools of *formosus* Sowerby, 1841. Color Plate 8.

1821 *Murex motacilla* Gmelin, 1791. Lesser Antilles. *Johnsonia*, vol. 1, no. 17, pl. 11. Uncommon. Color Plate 8.

1820

1822 *Murex consuelae* E. Vokes, 1963. Virgin Islands to Barbados. *Johnsonia*, vol. 1, no. 17, pl. 12. Synonym: *pulcher* A. Adams, 1853, not Sowerby, 1813. Uncommon. Verrill's mimeographed name *consuelae* of 1950 is probably not valid.

1823 *Murex senegalensis* Gmelin, 1791. Brazil. Common in shallow water. *Johnsonia*, vol. 1, no. 17, pl. 13.

1824 *Murex cailleti* Petit, 1856. Greater Antilles to Barbados. 54 to 200 fathoms. *M. kugleri* Clench and Farfante, 1945; *similis* Sowerby, 1841, not Schröter, 1805; *perelegans* E. Vokes, 1965; *elegans* Sowerby, 1841, not Donovan, 1804, are synonyms. Uncommon.

1825 *Murex chrysostoma* Sowerby, 1834. Lesser Antilles to Venezuela. Brazil. See *Johnsonia*, vol. 1, no. 17, pl. 5. Uncommon.

1826 *Murex articulatus* Reeve, 1845. Cuba and Jamaica. Synonyms are: *nodatus* Reeve, 1845, not Gmelin, 1791; *gundlachi* Dunker, 1883; *finlayi* Clench, 1955 (see E. Vokes, 1965, p. 196). Color Plate 8.

1827 *Murex reevei* E. Vokes, 1965. Cuba and the Gulf of Mexico. Synonym: *trilineatus* Reeve, 1845, not Sowerby, 1813.

Subgenus *Phyllonotus* Swainson, 1833

Large heavy shells with 3 to 6 varices and with a broad, short, upturned siphonal canal. Type: *margaritensis* Abbott, 1958.

Murex pomum Gmelin, 1791 1828
Apple Murex Color Plate 8

North Carolina to Florida and to Brazil. Bermuda.

2 to 4½ inches in length. Sturdy with a rough surface. No long spines. Colored dark-brown to yellowish tan. Aperture glossy, ivory, buff, yellow or orangish with a dark-brown spot on the upper end of the parietal wall. Outer lip crenulate and with 3 or 4 daubs of dark-brown. A very common shallow-water species. Bores holes into and feeds upon oysters, mainly *Crassostrea* (Radwin and Wells, 1968). A deep-water form is more delicately sculptured and has a red tinge. Synonyms are: *asperrimus* Lamarck, 1822; *mexicanus* Petit, 1852; *pomiformis* Mörch, 1852; and *oculatus* Reeve, 1845.

A southern Caribbean species, *M. margaritensis* Abbott, 1958 **(1829)**, (*imperialis* Swainson, 1831, not G. Fischer, 1807) has 4 to 6 varices per whorl and generally has a pink or yellow parietal wall. It hybridizes with *pomum* on Isla Margarita, Venezuela.

Other species:

1830 *Murex peratus* (Keen, 1960). Panama to Mexico. 15 to 40 fms. This is possibly *pomum* Gmelin, introduced some time ago

to the Pacific side through the Panama Canal, possibly in the form of an eggmass. Keen also described a subspecies, *decoris* in 1960 **(1831)**. See *The Nautilus*, vol. 73, p. 103.

Subgenus *Hexaplex* Perry, 1810

Shells large, with 5 or more spinose to foliaceous varices, and with a short, recurved siphonal canal. The type is *cichoreus* Gmelin, 1791, an Indo-Pacific species. Synonyms are: *Polyplex* Perry, 1810; *Centronotus* Swainson, 1833; *Muricanthus* Swainson, 1840; *Bassia* Jousseaume, 1880; *Truncularia* Monterosato, 1917; *Trunculariopsis* Cossmann, 1921; *Murithais* Grant and Gale, 1931; *Bassiella* Wenz, 1941; and *Aaronia* "Verrill" E. Vokes, 1968.

Murex fulvescens Sowerby, 1834 1832
Giant Eastern Murex Color Plate 8

North Carolina to Florida and to Texas.

5 to 7 inches in length. Characterized by the large shell, and the strong, straight, rather short spines. Exterior milky-white to dirty-gray. Aperture enamel-white. Thin spiral color lines are usually prominent on the whorls. Fairly common along the shallow areas of northeastern Florida and Texas where they are found abundantly during the breeding season. Well-known to shrimp fishermen, whose nets often ensnare them from 6 to 43 fathoms. Feeds on oysters (Radwin and Wells, 1960). *Murex burryi* Clench and Farfante, 1945, is the young of this species. *M. spinicostatus* Reeve, 1845, and *spinicostus* Kiener, 1843 are synonyms.
Murex (Hexaplex) strausi "Verrill" E. Vokes, 1968 (Tulane Studies in Geology, vol. 6, p. 105), reportedly from Dominica West Indies, probably originally was sent to Verrill from the Eastern Pacific and may be a synonym of some well-known species.

Murex erythrostomus (Swainson, 1831) 1833
Pink-mouthed Murex Color Plate 8

Gulf of California to Peru.

3 to 4 inches, having varices with blunt spines, and characterized by a beautiful pink aperture and parietal shield. An abundant, clam-eating species occurring from the low-tide mark to several fathoms. Synonyms are: *bicolor* Valenciennes, 1832; and *hippocastanum* Philippi, 1845, not Linné, 1758.

Murex brassica Lamarck, 1822 1834
Cabbage Murex Color Plate 8

Guaymas, Mexico, to Peru.

6 to 8 inches, whitish with 3 indistinct spiral bands of diffused brown. Old varices are marked by a narrow, frilly, pink ridge with tiny white teeth. Aperture yellow or white with weak pink border. No black marks present. Parietal shield very narrow. *Murex ducalis* Broderip, 1829, is a synonym. Moderately common in some areas, from tidal flats to a depth of about 150 feet.

Murex regius (Swainson, 1821) 1835
Regal Murex Color Plate 8

Gulf of California to Peru.

4 to 5 inches, similar to *erythrostomus*, but having the upper part of the parietal wall a glossy-brown to black, a color that also appears in the spire where the old posterior anal notches survive. Moderately common; tidal flats to 2 fathoms.

***Murex nigritus* Philippi, 1845**   **1835a**
Northern Radix Murex

Gulf of California.

4 to 6 inches, moderately elongate, with numerous, stout
black spines. Aperture white. Young are white. Common on
reefs where it feeds on other mollusks. The true Radish
Murex, *M. radix* Gmelin, 1791 (**1836**; Color Plate 8), is a
rounder, heavier shell with more spines and occurs from Costa
Rica to Ecuador. These two species are placed in the genus
Muricanthus Swainson, 1840, by many workers.

1835 a

Subgenus *Chicoreus* Montfort, 1810

The type of this subgenus is *ramosus* Linné, 1785. Synonyms are: *Triplex* Perry, 1810; *Cichoreum* Voigt, 1834;
Frondosaria Schlüter, 1838; *Euphyllon* Jousseaume, 1880;
Pirtus de Gregoria, 1885; *Torvamurex* Iredale, 1936; *Foveo-
murex* Wenz, 1941.

***Murex brevifrons* Lamarck, 1822**   **1837**
West Indian Murex   **Color Plate 8**

West Indies to Brazil.

3 to 6 inches in length. Numerous, stout, fairly long spines
on the varices which arch backwards and bear sharp fronds.
Raised, spiral lines prominent between the varices. Color variable from cream to dark-brown. Fairly common to abundant in
the West Indies in semibrackish, shallow waters where they
feed on barnacles and oysters. Synonyms are: *calcitrapa* Lamarck, 1822; ?*crassivaricosa* Reeve, 1845; ?*pudoricolor* Reeve,
1845; *purpuratus* Reeve, 1846; *toupiollei* Bernardi, 1860; *ap-
proximatus* Sowerby, 1879.

***Murex florifer dilectus* A. Adams, 1855**   **1838**
Lace Murex   **Color Plate 8**

South half of Florida to off South Carolina.

1 to 3 inches in length. Aperture small, nearly round. 8 to
10 crowded, frondose, scaly spines bordering the outer lip
and siphonal canal. Color light-brown, or whitish, and in
the latter case, the nuclear whorls at the spire are pinkish.
Usually 1 axial low ridge between each varix, although occasionally with more and smaller axial ribs.

The lace murex is one of Florida's most common species
in this genus. It lives in a wide variety of habitats from

mangrove, muddy areas to protected rocks and frequently
in clear, sandy areas. Feeds on various bivalves by boring a
hole in the shell. This species differs from the 4 to 5 inch-long
M. brevifrons in being smaller, in having closely crowded
scaly spines, and in having a round instead of elongate operculum. *M. arenarius* Clench and Farfante, 1945 (non Steuer,
1912), is a synonym. For many years this species was called
rufus Lamarck, 1822 (not *rufus* Montagu, 1803).

True *Murex florifer* Reeve, 1846 (**1839**) from the Bahamas
and West Indies is darker, larger, with the shoulder spine
much longer than the others, and with 1 large intervarical
node. Common; shallow water. *M. dilectus* is only a subspecies of this.

Other species:

1840 *Murex spectrum* Reeve, 1846. 1 to 6 fms. Jamaica; Lesser
Antilles to Brazil. Synonyms: *imbricatus* Higgins and Marrat,
1877 (non Brocchi, 1814); *argo* Clench and Farfante, 1945. Uncommon.

1841 *Murex (Siratus) tenuivaricosus* Dautzenberg, 1927. 15 to
30 meters, southern Brazil. Synonyms: *calcar* Kiener, 1843, not
Sowerby, 1823; *carioca* E. Vokes, 1968. Uncommon.

Subgenus *Murexiella* Clench and Farfante, 1945

Aperture almost circular; no anal notch; with 4 to 10
foliaceous varices with webs. The type of this feeble subgenus is *hidalgoi* Crosse, 1869. *Minnimurex* Woolacott,
1957, is a synonym. E. Vokes raised it to generic level and
gave a review of the West American species (1970, *The
Veliger*, vol. 12, p. 325).

***Murex hidalgoi* Crosse, 1869**   **1842**
Hidalgo's Murex

North Carolina to West Florida and to the Lesser Antilles.

1 to 1½ inches in length. Spines frondose and long, with
webbing in between which is exquisitely sculptured with
scalelike lamellations. Color grayish white to cream. Aperture small, subovate and white. 4 or 5 varices per whorl.
Uncommon; from 76 to 196 fathoms. The Pacific Gulf of
Panama counterpart is *diomedaeus* Dall, 1908 (**1843**).

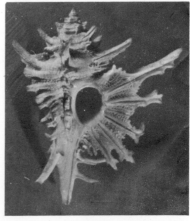

1842

***Murex glyptus* M. Smith, 1938**   **1844**
Carved Murex

West Florida to Yucatan, Mexico.

½ to 1 inch, elongate, spinose, whitish gray, with 5 to 8
axial rows of crowded, open spines or foliations per whorl.
The spines occur on squarish spiral cords. Nuclear whorls

1½ and bulbous. Uncommon from 4 to 17 fathoms. Also found fossil. Do not confuse with *leviculus* Dall.

Murex leviculus (Dall, 1889) 1845
Leviculus Murex

North Carolina to West Florida.

½ inch, somewhat elongate, with strong frondose varices bearing open fronds occurring in pairs. Brown to gray in color. Smoothish between the varices. 4 nuclear whorls, conical, not 1½ and bulbous as in the similar *glyptus* M. Smith. Uncommon; from 10 to 27 fathoms. Formerly placed in *Ocenebra* and *Favartia* by various authors. It is not a subspecies of *M. cellulosus* Conrad (see species 1953).

Murex macgintyi M. Smith, 1938 1846
Tom McGinty's Murex

Florida to the lower Caribbean. Brazil.

¾ inch, ovate to oval-elongate, with 6 or 7 spinose varices. 6 spines on each varix. Spiral rounded cords are present between the varices. Color of shell whitish to brown with darker spiral bands. The forma *facetus* E. Vokes, 1970, is quite ovate. It occurs fossil and living in Aruba, Dutch West Indies. Uncommon from 3 to 100 fathoms. Originally described from the Pliocene of Clewiston, Florida (*The Nautilus*, vol. 51, p. 88).

The Eastern Pacific counterpart of this species is the similar *humilis* Broderip, 1833 (**1847**) ranging from Mexico to Ecuador.

1846

Murex santarosanus (Dall, 1905) 1848
Santa Rosa Murex

Point Estero, California, to Baja California.

1½ inches in length, spire low, and with 6 curled-back, spined varices per whorl. Anterior surface of varices strongly fimbriated. Narrow intervarical space smooth. Color brownish white. Uncommon to rare on gravel bottom just offshore to 30 fathoms. Do not confuse with the common *Maxwellia gemma*.

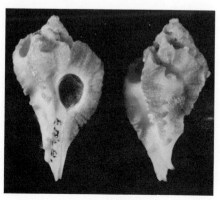

1848

Other *Murexiella*:

1849 *Murex humilis* Broderip, 1833. Gulf of California to Ecuador. (Synonym: *taeniatus* Sowerby, 1860.)

1850 *Murex laurae* (E. H. Vokes, 1970). Manzanillo, Colima, Mexico, 17 fms. *The Veliger*, vol. 12, p. 328.

1851 *Murex diomedaeus* (Dall, 1908). Gulf of Panama, 85 fms.

1852 *Murex keenae* (E. H. Vokes, 1970). Venado, Panama, Canal Zone. *The Veliger*, vol. 12, p. 328.

1853 *Murex lappa* (Broderip, 1833). Baja California to Ecuador, 20 fms. Uncommon. (Synonym: *minuscula* M. Smith, 1947.)

1854 *Murex peritus* Hinds, 1844. Baja California to Mexico, 7 to 52 fms.

1854a *Murex vittatus* Broderip, 1833. Gulf of California to Ecuador, intertidal to 11 fms.

Genus *Homalocantha* Mörch, 1852

Spire short; suture indented; varices up to 10 usually frilled with spines that expand at their ends. Operculum with a lateral nucleus. Type: *Murex scorpio* Linné, 1758.

Homalocantha oxycantha (Broderip, 1833) 1855
Oxycanth Murex   **Color Plate 8**

West Mexico to southern Ecuador

2 to 3 inches, with about 7 to 8 varices crossed by scaly, squarish spiral threads. Color whitish with brown and white ribs. Outer lip with 12 to 14 long, simple spines of various lengths. Moderately common in shallow water. Formerly misidentified as *M. melanamathos* Gmelin, which is from West Africa. *M. stearnsii* Dall, 1918, is a synonym.

Genus *Pterynotus* Swainson, 1833

Shells 3 inches or less in size, with 3 winglike or bladelike varices per whorl. Type: (*pinnatus* Swainson, 1822) *alatus* Röding, 1798 from the Indo-Pacific. *Pteronotus* Swainson, 1833, not Rafinesque, 1815, is the same, as is *Triplex* Newton, 1891; *Pterymurex* Rovereto, 1899; *Marchia* Jousseaume, 1880; *Timbellus* Gregorio, 1885. For a monograph of the Western Atlantic species, see E. Vokes, 1970, *Tulane Studies in Geol. and Paleo.*, vol. 8.

Pterynotus phaneus (Dall, 1889) 1856
Bright Wing Murex

Off South Carolina to Cuba.

½ inch, cream-white, thin and delicate, with about 6 whorls and an acute spire. Aperture very small. Previous siphonal canal exists as a long spine. 3 thin varices per whorl have delicately denticulated edges caused by spiral ridges extending upward on the back side of each varix. Between the varices are 4 low, inconspicuous ridges. Rare; 152 to 400

1856

fathoms. *Murex pygmaeus* Bush, 1893; *bushae* E. Vokes, 1970; *M. tristichus* Dall, 1889, not Beyrich, 1854; and *havanensis* E. Vokes, 1970, are doubtlessly synonyms.

Other species:

1857 *Pterynotus (Pterochelus) ariomus* (Clench and Farfante, 1945). 50 to 60 fms., off Hollywood, Florida. Rare. See *Johnsonia*, vol. 1, no. 17, pl. 20.

1858 *Pterynotus (Pterochelus) phillipsi* E. H. Vokes, 1966. Off Santa Barbara, California, near Anacapa Island, 100 fms. *The Veliger*, vol. 8, p. 165.

1860 *Pterynotus (Purpurellus* Jousseaume, 1880) *pinniger* (Broderip, 1833). Gulf of California to Ecuador. (Synonym: *inezana* Durham, 1950.) See Emerson and D'Attilio, 1969, *The Veliger*, vol. 12, p. 145.

1861 *Pterynotus (Purpurellus) macleani* Emerson and D'Attilio, 1968. Baja California to Panama, 12 to 25 fms. *The Veliger*, vol. 12, p. 147.

1862 *Pterynotus (Pterynotus) phyllopterus* (Lamarck, 1822). Lesser Antilles. (Synonym: *Murex rubridentatus* Reeve, 1846.) See *The Veliger*, vol. 14, no. 4, p. 350. 3 to 4 inches. Rare, 60 to 70 feet, on sand. Exterior brown; aperture cream with tiny purple teeth. Color Plate 8.

Genus *Pteropurpura* Jousseaume, 1880

The type of this genus is *macroptera* Deshayes, 1839. The chitinous operculum is similar to that of *Purpura* in that it is fan-shaped with the nucleus on the side facing the columella. *Centrifuga* Grant and Gale, 1906, is a synonym.

Pteropurpura macroptera (Deshayes, 1839)   **1863**
Frill-wing Murex

California.

The northern California form has a slender appearance because only the last varix is broadly developed. The spire is slightly longer, the anterior face of the last varix is beautifully frilled and its edge is only moderately scalloped. Color often coffee-brown. Moderately common; dredged offshore. Synonyms are: *carpenteri* Dall, 1899, and *petri* Dall, 1900. For nomenclatorial details see Emerson, 1964, *The Veliger*, vol. 6, p. 151.

1863

The southern California subspecies (**1863a**) (or form) is broader, with less frilled varices, but the last varix is very large, quite deeply scalloped and has a pair of flutes at the lower end. The earliest name available for this subspecies is *M. tremperi* Dall, 1910. *M. tremperi* Hemphill, 1911, is the same. I consider the albino form *alba* to be a variant commonly occurring in *trialata* found off Newport, California. Moderately common; dredged offshore.

Pteropurpura trialata (Sowerby, 1834)   **1864**
Western Three-winged Murex   **Color Plate 9**

Northern California to Baja California.

I believe this species to be only a form of *macroptera* Deshayes, 1839, but give here the opinion of some Californian workers in considering it separate.

2 to 3 inches in length, with 3 large wavy, winglike varices per whorl. Siphonal canal closed along its length. The body whorl between each varix is smoothish, with or without 1 low, rounded turbercle, and sometimes with 2 to 5 weak, spiral cords or threads. Anterior face of each varix with fine, crowded, axial fimbriations. Color grayish, dark- or light-brown, or with white spiral bands. Rarely albino, a form called *alba* Berry, 1908.

Typical *trialata* from southern California and Baja California reaches a length of 3 inches, is generally dark-chestnut to blackish brown with 4 to 6 narrow white bands and has very fine spiral threads which are sometimes scaled, beaded or smooth.

Pteropurpura centrifuga (Hinds, 1844)   **1865**
Centrifugal Murex   **Color Plate 9**

Baja California, the Gulf, to Panama.

2 inches, similar to *macroptera*, but with a single prominent node between the varices. Varices with 3 webbed spines, the one at the shoulder being very long. Color usually whitish. *Pterynotus swansoni* Hertlein and Strong, 1951, is a synonym (see Emerson, 1960, *Amer. Mus. Novitates*, no. 2009). Moderately common; 40 to 153 fathoms.

Pteropurpura erinaceoides (Valenciennes, 1832)   **1866**
Prickly Wing Murex

Baja California, including the Gulf of California.

½ to 2½ inches, solid, with a shouldered spire bearing about 7 pairs of blunt nodules connected by 2 spiral cords. Body whorl with 3 blunt, thick varices which have 5 short, backwardly turning, open spines. Intervarical area with 1 strong axial swelling crossed by 4 or 5 major cords. Color gray to white. Siphonal canal rather short and slightly twisted. Moderately common in the intertidal zones, especially in warm protected bays. *Murex californicus* Hinds, 1844, appears to be a synonym.

Pteropurpura vokesae Emerson, 1964   **1867**
Wrinkled Wing Murex   **Color Plate 9**

San Pedro to San Diego, California.

1½ inches, similar to *erinaceoides*, but more slender, with a longer spire, with thinner, flatter, brownish, fimbriated varices. Intervarical sculpture of numerous rasplike, fine spiral threads and 1 long, swollen axial ridge. Uncommon: 25 to 40 fathoms. This is *Murex rhyssus* Dall, 1919, not Tate, 1888. (Emerson, 1964, *The Veliger*, vol. 7, p. 5: "This species has been dredged in depths of 10 to 50 fathoms. Although beach

1865   1866   1867   1868

specimens are rarely found, specimens sometimes are found in kelp holdfasts that have been washed ashore. This may be a form of *erinaceoides*.") It may also be a young, more sculptured form of *trialata*.

Pteropurpura bequaerti (Clench and Farfante, 1945) 1868
Bequaert's Murex

North Carolina south to both sides of Florida.

1 to 2½ inches in length. Spire high. No spines. Each varix is a high, rounded, thin plate or web. Between these varical webs there is a single, low rounded knob. Color a uniform cream-white. A bizarre species which is the least spinose of our American forms. It is being collected in dredging operations along the west coast of Florida in increasing numbers, although it remains a rarity. It was named after one of our foremost malacologists now at the University of Arizona, Dr. Joseph C. Bequaert.

Subgenus *Shaskyus* Burch and Campbell, 1963

The type of this subgenus is *festivus* (Hinds, 1844).

Pteropurpura festiva (Hinds, 1844) 1869
Festive Murex   Color Plate 9

Morro Bay, California, to Baja California.

1½ to 2 inches in length, spire high, 3 varices per whorl. Color brownish cream with numerous fine, dark spiral lines. Varix, with its thin, fimbriated surface, curled backwards. 1 very large, rounded nodule between varices. Very common on rocks, on pilings or mud flats and down to 75 fathoms.

Subgenus *Calcitrapessa* S. S. Berry, 1959

The type of this subgenus is *Murex leeanus* Dall, 1890.

Pteropurpura leeana (Dall, 1890) 1870
Lee's Murex   Color Plate 9

Baja California and the Gulf of California.

2 to 3 inches, pale-brown, with 3 very long, round, curved spines on the last whorl. Between the varices is a very weak node. Uncommon; 40 to 60 fathoms.

Genus *Maxwellia* Baily, 1950

Shell solid, with an elongate, slightly opened siphonal canal and with 6 very broad varices per whorl. Operculum with a marginal nucleus. Type: *gemma* Sowerby, 1879. *The Nautilus,* vol. 64, p. 11.

Maxwellia gemma (Sowerby, 1879) 1871
Gem Murex   Color Plate 9

Santa Barbara, California, to Baja California.

1 to 1¼ inches in length, moderately high-spired, with 6 varices per whorl. The varices are swollen, roundish and smooth and connect with each other in the middle area of the whorl, but in the area of the suture, and again near the base of the shell, the varix is thin, elevated and curled back and may bear one or several small spines. There are several spiral low cords colored blackish blue which are more obvious on the middle or smoother part of the whorl. The spire appears to have squarish pits crudely dug out. Very common along rock areas under protective rubble and masses of worm tubes from low tide to 30 fathoms. Do not confuse with *Murex (Murexiella) santarosanus*.

Subfamily PURPURINAE Menke, 1828

The genera *Purpura* and *Thais* of the Western Atlantic were reviewed by W. J. Clench in 1947 in *Johnsonia*, vol. 2, no. 23, pp. 61–91. We consider *Nucella* as a good genus.

Genus *Morula* Schumacher, 1817

The type of this genus is *uva* (Röding, 1798) from the Indo-Pacific.

Subgenus *Trachypollia* Woodring, 1928

The type of this subgenus is *sclera* Woodring, 1928, a Miocene Jamaican species. *Morunella* Emerson and Hertlein, 1964, is a synonym.

Morula nodulosa (C. B. Adams, 1845) 1872
Blackberry Drupe

Off South Carolina to Brazil. Bermuda and Texas.

½ to 1 inch in length, elongate, grossly studded with round, black beads. Aperture purplish black. Outer lip thick, and with 4 to 5 relatively large, white beads. A common shallow-water species found under rocks. Formerly placed in the genus *Drupa* Röding, 1798.

Morula didyma (Schwengel, 1943) 1873
Twin Morula

South Florida to Texas and the West Indies. Brazil.

⅓ inch, ovate-fusiform, brown to peach-red, with small, dark-brown beads. Last whorl with 8 axial rows of 11 or 12 small, round, brown beads. In some specimens, the beads are spirally elongated. Spiral sculpture of microscopically fimbriated spiral threads. Columella raised and free on the left side; it has 2 or 3 pimples. Outer lip sometimes thickened by 4 or 5 internal spiral lirae. Not uncommon; from intertidal to 20 to 70 fathoms. Do not confuse with *Risomurex rosea* Reeve which has 3 spiral brown bands on the last whorl (see species 1926).

1872   1873   1874   1875

Morula lugubris (C. B. Adams, 1852) 1874

San Diego, California, to Panama.

⅓ to ½ inch, fusiform, color light-brown with dark-brown beads. Last whorl with 10 or 11 axial rows of round beads. The second and third beads below the suture are usually close together. Below them are 2 or 3 more beads. In fresh specimens there are numerous, fine spiral, microscopically fimbriated threads. Columella glossy-tan with 3 or 4 tiny pimples at the lower ⅓. Inside of outer lip with 3 or 4 spiral lirae. Common; intertidal rocks.

Subgenus *Evokesia* Radwin and D'Attilio, 1972

The type of this subgenus is *rufonotatum* Carpenter, 1864.

Morula ferruginosa (Reeve, 1846)   **1875**

Baja California to Sonora, Mexico.

¾ inch, fusiform, high spire; last whorl with 3 or 4 spiral rows of short black spines between which are microscopic spiral incised lines. Color black with white spots between the low spines. Columella whitish. Aperture slate-gray. Outer lip thin and scalloped and brown-bordered. Common; under intertidal rocks.

Genus *Acanthina* G. Fischer, 1807

Small, sturdy ovate rock shells with a small spine on the base of the outer lip. Type: *monoceros* Gmelin, 1791, or *calcar* Martyn from southern Argentina. *Monoceros* Sowerby, 1827 (not Lacépède, 1798), is a synonym.

Subgenus *Acanthinucella* Cooke, 1918

Type of this subgenus is *punctulata* (Sowerby, 1825).

Acanthina spirata (Blainville, 1832)   **1876**
Spotted Thorn Drupe

Puget Sound, Washington, to San Diego, California.

1 to 1½ inches in length, rather low-spired, solid, smoothish, except for numerous, poorly developed, spiral threads. Spine on lower, outer lip is strong, behind which on the base of the outside of the body whorl is a weak, spiral groove. Whorls slightly shouldered. Color bluish gray with numerous rows of small, red-brown dots. Aperture within is bluish white. A common southern species, locally known as the angular unicorn, found above high-tide mark on rocks; also on mussel beds. A yellow color form is named *aurantia* Dall, 1908.

Acanthina punctulata (Sowerby, 1825)   **1877**
Punctata Thorn Drupe

Monterey, California, to Baja California.

¾ to 1 inch, similar to *spirata,* but having a rounded shoulder with only a slight angulation. Surface rough with numerous small, square, red-brown spots. Formerly considered a form of *spirata,* but J. McLean (1969) separates them and says *spirata* occupies the lower intertidal zone, while *punctulata* is in the upper intertidal zone.

Acanthina paucilirata (Stearns, 1871)   **1878**
Checkered Thorn Drupe

San Pedro, California, to Baja California.

⅓ to ½ inch in length, characterized by about 6 spiral rows of small squares of black-brown on a cream-white background. Early whorls cancellate, later whorls smoothish except for 4 or 5 very small, smooth, raised, spiral threads. Top of whorl slightly concave. Spine at base of outer lip small and

1876   1878

needle-like. Aperture dentate, brownish with black squares on the outer lip. Siphonal canal short. Common above high-tide mark in southern California.

Acanthina lugubris (Sowerby, 1821)   **1879**

San Diego, California, to Baja California.

1 to 1½ inches, heavy, with a long, brown apertural spine, with an exterior dirty-gray to olive-brown and irregularly mottled with darker brown. Body whorl with 5 or 6 unevenly sized spiral rows of knobs. Aperture and columella light chocolate-brown. Inside of outer lip usually has several rows of elongate, glossy, cream bars. Common; rocks at low tide. Synonyms are *cymatum* Sowerby, 1825 (nude name): *armatum* and *denticulatum* Wood, 1828.

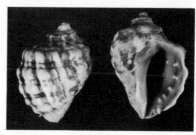

1879

Acanthina lugubris angelica Oldroyd, 1918   **1880**
Angelica's Rock-shell

Gulf of California.

Similar to and intergrading with *lugubris,* but more elongate, more angular shoulder; whitish columella, gray to purplish aperture and generally without the cream denticles on the inside of the outer lip. Common; intertidal rocks. A smooth-shouldered form with spiral rows of microscopic pinholes was named *tyrianthina* Berry, 1957. For egg capsules, see F. H. Wolfson, 1970, *The Veliger,* vol. 12, p. 375.

1880

Other species:

1881 *Acanthina* (*Neorapana* Cooke, 1918) *muricata* (Broderip, 1832) subspecies *tuberculata* (Sowerby, 1835). Gulf of California. See Keen, 1971, "Sea Shells of Tropical West America," fig. 1096. May belong to *Thais* (*Mancinella*).

1882 *Acanthina tyrianthina* S. S. Berry, 1957. Magdalena Bay and the Gulf of California.

Genus *Urosalpinx* Stimpson, 1865

Operculum purpuroid with a lateral nucleus. Shell fusiform, with rounded axial undulations. Type: *cinerea* (Say, 1822).

Urosalpinx cinerea (Say, 1822)   1883
Atlantic Oyster Drill

Nova Scotia to northeast Florida.
Washington to central California.

½ to 1 inch in length; without varices; outer lip slightly thickened on the inside and sometimes with 2 to 6 small, whitish teeth. Siphonal canal moderately short and straight. With about 9 to 11 rounded, axial ribs per whorl and with numerous, strong, spiral cords. Color grayish or yellowish white, often with irregular, brown, spiral bands. Aperture tan to dark-brown. This common species is very destructive to oysters. It occurs from intertidal areas down to about 25 feet or more. Females grow faster and hence are larger than males. They may reach an age of 7 years. The drills move inshore to spawn. Each female spawns once a year (May to September in Virginia; June to September in Canada and England). The female deposits 25 to 28 leathery, vase-shaped capsules, each containing 8 to 12 eggs. *U. aitkinae* Wheat, 1913 and *follyensis* B. Baker, 1951, are ecologic forms. The species was introduced to California prior to 1888. For shell growth, see D. Franz, 1971, *Biol. Bull.*, vol. 140, p. 63.

1883   1884

Urosalpinx perrugata (Conrad, 1846)   1884
Gulf Oyster Drill

Both sides of Florida (Panama?).

Similar to *cinerea*, but with 6 to 9 axial ribs which are quite large at the periphery of the whorl. The spiral cords are fewer and stronger. Aperture rosy-brown or yellow-brown. Outer lip more thickened on the inside and usually with 6 small, whitish teeth. This may be a subspecies of *cinerea*. Common on mud flats. Bores holes into its prey: mussels and young oysters (Radwin and Wells, 1968). Always compare with *Calotrophon ostrearum* Conrad which resembles this species very closely (see species 1965).

1885   1888

Urosalpinx tampaensis (Conrad, 1846)   1885
Tampa Drill

West central coast of Florida.

½ to 1 inch in length; the light-brown aperture is thickened and the outer lip has 6 small, white teeth. With about 9 to 11 sharp, axial ribs per whorl, crossed by about 9 to 10 equally strong spiral cords on the last whorl, thus giving the shell a cancellate sculpture. The whorls in the spire show only 2 spiral, nodulated cords. Exterior dark-gray. Operculum brown with 2 reddish spots. Common on mud flats. It bores a small hole in its favorite food, the mussel, *Brachidontes,* but also lives among oyster clumps.

Urosalpinx lurida (Middendorff, 1848)   1886
Lurid Dwarf Triton

Alaska to Catalina Island, California.

1 to 1⅓ inches in length; 5 to 6 whorls; moderately elongate spire whose whorls show the axial ribs more prominently than the numerous fine spiral threads. Suture well-impressed. Body whorl with 8 to 10 rounded ribs which are strongest and shouldered just below the suture but fade out below the periphery of the whorl. The smaller, smooth, spiral cords are elevated and prominent, often with numerous tiny axial lamellae between them. 6 to 8 small teeth on the inside of the outer lip. Color variable, whitish to rusty-brown, sometimes banded. Periostracum dark-brown and fuzzy. Siphonal canal usually sealed over. Very common from northern California north. Littoral to 30 fathoms.

This is a variable species which may be well-fimbriated (**1886a** *forma aspera* Baird, 1863, and *Coralliophila kincaidi* Dall, 1919), slender (form *munda* Carpenter, 1864), somewhat fat (*rotunda* Dall, 1919), or tall-spired with more cancellate sculpturing (*minor* Dall, 1919).

1886   1886a   1887

Urosalpinx circumtexta Stearns, 1871   1887
Circled Dwarf Triton

Moss Beach, California, to Baja California.

¾ to 1 inch in length, spire ⅓ the length of the shell. Characterized by very strong, rough spiral cords (15 on the body whorl, 6 on the whorls above). Under a lens the cords are seen to consist of arched, crowded, raised axial lamellae. The cords are often cream-white with the interspaces black-brown. Axial ribs wide, low, rounded and 7 to 9 per whorl. Some specimens are banded or have large, red-brown spots. A white form of this species was unnecessarily named *citrica* (Dall, 1919). An orange variety was named *aurantia* Stearns, 1895. A very abundant species on rocks at low tide to 30 fathoms.

Other species:

— *Urosalpinx carolinensis* Verrill, 1884, is a *Mohnia.*

1888 *Urosalpinx macra* Verrill, 1887. Off Cape Hatteras, North Carolina, to the Florida Keys, 85 to 928 fms.

1889 *Urosalpinx stimpsoni* Dall, 1927. Off Georgia, 440 fms.; off Fernandina, Florida, 294 fms.

1890 *Urosalpinx verrilli* Dall, 1927. Off Georgia, 440 fms.

1890a *Urosalpinx sclera* Dall, 1919. Graham Island, British Columbia, to Port Townsend, Washington. Puget Sound. 20 fms. Rare.

1890b *Urosalpinx subangulata* (Stearns, 1873). Washington to Santa Barbara Islands, California. (Synonym: *michaeli* Ford, 1888.)

Genus *Purpura* Bruguière, 1789

Shells with a very large aperture and a smooth narrow columella. *Haustrum* Perry, 1811; *Purpurella* Dall, 1871; *Patellapurpura* Dall, 1909; *Plicopurpura* Cossmann, 1903; and *Lepsia* Hutton, 1883, are synoyms. Type: *persica* (Linné, 1758).

Purpura patula (Linné, 1758)   **1891**
Wide-mouthed Purpura   **Color Plate 9**

Southeast Florida and the West Indies. Bermuda.

2 to 3½ inches in length; without an umbilicus. Exterior dull, rusty-gray. Columella salmon-pink. Inner borders of aperture with splotches of blackish brown. Common in the West Indies, uncommon in Florida. Rare in southeast Bermuda. The animal exudes a harmless liquid which stains the hands and collecting bag a permanent violet.

The subspecies *pansa* Gould, 1853 (**1892**) (West coast of Mexico south to Colombia), is similar in most respects, but the columella is colored a whitish cream. The subgenus *Patellapurpura* Dall, 1909, could be used for this group, but it seems little different from the Pacific *persica* or from *Plicopurpura* Cossmann, 1903.

Genus *Thais* Röding, 1798

Type of the genus is *nodosa* Linné, 1758, from West Africa.

Subgenus *Stramonita* Schumacher, 1817

The type of this subgenus is *haemastoma* Linné, 1758.

Thais haemastoma floridana (Conrad, 1837)   **1893**
Florida Rock-shell

North Carolina to Florida, and the Caribbean. Brazil.

2 to 3 inches in length, solid, smooth to finely nodulose. Color light-gray to yellowish with small flecks and irregular bars of brownish. Interior of aperture salmon-pink often with brown between the denticulations of the outer lip. Running inside the aperture high up on the body whorl above the parietal area there is a strong spiral ridge. Some specimens have a faint fold or plica on the base of the columella. This is a very common species, but quite variable in shape and color pattern. Feeds, by boring, on mussels, oysters, barnacles and clams (Butler, 1954, *Proc. Nat. Shellfish Assoc.*, vol. 45). Typical *haematoma* (Lamarck) occurs in the Mediterranean and West Africa. See additional remarks under *Thais rustica*.

Thais haemastoma canaliculata (Gray, 1839)   **1894**
Hays' Rock-shell

Northwest Florida to Texas.

This subspecies is characterized by its large size (up to 4½ inches in length), strongly indented suture and rugose sculp-

ture with a row of double, strong nodules on the shoulder of the whorls. M. D. Burkenroad (1931) has given a long account of the biology of this oyster pest. *T. haysae* Clench, 1927, is a synonym. (See Yen, 1941, *Proc. Malocol. Soc. London*, vol. 24, pl. 21.)

Thais haemastoma biserialis (Blainville, 1832)   **1895**
Two-row Rock-shell

Baja California to Chile.

1 to 2 inches, shoulder with 2 rows of 10 weak knobs. Exterior gray with purple-brown maculations and fine spiral threads. Columella straight and yellow-orange. Aperture strongly lirate within and yellow. Inside of outer lip stained with brown and bears white nodules or white barlike teeth. common; on rocks, intertidal.

Thais rustica (Lamarck, 1822)   **1896**
Rustic Rock-shell

Southeast Florida to Brazil. Bermuda.

1½ inches in length, irregularly sculptured with 2 spiral rows of blunt spines, one on the shoulder, the other at the center of the body whorl. Color dirty-gray to dull mottled brown. Interior of aperture whitish but generally margined with spots of dark-brown along the outer lip. Parietal wall glossy-white. This species is smaller and more nodulose than *floridana* and always has a white aperture. Erroneously called *Thais undata* Lamarck, which is an Indo-Pacific species.

Three confusing species are often found together in southern Florida. The young of *Pisania tincta* has the lower ⅓ of its white columella turned away (to the left) 20 degrees from the axis of the shell, and its early whorls in the spire are not shouldered. *Thais haemastoma floridana* is characterized by its almost straight, cream to orange columella, and by the numerous raised cream or white spiral ridges on the inside of the outer lip. *Thais rustica* has a stouter, white columella which is slightly twisted and purple-stained at the lower, inner corner. Its outer lip teeth occur in groups of 2 or 3, and near the edge of the lip are stained by heavy blotches of purple-brown.

Subgenus *Mancinella* Link, 1807

According to Clench, 1947, *Johnsonia*, vol. 2, p. 83, the type of this subgenus is *aculeata* Link, 1807.

Thais deltoidea (Lamarck, 1822)　　　　　　**1897**
Deltoid Rock-shell

Jupiter Inlet, Florida, to Brazil. Bermuda.

1 to 2 inches in length, heavy, and coarsely sculptured with two spiral rows of large, blunt spines. Parietal wall tinted with lavender, mauve or rose. Interior of aperture glossy-white. Exterior grayish white with mottlings of black or dull-brown. Columella with a small but distinct ridge at the base which forms the margin of the siphonal canal. This is an abundant species where intertidal rocks are exposed to the ocean surf.

Thais speciosa (Valenciennes, 1832)　　　　　　**1898**

Baja California to Peru.

1 to 1½ inches, solid, with 4 spiral rows of triangular spines, the top row of 9 being the largest. Exterior with numerous closely spaced spiral rows of red-brown color bars. Columella yellow; outer lip dotted with brown. An abundant, rock-dwelling species. *Purpura centiquadra* Duclos, 1832, is a synonym.

1895　　　　　1898　　　　　1900

Thais triangularis (Blainville, 1832)　　　　　　**1899**
Triangular Rock-shell

Baja California to Peru.

1899

1 inch, grayish white throughout, the slightly flattened columella, with its weak, oblique ridge, being enamel-white. Last whorl with 2 rows of 7 triangular equal-sized spines. Uncommon in the north, abundant southward. On rocks intertidally.

Subgenus *Thaisella* Clench, 1947

Suture with strong, curved lamellations. Type: *trinitatensis* Guppy, 1869.

Thais kiosquiformis (Duclos, 1832)　　　　　**1900**
　　　　　　　　　　　　　　　　　　　　Color Plate 9

Baja California to Peru.

1 to 2½ inches, with 8 very large triangular spines on the shoulder. Suture with numerous, short, sharp flutings. Exterior spirally threaded, dark-gray with 1 or 2 peripheral wide whitish bands. Columella white; aperture purplish to white, sometimes banded. Abundant in mangrove areas where it feeds on oysters.

Other species:

1901 *Thais trinitatensis* (Guppy, 1869). Guatemala to Brazil. Synonyms: *trinidadensis* M. Smith, 1939; *brujensis* M. Smith, 1946. Common. See *Johnsonia*, vol. 2, pl. 34.

1902 *Thais coronata* (Lamarck, 1822). Brazil to West Africa. Common. See *Johnsonia*, vol. 2, pl. 35.

Genus *Nucella* Röding, 1798

Boreal, cool-water species having solid white to orange shells, with spiral cords or axial fimbriations. Type: *lapillus* (Linné, 1758). See Rehder, 1962, *The Nautilus*, vol. 75, p. 109, who shows that *Polytropa* Swainson, 1840, is a synonym. For a review of the Pacific coast species, see Dall, 1915, *Proc. U.S. Nat. Mus.*, vol. 49, no. 2124.

Nucella lapillus (Linné, 1758)　　　　　　**1903**
Atlantic Dogwinkle

Southern Labrador to New York. Norway to Portugal.

1 to 2 inches in length, roughly sculptured or smoothish. Commonly with rounded, spiral ridges. Sometimes imbricated with small scales (form named *imbricata* Lamarck, 1822). Color usually dull-white, but sometimes yellowish, orange or brownish. Rarely with dark-brown or blackish spiral bands. This species has also been known as *Thais lapillus*. It gives off a tyrian purple dye once used as an indelible laundry marking ink. For feeding and habits, see R. W. Dexter, 1943, *The Nautilus*, vol. 57, p. 6; H. Agersborg, 1929, *The Nautilus*, vol. 43, p. 45; H. S. Colton, 1922, *Ecology*, vol. 3, p. 146.

1903

Nucella canaliculata (Duclos, 1832)   **1904**
Channeled Dogwinkle

Aleutian chain to Monterey, California.

1 inch in length, moderately globose, its spire higher than that of *emarginata*, but lower than that of *lamellosa*. Columella arched, flattened below. Characterized by about 14 to 16 low, flat-topped, closely spaced spiral cords on the body whorl. Suture slightly channeled. Color white or orange-brown, often spirally banded. Moderately common on rocks and mussel beds. Do not confuse with *lima* from Alaska. For feeding habits, see O. H. Paris, 1960, *The Veliger*, vol. 2, p. 41. Synonyms include *analoga* Forbes, 1850; **(1904a)** *decemcostata* Middendorff, 1849; and the forma *compressa* Dall, 1915 **(1904b)**.

1904   1904 a   1904 b

Nucella lamellosa (Gmelin, 1791)   **1905**
Frilled Dogwinkle

Bering Straits to Santa Cruz, California.

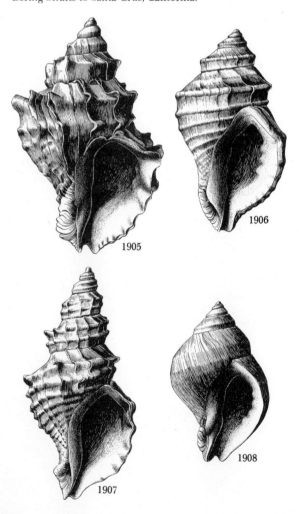

1905

1906

1908

1907

1½ to 5 inches in length, solid, usually with a fairly high, pointed spire. Columella almost vertical and straight, not flattened. Size, details of shape, sculpturing and color very variable. White, grayish, cream or orange, sometimes spirally banded. Smoothish or with variously developed foliated, axial ribs. Sometimes spinose. A very common rock-loving species. Synonyms include *Purpura crispata* Martyn; *rugosus* Perry, 1811; *ferrugineus* Eschscholtz, 1829; *lactuca* Eschscholtz, 1829; *septentrionalis* Reeve, 1846; *plicata* von Martens, 1872; *quillayutea* Reagan, 1909; and several forma described by Dall in 1915: *franciscana,* **(1906)** *hormica,* **(1907)** *neptunea, cymica* **(1908)** and *sitkana.*

Nucella emarginata (Deshayes, 1839)   **1909**
Emarginate Dogwinkle

Bering Sea to Mexico.

1 inch in length, with a rather short spire and with globose whorls. Aperture large. Columella strongly arched, and flattened and slightly concave below. Sculpturing variable, but characteristically with coarse spiral cords, usually alternatingly small and large. Cords often scaled or coarsely noduled. Exterior yellow-gray to rusty-brown, often with darker, narrow spiral bands. Interior and columella light- to chestnut-brown. Exceedingly common in many places along the coast where there are rocks. Synonyms include *conradi* Reeve, 1846; *ostrina* Gould, 1852 **(1910)**; and the forma *projecta* Dall, 1915, from Sitka, Alaska.

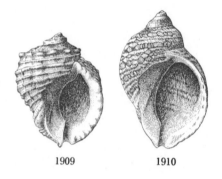

1909   1910

Nucella lima (Gmelin, 1791)   **1911**
File Dogwinkle

Alaska and Japan to northern California.

1 to 2 inches in length, very similar to *canaliculata*, but with 17 to 20 round-topped spiral cords, often smooth, sometimes minutely fimbriated. Cords often alternate in size. Color whitish or orange-brown, rarely banded. Common intertidally. Compare with *canaliculata*. Synonyms include *saxicola* Valenciennes, 1846, and *attenuata* Reeve, 1846 **(1912)**. Sometimes smoothish **(1913)**.

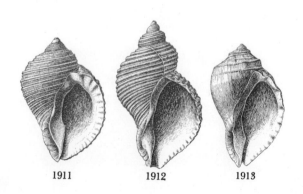

1911   1912   1913

Subfamily OCENEBRINAE Cossmann, 1908

The subfamily name Tritonaliinae Broderip, 1839, is a synonym.

Genus *Ocenebra* Gray, 1847

Tritonalia Fleming, 1828, may not be used as a name for this genus, according to Opinion 886 (*Bull. Zool. Nomen.*, vol. 26, p. 128; 1969). *Ocinebra* is a misspelling. Type: *erinacea* (Linné, 1758). For notes on Pacific coast species, see Bormann, 1946, *The Nautilus*, vol. 60.

Ocenebra interfossa Carpenter, 1864   1914
Carpenter's Dwarf Triton

Alaska to Baja California.

½ to ¾ inch in length, spire ½ the length of the shell; light-gray in color, delicately sculptured. 8 to 11 axial ribs on the body whorl crossed by about a dozen strong, microscopically scaled spiral cords. The surface is often fimbriated axially

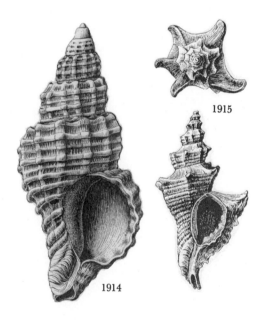

1915

1914

between the cords. Siphonal canal moderately long, usually sealed over. Littoral to several fathoms. Common. There are several named varieties of doubtful biological significance: *atropurpurea* Carpenter, 1865, and *clathrata* Dall, 1919, the latter having tabulate whorls and a clathrate sculpturing producing squarish pits. *O. fraseri* Oldroyd, 1920, is considered a subspecies by some and a synonym by others.

Ocenebra barbarensis (Gabb, 1865)   1915
Santa Barbara Dwarf Triton

British Columbia to Santa Barbara Islands, California.

½ to ¾ inch, brown, high spire, angular whorls, with 5 to 9 low-spined varices per whorl. Surface with numerous, strong, fimbriated spiral cords. Last varix fairly large and with a triangular extension on the shoulder. Aperture white; 5 or 6 tubercles on the inside of the outer lip. Uncommon; 10 to 50 fathoms.

Ocenebra beta (Dall, 1919)   1916
Beta Dwarf Triton

San Mateo County to San Diego, California.

⅓ to ½ inch, similar to *barbarensis*, but it is heavier, with a shorter spire and shorter siphonal canal. Color is tan to dark-brown, occasionally with white banding. Common; 8 to 35 fathoms on rocks and shale. *O. keenae* Bormann, 1946 (*The Nautilus*, vol. 60, p. 40), is a synonym.

Ocenebra foveolata (Hinds, 1844)   1917
Dim Dwarf Triton

San Mateo County to Baja California.

½ to 1 inch, 5 whorls tabulate in the spire; last whorl with 6 or 7 varices which may be sharply raised and slightly reflected or merely rounded. Spiral sculpture of numerous spiral fimbriated, rounded threads. Aperture fairly large, white. Siphonal canal short, closed. Color cream to brown, sometimes spirally banded. Common; 3 to 35 fathoms. North of San Pedro, particularly off Monterey, the form or subspecies *fusconotata* (Dall, 1919) (**1918**) is found from 3 to 15 fathoms. It is smaller and more delicately sculptured. Other synonyms: *squamulifera* Carpenter, in Gabb, 1869; and *epiphanea* Dall, 1919.

Ocenebra inornata (Récluz, 1851)   1919
Japanese Dwarf Triton

British Columbia and Washington.

1 to 1⅓ inches, whorls in spire squarish. Aperture large, round. Siphonal canal short, open along its length. Last whorl with 5 to 11 axial low, scalloped varices which occur at irregular intervals, sometimes being paired. Spiral sculpture of several uneven low cords which may form nodes as they cross the axial varices or axial riblike swellings. Outer lip has a wide varix. Color white to blue-gray to chestnut-brown. Aperture brown-stained. Introduced from Japan prior to 1929. Common on oyster beds. For habits and control, see *Olympia Oyster Bull.* for May 10, 1946 (Wash. State Oyster Lab.). *O. japonica* Dunker, 1860, is a synonym.

1916   1917   1919

Ocenebra gracillima Stearns, 1871   1920
Graceful Dwarf Triton

Monterey, California, to the Gulf of California.

⅓ inch in length, similar to *lurida* in shape; with 5 whorls, those in the spire weakly cancellate with the axial ribs the strongest. Last whorl without axial ribs, except for a rather strong, rounded varix behind the outer lip. Last whorl with about a dozen or so light-brown, spotted, spiral threads over a background of light yellowish gray. 3 to 5 fairly large teeth on the inside of the outer lip. Siphonal canal short, sealed over in adults. Periostracum thin, fuzzy, grayish with mauve-brown undertones. Interior of aperture light mauve-brown, usually with a whitish, spiral band on the middle of the body whorl. *O. stearnsi* Hemphill, 1911, and *obesa* Dall, 1919 is the same. Very common in rocky rubble and on wharf pilings.

Subgenus *Roperia* Dall, 1898

The type of this subgenus is *poulsoni* (Carpenter, 1864).

Ocenebra poulsoni (Carpenter, 1864)   1921
Poulson's Dwarf Triton

Santa Barbara, California, to Baja California.

1½ to 2 inches in length; a sturdy shell with a semigloss finish. 8 to 9 nodulated, rounded axial ribs per whorl crossed by numerous, very fine, incised spiral lines and 4 or 5 larger, rounded raised spiral cords. The latter make nodules on the ribs. Siphonal canal narrowly open. Periostracum thin, grayish or brownish and smoothish. When the periostracum is absent, the shell is glossy-white with numerous, fine spiral lines of dark- to yellow-brown. Aperture white. An exceedingly common species found in nearly every region on rocks and wharf pilings. *Roperia roperi* Dall, 1898, is a synonym.

1920   1921

Subgenus *Risomurex* Olsson and McGinty, 1958

The type of this subgenus is *schrammi* Crosse, 1863, from the West Indies. See *Bull. Amer. Paleontol.*, vol. 39, pl. 2, 1958. Nuclear whorl with a single keel.

Ocenebra muricoides (C. B. Adams, 1845)   1922
Adams' Dwarf Triton

Lower Florida Keys and the West Indies.

⅓ to ½ inch, fusiform, with a straight, high spire, with its whorls bearing 3 spiral noduled cords, the lower one (just above the suture) being cream white. Body whorl brown to black with 2 spiral rows of about 11 raspberry-red to pinkish white beads. Base with 2 rows of smaller white to cream beads. Outer lip very thick, brown, with 2 white spots, and the inner side with 4 to 5 large white denticles. Columella with 3 pimples and a sharp left edge. Aperture rose. Moderately common; under rocks; intertidal. Formerly in *Ocinebrina*. *Tritonalia caribbaea* Bartsch and Rehder, 1939, is a synonym.

Subgenus *Ocinebrina* Jousseaume, 1880

The type of this subgenus is *Murex aciculatus* Lamarck, 1822. *Corallina* Bucquot, Dautzenberg and Dollfuss, 1905, is a synonym.

Ocenebra minirosea Abbott, 1954   1923
Tiny Rosy Drill

South Florida, Bahamas to the Virgin Islands.

5 to 7 mm., resembling a miniature murex, orange to pink. Last whorl with 7 to 9 rounded, axial ribs bearing about 10 fluted spines. Nuclear whorl glassy-pink or white. Peristome complete and somewhat spoutlike. Inside of outer lip with 5 or 6 weak, elongate teeth. The ends of 3 or 4 siphonal canals

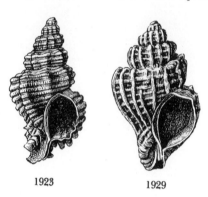

1923   1929

show at the left. Uncommon to rare; 10 to 40 fathoms. *Murex coccineus* A. Adams, 1853, non Lesson, 1844, is a synonym. *The Nautilus*, vol. 68, p. 43.

Ocenebra emipowlusi Abbott, 1954   1924
M. E. Powlus Drill

Off both sides of Florida.

7 to 10 mm., white to pinkish white, similar to *minirosea* Abbott, but more squat, and with 6 rounded axial varixlike ribs on the body whorl. Spiral sculpture of raised fimbriated cords or threads (17 to 20 on the last whorl). One nuclear whorl, smooth and with a carina. No anal fasciole present. Inside of outer lip with 5 or 6 weak, glossy, white lirae. Uncommon; offshore from 40 to 200 fathoms. *The Nautilus*, vol. 68, p. 41.

1922   1924

Other species:

1925 *Ocenebra* (*Risomurex*) *schrammi* (Crosse, 1863). Caribbean. See *Bull. Amer. Paleontol.*, vol. 39, pl. 2, 1958. Uncommon.

1926 *Ocenebra* (*Risomurex*) *rosea* (Reeve, 1856). Matagorda, Texas; Bahamas and West Indies. Nuclear whorls pink. Uncommon.

1929 *Ocenebra painei* (Dall, 1903). Dundas Bay, Alaska, to San Diego, California. 50 to 100 fms. Rare. 15 mm.

1930 *Ocenebra squamulifera* (Carpenter in Gabb, 1868). Santa Barbara to San Pedro, California. Dubious.

1931 *Ocenebra tracheia* Dall, 1919. 10 to 15 fms., Sitka, Alaska, to Del Monte, California. Rare.

1932 *Ocenebra crispatissima* S. S. Berry, 1953. Catalina Island, California, 15 to 33 fms. *Trans. San Diego Soc. Nat. Hist.,* vol. 11, p. 414.

1933 *Ocenebra micromeris* (Dall, 1890). Greater Antilles, living. Florida Pliocene. 7 mm.

1934 *Ocenebra buxea* (Broderip, 1833). Baja California to Chile. Common. This may be a *Cantharus.*

— *Ocenebra carmen* (Lowe, 1935). Gulf of California, 9 to 20 fms. Uncommon locally. Is an *Attiliosa* Emerson.

1935 *Ocenebra lugubris* (Broderip, 1833). Baja California to Ecuador; to 21 fms.

1936 *Ocenebra parva* (E. A. Smith, 1877). Gulf of California; Galapagos. Moderately common.

— *Ocenebra peasei* (Tryon, 1880), is a *Favartia.*

— *Ocenebra perita* (Hinds, 1844), is a *Murexiella.*

Genus *Ceratostoma* Herrmannsen, 1846

Type of this subgenus is *nuttalli* Conrad, 1837. *Spinostoma* Coen, 1943; *Microrhytis* Emerson, 1959; and *Cerostoma* Conrad, 1837, non Latreille, 1802, are synonyms.

Ceratostoma nuttalli (Conrad, 1837) **1938**
Nuttall's Purpura **Color Plate 9**

Point Conception, California, to Baja California.

1½ to 2 inches in length, similar to *foliatum* (and somewhat resembling *Ocenebra poulsoni*), but with much more poorly developed varices, and with one prominent, noduled rib between each varix. Spine on outer lip usually long and sharp. Exterior yellowish brown, sometimes spirally banded. Siphonal canal closed along its length. A common littoral species in the southern part of its range. A white color form has the name *albescens* Dall, 1919. A banded form is *albofasciata* Dall, 1919. *P. aciculiger* Valenciennes, 1846, is a synonym. Locally known as Nuttall's thornmouth.

Subgenus *Pterorytis* Conrad, 1862

Fusiform, 3 to 6 varices per whorl; varices bladelike. Peristome continuous. With a long spine at the base of the outer lip. Type: *umbrifer* Conrad, 1832, a Miocene fossil. *Neurarhytis* Olsson and Harbison, 1953, is a synonym.

Ceratostoma foliatum (Gmelin, 1791) **1937**
Foliated Thorn Purpura **Color Plate 9**

Alaska to San Pedro, California.

2 to 3 inches in length, with 3 large, thin, foliaceous varices per whorl which are finely fimbriated on the anterior side. Numerous spiral cords are rather prominent and of various sizes. Siphonal canal closed, its anterior tip turned up and to the right. Base of outer lip with a moderately strong spine. Aperture white. Exterior white to light-brown. Common in the Puget Sound area and to the north on rocks near shore. Also down to 35 fathoms. Appears in some books as *Purpura* or *Ceratostoma foliatum* Martyn.

Genus *Poirieria* Jousseaume, 1880

Shells with 6 long, low, leaflike varices on the body whorl and usually bearing an open spine at the shoulder of each varix. Type: *Murex zelandicus* Quoy and Gaimard, 1833, from New Zealand. *P. clenchi* (Carcelles, 1953) from Argentina is the only living American species in the nominate subgenus, *Poirieria.* The aperture lacks tiny denticles.

Subgenus *Paziella* Jousseaume, 1880

Shells small, with 6 spinose varices. Aperture with small denticles. The type is *pazi* Crosse, 1869. *Bathymurex* Clench and Farfante, 1945 is a synonym, as is *Dallimurex* Rehder, 1946, whose type is *nuttingi* Dall.

Poirieria pazi (Crosse, 1869) **1939**
Paz's Murex **Color Plate 9**

East Florida, the Bahamas to Cuba and Honduras.

1 to 2 inches, white, extremely spinose, with 5 or 6 long, almost straight spines which are open on their anterior edges. Base of shell with 2 or 3 spines, about ⅔ as long as those on the shoulder. 1½ nuclear whorls, rounded, glass-like, the first one bent to one side. Uncommon; 112 to 338 fathoms.

Poirieria hystricina Dall, 1889 **1940**
Atlantic Burr Murex

Off Cuba to the Lesser Antilles.

¾ inch, alabaster-white, spire turreted. Strongly sculptured with 4 or 5 rows of open spines on the last whorl with those on the shoulder being twice as long as long, scalelike spines.

Body whorl with 8 or 9 varices. 1½ nuclear whorls, glassy and very tiny. Rare; 148 to 254 fathoms.

Poirieria nuttingi (Dall, 1896) **1941**
Nutting's Murex

Off Key West, Florida.

1½ inches, white with a yellowish periostracum; elongate, with about 8 whorls, the last having about 8 denticulated varices bearing a short spine at the shoulder. Base of shell with about 6 strong spines. About 6 spiral low rounded ridges or cords between the varices. Uncommon; 110 to 220 fathoms.

Other species:

1942 *Poirieria (Paziella) atlantis* Clench and Farfante, 1945. 190 to 200 fms., off Bahia de Cochinos, Cuba. *Johnsonia*, vol. 1, no. 17, pl. 20. Rare.

1943 *Poirieria (Panamurex* Woodring, 1959) *carnicolor* (Clench and Farfante, 1945). 88 to 103 fms., off Barbados and Montserrat, West Indies. *Johnsonia*, vol. 1, no. 17, pl. 25. Gulf of Mexico, 25 to 162 fms.

1944 *Poirieria (Panamurex) valero* E. Vokes, 1970. 2 miles southwest of Cabo de Vela, Colombia, 10 to 22 fms. *Tulane Studies Geol. Paleontol.*, vol. 8, p. 47. May be a synonym of *mauryae* E. Vokes, 1970, a fossil.

1945 *Poirieria (Paziella) oregonia* (Bullis, 1964). Trinidad to Brazil, 105 to 275 fms. *Tulane Studies Zool.*, vol. 11, p. 106.

1946 *Poirieria (Paziella) actinophorus* Dall, 1889. Caribbean to Brazil, in 18 to 744 meters. See F. M. Bayer, 1971, p. 159.

Subgenus *Pazinotus* E. Vokes, 1970

4 to 7 varices, formed by lamellar flanges connecting varical spines. Aperture with denticles on both parietal and outer lip. Siphonal canal open, recurved. Type: *stimpsonii* Dall, 1889.

Poirieria stimpsonii (Dall, 1889)   **1947**
Stimpson's Drill

Florida to Barbados.

12 mm., thin, whitish, with 4 to 6 varices per whorl. Spiral sculpture of extremely fine faint striae and, on the last whorl, of 5 low cords, most prominent on the back of the varices. Spine at the shoulder of the varix is webbed on either side. Aperture round. Parietal lip trumpetlike. Columella with 3 teeth, the outer lip with 4. Siphonal canal slightly recurved, the end of the previous one projecting from it at the left. Rare; 50 to 100 fathoms. Formerly placed in *Eupleura*. See E. Vokes, 1970, *Tulane Studies Geol. Paleontol.*, vol. 8, p. 28.

1947

Subfamily ASPELLINAE Keen, 1971

Proposed in *The Veliger*, vol. 13, p. 296. Contains *Aspella, Eupleura, Calotrophon, Favartia, Attiliosa* and *Phyllocoma*.

Genus *Aspella* Mörch, 1877

The type of the genus is *anceps* (Lamarck, 1822). Synonym appears to be *Gracilimurex* Thiele, 1929.

Aspella anceps (Lamarck, 1822)   **1948**
Two-sided Aspella

North Carolina to the Caribbean. Brazil.
Indo-Pacific.

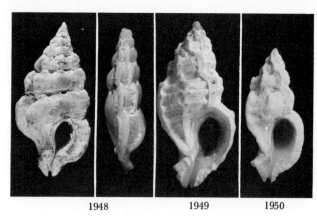

1948   1949   1950

½ inch, elongate, with an elevated spire. Color white or yellowish. Shell flattened, with a row of double or twin varices on each edge. There is usually a smaller secondary varix in the center of the space between the double varices. Spiral sculpture of a few threads. Suture deep and pitted. Aperture small, elliptical. Siphonal canal short, narrow, and recurved. Moderately common; under rocks from 6 to 20 feet. Rare in the lower Florida Keys. Rios (1970) reports it from 12 to 15 fathoms off Brazil.

Subgenus *Dermomurex* Monterosato, 1890

Synonym: *Poweria* Monterosato, 1884, non Bonaparte, 1840.

Aspella paupercula (C. B. Adams, 1850)   **1949**
Little Aspella

North Carolina to the Lesser Antilles.

½ to 1 inch. Outer surface covered with a thin layer of white cellulose lime which under the microscope appears to be finely vermiculated. Shell white. Axial sculpture of 3 to 5 strong varices between which run 3 to 5 spiral cords. Inside of outer lip with 3 to 6 denticles. Siphonal canal recurved, barely open along its length. Moderately common; under rocks in shallow water. Feed on attached bivalves. Egg capsules wafer-shaped and in rows. Larvae nonpelagic (R. C. Work, 1969, p. 669).

Aspella elizabethae McGinty, 1940   **1950**
Elizabeth's Aspella

Florida Keys to Panama; Bermuda.

⅓ to ½ inch, chalky-white, similar to *paupercula*, but smaller, more elongate. Nuclear whorl glossy, smooth, rounded, rapidly descending. Early postnuclear whorls with 6 axial varices, but later every other one becomes a short, elongate axial fold with 2 or 3 small beads. Uncommon to common; found under old coral slabs among the sandy tubes of worms in 3 to 18 feet of water. *The Nautilus*, vol. 53, pl. 10, vol. 54, p. 63.

Other species:

1951 *Aspella (Dermomurex) abyssicola* (Crosse, 1865). 270 fms., off Guadeloupe Island, Lesser Antilles. Rare. See *Johnsonia*, vol. 1, no. 17, pl. 20.

Subgenus *Gracilimurex* Thiele, 1929

Shell slender, only slightly flattened because of the inconspicuous varices. With a broad color band. Type: *Murex bicolor* Thiele, 1929 (not Risso, 1826), now *bakeri* Hertlein and Strong, 1951.

1952 *Aspella (Gracilimurex) bakeri* Hertlein and Strong, 1951. Gulf of California. 18 mm. (Synonym: *bicolor* Thiele, 1929, not Risso, 1826.)

Genus *Favartia* Jousseaume, 1880

Shells small, about 1 inch, with 5 to 7 low, poorly sculptured varices. The type is *breviculus* Sowerby, 1834.

Favartia cellulosa (Conrad, 1846)   **1953**
Pitted Murex

North Carolina to the Gulf of Mexico to Brazil. Bermuda.

1 inch in length. Shell rough, with 5 to 7 poorly developed fluted varices. It rarely develops spines, but when present they are short and stubby with a thin webbing connecting each spine in the varix. The siphonal canal strongly upturned. Aperture small, almost round. Color a dull grayish white.

1953   1955

This is one of the smallest and most compact species of muricids on the Atlantic Coast and is often found in shallow, intertidal waters, especially near oyster beds where it probably does moderate damage to young oysters. Its main food is mussels. Its identification is made difficult when the siphonal canal has been broken off completely. *Murex jamaicensis* Sowerby, 1879, is a synonym. An inch-long, chubby form (*nucea* Mörch, 1850) **(1954),** with a shorter and wider siphonal canal and heavily scaled varices, occurs in the West Indies, Brazil, Bermuda and Florida. For *leviculus* Dall, see under the *Murex* subgenus *Murexiella* (see species 1845).

Subgenus *Caribiella* Perrilliat, 1971

Type of the subgenus is *alveata* (Kiener, 1842).

Favartia alveata (Kiener, 1842)   **1955**
Frilly Dwarf Triton

Florida Keys and the West Indies to Brazil. Bermuda.

¾ to 1 inch, whitish, sometimes banded with brown. Body whorl with 6 or 7 frilled varices and a series of strong revolving ridges. Suture deep. Whorls shouldered. Outer lip delicately frilled. Siphonal canal rather short, nearly closed. Moderately common; just offshore in small crevices of reefs. *Murex intermedius* C. B. Adams, 1850, and *Aspella elegans* Perrilliat, 1971, are synonyms. See *Paleontologia Mexicana,* no. 32, p. 83.

Other species:

1956 *Favartia peasei* (Tryon, 1880). Baja California to Panama.

Genus *Muricopsis* Bucquoy, Dautzenberg and Dollfuss, 1882

The genus *Muricidea* Mörch, 1852, not Swainson, 1840, is the same as this genus. The type of *Muricopsis* is *blainvillei* (Payraudeau, 1826). *Jania* Bellardi, 1882, is the same.

Muricopsis oxytatus (M. Smith, 1938)   **1957**
Hexagonal Murex

Florida and the West Indies.

1 to 1½ inches in length, elongate, heavy, with a high spire, and sharply spinose on each of the 6 or 7 axial ribs on each whorl. Exterior chalk-white or tinted with orange-brown. Aperture white. A moderately common species from 1 to 30 fathoms on rocky bottom. *M. hexagonus* Lamarck, 1816, non Gmelin, 1791, is this species.

Muricopsis armatus (A. Adams, 1854)   **1958**
Armored Drill

Gulf of California.

1 inch, fusiform, bearing sharp, short, stout spines and having a high spire, thus resembling the Atlantic *oxytatus*. Color yellow to tan. Common; among intertidal rocks. Synonym: *squamulata* Carpenter, 1865.

1957   1958

Genus *Eupleura* H. and A. Adams, 1853

Shell dorso-ventrally compressed; with 2 lateral varices. Type: *caudata* (Say, 1822).

Eupleura caudata (Say, 1822)   **1959**
Thick-lipped Drill

South of Cape Cod to south half of Florida.

½ to 1 inch in length; apex pointed; siphonal canal moderately long, almost closed, coming to a sharp point below. Last varix large, rounded and with small nodules. Inside of outer lip with about 6 small, beadlike teeth. Whorls with spiral cords and strong axial ribs. There are 4 to 6 axial ribs between the last 2 varices. An abundant shallow-water species that feeds, by boring, upon young oysters. For sexual behavior studies, see Hargis and MacKensie, Jr., 1961, *The Nautilus,* vol. 75, p. 7. *E. etterae* B. B. Baker, 1951, is a large, ecologic form of this species.

Eupleura sulcidentata Dall, 1890   **1960**
Sharp-ribbed Drill

West coast of Florida. Bimini.

Similar to *E. caudata*, but more delicate, with the upper whorls more tabulate and spiny and with the spiral sculpture almost absent. There are only 2 or 3 axial ribs between the last 2 lateral varices. The varices are thin and sharp, and the entire shell is slightly compressed laterally (has a more oval outline from an apical or top view than does *caudata*). The axial ribs are often sharp and may bear a small spine at the top. Color gray, chocolate-brown, tan or rarely pinkish, and sometimes with narrow spiral brown bands. Moderately common; in shallow grass and mud areas.

Eupleura muriciformis (Broderip, 1833)   **1961**
Murex-shaped Drill

Gulf of California to Ecuador.

1 to 1½ inches, a variable species similar to *caudata* of the Atlantic. Color waxy-white to bluish gray or maculated with blackish. Common; from shore to offshore. Feeds on other small mollusks. Synonyms are: *plicata* Reeve, 1844; *triquetra* Reeve, 1844; *unispinosa* Dall, 1890; *limata* Dall, 1890.

1959   1960   1961

Other species:

1962 *Eupleura nitida* (Broderip, 1833). Mazatlan to Panama. Uncommon.

1963 *Eupleura pectinata* (Hinds, 1844). Pacific side of Central America. Rare.

— *Eupleura stimpsoni* Dall, see before under *Poirieria*, species 1947.

1964 *Eupleura grippi* Dall, 1911. San Diego, California, 15 fms. 21 mm.

1964

Genus *Calotrophon* Hertlein and Strong, 1951

Fusiform, somewhat chalky, whorls shouldered. Sculpture of axial ribs and spiral, imbricate ridges. Canal distinct, open; columella smooth, concave. Outer lip lirate within. 1½ nuclear whorls, rounded. Operculum with a terminal nucleus. Type: *bristolae* H. and S., which is *turritus* Dall, 1919. Synonyms are *Pseudosyrinx* Olsson and Harbison, 1958, and *Hertleinella* S. S. Berry, 1958. For an excellent review of this genus, see McLean and Emerson, 1970, *The Veliger*, vol. 13, no. 1, pp. 57–62.

Calotrophon ostrearum (Conrad, 1846)   **1965**
Mauve-mouth Drill

West coast of Florida to the Florida Keys.

1 inch in length, extraordinarily like *Urosalpinx perrugata* (species 1884), but more elongate, with a longer siphonal canal which is bent slightly back, and with a light- to dark-mauve aperture. Upper part of aperture angular, not rounded. Operculum dark-brown. Moderately common from low-tide area to 35 fathoms. Some colonies may have up to 25% albinos. *C. floridana* (Conrad, 1869), is this species, as are the fossils *attenuata* (Dall, 1890), and *perplexus* (Olsson and Harbison, 1935). Previously and erroneously placed in *Muricopsis* and *Cantharus*.

1965

Other species:

1966 *Calotrophon turritus* (Dall, 1919). Baja California, Mexico. 10 to 75 fms. (Synonyms: *bristolae* Hertlein and Strong, 1951, and *leucostephes* S. S. Berry, 1958.)

Subgenus *Attiliosa* Emerson, 1968

Shells about 10 to 40 mm., solid, whitish, with peripheral blunt knobs, slightly scaly spiral riblets. Outer lip lirate within; inner lip with a well-developed margin. Siphonal fasciole large, canal short. Nucleus of operculum apical. Type: *incompta* (S. S. Berry, 1960). See *The Veliger*, vol. 10, p. 380.

1967 *Calotrophon philippiana* (Dall, 1889). Gulf of Mexico; Key West to Yucatan, 20 to 30 fms. Jamaica and Cuba, 10 fms. Rare. 17 to 20 mm.

1968 *Calotrophon incompta* (S. S. Berry, 1960). Sonora, Mexico, to Secas Island, Panama, 18 to 80 meters. 20 to 35 mm.

1969 *Calotrophon rufonotata* (Carpenter, 1864). Gulf of California.

1970 *Calotrophon carmen* (Lowe, 1935). Gulf of California, 15 to 40 meters. Is *rufonotata?*

Subfamily TROPHONINAE Marwick, 1924

Genus *Boreotrophon* Fischer, 1884

Moderate-sized, fusiform, whitish shells with a long, slender siphonal canal. Varices usually lamellate, rarely spinose. Type: *clathratus* (Linné, 1758). A cold-water genus.

Boreotrophon clathratus (Linné, 1758) 1971
Clathrate Trophon

Arctic Seas to Maine.

1 to 2 inches in length, with rounded whorls, slightly flaring lip, and numerous axial, foliated ribs. Chalk-white in color. There are several forms described, and we have figured the subspecies *scalariformis* Gould, 1838. Commonly dredged on the Grand Banks, off New England. Synonyms: *multicostatus* (Eschscholtz, 1829); *lamellosa* (Gray, 1839); *gunneri* (Lovèn, 1846) (1971a).

Boreotrophon scitulus (Dall, 1891) 1972
Handsome Trophon

Alaska to San Diego, California.

1 to 1½ inches in length, rather fragile, pure-white in color, with a rather long siphonal canal. Characterized by 5 or 6 spiral rows of long, delicate, anteriorly hollowed spines. In the spire only 2 rows show. Operculum thin, light-brown, chitinous, ungulate. Rare, 50 to 250 fathoms.

Boreotrophon stuarti (E. A. Smith, 1880) 1973
Stuart's Trophon

Alaska to San Diego, California.

2 inches in length, waxy texture, pure-white to yellow-cream, with 9 to 11 strong, lamellalike, high-shoulder ribs per each of the 7 whorls. Whorls in spire cancellated by the 2 or 3 spiral raised cords. Body whorl with 5 very weak spiral rounded threads. *B. smithi* (Dall, 1902) is the same. Uncommon from low tide (in Alaska) to 25 fathoms.

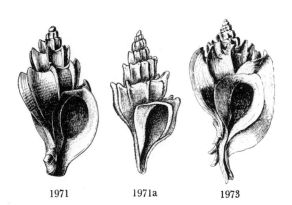

1971   1971a   1973

Boreotrophon pacificus (Dall, 1902) 1974
Northwest Pacific Trophon

Alaska to off Baja California. Arctic Canada.

¾ to 1 inch in length, similar to *multicostatus*, with the spire ⅖ the length of the shell, canal twice as long, with 12 to 20 ribs which are shouldered further below the suture. Suture slightly indented. Fairly common at low tide in Alaska. Occurs in very deep water farther south. *B. beringi* Dall, 1902, may be this species.

Boreotrophon dalli (Kobelt, 1878) 1975
Dall's Trophon

Arctic Ocean to Fuca Strait, Washington.

2 inches in length, spire ¼ and canal nearly ½ the length of the shell. 5 whorls globose, crowned at the shoulder with 12 to 20 short spines per whorl. The ribs over the whorl are moderately to obsoletely developed. Rare in about 30 to 50 fathoms.

Boreotrophon triangulatus (Carpenter, 1864) 1976
Triangular Trophon

Monterey to San Pedro, California.

Nearly 1 inch in length, 7 whorls, with 8 delicate axial ribs which bear rather short, erect protuberances on the shoulder. Nuclear whorl smooth, followed by squarish whorls which bear ribs from suture to suture, and which are smooth in between. Siphonal canal rather short. Grayish white with enamel-white aperture. *B. peregrinus* Dall, 1902, is the same. Do not confuse with *Trophon cerrosensis* (species 2006). The latter has longer spines, longer siphonal canal and the ribs are only on the periphery of the early whorls.

Boreotrophon orpheus (Gould, 1849) 1977
Orpheus Trophon

Alaska to Redondo Beach, California.

¾ to 1 inch in length, resembling *pacificus* but with a spire ¾ the length of the shell, and with 3 strong, but small, spiral cords showing in the spire, but with about the same number and type of axial ribs. It resembles an immature *stuarti*, but has more and smaller ribs and its whorls are not so sharply shouldered on top nor so strongly cancellate in the spire. Moderately common from 12 to 80 fathoms.

Boreotrophon macouni (Dall, 1910) 1978
Macoun's Trophon

Port Althorp, Alaska, to British Columbia.

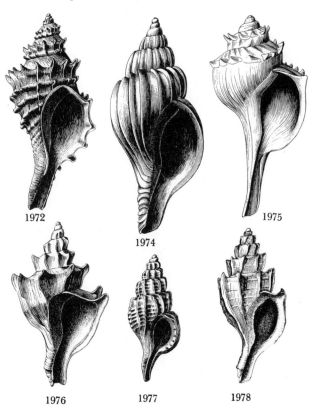

1972   1975

1974

1976   1977   1978

½ inch, background color dark-purple, ridges and varices white. 6 whorls, with thin sharp varices (9 on the last whorl) which are sharply angulated at the shoulder. Spiral sculpture of 2 to 7 low, flat ridges and microscopic lines. Aperture small, siphonal canal white, rather long and narrow, twisted slightly to the left. Dredged offshore; uncommon to rare.

Boreotrophon albospinosus (Willett, 1931)   1979
White-spined Trophon

Southern California.

½ to ¾ inch, dark purplish brown, with 8 sharply laminated varices on the last whorl. Similar to *macouni*, but larger, broader, with a shorter siphonal canal, more acute angles, longer spines and lacking the prominent spiral ridges. Not uncommon in gravel beds from 25 to 30 fathoms.

Boreotrophon avalonensis (Dall, 1902)   1980
Avalon Trophon

Monterey to San Diego, California.

½ inch, white, delicate, fusiform. Nucleus of 1½ whorls which are rounded, smooth and tilted. Whorls finely spirally striated, with 8 or 9 sharp, pressed-down varices rising into radiating narrowly grooved spines at the shoulder. Aperture subovate. Moderately uncommon; 40 to 150 fathoms.

Boreotrophon rotundatus (Dall, 1902)   1981
Rotund Trophon

Pribilof Islands, Bering Sea.

½ inch, with a short spire and with 5 rather rounded whorls. Last whorl with 14 angular, smoothish ribs. Aperture yellowish within. 70 fathoms; rare.

Other Pacific species:

1982 *Boreotrophon disparilis* (Dall, 1891) Fuca Strait, Washington, to off San Diego, California. 52 fms.

1983 *Boreotrophon elegantulus* (Dall, 1907) Aleutian Islands. 135 fms. 31 mm.

1984 *Boreotrophon beringi* (Dall, 1902) Arctic Ocean to Puget Sound. Also Japan. 5 fms. (Is *pacificus* Dall, 1902?) 40 mm.

1985 *Boreotrophon cepulus* (Sowerby, 1880) Bering Sea and Aleutian Islands.

1986 *Boreotrophon cymatus* (Dall, 1902) Pribilof Islands, Bering Sea.

1987 *Boreotrophon ithitomus* (Dall, 1919) Southeast Alaska. 253 fms.

1988 *Boreotrophon alaskanus* (Dall, 1902) Bering Sea, north of Unalaska; 225 fms. 32 mm.

1989 *Boreotrophon eucymatus* (Dall, 1902) Off San Diego and Catalina Island, California. 75 to 124 fms. 16 mm.

1990 *Boreotrophon bentleyi* (Dall, 1908) Southern California; 30 to 75 fms.

1991 *Boreotrophon staphylinus* (Dall, 1919) Alaska to southern California.

1992 *Boreotrophon calliceratus* (Dall, 1919) Monterey to San Diego, California. 120 to 131 fms.

Other Atlantic species:

1993 *Boreotrophon abyssorum* Verrill, 1885 (and form *limicola* Dall, 1902). Off Georges Bank to off North Carolina; 843 to 1,850 fms.

1994 *Boreotrophon aculeatus lacunellus* (Dall, 1889). Off Cape Fear, North Carolina; Gulf of Mexico; West Indies. 227 to 769 fms.

1995 *Boreotrophon clavatus* (G. O. Sars, 1878). Off Cape Sable, Nova Scotia to off Cape Hatteras, North Carolina; 843 to 2,033 fms.

1996 *Boreotrophon craticulata* (Fabricius, 1780) Hudson Strait to Nova Scotia. 30 to 80 fms. (Synonyms: *fabricii* Möller, 1842; *borealis* Reeve, 1845.)

1997 *Boreotrophon maclaini* (Dall, 1902). Off Greenland.

1998 *Boreotrophon truncatus* Strøm, 1768. Europe. Greenland to Georges Bank; 10 to 50 fms. Point Barrow, Alaska, 11 to 72 fms.

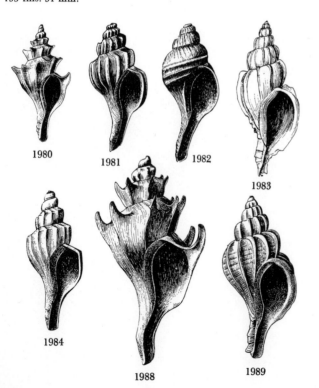

1980   1981   1982

1983

1984

1988   1989

1998

1999 *Boreotrophon verrillii* Bush, 1893. Off Cape Fear, North Carolina. 647 fms.

For Brazilian-Argentine species, see E. C. Rios, 1970, "Coastal Brazilian Seashells," p. 80, pl. 23.

Genus *Trophonopsis* Bucquoy, Dautzenberg and Dollfuss, 1882

Shells similar to *Boreotrophon* but with prominent spiral sculpture and secondary axial sculpture. Type: *muricatus* Montagu, 1803.

Trophonopsis lasius (Dall, 1919)   **2000**
Sandpaper Trophon

Bering Sea to Baja California.

1½ to 2 inches in length; spire ½ the length of the grayish white shell; 6 whorls with numerous, indistinct to moderately well-developed, rounded ribs and more numerous, small, frequently scaled, spiral cords—all of which gives the shell a rough, sandpaper feel. Whorls in spire shouldered slightly. Aperture enamel-white. Moderately common; 25 to 495 fathoms on gravel bottom. References in 1937 and earlier to *Trophon tenuisculptus* Carpenter are this species. The true *tenuisculptus* is an *Ocenebra*.

Other species:

2001 *Trophonopsis apolyonis* (Dall, 1919) Off Santa Barbara Islands, California, 216 to 339 fms.

2002 *Trophonopsis kamchatkanus* (Dall, 1902) Western Bering Sea, 100 fms. 24 mm.

2003 *Trophonopsis tripherus* (Dall, 1902) Washington to San Diego, California, 485 fms. 20 to 22 mm.

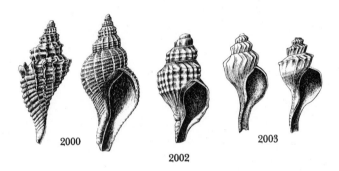

2004 *Trophonopsis tolomius* (Dall, 1919) San Miguel Island, California, 376 fms.

2005 *Trophonopsis keepi* (Strong and Hertlein, 1937) West end of San Nicolas Island, California, 30 to 50 fms. *Proc. Calif. Acad. Sci.,* 4th series, vol. 22.

Genus *Trophon* Montfort, 1810

Type of this genus is *geversianus* Pallas, 1774, from southern South America.

Subgenus *Austrotrophon* Dall, 1902

Type of this subgenus is *cerrosensis* Dall, 1891.

Trophon cerrosensis cerrosensis Dall, 1891   **2006**
Cerros Forreria

Off southern Baja California to Acapulco.

3 inches in length, with a rather long siphonal canal, and bearing 8 long axial ribs (or former varices) which are blade-like and are curled into long upswept, large spines. With or without small, low, spiral threads which may be more pronounced on the last whorl. Color yellowish to brownish white. Sometimes with subdued, wide brown lines. The form *pinnata* Dall, 1902 **(2007),** is a smaller, broader, 8 to 9 ribbed, spirally threaded variant from the same region. Uncommon from 21 to 74 fathoms. Do not confuse with species 1976.

Trophon cerrosensis catalinensis Oldroyd, 1927   **2008**
Catalina Forreria   **Color Plate 9**

Southern third of California.

2 to 3 inches in length, similar to the typical *cerrosensis,* but with 7 ribs, sturdier shell, less development of the blade-like ribs and often more brownish coloration. Formerly thought to be *Boreotrophon triangulatus* Carpenter. Moderately common offshore, sometimes cast ashore.

Subgenus *Zacatrophon* Hertlein and Strong, 1951

Trophon beebei Hertlein and Strong, 1948   **2009**
Beebe's Trophon

Gulf of California.

1 to 2 inches, lightweight, suture greatly tabulated, whorls somewhat loosely coiled. Color tan. Whorls with numerous small spines on the shoulder. Moderately common; 30 to 60 fathoms.

2009

Genus *Actinotrophon* Dall, 1902

Shell about 1 inch, with long fluted spines, wide umbilicus. Type: *Trophon actinophorus* Dall, 1889.

2010 *Actinotrophon actinophorus* Dall, 1889. West Indies, deep water.

2010

Subfamily *Typhinae* Cossmann, 1903

For reviews of this subfamily, see A. M. Keen, 1944, *Jour. Paleontol.,* vol. 18, and R. L. Gertman, 1969, *Tulane Studies in Geol. and Paleontol.,* vol. 7, pp. 143–191.

Genus *Typhis* Montfort, 1810

Small, white, with 4 spined varices per whorl, delicately sculptured shells with tiny, hollow tubes extending upward from the shoulder of the whorls. Aperture surrounded by a raised rim. Type: *tubifer* Bruguière, 1792. Synonyms: *Hirto-typhis* Jousseaume, 1880; *Monstrotyphis* Habe, 1961. No typical *Typhis* are known from the New World.

Subgenus *Rugotyphis* Vella, 1961

Has a large varical blade which is almost perpendicular (rather than parallel) to the axis of the shell. Tubes also perpendicular. Varices foliated or crenulated and midway between tubes. Type: *francescae* Finlay, 1924.

Atlantic species:

2011 *Typhis (Rugotyphis) puertoricensis* Warmke, 1964. Puerto Rico, 33 fms.; *The Nautilus*, vol. 78, p. 1.

2012 *Typhis (R.) cleryi* (Petit, 1840). Brazil (40 fms.) and Rio de Oro, Spanish Sahara. See Gertman, 1969, p. 154, fig. 2; and E. C. Rios, 1970, pl. 23.

Subgenus *Typhinellus* Jousseaume, 1880

Shell with 4 varices and tubes per whorl; tubes closer to the preceding than the succeeding varices, pointing away from the axis of the shell and away from the aperture. Usually with faint spiral sculpturing. Type: *sowerbii* Broderip, 1833.

Typhis sowerbii Broderip, 1833   **2013**
Sowerby's Typhis

West Florida to Texas and the Caribbean.
Mediterranean; also Pliocene of Italy.

20 mm., with 4 weakly crenulated, convex varices per whorl, and a spine at the top of each varix, pointing towards the apex. Tubes fairly long, closer to the preceding varices, pointing towards the apex. Anterior canal closed, narrow, curving sharply to the right and away from the aperture. Synonyms: *fistulatus* Risso, 1826 (non Schlotheim, 1820); *tetrapterus* Bronn, 1838; *syphonellus* Bonelli in Bellardi and Michelotti, 1841; *sowerbyi* H. and A. Adams, 1853; *fulva* Pallary, 1906; *minor* Pallary, 1906. Uncommon; 28 to 60 fathoms.

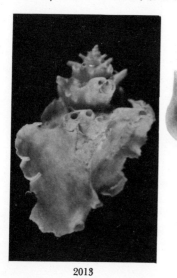

2013   2019

Other species:

2014 *Typhis (Talithyphis* Jousseaume, 1882) *expansus* Sowerby, 1874. Santo Domingo and Surinam to Brazil, 24 to 30 fms. See Gertman, 1969, p. 167. (Synonym: *melloleitaoi* Morretes, 1940.)

2015 *Typhis (Talityphis) latipennis* Dall, 1919. Gulf of California to Panama, in depths to 45 meters.

2016 *Typhis (Typhisopsis* Jousseaume, 1880) *clarki* Keen and Campbell, 1964. Northern end of the Gulf of California to Panama, intertidal to 16 meters.

2017 *Typhis (Typhisopsis) coronatus* Broderip, 1833. Magdalena Bay, Baja California, to Ecuador. (Synonyms: *quadratus* Hinds, 1843; *martyria* Dall, 1902.)

2018 *Typhis (Typhisopsis) grandis* A. Adams, 1855. Guaymas, Sonora, Mexico, 15 meters.

Genus *Siphonochelus* Jousseaume, 1880

With 4 varices and tubes per whorl. Tubes formed within the varices, pointing away from the aperture and away from the axis of the shell. Varix smooth and riblike, no spines or flanges. Spiral sculpture absent. Type: *arcuatus* Hinds, 1843. Synonyms: *Cyphonochelus* Jousseaume, 1882; *Trubatsa* Dall, 1889; *Choreotyphis* Iredale, 1936; *Eotyphis* Tembrock, 1963.

Siphonochelus longicornis (Dall, 1888)   **2019**
Long-horned Typhis

Florida Keys and the Caribbean.

1 inch, waxy white with pale rosy-brown hues between the varices. 4 varices per whorl. Tubes tapering, recurved, long (but often broken). Spiral sculpture of microscopic striae. Aperture small. Siphonal canal long and slender, nearly straight, with 3 former canals surrounding a chinklike umbilicus. Rare; 127 to 400 fathoms.

Other species:

2020 *Siphonochelus (Laevityphis* Cossmann, 1903) *bullisi* Gertman, 1969. Gulf of Darién, Atlantic Panama, 43 fms.; Pleistocene of Costa Rica. *Tulane Studies Geol. Paleontol.*, vol. 7, p. 178.

2021 *Siphonochelus tityrus* (F. M. Bayer, 1971). 60 meters, off Isla Margarita, Venezuela, to Trinidad. "Studies Tropical American Mollusks," p. 166.

Genus *Pterotyphis* Jousseaume, 1880

Shell with 3 varices and tubes per whorl. Siphonal canal with a narrow slit. Type: *pinnatus* (Broderip, 1833). *Trigonotyphis* Jousseaume, 1882, is a synonym.

Pterotyphis pinnatus (Broderip, 1833)   **2022**
Pinnate Typhis

Florida, Bahamas, Cuba to Panama.

15 to 20 mm., somewhat slender, yellowish buff, 6 whorls, suture deep. Spiral sculpture of fine raised threads. Tubes fairly short, bending upward and backward a little, and arising a short distance behind each varix, with which the base of the tube is connected by a low ridge. Anterior canal rather long and narrowly open in front. Uncommon; shallow water. *Typhis fordi* Pilsbry, 1943 (*The Nautilus*, vol. 57, p. 40), is a synonym.

A similar species **(2023)** *Pterotyphis fimbriatus* (A. Adams, 1854) from the Gulf of California is about 20 mm. in length, yellowish white with a blush of pink and brown on the body whorl and part of the lip varix.

Subgenus *Tripterotyphis* Pilsbry and Lowe, 1932

With 3 varices per whorl; tubes within the varices; shell sculptured. Type: *lowei* Pilsbry, 1931.

Atlantic species:

2024   *Pterotyphis* (*T.*) *triangularis* (A. Adams, 1856). Lower Florida Keys; Yucatan, Mexico; Bahamas to Panama. (Synonym: *cancellatus* Sowerby, 1841, non Gmelin, 1791.)

2022   2024

Pacific species:

2025   *Pterotyphis* (*T.*) *fayae* Keen and Campbell, 1964. West Mexico. *The Veliger*, vol. 7, p. 54. 20 mm.

2026   *Pterotyphis* (*T.*) *arcana* DuShane, 1969. West Mexico. 15 mm.

2027   *Pterotyphis* (*T.*) *lowei* Pilsbry, 1931. Gulf of California to Panama.

Family CORALLIOPHILIDAE Chenu, 1859

The family name Magilidae Thiele, 1925, is a synonym.

Genus *Coralliophila* H. and A. Adams, 1853

This group lacks a radula. They are usually associated with seafans and corals. The sturdy ovate shells are usually very variable in shape. Type: *violacea* Kiener, 1836, from the Indo-Pacific.

Coralliophila abbreviata (Lamarck, 1816)   **2028**
Short Coral-shell

Southeast Florida and the West Indies to Brazil. Bermuda.

¾ to 2 inches in length, solid, grayish white, rather misshapen, with rounded or squared shoulders, and with or without weak, rounded axial ridges. Spiral sculpture of crowded, variously sized cords which are made up of numerous microscopic scales. Aperture enamel-white, tinted with yellow, rounded above and constricted into a short siphonal canal below. Umbilicus small, shallow and funnel-shaped. Operculum chitinous, yellowish brown. A common species found living at the bases and on the branches of seafans and on live coral.

Coralliophila caribaea Abbott, 1958   **2029**
Caribbean Coral-shell

Off South Carolina to Florida and to Brazil. Bermuda.

2028   2029

½ to 1 inch, somewhat triangular in shape because of the angular shoulder of the body whorl and the rather flat-sided spire. Aperture elongate-triangular, usually white with a taint of purple within. Spiral ridges within the aperture usually absent. Operculum wine-red. It occurs commonly in nests under or in the holdfasts of seafans. This is *Coralliophila plicata* of authors, not Wood, 1818.

Other species:

2030   *Coralliophila aberrans* (C. B. Adams, 1850). Bermuda (R. Jensen); Massachusetts to the Gulf of Mexico (90 fms.) and the Caribbean. Brazil, 70 fms. to shore. (Synonyms: *Trophon lintoni* Verrill, 1882; *bracteata* of authors, non Brocchi, 1814; *profundicola* Haas, 1949; *Fusus lamellosus* Philippi, 1836, non Borson, 1821.)

2031   *Coralliophila lactuca* Dall, 1889. 352 fms., off Fernandina, Florida, to the Gulf of Mexico, 152 to 220 fms.; northwest of Little Bahama Bank, in 549 meters.

2030   2031

2032   *Coralliophila scalariformis* (Lamarck, 1822). Florida; Gulf of Mexico, to the Virgin Islands.

2033   *Coralliophila macleani* Shasky, 1970. Gulf of California, on the bases of gorgonians, from 2 to 10 meters.

2034   *Coralliophila* (*Pseudomurex*) *nux* (Reeve, 1846). Baja California to Ecuador. Intertidal, common.

2035   *Coralliophila* (*Pseudomurex*) *orcuttiana* Dall, 1919. Magdalena Bay to the Gulf of Tehuantepec, Mexico. (Synonyms: *Ocenebra sloati* and *hambachi* Hertlein, 1958.)

2036   *Coralliophila* (*Pseudomurex*) *parva* (E. A. Smith, 1877). Gulf of California to the Galapagos Islands.

—   [*Coralliophila incompta* Berry, 1960, from the Gulf of California, has been placed in *Attiliosa*, a subgenus of *Calotrophon*. See Emerson, 1968, *The Veliger*, vol. 10, p. 380.]

2037   *Coralliophila fax* F. M. Bayer, 1971. Off northwest end of Grand Bahama, in 494 meters. "Studies Tropical American Mollusks," p. 191.

Genus *Latiaxis* Swainson, 1840

Type of the genus is *mawae* Griffith and Pidgeon, 1854, of Japan.

Subgenus *Babelomurex* Coen, 1922

Shells bearing small or large upturned spines on the shoulder. Type: *babelis* (Requien, 1849) of the Mediterranean.

Latiaxis dalli Emerson and D'Attilio, 1963   **2038**
Dall's Latiaxis

Gulf of Mexico and the Caribbean. Brazil.

1 to 1½ inches, white; 8 whorls, with a sharply angular shoulder, bear about 10 broadly triangular spines, below which are 14 to 18 scaly ribs. Umbilicus wide and funnel-shaped. Operculum ovate, chitinous, with a lateral nucleus. Uncommon from 28 to 878 fathoms. Formerly called *deburghiae* (Reeve, 1857) which, however, is from Japan (see Emerson, 1963, *Amer. Mus. Novitates*, no. 2149).

2038

Subgenus *Pseudomurex* Monterosato, 1872

Sculpture reticulate, axial ribbing fairly strong, spiral sculpture scaly; outer lip with a rough outer margin; siphonal canal small. Type: *bracteata* (Brocchi, 1814).

Latiaxis costatus (Blainville, 1832)   **2039**
California Latiaxis

Point Conception, California, to Panama.

1 to 1¾ inches in length, variable in shape and the development of frills and spines. Deep-water forms (called *hindsi* Carpenter) bear triangular, flattened, upturned spines on the periphery of the whorl. Spiral cords are strongly scaled. Color light-gray with an enamel-white aperture. Moderately common offshore. A choice collector's item. This is probably *oldroydi* (Ida Oldroyd, 1929). *Purpura foveolata* C. B. Adams, 1852, is a synonym.

2039

Other species:

2040 *Latiaxis* (*Babelomurex*) *mansfieldi* (McGinty, 1940). Off Florida and the Gulf of Mexico to Panama. Brazil. 75 to 100 fms. Emerson, 1963, p. 6, fig. 4.

2041 *Latiaxis* (*Babelomurex*) *kincaidi* Dall, 1919. Puget Sound, Washington.

2040

2042 *Latiaxis* (*Tarantellaxis* Habe, 1970) *juliae* Clench and Aguayo, 1939. Barbados and off Aruba, 75 to 127 fms. 28 mm. *Mem. Soc. Cubana Nat. Hist.*, vol. 13, p. 194.

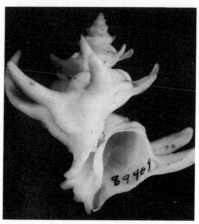

2042

Genus *Quoyula* Iredale, 1912

Aperture very large; anterior canal absent; lives attached to live coral branches. One species in the Eastern Pacific, none in the Western Atlantic. Type: *madreporarum* (Sowerby, 1834).

2043 *Quoyula madreporarum* (Sowerby, 1834). White externally; purplish pink within. 10 to 20 mm. Attaches to coral branches. Southern Gulf of California to Panama; Indo-Pacific.

Superfamily Buccinacea Rafinesque, 1815

Family COLUMBELLIDAE Swainson, 1840

Small shells, usually fusiform, solid. Aperture ovate to narrow. Parietal wall usually smooth; outer lips strong, often bearing denticles. No umbilicus. Nuclear whorls small, smooth, siphonal canal short, straight, but the fleshy siphon very long. Foot very narrow. Tentacles long and slender. Animal active. Operculum small, chitinous, smooth, sickle-shaped to oblong. Each radular row has 3 teeth—a flat, rectangular, noncuspid median tooth, flanked on either side by a bi- or tri-cuspid lateral tooth. Numerous genera. Also called the Pyrenidae. Carnivorous; feed on bivalves and crustaceans.

Genus *Columbella* Lamarck, 1799

Shells ½ inch, square-shouldered, spirally corded, with a thick, denticulated outer lip. Periostracum velvetlike. Type: *mercatoria* (Linné, 1758).

2044   2051   2049   2048

Columbella mercatoria (Linné 1758)   2044
Common Dove-shell

Northeast Florida and the West Indies to Brazil. Bermuda.

½ to ¾ inch in length, solid, squat, highly colored with white and brown, interrupted spiral bars over yellow, pink or orange background. Sometimes only maculated with one color (orange, brown or yellow). Outer lip thick, bearing about a dozen white teeth. Periostracum grayish, velvety. A common shallow-water species found under rocks. Not found on the west coast of Florida. Synonyms include *somersiana* Dall and Bartsch, 1911; *elongata* Usticke, 1959; *dysoni* Reeve, 1859; *variabilis* Schumacher, 1817; *rudis* Sowerby, 1844; and *fustigata* Kiener, 1841.

Columbella rusticoides Heilprin, 1887   2045
Rusty Dove-shell

South half of Florida to Key West; northwest Cuba.

Similar to *C. mercatoria*, but much more slender, smooth on the center of the body whorl, and with mauve-brown marks between the apertural teeth. Also more faintly colored and lacking spiral bars or lines of brown. Common down to 20 feet.

Other Pacific species:

2046 *Columbella aureomexicana* (Howard, 1963). Gulf of California to lower Mexico. (Synonym: *aureola* Howard, 1963, not Duclos in Chenu, 1846.)

Genus Anachis H. and A. Adams, 1853

Shell fusiform, with axial ribs, narrow aperture, with teeth on the inside of the thickened outer lip. Operculum elliptical, concentric, with an eccentric nucleus. Type: *scalarina* (Sowerby, 1832).

Subgenus Costoanachis Sacco, 1890

Shell fusiform, with prominent axial ribs on the body whorl and with spiral scratches between the ribs. Columella weakly denticulate. Type: *turrita* Sacco, 1890. *Glyptanachis* Pilsbry and Lowe, 1932, is a synonym.

Anachis sparsa (Reeve, 1859)   2047
Sparse Dove-shell

Southeast Florida; Caribbean to Brazil; Bermuda.

9 to 12 mm., broadly fusiform; spire high and slightly convex. Suture distinct. Axial ribs thin, rounded, strong, and with spiral incised lines between them on the base of the shell. Outer lip denticulate. Color ivory with scattered squares and flecks of orange-brown. Sandy bottoms; 13 to 30 fathoms.

Anachis lafresnayi (Fischer and Bernardi, 1856)   2048
Well-ribbed Dove-shell

Maine to east Florida; Yucatan.

Similar to *avara*, but with the aperture less than ½ the length of the shell, with flatter-sided whorls bearing about 15 (11 to 20) straight axial ribs which are crossed by spiral striae, the uppermost being the strongest. On the last whorl the ribs extend from the suture above to just below the slightly angulated periphery. Base of shell with spiral striae sometimes producing weak beading. Outer lip, when thickened, bears about 6 (3 to 8) indistinct, tiny denticles. Color dull-yellow to brownish, sometimes with spiral streaks of white. White posterior end of foot is small and blends into the mottlings of the rest of the foot. Abundant on shelly bottoms, rocks and pilings from low tide to 48 fathoms. (See A. H. Scheltema, 1968, *Breviora*, no. 304). Formerly *translirata* (Ravenel, 1861) according to G. Radwin *in thesis* 1970.

Anachis avara (Say, 1822)   2049
Greedy Dove-shell

Massachusetts Bay to east Florida and Texas.

10 to 22 mm. in length, moderately elongate, with the aperture about ½ the length of the entire shell, and with about a dozen, smooth, axial plications on the upper ½ of each whorl. Spiral, incised lines very weak or absent, but strong at the base. Aperture narrow, a little less than ½ the length of the shell. Weak, smooth varix present. The 3 nuclear whorls are smooth and translucent-white. Next few whorls with numerous axial riblets. Last whorl with 7 to 21 ribs (10 to 21 north of Cape Hatteras; 7 to 14 heavy ones south to east Florida). Color yellowish brown to dark gray-brown over which may be seen a faint pattern of irregular, white, large dots. Sometimes with dark-brown specklings. 4 to 12 (average 8) weak teeth inside inner lip. The top of the posterior end of the foot is white and sharply marked off from the rest of the mottled foot. A very common low-tide, eel-grass species from shore to 25 fathoms. *C. cleta* Duclos, 1846, is a synonym.

The subspecies *brasiliana* (von Martens, 1897) occurs from Brazil to Uruguay from shore to 10 fathoms (**2050**).

Anachis catenata (Sowerby, 1844)   2051
Chain Dove-shell

Bermuda, southeast Florida and the West Indies to Brazil.

¼ inch (6 mm.), white to yellow with 2 rows of large, squarish brown dots, one just below the suture, the other just below the periphery on the base of the shell. Axial ribs 16 to 19 strong. Spiral striae faint. Outer lip thickened, slightly notched above, toothed within. Columella with feeble pustules. Protoconch of 2 swollen, white whorls. Moderately common in shallow water in the West Indies.

Anachis floridana Rehder, 1939   2052
Florida Dove-shell

North Carolina to Florida and Texas.

2045   2053   2052

6 to 12 mm., resembling *avara* but the whorls in the spire are flat, although the spire is slightly convex-sided. Aperture less than ½ the length of the shell. Axial ribbing distinct on the body whorl only. Suture not indented strongly. Spiral sculpture absent above the base. Egg capsule with a wide, scalloped collar around the top. In *avara,* it is a simple volcano-shaped cone. In *lafresnayi,* it has circular wrinkles around the sides of the "volcano."

Anachis semiplicata Stearns, 1873   2053
Semiplicate Dove-shell

West Florida to Yucatan, Mexico.

½ inch, slender, fusiform (average width to length ratio is 0.36), spire flat-sided and acute. Last whorl with about 12 ribs on the upper ½. Spiral sculpture absent or very reduced, except on the base where there are a few spiral threads. Aperture narrow. Color gray, sometimes finely reticulated with browns. Abundant in shallow water in weeds. May be a subspecies of *lafresnayi. A. similis* Ravenel, 1861, is considered a dubious species.

Subgenus *Zafrona* Iredale, 1916

Type of the subgenus is *Colombella isomella* Duclos, 1840.

Anachis pulchella (Blainville, 1829)   2054
Beautiful Dove-shell

Florida Keys and the Caribbean to Brazil; Bermuda.

¼ to ⅓ inch (8 to 10 mm.), with numerous, weak axial riblets crossed by numerous spiral striae. Color yellowish white, variegated with light- to dark-brown. Columella callus raised into a little lip at its left edge, faintly nodulous. Moderately common in the West Indies. This is *subcostulata* (C. B. Adams, 1845), and *dicomata* Dall, 1889, according to Radwin *in thesis. A. pulchella* Sowerby, 1844, is the same.

2054

Subgenus *Suturoglypta* Radwin, 1968

Fusiform, with strong square-cut axial ribs, and squarely incised sutures. Type: *pretrii* Duclos, 1846.

Anachis hotessieriana (Orbigny, 1842)   2055
Hotessier's Dove-shell

South Florida and the West Indies.

4 to 6 mm., slender, 6 strongly plicate whorls, high spire, and gray to white with light-brown mottlings. Apical whorl smooth, brownish, later whorls glistening-smooth, with about 14 strong axial ribs per whorl with interstices of the same size. On the last whorl, the ribs disappear just below the

periphery. Varix swollen. Spiral sculpture strong or weak. Outer lip with brown-bordered posterior notch. 4 to 6 denticles within. Columella microscopically pimpled. The following may be synonyms, but Radwin (1968) believes them distinct: *pretrii* Duclos, 1846; *albella* C. B. Adams, 1850; *iontha* Ravenel, 1861; *acuta* Stearns, 1873; *samanensis* Dall, 1889; and *hotessieri* Orbigny, 1845. May belong to *Nassarina* Dall, 1889.

Subgenus *Parvanachis* Radwin, 1968

Small, stout, *Anachis*-like with inflated, ribbed whorls and thick apertural lips. Type: *obesa* (C. B. Adams, 1845).

Anachis obesa (C. B. Adams, 1845)   2056
Fat Dove-shell

Virginia to Florida, to Texas and to Uruguay. Bermuda.

3/16 to ¼ inch in length, moderately wide, dull-grayish with 1 or 2 subdued, spiral brown bands in some specimens. Small, sharp axial ribs are numerous; spiral incised lines numerous, not crossing ribs. There is a fairly strong, occasionally knobbed, spiral cord immediately below the suture. Varix large, smooth and rounded. Body whorl behind it usually smoothish. Parietal shield faintly developed, but with a sharp edge. Inner wall of outer lip with about 3 to 5 small teeth. The form *ostreicola* Melvill, 1881, from northwest Florida is dark-brown and with stronger spiral threads or wider incised lines. This is a common shallow-water species. Other synonyms: *cancellata* (Gaskoin, 1851); *ornata* (Ravenel, 1858); *sertularium* (Orbigny, 1841); *hermosa* von Ihering.

2055   2056   2060

Other Atlantic species:

2057 *Anachis (Parvanachis) rhodae* Radwin, 1968. Puerto Plata, Dominican Republic. *Proc. Biol. Soc. Wash.,* vol. 81, p. 147.

2058 *Anachis veleda* (Duclos, 1846). Brazilian coast. See E. C. Rios, 1970, pl. 25.

2059 *Anachis (Costoanachis) fenneli* Radwin, 1968. Rio de Janeiro, Brazil. *Proc. Biol. Soc. Wash.,* vol. 81, p. 148.

Anachis penicillata Carpenter, 1865   2060
Penciled Dove-shell

San Pedro, California, to Baja California.

⅕ inch in length, rather slender, with 6 whorls, of which the first 2 nuclear ones are smooth, the remainder with about 15 strong axial riblets per whorl, which are made slightly uneven by numerous, very fine spiral threads. Color translucent-cream with sparse spottings of brown. Common under rocks between tide marks.

Other Pacific species:

2061 *Anachis subturrita* Carpenter, 1866. San Pedro, California, to the Tres Marias Islands, Mexico.

2062 *Anachis adelinae* (Tryon, 1883). Baja California, the Gulf, to Sonoran coast of Mexico. 15 mm.

2063 *Anachis coronata* (Sowerby, 1832). Baja California, the Gulf, to Panama (subspecies *hannana* Hertlein and Strong, 1951). Common.

2064 *Anachis (Parvanachis) diminuta* (C. B. Adams, 1852). Baja California, near Pt. Abreojos, to Panama. Common.

2065 *Anachis fayae* Keen, 1971. Guaymas to Mazatlan, Mexico.

2066 *Anachis (Parvanachis) gaskoini* Carpenter, 1857. Southern Baja California to Manzanillo, Mexico. (Synonym: *bartschii* Dall, 1918.)

2067 *Anachis nigricans* (Sowerby, 1844). Gulf of California to Panama; Galapagos. Under rocks between tides; not common.

2068 *Anachis (Parvanachis) pygmaea* (Sowerby, 1832). Baja California, the Gulf, to Ecuador. Not uncommon. (Synonyms: *elegantula* Mörch, 1860, and *deshayesii* De Folin, 1867.)

2069 *Anachis sanfelipensis* Lowe, 1935. Northern end of the Gulf of California; intertidal; not common.

2070 *Anachis vexillum* (Reeve, 1858). Northern end of the Gulf of California to Acapulco, Mexico; intertidal; common.

2071 *Anachis (Glyptanachis* Pilsbry and Lowe, 1932) *hilli* Pilsbry and Lowe, 1932. Northern end of the Gulf of California to Nicaragua. Type of the subgenus.

2072 *Anachis (Bifurcium* Fischer, 1884) *bicanaliferum* (Sowerby, 1832). Gulf of California to Ecuador. (Synonym: *bicanaliculata* Duclos, 1835–40.)

Genus *Cosmioconcha* Dall, 1913

Shell delicate, smoothish, with spiral striae on the base. Usually with a spiral identation well below the suture. Outer lip may be thickened. Type: *Buccinum modestum* Powys, 1835. Four species from the tropical Eastern Pacific.

***Cosmioconcha calliglypta* (Dall and Simpson, 1901)   2073**
Flamed Dove-shell

Key West, Florida; Texas; Puerto Rico.

8 to 13 mm., with a high slender spire; columella smooth; outer lip denticulate. Slight anal sinus present. Weak, numerous axial riblets on early whorls. Color whitish to pinkish or yellow and with 2 rows of zigzag brown flammules. Uncommon; dredged offshore below 25 fathoms.

Other species:

2074 *Cosmioconcha nitens* (C. B. Adams, 1850). Cuba to Puerto Rico; Central America. (Synonym: *perpicta* Dall and Simpson, 1901.)

2074

2075 *Cosmioconcha palmeri* (Dall, 1913). Head of the Gulf of California to Acapulco, Mexico, to 24 fms. 18 mm.

2076 *Cosmioconcha parvula* (Dall, 1912). Off La Paz, Baja California. 112 fms. 15 mm.

2077 *Cosmioconcha pergracilis* (Dall, 1913). Gulf of California, 58 fms. 24 mm.

2078 *Cosmioconcha modesta* (Powys, 1835). El Salvador to Ecuador.

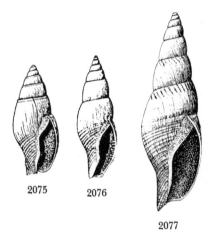
2075   2076   2077

Genus *Pyrene* Röding, 1798

Shell ovate, somewhat turnip-shaped, smoothish, aperture narrow. Type: *discors* (Gmelin, 1791) from the Indo-Pacific.

Subgenus *Conella* Swainson, 1840

Shell biconic, elongate-ovoid; columella smooth; apertural lip denticulate. Type: *picata* Swainson, 1840, which is *ovulata* Lamarck, 1822.

***Pyrene ovulata* (Lamarck, 1822)   2079**
Ovate Dove-shell

Florida Strait to Barbados; Bahamas.

12 to 20 mm., solid, ovoid with a rounded shoulder; spire concave with a papillate nucleus of 3 or 4 smoothish, milk-white whorls. Spiral sculpture of minute cords, except on the smooth periphery of the whorls. Color black to brown with sparse white splotches. Periostracum thick, axially laminate. Uncommon in the West Indies. Not recorded from the Florida Keys, as yet. Synonym: *picata* Swainson, 1840.

***Pyrene ovuloides* (C. B. Adams, 1850)   2080**
C. B. Adams' Dove-shell

Bahamas and Caribbean.

2079   2080

12 to 18 mm., similar to *ovulata*, but usually larger, always much more slender, and with a higher, concave spire. Color orange-brown with irregular blotches of white. Uncommon.

Other species:

2081 *Pyrene* (*Minipyrene* Coomans, 1967) *dormitor* (Sowerby, 1844). Southern Caribbean. See *Beaufortia*, vol. 14, no. 168, 1967.

2082 *Pyrene fuscata* (Sowerby, 1832). Magdalena Bay; the Gulf; to Peru. Common. (Synonyms: *gibbosa* Valenciennes, 1832; *meleagris* and *nodalina* Kiener, 1840; *pallescens* Wimmer, 1880.)

2083 *Pyrene haemastoma* (Sowerby, 1832). Magdalena Bay; the southern Gulf; to Ecuador; Galapagos Islands.

2084 *Pyrene major* (Sowerby, 1832). Southern Gulf of California to Peru. Common.

2085 *Pyrene strombiformis* (Lamarck, 1822). Gulf of California to Peru. Not uncommon. (Synonym: *bridgesii* Reeve, 1858.)

Genus *Parametaria* Dall, 1919

Shells conical with a rounded shoulder. They somewhat resemble cones of immature *Strombus*. Outer lip smooth; upper whorls rounded. Spiral threads on the base. Type of the genus is *dupontii* (Kiener, 1849). *Meta* Reeve, 1859, not Koch, 1835, is a synonym.

Parametaria dupontii (Kiener, 1849)   2086
Dupont's Dove-shell

Gulf of California to Tres Marias Islands, Mexico.

½ to 1 inch, resembling a smooth-shouldered cone. Shell thin, white, with numerous, irregular blotches of yellow and brown. Moderately common in shallow water.

Genus *Strombina* Mörch, 1852

Small, sturdy columbellids resembling miniature *Strombus* in that the spire is turreted and the aperture is surrounded by a thickened outer lip and parietal shield. There is only one species living in the lower Caribbean, but the eastern Pacific has about 24 recorded ones, 14 of which occur in the Gulf of California from depths of 1 to 60 fathoms. Type of the genus is *lanceolata* (Sowerby, 1832) from Ecuador.

2087 *Strombina pumilio* (Reeve, 1859). Venezuela and Isla Margarita.

2088 *Strombina angularis* (Sowerby, 1832). Gulf of California to Panama. (Synonym: *subangularis* Lowe, 1935.)

2089 *Strombina bonita* Strong and Hertlein, 1937. Gulf of California.

2090 *Strombina carmencita* Lowe, 1935. Gulf of California to Panama.

2091 *Strombina colpoica* Dall, 1916. Gulf of California to Panama.

2092 *Strombina dorsata* (Sowerby, 1832). Upper Gulf of California to Ecuador.

2093 *Strombina fusinoidea* Dall, 1916. Baja California to Panama. (Synonym: *fusiformis* Hinds, 1844, not Anton, 1839.)

2094 *Strombina gibberula* (Sowerby, 1832). Baja California and the Gulf to Peru.

2095 *Strombina hirundo* (Gaskoin, 1852). Baja California to Ecuador.

2096 *Strombina lilacina* Dall, 1916. Gulf of California.

2097 *Strombina maculosa* (Sowerby, 1832). Gulf of California to Panama.

2098 *Strombina marksi* Hertlein and Strong, 1951. Gulf of California.

2099 *Strombina paceana* Dall, 1916. Baja California and the Gulf.

2100 *Strombina recurva* (Sowerby, 1832). Baja California to Peru. (Synonym: *Drillia limonetta* Li, 1930.)

2101 *Strombina solidula* (Reeve, 1859). Southern Gulf of California.

Genus *Nitidella* Swainson, 1840

Shell fusiform, glossy, smooth. Type: *nitida* (Lamarck, 1822).

Nitidella nitida (Lamarck, 1822)   2102
Glossy Dove-shell

Southeast Florida and the West Indies to Brazil. Bermuda (rare).

½ inch in length, characterized by the long aperture (¾ that of the entire shell) and by the very glossy shell. Color whitish with heavy mottlings of light-yellow to mauve-brown. Outer lip with about 7 small teeth. Base of columella with 2 small, spiral, white plicae within. Common in the West Indies on rocks at low tide. Synonyms are: *nitidula* Sowerby, 1822, and *gracilis* Dillwyn, 1823.

Nitidella gouldi (Carpenter, 1857)   2103
Gould's Dove-shell

Alaska to San Diego, California.

½ inch in length, 7 whorls are smoothish and slightly convex. Spire almost flat-sided. Base of shell on exterior of canal with about 9 fine, incised spiral lines. Bottom of white columella with a single, low plait. Outer lip simple, sharp and often reinforced within by 4 or 5 weak pustules. Shell whitish with faint brown maculations, covered with a yellowish gray periostracum. Fairly common from just offshore to 300 fathoms.

Subgenus *Alia* H. and A. Adams, 1853

Type of this subgenus is *carinata* (Hinds, 1844).

Nitidella carinata (Hinds, 1844)   2104
Carinate Dove-shell

San Francisco, California, to Baja California.

⅓ inch in length, glossy, brightly variegated with orange, yellow, white and brown. Shoulder of last whorl usually

2102   2103   2104   2106

strongly swollen. Exterior of canal with about a dozen spiral, incised lines. Both ends of the aperture are stained dark-brown. Outer lip thickened, crooked, and with about a dozen small spiral threads or teeth inside. Fairly common in shallow water.

(2105) *N. gausapata* Gould, 1850 (California to Alaska), is similar, but without the swollen shoulder. *N. californiana* (Gaskoin, 1852) is a synonym. J. M. Crane, Jr., reports that the amphipod crustacean, *Pleustes*, lives near and closely resembles *carinata* in shape and coloration. (*The Veliger,* vol. 12, pl. 36, 1969).

Subgenus *Rhombinella* Radwin, 1968

Shell smooth, glossy with a thin, nondenticulate outer lip. Type: *laevigata* (Linné, 1758).

Nitidella laevigata (Linné, 1758) 2106
Smooth Dove-shell

Florida Keys and the West Indies. Bermuda.

½ to ¾ inch, glossy, smooth, whorls slightly shouldered. Outer lip smooth. Shell yellowish with axial, zigzag red, brown and white flames. Body whorl usually with 2 rows of brown spots, one just below the suture. Periostracum dull and buff-colored. Fairly common in shallow water. The subspecies *hendersoni* Dall, 1908, (2107) from Cuba has a higher, more acute spire.

Other Atlantic species:

2108 *Nitidella moleculina* Duclos, 1840. Florida Keys.

2109 *Nitidella parva* Dunker, 1847. Gulf of Mexico and West Indies.

Genus *Mitrella* Risso, 1826

Shell small, fusiform. Aperture less than ½ the length of the entire shell. Outer lip slightly sinuate. Sculpture smoothish. Type: *scripta* Linné, 1758.

Mitrella ocellata (Gmelin, 1791) 2110
White-spotted Dove-shell

Southeast Florida and the West Indies to Brazil. Bermuda. Baja California to Panama.

½ inch in length, smooth, characteristically dark black-brown with numerous small white dots which may be quite large just below the suture. Outer lip thick, with 5 or 6 small whitish teeth. Aperture short, narrow, purplish brown within. When beachworn, the color is reddish or yellowish brown. Common under rocks at low tide. Formerly known as *cribraria* Lamarck, 1822.

The west Mexico counterpart of this species is *guttata* Sowerby, 1832 (2111) differing only in having a lighter band just below the periphery. A subspecies was described from the Galapagos Islands, *baileyi* Bartsch and Rehder, 1939.

Mitrella argus Orbigny, 1842 2112
Argus Dove-shell

Southeast Florida and the West Indies.

¼ to ⅓ inch, fusiform, glossy, similar to *ocellata*, but instead of white polka dots, it has irregular axial streaks and diffused flames of white on a dark-brown background. Inside of outer lip with 7 to 9 small teeth. Moderately common; under stones at low-tide mark. Synonyms: *dichroa* Sowerby, 1844; *parvula* Dunker, 1847; *orphia* Duclos, 1846; *schrammi* Petit, 1853; *elegans* Dall, 1871.

Subgenus *Astyris* H. and A. Adams, 1853

Generally smooth except for spiral striae on base. Type: *rosacea* (Gould, 1840).

Mitrella rosacea (Gould, 1841) 2113
Rosy Northern Dove-shell

Greenland to New Jersey.
Alaska; Northern Europe.

6 to 7 mm., elongate, acutely conic, dingy-white to rose, with a rose-red apex. Smoothish, except for microscopic spiral lines. Suture faintly impressed. Aperture ⅖ the length of the shell and narrow-ovate. Outer lip sharp, slightly reflected, smooth within. Columella arching, slightly flattened, smooth and white. *M. holboellii* Möller, 1842 is a synonym. Common; 3 to 60 fathoms.

Mitrella lunata (Say, 1826) 2114
Lunar Dove-shell

Massachusetts to Florida, Texas and to Brazil. Bermuda.

³⁄₁₆ to ¼ inch in length, glossy, smooth, translucent-gray and marked with fine, axial zigzag brown to yellow stripes. Base of shell with fine, incised spiral lines. Aperture constricted, slightly sinuate. Outer lip with 4 small teeth on the inside. No prominent varix. Nuclear whorls very small and translucent. Color sometimes milky-white or mottled in brown. A very common species found at low tide. Probable synonyms: *duclosiana* Orbigny, 1842; *spirantha* Ravenel, 1861, the latter being a dark, stout, and stumpy form from the Carolina coast. Also *gouldiana* Stimpson, 1851; *wheatleyi* DeKay, 1843; *dissimilis* Stimpson, 1851; and possibly *zonalis* Gould, 1848.

Mitrella multilineata (Dall, 1889) 2115
Brown-banded Dove-shell

Chesapeake Bay to Florida Keys and the Gulf of Mexico.

⅓ inch, similar to *lunata*, but more slender and acute, the whorls less rounded, with a flat-sided spire, and a slightly angulated periphery. Body whorl with 5 or 6 pale-brown, narrow, even, spiral lines on a straw background. 4 or 5 den-

2110   2112   2114

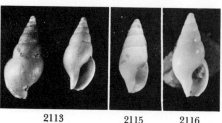

2113   2115   2116

ticles on inside of outer lip, the lowest being the largest and opposite a toothlike callus on the base of the columella. Dredged from 9 to 230 fathoms. Uncommon.

Mitrella raveneli (Dall, 1889)   **2116**
Ravenel's Dove-shell

North Carolina to both sides of Florida.

⅜ inch in length, resembling *lunata,* but translucent-whitish, without the mottlings, normally a slightly larger shell, more elongate, with a longer siphonal canal, and with a rather thin outer lip as in *lunata.* Commonly dredged from 5 to 90 (rarely to 200) fathoms. Rarely washed ashore.

Mitrella tuberosa (Carpenter, 1865)   **2117**
Variegated Dove-shell

Alaska to the Gulf of California.

¼ inch in length, slender, with a narrow, pointed, flat-sided spire. Shell smooth and usually glossy. Outer lip slightly thickened and with small teeth within. Color translucent-yellowish tan with opaque, light-brown flammules and maculations. Sometimes all-brown with tiny white dots. Early whorls in worn specimens have a lilac tinge. Periostracum thin and translucent. Common in shallow water; 7 to 30 fathoms. *M. variegata* Stearns may be this species.

Subgenus *Columbellopsis* Bucquoy, Dautzenberg and Dollfuss, 1882

Shell fusiform with a flat-sided spire. Type: *minor* (Scacchi 1836) from the Mediterranean. Synonym: *Tetrastomella* Bellardi, 1889.

Mitrella nycteis (Duclos, 1846)   **2118**
Fenestrate Dove-shell

South Florida to the Lesser Antilles.

6 to 8 mm., very slender, glossy-smooth. Its pointed, flat-sided, smooth spire is about ⅔ the entire length of the solid, thick shell. Varix large and swollen, inside with 1 strong and 4 or 5 weaker teeth. Columella with 4 or 5 tiny nodules caused by the 6 or 7 spiral cords on the base of the shell. Periphery of whorl slightly angular. Color white to grayish with large, irregular patches of light-brown above the periphery of the whorl, and with sparse flames of tan on the base. Uncommon; 1 to 3 fathoms in coral sand and eel-grass. Synonym: *fusiformis* Orbigny, 1842, non Anton, 1839, and *fenestrata* (C. B. Adams, 1850).

Other Atlantic species:

2119 *Mitrella (Astyris) diaphana* (Verrill, 1882). Martha's Vineyard, Massachusetts, to the Gulf of Mexico, 65 to 487 fms.

2117   2118   2121

2120 *Mitrella (Astyris) pura* (Verrill, 1882). Martha's Vineyard, Massachusetts, to Cape Florida, 70 to 487 fms.

2121 *Mitrella (Astyris) profundi* (Dall, 1889). Off North Carolina to Cuba, 34 to 805 fms. Gulf of Mexico, 100 to 220 fms.

2122 *Mitrella perlucida* (Dall, 1927). Off Fernandina, Florida, 294 fms.

2123 *Mitrella stemma* (Dall, 1927). Off Fernandina, Florida, 294 fms.

2124 *Mitrella (Fluella* Dall, 1924) *amphisella* (Dall, 1881). Off Georgia, 440 fms; West Indies, 413 fms. (Variety *rushii* (Dall, 1889) **(2125)** Off Georgia, 440 fms; off Fowey Rocks, Florida, 465 fms.)

2126 *Mitrella (Fluella) appressa* (Dall, 1927). Off Georgia, 440 fms.

2127 *Mitrella (Fluella) enida* (Dall, 1927). Off Georgia, 440 fms.; off Fernandina, Florida, 294 fms.

2128 *Mitrella (Fluella) vidua* (Dall, 1924). Off Fernandina, Florida, 294 fms. Type of the subgenus.

2129 *Mitrella (Plectaria* Dall, 1924) *crumena* (Dall, 1924). Off Fernandina, Florida, 294 fms.

2130 *Mitrella embusa* (Dall, 1927). Off Georgia, 440 fms.

2131 *Mitrella euribia* (Dall, 1927). Off Fernandina, Florida, 294 fms.

2132 *Mitrella projecta* (Dall, 1927). Off Fernandina, Florida, 294 fms.

2133 *Mitrella verrillii* (Dall, 1881). Off Fernandina, Florida; Florida Keys; West Indies, 310 to 805 fms. (Synonyms: *strix* Watson, 1882 and 1885; *subacta* Watson, 1882.)

2125   2133

2134 *Mitrella (Parasagena* Dall, 1924) *georgiana* (Dall, 1924). Off Georgia, 440 fms.

2135 *Mitrella (Parasagena) sagenata* (Dall, 1927). Off Georgia, 440 fms.

Other Pacific species:

2136 *Mitrella callimorpha* (Dall, 1919). San Diego, California, to Magdalena Bay, Baja California.

2137 *Mitrella casciana* (Dall, 1919). Off La Jolla, California, 110 to 199 fms.

2138 *Mitrella (Astyris) clementensis* Bartsch, 1927. San Clemente and Catalina Island, California, on rocks.

2139 *Mitrella aurantiaca* (Dall, 1871). Monterey, California, to Baja California.

2140 *Mitrella amiantis* (Dall, 1919). Kyska Harbor, Aleutian Islands, Alaska.

2141 *Mitrella permodesta* (Dall, 1890). Aleutians to off San Diego, California. 7 mm.

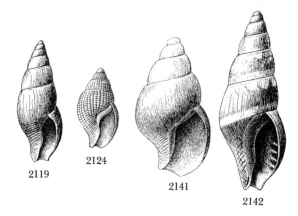

2119   2124   2141   2142

2142 *Mitrella hypodra* (Dall, 1916). Puget Sound, Washington, to Puerto Libertad, Mexico. 6.5 mm.

2143 *Mitrella lutulenta* (Dall, 1919). Washington to Coronado Islands (emended *luculenta* Dall, 1921).

2144 *Mitrella densilineata* (Carpenter, 1864). Gulf of California. Rare.

2145 *Mitrella granti* Lowe, 1935. San Felipe, Gulf of California.

2146 *Mitrella baccata* (Gaskoin, 1852). Magdalena Bay to Nicaragua. (Synonyms: *cervinetta* and *obsoleta* Carpenter, 1857.)

2147 *Mitrella caulerpae* Keen, 1971. Southern Baja California.

2148 *Mitrella dorma* Baker, Hanna and Strong, 1938. Gulf of California.

Genus *Nassarina* Dall, 1889

Shells small, ¼ inch, slender. Nucleus inflated, glassy white, smooth 1½ whorls. With axial noduled ribs. Anterior siphonal canal constricted; aperture narrow, interior of outer lip denticulate. Type: *glypta* Bush, 1885.

Nassarina glypta (Bush, 1885)   2149
Engraved Dove-shell

North Carolina to the Gulf of Mexico.

5 mm., semitransparent, tan, glossy, fusiform. Nucleus of 3½ white, glassy whorls. Upper whorls with 10 axial ribs crossed by 3 rows of spiral cords, forming whitish oblong nodules or beads. Body whorl with 5 or 6 spiral threads. Aperture a little more than ⅓ the length of the shell, narrow-ovate, pinched up anteriorly into a very narrow, short canal. Outer lip thick, with a varix, and with a thin, white edge and a shallow sinus close to the suture. About 5 denticles within the lip's edge. Columella's left edge raised and white, and on the inner side with 4 or 5 minute, white crenulations. Suture well-indented. Epidermis thin, dull, raised in folds along the lines of growth. Moderately common; 14 to 63 fathoms.

Subgenus *Steironepion* Pilsbry and Lowe, 1932

Anterior canal hardly discernible; siphonal fasciole weak. Type: *melanosticta* (Pilsbry and Lowe, 1932). *Psarostola* Rehder, 1943, is a synonym.

Nassarina monilifera (Sowerby, 1844)   2150
Many-spotted Dove-shell

Florida and the West Indies; Bermuda.

About ³⁄₁₆ inch in length, slender. Color white to yellow with 3 or 4 spiral rows of reddish brown spots on each whorl and about 6 rows at the base of the last whorl. Sculptured

2150   2152

with narrow axial ribs crossed by strong spiral cords which form nodules on crossing the ribs. Aperture small; interior of outer lip toothed; inner lip smooth. Common down to 42 fathoms. *N. sparsipunctata* (Rehder, 1943) is a color form (2151) with spots only on the upper 2 cords. A brown color form with white spots is also found in the West Indies.

Nassarina minor (C. B. Adams, 1845)   2152
Banded Dove-shell

West Indies; off West Florida.

About ¼ inch in length, with slightly rounded whorls. Color whitish, with a spiral brown band below the suture and another on the base of the shell. Sculptured with 9 to 13 rounded longitudinal ribs; these are cut by shallow, revolving grooves into low, irregular nodules. Otherwise resembling *N. monilifera*. Common from shallow dredgings to 120 fathoms.

Other species:

2153 *Nassarina bushiae* (Dall, 1889). Key West, Florida, 15 to 128 fms.; West Indies. (Originally *bushii*.)

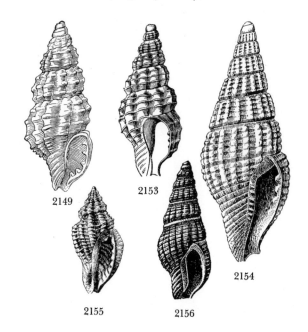

2149   2153   2154   2155   2156

2154 *Nassarina columbellata* Dall, 1889. Gulf of Mexico, 124 fms.

2155 *Nassarina grayi* Dall, 1889. Gulf of Mexico, and West Indies, 73 to 130 fms.

2156 *Nassarina metabrunnea* Dall and Simpson, 1901. Puerto Rico. 7 mm.

2157 *Nassarina xenia* (Dall, 1919). Cape San Lucas, Baja California. Rare.

2158 *Nassarina* (*Zanassarina* Pilsbry and Lowe, 1932) *anitae* Campbell, 1961. Gulf of California.

2159 *Nassarina* (*Z.*) *atella* Pilsbry and Lowe, 1932. Gulf of California.

2160 *Nassarina* (*Z.*) *pammicra* Pilsbry and Lowe, 1932. Gulf of California to Nicaragua.

2161 *Nassarina* (*Steironepion*) *tincta* (Carpenter, 1864). Gulf of California to Banderas Bay, Mexico. (Synonym: *Mangelia fredbakeri* Pilsbry, 1932.) See Keen, 1971, p. 595, species 1252.

Genus *Aesopus* Gould, 1860

Shell under ½ inch, solid, thick, subcylindrical, terminating in a blunt apex. Surface shiny. Sculpture of inconspicuous axial ribs and spiral striations; columella short. Type: *japonicus* Gould, 1860.

2162 *Aesopus stearnsii* (Tryon, 1883). Off Cape Fear, North Carolina, to Tampa Bay, Florida. Bermuda. Colombia. (Synonym: *filosa* Stearns, 1873, not Angas, 1867.) Common.

2175

2162

2163 *Aesopus* (*Ithiaesopus*) *eurytoideus* (Carpenter, 1864). San Diego to Panama.

2164 *Aesopus chrysalloideus* (Carpenter, 1864). Santa Monica to San Diego, California, 5 to 10 fms.

2165 *Aesopus sanctus* Dall, 1919. Santa Monica, California, to the Gulf of California, 14 to 20 fms. Uncommon.

2166 *Aesopus myrmecoon* Dall, 1916. Malaga Cove, Santa Monica Bay, California, shore to 3 fms.

2167 *Aesopus goforthi* Dall, 1912. Monterey, California (questionable locality data). 12 mm.

2167   2174

2168 *Aesopus babbi* (Tryon, 1882). Gulf of California.

2169 *Aesopus xenicus* Pilsbry and Lowe, 1932. Acapulco, Mexico. Rare.

2170 *Aesopus metcalfei* Reeve. West Indies to northeast Brazil. 7 to 16 fms.

2171 *Aesopus* (*Ithiaesopus* Olsson and Harbison, 1953) *arestus* Dall, 1919. Magdalena Bay, Baja California.

2172 *Aesopus* (*I.*) *fuscostrigatus* (Carpenter, 1864). Baja California.

2173 *Aesopus* (*I.*) *subturritus* (Carpenter, 1864). Southern California to Tres Marias Islands, Mexico. (Synonym: *petravis* Dall, 1908.)

Genus *Decipifus* Olsson and McGinty, 1958

Shell very small, ovate-elongate, beaded. Nuclear whorls relatively large, subcylindrical, 1½ smooth whorls with the apical tip inrolled. Sculpture of adult whorls is formed of low, narrow riblets finely beaded by spiral threads. Columella straight. Type: *sixaolus* Olsson and McGinty, 1958.

Decipifus sixaolus Olsson and McGinty, 1958   **2174**
Deceitful Whelk

Bahamas to the lower Caribbean.

2 to 5 mm., purple-brown, fusiform-ovate; axial beaded riblets 18 on the last whorl. Last whorl with 13 spiral beaded threads. Color sometimes whitish with scattered short, narrow bars of brown. Intertidal, in crevices. Uncommon. *Buccinum pulchellum* C. B. Adams, 1851 (not Blainville, 1829) is a synonym.

Other species:

2176 *Decipifus dictynna* (Dall, 1919). Cape San Lucas, Baja California.

2177 *Decipifus gracilis* McLean, 1959. Baja California.

2178 *Decipifus lyrta* (Baker, Hanna and Strong, 1938). Gulf of California.

2179 *Decipifus macleani* Keen, 1971. Gulf of California.

Genus *Amphissa* H. and A. Adams, 1853

This genus was formerly placed in the family Buccinidae. Type: *columbiana* Dall, 1916.

Amphissa versicolor Dall, 1871   **2180**
Joseph's Coat Amphissa

Vancouver Island, British Columbia, to Baja California.

½ inch in length, rather thin, but quite strong; surface glossy. 7 whorls. Suture well-impressed. Whorls in spire and upper ⅓ of body whorl with about 15 obliquely slanting, strong, rounded, axial ribs. Numerous spiral, incised lines are strongest on the base of the body whorl. Lower columella area with a

2180   2184   2187

small shield. Outer lip thickened within by about a dozen small white teeth. Color pinkish gray with indistinct mottlings of orange-brown. A common littoral to shallow-water species in 10 fathoms. Synonyms: *incisa* Dall, 1916, and *lineata* Stearns, 1872.

Amphissa columbiana Dall, 1916　　2181
Columbian Amphissa

Alaska to San Pedro, California.

1 inch in length, similar to *versicolor,* but characterized by its large size, numerous, weak, vertical, axial ribs (20 to 24 on the next to last whorl, and missing on the last part of the last whorl), and by the low, rounded varix behind the outer lip. Color yellow-brown with indistinct mauve mottlings. Periostracum thin, yellowish brown. Moderately common in shallow water from Oregon to Alaska. *A. altior* Dall, 1916 (**2182**) is a synonym.

Amphissa undata (Carpenter, 1864)　　2183
Carpenter's Amphissa

Monterey, California, to Baja California.

⅓ to ½ inch in length, similar to *versicolor,* especially in color, but with a much higher spire, stronger axial ribs, and much stronger, more acute, spiral cords. Moderately common from 25 to 265 fathoms.

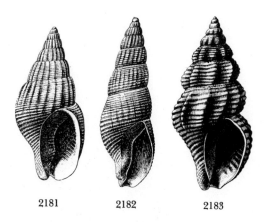

2181　　2182　　2183

Amphissa bicolor Dall, 1892　　2184
Two-tinted Amphissa

Farallon Islands, to San Diego, California.

½ to ⅝ inch in length, similar to *versicolor,* but thinner-shelled, usually with fewer ribs, glossy-white in color, but covered with a pale-straw periostracum, and without the small teeth inside the outer lip. Dredged commonly from 40 to 330 fathoms.

Other Pacific species:

2185 *Amphissa cymata* Dall, 1916. Monterey to Dana Point, Orange County, California.

2186 *Amphissa reticulata* Dall, 1916. Port Althorp, Alaska, to San Diego, California.

Amphissa haliaeeti (Jeffreys, 1867)　　2187
Atlantic Amphissa

Nova Scotia to off North Carolina.

5 to 12 mm., broadly fusiform, shell white with a yellow-tan periostracum .3 nuclear whorls, bulimoid, smooth, rounded. Postnuclear whorls with strong sinuous axial, low, smooth-topped ribs (10 to 20 per whorl) sometimes fading on the last.

Outer lip smooth within, with a low swollen varix behind. Spiral sculpture of numerous, fine, crowded threads of equal size. Peristome enamel-white, slightly flaring. Operculum ⅕ the size of the aperture, oval, tan-colored. Columella shield with a sharp left edge. Dredged from 30 to 640 fathoms; common.

Family BUCCINIDAE Rafinesque, 1815

Genus *Buccinum* Linné, 1758

Shells dull-gray to whitish, ovate, with a periostracum. Operculum relatively small, ovate, with the nucleus off to one side. Type: *undatum* Linné, 1758. Many species have been described from the arctic and subarctic seas, many of which are probably only ecologic forms.

Buccinum undatum Linné, 1758　　2188
Common Northern Buccinum

Arctic Seas to New Jersey. Europe.

2 to 4 inches in length, solid, chalky-gray to yellowish with a moderately thick, gray periostracum. Axial ribs 9 to 18 per whorl, low extending ¼ to ½ way down the whorl. Spiral cords small, usually about 5 to 8 between sutures. Outer lip slightly or well-sinuate and somewhat flaring. Aperture and parietal wall enamel-white. Anterior canal short. 1½ nuclear whorls fairly large, smooth and translucent-white. Operculum oval, concentric, chitinous and light yellow-brown. A very variable shell which sometimes lacks the axial ribs but may have numerous spiral threads. Common just offshore to several fathoms in cold water. *B. striatum* Pennant, 1777, is a synonym.

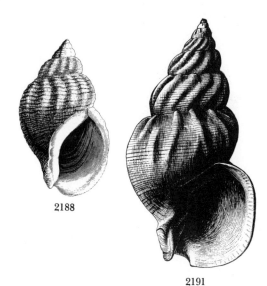

2188

2191

Buccinum scalariforme Möller, 1842　　2189
Silky Buccinum

Arctic Seas to Washington State.
Arctic Canada to the Gulf of Maine.

1½ to 2½ inches in length; aperture ½ the length of the shell. Outer lip slightly sinuate, thin and slightly flaring. Axial ribs small, numerous, intertwining and extending from suture to suture. Spiral sculpture of microscopic, beaded threads giving a silky appearance. Color light-brown. Common offshore 8 to 100 fathoms. Synonyms: *rhodium* Dall, 1919, *lyperum* Dall, 1919, *elatium* Tryon, 1880, *Buccinum tenue* Gray, 1839, non Schröter, 1805. Compare with *plectrum* Stimpson.

2189   2190   2192   2193

Buccinum totteni Stimpson, 1865   **2190**
Totten's Buccinum

Arctic Canada to northern Maine. Iceland.

1½ to 2 inches, thin-shelled, with 5 or 6 well-rounded
smoothish whorls. Aperture slightly less than ½ the length
of the shell. Yellow-brown with a thin straw-colored periostracum. Spiral sculpture of numerous, finely incised striations.
Outer lip thin and fragile. Moderately common; 8 to 50 fathoms

Buccinum plectrum Stimpson, 1865   **2191**
Plectrum Buccinum

Arctic Seas to Puget Sound.
Arctic Seas to the Gulf of St. Lawrence.

2 to 3 inches in length; aperture a little more than ⅓ the
entire length of the shell. Outer lip strongly sinuate, thickened
and flaring. Axial ribs small (but larger and fewer than those
in *scalariforme*), and limited to the upper ¼ of the whorl.
Spiral sculpture of numerous rough, but microscopic, incised
lines. Color grayish white. Common offshore in cold water
from 13 to 50 fathoms. Do not confuse with *B. scalariforme*.

Buccinum glaciale Linné, 1761   **2192**
Glacial Buccinum

Arctic Seas to Washington State.
Arctic Seas to the Gulf of St. Lawrence.

2 to 3 inches in length, fairly thick-shelled, but light in
comparison to its size. Characterized by its thick, flaring,
turned-back outer lip, and by the 2 wavy, strong spiral cords
on the periphery of the whorl. Spiral incised lines numerous.
Color mauve-brown. Aperture cream with a purplish flush
within. Common from low tide to several fathoms in the Arctic
region. *B. morchianum* Dunker, 1858, and *B. parallelum* Dall,
1919, are the carinate form of this species. Among the many
synonyms are *carinatum* Phipps, 1774; *donovani* Gray, 1839;
groenlandicum Hancock, 1846; *ekblawi* F. C. Baker, 1919; and
hancocki Mörch, 1857.

Buccinum baerii (Middendorff, 1848)   **2193**
Baer's Buccinum

Bering Sea; Aleutians to Kodiak Island.

1 to 2 inches in length, resembling a thin, beachworn *Nucella lima*. Aperture about ⅔ the length of the shell. Rather
thin, smoothish, except for microscopic, incised spiral lines.
Color drab-grayish with purplish to reddish undertones. Peri-

ostracum, when present, is thin, translucent and light-brown.
Operculum ⅕ the size of the aperture. Commonly found
washed ashore on the beaches of Alaska and the Aleutians. A
very unattractive shell. *Volutharpa morchiana* (Fischer, 1859),
is a synonym.

Other Atlantic species:

2194   *Buccinum cyaneum cyaneum* Bruguière, 1792. Labrador to
Cape Cod, Massachusetts, 26 to 471 fms. (Synonym: *groenlandicum* Mörch, 1857.)

2194   2206

2195   *Buccinum cyaneum perdix* "Beck" Mörch, 1868. Off Nova
Scotia and Cape Cod, Mass., 70 to 90 fms.

2196   *Buccinum cyaneum patulum* G. O. Sars, 1878. Greenland
to the Gulf of St. Lawrence.

2197   *Buccinum sandersoni* Verrill, 1882. East of Georges Bank
and south of Martha's Vineyard, Massachusetts, 156 to 524 fms.

2198   *Buccinum humphreysianum* Bennett, 1825. Labrador, 60
fms. Greenland. Circumboreal.

2199   *Buccinum hydrophanum* Hancock, 1846. Canadian Arctic
to Grand Banks, 60 fms.

2200   *Buccinum tanguaryi* Baker, 1919. Etah and Peeawahto
Point, Greenland.

2201 *Buccinum tumidulum* G. O. Sars, 1878. Etah, Foulke Fjord, Greenland, 2 to 10 fms.

2202 *Buccinum gouldii* Verrill, 1882 (*B. ciliatum* Gould, not Fabricius) Grand Banks (Gould). Le Have Bank, 60 fms. (Verrill).

2203 *Buccinum belcheri* Reeve, 1855. Etah, Greenland.

2204 *Buccinum inexhaustum* Verkrüzen, 1878. Newfoundland.

2205 *Buccinum ciliatum* O. Fabricius, 1780. Greenland to Gulf of St. Lawrence and Newfoundland Banks, 3 to 112 fms. Circumboreal. Bering Sea.

2206 *Buccinum abyssorum* Verrill, 1884. N. 40° to off North Carolina, 49 to 1,434 fms.

2208 *Buccinum tenebrosum* Hancock, 1846. Davis Strait. Circumboreal.

2209 *Buccinum micropoma* Thorson, 1944. Greenland and Arctic Canada. (See E. Macpherson, 1971, *Publ. Biol. Oceanogr.*, no. 3, Ottawa.)

Other Pacific species:

2210 *Buccinum hertzensteini* Verkrüzen, 1882. Siberia and Alaska.

2211 *Buccinum eugrammatum* Dall, 1907. Petrel Bank, Bering Sea, 42 to 54 fms. 55 mm.

2212 *Buccinum solenum* Dall, 1919. Bering Sea, Nunivak Island, to north of Unimak Island, 36 fms.

2213 *Buccinum oedematum* Dall, 1907. Pribilof and Sannakh Islands, to Tahwit Head, Washington. 56 mm.

2214 *Buccinum polare* Gray, 1839. Arctic Ocean and south in Bering Sea to Avacha Bay, Kamchatka, on the west, and to Alaska Peninsula on the east. Greenland and Siberia.

2215 *Buccinum chartium* Dall, 1919. Off Pribilof Islands, in 688 fms. Japan.

2216 *Buccinum pemphigus* Dall, 1907. Western Bering Sea, off Avacha Bay, 682 fms. 63 mm.

2217 *Buccinum pemphigus major* Dall, 1919. Western Bering Sea and south to Japan, 100 fms.

2218 *Buccinum pemphigus orotundum* Dall, 1907. Nunivak, Pribilof and Unimak Islands, Bering Sea. 60 mm.

2219 *Buccinum viridum* Dall, 1889. Both coasts of Bering Sea from the Pribilof Islands southward to Santa Barbara Islands, California, in 414 fms.

2220 *Buccinum planeticum* Dall, 1919. From the Pribilof Islands, Bering Sea, to the Queen Charlotte Islands, British Columbia.

2221 *Buccinum diplodetum* Dall, 1907. Off Sitka, Alaska, 1,569 fms., and Sea Lion Rock, Washington, in 877 fms. 38 mm.

2222 *Buccinum cnismatum* Dall, 1907. Bering Sea, north of Unalaska, in 300 fms. 38 mm.

2223 *Buccinum kadiakense* Dall, 1907. Kodiak Islands, Alaska. 21 mm.

2224 *Buccinum bulimuloideum* Dall, 1907. Near the Shumagin Islands, Alaska, in 159 fms. 16 mm.

2225 *Buccinum rondinum* Dall, 1919. Southeastern Bering Sea off the peninsula of Alaska, in 159 fms. 18 mm.

2226 *Buccinum castaneum* Dall, 1877. Sannakh and Shumagin Islands, 20 to 41 fms. Type of *Volutopsion* Habe and Ito, 1965. Venus, vol. 24, no. 1, p. 35.

2227 *Buccinum castaneum fluctuatum* Dall, 1919. Pribilof Islands to Unimak Pass and the Shumagin Islands, Alaska, in 30 to 56 fms.

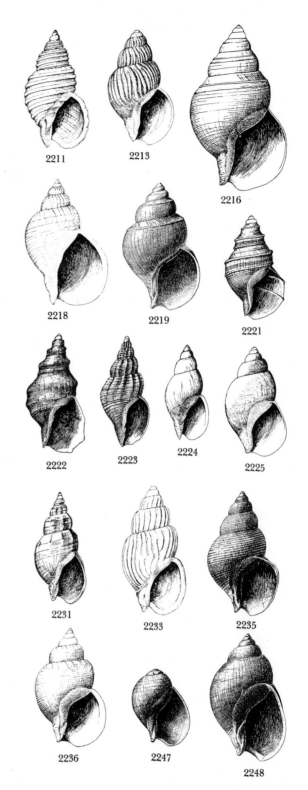

2228 *Buccinum castaneum triplostephanum* Dall, 1919. Kyska and Amchitka Islands, Aleutians, in 8 to 11 fms.

2229 *Buccinum castaneum incisulum* Dall, 1919. Western Bering Sea.

2230 *Buccinum picturatum* Dall, 1877. Aleutian Islands eastward to Bristol Bay and Kodiak Island, Alaska, 5 to 60 fms.

2231 *Buccinum simulatum* Dall, 1907. Petrel Bank, Bering Sea, 43 to 54 fms. Also northern Japan. 38 mm.

2232 *Buccinum ochotense* (Middendorff, 1848). Arctic Ocean north of Bering Strait and the Okhotsk Sea.

2233 *Buccinum sigmatopleura* Dall, 1907. Commander and Attu Islands, Bering Sea, in 135 fms. 59 mm.

2234 *Buccinum rossellinum* Dall, 1919. Southeast of Chirikoff Island, Alaska, in 695 fms.

2235 *Buccinum strigillatum* Dall, 1891. Fuca Strait (178 fms.) to San Diego (822 fms.) and Guadelupe Island.

2236 *Buccinum strigillatum fucanum* Dall, 1907. Straits of Fuca and Oregon coast. 45 mm.

2237 *Buccinum sericatum* Hancock, 1846. Arctic Ocean. Circumboreal.

2238 *Buccinum normale* Dall, 1885. Arctic Ocean from Point Barrow to Kotzebue Sound.

2239 *Buccinum physematum* Dall, 1919. Bernard Harbor, Arctic coast, west to Point Barrow and south to Bristol Bay, Alaska.

2240 *Buccinum angulosum* Gray, 1839. Bernard Harbor, Arctic coast, west to Point Barrow and south to vicinity of Bering Strait.

2241 *Buccinum angulosum subcostatum* Dall, 1885. Point Barrow, Alaska.

2242 *Buccinum angulosum transliratum* Dall, 1919. Point Barrow and southward to Bristol Bay.

2243 *Buccinum onismatopleura* Dall, 1919. Point Barrow and southward to Bristol Bay.

2244 *Buccinum fringillum* Dall, 1877. North end of Nunivak Island, Bering Sea.

2245 *Buccinum tenellum* Dall, in Kobelt, 1883. Sea Horse Islands, Arctic Ocean, south to the Aleutian Islands.

2246 *Buccinum fischerianum* Dall, 1871. St. George Island, Pribilof group, Bering Sea.

2247 *Buccinum ovulum* Dall, 1894. Amukhta Pass, Aleutians, in 248 fms.

2248 *Buccinum aleuticum* Dall, 1894. Near Unimak Island, 59 fms.

2249 *Buccinum percrassum* Dall, 1881. Arctic Ocean north of Bering Strait.

2250 *Buccinum chishimanum* Pilsbry, 1904. Bering Island, Bering Sea. Also Japan.

2251 *Buccinum striatissimum* Sowerby, 1899. Pribilof Islands; Lynn Canal, Alaska; Sea of Japan. 50 to 937 fms.

Genus *Volutharpa* Fischer, 1856

Shells about 1 inch, very thin and fragile, globular with a large aperture, short spire and well-developed, velvety periostracum. Operculum, when present, corneous, concentric. Type: *ampullacea* (Middendorff, 1848).

***Volutharpa ampullacea* (Middendorff, 1848)**   **2252**

Bering Sea to British Columbia.

2252

1 to 1½ inches, fragile, ovate, deep purplish brown. Outer lip flaring in adults. Suture canaliculate. Periostracum thin, grayish green and velvety. Moderately common; just offshore, *V. acuminata* Dall, 1871 (**2252a**) is a form with a slightly higher spire.

Other species:

2253 *Volutharpa perryi* (Jay, 1855). Japan; St. Paul Island, Bering Sea.

Genus *Volutopsius* Mörch, 1857

Shells large, with a large body whorl, a large aperture, a short anterior canal and a short, blunt spire. Type: *largillierti* Petit, 1851. Synonym: *Strombella* Gray, 1857, non Schlüter, 1838.

Subgenus *Volutopsius* Mörch, 1857

***Volutopsius castaneus* (Mörch, 1858)**   **2254**
Chestnut Buccinum

Aleutians and Kodiak Islands, Alaska.

2½ to 3½ inches in length, rather solid; with 4 whorls; aperture large and slightly flaring. Interior brownish white enamel. Columella slightly arched, white within, brown on the parietal wall. Exterior surface brownish and smoothish, except for coarse axial wrinkles appearing more as deformities in growth. Moderately common on rocks below low-tide mark.

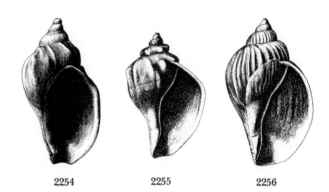

2254   2255   2256

Other species:

2255 *Volutopsius fragilis* (Dall, 1891). Bering Sea, 15 to 121 fms.; to Dutch Harbor, Aleutian Islands.

2256 *Volutopsius melonis* (Dall, 1891). Bering Sea, 227 fms. Type of *Harpofusus* Habe and Ito, 1965.

2257 *Volutopsius behringi* (Middendorff, 1849). Arctic Seas to the Pribilof Islands, 17 to 50 fms.; Kodiak Islands, Alaska. (Variety *kobelti* Dall, 1902.) A *Beringius?*

2258 *Volutopsius stefanssoni* Dall, 1919. Point Barrow, Alaska, to the Pribilof Islands. 110 mm.

2259 *Volutopsius rotundus* Dall, 1919. Kodiak Island to Cook's Inlet, Alaska.

2260 *Volutopsius middendorffii* (Dall, 1891). Bering Sea, 57 to 225 fms.

2261 *Volutopsius simplex* Dall, 1907. Off Bering Island, Bering Sea, in 72 fms. 101 mm.

2262 *Volutopsius attenuatus* (Dall, 1874). Arctic Ocean south to the Pribilof Islands and Bristol Bay.

2263 *Volutopsius trophonius* Dall, 1902. Bering Sea, in 81 fms., south of the Pribilof Islands. 67 mm.

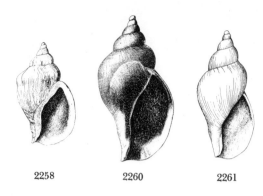

2258   2260   2261

2264 *Volutopsius filosus* Dall, 1919. From the Pribilof to the Aleutian Islands, Bering Sea. 64 mm.

2265 *Volutopsius callorhinus* (Dall, 1877). Pribilof Islands, Bering Sea.

2266 *Volutopsius callorhinus stejnegeri* (Dall, 1884). Bering Island. 60 mm. See *U. S. Nat. Mus. Bull.* 112, pl. 8, fig. 9.

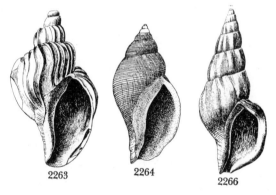

2263   2264   2266

2267 *Volutopsius regularis* (Dall, 1873). Pribilof, Aleutian and Sannakh Islands, Alaska.

2268 *Volutopsius norvegicus* (Gmelin, 1791). Off St. Lawrence Harbor, Newfoundland, 50 fms., to Bonaventure Island, Gulf of St. Lawrence. (Synonym: *largillierti* Petit, 1851.)

Subgenus *Pyrolofusus* Mörch, 1869

The type is *deformis* Reeve, 1847. *Pyrulofusus* is a misspelling. Inch-sized, hemispherical, egg capsule contains one snail. Synonym: *Heliotropis* Dall, 1873.

Volutopsius harpa (Mörch, 1858)   **2269**
Left-handed Buccinum

Pribilof Islands to Kodiak Island, Alaska.

3 to 4 inches in length, characteristically sinistral (left-handed), 4 whorls, with the smoothish nucleus indented. Sculpture of 6 to 12 very oblique, low, rounded axial ribs and numerous (often paired) raised spiral threads. Color ash-gray with a light brownish yellow periostracum. Interior of aperture tinted with tan. Operculum much smaller than the aperture. Dextral (right-handed) specimens are rarities (forma *dexius* Dall, 1907, (**2270**)). Fairly common in deep water, 74 to 140 fathoms. For egg capsules, see McT. Cowan, 1965, *The Veliger*, vol. 8, p. 1.

Other species:

2271 *Volutopsius deformis* (Reeve, 1847). Arctic Ocean; Alaska and northern Europe; Greenland. For egg capsule, see J. J. Gonor, 1964, *Arctic*, vol. 17, p. 48.

2268   2269

Genus *Beringius* Dall, 1886

The name *Jumala* Friele, 1882 was rejected by the International Commission (Opinion 469, 1957) on the quaint grounds that it was blasphemous. However, the word means Jehovah or Maker and does honor to God.

Shell large, volutid-shaped, with a rather large, bulbous nucleus. Type: *crebricostatus* Dall, 1877.

Beringius crebricostatus (Dall, 1877)   **2272**
Thick-ribbed Buccinum

Aleutians to the Shumagin Islands, Alaska.

4 to 5 inches in length, moderately heavy, with 5 to 6 whorls. Characterized by the very strong, rounded spiral cords (3 or 4 between sutures) which on the base of the shell tend to be flat-topped. Periostracum grayish brown, thin and semiglossy. A very handsome, but not commonly acquired species which occurs from 80 to 100 fathoms. Synonym: *undata* (Dall, 1919).

2272

Beringius kennicottii (Dall, 1907)   **2273**
Kennicott's Buccinum

Aleutians to Cook's Inlet, Alaska.

5 to 6 inches in length, not very heavy. Characterized by about 9 strong, arched, somewhat rounded axial ribs extending from suture to suture and, on the body whorl, extending ¾ the way down the whorl. Spiral sculpture of microscopic scratches, except on the base where there are a dozen or so weak threads. Periostracum light-brown, thin and usually flakes off in dried specimens. Shell chalky and whitish gray in color. Not uncommon in several fathoms of water; rarely in very shallow water. Synonym: *incisus* Dall, 1907.

Other species:

2274 *Beringius stimpsoni* (Gould, 1860). Arctic Ocean to the Aleutians.

2275 *Beringius malleatus* (Dall, 1884). Arctic Ocean and Bering Sea.

2276 *Beringius frielei* (Dall, 1894). Pribilof Islands, Bering Sea, 66 fms.

2277 *Beringius aleuticus* (Dall, 1894). Amukhta Pass, Aleutian Islands, 248 fms.

2278 *Beringius marshalli* (Dall, 1919). Pribilof Islands to Unalaska, Alaska, 78 fms.

2279 *Beringius indentatus* (Dall, 1919). Japan and Aleutian Islands.

2280 *Beringius brychius* (Verrill and Smith, 1885). Off Virginia, 2,574 fms.; off Georgia, 294 fms.

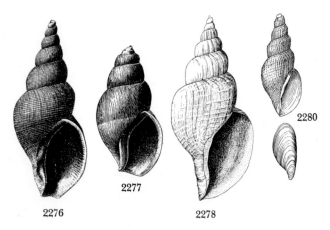

2276   2277   2278   2280

2281 *Beringius eyerdami* A. G. Smith, 1959. La Perouse Bank, 40 miles west of Cape Flattery, Washington, 100 fms. *The Nautilus*, vol. 73, no. 1 and 2; also *The Veliger*, 1960, vol. 2, p. 57, pls. 13, 14, and p. 95, pl. 23. For egg capsules, see *The Veliger*, vol. 7, p. 43.

2282 *Beringius turtoni* (Bean, 1834) form *ossiani* (Friele, 1879). Arctic Canada; Gulf of St. Lawrence; northern Norway. 10 to 700 fms. (illus. on p. 210).

Genus *Colus* Röding, 1798

Shell fusiform, slender, with numerous, moderately rounded whorls. Shell white to chalky under a brownish adhering periostracum. Sculpture of spiral threads. Type: *islandicus* (Gmelin, 1791). Numerous species described, many of which will probably prove to be only minor variants. *Sipho* Bruguière, 1792, is a synonym. The name *Colus* was considered masculine by Röding, Dall, C. W. Johnson and others. When meaning "distaff" it is feminine, but when meaning "spindle" as in this case, it is masculine.

Colus stimpsoni **(Mörch, 1867)**   **2283**
Stimpson's Colus

Labrador to off North Carolina.

3 to 5 inches in length, fusiform, moderately strong, chalky-white in color, but covered with a semiglossy, light- to dark-brown, moderately thin periostracum. Length of aperture about ½ the length of the entire shell. Sculpture of numerous incised spiral lines. Fairly common from 1 to 471 fathoms. Synonyms are: *brevis* Verrill, 1882; *liratulus* Verrill, 1882.

Subgenus *Anomalosipho* Dautzenberg and Fischer, 1912

Type of this subgenus is *dautzenbergii* Dall, 1916.

Colus pubescens **(Verrill, 1882)**   **2284**
Hairy Colus

Arctic Canada to off South Carolina.

2 to 2½ inches in length, very similar to *stimpsoni*, but the aperture is about ½ the entire length of the shell, the suture more abruptly impressed, the whorls slightly more convex and the siphonal canal usually, but not always, more twisted. Very commonly dredged from 18 to 640 fathoms.

Colus ventricosus **(Gray, 1839)**   **2285**
Ventricose Colus

Nova Scotia to Georges Bank, Massachusetts.

2 inches, very inflated for a *Colus*, about 5 whorls, with wide, faint spiral cords. Aperture large and slightly larger than the spire. Periostracum velvety, brownish and fairly thick. Uncommon; dredged offshore.

2281

2283   2284   2285   2286

Subgenus *Siphonellona* Wenz, 1941

Type of this subgenus is *pygmaeus* (Gould, 1841). Synonyms: *Siphonella* Verrill, 1879, non Hagenow, 1851; *Neptunella* Verrill, 1873, non Meek, 1864.

Colus pygmaeus (Gould, 1841)   2286
Pygmy Colus

Gulf of St. Lawrence to off North Carolina.

Less than 1 inch in length, with 6 to 7 fairly convex whorls, fairly fragile, chalk-white, with spiral raised lines, and covered with a light olive-gray, thin, velvety periostracum. Aperture slightly more than ½ the length of the entire shell. Commonly dredged from 1 to 640 fathoms. (Synonym: *planulus* Verrill, 1882.) Early whorls with 3 or 4 flat-topped spiral cords.

(2287) *Colus caelatus* (Verrill and Smith, 1880) (Massachusetts to North Carolina, deep water) is about the same size, but is characterized by about 12 strong axial ribs per whorl in addition to numerous fine spiral threads. It is chalky-white to gray. *C. hebes* Verrill, 1884, is a synonym.

2287

Subgenus *Aulacofusus* Dall, 1918

Canal short; spiral ribs prominent. Type: *spitzbergensis* (Reeve, 1855).

Colus spitzbergensis (Reeve, 1855)   2288
Spitzbergen Colus

Bering Sea to Washington State.
Arctic Seas to Maine.

2½ to 3 inches in length, rather light-shelled, and with 6 fairly well-rounded whorls. Spire long and of about 30 to 35

degrees. Siphonal canal short; columella almost straight. Outer lip flaring, slightly thickened. Sculpture of numerous (12 to 14 between sutures) low, flat-topped, small, equally sized spiral cords. Chalk-gray with a reddish to yellowish brown, thin periostracum. Commonly dredged from 1 to 142 fathoms. *Tritonium shantaricum* Middendorff, 1849, is a *nomen oblitum*.

2288

Other species (Atlantic):

2289 *Colus latericeus* (Möller, 1842). Gulf of St. Lawrence to off Newport, Rhode Island, 238 to 365 fms.

2290 *Colus islandicus* (Gmelin, 1791). Labrador to Norway. Arctic Canada.

2291 *Colus obesus* (Verrill, 1884). Martha's Vineyard, Massachusetts to off Fernandina, Florida, 102 to 843 fms.

2291

2292 *Colus perminutus* (Dall, 1927). Off Georgia, 440 fms. off Fernandina, Florida, 294 fms.

2293 *Colus rushii* (Dall, 1889). Off Fernandina, to Florida Strait, 193 to 294 fms.

2294 *Colus profundicola* (Verrill and Smith, 1884). Off Martha's Vineyard, Massachusetts, to off Delaware, 1,497 to 2,033 fms. (variety *dispar* Verrill, 1884).

2295 *Colus leptalens* (Verrill, 1884). Off Martha's Vineyard, Massachusetts, 452 fms.

Other species (s. g. *Aulacofusus*)

2296 *Colus hunkinsi* A. H. Clarke, 1962. North Canadian Basin; Alaska to Chukchi Rise. 290 to 1,208 fms. *Breviora,* no. 119, pl. 1, fig. 9.

2297 *Colus periscelidus* (Dall, 1891). Aleutians to Sannakh Islands, Alaska.

2298 *Colus herendeenii* (Dall, 1902). Southern Bering Sea, Nunivak Island to the Aleutians and eastward to the Shumagin Islands, Alaska, 41 to 284 fms. 70 mm.

2299 *Colus nobilis* (Dall, 1919). Near Pribilof Islands, 60 fms.

2300 *Colus calameus* (Dall, 1907). Western Bering Sea off Starichkoff Island, in 632 fms. 57 mm.

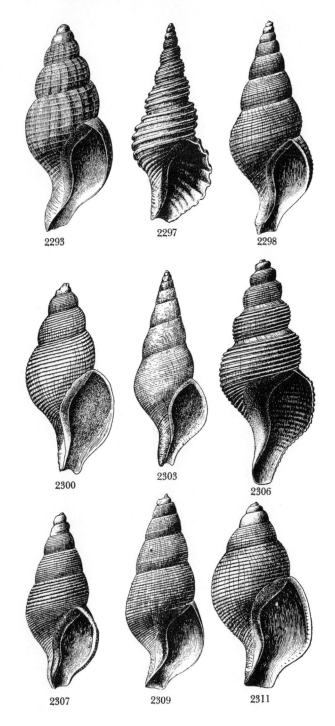

2293   2297   2298

2300   2303   2306

2307   2309   2311

2301 *Colus ombronius* (Dall, 1919). Eastern Bering Sea, from Nunivak Island, south to Bristol Bay and the Pribilof Islands.

2302 *Colus bristolensis* (Dall, 1919). Pribilof Islands north and east to Unimak Island, Alaska, 27 to 62 fms.

2303 *Colus esychus* (Dall, 1907). Point Barrow, Arctic Ocean, to Bering Island. 52 mm.

2304 *Colus roseus* (Dall, 1877). Off Cape Lisburne, Arctic Ocean, 10 to 15 fms.

2305 *Colus barbarinus* (Dall, 1919). Southern Bering Sea, off Khudubine Island, 53 fms.

2306 *Colus sapius* (Dall, 1919). Off Sitka, Alaska, in 1,569 fms.

2307 *Colus calathus* (Dall, 1919). Near Shumagin Islands, 159 fms. 27 mm.

2308 *Colus capponius* (Dall, 1919). Bering Strait near Port Clarence.

2309 *Colus acosmius* (Dall, 1891). Off Pribilof Islands, Bering Sea (688 fms.), to Unalaska (399 fms.).

2310 *Colus halidonus* (Dall, 1919). Pribilof Islands (81 fms.) to Monterey Bay, California (633 fms.).

2311 *Colus trophius* (Dall, 1919). Eastern Bering Sea south of Pribilof Islands (27 fms.) south to San Nicolas Island, California (1,100 fms.).

2312 *Colus tahwitanus* (Dall, 1918). Off Tahwit Head, Washington, in 178 fms. Type of *Limatofusus* Dall, 1918.

2313 *Colus morditus* (Dall, 1919). Gulf of Georgia, 60 to 200 fms.

2314 *Colus timetus* (Dall, 1919). Bering Sea, off Unalaska, in 19 fms.

2315 *Colus dimidiatus* (Dall, 1919). Off Tillamook Bay, Oregon, in 786 fms.

2316 *Colus severinus* (Dall, 1919). Off Pigeon Point and Monterey Bay, California, in 278 to 296 fms.

2317 *Colus pulcius* (Dall, 1919). Arctic Ocean, north of Bering Strait (Healy).

2318 *Colus georgianus* (Dall, 1920). Gulf of Georgia, in 60 to 200 fms. 40 mm.

2319 *Colus halimeris* (Dall, 1919). British Columbia to San Diego, California, 60 to 822 fms.

2320 *Colus trombinus* (Dall, 1919). Pribilof Islands, Bering Sea, in 36 fms.

Other species (s. g. *Anomalosipho*):

2321 *Colus martensi* (Krause, 1885). Plover Bay, Bering Strait, 20 fms. Arctic Canada, 18 fms.

2322 *Colus adonis* (Dall, 1919). Japan; Washington to San Diego, California, 175 to 822 fms.

2323 *Colus conulus* (Aurivillius, 1885). Arctic Seas. Deep water.

2324 *Colus dautzenbergii* (Dall, 1916). Greenland; Arctic Seas. New name for *Sipho verkruzeni* Dautz. and Fischer, 1912, non Kobelt, 1876.

2329

2282

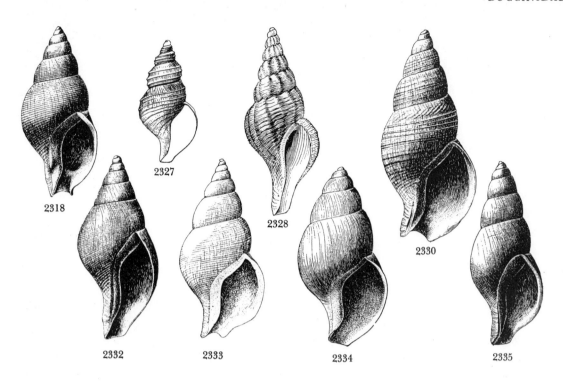

2318 2327 2328 2330

2332 2333 2334 2335

2326 *Colus sabinii* (Gray, 1824). Gulf of Maine.

2327 *Colus parvus* (Verrill and Smith, 1882). Off Martha's Vineyard, Massachusetts, 193 to 906 fms.

2328 *Colus glyptus* (Verrill, 1882). Off Martha's Vineyard, Massachusetts, 219 to 458 fms.

2329 *Colus lividus* (Mörch, 1862). Labrador to the Gulf of St. Lawrence.

Other species (s. g. *Latisipho* Dall, 1916)

2330 *Colus hypolispus* (Dall, 1891). Arctic Seas to Alaska. Type of the subgenus.

2331 *Colus errones* (Dall, 1919). Pribilof Islands, Bering Sea (18 fms.), to Strait of Fuca (308 fms.).

2332 *Colus halli* (Dall, 1873). Nunivak Island, Bering Sea, to San Diego, California, in 65 to 293 fms. 45 mm.

2333 *Colus jordani* (Dall, 1913). Bering Sea, 70 to 100 fms.; British Columbia, 67 to 142 fms.; Monterey Bay, California, 633 fms. 45 mm.

2334 *Colus aphelus* (Dall, 1889). Chirikoff Island, Alaska, to San Diego, California, 290 to 626 fms.

2335 *Colus halibrectus* (Dall, 1891). Southern Bering Sea, near Unalaska Island, 351 to 399 fms.

2336 *Colus clementinus* (Dall, 1919). Monterey Bay to San Diego, California, 339 to 704 fms.

2337 *Colus dalmasius* (Dall, 1919). Off British Columbia in 238 fms.

Genus *Liomesus* Stimpson, 1865

Shell 1 to 2 inches, solid, bucciniform, with spiral striations or threads. Columella short, somewhat twisted. Outer lip thickened. Operculum corneous, with a terminal nucleus. Periostracum usually well-developed. Type: *dalei* (J. de C. Sowerby, 1825).

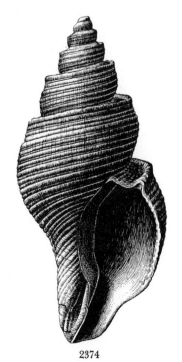

2374

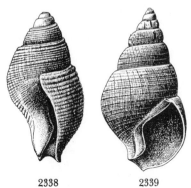

2338 2339

Liomesus stimpsoni Dall, 1889   2338
Stimpson's Liomesus

Off South Carolina to Florida.

1 to 1½ inches, solid, short-fusiform, whitish with pinkish to flesh suffusion. Nucleus minute and somewhat sunken and with 2½ smooth white whorls. Postnuclear whorls with numerous spiral threads crossed by fine growth lines. Columella obliquely truncate, arching and keeled on the anterior edge. Uncommon, 159 to 247 fms. Radula with arrow-shaped central and bicuspid lateral (*The Nautilus,* vol. 72, p. 100).

Other species:

2339 *Liomesus nassula* Dall, 1901. Pribilof Islands to Alaska, 34 to 121 fms. 43 mm.

2340 *Liomesus ooides* (Middendorff, 1848). Arctic Seas to Alaska. (Variety *canaliculata* Dall, 1874.) Type of *Pseudoliomesus* Habe and Ito, 1972.

2341 *Liomesus nux* Dall, 1877. Aleutian Islands to Shumagin Islands, Alaska.

Genus *Ptychosalpinx* Gill, 1867

Resembling *Liomesus* and possibly a subgenus of it. The columella has false plicae on it, caused by the sharp edge of the siphonal fasciole. The animal is pure-white, and without eyes. The tentacles are small, the proboscis very long. Operculum corneous, bluntly pointed, with a terminal nucleus. Dall reported 1 living species in deep water off our Atlantic coast. Type: *altilis* (Conrad), a Miocene Maryland species.

2342 *Ptychosalpinx globulus* (Dall, 1889). Florida Strait and the Bahamas, 338 to 966 fms. Off West Florida, 70 to 200 fms., uncommon. 22 mm.

2342

2343

Genus *Mohnia* Friele, 1878

Shells about 1 inch, resembling small *Colus,* but the operculum is paucispiral, and the early whorls are smooth. They lay their egg capsules singly. Type: *mohnii* Friele, 1878. Most are very deep water species:

2343 *Mohnia carolinensis* (Verrill, 1884). Off Cape Hatteras, North Carolina, to off the Florida Keys, 83 to 938 fms. See Radwin, 1972, *Trans. San Diego Soc. Nat. Hist.,* vol. 16, p. 339.

2344 *Mohnia caelatulus* (Verrill, 1884). Georges Bank, Massachusetts, to North Carolina, 326 to 1,356 fms.

2345 *Mohnia hispidulus* (Verrill, 1884). Off Delaware, 2,033 fms.

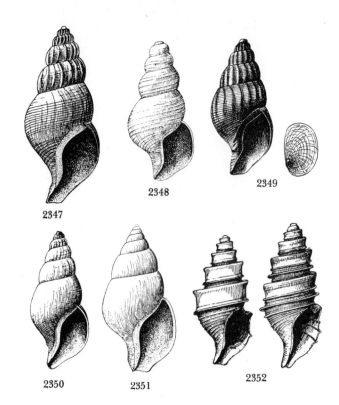

2347

2348

2349

2350   2351   2352

2346 *Mohnia simplex* (Verrill, 1884). Georges Bank, Massachusetts, to off Cape Hatteras, North Carolina, 95 to 843 fms.

2347 *Mohnia robusta* (Dall, 1913). Off the Pribilof Islands, 987 to 1,401 fms. 33 mm.

2348 *Mohnia corbis* (Dall, 1913). Off the Pribilof Islands; 1,771 fms. 30 mm.

2349 *Mohnia frielei* (Dall, 1891). Off the Queen Charlotte Island, British Columbia, 876 fms. 16 mm.

2350 *Mohnia vernalis* (Dall, 1913). Tillamook Bay, Oregon, to Monterey, California, 786 to 881 fms. 21 mm. A *Mohnia?*

2351 *Mohnia siphonoidea* (Dall, 1913). Off Pribilof Islands, 987 fms. 37 mm.

2352 *Mohnia exquisita* (Dall, 1913). Koniugi Island, Aleutians, 1,786 fms. 27 to 31 mm.

Genus *Neptunea* Röding, 1798

Shells 3 to 5 inches, sturdy, with 6 to 8 large whorls, moderately high spire, a distinct and usually twisted siphonal canal. Whorls in the nucleus small, smooth and swollen. Periostracum thin. Operculum large, irregularly triangular with a terminal nucleus. Arctic and boreal seas. Type: *antiqua* Linné, 1758 (by Monterosato, 1872). I largely follow A. N. Golikov, 1963, "The gastropod mollusk genus *Neptunea* Bolten," *Fauna of U.S.S.R.,* vol. 5, pt. 1, 218 pp.

Neptunea lyrata lyrata (Gmelin, 1791)   2353
Common Northwest Neptune   Color Plate 12

Arctic Ocean to off California.

4 to 5 inches in length, ¾ as wide, solid, fairly heavy. With 5 to 6 strongly convex whorls, bearing about 8 strongly to poorly developed, raised spiral cords (2 of which usually show in each whorl in the spire). Faint, quite small, spiral threads are also present. Exterior dull whitish brown. Aperture enamel-white with a tan tint. Fairly common in Alaska from shore to 50 fathoms. Off California and Oregon it occurs down

2353   2354   2355   2356

to 800 fathoms. This is *Chrysodomus lirata* Martyn. Among the varietal synonyms are: *phoenicea* Dall, 1891; *Cymatium pacificum* Dall, 1909; *obsoleta* Golikov, 1963.

Neptunea lyrata decemcostata (Say, 1826)   **2354**
New England Neptune   **Color Plate 12**

Grand Banks to off North Carolina.

3 to 4½ inches in length, rather heavy. Characterized by its grayish white, rather smooth shell which bears 7 to 10 very strong, reddish brown, spiral cords. The upper whorls show 2 to 3 cords. There is an additional band of brown just below the suture. A common cold-water species found offshore, but occasionally washed up on New England beaches. A form or possibly a subspecies with a short spire and broader shell, occurring between Grand Manan Island, New Brunswick, and Mount Desert Island, Maine, was named **(2355)** *turnerae* A. H. Clarke, 1956 (for *N. clenchi* Clarke, see under *despecta* (Linné)). (*The Nautilus*, vol. 69, p. 117.)

Neptunea ventricosa (Gmelin, 1791)   **2356**
Fat Neptune

Arctic Ocean and Bering Sea.

3 to 4 inches in length, heavy, with a large, ventricose body whorl. Axial ribs or growth lines coarse and indistinct, rarely lamellate. Shoulders sometimes weakly nodulated. Spiral cords absent or very weak. Color a dirty-brownish white. Aperture white or flushed with brownish purple. Moderately common offshore. This is *Chrysodomus satura* Martyn and its several poor varieties, such as *middendorffiana* MacGinitie, 1959.

Neptunea smirnia (Dall, 1919)   **2357**
Smirnia Neptune

Alaska to Washington.

2 to 4 inches, livid purple-brown, with an olivaceous periostracum. Early whorls with 4 or 5 low, flat spiral cords, but the last 2 whorls are smoothish, round-shouldered and have faint, fine, silky growth lines. Common; 60 to 100 fathoms. *N. fukueae* "Kuroda" (Kira, 1959, pl. 27, fig. 4) may be the Japanese synonym of this species.

Neptunea amianta (Dall, 1890)   **2358**

Bering Sea to off California.

3 inches, thin-shelled, globose, white, with a small, but prominent, inflated subglobular nucleus. Spiral sculpture of numerous, closely set, rounded, narrow cords. Siphonal canal short. Outer lip whitish, smooth, slightly thickened and flaring in adults. Uncommon; dredged as deep as 414 fathoms off the coast of California.

Neptunea despecta (Linné, 1758)   **2359**

Northern Europe to Greenland to Maine.

2 to 3 inches, fusiform, usually appearing worn and unattractive. Color dingy-white or brownish tan. 8 whorls, very convex, the last being ventricose. Sometimes spirally banded with light-chestnut. Spiral sculpture of fine, crowded threads. Upper whorls usually have 2 large, crude spiral cords which gradually disappear on the body whorl. Below these may be several others gradually diminishing in prominence. Aperture just less than ½ the length of the entire shell and sometimes dilated, the outer lip being angulated by the more prominent revolving ribs. Siphonal canal short and considerably recurved. Uncommon; offshore to 500 fathoms. A very variable species having a number of named varieties: *carinata* Pennant, 1777; *tornatus* Gould, 1840; *subantiquata* Dautzenberg and Fischer, 1911; *striata* Dautzenberg and Fischer, 1911; *fasciata* Jaeckel, 1952; *clenchi* A. H. Clarke, 1956 **(2360)** Golikov recognized a subspecies with spiral brown, wavy lines from the west side of Iceland: *fornicata* (Fabricius, 1780).

Neptunea pribiloffensis (Dall, 1919)   **2361**
Pribiloff Neptune

Bering Sea to British Columbia.

4 to 5 inches in length, similar to *N. lyrata,* but with a lighter shell, with weaker and more numerous spiral cords, and with more numerous and stronger secondary spiral threads. Outer lip more flaring and the siphonal canal with more of a twist to the left. Fairly commonly dredged from 50 to 100 fathoms.

Neptunea stilesi A. G. Smith, 1968   **2362**
Stiles' Neptune

Off Vancouver Island, British Columbia, to Washington.

3 to 4½ inches, somewhat resembling *lyrata,* but with a very large body whorl, unflaring lip when adult and a short canal. Color dingy-white to tan or yellowish and occasionally reddish brown. Major sculpture of widely spaced, distinct, rounded, low, spiral cords which are darker brown. Uncommon; 34 to 125 fathoms. *The Veliger*, vol. 11, p. 117.

2359

2360

2361

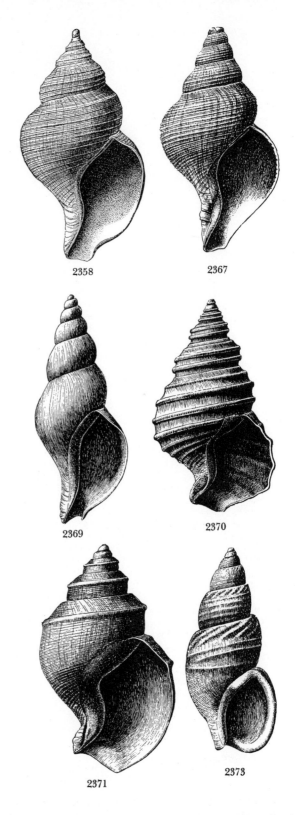

2358   2367

2369   2370

2371

2373

Other species:

2363 *Neptunea behringiana* (Middendorff, 1848). Siberia, Bering Sea to southeast Alaska. (Synonyms: *solutus* Dall, 1910; *nuceus* Dall, 1921; *cordatus* Dall, 1921.)

2357

2364 *Neptunea communis communis* (Middendorff, 1849). Arctic Seas: Bering Sea, Siberia, Faroes, west Greenland (forms: *borealis* Philippi, 1851; *elongata* Löyning, 1932, subspecies *clarki* Meek, 1923, in the northwest Pacific-Japan).

2365 *Neptunea vinosa* (Dall, 1919). Kamchatka to the Bering Sea. (Var. *umbonata* Golikov, 1963.) **(2366)**.

2367 *Neptunea insularis* (Dall, 1895). Sea of Okhotsk to the Bering Sea. (Var. *convexa* Golikov, 1963.) **(2368)**.

2362

2369 *Neptunea ithia* (Dall, 1891). Off Barkley Sound, British Columbia, to Monterey, California, 204 to 382 fms.

Subgenus *Ancistrolepis* Dall, 1894

Siphonal canal short and wide. Columella strongly incurved in the middle. Whorls with widely spaced spiral ribs. Type: *eucosmia* Dall, 1891.

Neptunea eucosmia (Dall, 1891)   **2370**
Channeled Neptune

Alaska to Oregon.

1½ inches in length, solid, outer lip sharp, strong and crenulated. Siphonal canal short, wide and slightly twisted. Spiral cords strong. Suture channeled. Shell chalk-white, but covered with a rather thick, yellow-brown to gray periostracum which is axially lamellate and bears minute, erect hairs. Aperture glossy-white. Not uncommonly dredged from 62 to 780 fathoms. *N. californica* (Dall, 1919), and *bicincta* (Dall, 1919), appear to be this species.

Other *Ancistrolepis*:

2371 *Neptunea (Ancistrolepis) magna* (Dall, 1895). Okhotsk and Bering Seas, 25 to 70 fms. Proc. U. S. N. Mus., vol. 17 p. 709.

2372 *Neptunea (Ancistrolepis) beringiana* (Dall, 1919). Bering Sea, 58 fms. (non Middendorff, 1848). Type of *Neancistrolepis* Habe and Ito, 1972.

2373 *Neptunea (Sulcosinus* Dall, 1895) *taphria* (Dall, 1891). Bering Sea, off Unalaska, in 351 fms. Type of the subgenus.

Subgenus *Sulcosipho* Dall, 1916

Shell like *Neptunea* but more slender and elongate and with a tabulated shoulder. Color white. Nuclear whorls inflated and slightly oblique. Type: *tabulata* (Baird, 1863).

Neptunea tabulata (Baird, 1863)   **2374**
Tabled Neptune   **Color Plate 12**

British Columbia to San Diego, California.

3 to 4 inches in length, moderately solid, with 8 whorls, colored white with a thin, brown periostracum. Characterized by the wide, flat channel next to the suture. It is bounded by

a raised, scaly or fimbriated spiral cord. Remainder of whorl with numerous sandpapery spiral threads. A choice collector's item, not uncommonly dredged from 30 to 218 fathoms.

Genus *Plicifusus* Dall, 1902

Large, cold-water shells resembling *Colus,* but strongly plicate axially, smooth or spirally sculptured, usually with an inconspicuous periostracum. Aperture large. Outer lip expanded. The columella has a built-up callus. These are deep-water Arctic species not likely to be encountered by shore collectors. Type: *kroyeri* (Möller, 1842). Synonym: *Parasipho* Dautzenberg and H. Fischer, 1912.

2375 *Plicifusus cretaceus* (Reeve, 1847). Labrador and the Gulf of St. Lawrence, 30 to 60 fms.

2376 *Plicifusus kroyeri* (Möller, 1842). Circumpolar Seas; Greenland; Bering Sea.

2376

2377 *Plicifusus arcticus* (Philippi, 1850). Circumpolar Seas; Greenland; Aleutians to Shumagin Islands, Alaska. 103 mm.

2378 *Plicifusus syrtensis* (Packard, 1867). Square Island, Labrador, 30 fms.

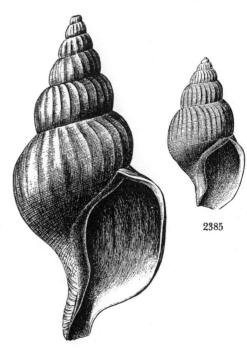

2385

2377

2379 *Plicifusus johanseni* Dall, 1919. Point Barrow to Icy Cape, Alaska.

2380 *Plicifusus verkruzeni* (Kobelt, 1876). Circumpolar Seas; Europe; Bering Sea.

2381 *Plicifusus* (*Retifusus* Dall, 1916) *virens* (Dall, 1877). Bering Sea; Kyska Harbor, 10 fms.; Middleton Island, Alaska.

2382 *Plicifusus* (*Retifusus*) *incisus* Dall, 1919. Bering Sea to Shumagin Islands, 38 to 54 fms.

2383 *Plicifusus* (*Retifusus*) *oceanodromae* Dall, 1919. Petrel Bank, Bering Sea; Aleutians to the Shumagin Islands.

2384 *Plicifusus* (*Microfusus* Dall, 1916) *brunneus* (Dall, 1877). Port Clarence, Bering Strait, to Nunivak and the Pribilof Islands.

2385 *Plicifusus* (*Latifusus* Dall, 1916) *griseus* (Dall, 1890). Bering Sea to San Diego, California, 27 to 414 fms.

2386 *Plicifusus* (*Helicofusus* Dall, 1916) *laticordatus* (Dall, 1907). Bering Sea to Washington, 33 to 559 fms.

Genus *Exilioidea* Grant and Gale, 1931

Shells small, about 1 inch, slender-fusiform, with numerous whorls, with a fairly large nucleus; siphonal canal elongate and straight. Periostracum well-developed and smooth. Sculpture of numerous fine, sinuous, axial riblets and spiral striations. Columella smooth. Outer lip thin and sharp. Operculum long, slightly arcuate, with a terminal nucleus. Type: *rectirostris* (Carpenter, 1865). Formerly confused with the fossil genus *Exilia* Conrad, 1860.

Exilioidea rectirostris (Carpenter, 1865)   **2387**

Behm Canal, Alaska, to Baja California.

¾ to 1 inch, slender, white, with a polished olive-brown periostracum. Aperture reddish brown to whitish within. The apex is usually eroded. Uncommon; in fine gravel and mud, 50 to 80 fathoms.

Other species:

2388 *Exilioidea kelseyi* (Dall, 1908). Crescent City, 30 fms., to off San Diego, 359 fms., California.

2388

Genus *Kelletia* Fischer, 1884

Type of the genus is *kelleti* (Forbes, 1850). Shell large, heavy, fusiform, with nodules on the shoulder.

Kelletia kelleti (Forbes, 1850)   **2389**
Kellet's Whelk   Color Plate 12

Santa Barbara, California, to San Quentin Bay, Mexico.

4 to 5 inches in length, characterized by its very heavy, white, fusiform shell, its fine, wavy suture and its sharp, crenulated outer lip. Whorls slightly concave between the suture and the shouldered periphery, which bears 10 strong, rounded knobs per whorl. Base with about 6 to 10 incised, spiral lines. Aperture glossy and white. Very commonly caught in traps from 10 to 35 fathoms. There are no other recent species in the genus. Often misspelled with two t's. Also found in northern Japan.

2389

Genus *Searlesia* Harmer, 1916

Shell about 1 inch, solid, fusiform; apex blunt; spirally and axially ribbed; siphonal canal rather short, slightly bent to the left. Type: *costifer* (S. V. Wood, 1848).

Searlesia dira (Reeve, 1846)   **2390**
Dire Whelk

Alaska to Monterey, California.

1 to 1½ inches in length, ½ as wide, with the brown aperture ½ the length of the dark-gray, fusiform shell. Outer lip

2390

thin but strong and with fine serrations which extend back into the shell as small spiral threads. Columella arched, chocolate-brown and glossy. Whorls in spire with 9 to 11 low, rounded, axial ribs, and all of the exterior with numerous fine, unequal-sized spiral threads. Siphonal canal short and slightly twisted to the left. A common shallow-water species commonly from northern California to the north.

Genus *Bartschia* Rehder, 1943

Shell about 2 inches, elongate, heavy, without an umbilicus. Nuclear whorls dome-shaped, 3½ smooth whorls. Parietal wall with a glossy callus. Outer lip with row of denticles on the inner edge. Type: *significans* Rehder, 1943.

Bartschia significans Rehder, 1943   **2391**
Significant Bartsch Whelk

Off both sides of southern Florida.

1½ to 2 inches, heavy, elongate-ovate, spire high. Nucleus bulbous, smooth, of 3½ whorls. Later 5 whorls with numerous, closely set spiral cords crossed by axial riblets, giving the upper whorls a latticed appearance. Color yellowish with maculations of light-brown. Outer lip roundly thickened; with denticles on the inner edge. Uncommon; 75 to 100 fathoms.

2391

Other species:

2392 *Bartschia agassizi* (Clench and Aguayo, 1941). Gulf of Mexico, off Florida, 180 fms.; off Cuba. *Mem. Soc. Cubana Hist. Nat.*, vol. 15, p. 179.

2393 *Bartschia canetae* (Clench and Aguayo, 1944). Off Havana, Cuba, 240 to 300 fms.; off Barbados, 288 fms. *Revista Soc. Malacol.*, p. 68.

2394 *Bartschia fusiformis* (Clench and Aguayo, 1941). Off south Cuba, 1,440 fms. *Mem. Soc. Cubana Hist. Nat.*, vol. 15, p. 179.

Genus *Bailya* M. Smith, 1944

Shell 15 mm., fusiform; sculpture subreticulated by axial riblets and spiral threads. Outer lip with strong varix. Inner margin of outer lip denticulate. Parietal wall with a sharp-edged, finely denticulate callus. Periostracum fairly thick. Operculum smaller than the aperture and with a terminal nucleus. Type: *Triton anomalus* Hinds, 1844.

Bailya intricata (Dall, 1884)   **2395**
Intricate Baily-shell

Southern half of Florida. Bahamas.

½ inch in length, fairly strong, pure-white in color and with cancellate sculpturing. Last whorl with 12 to 14 low axial ribs which are crossed by about a dozen spiral cords (between which there may be a much smaller thread). At their intersection there are small beads. Outer lip with a frilled, rounded varix. Columella smooth. Weak spiral cord present on inside of aperture on the upper parietal wall. Whorls slightly shouldered. No notch in lower part of outer lip. Nuclear whorl smooth, glassy and rounded. Uncommon from 1 to 50 fathoms.

2395   2396

Bailya parva (C. B. Adams, 1850)   **2396**

Bahamas and the West Indies.

¾ inch, elongate-fusiform. Flesh-colored with irregular pale-brown and white revolving stripes. Whorls rounded, deeply sutured. Sculptured with 10 to 12 narrow, rounded axial ribs (per whorl) which are crossed by less conspicuous spiral lines, and beaded at their intersections. Columella smooth and glossy. Similar to *intricata* but larger, has brown coloring, the axial ribs are more prominent and the whorls are not shouldered. Uncommon; shallow water, under rocks.

A similar species, *Bailya anomala* (Hinds, 1844) **(2397)** ranges intertidally from Guaymas, Mexico to Nicaragua. It is yellowish brown, about 15 mm. in length and has 10 to 12 axial ribs per whorl. *Fusus bellus* C. B. Adams, 1852, is a synonym.

Genus *Engina* Gray, 1839

Shells under 1 inch, broadly fusiform, solid, usually noduled. Outer lip thickened. Inner lip glazed and with curious, radially oriented, raised lirae. Central radular tooth quadrate and with 5 to 7 denticles. Lateral tricuspid. Type: *turbinella* Kiener, 1835 (*zonata* Gray, 1839, non Reeve, 1846). (See V. Orr, 1962, *The Nautilus*, vol. 75, p. 107.)

Engina corinnae Crovo, 1971   **2398**
Corinne's Engina

Southeast Florida.

10 to 13 mm., similar to *turbinella*, but differing in being ivory-white with sparse patches of brown lines between the 8 to 10 shoulder nodules, in having a white, instead of tan, nuclear whorl, in having 5 or 6 brown-spotted, spiral threads on the base (instead of 2 or 3 rows of white beads) and in having a light-pink to mauve aperture. Uncommon; rubble bottom and on *Spondylus*, 90 to 120 feet. Named after the veteran collector, Corinne E. Edwards (*The Veliger*, vol. 14).

Engina turbinella (Kiener, 1835)   2399
White-spotted Engina

Southeast Florida to Brazil.

⅓ to ½ inch in length, dark purple-brown with about 10 low, white knobs per whorl on the periphery. Base with 2 to 4 spiral rows of much smaller white knobs. Microscopic spiral threads numerous. Aperture thickened and constricted by 4 to 5 whitish teeth on the outer lip and by a twist of the columella just above the narrow siphonal canal. Do not confuse with some *Mitra* which has several columellar plications. Common under rocks at low tide. Scarce in Florida.

2399   2398

Other species:

2400 *Engina caribbaea* Bartsch and Rehder, 1939. Off West Florida and the Caribbean, 120 fms. 12 mm.

2401 *Engina fusiformis* Stearns, 1894. Gulf of California. (Synonyms: *solida* Dall, 1917; *senae* and *multa* Pilsbry and Lowe, 1932.)

2402 *Engina jugosa* (C. B. Adams, 1852). Head of the Gulf of California to Ecuador; Galapagos Islands.

2403 *Engina tabogaensis* Bartsch, 1931. Gulf of California to Panama.

Subfamily COLUBRARIINAE Dall, 1909

Genus *Colubraria* Schumacher, 1817

Shells resembling miniature *Charonia*. 2 nuclear whorls, small, smooth, white, rounded, descending rapidly. Postnuclear whorls axial and spirally ribbed, but becoming smoothish or weakly sculptured. Parietal wall raised into an elongate columellar shield. Type: *maculosa* (Gmelin, 1791). Formerly and erroneously placed in the Cymatiidae.

Colubraria lanceolata (Menke, 1828)   2404
Arrow Dwarf Triton

North Carolina to both sides of Florida to Brazil. Bermuda.

¾ to 1 inch in length, slender, with 7 whorls. Aperture long and narrow. Varix strong and curled back. Parietal shield elevated into a collar. Formerly distinct varices present every ⅔ of a whorl. Sculpture very finely cancellate and beaded. Nucleus brown, smooth and bulbous. Shell ash-gray with occasional orange-brown smudges. Uncommon; rocky bottom just offshore.

Colubraria obscura (Reeve, 1844)   2405
Obscure Dwarf Triton

Bermuda; southeast Florida to the Lesser Antilles. Brazil.

2404   2405   2406

½ to 1½ inches, similar to *lanceolata*, but fatter, with numerous, small beads and with wider varices. Uncommon; in shallow water under rocks on the reef. *C. testacea* (Mörch, 1852) is a synonym. Rare in Florida.

Subgenus *Monostiolum* Dall, 1904

The type of this subgenus is *swifti* (Tryon, 1881).

Colubraria swifti (Tryon, 1881)   2406
Swift's Dwarf Triton

Bermuda; Caribbean.

½ to ¾ inch, slender, elongate, whitish with sparse, light-brown irregular splotches. Nuclear whorls smooth, white, round, descending rapidly. Early whorls with axial riblets which disappear by the last whorl. Spiral sculpture of numerous spiral threads (12 between sutures) which may be microscopically beaded by axial growth lines. Parietal shield raised, glossy-white. Siphonal canal short, straight, constricted. Inside of outer lip may have 9 to 11 white short lirae. Upper corner of aperture U-shaped. Edge of outer lip with a swollen varix. Common in Bermuda, under rocks, shallow water. Uncommon in the West Indies.

Other species:

2407 *Colubraria monroei* McGinty, 1962. North coast of Cuba. *The Nautilus*, vol. 76, p. 42.

2408 *Colubraria* (*Monostiolum*) *biliratus* (Reeve, 1846). Gulf of California, 7 to 146 meters. Galapagos Islands. (Synonym: *Fusus apertus* Carpenter, 1857.)

2407

Subfamily PISANIINAE Tryon, 1880

1½ nuclear whorls, rapidly descending, smooth, translucent-brown. First few postnuclear whorls beaded.

Genus *Pisania* Bivona, 1832

Type: *striata* (Gmelin, 1791) (not *pusio* Linné).

Pisania pusio (Linné, 1758) 2409
Miniature Triton Trumpet **Color Plate 10**

Southeast Florida and the West Indies to Brazil. Bermuda.

1 to 1½ inches in length, sturdy, spirally striate to smooth, and usually with a glossy finish. Early whorls beaded. The outer lip is weakly toothed within, and the upper parietal wall has a small, white, swollen tooth near the top of the aperture. Color variable, but usually purplish brown with narrow spiral bands of irregular dark and light spots commonly chevron-shaped. Moderately common below low-water line in the region of coral reefs. Synonyms: *accinctus* (Born, 1778); *plumatum* (Gmelin, 1791); *fasciatum* (G. Fischer, 1807); *articulata* (Lamarck, 1822).

2409   2410   2411

Subgenus *Pollia* Gray in Sowerby, 1834

Type: *undosa* Linné, 1758 from the Indo-Pacific. *Tritonidea* Swainson, 1840, and *Gemophos* Olsson and Harbison, 1953, are synonyms.

Pisania auritula (Link, 1807) 2410
Gaudy Cantharus

Southeast Florida and the West Indies to Brazil. Bermuda.

Similar to *C. tinctus,* but broader, with shouldered whorls, with about 9 stronger axial ribs per whorl and with about 10 sharp spiral threads on the last whorl. Outer lip turned in as the varix is formed. Color brighter. Posterior canal longer. Common in the West Indies; intertidal. Formerly placed in *Cantharus.*

Pisania tincta (Conrad, 1846) 2411
Tinted Cantharus

North Carolina to Florida, Texas and the West Indies. Bermuda. Brazil.

¾ to 1¼ inches in length, heavy, spire evenly conic; aperture with a small canal at the top. Axial ribs low and weak. Spiral cords numerous and weak, forming weak beads as they cross the ribs. Inside of outer lip with small teeth which are strongest near the top. Color of shell variegated with yellow-brown, blue-gray and milky-white. Fairly common in shallow water. The young are easily confused with *Thais.* See remarks under *T. rustica.* Formerly placed in *Cantharus. Tritonidea bermudensis* Dall and Simpson, 1901, is a synonym. Rare in Texas.

Genus *Cantharus* Röding, 1798

Similar to *Pisania,* subgenus *Pollia,* but there is no posterior siphonal constriction due to parietal lirae. Type: *tranquebaricus* (Gmelin, 1791), from the Andaman Sea.

Cantharus cancellarius (Conrad, 1846) 2412
Cancellate Cantharus

West coast of Florida to Texas and Yucatan.

Similar to *P. tincta,* but with a lighter shell, higher spire, and with sharp, spiral threads and narrow, axial ribs which cross to make a beaded and cancellate sculpturing. Base of columella with fairly large, white spiral ridge. Posterior siphonal canal absent or weak. Varix very weak. Moderately common in shallow water. Dead shells not uncommon on Texas beaches.

2412   2421

Cantharus multangulus (Philippi, 1848) 2413
False Drill

North Carolina to Yucatan, Cuba and Bahamas.

1 to 1¼ inches in length, rather broad, with a short, fairly open siphonal canal. Outer lip sharp, finely crenulated. At the base of the columella there is a single, small fold. 8 to 9 short, rounded axial ribs on the periphery of the whorl. Spiral cords weak. Color gray with red-brown specklings; sometimes solid yellow-orange. Moderately common. (See Robertson, 1957, *Notulae Naturae,* no. 300). Feeds on barnacles.

The Pacific counterpart is *Cantharus panamicus* (Hertlein and Strong, 1951) (2414) from the southern part of the Gulf of California to Panama from 35 to 40 fathoms.

2413

Other species:

2415 *Cantharus massena* (Risso, 1826). Mediterranean. Bermuda and Bahamas?

2416 *Cantharus rehderi* S. S. Berry, 1962. Gulf of California to Panama. (Synonym: *elegans* Dall, 1908, non Griffith and Pidgeon, 1834.)

2417 *Cantharus shaskyi* S. S. Berry, 1959. Gulf of California, off-shore.

2418 *Cantharus* (*Gemophos* Olsson and Harbison, 1953) *elegans* (Griffith and Pidgeon, 1834). Magdalena Bay and the Gulf of California to Peru. Intertidal. (Synonyms: *Buccinum insignis* Reeve, 1846, and *Pisania aequilirata* Carpenter, 1857.)

2419 *Cantharus* (*Gemophos*) *sanguinolentus* (Duclos, 1833). Baja California, Gulf of California to Ecuador. (Synonyms: *Pollia haemastoma* Gray, 1839, and *Buccinum janelii* Valenciennes, 1846.)

2420 *Cantharus* (*Gemophos*) *vibex* (Broderip, 1833). Baja California to Panama. (Synonyms: *turbinelloides* Griffith and Pidgeon, 1834, and *marjoriae* M. Smith, 1944.)

Subgenus *Solenosteira* Dall, 1890

Shell solid, spirally corded, with weak axial ribs. Outer lip crenulate. Periostracum fuzzy and fairly thick. Radular buccinid. Type: *Cantharus pallidus* form *anomala* Reeve, 1847. No living Caribbean species. The similar genus *Hanetia* from Brazil is actually a muricid (see Emerson, 1968, *The Veliger,* vol. 11, p. 2).

Cantharus pallidus (Broderip and Sowerby, 1829)   2421
Pallid Cantharus

Baja California, the Gulf of Ecuador.

1 to 1½ inches, solid, heavy, white with a heavy, velvety yellowish periostracum. With numerous coarse spiral ribs and a varying number of low axial undulations. The latter may be obsolete in the form *anomala* Reeve, 1847. A very variable and very common species from intertidal regions to 20 fathoms. Synonyms are *Pyrula lignaria* Reeve, 1847; *Fusus turbinelloides* Reeve, 1848; and *capitaneus* Berry, 1957, the last having an extra well-developed periostracum with bristles.

Other species:

2422 *Cantharus berryi* J. H. McLean, 1970. Banderas Bay, Jalisco, Mexico. 10 to 15 fms. *The Veliger,* vol. 12, no. 3, p. 314.

2423 *Cantharus gatesi* (S. S. Berry, 1963). Off Mazatlan, Mexico, 15 fms. *Leaflets in Malacology,* no. 23, p. 144.

2424 *Cantharus mendozanus* (S. S. Berry, 1959). Magdalena Bay, Baja California, 10 to 25 fms. *Leaflets in Malacology,* no. 18, p. 111.

Genus *Antillophos* Woodring, 1928

Shell solid, elongate-oval, with a sharp spire, with 12 to 18 axial beaded ribs. 2 nuclear whorls, glossy and bearing a spiral carina just above the suture. Aperture spirally lirate within. Siphonal canal short, constricted a little and only slightly twisted. Type: *candei* (Orbigny, 1842).

Antillophos candei (Orbigny, 1842)   2425
Candé's Beaded Phos

Off both sides of Florida to the West Indies. Brazil.

½ to 1 inch, yellowish white, sometimes faintly banded with brown. Axial ribs variable in strength and number (13 to 20 per whorl). Sometimes the sculpturing appears to be openly reticulate; or may be finely beaded with crowded axial ribs. Upper and inner parietal wall has a strong spiral white ridge disappearing within. Columella with 2 to 4 small spiral lirae. Synonyms: *antillarum* Petit, 1853; *oxyglyptus* (Dall and Simpson, 1901); *virginiae* (Schwengel, 1942). Commonly found in sand from 1 to 100 fathoms.

2425

Other species:

2426 *Antillophos adelus* (Schwengel, 1942). West Indies; Santo Domingo.

2427 *Antillophos beauii* (Fischer and Bernardi, 1860). Yucatan to Cuba and the Lesser Antilles, 10 to 183 meters.

2427a *Antillophos veraguensis* (Hinds, 1843). Gulf of California to lower Mexico. (Synonyms: *Phos biplicatus* Carpenter, 1856; *alternatus* Dall, 1917.)

Genus *Engoniophos* Woodring, 1928

Shells similar to *Antillophos,* but the 2 nuclear whorls are very rounded, without a spiral carina, and descending rapidly. Axial ribs line up one under the other. Siphonal slot deep and oblique. Siphonal fasciole swollen. Base of columella with a strong, glossy spiral ridge. Stromboid notch present or absent. Type: *erectus* Guppy, 1873, a Miocene species.

Engoniophos unicinctus (Say, 1825)   2428

Greater Antilles; lower Caribbean.

½ to 1 inch, light-gray to brownish, spire acute; with 10 to 12 axial rounded ribs crossed by numerous subequal fine spiral threads. Interior of aperture stained and streaked with brown and spirally lirate. Common; 1 to 10 fathoms. *Nassa guadelupensis* Petit, 1852, is a synonym.

2428

Genus *Macron* H. and A. Adams, 1853

Shell heavy, ovate, smoothish or spirally corded. Periostracum usually thick. Outer lip with a toothlike projection anteriorly. Operculum corneous, with a terminal nucleus. Type: *aethiops* (Reeve, 1847).

Macron lividus (A. Adams, 1855)   **2429**
Livid Macron

Monterey, California, to Baja California.

¾ to 1 inch in length, ½ as wide, strong, with 5 whorls which are covered with a thick, feltlike, dark-brown periostracum. Shell yellowish to bluish white. Outer lip sharp, strong and near its base bearing a small, spiral thread. Columella strongly concave and white. Upper end of aperture narrow, with a small, short channel and with a white, toothlike callus on the parietal wall. Siphonal canal short and slightly twisted. Base of shell with a half-dozen incised spiral lines. Operculum chitinous, brown, thick, oval and with the nucleus at one end. Very common under stones at low tide.

Macron aethiops (Reeve, 1847)   **2430**
Aethiopian Macron

Baja California and (rare) the Gulf of California.

2 inches, ovate, solid, with a fairly sharp apex. Whorls rounded; suture impressed. Shell white with a dark greenish brown periostracum. Lower part of whorls with broad, flattish cords. Sculpturing quite variable. A common intertidal species on the outer coast of Baja California. Synonyms are: *kellettii* A. Adams, 1854; *orcutti* Dall, 1918.

2429   2430

Genus *Morrisonella* Bartsch, 1945

Shell 1 inch, elongate-fusiform, with 6 to 7 adult whorls. Aperture ⅖ the length of the shell. Siphonal canal short, wide, slightly bent. Columella simple. Sculpture reticulate. Operculum corneous, with an apical nucleus. Radula with a 3-cusped central and a 2-cusped marginal tooth. No eyes present at the base of 2 long tentacles. Penis large, with a tiny papilla at the end. Type: *Leucosyrinx pacifica* Dall, 1908. One deepsea species known:

2431 *Morrisonella pacifica* (Dall, 1908). 1,569 fms., southwest of Sitka, Alaska. See Bartsch, 1945, *The Nautilus,* vol. 59, p. 23, pl. 3, figs. 11–14.

Family MELONGENIDAE Gill, 1867

Genus *Melongena* Schumacher, 1817

Large, heavy, spined shells with a thick periostracum. Operculum corneous, unguiculate and with a terminal nucleus. Carnivorous, feeding mainly on other mollusks. Shallow-water inhabitants. Eggs laid in strings of horny, coin-shaped capsules. Larval veligers are free-swimming. Synonym: *Galeodes* Röding, 1798, not Olivier, 1791. Type: *melongena* Linné, 1758. A full review of the American *Melongena* is given by Clench and Turner, 1956, in *John-*

sonia, vol. 3, no. 35. For feeding habits and travel ability, see *The Nautilus,* vol. 70, p. 84, and vol. 72, p. 117.

Subgenus *Rexmela* Olsson and Harbison, 1953

Melongena corona (Gmelin, 1791)   **2432**
Common Crown Conch   **Color Plate 12**

Florida, the Gulf States and Mexico.

2 to 4 inches in length, very variable in size, color, shape and production of spines. Dirty-cream with wide, spiral bands of brown, purplish brown or dark bluish black. Pure white "albinos" are infrequent. Shoulder and base of shell with 1, 2, 3 or 4 rows of semitubular spines which may point upward or horizontally. Synonyms are: *belknapi* Petit, 1852; *subcoronata* Heilprin, 1887 (type of the subgenus); *aspinosa* Dall, 1890; *inspinata* Richards, 1933; *perspinosa* Pilsbry and Vanatta, 1934. Numerous varieties have been named. Some may be considered as subspecies, all of which are common:

2433 (Color Plate 12) subspecies *bicolor* (Say, 1827). Miami to Key West and Dry Tortugas. 1½ inches, dwarf, yellowish white with a few brown bands and dark spire. Shoulder with numerous, small, erect, fluted spines. Synonyms: *minor* Sowerby, 1878; *estephomenos* Melvill, 1881.

2434 (Color Plate 12) subspecies *johnstonei* Clench and Turner, 1956. Alabama to northwest Florida. 3 to 7 inches; shell elongate, shoulder tabulate or flattened, spines erect; color dark.

2435 (Color Plate 12) subspecies *bispinosa* (Philippi, 1844). Yucatan, Mexico. 2 to 3 inches, with 1 or 2 rows of blunt spines at the shoulder; scale-like spines on the siphonal fasciole; spirally threaded. Uncommon. Synonym: *martiniana* Philippi, 1844.

2436 subspecies *altispira* Pilsbry and Vanatta, 1934. St. Augustine to Marco Island. This may be only an ecologic form. 2 to 3 inches. Spire high; usually with only a single row of very short, erect spines on the sloping shoulder.

Melongena melongena (Linné, 1758)   **2437**
West Indian Crown Conch   **Color Plate 12**

West Indies.

3 to 6 inches in length, similar to *corona,* but heavier, with rounded shoulders; smaller, more solid spines, and with a distinct channel at the suture. Common in the Greater Antilles.

Melongena patula (Broderip and Sowerby, 1829)   **2438**
Pacific Crown Conch

Gulf of California to Ecuador.

6 to 10 inches, heavy, round-shouldered, brown to purplish brown with narrow bands of whitish. Aperture very large, yellowish to white. Periostracum smoothish, fairly thick. Fairly common; on intertidal mud flats.

Genus *Pugilina* Schumacher, 1817

Shell 3 to 6 inches, fusiform, noduled shoulder, smooth columella, heavy periostracum, corneous operculum. Type: *Murex morio* Linné, 1758. One species limited to east South American and West Africa.

Pugilina morio (Linné, 1758)   **2439**
Giant Hairy Melongena

Trinidad to Brazil. Central West Africa.

3 to 6½ inches, solid, bluntly spinose. Covered with a thick, brown, rough periostracum. Shell color dark-brown with one

or more narrow light bands at the periphery. Spire high, suture well-indented. Shoulder with strong blunt spines. Parietal wall glazed in chocolate-brown. Inner part of outer lip spirally lirate. Operculum claw-shaped, with an apical nucleus, corneous, black-brown. Common in shallow mangrove waters. *Fusus coronatus* Lamarck, 1803, and *fasciata* Schumacher, 1817, are synonyms.

Subfamily BUSYCONINAE Finlay and Marwick, 1937

Genus *Busycon* Röding, 1798

Shells 6 to 16 inches, subpyriform, heavy, dextrally or sinistrally coiled. Columella without strong folds. Periostracum thin to thick. Siphonal canal narrow and fairly long. 2 nuclear whorls, the first globose and folded over the second. Operculum corneous, unguiculate, with an apical nucleus. Central radular tooth with 5 or 7 cusps. Type: *carica* (Gmelin, 1791). *Fulgur* Montfort, 1810, and *Sycopsis* Conrad, 1867, are synonyms. An exhaustive review was made by S. C. Hollister, 1958, *Palaeontol. Amer.*, vol. 4, no. 28, which we follow in part. Also see Pulley, 1959, *Rice Inst. Pamphlet,* vol. 46, no. 1.

Busycon carica (Gmelin, 1791)   2440
Knobbed Whelk

South shore of Cape Cod to Cape Canaveral, Florida.

Adults 5 to 9 inches in length; characterized by having low tubercles on the shoulder of the whorl and in being right-handed. Aperture light orange-yellow, but sometimes brick-red. The young show axial streaks of brownish purple. Common in shallow water. Synonyms: *muricatum* Röding, 1758, and *aruanum* of Hollister, 1958, not Linné, 1758. Sinistral *carica* live offshore in southern New Jersey.

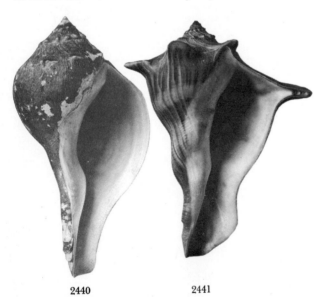

2440   2441

Busycon carica subspecies *eliceans* (Montfort, 1810)   2441
Kiener's **Whelk**

North Carolina to central east Florida.

Adults 5 to 8 inches, heavy, with a semiglossy surface, and a heavy swelling running around the center girth of the body whorl. The spines on the shoulder are generally prominent. Columella is usually enamel-white with yellowish or brownish suffusions. Presumed hybrids between this and the northern *carica* are not uncommon in the area of Virginia and North Carolina. Locally common.

Busycon candelabrum (Lamarck, 1816)   2442
Candelabrum Whelk   Color Plate 12

Banks off Texas to Yucatan, Mexico.

4 to 5 inches, dextral or right-handed, subpyriform to elongate. Siphonal canal somewhat narrow and long. Shoulders with numerous rather short spines. The anal or upper notch (where the outer lip meets the previous whorl) has spiral lirae, which are also present on the inside of the body whorl. Growth lines on the outer shell surface have reddish brown lines. Aperture salmon to yellow. Not uncommon at depths of 20 to 35 fathoms. This is not a dextral form of *perversum pulleyi.* The latter has a spire angle of about 90 degrees, while in *candelabrum* it is about 110 degrees.

Subgenus *Busycoarctum* Hollister, 1958

Turnip-shaped, with a long, thin canal. Operculum oval. Type: *coarctatum* (Sowerby, 1825).

Busycon coarctatum (Sowerby, 1825)   2443
Turnip Whelk   Color Plate 12

Bay of Campeche, Mexico.

Until 1950 this was considered a very rare species, but dredging activities of shrimp trawlers have brought a large number of them to light. Characterized by its turniplike shape, single row of numerous small, dark-brown spines, and by its golden yellow aperture. Often axially streaked with dark-brown. 5 inches in length.

Subgenus *Sinistrofulgur* Hollister, 1958

Busycon whelks which coil sinistrally. This character may be significant at the species level, but there is some doubt as to its phylogenetic significance at even a subgeneric level. Type: *contrarium* Conrad, 1840, recent forma *sinistrum* Hollister, 1958.

Busycon contrarium (Conrad, 1840)   2444
Lightning Whelk   Color Plate 12

Off New Jersey to Florida and the Gulf States.

4 to 16 inches in length, left-handed, with a row of moderately small, triangular knobs at the shoulder. Color grayish white with long, axial, wavy streaks of purplish brown which are blurred along their posterior edge. Albino shells are rare. A very common species in west Florida. Large white, left-handed specimens are sometimes recovered by scuba divers off the coast of central New Jersey.

The original fossil forms of *contrarium* have smoothish, rounded shoulders, a feature seen today in colonies found around Longboat Key, Sarasota, Florida. The living form with smooth shoulders has the name (**2445**) *aspinosum* Hollister, 1958. *B. sinistrum* Hollister, 1958 (**2446**) is the living, spinose form, although some workers may wish to consider it a separate species. Some authorities feel that the common lightning whelk should be called *perversum* Linné, an idea not without technical merit.

Busycon perversum (Linné, 1758)   2447
Perverse Whelk

Campeche Bay, Mexico.

4 to 8 inches in length, very heavy and with a glossy finish. This species should not be confused with the common *contrarium.* The name *B. perversum* or *Fulgur perversa* in most old popular books refers to *B. contrarium.* The perverse whelk

is an uncommon species. It is characterized by the heavy, polished shell and the swollen, rounded ridge around the middle of the whorl. Dredged from 4 to 10 fathoms. Synonyms: *kieneri* Philippi, 1848; *gibbosum* Conrad, 1853.

The subspecies *pulleyi* (Hollister, 1958) **(2448)** occurs from Breton Sound, Louisiana, to Texas and to the north Mexican coast. It is characterized by being much more slender, having a turreted spire and having more and smaller spines on the shoulders. It lacks the swelling around the girth of the body whorl. Intergrades exist in its southern range. In the northeast part of its range, this subspecies blends in with *contrarium*, giving rise to the possibility that this complex is one species, namely *perversum* Linné.

Subgenus *Busycotypus* Wenz, 1943

Suture deeply and squarely channeled; periostracum thick and velvety; central radular tooth with 3 cusps. Type: *canaliculatum* (Linné, 1758). *Sycofulgur* E. S. Marks, 1950, is a synonym. This is *Sycotypus* of authors, not Mörch, 1852.

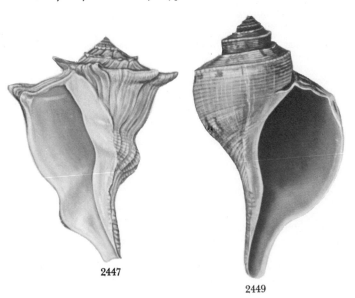

2447

2449

Busycon canaliculatum (Linné, 1758)   **2449**
Channeled Whelk

Cape Cod to St. Augustine, Florida.
Introduced to San Francisco Bay, California.

5 to 7½ inches in length, characterized by a deep, squarish, rather wide channel running along the suture and by the heavy, feltlike, gray periostracum. The shoulder has a strong carina or ridge which is weakly beaded in the early whorls. Common in shallow, sandy areas. Left-handed specimens are rare. Introduced to California about 1948 (R. Stohler, 1962, *The Veliger*, vol. 4, p. 211).

Subgenus *Fulguropsis* E. S. Marks, 1950

Suture channeled with a V-shaped groove, the inner face being almost vertical. *Pyrofulgur* Hollister, 1958, is a synonym. Both have the same type: *pyrum* Dillwyn, 1817, which is *spiratum* Lamarck, 1816.

Busycon spiratum spiratum (Lamarck, 1816)   **2450**
True Pear Whelk

Off south Texas to Yucatan, Mexico.

3 to 4 inches in length, pyriform, shoulder rounded or with a very weak carination. Suture V-shaped. Shell rather strong, heavy and its surface strongly spirally grooved. Interior of aperture strongly lirate. *B. pyrum* Dillwyn, 1817, is a synonym. Found offshore. Absent in Florida.

Busycon spiratum subspecies pyruloides (Say, 1822)   **2451**
Say's Pear Whelk   **Color Plate 6**

North Carolina to both sides of Florida.

4 to 6 inches, similar to typical *spiratum* from Mexico, but is more elongate, thinner-shelled, has very weak spiral threads on the outside; adults with the inside of the aperture smooth. Exterior sometimes with suffused spiral bands of purple-brown. Periostracum velvety-gray. Common; in shallow, sandy areas. Sinistral specimens very rare.

Busycon spiratum subspecies plagosum (Conrad, 1863)   **2452**
                                                    Color Plate 12

Alabama to Texas to Campeche Bay, Mexico.

Similar to typical *spiratum* and *pyruloides*, but the shell is longer, narrower, with a slimmer siphonal canal, with a deep suture which may be V-shaped or squarish. Most specimens have a strong carina on the shoulder which in the early whorls is beaded. *B. galvestonense* (Hollister, 1958), and *texanum* (Hollister, 1958), are minor colonial variants too difficult to distinguish consistently. Common offshore in several fathoms.

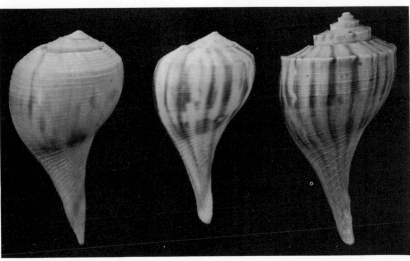

2450   2451   2352

Family NASSARIIDAE Iredale, 1916

Genus *Nassarius* Duméril, 1806

Shells less than 1 inch, ovate, with a sharp spire and broad, thickened parietal shield. Posterior end of foot with 2 short cirri. Animals are scavengers. Type: *arcularius* (Linné, 1758) of the Indo-Pacific. The genus *Nassa* Lamarck, 1799, non Röding, 1798, is a synonym.

Subgenus *Nassarius* Duméril, 1806

Nassarius vibex (Say, 1822)   2453
Common Eastern Nassa

Cape Cod to Florida, to Texas and the West Indies to Brazil.

½ inch in length, heavy, with a well-developed parietal shield. Last whorl with about a dozen, poorly developed, axial ribs which are coarsely beaded. Color gray-brown to whitish with a few splotches or broken bands of subdued, darker brown. A common sand or mud-flat species. Some specimens have numerous weak spiral cords. Parietal shield sometimes yellowish. Recently introduced (1967) to Bermuda (R. Jensen, *in litt.*).

Nassarius acutus (Say, 1822)   2454
Sharp-knobbed Nassa

Central Florida to Texas.

¼ inch in length, characterized by its glossy shell, its strong, pointed beads and in occasionally having a narrow, brown, spiral thread connecting the beads. Moderately common. Fossil specimens are twice as large.

2453   2554   2557

Nassarius insculptus (Carpenter, 1864)   2455
Smooth Western Nassa

Point Arena, California, to Baja California.

¾ inch in length, outer lip thickened, parietal wall thick, white but not very wide. Body whorl smoothish, except for weak, fine spiral threads. Axial ribs numerous only on early whorls. Color white; covered by a yellowish white periostracum. Moderately common; dredged from 20 to 200 fathoms. *N. eupleura* (Dall, 1916), is a synonym.

Nassarius tegula (Reeve, 1853)   2456
Western Mud Nassa

San Francisco to Baja California.

¾ inch in length, moderately heavy, with a heavy, whitish or brown-stained parietal callus. Body whorl smoothish around the middle, but with a spiral row of fairly large nodes below the suture. In the spire, the nodes are usually divided in two. Base of body whorl with a few weak, spiral threads. Outer lip thick. Color olive-gray to brownish, often with a narrow, whitish or purplish, spiral band. A common mud-flat species. *N. tegulus* is erroneous.

Subgenus *Reticunassa* Habe and Ito, 1965

Nassarius fraterculus (Dunker, 1860)   2457

Puget Sound, Washington.
Introduced from Japan.

8 to 13 mm., grayish, with a subdued spiral band, rather ovoid, 6 whorls, with about 16 rounded, curved axial ribs per whorl. Spiral sculpture of threads weak. Parietal wall raised, narrow and with a sharp left edge. Umbilicus chinklike. Well established at the Bayview State Park, Skagit County, by 1960.

Subgenus *Hinia* Gray, 1847

Type of this subgenus is *reticulatus* (Linné, 1758) from western Europe.

Nassarius albus (Say, 1826)   2458
Variable Nassa

North Carolina to Florida, Texas, and the West Indies to Brazil. Bermuda.

½ inch in length, relatively light-shelled, usually pure-white in color, but occasionally with 1 or 2 narrow, spiral bands of light yellowish brown. Number of strong, axial ribs per whorl varies from 8 to 12. Upper part of whorl shouldered. Numerous spiral, rounded cords are strong or weak. Parietal shield enamel-white, usually not well-developed. Common; 1 to 5 fathoms. *N. consensus* (Ravenel, 1861) is possibly only a form of this unusually variable species. Synonyms: *ambiguus* Pulteney, 1799, non Solander, 1766; *antillarum* Orbigny, 1842; *candei* Orbigny, 1842; *candidissimum* C. B. Adams, 1845; and *floridensis* Olsson and Harbison, 1953. Rare in Texas.

Nassarius trivittatus (Say, 1822)   2459
New England Nassa

Newfoundland to off northeast Florida.

¾ inch in length, rather light-shelled, 8 to 9 whorls; nuclear whorls smooth. Whorls in spire with 4 to 5 rows of strong, distinct beads. Parietal wall thinly glazed with white enamel. Outer lip sharp and thin. Whorls slightly channeled just below the suture. Color light-ash to yellowish gray. Common from shallow water to 45 fathoms. See R. S. Scheltema, 1964, *The Nautilus*, vol. 78, p. 49. They commonly lay their eggs on the underside of *Polinices* "sand collar" eggmasses.

2456

2459   2463

Nassarius perpinguis (Hinds, 1844)   2460
Western Fat Nassa

Puget Sound to Baja California.

¾ to 1 inch in length, fairly thin, with a rather fragile outer lip. Similar to *N. rhinetes* but with much finer sculpture (usually finely cancellate or minutely beaded), and yellowish white in color with 2 or 3 narrow, spiral bands of orange-brown, one of which borders the suture. The sculpture is variable with spiral threads often predominant. Very abundant along most of the coast. Intertidal flats to 50 fathoms.

Subgenus *Demondia* Addicott, 1965

Nassarius rhinetes S. S. Berry, 1953   2461
California Nassa

Squaw Creek, Oregon, to Baja California.

1 inch in length, without a thick parietal shield and the outer lip not thickened. Shell with numerous, rather coarse beads arranged in 20 to 30 axial, slanting ribs. 11 to 12 spiral threads on the last whorl; 5 to 7 on the whorls above. Color white with an ashy or yellow-gray periostracum. Moderately common just offshore to 35 fathoms. Compare with *perpinguis*. Alias *californianus* Conrad, 1856, a fossil. *Trans. San Diego Soc. Nat. Hist.*, vol. 11, p. 415.

Nassarius mendicus (Gould, 1849)   2462
Western Lean Nassa

Alaska to Baja California.

½ to ¾ inch in length, with a moderately high spire. Outer lip not thickened. Sculpture consists of numerous, small beads which are formed by the crossing of about a dozen small axial ribs and smaller spiral threads. Color yellowish gray. Common in shallow water in the north; shore to 25 fathoms.

(2462a) The subspecies or form *cooperi* (Forbes, 1850), has weaker spiral threads and about 7 to 9 strong, whitish smoother axial ribs which persist to the last of the body whorl. Color grayish yellow to whitish, often with fine, spiral, brown or mauve lines. Very common in the south; shore to 25 fathoms.

2458   2460   2462

Subgenus *Zaphon* H. and A. Adams, 1853

Type of the subgenus is *fossatus* (Gould, 1849).

Nassarius fossatus (Gould, 1849)   2463
Giant Western Nassa

Vancouver Island to Baja California.

1½ to 2 inches in length, orange-brown to brownish white in color. Early whorls coarsely beaded; last whorl with about a dozen coarse, variously sized, flat-topped spiral threads and with about a dozen short axial ribs on the top ⅓ of the last whorl. Outer lip with a jagged edge and constricted at the top. The largest and one of the common intertidal mud snails on the Pacific coast.

Other Atlantic species:

2464 *Nassarius nigrolabra* (Verrill, 1880). South of Martha's Vineyard, Massachusetts, 155 fms.

2465 *Nassarius hotessieri* (Orbigny, 1845). North Carolina to the West Indies, 35 to 85 fms. Common.

2466 *Nassarius scissuratus* (Dall, 1889). Florida Strait and the West Indies, 56 to 140 fms.

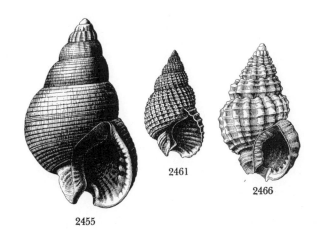

2461

2466

2455

Other Pacific species:

2467 *Nassarius angulicostis* (Pilsbry and Lowe, 1932). Gulf of California to Panama.

2468 *Nassarius catallus* (Dall, 1908). Baja California to Peru. (Synonym: *polistes* Dall, 1917.)

2469 *Nassarius cerritensis* (Arnold, 1903). Baja California, the Gulf of California. Also Pleistocene of southern California.

2470 *Nassarius gallegosi* Strong and Hertlein, 1937. Guaymas, Mexico, to Ecuador, offshore to 110 meters.

2471 *Nassarius guaymasensis* (Pilsbry and Lowe, 1932). Gulf of California.

2472 *Nassarius howardae* Chace, 1958. Gulf of California.

2473 *Nassarius insculptus* (Carpenter, 1864). California to the Gulf of California. (Synonym: *gordanus* Hertlein and Strong, 1951.)

2474 *Nassarius limacinus* (Dall, 1917). Gulf of California.

2475 *Nassarius miser* (Dall, 1908). Gulf of California to Panama, 150 to 590 meters.

2476 *Nassarius onchodes* (Dall, 1917). Baja California, the Gulf to Panama, in 18 to 55 meters.

2477 *Nassarius pagodus* (Reeve, 1844). Magdalena Bay to Ecuador. (Synonyms: *canescens* C. B. Adams, 1852; *acuta* Carpenter, 1857.)

2478 *Nassarius shaskyi* McLean, 1970. Gulf of California to Colombia.

2479 *Nassarius versicolor* (C. B. Adams, 1852). Magdalena Bay, Baja California, to Peru. (Synonyms: *rufocincta* A. Adams, 1852; *glauca*, *proxima*, *striata* and *striatula* C. B. Adams, 1852; *crebristriata* Carpenter, 1857; *lecadrei* De Folin, 1867.)

2480 *Nassarius* (*Arcularia* Link, 1807) *iodes* (Dall, 1917). Head of the Gulf of California to Mazatlan, Mexico.

2481 *Nassarius* (*A.*) *luteostoma* (Broderip and Sowerby, 1829). Gulf of California to Peru. (Synonym: *nodulifera* Carpenter, 1857.)

2482 *Nassarius (A.) moestus* (Hinds, 1844). Head of the Gulf of California.

2483 *Nassarius (A.) tiarula* (Kiener, 1841). Entire Gulf of California to Panama. (Synonym: *major* Stearns, 1894.)

Genus *Ilyanassa* Stimpson, 1865

Columella with a single strong spiral ridge near the base. Shell and periostracum darkly colored. Posterior end of foot without the cirri. Type: *obsoleta* (Say, 1822). The animal is omnivorous and possesses an internal crystalline style.

Ilyanassa obsoleta (Say, 1822)   2484
Eastern Mud Nassa

> Gulf of St. Lawrence to northeast Florida.
> Vancouver, British Columbia, to central California.

¾ to 1 inch in length, usually covered with mud and algae, and has its spire eroded at the tip. Color dark black-brown, rarely with a spiral white band. Sculpture of numerous rows of weak beads. Parietal wall thickly glazed with brown and gray. Columella with a single, strong spiral ridge near the base. Outer lip with half a dozen small grayish teeth which run back into the aperture. Very common on oozy, warm mud flats. Rarely, some colonies may have a color form with a spiral whitish band. Introduced to California prior to 1911. For biology and large bibliography, see S. C. Brown, 1969, *Malacologia,* vol. 9, no. 2, pp. 447–500.

2484

Family FASCIOLARIIDAE Gray, 1853

Shells usually large, with a well-developed siphonal canal. Interior of body whorl with numerous spiral lirae. Columella with a few weak folds near the base. Operculum horny, unguiculate. For feeding habits on worms and bivalves, see R. T. Paine, 1966, *The Veliger,* vol. 9, p. 17.

Subfamily PERISTERNIINAE Tryon, 1880

Genus *Leucozonia* Gray, 1847

Shell solid, ovate-fusiform. Anterior canal short. Columella truncate at the lower end and with 3 or 4 oblique plaits in the middle. Type: *nassa* (Gmelin, 1791). Shallow water.

Leucozonia nassa (Gmelin, 1791)   2485
Chestnut Latirus

> Florida to Texas and the West Indies to Brazil.

1½ inches in length, heavy, squat, with its whorls shouldered by about 9 large nodules. Characterized by its semiglossy, chestnut-brown color with a faint, narrow spiral band of whitish at the base of the shell which terminates into a

small, distinct spine on the outer lip. Columella with 4 weak folds at the base. Aperture yellowish tan within. Soft parts orange-red. Common among rocks at low tide. Alias *L. cingulifera* Lamarck, 1816; *fuscata* Gmelin, 1791; *angularis* Reeve, 1847; and *Turbinella knorrii* Deshayes, 1843. For anatomy and biology, see Marcus and Marcus, 1962, Bol. Fac. Fil. Univ. Sao Paulo, no. 261.

The subspecies, *leucozonalis* (Lamarck, 1822) **(2486)** a smoothshouldered, round, black or brown form occurs in Cuba, Haiti and Grand Cayman Island (see Abbott, 1958, Monograph 11, Acad. Nat. Sci. Philadelphia, p. 79, map 11). Both forms may rarely be colored light yellow-brown.

Leucozonia ocellata (Gmelin, 1791)   2487
White-spotted Latirus   Color Plate 11

> Southeast Florida and West Indies to Brazil.

¾ inch in length, ⅔ as wide, squat and heavy. Color darkbrown to blackish with a row of about 8 large, white, rounded nodules at the periphery and about 3 or 4 spiral rows of smaller white squares on the base of the shell. Base of columella with 3 small folds. Apex usually worn white. A common Caribbean intertidal species found under rocks. Rare in Palm Beach County, Florida.

2487

2485

Genus *Opeatostoma* S. S. Berry, 1958

Shell ovate, spirally corded and with a long tooth on the lower outer lip. Type: *pseudodon* (Burrow, 1815).

Opeatostoma pseudodon (Burrow, 1815)   2488
Thorn Latirus   Color Plate 11

> Gulf of California to Peru.

1 to 1½ inches, whorls shouldered and with numerous brown, fine, spiral, smooth ridges on a cream background. Periostracum brown. Outer lip with a long, white spine at the base. Common; among rocks at the low-tide mark. This is *Monoceros cingulatum* Lamarck, 1816, and *angulatum* Rogers, 1913.

Genus *Latirus* Montfort, 1810

Shell elongate-fusiform with a moderately long siphonal canal. Type: *gibbulus* (Gmelin, 1791) from the Indo-Pacific. *Lathyrus* is a misspelling.

Latirus cariniferus Lamarck, 1822   2489
Trochlear Latirus   Color Plate 11

> Southeast Florida, Gulf of Mexico, to the West Indies.

1 to 2½ inches, elongate, heavy, 8 to 10 whorls, with 7 to 9 low, rounded ribs which are noduled by 2 spiral cords in the

upper whorls and 4 cords on the wide periphery of the last whorl. Numerous fine spiral threads present. Umbilicus variable, but sometimes funnel-shaped. Color yellow, cream or yellow-brown, sometimes with brown axial bars between the ribs. Lower part of columella with 2 weak folds. Uncommon; in rocky reef areas in shallow water. Aperture bright-yellow. *L. trochlearis* Kobelt, 1876, is a synonym.

The form *macgintyi* Pilsbry, 1939 (**2490**) not uncommon in the Lower Florida Keys and off northwest Florida but rare in the West Indies, lacks the brown stripes between the axial ribs, and has a very open umbilicus.

Latirus infundibulum (Gmelin, 1791)   **2491**
Brown-lined Latirus   **Color Plate 11**

Off Palm Beach, Florida, and the West Indies to Brazil.

3 inches in length, heavy, resembling a *Fusinus* in shape, but characterized by the 3 weak folds on the columella, the light-tan to light-brown shell which bears small, darker brown, wavy, glossy, smooth spiral cords. 7 to 8 strong axial nodules per whorl. Umbilicus imperfect, sometimes funnel-shaped. Moderately common in the West Indies, rare in Florida. 1 to 30 fathoms.

Latirus angulatus (Röding, 1798)   **2492**
Short-tailed Latirus   **Color Plate 11**

South Florida and the West Indies to Brazil.

1 to 2½ inches in length, rather broad, with a short siphonal canal, with 8 to 9 rounded, long axial ribs crossed by numerous spiral threads. Color light-chestnut, reddish brown or dark-brown. Not so shouldered as, and less coarsely sculptured than, *cariniferus*. It is much stouter and not so elongate as *infundibulum,* but like that species may have narrow, brown spiral lines or threads. Moderately common in the West Indies. Rare in Florida. *L. cymatias* Schwengel, 1940, and *brevicaudatus* Reeve, 1847 are synonyms. *The Nautilus,* vol. 53, p. 110.

Other species:

2493 *Latirus (Polygona) virginensis* Abbott, 1958. Virgin Islands to Brazil. Monograph no. 11, Acad. Nat. Sci. Philadelphia, p. 76. Common. **Color Plate 11.**

2493a *Latirus (Latirus) varai* Bullock, 1970. Off Gibara, Oriente Province, Cuba, 100 fms. *The Nautilus,* vol. 83.

2493 a   2498

2494 *Latirus hartvigii* (Shuttleworth, 1859). Hartwig's Latirus. 1½ inches. Greater Antilles; Virgin Islands. Uncommon. **Color Plate 11.**

2495 *Latirus macmurrayi* Clench and Aguayo, 1941 (spelling emended herein from *mcmurrayi* in conformity with the I.C.Z.N. rules). Off Matanzas, Cuba, 190 fms. *Mem. Soc. Cubana Hist. Nat.,* vol. 15, p. 178.

2496 *Latirus hemphilli* Hertlein and Strong, 1951. Magdalena Bay, Baja California, to Panama Bay.

2497 *Latirus praestantior* Melvill, 1892. Gulf of California, in 10 to 20 meters. (Synonym: *Turbinella gracilis* Reeve, 1847.)

Genus *Dolicholatirus* Bellardi, 1884

Shell 2 inches, slender, fusiform, with swollen axial ribs crossed by spiral cords. Aperture small, oval internally lirate. 2 plaits on the columella. Type: *bronni* (Michelotti, 1897). Synonym: *Fusilatirus* McGinty, 1955.

Dolicholatirus cayohuesonicus (Sowerby, 1878)   **2498**
Key West Latirus

Florida; Bahamas to the Virgin Islands.

½ inch, fusiform, dark chocolate-brown with 7 or 8 large, rounded axial ribs per whorl; spiral sculpture of several raised, strong, smoothish major cords (5 or 6 in the spire whorls) between which may be a minor spiral thread. Aperture small, ovate, dark-brown with a sharply raised parietal wall bearing 3 strong, whitish teeth. Inside of body whorl with 4 or 5 raised, whitish spiral plicae. Periostracum dark-brown, fairly thick, and with fine axial threads. Operculum unguiculate, its exterior with crinkly, raised growth lines. Uncommon; in a few feet of water.

Other species:

2499 *Dolicholatirus pauli* (McGinty, 1955). 70 fms. off the Florida Keys. West coast of Florida, 70 to 200 fms. Rare. *Proc. Acad. Nat. Sci. Philadelphia,* vol. 107, p. 79.

2499

Subfamily FASCIOLARIINAE Gray, 1853

Genus *Fasciolaria* Lamarck, 1799

Shells large, fusiform, with rounded whorls, long siphonal canal. Operculum brown, chitinous, smooth-edged, unguiculate. Type: *tulipa* (Linné, 1758). For an ecologic study of the Florida species, see Fred E. Wells, Jr., 1970, *The Veliger,* vol. 13, no. 1, pp. 95–108.

Fasciolaria tulipa (Linné, 1758)   **2500**
True Tulip   **Color Plate 10**

North Carolina to south half of Florida to Texas and the West Indies. Brazil.

3 to 8 inches in length, with 2 or 3 small spiral grooves just below the suture, between which the shell surface is often crinkled. Sometimes with broken spiral color lines. A beautiful orange-red color variety is not uncommon on the Lower Florida Keys and Yucatan. Common. Giants reach a length of 10 inches.

Subgenus *Cinctura* Hollister, 1957

A prominent spiral ridge emerges from the aperture below the suture and extends across the parietal wall to the margin of the callus. Type: *hunteria* (Perry, 1811).

Fasciolaria lilium G. Fischer, 1807   2501
Banded Tulip

North Carolina to Yucatan, Mexico.

2 to 5 inches in length, whorls smooth, even at the sutures. The widely spaced, rarely broken, distinct, 7 to 11 spiral, purple-brown lines and cloudy background are characteristic. Rarely all-white. Formerly known as *distans* Lamarck, 1822. There are 4 distinct subspecies. Axial riblets (a probable ecologic dietary character), shape of shell and coloration distinguish these subspecies (see S. C. Hollister, 1957, *The Nautilus*, vol. 70, p. 73).

2501 (Color Plate 10) Subspecies *lilium* G. Fischer, 1807. Mississippi Delta west and southward to Yucatan, Mexico, and the northern side of Dry Tortugas, Florida, in 2 to 25 fathoms. Second nuclear whorl with 7 axial riblets, followed by 13 riblets per whorl on the next 2 whorls. Last whorl with 9 or 10 spiral brown lines. Background creamy white with mauve or yellowish axial flames. (Synonym: *distans* Lamarck, 1822.) 3 to 4 inches. Uncommon.

2502 (Color Plate 10) Subspecies *hunteria* (G. Perry, 1811). Cape Hatteras, North Carolina to Mobile Bay, Alabama, and all of Florida. No axial riblets on early whorls in shallow water colonies. Last whorl with 5 or 6 maroon spiral lines. Background ivory or bluish gray and with mauve axial flames. Upper whorls with 2 brown lines. 2½ to 3 inches. Common.

2503 (Color Plate 10) Subspecies or forma *branhamae* Rehder and Abbott, 1951. Gulf of Campeche, Mexico; south Texas. Protoconch followed by ¾ whorl of 15 axial riblets then ¾ whorl of 5 spiral threads. Siphonal canal long and orange-brown. Last whorl with 8 to 12 spiral purple-brown lines with 2 or 3 on upper third of canal. 4 to 5 inches. Uncommon.

2504 (Color Plate 10) Subspecies or forma *tortugana* Hollister, 1957. Whorl after the smooth protoconch has about 7 riblets. Last whorl with 6 primary spiral black-brown lines, 2 on above whorls. Background white with bright-red axial flames. 2 to 4 inches. Uncommon.

2505 Subspecies or forma *bullisi* Lyons, 1972. Deep water, off northwest Florida, 73 to 119 meters. Shell very elongate, with 10 to 12 thin spiral bands, between which may be smaller ones. Background color pale-yellow. No spiral ridge extending onto the upper part of the parietal wall. Uncommon. *The Nautilus*, vol. 85, p. 96.

Genus *Pleuroploca* P. Fischer, 1884

Shell fusiform, heavy and very large. Shoulder nodulose. Periostracum fairly thick. Type: *trapezium* (Linné, 1758) from the Indo-Pacific.

Pleuroploca gigantea (Kiener, 1840)   2506
Florida Horse Conch

North Carolina to Florida; Texas and Yucatan.

Almost 2 feet in length, although usually about 1 foot. Outer surface dirty-white to chalky-salmon, and covered with a fairly thick, black-brown periostracum which flakes off in dried specimens. The young (up to about 3½ inches) have a

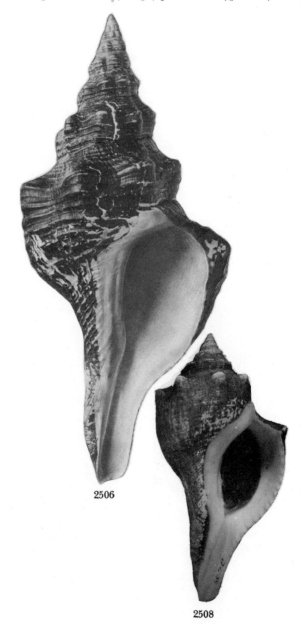

2506

2505

2508

thinner periostracum and the entire shell is a bright orange-red. Soft parts red. A form which lacks the nodules on the last whorl was named *reevei* Philippi, 1851. It occurs in northwest Sanibel Island. *P. papillosa* Sowerby, 1825 is insufficiently described to apply with any certainty to this species. This species was made the official shell for the state of Florida in 1969.

A similar, large species, *P. princeps* (Sowerby, 1825) **(2507)** (the Panama horse conch), occurs from the Gulf of California to Ecuador. Its operculum has deep, rounded grooves. Moderately common. Both of these horse conchs were previously put in the genus *Fasciolaria*.

Pleuroploca salmo (Wood, 1828)   2508
Granose Horse Conch

Gulf of California to Peru.

5 to 7 inches, solid, heavy, turnip-shaped, with a single row of blunt nodes or spines on the angular shoulder. Aperture, parietal wall glazed in orange or yellowish. 3 small oblique plica at the base of the columella. Periostracum either smoothish or minutely pustuled **(2509)** (form *granosa* Broderip, 1832). Uncommon to common, on intertidal mud flats.

Subfamily FUSININAE Swainson, 1840

Genus *Fusinus* Rafinesque, 1815

Shells spindle-shaped, elongate, with a long siphonal canal. Usually well-ribbed and with spiral threads. Columella lacks plicae. Operculum chitinous, with a terminal nucleus. Type: *colus* Linné, 1758, of the Indo-Pacific. This is *Fusus* of authors.

Subgenus *Fusinus* Rafinesque, 1815

Fusinus dupetitthouarsi (Kiener, 1846)   2510
Du Petit's Spindle

Baja California to Ecuador.

6 to 10 inches, spindle-shaped, white, with a fuzzy, greenish yellow periostracum. Early whorls rounded and with axial ribs, crossed by numerous somewhat equal-sized spiral threads.

Ribs gradually disappear and are only seen as knobs on the last whorls. Intertidal to several fathoms. Moderately common. Synonyms: *aplicatus* Grabau, 1904; *nodosus* Grabau, 1904; *funiculatus* Lesson, 1842.

Fusinus irregularis (Grabau, 1904)   2511
Irregular Spindle

Baja California.

4 to 6 inches, similar to *dupetitthouarsi*, but with angular whorls, strong spiral cord at the periphery, and a longer siphonal canal that is decidedly distorted. Rare on the Pacific coast of Baja California in 10 to 22 meters.

Subgenus *Heilprinia* Grabau, 1904

Similar to *Fusinus*, with a relatively short spire, with the protoconch sculptured with strong axial riblets over the entire surface of all of the nuclear whorls. Type: *caloosaensis* Heilprin, 1887.

Fusinus timessus (Dall, 1889)   2512
Turnip Spindle   Color Plate 11

Both sides of Florida to Texas.

About 3 to 5 inches in length, solid, pure-white to orange, with a thin, gray periostracum. Aperture round with a flaring, raised parietal wall which, like the inside of the outer lip, is enamel-white and bears numerous spiral threads. Each whorl with 10 to 12 low, short axial ribs at the periphery. Upper whorls with 8 to 9 small, but sharp and slightly wavy, smooth spiral cords. Last whorl and the long siphonal canal with a total of about 30 to 40 small cords between which is often a very fine one. Dredged uncommonly from 20 to 60 fathoms.

Fusinus eucosmius (Dall, 1889)   2513
Ornamented Spindle   Color Plate 11

Off both sides of Florida to Texas.

3 inches in length, with about 12 rounded whorls and with a small, roundish aperture located at the middle of the shell. Siphonal canal long, its diameter about equal to that of the aperture. Whorls with 8 to 10 large, rounded axial ribs which

2510   2511   2514   2516   2519

in the upper whorls are crossed by about 6 strong, sharp, slightly wavy spiral threads. Apex often leaning to one side. Color all-white to orange, with a rather heavy, grayish white to yellowish periostracum. Uncommonly dredged offshore 15 to 50 fathoms.

Fusinus dowianus Olsson, 1954   **2514**
Tom Dow's Spindle

Caribbean; Honduras.

3 to 5 inches, white with peach-brown between the ribs, especially in the spire. Whorls with 6 or 7 large, rounded, broad axial ribs per whorl crossed by alternating large and small, raised cords. Angle of spire about 40 degrees (60 degrees in *timessus*). Uncommon 20 to 50 fathoms. *The Nautilus,* vol. 67, p. 105. Very close to *eucosmius,* and may be only a southern subspecies of it.

Fusinus couei (Petit, 1853)   **2515**
Coue's Spindle   **Color Plate 10**

Gulf of Mexico; off southern Texas.

3 to 5 inches, narrowly elongate, heavy, white with the apex and siphonal tip suffused with yellow or orangish. Axial riblets on the first 4 or 5 whorls only; later whorls have numerous, strong, sharp, neatly raised spiral cords and threads. Outer lip finely denticulate. Do not confuse with the turrid, *Polystira tellea* (Dall). Dredged commonly by shrimp fishermen.

Fusinus helenae Bartsch, 1939   **2516**
Helen's Spindle

Gulf of Mexico to Venezuela.

3 to 4 inches, very similar to *couei,* but lighter in weight, whorls more evenly rounded, with much finer spiral threads, and usually white with axial streaks of yellow-brown. The first 3 whorls are brownish and have minute axial riblets. Uncommon; 17 to 55 fathoms.

Subgenus *Barbarofusus* Grabau and Shimer, 1909

Type of this subgenus is *harfordii* Stearns, 1871. *Harfordia* Dall, 1921, is a synonym.

Fusinus harfordii (Stearns, 1871)   **2517**
Harford's Spindle

Hope Island, British Columbia, to California.

2 inches in length, heavy, exterior dark, orange-brown, with 11 to 12 wide, rounded ribs crossed by small, sharply raised, finely scaled spiral cords. Rare in moderately deep water. The soft parts of living specimens give off a luminescent pinkish glow (see R. R. Talmadge, 1966, *The Veliger,* vol. 8, p. 199).

Fusinus kobelti (Dall, 1877)   **2518**
Kobelt's Spindle

Monterey to Catalina Island, California.

2½ inches in length, heavy, similar to *harfordii,* but with a longer siphonal canal, fewer and larger axial ribs (8 to 10 per whorl), colored white, except for several orange-brown spiral cords. Periostracum rather thick, opaque and light-brown. The spiral cords in *harfordii* are much larger and with squarish tops. Moderately common in shallow water to 35 fathoms.

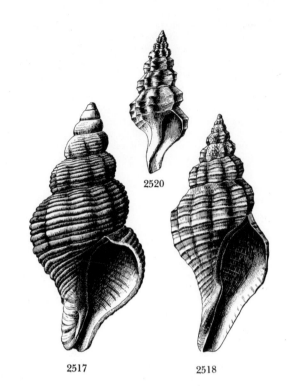

2520

2517   2518

Fusinus barbarensis (Trask, 1855)   **2519**
Santa Barbara Spindle

Oregon to San Diego, California.

4 to 5 inches in length, almost ¼ as wide, 9 to 10 rounded whorls, the early ones with about 10 low, axial ribs which are very weak or absent in the last 2 whorls. Spiral threads prominent and numerous. Color dirty gray-white, sometimes with a pinkish or yellowish cast. Dredged from 40 to 200 fathoms, and occasionally brought up in fishermen's nets. Common.

Other Atlantic species:

2520 *Fusinus amiantus* (Dall, 1889). Gulf of Mexico and Florida Strait, 805 fms.

2521 *Fusinus benthalis* (Dall, 1889). Florida Strait, 15 to 229 fms.

2522 *Fusinus bullatus* (Dall, 1927). Off Fernandina, Florida, 294 fms.

2523 *Fusinus aepynotus* (Dall, 1889). Gulf of Mexico and Florida Strait, 70 to 200 fms.

2524 *Fusinus alcimus alcimus* (Dall, 1889). Gulf of Mexico, 95 fms.

2525 *Fusinus alcimus rushii* (Dall, 1889). Off east Florida, 200 fms.

2526 *Fusinus amphiurgus* (Dall, 1889). Gulf of Mexico, 101 fms.

2527 *Fusinus halistreptus* (Dall, 1889). Florida Strait, 338 fms.

2528 *Fusinus schrammi* (Crosse, 1865). Off North Carolina to Florida, 294 to 440 fms.

2529 *Fusinus vitreus* (Dall, 1927). Off Georgia, 440 fms.

2530 *Fusinus ceramidus* (Dall, 1889). Off Barbados, 73 to 103 fms.

Other Pacific species:

2531 *Fusinus fredbakeri* (Lowe, 1935). San Felipe, Gulf of California to the Sonora coast; rare.

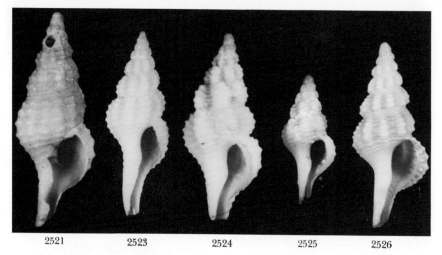

2521 2523 2524 2525 2526

2532 *Fusinus ambustus* (Gould, 1853). Baja California to Mazatlan, Mexico. Intertidal. Uncommon.

2533 *Fusinus monksae* (Dall, 1915). Banks Island, British Columbia, to Baja California. 20 to 100 fms. Common. (Synonym: *robustus* Trask, 1855, non Beyrich, 1853.)

Subgenus *Aptyxis* Troschel, 1868

Type of this subgenus is *syracusanus* (Linné, 1758) from the Mediterranean.

***Fusinus luteopictus* (Dall, 1877)**   2534
Painted Spindle

Monterey, California, to Baja California.

¾ inch in length, strong, with a thin outer lip. Color dark purplish brown with an indistinct, wide, spiral band of cream at the periphery. Common from low tide to 20 fathoms. Most workers now consider *Aptyxis* a subgenus of *Fusinus*.

***Fusinus cinereus* (Reeve, 1847)**   2535
Ashy Spindle

Gulf of California.

1 inch, high-spired, truncated short siphonal canal; with bulbous axial ribs crossed by brown spiral threads or cords.

2527   2530   2535

Color grayish with a brownish periostracum. Common; on rocks on intertidal mud flats. Synonyms: *taylorianus* (Reeve, 1848); *coronadoensis* Lowe, 1935; *sonoraensis* Lowe, 1935.

Other species:

2536 *Fusinus (Aptyxis) felipensis* Lowe, 1935. Upper Gulf of California. 19 mm.

2534

Superfamily Volutacea Rafinesque, 1815

Family OLIVIDAE Latreille, 1825

Shell elongate, very glossy, with a narrow aperture. With or without an operculum. Tentacles short, needlelike and set on thick, round pedicles of the same length.

Subfamily OLIVINAE Swainson, 1840

Shells usually over 1 inch; no operculum; radular ribbon long with about 100 transverse rows; central tooth tricuspidate. Tentacles and eyes present.

Genus *Oliva* Bruguière, 1789

Glossy shells, cylindrical in shape. Columella with several weak spiral lirae. Type: *oliva* Linné, 1758. For a more complete treatment, see "Olive Shells of the World," by R. F. Zeigler and H. C. Porreca, 1969.

***Oliva sayana* Ravenel, 1834**   2537
Lettered Olive   **Color Plate 13**

North Carolina to Florida and the Gulf States. Brazil.

2 to 2½ inches in length, moderately elongate, with a glossy finish and with rather flat sides. Color grayish tan with numerous purplish brown and chocolate-brown, tentlike markings. A common species found at night crawling in sand in shallow water. Formerly called *O. litterata* Lamarck, 1810, non Röding, 1798. Do not confuse with *O. reticularis* which is generally smaller, which has a much more shallow canal at the suture, whose apical whorls are slightly convex. Dead specimens buried for a long time in bay mud may take on an artificial black coloration. The yellow color (2537a; **Color Plate 13**) form has the name *citrina* Johnson, 1911, and is fairly rare. The lavender form is very rare, and does not hold its color for more than a few months after being collected. *O. circinata* Marrat, 1871, and *polita* Marrat, 1870, are also this species. For habits and eggs, see Olsson and Crovo, 1968, *The Veliger*, vol. 11, p. 31.

Oliva reticularis Lamarck, 1810   **2538**
Netted Olive   **Color Plate 13**

    Southeast Florida to Brazil. Bermuda.

1½ to 1¾ inches in length, similar to *sayana,* but smaller, more globose, with an oily finish and generally more lightly colored. Propodium and foot milk-white with light-brown lines. Sometimes yellowish, pure-white or very dark-brown in color. Rare in deep water off Miami, Florida (form *bollingi* Clench, 1934, (2539; **Color Plate 13**)). Common in Bermuda. A common West Indies species. A number of synonyms and color forms have been described: *memnonia* Duclos, 1844; *olivacea* Marrat, 1870; *sowerbyi* Marrat, 1870; *reclusa* Marrat, 1871; *bifasciata* Küster, 1878 (two brown bands); *greenwayae* Clench, 1937; *bollingi* Clench, 1934; *formosa* Marrat, 1870; *pattersoni* Clench, 1945 (solid dark-brown); *nivosa* Marrat, 1870; *olorinella* Duclos, 1835; *quersolina* Duclos, 1835; *pallida* Marrat, 1871; *tisiphona* Duclos, 1844.

Oliva scripta Lamarck, 1810   **2540**
Caribbean Olive   **Color Plate 13**

    Caribbean to Brazil.

1 to 2 inches, similar to *reticularis,* but more cylindrical, with a lower spire, a deeply channeled suture, a purplish interior to its aperture, and in general a more orange coloration. Synonyms include: *caribaeensis* Dall and Simpson, 1901; *trujilloi* Clench, 1938; *jamaicensis* Marrat, 1870, and possibly *graphica* Marrat, 1870 and *porcea* Marrat, 1870. Sometimes reddish, golden or solid-brown. Moderately common; usually offshore on sand bottoms.

Oliva spicata (Röding, 1798)   **2541**
Veined Olive   **Color Plate 13**

    Gulf of California to Panama.

1½ to 2½ inches, elongate, with a fairly high spire, resembling the Atlantic *reticularis.* Color variable but usually with numerous gray-brown zigzag fine axial lines. Aperture white within. Suture indented, and with feathery axial brown lines below. An abundant species with many forms: *violacea* Marrat, 1867 and *rejecta* Burch and Burch, 1962 (aperture violet); *cumingi* Reeve, 1850; *arachnoidea* Röding, 1798; *araneosa* Lamarck, 1811; *oriola* Duclos, 1835; *melchersi* Menke, 1851; *intertincta* Carpenter, 1857; *oblongata* Marrat, 1870; *fuscata* Marrat, 1871; *hemphilli* Johnson, 1911 (white form); *obesina* Duclos, 1835; *pindarina* Duclos, 1835; *subangulata* Philippi, 1848; *venulata* Lamarck, 1811; *punctata* Marrat, 1870. An abundant shallow-water species.

Oliva porphyria (Linné, 1758)   **2542**
Tent Oliva   **Color Plate 13**

    Gulf of California to Panama.

3 to 5 inches, characterized by triangular, tentlike white markings on a reddish brown or chestnut background. The peaks of the brown-bordered tents face towards the outer lip. Moderately common locally; in shallow water.

Oliva incrassata Lightfoot, 1786   **2543**
Angulate Olive   **Color Plate 13**

    Baja California, the Gulf, to Peru.

2 to 3½ inches, solid, heavy, angularly swollen at the shoulder. Color variable, but usually gray with dark-brown zigzag splotches. Outer lip and columella with pink to rose. Rarely one finds the golden form (*burchorum* Zeigler, 1969) at the northern end of the Gulf of California, and the rarer white form (*nivea* Pilsbry, 1910). Moderately common; low-tide mark in sandy areas. Synonyms: *angulata* Lamarck, 1811; *timorea* Marrat, 1871. Authorship of this species is Lightfoot, not Solander or Humphrey.

Oliva polpasta Duclos, 1835   **2544**
Polpast Olive   **Color Plate 13**

    Baja California, the Gulf, to Ecuador.

¾ to 1½ inches, similar to *spicata* but more pear-shaped, with numerous chocolate-brown spots on a gray-brown background. The lirations on the white columella are quite fine. Moderately common; in sandy shallow areas. *O. davisae* Durham, 1950, is a dark, shouldered form.

Oliva splendidula Sowerby, 1825   **2545**
Splendid Olive   **Color Plate 13**

    Gulf of California to Panama.

1 to 2 inches, oblong, with a fairly short spire. Base and apex purple to violet, rarely pink. Aperture yellowish white. Color creamy-flesh with 2 broad bands of brown which have minute triangular whitish spots. Uncommon; intertidal.

Other species:

2546a *Oliva ionopsis* S. S. Berry, 1969. Bahia de las Palmas, Baja California, 10 to 33 fms. 30 mm. *Leaflets in Malacology,* no. 26, p. 164.

2546 Fusiform Olive. *Oliva fulgurator* (Röding, 1798). 2 inches. Lower Caribbean. (Synonym: *fusiformis* Lamarck, 1816). **Color Plate 13.**

Subgenus *Stephonella* Dall, 1909

Oliva with a spiral incised line running around the lower middle of the shell, thus forming a wide band between it and the fasciole below that differs in coloration. Type: *undatella* Lamarck, 1810.

2547

Oliva undatella Lamarck, 1810   2547
Waved Olive

Baja California to Ecuador.

½ to ¾ inches, squat, small for an *Oliva,* yellowish white with hazy or cloudy streaks and blotches of bluish gray and brown. Basal zone yellowish, streaked with brown lines. Columella callus thick, narrow and with fine spiral lirae along its entire length, the 4 basal ones being the largest. Not uncommon; intertidal. Synonyms: *tenebrosa* Wood, 1828; *nedulina* Duclos, 1835.

Genus *Agaronia* Gray, 1839

Shells semiglossy, with an acuminate apex, and flaring lower outer lip. Operculum present. Type: *hiatula* (Gmelin, 1791).

Agaronia testacea (Lamarck, 1811)   2548
Panama False Olive   Color Plate 13

Gulf of California to Peru.

1 to 2 inches, light bluish gray to yellowish gray with brownish zigzag lines. Columella white. Interior of aperture violet-brown. Moderately common; burrow in sand at the midtide line. Synonyms: *reevei* Mörch, 1860; *griseoalba* von Martens, 1897: *philippi* von Martens, 1897.

Genus *Ancilla* Lamarck, 1799

Shell elongate to fusiform, solid, glossy, with an acuminate spire. Fasciole with 4 or 5 deep spiral grooves. Posterior parietal callus well-developed. Synonyms: *Anaulax* Roissy, in Montfort, 1805, and *Ancillaria* Lamarck, 1811. Type by monotypy: *cinnamonea* Lamarck, 1801, by being placed on the Official List of Generic Names in Zoology (I.C.Z.N. no. 170). *B. amplum* Gmelin is not the type. Three living species in American waters.

Subgenus *Eburna* Lamarck, 1801

Shell solid, fusiform, glossy, yellow, orange or white. Fasciole with 4 or 5 deep spiral grooves. Umbilical gutter deep and long. Columella smoothish, concave. Toothlike projection on the basal part of the outer lip. Type: *Buccinum glabratum* Linné, 1758.

Ancilla glabrata (Linné, 1758)   2548a
Golden Ancilla   Color Plate 13

Lower Caribbean.

2 to 3 inches, glossy, golden-yellow, with a white band at the glazed suture. Parietal callus above the deep umbilicus is white. Uncommon in sand offshore. *A. flavida* Lamarck, 1801, is a synonym.

Ancilla lienardi (Bernardi, 1858)   2548b
Lienard's Ancilla   Color Plate 13

Eastern Brazil.

1 to 1½ inches in length; dark orange, with a white, spiral fasciole groove above and below and orange spiral band on the base of the shell. Umbilicus deep and white. Aperture white. Differs from *glabrata* in being broader, smaller, and lacking the white sutural band. Uncommon; offshore.

Ancilla tankervillii (Swainson, 1825)   2548c
Tankerville's Ancilla   Color Plate 13

North coast of South America; Isla Margarita.

2 to 3 inches in length, lemon-yellow, without an umbilicus. Lower part of outer lip has a spiral internal ridge ending as a small tooth. Wide fasciole on the lower third of the last whorl is silky yellowish gray with numerous microscopic spiral scratches. Apex white. Interior of aperture stained with yellowish tan. Locally moderately common in sand, in 20 to 40 feet of water.

Subfamily OLIVELLINAE Troschel, 1869

Shell usually less than ¾ inch. Operculum chitinous, but sometimes absent. Radular ribbon wide, and with about 50 transverse rows of teeth, the central tooth being multicuspidate. A review of the American species of *Olivella* appears in A. Olsson, 1956, *Proc. Acad. Nat. Sci. Phila.,* vol. 108, pp. 155–225.

Genus *Jaspidella* Olsson, 1956

Operculum present, chitinous. Radular ribbon with about 100 rows of teeth, central tooth tricuspidate. No callus wash on the parietal wall. Columellar sculpture of a low, finely plaited fold at the end of the columella. Type: *jaspidea* (Gmelin, 1791).

Jaspidella jaspidea (Gmelin, 1791)   2549
Jasper Dwarf Olive   Color Plate 11

Southeast Florida to Brazil. Bermuda.

10 to 20 mm. in length, about 5 whorls, apex blunt, nuclear whorls large. Color variable, usually grayish white with small, dull maculations of purplish brown. Fasciole at base of columella wide, undivided and with irregular brown spots and bars. End of columella sharply keeled. A common West Indian species found in shallow water to 7 fathoms in sand. Compare with *nivea.*

Jaspidella blanesi (Ford, 1898)   2550
Blanes' Dwarf Olive

South half of Florida to Panama and Bahamas.

7 to 10 mm.; parietal callus wash is thin to absent. Columella short and deeply concave. Fasciole narrow and plain, slightly bulging in the middle. Suture narrowly channeled. Color translucent-white, usually with 3 rows of reddish brown spots, 1 at the suture, 1 on the middle of the last whorl, 1 at the base. Nucleus rather large. Albino forms named *alba* Ford, 1898, and *albata* Vanatta, 1915. Uncommon.

Other species:

2551 *Jaspidella miris* Olsson, 1956. Gulf of Mexico, 30 to 200 fms. *Proc. Acad. Nat. Sci. Phila.,* vol. 108, p. 214. Uncommon.

Genus *Olivella* Swainson, 1831

Distinguished from the genus *Oliva* by its much smaller shell (½ inch or less) and in possessing an operculum. Parietal callus extending to the upper suture. If the shell is rotated slightly on its axis, it will be seen that the columellar pillar is deeply excavated above the basal fold, this excavation being the result of secondary reabsorption. Type: *dama* (Mawe, 1828).

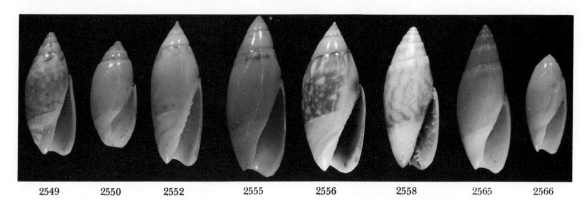

2549   2550   2552   2555   2556   2558   2565   2566

Olivella nivea (Gmelin, 1791)   **2552**
West Indian Dwarf Olive   **Color Plate 11**

Southeast Florida, the West Indies and Bermuda. Brazil.

20 to 34 mm. in length, about 7 whorls, apex sharply pointed; nucleus small, white, tan or purple. Suture channel is deep and fairly wide; with a strongly concave, etched, spiral indentation on the side of the preceding whorl. Color variable, usually cream-white with orange, tan or purple occurring in clumps in a spiral series just below the suture and just above the fasciole (that raised spiral ridge at the base of the shell). Fasciole lacks color. No operculum. Common from shore to 25 fathoms. Compare with *Jaspidella jaspidea* which has a more bulbous apex and an operculum.

A form from the Atlantic side of Panama, being 15 mm., with a deeper color and less attenuated nucleus was called *chiriquiensis* Olsson, 1956 (**2553**).

Olivella adelae Olsson, 1956   **2554**
Adele's Dwarf Olive

Florida.

10 to 13 mm., rather delicate, with a high body whorl and an elevated, evenly conic spire ½ the total length. Suture grooved. Parietal callus extending weakly beyond the end of the aperture about halfway to the suture. Color leaden-white, except for a band of brown spots below the suture and another at the edge of the white fasciole. Spire may be blotched with black. Wider and stubbier than *floralia*, and differing from small specimens of *nivea* in having small conic rather than drawn-out nuclear whorls. A common and widely distributed species in Florida; on sand bars in shallow water.

Olivella floralia (Duclos, 1853)   **2555**
Common Rice Olive

North Carolina to both sides of Florida and the West Indies. Bermuda. Brazil.

8 to 12 mm. in length, slender, fusiform and with a sharp apex. Color all white, but often with a dull-bluish undertone. Sometimes the body whorl is mottled brown. Apex white, orange, rose-pink or dull-purplish. Suture narrowly and deeply grooved. Fasciole double. Columella with numerous, very small folds. Common in shallow water.

Olivella dama (Wood, 1828)   **2556**
Dama Dwarf Olive   **Color Plate 13**

Gulf of California to Acapulco, Mexico.

18 to 20 mm., stout, spire almost as long as the aperture. Apex of spire usually violet. Parietal callus light violaceous. Outer lip brown within. Body whorl with brownish, faint zig-zag lines. Fasciole white or yellow. Common on the outer sides of sand spits at low tide. Synonyms are: *purpurata* Swainson, 1831, and *lineolata* Gray, 1839.

Subgenus *Dactylidia* H. and A. Adams, 1853

Columellar pillar structure smooth or weakly lirate, not terminating in an enlarged toothlike projection above. Outer lip smooth within. Type: *mutica* (Say, 1822). The single, 1-mm., dome-shaped egg capsules are laid in spring on intertidal broken shells. See R. T. Paine, 1962, *The Nautilus,* vol. 75, p. 139.

Olivella mutica (Say, 1822)   **2557**
Variable Dwarf Olive   **Color Plate 11**

North Carolina to Florida and the Bahamas.

¼ to ½ inch in length, ½ as wide, with a sharp apex. Strong, glossy callus is present on the parietal wall at the upper end of the aperture. Variable in color: ashy-grays and chocolate-browns to yellowish or whitish with wide bluish gray spiral bands. Sometimes brightly banded with white and browns. A very common species found in warm, shallow waters. Synonyms are: *rufifasciata* (Reeve, 1850) and *fimbriata* (Reeve, 1850).

Olivella dealbata (Reeve, 1850)   **2558**
Whitened Dwarf Olive

Florida to Texas and the Caribbean.

6 to 9 mm., moderately slender. Fasciole always white. Well-developed parietal callus extends halfway on to the penultimate whorl. Columella slightly concave, with 7 to 9 strong, narrow, slanting teeth. Suture deep and cutting into the previous whorls. Nuclear whorls small, white, tan or rose. Color of shell variable: whitish with sparse, axial flames or zigzag streaks of orange-brown or dark-brown. Common in shallow water in sand, especially on the west coast of Florida.

Olivella pusilla (Marrat, 1871)   **2559**
Very Small Dwarf Olive

Florida.

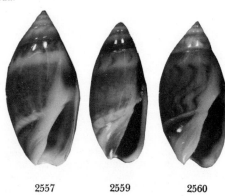

2557   2559   2560

6 to 8 mm., colored a rich mahogany-brown, or blue-gray, or either banded or entirely white. Subelliptical to bullet-shaped. Spire same length as aperture. Apical whorls very small. Sutures narrowly channeled. Parietal callus a thin wash only. Columella with 4 small ridges. Outer lip edged with white, internally a solid brown. Fasciole faintly divided. This is the commonest *Olivella* in Florida, particularly on sand flats along the west coast.

Subgenus *Niteoliva* Olsson, 1956

Sutures not covered by a callus. Type: *minuta* (Link, 1807).

Olivella minuta (Link, 1807)　　　　　**2560**
Minute Dwarf Olive　　　　　　　**Color Plate 11**

Southern Texas and the West Indies to Brazil.

10 to 15 mm., short, globose; suture open and grooved. Parietal callus strong, extending upward towards the suture. Fasciole wide, colored by a median brown band along the lower edge of an incised line. Nucleus small, glassy and somewhat elevated. Similar to *mutica* (Say) but fatter, with axial zigzag purplish brown color lines on the outer surface and on the inside of the inner lip, and in having its columella cutaway or narrowing sharply at the base. Common in the sandy surf of open beaches.

Olivella moorei Abbott, 1951　　　　　**2561**
Moore's Dwarf Olive

Off Key Largo to Key West, Florida.

¼ inch in length, apex bulbous. Characterized by its translucent shell with numerous, long, wavy, axial flammules of reddish brown on the sides of the whorls. Dredged from 115 to 144 fathoms. Named for Hilary B. Moore of the University of Miami, Florida.

2561

Subgenus *Callianax* H. and A. Adams, 1853

Columellar pillar structure is a simple fold at the base of the columella, sometimes lirate. Type: *biplicata* (Sowerby, 1825).

Olivella biplicata (Sowerby, 1825)　　　**2562**
Purple Dwarf Olive　　　　　　　**Color Plate 13**

Vancouver Island, British Columbia, to Baja California.

1 to 1¼ inches in length, globular to elongate, quite heavy. Upper columella wall with a heavy, low, white callus. Lower end of columella with a raised, spiral fold which is cut by 1, 2, or 3 spiral, incised lines. Color variable, but usually bluish gray or whitish brown with violet stains around the fasciole and lower part of the aperture. Brown and pure-white specimens are sometimes found. Abundant in summer months in

sandy bays and beaches. Sometimes dredged down to 25 fathoms on gravel bottom. Synonyms: *angelena* T. S. Oldroyd, 1918; *fucana* T. S. Oldroyd, 1921; and possibly *parva* T. S. Oldroyd, 1921. For bionomics and biology, see R. Stohler, 1959, *The Nautilus*, vol. 73, p. 65 and p. 95. For mating, egg capsule and veliger, see D. C. Edwards, 1968, *The Veliger*, vol. 10, p. 297. For growth of shell, see R. Stohler, 1969, *The Veliger*, vol. 11, p. 259.

Olivella pedroana (Conrad, 1856)　　　**2563**
San Pedro Dwarf Olive　　　　　**Color Plate 11**

Oregon to Baja California.

13 to 15 mm. in length, resembling *O. baetica*, but much heavier, much stouter, with a heavy callus, and colored light-buff to clouded, brownish gray with long, distinct, axial, zigzag stripes of darker brown. Fasciole and callus always white. The lowest columellar spiral ridge is single or rarely double. Moderately common from 1 to 15 fathoms. *O. pycna* S. S. Berry, 1935, is the same, and matches the neotype designated by Woodring in 1946. *O. intorta* Carpenter, 1857, is also this

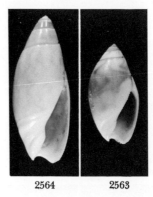

2564　　　　2563

species. Once used by Hupé Indians for necklaces (see *The Nautilus*, vol. 55, p. 92; 1942).

Olivella baetica Carpenter, 1864　　　**2564**
Beatic Dwarf Olive　　　　　　**Color Plate 11**

Kodiak Island, Alaska, to Baja California.

½ to ¾ inch in length, moderately elongate, rather light-shelled, glossy and colored a drab-tan with weak purplish brown maculations often arranged in axial flammules which may be more pronounced near the suture. Columellar callus weakly developed, the lower end with a double-ridged spiral fold. Fasciole white, often stained with brown. Early whorls usually purplish blue. *O. diegensis* and *mexicana* Oldroyd, 1921, are synonyms.

Subgenus *Olivina* Orbigny, 1839

Columellar pillar structure simple. Shell ovate. Parietal callus relatively thin. Type: *tehuelchana* (Orbigny, 1841).

Olivella bullula (Reeve, 1850)　　　　**2565**
Bubble Dwarf Olive

Southeast Florida and the Caribbean.

8 to 10 mm., subovate, rather high spire; 5 or 6 whorls; suture narrowly channeled. Parietal callus thin. 1 nuclear whorl, large, bulbous. Aperture ⁶/₁₀ the total length of the shell. Fasciole white, bounded by a finely raised thread. Columella below this thread is evenly and shallowly S-shaped. Lower ¼ of the columella thickened into a single, raised, wide, spiral plication. Sides of whorls with streaks and flames of light-brown. Operculum lanceolate, yellowish. Common; 25 to 200

fathoms. *O. bayeri* Abbott, 1951, is a synonym. This species is larger and thinner than *blanesi* and its pillar is nearly straight.

Subgenus *Macgintiella* Olsson, 1956

Distinguished by its bullet-shaped shell, with a relatively low spire and long, narrow canal at the top of the aperture opposite the parietal wall. Central tooth of radula is inversely hat-shaped, with a deep, central convex base. Type: *watermani* McGinty, 1940. *Mcgintiella* is a misspelling.

Olivella watermani McGinty, 1940   2566
Waterman's Dwarf Olive

Off the Florida Keys and Gulf of Mexico. Brazil.

9 to 12 mm., 4 to 6 whorls. Shaped like an *Oliva*. Nuclear whorls smooth, glassy-white, moderately bulbous. Sutural canal deep, not very narrow. The wall of the preceding whorl is strongly and concavely etched opposite the sutural canal. Upper series of plaits on columella 5 to 7, and separated from the lower series of 4 or 5 plaits on the base of the columella by a shallow U-shaped constriction. Inside of outer lip with a series of indistinct, raised, spiral cordlike thickenings. Common; 28 to 100 fathoms.

Olivella rosolina (Duclos, 1835)   2567
Rose-tipped Dwarf Olive

Bermuda (?) and the West Indies.

6 to 9 mm., white except for the pink end of the columella and sometimes with faint blotches of brown on the body whorl. Uncommon.

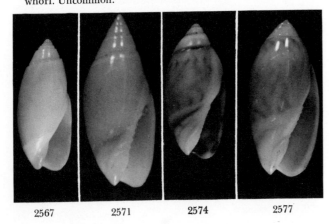

2567   2571   2574   2577

Other Atlantic species:

2568 *Olivella (Belloliva* Peile, 1922) *tabulata* Dall, 1889. Off Cape Canaveral, Florida, 540 fms.; off northeast Cuba, 220 to 225 fms. No operculum; radula of *Oliva*.

2569 *Olivella (Olivella) petiolita* (Duclos, 1835). Panama to Venezuela. Brazil.

2570 *Olivella (Olivella) macgintyi* Olsson, 1956. Off Palm Beach, Florida, 20 fms. *Proc. Acad. Nat. Sci. Phila.*, vol. 108.

2571 *Olivella (Olivella) stegeri* Olsson, 1956. Off Alacran Reef, Mexico, 28 fms. Gulf of Mexico, 10 to 90 fms.

2574 *Olivella (Niteoliva) verreauxii* (Duclos, 1857). West Indies. Brazil.

2575 *Olivella (Callianax) thompsoni* Olsson, 1956. 100 fms. Southwest of Sombrero Light, Marathon, Florida Keys.

2576 *Olivella (Macgintiella) rotunda* Dall, 1889. Gulf of Mexico, 70 fms. Puerto Rico to Barbados. Deep water. 20 mm.

2577 *Olivella (Macgintiella) fuscocincta* Dall, 1889. Lesser Antilles. Gulf of Mexico, 10 to 30 fms. Uncommon.

2578 *Olivella (Minioliva) perplexa* Olsson, 1956. South half of Florida. 4 mm. Type of the subgenus. Common.

2579 *Olivella (Minioliva) acteocina* Olsson, 1956. Bahamas to Panama.

Other Pacific species:

2580 *Olivella (Pachyoliva* Olsson, 1956) *semistriata* (Gray, 1839). Gulf of California to Peru.

2581 *Olivella alba* (Marrat in Sowerby, 1871). Baja California to Acapulco, Mexico. (Synonym: *miriadina* Duclos, 1835.)

2582 *Olivella fletcherae* S. S. Berry, 1958. Gulf of California.

2583 *Olivella gracilis* (Broderip and Sowerby, 1829). Guaymas, Mexico, to Panama.

2584 *Olivella sphoni* Burch and Campbell, 1963. Guaymas, Mexico, to Nicaragua. *The Nautilus*, vol. 76, no. 4.

2585 *Olivella steveni* Burch and Campbell, 1963. Upper Gulf of California to Guaymas, Mexico. *The Nautilus*, vol. 76, no. 4.

2586 *Olivella tergina* (Duclos, 1835). Magdalena Bay, Baja California, to Peru. (Synonym: *salinasensis* Bartsch, 1928.)

2587 *Olivella walkeri* S. S. Berry, 1958. Guaymas, Mexico.

2588 *Olivella (Callianax* H. and A. Adams, 1853) *intorta* Carpenter, 1857. Magdalena Bay to the Gulf of California.

2589 *Olivella (Dactylidella* Woodring, 1928) *anazora* (Duclos, 1835). Gulf of California to Peru. Type of the subgenus.

2590 *Olivella (D.) cymatilis* S. S. Berry, 1963. Magdalena Bay, Baja California.

2591 *Olivella (Dactylidia* H. and A. Adams, 1853) *zonalis* (Lamarck, 1811). Head of the Gulf of California to Acapulco, Mexico.

2592 *Olivella (Zanoetella* Olsson, 1956) *zanoeta* (Duclos, 1835). Upper Gulf of California to Ecuador.

Family MITRIDAE Swainson, 1831

An excellent account of the higher taxa is given by W. O. Cernohorsky, 1970, "Systematics of the Families Mitridae and Volutomitridae," Bull. no. 8, Auckland Institute and Museum. Classification is based largely on radular differences. Familial synonyms include Turritidae Gray, 1857; Strigatellacea Troschel, 1869; and Vexillidae Thiele, 1929.

Subfamily MITRINAE Swainson, 1831

Genus *Mitra* Lamarck, 1798

The shells are small to large, solid, elongate-fusiform, with spiral grooves or pits or smoothish and with 3 to 7 plicae on the lower part of the columella. Protoconch conical, paucispiral. Anterior canal not produced; siphonal notch prominent. Type of this genus is *Voluta mitra* Linné, 1758 (according to Opinion 885 (1969) of the I.C.Z.N.). Synonyms include: *Mitra* Röding, 1798; *Mitraria* Rafinesque, 1815; *Cucurbita* Scudder, 1882; *Papalaria* Dall, 1915; *Tiarella* Swainson, 1840; *Isara* H. and A. Adams, 1853; *Mutyca* H. and A. Adams, 1853; *Phaeomitra* von Martens, 1880; *Fuscomitra* Pallary, 1900; *Episcomitra* Monterosato, 1917; *Atrimitra* Dall, 1918; *Vicimitra* Iredale, 1929; *Volvariella* Coan, 1966.

Subgenus *Mitra* Lamarck, 1798

Mitra barbadensis (Gmelin, 1791)   **2593**
Barbados Miter

Southeast Florida and the West Indies to Brazil. Bermuda.

1 to 1½ inches in length, slender, with the aperture wide below and ½ the length of the entire shell. Characterized by its yellow-brown to fawn color which has an occasional fleck of grayish white. Aperture tan within. Columella with 5 slanting folds. The sides of the spire are almost flat. Weak spiral threads are commonly present especially in the earlier whorls. A common species under rocks on reefs facing the open ocean. Synonyms: *striatula* Schröter, 1804 and Lamarck, 1811.

Mitra swainsonii antillensis Dall, 1889   **2594**
Antillean Miter

North Carolina to Florida and the West Indies.

3 inches in length, about ¼ as wide, with the aperture ½ as long as the entire shell. 10 whorls smooth, except for 5 or 6 weak spiral threads on the upper ¼ of the whorl. Columella with 4 slanting, spiral folds, the largest being the uppermost. Color grayish white with a light-brown to olive periostracum. Short siphonal canal slightly recurved. Rare from 30 to 75 fathoms.

(2595) *Mitra swainsonii swainsonii* Broderip, 1836 (southern Gulf of California to Ecuador) is very similar to the Caribbean subspecies, grows to 5 inches in length and has a blackish olive periostracum. Uncommon; 20 to 70 fathoms. Synonyms: *mexicana* Dall, 1919; *zaca* Strong, Hanna and Hertlein, 1933 (Proc. Calif. Acad. Sci., vol. 21, no. 10); and *woodringi* Olsson, 1964 (fossil from Ecuador).

Subgenus *Atrimitra* Dall, 1918

Type of the subgenus is *idae* Melvill, 1893, from California.

Mitra fultoni E. A. Smith, 1892   **2596**
Fulton's Miter

San Diego, California, to Baja California and off Panama.

1 to 1½ inches, with moderately rounded whorls. Shell brown, covered with a shiny black periostracum. Interior of adult shell white. Surface of spire and entire body whorl with rows of evenly spaced, deep, microscopic pits. Rare; in shallow

water. Also fossil in southern California (see G. Sphon, *The Veliger*, vol. 4, p. 32).

Mitra idae Melvill, 1893   **2597**
Ida's Miter

Farallon Islands to San Diego, California.

2 to 3 inches in length, heavy, elongate. With 3 columella folds. Color mauve-brown, but usually covered with a thick, finely striate, black periostracum. The shell has microscopic pits only on the spire and upper part of the body whorl, while in the rarer *fultoni*, the pittings cover the spire and the entire body whorl. For mating behavior and egg capsule information, see J. M. Cate, 1968, *The Veliger*, vol. 10, p. 247. Uncommon offshore. Synonyms include: *montereyi* S. S. Berry, 1920; *coronadoensis* Baker and Spicer, 1930; *semiusta* S. S. Berry, 1957; *montereyensis* "Berry" Oldroyd, 1924. This is the type of the generic synonym, *Atrimitra* Dall, 1918.

Subgenus *Nebularia* Swainson, 1840

Spirally corded, rock-dwelling *Mitra*. Type: *contracta* Swainson, 1820, of the Indo-Pacific. Synonym: *Chrysame* H. and A. Adams, 1853.

Mitra nodulosa (Gmelin, 1791)   **2598**
Beaded Miter

North Carolina to Florida to Brazil. Bermuda.

¾ to 1 inch in length, solid, glossy, orange to brownish orange in color, and with about 17 long, axial riblets which are rather neatly beaded. Suture deep, with the whorls slightly shouldered. 3 columella folds, large and white. A common species frequently washed ashore or found under rocks at low tide. Synonyms include *granulosa* Lamarck, 1811; *granulata* Defrance in Blainville, 1824; *monilifera* C. B. Adams, 1852; *pallida* Nowell-Usticke, 1959.

Mitra straminea A. Adams, 1853   **2599**
Gulf Stream Miter

Florida to Puerto Rico.

20 to 28 mm., fusiform, white with some brown staining. Body whorl sculptured with 14 to 18 spiral cords. Narrow

2593   2597   2598

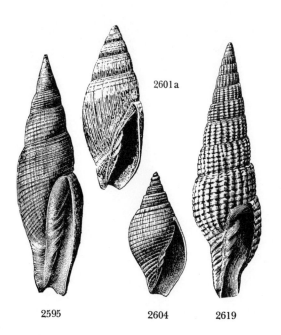

2601a

2595   2604   2619

aperture is more than ½ the length of the shell. 4 columellar folds strong. 2 to 180 fathoms; uncommon. Synonyms: *multilirata* A. Adams, 1853; *fluviimaris* Pilsbry and McGinty, 1949.

Subgenus *Dibaphimitra* Cernohorsky, 1970

Type of this subgenus is *florida* Gould, 1856. It is the only known living species.

Mitra florida Gould, 1856   **2600**
Royal Florida Miter   **Color Plate 10**

South half of Florida and the West Indies.

1½ to 2 inches in length, with about 6 whorls which bear on the last one about 16 spiral rows of evenly spaced, small, roundish dots of orange-brown. There are also odd patches of light orange-brown. 9 columella folds, the lower 7 being very weak. An uncommon species considered a choice collector's item; 1 to 30 fathoms. A live albino shell was collected by Mr. Fred Harper of Miami in 1968. Formerly known as *M. fergusoni* Sowerby, 1874, which is a synonym.

Subgenus *Strigatella* Swainson, 1840

Shells 5 to 50 mm., ovate, solid; sutures plain or coronate. Sculpture of spiral striae or smooth. Labial lip thick. Columella with 3 to 5 oblique folds. Type: *paupercula* (Linné, 1758). Mostly Indo-Pacific, no Caribbean species and one or two Eastern Pacific species:

2601   *Mitra (Strigatella) tristis* Broderip, 1836. Northern Gulf of California to Ecuador. Intertidal, under stones. Uncommon. (Synonyms: *olivacea* Anton, 1839; *albofasciata* Sowerby, 1874; *jousseaumiana* Mabille, 1898; *salinasensis* Bartsch, 1928.)

2601a   *Mitra (Strigatella) dolorosa* Dall, 1903. Gulf of California. Intertidal, under stones; uncommon. For eggs and spawning, see F. H. Wolfson, 1969, *The Veliger*, vol. 11, p. 282. Synonym of *tristis* Broderip, 1836?

Other *Mitra*:

2602   *Mitra (Nebularia) semiferruginea* Reeve, 1845. Bahamas and West Indies. (Synonyms: *clara* Sowerby, 1874; *fordi* Pilsbry and McGinty, 1949, *The Nautilus*, vol. 63, p. 12.)

2602   2599

2603   *Mitra (Nebularia) candida* Reeve, 1844. West Indies.

2604   *Mitra (Nebularia) crenata* Broderip, 1836. Guaymas, Mexico, to Ecuador. 17 mm. (Synonyms: *lowei* Dall, 1903; *loweana* Pilsbry, 1931; *sphoni* Shasky and Campbell, 1964.)

2605   *Mitra (?Mitra) zilpha* Dall, 1927. Off Georgia, 440 fms.

2606   *Mitra (?Mitra) grammatula* Dall, 1927. Off Georgia, 440 fms.

2607   *Mitra (Atrimitra) belcheri* Hinds, 1844. Mazatlan, Mexico, to Panama. Offshore to 20 or more fms. 3 to 4 inches, spirally corded, black. Uncommon.

2608   *Mitra (Mitra) catalinae* Dall, 1919. Crescent City, California, to Todos Santos Bay, Baja California. (Synonym: *diegensis* Dall, 1919.) Dwarf form of *idae* Melvill?

2609   *Mitra (Nebularia) brasiliensis* Oliveira, 1969. Northeast and east Brazil, 12 fms. (see E. C. Rios, 1970, pl. 33).

2610   *Mitra (Cancilla* Swainson, 1840) *larranagai* Carcelles, 1947. South Brazil to Argentina.

2611   *Mitra saldanha* Matthews and Rios, 1970. North Brazil, 53 fms. (see E. C. Rios, 1970, pl. 34).

Subfamily IMBRICARIINAE Troschel, 1867

Genus *Subcancilla* Olsson and Harbison, 1953

This genus is not represented in the Caribbean or the waters of the United States. There are 4 species in the tropical Eastern Pacific, 1 from Brazil. Spiral ribs are prominent. Type: *sulcata* "Swainson" Sowerby, 1825. This is *Tiara* of Woodring (1964), Keen (1958) and others, but not Swainson, 1831, which is a synonym of *Vexillum* Röding, 1798.

2612   *Subcancilla phorminx* (S. S. Berry, 1969). Off Rio Bolsa, Acapulco area, Guerrero, Mexico, 35 to 45 fms. 40 mm. *Leaflets in Malacology*, no. 26, p. 163.

2613   *Subcancilla attenuata* (Broderip, 1836). Gulf of California to Ecuador. (Synonyms: *hindsii* Reeve, 1844; *directa* S. S. Berry, 1960.)

2614   *Subcancilla gigantea* (Reeve, 1844). Panama to Ecuador. 70 mm. (Synonyms: *polystira* Pilsbry and Olsson, 1941.)

2615   *Subcancilla erythrogramma* (Tomlin, 1931). Nicaragua to Colombia. (Synonyms: *lineata* Broderip, 1836, non Schumacher, 1817; *calodinota* S. S. Berry, 1960.)

2616   *Subcancilla sulcata* ("Swainson" Sowerby, 1825). Gulf of California to Ecuador. (Synonyms: *funiculata* Reeve, 1844; *haneti* Petit, 1852.) Not *Voluta sulcata* Gmelin, 1791.

2617   *Subcancilla lopesi* (Matthews and Coelho, 1969). Northeast Brazil, 13 to 55 fms. (see E. C. Rios, 1970, pl. 33).

Family VEXILLINAE Thiele, 1929

Synonyms include Pusiinae Habe, 1961; Plesiomitrinae Bellardi, 1888. For anatomy and family placement, see W. F. Ponder, 1972, *Malacologia*, vol. 11, no. 2.

Genus *Vexillum* Röding, 1798

Typical *Vexillum* is confined to the Indo-Pacific. Type: *plicaria* (Linné, 1758). Synonyms include *Turricula* Fabricius, 1823; *Vulpecula* Defrance, 1824; *Tiara* Swainson, 1831; *Harpaeformis* Lesson, 1842.

Subgenus *Costellaria* Swainson, 1840

Fusiform to ovate, with axial ribs. Type: *semifasciata* (Lamarck, 1811), of the Indo-Pacific. Synonyms include *Uromitra* Bellardi, 1887; *Arenimitra* Iredale, 1929; *Pulchritima* Iredale, 1929; *Mitropifex* Iredale, 1929.

2618 2620 2621 2622 2623 2623a 2624 2632

Vexillum hendersoni (Dall, 1927)   2618
Henderson's Miter

Off Georgia to Florida and the West Indies.

½ to ¾ inch in length, fusiform in shape, with 8 whorls, each bearing a dozen sharp axial ribs which extend halfway down the whorl. Numerous microscopic, spiral cords present. Columella with 4 folds. Color drab pinkish gray with the upper ½ of the whorl bearing a wide, lighter, spiral band. Moderately common offshore down to 440 fathoms. This species was redescribed in 1943 by Rehder.

Vexillum styria Dall, 1889   2619
Dwarf Deepsea Miter

Both sides of Florida and the West Indies.

½ inch in length, fusiform in shape, moderately fragile and ashen-white in color. 10 whorls. Characterized by the numerous, very small, beaded, axial riblets and the thin, gray periostracum. Columella folds 5, the lower 2 being very weak. Nuclear whorls small, smooth and pointed. Commonly dredged from 30 to 333 fathoms.

Vexillum epiphanea Rehder, 1943   2620
Half-brown Miter

Southeast Florida, Bahamas and Bermuda.

½ to ¾ inches, moderately swollen, with 11 to 15 long, smooth, axial ribs on the body whorl, ending below in 2 or 3 beads. 4 whitish plicae on the chocolate columella. Color white to cream with the lower ½ of the last whorl brown to blackish and with a brown band at the suture which has flames extending downward between the whitish axial ribs. Aperture chocolate at the lower ⅓. Among dead coral at low tide to 8 feet. Uncommon.

Subgenus *Pusia* Swainson, 1840

Ovate, solid, with axial ribs and spiral threads or with nodes on the presutural ramp. Columella calloused and with 3 to 5 oblique folds. Type: *microzonias* (Lamarck, 1811). Synonyms: *Ebenomitra* Monterosato, 1917; *Pusiolina* Cossmann, 1921; *Idiochila* Pilsbry, 1921.

Vexillum hanleyi (Dohrn, 1862)   2621
Hanley's Miter

South half of Florida and the West Indies.

4 to 6 mm., resembling *gemmatum*, but with 13 to 16 ribs per whorl, narrower ribs, with numerous axial crinkles just below the suture, light-brown in color, with a broader peripheral cream band which bears 1 or 2 fine spiral lines of brown. Aperture tan with 2 broad darker-brown bands. Common; in shallow inshore grass flats near mud.

Vexillum gemmatum (Sowerby, 1871)   2622
Little Gem Miter

Florida Keys and the West Indies.

5 to 8 mm., 6 whorls, fusiform, black-brown with a broad white peripheral band, the latter covering most of the short, broad, axial nodules (11 to 13 per whorl). Suture well-indented, below which is a smooth, flattish black area which is rarely cream-spotted. Spiral sculpture obsolete or of weak broad spiral threads. Base of shell with 1 strong, raised, white-splotched cord, below which are 4 to 6 smaller, squarish black spiral cords. Aperture violet-black. Columella with 4 black teeth. 4 to 6 lirae inside body whorl. Moderately common in shallow water facing the open ocean. *Mitra exigua* C. B. Adams, 1845, may be an earlier name.

Vexillum albocinctum (C. B. Adams, 1845)   2623
Sulcate Miter

Southeast Florida and the West Indies. Brazil.

½ inch in length, rather fusiform in shape, with axial ribs as in *hendersoni*, but without spiral threads. 4 columella folds large and dark-brown. Color of shell dark chocolate-brown with a narrow, white, spiral band on the upper ½ of the whorl. Moderately common below low-water line under rocks in sand. Do not confuse with *Engina turbinella* which has no columella folds. *Voluta sulcata* Gmelin, 1791, p. 3455, not *sulcata* Gmelin, 1791, p. 3436, is this species. Possible synonyms are: *histrio* Reeve (2623a), 1844; *articulata* Reeve, 1845; *bifasciata* Mörch, 1852; *cruzana* Nowell-Usticke, 1959.

Vexillum arestum Rehder, 1943   2624
Pleasing Miter

West end of Cuba.

10 mm., narrowly ovate, whorls in spire rather flat, 1¾ nuclear whorls, brown and bulbous. Axial ribs weak and distantly spaced. Color yellowish to brown with a peripheral band of white which has a chocolate-brown squarish maculation between each rib. Base with 3 or 4 beaded cords. Columella has 4 plicae.

Vexillum puella (Reeve, 1845)   2625
Maiden Miter

North Carolina to the Bahamas and Caribbean.

10 mm., rotund, purple-brown with large, irregular white splotches below the suture, and with a strong reticulated sculpturing. Uncommon; just offshore. *Mitra albomaculata* Sowerby, 1874, is a synonym.

Other *Vexillum*:

2626 *Vexillum* (*Costellaria*) *laterculatum* (Sowerby, 1874). Southeast Florida, 30 fms., and the West Indies. (Synonyms: *oriflavens*

2625   2626   2627   2630   2633

Melvill, 1925; *olssoni* McGinty, 1955 (*Proc. Acad. Nat. Sci. Phila.*, vol. 107, p. 77).)

2627 *Vexillum (Costellaria) wandoense* (Holmes, 1860). North Carolina to the Florida Keys and the Gulf of Mexico, 12 to 440 fms. (Synonym: *rushii* Dall, 1887.)

2628 *Vexillum (Costellaria) styliolum* (Dall, 1927). Off Georgia and east Florida, 440 fms.

2629 *Vexillum (Costellaria) trophonium* (Dall, 1889). West Indies to Brazil, 50 fms.

2630 *Vexillum (Pusia) moisei* McGinty, 1955. 25 fms., off Palm Beach, Florida. West coast Florida, 20 to 40 fms., moderately common, *Proc. Acad. Nat. Sci. Phila.*, vol. 107, p. 77. Might be *exiguum* C. B. Adams, 1845.

2631 *Vexillum (Pusia) cubanum* (Aguayo and Rehder, 1936). Cuba and Grand Cayman Island. *Mem. Soc. Cubana Hist. Nat.*, vol. 9, no. 4, p. 266.

2632 *Vexillum (Pusia) dermestinum* (Lamarck, 1811). Caribbean. 20 mm. (Synonym: *albicostata* C. B. Adams, 1850.)

2633 *Vexillum (Pusia) variatum* Reeve, 1845. Caribbean. 20 mm.

Genus *Thala* H. and A. Adams, 1853

Shells small, usually about ½ inch or less, ovate, with cancellate or beaded sculpturing; aperture narrow and with strong lirations on the inside of the outer lip. With 3 or 4 columellar plicae. Early 3 or 4 whorls smooth, rounded. Type: *Mitra mirifica* Reeve, 1845. Synonyms: *Micromitra* Bellardi, 1888; *Mitromica* S. S. Berry, 1958.

Thala foveata (Sowerby, 1874)   **2634**
Beaded Florida Miter

South half of Florida. Bermuda.

6 to 8 mm., ovate, blackish brown to gray, with its surface cancellated by spiral and axial riblets, forming nodes where they cross. Nuclear whorls smooth, brown. Aperture

brown. Common; under stones, intertidal to 50 fathoms. *Mitra floridana* Dall, 1884, is a synonym.

Other species:

2635 *Thala gratiosa* (Reeve, 1845). San Diego, California; the Gulf of California to Panama. Galapagos. Pacific counterpart to *foveata* (Sby.). (Synonym: *nodocancellata* (Stearns, 1890).) See G. Sphon, 1969, *The Veliger*, vol. 12, p. 84, pl. 6.

2636 *Thala torticula* (Dall, 1889). Off Havana, Cuba.

2637 *Thala solitaria* (C. B. Adams, 1852). Jalisco, Mexico, to Panama; Galapagos. Synonym?: *Mitra orcutti* Dall, 1920.

2638 *Thala jeancateae* Sphon, 1969. Galapagos Islands, 50 to 60 fms. *The Veliger,* vol. 12, p. 85, pl. 6.

Family VOLUTOMITRIDAE Gray, 1854

Synonyms include Microvolutidae and Peculatoridae both Iredale and McMichael, 1962. A full treatment of this family is given in 1970 by W. O. Cernohorsky in Bull. No. 8, Auckland Institute and Museum, pp. 90–153.

Genus *Volutomitra* H. and A. Adams, 1853

Shell small, fusiform, with a periostracum. Sculpture smoothish. Columella with 3 or 4 folds. Siphonal canal short. Radula with a wishbone-shaped rachidian and rhomboidal laterals. Type: *groenlandica* (Beck in Möller, 1842). Some species with small operculum. One species reported from the Pacific coast, two from the Atlantic coast:

Volutomitra groenlandica (Möller, 1842)   **2639**
False Greenland Miter

Northern Canada to off Massachusetts.
Greenland to Norway.

10 to 30 mm., ovate-fusiform, white with a dark-brown periostracum. Teleoconch 5 whorls, nucleus of 2 smooth, rounded whorls. Sculptured with fine spiral striae and with axial costae on early whorls. Columella with 4 folds. Outer lip thin, simple. Uncommon; 15 to 280 fathoms.

Other species:

2640 *Volutomitra alaskana* Dall, 1902. Pribilof Islands, Alaska, south to off San Diego, California, in 822 fathoms. Hondo, Japan. 43 mm.

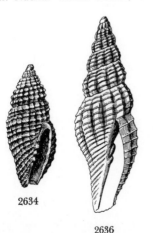

2634

2636

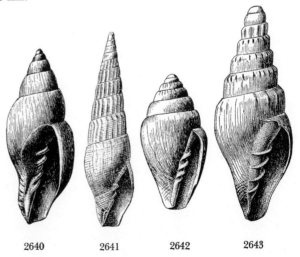

2640   2641   2642   2643

2641 *Volutomitra (Latiromitra* Locard, 1897) *bairdii* Dall, 1889. Off Cape Fear, North Carolina, 528 fms. 35 mm. Type illustrated here.

Genus *Microvoluta* Angas, 1877

Shell stout, broadly fusiform, with a large mammillate nucleus of 2 whorls. Outer lip smooth inside. Columella with 3 to 5 plicae. Surface smooth or axially plicate. Type: *M. australis* Angas, 1877. Two Caribbean Recent species reported:

2642 *Microvoluta blakeana blakeana* (Dall, 1889). Gulf of Mexico and West Indies, 80 to 640 fms. (var. *laevior* Dall, 1889). 10 mm.

2643 *Microvoluta intermedia* (Dall, 1889). Off St. Bartholomew, West Indies, 496 fms. 15.5 mm. (Synonym: *Mitra miranda* E. A. Smith, 1891, from Station 164B of the "Challenger", an Atlantic, not Australian dredge haul.)

Family TURBINELLIDAE Swainson, 1840

Shells heavy, with strong columellar folds. Operculum chitinous, heavy, unguilate. Xancidae and Vasidae are synonyms.

Subfamily TURBINELLINAE Swainson, 1840

Shells large, fusiform with columellar folds about at right angles to the axis of the shell.

Genus *Turbinella* Lamarck, 1799

Shells large, heavy, fusiform. Columella with 3 or 4 strong folds. Type: *pyrum* (Linné, 1758). *Xancus* Röding, 1798, was rejected by the I.C.Z.N., Opinion 489, 1957. Eggmass consists of chitinous, pill-shaped capsules closely packed in a long strand. Operculum clawlike, chitinous. Synonyms: *Turbinellus* Lamarck, 1801; *Turbinellarius* Duméril, 1806; *Buccinella* Perry, 1811; *Mazza* H. and A. Adams, 1853; *Turbofusula* Rovereto, 1900.

Turbinella angulata (Lightfoot, 1786)   2644
West Indian Chank

The Bahamas, northern Cuba; Yucatan to Panama.

7 to 14 inches in length, very heavy. Color cream-white with a thick, light-brown periostracum. Interior often tinged with glossy, pinkish cream or deep, brownish orange. Columella bears 3 strong, widely spaced folds. Middle of whorl on inside of aperture often with a spiral, weak ridge. A left-handed specimen of this species would be worth its weight in silver. Common in the Bahamas and Cuba. *T. scolyma* (Gmelin, 1791) is a synonym.

Turbinella laevigata Anton, 1839   2644a
Brazilian Chank

Northeast Brazil.

4 to 5 inches, solid ovoid, without axial ribs or knobs. Periostracum brown, fairly thick. Locally common. Synonym is *ovoidea* Kiener, 1841. See *Johnsonia,* vol. 2, no. 28, pl. 91.

2644

Genus *Surculina* Dall, 1908

Shell about 2 inches, slim-fusiform, with spiral and axial sculpture; siphonal canal elongate; columella without plications in the adult. The upper whorls (when broken open) show 2 slanting plicae on the columella. Rehder, 1967 (*Pacific Sci.,* vol. 21) has shown that this genus is best placed in the Turbinellidae and not in the turrids or volutes. *Phenacoptygma* Dall, 1918, is a synonym. One North American species:

2645 *Surculina cortezi* (Dall, 1908). Off Cortez Bank, 984 fms., and off San Diego, California.

Subfamily VASINAE H. and A. Adams, 1854

Shells medium-sized, with unequal-sized columellar folds set at a slightly oblique angle. Periostracum usually thick.

Genus *Vasum* Röding, 1798

Shells heavy, ovate-conic, with spiral cords or threads and axial nodes or blunt spines. Columella with 3 to 5 oblique folds. Type: *turbinellus* Linné, 1758. Synonyms: *Volutella* Perry, 1810; *Cynodonta* Schumacher, 1817; *Scolymus* Swainson, 1835.

Vasum muricatum (Born, 1778)   2646
Caribbean Vase

South half of Florida and the West Indies.

2½ to 4 inches in length, heavy. Blunt spines are at the shoulder and near the base. Shell chalk-white, covered by thick, black-brown periostracum. Aperture glossy-white and with a purplish tinge. Columella with 5 strong folds, the first and third being the largest. Rather common, often in pairs, in shallow water. Preys on worms and clams.

2646

The subspecies *coestus* (Broderip, 1833) **(2647)** (Panamanian vase) occurs from the Gulf of California to Panama, and differs only in having 4 (rarely 5) columella folds and in having heavier spiral cords. It is common.

Other species:

2648 *Vasum capitellum* (Linné, 1758). Puerto Rico to Trinidad and the north coast of South America.

2648

2648a *Vasum cassiforme* Kiener, 1841. Helmet Vase. Bahia, Brazil. 2½ inches. See *Johnsonia*, vol. 2, no. 28, pl. 93. (Synonym: *cassidiformis* Deshayes, 1845.

2648b *Vasum (Globivasum* Abbott, 1950) *globulus* Lamarck, 1816, subspecies *nuttingi* Henderson, 1919. Globe Vase. Lesser Antilles. 1 inch. See *Johnsonia*, vol. 2, no. 28, pl. 95.

Subgenus *Siphovasum* Rehder and Abbott, 1951

Differing from *Vasum* in having the rather long, narrow siphonal canal closed over, in having a strongly developed, elevated parietal wall and in having a high spire. Type: *latiriforme* Rehder and Abbott, 1951.

Vasum latiriforme Rehder and Abbott, 1951   **2649**
Latirus-shaped Vase

Off Yucatan, Mexico.

2 to 3 inches, solid, fusiform, spire $\frac{6}{10}$ that of the entire shell. Nuclear whorl bulbous, white. Axial sculpture of 6 ribs crossed by spiral cords. Color of shell peach to yellow. Siphonal canal long and almost closed along its length. Operculum horny, thick, unguiculate. Uncommon; 18 to 20 fathoms.

2649

Subfamily PTYCHATRACTINAE Stimpson, 1865

Genus *Ptychatractus* Stimpson, 1865

Shells about ¾ inch, fusiform, resembling a miniature *Colus* (*Buccinidae*) but having 2 or 3 weak plicae at the base of the columella. Spirally corded. Periostracum thin. Operculum with a terminal nucleus. Type: *ligatus* (Mighels and Adams, 1842).

Ptychatractus ligatus (Mighels and Adams, 1842)   **2650**
Ligate False Spindle

Gulf of St. Lawrence to Connecticut.

¾ inch, fusiform, rather thick, reddish brown, periostracum very thin. 6 whorls, well-rounded, with 6 or 7 spiral, raised cords. Aperture glossy, reddish brown. Canal short and straight. Outer lip thin, crenulated. Columella with 2 delicate, oblique folds above the beginning of the canal. Common; 15 to 60 fathoms.

2650

Other species:

2651 *Ptychatractus occidentalis* Stearns, 1873. Aleutians and Shumagin Islands. (May be *ligatus* Mighels and Adams.) 14 mm.

2652 *Ptychatractus californicus* Dall, 1908. Monterey Bay and off San Diego, California, 822 fms.

2651

Genus *Teramachia* Kuroda, 1931

Type of this genus, formerly considered a volute, is *tibiaeformis* Kuroda, 1931. The American species, *meekiana* Dall, 1889 (2653), was dredged in fresh condition in the Gulf of Mexico and off Cuba in 220 to 400 fathoms by the steamer *"Blake."* Dall referred it to *Mesorhytis*, a fossil genus. The shell is 15 mm. long, has 3 thin, elevated plicae on the columella (suggesting a volute or young marginellid). Nuclear whorl blunt, 1 whorl globose, followed by 3 whorls bearing 8 or 9 high, sharp axial ribs which disappear on the following 2 or 3 whorls. The shell is glossy yellow-tan to brown. A second species is *M. costatus* Dall, 1890 (2654), 14 mm. in length, dredged in 687 fms. off St. Kitts, Lesser Antilles. A third species (2655) *Teramachia chaunax* F. M. Bayer, 1971, was dredged in 201 to 589 meters off St. Lucia, Lesser Antilles.

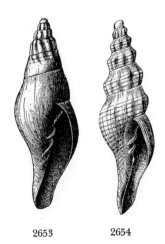

2653   2654

Genus *Metzgeria* Norman, 1879

Shells elongate-fusiform, axially plicate; columella weakly plicate; operculum ovate with a terminal nucleus. Type: *alba* Jeffreys.

Metzgeria californica Dall, 1903   2656
Californian False Spindle

Santa Barbara Channel, California.

½ inch, whitish, with a pale-yellow, wrinkled periostracum. Nucleus small, smooth, white, obliquely inclined, of 2 whorls. Suture deep. Whorls with 9 axial ribs. Spiral sculpture of

2656

numerous, subequal spiral threads. Offshore to 80 fathoms; uncommon to rare.

Metzgeria montereyana Smith and Gordon, 1948   2657
Monterey False Spindle

Monterey Bay, California.

½ inch, whitish, with a brown periostracum. Axial sculpture of 12 or 13 prominent, rounded, elevated ribs which, on the base, cross over the periphery to about the middle of the base. Spiral sculpture of 7 or 8 narrow spiral cords. Columella whitish, with 3 oblique plicae, the upper one being the smallest. Differs from *californica* in having 3 or 4 fewer axial ribs, in having less inflated whorls and in having a shorter siphonal canal. 15 fathoms; sand and broken shale. Rare.

Family VOLUTIDAE Rafinesque, 1815

Subfamily VOLUTINAE Rafinesque, 1815

For a full treatment of this group, see Weaver and du Pont, 1970, "Living Volutes," Monograph 1, Delaware Museum of Natural History, 375 pp., 79 colored pls.

Genus *Voluta* Linné, 1758

Shells stout, nodulose, with several strong plicae on the columella. Operculum chitinous, clawlike. Type: *musica* Linné, 1758. Synonyms: *Plejona* Röding, 1798; *Musica* Swainson, 1840; *Volutolyria* Crosse, 1877.

Voluta musica Linné, 1758   2658
Common Music Volute   Color Plate 10

Greater Antilles to Surinam.

2 to 2½ inches in length, heavy and with a polished finish. 3 nuclear whorls, bulbous and yellowish. 3 postnuclear whorls, plicate at the shoulder. Columella with about 9 evenly spaced folds. Characterized by the pinkish cream background and 2 to 3 spiral bands of fine lines which are dotted with darker brown (the "musical notes"). A moderately common West Indian species not found in the United States, but a favorite with collectors. A number of useless names have been applied to the numerous variations of this species. This is one of the few volutes to have an operculum. Among the many synonyms are: *lineata, chorea, incoronata, maculata, laevigata, reticulata, rosea, confusa, turbata, muta* all (Röding, 1798); *thiarella, violacea, nebulosa, guinaica, carneolata, laevigata* and *sulcata* all Lamarck, 1811; *guineensis* and *plicata* both Dillwyn, 1817; *polypleura* Crosse, 1876; *typica, damula* and *rugifera* all "Lamarck" Dall, 1907.

Voluta virescens **Lightfoot, 1786**   **2659**
Green Music Volute   **Color Plate 10**

Lower Caribbean.

2 inches in length, moderately heavy with the aperture ⅘ the total length of the shell. Whorls flat-sided and with weak, axial nodules high on the shoulder. Numerous spiral, incised lines and fine threads present. Columella with about a dozen folds of variable sizes. Exterior dull greenish brown with weak, narrow, spiral bands of lighter color dotted with black-brown. Aperture pale-cream to gray within. Erroneously reported from Florida, but not uncommon along the northern coast of South America. Shrimp boats bring specimens into United States ports. Synonyms are: *polyzonalis* and *fulva* both Lamarck, 1811; *pusio* Swainson, 1823.

Subfamily SCAPHELLINAE H. and A. Adams, 1858

Genus *Scaphella* Swainson, 1832

Shells 1 to 5 inches, fusiform, usually spotted with brown or purple-brown. Protoconch nipplelike, and may have a spurlike projection. Columella with several folds. Periostracum thin. Operculum absent. Radula consists of a Y-shaped central tooth. Type: *junonia* (Lamarck, 1804).

Scaphella junonia **(Lamarck, 1804)**   **2660**
The Junonia   **Color Plate 10**

North Carolina to both sides of Florida to Texas; Mexico.

5 to 6 inches in length, rather solid and smooth. 4 folds on the columella. Characterized by the cream background and the spiral rows of small reddish brown dots. Moderately common from 1 to 30 fathoms, but rarely washed ashore. A golden form occurs off Alabama **(2661)** (*johnstoneae* Clench, 1953) and specimens from Yucatan have a white background with smaller spots (*butleri* Clench, 1953) **(2662)**. About 50 or so

specimens a year are found on west Florida beaches, and many more are brought in by fishermen. The soft parts are mottled with purple. Shaw, 1808, also used the name *junonia*. Three sinistral specimens are known (1974).

Subgenus *Clenchina* Pilsbry and Olsson, 1953

The type of this dubious subgenus is *dohrni* (Sowerby, 1903) which is a synonym of *gouldiana* (Dall, 1887).

Scaphella gouldiana **(Dall, 1887)**   **2663**
Dohrn's Volute   **Color Plate 10**

Off the south half of Florida.

3 to 4 inches in length, similar to *junonia*, but much more slender, with a higher spire and with numerous exceedingly fine, incised (cut) spiral lines. Some specimens have the early whorls slightly angled and with short axial ribs. This is the form named *florida* (Clench and Aguayo, 1940). The typical *gouldiana* (Dall, 1887) has spiral bands of color or is all brownish orange and has numerous, short axial riblets or beads on the slightly carinate shoulder. Other synonyms and forms are *dohrni* (Sowerby, 1903) **(2664;** Color Plate 10); *robusta* (Dall, 1889) **(2665)** which has stronger spiral threads, a broader shell and heavier periostracum; *marionae* Pilsbry and Olsson, 1953 **(2666);** *cuba* Clench, 1946; *bermudezi* (Clench and Aguayo, 1940), a heavy specimen with stronger columellar plicate; *atlantis* Clench, 1946, from 210 fathoms off Camaguey, Cuba.

2667 *Scaphella (Clenchina) evelina* F. M. Bayer, 1971, is found in the southwestern part of the Caribbean Sea, in 77 to 641 meters.

Subgenus *Aurinia* H. and A. Adams, 1853

The type of this subgenus is *dubia* (Broderip, 1827). Synonyms are *Rehderia* Clench, 1946, and *Auriniopsis* Clench, 1953.

Scaphella dubia **(Broderip, 1827)**   **2668**
Dubious Volute

Off south half of Florida and the Gulf of Mexico.

3 to 9 inches in length, similar to *dohrni*, but more slender, much lighter in weight, usually with fewer rows of spots (5 to 8, instead of 8 to 12, rows of spots). Columellar plicate usually absent but there may be 2 to 4 weak ones in young, heavier forms. Spiral sculpture consists of numerous, silky spiral microscopic threads in small specimens, but absent in older ones. The following forms or possible subspecies were named: *kieneri* Clench, 1946 **(2669;** Color Plate 10) which comes from the northern Gulf of Mexico from mud bottoms, is up to 9 inches, smoothish and with a brown siphonal canal; *ethelae* (Pilsbry and Olsson, 1953) has a slightly longer spire and a cream siphonal canal; *georgiana* Clench, 1946, is about 3 inches, has a heavier shell, sometimes 2 weak plicae and has axial riblets; *schmitti* (Bartsch, 1931) **(2670)** is 2 to 4 inches, has a heavy, gray-brown "sugarlike" coating on the parietal wall. Young specimens have silky, spiral threads, and the nuclear whorls are sometimes small; another form is: *neptunia* Clench and Aguayo, 1940, from 322 fathoms off Jamaica. *Fusus tessellatus* Schubert and Wagner, 1829, appears to be a synonym.

Subgenus *Volutifusus* Conrad, 1863

Shells glossy because the mantle envelops part of the shell at times. Adults without columellar plicae. Periostracum absent. Type: *mutabilis* (Conrad, 1834), a Miocene species

2662

2664   2665   2666   2668   2670   2673   2674

Synonym: *Bathyaurinia* Clench and Aguayo, 1940. Three rare living species:

2671 *Scaphella* (*Volutifusus*) *aguayoi* (Clench, 1940). 400 fms., 164 miles east of St. Augustine, Florida.

2672 *Scaphella* (*Volutifusus*) *piratica* (Clench and Aguayo, 1940). 210 fms., off Punta Alegre, Camaguey, Cuba.

2673 *Scaphella* (*Volutifusus*) *torrei* (Pilsbry, 1937). 10 to 265 fms., off north and south coasts of Cuba. Above 3 species illustrated in *Johnsonia*, vol. 2, no. 22.

Genus *Arctomelon* Dall, 1915

Type: *stearnsii* (Dall, 1872). *Boreomelon* Dall, 1918, is a synonym.

Arctomelon stearnsii Dall, 1872   **2674**
Stearns' Volute

Alaska.

4 to 5 inches in length, strong; exterior chalky-gray with mauve-brown undertones. Aperture semiglossy, light-brown. Columella brownish with 2 moderately large folds and a weak one below. Nucleus bulbous, chalky-white. Uncommon offshore down to 100 fathoms.

Subfamily LYRIINAE Pilsbry and Olsson, 1954

Contains the genera *Lyria* and *Enaeta*, including about 22 living species, most of which occur in the Indo-Pacific.

Genus *Lyria* Gray, 1847

Axial ribs present on early whorls. Protoconch smooth. A single toothlike projection and serrations may be present on the inside of the outer lip. Columella with numerous plicae. Operculum present. Type: *Voluta nucleus* Lamarck, 1811.

2675 *Lyria* (*Lyria*) *beauii* (Fischer and Bernardi, 1857). Lesser Antilles. 50 to 70 mm. Uncommon, 50 to 80 feet on sand.

2676 *Lyria* (*Lyria*) *vegai* (Clench and Turner, 1967). Cabo Rojo, Dominica. 25 mm. *The Nautilus*, vol. 80, p. 83. Rare.

2677 *Lyria* (*Cordilyria* F. M. Bayer, 1971) *cordis* F. M. Bayer, 1971. 174 meters, 20 miles east southeast of Santo Domingo. Studies Tropical Amer. Mollusks, p. 204. See Abbott, 1972, "Kingdom of the Seashell," p. 138 for color photo of animal. Rare.

Genus *Enaeta* H. and A. Adams, 1853

Small, thick, conic-ovate shells, usually with weak axial ribs. Columella with 5 to 8 weak, irregular plicae. Outer lip has a small lirate swelling within at about the center. Type: *barnesii* (Gray, 1825) from the Tropical Eastern Pacific.

Enaeta cylleniformis (Sowerby, 1844)   **2678**
Guilding's Lyria

Southeast Florida (rare); Bahamas to Brazil.

½ inch, stout, thick-shelled, smoothish with weak axial ribs. Color whitish with light-yellow specks. Outer lip thickened within. Lower part of columella with 3 small teeth. Si-

2675   2676   2677

2678   2682

phonal canal slightly reflected. Not uncommon; in sand just offshore. Rare off Soldier Key, southeast Florida.

Enaeta barnesii (Gray, 1825)   2679
Barns' Lyria

Gulf of California to Peru.

1 to 1¼ inch, solid, ovate, pinkish gray to yellowish gray with spots and bands of brown. Lower columella with 5 to 7 weak folds. Uncommon; offshore to 20 fathoms. The spelling was emended to *barnesii* by Carpenter in 1864. This is *Voluta harpa* Barnes, 1824, not Mawe, 1823.

Enaeta cumingii (Broderip, 1832)   2680
Cuming's Lyria

Baja California to Peru.

1 inch, with a pointed spire. Axial ribs bear about 9 nodules at the shoulder. Color cream and purple-brown maculations. Moderately common; sandy mud at low tide. (Synonym: *pedersenii* Verrill, 1870.)

2679   2680

Other species:

2681 *Enaeta guildingi* (Sowerby, 1844). Antilles to Brazil. (See E. C. Rios, 1970, pl. 36.)

2682 *Enaeta reevei* (Dall, 1907). Cuba and east Honduras.

Family CANCELLARIIDAE Forbes and Hanley, 1853

Strong, biconic shells with several prominent spiral teeth, or lirae, on the columella. Sculpture usually cancellate. Many species are umbilicate. No operculum. For a review of various genera and subgenera, see J. G. Marks, 1949, *Jour. Paleontol.*, vol. 23, no. 5. The divisions in this family are very artificial.

I am indebted to Mr. Richard E. Petit of South Carolina, for assisting in the preparation of this family. He points out that the genera *Narona* H. and A. Adams, 1854, *Aphera* H. and A. Adams, 1854, and *Massyla* H. and A. Adams, 1854, are not represented in North American waters.

Genus *Cancellaria* Lamarck, 1799

Shells with strong cancellate sculpturing; slightly umbilicate; with a weak parietal callus; and with 2 strong columella folds. Type: *reticulata* (Linné, 1767).

Cancellaria reticulata (Linné, 1767)   2683
Common Nutmeg   Color Plate 10

North Carolina to both sides of Florida; Texas; Caribbean to Brazil.

1 to 1¾ inches in length, strong, with numerous spiral rows of small, poorly shaped beads which, with the weak axial and spiral threads, give a reticulate appearance. Columella with 2 folds, the uppermost being very strong and furrowed by 1 or 2 smaller ridges. Color cream to gray with heavy, broken bands and maculations of dark orange-brown. Uncommonly all white in western Florida. Common in shallow water in sand and grass from low-tide line to several fathoms. *C. conradiana* Dall, 1890, is a fossil species.

The subspecies *adelae* Pilsbry (**2684**) from the Lower Florida Keys is smoothish, except for incised lines on the body whorl. The aperture is faintly flushed with pink. Uncommon. (Adele's nutmeg.)

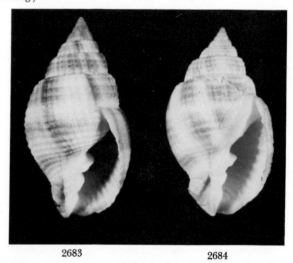

2683   2684

Other species:

2685 *Cancellaria (Cancellaria) decussata* Sowerby, 1832. Baja California to Ecuador. 20 fms. Uncommon.

2686 *Cancellaria (Cancellaria) obesa* Sowerby, 1832. Baja California to Ecuador. 20 fms. (Synonyms: *acuminata* Sowerby, 1832; *ovata* Sowerby, 1832.) This is the Pacific counterpart to *reticulata* (Linné).

2687 *Cancellaria (Cancellaria) gemmulata* Sowerby, 1832. Gulf of California to Panama; Galapagos Islands.

2688 *Cancellaria (Cancellaria) urceolata* Hinds, 1832. Magdalena Bay and the Gulf of California to Ecuador.

2689 *Cancellaria (Cancellaria) ventricosa* Hinds, 1843. Magdalena Bay, Baja California, to Panama. (Synonym: *affinis* C. B. Adams, 1852.)

2690 *Cancellaria (Pyruclia* Olsson, 1932) *solida* Sowerby, 1832. Gulf of California to Peru. Type of the subgenus.

2691 *Cancellaria (Mericella* Thiele, 1929) *corbicula* Dall, 1908. Santa Barbara Islands to the Coronado Islands, California. Deep water. 22 mm.

Subgenus *Crawfordina* Dall, 1919

The name *Crawfordia* Dall, 1918, is preoccupied by *Crawfordia* Pierce, 1908, an insect. Type: *crawfordiana* (Dall, 1891).

Cancellaria crawfordiana (Dall, 1891)   2692
Crawford's Nutmeg

Bodega to San Diego, California.

1 to 2 inches in length, heavy, white to pale-brown in color, but covered with a thick, rather fuzzy, gray-brown periostracum. Sculpture of 15 to 20 axial ribs crossed by numerous, flat-topped spiral cords. Columella with 2 slender folds and 1 obscure one at the bottom. Aperture enamel-white. Uncommon from 16 to 204 fathoms.

Other *Crawfordina*:

2693 *Cancellaria (Crawfordina) io* (Dall, 1896). San Diego, California, 650 fms., to the Gulf of Panama, 322 fms. 41 mm.

Subgenus *Progabbia* Dall, 1918

The type of this subgenus is *cooperi* Gabb, 1865.

Cancellaria cooperi Gabb, 1865   2694
Cooper's Nutmeg

Monterey, California, to Baja California.

2 to 3 inches in length, moderately heavy; columella with 2 small spiral folds. Whorls slightly shouldered, with about a dozen to 15 narrow axial ribs which at the top bear a single, low, sharp knob. Color brownish cream with a dozen or so narrow, brown spiral bands. Aperture orange-cream. Outer lip sometimes with numerous white, glossy, spiral cords on the inside. An uncommon, deep-water species, occasionally brought up in fish nets or caught on hook and line as deep as 300 fathoms. Said to grow to 7 inches in length.

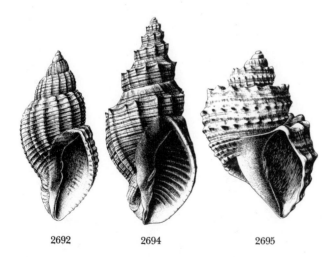

2692   2694   2695

Subfamily TRIGONOSTOMINAE Cossmann, 1899

Genus *Trigonostoma* Blainville, 1827

Whorls tabulate and loosely coiled, thus forming a broad umbilicus. Aperture usually triangular; columellar plicae

weak. Type: *trigonostoma* (Lamarck, 1822), now *pellucida* Perry, 1811. *Trigona* Perry, 1811, non Jurine, 1807. is a synonym.

Subgenus *Ventrilia* Jousseaume, 1887

Type: *tenerum* (Philippi, 1848). Synonym: *Emmonsella* Olsson and Petit, 1964.

Trigonostoma tenerum (Philippi, 1848)   2695
Philippi's Nutmeg

Southern half of Florida.

¾ to 1½ inches in length, fairly thin, but a quite strong shell. 4 whorls, strongly shouldered with the upper part of the whorl smooth and flat, and the sides with 3 to 5 spiral rows of strong nodules or blunt beads. Umbilicus very deep and funnel-shaped. Color light orangish brown. Uncommon just offshore. *T. stimpsoni* (Calkins, 1878), and *ventrilia* Jousseaume, 1887, are synonyms.

(2696) The Pacific Central American counterpart is *Trigonostoma bullatum* (Sowerby, 1832), 1 inch, whitish with indistinct brown bands, and with a range from Acapulco, Mexico, to Panama. Rare; offshore.

Subgenus *Bivetopsia* Jousseaume, 1887

Type of this subgenus is *chrysostoma* (Sowerby, 1832), from the Pacific side of Panama.

Trigonostoma rugosum (Lamarck, 1822)   2697
Rugose Nutmeg

Southeast Florida and the West Indies.

¾ to 1 inch, rather heavy, squat; color dirty-white with brown maculations; sculpture of strong axial ribs, about 14 on the body whorl, crossed by numerous smaller spiral cords. Aperture wide and expanded. Columella with 2 or 3 weak spiral teeth and 5 to 7 weak, raised, short lirae. Inside of outer lip strongly lirate. Umbilicus chinklike or absent. Locally common in the West Indies.

2697

Other species:

2698 *Trigonostoma (Bivetiella* Wenz, 1943) *pulchra* Sowerby, 1832. Guaymas, Mexico, to Ecuador. 28 fms. Type of the subgenus.

2699 *Trigonostoma (Ventrilia) goniostoma* (Sowerby, 1832). Head of the Gulf of California to Panama. Intertidal to offshore. Uncommon. (Synonym: *rigida* Sowerby, 1832.)

Genus *Agatrix* R. Petit, 1967

Shells small, without a large umbilicus. Periostracum tufted at the shoulders. Type: *agassizii* (Dall, 1889).

Agatrix agassizii (Dall, 1889)   2700
Agassiz's Nutmeg

North Carolina to Florida; Caribbean.

6 to 8 mm., brownish orange in color, somewhat squat, with a finely reticulated nucleus of 1½ whorls. Last whorl with about 14 axial rounded ribs crossed by about 8 spiral threads. Whorls somewhat shouldered. Periostracum brownish and tufted on the shoulder. Outer lip internally lirate, rather thin. Columella with 3 strong folds. Umbilicus chinklike. Moderately common; 18 to 50 fathoms.

Subgenus *Olssonella* Petit, 1970

Type of this subgenus is *smithii* (Dall, 1888). Sutures deep; spiral cords present. See *Tulane Studies Geol. Paleontol.*, vol. 8, p. 83.

Agatrix smithii (Dall, 1888)   2701
Smith's Nutmeg

Off the Carolinas.

8 to 10 mm., reddish brown, similar to *agassizii*, but much narrower, with a higher spire, with 8 or 9 axial ribs crossed by spiral threads, and with the columella bearing 2 folds. Periostracum fibrous, and dehiscent. Suture quite deep. Uncommon; 22 to 49 fathoms.

2700   2701

Other species:

2702 *Agatrix strongi* (Shasky, 1961). Gulf of California. Deep water. *The Veliger*, vol. 4, p. 18.

2703 *Agatrix* (*Olssonella*) *funiculata* (Hinds, 1843). Pacific Panama.

2704 *Agatrix* (*Olssonella*) *campbelli* (Shasky, 1961). Pacific Panama.

Subfamily ADMETINAE Troschel, 1865

Genus *Admete* Kröyer, 1842

Shell broadly fusiform, somewhat thin but strong; with well-rounded whorls bearing strong axial plications and weak spiral threads. Columella arched but bearing weak spiral folds. Periostracum fairly thick. Type: *couthouyi* (Jay, 1839) (synonym: *crispa* Möller, 1842).

Admete couthouyi (Jay, 1839)   2705
Common Northern Admete

Arctic Seas to Massachusetts.
Arctic Seas to San Diego, California.

½ to ¾ inch in length, moderately thick, with 6 whorls. Suture wavy, well-impressed. Sculpture coarsely reticulate, often beaded or with the axial cords the strongest. Columella

2705

strongly arched and bearing 2 to 5 very weak, spiral folds near the middle. Shell dull-white, covered with a fairly thick, gray-brown periostracum. Commonly dredged in cold waters. There are several other species on both of our coasts but they occur in very deep water. Synonyms include: *laevior* Leche, 1878; *undata* Leche, 1878; *middendorffiana* Dall, 1884; *crispa* Möller, 1842; *Murex costellifera* Sowerby, 1818 (a *nomen oblitum*); *buccinoides* Couthouy, 1838, non Sowerby, 1832; *borealis* A. Adams, 1855.

Other species:

2706 *Admete seftoni* S. S. Berry, 1956. 2½ miles north of Anacapa Island, California, 46 to 58 fms. *Jour. Wash. Acad. Sci.*, vol. 46, p. 156.

2707 *Admete* (*Microcancilla* Dall, 1924) *microscopica* Dall, 1889. Off Georgia, 440 fms.; off Fernandina, Florida, 294 fms. Gulf of Mexico, 200 to 780 fms.

2708 *Admete* (*Microcancilla*) *nodosa* Verrill and Smith, 1885. Off Nantucket, Massachusetts, to off Fernandina, Florida, 294 fms.

2709 *?Admete regina* Dall, 1911. Arctic Sea to the Pribilof Islands. Placement doubtful.

2710 *Admete californica* Dall, 1908. Prince William Sound, Alaska, to the Gulf of California, deep water. 16 mm.

2711 *?Admete gracilior* (Carpenter in Gabb, 1869). Akutan Island, Aleutian Islands, to San Diego, California, 10 to 40 fms.

Subgenus *Neadmete* Habe, 1961

Type of this subgenus is *japonica* (E. A. Smith, 1875).

2712 *Admete* (*Neadmete*) *unalaskensis* (Dall, 1873). Unalaska to Cape Blanco, Oregon. 20 mm. (young of *Admete modesta* Carpenter?).

2713 *Admete* (*Neadmete*) *circumcincta* (Dall, 1873). Unalaska to Departure Bay, British Columbia. 17 fms.

2714 *Admete* (*Neadmete*) *modesta* (Carpenter, 1865). Aleutian Islands to Puget Sound, Washington, 5 to 50 fms. (off San Clemente Island, Baja California).

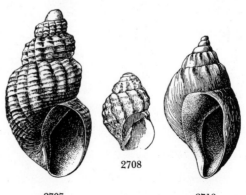

2707   2708   2718

2715 *Admete (Neadmete) rhyssa* (Dall, 1919). Point Pinos, California, to Todos Santo Bay, Baja California, 55 to 155 fms.

2716 *Admete (Neadmete) woodworthi* (Dall, 1905). Monterey to Santa Barbara Islands, California, 10 to 45 fms.

2717 *Admete (Neadmete) microsoma* (Dall, 1908). North Coronado Island, 656 fms., to Acapulco, Mexico, 660 fms.

Genus *Benthobia* Dall, 1889

Shell smooth, pale-brown, short-spired, resembling *Admete;* with a broad, concave columella without folds. Outer lip sinuous but not notched. Periostracum thin and smooth. Type: *tryonii* Dall, 1889. One species:

2718 *Benthobia tryonii* Dall, 1889. Off Cape Fear, North Carolina, 731 fms. 13 mm.

Family MARGINELLIDAE Fleming, 1828

Genus *Marginella* Lamarck, 1799

For a list of 1,300 names of Marginellidae, a classification, a list of the 580 recognized world living species and color illustrations and drawings of many species, consult Wagner and Abbott, 1967, "Van Nostrand's Standard Catalog of Shells." An excellent illustrated review of the Eastern Pacific species was made by E. Coan and B. Roth, 1966, *The Veliger*, vol. 8, p. 276. The type of *Marginella* is *glabella* (Linné, 1758). Synonyms: *Marginellarius* Duméril, 1806; *Marginellus* Montfort, 1810; *Marginilla* Swainson, 1831; *Cucumis* Deshayes, 1830; *Pseudomarginella* Maltzan, 1880; *Stanzzania* Sacco, 1890.

Subgenus *Dentimargo* Cossmann, 1899

Shells very glossy, spire rather high to medium; outer lip smooth or denticulate. Type: *M. dentifera* Lamarck, 1803. *Eburnospira* Olsson and Harbison, 1953, and *Volvarinella* Habe, 1951, are synonyms.

Marginella hematita Kiener, 1834   **2719**
Carmine Marginella   **Color Plate 11**

South Carolina to Florida and the West Indies. Brazil.

¼ inch in length, characterized by its glossy, bright and deep rose color, 4 strong columella teeth, pointed spire and thickened outer lip whose inner edge bears about 15 small, round teeth. Uncommon from 25 to 90 fathoms. *M. philtata* M. Smith and *M. jaspidea* Schwengel, 1940, are these species. *M. haematita* is a misspelling.

Marginella eburneola Conrad, 1834   **2720**
Tan Marginella

North Carolina to both sides of Florida and the West Indies.

⅜ inch in length, similar to *hematita*, but with a longer spire, only 7 to 9 teeth on the outer lip, with a shallow U-shaped notch at the top of the aperture, and the entire shell is yellow-tan to whitish. Uncommon from low tide to 600 fathoms. Synonyms are: *denticulata* Conrad, 1830, non Link, 1807; *opalina* Stearns, 1872; *destina* Schwengel, 1943.

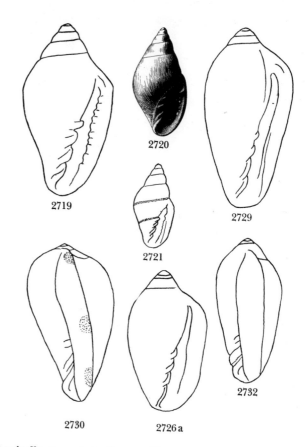

Marginella aureocincta Stearns, 1872   **2721**
Golden-lined Marginella

North Carolina to both sides of Florida and the West Indies. Brazil.

1⁄16 inch (4.0 mm.) in length; aperture ½ the length of the entire shell; spire pointed. Outer lip thickened, with about 4 very small teeth just inside the aperture. Columella with 4 strong folds or teeth. Color translucent-white, with 2 distinct, narrow spiral bands of light tan-orange on the body whorl (1 showing in the whorls of the spire). A very common species from low-water line to 90 fathoms. *M. smithii* Verrill, 1885, *macnairi* Bavay, 1922, and *virginiana* Verrill, 1885, are synonyms. A form without the color bands has the name *immaculata* Dall, 1890. It is possible that *M. idiochila* Schwengel, 1943, is this species. *M. perexilis* Bavay, 1922, May also be this species.

Other species:

2722 *Marginella (Dentimargo) eremus* Dall, 1919. (Synonym: *anticlea* Dall, 1919.) Galapagos Islands, 80 to 1,300 meters. Very rare. See Coan and Roth, 1966, *The Veliger*, vol. 8, no. 4, pl. 51.

2722a *Marginella (D.) incessa* Dall, 1927. Off Fernandina, Florida, deep water. 4.5 mm. May be immature.

2722b *Marginella (D.) esther* Dall, 1927. Off Fernandina, Florida, deep water. 5 mm. May be immature *eburneola* Conrad.

Subgenus *Prunum* Herrmannsen, 1852

Spire very low and almost covered over; aperture fairly narrow; the anterior siphonal canal is very shallow; outer lip thick and may be smooth or finely denticulate within. Type: *prunum* (Gmelin, 1791). *Leptegouana* Woodring, 1928, is a synonym.

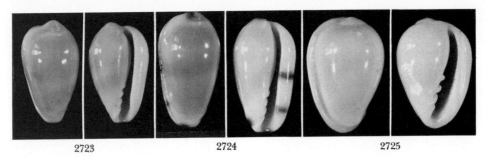

2723   2724   2725

***Marginella carnea* (Storer, 1837)**   **2723**
Orange Marginella   **Color Plate 11**

Southeast Florida and the West Indies.

¾ inch in length, very glossy; outer lip thickened, smooth and white. Apex half covered by a callus of enamel. Lower ⅓ of columella with 4 strong, slanting teeth. Shell bright-orange with a faint, narrow, whitish, spiral band on the middle of the whorl and one just below the suture. Moderately common in Florida in grassy areas to 6 fathoms.

***Marginella roosevelti* Bartsch and Rehder, 1939**   **2724**
Roosevelt's Marginella   **Color Plate 11**

The Bahamas and Caribbean.

1 inch in length, extremely close to *carnea*, differing only in being larger, and in having a brown spot on the apex and 2 large chocolate spots on the outer lip. There may be also 2 smaller spots at the anterior end of the shell. Apparently rare and possibly a color form of *carnea*.

***Marginella labiata* Kiener, 1841**   **2725**
Royal Marginella   **Color Plate 11**

Yucatan to Central America.

1 to 1¼ inches in length, similar to *carnea*, but stouter, lip orange-brown, body whorl whitish gray with 3 darker, subdued spiral bands. Outer lip with small teeth on its inner edge. Moderately common in east Mexico.

***Marginella hartleyanum* Schwengel, 1941**   **2726**
Hartley's Marginella

West coast of Florida.

7 to 8 mm., glossy, elongate, characterized by numerous small, irregularly placed, closely packed tan spots on a cream background, giving the shell a "measled" look. Outer lip thick, white, without teeth or spots. Lower columella with 4 squarish, spiral plicae, the next to the lowest being twice as large as the others. Very common; 20 to 40 fathoms. *The Nautilus*, vol. 55, p. 65.

***Marginella guttata* (Dillwyn, 1817)**   **2727**
White-spotted Marginella   **Color Plate 11**

Southeast Florida and the West Indies.

2726

½ to ¾ inch in length; outer lip smooth, white and with 2 or 3 brown spots on the lower ½. 4 columella teeth. Color of body whorl pale whitish with 3 obscure bands of light pinkish brown, and irregularly spotted with weak, opaque-white, roundish dots. Not uncommon in shallow water under rocks. *P. coniformis* Sowerby, 1850, and *longivaricosa* Lamarck, 1822, are synonyms.

***Marginella pruniosum* (Hinds, 1844)**   **2728**
Glowing Marginella   **Color Plate 11**

West Indies

⅓ to ½ inch, resembling *guttatum*, but with its white spots half as small. With numerous, weak, uneven, denticulations on the inner lip. With 3 weak, diffused spiral bands of yellowish brown (or absent). Without color spots on the outer lip, and with a slightly raised spire which is never covered by a labral callus. *Marginella nivea* (C. B. Adams) is a synonym. Common at depths of 5 to 20 feet on sand and grass bottoms.

***Marginella bella* Conrad, 1868**   **2728a**
La Belle Barginella

Off North Carolina to Key West.

¼ inch in length, glossy, white, sometimes with a bluish gray undertone. Sometimes with a rose tint on the body whorl. Spire moderately elevated. Outer lip thickened, without teeth. Lower ½ of columella with 4 strong, equally sized teeth. Commonly dredged from 1 to 200 fathoms. In 1890, Dall named a fossil variety of this species from Ballast Point, Tampa, Florida, as *inepta* (not *Hyalina inepta* Dall, 1927).

***Marginella amabilis* Redfield, 1852**   **2729**
Queen Marginella

Off North Carolina to Key West. Brazil.

⅜ inch in length, similar to *bella*, but with a shorter spire, more slanting columellar teeth, colored a translucent-tan with a heavy suffusion of orange on the shoulder of the whorl which becomes lighter on the lower part of the whorl. There is a fairly well-developed, white callus on the parietal wall. Uncommonly dredged from 25 to 125 fathoms.

***Marginella apicina* Menke, 1828**   **2730**
Common Atlantic Marginella   **Color Plate 11**

North Carolina to Florida, the Gulf States and the West Indies.

⅓ inch in length, glossy, with a dark nuclear whorl. Outer lip thickened, smooth, white, with 2 small, red-brown dots near the middle and a longer one at the very top and very bottom. Body whorl golden to brownish orange with 3 subdued, wide bands of darker color. A very common, shallow-water species. About 1 in every 5,000 specimens is sinistral. Synonyms are: *caribaea* Orbigny, 1842; *conoidalis* Kiener, 1841; *flavida* Redfield, 1846; *livida* Hinds, 1844; and *virginea* Jousseaume, 1875. Also called the coffee shell.

Marginella roscida Redfield, 1860   2731
Boreal Marginella

Massachusetts to South Carolina.

½ inch in length, similar to *apicina*, but has small, irregular, opaque-white spots, and with a higher spire, milky-cream color, with 3 faint, spiral bands of mauve or weak orange. Outer lip not sinuate, and is usually marked with 4 spots. Nucleus white, while in *apicina* it is usually bright-pink. Not uncommon from 18 to 132 fathoms. *M. limatula* Conrad, 1834, is evidently limited to the Miocene. Synonyms: *borealis* Verrill, 1884; *eulima* Dall, 1893; *beali* McGinty, 1940. See Abbott, *The Nautilus*, vol. 71, pl. 4, 1957.

Marginella virginiana Conrad, 1868   2732
Virgin Marginella

North Carolina to west Florida and Yucatan.

¼ inch in length, similar to *apicina*, but without spots on the thick varix; the third columella tooth is the largest; color of the last whorl whitish to cream, often with a faint curdling of darker orange-cream, and with a deeper, suffused band just below the suture and at the base of the shell. Moderately common, 14 to 56 fathoms.

Other Atlantic species:

2733 *Marginella (Prunum) evelynae* F. M. Bayer, 1943. Hillsborough Inlet, east Florida. *The Nautilus*, vol. 56, p. 114. Form of *carnea* Storer?

2734 *Marginella (Prunum) nobiliana* F. M. Bayer, 1943. Off Palm Beach, Florida, 70 fms. *The Nautilus*, vol. 56, p. 114. Form of *carnea* Storer?

Other species:

2735 *Marginella (Prunum) prunum* (Gmelin, 1791). Lower Caribbean to Brazil.

2736 *Marginella (Prunum) sapotilla* Hinds, 1844. Panama to Ecuador. See Coan and Roth, 1966, *The Veliger*, vol. 8, no. 4, pl. 48.

2737 *Marginella (Prunum) woodbridgei* Hertlein and Strong, 1951. San Jose, Guatemala. Pacific analogue of *apicina* Menke. See Coan and Roth, 1966, pl. 48.

2738 *Marginella (Prunum) albuminosa* Dall, 1919. West Mexico (locality correct?).

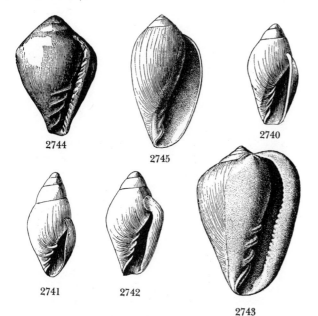

2744   2745   2740

2741   2742

2743

2739 *Marginella torticula* Dall, 1881. Gulf of Mexico and the West Indies, 152 to 229 fms.

2740 *Marginella yucatecana* Dall, 1881. Gulf of Mexico; Key West to Yucatan, Mexico, 125 to 640 fms.

2741 *Marginella fusina* Dall, 1881. Yucatan Strait, 640 fms.

2742 *Marginella seminula* Dall, 1881. Yucatan Strait, 640 fms.

2743 *Marginella cassis* Dall, 1889. Gulf of Mexico, 40 to 100 fms.; Cuba.

2744 *Marginella watsoni* Dall, 1881. Off Yucatan and Cuba, 480 to 805 fms.

2745 *Marginella cineracea* Dall, 1889 and 1890. Off Cape Fear, North Carolina, 14 mm.

2746 *Marginella (Glabella) reducta* Bavay, 1922. Cuba. (see *The Veliger*, vol. 16, p. 213).

2746a *Marginella (Prunum) canilla* Dall, 1927. Off Georgia and Fernandina, Florida, deep water. 9 mm.

Subgenus *Gibberula* Swainson, 1840

The type of this subgenus of minute white shells is *miliaria* (Linné, 1758). *Eratoidea* Weinkauff, 1879, is a synonym. *Giberula* is a misspelling.

Marginella lavalleeana Orbigny, 1842   2747
Snowflake Marginella

South half of Florida and the West Indies. Bermuda.

⅛ inch in length, resembling a miniature *apicina*, but pure-white in color. Like *Granulina ovuliformis*, but the aperture is not so long and has microscopic, spiral teeth inside the thin, curled-in outer lip. Columella with 3 or 4 oblique folds. Alias *minuta* Pfeiffer, 1840, not Gray, 1826. Common in shallow water to 40 fathoms.

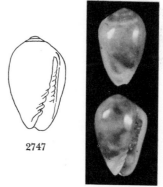

2747

2747

Other species:

2747a *Marginella (Gibberula) fernandinae* Dall, 1927. Off Fernandina, Florida; deep water. 9 mm.

Genus *Persicula* Schumacher, 1817

Spire low and usually covered, outer lip very thick and usually denticulate within. Type: *persicula* (Linné, 1758). *Rabicea* Gray, 1857 is a synonym.

Persicula catenata (Montagu, 1803)   2748
Princess Marginella

Southeast Florida and the West Indies. Brazil.

¼ inch in length, glossy; apex sheared off and sealed over by a weak callus. 7 columella teeth. Inside of outer lip with about 20 to 25 small teeth. Color translucent grayish with 7 spiral rows of teardrop-shaped, opaque-white spots and with 2 very subdued, wide spiral bands of light-brown. Uncommon in shallow water to 92 fathoms.

Persicula fluctuata (C. B. Adams, 1850)   2749
Fluctuating Marginella

Bahamas and Greater Antilles.

4 to 5 mm., ovuliform, glossy, light-tan to cream with 16 to 18 wavy longitudinal brown lines on the body whorl. The "waves" are darkest at the crests, and they point posteriorly or away from the outer lip. *P. catenata* has spiral rows of white dots, and the arrowlike brown lines point anteriorly or towards the outer lip. Outer lip sharp, with fine serrations within. 6 columellar plaits. Uncommon; in sand 1 to 6 fathoms.

Persicula pulcherrima (Gaskoin, 1849)   2750
Decorated Marginella

Southeast Florida and the West Indies. Brazil.

6 to 8 mm., beautifully decorated with alternating spiral bands of white dots on a buff background and bands of short, axial brown lines. These rows are separated by 4 spiral lines of brown dashes. Columella with 4 or 5 teeth below, and a few indistinct ones above. Uncommon in Florida; common in the Greater Antilles, 17 to 28 fathoms.

Other species:

2751 *Persicula interruptolineata* (Mühlfeld, 1816). Lower Caribbean. (Synonyms: *duchon* Jousseaume, 1875; *multilineata* Sowerby, 1846; *weberi* Olsson and McGinty, 1958.)

2752 *Persicula porcellana* (Gmelin, 1791). Panama, Pacific side. (Synonym: *tessellata* Lamarck, 1822.)

2753 *Persicula imbricata* (Hinds, 1844). Baja California to Ecuador; Galapagos. (Synonyms: *vautieri* Bernardi, 1853; *dubiosa* Dall, 1871; *adamsiana* Pilsbry and Lowe, 1932.)

2754 *Persicula bandera* Coan and Roth, 1965. Bahia de las Banderas, Jalisco, Mexico. 3 meters. Rare.

2755 *Persicula phrygia* (Sowerby, 1846). Baja California to Panama. See Coan and Roth, 1966, *The Veliger*, vol. 8, no. 4, pl. 50.

2756 *Persicula hilli* (M. Smith, 1950). Bahia Chamela, Jalisco, Mexico. *The Nautilus*, vol. 64, p. 61.

Genus *Hyalina* Schumacher, 1817

The type of this genus is *pallida* (Linné, 1767), from the West Indies.

Subgenus *Volvarina* Hinds, 1844

Shells small, elongate; low-spired; outer lip smooth or slightly denticulate. Type: *nitida* (Hinds, 1844).

Hyalina avena (Kiener, 1834)   2757
Orange-banded Marginella   Color Plate 11

North Carolina to Key West to Brazil. Bermuda.

11 to 13 mm., in length, slender; spire pointed but short. Outer lip curled in, white and smooth. Aperture narrow above, wide below. 3 to 4 slanting, columellar teeth. Color whitish, cream or yellowish with 4 to 6 spiral bands of subdued orange-tan. A moderately common, shallow-water species. The pink variety, especially common in Yucatan has been given the name *beyerleana* Bernardi, 1883. Do not confuse with the smaller *albolineata* Orbigny.

Hyalina veliei (Pilsbry, 1896)   2758
Velie's Marginella   Color Plate 11

South Carolina to Florida.

½ inch in length, somewhat like *amabiles* but with a higher, more pointed spire. Shell quite thin for a *Marginella*; color yellowish to whitish and somewhat translucent. Outer lip thickened, pushed in at the middle and white in color. Columella with 4 very distinct folds. Common in shallow water inside dead *Pinna* shells on mangrove mud flats.

Hyalina avenacea (Deshayes, 1844)   2759
Little Oat Marginella

North Carolina to both sides of Florida and to Brazil.

⅓ to ½ inch in length, slender, very similar to *H. avena*, but usually smaller, with a longer spire, more slender anterior end and pure, opaque-white in color, except for a very faint hint of straw color below the suture, again at the middle of the

body whorl and also near the base. Common from shallow water to 750 fathoms. This is *avenella* Dall, 1881.

Hyalina torticula (Dall, 1881)   2760
Knave Marginella

Off eastern Florida. Brazil.

⅓ inch in length, slender, fusiform, with a tall spire which is leaning to one side. Color opaque-white, glossy, and with a hint of straw-colored bands. *H. elusiva* (Dall, 1927), is a minor variant. Possibly a sport of *avenacea*. Uncommon in deep water to 38 fathoms.

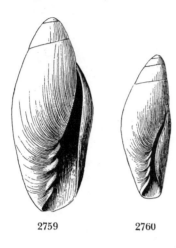

2759   2760

Hyalina albolineata (Orbigny, 1842)   2761
White-lined Marginella

Southeast Florida to Brazil. Bermuda.

5 to 8 mm., moderately elongate, with a short spire and white nucleus. Outer lip smooth, rather thin and white. Characterized by 3 very broad spiral bands of reddish brown, fawn, yellow or rose. Distinguished from *avena* by its smaller size, stouter shell, lower spire and the whitish bands between the color bands. Moderately common; 1 to 28 fathoms of water.

Hyalina taeniolata (Mörch, 1860)   2762
Californian Marginella

Santa Monica, California, to Mexico.

⅓ inch in length, slender, aperture ⅝ the length of the entire shell, with 4 whorls, and colored a grayish to bright-orange with 3 distinct or obscure, rather wide, spiral bands of white. Lower ⅓ of columella white and with 4 distinct spiral folds. Outer lip smooth, rounded, pushed in slightly, especially near the central portion. Moderately common in rocky rubble under stones at dead low tide. *H. californica* (Tomlin, 1916) and *parallela* (Dall, 1918) are synonyms. A pink subspecies, *rosa* Schwengel, 1938, occurs in the Galapagos.

Other species:

2763 *Hyalina styria* Dall, 1889 and var. *minor* Dall, 1927. Off Georgia and Fernandina, Florida, 294 to 440 fms.; West Indies.

2764 *Hyalina pallida* (Linné, 1758). Southeast Florida to the West Indies. *H. tenuilabra* (Tomlin, 1917), is the same.

2765 *Hyalina subtriplicata* (Orbigny, 1842). Southeast Florida to Cuba. 5 to 111 fms.

2766 *Hyalina lactea* (Kiener, 1841). West Indies. (Synonym: *abbreviata* C. B. Adams, 1850.)

2767 *Hyalina gracilis* (C. B. Adams, 1851). West Indies to Brazil, 26 fms.

Subfamily CYSTISCINAE Coan, 1965
(CYSTICINAE Stimpson, 1865)

Genus *Cystiscus* Stimpson, 1865

Shells minute; white; spire low; outer lip smooth within. Type: *cystiscus* Redfield, 1870, which is *capensis* Stimpson, 1865, non Krauss, 1848.

Cystiscus politulus (Dall, 1919)   2768
Polite Marginella

Santa Barbara, California, to Baja California.

3 mm., elongate; translucent; aperture widest anteriorly; outer lip thick, smooth within; columella with 3 strong plaits, 1 or 2 weaker ones above. Common; intertidal to 60 meters. "*C. regularis* Carpenter" of authors is this species. *C. myrmecoon* (Dall, 1919) may also be a synonym. See *The Veliger,* vol. 8, pl. 51.

Other species:

2769 *Cystiscus larva* (Bavay, 1922). Atlantic side of Panama. (Synonym: *Gibberula bocasensis* Olsson and McGinty, 1958).

Subgenus *Granula* Jousseaume, 1875

Kogomea Habe, 1951, is a synonym.

Cystiscus subtrigonus (Carpenter, 1864)   2770
Triangular Marginella

Monterey, California, to Baja California.

3 to 3.5 mm., in length; 2 to 2.5 mm. in breadth; ovate; spire moderately high; aperture slightly wider anteriorly; outer lip thickened, with strong denticles on its inner margin; columella with 4 folds. Common; intertidal to 100 meters. The name

2768   2771   2770

regularis Carpenter, 1864, because of conchological incompetence, falls into the synonymy of this species. *C. oldroydae* (Jordan, 1926), is also a synonym.

Cystiscus jewetti (Carpenter, 1857)   2771
Jewett's Marginella

Monterey, California, to Baja California.

¼ inch (5.0 mm) in length, snow-white, glossy, rather stout. Apex smoothed over and obscured. Outer lip smooth, slightly curled inward. Columella with 3 or 4 rather distinct, slanting, spiral folds with several smaller ones higher on the columella. Common from low tide to several fathoms. *C. nanella* (Oldroyd, 1925) is a synonym.

Other species:

2772 *Cystiscus (Granula) polita* (Carpenter, 1857). Southern California to Panama. Intertidal to 100 meters. (Synonyms: *coniformis* Mörch, 1860; *morchii* Redfield, 1870.)

2773 *Cystiscus (Granula) minor* (C. B. Adams, 1852). Costa Rica to Ecuador; Galapagos.

2773a *Cystiscus (Granula) ocellus* (Dall, 1927). Off Fernandina, Florida; deep water. 5 mm. Very similar to *subtrigonus* (Carpenter, 1864) from the Pacific coast.

Genus *Granulina* Jousseaume, 1888

Shell extremely small, apex covered over; aperture projecting higher than the apex; outer lip finely denticulate within. Type: *clandestina* (Brocchi, 1814). *Gibberulina* Monterosato, 1884 (a substitute name for *Bullata* Jousseaume, 1875) cannot be used for this genus. *Cypraeolina* Cerulli-Irelli, 1911, is a synonym.

Granulina ovuliformis (Orbigny, 1841)   2774
Teardrop Marginella

North Carolina to both sides of Florida and the West Indies.

⅛ inch (2.5 mm.) in length, globular, glossy, opaque-white. Aperture as long as the shell. Apex hidden under top of outer lip. Upper part of whorl slightly shouldered. Lower ⅓ of columella with 3 or 4 small, slanting teeth. Outer lip thickened. Do not confuse with *Marginella lavalleeana.* Common in shallow water to several fathoms. Alias *lacrimula* Gould, 1862; *hadria* Dall, 1889; and *amianta* Dall, 1889, and from Brazil, *clandestinella* Bavay, 1908, and 1913.

2774   2775

Granulina margaritula (Carpenter, 1857)   2775
Pear-shaped Marginella

Izhut Bay, Alaska, to Panama.

⅛ inch (2 to 3 mm.) in length; aperture as long as the shell. Glossy, translucent milk-white. Lower columella with 4 fairly strong folds with several microscopic teeth farther above. Outer lip curled in and with about 30 microscopic teeth. Animal black with the mantle spotted white. Very common all along the Pacific coast from low tide to 40 fathoms. On mud, gravel or backs of abalones. *Volutella pyriformis* Carpenter, 1864, is a synonym.

Other species:

2776 *Granulina tinolia* (Dall, 1927). Off Fernandina, Florida, 294 fms.

2777 *Granulina truncata* (Dall, 1922). Off Georgia, 440 fms. Off Fernandina, Florida, 294 fms.

Genus *Marginellopsis* Bavay, 1911

Shell minute; last whorl with a central band of granulations or microscopic dimples. Columella with 4 folds. Type: *serrei* Bavay, 1911.

Marginellopsis serrei Bavay, 1911   2778
Serré's Marginella

Lower Florida Keys and the Caribbean. Brazil.

1.5 to 2 mm., white, ovate, spire flat, outer lip thick and projecting above the spire. Sculpture on last whorl of numerous microscopic raised wavy ridges and elongate pustules. Parietal callus slightly raised. Base of columella with 1 large spiral ridge above which are 2 or 3 very small teeth. Inside of outer lip with numerous minute denticles. Uncommon; intertidal sand and rubble. Synonym: *fulva* Bavay, 1913.

2778

Superfamily Conacea Rafinesque, 1815

Sometimes referred to as the Toxoglossa because of the harpoonlike radular teeth and the poison gland found in many species.

Family CONIDAE Rafinesque, 1815

Genus *Conus* Linné, 1758

Shells inverted cone-shaped; aperture narrow. Outer lip thin and sharp. Type: *marmoreus* Linné, 1758, from the Indo-Pacific. Tropical Western Pacific species are highly venomous and the sting from the harpoonlike radula has been fatal to several humans. *Conus spurius atlanticus* has inflicted a beelike sting to at least one person in west Florida.

The radulae of 7 species of Caribbean cones were illustrated by G. L. Warmke, 1960, *The Nautilus,* vol. 73, p. 123. For radulae of West American Conidae, see J. Nybakken, 1970, *Amer. Mus. Novitates,* no. 2414, 29 pp.

Conus spurius atlanticus Clench, 1942   2779
Alphabet Cone   Color Plate 14

Florida and the Gulf of Mexico.

2780

2 to 3 inches in length; spire slightly elevated in the center. Top of whorls smooth, except for tiny growth lines. Color white with spiral rows of orange-yellow squares. Interior of aperture white. A rather common and attractive species found in shallow water. Typical *spurius spurius* Gmelin, 1791 (2779a; Color Plate 14) from the Bahamas and Antilles differs only in having the spots merging into occasional mottlings. Another race occurs off Yucatan in which the spots are sometimes smaller and a rather dark bluish purple.

Conus spurius aureofasciatus Rehder and Abbott, 1951   2780
Golden-banded Cone

Tortugas to off Yucatan, Mexico.

2 to 3 inches in length, similar to *spurius,* although sometimes more slender. Characterized by several spiral bands of light-yellow. Dredged in several fathoms of water. Uncommon to rare. It is probable that this species is only a freak color form of *spurius.*

Conus daucus Hwass, 1792   2781
Carrot Cone   Color Plate 14

Both sides of Florida and the West Indies. Brazil.

1 to 2 inches in length. Spire rather low, sometimes almost worn flat. Shoulder even and sharp. Spire with small, spiral threads. Color deep, solid orange to lemon-yellow, rarely with a lighter band. Spiral rows of minute brown dots sometimes present on sides. Interior of aperture pinkish white. Color of spire is orange with large white splotches. Uncommon; below 15 feet of water.

Conus juliae Clench, 1942   2782
Julia's Cone

Off North Carolina to off Texas and the West Indies.

1½ to 2 inches in length. Spire moderately high, flat-sided and with about 10 to 11 whorls. Shoulders of whorls slightly rounded; sides nearly flat. Color a pale pinkish brown to orangish with a moderate and indistinct band of cream or white at the mid area. This is overlaid with a series of fine spiral broken lines or dots of brown. Spire whitish with axial, zigzag, reddish brown streaks. Uncommon; offshore to 15 fathoms. Named after Mrs. William J. Clench, a great contributor to the cause of malacology.

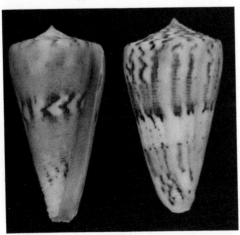

2782

Conus floridanus Gabb, 1868   2783
Florida Cone   Color Plate 14

North Carolina to both sides of Florida.

1½ to 1¾ inches in length. Spire well-elevated and slightly concave. Sides of whorls flat. The top of each whorl in the spire is concave and also has faint lines of growth. Color variable: usually white with elongate, rather wide patches of light orange-yellow to yellow. Spire with splashes of color. There is usually a white, spiral band around the middle of the whorl which may have small dots of yellowish brown. Moderately common in shallow water to 7 fathoms.

(2784; Color Plate 14) *Conus floridanus floridensis* Sowerby, 1870, pl. 14, is an extremely dark color form with spiral rows of reddish brown dots and heavier mottlings.

2785; Color Plate 14) *C. floridanus burryae* Clench, 1942, is another color form from off the Lower Florida Keys in which the spiral rows of brownish dots merge into solid lines. The lower end of the shell is very dark brown to deep brownish black. Uncommon.

Conus sennottorum Rehder and Abbott, 1951   2786
Sennott's Cone   Color Plate 14

Gulf of Mexico, from Tortugas to Yucatan.

1 inch in length, with a glossy, smooth finish. Slightly turnip-shaped. Color variable: white to bluish white with spiral rows of very small brown dots. Yellowish brown maculations may be present. Moderately common in 18 fathoms off Yucatan. Named after John and Gladys Sennott.

2786

Conus delessertii Récluz, 1843   2787
Sozon's Cone   Color Plate 14

South Carolina to Key West and the Gulf of Mexico. Bermuda.

2 to 4 inches in length. Spire elevated, slightly concave, with the top of each whorl also concave and with fine, arched lines of growth. There are 10 to 12 small spiral ridges at the lower end of the shell. Sides of whorls flat. Color as shown in the photograph, with the 2 whitish spiral bands being characteristic. Large and perfect specimens are collector's items, although individuals less than 2 inches in length are rather commonly dredged in 50 feet of water off both sides of Florida. Beach specimens have been collected on rare occasions. The type locality of Red Sea is evidently erroneous, and there is now little doubt that Récluz had a specimen from the American shores. *C. sozoni* Bartsch, 1939, is a synonym. I am retaining the popular name honoring the sponge diver, Sozon Vatikiotis.

Conus regius Gmelin, 1791   2788
Crown Cone   Color Plate 14

Southern Florida and the West Indies to Brazil.

1½ to 2½ inches in length. Spire low; shoulders of whorls with low, irregular knobs or tubercles. Color very variable even in the same locality. A rare yellowish color form (2788a; Color Plate 14) (*citrinus* Gmelin, 1791, not of Clench, 1942) occurs in the Lower Florida Keys, Cuba and the Antilles. The interior of the aperture of this species is white. Moderately common in Florida on outer reefs. *C. nebulosus* Hwass, 1792, is a synonym.

Conus cardinalis Hwass, 1792   2789
Cardinal Cone   **Color Plate 14**

Bahamas and the Caribbean.

¾ inch, similar to *regius,* but smaller, exterior reddish; aperture always tinted with rose, lavender or light-violet, while that of *regius* is generally white. Synonyms: *magellanicus* Hwass, 1792; *cinamomeus* Röding, 1798; *lubeckianus* Bernardi, 1861. Uncommon.

Conus mus Hwass, 1792   2790
Mouse Cone   **Color Plate 14**

Southeast Florida and the West Indies. Bermuda. Pacific side of Panama.

1 to 1½ inches in length. Spire elevated somewhat. Shoulders of whorls with low, irregular white knobs, between which are brown splotches. Color a dull bluish gray with olive-green or brown mottlings. Interior of aperture with 2 wide spiral bands of subdued brown. Periostracum thick, velvety and yellowish to greenish brown. The name *Conus citrinus* Gmelin (erroneously applied to this species in *Johnsonia* and elsewhere) is actually the yellow form of *regius.* The mouse cone is very common in intertidal, reef areas. Apparently it has been introduced to the Pacific side of the Panama Canal zone (*The Nautilus,* vol. 74, p. 81).

Conus bermudensis Clench, 1942   2791
Bermuda Cone   **Color Plate 14**

Bermuda, Florida and the West Indies.

1 to 2 inches, heavy, nearly smooth. Color porcelaneous white with faint pink irregular blotches. Rarely suffused with yellowish. Many specimens may have spiral rows of fine dark dots, especially at the somewhat carinate shoulder. Moderately common in Bermuda; 6 to 28 feet in sand. Rare elsewhere. Erroneously called *agassizi* Dall, 1889. *C. lymani* Clench, 1942, is a synonym.

Conus jaspideus Gmelin, 1791   2792
Jasper Cone

South half of Florida and the West Indies. Brazil.

½ to ¾ inch, somewhat fusiform, with a sharp spire about ⅓ the length of the shell. Sides of last whorl slightly convex. Color variable, usually brightly hued with reddish brown or yellowish mottlings. Rarely pink. Sculpture variable—sometimes smooth, but usually with about a dozen spiral cut lines. Some adult specimens bear spiral rows of pustules (form *verrucosus* Hwass, 1792). Nuclear whorls papilliform—either white, amber, rose or violet. Shoulder of last whorl usually rounded but may be sharp. Among the synonyms are: *branhamae* Clench, 1953; *havanensis* Aguayo and Farfante, 1947; *piraticus* Clench, 1945; *vanhyningi* Rehder, 1944; *pealii* Green, 1830; *anaglypticus* Crosse, 1865; *duvali* Bernardi, 1862. For others and more details see Abbott, 1958, Monograph 11, Acad. Nat. Sci. Phila.

Conus jaspideus subspecies stearnsi Conrad, 1869   2793
Stearn's Cone

North Carolina to both sides of Florida to Yucatan.

½ to ¾ inch in length. A small, slender, smooth, graceful cone with a high spire. Top of whorls concave. Sides almost flat. Color usually dull grayish to dark purplish brown with rows of tiny, white squares and with dull, yellowish brown

2792

2793

streaks or mottlings. Highly colored specimens may have rich reddish brown mottlings. Nuclear whorls brown. Moderately common from shallow water to 30 feet in sand. Intergrades with *jaspideus* in Caribbean waters in the Lower Florida Keys and well off the west coast of Florida.

Conus stimpsoni Dall, 1902   2794
Stimpson's Cone   **Color Plate 14**

Southeast Florida and the Gulf of Mexico.

1½ to 2 inches in length. A simple cone with a sharp, slightly concave, rather high spire, and with flat sides. It is usually smooth, but may have 15 to 20 cut spiral lines on the sides. Color is an even wash of yellowish white, but sometimes with 2 or 3 slightly darker, wide, yellowish spiral bands. Periostracum gray and rather thick. We have figured the holotype. Uncommon in rather deep water down to 30 fathoms.

Conus villepini Fischer and Bernardi, 1857   2795
Villepin's Cone   **Color Plate 14**

Southeast Florida to Yucatan.

1½ to 2½ inches in length. Spire rather well-elevated, very slightly concave. Each whorl in the spire is concave, with 3 to 4 spiral threads, and with fine, arched growth lines. Sides of shell smooth and slightly convex. There are about 9 indistinct spiral threads at the bottom end of the shell. Color of the thin periostracum is light yellowish brown. Shell light grayish white with a faint pinkish undertone. There are 3 or 4 long, irregular, axial streaks of dark reddish brown on the sides of

2794

2796

the last whorl. Interior of aperture blushed with rosy-white. We have illustrated the holotype of *amphiurgus* Dall in color which is a synonym. Uncommon in deep water.

The subspecies (or form) *fosteri* Clench and Aguayo (in Clench, 1942) (**2796**) has a much lower spire and the sides of the last whorl are slightly concave; otherwise it is indistinguishable from typical *villepini*. It is found off Bermuda and in deep water in the Caribbean to Brazil. Rare.

Conus mazei Deshayes, 1874   2797
Maze's Cone   **Color Plate 14**

Off Florida, the Gulf of Mexico; Caribbean.

1½ to 2 inches in length. A long, narrow and very handsome species which has rows of delicate beads on the very high spire. The sides of the whorls are smooth. Color cream with about 6 or 7 spiral rows of small red-brown spots. Rare.

The subspecies *macgintyi* Pilsbry, 1955 (**2798**), is similar but has numerous spiral incised lines around the last whorl. This subspecies occurs from off the Florida Keys to Yucatan and to Brazil, and is generally less than 1½ inches in length. Uncommon; 30 to 100 fathoms. *Conus rainesae* McGinty, 1953, is a synonym (*Notulae Naturae*, no. 249).

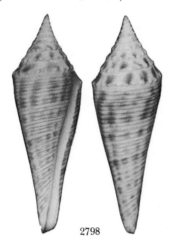

2798

Conus granulatus Linné, 1758   2799
Glory-of-the-Atlantic Cone   **Color Plate 14**

Southeast Florida and the West Indies.

1 to 1¾ inches in length. A fairly slender cone with rounded whorls in the spire which have spiral threads. Colored a brilliant orange-red to bright-red with flecks of brown and gold. Coarse spiral threads are usually present on the sides. Interior of aperture with a rosy-pink blush. A perfect specimen of this species is, indeed, a collector's item. It is very rare in Florida and not at all common in the West Indies. It lives in reefs just offshore.

Conus austini Rehder and Abbott, 1951   2800
Austin's Cone   **Color Plate 14**

Southeast Florida to Yucatan and West Indies.

2 to 2½ inches in length, pure white in color, although some may have a yellow-brown apex. Characterized by numerous odd-sized spiral threads on the sides. Sides of whorls flat to slightly rounded. Shoulders sharp to slightly rounded. Top of whorls slightly concave, with 1 smooth spiral carina and several much smaller threads. Shell sometimes with axial puckerings or riblike wrinkles. Periostracum velvety and grayish brown. Rare off Florida but common in 20 fathoms off Yucatan. (*Jour. Wash. Acad. Sci.*, vol. 41, p. 22.)

Conus clarki Rehder and Abbott, 1951   2801
Clark's Cone   **Color Plate 14**

Off Louisiana.

1 to 1½ inches in length, whitish in color and with small weak spots, rather turnip-shaped, similar to *austini*, but with 27 to 30 very strong, squarish spiral cords on the sides. The cords, and especially the one at the shoulder, are strongly beaded. Between the cords there are microscopic, axial threads. Periostracum gray. Apparently rare offshore in 29 fathoms. This and the preceding species were named after Austin H. Clark, scientist, author and gentleman. *C. frisbeyae* Clench and Pulley, 1952, is unquestionably this species.

Conus ermineus Born, 1778   2802
Turtle Cone   **Color Plate 14**

Gulf of Mexico and the Caribbean to Brazil.
West Africa.

2 to 3½ inches, characterized by its well-rounded shoulders and smooth, slightly convex whorls. Color usually grayish white with irregular, interlocking mottlings of brown, blue-black or dark-gray. With or without a few rows of minute dots. Spire with 4 or 5 spiral threads on each whorl. A deep-water Gulf of Mexico form is very large and orange-white or with only a few darker markings. Periostracum thin. Uncommon; offshore to 55 fathoms. Formerly called *ranunculus* Hwass, 1792, but that species has now proven to be based upon a specimen from the Indo-Pacific. *C. testudinarius* Hwass, 1792, is a synonym. (See Clench and Bullock, 1970, *Johnsonia*, vol. 4, p. 377.)

Conus californicus Hinds, 1844   2803
Californian Cone

Farallon Islands, California, to Baja California.

¾ to 1 inch in length. Spire moderately elevated and slightly concave. The shoulders of the shell are rounded, the sides very slightly rounded. The chestnut to pale-brown, velvety periostracum is rather thick. Shell grayish white in color. Interior whitish with a light-brown tint. Rather common in shallow water along certain parts of southern California. They sting and feed upon other mollusks, polychaete worms and fish (see P. R. Saunders and Fay Wolfson, 1961, *The Veliger*, vol. 3, p. 73, pl. 13).

Other Atlantic species:

2804 *Conus perryae* Clench, 1942. Off Sanibel Island, west Florida. Only one specimen found, and it looks very much like a young *ermineus*. (Synonym: *melvilli* L. Perry, 1939, non Sowerby, 1878.)

2808

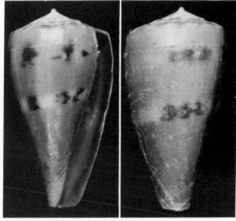

2805

2805   Conus abbotti Clench, 1942. Bahamas. *Johnsonia*, vol. 1, no. 6, pl. 4. 30 to 50 mm. Color Plate 14.

2806   *Conus flavescens* Sowerby, 1834. Southeast Florida and the Caribbean. 30 to 40 mm.

2811a   2806

2807   *Conus caribbaeus* Clench, 1942. Southeast Florida and the Caribbean. *Johnsonia*, vol. 1, no. 6, pl. 11. 20 to 30 mm.

2808   *Conus jaspideus* subspecies *pygmaeus* Reeve, 1844. Southern Caribbean. 20 mm.

2809   *Conus selenae* van Mol, Tursch and Kempf, 1967. Northeast Brazil. 15 mm. *Ann. Inst. Oceanogr.*, vol. 47, pl. 8. (Synonym: *yemanjae* van Mol, Tursch and Kempf, 1967.)

2810   *Conus ustickei* Miller in Usticke, 1959. 25 mm. Virgin Islands. "Check List Marine Moll. St. Croix," privately printed, p. 80.

2811   *Conus bajanensis* Usticke, 1968. Dredged south of Barbados, Lesser Antilles. "Caribbean Cones from St. Croix and the Lesser Antilles," privately printed, p. 28, fig. 1020D.

2811a   *Conus patae* Abbott, 1971. Southeast Florida and the West Indies. 13 mm. *The Nautilus*, vol. 85, p. 49. (Synonym: *C. rudiae* Magnotte, 1971, in mimeograph).

Other tropical Eastern Pacific species:

There are about 35 species and forms of *Conus* living from the Gulf of California to Ecuador. For their treatment, we recommend A. M. Keen's "Sea Shells of Tropical West America," and Hanna and Strong, 1949, *Proc. Calif. Acad. Sci.*, series 4, vol. 26, no. 9. The following species are found in the Gulf of California and west Central America:

2812   *Conus poormani* S. S. Berry, 1968. Off Morro Colorado, Sonora, Mexico, 24 to 26 fms. Uncommon.

2813   *Conus* (*Pyruconus* Olsson, 1967) *patricius* Hinds, 1843. 3 inches. Nicaragua to Ecuador. (Synonym: *pyriformis* Reeve, 1843.) Type of the subgenus. Color Plate 15.

—   *Conus chrysocestus* S. S. Berry, 1968. Off Morro Colorado, Sonora, Mexico, 30 to 45 fms. Both in Leaflets in Malacology, no. 26, p. 156 and 157 respectively. Uncommon. Is *fergusoni* Sowerby, 1873.

2814   *Conus bartschi* Hanna and Strong, 1949. Baja California. (Synonym: *andrangae* Schwengel, 1955.) Uncommon. Possibly a synonym of *brunneus* Wood, 1828.

2815   *Conus brunneus* Wood, 1828. Magdalena Bay, Baja California, to Ecuador. Common; intertidal. Color Plate 15.

2816   *Conus diadema* Sowerby, 1834. Revillagigedo Island, Baja California, to Ecuador. Subspecies *pemphigus* Dall, 1910, from the Tres Marias Islands (**2816a**) (Synonym: *prytanis* Sowerby, 1882). Color Plate 15.

2817   *Conus gladiator* Broderip, 1833. Central part of the Gulf of California to Ecuador. (Synonym: *tribunis* Crosse, 1865.) Color Plate 15.

2818   *Conus sponsalis nux* Broderip, 1833. Magdalena Bay, Baja California, to Ecuador. Color Plate 15.

2819   *Conus princeps* Linné, 1758. Head of the Gulf of California to Ecuador. Color form (**2819a**; Color Plate 15) *lineolatus* Valenciennes, 1832; and (**2819b**) form *apogrammatus* Dall, 1910. Common.

2820   *Conus perplexus* Sowerby, 1857. Magdalena Bay, Baja California, the Gulf, to Ecuador. Common. Color Plate 15.

2821   *Conus purpurascens* Sowerby, 1833. Both sides of Baja California to Ecuador. Common. (Synonyms: *regalitatis* Sowerby, 1834; *comptus* Gould, 1853; *rejectus* Dall, 1910.) Color Plate 15.

2821a   *Conus tornatus* Sowerby, 1833. Both sides of Baja California to Ecuador. Offshore to 20 fms. Common. (Synonym: *desmotus* Tomlin, 1937.)

2822   *Conus vittatus* Hwass in Bruguière, 1792. 2 inches. Gulf of California to Manta, Ecuador. Uncommon. Color Plate 15.

2823   *Conus ximenes* Gray, 1839. Throughout the Gulf of California to Panama. Offshore to 50 fms. Common. (Synonym: *mahogani* Reeve, 1843.) Color Plate 15.

2824   *Conus* (*Cylindrus*) *dalli* Stearns, 1873. Southern Gulf of California to Panama; Galapagos. Locally common. Color Plate 15.

2825   *Conus lucidus* Wood, 1828. Baja California to Ecuador. Locally common. (Synonym: *reticulatus* Mawe, 1832.) Color Plate 15.

2826   *Conus* (*Lithoconus*) *archon* Broderip, 1833. Gulf of California to Panama. Offshore from 14 to 220 fms. Common. (Synonym: *sanguineus* Kiener, 1849.) Color Plate 15.

2827   *Conus* (*L.*) *arcuatus* Broderip and Sowerby, 1829. Gulf of California to Panama. Moderately common; offshore. Color Plate 15.

2828   *Conus* (*L.*) *dispar* Sowerby, 1833. Gulf of California. Uncommon.

2829   *Conus* (*L.*) *fergusoni* Sowerby, 1873. Gulf of California to Ecuador. Common. (Synonyms: *xanthicus* Dall, 1910; *chrysocestus* S. S. Berry, 1968.) Color Plate 15.

2830   *Conus* (*L.*) *gradatus* Wood, 1828. Both sides of Baja California to Clipperton Island. (Synonym: *helenae* Schwengel, 1955).

2831   *Conus* (*L.*) *recurvus* Broderip, 1833. Both sides of Baja California to Colombia; 20 to 80 fms. Uncommon. (Synonyms: *scariphus* Dall, 1910; *magdalenensis* Bartsch and Rehder, 1939.) Color Plate 15.

2832   *Conus* (*L.*) *regularis* Sowerby, 1833. Both sides of Baja California to Panama. Common. (Synonyms: *syriacus* Sowerby, 1833; *angulatus* A. Adams, 1854; *monilifer* Broderip, 1833; *thaanumi* Schwengel, 1955.) Color Plate 15.

2833 *Conus (L.) scalaris* Valenciennes, 1832. Both sides of Baja California to Acapulco, Mexico. Not uncommon. Offshore.

2834 *Conus (L.) virgatus* Reeve, 1849. Baja California to Ecuador. Uncommon. (Synonym: *signae* Bartsch, 1937.) Color Plate 15.

Family TEREBRIDAE H. and A. Adams, 1854

Genus *Terebra* Bruguière, 1789

I am indebted to Twila Bratcher and R. D. Burch for supplying the information on the Pacific coast species. Type of the genus *Terebra* is *subulata* (Linné, 1758), from the Indo-Pacific. We have not attempted to assign all of the species to the numerous subgenera available.

Terebra taurinus Lightfoot, 1786	**2835**
Flame Auger	**Color Plate 10**

Southeast Florida, to south Texas and the West Indies to Brazil.

4 to 6 inches in length, heavy, rather slender. Characterized by a cream color with 2 spiral rows of axial, red-brown bars, the upper series being twice as long as the lower one. Upper whorls faintly and axially ribbed. Upper ½ of each whorl swollen and with a single incised line. *T. flammea* Lamarck, 1822, and *T. feldmanni* Röding, 1798, are this species. Formerly considered quite rare, but now not infrequently dredged in the Gulf of Mexico and Lake Worth, east Florida, 2 to 40 fathoms.

Terebra dislocata (Say, 1822)	**2836**
Common American Auger	

Maryland to Florida, Texas and the West Indies. Brazil. Redondo Beach, California, to Panama.

1½ to 2 inches in length, slender. Whorls with about 25 axial ribs per whorl which are divided ⅓ to ½ their length by a deep, impressed, spiral line. Many specimens show prominent, squarish, raised spiral cords between the ribs. Columella with 2 fused spiral folds near the base. Color a dirty, pinkish gray, but sometimes orangish. A common shallow-water species. *T. rudis* Gray, 1834, and *petiti* Kiener, 1939, are synonyms. Fossil specimens are found occasionally on New Jersey beaches.

Subgenus *Myurella* Hinds, 1845

Type of this subgenus is *myuros* Lamarck, 1822, from the Indo-Pacific.

Terebra floridana Dall, 1889	**2837**
Florida Auger	

Off South Carolina to south Florida. Brazil.

2 to 3 inches in length, very long and slender. Color light-yellow to yellowish white. Each whorl has just below the suture a row of about 17 to 23 oblong, slightly slanting, smooth axial ribs. Below this, and separated from it by an impressed line, is a similar row of much shorter, axial ribs. The lower ⅓ of the whorl is marked by 3 or 4 raised, spiral threads only. Columella with a single, strong fold near the bottom. Operculum horny, translucent brown, quadrate ungulate. A fairly rare species; 5 to 118 fathoms.

The Central American subspecies *stegeri* Abbott, 1954 (**2838**) from off the Yucatan Peninsula, differs in being stouter (68 × 13 mm.), bright-orange, lemon-yellow to waxy white, in having the siphonal canal considerably more twisted to the left, in having weaker and more numerous axial riblets (28 to 39)

2836   2839

2842   2846

which increase in number in later whorls, and in lacking the upper fold on the columella. Uncommon; 12 fathoms.

Subgenus *Strioterebrum* Sacco, 1891

Type of this subgenus is *basteroti* Nyst, 1845, from the Miocene of Europe.

Terebra concava Say, 1827	**2839**
Concave Auger	

North Carolina to Florida and Texas to Brazil.

¾ inch in length, slender, about 12 whorls, semiglossy, and with slightly concave whorls. Whorls in spire with a large, heavily nodulated or beaded, swollen spiral cord just below the suture. Above the suture there is a spiral series of 20 very small beads per whorl. The concave middle of the whorl bears about 5 microscopic, incised spiral lines. Color yellowish gray. Uncommon; offshore in shallow water. Fossil specimens are occasionally found on New Jersey beaches. Do not confuse with the larger yellow *T. floridana* which has 2 spiral rows of elongate beads just below the suture. *T. vinosa* Dall, 1889, is a synonym.

Terebra arcas Abbott, 1954	**2840**
Arcas Cays Auger	

West coast of Florida to Mexico.

15 to 27 mm., its width ¼ of its length, semiglossy white with yellowish orange early whorls; rarely with a wide band

2837   2845

of orange. With 12 to 16 axial ribs per whorl with the interstices concave. Spiral sculpture of 5 to 7 incised lines, with the topmost one cutting across the ribs. Siphonal fasciole bordered above by a small, sharp thread. Columella with 1 weak, spiral fold at the base. Uncommon; 2 to 50 fathoms. Differs from *glossema* Schwengel which has 27 arching ribs on the last whorl.

Terebra glossema Schwengel, 1940   **2841**

Southeast Florida and Cuba; Bahamas.

1 inch, ¼ as wide and long. 1½ nuclear whorls, smooth, pinkish. 14 adult whorls white, with small, retractively curved ribs (27 on the last whorl). Spiral cut grooves 5 or 6 between ribs. Spiral lines present between the grooves on the last 3 whorls. Siphonal fasciole convex. Columella smooth and recurved. Sometimes buff-colored with a paler band on the base. Not uncommon; 1 to 8 fathoms. *The Nautilus*, vol. 53 (1940), pl. 12, and vol. 56 (1942), p. 65, pl. 6.

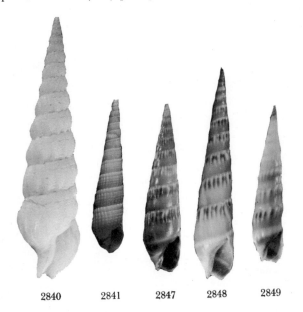

2840   2841   2847   2848   2849

Terebra protexta Conrad, 1845   **2842**
Fine-ribbed Auger

North Carolina to Florida and Texas to Brazil.

20 to 25 mm. in length, about 13 whorls, dull-white in color and with a well-indented suture. Whorls in spire slightly concave with about 22 fine axial ribs running from suture to suture, but which are broken weakly by 7 to 9 incised spiral lines. The upper line is about ¼ the way down the whorl.

Several forms exist which have been given names: form *lutescens* Dall, 1889 **(2843)** has about 30 to 32 finer axial riblets per whorl which are made slightly beaded by the spiral lines; in the form *limatula* Dall, 1889 **(2844)** the ribs and the spiral threads are about equal in size and give a reticulated pattern. All occur together in fairly deep water and are common.

Terebra rushii Dall, 1889   **2845**
Rush's Auger

Off south Florida.

15 mm., brilliant, polished white, columella not keeled, sides of whorls flat and with 4 strong, flat spiral cords. Base with fine spiral threads. Growth lines on the earlier whorls. Uncommon; 8 fathoms.

Subgenus *Hastula* H. and A. Adams, 1853

Type of this subgenus is *strigilata* (Linné, 1758) from the Indo-Pacific.

Terebra hastata (Gmelin, 1791)   **2846**
Shiny Atlantic Auger

Southeast Florida and the West Indies to Brazil. Bermuda.

1¼ to 1½ inches in length. Characterized by its smooth, highly glossy finish, its numerous axial ribs which extend from suture to suture, and by its bright yellowish color and white band below the suture. Columella smoothish and white. This is the "fattest" species in the Western Atlantic, and is fairly common in the West Indies.

Terebra cinerea (Born, 1778)   **2847**
Gray Atlantic Auger

East Florida and the West Indies and Brazil.
Sonora, Mexico, to Ecuador.

1 to 2 inches in length, slender, with flat-sided whorls and a sharp apex. Numerous small riblets extend halfway down the whorls (about 45 to 50 per whorl). Color all-cream or bluish brown; sometimes with darker spots below the suture. Surface with exceedingly fine, numerous rows of pinpricks which give the shell a silky appearance under the lens. Moderately common in the intertidal zone of beaches. Compare with *salleana* Deshayes. Synonyms include: *livida* Dillwyn, 1817; *aciculina* Lamarck, 1822; *castanea* Kiener, 1839; *jamaicensis* C. B. Adams; and *acuminata* Reeve, 1860. Animal paralyzes with its poison radula and swallows whole small polychaete worms (see Marcus and Marcus, 1960, *Bol. Fac. Fil. Cien. Letr. Univ. Sao Paulo*, no. 260, *Zool.* no. 23).

Terebra salleana Deshayes, 1859   **2848**
Salle's Auger

Northwest Florida to Texas and Vera Cruz, Mexico. Brazil.

1 to 1⅓ inches in length, similar to *cinerea*, but always a dark bluish gray or brownish, with fewer, larger punctations, with about 30 ribs per whorl, and with a purple, not white, nucleus. The spiral rows of pinpoint pits are more widely spaced than those of *cinerea*. Common in shallow water.

Terebra maryleeae R. D. Burch, 1965   **2849**
Marylee's Auger

Texas to British Honduras.

1 inch, with flat to slightly concave whorls; color dark-brown with an obscure white band at the periphery of the body whorl; sculpture of low, thin axial ribs on the posterior ⅓ of each whorl. Body whorl with a series of low, thin, weak nodules at the periphery. No pinpoint pits. Rarely all-white in color. Moderately common on intertidal sand flats. *The Veliger*, vol. 7, p. 242. See *Texas Conchologist*, vol. 4, no. 9, May 1968, for radulae.

Terebra pedroana Dall, 1908   **2850**
San Pedro Auger

Redondo Beach, California, to Baja California.

1 to 1¼ inches in length, strong, slender, with about 12 whorls and colored grayish to whitish yellow or brownish. Sculpture between sutures of first a fairly broad row of well to poorly developed nodules (about 15 to 18 per whorl), followed below by a flat area which is weakly and axially wrinkled or ribbed and with numerous, fine, spiral, incised lines. Siphonal canal bounded by a sharp spiral line on the outer shell. Fairly common in shallow water from sandy shores to 30 fathoms.

2850   2851   2852   2853   2854

Terebra crenifera Deshayes, 1859 — 2851
Western Crenate Auger

California to Ecuador.

1 to 1½ inches in length, with about 15 whorls. Color varies from white to tan. There are sharp, widely spaced, straight axial ribs, usually starting from node on the subsutural band. The interspaces are filled with fine spiral lines. The aperture is quadrate and the columella straight with no plication. Uncommon; in shallow water to 60 fathoms. Synonym: *ligyrus* Pilsbry and Lowe, 1932.

Terebra danai Berry, 1958 — 2852
Dana Auger

Southern California to the West Coast of Baja California.

1 to 1¼ inches in length, about 10 slightly convex whorls, usually of a dull brownish color with whitish subsutural band. Early whorls have axial ribs, narrower than interspaces, which often fade in later whorls. Subsutural band, which is set off by inconspicuous groove, is weakly noded though occasional specimens have strong nodes. Spiral sculpture of shallow, unevenly spaced spiral grooves. Aperture quadrate; columella curved with 1 weak plication at anterior end, laminated. Formerly fairly common in shallow water, uncommon in recent years. Synonyms: *simplex* Carpenter, 1856 (non Conrad, 1830); *philippiana* Dall, 1921 (non Deshayes, 1859).

Terebra hemphilli Vanatta, 1924 — 2853
Hemphill's Auger

West coast of Baja California.

½ to ¾ inch, about 10 whorls. This species, although smaller and more slender, resembles *T. danai* in color and sculpture. The columella is straighter and the aperture semi-elongate. Fairly common in the Scammon's Lagoon area.

Other Atlantic species:

2854 *Terebra nassula* Dall, 1889. Yucatan, Mexico, to the Lesser Antilles, 95 to 640 fms.

2855 *Terebra texana* Dall, 1898. Matagorda Island, Texas.

2856 *Terebra* (*Fusoterebra* Sacco, 1891) *benthalis* Dall, 1899 Florida Strait, 100 to 400 fms.

2857 *Terebra evelynae* Clench and Aguayo, 1939. Northern Cuba, 145 to 230 fms. *Mem. Soc. Cubana Hist. Nat.*, vol. 13 p. 196.

2858 *Terebra juanica* Dall and Simpson, 1901. Puerto Rico 5 mm.

2856   2858

Other Gulf of California species:

2859 *Terebra adairensis* Campbell, 1964. Sonora, Mexico to Sinaloa, Mexico.

2860 *Terebra affinis* Gray, 1834. San Luis Gonzaga Bay, Baja California.

2861 *Terebra allyni* Bratcher and Burch, 1970. Baja California to Jalisco, Mexico.

2862 *Terebra armillata* Hinds, 1844. West coast of Baja California to Peru. (Synonyms: *albicostata* Adams and Reeve, 1850; *marginata* Deshayes, 1857.)

2863 *Terebra balaenorum* Dall, 1908. West coast of Baja California to Jalisco, Mexico. (Synonym: *pulchella* Deshayes, 1857 (not Röding, 1798).)

2864 *Terebra berryi* Campbell, 1961. Baja California to Costa Rica.

2865 *Terebra brandi* Bratcher and Burch, 1970. Gulf of California to Peru.

2866 *Terebra bridgesi* Dall, 1908. Gulf of California to Panama. (Synonym: *dushanae* Campbell, 1964.)

2867 *Terebra churea* Campbell, 1964. Sonora, Mexico.

2868 *Terebra corintoensis* Pilsbry and Lowe, 1932. Baja California to Ecuador.

2869 *Terebra dorothyae* Bratcher and Burch, 1970. Gulf of California to the Galapagos.

2870 *Terebra elata* Hinds, 1844. Baja California to Ecuador. (Synonyms: *aspera* Hinds, 1844 (not Bosc, 1801); *radula* Hinds, 1844 (not Gravenhorst, 1807); *petiveriana* Deshayes, 1857.)

2871 *Terebra grayi* E. A. Smith, 1877. Sonora, Mexico, to Guatemala. (Synonym: *gracilis* Gray, 1834 (not I. Lea, 1833).)

2872 *Terebra hindsi* Carpenter, 1857. Baja California to Colombia.

2873 *Terebra intertincta* Hinds, 1844. Baja California to Ecuador.

2874 *Terebra iola* Pilsbry and Lowe, 1932. Outer coast of Baja California to Mazatlan, Mexico.

2875 *Terebra larvaeformis* Hinds, 1844. West coast of Baja California to Ecuador. (Synonym: *isopleura* Pilsbry and Lowe, 1932.)

2876 *Terebra ornata* Gray, 1834. Baja California to Ecuador.

2877 *Terebra panamensis* Dall, 1908. Baja California to Panama.

2878 *Terebra puncturosa* Berry, 1959. Baja California to Panama. *Leaflets in Malacology,* no. 18. p. 112.

2879 *Terebra robusta* Hinds, 1844. West coast of Baja California to the Galapagos. (Synonyms: *lingualis* Hinds, 1844; *loroisi* Guerin, 1854; *insignis* Deshayes, 1857; *macrospira* Li, 1930; *dumbauldi* Hanna and Hertlein, 1961.)

2880 *Terebra roperi* Pilsbry and Lowe, 1932. Conception Bay, Baja Cailfornia to Ecuador.

2881 *Terebra rufocinerea* Carpenter, 1857. West coast of Baja California to Tartar Shoals.

2882 *Terebra shyana* Bratcher and Burch, 1970. West coast of Baja California to Manzanillo, Mexico.

2883 *Terebra specillata* Hinds, 1844. West coast of Baja California to Ecuador.

2884 *Terebra stohleri* Bratcher and Burch, 1970. Baja California to Socorro Island, Mexico.

2885 *Terebra tiarella* Deshayes, 1857. West coast of Baja California.

2886 *Terebra tuberculosa* Hinds, 1844. West coast of Baja California to Guatamala.

2887 *Terebra variegata* Gray, 1834. West coast of Baja California to Ecuador. (Synonyms: *africana* Griffith and Pidgeon, 1834 (ex-Gray ms); *hupei* Lorois, 1857.)

(Above list supplied by Twila Bratcher and R. D. Burch.)

Family TURRIDAE Swainson, 1840

The family Turridae is a very large and diverse group of toxoglossate gastropods which are very difficult to classify. Most of them have a slit, notch or U-shaped canal at the top of the outer lip, known as the "turrid notch." A book of this size cannot do justice to the many interesting species found in our waters. The family probably contains no less than 500 genera and subgenera and several thousand species. A valuable review of the family is given by A. W. Powell in the *Bulletin of the Auckland Institute and Museum,* no. 5, pp. 1–184, 1966. Those interested should consult the works of Grant and Gale, Bartsch, Dall, Rehder and Woodring. The family will continue to undergo many major changes for several years. Mrs. Virginia Maes kindly suggested many improvements in the manuscript.

The tropical west American species are treated in detail by James McLean in Keen's 1971 second edition of "Sea Shells of Tropical West America," Stanford University Press.

Subfamily PSEUDOMELATOMINAE J. H. McLean, 1971

Posterior sinus on the shoulder; operculum leaf-shaped, with a terminal nucleus. Rachidian tooth of radula rectangular in shape, with a large projecting cusp; marginal tooth massive, tapered to a sharp point; poison gland present.

Genus *Hormospira* Berry, 1958

Shells 1½ inches, elongate-fusiform, with oblique nodules on the shouldered periphery of the whorls. Periostracum

thin. Operculum leaf-shaped with a terminal nucleus. Radula is unusual for a turrid (see Powell, 1966, p. 33). Type: *maculosa* (Sowerby,1834).

Hormospira maculosa (Sowerby 1834)   **2888**
Western Maculated Turrid

Gulf of California to Ecuador.

1 to 1½ inches, high-spired, narrowly fusiform, with nodes midway between sutures. Color cream to bluish with the nodes white and with dots, flecks and axial flames of brown. Moderately common; intertidal to 16 fathoms.

2888

Other species:

2889 *Hormospira* (*Tiariturris* S. S. Berry, 1958) *libya* (Dall, 1919). Baja California to Mazatlan, Mexico. 20 to 66 fms. Uncommon.

2890 *Hormospira* (*Tiariturris*) *spectabilis* Berry, 1958. Isla Angel de la Guarda, Baja California, 67 fms. Off Cape Tepopa, Baja California, 6 fms.

Genus *Pseudomelatoma* Dall, 1918

Shells about 1½ inches, solid, compact, dark, fusiform, with slanting axial ribs above which is a subsutural spiral row of beads. Periostracum brownish, opaque. Operculum leaf-shaped, with a terminal nucleus. Pacific coast only. *Laevitectum* Dall, 1919, is probably a synonym. Type: *penicillata* (Carpenter, 1865).

Pseudomelatoma moesta (Carpenter, 1865)   **2891**
Doleful Turrid

Southern California to Baja California.

1 to 1½ inches, dark-brown, with a beaded band just below the suture, with 9 or 10 slightly curved ribs per whorl, and with faint incised lines. In some specimens the axial ribs are very weak or obsolete. Moderately common; under rocks intertidally.

Other species:

2892 *Pseudomelatoma torosa* (Carpenter, 1865). Oregon to Scammon Lagoon, Baja California. (Synonym: *aurantia* Carpenter, 1865.) Scarce in shallow water; common at 35 fms.

2893 *Pseudomelatoma penicillata* (Carpenter, 1865). Magdalena Bay to Gulf of California. May be form of *moesta* (Carpenter, 1864).

2891   2892

2894 *Pseudomelatoma grippi* (Dall, 1919). San Diego, California. Formerly in *Bela*. (Synonym: *Crassispira martinensis* Dall, 1919,)

2895 *Pseudomelatoma sticta* S. S. Berry, 1956. Anacapa Island, California, 26 fms. *Jour. Wash. Acad. Sci.*, vol. 46, p. 156.

2896 *Pseudomelatoma* (*Laevitectum* Dall, 1919) *eburnea* (Carpenter, 1865). Gulf of California. Rare. Type of the subgenus.

Subfamily TURRICULINAE Powell, 1942

Members of this subfamily have shells of moderate (2 to 3 inches) size, narrowly fusiform, with a tall spire and long siphonal canal. The sinus is usually deep, rounded and always on the shoulder slope. Operculum with a mediolateral nucleus or, in the case of most American forms, with a terminal nucleus.

Genus *Leucosyrinx* Dall, 1889

Shells 1 to 3 inches, fusiform, with a tall spire. Protoconch small, globular, 2 smooth whorls. Adult whorls with axial nodes on the periphery and with spiral cords below. Sinus fairly deep, broadly concave. Outer lip arching forward. Operculum leaf-shaped with a terminal nucleus. Type: *Pleurotoma* (*Pleurotomella*) *verrillii* Dall, 1881.

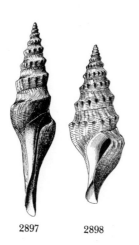

2897   2898

Leucosyrinx tenoceras Dall, 1889
Tenoceras Turrid

2897

Off North Carolina to Mississippi and Lesser Antilles.

2½ inches, aperture half as long as the slender, grayish white shell. Shoulder of whorl angular, those in the spire bearing about 14 nodes. Spiral sculpture of numerous, flattened indistinct threads. Columella smooth and white. Operculum pear-shaped, pointed, with a thin marginal rib, horn-colored. Rare; 478 to 724 fathoms.

2892   2900

Other species:

2898 *Leucosyrinx verrillii* (Dall, 1881). North Carolina, Gulf of Mexico and the West Indies. 150 to 940 fms.

2899 *Leucosyrinx sigsbei* (Dall, 1881). Yucatan Straits, 640 fms.; off Bequia, West Indies, 1,591 fms.

2899

2900 *Leucosyrinx subgrundifera* (Dall, 1888). North Carolina; Gulf of Mexico, 528 to 940 fms.

2900a *Leucosyrinx kincaidi* Dall, 1919. Shelikoff Strait, Alaska. *Proc. U. S. Nat. Mus.*, vol. 56, p. 6.

— *Leucosyrinx persimilis* Dall, 1890 (and varieties *blanca* Dall, 1919; *leonis* Dall, 1908) are *Aforia goodei* (Dall, 1890).

Genus *Knefastia* Dall, 1919

Shells 2 to 3 inches, robust, resembling *Latirus*, with angular axial nodes crossed by coarse spiral cords. Periostracum thick. Sinus deep, U-shaped, occupying the middle of the shoulder slope. Operculum leaf-shaped with a terminal nucleus. Living only in the Eastern Tropical Pacific. Type: *olivacea* Sowerby, 1833.

Knefastia olivacea (Sowerby, 1833)   2901
Olivaceous Knefastia

Gulf of California to Ecuador.

1 to 2 inches, heavy, broadly fusiform, with about 12 irregular, axial knobs on the angular shoulder of the last whorl. Spiral sculpture of numerous distinct spiral raised threads. Outer lip lirate within. Shell orangish to purple-brown with olive-brown periostracum. Moderately common; offshore. Operculum thin, brown, leaf-shaped.

2901   2902

Knefastia dalli Bartsch, 1944   2902
Dall's Knefastia

Gulf of California.

2 inches, similar to *olivacea*, but much slimmer, with lower but longer, rounded axial nodules, crossed by strong spiral threads. Pinkish white with brown variegations. Periostracum brownish to olive. Inside of outer lip smoothish. Uncommon; offshore.

Other species:

2903 *Knefastia tuberculifera* (Broderip and Sowerby, 1829). Gulf of California to Nicaragua. Rare.

2904 *Knefastia walkeri* S. S. Berry, 1958. Off Angel de la Guarda Island, Gulf of California. Uncommon.

2905 *Knefastia funiculata* Kiener, 1839–40. Mexico to Panama. Uncommon.

2906 *Knefastia nigricans* (Dall, 1919). Dredged off Baja California. *Proc. U.S. Nat. Mus.*, vol. 56, p. 3.

2907 *Knefastia princeps* S. S. Berry, 1953. Cedros Island, Baja California. *Trans. San Diego Soc. Nat. Hist.*, vol. 11, p. 420.

Genus *Fusiturricula* Woodring, 1928

Sculpture of stout, foldlike oblique axials crossed by spiral threads. The outer lip is produced greatly forward, thus giving the anal sinus a very deep appearance. Type: *fusinella* (Dall, 1908). Considered a subgenus of *Turricula* by

some workers. The subgenus *Fusisyrinx* Bartsch, 1934, is probably a synonym, as may be *Cruziturricula* Marks, 1951, although McLean (in Keen, 1971) accepts the latter as a genus in the subfamily Borsoniinae.

Main species:

2908 *Fusiturricula armilda* (Dall, 1908). Gulf of California to Panama, 35 to 58 fms.

2909 *Fusiturricula dolenta* (Dall, 1908). Panama Bay, 47 to 200 fms.

2910 *Fusiturricula fusinella* (Dall, 1908). Gulf of California to Panama, 58 to 153 fms.

2911 *Fusiturricula enae* Bartsch, 1934. Puerto Rico, 190 to 350 fms. *Smithsonian Misc. Coll.*, vol. 91, p. 13.

2912 *Fusiturricula howelli* Hertlein and Strong, 1951. Off Costa Rica, 42 to 61 fms.

2913 *Fusiturricula* (*Cruziturricula* Marks, 1951) *lavinia* (Dall, 1919). West coast of Mexico.

2914 *Fusiturricula notilla* (Dall, 1908). Gulf of California, 58 fms.

2916 *Fusiturricula* (*Fusisyrinx* Bartsch, 1934) *fenimorei* Bartsch, 1934 (type of subgenus). Puerto Rico, 80 to 180 fms. Deep.

2917 *Fusiturricula* (*Cruziturricula* Marks, 1951) *arcuata* (Reeve, 1843). Panama. Rare. (Synonym: *Turricula panthea* Dall, 1919, pl. 1, fig. 6.)

2918 *Fusiturricula jaquensis* (Sowerby, 1850). Off Paramaribo, Surinam, 30 fms.; also Miocene West Indies.

Genus *Megasurcula* Casey, 1904

Shells up to 4 inches, biconic-ovate, with a tall spire and a long, large, body whorl that gradually tapers into a short deeply notched anterior siphonal canal. The fasciole is ridged. Protoconch of 2 smooth, small whorls. Anal sinus broad and shallow. Operculum leaf-shaped with a terminal nucleus. Type: *carpenteriana* (Gabb, 1865).

Megasurcula carpenteriana (Gabb, 1865)   2919
Carpenter's Turrid

Central California to Baja California.

2 to 3 inches, fusiform, golden-tan with numerous, narrow, reddish brown bands. Anal notch shallow. Shoulder of whorls smoothish or slightly angular. Axial sculpture of fine, microscopic growth lines. Common; 5 to 240 fathoms on mud bottom. A shouldered form with small nodes was given the name *tryoniana* (Gabb, 1866). Once erroneously placed in the Eocene

2919

genus *Cryptoconus* Koenen, 1867. Other synonymic forms include *granti* Bartsch, 1944, and *tremperiana* (Dall, 1911) (2920).

Other species:

2921 *Megasurcula remondii* (Gabb, 1866). Monterey, California, to Todos Santos Bay, Baja California. Gravel bottoms, 20 to 55 fms. Synonym: *stearnsiana* (Raymond, 1904). Also Pliocene.

Genus *Rhodopetoma* Bartsch, 1944

Shells 1 inch, deepsea, very similar in appearance to *Borsonella* but without a columella fold. Spire high, body whorl short. Sculpture of rounded subsutural folds, a shallow shoulder sulcus, below which is a rounded peripheral bulge. Sinus broadly rounded and subsutural. Operculum lanceolate, with a terminal nucleus. Type: *rhodope* (Dall, 1919). Only 2 American species known:

2922 *Rhodopetoma amycus* (Dall, 1919). 871 fms., off Monterey, California. (Formerly in *Leucosyrinx*.)

2923 *Rhodopetoma rhodope* (Dall, 1919). Off Santa Rosa Island, 82 fms.; off San Diego, 633 fms. (Formerly in *Borsonella*.)

Genus *Cochlespira* Conrad, 1865

Shells white, up to 2 inches, fusiform, with a strong spiny carina on the upper part of the whorl. Siphonal canal narrow and long. Operculum leaf-shaped, with a terminal nucleus. Synonyms are *Ancistrosyrinx* Dall, 1881; *Coronasyrinx* Powell, 1944; and *Pagodasyrinx* Shuto, 1969. Type: *cristata* (Conrad, 1847).

Cochlespira radiata (Dall, 1889)   2924
Common Star Turrid

Off North Carolina to Florida, the Gulf of Mexico and the West Indies.

½ inch in length. A delicate, glossy, translucent and highly ornamented species. Anterior canal very long. Shoulders keeled, with numerous, small, sharp, triangular spines. Commonly dredged from 30 to 170 fathoms.

Cochlespira elegans (Dall, 1881)   2925
Elegant Star Turrid

Off both sides of Florida.

1 to 2 inches, similar to *radiata,* but larger, more elongate and with 2 spiral rows of more numerous and duller spines on the sharp shoulder. Very rare; 30 to 200 fathoms.

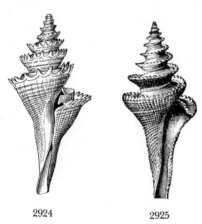

2924   2925

Other species:

2926 *Cochlespira cedonulli* (Reeve, 1843). Guaymas, Mexico, to Panama, 30 to 150 fms. Uncommon. ¾ inch.

Genus *Aforia* Dall, 1889

Shells 2 to 3 inches, chalky-gray, light build, fusiform, with spiral lirae and carinae. Periostracum greenish. Operculum ovate, with a subterminal nucleus. Type: *circinata* (Dall, 1873). *Irenosyrinx* Dall, 1908, is a synonym.

Aforia circinata (Dall, 1873)   2927
Keeled Aforia

Bering Sea; Alaska and Japan.

2 to 3½ inches, light build, with a tall carinated spire. 25 to 40 spiral cords from the peripheral carina to the end of the long, slender siphonal canal. Common; 6 to 300 fathoms. Synonyms: *insignis* Jeffreys, 1883; *nojimensis* Yokoyama, 1920; *hondoana* (Dall, 1925); *diomedea, okhotskensis, sakhalinensis* and *chosenensis* all Bartsch, 1945; *minatoensis* Otuka, 1949; and *otohimei* Ozaki, 1958.

2927

Other species:

2928 *Aforia pacifica* (Dall, 1908). Off Sitka, Alaska, 1,569 fms.

2929 *Aforia crebristriata* (Dall, 1908). Off Sitka, Alaska, 1,569 fms. Both *Bull. Mus. Comp. Zool.*, vol. 43, p. 272, 270.

2930 *Aforia goodei* (Dall, 1890). Queen Charlotte Sound, British Columbia, to southern Chile, 1,220 to 1,950 meters. (Synonyms: *amycus* Dall, 1919; *persimilis* Dall, 1890; *leonis* Dall, 1908; *blanca* Dall, 1919.)

Genus *Pyrgospira* J. H. McLean, 1971

Shell of medium size, high-spired, resembling a *Drillia,* with tabulate whorls, shoulder concave, periphery nodulated. Anterior canal short, turrid notch deep. Radula with marginal teeth only; main arm of the tooth massive, distal limb small. Type: *P. obeliscus* (Reeve, 1843).

Pyrgospira ostrearum (Stearns, 1872)   2931
Oyster Turrid

North Carolina to south half of Florida. Cuba.

⅓ to ⅔ inch in length; light yellow-brown to chestnut. Sinus U-shaped. About 20 weakly beaded axial ribs per whorl. Just below the suture there is a single, smooth, strong spiral cord. Below this on the shoulder concavity there are 2 to 5 spiral threads crossed by axial curved growth lines. Spiral threads moderately strong to weak (16 to 20 on the last whorl, 4 between sutures). Lower part of outer lip thin and strongly crenulate or wavy. Ribs on last whorl are beaded on the base. Common from low water to 90 fathoms. *Pyrgospira tampaensis*

Bartsch and Rehder, 1939, is very similar, and probably is this species.

Other species:

2932 *Pyrgospira obeliscus* (Reeve, 1843). Head of the Gulf of California to Colombia. (Synonyms: *Clathrodrillia aenone* Dall, 1919; *C. nautica* Pilsbry and Lowe, 1932; *Crassispira tomliniana* Melvill, 1927.)

Subfamily TURRINAE Swainson, 1840

Fairly large, fusiform shells distinguished from the Turriculinae by their sinus which is either on the peripheral carina or on a minor spiral cord immediately above the periphery. The sinus may be broadly V-shaped or a deep narrow slit. Operculum leaf-shaped with a terminal nucleus.

Genus *Gemmula* Weinkauff, 1875

Shells white, solid, with 1 or 2 beaded peripheral carinae, giving a coglike appearance. Sinus deep and narrow. Nuclear whorls tall, conical, and axially costate. Type: *hindsiana* Berry, 1958.

Gemmula periscelida (Dall, 1889)   **2933**
Atlantic Gem Turrid

North Carolina to Tortugas, Florida.

1½ to 2 inches in length, heavy and with the sinus or anal notch well below the suture. With numerous spiral lirae, the one at the top of the whorl bearing numerous, coglike, axial riblets. Color ash-gray with a straw-colored periostracum. Rare in 100 fathoms.

2933

Gemmula hindsiana S. S. Berry, 1958   **2934**
Hinds' Gem Turrid

Magdalena Bay to the Gulf of California.

¾ inch, fusiform, with coglike nodes on the angular carina of the shoulder. Color brown with white nodes. Uncommon; offshore. Formerly *gemmata* Reeve, 1843, not Conrad, 1835.

Genus *Ptychosyrinx* Thiele, 1925

Similar to *Gemmula*, but with a broadly open U-shaped sinus, not a narrow slit. Mainly an Indo-Pacific group. Type: *bisinuata* von Martens, 1901. *Bathybermudia* Haas, 1949, is a synonym. The latter has only one deep water species, *P. carynae* Haas, 1949 (**2657**) (1,700 fathoms, off Bermuda). See *Bull. Inst. Catalana Hist. Nat.*, vol. 37, p. 70.

Genus *Carinoturris* Bartsch, 1944

Shells ¾ inch, smooth, except for a strong carina between sutures. White with an olive periostracum. Protoconch of 1 smooth rounded whorl. Operculum ovate, with a terminal nucleus. Deep water. Type: *adrastia* (Dall, 1919).

2935 *Carinoturris adrastia* (Dall, 1919). Monterey to San Diego, California. 300 to 581 fms.

2936 *Carinoturris polycaste* (Dall, 1919). Off Oregon to Gulf of California. 786 fms.

2937 *Carinoturris fortis* Bartsch, 1944. Monterey Bay, California. 298 fms., sand and mud. Rare.

Genus *Polystira* Woodring, 1928

Large white to brown, fusiform, with strong, smooth spiral cords. Protoconch stout, cylindrical, of almost 2 whorls, the last ¼ whorl being axially ribbed. Peripheral sinus broadly V-shaped. Operculum leaf-shaped, with a terminal nucleus. Type: *albida* Perry, 1811. *Oxytropa* Glibert, 1955, is a synonym. *Pleuroliria* Gregorio is not this genus.

Polystira albida (Perry, 1811)   **2938**
White Giant Turret   Color Plate 10

South Florida to Texas and the West Indies.

3 to 4 inches in length, pure-white in color. With about 5 to 7 spiral smooth cords of unequal size between the well-impressed sutures. Largest squarish cord behind the slotlike, deep sinus. *P. virgo* Lamarck, 1822, and "Wood" are this species. Not uncommonly dredged in the Gulf of Mexico; 26 to 125 fathoms.

Polystira tellea (Dall, 1889)   **2939**
Delicate Giant Turret

Off Florida to Louisiana.

3 to 3½ inches in length. Grayish white. Sculpture not so distinct nor so smooth as in *albida*. Sinus higher on the shoulder and broadly and shallowly V-shaped. Axial minute fimbriations cover the exterior. Not uncommonly dredged off Key West. Do not confuse this and the preceding species with *Fusinus couei*.

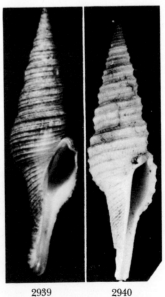

2939   2940

Other species:

2940 *Polystira vibex* (Dall, 1889). Gulf of Mexico to the West Indies, 40 to 200 fms. Moderately common.

2941 *Polystira artia* (S. S. Berry, 1957). Off Angel de la Guarda Island, Gulf of California, 67 fms.

2942 *Polystira nobilis* (Hinds, 1843). Gulf of California to Acapulco, Mexico, 20 fms. Common.

2943 *Polystira oxytropis* (Sowerby, 1834). Guaymas to Colombia, 7 to 20 fms. (Synonym: *albicarina* Sowerby, 1870.) Not uncommon. 1 to 1½ inches.

2944 *Polystira picta* (Reeve, 1843). Guaymas to Colombia, 6 to 14 fms. Uncommon. (Synonym: *rombergii* Mörch, 1857.)

2945 *Polystira parthenia* (S. S. Berry, 1957). Gulf of Nicoya, Costa Rica, 10 fms. Rare?

2946 *Polystira formosissima* (E. A. Smith, 1915). Brazil, 7 fms.

Genus *Antiplanes* Dall, 1902

Shell always sinistral in this subgenus (type: *voyi* Gabb, 1866); dextral in the subgenus *Rectiplanes* Bartsch, 1944. Type: *santarosana* Dall, 1902. Whorls smoothish, with microscopic spiral striations. Periostracum thick, grayish green. Sinus broadly open. Cold-water genus. *Rectisulcus* Habe, 1958, is a synonym.

Antiplanes voyi (Gabb, 1866)   **2947**
Perverse Turrid

Forrester Island, Alaska, to off California.

2 inches, sinistral, whitish shell with a greenish brown periostracum. Sinus shallowly V-shaped. Protoconch smooth, globular, and with only 2 whorls. Uncommon; 30 to 300 feet.

2947

Other species:

2948 *Antiplanes vinosa* (Dall, 1874). Bering Sea to San Diego, California.

2949 *Antiplanes kamchatica* Dall, 1919. Southwest Bering Sea, 48 to 100 fms.

2950 *Antiplanes catalinae* (Raymond, 1904). Esteros Bay to San Diego, California.

2951 *Antiplanes rotula* Dall, 1921. Forrester Island, Alaska, to San Diego, California. (Synonym: *smithi* Arnold, 1903, not Forbes, 1840.)

2952 *Antiplanes diaulax* (Dall, 1908). Off Coronado Islands.

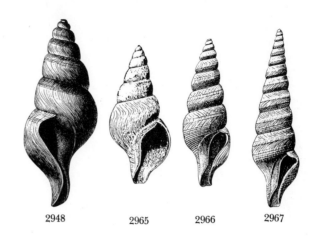

2948   2965   2966   2967

2953 *Antiplanes major* Bartsch, 1944. Monterey Bay to Santa Cruz, California. 43 to 278 fms., but commonest between 50 to 75.

2954 *Antiplanes profundicola* and *diomedia* (**2955**) Bartsch, 1944. 659 fms. and 328 fms. respectively, both off Point Sur, Monterey, California.

2956 *Antiplanes hyperia* Dall, 1919. Off Drake's Bay and the Coronado Islands.

2957 *Antiplanes abarbarea* Dall, 1919. Cape Martin to Cortez Bank, California.

2958 *Antiplanes briseis* Dall, 1919. Drake's Bay to Coronado Islands.

2959 *Antiplanes bulimoides* Dall, 1919. Bowers Bank, Bering Sea.

2960 *Antiplanes beringi* (Aurivillius, 1885). Bering Sea south of St. Lawrence Island to the Aleutians and eastward to the Shumagin Islands.

2961 *Antiplanes litus* Dall, 1919. Coast of Washington to Esteros Bay, California.

— *Antiplanes amphitrite* Dall, 1919, see *Borsonella*.

2962 *Antiplanes antigone* Dall, 1919. Off San Diego, California, in 822 fms.

2963 *Antiplanes agamedea* Dall, 1919. Off Cape San Quentin, in 359 fms.

2964 *Antiplanes amycus* Dall, 1919. Monterey Bay, in 581 fms. All *Proc. U.S. Nat. Mus.*, vol. 56, no. 2288.

2965 *Antiplanes piona* (Dall, 1902). Bering Sea. 40 mm.

Other _Rectiplanes_ species:

2966 *Antiplanes santarosana* (Dall, 1902). Off Santa Rosa Island. Point Sur to off San Diego, California.

2967 *Antiplanes thalaea* (Dall, 1902). Unimak Pass, Aleutians, to off San Diego, California, 252 fms. 40 mm.

2968 *Antiplanes willetti* S. S. Berry, 1953. Off Forrester Island, southwest Alaska, 50 fms. *Trans. San Diego Soc. Nat. Hist.*, vol. 11, p. 419.

Subfamily BORSONIINAE Bellardi, 1875

Shell biconic or fusiform in shape; with columellar plicae; sinus on the shoulder, poorly developed. Operculum present or absent. Radula with 2 slender marginals. Shell usually with columella plications. We have not included the genus *Cordieria* Rouault, 1848, for which see *C. rouaultii* Dall, 1889, p. 98, from Barbados, 76 fms.

Genus *Taranis* Jeffreys, 1870

Shells white, small, 3 to 6 mm., ovate-biconic, angular whorls, spirally keeled, with axial lamellations giving a fenestrate sculpture. Aperture large. Protoconch paucispiral, 2 whorls, papillate, with microscopic stippled spiral lirae. Sinus shallow. Columella abruptly twisted. Type: *moerchii* Malm, 1863. Synonyms: *Allo* Lamy, 1934; *Feliciella* Lamy, 1934; *Fenestrosyrinx* Finlay, 1926. Atlantic species:

2969 *Taranis cirrata* (Brugnone, 1862). Off Cape Hatteras, North Carolina, 124 fms. Florida Strait, 150 to 200 fms. (Synonym: *moerchii* Malm, 1863, *fide* Dall, 1889, p. 128.)

2969

2970 *Taranis cirrata tornatus* Verrill, 1884. Off Cape Sable, Nova Scotia, 1,255 fms.

— *Taranis pulchella* Verrill, 1880, is a *Microdrillia*, *fide* V. Maes, *in litt.*

Genus *Borsonella* Dall, 1908

Shells 1 inch, fusiform, of a rather simple, graceful outline, with a tall spire of bluntly angulated whorls. Protoconch small, bluntly rounded, of 1 or 2 smooth whorls. Adult whorls with a wide, concave, sunken shoulder slope, below which is a rounded peripheral fold which may be smooth or spirally lirate. Periostracum light-olive. Sinus deep, broadly U-shaped, occupying the whole of the shoulder sulcus. Columella with a nearly horizontal strong plica. Operculum absent. Type: *angelana* Hanna, 1924. Living off the Pacific coast and Galapagos Islands.

2971 *Borsonella angelana* Hanna, 1924. Drake's Bay to Coronado Islands, Baja California. (Synonym: *dalli* Arnold, 1903.)

2972 *Borsonella agassizii* (Dall, 1908). Gulf of Panama, deep water.

2973 *Borsonella omphale* Dall, 1919. Off Point Loma, California. Deep water.

2974 *Borsonella coronadoi* (Dall, 1908). San Clemente Island, California, to Coronado Islands.

2975 *Borsonella diegensis* (Dall, 1908). Off San Diego to Panama; Galapagos Islands. Deep water.

2976 *Borsonella barbarensis* Dall, 1919. Off Santa Barbara Islands, in 414 fms.

2977 *Borsonella bartschi* (Arnold, 1903). San Pedro to San Diego, California, 75 fms.

2978 *Borsonella nicoli* Dall, 1919. Off Point Loma, California. Deep water.

2979 *Borsonella civitella* Dall, 1919. Off Point Loma, California. Deep water.

2980 *Borsonella nychia* Dall, 1919. Off Point Loma, California, 101 fms. All *Proc. U.S. Nat. Mus.*, vol. 56, no. 2288.

2981 *Borsonella pinosensis* Bartsch, 1944. Off Point Pinos, Monterey, California, 40 to 50 fms. Rare.

Subgenus *Borsonellopsis* J. H. McLean, 1971

Shell broad and angulate, with nodes at periphery; columella fold weak or lacking; operculum vestigial. Deep water. Type: *Leucosyrinx erosina* Dall, 1908, from the Gulf of Panama.

2982 *Borsonella (Borsonellopsis) callicesta* (Dall, 1902). Off Santa Barbara, California, to Panama; Galapagos Islands, in 700 to 4,000 meters. (Synonyms: *Gemmula esuriens* and var. *pernodata* Dall, 1908; *Antiplanes amphitrite* and var. *beroe* Dall, 1919; *Cryptogemma cymothoe* and *eidola* Dall, 1919.) 20 mm.

2982

Genus *Microdrillia* Casey, 1903

Shells ½ inch or less, fusiform, strong, tall-spired, whitish, with strong, smooth, even-sized spiral keels. Axial growth threads between ribs. Protoconch of 3 to 5 whorls, the tip smooth, the remainder axially costate. V-shaped sinus at the top of the outer lip. Siphonal canal short. Type: *cossmanni* (Meyer, 1887) from the U.S. Oligocene.

2983 *Microdrillia comatotropis* (Dall, 1881). Off West Florida, 26 to 100 fms. 3 mm.; yellowish. 3 or 4 cords between sutures. Moderately common. *Pleurotoma tiara* Watson, 1881 (Aug.), 390 fms., West Indies is a synonym.

2984 *Microdrillia pulchella* (Verrill, 1880). Off Martha's Vineyard, Massachusetts, 484 fms.

2985 *Microdrillia tersa* Woodring, 1928. Panama Bay, Pacific coast, in 22 meters; also Miocene of Jamaica.

2983   2984

Genus *Darbya* Bartsch, 1934

Known only from the type species, *lira* Bartsch, 1934 **(2985a),** from 80 to 360 fathoms in the Puerto Rican Deep. Elongate-fusiform, 1 inch and characterized by a strong, broad, blunt fold on the middle of the columella. (*Smithsonian Misc. Coll.*, vol. 91, no. 2, p. 22).

Genus *Bathytoma* Harris and Burrows, 1891

Shells large, 3 inches, biconical, with a conic spire. Protoconch of 3 whorls, the last with arching riblets. Adult sculpture of spiral beaded threads. Columella with an oblique plica on the lower ½. Type: *Murex cataphractus* (Brocchi, 1814).

Bathytoma viabrunnea **(Dall, 1889)** **2986**
Brown-banded Turrid

Southeast Florida and the West Indies.

1½ to 2 inches in length, heavy and thick-shelled. Nuclear whorl smooth and brown. Sculpture of numerous microscopically beaded spiral rows of fine threads. Color yellowish to orangish white with a spiral, suffused band of light-brown well below the suture. Nucleus dark-brown and with tiny arched, smooth ribs. Anal sinus very wide. Operculum ⅙ the length of the aperture; quadrate, light-brown. Rare from 100 to 350 fathoms. Erroneously placed in *Genota*.

Other species:

2987 *Bathytoma mitrella* (Dall, 1881). Yucatan Strait, Mexico, 640 fms.; Sombrero, Puerto Rico, 450 fms. (*Genota*).

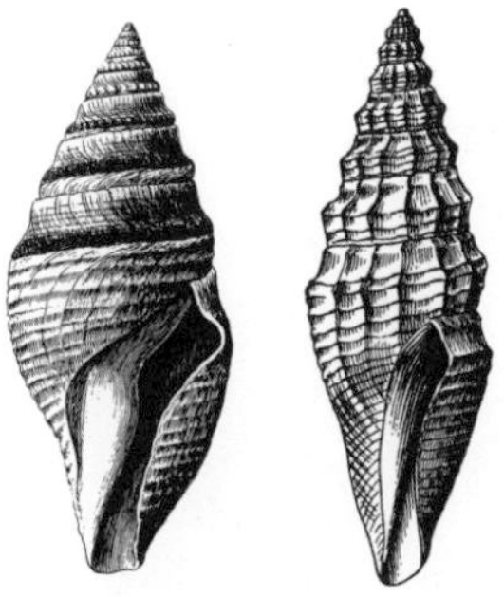

2986 2987

Subfamily MITROLUMNINAE J. H. McLean, 1971

Genus *Mitromorpha* P. P. Carpenter, 1865

Shells small, 8 mm., solid, biconic. Whorls convex but overlapping so that the spire profile is straight-conic. Protoconch 2 whorls, smooth, depressed-papillate with the tip flattened and inrolled. Adult whorls spirally corded. Sinus obsolete. Inner lip without folds. Type: *filosa* (Carpenter, 1865).

Mitromorpha filosa **(Carpenter, 1865)** **2988**
Filose Turrid

Monterey, California, to the Gulf of California.

⅜ inch in length, solid, light orange-brown in color. Spiral cords may be slightly beaded in some specimens. Microscopic axial threads present between ribs, which cross the parietal wall. Outer lip thickened within by a dozen short lirae. Uncommon offshore. *M. carpenteri* Gilbert, 1954, is a synonym.

Mitromorpha aspera **(Carpenter, 1864)** **2989**
Beaded Turrid

Monterey, California to the Gulf of California.

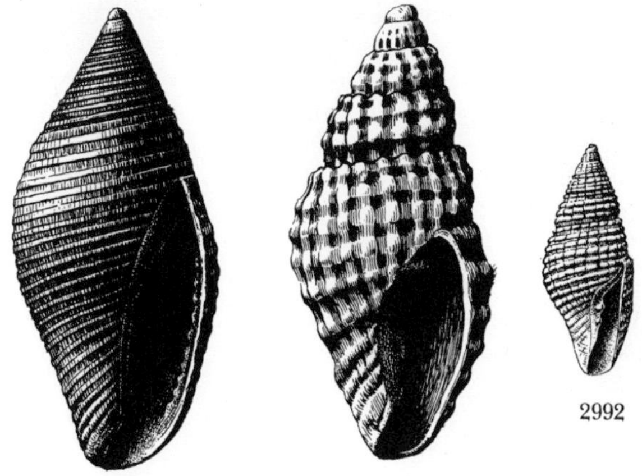

2988 2989 2992

⅜ inch in length, strongly beaded and somewhat cancellate, with a glossy finish and light orange-brown in color. Moderately common offshore.

Other species:

2990 *Mitromorpha gracilior* "Hemphill" Tryon, 1884. Forrester Island, Alaska, to San Diego, California. (Synonym: *intermedia* Arnold, 1903.)

2991 *Mitromorpha dormitor* (Sowerby, 1844). West Indies. ¼ inch, lavender color. Locally common.

Genus *Mitrolumna* Bucquoy, Dautzenberg and Dollfus, 1882

Shell broadly spindle-shaped, without a sinus, with axial riblets and spiral threads. Columella with 2 or 3 weak folds. Type: *olivoidea* (Cantraine, 1835).

Mitrolumna biplicata **Dall, 1889** **2992**
Biplicate Mitralike Turrid

Bermuda; both sides of Florida to Barbados.

6 to 8 mm., biconic, cancellated, yellowish with brown flames. Nucleus white, glassy, globose, 1½ whorls. Outer lip lirate within. Inner lip with 2 strong plications near its middle, the upper one the largest. Moderately common; 20 to 200 fathoms. *Mitra haycocki* Dall and Bartsch, 1911, is a synonym. This species is placed in *Mitrolumna* on the advice of Virginia Maes, *in litt.*

Subgenus *Arielia* Shasky, 1961

Shell small, 12 mm., narrowly biconic, with a flat-sided high spire. Long axial, foldlike ribs are crossed by rounded cords. Protoconch of 1½ smooth whorls. Outer lip lirate within. With a shallow sutural sinus. Inner lip without a parietal callus pad, but with two distinct, medially located, columella folds. Color white with broad, brown bands. Known only from the type, *Mitrolumna mitriformis* (Shasky, 1961) **(2993)**, from the Gulf of California in 40 to 90 fathoms (*The Veliger,* vol. 4, no. 1, p. 20).

Subfamily CLAVINAE Powell, 1942

Shells between ¼ and ½ inch; spire tall and the anterior canal short. Sinus on the shoulder, moderately to deeply

U-shaped, often rendered subtubular by a parietal tubercle. Operculum with an apical nucleus. In some forms the radula has a central tooth, broad comblike laterals and a pair of long marginals, but many others have quite different radulae. This is a heterogeneous grouping in need of revision. Drilliinae Morrison, 1966, is a synonym.

Genus *Clavus* Montfort, 1810

Sculpture of prominent peripheral axials which may be produced into pointed, solid tubercles. Spiral sculpture absent. Sinus wide. Outer lip thin, with a well-defined stromboid notch. Siphonal canal short, widely open. Type: *flammulatus* Montfort, 1810. Synonyms: *Eldridgea* Bartsch, 1934; *Clavicantha* Swainson, 1840; *Aliceia* Dautzenberg and Fischer, 1897; *Tylotia* Melvill, 1917. There are 2 deep water Caribbean species in the typical subgenus *Clavus*: (2994) *cadenasi* (Clench and Aguayo, 1939) from off Cuba; and (2995) *johnsoni* (Bartsch, 1934) from off Puerto Rico.

Genus *Drillia* Gray, 1838

Shells fusiform, white or pink-banded, with strong axial ribs. Spiral sculpture of weak threads. Anterior canal short and straight. Sinus spoutlike. 2½ nuclear whorls, smooth and small. Type: *umbilicata* Gray, 1838, from West Africa. The genus *Douglassia* Bartsch, 1934, is a synonym and it has as its type *enae* Bartsch, 1934, from off Puerto Rico.

Subgenus *Clathrodrillia* Dall, 1918

Fusiform, with a heavy, rounded varix at about ⅓ whorl back from the aperture. Protoconch of 2 smooth whorls. Type: *gibbosa* (Born, 1778).

Drillia acurugata (Dall, 1890)   2996
Roughed Drillia

South Florida.

½ to ⅝ inch, narrowly fusiform, uniform light-brown to gray-brown, about 9 whorls. 2 nuclear whorls, smooth, brown. Postnuclear whorls with 13 to 15, short, slightly slanting, rounded axial riblets. Between these are 5 to 8 spiral threads. Just below the suture is a single, sharp spiral cord, and below that (on the upper shoulder) is a wide concave area crossed by cresent axial numerous growth lines. Common; low tide to 2 fathoms. Do not confuse with *Crassispira cubana* (Melvill, 1923) which has white-spotted, much longer axial riblets and spiral threads in the shoulder concavity.

Drillia solida C. B. Adams, 1830   2997
Solid Drillia

Southeast Florida and the Caribbean.

½ inch, pitch-black (brown when beachworn), broadly fusiform, with an oily sheen. About 15 sharp, smooth-topped ribs per whorl, on the last extending down ¾ the way, below which are 7 or 8 spiral threads on the base. 8 or 9 fine spiral threads between the ribs on the upper whorls. Suture bounded below by a single smooth cord. Below this is a broad concave gutter bearing 3 to 5 spiral, sometimes beaded, threads. Sinus deep and spoutlike. Moderately common; shallow water. *D. ebenina* Dall, 1890 is a synonym. The similar *cubana* and *fulvescens* are in the genus *Crassispira*.

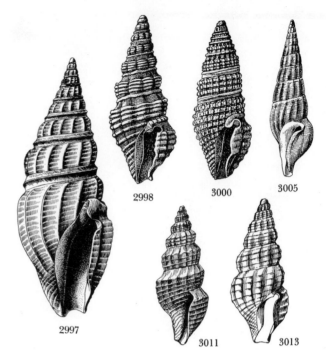

2998   3000   3005

2997

3011   3013

Other *Drillia*:

2998 *Drillia (Clathrodrillia) interpleura* Dall and Simpson, 1901. Puerto Rico. 10 mm.

2999 *Drillia (Clathrodrillia) gibbosa* (Born, 1778). North coast of South America, 1 to 6 fms. 1 to 2 inches.

2999

3000 *Drillia (Clathrodrillia) ponciana* Dall and Simpson, 1901. Puerto Rico. 6.5 mm.

3001 *Drillia (Clathrodrillia) fancherae* (Dall, 1903). Avalon, Catalina, to Redondo Beach, California, 25 fms. 10 mm.

3002 *Drillia (Imaclava* Bartsch, 1944) *unimaculata* (Sowerby, 1834). Head of the Gulf of California, in 20 to 70 meters. Type of the subgenus. (Synonyms: *Clavus pembertoni* Lowe, 1935; *ima* Bartsch, 1944.)

3003 *Drillia (Imaclava) asaedai* (Hertlein and Strong, 1951). Baja California to Ecuador, down to 82 meters.

3004 *Drillia (Imaclava) pilsbryi* (Bartsch, 1950). Sonora, Mexico, to Panama, in 10 to 70 meters.

Other Atlantic species of doubtful placement:

3005 *Drillia albicoma* (Dall, 1889). Gulf of Mexico, 84 to 804 fms.

— [*Clathrodrillia ebur* (Reeve, 1845) is an Indo-Pacific species, fide Tomlin, 1934.]

3006 *Drillia inimica* Dall, 1927. Off Georgia, 440 fms.

3007 *Drillia orellana* Dall, 1927. Off Georgia, 440 fms.

3008 *Drillia dolana* Dall, 1927. Off Fernandina, Florida, 294 fms.

3009 *Drillia fanoa* Dall, 1927. Off Fernandina, Florida, 294 fms. Above 4 in *Proc. U.S. Nat. Mus.*, vol. 70, no. 2667.

3010 *Drillia amblytera* (Bush, 1893). Off Cape Hatteras, North Carolina, 142 fms.

3011 *Drillia detecta* (Dall, 1881). Gulf of Mexico, 339 fms.

3012 *Drillia canna* (Dall, 1889). Off Cape Lookout, North Carolina, 52 fms; Gulf of Mexico, 50 fms.

3013 *Drillia pharcida* (Dall, 1889). Off Florida, both sides, 150 to 229 fms. (Synonym: *exasperata* Dall, 1881, non Reeve.)

3014 *Drillia acrybia* (Dall, 1889). Off east Florida, 136 to 294 fms.

Genus *Kylix* Dall, 1919

Second whorl of protoconch is carinate; siphonal canal elongate; back of the last whorl smooth; axial ribs sinuous. Usually with lighter-colored beads adjacent to the suture. Type: *alcyone* Dall, 1919.

3015 *Kylix paziana* (Dall, 1919). Guaymas to La Paz, Mexico, 20 fms.

3016 *Kylix alcyone* Dall, 1919. Type of the genus. Off Cape Lobos, Baja California, 76 fms.

3017 *Kylix halocydne* (Dall, 1919). Ventura County, California, to Magdalena Bay, Baja California, 30 to 60 meters.

3018 *Kylix hebuca* (Dall, 1919). Head of the Gulf of California to Sonora, Mexico, in 10 to 25 meters.

3019 *Kylix ianthe* (Dall, 1919). West Mexico, 13 to 50 meters.

Genus *Neodrillia* Bartsch, 1943

Solid, elongate-conic, with a tall spire. Protoconch of 2½ whorls, the first 1½ smooth, succeeded by very closely spaced hairlike threads, followed by distinct axials, later crossed by weak spiral striae. Adult sculpture of distinct, bluntly rounded strong axials, the whole surface crowded with strong spiral threads. Type: *cydia* Bartsch, 1943. Two species living in the Caribbean:

Neodrillia cydia **Bartsch, 1943**   **3020**
Glorious Drillia

Both sides of Florida and the West Indies.

10 to 20 mm., stout, white, with a row of brown spots on the base of the strong 8 or 9 axial rounded ribs per whorl. Spiral sculpture of crowded, fine, raised threads. Sinus deep and somewhat tubular. Moderately common; 1 to 70 fathoms. Rarely with 2 spiral bands of brown. (Synonyms: *antiguensis encia;* and *barbadensis* and *jamaicensis* all Bartsch, 1943.) Type of the genus. See *Mem. Soc. Cubana Hist. Nat.*, vol. 17, p. 83, 1943.

Other species:

3021 *Neodrillia euphanes* (Melvill, 1923). Cuba, 100 to 150 fms.

Genus *Cerodrillia* Bartsch and Rehder, 1939

Similar to *Neodrillia*. Protoconch of 2 smooth rounded whorls. The type of the genus is *clappi* Bartsch and Rehder, 1939.

Cerodrillia perryae **Bartsch and Rehder, 1939**   **3022**
Perry's Drillia

Both sides of Florida.

½ inch in length, flesh-colored, with a broad, golden-brown band around the periphery. 8 to 9 axial ribs per whorl. Faint spiral lines present. Not uncommon. *C. clappi* Bartsch and Rehder, 1939 (**3022a**) from the Lower Florida Keys is probably only a variant of *perryae. C. thea* has shorter axial ribs and is uniform chocolate-brown.

Cerodrillia thea **(Dall, 1883)**   **3023**
Thea Drillia

West coast of Florida.

3023

½ inch in length, thick-shelled, with a glossy-brown to bluish finish, and with 8 to 10 short, slanting ribs cream in color. Outer lip prominent. Sinus deep and U-shaped. Common in shallow water to 111 fathoms. Usually with weeds. *C. carminura* Dall, 1889, is a synonym.

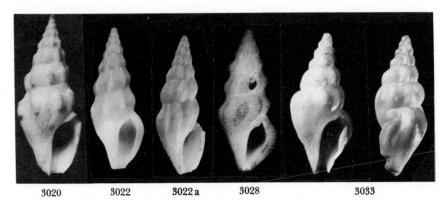

3020   3022   3022a   3028   3033

Other species:

3024 *Cerodrillia* (*Lissodrillia* Bartsch and Rehder, 1939) *schroederi* Bartsch and Rehder, 1939. Off west Florida, 40 to 60 fms. Type of the subgenus. *Proc. U.S. Nat. Mus.*, no. 3070, p. 129. Rose-stained. 4 mm.

3025 *Cerodrillia* (*Lissodrillia*) *simpsoni* Dall, 1887. North Carolina to the Gulf of Mexico, 15 to 18 fms. Common.

3026 *Cerodrillia* (*Viridrillia* Bartsch, 1943) *williami* Bartsch, 1943. Type of the subgenus. Off southeast Florida, 15 to 90 fms. Common. *Viridrillina* Bartsch, 1943 is a synonym of *Viridrillia*.

3027 *Cerodrillia* (*Viridrillia*) *cervina* Bartsch, 1943. Off Cape Hatteras, to Frying Pan Shoals, North Carolina, 12 to 63 fms.

3028 *Cerodrillia* (*Viridrillia*) *bahamensis* Bartsch, 1943. Bimini and Nassau, Bahamas. Shore.

3029 *Cerodrillia* (*Viridrillia*) *hendersoni* Bartsch, 1943. Off Key West, Florida, 35 to 90 fms. Type of *Viridrillina* Bartsch, 1943.

3030 *Cerodrillia* (*Cerodrillia*) *verrilli* (Dall, 1881). Off west Florida, 40 to 50 fms. 4 mm. Glistening, all-white.

3030

3031 *Cerodrillia abdera* (Dall, 1919). Baja California.

3032 *Cerodrillia bealiana* Schwengel and McGinty, 1942. South Florida, 20 to 50 fms. West Florida, 14 to 40 fms. *The Nautilus*, vol. 56, p. 15.

3033 *Cerodrillia cybele* (Pilsbry and Lowe, 1932). Gulf of California to Ecuador.

3033a *Cerodrillia girardi* Lyons, 1972. Off west Florida. 49 to 55 meters. *The Nautilus*, vol. 86, p. 4.

Genus *Crassispira* Swainson, 1840

Shells 1 to 2 inches, solid, fusiform, usually brown or black, with a short, truncated anterior end. Sculptured with a broad, flat subsutural fold, usually smooth, sometimes nodulose, bounded below by a sharp, raised edge. Sinus moderately deep, U-shaped, occupying the shoulder concavity between the subsutural margin and the shoulder angle. Outer lip not variced, but with a very shallow stromboidlike notch. Operculum ovate, with a terminal nucleus. Protoconch paucispiral, smooth at first, then developing axial riblets. Type: *bottae* (Kiener, 1839), which is now *incrassata* (Sowerby, 1834).

There are about 70 species in the Tropical Eastern Pacific. See Myra Keen's "Sea Shells of Tropical West America." The following are reported from the Gulf of California and/or Baja California, and Florida.

3034 *Crassispira* (*Crassispira*) *incrassata* (Sowerby, 1834). Gulf of California, 20 fms. Uncommon. (Synonym: *bottae* Kiener, 1839.)

3035 *Crassispira* (*Crassispira*) *montereyensis* Stearns, 1871. Monterey, California, to Mazatlan, Mexico. Below low tide in rock crevices; uncommon.

3036 *Crassispira* (*Crassispira*) *maura* (Sowerby, 1834). Gulf of California to Ecuador. (Synonyms: *nigricans* Dall, 1919; *inaequistriata* Li, 1930; *perla* M. Smith, 1947.)

3038   3054   3070   3071

3037 *Crassispira* (*Crassispira*) *phasma* Schwengel, 1940. Off Palm Beach, Florida, in 12 fms. *The Nautilus*, vol. 54, p. 50.

3038 *Crassispira gundlachi* (Dall and Simpson, 1901). Puerto Rico.

3039 *Crassispira* (*Nymphispira* J. H. McLean) *nymphia* Pilsbry and Lowe, 1932. Gulf of California. Uncommon.

3040 *Crassispira* (*Nymphispira*) *bacchia* Dall, 1919. La Paz, Baja California. 8 fms. Rare.

3041 *Crassispira* (*Strictispira* J. H. McLean, 1971) *ericana* Hertlein and Strong, 1951. Gulf of California, 4 to 13 fms.

3042 *Crassispira* (*Strictispira*) *stillmani* Shasky, 1971. Gulf of California to Panama.

3043 *Crassispira* (*Striospira* Bartsch, 1950) *kluthi* E. K. Jordan, 1936. Baja California and the Gulf to Ecuador. (Synonyms: *Clavatula luctuosa* Hinds, 1843, not Orbigny, 1842; *lucasensis* and *tabogensis* Bartsch, 1950.) Type of the subgenus.

3044 *Crassispira* (*Striospira*) *nigerrima* (Sowerby, 1834). Guaymas, Mexico, to Ecuador. Common. (Synonyms: *cornuta* Sowerby, 1834; *thiarella* Kiener, 1839.)

3045 *Crassispira* (*Striospira*) *tepocana* Dall, 1919. Sonoran Coast, Mexico, to Ecuador. 36 fms. Rare.

3046 *Crassispira* (*Striospira*) *xanti* Hertlein and Strong, 1951. Sonora, Mexico, to Ecuador, 4 to 30 fms. Uncommon.

3047 *Crassispira* (*Gibbaspira* J. H. McLean) *rudis* (Sowerby, 1834). Mexico to Ecuador. Uncommon. Type of the subgenus.

3048 *Crassispira* (*Crassiclava* J. H. McLean, 1971) *turricula* (Sowerby, 1834). Baja California to Ecuador; uncommon. (Synonyms: *corrugata* Sowerby, 1834; *sowerbyi* Reeve, 1843.) Type of the subgenus.

3049 *Crassispira* (*Crassiclava*) *cortezi* Shasky and Campbell, 1964. Gulf of California.

3050 *Crassispira* (*Burchia* Bartsch, 1944) *semiinflata* (Grant and Gale, 1931). Santa Barbara, California, to Baja California. Type of the subgenus. *The Nautilus*, vol. 57, p. 115, and vol. 52, p. 21. (Synonym: *redondoensis* T. Burch, 1938.)

3051 *Crassispira* (*Burchia*) *unicolor* (Sowerby, 1834). Head of the Gulf of California to Ecuador, intertidal to 20 meters. (Synonyms: *erebus* Pilsbry and Lowe, 1932; *tangolaensis* Hertlein and Strong, 1951.)

Subgenus *Crassispirella* Bartsch and Rehder, 1939

Sinus deep, laterally directed, parietal callus directed downward. Shoulder with subsutural thread. Axially ribbed and spirally threaded. Type of this subgenus is *rugitecta* (Dall, 1918), from Baja California.

Crassispira cubana Melvill, 1923   3052
Cuban Turrid

South Florida and the West Indies. Yucatan.

¾ inch in length; a solid brown-black in color and with a slight sheen. 14 to 16 short, white-beaded axial ribs per whorl. Spiral threads numerous and fine. Subsutural cord is white-spotted. Sinus small, its posterior end round, its opening narrow. Not uncommon below low water under rocks. *C. mesoleuca* Rehder, 1943, is a synonym.

Crassispira sanibelensis Bartsch and Rehder, 1939   3053
Sanibel Turrid

West coast of Florida; Bahamas.

1 inch, similar to *Clathrodrillia solida,* but with 7 longer and wider axial ribs, with a large sinus, and colored orange-chestnut with white between the ribs. The chestnut markings appear as long axial flames on the body whorl. Uncommon; just offshore.

Other Atlantic species:

3054 *Crassispira (Crassispirella) tampaensis* Bartsch and Rehder, 1939. West Florida. (Synonym: *bartschi* L. Perry, 1955.)

3055 *Crassispira (Crassispirella) nigrescens* (C. B. Adams, 1845). Grand Cayman to Barbados.

3056 *Crassispira (Crassispirella) drangai* Schwengel, 1951. Barbados, West Indies. (Very close to *Clathrodrillia solida* C. B. Adams.)

3057 *Crassispira (Crassispirella) candace* Dall, 1919. Probably West Indies, not the Gulf of California. Rare.

3058 *Crassispira (Crassispirella) rhythmica* Melvill, 1927. South Florida.

3059 *Crassispira (Crassispirella) quadrifasciata* (Reeve, 1845). Greater Antilles to Brazil.

3060 *Crassispira polytorta* (Dall, 1881). Gulf of Mexico and off Cuba. 413 fms. 33 mm.

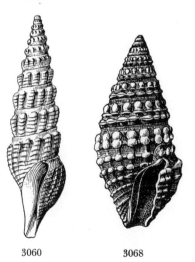

3060   3068

Other Pacific species:

3061 *Crassispira (Crassispirella) rugitecta* (Dall, 1918). Outer Baja California and the Gulf of California. Type of the subgenus.

3062 *Crassispira (Crassispirella) brujae* Hertlein and Strong, 1951. Gulf of California, 45 fms. Rare.

3063 *Crassispira (Crassispirella) chacei* Hertlein and Strong, 1951. Gulf of California, 40 to 60 fms. Uncommon.

3064 *Crassispira (Crassispirella) discors* (Sowerby, 1834). Gulf of Ecuador. Uncommon.

3065 *Crassispira (Crassispirella) rustica* (Sowerby, 1834). Upper Gulf of California to Panama, offshore to 20 meters.

Subgenus *Monilispira* Bartsch and Rehder, 1939

Fusiform, high-spired, with light-colored beads on the periphery. Subsutural cord weak. Surface covered with fine spiral striae. Type of this subgenus is *Drillia monilifera* (Carpenter, 1864).

Crassispira albomaculata (Orbigny, 1842)   3066
White-banded Drillia

Florida to Texas and the West Indies.

½ inch in length, resembling a *Cerithium* in shape; color dark blackish brown with a yellow-white band bearing about 13 knobs per whorl. Last whorl with 2 or 3 spiral white bands. 2 smooth cords and numerous fine striae are on the black portions. Fairly common in shallow water under rocks. Synonyms: *zebra* Kiener, 1846, non Perry, 1811; *ornata* Orbigny, 1842, not Defrance, 1826; *albocincta* C. B. Adams, 1845. *C. (Pilsbryspira) albinodata* (Reeve, 1843) is from the tropical Eastern Pacific.

Crassispira leucocyma Dall, 1883   3067
White-knobbed Drillia

Florida to Texas and the West Indies.

⅓ inch in length. 1 nuclear whorl smooth. Shell dark to light grayish brown with pairs of large, round, white nodules connected by 2 fine threads on the periphery. 1 small and 1 large smooth, spiral lirae below the suture. Deep concavity below this has crescent-shaped growth lines. An abundant shallow-water, intertidal species found in rubble and near eel-grass.

Other *Monilispira:*

3068 *Crassispira (Monilispira) melonesiana* (Dall and Simpson, 1901). Puerto Rico.

3069 *Crassispira (Monilispira) monilis* (Bartsch and Rehder, 1939). West coast of Florida. Uncommon.

3073

3074

3072

3070 *Crassispira (Monilispira) greeleyi* (Dall, 1901). Off Rio Goyanna, Brazil.

3071 *Crassispira (Monilispira) appressa* (Carpenter, 1864). Cape San Lucas. Rare.

3072 *Crassispira (Monilispira) pluto* Pilsbry and Lowe, 1932. Gulf of California. Common.

3073 *Crassispira (Monilispira) monilifera* Carpenter, 1864. Type of the subgenus. Gulf of California.

3074 *Crassispira (Monilispira) trimariana* Pilsbry and Lowe, 1932. Tres Marias Islands, Mexico. Rare.

Subgenus *Buchema* Corea, 1934

Fusiform, axially ribbed and spirally threaded. Sinus deep, U-shaped, spoutlike. Subsutural cord present. Base truncate. Radula of the duplex type. Type: *Carinodrillia tainoa* Corea, 1934.

3075 *Crassispira (Buchema) tainoa;* **(3076)** *suimaca;* **(3077)** *mamona;* **(3078)** *apitoa;* and **(3079)** *liella* all Corea, 1934, all off north Puerto Rico, 30 to 120 fms. *Smithsonian Misc. Coll.*, vol. 91, no. 16.

3080 *Crassispira (Buchema) granulosa* (Sowerby, 1834). Gulf of California to Ecuador, in 20 to 55 meters. (Synonym: *Clathrodrillia callianira* Dall, 1919.)

Subgenus *Carinodrillia* Dall, 1919

3081 *Crassispira (Carinodrillia* Dall, 1919) *adonis* Pilsbry and Lowe, 1932. Gulf of California to Ecuador; Galapagos Islands.

3082 *Crassispira (Carinodrillia) dichroa* Pilsbry and Lowe, 1932. Gulf of California to Ecuador, in 10 to 40 meters.

3083 *Crassispira (Carinodrillia) halis* Dall, 1919. Head of the Gulf of California to Ecuador, in 20 to 55 meters. Type of the subgenus.

3084 *Crassispira (Carinodrillia) hexagona* (Sowerby, 1834). Head of the Gulf of California to Ecuador, intertidal to 40 meters. (Synonym: *Clathrodrillia pilsbryi* Lowe, 1935.)

Genus *Suavodrillia* Dall, 1918

Shell 1 inch, solid, fusiform, tall spire; siphonal canal twisted and deeply notched. Spiral sculpture of strong, narrow, rounded keels. Shoulder slope wide, flat and steep. Protoconch 2½ smooth whorls, the tip slightly inrolled. Outer lip thin; sinus moderately deep and U-shaped. Perios-

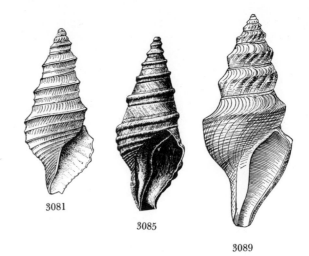

3081

3085

3089

tracum yellowish. Operculum leaf-shaped and with a terminal nucleus. Type: *kennicottii* (Dall, 1871). Cold-water species:

3085 *Suavodrillia kennicottii* (Dall, 1871). Bering Sea to Aleutians.

3086 *Suavodrillia willetti* Dall, 1919. Forrester Island, Alaska.

3087 *Suavodrillia textilia* Dall, 1927. Off Georgia, 440 fms.; off Fernandina, Florida, 294 fms.

3088 *Suavodrillia (Typhlomangelia* G. O. Sars, 1878) *nivalis* (Lovén, 1846). North Atlantic.

3089 *Suavodrillia (Typhlomangelia) tanneri* (Verrill and Smith, 1884). Off Nantucket, Massachusetts, 1290 fms.

Genus *Inodrillia* Bartsch, 1943

Shells white, about 1½ inch, fusiform, high spire, curved axial ribs, short siphonal canal; sinus deep U-shaped, and somewhat tubular. Weak sinuation of the lower outer lip. Type: *nucleata* (Dall, 1881). Synonyms: *Inodrillara* Bartsch, 1943; *Inodrillina* Bartsch, 1943. See *Mem. Soc. Cubana Hist. Nat.*, vol. 17, p. 104. Most of Bartsch's species are synonyms of each other.

Inodrillia aepynota **(Dall, 1889)**   **3090**
Tall-spired Turrid

North Carolina to southeast Florida.

3090   3092   3095   3100   3104

½ inch in length; chalk-white to pinkish white. Axial ribs very strong. Large varix behind thin, sharp outer lip. Sinus spoutlike. Moderately common from 63 to 120 fathoms.

Other species:

3091 *Inodrillia dalli* (Verrill and Smith, 1882). Martha's Vineyard, Massachusetts, to the Gulf of Mexico, 120 to 196 fms. (Synonyms or forms: *acloneta* Dall, 1889; *cestrota* Dall, 1889, p. 92.)

3092 *Inodrillia nucleata* (Dall, 1881). Gulf of Mexico to the Lesser Antilles, 229 to 464 fms. Type of the genus. (Synonym: *amblia* Watson, 1882.)

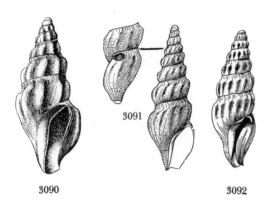

3091

3090   3092

3093 *Inodrillia hesperia* Bartsch, 1943. Off Tortugas and Key West, Florida, 110 to 144 fms.

3094 *Inodrillia acova* Bartsch, 1943. Off southeast Florida, 75 to 135 fms.

3095 *Inodrillia avira* Bartsch, 1943. Off southeast Florida, 100 to 209 fms.

3096 *Inodrillia gibba* Bartsch, 1943. Off Western Dry Rock, southeast Florida, 80 fms.

3097 *Inodrillia carpenteri* (Verrill and Smith, 1880). Off Martha's Vineyard, Massachusetts, 86 to 155 fms. (a *Cymatosyrinx?*).

3098 *Inodrillia martha* Bartsch, 1943. Off Martha's Vineyard, Massachusetts, to off Atlantic City, New Jersey, 102 to 146 fms.

3099 *Inodrillia hatterasensis* Bartsch, 1943. Off Cape Hatteras, North Carolina, 63 to 142 fms.

3100 *Inodrillia miamia* Bartsch, 1943. Off southeast Florida, 23 to 100 fms.

3101 *Inodrillia vetula* Bartsch, 1943. Off Key West, Florida, 90 to 100 fms.

3102 *Inodrillia ino* Bartsch, 1943. Off Sambo Reef, Florida, 100 to 135 fms.

3103 *Inodrillia dido* Bartsch, 1943. Off Key West, Florida, 80 to 90 fms.

3104 *Inodrillia hilda* Bartsch, 1943. Off Sambo Reef, Florida, 118 to 135 fms.

Genus *Splendrillia* Hedley, 1922

Type of this genus is *woodsi* (Beddome, 1883), from Australia.

Subgenus *Syntomodrillia* Woodring, 1928

Shells ½ inch, fusiform, with slanting axial ribs, without a subsutural cord or fold. Ribs extend from suture to suture. Protoconch of 2 smooth whorls. Type: *lissotropis* (Dall, 1881).

American species are:

3105 *Splendrillia lissotropis* (Dall, 1881). Gulf of Mexico to the West Indies, 73 to 248 fms. (var. *scissurata* Dall, 1890, from off Florida).

3106 *Splendrillia carolinae* (Bartsch, 1934). Caribbean, 160 fms.

3107 *Splendrillia tantula* (Bartsch, 1934). Caribbean, 95 fms. Smiths. Misc. Coll., vol. 91, p. 28.

3108 *Splendrillia woodringi* (Bartsch, 1934). Off west coast of Florida, 20 to 50 fms., to Barbados, 103 fms. Moderately common. 12 mm.

3109 *Splendrillia fucata* (Reeve, 1845). North Carolina to Yucatan and the West Indies, 14 to 50 fms. 40 mm. (Synonym: *paria* Reeve, 1846.)

3110 *Splendrillia* (*Fenimorea*) *halidorema* Schwengel, 1940. Southeast Florida to Jamaica. 10 to 12 fms. *The Nautilus*, vol. 54, p. 50.

3111 *Splendrillia janetae* Bartsch, 1934. Gulf of Mexico, 60 to 70 fms.; West Indies.

3112 *Splendrillia moseri* (Dall, 1889). North Carolina to the Gulf of Mexico, 3 to 49 fms., to Barbados.

3113 *Splendrillia moseri* subspecies *brunnescens* Rehder, 1943. Off west Florida.

3114 *Splendrillia bratcherae* McLean and Poorman, 1971. Gulf of California, 10 to 40 meters.

3115 *Splendrillia lalage* (Dall, 1919). Baja California. (Synonym: *Elaeocyma baileyi* S. S. Berry, 1969.)

Genus *Cymatosyrinx* Dall, 1889

Shells polished, 1 to 1½ inches, tall-spired, truncated body whorl, terminated in a short, twisted, very deeply notched anterior canal. Whorls with a strong subsutural cord or fold which is usually nodulated. Axial ribs oblique. Outer lip with a distinct stromboid notch. Sinus deep, U-shaped, bounded by a thick parietal pad. Second nuclear whorl with a peripheral carina. Type: *lunata* (Lea, 1843),

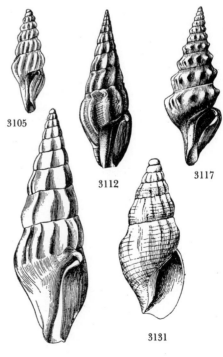

3105

3112

3117

3131

3118

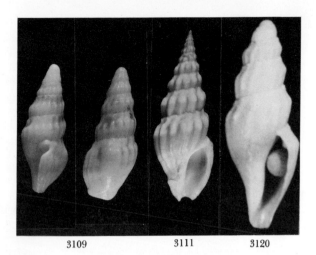

3109   3111   3120

Miocene. This group is in need of revision and for the present I am considering the following to be subgenera: *Elaeocyma* Dall, 1918; *Leptadrillia* Woodring, 1928.

Atlantic species:

3116 *Cymatosyrinx bartschi* Haas, 1941. Bermuda, 30 fms. *Publ. Field Mus. Nat. Hist.,* vol. 24, p. 172.

3117 *Cymatosyrinx centimata* (Dall, 1889). Off North Carolina to Gulf of Mexico, 731 to 1,920 fms.

3118 *Cymatosyrinx pagodula* (Dall, 1889). Florida and the West Indies, 50 to 175 fms.

3119 *Cymatosyrinx* (*Leptadrillia*) *splendida* (Bartsch, 1934). Gulf of Mexico, off Florida, to the Greater Antilles, 50 to 230 fms.

3120 *Cymatosyrinx* (*Leptadrillia*) *fritellaria* Dall, 1927. Off Fernandina, Florida, 294 fms.

Pacific coast species:

3121 *Cymatosyrinx* (*Elaeocyma*) *johnsoni* Arnold, 1903. San Pedro, California.

3122 *Cymatosyrinx* (*Elaeocyma*) *empyrosia* (Dall, 1899). San Pedro to San Diego, California. Type of the subgenus.

3123 *Cymatosyrinx* (*Elaeocyma*) *hemphilli* (Stearns, 1871). Santa Barbara to the Gulf of California.

3124 *Cymatosyrinx* (*Elaeocyma*) *halocydne* (Dall, 1919). San Pedro to Santa Barbara Islands.

3125 *Cymatosyrinx allynianus* Hertlein and Strong, 1951. Southern Gulf of California, 50 to 55 fms.

3126 *Cymatosyrinx arenensis* Hertlein and Strong, 1951. Southern Gulf of California, 50 to 55 fms.

3127 *Cymatosyrinx* (*Elaeocyma*) *ricaudae* (S. S. Berry, 1969). Boca Soledad, Baja California, 24 fms. 29 mm. *Leaflets in Malacology,* no. 26.

Genus *Bellaspira* Conrad, 1868

Shell about ⅓ inch, ovate-biconic, spire elevated, short siphonal canal. 2 nuclear whorls, smooth. Sinus narrow and pointing upwards. Outline below suture is concave. Glossy smooth axial ribs are lined up one under the other. Radula with a vestigial rachidian, comblike laterals and long, narrow marginals. Operculum leaf-shaped, nucleus terminal. See J. H. McLean and Poorman, 1970, *Contrib. in Sci., Los Angeles,* no. 189. Type: *virginiana* Conrad, 1862.

Bellaspira pentagonalis (Dall, 1889)   3128
Pentagonal Bellaspire

North Carolina to Florida to the Lesser Antilles.

¼ inch, fusiform, solid, glossy, cream with a large orange spot on each rib just above the suture. Each whorl with 5 smooth axial ribs which are lined up one under the other. Outer lip thin, sharp. Siphonal canal short. Microscopic spiral threads on last whorl. Not uncommon; 34 to 200 fathoms. *Fenimorea pentapleura* Schwengel, 1940, is a synonym.

Other species:

3129 *Bellaspira brunnescens* (Rehder, 1939). Cuba.

3128   3129

3130 *Bellaspira margaritensis* J. H. McLean and Poorman, 1970. Isla Margarita, Venezuela, 17 to 21 fms. *Contrib. in Sci.,* no. 189, p. 5. 15 mm.

3131 *Bellaspira grippi* (Dall, 1908). Redondo Beach, California, to Asuncion Island, Baja California, down to 25 fms. 9 mm.

3132 *Bellaspira clarionensis* J. H. McLean and Poorman, 1970. Clarion Island, Revillagigedo Islands, 28 to 45 fms. 13 mm. May be *grippi* (Dall). *Contrib. in Sci.,* no. 189, p. 9.

3133 *Bellaspira acclivicosta* J. H. McLean and Poorman, 1970. Off Guaymas, Mexico, 15 to 20 fms. *Contrib. in Sci.,* no. 189, p. 7.

3134 *Bellaspira melea* Dall, 1919. Baja California to Colombia, 10 to 40 fms. 13 to 19 mm.

Genus *Compsodrillia* Woodring, 1928

Shell about ½ inch, elongate-fusiform, tall spire, inflated body whorl, constricted siphonal canal, angular whorls. Short siphonal canal. Protoconch 2 whorls smooth, followed by axial riblets. Postnuclear whorls with heavy, rounded axial ribs crossed by spiral cords and threads. Sinus subsutural deep and U-shaped. Outer lip sharp, with a varix behind. Type: *urceola* Woodring, 1928.

3137

3137 *Compsodrillia tristicha* (Dall, 1889). Off Mississippi, 111 to 210 fms.; southwest Florida, 100 fms.

3138 *Compsodrillia acestra* (Dall, 1889). Florida Straits, 400 fms.

3139 *Compsodrillia disticha* Bartsch, 1934. Off west Florida, 130 fms.

3140 *Compsodrillia albonodosa* (Carpenter, 1857). Gulf of California, intertidal to 20 meters. (Synonym: *soror* Pilsbry and Lowe, 1932.)

3141 *Compsodrillia duplicata* (Sowerby, 1834). Gulf of California to Ecuador, in 20 to 45 meters.

3142 *Compsodrillia haliplexa* (Dall, 1919). Gulf of California to Ecuador, in 10 to 55 meters.

3143 *Compsodrillia opaca* McLean and Poorman, 1971. Baja California and the Gulf of California, in 95 to 140 meters.

3144 *Compsodrillia thestia* (Dall, 1919). Gulf of California.

3145 *Compsodrillia alcestis* (Dall, 1919). Gulf of California to Colombia, in 20 to 90 meters.

3146 *Compsodrillia bicarinata* (Shasky, 1961). Gulf of California to Ecuador, in 40 to 110 meters.

3147 *Compsodrillia excentrica* (Sowerby, 1834). Gulf of California to Ecuador, intertidal to 30 meters.

3148 *Compsodrillia olssoni* McLean and Poorman, 1971. Gulf of California to Ecuador, 10 to 70 meters.

Compsodrillia eucosmia (Dall, 1889)   **3135**
Eucosmia Drillia

Gulf of Mexico to the West Indies.

½ to ¾ inch, resembling a miniature *Fusinus*. Color soft brown with short rounded white axial ribs crossed by 4 strong spiral threads. Top ¼ of shell dark-brown. Moderately common; 10 to 60 fathoms.

Other species:

3136 *Compsodrillia haliostrephis* (Dall, 1889). Gulf of Mexico, off Florida, 40 to 50 fms. Common. 17 mm.

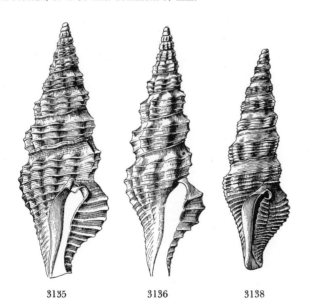

3135   3136   3138

Genus *Ophiodermella* Bartsch, 1944

Shells 1 to 2 inches, narrowly fusiform, with a tall straight-sided spire. With distinct riblets on the anal fasciole. Protoconch of 2 smooth, loosely wound whorls. Sinus weak, widely open. Suture not margined. Aperture long and narrow. Outer lip thin. Operculum with a terminal nucleus. Type: *inermis* (Hinds, 1843). Pacific coast species were previously placed in *Moniliopsis* Conrad, 1865:

3149 *Ophiodermella grippi* (Dall, 1919). San Diego, California.

3150 *Ophiodermella rhines* (Dall, 1908). Puget Sound to San Diego, California. (Synonym: *halcyonis* Dall, 1908.)

3151 *Ophiodermella inermis* (Hinds, 1843). California to Baja California. (Synonym: *ophioderma* Dall, 1908.)

3152 *Ophiodermella incisa* (Carpenter, 1865). Puget Sound to San Pedro, California. (Subsp. *fancherae* Dall, 1903 **(3153)**, Santa Rosa Island, to Baja California.) 10 mm.

3154 *Ophiodermella montereyensis* Bartsch, 1944. Monterey Bay, California, 10 to 50 fms., sand and mud. Uncommon.

3153   3155

Genus *Hindsiclava* Hertlein and Strong, 1955

Shell 1 to 2 inches, fusiform, narrow, tall-spired, with a long, tapering siphonal canal. Axial riblets and rows of beads numerous. Outer lip thin, no varix behind. Sinus high on shoulder, deep, U-shaped. Subsutural cord present. Type: *militaris* (Reeve, 1843). *Turrigemma* S. S. Berry, 1958, is a synonym.

3155 *Hindsiclava alesidota* (Dall, 1889). Off Cape Hatteras, North Carolina, to both sides of Florida to Barbados. 27 to 75 fms. (Synonym: *macilenta* Dall, 1889.) 1½ inches, yellow-white.

3156 *Hindsiclava chazaliei* (Dautzenberg, 1900). Lower Caribbean.

3157 *Hindsiclava andromeda* (Dall, 1919). Baja California. (Synonym: *Turrigemma torquifer* S. S. Berry, 1958.)

3158 *Hindsiclava militaris* (Reeve, 1843). Gulf of California to Colombia, in 20 to 55 meters. (Synonyms: *Turricula dotella* and *notilla* Dall, 1908.)

Subfamily MANGELIINAE Fischer, 1887

Shell small, ovate or fusiform, with a short canal and without an operculum in most genera, except *Propebela* and *Oenopota*. Sinus on shoulder usually very shallow, sometimes tubular. Radula with 2 slender marginals. Cytharinae is a synonym.

Genus *Mangelia* Risso, 1826

Operculum absent. Shells ½ inch or less, narrowly fusiform, spire tall, whorls convex, sculpture variable (ribbed or reticulated), usually finely spirally striate. Sinus usually deep and fluted. Inner and outer lip may have denticulations. Type: *attenuatus* Montagu, 1803. *Mangilia* is a misspelling. This is *Cythara* of authors.

There are many dozens of species assigned to this genus, many of which may belong to other genera. About 50 are found along our Atlantic coast, and perhaps an equal number along the Pacific coast from Alaska to Mexico.

Mangelia bartletti (Dall, 1889)   3159
Bartlett's Turrid

South Florida and the West Indies, Bermuda.

10 mm., oval, spire acute; nucleus glassy, dark-grown, inflated. Color pale-yellowish with touches of pale-brown on the varix. Spiral sculpture of fine subequal threads. 20 axial

rounded riblets on the whorls of the spire. Moderately common; 3 to 450 fathoms. Placed in the subgenus *Tenaturris* Woodring, 1928, by some workers.

Mangelia stellata (Stearns, 1872)   3160
Stellate Turrid

South half of Florida; Bahamas; Mexico.

4 to 5 mm., broadly fusiform, brown or whitish, angulate whorls with about 12 strong, axial, rounded, narrow ribs crossed by 3 to 5 spiral smaller cords. 2 nuclear whorls, increasing in size rapidly; smooth, brown or white. Sinus V-shaped, subsutural, large, deep. Common; grass flats; shallow water. This is the type of the subgenus *Stellatoma* Bartsch and Rehder, 1939.

3159   3160   3170

Atlantic "Mangelia":

3161 *Mangelia acloneta acloneta* (Dall, 1889). Off Fernandina, Florida, 294 fms. Variety *cestrota* (Dall, 1889). Gulf of Mexico, 196 fms.

3162 *Mangelia acrocarinata*, *areia* (**3163**), *chasmata* (**3164**), *christina* (**3165**), *crossata* (**3166**), *cratera* (**3167**), *cryera* (**3168**), *ischna* (**3169**), *sagena* (**3170**), *lastica* (**3171**), *loraeformis* (**3172**), *percompacta* (**3173**), *rhabdea* (**3174**), *sericifila* (**3175**), *strongyla* (**3176**), *subcircularis* (**3177**) and *tachnodes* (**3178**) all Dall, 1927, all off Fernandina, Florida. 294 fms. *Proc. U.S. Nat. Mus.*, vol. 70, no. 2667.

3179 *Mangelia astricta* Reeve, 1846. Gulf of Mexico and Florida Keys.

3180 *Mangelia (Benthomangelia) bandella* (Dall, 1881). Off east Florida and the Gulf of Mexico, 100 to 1,200 fms. Common. May be a *Pleurotomella*. 9 mm.

3181 *Mangelia ceroplasta* Bush, 1885. Off Cape Hatteras, North Carolina, 10 to 17 fms.

3182 *Mangelia (Glyphoturris* Woodring, 1928) *rugirima* Dall, 1889. South Florida, 1 to 200 fms. Moderately common.

3183 *Mangelia (Glyphoturris) quadrata* (Reeve, 1845). Bermuda. North Carolina to the Gulf of Mexico and the West Indies. (Subspecies: *diminuta* C. B. Adams, 1850 (**3184**); *eritima* Bush, 1885. 7 mm. (**3185**)).

3186 *Mangelia (Saccharoturris* Woodring, 1928) *monocingulata* Dall, 1889. Barbados. 100 fms. Gulf of Mexico, off Florida, 220 fms. 6 to 8 mm.

3187 *Mangelia (Brachycythara* Woodring, 1928) *biconica* C. B. Adams, 1850. North Carolina to the West Indies. Gulf of Mexico to Yucatan, 30 fms.

3188 *Mangelia (Stellatoma) antonia* (Dall, 1881). Campeche Banks, Mexico. Off Fernandina, Florida, 640 fms. 18 mm.

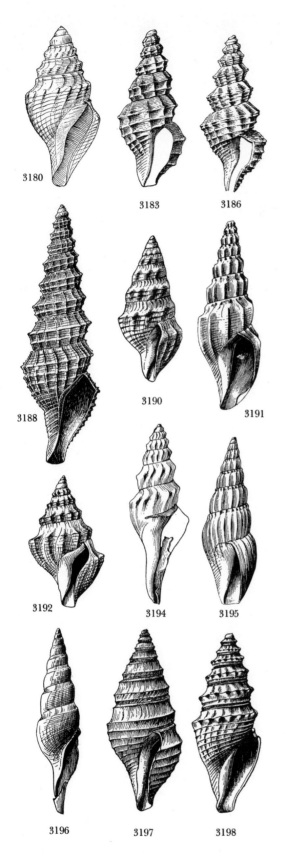

3180

3183   3186

3188   3190   3191

3192   3194   3195

3196   3197   3198

3189 *Mangelia dalli* (Verrill, 1882). South of Martha's Vineyard, Massachusetts, to off Delaware, 94 to 143 fms. An *Inodrillia?*

3190 *Mangelia elusiva* (Dall, 1881). Gulf of Mexico, 640 fms.; West Indies. (Synonym: *perpauxilla* Watson, 1881.) 9 mm.

3191 *Mangelia exsculpta* (Watson, 1881). Gulf of Mexico. 640 fms.

— *Mangelia glypta* Bush, 1885, is a *Nassarina*.

3192 *Mangelia ipara* (Dall, 1881). Off Fernandina, Florida, 294 fms., Gulf of Mexico, 640 fms. (Synonym: *chyta* Watson, 1881.)

3193 *Mangelia leuca* Bush, 1893. Off Cape Lookout, North Carolina, 603 fms.

— *Mangelia monilifera* (Sowerby, 1844), is a columbellid *Nassarina*.

3194 *Mangelia pelagia* (Dall, 1881). Off Georgia, 440 fms.; Gulf of Mexico, 539 fms. 10 mm.

3195 *Mangelia pourtalesii* (Dall, 1881). Off Fernandina and Florida Straits, 294 to 447 fms. 17 mm.

3196 *Mangelia scipio* Dall, 1889. Off Georgia and the West Indies, 124 to 982 fms. 14 mm.

3197 *Mangelia subsida* (Dall, 1881). Gulf of Mexico, 339 fms.

3198 *Mangelia* (*Anticlinura* Thiele, 1934) *toreumata* Dall, 1889. Off Fernandina, Florida, and the West Indies, 294 to 391 fms. Gulf of Mexico, 220 fms. 10 mm.

Pacific species:

The tropical Eastern species are listed and illustrated by Myra Keen in her "Sea Shells of Tropical West America." Those from Alaska to California are listed below. Many may belong to *Kurtzia* and *Kurtzina*.

3199 *Mangelia aleutica* Dall, 1871. Cape Sabine, Arctic Ocean, to Fuca Strait.

3200 *Mangelia* (*Glyptaesopus* Pilsbry and Olsson, 1941) *cetolaca* Dall, 1908. Baja California to Salina Cruz, Mexico. Pleistocene of San Pedro and Santa Monica, California.

3201 *Mangelia* (*Crockerella* Hertlein and Strong, 1951) *crystallina* (Gabb, 1865). Southern California.

3202 *Mangelia carlottae* Dall, 1919. Off Queen Charlotte Islands, British Columbia, 876 fms.

3203 *Mangelia nunivakensis* Dall, 1919. Nunivak Island, Bering Sea.

3204 *Mangelia eriopis* Dall, 1919. Forrester Island, Alaska, and the Queen Charlotte Islands.

3205 *Mangelia oldroydi* Arnold, 1903. Pleistocene, San Pedro, California.

3206 *Mangelia constricta* Gabb, 1865. Catalina Island, 80 fms.

3207 *Mangelia granitica* Dall, 1919. Granite Cove, Port Althorp, Alaska.

3208 *Mangelia althorpi* Dall, 1919. Granite Cove, Port Althorp, Alaska.

3209 *Mangelia sculpturata* Dall, 1886. Chiachi Islands to Port Etches, Alaska.

3210 *Mangelia hooveri* Arnold, 1903. San Pedro to San Diego, California. Also Pleistocene.

3211 *Mangelia eriphyle* Dall, 1919. Off Esteros Bay, California, to Coronado Islands.

3212 *Mangelia cesta* Dall, 1919. "California" (Mrs. Blood).

3213 *Mangelia evadne* Dall, 1919. Off Santa Rosa Island, California, in 53 fms.

3214 *Mangelia painei* Arnold, 1903. San Pedro to San Diego, California. Also Pleistocene.

3215 *Mangelia philodice* Dall, 1919. Point Ano Nuevo, California, to Coronado Islands, 65 to 71 fms.

3216 *Mangelia perattenuata* Dall, 1905. Monterey Bay, California. Rare.

Mangelids of the subgenus *Agathotoma* Cossmann, 1899:

3217 *Mangelia densilineata* (Dall, 1921). San Pedro, California, to the Gulf of California.

3218 *Mangelia janira* (Dall, 1919). San Diego to Baja California.

3219 *Mangelia pomara* (Dall, 1919). San Pedro, California.

3220 *Mangelia stellata* (Mörch, 1860). San Diego, California, Gulf of California to Ecuador. (Synonyms: *Cytharella hippolita* Dall, 1919; *C. taeniornata* Pilsbry and Lowe, 1932.)

3221 *Mangelia quadriseriata* (Dall, 1919). Gulf of California.

3222 *Mangelia alcippe* (Dall, 1918). Head of the Gulf of California to Ecuador, intertidal to 20 meters. (Synonyms: *Pleurotoma parililis* E. A. Smith, 1888, not Edwards, 1860; *Cytharella euryclea* and *pyrrhula* Dall, 1919.)

3223 *Mangelia aculea* (Dall, 1919). San Diego, California, to Baja California.

Cytharella mangelids:

The following Pacific coast species are placed in the genus *Cytharella* Monterosato, 1875, by some workers, and in the genus *Tenaturris* Woodring, 1928, by others.

3224 *Mangelia hexagona* Gabb, 1865. Monterey to San Diego, California. 10 to 50 fms., sand. Rare. Baja California.

3225 *Mangelia branneri* Arnold, 1903. California Pleistocene. Living in Panama.

3226 *Mangelia victoriana* (Dall, 1897). Victoria, Vancouver Island.

3227 *Mangelia amatula* (Dall, 1919). Monterey to San Diego, California.

3228 *Mangelia merita* Hinds, 1843. Southern California to Gulf of Nicoya, Central America. (Synonyms: *fusconotata* Carpenter, 1864, and *nereis* Pilsbry and Lowe, 1932.)

3229 *Mangelia louisa* (Dall, 1919). San Luis Obispo, California.

3230 *Mangelia verdensis* (Dall, 1919). Gulf of California. Rare. (Synonym: *Cytharella burchi* Hertlein and Strong, 1951.)

3231 *Mangelia* (*Notocytharella* Hertlein and Strong, 1955) *striosa* (C. B. Adams, 1852). Cape San Lucas and the Gulf of California to Panama. (Synonyms: *Pleurotoma exigua* C. B. Adams, 1852; *Cytharella niobe* Dall, 1919; *hastula* Pilsbry and Lowe, 1932.) Type of the subgenus.

Genus *Cryoturris* Woodring, 1928

Small, biconic, angulate whorls, axial ribs crossed by spiral threads, giving a frosty surface. Canal short. Sinus weak. Type of the genus is *engonia* Woodring, 1928, a Miocene species.

Cryoturris cerinella (Dall, 1889)   3232
Little Waxy Mangelia

South Carolina to Florida and Texas.

8 to 10 mm., narrowly elongate, yellowish white, aperture 1/3 length of the shell. 7 to 9 adult whorls, angular, with 7 or 8 rounded axial ribs usually lined up under each other, crossed by a strong carinal thread midway between sutures and by numerous granular, microscopic threads above and below. Outer lip sharp, behind it a long, rounded, riblike varix. Columella straight, white. Siphonal canal short. 2 nuclear white whorls, smooth. Common; 2 to 40 fathoms. Formerly placed in *Kurtziella*.

3232   3234   3238   3242

Cryoturris citronella (Dall, 1889)   3233
Little Yellow Mangelia

Off both sides of Florida.

6 to 10 mm., elongate-fusiform, yellowish white, aperture ⅔ the length of the shell. 2 nuclear whorls, small, brown, smooth, followed by 2 whorls of numerous axial riblets. Adult whorls with 18 or 19 narrow, axial riblets crossed between sutures, by 3 spiral cords, and numerous microscopic threads, giving a frosty surface. Fairly large axial varix behind outer lip is rounded. Subsutural sinus U-shaped, shallow. Common; 30 to 230 fathoms.

Cryoturris fargoi McGinty, 1955   3234
Fargo's Mangelia

Off west Florida.

10 to 12 mm., similar to *citronella*, but stubbier, finer ribbing, light reddish brown in color, with about 8 (instead of 3) spiral threads between sutures. 2½ nuclear whorls, brown, smooth, the last bulging at the periphery where it is slightly nodulose. Uncommon; 10 to 30 fathoms.

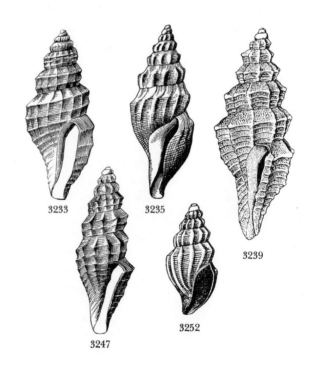

3233   3235   3239
3247   3252

Other species:

3235 *Cryoturris filifera* (Dall, 1881). Off Mississippi and west Florida. 30 to 180 fms. 12 mm., white.

3236 *Cryoturris elata* (Dall, 1889). Off North Carolina, 15 to 22 fms.

3237 *Cryoturris quadrilineata* (C. B. Adams, 1850). West Indies; southeast Florida. Uncommon.

3238 *Cryoturris trilineata* (C. B. Adams, 1845). West Indies.

Genus *Kurtzia* Bartsch, 1944

Fusiform, with latticed sculpturing. Protoconch 1½ smooth whorls, then 2 whorls of curved axials over 2 spiral threads. Adult whorls fenestrate with crisp spiral cords and long axial ribs. Operculum oval and with a terminal nucleus. Type: *arteaga* (Dall and Bartsch, 1910).

3239 *Kurtzia arteaga* (Dall and Bartsch, 1910). Vancouver Island, British Columbia, to the Gulf of California (and subspecies *roperi* Dall, 1919). 10 mm. (Synonym: *gordoni* Bartsch, 1944.)

3240 *Kurtzia granulatissima* (Mörch, 1860). Magdalena Bay and the Gulf of California to Costa Rica, in 20 to 40 meters. (Synonym: *Philbertia aegialea* Dall, 1919.)

3241 *Kurtzia aethra* (Dall, 1919). Sonora, Mexico, 20 to 70 meters.

Genus *Kurtziella* Dall, 1918

Shells white, ¼ to ½ inch, elongate-biconic, with a tall spire and a long, narrow, angulated body whorl, with a widely open, unnotched anterior canal. Protoconch of 2 whorls, the last reticulated. Adult sculpture of regular, prominent but rather narrow axial ribs, somewhat thickened at the shoulder, overridden by dense spiral lirae. Sinus a broad open **V**. Operculum absent. Type: *cerina* (Kurtz and Stimpson, 1851).

Kurtziella limonitella (Dall, 1883)   **3242**
Punctate Mangelia

North Carolina to both sides of Florida.

⅜ inch in length, semitranslucent and yellowish white. Sinus widely V-shaped. Between the strong, rounded, axial ribs there are numerous rows of microscopic opaque-white punctations. Uncommon from a few to 48 fathoms. Placed in *Stellatoma* by some workers.

Kurtziella atrostyla (Tryon, 1884)   **3243**
Brown-tipped Mangelia

North Carolina to Florida and Texas; West Indies.

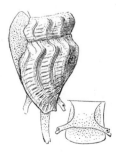

3243   3248   3248

7 mm., milky-white with a brown-stained columella, a brown subsutural band and sometimes brown on the thickened outer lip. Rarely all-brownish. 6 whorls, keeled at the shoulder; 8 to 10 axial riblets and fine spiral striations. Aperture and short anterior canal slightly oblique; outer lip with a shallow, rounded, posterior turrid notch. Moderately common; from intertidal to 48 fathoms. *K. ephamilla* Bush, 1885, is a synonym. *Mangilia* (sic) *atrostyla* Dall, 1889, is the same.

Subgenus *Rubellatoma* Bartsch and Rehder, 1939

Shells ⅓ inch, fusiform, tall spire, bluntly angulate whorls. Siphonal canal strongly flexed, short and unnotched. 2 nuclear whorls, smooth, followed by a stage of curved slender axials. Axial ribs broadly rounded. Surface densely and spirally threaded. Outer lip thin but variced behind. Type: *rubella* Kurtz and Stimpson, 1851.

Kurtziella rubella (Kurtz and Stimpson, 1851)   **3244**
Reddish Mangelia

North Carolina to southeast Florida; Texas.

¼ inch in length. Sinus shallow and U-shaped. Axial ribs long and rounded (about 9 per whorl). Spiral sculpture of numerous incised lines. Color grayish cream with light-reddish between the ribs. Commonly dredged from 9 to 80 fathoms.

(3245) *K. diomedea* Bartsch and Rehder, 1939, from off west Florida is extremely similar, but is more brightly colored with a wide spiral band of reddish brown. Uncommon to rare.

3244   3245

3246 *Kurtziella perryae* Bartsch and Rehder, 1939. West coast of Florida. *Proc. U.S. Nat. Mus.*, no. 3070, p. 134, pl. 17. Uncommon.

3247 *Kurtziella serga* (Dall, 1881). Florida Strait; West Indies; Bermuda; Yucatan. (Synonym: *acanthodes* Watson, 1881.)

3248 *Kurtziella cerina* (Kurtz and Stimpson, 1851). Massachusetts to New Jersey (Pleistocene?) to Florida; Yucatan, 20 to 30 fms.

3249 *Kurtziella accincta* (Montagu, 1808). Bermuda; southeast Florida and the Caribbean. (Synonym: *dorvilliae* Reeve, 1845.)

3250 *Kurtziella newcombei* (Dall, 1919). Vancouver Island, to Drake's Bay, California.

3251 *Kurtziella plumbea* (Hinds, 1843). British Columbia to Mazatlan, Mexico, in 10 to 50 meters. 8 to 11 mm. (Synonyms: *Mangelia alesidota, hebe, oenoa* and *tersa* Dall, 1919; *barbarensis* Oldroyd, 1924; *wrighti* Jordan, 1936; *angulata* Carpenter, 1864; *sulcata* Carpenter, 1865; *subangulata* Carpenter, 1857; *hecetae* Dall and Bartsch, 1910.)

3252 *Kurtziella antiochroa* (Pilsbry and Lowe, 1932). Head of the Gulf of California to Ecuador, in 10 to 50 meters. (Synonym: *Mangelia cymatias* Pilsbry and Lowe, 1932.)

3253 *Kurtziella* (*Rubellatoma*) *powelli* Shasky, 1971. Head of the Gulf of California to Ecuador, intertidal.

3254 *Kurtziella* (*Granoturris* Fargo, 1953) *antipyrgus* (Pilsbry and Lowe, 1932). Gulf of California to Colombia, in 20 to 70 meters.

3255 *Kurtziella* (*Kurtzina* Bartsch, 1944) *beta* (Dall, 1919). Farallon Islands, California, to Santa Maria Bay, Baja California, in 45 to 110 meters. Type of the subgenus.

3255a *Kurtziella* (*Kurtzina*) *cyrene* (Dall, 1919). Head of the Gulf of California to Ecuador, in 10 to 70 meters.

3256 *Kurtziella* (*Kurtzina*) *variegata* (Carpenter, 1864). Alaska to southern California. (Synonyms: *pulchrior* Dall, 1919; *nitens* Carpenter, 1864.)

Genus *Ithycythara* Woodring, 1928

Shells 9 mm., white, elongate, tall spire, and with prominent, concave-sided axial ribs which are aligned from whorl to whorl. Inner lip with a deeply set denticle just above the middle and with some smaller denticles below. Type: *psila* (Bush, 1885).

Ithycythara lanceolata (C. B. Adams, 1850)   **3257**
Spear Mangelid

South Florida and the West Indies.

6 to 12 mm., elongate; slender. Color tan with spiral bands of opaque-white and brown, the darkest band being at the suture. Whorls 8; suture slightly impressed. With 6 or 7 prominent, narrow, axial ribs per whorl. Ribs have a slight nodule at the middle of the whorl. Aperture narrow, finely toothed within. 10 to 40 fms; moderately common.

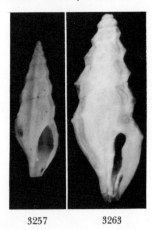

3257   3263

Ithycythara parkeri Abbott, 1958   **3258**
Parker's Mangelid

West Florida and the Caribbean.

5 to 6 mm., slender, with 7 flattened whorls bearing 5 slanting, straight, strong, sharp, narrow ribs which are slightly shouldered. Spiral sculpture of about 26 fine threads between sutures. Color chalk-white with a large, faint, orange-brown splotch near the center of the whorl between each rib. Common; 1 to 30 fathoms.

Other species:

3259 *Ithycythara muricoides* (C. B. Adams, 1850). West Indies.

3260 *Ithycythara hyperlepta* Haas, 1953. Brazil. Fieldiana, vol. 34, p. 207.

3261 *Ithycythara pentagonalis* (Reeve, 1845). South Florida and the West Indies. Brazil.

3262 *Ithycythara cymella* (Dall, 1889). Off West Florida to the Lesser Antilles, 70 to 200 fms. 13 mm.

3263 *Ithycythara psila* (Bush, 1885). Off North Carolina to Puerto Rico.

3264 *Ithycythara penelope* (Dall, 1919). Guaymas, Mexico, to Panama; Galapagos Islands.

Genus *Nannodiella* Dall, 1919

Fusiform with a high spire and truncate base. Protoconch of 4 whorls, tip minute, globular, last 1 or 2 nuclear whorls with an angular keel. Aperture small with a conspicuous laterally projecting spoutlike anal sinus. Adult whorls cancellate and angulate. Type: *nana* (Dall, 1919) from the Pacific Panamic province.

3265 *Nannodiella melanitica* (Bush, 1885). North Carolina to Florida and Texas, 7 to 47 fms. (same as *oxia?*).

3266 *Nannodiella oxia* (Bush, 1885). North Carolina to Florida, 14 to 15 fms. Gulf of Mexico, 40 to 50 fms. Yucatan, common. 5 mm.

3267 *Nannodiella vespuciana* (Orbigny, 1842). Florida and the West Indies. (Synonyms: *oxytata* Bush, 1885 and *melanitica* Dall, 1889.) Intertidal to 30 fms., common. 5 mm.

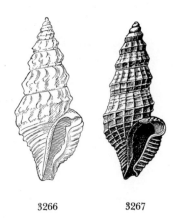

3266   3267

3268 *Nannodiella fraternalis* (Dall, 1919). Gulf of California to Colombia, in 20 to 70 meters.

3269 *Nannodiella nana* (Dall, 1919). Gulf of California to Colombia, in 20 to 70 meters.

Genus *Glyphostoma* Gabb, 1872

Shells ¾ inch, fusiform-biconic, solid, whorls bulging. Aperture long and narrow. Outer lip strongly variced. Both outer and inner lips heavily sculptured with denticles and ridges. Sinus wide, deep, strongly rimate. Fasciole with strong growth wrinkles. Sculpture of strong axial threads. Protoconch of 3 whorls, the second one having a strong keel. Type: *dentifera* Gabb, 1872. *Rhiglyphostoma* Woodring, 1970.

Glyphostoma gabbii Dall, 1889   **3270**
Gabb's Mangelia

Florida, the Gulf of Mexico and the West Indies.

10 to 17 mm., in length. The 3 nuclear whorls are brownish, smooth and with a single, strong carina at the periphery. Shell white with 2 wide spiral bands of rose-brown on the whorl. The upper one is interrupted by about 15 short white ribs per

Propebela gouldii (Verrill, 1882)   **3317**
Gould's Northern Turrid

Nova Scotia to off Cape Cod, Massachusetts.

10 to 15 mm., white to greenish white; 6 or 7 whorls, squarely carinate-shouldered. 15 sharp ribs on the last whorl with well-marked, raised, spiral lines between. Sinus shallow, broadly concave. Radular teeth broad and short. Very common in Massachusetts Bay and the Gulf of Maine, in 15 to 115 fathoms. Rarely to 122 fathoms; on mud or shelly bottom. (*Trans. Conn. Acad.*, vol. 5, p. 465.)

Other Atlantic species (probably *Propebela*):

3318 *Propebela hebes* (Verrill, 1880). Off Newport, Rhode Island, 282 to 500 fms.

3319 *Propebela concinnula* (Verrill, 1882). Labrador to Cape Cod, Massachusetts, 16 to 118 fms.

3320 *Propebela pingelii* (Möller, 1842). Nova Scotia to Georges Bank, 20 to 90 fms. Arctic Seas.

3321 *Propebela pygmaea* (Verrill, 1882). Off Martha's Vineyard, Massachusetts, 312 to 487 fms.

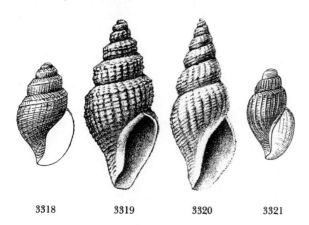

3318   3319   3320   3321

3322 *Propebela rosea* (G. O. Sars, 1878). Halifax, Nova Scotia 15 to 57 fms. *P. rosea* (Lovén, 1846)?

3323 *Propebela rathbuni* (Verrill, 1884). Off Delaware, 1,395 fms.

3323

3324 *Propebela sarsii* (Verrill, 1880). Labrador and the Gulf of St. Lawrence, 10 to 20 fms. (Synonym: *cancellata* Sars, non Couthouy.)

3325 *Propebela subturgida* and *subvitrea* (**3326**) both (Verrill, 1884). Off North Carolina, 843 fms.

3327 *Propebela tenuicostata* (G. O. Sars, 1878). Eastport, Maine to off southeast Nantucket, Massachusetts, 843 to 1,290 fms. Bering Sea and Alaska. Europe.

3328 *Propebela tenuilirata* (Dall, 1871). Off Martha's Vineyard, Massachusetts, 365 fms. (Synonym: *simplex* Verrill, non Mid-

dendorff, 1849.) Northern Alaska. Variety *cymata* Dall, 1919 (**3329**) Bering Sea.

3330 *Propebela trevelliana* (Turton, 1834). Gulf of St. Lawrence and the Gulf of Maine. Behm Canal, Alaska. (*trevelyana* is a misspelling.)

3331 *Propebela exarata* (Möller, 1842). Greenland to off Martha's Vineyard, Massachusetts, 5 to 487 fms.

3332 *Propebela nobilis* (Möller, 1842). Greenland to Maine, 7 to 15 fms.; Bering Sea, Alaska, 10 to 60 fms.

3333 *Propebela angulosa* (G. O. Sars, 1878). Off Metis, Quebec.

3334 *Propebela blakei* (Verrill, 1885). Off Chesapeake Bay, 2,021 fms.

3335 *Propebela blaneyi* (Bush, 1909). Frenchmans Bay, Maine.

3336 *Propebela elegans* (Möller, 1842). Greenland and the Gulf of the St. Lawrence. Northern Europe.

Other Pacific *Propebela*:

3337 *Propebela fidicula* (Gould, 1849). Aleutian Islands to Bellingham Bay, Alaska.

3338 *Propebela inequita* (Dall, 1919). Plover Bay to Boca de Quadra, Alaska.

3339 *Propebela miona* (Dall, 1919). Boca de Quadra, Alaska, to Point Reyes, California.

3340 *Propebela pribilova* (Dall, 1919). Arctic Ocean to Esteros Bay, California.

3341 *Propebela monterealis* (Dall, 1919). Monterey, California, 581 fms. Rare.

3342 *Propebela* (*Turritoma* Bartsch, 1944) *diomedea* Bartsch, 1944. Monterey, California, 581 fms.

3343 *Propebela tabulata* (Carpenter, 1865). Sitka, Alaska to Washington. California records doubtful.

3344 *Propebela casentina* (Dall, 1919). Off Point Pinos, California, 871 fms. Rare.

3345 *Propebela* (*Nodotoma* Bartsch, 1941) *impressa* (Mörch, 1869). Arctic Ocean to Kodiak Island, Alaska. Norway. Type of the subgenus.

3346 *Propebela profundicola* Bartsch, 1944. Point Pinos, California, 871 fms.

3347 *Propebela popovia* (Dall, 1919). Bristol Bay, Bering Sea.

3348 *Propebela* (*Turritoma*) *smithi* Bartsch, 1944. Off Point Pinos Light, California, 293 fms. Proc. Biol. Soc. Wash., vol. 57, p. 67.

Pacific "Lora" (*Propebela?*):

— *Lora kyskana* (**3350**), *quadra* (**3351**), *althorpensis* (**3352**), *fiora* (**3353**), *tenuissima* (**3354**), *galgana* (**3355**), *luetkeni* "Krause, 1885" (**3356**), *lutkeana* (**3357**), *nazanensis* (**3358**), *amiata* (**3359**), *pavlova* (**3360**), *alitakensis* (**3361**), *rassina* (**3362**), *healyi* (**3363**), *chiachiana* (**3364**), *mitrata* (**3365**), *lawrenciana* (**3366**) all Dall, 1919, all Bering Sea and Alaska, Proc. U.S. Nat. Mus., vol. 56.

3367 *Lora babylonia* Dall, 1919. Off Tillamook, Oregon. *Proc. Biol. Soc. Wash.*, vol. 32, p. 250.

3368 *Lora maurellei* (Dall and Bartsch, 1910). Barkley Sound, Vancouver Island, British Columbia. 8.5 mm.

— *Lora grippi* Dall, 1908, see under *Bellaspira*.

3369 *Lora regulus* Dall, 1919. Off Point Reyes, California, 61 fms.

3370 *Lora lotta* Dall, 1919. Queen Charlotte Islands, British Columbia, to San Diego, California.

3371 *Lora sixta* Dall, 1919. Off San Diego, California, in 359 to 822 fms.

3372 *Lora albrechti* Krause, 1885. Plover Bay, Bering Strait, to Port Etches, Alaska, 6 to 17 fms.

3373 *Lora metschigmensis* Krause, 1885. Metchigme Bay, Bering Strait.

3374 *Lora morchi* Leche, 1878. Novaia Zemlia, Arctic Ocean, to Bering Strait.

3375 *Lora elegans* (Möller, 1842). Blizhni Islands to St. Lawrence Island, Bering Sea (Aurivillius). Circumboreal.

3376 *Lora novaiasemliensis* (Leche, 1878). Kara Sea, Arctic Ocean to north of Bering Strait.

3377 *Lora woodiana* (Möller, 1842). White Island, Arctic Ocean to St. Lawrence Island, Bering Strait. Circumboreal (Aurivillius).

3378 *Lora sancta-monicae* Arnold, 1903. Santa Monica, California. Pleistocene.

3379 *Lora solida* (Dall, 1887). Bering Strait to Puget Sound.

3380 *Lora murdochiana* (Dall, 1885). Point Barrow, Arctic Ocean, to the Pribilof Islands, Bering Sea.

3381 *Lora simplex* (Middendorff, 1849). Point Barrow, Arctic Ocean, to the Pribilof Islands and the Okhotsk Sea.

3382 *Lora arctica* (A. Adams, 1855). Shumagin Islands to Chirikoff Island, Alaska.

3383 *Lora laevigata* (Dall, 1871). Kotzebue and Norton Sounds and south to Chirikoff Island, Alaska.

3384 *Lora nodulosa* (Krause, 1885). St. Lawrence Bay, Bering Strait, to the Aleutians and Cook's Inlet.

3385 *Lora beckii* (Möller, 1842). Bernard Harbor, Arctic coast, and eastward.

Genus *Oenopota* Mörch, 1852

Resembles *Propebela* in shape and sculpture, but lacks shouldered whorls and the sculpturing is weaker on the shoulders. Protoconch narrowly papillate of about 2 to 2½ whorls, initially smooth, then developing 3 spiral cords crossed by spaced axials. Periostracum thin, pale-brown. Type: *pyramidalis* (Ström, 1788); (*pleurotomaria* Couthouy, 1835, is a synonym).

Oenopota harpularia (Couthouy, 1838)   3386
Harp Lora

Arctic Seas to Rhode Island.
Alaska to Puget Sound, Washington.

12 mm., ovate-oblong, turreted, brownish flesh color. 6 to 8 whorls, angulate, flattened above the angulation, with a slightly sloping shoulder. Whorls with about 18 oblique, rounded riblets becoming obsolete at the base of the shell; spiral threads fine and numerous. Outer lip sharp. Inner lip white, smooth and slightly arched. Do not confuse with *Propebela turricula*.

Oenopota pyramidalis (Ström, 1788)   3387
Pyramid Lora

Arctic Canada to Massachusetts. Norway.
Bering Sea to Puget Sound, Washington.

11 to 20 mm., elongate-fusiform, reddish brown. 8 whorls, slightly convex, with numerous obliquely axial riblets (18 to 22 per whorl). Interspaces as wide as the riblets. Ribs extend halfway down the last whorl. Spiral sculpture of fine lines, more conspicuous at the base of the shell. Slightly shouldered at the suture. Turrid notch weak. Outer lip sharp. Columella smooth. Siphonal canal short. *Bela pleurotomaria* (Couthouy, 1838) is a synonym. Common; offshore from 5 to 255 fathoms.

3386   3388

Oenopota bicarinata (Couthouy, 1838)   3388
Two-corded Lora

Arctic Canada to New York.
Arctic Seas to Chirikoff Islands, Alaska.

5 to 10 mm., fusiform, turreted, dusky-white to purplish gray. 6 whorls, convex. Spiral sculpture prominent. Periphery of whorl with a deep spiral groove, with a prominent ridge above and below. Growth lines microscopic. Outer lip sharp and crenulated. Turrid notch weak. Common; 6 to 100 fathoms. Also found in the stomachs of fish. C. W. Johnson considers *violacea* Mighels and Adams, 1842, as a synonym. *O. cylindracea* Möller, 1842, is a synonym, as is the Bering Sea variety, *exserta* Aurivillius, 1885.

Oenopota decussata (Couthouy, 1839)   3389
Decussate Lora

Arctic Canada to Massachusetts.

5 to 7 mm., ovate, white to flesh-colored, with an olive periostracum. 6 whorls, with 25 to 30 inconspicuous, oblique riblets on the upper ½ of the whorls. Numerous spiral threads make a decussate network over the whorl shell. Suture indented. Outer lip sharp, with a shallow notch as it joins the whorl above. Uncommon; 10 to 75 fathoms. Verrill, in 1882, described a small variety, *pusilla*. *Bela tenuicostata* M. Sars, 1868, and *conoidea* G. O. Sars, 1878, are synonyms.

Subgenus *Funitoma* Bartsch, 1941

Type of the subgenus is *areta* Bartsch, 1941, from Japan.

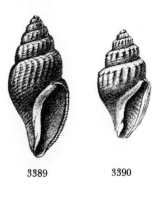

3389   3390

Oenopota incisula (Verrill, 1882) 3390
Incised Northern Turrid

Arctic Canada to off Rhode Island; Greenland.

6 to 8 mm., greenish white to pinkish white, covered by a thin, greenish periostracum. Last whorl with about 20 broad, flattened, straight ribs which are slightly nodulose at the shoulder and separated by concave grooves. Characterized by fine, incised, narrow spiral grooves, 3 to 5 on the penultimate whorl, and 20 to 28 on the last. Sinus deep, rounded; canal short. One of the most common and widely distributed New England turrids; on sand, mud or gravel; 5 to 500 fathoms. 30 to 40 radular teeth are narrowly lanceolate. Synonym: *impressa* Verrill, 1880.

Other species:

3391 *Oenopota (Granotoma) excurvata* (Carpenter, 1865). Bering Sea to Puget Sound, Washington.

3392 *Oenopota (Granotoma) krausei* (Dall, 1886). Port Etches, Alaska.

3393 *Oenopota alaskensis* (Dall, 1871). Bering Strait to Puget Sound, Washington, 3 to 4 fms.

3394 *Oenopota harpa* (Dall, 1885). Arctic Seas; Alaska to Queen Charlotte Islands, British Columbia.

3395 *Oenopota reticulata* (Brown, 1827). Arctic Seas to Baja California. Arctic Canada. (Synonyms: *leucostoma* Reeve, 1845; *surana, colpoica, pitysa* and *diegensis* all Dall, 1919.)

3396 *Oenopota rosea* (Lovén, 1846). Alaska to Washington.

Subfamily DAPHNELLINAE Casey, 1904

Shell fusiform or ovate, canal short. Operculum absent. Sinus adjoining the suture. The protoconch has diagonally cancellate sculpturing. Radula with 2 slender, curved marginals only.

Genus *Daphnella* Hinds, 1844

Shells 10 to 30 mm., elongate-ovate, with a long body whorl usually more than ½ the length of the shell. Siphonal canal short. Protoconch of several cancellated whorls. Sinus at the suture is small and L-shaped. Adult sculpture of fine reticulated threadings. Type: *lymneiformis* (Kiener, 1840). Synonyms: *Eudaphne* Bartsch, 1931 (non Reuss, 1922); *Eudaphnella* Bartsch, 1933; *Paradaphne* Laseron, 1954.

Daphnella lymneiformis (Kiener, 1840) 3397
Volute Turrid

Southeast Florida and the West Indies.

⅓ to ½ inch in length; resembles a miniature, elongate *Scaphella* volute shell. With about 8 whorls, the nuclear ones smoothish, the next 4 with strong, axial ribs, but the last 2 whorls with only numerous fine spiral threads crossed by exceedingly fine growth lines. Aperture elongate, rather expanded and a little flaring below. Sinus moderately large and simple. Color cream with yellowish brown maculations. Uncommon from shallow water to 25 fathoms. *D. decorata* (C. B. Adams, 1850) is a synonym.

Daphnella stegeri McGinty, 1955 3398
Steger's Turrid

Southeast Florida to Yucatan, Mexico.

½ inch, white with smear of tan; nucleus of 3 rounded, reticulated, brown whorls. Early whorls axially ribbed. Last 2 whorls with strong, raised cords (3 on the body whorl, 1 between sutures). Entire surface has a frosted appearance under the high-power lens. Sinus deep and U-shaped. Uncommon; 25 to 70 fathoms.

3398

Daphnella morra (Dall, 1881) 3399
Morro Turrid

Off North Carolina to both sides of Florida; Texas.

¼ inch in length, yellowish tan to glistening brown. Anal notch deep. 16 to 450 fathoms. Common.

Other Atlantic species:

3400 *Daphnella leucophlegma* (Dall, 1881). Gulf of Mexico and the West Indies, 805 fms.

3401 *Daphnella corbicula* Dall, 1889. Off North Carolina to both sides of Florida to Barbados, 25 to 130 fms. 11 mm.

3397   3399   3406   3408   3408a   3411

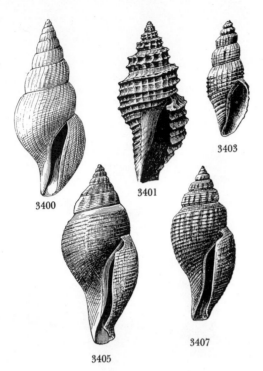

3400   3401   3403   3405   3407

3402 *Daphnella retifera* Dall, 1889. Off North Carolina, 49 to 63 fms.

3403 *Daphnella elata* Dall, 1889. Off North Carolina to Florida to Yucatan, 10 to 40 fms. Common. 5 mm.

3404 *Daphnella peripla* (Dall, 1881). Southwest of Egmont Key, to Yucatan, 220 to 640 fms.

3405 *Daphnella reticulosa* Dall, 1889. Barbados, Lesser Antilles, 76 fms. 11.5 mm.

3406 *Daphnella pompholyx* Dall, 1889. Off Fernandina, Florida, 294 fms. 12.5 mm.

3407 *Daphnella eugrammata* Dall, 1902. Florida Strait, 203 fms. 9 mm.

3408 *Daphnella grundifera* (Dall, 1927). Off Fernandina, Florida. Type illustrated.

3408a *Daphnella margaretae* Lyons, 1972. Off Palm Beach Co. to off Key West, Florida. 35 to 85 meters. 10 mm. *The Nautilus*, vol. 86, p. 6.

Other Pacific species:

3409 *Daphnella clathrata* (Gabb, 1865). San Miguel Island to San Diego and Cortez Bank, California.

3410 *Daphnella fuscoligata* (Dall, 1871). Monterey to San Diego, California, 5 to 15 fms. in crevices and kelp. (Synonym: *Mangelia crassaspera* Grant and Gale, 1931.)

3411 *Daphnella bartschi* Dall, 1919. Magdalena Bay and the Gulf of California; Galapagos Islands.

3412 *Daphnella mazatlanica* Pilsbry and Lowe, 1932. Gulf of California to Ecuador; Galapagos Islands.

Genus *Eubela* Dall, 1889

Shell smooth, glossy, with a sutural band, the outer lip sharp, with a shallow anal sinus, and having a short angular canal. Type: *Daphnella limacina* Dall, 1881.

3413 *Eubela limacina* (Dall, 1881). Off Martha's Vineyard, Massachusetts, 368 fms., to the Gulf of Mexico to Brazil, 350 fms. (Synonym: *hormophora* Watson, 1881.)

3414 *Eubela macgintyi* Schwengel, 1943. Off Lake Worth, east Florida, 80 fms. West coast of Florida, 30 to 50 fms. Uncommon.

3415 *Eubela calyx* (Dall, 1889). 36 miles southwest of Cape Hatteras, North Carolina, 124 fms.

3416 *Eubela sofia* variety *hyperlissa* (Dall, 1889). Off Cape Fear, North Carolina, 731 fms.

Genus *Pleurotomella* Verrill, 1873

Shells white, broadly fusiform, with well-rounded whorls, bearing a broad, smooth, slightly concave subsutural band, below which are numerous, fine but distinct, curved, axial ribs. Sinus deeply indented. Protoconch of 3 to 4 whorls, narrowly turbinate, with cancellate sculpture or axial riblets crossed by spiral threads. Type: *packardii* Verrill, 1873. This group is blind and without operculum. F. Nordsieck, 1968, Die europ. Meeres-Gehäuseschnecken (Prosobranchia), created the subfamily Pleurotomellinae for this group of genera.

Other species:

3417 *Pleurotomella* (*Theta* Clarke, 1959) *lyronuclea* A. H. Clarke, 1959. 185 miles west of Bermuda; 2,843 fms. *Proc. Malacol. Soc. London*, vol. 33, p. 234, pl. 13.

3418 *Pleurotomella packardii packardii* (Verrill, 1872). Gulf of Maine, 85 to 110 fms. Off Cape Cod, 96 fms. 20 mm.

3419 *Pleurotomella packardii benedicti* Verrill and Smith, 1884. Southeast of Nantucket, 1,290 fms. 20 mm.

3420 *Pleurotomella packardii formosa* Jeffreys, 1883. Off Massachusetts, 906 fms.

3421 *Pleurotomella bruneri* Verrill and Smith, 1884. Off Delaware Bay, 1,608 to 2,033 fms.

3422 *Pleurotomella leucomata* (Dall, 1881). Gulf of Mexico, 533 to 940 fms.

3423 *Pleurotomella catherinae* Verrill and Smith, 1884. South of Martha's Vineyard, Massachusetts, to North Carolina, 843 to 2,033 fms.

3424 *Pleurotomella agassizii agassizii* Verrill and Smith, 1880. South of Martha's Vineyard, Massachusetts, to the West Indies, 39 to 1,608 fms. 30 mm. Type illustrated.

3425 *Pleurotomella sandersoni* Verrill, 1884. South of Martha's Vineyard, Massachusetts, to Delaware Bay, 1,290 to 2,033 fms.

3426 *Pleurotomella pandionis* Verrill, 1880. South of Martha's Vineyard, Massachusetts, 238 to 312 fms. 40 mm.

3427 *Pleurotomella emertonii* Verrill and Smith, 1884. Off Virginia, 1,917 fms. 25 mm.

3428 *Pleurotomella tincta* Verrill, 1885. Off North Carolina, 2,512 fms.

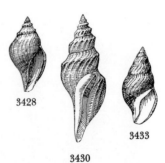

3428   3430   3433

3429 *Pleurotomella jeffreysii* Verrill, 1885. Off Delaware and Chesapeake Bays, 984 to 1,525 fms. In subgenus *Majox* Nordsieck, 1968.

3429   3434

3430 *Pleurotomella chariessa phalera* Dall, 1889. Off Cape Fear, North Carolina, 731 fms. (*P. chariessa* Watson, 1879, and its varieties are in *Majox* Nordsieck, 1968.)

3431 *Pleurotomella chariessa aresta* Dall, 1889. Off Cape Fear, North Carolina, 731 fms.

3432 *Pleurotomella chariessa tellea* Dall, 1889. Off Cape Fear, North Carolina, 731 fms.

3433 *Pleurotomella frielei* Verrill, 1885. Off Delaware Bay, 1,178 fms.

3434 *Pleurotomella bairdii* Verrill and Smith, 1884. Off Delaware and Chesapeake Bays, 1,608 to 1,721 fms. 35 mm. Type of subgenus *Majox* Nordsieck, 1968.

3435 *Pleurotomella lottae* Verrill, 1885. South of Martha's Vineyard, Massachusetts, 1,525 fms.

3436 *Pleurotomella hadria* Dall, 1889. Off Cape Fear, North Carolina, 407 to 731 fms. Gulf of Mexico, 1,181 fms.

3437 *Pleurotomella vaginata* Dall, 1927. Off Georgia, 440 fms.

3438 *Pleurotomella lineola* Dall, 1927. Off Fernandina, Florida, 294 fms.

3439 *Pleurotomella stearina* Dall, 1927. Off Georgia, 440 fms.

3440 *Pleurotomella corrida* Dall, 1927. Off Georgia, 440 fms. Off Fernandina, Florida, 294 fms.

3441 *Pleurotomella aperta* Dall, 1927. Off Georgia, 440 fms.

3442 *Pleurotomella atypha* Bush, 1893. Off Cape Fear, North Carolina, 464 to 647 fms.

3443 *Pleurotomella sulcifera* Bush, 1893. Off Cape Fear, North Carolina, 647 fms.

3444 *Pleurotomella leptalea* Bush, 1893. Off Cape Fear, North Carolina, 647 fms.

3445 *Pleurotomella dalli* Bush, 1893. Off Cape Fear, North Carolina, 647 fms.

3445a *Pleurotomella thalassica* Dall, 1919. Off Tillamook, Oregon, 786 fms.

3446 *Pleurotomella herminea* Dall, 1919. Off Catalina Island, California, 334 to 600 fms.

Subgenus *Gymnobela* Verrill, 1884

Shells more rotund than *Pleurotomella*, without a pronounced siphonal canal, and without the subsutural concave band. Type: *engonia* Verrill, 1884.

Pleurotomella blakeana (Dall, 1889)   **3447**
Blake's Turrid

Off Massachusetts to the Lower Florida Keys and Gulf of Mexico.

8 mm., glistening, whitish. Axial riblets are sharp-edged; spiral threads weak, spire low. Turrid notch quite shallow. Deep water, from 100 to 1608 fathoms. Uncommon. The form *agria* (Dall, 1889) has a higher spire, stronger spiral threads and a dark periostracum. *Gymnobela brevis* Verrill, 1885, is a synonym.

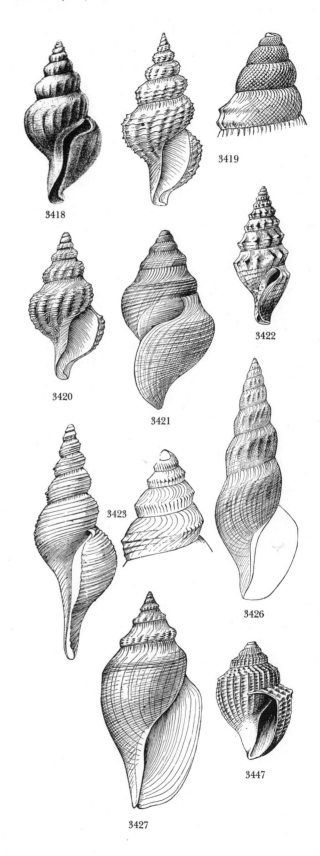

3418   3419   3420   3421   3422   3423   3426   3427   3447

Other species:

3448 *Pleurotomella curta curta* (Verrill, 1884). Southeast of Nantucket, 1,290 fms.

3449 *Pleurotomella curta subangulata* (Verrill, 1884). With the typical form.

3448

3450 *Pleurotomella engonia* (Verrill, 1884). South of Martha's Vineyard, Massachusetts, 1,290 to 1,608 fms.

3451 *Pleurotomella vitrea* Verrill, 1885. Off Delaware Bay, 384 to 428 fms.

3452 *Pleurotomella extensa* (Dall, 1881). Gulf of Mexico and West Indies, 640 to 1,000 fms.

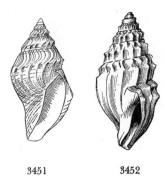

3451　　　　3452

3453 *Pleurotomella tornata tornata* (Verrill, 1884). Off Nova Scotia, 1,255 fms.

3454 *Pleurotomella tornata malmii* Dall, 1889. Gulf of Mexico, 805 fms.

3455 *Pleurotomella imitator* (Dall, 1927). Off Fernandina, Florida, 294 fms.

3456 *Pleurotomella illicita* (Dall, 1927). Off Fernandina, Florida, 294 fms.

3457 *Pleurotomella lanceata* (Dall, 1927). Off Fernandina, Florida, 294 fms.

Subclass *Euthyneura* Spengel, 1881

Univalves, including the air-breathing land snails, the sea slugs, opisthobranchs, nudibranchs and many nonoperculate fresh-water snails, all having a nervous system that is not crossed by the torsion action in the embryonic young.

Order *Pyramidelloida* Gray, 1840

Superfamily Pyramidellacea Gray, 1840

Family PYRAMIDELLIDAE Gray, 1840

This is a well-known family of very small gastropods which are extremely baffling to novices attempting to identify any one of the several hundred so-called species. Even among the experts there is not always agreement on what constitutes a species, subgenus or genus in this group. It would be impossible to present in a book this size even an account of only the most common species. Those interested in delving into this interesting maze of species are referred to the works of P. Bartsch, W. H. Dall and K. Bush. See P. Bartsch, 1909, "Pyramidellidae of New England and the adjacent region," *Proc. Boston Soc. Nat. Hist.*, vol. 43, pp. 67–113.

Many species of pyrams are ectoparasites, feeding on tubicolous polychaete worms, gastropods and bivalves. The pyrams attach themselves to the host by means of an oral sucker, and pierce the body wall of the host with a buccal stylet. They suck the host's blood by means of a buccal pump. Embryological and other data have shown that this family of mollusks is closely related to the tectibranch mollusks, rather than to the prosobranchs with which they have been formerly placed. (See *Journal of the Marine Biological Association*, vol. 28, pp. 493–532, 1949.)

Genus *Pyramidella* Lamarck, 1799

Shells up to 1 inch, glossy, solid, elongate-conic. Columella with 3 strong spiral folds. Type: *dolabrata* (Linné, 1758). *Pyramidellus* Montfort, 1810, is a synonym.

Pyramidella dolabrata (Linné, 1758)　　　　**3459**
Giant Atlantic Pyram　　　　**Color Plate 3**

Southeast Florida and the West Indies. Brazil.

¾ to 1 inch in length, solid and glossy-smooth. Columella large, and with 2 or 3 strong, spiral plicae. Color opaque-white with 3 fine, spiral lines of brown, 1 of which is just above the suture. Common in the West Indies and possibly present in the Lower Florida Keys. This species is a sand-dweller; from 1 to 31 fathoms.

(3460) A form or separate species is *subdolabrata* Mörch, 1854, from the West Indies. It resembles *dolabrata,* but has duller color, weaker suture, less strongly developed columellar folds and the color bands are wider and less distinct. Uncommon.

3459　　　　3461　　　　3462　　　　3470

Subgenus *Longchaeus* Mörch, 1875

Shell elongate-conic, not umbilicate, having 3 columellar folds, a basal fasciole and a peripheral sulcus. Entire surface with fine growth lines and microscopic spiral striations. Type: *punctata* (Schubert and Wagner, 1829). Operculum thin, paucispiral and notched to accommodate the folds on the shell's columella.

Pyramidella candida Mörch, 1875   3461
Brilliant Pyram

North Carolina to Florida; West Indies to Brazil.

12 to 14 mm., elongate-conic, flat-sided, glossy-white, with about 12 flat-sided whorls. Suture channeled and crenulated. Columella with 3 strong, spiral plaits. Operculum thin, light-brown. Uncommon; just offshore to 18 fathoms.

Pyramidella crenulata (Holmes, 1859)   3462

South Carolina to Texas; West Indies.

10 to 12 mm., elongate-conic, flat-sided, glossy-tan to pink, but frequently with eroded spots. 13 whorls, flat-sided; suture horizontally channeled and crenulated. Sometimes the body whorl has 2 or 3 spiral lines. Common; shallow bays on sand, mud and grass. Synonyms: *arenosa* Tuomey and Holmes, 1857 (not Conrad, 1843); *floridana* (Mörch, 1875); *suprapulchra* Gregorio, 1890.

Pyramidella adamsi Carpenter, 1864   3463
Adams' Pyram

Monterey, California, to southern Mexico.

12 to 19 mm., early whorls white, the later ones brown with axial whitish variegations. 2 nuclear whorls, smooth, planorboid and set at right angles. About 11 postnuclear whorls, flattish, and with a light-colored spiral sulcus on the periphery. Entire surface with growth lines and microscopic spiral striations. Aperture brown and white-banded within, and with 5 spiral cords. Columellar fasciole with 3 folds, the upper one being the largest. Common; mud and sand flats, intertidal. *P. variegatus* Carpenter, 1864, is a synonym.

3463

3464

Pyramidella mexicana Dall and Bartsch, 1909   3464

San Pedro, California, to Baja California.

12 to 19 mm., similar to *adamsi,* but solid dull-brown, except for a narrow white band at the periphery. The whorls in the spire are minutely crenulated at the top. Uncommon; shallow water.

Other Pacific *Longchaeus*:

3465 *Pyramidella bicolor* Menke, 1854. Guacamoyo, Mexico to Costa Rica. 10 mm.

3466 *Pyramidella mazatlanicus* Dall and Bartsch, 1909. San Diego, California, to Baja California and southern Mexico.

Subgenus Eulimella Jeffreys, 1847

Elongate-conic, surface polished, with faint growth lines and microscopic spiral striations; base without a fasciole;

columella with 2 folds. Type: *crassula* Forbes, 1843. *Belonidium* Cossmann, 1846, and *Loxoptyxis* Cossmann, 1888, are synonyms. The name was first validated in May 1847 in the *Annals Mag. Nat. Hist.,* vol. 19, p. 311 by Jeffreys.

3467 *Pyramidella crassula* Forbes, 1843. North Carolina to the West Indies, 6 to 25 fms. (Synonym: *scillae* Scacchi, 1835.)

3468 *Pyramidella unifasciata* Forbes, 1843. Off North Carolina to Gulf of Mexico, 80 to 120 fms.

3469 *Pyramidella ventricosa* Forbes, 1843. Eastport, Maine, 100 fms.

3470 *Pyramidella lissa* Verrill, 1884. Off Cape Hatteras, North Carolina, 142 fms.

(For the species *charissa, lucida* and *nitida* all Verrill, 1884, see under the Melanellidae.)

Subgenus Sulcorinella Dall and Bartsch, 1904

Columella with 1 large fold. Shell umbilicate, surface polished, with microscopic striations. One peripheral sulcus. Type: *dodona* Dall and Bartsch, 1904, a Florida Oligocene species.

3471 *Pyramidella camara,* Bartsch in Dall, 1927. Off Georgia, 440 fms.

Subgenus Stylopsis A. Adams, 1860

Shell less than 4 mm., fragile, elongate, with swollen but laterally flattened whorls (8 to 10), white; no umbilical chink, no fold on columella. Spiral sculpture of fine grooves. Type: *typica* A. Adams, 1860. Doubtfully placed in the genus *Pyramidella.*

3472 *Pyramidella resticula* (Dall, 1889). Key West, and Florida Key, Florida, intertidal.

Subfamily ODOSTOMIINAE Pelseneer, 1928
Genus Odostomia Fleming, 1813

Small, elongate-conic shells, white, some smooth, some cencellate; all with a small fold on the upper part of the columella. Several subgenera have been proposed, but many species straddle these artificial groupings. Type: *plicata* (Montagu, 1803). Synonyms: *Odontostomia* Jeffreys, 1839; *Ptychostomon* Locard, 1886; *Turritodostomia* Sacco, 1892; *Heida* Dall and Bartsch, 1904. Many, if not most, species are ectoparasites feeding on other mollusks and marine worms. They are not host-specific.

Subgenus Odostomia Fleming, 1813

Axial and spiral sculpture absent, usually. Columellar fold present. Type: *plicata* (Montagu, 1803). *Heida* Dall and Bartsch, 1904, is a synonym.

Odostomia laevigata (Orbigny, 1842)   3473
Ovoid Odostome

North Carolina to Florida and the West Indies. Brazil.

3 to 5 mm. long, elongate-ovate, grayish white (alive) to brown (dead), with 4 to 6 slightly rounded postnuclear whorls;

nucleus small and deeply and obliquely buried in the apex. Umbilicus obsolete to chinklike. Microscopic spiral scratches absent or present, rarely with 1 or 2 prominent one just below the suture. Columellar tooth weak. A very variable species given many names: *ovuloides* (C. B. Adams, 1850); *caloosaensis* (Dall, 1893); *pomeroyi, cooperi, stearnsi, pinellasensis, bassleri, matsoni, gunteri, schwengelae, stephensoni, conradi* and *hielprini* all Bartsch, 1955.

3473

Odostomia gibbosa Bush, 1909   3474

Maine to southern Massachusetts.

3 mm., solid ovate, yellowish white, shiny. Oblique small nuclear whorls completely immersed. Postnuclear whorls inflated. Periphery of the last whorl obscurely angulated. Sculpture of fine growth lines and very faint spiral striations. Outer lip thick within. Columella stout, curved and with a strong oblique fold at the top. Uncommon. This is what Bartsch in 1909 erroneoulsy called *modesta* Stimpson.

Odostomia conoidea (Brocchi, 1814)   3475
Conoidal Odostome

North Carolina to west Florida.
British Isles. Mediterranean.

3 to 4 mm., white, elongate-ovate, with slightly flattened whorls, the last being slightly angular at the low periphery. Nuclear whorl set at right angles and half-immersed. Aperture minutely spirally grooved inside the outer lip. Upper part of the columella with a weak fold. The form or subspecies *acutidens* Dall, 1884, from west Florida, is less elevated, more angular at the base and has a stronger columellar fold. Moderately common; shallow water, 14 to 22 fathoms. It is likely that *O. modesta* Stimpson, 1851, is a synonym.

Odostomia dealbata (Stimpson, 1851)   3476
White New England Odostome

Massachusetts to Connecticut.

4 mm., slender, high-spired, white, smooth, shiny. 6 whorls, moderately convex and slightly increasing in size. Base rounded. Aperture ovate. Columella with a small fold at the top. Uncommon; 3 fathoms, on shelly bottoms.

Odostomia dinella Dall and Bartsch, 1909   3477
Dinella Odostome

Los Angeles County to San Diego, California.

2 to 2.5 mm., ovate, white, shiny, semitransparent. Nuclear whorl almost vertically deeply embedded by the following whorl. Postnuclear whorls rounded, obscurely angulated at the periphery. Narrowly umbilicate. Columella with a weak fold at the top. Parietal wall covered with a callus. Moderately common; 25 to 35 fathoms. Synonyms: *farella* Dall and Bartsch, 1909; *coronadoensis* Dall and Bartsch, 1909; and *orcutti* Bartsch, 1917, the last being a dead dwarf specimen.

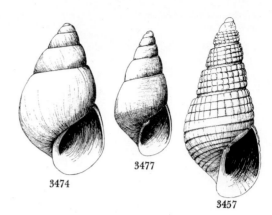

3474   3477

3457

Other Atlantic *Odostomia* (s. s.):

3478 *Odostomia modesta* Stimpson, 1851. Eastport, Maine, to off Martha's Vineyard, Massachusetts, 6 to 115 fms. (is *conoidea* Brocchi, 1814?).

3479 *Odostomia disparilis* Verrill, 1884. Off Cape Hatteras, North Carolina, 142 fms.

3480 *Odostomia cancellata* Orbigny, 1842. Off Cape Hatteras, North Carolina, 14 fms.; Cuba.

3481 *Odostomia tornata* Verrill, 1884. Off Cape Hatteras, North Carolina, to Florida Keys, 15 to 142 fms.

3482 *Odostomia engonia* Bush, 1885. Off Carolina to the Florida Keys, 16 to 200 fms.

3483 *Odostomia teres* Bush, 1885. Off North Carolina and South Carolina, 14 to 22 fms.

3484 *Odostomia unidentata* Fleming, 1813. Off North Carolina to the Florida Keys, 63 to 200 fms.

— *Odostomia barretti* Morrison, 1965, non Morlet, 1885, is *Hydrobia booneae* Morrison, 1973, from Texas (*The Nautilus*, vol. 87).

Other Pacific *Odostomia* (s. s.):

3485 *Odostomia* (*?Heida*) *kelseyi* Bartsch, 1912. San Diego, California.

3486 *Odostomia* (*Odostomia*) *cassandra* Bartsch, 1912. Southwest Alaska to British Columbia. (Synonym: *O.* (*Evalea*) *cassandra* Dall and Bartsch, 1913.) See J. X. Corgan, *The Nautilus*, vol. 83, p. 72.

Subgenus *Chrysallida* Carpenter, 1857

Odostomias having strong axial ribs crossed by equally strong spiral keels between the sutures, thus forming small nodules. Base with strong spiral cords or threads, but weak axial sculpture. Type: *torrita* Dall and Bartsch, 1909 (*communis* Carpenter 1857, not C. B. Adams).

Odostomia seminuda (C. B. Adams, 1837)   3487
Half-smooth Odostome

Nova Scotia to Florida to Texas.

4 to 5 mm., milk-white. 2 nuclear whorls, smooth, half-obliquely immersed in the whorl below. Suture channeled. Upper whorls with 4 strong spiral cords bearing squarish nodules. Last whorl with 5 or 6 spiral cords, the upper 2 or 3 cut into nodules by deep incised axial lines. Base with 3 or 4 spiral incised lines. Aperture slightly flaring below. Upper part of columella with an oblique fold which is stronger within the aperture. Common; shallow water. *O. granatina* Dall, 1883, is a synonym. *O. willisi* Bartsch, 1909, from Prince Edward Island, 2.3 mm. long, is probably a dwarf *seminuda*.

Feeds on snails (*Crepidula*) and bivalves (Robertson, 1957, *The Nautilus*, vol. 70, p. 96). Do not confuse with *diantho-phila* and *dux*.

Odostomia bushiana Bartsch, 1909   3488
Bush's Odostome

Cape Cod, Massachusetts, to Rhode Island; Texas.

2 to 3 mm., chalk-white. Nuclear whorl completely immersed in the following whorl. Suture channeled. Upper whorls with 3 or 4 spiral, nodulated or beaded cords. The spiral square grooves between these cords bear 20 to 26 axial riblets. Base with 5 spiral cords. Columella strong, curved and with a stout oblique fold above. Rare; 10 fathoms.

Odostomia dianthophila Wells and Wells, 1961   3489
Serpulid Odostome

Southern Massachusetts to North Carolina.

White, minute, only 1 to 1.8 mm., ovately fusiform, like *seminuda*, but with 5 (not 4) spiral rows of tubercles; axial ribs may extend on to the base. Aperture somewhat constricted and the outer lip sinuate. Nuclear whorl immersed. Feeds

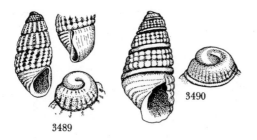

3489

on and lives in the serpulid annelid, *Hydroides dianthus* Verrill, shore to 6 fathoms. *The Nautilus*, vol. 74, p. 152, vol. 81, p. iii.

Odostomia dux Dall and Bartsch, 1906   3490
Dux Odostome

Massachusetts to North Carolina.

White, 3 mm., 5 whorls, similar to *seminuda*, but with 3 spiral rows of tubercles (not 4) and smooth spiral keel just above the suture. Nuclear whorls deeply immersed. Uncommon; subtidal to 22 fathoms. Japan records are erroneous. See *The Nautilus*, vol. 74, p. 151.

Odostomia helga Dall and Bartsch, 1909   3491
Helga's Odostome

Redondo Beach, California, to Baja California.

3 to 4.5 mm., milk-white; upper whorls with 4 or 5 spiral, flat-topped cords of unequal size, the uppermost one being weakly nodulated. Suture well-impressed. Last whorl and base with about a dozen spiral incised lines. Common; shore to 25 fathoms. Sometimes found on *Haliotis*.

Odostomia astricta Dall and Bartsch, 1907   3492
Latticed Odostome

Monterey, California, to Baja California.

3 to 4 mm., elongate-conic, white; nuclear whorls obliquely immersed. Upper whorls covered with a lattice work of small rounded beads connected by 4 or 5 spiral threads and 16 to 26 axial riblets. Base with 5 to 7 narrow, spiral cords, between which are numerous, microscopic axial threads. Aperture oval. Outer lip moderately strong. Columella twisted and provided

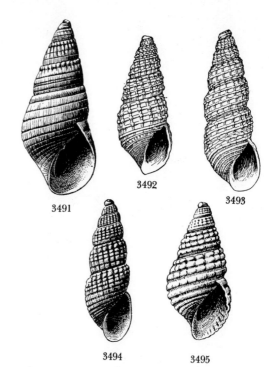

3491   3492   3493

3494   3495

above by a fairly strong fold. Parietal wall with a strong to weak callus. A variable species with several synonyms: *cooperi* and *montereyensis*, both Dall and Bartsch, 1907, and *lucca* and *oldroydi* Dall and Bartsch, 1909. Common; shore to 25 fathoms.

Odostomia oregonensis Dall and Bartsch, 1907   3493
Oregon Odostome

British Columbia to central California.

3 to 4 mm., slender, translucent whitish, similar to *astricta*. Nuclear whorls immersed except for the last one which has 3 spiral reels with axial threads between them. Postnuclear whorls not increasing rapidly, and with 4 to 7 spiral rounded threads crossed by 16 to 22 axial threads, the intersection of which form small beads or nodules. This replaces *astricta* north of Monterey Bay. Moderately common; shore to 15 fathoms.

Odostomia virginalis Dall and Bartsch, 1909   3494
Virginal Odostome

Redondo Beach, California, to Baja California.

3 mm., semitranslucent, slender, similar in shape to *oregonensis*, with a high spire. The sculpture is not openly latticed but rather of closely packed beads or squarish nodules formed by 4 to 6 incised lines crossing 18 to 20 axial ribs per whorl. Nodules strongest at the shoulders. Base with 4 spiral cords and about 7 incised spiral grooves, within which are microscopic axial threads. Columella slender, with a deep-seated fold at the top. Common shore to 25 fathoms.

Odostomia cincta (Carpenter, 1864)   3495
Santa Barbara to San Diego, California.

3 mm., ovate, shiny white; whorls slightly flattened. With 18 to 20 axial riblets per whorl crossed by 4 spiral grooves which form nodules on the ribs. Nodules becoming obsolete on the spiral cords on the base. Suture quite well-channeled. Aperture ovate, flaring slightly at the base. Fold on upper part of columella weak. Common; shallow water. *O. vincta* Dall and Bartsch, 1909, is probably only a variety.

Other Atlantic *Chrysallida:*

3496 *Odostomia toyatani* Henderson and Bartsch, 1914. Chinco-teague Bay, Virginia, to Beaufort, North Carolina.

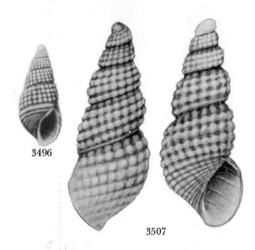

3496

3507

3497 *Odostomia jadisi* Olsson and McGinty, 1958. Atlantic Panama to Brazil. *Bull. Amer. Paleontol.,* vol. 39, p. 43.

Other Pacific *Chrysallida:*

3498 *Odostomia catalinensis* Bartsch, 1927. On abalones, Catalina Island, California.

3499 *Odostomia clementensis* Bartsch, 1927. Monterey Bay, 5 to 15 fms., to San Clemente Island, California.

3500 *Odostomia clementina* Dall and Bartsch, 1909. Santa Cruz to South Coronado Island, California.

3501 *Odostomia cumshewaensis* Bartsch, 1921. Cumshewa Inlet, British Columbia. Proc. Biol. Soc. Wash., vol. 34, p. 34.

3502 *Odostomia dicella* Bartsch, 1912. La Jolla to San Diego, California.

3503 *Odostomia eugena* Dall and Bartsch, 1909. Redondo Beach, California, to San Hipolite Point, Baja California.

3504 *Odostomia euglypta* E. Jordan, 1920. Crescent City. to Trinidad, California.

3505 *Odostomia fia* Bartsch, 1927. Todos Santos Bay, Baja California.

3506 *Odostomia heterocincta* Bartsch, 1912. San Diego, California.

3507 *Odostomia ornatissima* (Haas, 1943). Monterey, California. *Zool. Series Field Mus. Nat. Hist.,* vol. 29, p. 4. 2.7 mm. White.

3508 *Odostomia promeces* Dall and Bartsch, 1909. Baja California.

3509 *Odostomia pulcherrima* Dall and Bartsch, 1909. San Pedro to San Diego, California. 5 mm.

3510 *Odostomia pulcia* Dall and Bartsch, 1909. San Pedro, California, to San Martin Island, Baja California. On *Homalopoma bacula* Carpenter. 3 mm.

3511 *Odostomia ritteri* Dall and Bartsch, 1909. Catalina Island to San Diego, California, 25 to 40 fms. 4.5 mm.

3512 *Odostomia sanctorum* Dall and Bartsch, 1909. Baja California.

3513 *Odostomia sapia* Dall and Bartsch, 1909. San Diego, California.

3514 *Odostomia thalia* Bartsch, 1912. Coronado Islands.

3515 *Odostomia trachis* Dall and Bartsch, 1909. Los Angeles County to Baja California. May be form of *cincta* Carpenter. Common; shore to 35 fms.

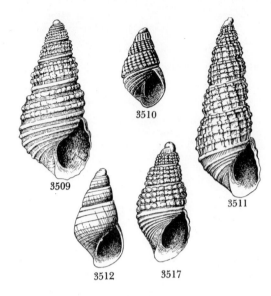

3509

3510

3511

3512   3517

3516 *Odostomia tremperi* Bartsch, 1927. San Clemente Island, California.

3517 *Odostomia vicola* Dall and Bartsch, 1909. Monterey to San Pedro, California. Above species by Dall and Bartsch, 1909 and 1907 appeared in *Proc. U. S. Nat. Mus.,* vol. 33 and Bull. 68, *U. S. Nat. Mus.,* respectively.

3518 *Odostomia acrybia* Dall and Bartsch, 1909. Point Abreojos, Baja California.

3519 *Odostomia audoax* Baker, Hanna and Strong, 1928. Baja California.

3520 *Odostomia contrerasi* Baker, Hanna and Strong, 1928. Gulf of California.

3521 *Odostomia vizcainoana* Baker, Hanna and Strong, 1928. Gulf of California.

Subgenus *Menestho* Möller, 1842

Odostomias lacking strong axial riblets and having few to many spiral cords. Type: *Turbo albulus* Fabricius, 1780.

***Odostomia impressa* (Say, 1821)**   **3522**
Impressed Odostome

Massachusetts to the Gulf of Mexico.

4 to 6 mm., elongate-conic, milk-white, 6 to 7 whorls almost flat-sided, and bearing 4 spiral, smooth-topped, strong cords; the grooves between are crossed by crowded microscopic, axial threads. Nuclear whorls small, obliquely immersed in the following turns. Base with 5 or 6 weaker spiral cords. Aperture elongate-ovate, roundly pointed below. Upper part of columella has a strong oblique fold. Moderately common; shallow water. Feeds on the various mollusks and the ascidian sea squirt, *Molgula* (J. F. Allen, 1958, *The Nautilus,* vol. 72, p. 14). This species normally lives but one year, being spawned the first summer, and spawning and dying the second (H. W. Wells, 1959, *The Nautilus,* vol. 72, p. 140).

***Odostomia trifida* (Totten, 1834)**   **3523**
Three-toothed Odostome

Maine to New Jersey.

4 to 5 mm., shiny-white. Whorls with 3 or 4 spiral grooves of various widths. Suture slightly impressed. Base with about 10 or 12 spiral scratches. Grooves have axial threads in them. Columella fold present. Synonyms: *bedequensis* Bartsch, 1909;

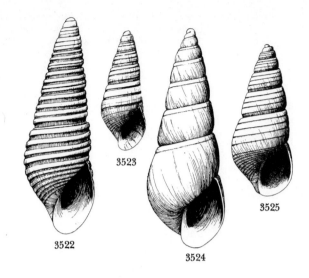

3523

3525

3522

3524

insculpta DeKay, 1847. Common; shallow water. May be form of bisuturalis.

Odostomia bisuturalis (Say, 1821)   3524
Two-sutured Odostome

Gulf of St. Lawrence to Delaware.

4 to 6 mm., elongate-conic, milk-white. 2 nuclear whorls, helicoid, half obliquely immersed. Postnuclear whorls smooth, with 1 deeply incised channel below the impressed suture. Microscopic wavy spiral scratches present. Anterior end of aperture lip free. Oblique columella fold well within aperture. *O. ovilensis* Bartsch, 1909, is a synonym. Common; below mean low tide clinging to stones. Feeds on oysters and herbivorous gastropods. (A H. Scheltema, 1965, *The Nautilus*, vol. 79, p. 7.)

Odostomia fetella Dall and Bartsch, 1909   3525
Fetella Odostome

Santa Monica Bay, California, to Baja California.

4 to 5 mm., elongate-conic, milk-white. Nuclear whorls obliquely ⅔ immersed. Postnuclear whorls moderately rounded and with 3 or 4 deep, narrow, incised grooves between sutures. Base with about 10 fine spiral threads. Very similar to the Atlantic *trifida*, and may have been introduced, but only careful observations on living specimens will decide this. Found on oyster beds. Commonest *Menestho* on the West Coast.

Other Atlantic *Menestho:*

3526 *Odostomia bruneri* Verrill, 1882. Off Rhode Island, 487 fms.

3527 *Odostomia beauforti* Jacot, 1921. Piver's Island, Beaufort, North Carolina. *Jour. Elisha Mitchell Sci. Soc.*, vol. 36, p. 129.

3528 *Odostomia sulcosa* (Mighels, 1843). Georges Bank and off Martha's Vineyard, Massachusetts, 45 to 365 fms. (Synonyms: *O. sulcata* Verrill, 1880, non A. Adams, 1860; *morseana* Bartsch, 1909, *Proc. Boston Soc. Nat. Hist.*, vol. 34, p. 104.)

3528

Other Pacific *Menestho:*

3529 *Odostomia amilda* Dall and Bartsch, 1909. San Diego to Baja California. 2.6 mm.

3530 *Odostomia enora* Dall and Bartsch, 1909. San Pedro, California. 2.8 mm.

3531 *Odostomia exara* Dall and Bartsch, 1907. Monterey to San Diego, California, 5 to 15 fms. 4 mm.

3532 *Odostomia excisa* Bartsch, 1912. Catalina Island, California. 4. mm.

3533 *Odostomia farma* Dall and Bartsch, 1909. Catalina Island to La Jolla, California. 2.4 mm.

3534 *Odostomia gloriosa* Bartsch, 1912. San Diego, California, to Baja California. 3.1 mm.

3535 *Odostomia harfordensis* Dall and Bartsch, 1907. Port Harford, California. 3.2 mm.

3536 *Odostomia hypocurta* Dall and Bartsch, 1909. Bristol Bay, Bering Sea, Alaska, 33 fms. 4.3 mm.

3537 *Odostomia pharcida* Dall and Bartsch, 1907. Queen Charlotte Islands, British Columbia. 2.2 mm.

3538 *Odostomia aequisculpta* Carpenter, 1864. San Diego, California, to Baja California.

3539 *Odostomia churchi* A. G. Smith and M. Gordon, 1948. Off Cabrillo Point, Monterey Bay, California. 1.8 mm. *Proc. Calif. Acad. Sci.*, series 4, vol. 26, p. 224.

3540 *Odostomia ciguatonis* Strong, 1949. Gulf of California. Bull. 48, So. Calif. Acad. Sci., pt. 2.

3541 *Odostomia grijalvae* and *navarettei* (3542) Baker, Hanna and Strong, 1928. Both Gulf of California.

Subgenus *Evalea* A. Adams, 1860

Odostomias smoothish, except for numerous, microscopic, spiral incised lines. Type: *elegans* A. Adams, 1860.

Odostomia angularis Dall and Bartsch, 1907   3543

Sitka, Alaska, to San Diego, California.

3 to 6 mm., elongate-conic, white. Whorls with microscopic, slender, wavy, subequal, spiral striations, of which about 33 occur upon the last whorl between the summit and the periphery. Periphery of the last whorl marked by a slender raised keel, decidedly angulated. Common; shore to 35 fathoms. A variable species; some of the synonyms are *minutissima*, *notilla* and *raymondi* all Dall and Bartsch, 1909.

Odostomia gravida Gould, 1852   3544

Monterey, California, to Baja California.

3 to 6 mm., white to bluish white, similar to *angularis*, but the somewhat angulate periphery of the last whorl lacks a raised thread. Rarely with chinklike umbilicus. Moderately common; shore to 30 fathoms. Has been found on *Mytilus* on wharf pilings. Synonyms include: *californica, donilla* both Dall and Bartsch, 1909; *io* Dall and Bartsch, 1903.

Odostomia tenuisculpta Carpenter, 1864   3545
Fine-sculptured Odostome

Alaska to Baja California.

3 to 6 mm., elongate-ovate, the last whorl being ¾ the size of the entire shell. Color yellow to white. Base gently rounded. Aperture fairly large and ovate. Surface with numerous, microscopic spiral striations. Similar to *gravida* and *angularis* but lacks the peripheral angulation and spiral keel. Some-

times quite smooth. Shape variable and malformed specimens not uncommon. Common; shore to 30 fathoms. Commonly found on live oysters and abalones. It has many synonyms, most of them created by Bartsch who described specimens: *straminea* Carpenter, 1865 (smoothish form); *sitkaensis* Clessin, 1900; *tillamookensis, jewetti, deliciosa, phanea, valdezi* all Dall

and Bartsch, 1907, *Proc. U.S. Nat. Mus.*, vol. 33; *socorroensis, clesseni, kadiakensis, amchitkana, stephensae, hagemeisteri, obesa, santarosana, baranoffensis* all Dall and Bartsch, 1909, U.S. Nat. Mus. Bull. 68.

Other Atlantic *Evalea:*

3546 *Odostomia virginica* Henderson and Bartsch, 1914. Chincoteague Island, Virginia.

3547 *Odostomia fernandina* Bartsch in Dall, 1927. Off Fernandina, Florida, 294 fms. Also see species 3645.

3548 *Odostomia ryclea* Bartsch in Dall, 1927. Off Georgia, 440 fms.

3549 *Odostomia ryalea* Bartsch in Dall, 1927. Off Georgia, 440 fms.

3550 *Odostomia pocahontasae* Henderson and Bartsch, 1914. Chincoteague Bay, Virginia. 2.4 mm.

3546   3550

Other Pacific *Evalea:*

3551 *Odostomia aleutica* (Alaska), *altina* (**3552**) (San Diego), *capitana* (**3553**) (Alaska), *esilda* (**3554**) (off San Diego), *herilda* (**3555**) (San Diego), *killisnooensis* (**3556**) (Alaska), *movilla* (**3557**) (off San Diego), *nemo* (**3558**) (San Diego), *nunivakensis* (**3559**) (Bering Sea), *phanella* (**3560**) (San Pedro), *pratoma* (**3561**) (off Santa Rosa), *profundicola* (**3562**) (off San Diego), *resina* (**3563**) (Arch Beach), *septentrionalis* (**3564**) (Unalaska), *unalaskensis* (**3565**) (Unalaska) all Dall and Bartsch, 1909.

3566 *Odostomia columbiana* (Alaska to Washington), *tacomaensis* (**3567**) (Tacoma, Washington) both Dall and Bartsch, 1907.

3568 *Odostomia atossa* Dall, 1908. San Pedro, California. 6.5 mm.

3569 *Odostomia inflata* Carpenter, 1864. Alaska to Monterey, California. 6.2 mm.

3570 *Odostomia baldridgae* (San Pedro), *calcarella* (**3571**) (off Santa Rosa), *callimene* (**3572**) (San Pedro), *calliope* (**3573**) (off La Jolla), *cookeana* (**3574**) (Alaska), *hypatia* (**3575**) (British Columbia), *skidegatensis* (**3576**) (Alaska to California), *thea* (**3577**) (San Pedro) all Bartsch, 1912, *Proc. U.S. Nat. Mus.*, vol. 42.

3578 *Odostomia martinensis* Strong, 1938. San Martin Island, Baja California.

3579 *Odostomia valeroi* Bartsch, 1917. Baja California.

3580 *Odostomia bachia* (San Clemente Island), *eyerdami* (**3581**) (Alaska) *strongi* (**3582**) (Catalina Island), *whitei* (**3583**) (California), *chinooki* (**3584**) (Olga, Washington) all Bartsch, 1927, *Proc. U.S. Nat. Mus.*, vol. 70.

— *Odostomia cassandra* Bartsch, 1912, see under *Odostomia* (*Odostomia*).

3543   3544   3545

3551   3552   3553

3554   3555   3556

3557   3566   3567

3568   3569   3585

3585 *Odostomia barkleyensis* (Vancouver Island, British Columbia) 3 mm., *spreadboroughi* (**3586**) (Vancouver Islands, British Columbia) 3.8 mm., *vancouverensis* (**3587**) (Alaska to Vancouver) 4.7 mm., *youngi* (**3588**) (Vancouver Island, British Columbia) 6.5 mm., all Dall and Bartsch, 1910, *Canada Geol. Surv. Mem.*, vol. 14-N, no. 1143.

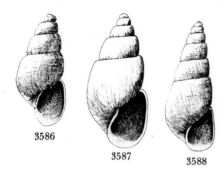

3586   3587   3588

3589   *Odostomia edmondi* E. Jordan, 1920. Trinidad, California.

3590   *Odostomia franciscana* (San Francisco Bay, California), *willetti* (**3591**) (Alaska to Queen Charlotte Islands, British Columbia), both Bartsch, 1917.

3590   3591

3592   *Odostomia quadrae* Dall and Bartsch, 1910. Drier Bay, Alaska, to Vancouver Island, British Columbia. 6.2 mm.

3593   *Odostomia cypria* Dall and Bartsch, 1912. Skidegate, Queen Charlotte Islands, British Columbia, *Proc. U.S. Nat. Mus.*, vol. 42, p. 282.

Subgenus *Amoura* de Folin, 1876

Shells large for the genus, up to 13 mm., usually inflated or ovate; sculpture of fine axial growth lines and still finer, wavy, closely placed, spiral striations. Formerly called *Amaura* Möller, 1842, non Geyer, 1837. Type of the subgenus is *A. candida* Möller, 1842.

Odostomia satura Carpenter, 1865   3594
Full Odostome

Puget Sound, Washington to San Diego, California.

6 mm., broadly conic, yellowish white, umbilicate. Base short and decidedly rounded. Growth lines coarse, suggesting axial folds. Suture well-impressed. Columella with a toothlike, oblique fold at about the middle. *O. gouldii* Carpenter, 1865, is a form having a smaller umbilicus. Uncommon; shallow water.

Odostomia nuciformis Carpenter, 1865   3595
Nut-shaped Odostome

Alaska to San Diego, California.

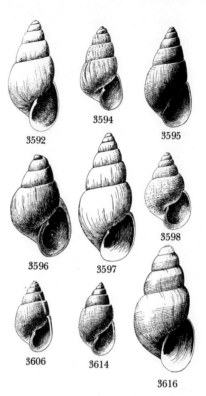

3592   3594   3595

3596   3597   3598

3606   3614

3616

6 to 8 mm., pupoid in shape with an ovoid, large last whorl. White to yellow. Well-impressed suture. Spiral sculpture of microscopic wavy striations. Common; shallow water, near roots of eel-grass. Variable in shape, hence several synonyms: *avellana* Carpenter, 1865; *orcia* Dall and Bartsch, 1909; *talpa* Dall and Bartsch, 1909; *nota* Dall and Bartsch, 1909; *pesa* Dall and Bartsch, 1909; *elsa* Dall and Bartsch, 1909.

Other Pacific *Amoura:*

3596   *Odostomia beringi* Dall, 1871. Norton Sound, Alaska. 5.7 mm.

3597   *Odostomia krausei* Clessin, 1900. Kodiak to Killisnoo, Alaska. 8 to 10 mm.

3598   *Odostomia kennerleyi* Dall and Bartsch, 1907. Afognak Island, Alaska to Nanaimo, British Columbia, to Monterey, California. 10 mm.

3599   *Odostomia lastra* (Catalina Island to San Diego), *elsa* (**3600**) (Alaska), *farallonensis* (**3601**) (Farallones Islands), *sillana* (**3602**) (Unalaska), *arctica* (**3603**) (Bering Sea), *moratora* (**3604**) (off Point Reyes, California), *iliuliukensis* (**3605**) (Unalaska), *subturrita* (**3606**) (Santa Barbara to Baja Calif.) all Dall and Bartsch, 1909, U.S. Nat. Mus. Bull. 68.

3607   *Odostomia subglobosa* (San Diego), *helena* (**3608**) (San Pedro), *grippiana* (**3609**) (Nanaimo, British Columbia), *eldorana* (**3610**) (Kodiak, Alaska) all Bartsch, 1912, *Proc. U.S. Nat. Mus.*, vol. 42.

3611   *Odostomia engbergi* (San Juan Islands, Puget Sound, Washington), *sanjuenensis* (**3612**) (Alaska to San Juan Islands), *washingtonia* (**3613**) (San Juan Islands) all Bartsch, 1920, *Jour. Wash. Acad. Sci.*, vol. 10.

3614   *Odostomia canfieldi* Dall, 1908. Barkley Sound, British Columbia, to San Diego, California.

3615   *Odostomia martensi* Dall and Bartsch, 1906. Killisnoo, Alaska.

Other Atlantic *Amoura:*

3616   *Odostomia hendersoni* Bartsch, 1909. South Cape Cod, Massachusetts. Originally placed in the subgenus *Iolaea.*

Subgenus *Iolaea* A. Adams, 1867

Shell ovately turreted, about 4 mm., umbilicate, with 3 strong, smooth spiral keels between which are numerous axial riblets. *Iole* A. Adams, 1860, non Blyth, 1844, and *Iolea* Pascoe, 1858 are synonymous. Type of *Iolaea* is *Iole scitula* A. Adams, 1860. *Iolina* Baily, 1948, was unnecessary.

Odostomia eucosmia Dall and Bartsch, 1909   3617
Graceful Odostome

Palos Verdes, California, to Baja California.

2.5 mm., milk-white, narrowly umbilicate. Nuclear whorls deeply immersed. Postnuclear whorls with 3 strong spiral keels between which are retractively slanting numerous riblets. Suture V-shaped. Base with 3 spiral cords. Aperture large. Outer lip angulated by the keels. Columella fold high and weak. Common; shallow water. Synonym: *insculpta* Carpenter, 1864, non DeKay, 1843.

Odostomia amianta Dall and Bartsch, 1907   3618

Moss Beach, California, to Baja California.

2.5 mm., yellowish white, similar to *eucosmia*, but much broader, with the last whorl being equal in length to ½ that of the entire shell. Spiral keels 3 on early whorls, 4 on last 2 whorls. Base with 8 spiral cords. Common; shore to 40 fathoms.

Other species of *Iolaea*:

3619 *Odostomia delicatula* Carpenter, 1864. Baja California. 2.3 mm.

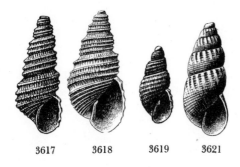

3617   3618   3619   3621

Subgenus *Evalina* Dall and Bartsch, 1904

Odostomias having feebly developed axial ribs only near the top of the whorls. Spiral sculpture of numerous fine lirations. Type: *americana* Dall and Bartsch, 1904. Two North American species:

3620 *Odostomia winkleyi* Bartsch, 1909. Massachusetts to Branford, Connecticut. 3.0 mm. Whitish, faint spiral lines crossing growth lines.

3621 *Odostomia americana* Dall and Bartsch, 1904. San Pedro, California, to Coronado Islands, 7 to 10 fathoms.

Subgenus *Ivara* Dall and Batsch, 1903

Top of the whorls strongly tabulate. Spiral sculpture of unequal fine lirations. Axial riblets feeble and only at the top of the whorls. Type: *turricula* Dall and Bartsch, 1903.

Odostomia turricula Dall and Bartsch, 1903   3622

Monterey, California, to Baja California.

4 mm., milk-white. Postnuclear whorls moderately rounded, but very broadly tabulated on the shoulders where there are very feeble axial riblets. Spiral sculpture of 7 to 9 incised grooves, sometimes punctate. Common; 3 to 25 fathoms. An alternate and rejected misspelling is *terricula*.

Other Atlantic *Ivara*:

3623 *Odostomia terryi* Olsson and McGinty, 1958. Atlantic Panama. *Bull. Amer. Paleontol.*, vol. 39, no. 177, p. 43.

Subgenus *Miralda* A. Adams, 1864

Odostomias with very strong spiral keels between the sutures and on the base. Upper keels usually nodulated. Type: *diadema* A. Adams, 1864. Synonym: *Lia* Folin, 1870; *Ividia* Dall and Bartsch, 1904.

Odostomia aepynota Dall and Bartsch, 1909   3624

Palos Verdes, California, to Baja California.

2 mm., translucent-white; nuclear whorls small, obliquely immersed in the first postnuclear whorls which have 4 spiral cords. Postnuclear whorls with 2 very strong spiral keels, the uppermost beaded, the lower smooth. Last whorl with 3 smooth spiral keels. Columella fold greatly reduced to absent. Moderately common; shore to 12 fathoms. Synonym: var. *planicosta* Baker, Hanna and Strong, 1928.

Other Pacific *Miralda*:

3625 *Odostomia hemphilli* Dall and Bartsch, 1909. San Pedro, California, to Baja California. 3.5 mm.

3626 *Odostomia porteri* Baker, Hanna and Strong, 1928. Gulf of California.

Other Atlantic *Miralda*:

3627 *Odostomia havanensis* Pilsbry and Aguayo, 1933. Southern Florida and the Caribbean. Brazil. 2 mm. *The Nautilus*, vol. 46, p. 118.

3628 *Odostomia abbotti* Olsson and McGinty, 1958. Western Caribbean. *Bull. Amer. Paleontol.*, vol. 39, no. 177.

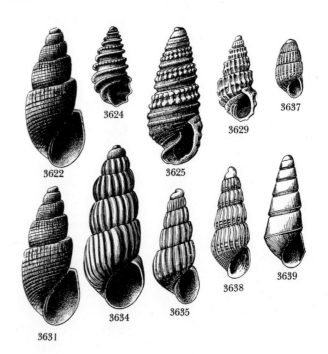

3624   3637   3629   3622   3625   3631   3634   3635   3638   3639

Subgenus *Ividella* Dall and Bartsch, 1909

Odostomias with lamellar spiral ridges and equally strong lamellar axial ribs on the spire and base. Type: *navisa* Dall and Bartsch, 1907. This is *Ividia* Dall and Bartsch, 1904; and *Funicularia* Monterosato, 1884, non Lamarck. If *Ividia* is valid, the type is *armata* Carpenter.

Odostomia navisa Dall and Bartsch, 1907   **3629**
Navisa Odostome

Monterey, California, to Baja California.

2.7 mm., translucent milk-white. Whorls shouldered, bearing 2 strong axial ribs crossed by 18 to 20 axial ribs, giving a beaded cancellated surface (with concave quadrangular depressions). Umbilicus chinklike. Suture deeply channeled. Columella fold conspicuous. A form, *delmontensis* Dall and Bartsch, 1907 (**3630**), is more elongate, and with a larger umbilicus. Not uncommon; offshore to 70 fathoms.

Other *Ividella*:

— *Odostomia pedroana* Dall and Bartsch, 1909, is now in the genus *Peristichia*.

3631 *Odostomia quinquecincta* (Carpenter, 1856). Mazatlan, Mexico. 1.8 mm.

3632 *Odostomia mendozae* Baker, Hanna and Strong, 1928. Baja California.

3633 *Odostomia ulloana* Strong, 1949. La Paz, Baja California.

Subgenus *Salassiella* Dall and Bartsch, 1909

Shells 3 to 4 mm., pupiform, whorls inflated, marked by axial ribs which extend from the top of the last whorl to the base. Varices strong, irregularly distributed. Type: *laxa* Dall and Bartsch, 1909. Resemble short, squat *Turbonilla*.

Odostomia laxa Dall and Bartsch, 1909   **3634**
Lax Odostome

Catalina Island to Baja California.

4 to 5 mm., white, resembling a stubby *Turbonilla;* whorls inflated, contracted at the sutures, and with 18 to 28 lamellar, flexuous crowded axial ribs per whorl, those on the last whorl going down to the chinklike umbilicus. Columella fold high up and weak. Uncommon; 30 to 70 fathoms.

Other *Salassiella*:

3635 *Odostomia richi* Dall and Bartsch, 1909. San Pedro, California; 3 mm.

3636 *Odostomia heathi* A. G. Smith and M. Gordon, 1948. Off Cabrillo Point, Monterey Bay, California. 2.8 mm. *Proc. Calif. Acad. Sci.*, series 4, vol. 26, p. 223.

3636a *Odostomia gabrielensis* Baker, Hanna and Strong, 1928. Gulf of California.

Subgenus *Besla* Dall and Bartsch, 1904

Very small, 2 mm., with axial ribs and 3 strong, spiral raised threads, 1 at and 2 posterior to the periphery between the sutures; base marked by raised spiral threads. Type: *convexa* Carpenter, 1856. A subgenus of dubious distinctiveness. Three known species:

3637 *Odostomia callimorpha* Dall and Bartsch, 1909. San Pedro, California. 1.5 mm.

3638 *Odostomia convexa* Carpenter, 1856. Gulf of California, 25 fms., to Mazatlan, Mexico (on *Spondylus*). 2.4 mm.

3639 *Odostomia excolpa* Bartsch, 1912. Gulf of California.

Subgenus *Eulimastoma* Bartsch, 1916

2 to 5 mm., high-spired, with a single columellar plication. Base of whorl with angulation or keel. 2 nuclear whorls, almost planorboid, partially immersed. Without axial sculpture. With or without a small umbilicus. Type: *dotella* Dall and Bartsch, 1909. Synonyms: *Telloda* Hertlein and Strong, 1951; *Parodostomia* Laseron, 1959. Two known North American species: (**3640**) *dotella* Dall and Bartsch, 1909, and (**3641**) *subdotella* Hertlein and Strong, 1951, from the tropical Eastern Pacific.

Subgenus *Syrnola* A. Adams, 1860

Odostomias with rather narrow, smooth shells with polished surfaces. Columella with a single fold. Some species are brown and have a thin brownish periostracum. Operculum very thin and paucispiral. Type: *gracillima* A. Adams, 1860. In 1892, Dall placed this as a subgenus of *Odostomia,* but later with several papers co-authored with Bartsch, he unwisely, I believe, put it under *Pyramidella*. Some of the species assigned here may be typical *Odostomia*. Others may belong to the genus *Sayella*.

Odostomia producta (C. B. Adams, 1840)   **3642**
Produced Odostome

Massachusetts Bay to New Jersey.

5 mm., similar to *Sayella fusca*, but much more slender. Periostracum dusky, horn-colored and often eroded. Apex usually eroded away. Aperture ¼ the length of the shell; ovate. Columella fold rather strong. Moderately common; intertidal. This species may be a *Sayella*.

Other Atlantic *Syrnola*:

3643 *Odostomia smithii* Verrill, 1880. South of Martha's Vineyard, Massachusetts, 85 to 146 fms.

3644 *Odostomia attenuata* (Dall, 1892). Off Rebecca Shoals, Florida, 430 fms.

3645 *Odostomia fernandina* (Bartsch in Dall, 1927). Off Georgia, 440 fms.; off Fernandina, Florida, 294 fms. Also see species 3547.

3646 *Odostomia floridana* (Bartsch in Dall, 1927). Off Fernandina, Florida, 294 fms.

3647 *Odostomia georgiana* (Bartsch in Dall, 1927). Off Georgia, 440 fms.; off Fernandina, Florida, 294 fms.

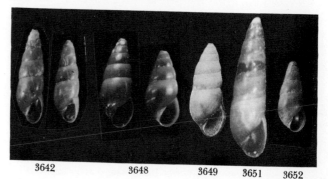

3642   3648   3649   3651   3652

Genus *Sayella* Dall, 1885

Shells elongate-pupoid, fragile, with convex whorls. Tentacles flat, triangular. Periostracum brownish. Intertidal mud flats. Type: *Leuconia hemphillii* Dall, 1884. See J. P. E. Morrison, 1939, *The Nautilus,* vol. 53, pp. 43–45.

Sayella fusca (C. B. Adams, 1839)   **3648**
Brown Sayella

Prince Edward Island to Florida.

5 mm., light-brown, translucent, elongate-conic, with the apex usually knocked or worn off. Periostracum glossy violet-brown. A whitish band appears below the suture. Whorls with irregular growth lines and microscopic spiral striations. Sutures well-marked. The preceding whorl shining through the shell at the top of each whorl lends this the aspect of having a double suture. Aperture ear-shaped, flaring anteriorly. Oblique fold present well within the aperture. Operculum paucispiral. Common; intertidal. *Pyramidella winkleyi* Bartsch, 1909, is a synonym. Also see *Odostomia producta* (C. B. Adams).

Sayella livida Rehder, 1935   **3649**
Livid Sayella

Texas.

3 to 4 mm., elongate-ovate, straw-yellow, with a rather wide, subsutural, white band. 1¼ nuclear whorls, inverted, colorless, glassy-smooth. Remaining 6 whorls convex, smoothish and with crowded microscopic spiral lines. Last whorl ½ the length of the shell. Periostome red-brown. *The Nautilus,* vol. 48, p. 129.

Other species:

3650   *Sayella hemphilli* (Dall, 1889). Cedar Keys, Florida. 4 mm.

3650

3651   *Sayella crosseana* (Dall, 1885). Egmont Key, west Florida; Texas; West Indies. 4 mm.

3652   *Sayella chesapeakea* Morrison, 1939. Chesapeake Bay, Maryland. 3 to 4 mm. Apex eroded. *The Nautilus,* vol. 53, p. 44. Mud flats.

Genus *Triptychus* Mörch, 1875

Type of this genus is *niveus* Mörch, 1875, Columella with 2 cords.

Triptychus niveus Mörch, 1875   **3653**
Three-corded Pyram

Southeast Florida to the West Indies.

6 to 8 mm., slender, pure-white, nucleus of 1 large, bulbous smooth whorl. Suture broad and indented. Spiral sculpture of 3 raised, rounded cords on the upper whorls, the 2 uppermost being nodulated, the lower smooth; last whorl with 6 cords, the lower 4 being smooth. Columella with 2 strong spiral cords. Aperture produced below. Common in shallow-water dredgings. *Pyramidella vincta* Dall, 1884, is the same. A Plio-

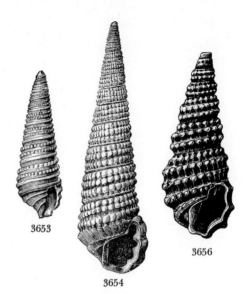

3653   3656

3654

cene relative, *T. biseriatus* (Gabb, 1874) occurs in Costa Rica fossil beds.

Genus *Peristichia* Dall, 1889

Shells elongate, small, white, spirally corded, similar to *Triptychus,* but with only 1 basal spiral cord (instead of 2) and lacking columella folds. Type: *toreta* Dall, 1889. Rehder, 1943, placed the following uncommon species in the genus:

3654   *Peristichia toreta* Dall, 1889. North Carolina to west Florida, 2 to 22 fms. 11 mm.

3655   *Peristichia agria* Dall, 1889. North Carolina to Florida Keys, 2 to 63 fms. See Rehder, *Proc. U.S. Nat. Mus.,* vol. 93, p. 195, pl. 20. 6 mm.

3655

Peristichia pedroana (Dall and Bartsch, 1909)   **3656**
San Pedro Corded Pyram

San Pedro to central Baja California.

6 to 7 mm., elongate, chocolate-brown, glistening, with beaded cancellate sculpturing caused by 3 spiral cords between well-impressed sutures and by slightly slanting riblets (14 to 24 per whorl). Irregular varices may be present. Columella strong, almost straight. Moderately common; in gravel in subtidal shallow water.

Other species:

3657   *Peristichia hermosa* (Lowe, 1935). San Felipe, Baja California.

Genus *Cingulina* A. Adams, 1860

Shell small, with spiral cords or threads; axial sculpture of microscopic riblets. Varices absent. Type: *cingulata* Dunker.

Cingulina babylonia (C. B. Adams, 1845)   3658
Babylon Pyram

Bermuda and the West Indies.

2 mm., dull-white, nuclear whorls smooth, remaining 4 with 3 or 4 prominent spiral cords on the body whorl and 2 similar cords on the above whorls. Space between cords concave. *Odostomia judithae* Usticke, 1959, is this species. Moderately common in offshore dredgings.

Other species:

3659 *Cingulina evermanni* Baker, Hanna and Strong, 1928. Gulf of California.

3660 *Cingulina urdeneta* (Bartsch, 1917). Gulf of California.

Subfamily TURBONILLINAE Simroth, 1907

Genus *Turbonilla* Risso, 1826

Small (less than ⅓ inch), elongate, many-whorled, with a single columellar fold sometimes not visible in the aperture. Nuclear whorl heterostrophic. Sculpture both axial riblets and spiral lirae, sometimes reduced or absent. This is an overly named group having no less than 24 subgenera, most of which are based on transient and variable characters. Many of the species described by Dall and Bartsch will probably fall into synonymy when more careful biological studies are made.

Subgenus *Turbonilla* Risso, 1826

Small and slender, bluish to milk-white, without spiral sculpture. Prominent axial ribs extend from the top of the whorl to the umbilical region. Columella straight or slightly twisted. Type: *typica* Dall and Bartsch, in Arnold, 1903.

Turbonilla nivea (Stimpson, 1851)   3661
Milky Turbonille

Maine to Connecticut.

5 to 7 mm., regularly elongate, sides of the spire straight; milk-white. 2½ nuclear whorls, helicoid, set at right angles. Postnuclear whorls moderately rounded, bearing numerous, low, rounded, protractively slanting ribs (20 to 22 on the last eighth or ninth whorl). Ribs absent on base. Uncommon; shallow water.

Turbonilla stricta Verrill, 1873   3662
Constricted Turbonille

South Massachusetts to North Carolina.

4 mm., similar to *nivea*, but only 14 to 17 riblets on the last sixth or seventh whorl. Intercostal spaces shallow and less than ½ the width of the riblets. Entire surface crossed by exceedingly fine striations. Uncommon; shallow water.

Turbonilla diegensis Dall and Bartsch, 1909   3663
San Diego Turbonille

San Pedro to San Diego, California.

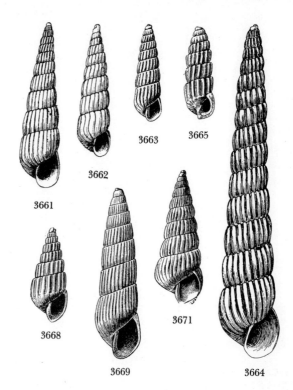

5 mm., dingy-white, 10 postnuclear whorls, rounded and somewhat overhanging, the greatest convexity being on the lower ⅓ of the exposed portion of the whorls. 18 riblets on the last whorl. Intercostal dug-out grooves extend down to the periphery but not on to the base of the shell. Moderately common; sublittoral.

Turbonilla acra Dall and Bartsch, 1909   3664
Acra Turbonille

Catalina Island to the Coronado Islands.

10 mm., long and slender, milk-white, 17 whorls. Axial ribs strong, 14 to 20 on the first 15 whorls; 30 weak irregular ones on the last whorl. Uncommon; 30 to 50 fathoms.

Other *Turbonilla* (s. s.):

3665 *Turbonilla gilli* Dall and Bartsch, 1907. Catalina Island to San Diego, California, 12 to 40 fms. 3 mm. (And variety *delmontensis* Dall and Bartsch, 1907 **(3666)**, Monterey, California, 5 to 25 fms. Rare. 3.4 mm.)

3667 *Turbonilla fackenthallae* A. G. Smith and M. Gordon, 1948. Off Cabrillo Point, California, 20 to 30 fms. *Proc. Calif. Acad. Sci.*, series 4, vol. 26, p. 220. 7.7 mm.

3668 *Turbonilla centrota* and **(3669)** *lucana* both Dall and Bartsch, 1909. Cape San Lucas, Baja California.

Subgenus *Chemnitzia* Orbigny, 1839
(in Webb and Berthelot)

Turbonillas without spiral sculpturing, having prominent axial ribs which fuse or terminate at the periphery. Intercostal areas sunken. Base smooth. Type: *campanellae* Philippi, 1836.

Turbonilla aequalis (Say, 1827)   3670
Equal Turbonille

Southern Massachusetts.

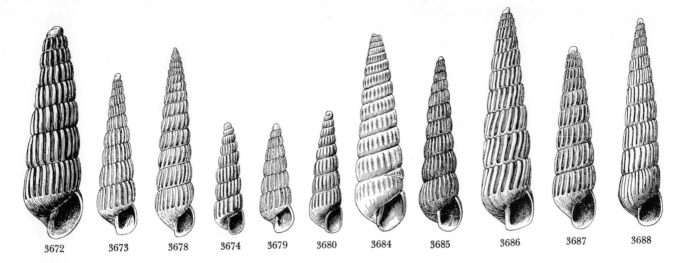

3672   3673   3678   3674   3679   3680   3684   3685   3686   3687   3688

4.5 mm., slender, with 9 somewhat flattened whorls bearing axial ribs which are a little oblique, especially at the upper ends, close to the sutures. Body whorl and base of shell smooth. Intercostal areas sunken and smooth. Aperture ovate. Rare, 6 to 8 fathoms.

Turbonilla curta Dall, 1889   3671
Short Turbonille

North Carolina to west Florida.

8 mm., not very slender, waxen-white, with 9 or 10 slightly inflated whorls. Last whorl with about 25 closely set rounded ribs, only very slightly curved. Base smooth, surface polished. Suture distinct. Columella with a faint spiral fold. Common; shore to 640 fathoms.

Turbonilla kelseyi Dall and Bartsch, 1909   3672
Kelsey's Turbonille

Santa Barbara, California, to the Gulf of California.

5 mm., white, 8 to 10 whorls with slightly sinuous, protractive, rounded axial riblets (14 to 24 per whorl). Intercostal spaces as wide as the ribs. Base smooth. Common; shore to 30 fathoms.

Turbonilla raymondi Dall and Bartsch, 1909   3673
Raymond's Turbonille

Redondo Beach to San Diego, California.

6 mm., narrow and pointed, with 11 rounded whorls bearing 18 to 22 riblets per whorl. Intercostal spaces deeply grooved and 1½ times as wide as the ribs. Suture strongly impressed and wavy. Base smooth. Moderately common.

Other *Chemnitzia:*

3674 *Turbonilla aepynota* Dall and Bartsch, 1909. San Pedro, California to San Martin Island, Mexico, tidepools to 70 fms. 3.2 mm.

3675 *Turbonilla clarinda* Bartsch, 1912. Redondo Beach to San Diego, California, 12 to 75 fms.

3676 *Turbonilla engbergi* Bartsch, 1920. Puget Sound, Washington.

3677 *Turbonilla gabbiana* Cooper, 1870. Monterey, California. (Synonyms: *gracillima* Gabb; *cayucosensis* Willett, 1929.)

3678 *Turbonilla hypolispa* Dall and Bartsch, 1909. Santa Monica, California, to San Martin Island, Mexico, 12 to 35 fms.

3679 *Turbonilla muricatoides* Dall and Bartsch, 1907. Monterey to San Pedro, California, 15 to 25 fms.

3680 *Turbonilla santarosana* Dall and Bartsch, 1909. Monterey to South Coronado Island, California, 7 to 53 fms.

3681 *Turbonilla amortajadensis* Baker, Hanna and Strong, 1928. Gulf of California.

Subgenus *Strioturbonilla* Sacco, 1892

Ribbed turbonillas with finely and closely packed spiral striations on the spire and base. Type: *alpina* Sacco, 1892. White to yellowish. About 40 North American species described.

Turbonilla hemphilli Bush, 1899   3682
Hemphill's Turbonille

West coast of Florida

10 to 12 mm., slender, buff to pale-brown with a waxy luster, often with a wide, dark band below the suture. Apex acute with 2 obliquely tilted nuclear whorls. 12 to 14 postnuclear whorls, only slightly convex, flattened at the pe-

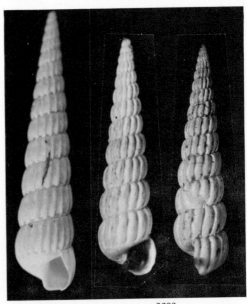

3682   3683

riphery. Suture distinct. Axial ribs 18 or 19 with the intercostal spaces about the width of the ribs; spirally striate.

Turbonilla dalli Bush, 1899   3683
Dall's Turbonille

North Carolina to both sides of Florida.

8 to 9 mm., bluish white, translucent with a dull luster. Suture deep. 12 whorls, very convex. About 16 ribs per whorl, opaque, very strong, a little oblique. Wide spaces between the ribs are concave, with squarish ends just above the suture. Surface covered with fine microscopic striae. Common; shore to several fathoms.

Turbonilla belotheca Dall, 1889   3684
Belotheca Turbonille

South half of Florida; West Indies.

12 to 14 mm., polished-ivory in color with a faint broad yellowish band above the periphery (or all yellow-brown). 15 whorls with a slightly malleated aspect. 20 ribs per whorl, more distinct on the upper whorls. The ribs fade out at their ends, thus not reaching the sutures. Spiral striations neither prominent nor constant. Columellar fold distinct. Uncommon; 20 to 95 fathoms.

Turbonilla stylina (Carpenter, 1865)   3685
Many-named Turbonille

Monterey, California, to the Gulf of California.

4 to 8 mm., slender, white, well-impressed sutures with 9 to 13 postnuclear, somewhat flat-sided whorls bearing 16 to 25 slightly slanting, rounded ribs which on the last whorl descend only to the periphery. Intercostal spaces well-impressed and as wide as the ribs. A very variable, widely distributed and common species, usually identified as *buttoni* by Western collectors. The number, strength and size of the ribs varies, as does the size and proportion of the whole shell. Synonyms include: *buttoni* (**3686**), *asser, profundicola, humerosa, nicholsi, calvini, galianoi* all Dall and Bartsch, 1909, Bull. 68, U.S. Nat. Mus.

Turbonilla torquata (Gould, 1852)   3687
Vancouver Turbonille

Alaska to San Diego, California.

6 to 9 mm., not as slender as *stylina*, white; 10 postnuclear whorls which have their greatest convexity on the lower ½. Axial ribs 10 to 18 per whorl. The ribs terminate well above the suture leaving a smooth band around the whorl above the suture. Entire surface marked with faint, wavy, spiral striations. Intercostal spaces deep and narrower than the ribs. Common; 2 to 75 fathoms. *T. vancouverensis* (Baird, 1863) is a synonym.

Turbonilla attrita Dall and Bartsch, 1909   3688
Attrita Turbonille

San Pedro, California, to Baja California.

7 to 8 mm., slender, characterized by high, flat-sided whorls bearing closely set, feeble ribs (18 to 22 on the first 9 postnuclear whorls). Last whorl with feeble to obsolete ribs. Intercostal spaces shallow and ½ the width of the ribs. Entire surface with spiral striations. Common; from pilings to 35 fathoms.

Other Atlantic *Strioturbonilla:*

3689 *Turbonilla bushiana* Verrill, 1882. Off Georges Bank to Long Island, New York, 365 to 1525 fms. (Synonym: *formosa* Verrill and Bush, 1880, non Klipst., 1856.) 11 mm. (var. *abyssicola* Bartsch, 1909). 13 mm.

3690 *Turbonilla nema, pyrrha* (**3691**), *theona* (**3692**), *electra* (**3693**), *rhea* (**3694**), *sirena* (**3695**), *leta* (**3696**), *myia* (**3697**), *enna* (**3698**), *idothea* (**3699**), *nonica* (**3700**) all Bartsch in Dall, 1927, *Proc. U.S. Nat. Mus.,* vol. 70, all off Fernandina, Florida, and Georgia, 294 to 440 fms. [*enna* is a homonym of species 3865].

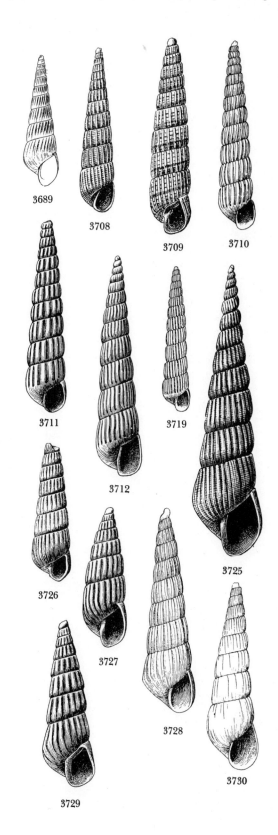

3689

3708

3709

3710

3711

3719

3712

3725

3726

3727

3728

3729

3730

3701 *Turbonilla exilis* (C. B. Adams, 1850). North Carolina to the West Indies, 3 to 63 fms.

3702 *Turbonilla perlepida* Verrill, 1885. Off Chesapeake Bay, 70 fms.

3703 *Turbonilla grandis* Verrill, 1885. Off Virginia, 1,582 fms.

3704 *Turbonilla virgata* Dall, 1892. North Carolina to Florida, 12 to 80 fms.

3705 *Turbonilla louiseae* A. H. Clarke, 1954. Off Cape Ann, Massachusetts. From haddock stomach, 52 fms. *The Nautilus*, vol. 67, p. 118.

3706 *Turbonilla textilis* (Kurtz, 1860). Fort Johnson, South Carolina.

3707 *Turbonilla pusilla* (C. B. Adams, 1850). North Carolina to the West Indies, 16 to 294 fms. Brazil.

3708 *Turbonilla insularis* Dall and Simpson, 1901. Mayaguez, Puerto Rico. 7 mm.

3709 *Turbonilla portoricana* Dall and Simpson, 1901. Mayaguez, Puerto Rico. 5 mm.

Other Pacific *Strioturbonilla:*

3710 *Turbonilla aresta* (Santa Rosa Island to San Diego); **(3711)** *simpsoni* (Redondo Beach to San Diego), **(3712)** *carpenteri* (Redondo Beach to Coronado Islands) all Dall and Bartsch, 1909, Bull. 68, U.S. Nat. Mus.

3713 *Turbonilla dinora* (San Diego); **(3714)** *encella* (San Pedro); **(3715)** *bakeri* (Redondo Beach to Baja Calif.); **(3716)** *dracona* (San Diego); **(3717)** *cookeana* (Gulf of California) all Bartsch, 1912, *Proc. U.S. Nat. Mus.*, vol. 42.

3718 *Turbonilla canadensis* Bartsch, 1917. Forrester Island, Alaska.

3719 *Turbonilla serrae* Dall and Bartsch, 1907. Monterey, California, to Coronado Islands.

3720 *Turbonilla barcleyensis* Bartsch, 1917. Barcley Sound, British Columbia.

3721 *Turbonilla kincaidi* Bartsch, 1921. Puget Sound, Washington. *Proc. Biol. Soc. Wash.*, vol. 34, p. 33.

3722 *Turbonilla asuncionis* Strong, 1949. Baja California. Bull. 48, So. Calif. Acad. Sci., pt. 2.

3723 *Turbonilla chalcana* and **(3723a)** *nahuana* Baker, Hanna and Strong, 1928. Gulf of California.

3724 *Turbonilla schmitti* Bartsch, 1917. Pt. Abreojos, Baja California. 6.3 mm.

3718   3720   3724

Subgenus *Pyrgolampros* Sacco, 1892

Brown to yellow turbonillas with low, broad axial ribs that gradually disappear over the periphery and base of the

last whorl. Spiral striations present. Surface covered with a thin periostracum. Intercostal spaces not grooved out or sunken. Type: *mioperplicatulus* Sacco, 1892. This subgenus has not been recognized among the Atlantic coast species of *Turbonilla*.

Turbonilla chocolata (Carpenter, 1865)   **3725**
Chocolate Turbonille

Redondo Beach to San Diego, California.

8 to 14 mm., pointed-elongate, wax-yellow to dark golden-brown, variously banded. 2½ nuclear whorls, set at right angles, and usually not immersed. Long axial ribs slightly slanting retractively and extending over the base of the last whorl where they fade out. Ribs on last whorl may become irregular and enfeebled. Ribs per whorl 20 to 28. Wavy, fine, spiral striations present. A variable species having several synonyms: *halia, painei, keepi* all Dall and Bartsch, 1909; *lowei, pedroana* Dall and Bartsch, 1903; *berryi* Dall and Bartsch, 1907. Common; 5 to 20 fathoms.

Turbonilla aurantia (Carpenter, 1865)   **3726**
Golden Turbonille

British Columbia to Monterey, California.

5 to 7 mm., rather stubby, the sides of the whorls only slightly convex. Suture very well-indented, sometimes crenulated by the ribs below. Color golden-yellow, light-brown to reddish brown. 20 to 22 broad ribs per whorl, becoming obsolete as they pass over the base of the shell. Spiral striae present. Synonyms include: *valdezi* and *victoriana* Dall and Bartsch, 1907; *gouldi* Dall and Bartsch, 1909. Common; shore to 20 fathoms.

Other *Pyrgolampros:*

3727 *Turbonilla ridgwayi* (San Diego); *halibrecta* **(3728)** (Catalina Island); *alaskana* **(3729)** (Kodiak to Sitka); *halistrepta* **(3730)** (Newport, California), *lituyana* **(3731)** (Alaska) all Dall and Bartsch, 1909, Bull. 68, U. S. Nat. Mus.

3732 *Turbonilla newcombei* (British Columbia), *taylori* **(3733)** (British Columbia), *lyalli* **(3734)** (British Columbia), *oregonensis* **(3735)** (Washington and Oregon) all Dall and Bartsch, 1907, *Proc. U. S. Nat. Mus.*, vol. 33.

3736 *Turbonilla talma* (Vancouver Island) 9 mm., *macouni* **(3737)** (Vancouver Island) 14 mm., *pesa* **(3738)** (Vancouver Island) 6 mm., *rinella* **(3739)** (Vancouver Island) 8 mm. all Dall and Bartsch, 1910, *Canada Geol. Surv., Memoir*, vol. 14-N.

3740 *Turbonilla pugetensis* (Puget Sound), *tremperi* **(3741)** (San Diego Bay), *franciscana* **(3742)** (San Francisco Bay) all Bartsch, 1917. *Proc. U. S. Nat. Mus.*, vol. 52.

3743 *Turbonilla gloriosa* Bartsch, 1912. Off San Diego, California, in 12 fms.

3744 *Turbonilla shuyakensis* Bartsch, 1927. Afognak Island, Alaska. *Proc. U. S. Nat. Mus.*, vol. 70.

3745 *Turbonilla stelleri* Bartsch, 1927. Afognak Island, Alaska. 8.3 mm.

3746 *Turbonilla strongi* Willett, 1931. Catalina Island, California, 25 fms. *The Nautilus*, vol. 45.

3747 *Turbonilla ilfa* Bartsch, 1927. San Pedro, California. 12.5 mm.

3748 *Turbonilla middendorfii* and **(3748a)** *eyerdami* Bartsch, 1927. Both from Afognak Island, Alaska.

3749 *Turbonilla stillmani* A. G. Smith and M. Gordon, 1948. Monterey Bay, California. 3.5 mm. *Proc. Calif. Acad. Sci.*, series 4, vol. 26, p. 221.

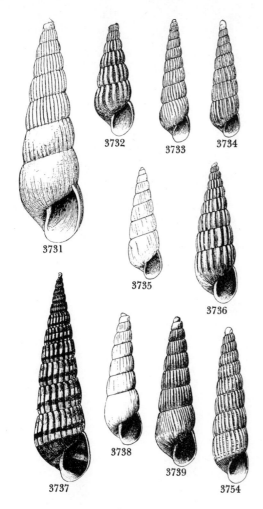

3732 3733 3734

3731

3735

3736

3737 3738 3739 3754

3750 *Turbonilla willetti* A. G. Smith and M. Gordon, 1948. Monterey Bay, California. 5.8 mm.

3751 *Turbonilla francisquitana;* **(3752)** *gonzagensis;* and **(3753)** *pazensis* all Baker, Hanna and Strong, 1928. Gulf of California.

Subgenus *Pyrgiscus* Philippi, 1841

Turbonillas having strong spiral incised grooves as well as axial ribs. Summits of the whorls not strongly shouldered. Type: *Melania rufa* Philippi, 1841. About 75 so-called species have been described from North American water, 42 of these from the Atlantic coast. Have no varices or lirations on the inner lip (see subgenus *Mormula*).

Turbonilla interrupta (Totten, 1835)   3754
Interrupted Turbonille

Gulf of St. Lawrence, Canada, to the West Indies. Brazil.

6 to 8 mm., slender, wax-yellow. 2½ nuclear whorls, depressed helicoid, set at right angles. 9 to 11 postnuclear whorls, slightly convex, with 20 to 28 broad, low axial ribs (the intercostal spaces being not as wide). Spiral incised lines cross the intercostal spaces, 4 to 8 on the early whorls, 8 to 14 on the remaining whorls. Rarely with 30 to 38 enfeebled ribs on the last whorl. Base with the disappearing ends of the axial riblets and with 12 to 15 microscopic spiral striations. A common shallow-water species which is highly variable and has several synonyms: *winkleyi* Bartsch, 1909; *senilis* Bartsch, 1909; *buteonis* Bartsch, 1909; *areolata* Verrill, 1873; *vineae* Bartsch, 1909; *cascoensis* Bartsch, 1909; *fulvocincta* Jeffreys, 1884.

3740   3741   3742

Turbonilla conradi Bush, 1899   3755
Conrad's Turbonille

West Florida.

8 to 10 mm., slender, waxen-gray to yellowish. Nucleus 2 whorls set at right angles. 12 postnuclear whorls, regularly coiling. Suture distinct, slightly undulating. 18 to 24 axial ribs, broad, rounded, slightly oblique. Intercostal spaces shallow and wider than the ribs. Spaces crossed by conspicuous incised lines which become deep grooves above the suture. Base rounded, with 3 incised spiral grooves and numerous smaller ones. Columella straight, thickened. Common; intertidal to 2 fathoms.

Turbonilla protracta Dall, 1892   3756
Drawn-out Turbonille

North Carolina to the West Indies.

4 mm., pale yellow-brown with the spiral lines darker. 6 postnuclear whorls. Sculpture of numerous, small, narrow, closely set slightly oblique riblets, extending from the suture nearly to the umbilical region; interspaces spirally striate, the stronger one appearing to be punctate. Whorls gently rounded, axially drawn out. Columella slender and twisted. Moderately common; shallow water. Also occurs in tertiary beds of the Carolinas.

Turbonilla tenuicula (Gould, 1853)   3757
Slight Turbonille

Monterey, California, to Baja California.

5 to 7 mm., somewhat stubby and inflated, varying from white to waxy-yellow to dark-brown, and sometimes spirally banded. 3 nuclear whorls, planorboid, slightly slantingly im-

3755   3759   3761   3777   3781

mersed. Postnuclear whorls slightly convex to flattened, strongly shouldered. 18 to 28 axial ribs per whorl, curved, thick and connected at the tops, which appear beaded. Ribs extend feebly over the rounded base of the last whorl. Spiral sculpture of 10 to 16 incised lines between sutures. Suture deep, subchanneled and wavy. An abundant species from shore to 25 fathoms. Synonyms: *terebralis* Carpenter, 1857; *unifasciata* Carpenter, 1857; *subcuspidata* Carpenter, 1864; *crebrifilata* Carpenter, 1864; *antemunda* Dall and Bartsch, 1909; *macra* Dall and Bartsch, 1909.

Other Atlantic *Pyrgiscus*:

3758 *Turbonilla edwardensis* Bartsch, 1909. Prince Edward Island, Canada.

3759 *Turbonilla elegantula* Verrill, 1882. Vineyard Sound to off New Haven, Connecticut. (Synonym: *elegans* Verrill, 1872, non Wood, 1842.) (Var. *branfordensis* Bartsch, 1909, Massachusetts to Connecticut.)

3760 *Turbonilla conoma* Bartsch in Dall, 1927. Off Fernandina, Florida, 294 fms.

3761 *Turbonilla mighelsi* Bartsch, 1909. Vineyard Sound and Woods Hole, Massachusetts, 1 to 5 fms. Near New Haven, Connecticut.

3762 *Turbonilla miona* Bartsch in Dall, 1927. Off Georgia, 440 fms.

3763 *Turbonilla hecuba* Dall and Bartsch, 1913. Barrington Passage, Nova Scotia, 19 fms.

3764 *Turbonilla pocahontasae* Henderson and Bartsch, 1914. Chincoteague Island, Virginia.

3765 *Turbonilla powhatani* Henderson and Bartsch, 1914. Chincoteague Island, Virginia.

3766 *Turbonilla rathbuni* Verrill and Smith, 1880. Woods Hole, Massachusetts, south of Martha's Vineyard to North Carolina, 64 to 365 fms. Off Nantucket Shoals, 250 to 547 fms. Newport, Rhode Island.

3767 *Turbonilla sumneri* Bartsch, 1909. Woods Hole, Massachusetts.

3768 *Turbonilla toyatani* Henderson and Bartsch, 1914. Chincoteague Island, Virginia.

3769 *Turbonilla verrillii* Bartsch, 1909. Martha's Vineyard and Woods Hole, Massachusetts.

3770 *Turbonilla virginica* Henderson and Bartsch, 1914. Chincoteague Island, Virginia.

3771 *Turbonilla whiteavesi* Bartsch, 1909. Shediac Bay, New Brunswick, Prince Edward Island.

3772 *Turbonilla multicostata* (C. B. Adams, 1850). North Carolina to the West Indies.

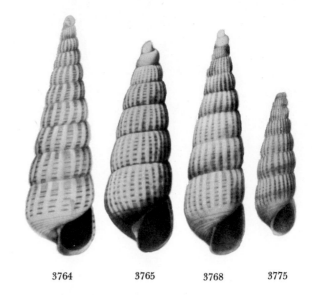

3764   3765   3768   3775

3773 *Turbonilla obeliscus* (C. B. Adams, 1850). North Carolina to the West Indies, 1 to 63 fms.

3774 *Turbonilla reticulata* (C. B. Adams, 1850). (Synonym: *T. cancellata* Holmes.) North Carolina to the West Indies.

3775 *Turbonilla puncta* (C. B. Adams, 1850). North Carolina to the West Indies, 12 to 15 fms.

3776 *Turbonilla virga* Dall, 1884. North Carolina to the Florida Keys and west Florida, 2 to 15 fms.

3777 *Turbonilla punicea* Dall, 1884. North Carolina to Florida and Gulf of Mexico, 1 to 31 fms.

3778 *Turbonilla kurtzii* Mazÿck, 1913. Sullivans Island, South Carolina.

3779 *Turbonilla textilis* (Kurtz, 1860) North Carolina to Gulf coast of Florida, 2 to 14 fms.

3780 *Turbonilla subulata* (C. B. Adams, 1850). North Carolina to the West Indies, 1 to 63 fms.

3781 *Turbonilla incisa incisa* Bush, 1899. West coast of Florida.

3782 *Turbonilla incisa constricta* Bush, 1899. West coast of Florida.

3783 *Turbonilla emertoni* Verrill, 1882. Off Martha's Vineyard, Massachusetts, 238 fms.

3784 *Turbonilla stimpsoni* Bush, 1899. South Carolina.

3785 *Turbonilla unilirata* Bush, 1899. Off Cape Hatteras, North Carolina, 16 fms.

3786 *Turbonilla laevis* (C. B. Adams, 1850). North Carolina to the West Indies, 15 to 107 fms.

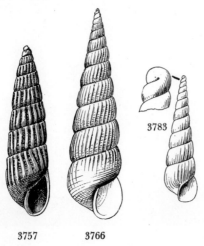

3757   3766

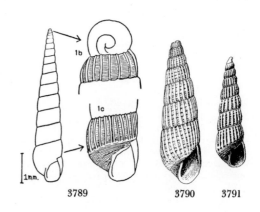

3789   3790   3791

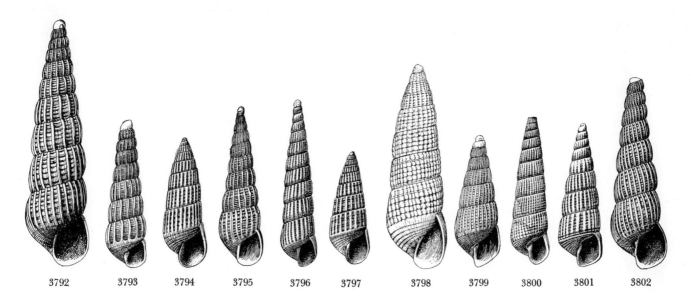

3792 3793 3794 3795 3796 3797 3798 3799 3800 3801 3802

3787 *Turbonilla viridaria* Dall, 1884. South Florida.

3788 *Turbonilla haycocki* Dall and Bartsch, 1911. Bermuda; West Indies; north Brazil, 42 fms.

3789 *Turbonilla alfredi* Abbott, 1958. Grand Cayman Island.

Other Pacific *Pyrgiscus:*

3790 *Turbonilla adusta* (Redondo to San Diego), *almo* (**3791**) (Monterey to Baja California), *aragoni* (**3792**) (Monterey to Redondo), *callia* (**3793**) (San Diego), *jewetti* (**3794**) (San Pedro to Baja California), *nereia* (**3795**) (Redondo to San Diego), *nuttingi* (**3796**) (San Pedro to San Diego), *obesa* (**3797**) (Redondo to San Diego), *pluto* (**3798**) (San Pedro to San Diego), *recta* (**3799**) (San Diego to Baja California), *signae* (**3800**) (Redondo to San Pedro), *vexativa* (**3801**) (San Pedro to San Diego), *weldi* (**3802**) (Southern California, *arata* is a synonym), *wickhami* (**3803**) (Catalina Island) all Dall and Bartsch, 1909, Bull. 68, U. S. Nat. Mus.

3804 *Turbonilla antestriata* (Esteros Bay to San Diego), *canfieldi* (**3805**) (Monterey to Baja California), *eucosmobasis* (**3806**) (Santa Barbara to San Diego), *morchi* (**3807**) (Monterey to San Diego) all Dall and Bartsch, 1907, *Proc. U. S. Nat. Mus.*, vol. 33.

3808 *Turbonilla auricoma* Dall and Bartsch, 1903. Santa Monica, California, to Baja California.

3809 *Turbonilla burchi* Gordon, 1938. Off Redondo Beach, California, 25 fms. *The Nautilus*, vol. 52.

3810 *Turbonilla callimene* (San Diego), *grippi* (**3811**) (San Diego) both Bartsch, 1912. *Proc. U. S. Nat. Mus.*, vol. 42.

3812 *Turbonilla castanea* Keep, 1888. Monterey, California, to Baja California. 10 mm.

3813 *Turbonilla castanella* Dall, 1908. Monterey to Redondo Beach California, 10 to 40 fms. 13 mm.

3814 *Turbonilla ceralva*; (**3815**) *halidoma*; (**3816**) *histias*; (**3817**) *lara*; (**3818**) *larunda*; (**3819**) *lepta*; (**3820**) *macbridei*; (**3821**) *marshalli*; (**3822**) *pequensis*; (**3823**) *sanctorum*; (**3824**) *superba* all Dall and Bartsch, 1909. All Gulf of California.

3825 *Turbonilla delmontana* Bartsch, 1937. Off Del Monte, California, 10 fms. (Synonym: *delmontensis* Bartsch, 1927, non Bartsch, 1907.) *The Nautilus*, vol. 50, p. 100.

3826 *Turbonilla dora* (San Diego); (**3827**) *eva* (San Diego); (**3828**) *ina* (Long Beach to San Diego); (**3829**) *ista* (Pt. Vincent to San Diego) all Bartsch, 1917, *Proc. U. S. Nat. Mus.*, vol. 52.

3830 *Turbonilla virgo* (Carpenter, 1864). Santa Barbara, California.

3831 *Turbonilla alarconi*; (**3832**) *aripana*; (**3833**) *cochimana*; (**3834**) *guaicurana*; (**3835**) *kaliwana*; (**3836**) *pericuana* all Strong, 1949. All Gulf of California.

3837 *Turbonilla almejasensis*; (**3838**) *baegerti*; (**3839**) *bartolomensis*; (**3840**) *cabrilloi*; (**3841**) *corsoensis*; (**3842**) *cortezi*; (**3843**) *lamna*; (**3844**) *lazaroensis*; (**3845**) *mariana*; (**3846**) *tecalo*;

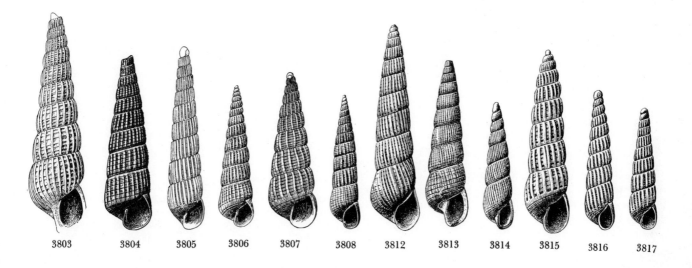

3803 3804 3805 3806 3807 3808 3812 3813 3814 3815 3816 3817

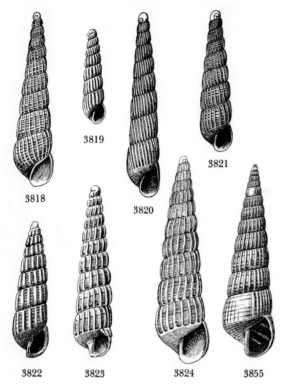

3819

3818

3821

3820

3822   3823   3824   3855

(**3847**) *ulloa* all Bartsch, 1917, all Baja California or the Gulf of California.

3848 *Turbonilla azteca;* (**3849**) *johnsoni;* (**3850**) *mayana;* (**3851**) *tolteca* all Baker, Hanna and Strong, 1928. Gulf of California.

3852 *Turbonilla domingana;* (**3853**) *vivesi;* (**3854**) *yolettae* all Hertlein and Strong, 1951. All Gulf of California.

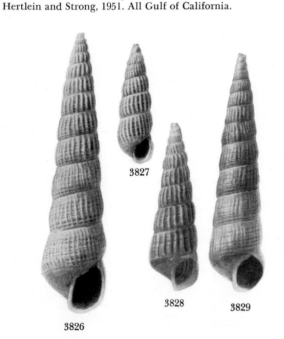

3827

3828   3829

3826

Subgenus *Mormula* A. Adams, 1864

Turbonillas with irregularly disposed varices on the outer surface, which usually mark internal lirations on the outer lip. Axial ribs and deeply incised spiral lines present. Sculpture never nodulose. Type: *rissoina* A. Adams, 1864. About 10 Pacific and 1 Atlantic species.

Turbonilla tridentata (Carpenter, 1864)   3855
Three-toothed Turbonille

Monterey, California, to Baja California.

9 to 11 mm., long and slender, chestnut-brown, obscurely banded, characterized by several irregularly placed, whitish broad varices, and 3 strong internal axial lirations within the outer lip. Periostracum dark-brown. 20 to 24 axial ribs per whorl, becoming obsolete on the last whorl of old specimens. Spiral grooves over the intercostal spaces and over the ribs. Common; intertidal to 35 fathoms. Synonyms: *ambusta* Dall and Bartsch, 1909; *catalinensis* Dall and Bartsch, 1909.

Other *Mormula:*

3856 *Turbonilla anira* Bartsch in Dall, 1927. Off Fernandina, Florida, 294 fms.

3857 *Turbonilla lordi* (E. A. Smith, 1880). Sitka, Alaska, to Puget Sound, Washington. 20 mm. Largest known species.

3858 *Turbonilla regina* (southern California), (**3859**) *santosana* (Baja California), (**3860**) *heterolopha* (San Diego to Baja California), (**3861**) *periscelida* (Santa Rosa Island), all Dall and Bartsch, 1909.

3862 *Turbonilla eschscholtzi* Dall and Bartsch, 1907. Alaska to British Columbia. 18 mm.

3863 *Turbonilla pentalopha* Dall and Bartsch, 1903. San Pedro, California, to Todos Santos Bay, Baja California.

3864 *Turbonilla clementina* Bartsch, 1927. San Clemente Island, California.

3865 *Turbonilla enna* Bartsch, 1927. San Pedro, California, 6 fms. Also see species 3698.

3866 *Turbonilla coyotensis* Baker, Hanna and Strong, 1928. Gulf of California.

3867 *Turbonilla ignacia* Dall and Bartsch, 1909. Baja California.

3868 *Turbonilla scammonensis* Bartsch, 1912. Baja California.

3869 *Turbonilla sebastiani and* (**3869a**) *viscainoi* both Bartsch, 1917. Magdalena Bay, Baja California.

Subgenus *Bartschella* Iredale, 1917

Turbonillas with well-rounded whorls, marked with strong axial ribs and strong spiral cords, the junctions of which are usually subnodulous. Type: *subangulata* (Carpenter, 1857). This is *Dunkeria* Dall and Bartsch, not Carpenter, 1865.

Turbonilla laminata (Carpenter, 1865)   3870
Laminated Turbonille

Cayucos, California, to the Coronado Islands.

6 to 7 mm., broadly conic, wax-yellow at the tip, chestnut-brown on the last few whorls; rarely banded. Columellar area white. Postnuclear whorls inflated, rounded, with strong, vertical axial ribs (18 to 28 on the upper whorls, 30 to 40 on the last). Intercostal areas strongly pitted. Ribs tuberculate. Base with fading riblets and about 10 spiral lirations. Common; low tide to 25 fathoms.

Other *Bartschella:*

3871 *Turbonilla pauli* A. G. Smith and Gordon, 1958. Monterey Bay, California. *The Nautilus,* vol. 71, p. 151. (Synonyms: *bartschi* Smith and Gordon, 1948; *bartschiana* Smith and Gordon, 1950, *The Nautilus,* vol. 62, p. 105.)

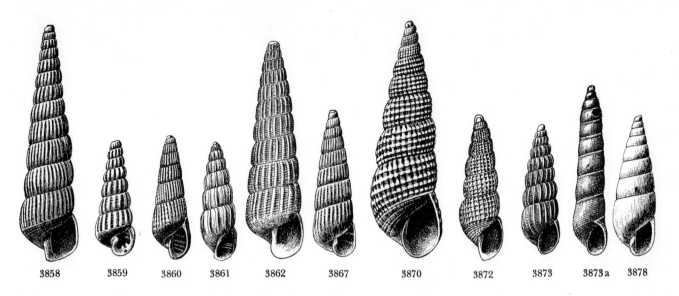

3858   3859   3860   3861   3862   3867   3870   3872   3873   3873a   3878

3872 *Turbonilla excolpa* Dall and Bartsch, 1909. Gulf of California.

Subgenus *Dunkeria* Carpenter, 1856

Whorls tabulate at the top, with the suture considerably indented. Type: *paucilirata* Carpenter, 1856. Synonym: *Pyrgisculus* Dall and Bartsch, 1909. Only 1 species was reported from California (**3873**) (*swani* Dall and Bartsch, 1909), but it is closest in characters to *Pyrgiscus*. Dall and Bartsch, 1909, reported 4 species from the Gulf of California.

Subgenus *Careliopsis* Mörch, 1875

Shell minute, slender, thin, white, with very fine, reticulated sculpturing. Type and only species reported from the Atlantic (Florida to Virgin Islands): (**3873a**) *Turbonilla styliformis* Mörch, 1875. One species was reported from the Gulf of California (**3873b**): *T. stenogyra* Dall and Bartsch, 1909.

Subgenus *Ptycheulimella* Sacco, 1892

Turbonillas smoothish, elongate-conic, white. Axial sculpture consisting of obsolete ribs shown only on the first few postnuclear whorls. Spiral sculpture of microscopic striations. Type: *pyramidata* (Deshayes, 1833). 3 west Mexican and 4 deep-water Atlantic species:

3874 *Turbonilla hespera* Bartsch in Dall, 1927. Off Fernandina, Florida, 294 fms.

3875 *Turbonilla melea* Bartsch in Dall, 1927. Off Fernandina, Florida, 294 fms.

3876 *Turbonilla polita* (Verrill, 1872). Eastport, Maine, 20 fms.

3877 *Turbonilla emertoni* Verrill, 1882. Off Martha's Vineyard, Massachusetts, 238 fms.

3878 *Turbonilla abreojensis* Dall and Bartsch, 1909. Baja California.

3879 *Turbonilla magdalinensis* Bartsch, 1927. Magdalena Bay, Baja California.

3880 *Turbonilla penascoensis* Lowe, 1935. Sonora, Mexico.

Subgenus *Ugartea* Bartsch, 1917

3881 *Turbonilla juani* Bartsch, 1917. Magdalena Bay, California, 13.5 fms. 4.2 mm.

3881

Genus *Kleinella* A. Adams, 1860

Shell ovate, umbilicate, surface cancellated; spire produced; aperture elongated, angular behind, produced in front. Type of the genus is *cancellaris* A. Adams, 1860.

Subgenus *Euparthenia* Thiele, 1931

Type of the subgenus is the Mediterranean *bulinea* Lowe, 1840. Synonym: *Parthenia* Lowe, 1840, non Robineau-Desvoidy, 1830. One Western Atlantic species:

3882 *Kleinella cedrosa* (Dall, 1884). Cedar Keys, west Florida.

Subfamily CYCLOSTREMELLINAE Moore, 1966

For details, see page 82. R. Robertson (1973, *The Nautilus*, vol. 87, p. 88) has shown that the genus *Cyclostremella* Bush, 1897, is not a prosobranch, and belongs here.

Order *Entoconchida* Fischer, 1883 (Abbott, 1973)
(Parasita Fischer, 1883)

Vermiform snails with most organ systems lacking; shell present in juvenile stages only and with only a slight tendency to spiral. Hermaphroditic; pseudopallium well-developed. Ctenidium and radula missing. Parasites of echinoderms. See "The classification of endoparasitic gastropods," by E. S. Tikasingh and I. Pratt, 1961, *Systematic Zoology*, vol. 10, pp. 65–69.

Family ENTOCONCHIDAE Gill, 1871

Larval shells with a tendency to spiral. "Dwarf males" present in the pseudopallium. Intestine present.

Genus *Entocolax* Voigt, 1888

Small, shell-less, saclike snails without radula or gills. They live embedded in the intestine of holothurian sea cucumbers. Type: *ludwigii* Voigt, 1888.

3883 *Entocolax ludwigii* Voigt, 1888. Bering Sea, internal parasite of *Myriotrochus*, a holothurian.

Genus *Entoconcha* J. Müller, 1852

Larval shell minute, ovate, smooth; spire short and obtuse. Aperture semilunar, angulated above. Soft parts very long and wormlike. Internal parasite of the red sea cucumber, *Stichopus*. Type: *mirabilis* J. Müller, 1852.

3884 *Entoconcha mirabilis* J. Müller, 1852. Puget Sound, Washington. Embedded in the tissues at the oral end of the sea cucumber, *Parastichopus californicus*. (Synonym: *Helicosyrinx parasitica* Baur, 1864.) See also *Comenteroxenos parastichopoli*.

Family ENTEROXENIDAE Heding and Mandahl-Barth, 1938

Wormlike parasitic snails. Permanently hermaphroditic. No tendency of larval shells to spiral. Intestine absent.

Genus *Comenteroxenos* Tikasingh, 1961

Ovaries consist of a system of complexly branching and anastomosing tubules, while the ovary of *Enteroxenos* (of Europe) is made up of a single axial canal with many side branches. Testis single. Type: *parastichopoli* Tikasingh, 1961.

Comenteroxenos parastichopoli **Tikasingh, 1961**   **3885**

Puget Sound, Washington.

4 to 5 inches in length, tubular, wormlike parasite of the sea cucumber, *Parastichopus californicus* (Stimpson) where it

lives attached to the anterior end of the intestine of the host. In life, the mollusk is transparent-yellow to opaque-orange. Pseudopallial cavity present with a ciliated tubule at the proximal end. (This may be what earlier Puget Sound workers called *Entoconcha mirabilis*.) See *Jour. Parasitol.*, 1961, vol. 47, p. 268. Illustration shows young and adult attached to holothurian intestine (HI). EC are egg capsules.

Genus *Thyonicola* Mandahl-Barth, 1946

Long, vermiform parasites of holothurians, with a marked tendency to coiling. True body wall of the snail is covered by the peritoneum of the host. Most organ systems absent. Reproductive system well-developed. Type: *mortenseni* Mandahl-Barth, 1946, from northern Europe.

Thyonicola americana **Tikasingh, 1961**   **3886**

Puget Sound, Washington.

Up to 12 inches in length, vermiform, greatly coiled in parts, width up to 2 mm.; both ends rounded. Skin smooth and varies from a light-yellow to orange, sometimes somewhat greenish. Hermaphroditic. 2 to 5 testes at proximal end in the

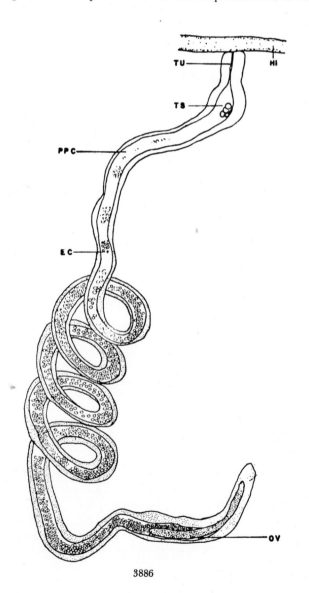

3886

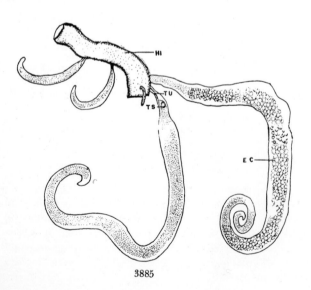

3885

pseudopallial cavity. Ovary occupying ⅓ of pseudopallial cavity at distal end and complexly branching and anastomosing. Live within the coelom and attached to and covered by the peritoneum of the white, frilly sea cucumber, *Eupentacta quinquesemita* (Selenka) and *E. pseudoquinquemita* Deichmann. Illustration from Tikasingh, 1961.

Order *Cephalaspidea* P. Fischer, 1883

Popularly known as the bubble shells. Shell usually present; gills present; head with a tentacular shield. Jaws feeble or none. Stomach usually with gizzard plates. With or without an operculum. These are the so-called Tectibranchia *sensu stricto*.

Superfamily Acteonacea Orbigny, 1842

Family ACTEONIDAE Orbigny, 1842

Strong shell external; foot with a chitinous operculum. Larval shell heterostrophic. Radula without median teeth, laterals few to numerous. Vas deferens closed. Penis external, nonretractile. Cephalic shield divided in front into 2 lobes. The name Pupidae Kuroda, 1941, is a synonym. The family name Tornatellidae Fleming, 1828, cannot be used.

Genus *Acteon* Montfort, 1810

External shell with a prominent spire; columella with 1 fold. Cephalic disk divided; operculum thin, corneous. Erroneously spelled *Actaeon*. Type: *tornatilis* (Linné, 1758). Synonyms: *Tornatella* Lamarck, 1816; *Speo* Risso, 1826; *Kanilla* Sowerby, 1838; *Myosota* Gray, 1847.

Acteon punctostriatus (C. B. Adams, 1840)   3887
Adams' Baby-bubble

Cape Cod to the Florida Keys to Texas and to Argentina. Bermuda.

3 to 6 mm. in length, solid, moderately globose, with a rather high spire. Columella with a single, twisted fold. Lower ½ of body whorl with numerous spiral rows of fine, punctate dots. Color white. Commonly found from low tide to 60 fathoms. *A. punctata* Orbigny, 1842, is a synonym.

3887

Acteon candens Rehder, 1939   3888
Rehder's Baby-bubble

North Carolina to southeast Florida and Cuba.

7 to 10 mm. in length, very similar to *punctostriatus,* but larger, very much thicker-shelled, glossy, opaque milk-white with light orange-brown suffusions on the body whorl. Commonly dredged in a few fathoms of water. *The Nautilus,* vol. 53, p. 21.

Other Atlantic species:

3889 *Acteon finlayi* McGinty, 1955. Cuba and off Palm Beach, Florida, 40 fms. Gulf coast of Florida, 30 to 200 fms. Brazil, 22 to 44 fms. *Proc. Acad. Nat. Sci. Phila.,* vol. 107, p. 81.

3888   3889   3912

3891 *Acteon pusillus* (Forbes, 1844). Off Florida Keys, 110 to 450 fms. Europe. (Not Macgillivray, 1843.)

3892 *Acteon cumingii* A. Adams, 1854. Off Cape Florida, 8 fms.; West Indies.

3893 *Acteon melampoides* (220 to 2,574 fms. Off Virginia to Florida Keys); *perforatus* (**3894**) (Gulf of Mexico, 805 fms.); *danaida* (**3895**) (off east Florida, 294 to 339 fms.); *incisus* (**3896**) (east Florida, 294 fms., Gulf of Mexico, 640 fms.) all Dall, 1881, Bull. 9, Mus. Comp. Zool.

3897 *Acteon delicatus* Dall, 1889. Florida Strait and Gulf of Mexico, 250 to 400 fms.

3898 *Acteon juvenis, lacunatus* (**3899**) *liostracoides* (**3900**) *parallelus* (**3901**) *particolor* (**3902**) *propius* (**3903**) *semicingulata* (**3904**) (all off Fernandina, Florida, 294 fms.) all Dall, 1927.

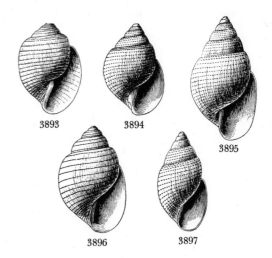

3893   3894

3895

3896   3897

Other Pacific species:

3905 *Acteon traskii* Stearns, 1897. San Diego, California, to Panama.

Genus *Rictaxis* Dall, 1871

Columella without a fold. Type: *punctocaelatus* (Carpenter, 1864).

Rictaxis punctocaelatus (Carpenter, 1864)   3906
Carpenter's Baby-bubble

British Columbia to Baja California.

10 to 20 mm. (¾ inch) in length, solid, oblong, 4 to 5 whorls, with 2 broad, ashy or brown spiral zones and about 26 spiral grooves on the body whorl. Columella obliquely truncated at base, and with 1 spiral fold. Base stained orange. Commonly found in shallow water in sand. *R. vancouverensis* Oldroyd, 1927, and *coronadoensis* Stearns, 1898, are the same species.

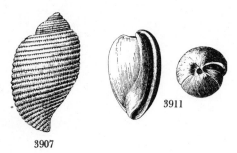

3907

Other species:

3907 *Rictaxis painei* Dall, 1903. Catalina Island, to San Diego, California, 30 to 50 fms.

Genus *Crenilabium* Cossmann, 1889

Shell 5 mm., white, elongate-fusiform, translucent, glossy. Spiral striae numerous, not punctate. First whorl mammilate. Columella folded and strong. Type: *aciculatum* Cossmann, 1889. *Lissacteon* Monterosato, 1890, is a synonym.

Crenilabium exilis (Jeffrey, 1870)   3908

Massachusetts to Florida.

3 to 5 mm., elongate-fusiform, white; spiral striae numerous. Apex rounded, bulbous. Uncommon. 150 to 487 fathoms. Synonyms: *insculptus* Verrill, 1880; *nitidus* Verrill, 1882.

Genus *Microglyphis* Dall, 1902

Spire very short, shell globose. Columella with 2 folds. Columella truncated below and has a sulcus behind it and is produced laterally into a rather wide spiral flange. No operculum. Type: *Acteon curtulus* Dall, 1890 from Patagonia. Steinberg (1963, *The Veliger*, vol. 5, p. 114) places this genus in the family Ringiculidae. Two west coast species:

3909 *Microglyphis breviculus* (Dall, 1902). Monterey, California, 66 to 200 fms.

3910 *Microglyphis estuarinus* (Dall, 1908). Washington to the Gulf of California, 92 fms.

Genus *Ovulacteon* Dall, 1889

Shell about 5 mm., globular, resembling a marginellid, involute, with an apical perforation (as in *Bulla*). Columella

simple, without folds. Aperture narrow, with a continuous, thickened edge. Type and only known species:

3911 *Ovulacteon meekii* Dall, 1889. Off Fernandina, Florida, 294 fms.; North Bimini, 200 fms.; north Cuba, 450 fms. Color white; length 5.5 mm.

Genus *Bullina* Férussac, 1822

Synonyms: *Bullinula* Swainson, 1840; *Perbullina* Iredale, 1929. Type: *scabra* (Gmelin, 1791).

Bullina exquisita McGinty, 1955   3912
Exquisite Bubble

Off Palm Beach, Florida.

8 mm., thin-shelled, perforated, color white with blotches of rose. Sculpture of rounded spiral threads about as wide as their internals. Spire obtuse, about 3 whorls, not including a large, smooth and distorted nucleus of 1 whorl. Last whorl spirally grooved and with minute, oblong puncture holes. Umbilicus nearly closed. Very rare; 60 fathoms.

Other species:

3913 *Bullina torrei* Aguayo and Rehder, 1936. Cuba.

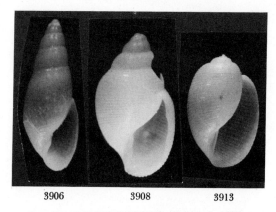

3906   3908   3913

Family RINGICULIDAE Philippi, 1853

Genus *Ringicula* Deshayes, 1838

Shells minute, white, glossy, resembling a miniature helmet shell (*Cassis* or *Phalium*). With globose whorls, thick outer lip, and 3 folds on the columella. Type: *ringens* (Lamarck, 1804).

Ringicula semistriata Orbigny, 1842   3914
Orbigny's Helmet-bubble

North Carolina to southeast Florida and the West Indies.

2 to 3 mm. in length, thick-shelled, resembling a miniature *Phalium*. 4 globose whorls, spire elevated. Aperture oblong;

3914

columella thickened by 3 folds, 1 above, 2 below. Outer lip very thick, swollen in the middle by a large tooth. Whorls white, smooth, except for fine striations on the base. Not uncommonly dredged from 34 to 107 fathoms.

Ringicula nitida Verrill, 1873   3915
Verrill's Helmet-bubble

Maine to the Gulf of Mexico. Bermuda.

2 to 4.5 mm., whorls 5, exterior smooth, with a simple, thickened outer lip, and with 2 smaller spiral ridges on the columella. Not uncommon; 100 to 447 fathoms.

Other species:

3916 *Ringicula peracuta* Watson, 1884. Off Bermuda, 1075 fms.

Family ACTEOCINIDAE Pilsbry, 1921

Nuclear whorls heterostrophic, visible, not enclosed by later whorls. With or without a radula. Shell strong.

Genus *Acteocina* Gray, 1847

Synonyms: *Tornatina* A. Adams, 1850; *Didontoglossa* Annandale, 1924; *Neacteocina* Kuroda and Habe, 1952. Type: *wetherelli* (I. Lea, 1833). This is *Retusa* of authors, not T. Brown, 1827.

Acteocina culcitella (Gould, 1852)   3917
Western Barrel-bubble

Kodiak Island, Alaska, to Baja California.

½ to ¾ inch in length, moderately solid, oblong but more constricted at the upper portions. Spire of 5 whorls, elevated, pointed and with a tiny, pimplelike nucleus (usually eroded in northern specimens). Suture narrowly and deeply channeled. Body whorl swollen at the lower ½. With numerous microscopic, wavy, incised spiral lines. Color yellowish, sometimes with numerous golden-yellow, fine spiral lines. Columella is a single, raised spiral cord. Common in shallow water down to 25 fathoms. *A. cerealis* Gould, 1853, is probably the same species.

The northern subspecies or form *eximia* (Baird, 1863) **(3918)** (Alaska to Monterey) has its shorter spire set in the circular depression at the top of the shell, and its columellar fold is very weak. Fairly common; 10 to 158 fathoms.

Acteocina candei (Orbigny, 1842)   3919
North Carolina to Argentina.

4 to 6 mm. in length, solid, white, somewhat spindle-shaped; with a projecting 1½ whorled, heterostrophic protoconch. A spiral carina is present on the shoulder just below the suture. Columella with a single spiral ridge. Radula minute, the lateral tooth having 5 to 7 denticles (while in *Acteocina canaliculata* there are 16 to 20 tiny ones). Lives offshore in oceanic waters. Do not confuse with the estuarine *Acteocina canaliculata* whose one-whorled protoconch is partially submerged. Common. Larvae pelagic. See H. W. and Mary Wells, 1962, *The Nautilus*, vol. 75, p. 87.

Other Atlantic species:

3920 *Acteocina bullata* (Kiener, 1834). Florida Keys; Bermuda; West Indies.

3921 *Acteocina rectus* (Orbigny, 1842). Florida and the West Indies.

3922 *Acteocina (Coleophysis* Fischer, 1883) *perplicatus* Dall, 1889. Florida Strait, 220 fms.; West Indies.

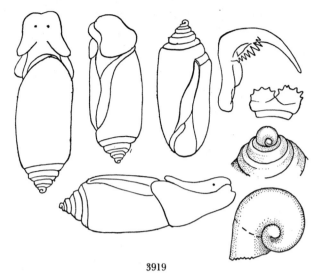

3919

3923 *Acteocina (Coleophysis) eburnea* Verrill, 1885. Off Cape Hatteras, North Carolina, to the Florida Keys, 10 to 70 fms.

3924 *Acteocina bermudensis* (Vanatta, 1901). Bermuda. *Proc. Acad. Nat. Sci. Phila.*, for 1901, p. 183.

Other Pacific species:

3925 *Acteocina smirna* Dall, 1919. San Diego, California, to Costa Rica.

3926 *Acteocina inculta* (Gould and Carpenter, 1857). Monterey, California, to the Gulf of California. Common; mud flats. 4 mm. (Synonym: *planulata* Carpenter, 1865.)

3927 *Acteocina intermedia* Willett, 1928. Monterey, California, to Todos Santos Bay, Baja California. 10 to 30 fms., common. *The Nautilus*, vol. 42, p. 37.

3928 *Acteocina magdalensis* Dall, 1919. Newport Bay, California, to Baja California.

3929 *Acteocina oldroydi* Dall, 1925. Departure Bay, British Columbia.

3930 *Acteocina angustior* Baker and Hanna, 1927. Gulf of California, to Acapulco, Mexico. *Proc. Calif. Acad. Sci.*, ser. 4, vol. 16, p. 124.

3931 *Acteocina (Coleophysis* Fischer, 1883) *carinata* (Carpenter, 1857). Redondo Beach, California, to Panama.

3932 *Acteocina (Coleophysis) harpa* (Dall, 1871). Forrester Island, Alaska, to San Diego, California.

3933 *Acteocina infrequens* (C. B. Adams, 1852). Santa Monica, California, to Panama.

Subgenus *Utriculastra* Thiele, 1925

Type: *canaliculata* (Say, 1822).

Acteocina canaliculata (Say, 1822)   3937
Channeled Barrel-bubble

Nova Scotia to Florida, Texas and the West Indies.

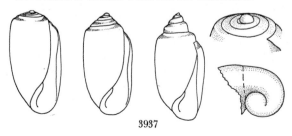

3937

4 to 6 mm. in length, solid, oblong with its spire moderately elevated, but almost invariably eroded. Glossy smooth, except for microscopic growth lines. Outer lip thin, advanced above. Columella of a single, raised, strong spiral ridge. Suture slightly channeled. Nucleus (when present) very small and pimplelike. Color white to cream, commonly with dark, rust-brown staining. Formerly placed in the genus *Retusa*. It is primarily an estuarine inhabitant, lays a round, stalked egg-mass, has a minute radula, and has no free-swimming veliger. Do not confuse with the similar-looking *Acteocina candei* (see H. W. and Mary Wells, 1962, *The Nautilus*, vol. 75, p. 87). For development, see D. Franz, 1971, *Trans. Amer. Micros. Soc.* vol. 90, p. 174.

Family APLUSTRIDAE Gray, 1847

Shell globose or oval, thin, with exposed spire of several whorls and a minute, uptilted, nearly immersed nucleus. Last whorl large, conspicuously banded with color. Animal voluminous, the foot large and flat. Head disc bearing 2 or 4 tentacular processes in front, and with 2 lobes behind. Radula lacks central teeth. Hydatinidae Pilsbry, 1895, is a synonym.

Genus *Hydatina* Schumacher, 1817

Shell globose, thin, smooth, spirally banded with color. Head with 4 flat tentacular processes. Epipodial lobes lacking. Edges of large foot reflexed over the shell. Type: *physis* (Linné, 1758).

Hydatina vesicaria (Lightfoot, 1786)   3938
Brown-lined Paper-bubble   **Color Plate 10**

South half of Florida to Brazil. Bermuda.

1 to 1½ inches in length, very thin, fragile, globose. Periostracum thin, buff to greenish. Shell characterized by many close, wavy, brown spiral lines. Animal large and colorful. Foot very broad. Moderately common in certain shallow, warm-water areas where they burrow in silty sand. Formerly called *H. physis* Linné which, however, is believed to be limited to the Indo-Pacific. Synonym: *H. verrilli* Pilsbry, 1949 (*The Nautilus*, vol. 63, p. 15).

Genus *Micromelo* Pilsbry, 1895

Animal not completely retractile into the oval, thin, flat-spired shell which has a minute, uptilted, subimmersed nucleus. Head disc bearing 2 flat tentacular processes in front, produced behind into 2 large lobes partly covering the shell. Last whorl spirally striate and punctate. Type: *undatus* (Bruguière, 1792).

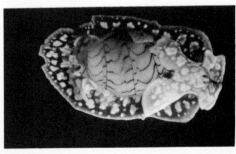

3939

Micromelo undatus (Bruguière, 1792)   3939
Miniature Melo

Southeast Florida to Brazil. Bermuda. Ascension Island.

½ inch in length, oval, rather thin and moderately fragile. Characterized by its whitish to cream color overlaid by 3 widely spaced, fine spiral lines of red and by many or few axial, wavy, lighter red flammules or lines. Uncommon. Found at low tide among delicate, feathery green algae. Body and foot translucent blue-gray with pinkish white, opaque splotches.

Superfamily Cylichnacea A. Adams, 1850

Parapodia large; shell external and cylindrical.

Family CYLICHNIDAE A. Adams, 1850

Parapodia thick; stomach with 3 flat gizzard plates. Scaphandridae is a synonym.

Genus *Cylichna* Lovén, 1846

Small, cylindrical shells with the lower end truncate and with the spire involute, leaving a small apical concavity. Columella with 1 oblique fold. Synonyms: *Cyclina* Gray, 1857; *Bullinella* Newton, 1891. Type: *cylindracea* (Pennant, 1777).

Cylichna gouldi (Couthouy, 1839)   3940
Gould's Barrel-bubble

Massachusetts Bay to off Cape Cod. Arctic Seas.

⅜ inch (9 mm.) in length, fragile, chubby, with the spire usually sunk in and consisting of 4 or 5 whorls. Color dirty-white with a yellowish periostracum. The whorls are much more globose and the anterior end more constricted than in the much smaller species, *Retusa obtusa*. Formerly placed in the genus *Retusa*. Uncommon from 26 to 34 fathoms.

Cylichna alba (Brown, 1827)   3941
Brown's Barrel-bubble

Arctic Seas to North Carolina.
Bering Sea to San Diego, California.

¼ inch (5 mm.) in length, fragile, narrowly oblong with flat sides. Apex with a dished, shallow depression. Upper ⅔ of aperture narrow; below it is wide. Columella short, rounded, slightly raised. Shell white, smoothish, except for microscopic, spiral scratches. Periostracum thin, shiny, yellowish, but often darkly stained with brown. Commonly dredged from 1 to 1,000 fathoms in cold water. *C. corticata* Beck, 1842, is a synonym.

Other Atlantic species:

3942 *Cylichna occulta* Mighels and Adams, 1842. Greenland to Maine; Alaska.

3943 *Cylichna vortex* (Dall, 1881). Massachusetts to off Chesapeake Bay, 326 to 1,356 fms. Off Dry Tortugas, Florida. (Synonym: *dalli* Verrill, 1882.)

3944 *Cylichna verrillii* Dall, 1889. Off North Carolina to the West Indies, 50 to 124 fms.

3945 *Cylichna eburnea* Verrill, 1885. Off Cape Hatteras, North Carolina, 70 fms.

Other Pacific species:

3946 *Cylichna nucleola* (Reeve, 1855). Arctic Seas to Kodiak Island, Alaska.

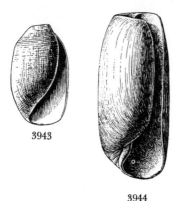

3943

3944

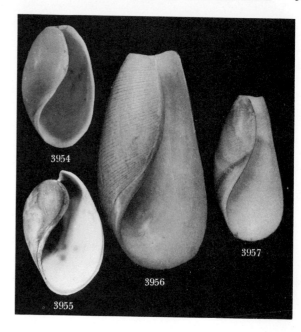

3954

3955

3956

3957

3947 *Cylichna diegensis* (Dall, 1919). Santa Monica, California, to Baja California.

3948 *Cylichna attonsa* (Carpenter, 1864) . Kodiak Island, Alaska, to Baja California, 10 to 40 fms.

3949 *Cylichna fantasma* (Baker and Hanna, 1927). Gulf of California.

Genus *Cylichnella* Gabb, 1873

Shell oblong-oval, capable of containing the entire body. Head shield with 2 rear flaps. Radula without marginals. Columella with 2 folds. Type: *bidentata* (Orbigny, 1841).

Cylichnella bidentata (Orbigny, 1841)   3950
Orbigny's Barrel-bubble

North Carolina, Florida to Texas and to Brazil.

3 to 4 mm. in length, somewhat resembling *alba,* but its columella has a spiral, callous fold and an indistinct nodule below. The shell is more oval. Glossy-white. Periostracum yellowish or partly orangish. Cephalic shield with 2 black eyes and with 2 long pointed extensions posteriorly. Eggmass pear-shaped, 2 mm., on a stalk 2 mm., and containing about 400 eggs. Commonly found from shallow water to 200 fathoms. This is *C. biplicata* of authors, not A. Adams. For anatomical details, see Marcus, 1958, *Amer. Mus. Novitates,* no. 1906.

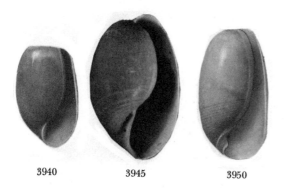

3940   3945   3950

Other species:

3951 *Cylichnella oryza* (Totten, 1835). Maine to Connecticut, 2 to 4 fms.

3952 *Cylichnella defuncta* and **(3953)** *gonzagensis* both (Baker and Hanna, 1927). Gulf of California.

Genus *Scaphander* Montfort, 1810

Synonyms: *Bullaria* Rafinesque, 1815; *Assula* Schumacher, 1817. Type: *lignarius* (Linné, 1767).

Scaphander punctostriatus Mighels, 1841   3954
Giant Canoe-bubble

Arctic Seas to Florida and the West Indies. Northern Europe.

1 to 1½ inches in length, very lightweight, but moderately strong. Ovate-oblong. Apex with a slightly sunken area. Aperture constricted above, roundly open below. Columella simple, rounded. Shell smoothish, except for numerous, spiral rows of microscopic, elongate punctations. Color chalk-white, with a straw periostracum. Fairly common from 20 to 1,000 fathoms.

(3955) *Scaphander nobilis* Verrill, 1884 (Noble Canoe-bubble) from off New England is the same size or smaller, has a proportionately much larger aperture, its outer lip is wing-like above, and the microscopic punctations are round. It is uncommon, from 996 to 1,309 fms.

Other species:

3956 *Scaphander pilsbryi* McGinty, 1955. 20 fms., off Pensacola, Florida. *Proc. Acad. Nat. Sci. Phila.,* vol. 107, p. 82.

3957 *Scaphander watsoni* Dall, 1881. Off Cape Hatteras to Florida, 63 to 324 fms., to Barbados and Venezuela. Subspecies *rehderi* Bullis, 1956. **(3958)** *Bull. Marine Sci. Gulf and Caribbean,* vol. 6, pp. 1–17.

3959 *Scaphander (Sabatia* Bellardi, 1876) *bathymophila* Dall, 1881. 100 miles east of Delaware, 554 fms., to Yucatan Straits, 640 fms.

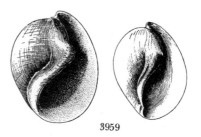

3959

3960 *Scaphander willetti* Dall, 1919. Forrester Island, Alaska.

3961 *Scaphander darius* Marcus and Marcus, 1967. Lower Caribbean, 49 to 174 meters. *Bull. Marine Sci.,* vol. 17, p. 603.

Family PHILINIDAE Gray, 1850

Related to the Scaphander canoe-shells, but different in having the mantle reflexed and closed over the shell, in lacking central teeth in the radula and in having a much more degenerate shell. With 3 thick gizzard plates. Parapodia thick.

Genus *Philine* Ascanius, 1772

Shell thin, with a very large aperture. 3 solid flat gizzard plates. Synonyms: *Lobaria* O. F. Müller, 1776; *Bullaea* Lamarck, 1801. Type: *aperta* (Linné, 1767).

Subgenus *Ossiania* Monterosato, 1884

Type of this subgenus is *quadrata* (S. Wood, 1839).

Philine quadrata (S. Wood, 1839)   3962
Quadrate Paper-bubble

Arctic Seas to North Carolina.

⅓ inch in length, moderately fragile, semi-transparent, white, squarish-oval and more constricted toward the top. Aperture large, flaring, and rounded below. Early whorls very small. Sculpture of numerous spiral rows of microscopic oval punctations. Suture deep. The narrow top of the aperture is slightly higher than the apex. Commonly dredged off the New England states from 20 to 400 fathoms. *P. formosa* Stimpson, 1850, is a synonym.

Philine lima (Brown, 1827)   3963
File Paper-bubble

Arctic Seas to Cape Cod, Massachusetts.

⅓ inch in length, much more oblong than *quadrata*, with the top of the aperture well below the apex, and sinuate from a top view. Columella fairly strong. Sculpture of spiral rows of scalloped lines forming chains, between which are a single scalloped line. Moderately common in fairly shallow but cold water. Alias *P. lineolata* (Couthouy, 1839). This is the type of the subgenus *Retusophiline* Nordsieck, 1972.

Philine sinuata Stimpson, 1850   3964
Sinuate Paper-bubble

Maine to Connecticut.
Bering Straits, Alaska.

Shell minute, 2 mm., ovate, white, translucent, fragile, axially striate. Spire conspicuous. Aperture large, anteriorly dilated. Animal 5 mm. in length, oblong, elongated, convex posteriorly; yellowish in color, darkest behind, with dots and patches of white. The reflected pedal lobes are rather narrow, and terminate near the middle of the part occupied by the shell. Eggs laid in globular, yellowish, gelatinous, 6 mm., masses, usually in August. Uncommon; 4 to 7 fathoms on sand bottom.

Philine sagra Orbigny, 1841   3965
Crenulated Paper-bubble

North Carolina to Florida and to Brazil.

⅛ to ¼ inch in length, oblong, fragile, white, with a large aperture, with numerous spiral lines of small oblong rings placed end to end, and characterized by the finely crenulated lip. Top of the aperture the same height as the apex. Not uncommon from 15 to 47 fathoms.

Philine bakeri Dall, 1919   3966
Baker's Paper-bubble

Santa Barbara to South Coronado Island, Baja California.

Living animal pale-cream, massive, 35 mm. long; shell buried within. Shell minute, 2 mm., translucent, of 2 or more whorls enfolded (except for the blunt subglobular nucleus) by the last whorl. Last whorl narrow, obliquely expanded in front. Sculpture of numerous, fine, incised, punctate spiral lines with wider interspaces; axis gyrate, previous; aperture as long as the shell, narrow behind with a very slight sulcus, but widely expanded in front; outer lip thin, sharp, straight, inner lip hardly glazed. Rare to uncommon.

Other Atlantic species:

3967 *Philine amabilis* Verrill, 1880. Off Martha's Vineyard, Massachusetts, 120 to 130 fms.

3968 *Philine angulata* Jeffreys, 1867. Casco Bay, Maine.

3969 *Philine cingulata* G. O. Sars, 1878. Off Cape Sable, Nova Scotia, 90 fms.

3970 *Philine finmarchia* M. Sars, 1878. Nova Scotia to off south Cape Cod, Massachusetts, 16 to 90 fms. (Synonyms: *cingulata* G. O. Sars, 1878; *fragilis* Dautzenberg, 1911.)

3971 *Philine fragilis* G. O. Sars, 1878. Off Nova Scotia, 90 fms.; Gulf of Maine, 88 to 92 fms.

3972 *Philine infundibulum* Dall, 1889. Florida Strait, Cuba and Barbados, 118 to 372 fms. (Bermuda record an error.)

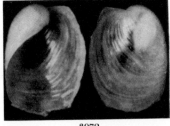

3972

3973 *Philine tincta* Verrill, 1882. South of Martha's Vineyard, Massachusetts, 65 fms.

3974 *Philine (Hermania) scabra* Müller, 1784. Greenland.

Other Pacific species:

3975 *Philine polaris* Aurivillius, 1885. Arctic Seas to Nanaimo, British Columbia.

3976 *Philine alba* Mattox, 1958. Catalina Island to San Benito Island, Baja California. 46 to 247 meters (*Bull. So. Calif. Acad.*

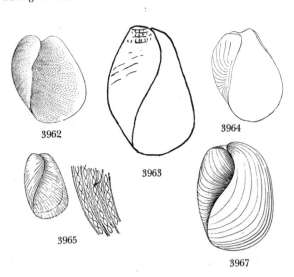

3962

3963

3964

3965

3967

Sci., vol. 57, p. 98). Gulf of Mosquitoes, Atlantic side of Panama (Marcus and Marcus, 1967, p. 607).

3977 *Philine californica* Willett, 1944. Off Redondo Beach, California, 50 fms. 5.5 mm. *Bull. So. Calif. Acad. Sci.*, vol. 43, p. 72.

Genus *Woodbridgea* S. S. Berry, 1953

*Haminoea*like, fragile, minute (1 to 2 mm.) shells, with a narrow umbilical chink. Type: *williamsi* S. S. Berry, 1953.

3978 *Woodbridgea* (Berry, 1953) *williamsi* S. S. Berry, 1953. Cedros Island, Baja California. *Trans. San Diego Soc. Nat. Hist.*, vol. 11, p. 422. Family placement unknown. Possibly Diaphanidae. *Brocktonia* of Dall, 1921, not Iredale, 1915, is the same.

3979 *Woodbridgea polystrigma* (Dall, 1908). San Diego, California.

Family DORIDIIDAE Gray, 1847

Body oblong, with 2 dorsal shields separated by a transverse furrow, the head shield having narrow, free, lateral and hind margins; posterior shield or mantle produced backward in 2 lobes, or wings. Shell internal, posterior, consisting of a flat, spiral whorl and a minute spire. Ctenidium posterior, on right side, large, bipinnate. The family name Aglajidae is synonymous.

Genus *Doridium* Meckel, 1809

Body cylindrical, with a cephalic shield bearing the eyes, a posterior shield under which the shell lies and 2 parapodia on the sides. Foot truncate anteriorly. No tentacles. *Aglaia* Renier, 1804 is a *nomen nudum*. Synonyms: *Posterobranchaea* Orbigny, 1837. Type: *tricolorata* Renier, 1807. For an annotated catalog of 43 known species, see E. and E. Marcus, 1966, *Studies in Trop. Oceanogr. Miami*, no. 4, pt. 1, pp. 164–172. The name *Aglaja* may be conserved someday.

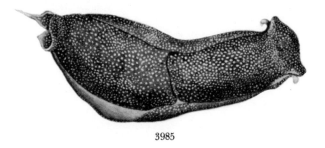

3985

Doridium pusum Marcus and Marcus, 1967 **3980**
Pusa Aglaja

Southeast Florida.

45 mm., grayish black with specklings of yellow to white. Head shield rectangular and ½ the length of the body. No posterior flagellum on mantle. Gills colorless. Internal shell 15 mm., yellow with brown lines of growth, thin, calcified and with 3 spinelike dorsal processes in the inner margin. Rare. Described from one specimen. *Studies Trop. Oceanogr. Miami*, no. 6, p. 18.

Doridium diomedeum (Bergh, 1894) **3981**
Diomedia Aglaja

Alaska to Monterey, California.

10 to 15 mm. long, animal uniformly black with yellowish flecks in some specimens. Head shield narrowed and rounded in front and nearly as long as the mantle shield. Between the head shield and foot a small fold occurs on either side of the mouth. The posterior mantle lobes are short and broad; without a flagellum. Shell delicate, light-brown, longer than broad, large; the larval shell projects behind. Columella distinct. No jaws or radula. The small ctenidium is yellow. Penial papilla as cuticularized tip. Common; mud flats, intertidally. It spawns from June to August in California. Feeds on nematodes.

Doridium nanum Steinberg and Jones, 1960 **3982**

San Francisco Bay, California.

3 to 12 mm., similar to *diomedeum*, but translucent grayish white with irregular black flecks and small yellow-brown dots on the dorsum, foot and both inner and outer surfaces of the parapodia. No flagellum. Brownish yellow organs show through the body wall. Shell is only slightly calcified and not as produced at the right posterior margin. Common; on mud bottoms, in 5 to 371 feet of water. *The Veliger*, vol. 2, p. 73, pl. 16.

Other species:

3983 *Doridium adellae* (Dall, 1894). Eagle Harbor, Puget Sound, 20 fms. Dark-plum, mottled with fine wiggly spots of golden yellow.

3984 *Doridium purpureum* (Bergh, 1894). Southern California. No flagellum.

3985 *Doridium ocelligerum* (Bergh, 1894). Sitka, Alaska, to Santa Barbara, California. Uncommon; mud flats to 10 fathoms. Black with blue and yellow dots. Flagellum on mantle.

3987 *Doridium gemmatum* (Mörch, 1863). Virgin Islands. Yellow or dull fleshy with close, longitudinal lines on head shield, the latter with green, shining, convex spots. *D. punctilucens* (Bergh, 1893) may be a synonym.

Genus *Navanax* Pilsbry, 1895

The parapodia envelop the cylindrical body, leaving the rectangular head and bifid tail free. Synonyms: *Navarchus* Cooper, 1863, non Filippi and Verany, 1859; *Strategus* Cooper, 1863, non Kirby and Spence, 1828. Type: *inermis* (Cooper, 1863).

Navanax inermis (Cooper, 1863) **3988**
Californian Navanax

Elkhorn Slough, California, to Baja California.

Up to 7 inches long (living). The parapodia envelop the cylindrical body leaving the rectangular head and bifid tail free. Body brown or purplish black with elongate yellow dashes and dots. Outer edge of parapodia orange with rows of yellow and blue spots. Black eyes surrounded by white haloes. Internal vestigial shell calcareous, brownish, distorted and with a point. Pharynx ⅓ the size of the body, but lacks jaws and radulae. Rapaciously swallows *Bulla, Haminoea* and its own young. Gives off a yellow fluid when irritated. Has commensal copepods and grapsoid crabs living on it. Common in shallow bays and estuaries and to a depth of 10 fathoms. For its biology, see R. T. Paine, 1963, *The Veliger*, vol. 6, pp. 1–9. *Aglaja bakeri* MacFarland, 1924, is a synonym.

3988

Genus *Chelidonura* A. Adams, 1850

Philinelike body; cephalic shield bearing 3 or 4 short nodules with microscopic filaments. Posterior appendages 2 and long. Eyes sometimes visible. Shell internal, slightly enrolled, without a columella, right edge arched. Type: *hirundinina* (Quoy and Gaimard, 1833). *Chelinodura* is a misspelling.

Chelidonura phocae Marcus, 1961   3989
Phoca Chelidone

Tomales Bay, central California.

10 to 20 mm. long (living), dark-red, or olive with white dots. Head shield truncate anteriorly and longer than the mantle shield. Corners of the foot are finger-shaped. Sole of foot, with red dots, is as long as the body. The triangular posterior mantle lobes are of equal or subequal size, sometimes the left one bearing a filiform prolongation. Calcified internal shell almost circular. The outer lip extends into the right mantle lobe. No jaws or radulae. The pleurembolic penis has cuticular conical warts on its base, and the prostate is enclosed in a sac. Uncommon; tidepools on mud flats down to 9 feet.

Chelidonura evelinae Marcus, 1955   3990
Evelina's Chelidone

Southeast Florida to Brazil.

40 to 60 mm. (living). A row of metallic-blue spots bordered with black rings on the margins of the parapodia characterize this uncommon species found among algae in shallow water to 10 feet.

Other species:

3991 *Chelidonura hirundinina* (Quoy and Gaimard, 1833). Southeast Florida on *Halimeda;* Curaçao on *Thalassia;* Indo-West Pacific.

Family GASTROPTERIDAE Swainson, 1840

Genus *Gastropteron* Meckel (in Kosse), 1813

Shell entirely internal and consisting of a minute, nautiloid, calcareous spire. Body sack-shaped, with 2 large, winglike, fleshy flaps, one on each side of the body. These peculiar, small sea-slugs swim through the water in a batlike manner. Type: *meckeli* Kosse, 1813. Synonyms: *Gastroptera* Blainville, 1825; *Sarcopterus* Rafinesque, 1814.

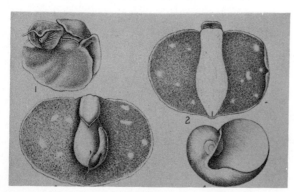

3992

Gastropteron rubrum (Rafinesque, 1814)   3992
Bat-wing Sea-slug

West coast of Florida to Texas to Brazil.
Mediterranean.

⅓ inch to 1 inch in length. General color varying from red-purple to pale-rose, sometimes with bluish white spots. There is a vivid, iridescent-blue border on the head disc and the "wings." This is probably *G. meckeli* "Dall." Moderately common; found swimming in quiet bays, but also crawls on the bottom.

Gastropteron pacificum Bergh, 1894   3993
Pacific Bat-wing Sea-slug

Alaska to the Gulf of California.

Similar to *rubrum,* but yellowish with red flecks. There are 12 to 20 gill leaflets. Margin of mantle without a flagellum, as in *rubrum.* Uncommon from 9 to 134 fathoms.

G. cinereum Dall, 1925 (British Columbia) is 11 mm. in length, and a uniform dusky-slate color. It also lacks a posterior flagellum on the mantle. Uncommon. See Hans Bertsch, 1964, *The Veliger,* vol. 11, p. 431. Erroneously reported from the Galapagos.

3993

Family RUNCINIDAE H. and A. Adams, 1854

Very small animals (3 to 6 mm.) without a shell, without a distinct cephalic shield, without a trace of parapodia. Body lanceolate. 2 eyes visible up front. Tail extends beyond the notum where leaflets of gills are located. Jaws of slivers. Gizzard with 4 solid plates, grossly denticulate on their internal faces.

Genus *Runcina* Forbes and Hanley, 1853

Characters of the family. Radula 1·1·1, the median one large and denticulate, and the lateral ones hooked. Type: *coronata* (Quatrefages, 1844).

Other species:

3994 *Runcina inconspicua* Verrill, 1901. Bermuda. *Trans. Conn. Acad. Arts Sci.,* vol. 11.

3995 *Runcina prasina* (Mörch, 1863). Virgin Islands.

Superfamily Diaphanacea Odhner, 1922

Family DIAPHANIDAE Odhner, 1922

Genus *Diaphana* Brown, 1837

Shell small, thin, corneous brown, umbilicated, swollen, the last whorl shouldered or globose; spire low or sunken in an apical umbilicus. Aperture as long as the shell, rising above the spire. Outer lip thin. Type: *minuta* (T. Brown, 1827). This is *Roxania* Turton, 1834, a preoccupied name.

Diaphana minuta (Brown, 1827) 3996
Arctic Paper-bubble

Arctic Seas to Connecticut. Europe.
Bering Strait, Alaska.

3 to 5 mm. in length, globose, thin, fragile and transparent-tan in color. Last whorl globose below, constricted somewhat above. Apex large, globose, obliquely and mammillarly projecting. Suture deep. Columella long, straight, not thickened, the edge partly closing the narrow umbilicus. Moderately common from 6 to 16 fathoms. *Diaphana debilis* Gould, 1840; *hyalina* Turton, 1834; *D. hiemalis* Couthouy, 1839; and *D. globosa* Lovén, 1846, are considered synonyms of this species by Lemche (1948) and other modern workers.

3996

Other species:

3997 *Diaphana lottae* Bush, 1893. Off Cape Lookout, North Carolina, 603 fms.

3998 *Diaphana brunnea* Dall, 1919. Kodiak Island, Alaska.

3999 *Diaphana californica* Dall, 1919. San Pedro, California, to South Coronado Island. Both *Proc. U.S. Nat. Mus.*, vol. 56, no. 2295.

Superfamily Bullacea Rafinesque, 1815

Family BULLIDAE Rafinesque, 1815

Genus *Bulla* Linné, 1758

Shell oval, compactly involute, solid, with a mottled pattern, smooth; spire sunken, umbilicated. Eyes small, wide apart, about halfway back on the animal's shield. The names *Vesica* Swainson, 1840, and *Bullaria* Rafinesque, 1815, have been ill-advisedly used for this genus. Fortunately, the name *Bulla* has been conserved for this group of bubble-shells by the International Commission for Zoological Nomenclature (Opinion 196). Type: *ampulla* (Linné, 1758). Synonyms include: *Bullus* Montfort, 1810; *Quibulla* Iredale, 1929.

Bulla striata Bruguière, 1792 4000
Common Atlantic Bubble   Color Plate 10

Florida to Texas and to Brazil. Bermuda.

½ to 1 inch in length, fragile or heavy, light or dark, without or with spiral grooves well-marked toward the base of the shell and within the apical perforation. The whorls may be compressed at the apical end. Columella usually with a brown-stained callus. Locally common. *B. amygdala* Bruguière, 1792, is probably a smooth form of this species. Typical *striata* comes from the Mediterranean. The Caribbean form may take the subspecies name *umbilicata* Röding, 1798 (synonym: *occidentalis* A. Adams, 1850) if one wished to recognize it as distinct.

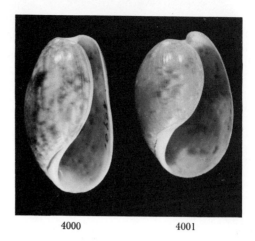

4000   4001

Bulla gouldiana Pilsbry, 1895 4001
California Bubble

Santa Barbara, California, to Ecuador.

1½ to 2 inches in length, rotund, fragile. Grayish brown with darker, streaked mottlings which are bordered posteriorly with cream. Periostracum dark-brown and microscopically crinkled. Collected abundantly at night, intertidal on mud flats. *Bulla punctulata* A. Adams from Baja California south is much heavier and constricted or narrowed at the top ⅓ of the shell.

Other species:

4002 *Bulla clausa* Dall, 1889. Florida.

4003 *Bulla gemma* Verrill, 1880. South of Martha's Vineyard, Massachusetts, 100 to 115 fms.

4004 *Bulla eburnea* Dall, 1881. North Carolina to Florida and the West Indies, 107 to 337 fms.

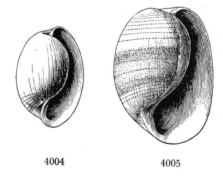

4004   4005

4005 *Bulla abyssicola* Dall, 1881. Yucatan Strait, 640 fms.

4006 *Bulla punctulata* A. Adams in Sowerby, 1850. Baja California to Peru. (Synonyms: *adamsi* Menke, 1850; *punctata* A. Adams, 1850; *quoyii* A. Adams, 1850, non Gray, 1843; *quoyana* Dall, 1919.)

4007 *Bulla bermudae* Verrill and Bush, 1900. Bermuda. *Trans. Conn. Acad. Sci.*, vol. 10, p. 523. *Bulla?*

Family HAMINOEIDAE Pilsbry, 1895

Genus *Atys* Montfort, 1810

Shell globose-oval to subcylindrical, involute, with the spire concealed. Aperture as long as the shell, produced above the apex. Lip rising from the center of the spire, and having an angular fold there. Columella short, with a fold-like truncation, or arcuate; umbilicus generally not wholly closed. Type: *naucum* (Linné, 1758). The family Atyidae Thiele, 1926, is the same.

Atys riiseana Mörch, 1875   **4008**
Riise's Paper-bubble

Southeast Florida and the West Indies. Brazil.

7 to 11 mm., oblong-oval, somewhat more compressed above than below, moderately thick, translucent-white in color. Center of shell smoothish, the ends with about a dozen fine, spiral, incised grooves. From the center of the narrow, concave vertex there arises a distinct plicated, slightly twisted lip. Columella vertical, thickened but not toothed, the edge reflexed, partly concealing the narrow but distinct umbilicus. Common; 1 to 50 fathoms. *A. caribaea* is smaller, narrower, with spiral lines over all of the last whorl and sometimes with axial threads at the upper end.

Atys sandersoni Dall, 1881   **4009**
Sanderson's Paper-bubble

North Carolina to southeast Florida and the West Indies. Brazil.

¼ to ⅓ inch in length, similar to *riiseana*, but thicker-shelled, with flatter sides, deeper and wider umbilicus and with more numerous and finer spiral lines at each end. Fairly common from shallow water to over 100 fathoms.

Atys caribaea (Orbigny, 1841)   **4010**
Sharp's Paper-bubble

Southeast Florida and the West Indies. Brazil.

6 to 8 mm., subcylindrical, solid, porcelaneous, glossy, translucent bluish white, very finely spirally striate over the last whorl, but strongest at the ends. Apex with an extremely small perforation. Base umbilicate. Lip rises from the right side of the apical perforation, and lacks a strong twist. Columella concave with a slight twist. Rarely there are microscopic axial threads at the upper end. Common; 2 to 100 fathoms. "*A. lineata* Usticke" and *sharpi* Vanatta, 1901, are synonyms.

4009   4010

Other Pacific species:

4011 *Atys casta* Carpenter, 1864. Catalina Island, California, to the Gulf of California.

4012 *Atys nonscripta* (A. Adams, 1850). San Diego, California.

4013 *Atys chimera* Baker and Hanna, 1927. Gulf of California.

4014 *Atys liriope* Hertlein and Strong, 1951. Gulf of California, in 91 meters.

Genus *Haminoea* Turton and Kingston, 1830

Shell thin and fragile, oval, with the spire concealed. Synonym: *Haminea* Gray, 1847. Type: *hydatis* (Linné, 1758).

Key to the Atlantic Species

(To determine on which side of the apical perforation the lip arises, hold the shell with the apex toward you and the apertural lip facing to the right.)

a. Apertural lip arising on the left side of the perforation, and angled near its insertion:
   b. Shell with numerous fine spiral grooves; ⅓ inch; yellowish to whitish; southeast Florida to Texas to Brazil; Bermuda   **(4015)** *elegans* (Gray, 1825)
   bb. Shell smooth; ⅜ inch; West Indies
                          (4016) *glabra* (A. Adams, 1850)
aa. Apertural lip arising on right side; not angled:
   c. Well-grooved spirally:
      d. Sides of whorls globose; ½ inch; amber to whitish; Cape Cod to North Carolina; common   **(4017)** *solitaria* (Say, 1822)
      dd. Sides of whorls flattish; ⅓ to ½ inch, translucent-white; west Florida to Texas; Bermuda. Common
                          (4018) *succinea* (Conrad, 1846)
   cc. Spiral striae absent or excessively fine; ½ inch; translucent greenish yellow; globose; Gulf to Brazil; Bermuda
                          (4019) *antillarum* (Orbigny, 1841)

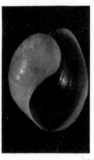

4015

Haminoea petiti (Orbigny, 1841)   **4020**
Petit's Paper-bubble

West Indies.

Fragile, 7 to 10 mm., cylindric-oval, translucent greenish yellow in color; smoothish, except for microscopic growth lines; glistening surface. Vertex wide, imperforate, usually with a chalky yellow callus. Columella thick, whitish, slightly reflected. Uncommon to locally common; 1 to 2 fathoms in grass and sand in protected bays and lagoons.

Haminoea virescens (Sowerby, 1833)   **4021**
Sowerby's Paper-bubble

Puget Sound, Washington, to the Gulf of California.

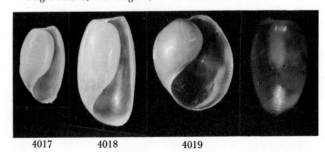

4017   4018   4019

½ inch in length, very fragile, a translucent greenish yellow in color. Aperture very large and open. Upper part of outer lip high and narrowly winged. No apical hole. The animal is dark-green with yellowish marks: dots on the head shield, mottlings on the parapodia. A common, littoral species on the open coast. *H. cymbiformis* Carpenter, 1857, *H. olgae* Dall, 1919, *H. strongi* Baker and Hanna, 1927, and *rosacea* Spicer, 1933, are the same.

Haminoea vesicula Gould, 1855   4022
Gould's Paper-bubble

Alaska to the Gulf of California.

¾ inch in length, very fragile, similar to *virescens*, but with a barrel-shaped whorl (from an apertural view), proportionately smaller aperture, with a tiny apical perforation, and with a lower, more rounded wing on the upper part of the outer lip. Shell color much the same, but the thin periostracum is often rusty-brown or yellowish orange. A common, littoral, bay species.

Other species:

4023   *Haminoea angelensis* Baker and Hanna, 1927. Baja California.

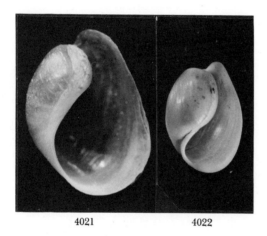

4021   4022

Family RETUSIDAE Thiele, 1926

Genus *Retusa* Brown, 1827

Type: *obtusa* (Montagu, 1808). Synonym: *Utriculus* T. Brown, 1844, non Schumacher, 1817. For the subgenus *Sulcularia* Dall, 1921, see *Sulcoretusa* Smith and Gordon, 1949, under *Cylichnina* below. Radula absent.

Retusa obtusa (Montagu, 1807)   4024
Arctic Barrel-bubble

Arctic Seas to off North Carolina. Alaska.

3 mm. in length, fairly fragile, smooth, stubby and with the spire commonly slightly sunken or only a little elevated. Columella smooth. A chinklike umbilicus is present. Color translucent-white with yellowish brown staining. Common from shore to 290 fathoms. *R. pertenuis* Mighels, 1843 and *R. turrita* Möller, 1842, are this species.

Retusa sulcata (Orbigny, 1842)   4025
Sulcate Barrel-bubble

North Carolina to southeast Florida and the West Indies.

2 mm. in length. Characterized by its small size, fine axial threads, white color, oblong shape, flat sides and deeply sunken spire. Moderately common from 3 to 95 fathoms.

Subgenus *Cylichnina* Monterosato, 1884

The type of this subgenus is *umbilicata* Montagu, 1803, from Europe. *Sulcularia* Dall, 1921, and *Sulcoretusa* Smith and Gordon are synonyms.

4025a   *Retusa xystrum* Dall, 1919. California.

4025b   *Retusa montereyensis* Smith and Gordon, 1948. Monterey, California.

Genus *Pyrunculus* Pilsbry, 1894

Type: *pyriformis* (A. Adams, 1850).

Pyrunculus caelatus (Bush, 1885)   4026
Bush's Barrel-bubble

North Carolina to southeast Florida. Texas.

3 mm. in length, pyriform in shape, rather thick and opaque-white. Spire concealed within a very deep pit. Rather rare from 15 to 43 fathoms.

Genus *Volvulella* Newton, 1891

Shells up to 9 mm., translucent to white, ovate-cylindrical, with a pointed vertex. The 2-coiled sinistral shell is buried deeply within. No radula present. Type: *acuminata* (Bruguière, 1792). Synonyms: *Volvula* A. Adams, 1862, non Gistel, 1848; *Rhizorus* Montfort of authors, not of Montfort, 1810. For a detailed review of this genus, see H. W. Harry, 1967, *The Veliger*, vol. 10, p. 133. Marcus and Marcus (1960) place this genus in the Retusidae. Do not confuse the genus name with *Volvatella* Pease, 1860, which are Panamanian and Indo-Pacific sacoglossids.

Volvulella persimilis (Mörch, 1875)   4027
Southern Spindle-bubble

North Carolina to southeast Florida and Texas to Brazil. Bermuda.

3 to 4 mm. in length, fragile, translucent-white, spindle-shaped, with a sharp, long spikelike apex. Glossy and with 4 or 5 very fine, indistinct, punctate, spiral lines at each end. Outer lip thin, following the curvature of the body whorl to just below the middle where it continues in a straight line. Umbilicus chinklike. Periostracum thin and pale-straw. Common from 5 to 209 fathoms. Synonyms: *oxytata* (Bush, 1885); *mörchi* Dall, 1927; and *ischnatracta* (Pilsbry, 1931). For anatomy, see Marcus and Marcus, 1960 (*Bull. Marine Sci. Gulf and Caribbean,* vol. 10, no. 2).

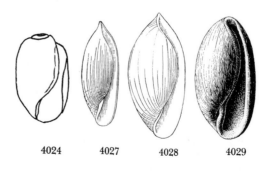

4024   4027   4028   4029

4026

Volvulella recta (Mörch, 1875) 4028
Spined Spindle-bubble

North Carolina to Florida and Alabama, and the West Indies.

2 to 3 mm. in length, spindle-shaped, ovate-oblong, fragile, smoothish, except for axial striae at the apical end and minute spiral lines at each end. Upper end of aperture ends in a very sharp, rather prolonged spike. Umbilicus rarely, if ever, present. Commonly dredged from 12 to 150 fathoms. *Bulla acuta* Orbigny, 1841, non Grateloup, 1828, is a synonym and a preoccupied name. Synonyms also include *minuta* (Bush, 1885), and *bushii* (Dall, 1889).

Volvulella paupercula (Watson, 1883) 4029
Spineless Spindle-bubble

North Carolina to Florida and the Caribbean.

3 to 4 mm., very similar to *recta*, but more ovate and barely produced at the top. The shell is unusually thick for the genus. Axial striae are absent at the apical end, although finer growth lines are present. Coarse spiral striae are prominent at both ends. The apical spine is small or usually absent, and there may be a minute umbilicus at the apex in spineless shells. Moderately common; 75 to 190 fathoms. *V. aspinosa* (Dall, 1889), is a synonym.

Volvulella cylindrica (Carpenter, 1864) 4030
Pacific Coast Spindle-bubble

Vancouver Island, British Columbia, to Panama.

4 to 10 mm., elongate, with very slightly convex or concave sides. Apical spine well to poorly produced. Spiral striae weak at both ends, and weaker ones sometimes evident on the midportion of the whorl. Dredged offshore; locally common. Spineless forms were named *cooperi* Dall, 1919, and *lowei* (Strong and Hertlein, 1937). *V. callicera* Dall, 1919, is a young specimen.

Other *Volvulella*:

4031   *Volvulella californica* Dall, 1919. Santa Cruz, California, to Todos Santos Bay, Baja California. May be "fat" growth form of *cylindrica*. 4 to 5 mm.

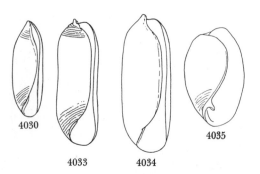

4030   4033   4034   4035

4032   *Volvulella catharia* Dall, 1919. Panama Bay, 62 fms.; Galapagos Islands, 40 fms. 2.8 mm., like *paupercula* of the Atlantic.

Subgenus *Paravolvulella* Harry, 1967

Elongate-cylindrical, flat-sided, rounded shoulder; with a deep sinus at the apical end of the outer lip. Parietal lip curved upward medially to the sinus to form a short, but acute, spine. Type: *texasiana* Harry, 1967.

Volvulella texasiana Harry, 1967 4033
Texas Spindle-bubble

Off Galveston, Texas.

4 mm., translucent gray, flecked with opaque-white and often stained with rust. Spine short (bilobed in side view) and with a minute deep pit at its apex. Spiral lines on the basal and apical ends. Microscopic wavy spiral striae on the middle part of the flat-sided whorls. Uncommon locally; 7 to 10 fathoms. *The Veliger*, vol. 10, p. 141.

Volvulella panamica Dall, 1919 4034

Off Redondo, California, to Panama.

3 to 4.5 mm., very similar to *texasiana*, with spiral lines over all of the shell, the stronger ones being at the ends. Apical end abruptly rounded; spine blunt and short. Apical end of lip deeply sinuate. Spine without a ridge. Brown-stained at both ends. Uncommon; down to 75 fathoms. *V. tenuissima* Willett, 1944, is a synonym.

Genus *Micraenigma* S. S. Berry, 1953

Shell 2 mm., white, translucent, *Bulla*-shaped, imperforate at both ends, columella with a downward-slanting toothlike plate. Type: *oxystoma* S. S. Berry, 1953. Placement of this genus is uncertain. One species, possibly malformed:

4035   *Micraenigma oxystoma* S. S. Berry, 1953. Cedros Island, Baja California. *Trans. San Diego Soc. Nat. Hist.*, vol. 11, p. 423.

Family VOLVATELLIDAE Pilsbry, 1895

Genus *Cylindrobulla* P. Fischer, 1856

Shell thin and fragile, cylindrical, with a sunken spire. Aperture long, straight-sided, produced at the vertex in a deep slit following the suture. Base obliquely truncated, entirely open, showing the whole interior of the body whorl as well as the spirally ascending columella, when viewed from below. Type: *beauii* P. Fischer, 1856. The family name Cylindrobullidae Thiele, 1926, is a synonym of Volvatellidae, as is Arthessidae Evans, 1950.

Cylindrobulla beauii P. Fischer, 1856 4036
Beau's Paper-bubble

South Florida and the West Indies. Bermuda.

Shell 7 to 14 mm., paper-thin, elastic, cylindrical. Surface smooth, pale-straw to white. Slit at the suture is ⅔ of a whorl

4036

in length. Not uncommon; shallow water to 95 fathoms. Reported in grassy shallows in Bermuda by R. Jensen (*in litt.*).

Other species:

4037 *Cylindrobulla californica* Hamatani, 1971. Gulf of California, on *Caulerpa* weed. *Publ. Seto Marine Biol. Lab.*, vol. 19, p. 112.

The Sand Nudibranchs

Order *Acochlidioidea* Küthe, 1935
(Hedylopsoidea Bergh, 1896)

Opisthobranchs with small bodies, without a shell, without gills, without a cephalic shield. Intestinal tract distinct from the foot. Rhinophores usually present. Oval lobes usually present. Denticulate radular present. No jaws. No gizzard plates. Lives in the interstitial waters of coarse sand, usually near estuaries. For a key to the world species, see Bertil Swedmark, 1971, *Smithsonian Contrib. Zool.*, no. 76, p. 42.

Family MICROHEDYLIDAE Odhner, 1938

Visceral sac elongated, detached from the foot which is very short in back. Liver in the form of an elongated tube. Animal moneiceous or dioecious. Without a penis. The genus *Ganitus* Marcus, 1953, with its type *evelinae* Marcus, 1953, lacks rhinophores, and is known only from Brazil.

Genus *Unela* Marcus, 1953

With flattened tentacles and cylindrical rhinophores. Radular formula: 1·1·1. No jaws. Epidermal glands voluminous. Sexes separate. No copulatory organ. Female with a ciliated furrow from the genital opening to the base of the right rhinophore. Type: *remanei* Marcus, 1953.

Unela remanei Marcus, 1953   **4038**
Remane's Sand Nudibranch

Southeast Florida to Brazil.

2 to 5 mm. White. Labial tentacles broader than high. Eyes not present. Radula has median tooth with 7 denticles, and an oblong, simple lateral on each side. 40 to 50 rows of teeth. Found in coarse sand of the lower tidal zone.

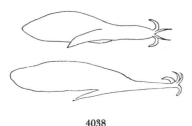

4038

The Pteropods or Sea Butterflies

These small, pelagic gastropods are very abundant in the open seas in nearly every part of the world. They are occasionally washed ashore, but more commonly their shells are found in dredge hauls. The identification of pteropods is important to many types of oceanographic studies. There are two unrelated orders, *Thecosomata* most of which have shells, and the *Gymnosomata*, without shells. Every known American species (Eastern Pacific and Western Atlantic) of the shelled *Thecosomata* has been included and figured.

The most exhaustive and useful work is that of S. van der Spoel, 1967, "Euthecosomata—a group with remarkable developmental stages (Gastropoda, Pteropoda)," 375 pp. 366 figs., J. Noorduijn en Zoon, Gorinchem, Netherlands. The book deals with the taxonomy, anatomy and ecology of the shelled or Thecosomata pteropods. For Western North Atlantic distributions, see Chen and Bé, 1964, *Bull. Marine Sci.*, vol. 14, no. 2, pp. 185–220.

An illustrated zoogeographical account of "The Thecosomata and Gymnosomata of California" (none is endemic) is given by J. A. McGowan, 1968, Supplement to vol. 3, *The Veliger*, pp. 103–130, 20 pls., from which some of our illustrations are taken. For a recent study of shell variation of the Cavoliniidae, see van der Spoel, 1970, *Basteria*, vol. 34, pp. 103–151.

Order *Thecosomata* Blainville, 1824

Foot reduced, but the epipodia greatly expanded and used for swimming. Gizzard armed with 4 hard plates. Jaws and a small, triserial radula present. Only one genus (*Peraclis*) has a true gill. Penis unarmed. All have calcareous shells, except for the Cymbulidae which have a transparent pseudoconch.

Suborder *Euthecosomata* Meisenheimer, 1905

Swimming fins laterally separated and dorsal to the mouth. Tentacles not paired and symmetrical. Shell composed of calcareous material; proboscis and rostrum absent. Mouth and wings in the same level.

Family LIMACINIDAE Blainville, 1823

Shells sinistrally coiled, trochoid or planorboid in shape, fragile, small. Operculum thin, chitinous, eccentrically spiraled. Paired fins large, their outer borders not subdivided. Right tentacle much larger than the left. Mantle cavity at the dorsal side. No ctenidium. Produces floating, gelatinous egg ribbons about 4 mm. in length. Spiratellidae Dall, 1921 is a synonym, as is Spirialidae Chenu, 1859.

Genus *Limacina* Bosc, 1817

Small sinistral shells with the characters shown in the family above. Type: *helicina* (Phipps, 1774). 8 living species and 8 subspecies or forms. Synonyms include: *Spiratella* Blainville, 1817; *Heterofusus* Fleming, 1823; *Helicophora* Gray, 1842; *Scaea* Philippi, 1844; *Spirialis* Lovén, 1847; *Heliconoides* A. Adams, 1858. I follow van der Spoel in abandoning *Spiratella*.

Subgenus *Limacina* Bosc, 1817

Relatively large shells with the last whorl ½ to ⅘ of the entire shell, and usually with striations. Umbilicus wide or narrow and very deep. Cold water forms.

Limacina helicina (Phipps, 1774)   **4039**
Helicid Pteropod

> Arctic Seas to the Gulf of Maine (*helicina*).
> Japan to Alaska (*acuta*).
> Alaska to Baja California (*pacifica*).

2 to 6 mm. in length, 6 whorls, spire short, shell wider than long. Surface smooth or with relatively large, axial threads. Spire flattened; last whorl much larger than the preceding ones. Wings of animal with a protrusion to the anterior border. Abundant enough in the Arctic seas to serve as an important source of food for whales. 3 forms border North America:

(4040) Forma *helicina* (Phipps, 1774). (North Europe; Greenland and north Canada to the Gulf of Maine.) Up to 6 mm., umbilical keel present; protrusion on wings; transverse striae present.

(4041) Forma *pacifica* Dall, 1871. (South Alaska to off Baja California.) Up to 2 mm. Spire flat; umbilical keel absent; protrusion on wings absent; axial striae absent. McGowan's 1963 var. B.

(4042) Forma *acuta* van der Spoel, 1967. (Japan to south Alaska.) Up to 2.5 mm. Spire higher; umbilical keel absent; protrusion on wing present; axial striae present.

Limacina retroversa (Fleming, 1823)   **4043**
Retrovert Pteropod

> Arctic Seas to Cape Cod, Massachusetts.

Up to 5 or 6 mm. in length, spire slightly elevated, umbilicus distinct, shell higher than wide. Entire surface covered with fine, spiral lines. 5 to 7 whorls. *Limacina balea* Möller, 1841, *Spirialis gouldii* Stimpson, 1851, and *macandrewi* Forbes and Hanley, 1853, are this species. For biology, see S. Hsiao, 1939, *Biol. Bull.*, vol. 76, pp. 7 and 280; A. C. Redfield, 1939, *Biol. Bull.*, vol. 76, p. 26; J. E. Morton, 1954, *Jour. Marine Biol. Assoc. U.K.*, vol. 33, p. 297.

Subgenus *Thilea* Strebel, 1908

Last whorl ⅘ or more of the entire shell; smooth; umbilicus narrow. No protrusion on anterior border of wing. Tropical and semitropical waters. Type: *helicoides* Jeffreys,

1877, from the eastern North and South Atlantic. *Embolus* Jeffreys, 1869, is a synonym. *Thielea* is a misspelling.

Limacina inflata (Orbigny, 1836)   **4044**
Planorbid Pteropod

> North Atlantic to Florida to Brazil.
> World seas; Polynesia.

1.3 to 1.5 mm. in diameter, spire depressed, with globose whorls in one plane to give a planorboid shape. Outer lip with a toothlike projection at the periphery. Umbilicus deep. Operculum thin, horny, coiled sinistrally. Synonyms: *rostralis* Eydoux and Souleyet, 1840; *scaphoides* Gould, 1852; *elata* Costa, 1861. Illustration from S. van der Spoel, 1967.

Limacina lesueurii (Orbigny, 1836)   **4045**
Lesueur's Pteropod

> Off Massachusetts to Brazil to Europe.
> Eastern Pacific.

1.0 to 1.5 mm. in length, spire elevated somewhat; 4 to 5 whorls; umbilicus distinct. Shell not as high as wide. Spiral lines only around the umbilicus.

Subgenus *Munthea* van der Spoel, 1967

Shells small; last whorl not proportionately very large. Surface smooth. Umbilicus narrow. Anterior border of wings without a protrusion. Type: *trochiformis* (Orbigny, 1836).

Limacina trochiformis (Orbigny, 1836)   **4046**
Trochiform Pteropod

> Massachusetts to the Gulf of Mexico to Brazil.
> Baja California southward.

1 mm. in length, very close in characters, except shape, to *L. retroversa*. Color of shell white, with the thicker parts purple-brown. Umbilicus narrow and deep. Columellar aperture border concave. Synonyms: *naticoides* Souleyet, 1852; *contorta* Sykes, 1905.

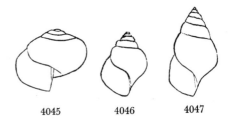

4045   4046   4047

Limacina bulimoides (Orbigny, 1836)   **4047**
Bulimoid Pteropod

> New York to West Florida and southern Brazil.
> Far off southern California southward.

2 mm. in length, spire high, shell twice as long as wide. Umbilicus very indistinct. Lip fragile and often broken. 6 to 7 whorls with a distinct brown suture. Thickened inner aperture is chestnut-brown.

Family CUVIERIDAE Gray, 1840

Shell symmetrical (not coiled), fragile, white to brown, and of various shapes—needlelike, cylinder-shaped, flattened triangular or bulbous. No operculum. Anus opens on the left.

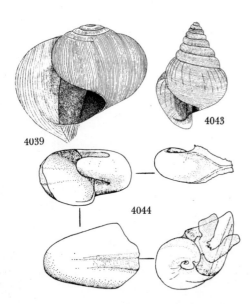

4039   4043

4044

Subfamily CLIONAE Jeffreys, 1869

Genus *Creseis* Rang, 1828

Shell a long cone, almost circular in cross-section, needle-like. Type: *virgula* Rang, 1828. Synonym: *Boasia* C. W. Johnson, 1934.

Creseis acicula (Rang, 1828)   **4048**
Straight Needle-pteropod

> Atlantic and Pacific, pelagic (50°N to 40°S).
> Baja California southward.

20 to 33 mm. (about an inch) in length. A long, straight, slender cone tapering to a sharp point. The shape of this species varies greatly, and some authors consider *virgula* to be an extreme form connected by the formae *clava* (Rang, 1828) and *conica* (Eschscholtz, 1829). All forms occur in our Atlantic offshore waters. Other synonyms: *acus* Eschscholtz, 1829; *aciculata* Orbigny, 1836; *recta* Gray, 1850; *vitrea* Verrill, 1872; *conoidea* Costa, 1873; *unguis* Eschscholtz, 1829; *cornucopiae* and *caligula* Eschscholtz, 1829; *corniformis* Orbigny, 1836; *falcata*, *placida* and *munda* Gould, 1852; *flexa* Pfeffer, 1879; *africana* Bartsch and *constricta* Chen and Bé, 1964. For behavioral biology, see L. S. Kornicker, 1959, *Bull. Marine Sci.*, vol. 9, p. 331.

Creseis virgula (Rang, 1828)   **4049**
Curved Needle-pteropod

> Atlantic and Pacific, pelagic (45°N to 40°S).
> Central California southward.

8 to 10 mm. in length. A drawn-out, slender shell similar to *acicula*, but with its narrow end hooked to one side. The amount of bend of hook is variable.

Genus *Styliola* Lesueur, 1825

Conical shell with a finely pointed posterior. A single longitudinal rib runs from the anterior dorsal edge of the aperture, angling to the left, back to the small tip. Type: *subula* (Quoy and Gaimard, 1827). *Styliola* Gray, 1850, is the same.

Styliola subula (Quoy and Gaimard, 1827)   **4050**
Keeled Clio

> Worldwide in warm seas, pelagic.
> Massachusetts to the Gulf of Mexico to Brazil.
> Oregon to Baja California.

10 mm. in length, conical, straight, considerably elongated. The surface is smooth, and with a dorsal groove not parallel to the axis of the shell, but slightly oblique, turning from left to right, with only the anterior extremity (which ends in a rostrum) in the median line. There is only one species in the genus and it is worldwide in distribution. Synonyms: *spinifera* Rang, 1829; *subulata* A. Adams, 1853.

Genus *Hyalocylis* Fol, 1875

Conical shell without protoconch, posterior tip truncate and open. Raised transverse annulations over fragile shell. *Hyalocylix* is a misspelling. Type and only species *striata* (Rang, 1828).

Hyalocylis striata (Rang, 1828)   **4051**
Striate Clio

> Worldwide in warm seas, pelagic.
> Nova Scotia to Gulf of Mexico to Brazil.

8 mm. in length, conical, slightly compressed dorso-ventrally (oval in cross-section); apex slightly recurved dorsally; surface with transverse grooves; embryonic shell small, smooth, bulbous and separated from the main shell by a constriction. Synonyms: *compressa* Eschscholtz, 1829; *zonata* Chiaje, 1841; *annulata* Deshayes, 1853; *phaeostoma* Troschel, 1854, *chierchiae* Boas, 1886. It has not been recorded from the Pacific coast of United States.

Genus *Clio* Linné, 1767

Shell of a somewhat angular form, colorless, compressed dorso-ventrally, and with lateral keels. A cross-section of the anterior or open portion is thus always angular at the sides. There is generally a crest or ridge extending longitudinally along the back. Embryonic shell varies in form, but is always definitely separated from the rest of the shell. *Euclio* Bonnevie, 1913; *Balantium* Children, 1823; *Proclio* Hubendick, 1951; and *Cleodora* Péron and Lesueur, 1810, are the same genus. Type: *pyramidata* Linné, 1767.

Clio pyramidata Linné, 1767   **4052**
Pyramid Clio

> Worldwide, pelagic.
> East and West United States.

16 to 21 mm. in length. No lateral keels on the posterior portion; without lateral spines. Lateral margins very divergent. No posterior transverse grooves. Dorsal ribs undivided. Common. The shell exhibits considerable variation in form. The following forms are recognized: 1, *lanceolata* (**4053**) Lesueur, 1813 (synonyms: *caudata* Lamarck, 1819; *exacuta* Gould, 1852; *occidentalis* Dall, 1871; *lobata* Sowerby, 1878; *lata* Boas, 1886); 2, *convexa* (**4054**) (Boas, 1886); 3, *martensii* (Pfeffer, 1880); 4, *excisa* van der Spoel, 1963; 5, *sulcata* (Pfeffer, 1879); 6, *antarctica* Dall, 1908. The last 3 are found only in antarctic waters. See van der Spoel, 1969, *Vidensk. Meddr. Dansk Naturh. Foren.*, vol. 132, pp. 95–114.

Clio cuspidata (Bosc, 1802)   **4055**
Cuspidate Clio

> Atlantic and Eastern Pacific, pelagic.
> Indo-Pacific.

Up to 20 mm. Without lateral keels on the posterior portion. Lateral spines very long. Hyaline glassy. Common. Synonym: *tricuspidata* Bowdich, 1820.

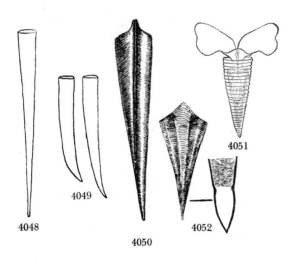

4048   4049   4050   4051   4052

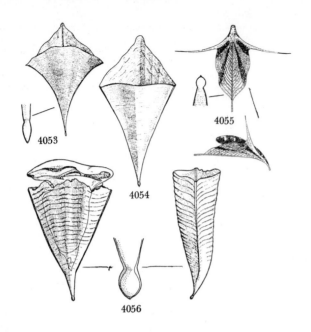

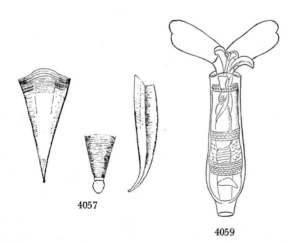

Clio recurva (Children, 1823)   4056
Wavy Clio

Worldwide. Warm water, pelagic.
New York to the Gulf of Mexico. California.

1 inch in length, with lateral keels over its entire length. 3 dorsal ribs markedly projecting. A large, fragile, transparent and very exquisite species. This is *C. balantium* Rang, 1834. Rare in the Californian current. Found at various depths. Illustration from S. van der Spoel, 1967.

Clio polita (Pelseneer, 1888)   4057
Two-keeled Clio

Atlantic and Eastern Pacific; bathypelagic.

Up to 14 mm. With lateral keels over its entire length. Dorsal ribs very slightly projecting. The posterior portion of the shell is narrow. Rare. *C. falcata* Pfeffer, 1880, is the same species. Lives at depths of 135 to 270 meters.

Clio chaptalii Gray, 1850   4058
Chaptal's Clio

Florida to West Africa; bathypelagic.
Gulf of Panama; Indo-Pacific.

Up to 19 mm. Outline of shell in ventral view even more perfectly triangular than *Clio recurva*, and the lateral ribs are sharp and never flattened like those in *recurva*. Wavy transverse striations on the ventral and dorsal sides. 3 dorsal ribs, none underneath. Width of aperture ¾ length of shell.

Subfamily CUVIERININAE Gray, 1840

Genus *Cuvierina* Boas, 1886

Shell cylindrical, shaped somewhat like a fat cigar. Surface smooth. A cross-section is almost circular. Behind the aperture the shell is slightly constricted. There is only one species in the genus. The genera *Cuvieria* Rang, 1827 (non Péron and Lesueur, 1807) and *Herse* Gistel 1848 (non Oken 1815) are synonyms. Type and only species: *columnella* Rang, 1827.

Cuvierina columnella (Rang, 1827)   4059
Cigar Pteropod

Worldwide, pelagic (50°N to 42°S).

4 to 10 mm. in length. See generic description and figure. The shell varies somewhat in shape. *C. oryza* Benson, 1835, *C. urceolaris* (Mörch, 1850) of the Indo-Pacific, and *cancellata* Pfeffer, 1880, *obtusa* Orbigny, 1839; *rosea* Gray, 1847, and *caliciformis* Meisenheimer, 1905, are the same. Common in the Atlantic; rare off California. The Atlantic Ocean form (forma *atlantica* van der Spoel, 1970, *Basteria*, vol. 34, p. 115) is about 8 to 10 mm. *Styliola sinecosta* F. E. Wells, 1974, from the West Indies (*Veliger*, vol. 16, p. 293) is probably a synonym.

Subfamily CAVOLINIINAE H. and A. Adams, 1854

Genus *Diacria* Gray, 1847

Similar to *Cavolinia s.s.*, but the dorsal lip of the shell is thickened into a pad, and not thin as in the true Cavoline. *Pleuropus* Pfeffer, 1879, is a synonym. Type: *trispinosa* Blainville, 1821.

Diacria trispinosa (Blainville, 1821)   4060
Three-spined Cavoline

Worldwide, pelagic (60°N to 41°S).
Nova Scotia to Brazil.
Off California.

About 5 to 13 mm. in length. Dorsal lip thickened into a pad. Shell with a long lateral spine on each side, and a very long terminal one which falls off in later life. Aperture scarcely discernible. Ventral side of shell very slightly convex. *C. mucronata* Quoy and Gaimard, 1827, *C. cuspidata* Delle Chiaje, 1841, and *C. major* (Boas, 1886) and *C. reeviana* Dunker, 1853, are this species. Very common.

Diacria quadridentata (Blainville, 1821)   4061
Four-toothed Cavoline

Worldwide, pelagic (40°N to 30°S).

2 to 4 mm. in length. Dorsal lip thickened into a pad. Without prominent lateral spines. Aperture well-developed. Ventral side greatly inflated. Upper lip longer than the bottom one. *O. inermis* Gould, 1852; *minuta* Sowerby, 1878; *orbignyi* Souleyet, 1852; *intermedia* Sowerby, 1878; *costata* Pfeffer, 1880; and

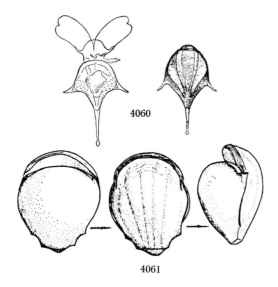

4060

4061

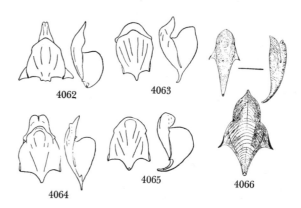

4062   4063

4064   4065   4066

quadrospinosa Orbigny, 1836, are synonyms. Quite common; both sides of United States. The Atlantic forma was named *danae* van der Spoel, 1969 (*Basteria,* vol. 33, p. 105). Illustration from S. van der Spoel, 1967.

Genus *Cavolinia* Abildgaard, 1791

Shell squat, bulbous, horny-brown in color, characterized by a much-constricted aperture, which is, however, very broad transversely. Sides of shell often prolonged into spine-like projections. *Cavolinia* is a conserved name. The original spelling was *Cavolina* (non Bruguière, 1792). Type: *tridentata* (Niebuhr, 1775). *Hyalaea* Lamarck, 1801 is a synonym of this genus.

Cavolinia longirostris (Blainville, 1821)   4062
Long-snout Cavoline

Worldwide, pelagic (50°N to 50°S).

5 to 9 mm. in length. Dorsal lip with a thin margin. Posterior portion of the ventral lip markedly projecting laterally. Common. *Hyalaea limbata* Orbigny, 1836; *ecaudata* Blainville, 1821; *femorata* Gould, 1852; *fissilabris* Benson, 1861; *strangulata* Deshayes, 1823; *angulosa* Gray, 1850; *angulata* Souleyet, 1852; *obtusa* Sowerby, 1878; and *couthouyi* Dall, 1908, are synonyms.

Cavolinia gibbosa (Orbigny, 1836)   4063
Gibbose Cavoline

Worldwide, pelagic (45°N to 38°S).

About 8 to 10 mm. in length. Dorsal lip with a thin margin. Shell without appreciable lateral points. Ventral lip not more developed than the dorsal. Ventral surface with an anterior transverse keel. *C. hargeri* Verrill, 1882; *flava* Orbigny, 1836; *plana* Meisenheimer, 1905; and *gegenbauri* Pfeffer, 1880, are synonyms. Common.

Cavolinia tridentata (Niebuhr, 1775)   4064
Three-toothed Cavoline

Worldwide, pelagic (50°N to 45°S).
Newfoundland to the Gulf of Mexico.
Off California.

10 to 20 mm. in length. Dorsal lip with a thin margin. Ventral lip not more developed than the dorsal one. Shell without appreciable lateral points. The shell is as broad at the end of the lips as it is at the anterior end. *C. gibbosa* is narrower at the ends of the lips. *Hyalaea affinis* Orbigny, 1836, is merely a form of this species. *C. telemus* Linné, 1758, might possibly be this species. There are numerous synonyms and several forms.

Cavolinia uncinata (Rang, 1829)   4065
Uncinate Cavoline

Worldwide, pelagic (45°N to 45°S).
Newfoundland to Brazil.
Costa Rica to Ecuador.

6 to 11 mm. in length. Brownish. Dorsal lip with a thin margin. Ventral lip not more developed than the dorsal one. Shell with distinct lateral points. Upper lip flattened posteriorly. Generally a warm-water species. *C. uncinatiformis* Pfeffer, 1880, and *roperi* and *pulsata* van der Spoel, 1967, are synonyms.

Cavolinia inflexa (Lesueur, 1813)   4066
Inflexed Cavoline

Worldwide; warm waters (55°N to 45°S).

3 to 8 mm. in length, similar to *uncinata*, but the upper lip is directed straight forward, instead of flattened posteriorly; and the' ventral side of the shell is weakly, instead of strongly convex. *C. labiata* Orbigny, 1836; *C. imitans* Pfeffer, 1879; *C. elongata* Blainville, 1821; *pellucida* Eschscholtz, 1825; *vaginellina* Cantraine, 1835; *depressa* Orbigny, 1836; *pleuropus* Souleyet, 1852; *longa* Boas, 1886; *curvata* Grieg, 1924; *lata* Boas, 1886; and *labiata* Tesch, 1913, are this species. Abundant from 88 meters in the day to 98 meters at night. See Wormelle, 1962, *Bull. Marine Sci.,* vol. 12, p. 93.

Other species:

— *Cavolinia globulosa* (Gray, 1850). Southwest Pacific.

Family PERACLIDIDAE C. W. Johnson, 1915

Shell fragile, with sinistral or left-handed whorls (resembling the fresh-water pond snail, *Physa*); aperture very large and elongated; columella prolonged into an elongate rostrum; no umbilicus. A noncalcareous hexagonal mesh-work covers much of the shell. Operculum thin, paucispiral, sinistral and subcircular in outline.

Genus *Peracle* Forbes, 1844

Characters of the family. 3 species known. *Peraclis* Pelseneer is an emendation. *Procymbulia* Meisenheimer, 1905, is a synonym.

Peracle reticulata (Orbigny, 1836)　　　　4067
Reticulate Pteropod

Worldwide, pelagic (40°N to 20°S).
Off California southward.

4 mm. in length, brownish yellow, sinistral and with 4 whorls. Suture deep. The surface exhibits a raised hexagonal reticulation, the sides of the hexagons bearing a regular row of minute teeth. *P. physoides* Forbes, 1844, *P. clathrata* Eydoux and Souleyet and *diversa* Monterosato, 1875, are the same.

Peracle bispinosa Pelseneer, 1888　　　　4068
Two-spined Pteropod

Atlantic, pelagic (38°N to 28°S).
Eastern Pacific, Baja California south.

7 mm. in length, milky-white, similar to *reticulata*, but with a wide, shallow suture bearing axial ridges, and with the shoulder of the outer lip bearing a small, triangular projection. Uncommonly collected.

Other species:

4069 *Peracle apicifulva* Meisenheimer, 1906. Eastern Pacific; California southward. Bermuda.

4070 *Peracle tricantha* Fischer, 1882. North Atlantic; off Bermuda.

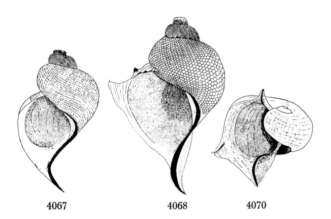

4067　　　　4068　　　　4070

Family CYMBULIIDAE Gray, 1840

Animal translucent, with a large, boat-shaped, transparent, gelatinous pseudoconch or "soft shell." Fins united to form 1 large swimming plate.

Genus *Cymbulia* Péron and Lesueur, 1810

Soft shell, spinose, elongate, pointed sharply in front and concavely squared off posteriorly. Swimming plate bilobed; centrally located. Mouth furnished with minute tentacles.

Cymbulia peroni Blainville, 1818　　　　4071
Péron's Cymbulia

Tropical Eastern Pacific, Indo-Pacific.

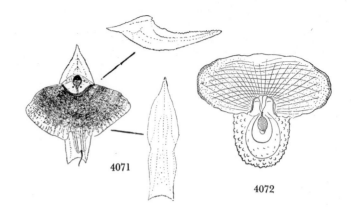

4071　　　　4072

½ to ¾ inch, characterized by its oblong, boat-shaped gelatinous, spinose pseudoconch and the 2 centrally located oval, transparent swimming lobes. Uncommon.

Genus *Corolla* Dall, 1871

Animal milky-translucent. Gelatinous, tuberculate shell ovoid-slipper shaped. Swimming plate large oval, with reticulated muscle bands showing, and with a deep sinus near the mouth region. Type: *spectabilis* Dall, 1871. *Cymbuliopsis* Pelseneer, 1888, is a synonym.

Corolla calceola Verrill, 1880　　　　4072
Atlantic Corolla

Northwest Atlantic; pelagic at surface.

1½ inches, gelatinous shell tuberculate. Swimming lobe large, oval. No small foot lobes present in the region of the eyes and mouth. Uncommon.

Corolla spectabilis Dall, 1871　　　　4073
Spectacular Corolla

Off west North America; pelagic at surface.

2 inches, similar to the Atlantic *calceola*, but has 2 small, but prominent, flaplike foot lobes. Moderately common in cool waters at the surface. Patches of them occur infrequently southward. *Cymbuliopsis vitrea* Heath and Spaulding, 1901, from Monterey Bay, California, may be a synonym.

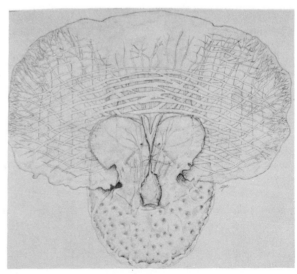

4073

Family DESMOPTERIDAE Dall, 1921

Without shell pseudoconch or proboscis. Swimming plate attached to the cylindrical body at a small area of its midpoint. Radula small. One genus:

Genus *Desmopterus* Chun, 1889

Body relatively small, cylindrical, stubby, with vestigial tentacles. Swimming plate broad oval, thin, with 3 lobes along the posterior edge and with 2 small short epipodial tentacles. One Pacific species:

4074 *Desmopterus pacificus* Essenberg, 1919. Off southern California. Width of swimming plate is 3 mm. *Univ. Cal. Publ. Zool.*, vol. 19, no. 2, p. 85. (Illustration from J. A. McGowan, 1968, *The Veliger*, supplement to vol. 3.)

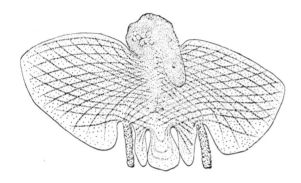

4074a

4074a *Desmopterus papilio* Chun, 1889. 5 mm. 5 lobes posteriorly. Newfoundland to Cape Hatteras, North Carolina; western Europe, open Atlantic. Illus. from van der Spoel, 1972, Conseil Internat. Explor. de la Mer, Zooplankton sheet 140–142. *D. pacificus* may be a synonym.

Order *Gymnosomata* Blainville, 1824

Pteropods characterized by the absence of shell, pallial cavity and mantle-skirt; by the presence of a well-developed head, bearing 2 pairs of tentacles, of which the 2 posterior bear rudimentary eyes. Jaws and radula present. Penis on the right side of the foot, anus on the right side. Found pelagic in all seas, and sometimes in great abundance. Rarely exceed 1 inch in length. They are carnivorous, feeding on shelled pteropods. Ascend to the surface at night, and sink to a lower level in the daytime. When Mother Nature had finished making the marine mollusks and had stamped out the familiar types, she used the leftover scraps to make this group of odd-looking creatures. So ashamed were they at the results, they took up a hidden life in the open seas, coming near the surface only at night.

Family CLIONIDAE Rafinesque, 1815

Body cylindrical, about 1 inch, tapering at the posterior end. Anterior end with 2 short tentacles and 2 or 3 pairs of conical, buccal appendages. Fin lobes relatively small, oval, 1 on each side of the constricted area in back of the head. No jaws; no gills. Skin not pigmented.

Genus *Clione* Pallas, 1774

Head with 2 eye-tubercles and 2 simple tentacula. Mouth with lateral lobes, each supporting 2 or 3 conical retractile processes, furnished with microscopic suckers. Fins ovate; foot lobed. Radula with 1 central and 12 laterals on each side. Fins flap 180 degrees when used in swimming. Food of whales.

Clione limacina (Phipps, 1774)   **4075**
Common Clione

Arctic Seas to North Carolina.
Alaska–Canada–northern Europe.

C. papillonacea Verrill, 1873, *elegantissima* Dall, 1871, and *kincaidi* Agersborg, are synonyms. Seasonally abundant. Sometimes cast ashore (*The Nautilis*, vol. 73, p. 76). For biology, see M. V. Lebour, 1931, *Jour. Marine Biol. Assoc. U. K.*, vol. 17, p. 785, and J. E. Morton, 1958, *loc. cit.*, vol. 37, p. 287.

4075

Genus *Clionina* Pruvot-Fol, 1924

Similar to *Clione* but with 2 pairs of buccal cones instead of 3. Central radular plate with a single, long cusp. Type: *longicaudata* (Souleyet, 1852).

4076 *Clionina longicaudata* (Souleyet, 1852). 3 to 10 mm. Widely distributed in the tropical Atlantic, Indian Ocean and Western Pacific. See A. Franc, 1951, *Bull. Soc. France*, vol. 76, p. 5.

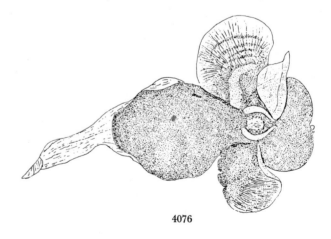

4076

Genus *Paedoclione* Danforth, 1907

Resembling *Clione limacina*, but there is only 1 cephalocone on the left, and 2 on the right. Type: *doliiformis* Danforth, 1907. Could be an aberrant *Clione*.

4077 *Paedoclione doliiformis* Danforth, 1907. Gulf of Maine, pelagic. *Proc. Boston Soc. Nat. Hist.*, vol. 34, pp. 1–19. Rarely cast ashore.

Family CLIOPSIDAE Costa, 1873

With a long, protrusible proboscis, but lack buccal appendages. Jaw present. Not pigmented.

Genus *Cliopsis* Troschel, 1854

Type: *krohni* Troschel, 1854. *Clionopsis* is a misspelling.

4078 *Cliopsis krohni* Troschel, 1854. Off Cape Hatteras; warm-water Atlantic; Indo-Pacific; off Pacific Panama. Mediterranean. Synonyms: *grandis* Boas, 1886; *modesta* Pelseneer, 1887; *microcephalus* Tesch, 1904.

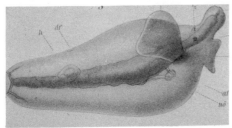

Family THLIPTODONTIDAE Kwietniewski, 1910

Head very large, almost ½ the mass of the entire body, and not set off by a constriction of the body. Pharynx large but there are no jaws. Radula large, triseriate, well-developed. No external "gills."

Genus *Thliptodon* Boas, 1886

Pharynx contains large, saclike spaces containing numerous, minute hooks. Type: *gegenbauri* Boas, 1886. *Pteroceanis* Meisenheimer, 1903, is a synonym.

4079 *Thliptodon diaphanus* (Meisenheimer, 1903). Worldwide warm to boreal seas. Common.

Family PNEUMODERMATIDAE Latreille, 1825

With suckers on the ventral side of the protrusible proboscis. They have a lateral gill-like structure on their right side. Jaw present. Skin pigmented.

Genus *Pneumoderma* Cuvier, 1805

With 2 lateral acetabuliferous arms and elaborately fringed lateral and hind gills. Two recognized species. Type:

atlanticum (Oken, 1816). *Pneumodermon, Pneumonoderma* and *Pneumodermis* are illegal emendations of the name.

Other species:

4080 *Pneumoderma atlanticum* (Oken, 1815). Abundant. World-wide warm seas. Caribbean; Bay of Panama. Pelagic. Synonyms: *violaceum* Orbigny, 1835; *peronii* Lamarck, 1819; *boasi* Pelseneer, 1887; *pacificum* Dall, 1872; *souleyeti* Pelseneer, 1887; and probably *eurycotylum* Meisenheimer, 1905, *heterocotylum* Tesch, 1904, and *meisenheimeri* Pruvot-Fol, 1926.

4081 *Pneumoderma mediterraneum* (van Beneden, 1838). Tropical Atlantic; Mediterranean; Indian Ocean.

Genus *Crucibranchaea* Pruvot-Fol, 1942

Resembling *Pneumodermopsis*, but has no lateral gill, only a posterior one with 4 branches. With very strong, thick lateral arms bearing 2 or 3 rows of small suckers, and having a very large one at each end. One species, the type: *macrochira* (Meisenheimer, 1905). Van der Spoel, *Basteria,* vol. 36, 1972, considers this a subgenus of *Pneumodermopsis*.

4082 *Crucibranchaea macrochira* (Meisenheimer, 1905). ½ inch. North Atlantic from the West Indies to England. Indo-Pacific. Worldwide pelagic.

Genus *Pneumodermopsis* Keferstein, 1862

Lateral gill generally present; posterior gill mostly absent. One median acetabuliferous arm, and at either side of its base lateral groups of suckers carried on very short arms, or sessile on walls of the buccal cavity. Type: *ciliata* (Gegenbaur, 1855). About 7 species. A key is given by Tesch. 1950, p. 32. For ecology, see G. A. Cooper and D. Forsyth, 1963, *Bull. Marine Ecology,* vol. 6, pp. 31–38.

4083 *Pneumodermopsis ciliata* (Gegenbaur, 1855). Greenland and Iceland to West Africa; North Pacific; Indo-Pacific. Common.

Family NOTOBRANCHAEIDAE Pelseneer, 1886

Genus *Notobranchaea* Pelseneer, 1886

2 pairs of conical buccal appendages associated with the proboscis. Posterior "gill" made up of 3 radiating crests

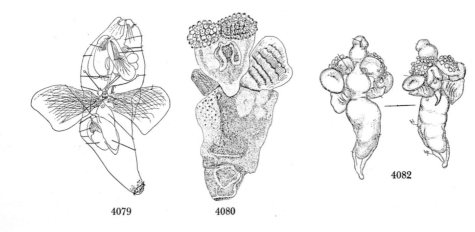

4079   4080

4082

4083

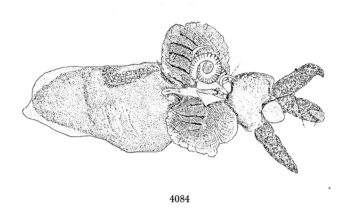

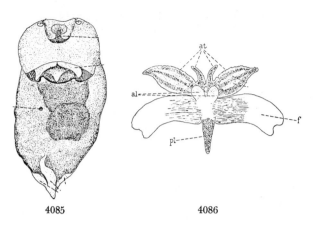

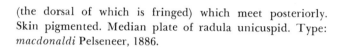

4084   4085   4086

(the dorsal of which is fringed) which meet posteriorly. Skin pigmented. Median plate of radula unicuspid. Type: *macdonaldi* Pelseneer, 1886.

4084 *Notobranchaea macdonaldi* Pelseneer, 1886. Worldwide; West Indies; Pacific Panama; Eastern Atlantic; Indo-Pacific. 1 inch. Synonyms: *inopinata* Pelseneer, 1887; *pelseneeri* Pruvot-Fol, 1942; *grandis* Pruvot-Fol, 1942.

Genus *Prionoglossa* Tesch, 1950

Similar to *Notobranchaea*. No buccal cones. Hind gill tetraradiate. Median plate of radula sawlike (with many small denticles on the concave side). Type: *tetrabranchiata* (Bonnevie, 1913). One species recognized.

4085 *Prionoglossa tetrabranchiata* (Bonnevie, 1913). Western and Eastern Atlantic; Indo-Pacific; pelagic.

Suborder *Gymnoptera* Van der Spoel, 1972

Family HYDROMYLIDAE Pruvot-Fol, 1942

Animal short, rounded; head distinct, retractile into a pouch formed by a thickening of the mantle; swimming lobes long and narrow. Radula present. Eurybiidae Troschel, 1856, and Halopsychidae Cooke, 1895, are synonyms. The family Anopsiidae Hoffmann, 1931 (and Thiele, 1931) is synonymous.

Genus *Hydromyles* Gistel, 1848

Body small, globular; with 2 elongate-oval swimming lobes. No tentacles. *Psyche* Rang, 1825, non Linné, 1758; *Euribia* Rang, 1827 (non Meigen, 1800); *Halopsyche* Bronn, 1862; *Hydromyles* Gistel, 1848; and *Verrillopsyche* Cossmann, 1910, are synonyms. *Eurybia* is a misspelling. *Anopsia* Gistel, 1848, is a synonym.

4086 *Hydromyles globulosa* (Rang, 1825). North Atlantic records may be erroneous. Found in North Pacific and Indo-Pacific abundantly. *Anopsia gaudichaudii* Souleyet, 1852, is a synonym. Figure symbols: al = anterior footlobe; at = anterior tentacles; f = fins; pl = posterior footlobe.

Order *Basommatophora* A. Schmidt, 1855

Without ctenidial gills; lung with a small pneumostome or spoutlike opening. Head with 1 pair of tentacles. Usually without an operculum. No free-swimming larvae. Includes the fresh-water lymnaeid, physid and ancylid snails, as well as the melampids and siphonariids.

Superfamily Melampidacea Stimpson, 1851

The superfamily Ellobiacea is synonymous.

An account of the ecological and habitat preferences of the ellobiids of the Florida Keys is given by J. P. E. Morrison, 1958, *The Nautilus*, vol. 71, pp. 118–124. Most live in shady, muddy areas under mangroves in the intertidal area or under logs, coconut husks and rocks at the high-tide debris zone. The genus *Sayella* is now placed in the Pyramidellidae.

Family MELAMPIDAE Stimpson, 1851

Small brown, high-spired snails living in shady areas above low-tide line in marine or brackish areas. Aperture armed with teeth. Sperm groove (between female and male openings) open and ciliated. The family names Auriculidae and Ellobiidae are synonyms.

Genus *Melampus* Montfort, 1810

Shells ovate-fusiform, small, brown, yellow or with narrow bands. Columella with several strong teeth. Synonyms: *Conovulus* Lamarck, 1816; *Melampa* Schweigger, 1820. Type: *coffeus* Linné, 1758. These vegetarian airbreathers lay their gelatinous eggmasses on marsh grasses.

***Melampus bidentatus* Say, 1822**   **4087**
Eastern Melampus

Southern Quebec to Texas; West Indies; Bermuda.

10 to 15 mm., 5 or 6 whorls, the last one being ¾ the length of the shell. Spire short, blunt. Shell ovate-conical, broadest at about the upper ⅓ where there is a faint angulation. Color of fresh specimens, brownish horn, smooth, shining, with 3 or 4 darker narrow bands. Often eroded and gray with longitudinal wrinkles. In adults, there are 1 to 4 elevated lirae within the outer lip; inner lip has 2 folds. Abundant in salt marshes on stems of grasses. Animal reddish brown. Synonyms: *corneus* (Deshayes, 1830); *lineatus* Say,

4087   4092

1822; *redfieldi* Pfeiffer, 1854 (from Bermuda). The upper shoulders of the whorls in this species have spiral incised lines, while in *coffeus* they are smooth. For a life history study, see P. A. Holle, 1957, *The Nautilus*, vol. 70, p. 90, but for correct name see Morrison, 1964, *The Nautilus*, vol. 77, p. 119. Free-swimming veligers hatch at spring high tides (see *Biol. Bull.*, vol. 139, p. 434, (1970)).

Melampus coffeus (Linné, 1758)   4088
Coffee Melampus

South half of Florida; the West Indies to Brazil. Bermuda.

12 to 20 mm. Restricted to areas of mangroves where they live on mud flats. The upper shoulders of the shell are lacking spiral incised lines. Synonym: *olivula* (Moricand, 1844) from Brazil.

Melampus olivaceus Carpenter, 1857   4089
Californian Melampus

Mugu Lagoon, California, and the Gulf of California.

10 to 15 mm., ovoid, solid, low-spired, with rounded shoulder. Color brown with 2 or 3 spiral whitish bands. Periostracum brownish. Columella short, whitish and with a white fold at the middle. Outer lip sharp, bordered with about a dozen fine white teeth. Abundant in lagoons and back bays at the high-tide line where debris gives shade and moisture. A subspecies was described from Mission Bay, Pacific Beach, California—*californianus* S. S. Berry, 1964 (*Leaflets in Malacology*, no. 24, p. 153).

Subgenus *Pira* H. and A. Adams, 1855

Type: *fasciata* (Deshayes, 1830) fide Morrison, 1964.

Melampus monilis (Bruguière, 1789)   4090
Caribbean Melampus

Florida; West Indies to Brazil. Bermuda.

Characterized by a single spiral row of epidermal setae (or pit-scars if the setae have rubbed off) in the middle of the whorls on the spire. Common on mud and among rocks near

mangroves rather near the low-tide line. Misnamed as "*flavus* Gmelin" by Holle and Dineen, 1957 and 1959.

Other species:

4091 *Melampus mousleyi* S. S. Berry, 1964. Cholla Cove, Bahia de Adair, Sonora, Mexico. *Leaflets in Malacology*, no. 24, p. 152.

Genus *Detracia* Gray, in Turton, 1840

Shells brown, ovate-fusiform. Type: *bullaoides* (Montagu, 1808). Synonyms: *Tifata* H. and A. Adams, 1855; *Eusiphorus* Conrad, 1863. Eggs laid in small gelatinous masses. Larvae pelagic.

Detracia bullaoides (Montagu, 1808)   4092
Bubble Melampus

South half of Florida; upper Caribbean; Bermuda.

10 to 12 mm., elongate, widest part is at the middle of the last whorl. Spire obtuse, with 9 whorls and a minute white pimplelike apex. Color glossy, coffee-brown with white axial flames in the spire. Base of columella brown, with single prominent tooth set at right angles. Common in shady areas near mangroves in intertidal areas.

Detracia floridana (Pfeiffer, 1856)   4093
Florida Melampus

Delaware to Louisiana.

6 to 8 mm., ovate-conic to fusiform, dark-brown with a few narrow bands of whitish. Columella with a moderately large tooth (not upturned). Above it on the parietal wall is a smaller horizontal lamella or tooth. Behind the outer lip there are about 10 fine, spiral, white, raised lirae. An abundant species in salt marshes (according to Morrison, 1951, 4 billion individuals per square mile). Gelatinous, ovoid, dome-shaped eggmasses 3 mm. in diameter, usually covered by detritus, and containing 20 to 50 eggs. Larvae free-swimming (J. P. E. Morrison, 1953, *Ann. Rep. Amer. Malacol. Soc.*, p. 15).

Detracia clarki Morrison, 1951   4094
Clark's Melampus

Southeast Florida and north Cuba.

6 to 13 mm. long, subcylindrical, smoothish, with 10 to 12 whorls. A few spiral incised lines are present above the shoulder and near the base. Lower columella with 2 strong white, dished, upturned teeth, the lower one extending halfway over the outer lip. Within the outer lip there are several spiral, white lirae. Color brown with irregularly sized spiral bands of whitish. Do not confuse with young or dwarf *Melampus bidentatus*. Common on intertidal mud flats in the shade of mangroves, associated with *Melampus coffeus*.

Genus *Tralia* Gray in Turton, 1840

Type of the genus is *ovula* (Bruguière, 1789).

Tralia ovula (Bruguière, 1789)   4095
Egg Melampus

Southeast Florida to Barbados; Bermuda.

10 to 15 mm., elongate-ovoid, glossy, rich deep chestnut color, sometimes blackish; 3 whitish folds on the inner lip (1 on the columella, 2 on the parietal wall). Outer lip thickened, inflected in the center with a single revolving ridge on its inside. Moderately common. *Voluta pusilla* Gmelin, 1791, is a synonym.

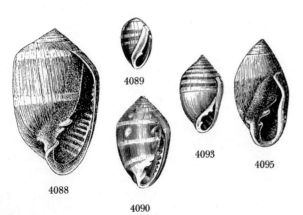

4089

4088

4090

4093

4095

Subfamily PEDIPEDINAE Crosse and Fischer, 1880

Genus *Pedipes* Bruguière, 1792

There are only 2 species in North America of this genus of small land snails living in isolated colonies in damp, brackish-water areas. Seldom over 8 mm. in length; sculptured with incised spiral grooves; rounded-ovate; no umbilicus; 3 plicae on inner lip, only 1 on outer lip. Young are very thin-shelled and lack tooth on outer lip. No operculum. *Carassa* Gistel, 1847, is a synonym. Type: *afer* Gmelin, 1791.

Pedipes mirabilis (Mühlfeld, 1816)   4096
Miraculous Pedipes

Bermuda; southeast Florida to Texas and to Brazil.

3 to 5 mm., solid, 4 to 5 whorls; brown; suture distinct; with fine spiral grooves; upper parietal plica long, lower 2 shorter; 1 plica on outer lip of adults. Colonies are sporadic in occurrence and differ in size of individuals. Common near brackish warm water. Under rocks above tide line. The synonyms are: *quadridens* Pfeiffer, 1839; *globulosus* C. B. Adams, 1845; *ovalis* C. B. Adams, 1849; *tridens* Pfeiffer, 1854; *globulus* Pfeiffer, 1856; *naticoides* Stearns, 1869 (from Tampa, Florida); and *insularis* Haas, 1950 (from Bermuda).

Pedipes angulatus C. B. Adams, 1852   4097
Angular Pedipes

Southern California to Panama.

5 to 8 mm., globose with slightly angled shoulder, and hence an expanded, bell-shaped outer lip; chestnut-brown; lower columella broad and with 3 teeth. Outer lip with 1 large, low tooth. Colonies appear to be rare; individuals abundant. *P. lirata* Binney, 1860, is a synonym.

Pedipes unisulcatus Cooper, 1866   4098
One-grooved Pedipes

Palos Verdes, California, and the Gulf of California.

5 to 7 mm., light-brown, ovoid, with a rather high, slightly convex spire. Nuclear whorls spirally striate, last whorl smooth. Aperture large; the outer lip thickened on the inner rim; columella wall with 3 very strong plicae, the uppermost projecting downward. Common; under slabs of rocks at the upper tide line. *P. biangulatus* Jaeckel, 1927, is a synonym.

Genus *Marinula* King and Broderip, 1832

Resembling *Laemodonta,* up to 10 mm., but smoothish, 3 plicae on inner lip, but none on outer. Synonyms: *Cremno-*

bates Swainson, 1855; *Maripythia* Iredale, 1936. Type: *pepita* King, 1832.

Marinula rhoadsi Pilsbry, 1910   4099
Rhoads' Marinula

Head of Gulf of California to Sonora, Mexico.

5 to 6 mm., ovate-conic, pale-yellowish with a narrow brownish band at the shoulder, below which the shell is brownish yellow. Colonies uncommon.

Marinula succinea (Pfeiffer, 1854)   4100
Pfeiffer's Marinula

Georgia, Florida, Bahamas, Cuba.

4 mm., light-brown, ovate with a broad last whorl and rather straight-sided apex. 3 teeth strong on the columella, with the middle one being at right angles to the axis of the shell. Uncommon; the only species of *Marinula* in the Floridian region. Found in marshy areas. *M. elongata* (Dall, 1885) is a synonym.

Genus *Laemodonta* Philippi, 1846

This genus of very small, high-spired snails is well-represented in the South Pacific, but only one species occurs in North America, namely *cubensis* Pfeiffer which is placed in the subgenus *Bullapex* Haas, 1950, because of its large white, inflated apex which is produced at nearly right angles to the axis of the shell. *Plectotrema* H. and A. Adams, 1853, and *Enterodonta* Sykes, 1894 are synonyms of *Laemodonta.* Type: *striata* Philippi, 1846.

Laemodonta cubensis (Pfeiffer, 1854)   4101
Cuban Dwarf Pedipes

Bermuda; south Florida Keys to Barbados.

2 to 3.3 mm.; ovate-conic with a high spire; thin-shelled; straw-yellow with large, white, swollen, tilted apex. Suture indistinct; spiral scratches microscopic, crossed by finer growth lines. Umbilicus minute. Outer lip with 1 or 2 teeth, inner with 3 larger ones. Abundant along the upper strand line under broken coral and weedy debris.

Genus *Apodosis* Pilsbry and McGinty, 1949

Shell 3 mm., thick, oblong-conic, spirally striate. Aperture ½ length of shell. Columella with a single entering fold. Outer lip and columella very thick, plain within. The Nautilus, vol. 63, p. 9. Type: *novimundi* Pilsbry and McGinty, 1949.

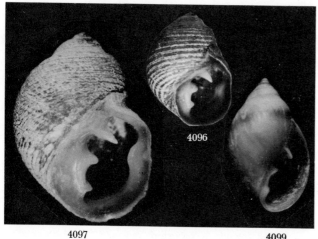

4097   4099

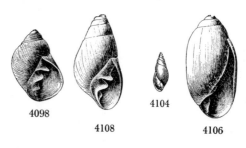

4098   4104   4108   4106

4101   4102

4102 *Apodosis novimundi* Pilsbry and McGinty, 1949. Florida Keys, Bahamas, Jamaica. 3 mm. Pale-brown.

Genus *Ovatella* Bivona, 1832

Type of the genus is *firminii* (Payraudeau, 1826).

Subgenus *Myosotella* Monterosato, 1906

Type of this subgenus is *myosotis* Draparnaud, 1801. Synonyms: "*Phytia*" Gray, 1821, a mistake for *Pythia* Röding, 1798; *Alexia* Gray, 1847 (non Stephens, 1835); *Kochia* Pallary, 1900 (non Frech, 1888); *Nealexia* Wenz, 1920.

Ovatella myosotis (Draparnaud, 1801)   4103

Nova Scotia to the West Indies. Bermuda.
Puget Sound, Washington, to California.

5 to 7 mm., ovate-fusiform, dark-brown to purplish brown, semiglossy. Spire elevated, with 7 or 8 slightly convex whorls, with a distinct suture and below which may be a spiral marginal line. Inner lip of adult with 3 white folds, the lowest one formed by the turning of the lip within the aperture; second thin, transverse, prominent tooth a little below the middle on the inner lip; the third minute one is a little above. Umbilicus minute. 2 or 3 teeth within the outer lip. *O. denticulata* (Montagu, 1808), and *Carychium personatum* Michaud, 1831, are synonyms. The subspecies *bermudensis* (H. and A. Adams, 1855), occurs in Bermuda only.

4103   4105

Genus *Blauneria* Shuttleworth, 1854

Shell minute, 3 mm., slender, smooth, sinistral, smooth, with a single columellar fold. Type: *heteroclita* (Montagu, 1808).

4104 *Blauneria heteroclita* (Montagu, 1808). Florida, Alabama and the West Indies to Brazil. Bermuda. 3 mm.

Subfamily CASSIDULINAE Odhner, 1925

Genus *Microtralia* Dall, 1894

Type of the genus is *minusculus* (Dall, in Simpson, 1889).

4105 *Microtralia occidentalis* (Pfeiffer, 1854). Bermuda; Florida and the Greater Antilles.

Subfamily ELLOBIINAE H. and A. Adams, 1855

Genus *Ellobium* Röding, 1798

Synonyms: *Auricula* Lamarck, 1799; *Marsyas* Oken, 1815; *Geovula* Swainson, 1840. Type: *aurismidae* (Linné, 1758).

4106 *Ellobium pellucens* (Menke, 1830). Florida and the West Indies. Brazil.

Superfamily Siphonariacea Gray, 1840
(Patelliformia)

Family TRIMUSCULIDAE Zilch, 1959

These low, conic, limpetlike, air-breathing snails have white shells, radiating riblets and a weak siphonal groove on the inner right side. Gadiniidae, Gray, 1840, is a synonym.

Genus *Trimusculus* Schmidt, 1818

Shell limpetlike, oval, radially striate, white. An air-breathing pulmonate snail. Muscle scar horseshoe-shaped. Rock-dwelling, intertidal limpetlike snails. Type: *mammillaris* (Linné, 1758). Synonyms: *Gadinia* Gray, 1824; *Clypeus* Scacchi, 1833; *Mouretia* Sowerby, 1842; *Muretia* Orbigny, 1843; *Gardinia* Pictat, 1855; *Rowellia* Carpenter, 1864; *Gadinalea* Iredale, 1940.

Trimusculus goesi (Hubendick, 1946)   4107
Goes' Gadinia

West Indies; Bermuda.

¼ to ½ inch, white, conic, base almost circular. Nuclear whorl white, glossy-smooth, raised, pointing backwards. Shell irregular. Radial riblets numerous, crowded, fairly coarse. Apex near the center. Uncommon; on rocks in intertidal area. *T. carinatus* (Dall, 1870), which may actually be from the Pacific side of Panama, is larger, flatter, with finer riblets and with its apex near the back end.

4107

Trimusculus reticulatus (Sowerby, 1835)   4108
Reticulate Gadinia

Trinidad, California, to the Gulf of California.

12 to 25 mm. in diameter, white, base almost circular, apex central and smooth, below which radiate 40 to 50 minutely

fimbriated riblets. Edge of shell thick, but minutely crenulated. Siphonal groove on the inside on the right is very weak. Moderately common; in colonies in protective overhanging rock crevices at about the mid-tide level. For detailed anatomy, see Dall, 1870, *Amer. Jour. Conchol.*, vol. 6, p. 11. Synonym: *Rowellia radiata* Cooper, 1865.

Trimusculus stellatus (Sowerby, 1835).   **4109**
Stellate Gadinia

Gulf of California to Nicaragua.

10 to 15 mm., base round to slightly oblong and denticulate. Spire low, usually eroded. Radial riblets numerous and distinct. Thin and whitish. Intertidal on rocks; uncommon.

Other species:

4110 *Trimusculus carinatus* (Dall, 1870). Carinate Gadinia. South Florida Keys to the West Indies. Pacific side of Panama? See species 4107.

Family SIPHONARIIDAE Gray, 1840

Genus *Siphonaria* Sowerby, 1823

Shells closely resembling the true limpets, *Acmaea,* but at once distinguished by the nature of the muscle scars on the inside. In both, the long, narrow scar is horseshoe-shaped, but in *Siphonaria* the gap between the ends is located on one side of the shell, while in *Acmaea* it is located at the front end. In some *Siphonaria,* the area near the gap is trough-shaped. These are air-breathers and are more closely related to the land garden snails than to the gill-bearing, water-breathing limpets. Type: *laciniosa* (Linné, 1758).

Subgenus *Patellopsis* Nobre, 1886

Type of the subgenus is *pectinata* (Linné, 1758).

Siphonaria pectinata (Linné, 1758)   **4111**
Striped False Limpet

Eastern Florida, Texas, Mexico and the Caribbean.

1 inch in length, rather high, with an elliptical base. Exterior with numerous, fine, radial threads or rather smooth-ish. Color whitish with numerous, brown, bifurcating, radial lines. Interior glossy, similarly striped. Center cream to brown. Muscle scar with 3 swellings, the gap occurring between the 2 at the side. Do not confuse with *Acmaea leucopleura* which commonly has a blackish owl-shaped figure inside. Common along the shores on rocks close to the low-water mark, so as to remain wet. Eggs are laid in soft, gelatinous strings on hard surfaces. Their larvae pass through a pelagic stage after hatching. This is *S. naufragum* Stearns, 1872, and *S. lineolata* Orbigny, 1842.

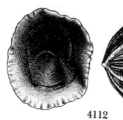

4111   4112

Subgenus *Liriola* Dall, 1870

Shell thin, strong; smooth, or furnished with fine radiating lines which do not interrupt the margin. Apex marginal, twisted to the left of the median line in most of the species. The gill passes behind the heart and lung. The jaw is simple and arcuate. Rhachidian tooth is moderate, with a simple, pointed cusp. Inner laterals are long, narrow and strongly bidentate. Outer laterals are broad and tridentate with short cusps. Type: *thersites* Carpenter, 1864.

Siphonaria alternata Say, 1826   **4112**
Say's False Limpet

North Carolina to Florida, the Bahamas and Bermuda.

½ to ¾ inch in length, with about 20 to 25 small, white, radial ribs between which are smaller riblets. Background gray to cream. Interior glossy-tan, sometimes striped or mottled with dark-brown. Fairly common on rocks near the shore-line usually between the high-tide and mid-tide line. They occur in isolated, well-populated colonies. *S. brunnea* Hanley, 1858, and *picta* Orbigny, 1842 are synonyms.

Siphonaria thersites Carpenter, 1864   **4113**
Carpenter's False Limpet

Alaska to Vancouver Island, British Columbia.

Shell 7 to 10 mm. in length, subconical, apex to one side, periostracum thin, light-colored, smooth, exterior of shell dark reddish brown; with a few ribs. Siphonal groove strong and forms an extension of the edge of the shell. Interior dark-brown, with a light border. Most northerly siphonariid known. Common; on shore rocks.

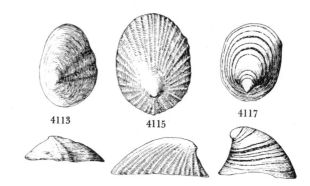

4113   4115   4117

Subgenus *Kerguelenia* Rochebrune and Mabille, 1887

Shell small, thin, cap-shaped with the apex near the posterior end. Type: *lateralis* (Gould, 1846).

Siphonaria henica (Verrill and Bush, 1900)   **4114**
Henica False Limpet

Bermuda.

7 mm., as wide as long. Apex median and rather near the posterior end. Siphon opens out far in front. Nucleus with 1½ whorls, pointing a little to the left. Upper side with numerous, crowded, fine radial striations and concentric growth lines. Internal color glossy-white. Animal white with a tinge of yellow. Rare; Shelly Bay.

Other species:

4115 *Siphonaria (K.) brannani* Stearns, 1872. Santa Barbara to Santa Catalina Island, California. 9 mm.

4116 *Siphonaria (K.) williamsi* S. S. Berry, 1969. Natividad Island, Baja California. *Leaflets in Malacology,* no. 26, p. 166. 7 mm.

Genus *Williamia* Monterosato, 1884

Shell limpetlike, thin, usually horn-colored, conical, with the apex hooked over towards the posterior end. Internal horseshoe scar opens to the right side. Synonyms: *Scutulum* Monterosato, 1877, non Tournouer, 1869; *Allerya* Mörch, 1877, non Bourguignat, 1876; *Aporemodon* Robson, 1913; *Parascutum* Cossmann, 1890. Type: *gussoni* (O. G. Costa, 1829).

Williamia peltoides (Carpenter, 1864)   **4117**
Pelta False Limpet

Crescent City, California, to the Gulf of California and to Panama.

7 to 12 mm., limpetlike, high-domed with the curved apex ⅔ the way from the anterior end and pointing backward. Base oval in outline. Shell moderately thin, semitranslucent, reddish brown with about 12 to 15 narrow, radial greenish white rays. Moderately common locally; under rock slabs in the intertidal area. *W. vernalis* (Dall, 1870) is a synonym. See *The Veliger,* vol. 8, p. 19.

Williamia krebsii (Mörch, 1877)   **4118**
Krebs' False Limpet

South Florida, off Texas and the West Indies. Bermuda.

4 to 6 mm. in length, rather fragile, smoothish, with a fairly high arching back ending in a whitish smooth apex curling slightly to the right at the anterior end of the shell. Color brown, rayed with tan. Moderately common in shallow water to 20 fathoms.

4118

False Sea-hares

Order *Sacoglossa* von Ihering, 1876

Radula with a single transverse row of teeth, and with a frontal sac or caecum enclosing worn teeth. No jaws. Right and left "liver" usually of equal size. Nerve ring behind the pharynx. *Saccoglossa* is a misspelling. Also called *Ascoglossa* Bergh, 1877, and (in part) *Monostichoglossa* Pagenstecher, 1875.

Superfamily Oxynoacea H. and A. Adams, 1854

Family OXYNOIDAE H. and A. Adams, 1854

With a thin, bubblelike shell partially covered by the side parapodia. Head with 2 inrolled tentacles in front of the eyes. Without cerata, but the skin is pimpled. Nonswimmers.

Genus *Oxynoe* Rafinesque, 1819

Fragile shell of 1 large whorl with a large aperture. It covers the gills. With long slender tail grooved medially. 2 inrolled tentacles at the front of the head. Synonyms: *Icarus* Forbes, 1844; *Lophocercus* Krohn, 1847. Type: *olivacea* Rafinesque, 1819.

Oxynoe antillarum Mörch, 1863   **4119**
Antillean Oxynoe

West Florida and the West Indies.

Shell 6 to 10 mm. in length, fragile, vitreous. Aperture large, narrow at the top. Smooth except for a few growth lines. Living animal is 1 inch, light-brown to greenish with white spots. Parapodia and tail bear minute white papillae, and the outer or ventral surface of the parapodia has yellowish dots. Edge of foot with alternating bars of frosty-white and pea-green. Eggs are yellow-orange and laid in a coiled, wide, jelly band. Common on the green alga, *Caulerpa racemosa.*

4119

Other species:

4120 *Oxynoe aguayoi* Jaume, 1945. Havana, Cuba. Lives in 1 to 3 feet, in algae. *Revista Soc. Mal. Carlos de la Torre,* vol. 3, p. 22, pl. 2.

4121 *Oxynoe panamensis* Pilsbry and Olsson, 1943. Gulf of California. Figured by Keen, 1971, pl. XIX.

Genus *Lobiger* Krohn, 1847

External, large-mouthed shell, with its apex on the side, is fragile and sits on top of the body. To each side are 2 long, flat, winglike lobes which flap up and down to aid swimming. Front of head with 2 small, inrolled tentacles. Type: *serradifalci* (Calcara, 1840). *Dipterophysis* Pilsbry, 1896, is believed to be a malformed *Lobiger.*

Lobiger souverbii P. Fischer, 1857   **4122**
Souverbie's Lobiger

South half of Florida to Brazil; Indo-Pacific. West Mexico.

Animal elongate, bright lettuce-green, reticulated with reddish purple, and foot with pale, elevated papillae. Epipodial

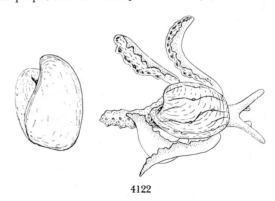

4122

extensions of 2 winglike lobes on each side which flap back and forth. Shell external, mantle only narrowly covering the edges, 12 mm. long, 8.5 mm. wide, oval, thin, transparent, with a huge aperture and with the columellar margin reflected. Uncommon; 6 fathoms, in the green seaweed, *Caulerpa crassifolia.* Synonyms: *pilsbryi* Schwengel, 1941; *viridis* Pease, 1863; *picta* Pease, 1863; *nevilli* Pilsbry, 1895.

Superfamily Plakobranchiacea Gray, 1840

Family PLAKOBRANCHIDAE Gray, 1840

Body sluglike, swollen in the middle, with a pair of slotted rhinophores in front. No cerata. With parapodia marked off from the back, used in swimming. Penis under the right rhinophore. The family name Elysiidae H. and A. Adams, 1854, is synonymous. Placobranchidae is a misspelling.

Genus *Elysia* Risso, 1818

Body sluglike, characters of the family. Coloration usually green with variations in yellows, orange and black markings. Synonyms: *Laplysia* Montagu, 1810, Bosc, 1802; *Notarche* Risso, 1818; *Elisia* Cantraine, 1840. Type: *timida* (Risso, 1818).

Elysia chlorotica (Gould, 1870) 4123
Eastern Emerald Elysia

Nova Scotia to North Carolina.

20 to 30 mm., characters of the genus; emerald-green, dotted with white and red spots. 2 rhinophores delicate, lanceolate and furrowed beneath. Eyes prominent behind the rhinophores. Penis as large as a rhinophore. Abundant in brackish, shallow water. Found on the marine plants, *Zostera* and *Ruppia.*

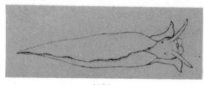

4123

Elysia evelinae Marcus, 1957 4124
Evalina's Elysia

Southeast Florida to Brazil.

3 to 5 mm.; a dark circumanal band and the thickened tip of the radular tooth characterize this species. Among algae from sheltered seawalls. Uncommon. *Jour. Linn. Soc.,* vol. 43, p. 412.

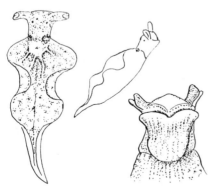

4124

Elysia hedgpethi Marcus, 1961 4125
Hedgpeth's Elysia

San Juan Island, Washington, to the Gulf of California.

10 to 30 mm. in length, bright-green with blue flecks, a white line on the rhinophores and center of the foot. Common; intertidal to subtidal. *Elysia bedeckta* MacFarland, 1966, is a synonym.

4125

Subgenus *Elysiella* Verrill, 1872

Type of this subgenus is *verrilli* Thiele, 1931 (synonym: *catula* Verrill, 1872, non Gould, 1870).

Elysia catula Gould, 1870 4126
Kitty Cat Elysia

Massachusetts to Virginia.

6 mm. long, brownish sea-green, with a whitish spot between the tentacles, another running obliquely inwards and backwards from the outer base of each rhinophore halfway to the median line, a small one at the top of the parapodia. Foot paler green. Rhinophores like elongated cat's ears. Dorsum with fine longitudinal folds. Common; on the marine plant, *Zostera.*

4126

Elysia papillosa Verrill, 1901 4127
Papillose Elysia

Southeast Florida to lower Caribbean. Bermuda.

8 to 12 mm., color gray with 1 or 2 transverse brown bands on the rhinophores and with numerous conical, white papillae sprinkled over the surfaces. Rhinophores large. Radula ribbon with about 13 teeth, with the edge of the pointed tooth bearing coarse serrations. Uncommon; found in shallow water on the alga, *Halimeda.*

Other species:

4128 *Elysia cauze* Marcus, 1957. Virginia Key, Florida, to Sao Paulo, Brazil, on the green alga *Caulerpa.* 40 mm. Greenish gray; foot green. *Jour. Linn. Soc.,* vol. 43, p. 405. (Synonym: *scops* Marcus and Marcus, 1967.)

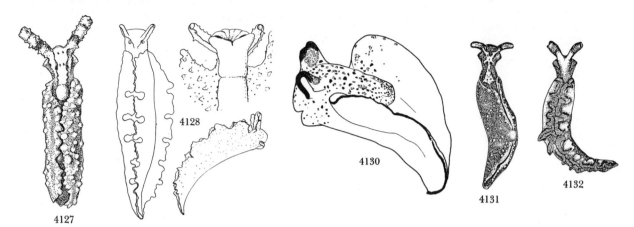

4127 4128 4130 4131 4132

4129 *Elysia subornata* Verrill, 1901. Bermuda. *Trans. Conn. Acad. Sci.*, vol. 11, p. 29.

4130 *Elysia ornata* (Swainson, 1840). Bermuda; West Indies.

4131 *Elysia tuca* Marcus and Marcus, 1967. Southeast Florida and southern Caribbean. *Studies Trop. Oceanogr. Miami*, no. 6, p. 29. Dark-green, clouded with yellowish. White on neck. Parapodia straight-edged. 10 mm.

4132 *Elysia duis* Marcus and Marcus, 1967. Southeast Florida. *Studies Trop. Oceanogr. Miami*, no. 6, p. 31. Light-green with black, red and bright-blue bands. 10 mm.

4133 *Elysia picta* Verrill, 1901. Bermuda. *Trans. Conn. Acad. Sci.*, vol. 11, p. 30. Head and back reddish brown; neck bears a yellow cross; purplish spot between the parapodia in front.

4134 *Elysia flava* Verrill, 1901. Bermuda. *Trans. Conn. Acad. Sci.*, vol. 11, p. 30.

Genus *Tridachia* Deshayes, 1857

Head rounded in front, tentacles large and canaliculate. Parapodialike borders of the mantle are folded and undulating along the length of the animal. Type: *schrammi* Deshayes, 1857, a *nomen nudum* and "synonym" of *crispata* (Mörch, 1863). *Thridachia* is an emendation.

Tridachia crispata (Mörch, 1863)   4135
Common Lettuce Slug

South Florida and the Caribbean. Bermuda. ·

20 to 40 mm. long, green or bluish with white spots on the back and the edges of the undulating, folded-up, "parapodia." Rhinophores simple and fairly thick. The oesophagus is without a loop. A common West Indian shallow-water species. The

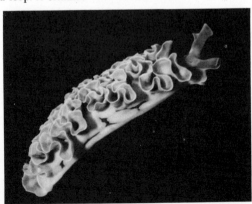

4135

similar *T. schrammi* Deshayes, 1857 (*whiteae* Marcus, 1957) from the Florida Keys and West Indies has an orange parapodial band containing zooxanthellae algae, and has an oesophagus with a loop. Synonyms: *verrilli* Pruvot-Fol, 1946; *pruvotfolae* Marcus, 1957. *T. crispa* Verrill is the same. For zooxanthellae symbiosis, see Yonge and Nicholas, 1940, *Carnegie Inst. Wash.*, Publ. 517, p. 287.

Other species:

4136 *Tridachia* (*Tridachiella* MacFarland, 1924) *diomedea* Bergh, 1894. 34 mm. Gulf of California to Panama. Type of the subgenus.

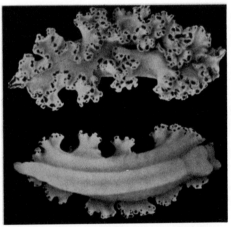

4136

Family HERMAEIDAE H. and A. Adams, 1854

With dorsal cerata, without oval tentacles, but with auriculate rhinophores. Foot usually rounded in front. *Stiligeridae* Iredale and O'Donoghue, 1923, is a synonym.

Genus *Aplysiopsis* Deshayes, 1835

Rhinophores large, auriculate. 2 tentacles at the front of the head. Foot with a marked gutter in front. Back with 2 rows of cerata with the liver ramifications seen inside. Radula long and with 2 turns inside the radular sac. Type: *elegans* Deshayes, 1835. Synonym: *Hermaeina* Trinchese, 1873. One Californian species:

4137 *Aplysiopsis* (*Phyllobranchopsis* Cockerell and Eliot, 1905) *enteromorphae* (Cockerell and Eliot, 1905). Monterey to San Pedro, California.

Genus *Hermaea* Lovén, 1844

Body translucent, elongate, with 2 inrolled rhinophores which are bifurcate at the ends. Cerata numerous down the sides of the dorsum. No head tentacles. Radula of a single row of 30 or so bladelike teeth in the sac. Type: *bifida* (Montagu, 1816). One New England species:

4138 *Hermaea cruciata* Gould, 1870. Naushon Island, Massachusetts. Figured in Gould and Binney, 1870, *Invert. Mass.,* pl. 17, fig. 256. (?Synonym: *coirala* Marcus, 1955, from Brazil.)

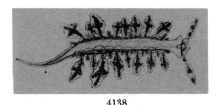

4138

4139 *Hermaea hillae* Marcus and Marcus, 1967. Puerto Peñasco, Sonora, Mexico, among intertidal hydroids. *Studies Trop. Oceanogr. Miami,* no. 6, p. 151.

Genus *Placida* Trinchese, 1876

Type of this genus is *brevirhinum* Trinchese, 1876.

Placida dentritica (Alder and Hancock, 1843)   4140

North Atlantic to Connecticut to North Carolina; Caribbean. Monterey to Carmel Bay, California. Japan.

8 to 14 mm. long, greenish, elongate, with numerous greenish finger-shaped cerata occurring in clumps of 3 to 5 down each lateral margin of the dorsum. Tail long, flat and wide. Head with 2 long, earlike smooth rhinophores which are channeled along their outer sides. No mandibles. Radula with a single series of about 40 bladelike teeth. Found in outer tidepools feeding on *Bryopsis* and *Codium*. Synonym: *Hermaea ornata* MacFarland, 1966.

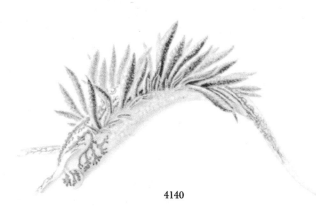

4140

Genus *Costasiella* Pruvot-Fol, 1951

Hermaeid sacoglossids with large eyes which are contiguous and lie between, not behind, the smooth rhinophores; anterior angles of the foot more or less produced. Cerata fusiform, containing branches of the albumen gland as well as the diverticula of the "liver" or digestive gland. Penis unarmed. Radula not denticulate. Type: *virescens* Pruvot-Fol, 1951.

4141 *Costasiella nonatoi* Marcus and Marcus, 1960. Ubatuba, Brazil. 2 to 5 mm. *Bull. Marine Sci. Gulf and Caribbean,* vol. 10, p. 149. Marcus and Marcus, 1969, *Beitr. Neotrop. Fauna,* vol. 6, place this in *Placida*.

Genus *Alderia* Allman, 1845

Foot large and fairly broad. Head without tentacles (rhinophores), but with frontal veil with a triangular protuberance on either side with an eye. Center of back smooth, but there are 1 or 2 rows of a dozen or so club-shaped cerata on each side. Anus on the midline of the back at the posterior end. The cerata contain the lobes of the liver. Penis with a stylet. Teeth smooth. Type: *modesta* (Lovén, 1844). Synonym: *Canthopsis* Agassiz, 1850.

Alderia modesta (Lovén, 1844)   4142
Modest Alderia

Nova Scotia to New Jersey. Europe.
Puget Sound, Washington, to Monterey, California.

6 to 8 mm., characters of the genus, beige to greenish, with the liver dark-green. Dorsum marbled in yellow-brown and blackish. Cerata cylindrical, arranged longitudinally on either side of the posterior ¾ of the body (13 to 16 on each side), speckled with numerous black pigment spots and a few opaque white dots distally. Radula uniserial with 11 to 14 spoon-shaped teeth and a sac containing discarded teeth. A common species preferring slightly brackish water and marsh conditions. *A. harvardiensis* Gould, 1870; *amphibia* Allman, 1845; *scaldiana* Nyst, 1855; are synonyms. For biology and habits, see Hand and Steinberg, 1955, *The Nautilus,* vol. 69, p. 22.

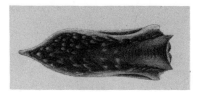

4142

Genus *Stiliger* Ehrenberg, 1831

Animal sluglike; 2 rhinophores simple, cylindro-conical. Cerata large, few in number, with the ramified liver ends inside. Head not separated from the body. Anus anterior ⅓ of body. Eyes behind the bases of the rhinophores. Teeth in the form of a saber. Penis armed with a chitinous stylet. Eggs in jellylike, elongate masses. Type: *ornatus* Ehrenberg, 1831. Synonyms: *Calliopaea* Orbigny, 1837; *Custiphorus* Deshayes, 1864.

Stiliger fuscatus (Gould, 1870)   4143
Dusky Stiliger

New Hampshire to Virginia.

6 to 8 mm., with the characters of the genus; dark slate-colored above, with a light streak extending from the light-colored rhinophores. 2 eye spots are black. Cerata long, club-shaped, slender at the base, black, and white at both ends. An upper row of 4 on each side begins about ⅓ down the body. A lower series of 5 or 6 smaller ones alternate with the upper large ones. Foot pale-yellowish. Eggs in oval jelly masses. Uncommon in salt marshes. For anatomical details, see Marcus, 1958, *Amer. Mus. Novitates,* no. 1906.

4143

Stiliger vossi Marcus and Marcus, 1960   4144
Voss' Stiliger

Florida.

1 to 2 mm., white with sparse black dots scattered on the
back and sides. Rhinophores smooth. Cerata form 2 groups
on either side, the anterior with 2, the posterior with 1 or 2
cerata. Digestive gland seen inside cerata. Labial disc bears a
thick brown cuticle. Radula of 30 teeth, saber-shaped with
minute denticulations.

Stiliger fuscovittatus Lance, 1962   4145
Brown-streaked Stiliger

Washington to the Gulf of California.

10 mm., eolidiform. Cerata deciduous. Color translucent
with brown liver material. Surface spotted with white opaque
dots and sparse brown streaks. Rhinophores simple, smooth,
nonretractile, with black eye spots just behind them. Radula
of 7 saber teeth in the ascending series, and about 23 on the
posterior descending series. Seasonally common from April
through June in bays on the red alga, *Polysiphonia*, in the
northern half of its range.

Other species:

4146 *Stiliger (Ercolania* Trinchese, 1872) *costai* Pruvot-Fol, 1951.
Key Largo, Florida, among *Padina*; Mediterranean (see Marcus
and Marcus, 1960, p. 146).

4147 *Stiliger vancouverensis* O'Donoghue, 1924. Vancouver,
British Columbia. Removed from *Hermaea* by Marcus and
Marcus, 1967, p. 153.

4148 *Stiliger niger* Lemche, 1935. Northwest Europe. 8 mm.
Vidensk. Medd. Dansk. naturh. Foren., vol. 99, p. 135.

4149 *Stiliger bellulus* (Orbigny, 1837). Western Europe. (Synonym: *mariae* Bergh, 1872.)

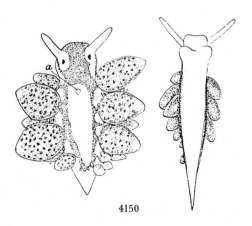

4150

4150 *Stiliger vanellus* Marcus, 1957. Southeast Florida to Brazil.

Family CALIPHYLLIDAE Tiberi, 1880

Genus *Cyerce* Bergh, 1871

Type of the genus is *elegans* Bergh, 1871. Two species
reported from the Atlantic:

4151 *Cyerce antillensis* Engel, 1927. South Florida and the
Caribbean. 35 mm.

4152 *Cyerce cristallina* (Trinchese, 1881). South Florida and
the lower Caribbean. See Marcus and Marcus, 1967, pl. 1, fig. 5
(in color). *C. iheringi* Pelseneer, 1892, is a synonym.

4152

Family LIMAPONTIIDAE Gray, 1847

Body sluglike, less than 10 mm., without cerata, without
parapodia. Eyes prominent in a clear area behind the 2
rudimentary tentacles. Anus centrally dorsal. Liver poorly
ramified. Radula in the form of sabres.

Genus *Limapontia* Johnston, 1836

Body sluglike, characters of the family. Without tentacles.
Type: *nigra* Müller, 1773. Synonyms: *Chalidis* Quatrefages,
1844; *Pontolimax* Crephlin, 1848.

4153 *Limapontia zonata* Girard, 1852. Massachusetts. 2 mm.,
pale-red with bands of white.

Family OLEIDAE Thiele, 1926

Genus *Olea* Agersborg, 1923

7 to 10 mm., limaciform, back with 2 rows of papillae on
each side. 2 small "eye" pigment spots on the sides of the
neck. Oral and lateral labial lobe somewhat fused. Radula
and jaws absent. Type: *hansineensis* Agersborg, 1923. *The
Nautilus*, vol. 36, p. 133.

Olea hansineensis Agersborg, 1923   4154
Hansine's Sea Slug

Puget Sound, Washington.

7 to 10 mm., dark-brown and studded with lighter spots of
various sizes. Tips of the papillae are lighter. Ventral lip is
bilobed. Live on *Zostera* eelgrass. Common in bays. Eggs are
laid in a simple coiled string, 1 mm. in diameter.

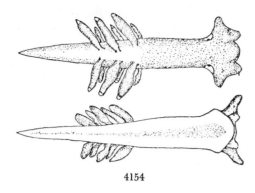

4154

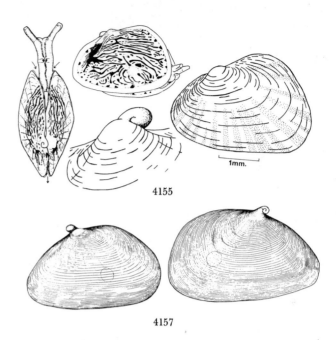

4155

4157

Superfamily Juliacea E. A. Smith, 1885

Family JULIIDAE E. A. Smith, 1885

Small sacoglossate snails having 2 large external shelly valves. Shell and soft parts usually greenish to yellow. Shell thick or thin.

Subfamily BERTHELINIINAE Beets, 1949

Shells lenticular in shape, ovate to quadrate, with a weak hinge. Spiral protoconch on 1 valve retained in the adult. Adductor muscle scar central, undivided, circular. Shell rather fragile.

Genus *Berthelinia* Crosse, 1875

Left valve with a spiral protoconch. Right valve slightly smaller, without a spiral apex. Outline of shell somewhat quadrate. Color of shell greenish to yellowish. Barbed penial stylet present. Associated with the green alga, *Caulerpa*. Type: *elegans* Crosse, 1875, from the Paris Basin Eocene. A review of West American species was made by Keen and A. G. Smith, 1961, *Proc. Calif. Acad. Sci.*, vol. 30, pp. 47–66.

Subgenus *Edenttellina* Gatliff and Gabriel, 1911

The type is *typica* Gatliff and Gabriel, 1911. *Tamanovalva* Kawaguti and Baba, 1959, is a synonym.

Berthelinia caribbea Edmunds, 1963   4155
Caribbean Bivalved Snail

Jamaica and Puerto Rico.

Shells 3 to 4 mm. long, resembling a tiny clam; quadrate, anterior end slightly spathate; height 60 to 73% of the length; green or grayish green, often tinged with brown, and with faint radial rays of yellow. Protoconch in left valve of 1¼ whorls. Hinge with a strong posterior cardinal in each valve, and a weak anterior cardinal in the right valve. Mantle with dark reddish brown narrow stripes. Fine brown line on dorsal side of neck. Rhinophores tubelike. Oval tentacles greatly reduced. Radula with about 24 to 30 hooklike teeth with a row of about 60 fine denticles on each side. Animal greenish with white specks. Anterior margin of foot sometimes blue.

Lives and feeds in the green alga *Caulerpa verticillata* Agardh in salt-water channels of mangrove swamps. For eggs, larvae and ecology, see John Grahame, 1970, *Bull. Marine Sci.*, vol. 19, p. 868.

Other species:

4156 *Berthelinia chloris* (Dall, 1918). See A. M. Keen and A. G. Smith, 1961, *Proc. Calif. Acad. Sci.*, vol. 30, pp. 47–66. Magdalena Bay, Baja California. 9 mm.

4157 *Berthelinia chloris* subspecies *belvederica* Keen and A. G. Smith, 1961. Puerto Ballandra Bay, La Paz, Baja California. Shell 8.5 mm. On *Caulerpa sertularioides* from 5 to 8 feet.

Subfamily JULIINAE E. A. Smith, 1885

Genus *Julia* Gould, 1862

Shells minute, heart-shaped, with a heavy hinge, a toothlike ridge in one valve and a fossettelike fold in the other. Greenish in color, posterior end sharp and pointed. Type: *exquisita* Gould, 1862. *Prasina* Deshayes, 1863, is a synonym. One Eastern Pacific species:

4158 *Julia thecaphora* (Carpenter, 1857). La Paz, Gulf of California, to Peru. 3 to 4 mm. (Synonym: *equatorialis* Pilsbry and Olsson, 1944.)

True Sea-hares

Order *Aplysiacea* Rafinesque, 1815

2 flaplike skin extensions, or parapodia, extend from the foot up over the back where there is a plicate gill on the right side. In some, a mantle shelf covers the gill and, when present, a thin chitinous shell. Radula with several rows of teeth. Stomach with large, corneous, angular, grinding teeth. Hermaphrodites. Eggs laid in long, thin, jelly strands. Most adults are capable of swimming by flapping the parapodia. The name of a fish order, Anaspidea, is used by some workers for this molluscan order. For a review of Pacific coast species, see R. D. Beeman, supplement to vol. 3, *The Veliger*, 1968.

Family AKERIDAE Pilsbry, 1893

In recent readjustments in the classification of the cephalaspid marine mollusks, the family Akeridae has been reduced to contain only the genus *Akera* O. F. Müller, 1776. Type: *A. bullata* Müller, 1776. Synonyms: *Vitrella* Swainson, 1840; *Eucampe* Gray, 1847; *Aceras* Locard, 1886. Formerly placed in this family were *Haminoea* Turton and Kingston, 1830 (see family Atyidae) and *Cylindrobulla* P. Fischer, 1856 (see family Volvatellidae). The family name was preserved in Opinion 539, I.C.Z.N.

Akera has a cephalic shield, a thin, inrolled, external bulloid shell and lacks tentacles. Parapodia used for swimming. Two species reported from the Caribbean:

4159 *Akera bayeri* Marcus and Marcus, 1967. Off Colombia, Caribbean, 87 meters. *Bull. Marine Sci.,* vol. 17, p. 610.

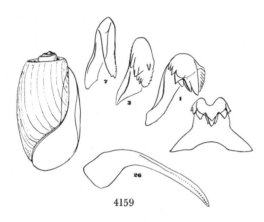

4159

4160 *Akera thompsoni* Olsson and McGinty, 1951. Florida. *The Nautilus*, vol. 65, no. 2, pl. 3.

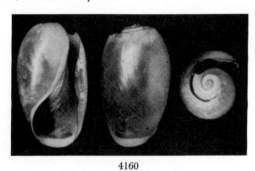

4160

Family APLYSIIDAE Rafinesque, 1815

Popularly known as "sea-hares" because the erect, anterior 2 rhinophores resemble ears. In front are 2 broad cephalic tentacles. Eyes small, embedded in front of the rhinophores. Pair of parapodia skinflaps used in swimming, and cover the plicate gill, anus, anal siphon and opaline gland pores. When irritated, most species give off a purple, maroon or blackish viscous fluid. (See Tobach, Gold and Ziegler, 1965, *The Veliger*, vol. 8, p. 16.) Shell thin, chitinous, flat and under the skin. An excellent account of the 4 Californian species is given by R. D. Beeman, 1968, Supplement to vol. 3 of *The Veliger*, pp. 87–102. It also contains a valuable bibliography.

Subfamily APLYSIINAE Rafinesque, 1815

Parapodia about the same size, widely separated, especially anteriorly, and used for swimming. Foot relatively narrow. Shell large, flat, chitinous, sometimes weakly calcified. Skin smooth, without filaments. Penis without spines.

Nervous system has very long visceral connectives. Some giant nerve cell bodies are over 800 μ in diameter. For a review of recent neuro-physiological research, see L. Simpson, H. Bern and R. I. Nishioka, 1966, *Amer. Zool.,* vol. 6, pp. 123–138.

Genus *Aplysia* Linné, 1767

Dorsal lobes free, well-separated and used for swimming. Shell internal, thin, flat, horny, with little or no lime, and colored amber. Skin smoothish. They give off a harmless purple ink. All are vegetarians. Eggs are deposited in long, tangled gelatinous strings of variously colored eggs among rocks and seaweeds, usually intertidally. One adult may produce over 86 million eggs in one season. The largest known living gastropod is *Aplysia vaccaria* from California, weighing 35 pounds and reaching 30 inches in length. *Tethys* is a name which was for a long time applied to this group, but it is now restricted to a nudibranch genus. *Aplysia* is a conserved name (Opinion 200, I.C.Z.N.). Type: *depilans* (Linné, 1758).

Aplysia willcoxi Heilprin, 1886   **4161**
Willcox's Sea-hare

Cape Cod to both sides of Florida; Texas; West Indies. Bermuda.

5 to 9 inches in length. Mantle under the lobes with a minute perforation or fleshy tube above the area of the shell. Color dark-brown with slight maculations on the swimming lobes, head and neck. There are large, rounded, fairly regular, yellowish scallopings along the inner border of the lobes. Mantle and gills light-purple and yellow. Common. The form *perviridis* Pilsbry, 1895, is clear-green on the head and tentacles, the lobes olive-green with a coarse-meshed reticulation of black, subdivided by fine veins; irregularly maculated all over with light-green, with an occasional clumping of white dots. Rare in New England (D. Merriman, 1937, *The Nautilus,* vol. 50, p. 95).

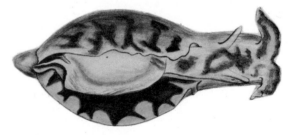

4161

Other species:

4162 *Aplysia donca* Marcus and Marcus, 1959. Port Aransas, Texas. 65 mm. (*Publ. Inst. Marine Sci. Univ. Texas*, vol. 6, p. 251).

Aplysia vaccaria Winkler, 1955   **4163**
Giant Black Sea-hare

Morro Bay, California, to the Gulf of California.

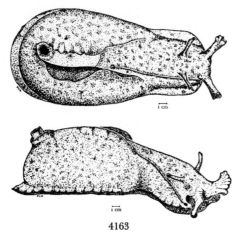

4163

Up to 30 inches in length (35 lbs.); parapodia overlapping, frilled and posteriorly joined high up near the anal siphon. Body firm. Shell foramen distinct. Shell without projecting rectangular plate. Ground color deep purplish black, usually with fine gray or white markings; foot uniformly deep bluish black. Rhinophores ruffled. Common; only along rock coasts and in kelp beds, intertidally to about 60 feet. Feeds only on the brown alga, *Egregia*. Does not emit an ink. (Illustration from R. D. Beeman, 1968, p. 92.)

Aplysia reticulopoda Beeman, 1960   4164
Net-footed Sea-hare

Laguna Beach to San Clemente Island, California.

6 to 15 inches, pale yellowish white, with purplish black network in grooves of foot. Parapodia short, not overlapping dorsally, edges smooth. Rhinophores smooth, bluntly tapering. Uncommon; subtidally. (Illustration from R. D. Beeman, 1968, p. 93.)

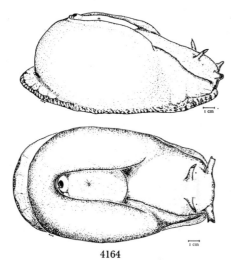

4164

Other species:

4165 *Aplysia (Aplysia) cedrosensis* Bartsch and Rehder, 1939. Cedros Island, Gulf of California. *Smithsonian Misc. Coll.*, vol. 98, pp. 1–18.

Subgenus *Varria* Eales, 1960

Aplysia dactylomela Rang, 1828   4166
Spotted Sea-hare

South half of Florida to Brazil. Bermuda.
West Africa to South Africa.

4166

4 to 5 inches in length, characterized by its pale-yellow to yellowish green color and the fairly large, usually irregular circles of violet-black scattered over the body. Common in some grassy localities. Sometimes digs into the mud. *A. protea* Rang, 1828; *panamensis* Pilsbry, 1895; and *aequorea* Heilprin, 1888, are synonyms. Has a strong, musky odor.

Aplysia brasiliana Rang, 1828   4167
Sooty Sea-hare

Lower Florida Keys. The West Indies?

4 inches in length. Color deep purple-black, the inside of the swimming lobes slightly lighter, and with blotches of black at the edges. Mantle purple-black, spotted irregularly with lighter purple. Uncommon. *A. floridensis* (Pilsbry, 1895) is a synonym.

4167

Aplysia morio Verrill, 1901   4168
Giant Black Sea-hare

Rhode Island to Texas; Bermuda and the West Indies.

8 to 12 inches, dark reddish brown to black, with black stripes on the head and with very broadly overlapping purple-edged lateral flaps. Obtuse posteriorly. Flaps entirely free posteriorly and extending to the end of the short foot. Rhinophores small; tentacles large and broad, foliaceous. No mantle pore. Shell very thin, pale-yellow, oblong-ovate, with the posterior sinus long. Uncommon. *A. modesta* Thiele, 1910, is a synonym.

Subgenus *Tullia* Pruvot-Fol, 1933

Synonym: *Metaphysia* Pilsbry, 1951.

Aplysia juliana Quoy and Gaimard, 1832   4169
Walking Sea-hare

Biscayne Bay, Florida, to Brazil.
Gulf of California.

2 to 4 inches in length. Mantle under the lobes with large perforation. Sole of foot with a characteristic, muscular disk at each end. Exterior of animal dark-olive, indistinctly mottled

with irregular spots of dusky buff, and having small, sparsely scattered, ragged black spots. Sole of foot yellowish olive. Found under rock ledges at low tide. For details see *Notulae Naturae*, Philadelphia, no. 240, pp. 1–6, illustrated. *A. badistes* Pilsbry, 1951, is a synonym, as is *parva* Pruvot-Fol, 1953. The parapodia joined low down posteriorly distinguish *cervina* from *juliana*.

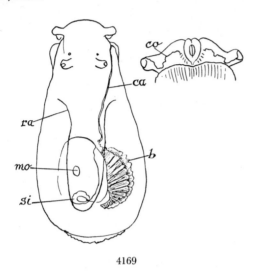

4169

Subgenus *Neaplysia* Cooper, 1863

Type of the subgenus is *californica* Cooper, 1863.

Aplysia californica Cooper, 1863   **4170**

Humboldt Bay, California, to the Gulf of California.

5 to 12 inches, bulky. Parapodia joined posteriorly near the top of the foot or only slightly separated; widely separated anteriorly. Body soft. Gives off a purple-colored secretion. Shell foramen minute or closed. Shell with a projecting rectangular "accessory" plate. Body variously colored, mainly sooty-brown to light-greenish with sooty spots. Young specimens are crimson-red. Common in rocky coast areas, in kelp beds, in bays and estuaries. Intertidal to 60 feet. Feeds mainly on red and green algae. *A. nettiae* Winkler, 1959, is a synonym.

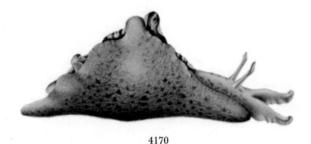

4170

Other species:

4171 *Aplysia tarda* Verrill, 1901. Bermuda. *Trans. Conn. Acad. Sci.*, vol. 11, p. 26.

4172 *Aplysia megaptera* Verrill, 1900. Bermuda. *Trans. Conn. Acad. Sci.*, vol. 10, p. 545.

4173 *Aplysia (Pruvotaplysia* Engel, 1936) *parvula* Mörch, 1863. West Indies and Gulf of California. 40 to 60 mm. Type of the subgenus.

4174 *Aplysia (Varria) cervina* (Dall and Simpson, 1901). South Carolina to Florida to Brazil. 2 inches. *Aplysia pilsbryi* Letson, 1898, may be this species.

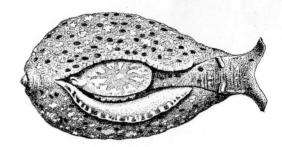

4174

4174a *Aplysia (Syphonota* H. and A. Adams, 1854) *geographica* (Adams and Reeve, 1850). An Indo-Pacific species reported from Dry Tortugas, Florida by K. White, 1952 (*Proc. Malacol. Soc. London*, vol. 29).

Subfamily DOLABRIFERINAE Pilsbry, 1895

Sea-hares with the parapodia separated anteriorly only by the genital groove, broadly joined posteriorly, small, not freely movable, and displaced to the right. Shell small, flat, calcareous or entirely missing. Body rather flattened. Foot as wide as, or wider than, the body.

Genus *Dolabella* Lamarck, 1801

Body plump, somewhat warty. Posterior end truncated and circular. Rhinophores and tentacles present. Mantle foramen large, elongate, its edges very thin, exposing the yellowish shell below. Shell flattish, calcareous apex. Type: *callosa* Lamarck, 1801. One Western American species:

4175 *Dolabella auricularia* (Lightfoot, 1786). Gulf of California to Ecuador; Indo-Pacific. 4 to 5 inches. (Synonyms: *californica* Stearns, 1877; *scapula* "Martyn" Engel, 1942; *agassizi* MacFarland, 1918.)

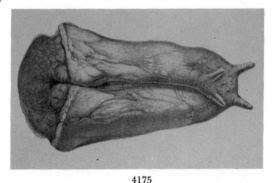

4175

Genus *Dolabrifera* Gray, 1847

Shell internal, solid, triangular with a thick nucleus posterior. One species common in the West Indies, the type of the genus, *D. dolabrifera* (Rang, 1828) of which *ascifera* (Rang, 1828) is a synonym. Rarely found in Florida Keys.

Dolabrifera dolabrifera (Rang, 1828)   **4176**
Warty Sea-cat

Southeast Florida and the West Indies to Brazil. Bermuda. Baja California.

PLATE 1: PLEUROTOMARIA AND ABALONES

3 ADANSON'S PLEUROTOMARIA. *Perotrochus adansonianus* (Crosse and Fischer, 1861). 6 inches. Bermuda and Florida to Brazil.

21 BLACK ABALONE. *Haliotis cracherodii* Leach, 1814. 6 inches. Coos Bay, Oregon, to Baja California.

23 RED ABALONE. *Haliotis rufescens* Swainson, 1822. 12 inches. Oregon to Baja California.

24 PINK ABALONE. *Haliotis corrugata* Wood, 1828. 6 inches. Monterey, California, to Baja California.

25 GREEN ABALONE. *Haliotis fulgens* Philippi, 1845. 7 inches. Point Conception, California to Baja California. Interior and exterior view.

26 NORTHERN GREEN ABALONE. *Haliotis walallensis* Stearns, 1899. 4 inches. British Columbia to La Jolla, California.

27 WHITE ABALONE. *Haliotis sorenseni* Bartsch, 1940. 10 inches. Point Conception, California to Baja California.

28 JAPANESE ABALONE. *Haliotis kamtschatkana* Jonas, 1845. 5 inches. Japan; Alaska to California.

29 THREADED ABALONE. *Haliotis assimilis* Dall, 1878. 5 inches. Point Conception, California to Baja California.

PLATE 2

PLEUROTOMARIA, KEYHOLE LIMPETS AND TURBANS

1 QUOY'S PLEUROTOMARIA. *Perotrochus quoyanus* (Fischer and Bernardi, 1856). 2 inches. Gulf of Mexico and the West Indies; Bermuda.

105 ROUGH KEYHOLE LIMPET. *Diodora aspera* (Rathke, in Eschscholtz, 1833). 2 inches. Alaska to Baja California.

114 GREAT KEYHOLE LIMPET. *Megathura crenulata* (Sowerby, 1825). 3 inches. Monterey, California, to Baja California.

130 BARBADOS KEYHOLE LIMPET. *Fissurella barbadensis* (Gmelin, 1791). 1 inch. Southeast Florida, Bermuda and the West Indies.

263 BAIRD'S SPINY MARGARITE. *Lischkeia bairdii* (Dall, 1889). 2 inches. Bering Sea to off Chile.

310 CHOCOLATE-LINED TOP SHELL. *Calliostoma javanicum* (Gmelin, 1791). ¾ inch. Southeast Florida and West Indies.

312 JUJUBE TOP SHELL. *Calliostoma jujubinum* (Gmelin, 1791). ¾ inch. North Carolina to the West Indies.

355 PACIFIC RINGED TOP SHELL. *Calliostoma annulatum* (Lightfoot, 1786). 1 inch. Alaska to southern California.

356 CHANNELED TOP SHELL. *Calliostoma canaliculatum* (Lightfoot, 1786). 1 inch. Alaska to southern California.

360 WESTERN RIBBED TOP SHELL. *Calliostoma ligatum* (Gould, 1849). ½ inch. Alaska to southern California.

375 SUPERB GAZA. *Gaza superba* (Dall, 1881). 1 inch. Gulf of Mexico and the West Indies; deep water.

388 QUEEN TEGULA. *Tegula regina* (Stearns, 1892). 2 inches. Catalina Island, California, to Baja California.

474 CHESTNUT TURBAN. *Turbo castanea* Gmelin, 1791. 1½ inches. North Carolina to Brazil.

475 CHANNELED TURBAN. *Turbo canaliculatus* Hermann, 1781. 3 inches. Southeast Florida and the West Indies; Brazil.

476 FILOSE TURBAN. *Turbo cailletii* Fischer and Bernardi, 1856. ¾ inch. Southeast Florida and the West Indies.

479 LONG-SPINED STAR SHELL. *Astraea phoebia* Röding, 1798. 1 inch. Southeast Florida and the West Indies. Also may have short spines.

480 SHORT-SPINED STAR SHELL. *Astraea brevispina* (Lamarck, 1822). 1 inch. Southern Caribbean. Red blotch is characteristic.

481 AMERICAN STAR SHELL. *Astraea americana* (Gmelin, 1791). 1 inch. Southeast Florida and Bahamas.

482 SHINGLED STAR SHELL. *Astraea tecta* (Lightfoot, 1786). 1½ inches. Eastern Caribbean.

483 CARVED STAR SHELL. *Astraea caelata* (Gmelin, 1791). 2 inches. Southeast Florida and and the West Indies.

484 GREEN STAR SHELL. *Astraea tuber* (Linné, 1758). 1½ inches. Southeast Florida and the West Indies.

488 BLOOD-SPOTTED STAR SHELL. *Astraea olivacea* (Wood, 1828). 2 inches. Western Mexico to Ecuador.

PLATE 3

NERITES, PERIWINKLES AND JANTHINAS

519 BLEEDING TOOTH. *Nerita peloronta* Linné, 1758. 1 inch. Southeast Florida to Brazil.

520 FOUR-TOOTHED NERITE. *Nerita versicolor* Gmelin, 1791. ¾ inch. Southeast Florida to Brazil.

521 TESSELLATE NERITE. *Nerita tessellata* Gmelin, 1791. ½ inch. Florida to Texas and the Caribbean.

522 ANTILLEAN NERITE. *Nerita fulgurans* Gmelin, 1791. 1 inch. Southeast Florida to Brazil.

523 ORNATE NERITE. *Nerita scabricosta* Lamarck, 1822. 1 inch. Baja California to Ecuador.

525 FUNICULATE NERITE. *Nerita funiculata* Menke, 1851. ½ inch. Baja California to Peru; Galapagos Islands.

526 ZEBRA NETITE. *Puperita pupa* (Linné, 1758). ⅓ inch. Southeast Florida and the Caribbean.

527 VIRGIN NERITE. *Neritina virginea* (Linné, 1758). ⅓ inch. Florida to Texas and the Caribbean.

528 OLIVE NERITE. *Neritina reclivata* (Say, 1822). ½ inch. Florida to Texas and the Caribbean.

531 YELLOW-STRIPED NERITE. *Theodoxus luteofasciatus* Miller, 1879. ⅓ inch. Gulf of California to Peru.

550 NORTHERN YELLOW PERIWINKLE. *Littorina obtusata* (Linné, 1758). ⅓ inch. Labrador to New Jersey.

556 ZIGZAG PERIWINKLE. *Littorina ziczac* (Linné, 1758). ½ inch. Southeast Florida and the Caribbean.

561 ASPER PERIWINKLE. *Littorina aspera* Philippi, 1846. ½ inch. Baja California to Ecuador.

567 FASCIATE PERIWINKLE. *Littorina fasciata* Gray, 1839. 1¼ inch. Baja California to Ecuador.

938 COMMON SUNDIAL. *Architectonica nobilis* Röding, 1798. 1½ inches. North Carolina to Texas; Caribbean.

1178 COMMON JANTHINA. *Janthina janthina* (Linné, 1758). 1 inch. Worldwide on the open, tropical seas.

1179 ELONGATE JANTHINA. *Janthina globosa* (Swainson, 1822). 1 inch. Worldwide on the open, tropical seas.

1180 PALLID JANTHINA. *Janthina pallida* (Thompson, 1840). 1 inch. Worldwide on the open, tropical seas.

1181 DWARF JANTHINA. *Janthina exigua* Lamarck, 1816. ¼ inch. Worldwide on the open, tropical seas.

1182 RECLUZIA SNAIL. *Recluzia rollandiana* Petit, 1853. 1 inch. Florida to Texas to Brazil; open, tropical seas.

1410 HENDERSON'S NISO. *Niso hendersoni* Bartsch, 1953. 1 inch. Southeast United States.

1623 COFFEE-BEAN TRIVIA. *Trivia pediculus* (Linné, 1758). ½ inch. Florida and the West Indies.

1652 PUSTULED COWRIE. *Jenneria pustulata* (Lightfoot, 1786). ¾ inch. Gulf of California to Ecuador.

1662 VIDLER'S SIMNIA. *Simnia aequalis vidleri* (Sowerby, 1881). ¾ inch. Monterey, California to Panama.

1664 FLAMINGO TONGUE. *Cyphoma gibbosum* (Linné, 1758). 1 inch. North Carolina to Florida and to Brazil. Bermuda.

1668 EMARGINATE CYPHOMA. *Cyphoma emarginatum* (Sowerby, 1830). ¾ inch. Western Mexico.

1669 INTERMEDIATE CYPHOMA. *Cyphoma intermedium* (Sowerby, 1828). 1¼ inches. Florida to Brazil; Bermuda.

3459 GIANT ATLANTIC PYRAM. *Pyramidella dolobrata* (Linné, 1758). 1½ inches. Florida to Brazil.

519   521   520   522

525   523   527   528   526

531   567   561   556   550

1178   1181   1179   1180

938   1664   1182

1669   1662   1652   1623   1410   1668   3459

PLATE 4

CARRIER SHELLS, CONCHS AND MOON SNAILS

1572 ATLANTIC CARRIER SHELL. *Xenophora conchyliophora* (Born, 1780). 3 inches. South Florida to Brazil.

1579 WEST INDIAN FIGHTING CONCH. *Strombus pugilis* Linné, 1758. 3 inches. Southeast Florida to Brazil.

1580 FLORIDA FIGHTING CONCH. *Strombus alatus* Gmelin, 1791. 3 inches. Florida to Yucatan, Mexico.

1580a Immature specimen of Florida Fighting Conch in which the faring outer lip has not been formed.

1581 PANAMANIAN FIGHTING CONCH. *Strombus gracilior* Sowerby, 1825. 3 inches. Gulf of California to Peru.

1582 GRANULATED CONCH. *Strombus granulatus* Swainson, 1822. 3 inches. Gulf of California to Ecuador.

1584 MILK CONCH. *Strombus costatus* Gmelin, 1791. 5 inches. South Florida and the West Indies; Bermuda.

1585 HAWK-WING CONCH. *Strombus raninus* Gmelin, 1791. 3 inches. Southeast Florida and the West Indies.

1586 ROOSTER-TAIL CONCH. *Strombus gallus* Linné, 1758. 4 inches. Southeast Florida and the West Indies.

1587 PERUVIAN CONCH. *Strombus peruvianus* Swainson, 1823. 5 inches. West Mexico to northern Peru.

1673 BROWN MOON SNAIL. *Polinices hepaticus* (Röding, 1798). 1 inch. Florida and the West Indies.

1677 ATLANTIC MOON SNAIL or SHARK EYE. *Polinices duplicatus* (Say, 1822). 2 inches. Cape Cod, Massachusetts to Texas.

1715 COLORFUL ATLANTIC NATICA. *Natica canrena* (Linné, 1758). 1 inch. North Carolina to Brazil; Bermuda.

1717 CAYENNE NATICA. *Natica cayennensis* Récluz, 1850. 1 inch. Caribbean to Brazil.

1729 ELENA NATICA. *Stigmaulax elenae* (Récluz, 1844). 1 inch. Baja California to Ecuador.

1582

1572

1585

1584

1586

1587

1729

1715

1717

1673

1677

1579

1580

1580

1581

PLATE 5

COWRIES

1638 MEASLED COWRIE. *Cypraea zebra* Linné, 1758. 3 inches. Southeast Florida to Brazil.

1639 ATLANTIC DEER COWRIE. *Cypraea cervus* Linné, 1771. 4 inches. Off North Carolina to Florida; northern Cuba; Yucatan, Mexico. Bermuda.

1639a Immature shell of Atlantic Deer Cowrie.

1641 ATLANTIC GRAY COWRIE. *Cypraea cinerea* Gmelin, 1791. 1 inch. Southeast Florida to Brazil.

1642 ANNETTE'S COWRIE. *Cypraea annettae* Dall, 1909. 1½ inches. Gulf of California to Peru.

1643 ATLANTIC YELLOW COWRIE. *Cypraea spurca acicularis* Gmelin, 1791. ¾ inch. South Carolina to Brazil.

1644 MOUSE COWRIE. *Cypraea mus* Linné, 1758. 1¼ inches. Southern Caribbean.

1645 SURINAM COWRIE. *Cypraea surinamensis* Perry, 1811. 1 inch. Southeast Florida to Brazil.

1646 CHESTNUT COWRIE. *Cypraea spadicea* Swainson, 1823. 1½ inches. Monterey, California, to Baja California.

1649 DWARF ARABIAN COWRIE. *Cypraea arabicula* Lamarck, 1811. 1 inch. Gulf of California to Peru; Galapagos.

1641

1645

1643

1649

1642

1646

1638

1644

1638

1639

1639a

1639

PLATE 6

BONNETS, HELMETS AND FIG SHELLS

1730 ROYAL BONNET. *Sconsia striata* (Lamarck, 1816). 1½ inches. Gulf of Mexico and Caribbean; deep water.

1734 MATTHEWS' MORUM. *Morum matthewsi* Emerson, 1967. ½ inch. Northeast Brazil.

1736 TUBERCULATE MORUM. *Morum tuberculosum* (Reeve, 1842). ¾ inch. Baja California to Peru.

1737 SCOTCH BONNET (typical form). *Phalium granulatum* (Born, 1778). 3 inches. North Carolina to Texas to Brazil.

1738 SCOTCH BONNET (smooth form). *Phalium granulatum* forma *cicatricosum* (Gmelin, 1791). 3 inches. Florida to Brazil.

1739 PANAMANIAN SCOTCH BONNET. *Phalium granulatum* subspecies *centiquadrata* (Valenciennes, 1832). 3 inches. Gulf of California to Peru; Galapagos.

1741 ATLANTIC CASMARIA. *Casmaria ponderosa* subspecies *atlantica* Clench, 1944. 1 inch. Southeast Florida and the Caribbean.

1747 RETICULATED COWRIE-HELMET. *Cypraecassis testiculus* (Linné, 1758). 2 inches. North Carolina to Texas to Brazil; Bermuda.

1748 THIN-SHELLED COWRIE-HELMET. *Cypraecassis tenuis* (Wood, 1828). 5 inches. Southern Baja California and Galapagos Islands.

1749 CONTRACTED COWRIE-HELMET. *Cypraecassis coarctatus* (Sowerby, 1825). 2 inches. Gulf of California to Ecuador.

1784 ATLANTIC PARTRIDGE TUN. *Tonna maculosa* (Dillwyn, 1817). 3 inches. South Florida to Brazil; Bermuda.

1787 CROSSE'S FALSE TUN. *Eudolium crosseanum* (Monterosato, 1869). 1½ inches. New Jersey to Lesser Antilles; Mediterranean to South Africa.

1796 COMMON FIG SHELL. *Ficus communis* Röding, 1798. 3 inches. North Carolina to Mexico. Shell light-weight; surface reticulated.

1800 VENTRICOSE FIG SHELL. *Ficus ventricosa* (Sowerby, 1825). 3½ inches. California to northern Peru.

2451 SAY'S PEAR WHELK. *Busycon spiratum* subspecies *pyruloides* (Say, 1822). North Carolina to west Florida. Shell solid; surface smoothish.

1749

1739

1738

1737

1747

1748

1730

1784

1736

1736

1787

1796

1734

2451

1741

1800

PLATE 7

TRITONS AND FROG SHELLS

1751 ANGULAR TRITON. *Cymatium femorale* (Linné, 1758). 5 inches. Southeast Florida to Brazil.

1753 ATLANTIC HAIRY TRITON. *Cymatium pileare* (Linné, 1758). 2 inches. Southeast United States to Brazil.

1757 KREBS' HAIRY TRITON. *Cymatium krebsii* (Mörch, 1877). 1½ inches. Southeast Florida and the Caribbean.

1760 GOLD-MOUTHED HAIRY TRITON. *Cymatium nicobaricum* (Röding, 1798). 2 inches. Southeast Florida to Brazil; Bermuda.

1761 POULSON'S TRITON. *Cymatium cingulatum* (Lamarck, 1822). 2 inches. North Carolina to Texas to Brazil.

1764 KNOBBED TRITON. *Cymatium muricinum* (Röding, 1798). 1½ inches. Southeast Florida to Brazil.

1765 CARIBBEAN TRITON. *Cymatium moritinctum* subspecies *caribbaeum* Clench and Turner, 1957. 2 inches. South Carolina to Brazil; Bermuda.

1770 ATLANTIC DISTORSIO. *Distorsio clathrata* (Lamarck, 1816). 2 inches. North Carolina to Texas; Caribbean to Brazil.

1776 ATLANTIC TRITON'S TRUMPET. *Charonia variegata* (Lamarck, 1816). 10 inches. Southeast Florida and the West Indies; Bermuda.

1777 ST. THOMAS FROG SHELL. *Bursa thomae* (Orbigny, 1842). ½ inch. South Carolina to Brazil.

1780 HEAVENLY FROG SHELL. *Bursa caelata* (Broderip, 1833). 2 inches. Baja California to Peru; Southeast Florida to Brazil.

1781 GRANULAR FROG SHELL. *Bursa granularis* Röding, 1798. 1 inch. Southeast Florida and the Caribbean; Indo-Pacific.

1782 CHESTNUT FROG SHELL. *Bursa bufo* (Bruguière, 1792). 1½ inches. Southeast Florida and the West Indies; Brazil.

1783 CALIFORNIAN FROG SHELL. *Bursa californica* (Hinds, 1843). 3 inches. Monterey, California, to the Gulf of California.

1751

1776

1765

1770

1753

1760

1764

1761

1782

1757

1780

1781

1777

1783

PLATE 8

MUREXES

1807 CABRIT'S MUREX. *Murex cabritii* Bernardi, 1858. 1½ inches. South Carolina to the Gulf of Mexico; Caribbean.

1816 BENT-SNOUT MUREX. *Murex recurvirostris* Broderip, 1833. 2 inches. Gulf of California to Ecuador. 1816a is the color form *lividus* Carpenter, 1857.

1817 BEAU'S MUREX. *Murex beaui* Fischer and Bernardi, 1857. 4 inches. South Florida and Gulf of Mexico to Brazil.

1818 ANTILLEAN MUREX. *Murex formosus* Sowerby, 1841. 3 inches. Southeast Florida to Brazil.

1820 CIBONEY MUREX. *Murex ciboney* Clench and Aguayo, 1945. 2 inches. Caribbean. (form *springeri* Bullis, 1964).

1821 MOTACILLA MUREX. *Murex motacilla* Gmelin, 1791. 3 inches. Lesser Antilles.

1824. CAILLET'S MUREX. *Murex cailleti* Petit, 1856. 3 inches. West Indies.

1826 ARTICULATE MUREX. *Murex articulatus finlayi* Clench, 1959. 2½ inches. Cuba and Jamaica.

1828 APPLE MUREX. *Murex pomum* Gmelin, 1791. 2 inches. North Carolina to Brazil.

1832 GIANT ATLANTIC MUREX. *Murex fulvescens* Sowerby, 1834. 5 inches. North Carolina to Texas.

1833 PINK-MOUTHED MUREX. *Murex erythrostomus* Swainson, 1831. 3 inches. Gulf of California to Peru.

1834 CABBAGE MUREX. *Murex brassica* Lamarck, 1822. 6 inches. West Mexico to Peru.

1835 REGAL MUREX. *Murex regius* (Swainson, 1821). 3½ inches. Gulf of California to Peru.

1836 RADISH MUREX. *Murex radix* Gmelin, 1791. 4 inches. Panama to southern Ecuador.

1837 WEST INDIAN MUREX. *Murex brevifrons* Lamarck, 1822. 2 inches. Lower Florida Keys and the Caribbean.

1838 LACE MUREX. *Murex florifer dilectus* A. Adams, 1854. 2 inches. South half of Florida.

1855 OXYCANTH MUREX. *Homalocantha oxyacantha* (Broderip, 1833). 2 inches. West Mexico to Ecuador.

1862 MARTINIQUE MUREX. *Pterynotus phyllopterus* (Lamarck, 1822). 3½ inches. Lesser Antilles.

PLATE 12

NEPTUNES, MELONGENAS AND WHELKS

2353 COMMON NORTHWEST NEPTUNE. *Neptunea lyrata lyrata* (Gmelin, 1791). 4 inches. Arctic Seas to north California.

2354 NEW ENGLAND NEPTUNE. *Neptunea lyrata decemcostata* (Say, 1826). Nova Scotia to off Massachusetts.

2374 TABLED NEPTUNE. *Neptunea tabulata* (Baird, 1863). 3½ inches. British Columbia to off San Diego, California.

2389 KELLET's WHELK. *Kelletia kelleti* (Forbes, 1850). 4 inches. Santa Barbara, California, to Baja California, Mexico.

2432 COMMON CROWN CONCH. *Melongena corona* (Gmelin, 1791). 3 inches. Florida.

2433 DWARF CROWN CONCH. *Melongena corona bicolor* (Say, 1827). 1 inch. Miami to Key West, Florida; Dry Tortugas.

2434 JOHNSTONE's CROWN CONCH. *Melongena corona johnstonei* Clench and Turner, 1956. 5 inches. Alabama and northwest Florida.

2435 YUCATAN CROWN CONCH. *Melongena corona bispinosa* Philippi, 1844. 2 inches. Yucatan, Mexico.

2437 WEST INDIAN CROWN CONCH. *Melongena melongena* (Linné, 1758). 6 inches. Greater and Lesser Antilles to Panama.

2439 GIANT HAIRY MELONGENA. *Pugilina morio* (Linné, 1758). 5 inches. Trinidad to Brazil; central West Africa.

2442 CANDELABRUM WHELK. *Busycon candelabrum* (Lamarck, 1816). 5 inches. Texas banks to Yucatan, Mexico.

2443 TURNIP WHELK. *Busycon coarctatum* (Sowerby, 1825). 5 inches. Bay of Campeche, Mexico.

2444 LIGHTNING WHELK. *Busycon contrarium* (Conrad, 1840). 8 inches. Off New Jersey to Florida and to Texas.

2452 TEXAS PEAR WHELK, *Busycon spiratum* subspecies *plagosum* (Conrad, 1863). Alabama to Mexico.

2353

2354

2374

2437

2389

2439

2434

2432

2433

2435

2444

2452

2442

2443

PLATE 14

WESTERN ATLANTIC CONES

2779 ALPHABET CONE (Florida form). *Conus spurius* Gmelin, 1791, form *atlanticus* Clench, 1942. Florida and Bahamas.

2779a ALPHABET CONE (West Indian form). *Conus spurius* Gmelin, 1791. 2½ inches. West Indies.

2781 CARROT CONE. *Conus daucus* Hwass, 1792. 1½ inches. Both sides of Florida and the West Indies to Brazil.

2783 FLORIDA CONE. *Conus floridanus* Gabb, 1868. (Typical form). 1½ inches. North Carolina to both sides of Florida.

2784 FLORIDENSIS CONE. *Conus floridanus* form *floridensis* Sowerby, 1870. 1½ inches. Florida.

2785 BURRY'S CONE. *Conus floridanus* form *burryae* Clench, 1942. 1½ inches. Eastern Florida.

2786 THE SENNOTT'S CONE. *Conus sennottorum* Rehder and Abbott, 1951. 1 inch. Off Tortugas, Florida, and off Yucatan, Mexico.

2787 SOZON'S CONE. *Conus delessertii* Récluz, 1843. 4 inches. South Carolina to the Gulf of Mexico; Bermuda.

2788 CROWN CONE. *Conus regius* Gmelin, 1791. 2 inches. Southern Florida and the West Indies to Brazil.

2788a YELLOW FORM OF CROWN CONE. *Conus regius* form *citrinus* Gmelin, 1791. 2 inches. Southern Florida and the West Indies.

2789 CARDINAL CONE. *Conus cardinalis* Hwass, 1792. 1 inch. Bahamas and the Caribbean.

2790 MOUSE CONE. *Conus mus* Hwass in Bruguière, 1792. 1 inch. Southeast Florida and the West Indies; Bermuda.

2791 BERMUDA CONE. *Conus bermudensis* Clench, 1942. 2 inches. Bermuda, Florida and the West Indies.

2794 STIMPSON'S CONE. *Conus stimpsoni* Dall, 1902. 2 inches. Southeast Florida and the Gulf of Mexico; deep water.

2795 VILLEPIN'S CONE. *Conus villepini* Fischer and Bernardi, 1857. 2½ inches. Southeast Florida and off Yucatan, Mexico.

2797 MAZE'S CONE. *Conus mazei* Deshayes, 1874. 2 inches. Off Florida, the Gulf of Mexico; Caribbean; deep water.

2799 GLORY-OF-THE-ATLANTIC. *Conus granulatus* Linné, 1758. 1¾ inches. Southeast Florida and the West Indies.

2800 AUSTIN'S CONE. *Conus austini* Rehder and Abbott, 1951. 2 inches. Southeast Florida to Yucatan, Mexico, and the West Indies; deep water.

2801 CLARK'S CONE. *Conus clarki* Rehder and Abbott, 1951. 1 inch. Off Louisiana; deep water.

2802 TURTLE CONE. *Conus ermineus* Born, 1778. 3 inches. Gulf of Mexico, the Caribbean to Brazil.

2805 ABBOTT'S CONE. *Conus abbotti* Clench, 1942. 2 inches. Bahamas.

2781

2794

2789

2799

2800

2790

2801

2786

2797

2787

2802

2795

2785

2784

2783

2805

2791

2788

2788a

2779a

2779

PLATE 15

TROPICAL WEST AMERICAN CONES

2813 PEAR-SHAPED CONE. *Conus patricius* Hinds, 1843. 4 inches. Nicaragua to Ecuador.

2815 BROWN CONE. *Conus brunneus* Wood, 1828. 1½ inches. Baja California to Ecuador; Galapagos Islands.

2816 DIADEM CONE. *Conus diadema* Sowerby, 1834. 1 inch. Baja California to Ecuador; Galapagos Islands.

2817 GLADIATOR CONE. *Conus gladiator* Broderip, 1833. 1 inch. Baja California to Ecuador; Galapagos Islands.

2818 NUT CONE. *Conus sponsalis* subspecies *nux* Broderip, 1833. ½ inch. Baja California to Ecuador; Galapagos Islands.

2819 PRINCE CONE. *Conus princeps* Linné, 1758. 2 inches. California to Ecuador. Typical color form.

2819a FINE-LINED PRINCE CONE. *Conus princeps* form *lineolatus* Valenciennes, 1832. Gulf of California to Ecuador.

2819b GOLDEN PRINCE CONE. *Conus princeps* form *apogrammatus* Dall, 1910.

2820 PUZZLING CONE. *Conus perplexus* Sowerby, 1857. ¾ inch. Baja California to Ecuador.

2821 PURPLISH CONE. *Conus purpurascens* Sowerby, 1833. 2 inches. Baja California to Ecuador.

2822 VITTATE CONE. *Conus vittatus* Hwass in Bruguière, 1792. 1 inch. Gulf of California to Ecuador.

2823 XIMENES CONE. *Conus ximenes* Gray, 1839. 1 inch. Gulf of California to Panama.

2824 DALL'S CONE. *Conus dalli* Stearns, 1873. Gulf of California to Panama; Galapagos.

2825 LUCID CONE. *Conus lucidus* Wood, 1828. ¾ inch. Baja California to Ecuador.

2826 ARCHON CONE. *Conus archon* Broderip, 1833. 2 inches. Gulf of California to Panama.

2829 FERGUSON'S CONE. *Conus fergusoni* Sowerby, 1873. 3 inches. Gulf of California to Ecuador.

2831 RECURVED CONE. *Conus recurvus* Broderip, 1833. 2 inches. Gulf of California to Colombia.

2832 REGULAR CONE. *Conus regularis* Sowerby, 1833. 1½ inches. Baja California to Panama.

2834 BRANCHING CONE. *Conus virgatus* Reeve, 1849. 2 inches. Baja California to Ecuador.

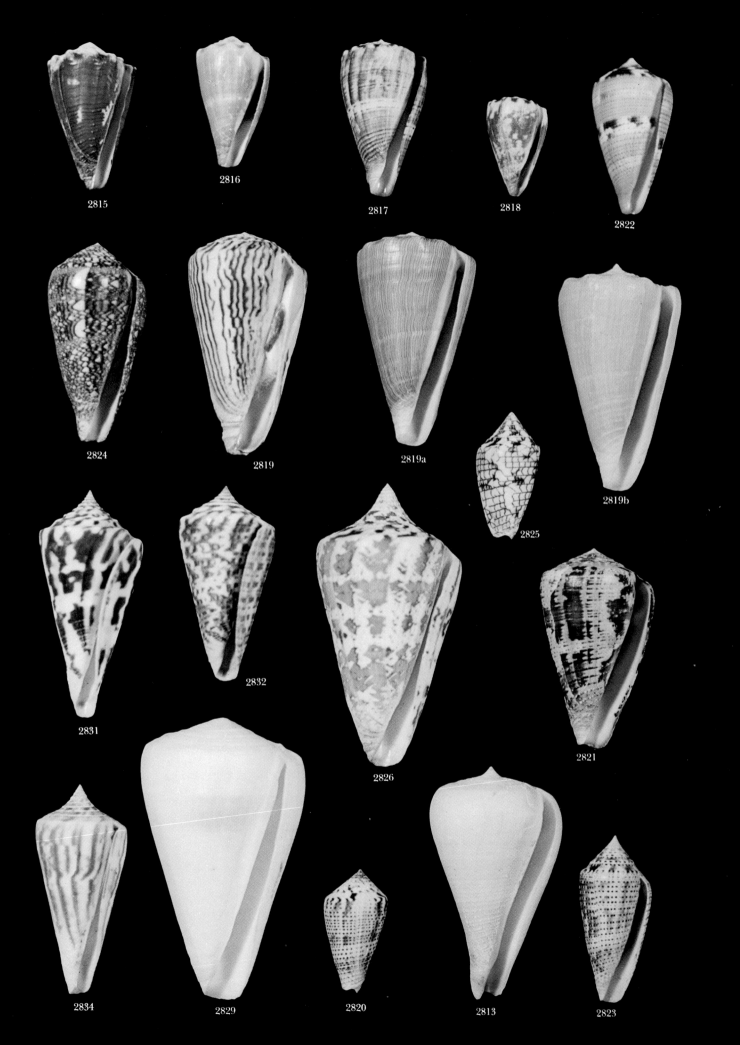

2815

2816

2817

2818

2822

2824

2819

2819a

2825

2819b

2831

2832

2826

2821

2834

2829

2820

2813

2823

PLATE 16

NEW ENGLAND NUDIBRANCHS

4305 PILOSE DORIS. *Acanthodoris pilosa* (Müller). 1 inch. Arctic to Connecticut; Alaska.

4325 YELLOW FALSE DORIS. *Adalaria proxima* (Alder and Hancock). ½ inch. Arctic to Maine.

4335 ATLANTIC ANCULA. *Ancula gibbosa* (Risso). ½ inch. Arctic to Massachusetts.

4357 FROND EOLIS. *Dendronotus frondosus* (Ascanius). 2 inches. Arctic to Rhode Island; to Washington State.

4408 RED-FINGERED EOLIS *Coryphella verrucosa* subspecies *rufibranchialis* (Johnston). 1 inch. Arctic to New York.

4431 PAINTED BALLOON EOLIS. *Eubranchus pallidus* (Alder and Hancock). ½ inch. Arctic to Boston.

4433 DWARF BALLOON EOLIS. *Capellinia exigua* (Alder and Hancock). ⅕ inch. Arctic to Massachusetts.

4436 JOHNSTON'S BALLOON EOLIS. *Tergipes tergipes* (Forskål). ⅓ inch. Arctic to New York.

4438 ORANGE-TIPPED EOLIS. *Catriona aurantia* (Alder and Hancock). ½ inch. Arctic to Connecticut.

4470 PAPILLOSE EOLIS. *Aeolidia papillosa* (Linné). 2 inches. Arctic to Maryland; to California.

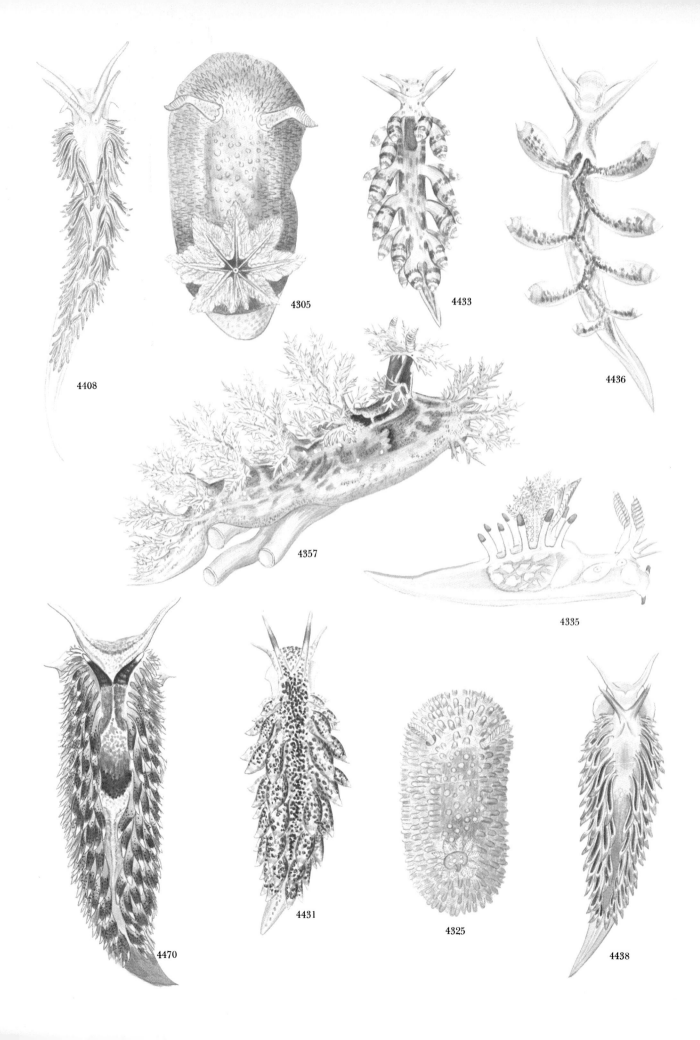

4305

4433

4436

4408

4357

4335

4470

4431

4325

4438

PLATE 17

PACIFIC COAST NUDIBRANCHS

4211 MONTEREY DORIS. *Archidoris montereyensis* (Cooper). 1½ inches. California.

4215 NOBLE PACIFIC DORIS. *Montereina nobilis* (MacFarland). 4 inches. California.

4220 HEATH'S DORIS. *Discodoris heathi* MacFarland. 1 inch. California.

4234 SAN DIEGO DORIS. *Diaulula sandiegensis* (Cooper). 2½ inches. Alaska to California.

4238 MACFARLAND'S PRETTY DORIS. *Rostanga pulchra* MacFarland. ¾ inch. California.

4252 PORTER'S BLUE DORIS. *Glossodoris porterae* (Cockerell). ½ inch. California.

4282 LAILA DORIS. *Laila cockerelli* MacFarland. ¾ inch. California.

4283 ORANGE-SPIKED DORIS. *Polycera atra* MacFarland. ¾ inch. California.

4295 CARPENTER'S DORIS. *Triopha carpenteri* (Stearns). 1 inch. California.

4296 MACULATED DORIS. *Triopha maculata* MacFarland. 1 inch. California.

4297 MACFARLAND'S GRAND DORIS. *Triopha grandis* MacFarland. 3 inches. California.

4339 HOPKINS' DORIS. *Hopkinsia rosacea* MacFarland. 1 inch. Monterey to San Pedro, California.

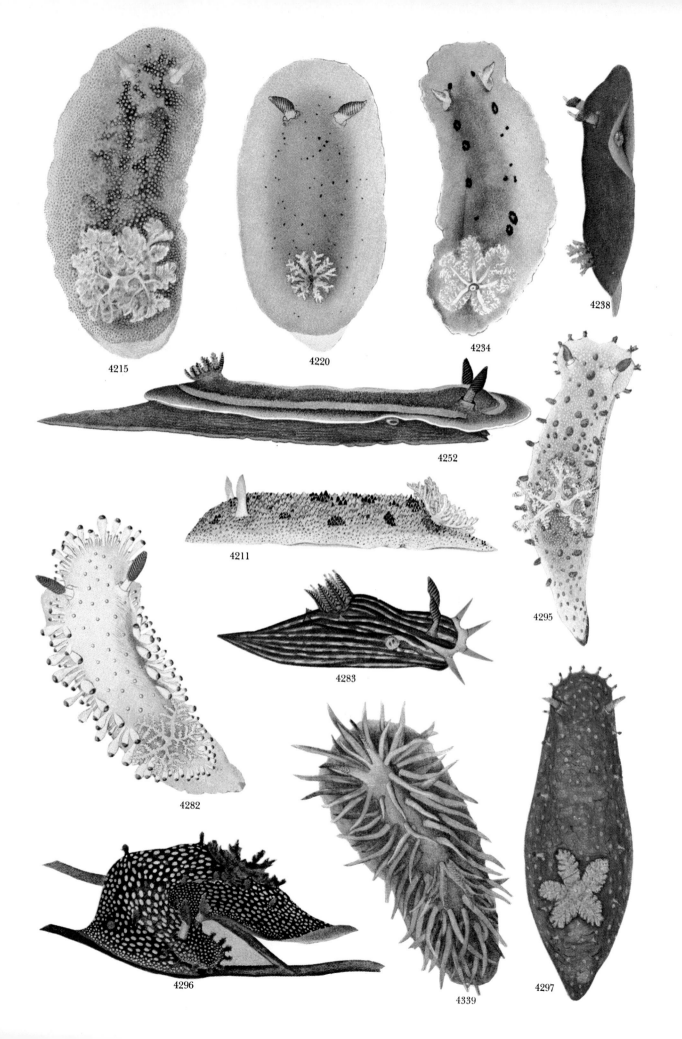

4215

4220

4234

4238

4252

4295

4211

4282

4283

4296

4339

4297

PLATE 18

SCALLOPS

5133 Zigzag Scallop. *Pecten ziczac* (Linné, 1758). 3 inches. North Carolina to the West Indies; Bermuda.

5134 Ravenel's Scallop. *Pecten raveneli* Dall, 1898. 2 inches. North Carolina to Florida and the Caribbean.

5136 Paper Scallop. *Amusium papyraceum* (Gabb, 1873). 2 inches. Gulf of Mexico to the West Indies; deep water.

5137 Laurent's Scallop. *Amusium laurenti* (Gmelin, 1791). 3 inches. Atlantic side of Central America to the Greater Antilles; deep water.

5153 Tryon's Scallop. *Aequipecten glyptus* (Verrill, 1882). 2 inches. Off Rhode Island to Florida to Texas; deep water.

5194 Atlantic Bay Scallop (New England subspecies). *Argopecten irradians irradians* (Lamarck, 1819). 3 inches. Cape Cod to New Jersey.

5195 Atlantic Bay Scallop (Florida subspecies). *Argopecten irradians concentricus* (Say, 1822). 3 inches. New Jersey to Georgia; West Florida to Louisiana.

5198 Calico Scallop (6 color varieties). *Argopecten gibbus* (Linné, 1758). 2 inches. Off Maryland to Florida to south Texas and Brazil.

5200 Pacific Calico Scallop. *Argopecten circularis* (Sowerby, 1835). 2½ inches. Santa Barbara, California, to Peru.

5202 Atlantic Deep-sea Scallop. *Placopecten magellanicus* (Gmelin, 1791). 8 inches. Labrador to off Cape Hatteras, North Carolina.

5205 Lion's Paw. *Lyropecten nodosus* (Linné, 1758). 5 inches. North Carolina to Texas and to Brazil; Ascension Island.

PLATE 19

SCALLOPS

5138 SENTIS SCALLOP (3 color varieties). *Chlamys sentis* (Reeve, 1853). 1½ inches. North Carolina to southeast Florida to Brazil.

5140 ORNATE SCALLOP. *Chlamys ornata* (Lamarck, 1819). 1½ inches. Southeast Florida to Brazil; Bermuda.

5141 MANY-RIBBED SCALLOP. *Chlamys multisquamata* (Dunker, 1864). 2 inches. Southeast Florida and the West Indies; Bermuda.

5143 LITTLE KNOBBY SCALLOP. *Chlamys imbricata* (Gmelin, 1791). 1¾ inches. Southeast Florida and the West Indies; Bermuda.

5145 PACIFIC PINK SCALLOP. *Chlamys hastata hericia* (Gould, 1850). 2¾ inches. Alaska to California.

5146 HINDS' SCALLOP. *Chlamys rubida* (Hinds, 1845). 2½ inches. Alaska to off San Diego, California.

5147 ICELAND SCALLOP. *Chlamys islandica* (Müller, 1776). 3½ inches. Arctic seas to Cape Cod, Massachusetts; and Alaska to Puget Sound, Washington.

5156 ROUGH SCALLOP. *Aequipecten muscosus* (Wood, 1828). 1¾ inches. North Carolina to Texas to Brazil; Bermuda.

5187 KELP SCALLOP. *Leptopecten latiauratus* (Conrad, 1837). 1 inch. Santa Barbara to Baja California.

5138

5140

5141

5143

5156

5145

5146

5147

5187

PLATE 20

MUSSELS, PEARL OYSTERS AND LIMA FILE CLAMS

5043 YELLOW MUSSEL. *Brachidontes modiolus* (Linné, 1767). 1 inch. Southern Florida and the West Indies.

5044 SCORCHED MUSSEL. *Brachidontes exustus* (Linné, 1758). ¾ inch. North Carolina to Uruguay.

5090 TULIP MUSSEL. *Modiolus americanus* (Leach, 1815). South Carolina to Brazil; Bermuda.

5105 ATLANTIC RIBBED MUSSEL. *Geukensia demissa* (Dillwyn, 1817). 3 inches. Quebec to South Carolina; California.

5117 ATLANTIC WING OYSTER. *Pteria colymbus* (Röding, 1798). 2 inches. North Carolina to the West Indies.

5122 ATLANTIC PEARL OYSTER. *Pinctada imbricata* Röding, 1798. 3 inches. South half of Florida to Brazil; Bermuda.

5124 FLAT TREE OYSTER. *Isognomon alatus* (Gmelin, 1791). 3 inches. Florida to Texas and the West Indies to Brazil; Bermuda.

5125 LISTER'S TREE OYSTER. *Isognomon radiatus* (Anton, 1839). 1½ inches. Florida to Texas to Brazil.

5126 TWO-TONED TREE OYSTER. *Isognomon bicolor* (C. B. Adams, 1845). 1½ inches. Florida Keys and the Caribbean; Bermuda.

5216 KITTEN'S PAW. *Plicatula gibbosa* Lamarck, 1801. 1 inch. North Carolina to Texas and the West Indies; Bermuda.

5232 COMMON JINGLE SHELL. *Anomia simplex* Orbigny, 1842. 1 inch. Cape Cod, Massachusetts, to Texas to Brazil; Bermuda. Has yellow, orange and white color varieties or stained black.

5237 FALSE PACIFIC JINGLE. *Pododesmus macroschisma* (Deshayes, 1839). 3 inches. Alaska to Baja California; Japan.

5240 SPINY LIMA. *Lima lima* (Linné, 1758). 1½ inches. Southeast Florida to Brazil; Bermuda.

5242 ROUGH LIMA. *Lima scabra* (Born, 1778). 2 inches. Off South Carolina to Texas to Brazil.

PLATE 21

THORNY OYSTERS, JEWEL BOXES AND CARDITAS

5221 ATLANTIC THORNY OYSTER. *Spondylus americanus* Hermann, 1781. 4 inches. Off North Carolina to Texas to Brazil.

5222 DIGITATE THORNY OYSTER. *Spondylus ictericus* Reeve, 1856. 2½ inches. Florida to Brazil.

5223 PACIFIC THORNY OYSTER. *Spondylus princeps* Broderip, 1833. 5 inches. Gulf of California to Panama.

5384 LEAFY JEWEL BOX. *Chama macerophylla* (Gmelin, 1791). 3 inches. North Carolina to Brazil; Bermuda. (Two color forms shown.)

5385 LITTLE CORRUGATED JEWEL BOX. *Chama congregata* Conrad, 1833. 1 inch. North Carolina to Texas, to Brazil.

5386 WHITE SMOOTH-EDGED JEWEL BOX. *Chama sinuosa* Broderip, 1835. 3 inches. South half of Florida and the West Indies; Bermuda.

5387 CHERRY JEWEL BOX. *Chama sarda* Reeve, 1847. 1 inch. Florida Keys to Brazil.

5388 CLEAR JEWEL BOX. *Chama pellucida* Broderip, 1835. 2 inches. Oregon to Chile.

5395 ATLANTIC LEFT-HANDED JEWEL BOX. *Pseudochama radians* (Lamarck, 1819). 2 inches. North Carolina to Texas and the West Indies; Bermuda.

5400 FLORIDA SPINY JEWEL BOX. *Arcinella cornuta* Conrad, 1866. 1½ inches. North Carolina to Florida to Texas.

5478 BROAD-RIBBED CARDITA. *Carditamera floridana* Conrad, 1838. 1½ inches. Southern half of Florida and east Mexico.

5479 WEST INDIAN CARDITA. *Carditamera gracilis* (Shuttleworth, 1856). 1 inch. Mexico to Puerto Rico.

5480 PANAMANIAN CARDITA. *Carditamera affinis* (Sowerby, 1833). 1½ inches. Gulf of California to Peru.

PLATE 22

TRUE OYSTERS, LUCINES AND COCKLES

5279 FRONS OYSTER (oval shape). *Lopho frons* (Linné, 1758). 1½ inches. Florida, Louisiana to Brazil.

5279a FRONS OYSTER (elongate shape). 1½ inches. Attaches to gorgonian stems. Florida to Brazil.

5283 PENNSYLVANIA LUCINA. *Linga pensylvanica* (Linné, 1758). 2 inches. North Carolina to Florida and the West Indies.

5297 TIGER LUCINA. *Codakia orbicularis* (Linné, 1758). 3 inches. Florida to Texas, to Brazil; Bermuda.

5305 THICK LUCINA. *Lucina pectinatus* (Gmelin, 1791). North Carolina to Texas, to Brazil.

5320 WESTERN RINGED LUCINA. *Lucinoma annulata* (Reeve, 1850). 2 inches. Alaska to southern California.

5326 BUTTERCUP LUCINA. *Anodontia alba* Link, 1807. 2 inches. North Carolina to Texas and the West Indies; Bermuda.

5327 CHALKY BUTTERCUP. *Anodontia philippiana* (Reeve, 1850). 4 inches. North Carolina to east Florida, Cuba and Bermuda.

5329 FLORIDA LUCINA. *Pseudomiltha floridana* (Conrad, 1833). 1½ inches. West Florida to Texas.

5332 CROSS-HATCHED LUCINA. *Divaricella quadrisulcata* (Orbigny, 1842). ½ inch. Massachusetts to Brazil.

5536 GIBB'S CLAM. *Eucrassatella speciosa* (A. Adams, 1852). 2 inches. North Carolina to the Gulf of Mexico and West Indies.

5546 PRICKLY COCKLE. *Trachycardium egmontianum* (Shuttleworth, 1856). 2 inches. North Carolina to Florida.

5549 YELLOW COCKLE. *Trachycardium muricatum* (Linné, 1758). 2 inches. North Carolina to Texas and the West Indies.

5555 SPINY PAPER COCKLE. *Papyridea soleniformis* (Bruguière, 1789). 1½ inches. North Carolina to Brazil.

5559 ATLANTIC STRAWBERRY COCKLE. *Americardia media* (Linné, 1758). 1½ inches. North Carolina to Brazil; Bermuda.

5572 COMMON EGG COCKLE. *Laevicardium laevigatum* (Linné, 1758). 2 inches. North Carolina to both sides of Florida and the West Indies; Bermuda.

5279

5279a

5536

5297

5327

5326

5305

5283

5329

5320

5332

5555

5572

5559

5546

5549

PLATE 23

SANGUIN CLAMS AND TELLINS

5636 SUNRISE TELLIN. *Tellina radiata* Linné, 1758. 3 inches. South Carolina to Florida and the West Indies; Bermuda. (rayed and plain form shown).

5637 SMOOTH TELLIN. *Tellina laevigata* Linné, 1758. 3 inches. North Carolina to the Caribbean; Bermuda.

5639 GREAT TELLIN. *Tellina magna* Spengler, 1798. 4 inches. North Carolina to the south half of Florida, and the West Indies.

5640 SPECKLED TELLIN. *Tellina listeri* Röding, 1798. 3 inches. North Carolina to the south half of Florida to Brazil; Bermuda.

5643 FAUST TELLIN. *Tellina fausta* Pulteney, 1799. 3 inches. North Carolina to Florida and the West Indies.

5660 ROSE PETAL TELLIN. *Tellina lineata* Turton, 1819. 1½ inches. All of Florida to Texas and to Brazil.

5661 ALTERNATE TELLIN. *Tellina alternata* Say, 1822. 2½ inches. North Carolina to Texas.

5662 TAYLOR'S TELLIN. *Tellina alternata* subspecies *tayloriana* Sowerby, 1867. 2½ inches. Texas to Tampico, Mexico.

5696 CANDY STICK TELLIN. *Tellina (Scissula) similis* Sowerby, 1806. 1 inch. South half of Florida and the Caribbean; Bermuda.

5704 LARGE STRIGILLA. *Strigilla carnaria* (Linné, 1758). 1 inch. South half of Florida to Argentina.

5749 ATLANTIC GROOVED MACOMA. *Psammotreta intastriata* (Say, 1827). 2½ inches. South Carolina to Florida and the Caribbean; Bermuda.

5766 ATLANTIC SANGUIN. *Sanguinolaria sanguinolenta* (Gmelin, 1791). 2 inches. Southeast Florida to Texas and the Caribbean.

5767 OPERCULATE SANGUIN. *Sanguinolaria cruenta* (Lightfoot, 1786). 3 inches. Caribbean to Brazil.

5770 NUTTALL'S MAHOGANY CLAM. *Sanguinolaria (Nuttallia) nuttalli* Conrad, 1837. 3 inches. Southern California to Baja California.

5771 GAUDY ASAPHIS. *Asaphis deflorata* (Linné, 1758). 2 inches. Southeast Florida to Brazil; Bermuda.

5771

5766

5643

5639

5770

5660

5661

5662

5704

5696

5640

5749

5636

5637

5767

PLATE 24

VENERID CLAMS

5852 PRINCESS VENUS. *Periglypta listeri* (Gray, 1838). 3 inches. Southeast Florida and the West Indies.

5855 RIGID VENUS. *Ventricolaria rigida* (Dillwyn, 1817). 3 inches. Florida Keys and West Indies to Brazil.

5857 EMPRESS VENUS. *Circomphalus strigillinus* (Dall, 1902). 1½ inches. Off South Carolina to Brazil.

5865 CROSS-BARRED VENUS. *Chione cancellata* (Linné, 1758). 1 inch. North Carolina to Texas, to Brazil.

5867 LADY-IN-WAITING VENUS. *Chione intapurpurea* (Conrad, 1849). 1½ inches. North Carolina to Texas, to Brazil.

5873 KING VENUS. *Chione paphia* (Linné, 1767). 1½ inches. Lower Florida Keys and the West Indies.

5874 IMPERIAL VENUS. *Chione latilirata* (Conrad, 1841). 1 inch. North Carolina to Texas, to Brazil.

5888 WEST INDIAN POINTED VENUS. *Anomalocardia brasiliana* (Gmelin, 1791). 1 inch. West Indies to Brazil.

5911 TRIGONAL TIVELA. *Tivela mactroides* (Born, 1778). 1½ inches. West Indies to Brazil.

5915 PISMO CLAM. *Tivela stultorum* (Mawe, 1823). 4 inches. San Mateo County, California, to Baja California.

5930 LIGHTNING VENUS. *Pitar fulminata* (Menke, 1828). 1 inch. North Carolina to Florida and the West Indies.

5934 SCHWENGEL'S VENUS. *Pitar cordatus* (Schwengel, 1951). 1½ inches. Off southwest Florida to Texas to Brazil.

5936 ROYAL COMB VENUS. *Pitar dione* (Linné, 1758). 1½ inches. East Mexico to Panama and the West Indies.

5949 SUNRAY VENUS. *Macrocallista nimbosa* (Lightfoot, 1786). 5 inches. North Carolina to Florida to Texas.

5950 CALICO CLAM. *Macrocallista maculata* (Linné, 1758). 2½ inches. North Carolina to Brazil; Bermuda.

5959 ELEGANT DOSINIA. *Dosinia elegans* Conrad, 1846. 3 inches. North Carolina to Texas; Caribbean.

5852

5855

5857

5911

5915

5930

5936

5934

5950

5959

5949

5888

5873

5867

5865

5874

2 to 4 inches, ovate-oblong, tapering toward the head. Tentacles and rhinophores slit and expanded distally, the latter nearer to the front margin than to the dorsal slit. Pleuropodial lobes arising far behind the middle of the length; scarcely mobile, united behind, and enclosing a large gill cavity; dorsal slit short. Mantle small, not perforated over the shell, nor covering much of the gill. Genital pore in front of the gill, under the mantle. Shell small, solid, calcareous, the apex projecting. Animal greenish to yellow-brownish, with numerous short, pointed, fleshy projections; moderately common; shallow water, 1 to 4 feet. Synonyms: *ascifera* (Rang, 1828); *virens* Verrill, 1901.

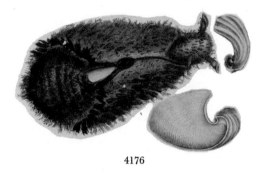

4176

Genus *Petalifera* Gray, 1847

Type of the genus is *virescens* Risso, 1826. *Aplysiella* P. Fischer, 1872, is a synonym.

Petalifera ramosa Baba, 1959   4177

Southeast Florida; southern Japan.

30 mm., translucent with a yellowish green viscera, and with a mid-dorsal white streak and a delicate pattern of white spots and lines on the slightly ragged tentacles and rhinophores. Most of the fringed papillae are wartlike and have a white center with a white ring around the base. Head broad; dorsal slit extends more than halfway back from the front end. Shell resembling that of a dolabriferid. Rare.

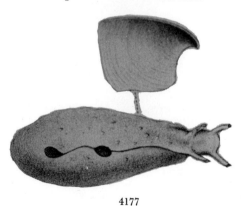

4177

Genus *Phyllaplysia* P. Fischer, 1872

Small (2 to 3 inches), flattened, smooth-skinned, or with small projections; foot broad. Body usually green. Shell, if present, is thin, fragile, calcareous, dorsally concave, with a secondary structure adhering to the dorsal surface of the mantle shelf in the parapodial cavity. Type: *lafonti* (Fischer, 1870). There is one Caribbean to Brazil species, *engeli* Marcus, 1955.

Phyllaplysia taylori Dall, 1900   4178
Taylor's Zebra Sea-hare

British Columbia to San Diego, California.

2 to 3 inches (living), head with a pair of broad, flat cephalic tentacles and a pair of smaller, tubular, nonstriped rhinophores. Body bright-green with numerous fine, irregular gray or black lines of various patterns. Common in shallow water on *Zostera* eel-grass where it grazes on sessile diatoms. Eggmasses are laid in flat packets on the grass blades. They hatch in 3 weeks with a limpetlike larval shell, 2 mm. long, protecting them for a few days. *P. zostericola* McCauley, 1960, is a synonym. See R. D. Beeman, 1963, *The Veliger*, vol. 6, p. 43.

4178

Other species:

4179 *Phyllaplysia engeli* Marcus, 1955. Southeast Florida to Brazil.

Subfamily NOTARCHINAE Eales, 1925

Sea-hares with filamentous appendages, frequently with brilliant "eye spots" in the skin. Parapodia almost completely joined. Shell usually absent, sometimes minute and orbicular. Marcus and Marcus (1962, *Bull. Marine Sci.*, vol. 12, p. 457), reported a Japanese species from off Islamorada, Florida Keys (*Notarchus punctatus armatus* Baba, 1938).

Genus *Bursatella* Blainville, 1817

Type: *leachii* Blainville, 1817. Synonyms: *Ramosaclesia* Iredale, 1929.

Bursatella leachii pleii Rang, 1828   4180
Ragged Sea-hare

North Carolina to Florida, West Indies to Brazil.

4 inches in length, elongate-oval, plump, soft and flabby. Greenish gray to olive in color, sometimes with white flecks. Surface covered with numerous, ragged filaments. Shell absent in adults. Commonly found in grassy, mud-bottom areas at low tide. The east side of Sanibel Island is a good collecting spot. This is the only Western Atlantic species known in this genus. Formerly placed in another genus, *Notarchus* Cuvier, 1817. *B. lacinulata* Gould, 1852, is a synonym.

4180

Genus *Aclesia* Rang, 1828

Parapodial cavity not large, extending a short distance behind the union of the parapodia. Body plump, somewhat compressed laterally. Covered with numerous tubercles, some of which are conical or filiform. Parapodial lobes broadly united behind, closing the gill cavity. Tentacles and grooved rhinophores present. The common genital aperture is slightly to the right of the median line of the body and well in front of the anterior ends of the parapodial lobes, not between them as in *Notarchus* or *Bursatella*. Type: *savignana* Audouin, 1826. One western Central American species:

4181 *Aclesia rickettsi* MacFarland, 1966. Espirita Santo Island, Gulf of California, and La Paz Bay, Baja California. 33 mm. *Mem. Calif. Acad. Sci.*, vol. 6, p. 27. (May be *Stylocheilus rickettsi* MacFarland, 1966.)

Genus *Stylocheilus* Gould, 1852

Type of the genus is *lineolatus* Gould, 1852.

Stylocheilus longicauda (Quoy and Gaimard, 1824)   **4182**

Circumtropical; West Indies.
Baja California (abundant). Brazil.

½ to 1 inch, with small, sparse filaments; greenish with numerous gray pencil lines and occasional very small blue ocellar spots surrounded by orange circles. Penis and its sheath are spiny. Synonym: possibly *Aclesia rickettsi* MacFarland, 1966.

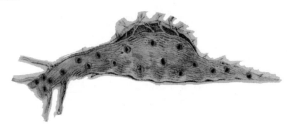

4181

Other species:

4183 *Stylocheilus citrinus* (Rang, 1828). Circumtropical. Florida to Brazil. East Indies. May be a yellow to reddish color form of *longicauda*. Has no "eye spots."

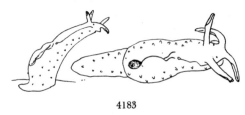

4183

Umbrella Shells and Pleurobranchs

Suborder *Notaspidea*

With flat calcareous, internal or external shells, and with a set of gill plumes usually on the right side in the region of the anus. This group stands somewhere between the tectibranchs (Bullacea, Aplysiacea, etc.) and the nudibranchs.

Superfamily Tylodinacea Gray, 1847

Family TYLODINIDAE Gray, 1847

Shell limpetlike, external and with a sinistral nucleus. Gills on the posterior right side. 2 or 3 pairs of tentacles.

Genus *Tylodina* Rafinesque, 1819

Shell smaller than the animal and covered with a fuzzy periostracum. Type: *punctulata* Rafinesque, 1814. Synonym: *Joannisia* Monterosato, 1884.

Tylodina fungina Gabb, 1865   **4184**
Yellow Umbrella Shell

Morro Bay, California, to Costa Rica. Galapagos.

15 to 30 mm., limpetlike, high-spired with the apex central. Base oval in outline. Brown periostracum extends over the margin of the shell. Exterior with weak, undulating radial ribs. Interior glossy-white and irregular. Muscle scar horseshoe-shaped with the opening at the right. Mantle and foot are bright-yellow, matching the color of the sublittoral sponges upon which it browses. Moderately common; intertidal under rocks.

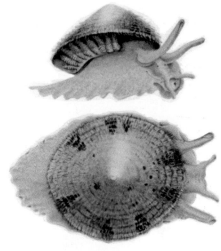

4184

Other species:

4185 *Tylodina americana* Dall, 1890. Northern Gulf of Mexico to off Cuba. Bermuda. Rare; 26 to 80 fms. (Synonym: *Umbraculum bermudense* of Dall, 1889, pl. 14, figs. 9–10, not an *Umbraculum*.)

Subfamily UMBRACULINAE Dall, 1889

Shell external, dorsal, a flat shield, with a small, subcentral, sinistral nucleus. Foot massive and warty. Head has 2 auriculate tentacles with eyes at the inside base. Gill plume large, at the right posterior side and between the mantle and foot. The family name Umbrellidae Gray, 1840, is synonymous.

Genus *Umbraculum* Schumacher, 1817

Animal with a large thick foot above which is a flat, calcareous, limpetlike shell, covered dorsally by a thin periostracum. Penis large, resembling an enrolled siphon, nonretractile and jutting out between the 2 tentacles. Type: *Patella sinica* Gmelin, 1791. Synonyms: *Gastroplax* Blainville, 1819; *Umbrella* Lamarck, 1819; *Ombrella* Blainville, 1824; *Operculatum* H. and A. Adams, 1854.

Umbraculum umbraculum (Lightfoot, 1786)   4186
Atlantic Umbrella Shell

> South Florida to off Texas and the Caribbean. Bermuda.

Animal 4 to 5 inches, oval and ½ as thick as long, uniform dull-yellow. Shell sits on top and is almost the size of the soft parts. Shell 1 to 2 inches across, flat, calcareous, thick, glossy-brown to whitish on the underside. Top with a weak yellowish periostracum and feltlike in a marginal rim. Apparently feeds on sponges. Uncommon; 1 to 40 fathoms. *U. bermudense* (Mörch, 1875), is probably a synonym, but topotype material has not been dissected. Might be *umbraculum* (Gmelin, 1791). *U. plicatulum* (von Martens, 1881) is a synonym.

4186   4187

Umbraculum ovale (Carpenter, 1856)   4187
Pacific Umbrella Shell

> Gulf of California to Panama.

Shell 30 to 50 mm., oval, low and covered with a thin, light-brown periostracum. Underside of shell glossy, white with a brown center. Found offshore; rare. Might be *umbraculum* (Lightfoot, 1786).

Subgenus *Hyalopatina* Dall, 1889

Shell oval, translucent-white, flat, very thin. Nucleus sinistral, of 1 whorl, half-immersed. Concentrically faintly undulated surface. Type and only known species:

4188 *Umbraculum (Hyalopatina) rushii* Dall, 1889. Off Great Isaac Light, Bahamas, 30 fathoms. Soft parts unknown.

4188

Superfamily Pleurobranchacea Menke, 1828
(Odhner, 1939)

Family PLEUROBRANCHIDAE Menke, 1828

Tough mantle dorsum as large as the foot and covering an internal thin, calcareous, oblong shell. Gill-plume located externally on the right side of the body. Rhinophores and a buccal veil present at the anterior end. For a key, see Odhner, 1926, *Results Swedish Antarctic Expedition*, vol. 2, no. 1, pp. 1–100.

Subfamily PLEUROBRANCHINAE Menke, 1828

Usually with an internal shell. Mantle edge very distinct from foot below. Anterior end of notum round or notched and distinct from the frontal veil. Shell internal. Rhinophores connected medianly under the border of the mantle.

Genus *Gymnotoplax* Pilsbry, 1896

Foot without a pedal gland. Radular teeth long, denticulate at the ends. Sections of the jaw simple or with a single denticle. Type: *americanus* (Verrill, 1885). *Berthellina* Gardiner, 1936, is a synonym.

Other species:

4189 *Gymnotoplax americanus* (Verrill, 1885). Off Martha's Vineyard, Massachusetts, 250 fms. Bermuda.

4190 *Gymnotoplax engeli* (Gardiner, 1936). Mediterranean. San Diego, California, to Guaymas, Mexico. Japan? Society Islands? (Synonym: *Berthella plumula* of Bergh.)

4191 *Gymnotoplax quadridens* (Mörch, 1863). Greater and Lesser Antilles; western Caribbean. Pacific side of Panama.

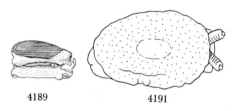

4189   4191

4192 *Gymnotoplax amarillius* (Mattox, 1953). Puerto Rico. *The Nautilus*, vol. 66, p. 109.

Genus *Pleurobranchus* Cuvier, 1804

Characters of the subfamily. The various genera are open to question as to their various values. Synonyms: *Oscaniella* Bergh, 1897; *Westernia* Blainville, 1827; *Gervisia* Blainville, 1827; *Cleanthus* Leach (in Gray) 1847. Type: *peronii* Cuvier, 1804.

Pleurobranchus areolatus Mörch, 1863   4193
Atlantic Pleurobranch

> Southeast Florida to Barbados.
> Gulf of California to Panama.

1½ to 2 inches in length. Mantle with U-shaped notch in front where 2 tubelike rhinophores protrude up. Dorsum or back with numerous small rounded warts. Color yellowish orange with irregular splotches of deep maroon-brown. Largest warts translucent pale-yellow to orangish. Gill-plume on right side of body, with 20 to 22 primary leaflets on each side, with a nodule on the main stem where they originate. Primary leaflets with 15 smaller leaflets, each of which has 5 to 10 microscopic plates. Shell small, calcareous, pinkish white, flat with a small spire, and located under the dorsum. Moderately common in shallow water in winter on Soldier Key, near Miami. *P. gardineri* White, 1952; *crossei* Vayssière, 1896; and *atlanticus* Abbott, 1949, are synonyms.

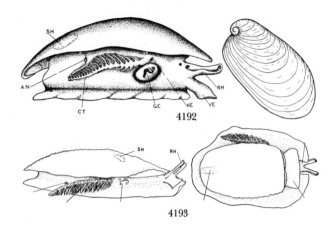

4192

4193

Other species:

4194 *Pleurobranchus digueti* Rochebrune, 1895. San Pedro, California, to the Gulf of California. 20 to 35 mm.

Pleurobranchus californicus (Dall, 1900)   **4195**
Californian Pleurobranch

Crescent City to San Diego, California.

20 to 50 mm., elongate oval, somewhat arched, waxen-white to cream, sprinkled with opaque-white flecks. Shell white, thin, quadrate, with a spiral nucleus. Gill-plume of about 25 pinnules on each side of the rhachis which bears a double row of alternating tubules at the base of the pinnules. Radula hooked, in 110 rows, a very large number in each row. Under rocks at low tide; uncommon. "*Pleurobranchus chacei* Burch, 1944" a mimeographed pseudo-species, is this species. *Pl. denticulata* MacFarland, 1966, is a synonym. This species, with a tuberculate rhachis, has been erroneously placed in *Berthella* by some workers.

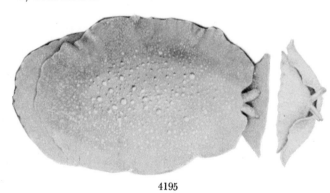

4195

Other species:

4196 *Pleurobranchus strongi* MacFarland, 1966. Monterey to Santa Barbara, California. 20 mm. Yellow. Rare. Erroneously placed in *Berthella* because of tubercles on the rhachis.

4197 *Pleurobranchus reesi* White, 1952. Bird Key Reef, Florida Keys. *Proc. Malacol. Soc. London*, vol. 29, p. 106.

Subgenus *Pleurobranchopsis* Verrill, 1900

Type of the subgenus is *verrilli* Thiele, 1931.

4198 *Pleurobranchus (Pleurobranchopsis) verrilli* Thiele, 1931. Bermuda. 1¾ inches. (Synonym: *aurantiaca* Verrill, 1900, non Risso, 1818.)

Genus *Berthella* Blainville, 1825

With a smooth gill rhachis, hook-shaped radular teeth, and a shell at least ½ the body length. Type: *plumula* (Montagu, 1803). Synonym: *Bouvieria* Vayssière, 1896.

Other species:

4199 *Berthella sideralis* (Lovén, 1847). Unalaska, Alaska (record doubtful); Europe; Scandinavia. For a description, see Odhner, 1939, p. 21, and MacFarland, 1966, p. 69.

4200 *Berthella tupala* Marcus, 1957. Southeast Florida to Brazil. *Jour. Linn. Soc. (Zool.)*, vol. 43, p. 416.

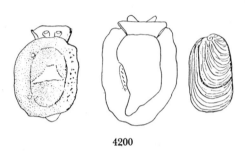

4200

Subfamily PLEUROBRANCHAEINAE Pilsbry, 1896

Anterior head veil contiguous with the notum and having tentaclelike angulations. Radular teeth bicuspid. Rhinophores fastened laterally between the borders of the velum and mantle. No shell.

Genus *Pleurobranchaea* Leue, 1813

Characters of the subfamily. Type: *meckelii* Leue, 1813. Gill rhachis smooth. Anal opening in front of or behind middle of the gill. Radula teeth with long lateral cusps on the upper part of the tooth. Jaw platelets are long prisms. Prostate present.

Pleurobranchaea hedgpethi Abbott, 1952   **4201**

North Carolina to Texas to Brazil.

20 to 50 mm., sepia-brown with lighter mottlings. Dorsal tail-spur black. Gill attached along most of its length; with 26 pairs of pinnules per gill with 35 leaflets on each. Branchial membrane extends to the thirteenth pinnule. Anus lies over the seventh pinnule. Radula of 36 rows, with 56 teeth in each half-row. Border of genital aperture with a dorsal flap pointing upward or forward. Moderately common; 14 to 50 fathoms. *P.*

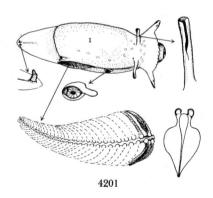

4201

hamva Marcus and Marcus, 1957, is a synonym according to Marcus and Marcus, 1967, p. 200.

Other species:

4202 *Pleurobranchaea tarda* Verrill, 1880. Off Martha's Vineyard, Massachusetts, 28 to 310 fms. Off Delaware, 31 to 300 fms.

4203 *Pleurobranchaea occidentalis* Bergh, 1897. Southeast Florida, Cuba to Martinique, 50 to 92 fathoms.

4204 *Pleurobranchaea agassizii* Bergh, 1897. Southeast Florida and the Grand Bahama Bank, 125 to 310 fathoms.

4205 *Pleurobranchaea californica* MacFarland, 1966. San Francisco to Monterey, California. 1 to 4 inches, light-brown with white mottlings. Dredged offshore to 110 fathoms; common. *Mem. Calif. Acad. Sci.*, vol. 6, p. 94.

Genus *Koonsia* Verrill, 1882

Same as *Pleurobranchaea* but the penis has small hooks. *Pleurobranchillus* Bergh, 1893, is a synonym. Type: *obesa* Verrill, 1882. One species:

4206 *Koonsia obesa* Verrill, 1882. Off Martha's Vineyard, Massachusetts, 192 to 258 fms. Off Delaware Bay, 312 fms.

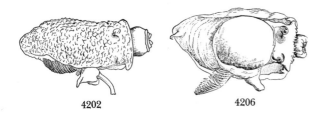

4202   4206

Order *Nudibranchia* Cuvier, 1817

The Nudibranchs

Shell-less, crawling sea slugs without true ctenidial gills. Rhinophores pillarlike, often bearing leaflike platelets, but not inrolled or auriculate. Jaws, when present, are paired, horny plates, not made up of a series of interlocking platelets.

Mother Nature was possibly on a drug spree and feeling very gay when she manufactured this beautiful and amazingly diversified group of sea fairies. Weird shapes, startling ornamentations, riotous colorations, intertwining evolutionary patterns and confusing, manmade nomenclature are lures that lead new students into this fascinating group of animals. I believe the classification will not be settled for several hundred years. About 275 species are reported from North America.

The pigeonholing of various groups is aided by a knowledge of the basic anatomy. *Holohepatic* means the "liver" or digestive gland is compact. *Cladohepatic* means the liver has been broken up and dispersed throughout the body and often into the club-shaped cerata or fleshy fingers on the backs of the animals. The condition of cladohepatic originated independently in various groups (like fins in whales, fish, skin-divers, and airplanes), and can be used only for identification, not classification.

The presence or absence of collarlike sheaths at the bases of the rhinophores is an important and useful key to sorting out certain suborders and families. Unfortunately for the person who does not have the time and training, the identification of many genera depends upon minute dissections and the examination of the radulae under high-power magnification.

There are many useful publications dealing with nudibranchs. Serious students should consult Nils Odhner, 1939, *Det Kongel. Norske Vidensk. Selsk.*, Skrifter, 1939, Trondheim, pp. 3–93 (in English); and Alice Pruvot-Fol, 1954, *Faune de France*, vol. 58, pp. 1–460 (in French), for higher classifications. Species-level material is found in *The Veliger*, occasionally in *The Nautilus* and in MacFarland's pretty 1966 tome (*Mem. Calif. Acad. Sci.*, vol. 6, 546 pp.). The spawn of the Pacific coast species were illustrated by Anne Hurst, 1967, *The Veliger*, vol. 9, p. 255. Caribbean species occur in the publications of the Institute of Marine Sciences, Miami, Florida (Marcus and Marcus, 1967, *Studies Trop. Oceanogr.*, no. 6). We are indebted to them for many illustrations used in this book.

For the use of an antioxidant (Ionol C. P.-40) (0.3% by volume in a solution of 5% formalin in seawater) to preserve the coloration of opisthobranchs, see *The Veliger*, 1969, vol. 11, p. 289.

As E. Marcus has pointed out (1957, *Jour. Linn. Soc.*, vol. 43, p. 465) many species have probably been transported to various parts of the world on the bottoms of algae-encrusted ships.

Suborder *Doridoida* Rafinesque, 1815

Holohepatic (liver compact, not scattered) nudibranchs with the right liver absent or reduced to a mere coecal appendix. Blood gland and (as a rule) 2 vesiculae seminales present. Feed on sponges; lay eggs in coiled ribbons. May be divided for convenience into four superfamilies:

1. True Dorids (Cryptobranchs)
Superfamily Dorididacea Rafinesque, 1815

Gills and rhinophores retractile into hollows. The rhinophores with a collar or sheath at the base. Mantle generally well-developed all around the body. Jaws present or absent. Includes such genera as *Doris*, *Cadlina*, *Rostanga*, *Aldisa*, *Archidoris*, *Discodoris*, *Anisodoris* and *Diaulula*.

2. False Dorids (Phanerobranchs)
Superfamily Pseudodoridacea Eliot, 1922

Gills and rhinophores not retractable into hollows, although they will contract when disturbed or preserved. Jaws not present. Includes such genera as *Aegires*, *Laila*, *Polycera*, *Triopha*, *Crimora*, *Acanthodoris* and *Lamellidoris*.

3. Swimming Dorids
Superfamily Hexabranchacea

Lively swimmers, by undulating the body. Usually bright red and yellow. No mandibles. Gills, usually 6-pinnate (but also 5- or 7-pinnate) nonretractile, with the wide anal opening in the middle. Rhinophores contractile only, with leaflets, smooth shafts and pockets with smooth rims. With numerous uniform lateral teeth and a small quadrate rhachial thickening. Nervous system highly concentrated. The genus *Hexabranchus* Ehrenberg, 1831, is mainly Indo-Pacific, but Marcus and Marcus, 1962 (*Bull. Marine Sci.*, vol. 12, p. 468) report *H. morsomus* Marcus and Marcus, 1962, from the Virgin Islands.

4. Pore-mouth Dorids
Superfamily Dendrodoridacea Pruvot-Fol, 1935

Dorids with the mouth a simple, porelike opening (hence the name Porostomata used by some workers for this superfamily) followed by a muscular buccal vestibule, then by a long digestive tube. No jaws, no radula. Contains families Phyllidiidae and the Dendrodorididae.

Family DORIDIDAE Rafinesque, 1815

Branchial plumes in an arc or circle usually joined together at their bases, usually retractile into a cavity. Rhinophores always with a perfoliate club. Pharyngeal bulb never suctorial and without jaws. No genital spines or platelets. Oral veil or palps produced into a fingerlike tubercle on each side. Radula teeth numerous, and basically of the same type (the marginals being only foreshortened modifications of the more central hook-shaped ones). Prostate distinct.

Subfamily DORIDINAE Rafinesque, 1815

Genus *Doris* Linné, 1758

Mantle (or back) with strong tubercles which are claviform and with small bases. Tubercles surround the branchial opening, and there are 2 large tubercles on either side of the rhinophoral openings. Branchial gills numerous but simply or singly pinnate. Oral palps digitiform. No glans penis. Type: *verrucosa* Linné, 1758. *Doridigitata* Orbigny, 1839, and *Staurodoris* Bergh, 1878, are synonyms.

Doris verrucosa Linné, 1758   4207

South Carolina to Florida and Brazil.
British Isles to South Africa.

40 to 50 mm., mantle with numerous globular and claviform tubercles. 10 to 18 branchial gills, pinnate. Color grayish or uniform yellowish, sometimes with a darker zone in the mid-

dle of the back. Radula of about 66 rows of hooklike teeth (45 to 60 per row). Uncommon in eastern United States. Synonyms: *Staurodoris januarii* Bergh, 1878; *Doridigitata derelicta* Fischer, 1867.

Other species:

4208 *Doris odonoghuei* Steinberg, 1963. Vancouver Island and San Juan Island, Washington. Synonyms: *Doris echinata* O'Donoghue, 1922, non Lovén, 1846; *Doris maculata* Iredale and O'Donoghue, 1923, non Garstang, 1896. Renamed by Steinberg in *The Veliger*, vol. 6, no. 2, p. 63.

4209 *Doris pickensi* Marcus and Marcus, 1967. Sonora coast of west Mexico. 25 mm., whitish and speckled with brown. *Studies Trop. Oceanogr. Miami*, no. 6, p. 184.

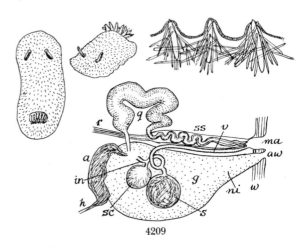

4209

Genus *Siraius* Marcus, 1955

Similar to *Doris,* but the notal warts are stiffened by internal tufts of hard spicules, and there are 4 lobes around each rhinophoral pit. Type: *ilo* Marcus, 1955.

Siraius kyolis Marcus and Marcus, 1967   4210
Kyolis Doris

Southeast Florida.

12 mm. in length, ½ as wide, dirty-yellow with minute blackish spots to tan with greenish black spots. Body rigid. Center of notum with low warts stiffened by internal tufts of spicules. 4 lobes surround the pits of the rhinophores, whose clubs bear about 15 foliations. Up to 11 unipinnate gills. Anal papilla in the center of the plumes. Foot bilabiate in front. Radula with 47 rows with 2·35·0·35·2 tooth formula. Marginal teeth pectinate. Shallow water; uncommon.

4207

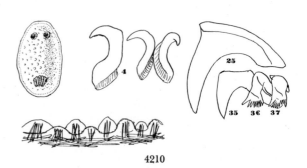

4210

Subfamily ARCHIDORIDINAE Bergh, 1891

Oral palps short and thick. Without a distinct prostate. Penis with a gland. Bergh later considered this subfamily not worth recognizing.

Genus *Archidoris* Bergh, 1878

Body not hard, dorsum granular or tubular; tentacles short, thick, with an external, longitudinal sulcus. No labial armature. Branchial plumes not numerous, 3- to 4-pinnate. Center of radula naked, marginal teeth hooked and bearing minute denticles. Penis and vagina unarmed (without hooks). Type: *tuberculata* (Cuvier, 1804) (non Müller, 1778).

Archidoris montereyensis (Cooper, 1862)   **4211**
Monterey Doris   **Color Plate 17**

Alaska to San Diego, California.

1 to 2 inches in length. Notum with low, bluntly conical, small tubercles of uniform size. Rhinophore stalks conical, the clavus slightly dilated, conical, perfoliate with 24 to 30 leaves on each side. Each of the 7 branchial plumes large, spreading and 3- to 4-pinnate. Radula with 33 rows; center naked; with 42 to 49 strongly hooked, denticulate pleural teeth. Common in tide pools to 26 fathoms. *A. nyctea* Bergh, 1900, is probably a synonym.

Other species:

4212 *Archidoris odhneri* (MacFarland, 1966). Cypress Point to Point Conception, California. 90 mm., all-white, with large and small tubercles. Rare; 80 ft. Erroneously placed in *Austrodoris* Odhner by MacFarland. See *The Veliger*, vol. 11, p. 90.

4213 *Archidoris tuberculata* (Cuvier, 1804). Bare Island, British Columbia to Baja California, 10 fms.

Genus *Atagema* Gray, 1850

Characterized by strongly hooked teeth, with a central tooth. Penis and vagina unarmed. Prebranchial hump and dorsal ridge present. 18 to 30 rows of teeth. Spicules in notum. Type: *carinata* (Quoy and Gaimard, 1832) from New Zealand. One American species.

Atagema quadrimaculata Collier, 1963   **4214**
Four-spotted Dorid

Monterey Bay to San Diego, California.

4214

30 to 64 mm. in length, oblong-elongate. Notum much larger than the foot, coarsely textured with numerous small papillae and fusiform minute spicules; tan to white. Sides of foot with brownish spots. 2 pairs of laterally placed dark-brown to black spots in depressions in the notum. Retractile rhinophores 2 mm. high, with 16 leaves. Anterior end of foot indented. Minute oral tentacle on each lip. Labial cuticle weak, without rods. *Petelodoris spongicola* MacFarland, 1966, is a synonym. Uncommon; lives on sponges under rocks.

Subfamily DISCODORIDINAE Bergh, 1891

Mouth armed with a simple cuticle which may have microscopic, chitinous slivers. Branchial opening sometimes lobed. Foot much smaller than the granular mantle. Teeth of the same type. The genus *Nuvuca* Marcus and Marcus, 1967, is reported from the lower Caribbean.

Genus *Montereina* MacFarland, 1905

Body firm, mantle (or dorsum) tuberculate. Oral palps (tentacles) slender and conical. Vagina and penis unarmed. Prostate gland large. Type: *nobilis* (MacFarland, 1905). Synonym: *Anisodoris* of authors, not Bergh, 1898. Do not confuse the name with *Peltodoris* Bergh, 1880, a Mediterranean genus.

Montereina nobilis (MacFarland, 1905)   **4215**
Noble Pacific Doris   **Color Plate 17**

Vancouver Island, British Columbia, to west Baja California.

4 inches in length. Rhinophore stalk stout, conical, the clavus perfoliate, with about 24 leaves, and the stalk deeply retractile within low sheaths, the margins of which are tuberculate. Each of the 6 branchial plumes large and spreading, 3- to 4-pinnate. A thin, membrane-like expansion joins the bases of the plumes. Radula with 26 rows; center naked; with 55 to 62 strongly hooked pleural teeth. The orange egg-ribbon contains about 2 million eggs. Moderately common in tide pools to 5 fathoms.

Genus *Anisodoris* Bergh, 1898

Similar to *Discodoris*, but the labial cuticle is smooth, lacking the labial rodlets. Notal papillae granular. Type: *punctuolata* (Orbigny, 1837).

4216 *Anisodoris worki* Marcus and Marcus, 1967. Biscayne Bay, Florida. Under rocks of causeway. 26 mm. Yellowish. *Studies Trop. Oceanogr. Miami*, no. 6, p. 66.

4217 *Anisodoris prea* Marcus and Marcus, 1967. Biscayne Bay, Florida. 28 mm. Creamy-white. *Loc. cit.*, p. 70.

Genus *Peltodoris* Bergh, 1880

Rigid notum is velvety because of densely standing, uniform caryophyllidia. Labial cuticle without rodlets. Type: *atromaculata* Bergh, 1880.

4218 *Peltodoris greeleyi* MacFarland, 1909. Southeast Florida to São Paulo, Brazil. See Marcus and Marcus, 1967, p. 72, fig. 94.

4219 *Peltodoris hummelinckei* Marcus, 1963. Aruba and Curaçao, lower Caribbean.

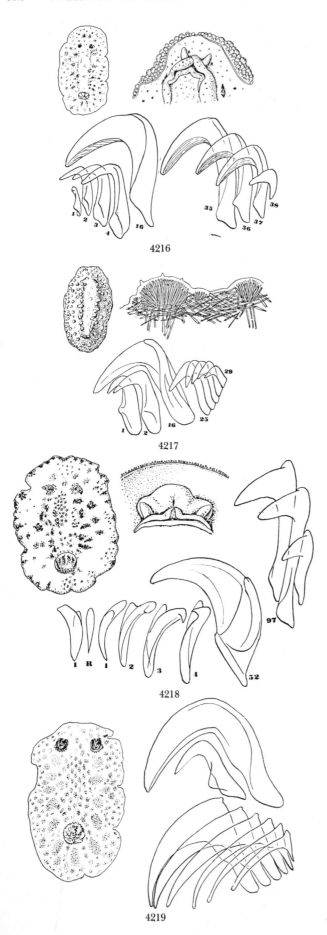

4216

4217

4218

4219

Genus *Discodoris* Bergh, 1877

Body rather soft, oval in outline; branchial aperture slightly crenulate, stellate or bilabiate; anterior margin of the foot bilabiate, the upper lip more or less notched. Mouth with chitinous, microscopic slivers. Oral tentacles long. Prostate gland large. Genital organs unarmed. Living slugs can drop off pieces of the notum brim at will. Type: *pardalis* (Alder and Hancock, 1864).

Discodoris heathi MacFarland, 1905
Heath's Doris

4220
Color Plate 17

Vancouver Island, British Columbia, to west Baja California.

1 inch in length. Mantle thick, densely spiculate. Rhinophores cylindro-conical, the stalk stout, the clavus with 10 to 15 leaves, and wholly retractile. Each of the 8 to 10 branchial plumes are tripinnate. Radula colorless, with 20 rows of teeth; center naked. 36 to 42 strongly hooked pleural teeth. Rather common in rock pools in the summer; rare otherwise to 5 fathoms. *Discodoris fulva* O'Donoghue, 1924, is a synonym.

Discodoris hedgpethi Marcus and Marcus, 1959
Hedgpeth's Doris

4221

Northern Texas.

2 to 4 inches; pale olive-green with black spots and blotches up to 3 mm. in diameter; sole of foot less spotted. Surface with numerous small papillae stiffened by spicules. Foot broad, bilabiate and notched in front, and undulate on the sides. Conical oral tentacles not grooved. Rhinophores set well-apart. The rims of the rhinophorial grooves are beset with papillae. With 25 leaflets on either side. Gill plume 6 and are tripinnate or quadripinnate. Lips of mouth bear 2 triangular areas of stratified rodlets. Radula of 26 rows; no median tooth, and 40 hooked laterals. Withdrawable penis armed with spines. Uncommon. (*Publ. Inst. Marine Sci. Univ. Texas*, vol. 6, p. 254). *D. spetteda* Marcus and Marcus, 1966, is a synonym.

Other species:

4222 *Discodoris phoca* Marcus and Marcus, 1967. Key Biscayne, Florida, among mangroves, feeding on sponges. *Studies Trop. Oceanogr. Miami*, no. 6, p. 78. 55 mm. Purplish brown with a net of white dots.

4223 *Discodoris alba* White, 1952. Dry Tortugas, Florida. *Proc. Malacol. Soc. London*, vol. 29, p. 113.

4224 *Discodoris voniheringi* MacFarland, 1909. Alagoas, Brazil.

4225 *Discodoris purcina* Marcus and Marcus, 1967. Key Biscayne, Florida. 25 mm. *Studies Trop. Oceanogr. Miami*, no. 6, p. 81.

4226 *Discodoris pusae* Marcus, 1955. Biscayne Bay, Florida, to Bahia Blanca, Argentina, 34 fms. See Marcus and Marcus, 1967, p. 82; 1969, p. 21.

4227 *Discodoris aurila* Marcus and Marcus, 1967. Deale Beach, Pacific side of Canal Zone. 20 mm. Light-brown, sprinkled with white, and a dark band on the midline. *Studies Trop. Oceanogr. Miami*, no. 6, p. 85.

4228 *Discodoris notha* Bergh, 1877 and 1904. Antilles.

4229 *Discodoris mavis* Marcus and Marcus, 1967. Puerto Peñasco, Sonora, Mexico. 24 mm. Orange-pink with brownish spots. *Loc. cit.*, p. 187.

Genus *Taringa* Marcus, 1955

Similar to *Discodoris*, but with an armed penial papilla. Type: *telopia* Marcus, 1955.

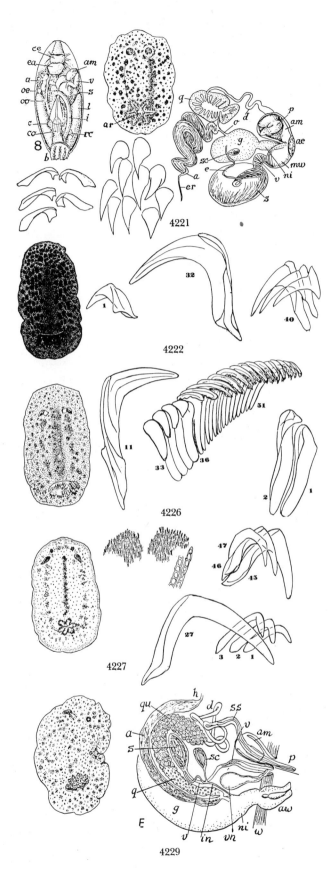

4221

4222

4226

4227

4229

4230 *Taringa telopia telopia* Marcus, 1955. Brazil.

4231 *Taringa telopia disa* Marcus and Marcus, 1967. Biscayne Bay, Florida. *Studies Trop. Oceanogr. Miami,* no. 6, p. 87. 18 mm. Brown with black dots.

4232 *Taringa aivica* Marcus and Marcus, 1967. Deale Beach, Pacific side of the Canal Zone. *Loc. cit.,* p. 89. Subspecies *timia* Marcus and Marcus, 1967, *loc. cit.* Sonora, Mexico.

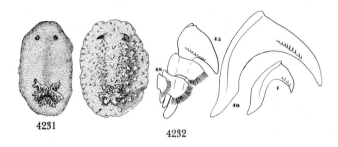

4231 4232

Genus *Tayuva* Marcus and Marcus, 1967

A discodorid with pointed tentacles, labial plates with rodlets, hook-shaped radular teeth, stout penial papilla, large vestibule (atrium) stiffened by spicules and containing the penial papilla and the vaginal aperture. Nidamental opening independent from that of the atrium. Type: *ketos* Marcus and Marcus, 1967.

4233 *Tayuva ketos* Marcus and Marcus, 1967. Sonora coast of west Mexico. Under rocks in large tide pools. 47 mm. Whitish with brown blotches and brown papillae. *Studies Trop. Oceanogr. Miami,* no. 6, p. 192.

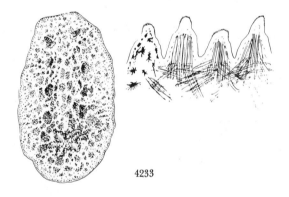

4233

Genus *Diaulula* Bergh, 1880

Body fairly soft; back silky finish; branchial aperture round and crenulate; branchial plumes tripinnate. Head covered by the mantle notum. Radula with 50 to 60 hooked pleural teeth. Penis unarmed. Type: *sandiegensis* (Cooper, 1862).

4234
Color Plate 17

Diaulula sandiegensis (Cooper, 1862)
San Diego Doris

Alaska to the Gulf of Japan. California to the Gulf of California.

2 to 3 inches in length. Body soft; back velvety. Rings of black varying greatly in number and clarity (2 or 3 to 30). Darker specimens are commonest in the northern half of its range. Rhinophores conical, the clavus with 20 to 30 leaves, deeply retractile into a conspicuous sheath with a crenulate margin. 6 branchial plumes tripinnate. Radula broad, with 19 to 22 rows, each row with 26 to 30 falcate teeth on each side of the naked center. Moderately common in rock pools of the *Fucus* seaweed zone at all seasons and down to 10 fathoms.

The broad, white, spiral egg bands are commonly laid from June to August.

Subfamily ASTERONOTINAE Thiele, 1935

Genus *Aphelodoris* von Ihering, 1886

Dorids with denticles on the outer side of the lateral teeth. Penis unarmed. Prostate voluminous. Type: *millegrana* (Alder and Hancock, 1854).

4235 *Aphelodoris antillensis* (Bergh, 1879). Southeast Florida, and the Caribbean. See Marcus and Marcus, 1967, p. 92.

Subfamily THORUNNINAE Odhner, 1926

Genus *Aldisa* Bergh, 1878

Tuberculated mantle; oral palps tentaclelike. Radula characterized by long spathate teeth with the ends finely serrated or denticulate. Oral lips smooth, not armed. Type: *zetlandica* (Alder and Hancock, 1854).

Aldisa sanguinea (Cooper, 1862)   **4236**
Blood-red Doris

Bodega Bay to Baja California. Japan.

½ inch in length. In form, superficially resembling our figure of *Archidoris montereyensis*, but bright-scarlet to light-red, sprinkled everywhere with very minute, black spots. Characterized by 2 or 3 very large, oval spots of black on the back. Usually with 3 squarish blotches of yellow. Rhinophores similar in form, with 12 to 15 leaves in the clavus. Branchial plumes 8 to 10, simply or irregularly bipinnate. Radula with 70 rows of teeth, each row with 70 to 100 teeth which are long, slender and with small, swollen bases. Not uncommon in rock pools and down to 5 fathoms.

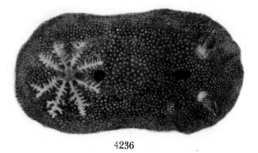

4236

4237 *Aldisa zetlandica* (Alder and Hancock, 1854). Point Barrow, Alaska to Iceland and northern Europe. Azores and Cape Verde Islands.

Subfamily ROSTANGIDAE Pruvot-Fol, 1954

Labial armature in 2 plates. Radula of 3 kinds of nondenticulate, hooked teeth. Mantle tubercles spiculose. Prostate present.

Genus *Rostanga* Bergh, 1879

Back covered with minute, spiculose or stiff papillae; branchiae of simple-pinnate leaves. Type: *rubra* (Risso, 1818).

Rostanga pulchra MacFarland, 1905   **4238**
MacFarland's Pretty Doris   Color Plate 17

Vancouver Island, British Columbia, to Argentina. Japan.

¾ inch in length. Rhinophores short, stout, translucent-pink, stalk stout, prolonged beyond the 20- to 24-leaved clavus as a blunt, cylindrical process which is ¼ the length of the entire rhinophore. 10 to 12 erect, separate, retractile branchial plumes. Radula with 65 to 80 rows of 80 denticulate pleural teeth. Lives on red sponges. The egg-ribbon is orange-red and often laid on the sponge. Common. Not uncommon in the Gulf of California (Lance, 1966, *The Veliger*, vol. 9, p. 72).

Subfamily GLOSSODORIDINAE Odhner, 1939

Back smooth, body elongate. Labial armature strong, of very minute hooks. Center of radula very narrow, often with minute, compressed, spurious teeth which are denticulate. The subfamily Chromodoridinae is synonymous.

Genus *Glossodoris* Ehrenberg, 1831

Radula large, without median teeth, the laterals being denticulate. Genital organs unarmed. Type: *gracilis* (Rapp, 1827). *Chromodoris* Alder and Hancock, 1855, is a synonym, as are *Hemidoris* Stimpson, 1855; *Goniobranchus* Pease and *Actinodoris* Ehrenberg, 1831. For a catalog of 233 *Glossodoris* of the world, see Pruvot-Fol, 1951 (*Jour. de Conchyl.*, vol. 91, pp. 76–164).

Glossodoris macfarlandi (Cockerell, 1902)   **4239**
MacFarland's Blue Doris

Monterey, California, to Baja California.

¼ inch in length, like a small *porterae*, but with a ground color of reddish purple (not dissolving out at death); mantle with a yellow-orange margin and 3 longitudinal yellow stripes. End of foot with an orange stripe. Rare; subtidal to 5 fathoms.

4239

Glossodoris aila (Marcus, 1961)   **4240**
Aila Blue Doris

North Carolina.

12 mm. long, 7 mm. wide, 6 mm. high; red with blue and yellowish marks. Notum smooth, broad behind, without glands. Spicules absent. Anterior border of foot bilabiate. Tentacles broad, grooved. Rhinophores with 15 to 18 leaflets. Branchial pouch broad, with a smooth rim. 13 unipinnate gills. Labial ring with bifid hooklets. Radula with 57 rows, of 1 median and 43 laterals, the latter hooked and denticulate. Rare.

Glossodoris neona (Marcus, 1955)   **4241**
Neona Blue Doris

Southeast Florida; Caribbean to Brazil.

10 mm., anterior part bluish white with red and opaque-cream meshes, and a red marginal line. Rhinophores violet, with 16 leaflets. 9 to 12 unipinnate gills near the posterior end. Notum broad, flat and smooth. Head triangular with a

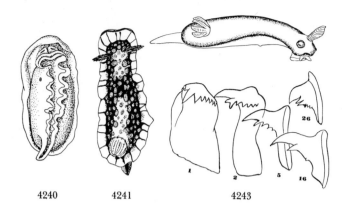

4240   4241   4243

longish mouth; grooved tentacles; pointed tail projecting beyond the notum. Some spicules in skin. Labial armature of rodlets with bifid tips. Radula of 33 rows of 66 denticulate teeth. Uncommon; shallow water. *Glossodoris clenchi* (Russell, 1935) is probably an earlier name.

Other species:

4242 *Glossodoris dalli* (Bergh, 1879). Puget Sound, Washington.

4243 *Glossodoris nyalya* (Marcus and Marcus, 1967). South Florida. Brilliant-blue with bright-red marginal band on the notum. 10 mm. *Studies Trop. Oceanogr. Miami*, no. 6, p. 53.

4244 *Glossodoris tura* (Marcus and Marcus, 1967). Pacific side of the Canal Zone. *Studies Trop. Oceanogr. Miami*, no. 6, p. 55.

4245 *Glossodoris roseopicta* (Verrill, 1900). *Trans. Conn. Acad.*, vol. 10, p. 549; vol. 11, p. 33.

4246 *Glossodoris norrisi* (Farmer, 1963). Pacific coast of Baja California to the Gulf of California. *Trans. San Diego Soc. Nat. Hist.*, vol. 13, p. 81.

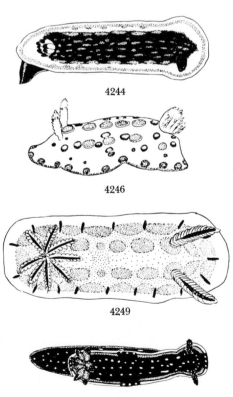

4244

4246

4249

4250

4247 *Glossodoris banksi banksi* (Farmer, 1963). Gulf of California. (Synonym: *sonora* (Marcus and Marcus, 1967, p. 173).)

4248 *Glossodoris sedna* (Marcus and Marcus, 1967). Gulf of California; intertidal to 16 meters. (Synonym: *fayeae* Lance, 1968.)

4249 *Glossodoris binza* Marcus, 1963. Curaçao, Lower Caribbean.

Subgenus *Hypselodoris* Stimpson, 1855

Type of the subgenus is *obscura* (Stimpson, 1855).

***Glossodoris californiensis* (Bergh, 1879)**   **4250**
Californian Blue Doris

Monterey to Gulf of California; Panama.

2 inches in length. Like *G. porterae*, but with numerous, bright, orange, oblong spots in 2 rows on the mantle, another row down each side of the foot, and a group of round spots on the anterior end. Common in tide pools and to 10 fathoms. *G. universitatis* Cockerell, 1901, is a synonym. Common intertidally in the Gulf of California.

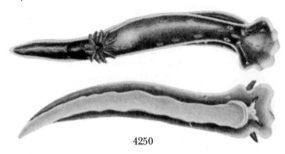

4250

***Glossodoris agassizi* (Bergh, 1894)**   **4251**
Agassiz's Blue Doris

Gulf of California to Panama.

Similar to *G. californiensis* in being dark-blue in color, but *agassizi* has many more and much smaller yellow dots or streaks. Along the mantle border there occur 2 stripes. The inner is light-green and the outer one is the same yellow shade as the dots. Between the 2 stripes is an area of navy-blue or black. There is a golden stripe down the posterior side of the blue-black rhinophores. Gills white, tipped with black. See G. G. Sphon, 1971, *The Veliger*, vol. 14, p. 214.

***Glossodoris porterae* (Cockerell, 1902)**   **4252**
Porter's Blue Doris   **Color Plate 17**

Monterey to Baja California.

½ inch in length, characterized by its deep ultramarine blue (dissolves out at death) and by the 2 orange stripes. Foot without orange marks. Fairly common in rocky tide pools and to 10 fathoms.

Other *Glossodoris* (*Hypselodoris*):

4253 *Glossodoris edenticulata* (White, 1952). Southeast Florida and west Florida, 8 to 180 ft. 50 mm. *Proc. Malacol. Soc. London*, vol. 29, p. 113. Blue-black with yellow lines. See *The Nautilus*, vol. 88, no. 2 (1974).

4254 *Glossodoris acriba* Marcus and Marcus, 1967. Lower Caribbean. *Studies Trop. Oceanogr. Miami*, no. 6, p. 60.

4255 *Glossodoris aegialia* (Bergh, 1905). Gulf of California.

4253

4256 *Glossodoris sycilla* (Bergh, 1890). Caribbean. Yucatan, Mexico.

4257 *Glossodoris clenchi* (Russell, 1935). Harrington Sound, Bermuda. *The Nautilus*, vol. 49, p. 59. May be an earlier name for *neona* Marcus and Marcus, 1967.

Subgenus *Felimare* Marcus and Marcus, 1967

Glossodorids with a rhachidian radular plate nearly as high as the neighboring teeth and provided with a smooth cusp. Lateral teeth bicuspidate. Labial armature of simple hooklets. Penis unarmed. Spermatheca with a single duct and with the spermatocyst located distally to it. Type: *bayeri* Marcus and Marcus, 1967 (is *zebra* Heilprin?). One known species:

4258 *Glossodoris (Felimare) zebra* (Heilprin, 1889). Bermuda. The Bermuda Islands, p. 176 and 187. For biology see Crozier and Arey, 1914 to 1922. For anatomy, see Smallwood and Clark, 1912, *Jour. Morphol.*, vol. 23, p. 625.

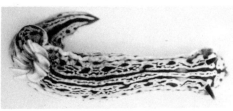

4258

4259 *Glossodoris (Felimare) bayeri* Marcus and Marcus, 1967. Biscayne Bay, Florida. 21 mm. Blue with orange-yellow lines edged in black. *Studies Trop. Oceanogr. Miami,* no. 6, p. 62. *Glossodoris zebra* Heilprin, 1889 is probably an earlier name for this species.

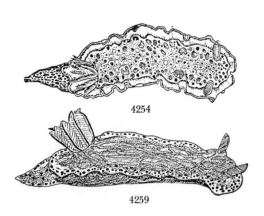

4254

4259

Subfamily INUDINAE Marcus and Marcus, 1967

The spelling Inudiinae is incorrect. Characters of the genus *Inuda*.

Genus *Inuda* Marcus and Marcus, 1967

Dorids with cleft hooks on the labial cuticle, multidentate radula with bicuspid rhachidian tooth and denticulate lateral teeth. Prostate well-developed; male copulatory organ acrembolic, with spines; vagina unarmed. Type: *luarna* Marcus and Marcus, 1967.

4260 *Inuda luarna* Marcus and Marcus, 1967. Puerto Peñasco, Sonora, Mexico. 27 mm. Whitish mottled with brown. *Studies Trop. Oceanogr. Miami,* no. 6, p. 182.

Subfamily CADLININAE Bergh, 1891

Genus *Cadlina* Bergh, 1879

Similar to *Glossodoris*, but body is broader and flat, the branchial gills tri- or bipinnate and the radula has a median tooth. Animal usually small, generally white, and with spicules in the skin. Vas deferens and penis studded with microscopic, chitinous hooks. Type: *laevis* (Linné, 1767). Synonyms: *Acanthochila* Mörch, 1868; *Cadlina* Bergh, 1879; *Juanella* Odhner, 1921; *Echinochila* Mörch, 1868, was rejected by the I.C.Z.N. opinion 812 (1967).

For key to the Pacific coast species see Marcus and Marcus, 1967, *Studies Trop. Oceanogr. Miami,* no. 6, p. 169.

***Cadlina laevis* (Linné, 1767)**   4261
White Atlantic Doris

Arctic Seas to Massachusetts. Europe.

1 inch in length, similar to *nobilis*, but a pure, waxy, semitransparent white. Back with numerous very small, obtuse, opaque-white tubercles. An irregular row of white or sulfur-yellow, angular spots located down each side near the margin of the back. Rhinophores opaque-white or yellowish, with 12 or 13 leaflets, surmounted by a short, blunt point. Branchial plumes of 5 imperfectly tripinnate, transparent-white plumes. Radula with 50 to 70 rows of teeth. 29 to 30 pleural teeth on each side of the central tooth, the latter with 3 to 4 denticles on each side of the center hook. Locally uncommon. *C. repanda* (Alder and Hancock, 1842), *C. obvelata* (Müller, 1776), and *C. planulata* (Stimpson, 1853), are this species.

***Cadlina flavomaculata* (MacFarland, 1905)**   4262
Yellow-spotted Doris

Vancouver Island, British Columbia, to Baja California.

¾ inch in length. Characterized by the 2 rows of 6 to 10 lemon-yellow spots borne upon low tubercles. Rhinophores

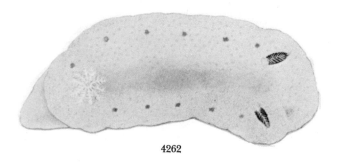

4262

with 10 to 12 leaves in its club. Branchial plumes small, 10 to 11, either simple pinnate or bipinnate. Foot is white. Radula with about 77 rows of teeth, with 23 pleural teeth on each side of the central tooth which has 4 to 6 equal-sized denticles. All times of the year in small numbers in rocky tide pools and down to 10 fathoms.

Cadlina marginata (MacFarland, 1905) 4263
Yellow-rimmed Doris

British Columbia to Baja California.

1½ inches in length, similar to *nobilis*, but covered everywhere with low, yellow-tipped tubercles surrounded by a narrow ring of white and forming the center of a clearly marked polygonal area. Ground color a translucent yellowish white. There is a distinct narrow band of lemon-yellow around the margins of the mantle and the lateral and posterior edges of the foot. Rhinophores with 16 to 18 leaves in the clavus. Branchial plumes 6, bipinnate, sheath with yellow-tipped tubercles on the margin. 90 rows of teeth, with about 47 pleural teeth on each side of the central tooth which has 4 to 6 even-sized denticles. Common in rock pools and subtidally. The name *luteomarginata* MacFarland, 1966, is unnecessary for this species.

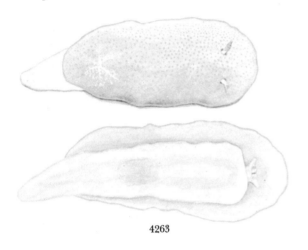

4263

Other species:

4264 *Cadlina sparsa* Odhner, 1921. Santa Barbara to San Diego, California, and Chile. (Marcus, 1961, *The Veliger*, vol. 3, supplement, p. 15.)

4265 *Cadlina modesta* MacFarland, 1966. Moss Beach to La Jolla, California. *Mem. Calif. Acad. Sci.*, vol. 6, p. 147. Intertidal to 5 fms.

4266 *Cadlina pacifica* Bergh, 1879. Unalaska and the Shumagin Islands.

4267 *Cadlina evelinae* Marcus, 1958. Santos, Brazil. *Amer. Mus. Novitates*, no. 1906. Gulf of California.

4268 *Cadlina limbaughi* Lance, 1962. Santa Barbara County, California. 27 mm. *The Veliger*, vol. 4, p. 155, pl. 38.

4269 *Cadlina rumia* Marcus, 1955. Southeast Florida, Caribbean, to Brazil. *Bol. Fac. Filos. Cienc. Univ. S. Paulo, Zoologia*, vol. 20, p. 119.

Genus *Geitodoris* Bergh, 1891

Flat, oval dorids with a tuberculate notum and with a marbled coloration. Branchial gills 5, small and tripinnate. Labial armature of slivers weak and sometimes absent. Marginal radular teeth spathate and serrated along the interlocking edges. A poorly defined group with only one American species. Type: *planata* (Alder and Hancock, 1846).

4270 *Geitodoris complanata* (Verrill, 1880). South of Martha's Vineyard, Massachusetts, 85 to 146 fathoms.

Genus *Thordisa* Bergh, 1877

A dorid with granular or villose mantle. Branchial gills few in number and sparse. No armature in the lips or penis. Teeth simple and hooked with the laterals prominently pectinate. Oral palps elongate. Type: *villosa* (Alder and Hancock, 1864).

Thordisa bimaculata Lance, 1966 4271
Two-spotted Dorid

Santa Barbara to La Jolla, California.

About 28 mm. long, 12 mm. broad, 6.5 mm. high; doridiform. Branchiae completely retractible. Numerous papillae on dorsum have microscopic spicules around the base. Foot ⅔ width of body. Oral tentacles arise behind lateral margins of the oral lobes. Notum orange to yellow, with one brown spot behind the rhinophores, and another in front of the 6 tripinnate branchiae. Dark rhinophores bear 14 to 16 horizontal lamellae. Middle intertidal zone to 15 feet. A similar unnamed species was recorded off Panama on the Atlantic side by Marcus and Marcus, 1967. Figures from *The Veliger*, vol. 9, p. 73.

Other species:

4272 *Thordisa diuda* Marcus, 1955. Brazil.

Subfamily PLATYDORIDINAE Bergh, 1891

Body flat and of a very firm consistency. Mantle granular or almost smooth, without tubercles and much larger than the foot. Penis, and sometimes the vagina, armed with spiny plaques.

Genus *Platydoris* Bergh, 1877

Flat, oval, rounded, firm-skinned dorids with the ridged or coriaceous mantle finely roughed and larger than the foot. 6 to 8 tripinnate branchial gills. Oral palps digitiform. Vagina and vas deferens canal armed with round, chitinous platelets which have a central spine. Large prostate. No labial armature. Teeth hooked and simple; marginals serrated or pectinate. Type: *argo* (Linné, 1758).

Platydoris macfarlandi Hanna, 1951 4273
MacFarland's Flat Doris

California.

30 mm. in length flattened, broadly oval. Deep dark-red above when alive. Mantle velvety, with broad, sharply edged margins. Rhinophores retractile. Gills retractile. Labial palps well-developed. Uncommon; 1 to 110 fathoms. *The Nautilus*, vol. 65, p. 1.

Other species:

4274 *Platydoris angustipes* (Mörch, 1863). Southeast Florida; Jamaica; Virgin Islands to Brazil. 40 mm. Orange-red with darker mottlings. See Marcus and Marcus, 1967.

Subfamily CONUALEVIINAE Collier and Farmer, 1964

Genus *Conualevia* Collier and Farmer, 1964

The smooth, retractile rhinophores and the absence of a ridge surrounding the rhinophores and branchial pits are characteristic. Body soft. Type: *marcusi* Collier and Farmer, 1964. *Trans. San Diego Soc. Nat. Hist.,* vol. 13, no. 19, p. 281.

***Conualevia alba* Collier and Farmer, 1964   4275**
White Smooth-horned Dorid

Newport to San Diego, California.

20 to 24 mm., dorid in shape, soft notum covers the entire foot. Animal white. Retractile rhinophores smooth, long, slender and with microscopic black dots arranged in rings. No ridge around the rhinophoral or branchial chambers. 8 white, retractile, tripinnate gills in a circle around the anus. 34 rows of radulae with 56 simple hooklike teeth on each side. No central. Common; shallow water.

Other species:

4276 *Conualevia marcusi* Collier and Farmer, 1964. Gulf of California. Light-orange, papillose, 16 unipinnate gills. 15 mm.

4277 *Conualevia mizuna* Marcus and Marcus, 1967. Puerto Peñasco, Sonora, Mexico. Synonym of *alba?* 16 mm., white with yellow rhinophores and gills.

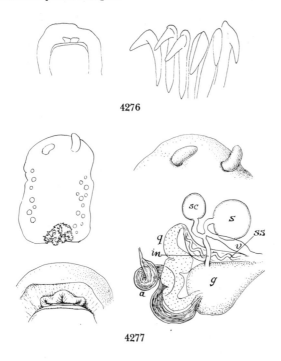

4276

4277

Superfamily Pseudodoridacea Eliot, 1922
False Dorids or Phanerobranchs

Dorid-like. Rhinophores and branchial gills not retractile, but will contract somewhat when disturbed or preserved. No oval palps or tentacles. Replaced by a large buccal veil, sometimes bilobed. No jaws. Feed on tunicates and bryozoans, instead of sponges.

The so-called **Nonsuctoria** dorids do not have a crop in the pharynx. Radula teeth numerous; in some groups they are all the same type, in others they are not. This group includes the families Aegiretidae, Polyceridae and Triophidae.

The so-called **Suctoria** dorids have a side crop in the pharynx. Radula teeth differentiated into 2 or 3 types. This group includes the Lamellidorididae (Onchidorididae of some authors), Acanthodorididae, Corambidae, Goniodorididae.

Superfamily Polyceratacea Alder and Hancock, 1845
The Nonsuctorial False Dorids

Family AEGIRETIDAE Fischer, 1883

Elongate, slug-shaped dorids. Branchial gills bi- or tripinnate. With numerous papillae on the mantle. All radula teeth simple and hook-shaped. Notodorididae Thiele, 1931, is a synonym.

Genus *Aegires* Lovén, 1844

Rhinophoral apertures with papillated sheaths. No mantle brim; back without crista and with numerous papillae. *Aegirus* is a misspelling. Type: *punctilucens* Orbigny, 1837.

***Aegires albopunctatus* MacFarland, 1905   4278**
White-spotted Doris

Vancouver Island, British Columbia, to the Gulf of California.

13 to 21 mm., long, pale yellowish white, and often with scattered small brown spots. Thickly set everywhere with short blunt tubercles. Anterior end round and beset with large blunt tubercles. Club of rhinophore lemon-yellow and smooth. Rhinophores retractile within a tuberculate sheath which has 2 quite large tubercles on the outer side. Branchial plumes 3 tripinnate, small. Mouth lined with a thick cuticle, and has a thick mandibular plate. Radula with 16 to 22 rows, 17 pleural hamate teeth on either side. Abundant in summer in southern California. Under stones, intertidally, and down to 10 fathoms.

Family GYMNODORIDIDAE Odhner, 1941

Genus *Nembrotha* Bergh, 1877

Body limaciform, almost smooth. Rhinophores retractible and foliated. Gills with several leaves. Tentacles short and lappet-formed. Distinct jaws absent. Radula small. *Angasiella* Crosse, 1864, may be an earlier name. Type: *nigerrima* Bergh, 1877. Worldwide in warm seas. The following American species are known:

4279 *Nembrotha gratiosa* Bergh, 1890. Off Key West, Florida, 26 fms.

4280 *Nembrotha divae* Marcus, 1958. Cabo Frio, Brazil. *Bol. Inst. Oceanogr. S. Paulo,* vol. 7, p. 47.

4281 *Nembrotha eliora* Marcus and Marcus, 1967. Puerto Lobos, Sonora, Mexico. 80 mm.

4281

Family POLYCERATIDAE Alder and Hancock, 1845

Body limaciform (sluglike); branchial plumes not retractile. With ceratalike appendages on the front edge of the frontal veil. 2 types of radula teeth.

Genus *Laila* MacFarland, 1905

With club-shaped papillae; rhinophores perfoliate, retractile. Branchial plumes tripinnate. Tentacles blunt, canaliculate. No mandibles. Radula uniserial, of hooklike teeth. Penis armed with hooks. Type: *cockerelli* MacFarland, 1905.

Laila cockerelli MacFarland, 1905   **4282**
Laila Doris   **Color Plate 17**

Vancouver Island, British Columbia, to Baja California.

¾ inch in length. Rhinophores with 13 leaves in the clavus. 5 branchial plumes tripinnate, nonretractile into the cavity. 76 to 82 rows of radula; center with a series of rectangular, flattened plates; on the side are 2 pleural teeth, then 10 to 13 closely set pavementlike uncinal teeth. Glans penis long, armed with 10 to 12 irregular rows of minute, thornlike hooks. Found under shelving rocks in tide pools and down to 5 fathoms. Commoner in winter and spring.

Genus *Polycera* Cuvier, 1817

Frontal margin with fingerlike processes. Rhinophores nonretractile and without sheaths. Fingerlike processes bordering branchial plumes. Center of radula naked, flanked by 2 lateral teeth and several uncini. *Themisto* Oken, 1815, non 1807, is a synonym. Type: *quadrilineata* Müller, 1776. For a key to American *Polycera,* see Marcus and Marcus, 1967, *Studies Trop. Oceanogr. Miami,* no. 6, p. 197.

Polycera atra MacFarland, 1905   **4283**
Orange-spiked Doris   **Color Plate 17**

San Francisco, California, to Sonora, Mexico.

15 to 30 mm. in length. The blue-black lines shown in our figure are usually thinner and less conspicuous. With 4 to 6 velar appendages. 8 or 9 gill plumes, with 1 to 4 extrabranchial tubercles. Common on brown algae, intertidal to 25 fathoms. 9 or 10 rows of radular teeth, dark-amber; 2 pleurals, 3 or 4 uncinal teeth.

Polycera hedgpethi Marcus, 1964   **4284**
Hedgpeth's Western Polycera

Central California to the Gulf of California.

30 to 50 mm. long, gray with small black dots, with yellow-orange on the rhinophores, on the corners of the foot and on the 4 to 6 velar appendages. Color spots on the low tubercles all over the body. Rhinophores with 12 leaves. Oral tentacle

4284

short. Tail pointed. 3 extrabranchial appendages on either side of the 9 tripinnate gills. They are shorter than the gills. Jaws with wings. Radula red, 17 rows of teeth. Abundant on the branching bryozoan, *Bugula* (*The Nautilus,* vol. 77, p. 128). *P. gnupa* Marcus and Marcus, 1967, is a synonym.

Polycera zosterae O'Donoghue, 1924   **4285**
Eelgrass Polycera

Vancouver Island, British Columbia, to San Juan Island, Washington.

More than 1 pair of extrabranchial appendages on either side. Jaws with upper winglike expansion. Three gills. Rare; intertidal.

Polycera hummi Abbott, 1952   **4286**
Humm's Polycera

North Carolina to northwest Florida.

On either side are 3 or 4 extrabranchial processes which are cylindrical with pointed tips and longer than the 7 to 9 gills. Rhinophores with 14 or 15 leaflets. Foremost gill especially long. Moderately common. *Florida State Univ. Studies,* vol. 7.

4286

Polycera aurisula Marcus, 1957   **4287**
Marcus' Polycera

North Carolina to Florida and to Brazil.

10 mm. in length; gray with dull-blue and orange markings. 7 to 9 gills. Extrabranchial processes carrot-shaped and longer than the gills. Velum bears 2 or 3 processes on either side.

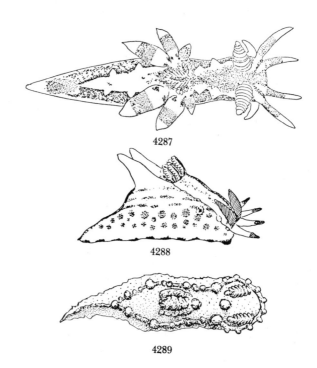

4287

4288

4289

Rhinophores have up to 15 leaflets. The edges of the outer marginal radular teeth are smooth, while in *hummi* they have 5 denticles. Uncommon; shallow water, in algae.

4288 *Polycera chilluna* Marcus, 1961. Beaufort, North Carolina, 18 mm. *Jour. Elisha Mitchell Sci. Soc.*, vol. 77, p. 143. 18 mm.

4289 *Polycera herthae* Marcus and Marcus, 1963. Curaçao and Antigua, West Indies. *Studies Fauna Curaçao and other Caribbean Ids.*, vol. 19, p. 34.

4290 *Polycera alabe* Collier and Farmer, 1964. Gulf of California. *Trans. San Diego Soc. Nat. Hist.*, vol. 13.

Subgenus *Palio* Gray, 1857

Appendices on the top of the back are small and numerous. Jaws triangular but without winglike extensions. Gill plumes tripinnate. Type: *dubia* Sars, 1829. (See Bergh, 1941, *Meddel. Goteborgr K. Vet. Handl.*, ser. B, vol. 1, pp. 1–20.)

4291 *Polycera (Palio) dubia* (Sars, 1829). Greenland, Labrador to Connecticut. 3 to 20 fms. Northern Europe. (Synonyms: *lessoni* Orbigny, 1837; *modesta* Lovén, 1844; *illuminata* Gould, 1841.)

4291

Genus *Polycerella* Verrill, 1881

Form similar to *Polycera*, but the dorsum is warty and the rhinophores are not retractible and have a simple club. Type: *emertoni* Verrill, 1881. Four Western Atlantic species reported:

4292 *Polycerella emertoni* Verrill, 1881. Woods Hole, Massachusetts, to Barnegat Bay, New Jersey. On hydroids. (For anatomy, see Bergh, 1883.) See *The Veliger*, vol. 14, p. 265, 1972.

4293 *Polycerella davenportii* Balch, 1899. Cold Spring Harbor, New York, on hydroids, 3 fms. May be *emertoni*.

4294 *Polycerella conyna* Marcus, 1957. Jarvis Sound, Cape May, New Jersey; Virginia Key, Florida, to São Paulo, Brazil. 3 mm. (*Bull. Marine Sci. Gulf and Caribbean*, vol. 10, p. 159). Also see D. Franz, 1968, *The Nautilus*, vol. 82, p. 8. Feeds on bryozoans.

— *Polycerella zoobotryon* Smallwood, 1910, see in Goniodorididae species 4341.

Subfamily TRIOPHINAE Odhner, 1941

Similar to the Polycerinae but the frontal and lateral appendices are ramified. Radula with smooth central or rachidian teeth, as well as 2 types of laterals. Jaws poorly developed. The subfamily Caloplocaminae Pruvot-Fol, 1954, is the same.

Genus *Triopha* Bergh, 1880

Gills with 5 branches. Type: *modesta* Bergh, 1880 (*carpenteri* Stearns, 1873).

Triopha carpenteri (Stearns, 1873)
Carpenter's Doris

4295
Color Plate 17

British Columbia to San Diego, California. Japan.

1 to 4 inches in length. Rhinophores with 20 to 30 leaves in the club. 5 branchial plumes, large, tripinnate. 30 to 33 rows of radulae, with 4 teeth on the center part (the rhachis); pleural teeth 9 to 18, strongly hooked. Uncinal teeth 9 to 18, quadrangular in outline. Very common in rock pools and to 10 fathoms. Bergh, 1894, placed his *modesta* in the synonymy of *carpenteri*.

Triopha maculata MacFarland, 1905
Maculated Doris

4296
Color Plate 17

Bodega to west Baja California.

1 to 2 inches in length. Rhinophore stalk and club same length, the latter with 18 leaves. Branchial plumes 5, tripinnate 14 rows of teeth. Blunt glans penis armed with minute hooks. Abundant in summer in rock pools, in winter uncommon; also to 10 fathoms.

Triopha grandis MacFarland, 1905
MacFarland's Grand Doris

4297
Color Plate 17

Santa Cruz, California, to Baja California.

3 to 6 inches in length. With 8 to 12 tuberculate processes in front of head, and 6 or 7 more down the sides of the back. Back yellowish brown, often flecked with bluish spots. Tips of yellow processes, tip of tail and tips of branchial plumes with yellowish red. Rhinophores set in conspicuous sheaths, club yellow with 20 leaves. Branchial plumes 5, bushy, tri- and quadripinnate. 18 rows of radular teeth; 4 centrals, 8 pleurals and 8 uncinal teeth. Found on brown kelp off the rocky coastline to 4 fathoms. Fairly common.

Other species:

4298 *Triopha catalinae* (Cooper, 1863). Santa Catalina Island, California. May be earlier name for *carpenteri*, but recent workers consider it a *nomen dubium*.

4299 *Triopha aurantiaca* Cockerell, 1908. Vancouver, British Columbia, to La Jolla, California. (Synonym: *T. elioti* O'Donoghue, 1922.)

4300 *Triopha scrippsiana* Cockerell, 1915. California *Jour. Ent. Zool. Pomona College*, vol. 7, no. 4.

Genus *Issena* Iredale and O'Donoghue, 1923

Body slug-shaped; tail with bordering papillae. Edge of mantle with sparse, simple papillae which may also be on the center of the back. Oral palps auriculate. Rhinophores retractile into a prominent oblique sheath. Branchial plumes bi- or tripinnate. Jaws triangular. Central tooth flat, simple. 2 laterals large and hooked; numerous marginals small, simple, flat. *Issa* Bergh, 1880, non Walker, 1867, and *Colga* Bergh, 1880, non *Kolga* Koren and Danielssen, 1848, are synonyms. Type: *pacifica* (Bergh, 1894).

4301 *Issena pacifica* (Bergh, 1894). Northern Europe; off Nova Scotia, 90 to 92 fms.; Gulf of Maine to off Cape Cod, Massachusetts, 25 to 80 fms. (Synonyms: *lacera* Abildgaard, 1806, non Cuvier, 1804; *abildgaardi* Pruvot-Fol, 1934.)

4302 *Issena ramosa* (Verrill and Emerton, 1881). South of Martha's Vineyard, Massachusetts, 100 to 130 fms. May be *pacifica* Bergh, 1894.

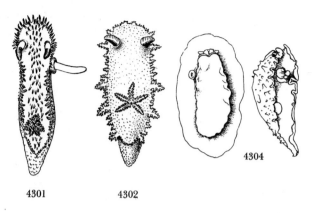

4301   4302   4304

Genus *Crimora* Alder and Hancock, 1862

3 branchial gills bi- or tripinnate; border of mantle with fairly large papillae and filaments, sometimes ramified. Rhinophores with sheaths and short stems. No jaws. Radula of 3 types, the outer marginals being very long, narrow and minutely serrated or smooth. Oral palps blunt. Type: *papillata* Alder and Hancock, 1862.

Crimora coneja Marcus, 1961   4303

Point Loma, California.

7 mm., narrow with a long tail; white in color. 25 papillae on either side and those on the notum behind the gills are orange, with black caps. Some papillae white. 7 small papillae on the frontal veil edge. Tail without knobs. Rhinophores with 9 leaflets. Anterior end of foot bilabiate. Gills 3 and tripinnate. Labial cuticle present. Radula 53 rows with a formula of 9·6·2·(1)·2·6·9. Penis bears cuticular hooklets on almost its entire length.

Family HETERODORIDIDAE Fischer, 1883

Dorid in shape; anus lateral. Median radula tooth simple, hamate, smooth, the laterals similar. No gills or accessory lamellae. Body elevated. Velum well-developed, free. Liver divided into a few large masses.

Genus *Heterodoris* Verrill and Emerton, 1882

Characters of the family. Type: *robusta* Verrill and Emerton, 1882. *Atthila* Bergh, 1899, is a synonym.

4304 *Heterodoris robusta* Verrill and Emerton, 1882. Deep water; North Atlantic.

Superfamily Onchidoridacea Gray, 1853

The Suctorial False Dorids

A pocketlike crop used in sucking is present as an offshoot, in the pharnyx. They feed mainly on tunicates and bryozoans. Gills nonretractile. No jaws. Penis usually spinose.

Family ONCHIDORIDIDAE Gray, 1854

Rhinophores without sheaths, but retractile into holes. Mantle well-developed all around. Velum replacing the oral tentacles. Gill plumes contractile, but not disappearing into

the cavity. Mantle spiculose. Genital organs usually unarmed.

Subsequent family names for this group, such as Lamellidorididae Altena 1937, and potentially Villiersiidae, cannot replace the old, well-known Onchidorididae according to article 40 of the I.C.Z.N.

Genus *Acanthodoris* Gray, 1850

Body doridlike with a furry back. Foot large. Rhinophoral openings minutely lobated. Labial disc armed with minute hooks. Center of radula naked; first pleural tooth large, external pleurals 4 to 8, small. Glans penis armed. Vagina very long. Oral veil produced laterally. Radula without central tooth; lateral very large; marginals small, about 5. Type: *pilosa* (Müller, 1776).

Acanthodoris pilosa (Müller, 1776)   4305
Pilose Doris   Color Plate 16

Arctic Seas to Ocean City, Maryland.
Alaska to Vancouver, British Columbia.
Japan. Europe.

½ to 1¼ inches in length. Semitransparent. Color variable, ranging from pure-white to yellowish white, canary-yellow, yellowish brown, gray-speckled, purple-brown and black. Back covered with soft, slender, conical, pointed papillae, which give it a hairy appearance. Rhinophores long, its club bent backwards and with 19 or 20 leaves. Sheath denticulate. Branchial plumes 7 to 9, large and spreading, tripinnate, transparent. A number of color forms have been described from Alaska by Bergh and from New England by A. E. Verrill. Radula with about 27 rows. No central tooth, 4 pleurals on each side. Moderately common at low tide, sometimes found out of water or down to 30 fathoms. *A. pilosa* (Abildgaard, 1789) is the same species, as is Verrill's 1879 variety, *ornata*.

Acanthodoris brunnea MacFarland, 1905   4306
Pacific Brown Doris

British Columbia to Point Conception, California.

¾ inch in length, somewhat like *A. pilosa*. Somewhat broader at the anterior end. Brown tubercles on back rounder, fewer, not as pointed. Back brown with flecks of black and with small spots of lemon-yellow between the tubercles. Rhinophores deep blue-black, tipped with yellowish white. Club with 20 to 28 obliquely slanting leaves. 7 branchial plumes, wide-spreading, bipinnate. About 10 tubercles are included within the rosette, 4 or 5 of them large and enclosing the anal papilla. Anterior margin of back is yellow. 24 to 28 rows of radular teeth. No centrals. First pleural large, with 14 to 19 denticles on the inner border. 6 to 7 other smaller pleural teeth. Intertidal to 60 fathoms.

4306

Other species:

4307 *Acanthodoris rhodoceras* Cockerell and Eliot, 1905. Dillon Beach to west Baja California. (Marcus, 1961, *The Veliger*, vol. 3, supplement.) Intertidal to 12 fms.

4308 *Acanthodoris atrogriseata* O'Donoghue, 1927. Vancouver Island, British Columbia.

4309 *Acanthodoris armata* O'Donoghue, 1927. Vancouver Island, British Columbia.

4310 *Acanthodoris hudsoni* MacFarland, 1905. Vancouver Island, British Columbia, to Monterey Bay, California. Rare; intertidal.

4311 *Acanthodoris lutea* MacFarland, 1925. Dillon Beach to Point Loma, San Diego, California, and west Baja California. Most common in August. Intertidal to 30 fms.

4312 *Acanthodoris nanaimoensis* O'Donoghue, 1921. Vancouver, British Columbia, to San Luis Obispo County, California. Common in summer; intertidal. (Synonym: *columbina* MacFarland, 1926.)

4313 *Acanthodoris pina* Marcus and Marcus, 1967. Puerto Peñasco, Sonora, Mexico. Intertidal rocks. 18 mm. Black with red and white papillae. Rhinophores with 20 leaves. *Studies Trop. Oceanogr. Miami*, no. 6, p. 201.

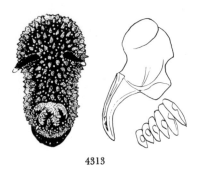

4313

Genus *Lamellidoris* Alder and Hancock, 1855

Small dorids. Lips of mouth armed with a ring of papillae. Rhinophores retractile into holes. Penis unarmed. Radula with 2 or 3 unequal lateral teeth, with or without a median tooth. Gill plumes simple, very small and somewhat contractile. Mantle tuberculate. Buccal veil large. *Villiersia* Orbigny, 1837, is an earlier *nomen oblitum*, according to Pruvot-Fol, 1954. *Onchidorus* (emended by some to *Onchidoris*) Blainville, 1816, is a *nomen dubium* according to Pruvot-Fol. The type of the genus is *Doris bilamellata* Linné, 1767.

Lamellidoris fusca (Müller, 1776)   **4314**
Dusky Doris

Arctic Seas to Massachusetts; Europe.
Japan and Alaska to Bodega, California.

7 to 35 mm., elliptical; pale-rust to flesh, or marbled with the two, with brownish blotches concentrated along the back in 2 or 3 longitudinal bands. Surface covered with rather large, numerous, unequal wartlike to clavate protuberances, the tips of the larger ones cream-colored. Rhinophores with 15 to 20 leaflets on either side; the rim of the pit is smooth. Gills brownish, unipinnate, from 16 to 32 or more. A cluster of branchial glands is situated between every 2 neighboring gills. Spicules may stand out of the surface of the notal tubercles, and they are slightly elbowed, rounded at the ends and some with a spine on the elbow. A narrow girdle of minute rods forms the labial armature. Radula with about 30 rows of 5 teeth.

This is *bilamellata* Linné, 1767 (not his *Limax* of 1761) and *coronata* "Agassiz" Gould, 1870. A very common, variable and widely distributed doridlike nudibranch. Shore or floating logs to several fathoms. Feeds on crustaceans.

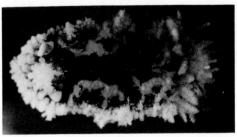

4314

Lamellidoris muricata (Müller, 1776)   **4315**
Muricate Doris

Arctic Seas to Rhode Island. Europe.
Alaska to central California.

6 to 10 mm., white to bluish white with the pinkish digestive gland showing through the skin which is covered with numerous small, claviform papillae up to 0.3 mm. in diameter and packed with tiny spicules, the ends of which may project above the truncate ends of these tubercles. Nonretractile gills small, 10 or 11. It is probable that *hystricina* (Bergh, 1878) and *varians* (Bergh, 1878) are Pacific synonyms of this widely distributed species (from which our Pacific range originates). Rare; down to 3 fathoms.

Other species:

4316 *Lamellidoris diaphana* Alder and Hancock, 1845. Eastport, Maine. Europe.

4317 *Lamellidoris aspera* Alder and Hancock, 1842. Eastport, Maine, to Newport, Rhode Island, 1 to 30 fms. On seaweed, *Chondrus*.

4318 *Lamellidoris diademata* Gould, 1870. Beverly to Boston, Massachusetts.

4319 *Lamellidoris grisea* Gould, 1870. Eastport, Maine, to Boston, Massachusetts.

4320 *Lamellidoris tenella* Gould, 1870. Near Eastport, Maine, to Beverly, Massachusetts. Until the anatomy is known of the above 3 species, their placement is uncertain.

4321 *Lamellidoris hystricina* (Bergh, 1878). Alaska to west Baja California.

4322 *Lamellidoris aureopuncta* and **(4323)** *miniata* both Verrill, 1901. Bermuda. *Trans. Conn. Acad. Sci.*, vol. 11, p. 31 and p. 32.

4324 *Lamellidoris lactea, quadrimaculata* and *olivacea* Verrill, 1900. Bermuda. *Trans. Conn. Acad. Sci.*, vol. 10, pp. 548 and 549.

Genus *Adalaria* Bergh, 1879

Rhinophoral openings smooth. Papillae on back claviform. Lips and penis smooth. Median tooth present. Laterals smooth. Type: *proxima* Alder and Hancock, 1845. Pruvot-Fol, 1954, makes this a subgenus of *Lamellidoris*. Do not confuse the genus name with *Alderia* Allman, which is a stiligerous sacoglossate.

Adalaria proxima (Alder and Hancock, 1854)   **4325**
Yellow False Doris   **Color Plate 16**

Arctic Seas to Eastport, Maine. Europe.

½ inch in length; deep yellow, white or yellow-orange. Back covered with stout, subclavate or elliptical bluntly pointed tubercles, set at a little distance apart, and mixed with smaller ones. Calcareous spicules appear through the skin, radiating from the tubercles. Rhinophores with 15 leaves reaching almost to the base. Margin of sheath smooth. Branchial plumes 11. 40 rows of radular teeth. No central tooth. First pleural large, sickle-shaped, other 11 small and platelike. Uncommon (?) in New England.

Family GONIODORIDIDAE H. and A. Adams, 1854

Rhinophores not retractile. Oral tentacles replaced by a velum. Mantle brim narrow or absent. Penis armed with hooks. Feed on tunicates and ascidians. Okeniidae Odhner, 1926, is the same.

Genus *Okenia* Menke, 1830

Rhinophores long, with numerous oblique leaflets, sometimes cup-shaped. Back with 4 to 7 filaments. Labial ring of hooks all around the mouth. Type: *elegans* Bronn, 1826. *Idalia* Leuckart, 1828, is a synonym (non Huebner, 1819), as is *Okenia* Leuckart, 1826, a *nomen nudum*, but *Okenia* Menke, 1830, although proposed in synonymy, is available according to the 1964 code. *Cargoa* Vogel and Schultz, 1970, is a synonym.

Okenia sapelona (Marcus and Marcus, 1967)   **4326**
Sapelo Okenia

Georgia.

Length while alive is 5 to 7 mm. General color an iridescent pale-blue; rhinophores and gills maroon. Yellow dots present on lateral margins. Tips of lateral appendages (la) and gills (g) are pale-yellow. 5 dorso-median tubercles (tu) are pale-blue. The tentacles (t) are triangular flaps. 7 to 9 gills (g) are unipinnate. Labial cuticle composed of smooth, polygonal elements which are conical in side view. Radula of 12 rows with a formula of 1·1·0·1·1. See *Malacologia*, vol. 6, p. 203.

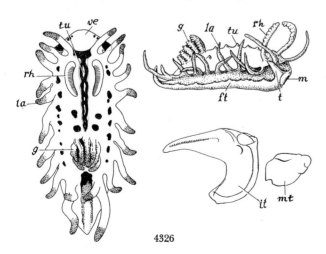

4326

Okenia angelensis (Lance, 1966)   **4327**
Angeles Okenia

Santa Barbara, California, to the Gulf of California.

When alive, 10 mm. long, 2.5 mm. broad. 1 pair of papillae anterior to the nonretractile rhinophores, which have 1 to 3 incomplete lamellae. Body translucent-white, flecked with yellow and white granules, and with clumps of reddish brown dots. 5 to 7 nonretractile gills separated at their bases. Penis unarmed. Nodulose spicules are elongate rods and more numerous on sides of the body.

Other species of *Okenia*:

4328 *Okenia evelinae* Marcus, 1957. Virginia Key, Florida, to São Paulo, Brazil.

4329 *Okenia vancouverensis* (O'Donoghue, 1921). Graham Island, Queen Charlotte Island; Vancouver Island, British Columbia.

4330 *Okenia plana* Baba, 1960. San Francisco Bay, California, mud flats, common; Japan also.

4331 *Okenia impexa* (Marcus, 1957). Beaufort, North Carolina, to São Paulo, Brazil. *Jour. Linn. Soc. London, Zool.*, vol. 43, p. 435.

4332 *Okenia cupella* (Vogel and Schultz, 1970). Mouth of York River, Virginia. 2 mm. Whitish with minute white and brown spots. *The Veliger*, vol. 12, no. 4, p. 388.

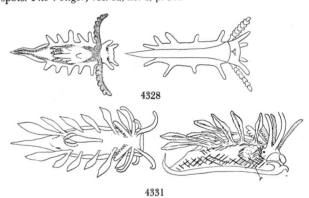

4328

4331

Subgenus *Idalla* Ørsted, 1844

Buccal veil large; borders of mantle with a single row of very long filaments. Gill plumes simple, long and numerous. Without dorsal filaments in the middle of the back. Radula without median; with a very curved, large lateral and with 1 small, denticulate marginal. Labial ring of hooks incomplete. Type: *caudata* Ørsted, 1844. *Idaliella* Bergh, 1881, is a synonym. Two New England species:

4333 *Okenia pulchella* Alder and Hancock, 1854. Arctic Seas to Salem, Massachusetts. Europe (*fusca* Odhner, 1907, is a color form from Sweden).

4334 *Okenia modesta* Verrill, 1875. Massachusetts to Long Island Sound, New York, 19 to 192 fms. (*velifer* Sars, 1878, is a synonym).

Genus *Ancula* Lovén, 1846

Body slug-shaped, without a distinct mantle-notum, but the back crests up on either side of 3 tripinnate gills and bears 5 or 6 simple, cylindrical fingers on either side. Rhinophores with a dozen oblique leaflets, and at the base there are 2 horizontal anterior filaments. Small buccal veil with 2 lateral lobes. Labial ring of chitinous scales. No median radula tooth. Type: *gibbosa* (Risso, 1818). *Miranda* Alder and Hancock, 1846, is a synonym. Do not confuse with *Polycera*.

Ancula gibbosa (Risso, 1818)   4335
Atlantic Ancula   Color Plate 16

Arctic Seas to Massachusetts. Europe.

½ inch in length, of a transparent watery-white, smooth. Rhinophores with 8 to 10 leaves. 3 branchial plumes, tripinnate. Labial armature of rows of imbricated hooks. Radula narrow, center naked, 25 to 27 rows of teeth; inner pleural large, denticulate on the inner margin. Outer pleural tooth small, smooth. Moderately common. Synonyms: *cristata* Alder, 1841; *cristata* Alder and Hancock, 1846; *sulphurea* Stimpson, 1853.

Ancula pacifica MacFarland, 1905   4336
Pacific Ancula

Moss Beach to San Diego, California.

9 to 16 mm. in length (living) very similar to *A. cristata*. Color translucent-yellow with 3 narrow, orange lines on the anterior ½ of the back, and 1 down the center of the back ½. Rhinophores with 9 yellowish leaves. 3 branchial plumes. 4 (not 6) fingerlike processes on each side of the plumes. 35 rows of teeth in the radula. Center with a small quadrangular plate, flanked by 1 large and 1 small pleural tooth. Uncommon; intertidal in the summer. Common in San Francisco Bay in early summer.

4336

Ancula lentiginosa Farmer and Sloan, 1964   4337
Freckled Ancula

Monterey to Santa Barbara, California; to Baja California.

21 mm., with 3 nonretractile, tripinnate branchiae. 1 pair of extra-branchial processes present. Rhinophores with 7 leaves, without sheaths, and with a short anterior process near the base. Color of animal tan to white with reddish brown freckles. Uncommon. *The Veliger*, vol. 6, p. 148.

Other species:

4338 *Ancula evelinae* Marcus, 1961. Beaufort, North Carolina. 3 mm. cream with black blotches. Rhinophores with 7 leaflets. (*Jour. Elisha Mitchell Sci. Soc.*, vol. 77, p. 144.)

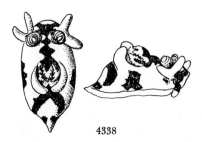

4338

Genus *Hopkinsia* MacFarland, 1905

Rhinophores nonretractile, without sheaths and with about 20 leaflets. Anterior end of foot strongly bilobed. Branchial plumes 7 to 14, simple, nonretractile. Mantle with

numerous, long, soft papillae. Dorsum firm, with closely packed spicules rendering it almost calcareous. Labial ring of rodlets. Penis with hooks. Radula of 16 rows (formula: 1·1·0·1·1). Type: *rosacea* MacFarland, 1905. *Hopkinsea* O'Donoghue, 1927, is a misspelling.

Hopkinsia rosacea MacFarland, 1905   4339
Hopkins' Doris   Color Plate 17

Coos Bay, Oregon, to west Baja California.

15 to 30 mm. in length; deep rose-pink color throughout. Rhinophores long and tapering, the anterior side smooth along the entire length. ¾ of the posterior side bears about 20 pairs of oblique plates. Branchial plumes 7 to 14, entirely narrow and naked. Radula with 1 large pleural tooth on each side, flanked by a tiny, triangular pleural. Spiral egg ribbon rosy. Moderately common at all times of year under shelving rock between tide marks.

Genus *Trapania* Pruvot-Fol, 1931

Similar to *Ancula* but there is an appendix on the back of the base of the rhinophore, and one on each back side of the branchial gills. Front foot arrowhead-shaped. Radula formula 1·0·1. Buccal armature and hooks on penis. Type: *fusca* Lafont, 1874. *Drepania* Lafont, 1874, non Hübner, 1816, and *Drepanida* MacFarland, 1931, are synonyms. One Pacific coast species:

4340 *Trapania velox* (Cockerell, 1901). Santa Barbara to San Diego, California. Rare; intertidal to 3 fathoms.

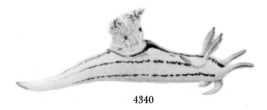

4340

Genus *Bermudella* Odhner, 1941

Type of the genus is *Polycerella zoobotryon* Smallwood, 1910 (*Meddel. Göteborg Mus. Zool. Avdel.*, vol. 91, p. 15). One Western Atlantic species:

4341 *Bermudella zoobotryon* (Smallwood, 1910). Bermuda. *Proc. Zool. Soc. London*, for 1910, p. 143. 5 to 6 mm.

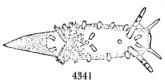

4341

Family CORAMBIDAE Bergh, 1869

Small dorids with the anus and the branchial plumes located posteriorly behind the foot and mantle. Genital orifice on the right. Branchial plumes simple, pinnate. With a sucking crop. Hypobranchiaeidae Fischer, 1887, is the same.

Genus *Corambe* Bergh, 1869

Characterized by a notch in the posterior margin of the notum on the midline. Type: *sargassicola* Bergh, 1872. Two American species:

4342 *Corambe pacifica* MacFarland and O'Donoghue, 1929. Vancouver Island, British Columbia, to Baja California. 10 to 14 mm. in length. Common in the summer on bryozoan colonies on offshore kelp.

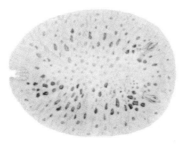

4342

4343 *Corambe evelinae* Marcus, 1956. Ubatuba, northeast Brazil. Low-tide mark. *Bol. Inst. Oceanogr. São Paulo*, vol. 7, p. 53.

Genus *Doridella* Verrill, 1870

Similar to *Corambe,* but without the posterior notch in the notum. Type: *depressa* Balch, 1899. *Corambella* Balch, 1899, is a synonym.

Doridella obscura Verrill, 1870   **4344**
Obscure Corambe

Massachusetts to Florida to Texas.

2 to 4 mm. (rarely 7 mm.) long, ¾ as wide; with numerous dendritic black pigment cells deep within the notum and with a reticulated yellow to brown pattern. Feeds on encrusting bryozoan, *Membranipora*. Lays eggs in a jellylike spiral 2 to 3 mm. across. Moderately common; intertidal to 25 feet on oyster beds in brackish water. *Corambella depressa* Balch, 1899, and *C. baratariae* Harry, 1953, are synonyms (see D. Franz, 1967, *The Nautilus,* vol. 80, no. 3, p. 73).

Doridella burchi Marcus and Marcus, 1967   **4345**
Burch's Corambe

Georgia.

7 to 8 mm. long (living). Interior organs visible through the translucent-whitish round notum which has numerous round brown spots with yellow marks between. Foot cordate, bilabiate in front. Rhinophores with 2 lamellae on either side. Border of rhinophoral pit with a broad collar, smooth outside, slightly radially folded on the inner edge. Pair of gills at the posterior end, each gill with 2 plates, one with 8 leaves, the other with 9. Anus opens between the gills, not on a papilla. Radula with 40 rows. Lives and feeds on the bryozoan, *Alcyonidium*. Common; 2 to 8 fathoms. (*Malacologia,* 1967, vol. 6, p. 206.)

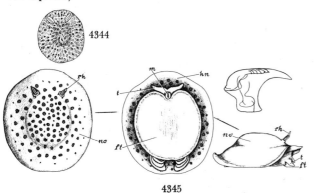

4345

Doridella steinbergae (Lance, 1962)   **4346**
Joan Steinberg's Corambe

Washington to the Coronado Islands.

5 to 8 mm. in length, body translucent with a reticulation of white lines on the back. Irregular splotches of rust-red present amid the reticulations. 3 to 6 gills present on either side at the posterior end of the foot. 2 oral tentacles project just beyond the front of the notum. Retractile rhinophores about 1 to 2 mm., smooth, blunt-ended. Sheaths at base smooth-margined and ⅕ length of rhinophore. 41 to 52 rows of radular teeth. No central. First large lateral with 5 to 7 minute denticles. 4 other laterals trapezoidal, without denticles. Common during the summer on its food, bryozoan colonies on offshore kelp. *C. bolini* MacFarland, 1966, is a synonym. For feeding, see J. W. McBeth, *The Veliger,* vol. 11, p. 145.

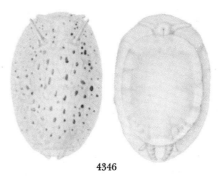

4346

Other species:

4347 *Doridella carambola* Marcus, 1955. São Sebastao, Brazil. *Zoologia,* vol. 20, no. 207, p. 89.

Superfamily Dendrodoridacea Pruvot-Fol, 1935

Dorids with the mouth a simple porelike opening (hence the name Porostomata used by some workers for this superfamily), followed by a muscular buccal vestibule, then by a long digestive tube. No jaws, no radula. Contains the families Phyllidiadae and Dendrodorididae.

Family DENDRODORIDIDAE Pruvot-Fol, 1935

Body dorid-shaped, soft. Pharyngeal bulb and elongated sucking tube, destitute of mandibles and radulae. Penis armed with a series of hooks.

Genus *Dendrodoris* Ehrenberg, 1831

Characterized by the branchial plumes being far back and the rhinophores set quite near the front, in having no buccal palps on either side of the mouth and in having a soft translucent skin. Prostate voluminous. Penis is spiculose. Type: *limbata* (Cuvier, 1804). *Doridopsis* Alder and Hancock, 1865, is a synonym, but not *Doriopsis* Pease. *Rhacodoris* Mörch, 1863, and *Haustellodoris* Pease, 1871, are synonyms.

Dendrodoris krebsii (Mörch, 1863)   **4348**
Krebs' Doris

Georgia to Florida and to Brazil.
Baja California to Mazatlan, Mexico.

30 to 70 mm. long, half as wide, dark brick-red, peppered with yellow and black spots, with a yellow border. Underside with orange blotches. Rhinophores darker, but with white tips. Notum and foot have frilled borders. Rhinophore with about 20 leaves. There are 6 tripinnate gills. Surface of notum and the borders of the rhinophoral and branchial pockets are smooth. Synonyms: var. *pallida* Bergh, 1879; *atropos* Bergh, 1879.

Genus *Doriopsilla* Bergh, 1880

Differing from *Dendrodoris* by its very spiculose, firm texture. The "ptyaline" gland which empties into the buccal cavity is absent. Porelike mouth is flanked by minute tentacles. Type: *areolata* Bergh, 1880.

Doriopsilla areolata Bergh, 1880   4349
Areolate Doris

West Indies; Mediterranean; West Africa.

15 to 30 mm., elongate-oval, with the notum wider than the foot. Color light-brown with darker-brown specklings in the middle zone, whose center is lighter again. Usually, there is a white, silky netting just under the surface of the notum, which bears 2 longitudinal rows of low, swollen nodules which have silky spicules. Elsewhere smaller tubercles are scattered all over the notum. Rims of retractile, 27-leafed rhinophores are smooth; edge of branchial pit is ragged. 5 gills present. Probable synonyms are *D. pelseneeri* Oliviera, 1895, and *fedalae* Pruvot-Fol, 1953. Dredged from 79 to 140 meters, and found in shallower water.

Doriopsilla pharpa Marcus, 1961   4350

North Carolina to Georgia.

20 mm., yellow with strong or subdued dark-brown specks. The notum is firm and slightly bossed. Rhinophores with 12 leaves. 4 gills. Papillae stiffened with bundles of spicules. Eversible penis with hooks. Moderately common; 9 to 10 fathoms.

Doriopsilla albopunctata (Cooper, 1863)   4351
Common Yellow Doris

Bolinas Bay, California, to the Gulf of California.

2 inches in length. Back soft, with low, papillalike elevations tipped with white. Rhinophores with 18 to 20 leaves in the clavus which is ⅔ the length of the entire rhinophore. It is completely retractile. 5 branchial plumes are tripinnate. No mandibles or radula. Young specimens are nearly always bright-yellow, but in southern California many adults tend towards a soft-brown, especially in the center of the notum. One of the commonest Pacific Coast species. In tide pools at all times of the year, especially common in summer. Coiled egg band is yellow. Synonyms are *D. fulva* (MacFarland, 1905) and *Doridopsis reticulata* Cockerell and Eliot, 1905 (see Steinberg, 1961, *The Veliger*, vol. 4, p. 57). Down to 20 fathoms.

4351

Other Pacific species:

4352   *Doriopsilla nigromaculata* (Cockerell and Eliot, 1905). La Jolla to San Diego, California. Rare.

4353   *Doriopsilla rowena* Marcus and Marcus, 1967. Puerto Peñasco, Sonora, Mexico. 8 to 10 mm. Yellow to pink to orangish with dark-brown speckles.

Other Atlantic species:

4354   *Doriopsilla leia* Marcus, 1961. Beaufort, North Carolina. 7 mm. (*Jour. Elisha Mitchell Sci. Soc.*, vol. 77, p. 144).

Other Pacific species:

4355   *Doriopsilla janaina* Marcus and Marcus, 1967. Sonora, Mexico, to Pacific side of the Canal Zone, Panama. *Studies Trop. Oceanogr. Miami*, no. 6, p. 96.

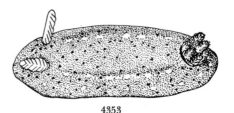
4353

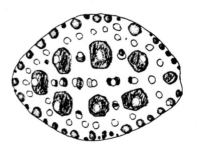
4356

Family PHYLLIDIIDAE Rafinesque, 1815

No dorsal branchial gills but with lamellae under the sides of the mantle. Body oval; the anus on or under the mantle. Rhinophores on the dorsal side of the mantle. Mantle of large pustulose tubercles, usually white or clear on a background of black.

Genus *Phyllidiopsis* Bergh, 1876

Body stiff, with numerous spicules, and conical tubercles on mantle. Buccal palps absent. Anterior border of foot notched. Up to 100 branchial leaflets on either side of foot. Type: *cardinalis* Bergh, 1876. One American species:

4356   *Phyllidiopsis papilligera* Bergh, 1890. Gulf of Mexico, 400 km. west of Cape Sable, Florida; Grand Bahama Island; Virgin Islands (Marcus and Marcus, 1962, *Bull. Marine Sci.*, vol. 12, p. 475). Jamaica.

Suborder *Arminoidea* Iredale and O'Donoghue, 1923

Contains some forms resembling those of dorids, and some like the aeolids.

Superfamily Dendronotacea Gray, 1857

The liver is mostly compact, but is in part ramified (so-called Heterohepatica) and extends into the greatly frilled appendices. No cnidosacs present. Anal papilla on the right side midway between first and second arborescent ceras. Rhinophores with a basal sleeve developed from the mantle.

Family DENDRONOTIDAE Gray, 1857

Only one genus is known in the family.

Genus *Dendronotus* Alder and Hancock, 1845

Body limaciform; 2 rhinophores laminated, with arborescent sheath; numerous cerata ramose. Arrow-shaped central tooth with a denticulate margin; about 9 elongate laterals on each side. About 40 rows of teeth. Penis unarmed. Type: *frondosus* (Ascanius, 1774). A detailed review of the 6 known species is given by Gordon A. Robilliard, 1970, in *The Veliger,* vol. 12, no. 4, pp. 433–479. *Campaspe* Bergh, 1863, is a synonym.

Dendronotus frondosus (Ascanius, 1774)   **4357**
Frond Eolis   **Color Plate 16**

Arctic Seas to New Jersey. Europe.
Alaska to southern California. Japan.

10 to 60 mm. in length. Grayish white, overlaid with browns, yellows and whites; conical papillae on dorsum yellow- or white-capped. Rhinophores with 8 to 12 large leaves, interspaced by about 15 smaller ones. Radula with 29 to 48 rows. Median tooth with strong denticles. 7 to 14 laterals with 2 to 5 denticles. Rhinophores as long as the first pair of cerata. *D. arborescens* Müller, 1776; *reynoldsii* Couthouy, 1837; *rufus* O'Donoghue, 1921; *venustus* MacFarland, 1966; *cervina* Gmelin, 1791; *lactea* Thompson, 1840; *pulchella* Alder and Hancock, 1842; *fabricii* "Beck" Mörch, 1857; *luteolus* Lafont, 1871; *purpureus* Bergh, 1879; *aurantiaca* Friele, 1879; *pusilla* Bergh, 1863; *major* Bergh, 1886, are synonyms. Common from shore to 200 fathoms. They reproduce during cold weather. They can mature by the second year. Feed on various hydroids. They are able swimmers and are usually found on rocky bottoms.

Dendronotus iris Cooper, 1863   **4358**
Giant Frond Eolis

Alaska; British Columbia to the Coronados Islands, Mexico.

5 to 8 inches in length; limaciform, heavy and tapering posteriorly; grayish, brown, or orange-red. 16 to 31 brown-colored leaves, all told, in the club of the thick-stalked rhinophore. Veil with 2 to 4 pairs of stout papillae. Dorsum skin smooth. Distinguished from *frondosus* by the 2 to 6 small but well-marked dendriform papillae on the posterior edge of the rhinophore sheath, by the thick, cylindrical, untapered penis (which is attenuated in *frondosus*), and by the opaque white line around the foot. The esophagus is very long and convoluted in this species. Radula with 34 to 61 rows; median tooth pointed and with 9 to 18 denticles on each concave edge; 11 to 21 simple lateral teeth on each side. For swimming habits, see Agersborg, 1922, *Biol. Bull.,* vol. 42, p. 257. Usually found down to 25 fathoms, over muddy bottoms. Feeds on the anemone, *Cerianthus*. Probably the largest of the American nudibranchs. *D. giganteus* O'Donoghue, 1921, is a synonym. See Agersborg, 1922, *Biol. Bull.,* vol. 42, pp. 257–266 for locomotion.

Dendronotus robustus Verrill, 1870   **4359**
Robust Frond Eolis

Greenland to Cape Cod, Massachusetts.
Norway to Siberia.

1 to 3 inches, stouter, less compressed laterally and less acutely tapered posteriorly than is *frondosus*. The anus is located nearer the second right ceras than in other species. No prominent cardiac prominence. Dorsum skin smooth or with scattered, small tubercles. Color pale-gray with numerous yellow dots, or a reddish body with white dots. Veil enormous, with 8 to 10 sulfur-yellow papillae, the outer 2 being largest. Foot as wide as the body. 6 or 7 pairs of short, stout, slightly arborescent cerata. No lateral papillae on the stout, round rhinophore. Clavus with 10 to 12 leaves and surrounded by 5 papillae. *D. velifer* Sars, 1878, is a synonym. Uncommon; down to 200 meters.

4359

Dendronotus dalli Bergh, 1879   **4360**
Dall's Frond Eolis

Siberia to Alaska to Vancouver Island, British Columbia.

3 to 5 inches, laterally compressed, limaciform. Cardiac hump is slight. Skin usually smooth. Sole of foot white or pale-pink. Body translucent-white to salmon-pink or rarely mauve. Rarely white-dotted. Cerata in 4 to 8 pairs, short, extensively branched and "fuzzy." Rhinophore short, stout, with many small branches, and the sheath with about 5 variable crown papillae. Veil with 4 or 5 pairs of stout, branched papillae. Yellow clavus with 16 to 33 leaves. Radula with 37 to 48 rows; median broadly pointed and smooth-edged; denticulated laterals 9 to 14 on each side. Common on hydroids on rocks from 20 to 30 meters.

Dendronotus rufus O'Donoghue, 1921   **4361**
Red Frond Eolis

British Columbia to Seattle, Washington.

3 to 7 inches, limaciform and fairly tubby, with a high, rounded dorsum. Color from pink to deep brick-red, with a red line around the foot. Sole of foot white. Cerata 6 to 9 pairs, magenta, white-dotted and the largest and most dendritic of all species. Rhinophore branched and as high as the first ceras. Subconical, yellowish to purple-spotted clavus small and with 19 to 24 shallow-cut leaves. Rhinophore sheath with 5 radiating long papillae. 5 pairs of veil papillae are very dendritic. Radula with 32 to 35 rows; 6 to 8 laterals on each side. Common on rocks from 7 to 40 meters.

Dendronotus subramosus MacFarland, 1966   **4362**
Subramose Frond Eolis

Washington to Los Coronados, Mexico.

1 to 2½ inches. Yellow- or white-tipped, conical papillae often present on skin. 4 brown lines run the length of the

dorsum. Body white to brown or rarely yellow to orange. 3 to 6 pairs of stout, upright cerata are branched in a rosette pattern. Rhinophore without lateral papillae. Clavus conical and with 9 to 14 leaves. Radula with 54 to 72 rows. Laterals 5 to 7, bearing 4 to 7 denticles. Usually lives offshore from 15 to 120 meters on gravel and shell bottoms.

Other species:

4363 *Dendronotus albus* MacFarland, 1966. British Columbia to Monterey Bay to Santa Barbara, California; Mexico. Rare; subtidal to 25 meters. White; cerata 4 to 8 pairs.

4364 *Dendronotus nanus* Marcus and Marcus, 1967. Puerto Peñasco, Sonora, Mexico. 13 mm. Reddish gray with white, black and orange appendages. *Studies Trop. Oceanogr. Miami,* no. 6, p. 210.

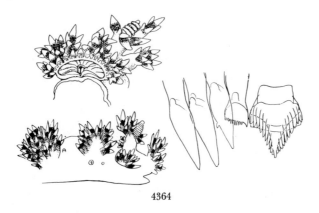

4364

4365 *Dendronotus diversicolor* Robilliard, 1970. British Columbia and San Juan Island, Washington. 1 to 3 inches. White or lilac. 4 pairs cerata. *The Veliger,* vol. 12, no. 4, p. 470. Same as *albus* MacFarland?

Family TRITONIIDAE Menke, 1828

Also called the Duvauceliidae Odhner, 1926, and Tritoniadae. No dorsal branchiae.

Genus *Tritonia* Cuvier, 1797

Body slug-shaped; no dorsal branchial plumes. A ring of frilly appendages present around the edges of the back. Rhinophores retractile within high sheaths with rather thick walls. Outer surface of the sheaths covered with small frilly appendages. Frontal veil fimbriated. Radula with a median and numerous smooth, hooked laterals. Jaws present. No prostate. Feeds on Alcyonarians. Type: *hombergi* Cuvier, 1802. Synonym: *Sphaerostoma* MacGillivray, 1843; *Candellista* Iredale and O'Donoghue, 1923.

For neural behavior and giant brain cell studies, see A. O. D. Willows, Feb. 1971, *Scientific American,* vol. 224, no. 2.

Subgenus *Duvaucelia* Risso, 1818

Body slender, with 4 to 6 digitations on the frontal veil. Small number of lateral teeth. Dorsal appendages distinct and not numerous. Type: *gracilis* Risso, 1818. *Candiella* Gray, 1857, is a synonym.

Tritonia festiva (Stearns, 1873)   **4366**
Festive Tritonia

Vancouver Island, British Columbia to the Coronado Islands, Mexico.
Central Japan.

10 mm., elongate, broad anteriorly where the cephalic veil bears 9 to 11 simple, prominent papillae. Rhinophores laterally placed in prominent smooth collars. Ends of rhinophores with clusters of pinnate plumes. 2 spoon-shaped tentacles in front underneath and on either side of the mouth. Notum tapers posteriorly to a point and bears 8 or 9 tufted small gills along each border. Pharynx $\frac{1}{3}$ length of body. Jaws and radula present (formula: $33 \cdot 1 \cdot 1 \cdot 1 \cdot 33$). Dorsal white circles are filled with orange. Anatomical details in Marcus, 1961, *The Veliger,* vol. 3, supplement, p. 31. Rare; shallow water. Synonyms: *reticulata* Bergh, 1881; *undulata* (O'Donoghue, 1924).

4366

Tritonia exsulans Bergh, 1894   **4367**
Rosy Tritonia

Japan and Alaska to Baja California.
West coast of Florida (fide Marcus, 1961).

40 to 60 mm., bright salmon to rose on the back, with the sides and sole of the foot translucent-white. Cephalic veil bears about 22 simple papillae, larger ones on the sides. Veil flanked by conical tentacles. Rhinophore rims crenulate. Rhinophores with terminate pinnate plumes. 28 tripinnate gills are arranged in 2 rows on either side of the notum. Jaws and radula present (formula: $52 \cdot 1 \cdot 1 \cdot 1 \cdot 52$). Uncommon; 2 to 150 fathoms. The Florida record is surprising, if it is this species.

4367

Tritonia bayeri Marcus, 1967   **4368**
Bayer's Tritonia

Georgia to southeast Florida.

4 to 12 mm., diaphanous with an opaque-white diffused network on the notum. Branchiae white. Back smooth. Radula with 34 rows. Second inner lateral tooth is denticulated. Male copulatory organ conic. Lives and feeds on *Briareum asbestinum* and *Pseudopterogorgia*.

(4369) A subspecies (*misa* Marcus, 1967) was described from 35 to 44 fathoms off Sapelo Island, Georgia. It differs in having 4 instead of 2 velar appendages and in minor radular characters. It may not be a good subspecies.

Other species:

4370 *Tritonia palmeri* (Cooper, 1862). San Diego to San Pedro, California. Placement to subgenus dubious.

4371 *Tritonia (Duvaucelia) diomedea* (Bergh, 1894). Alaska to British Columbia. For brain studies, see *Scientific American,* Feb. 1971.

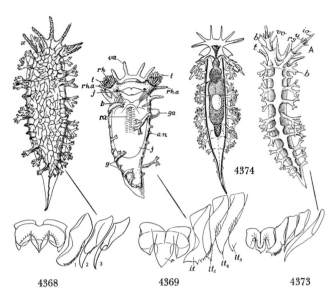

4368   4369   4373

4372 *Tritonia (Duvaucelia) gilberti* MacFarland, 1966. San Francisco Bay entrance, 20 fms., to Drakes Bay, California, 20 to 23 fms. (*Calif. Acad. Sci.*, Memoir 6, p. 235.)

4373 *Tritonia (Tritonidoxa* Bergh, 1907) *wellsi* Marcus, 1961. Beaufort, North Carolina to southern Brazil. 5 to 7 mm. (*Jour. Elisha Mitchell Sci. Soc.*, vol. 77, p. 146.)

4374 *Tritonia (Candiella) pickensi* Marcus and Marcus, 1967. Puerto Peñasco, Sonora, Mexico. 7 mm. White with bluish sides. *Studies Trop. Oceanogr. Miami*, no. 6, p. 207.

Genus *Tochuina* Odhner, 1963

Body chunky, not elongate. Anus behind the mid-body section. Penis flagelliform. Animal reddish in life. Radula broad with about 250 elongate smooth teeth on each side of the median tooth. No stomachal plates. Liver fused into a single mass, covering upper and left sides of the stomach, right liver thus indistinctly marked off. Branchial gills ramose in lateral dorsal margins. (Copied from Odhner, 1963, *The Veliger*, vol. 6, p. 51.) Type: *tetraquetra* (Pallas, 1788).

4375 *Tochuina tetraquetra* (Pallas, 1788). Alaska to Monterey Bay, California. Japan. Intertidal to 15 fms. 6 to 12 inches. (Synonyms: *gigantea* Bergh, 1881; *aurantia* Mattox, 1955.) Once assigned to *Tritoniopsis* and *Duvaucelia*.

Family HANCOCKIIDAE MacFarland, 1923

Body slug-shaped, oval veil prolonged into digitate lobes at the corners. Foot rounded in front. Cerata forming digitate lobes. Rhinophores with leaflets, retractile into sheaths. Cnidocysts and "liver" present in the cerata and rhinophore sheaths. Second stomach with a cuticular armature.

Genus *Hancockia* Gosse, 1877

Characters of the family. Rhinophores bulbous at the base, with vertical plications, set in long sheaths. 5-lobed processes on the dorsal margins. Jaws denticulate. Radula triseriate. Genitalia unarmed. Cnidocysts present. Type: *uncinata* Hesse, 1872 (synonym: *eudactylota* Gosse, 1877).

Govia Trinchese, 1885, is a synonym. Eastern Europe; India; western America.

Hancockia californica MacFarland, 1923   **4376**
Hancock Nudibranch

Dillon Beach, California, to Baja California.

20 mm. long, 3 mm. wide (living), reddish brown to greenish with blotches of brown. Cerata, rhinophore shafts white or sprinkled with opaque-white. Lateral velar lobe on each side of head with 6 to 10 subequal fingers. Retractile rhinophores with 6 to 8 vertical leaflets on either side, and with 6 to 9 slanting, nodulose ridges (cnidosacs) on the funnel-shaped sheaths. Lateral margins of notum with 4 to 7 cerata, each of which has 4 to 16 fingerlike processes arranged in a horseshoe. Eggs in a pale-greenish coiled ribbon on brown algae. Uncommon; tide pools.

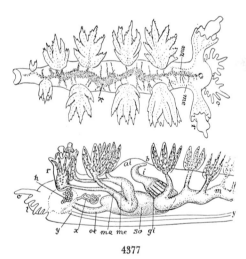

4376

Other species:

4377 *Hancockia ryrca* Marcus, 1957. Among *Padina*, São Paulo, Brazil; low tide.

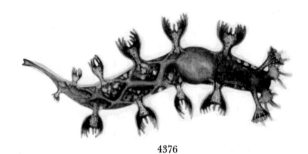

4377

Family TETHYIDAE Rafinesque, 1815

With an enormous buccal vestibule. Without radula. With peculiar, flat, leaf-shaped papillae on the sides of the back (have been described as parasites and planarian worms by early workers). The family name Fimbriidae is based upon a nonbinomial name. The accepted genus *Fimbria* is a lucinoid bivalve. *Tethys fimbria* Linné, 1758, (type of *Tethys* Linné, 1758) is abundant in western Europe. A subspecies was described from Santo Domingo: *dominguensis* Pruvot-Fol, 1954 (*Faune de France*, vol. 58, p. 359) **(4377a)**.

Genus *Melibe* Rang, 1829

Characters of the family. Pair of strong mandibles present. No radula present. Type: *rosea* Rang, 1829. Limited to the Indo-Pacific Ocean.

Subgenus *Chioraera* Gould, 1852

Radula and mandibles absent. Type: *leonina* Gould, 1853. The genus *Filurus* De Kay, 1843 (with its east North American type, *dubius* De Kay, 1843) may be this subgenus, but its anatomy is unknown.

Melibe leonina (Gould, 1853)   4378
Lion Nudibranch

Dall Island, Alaska, to the Gulf of California.

30 to 70 mm., translucent-yellow, body slim, but the enormous round head is separated by a narrow neck. Edge of concave underhead bears long outer and short inner row of cirri. Back bears 4 or 5 huge, flat, leaflike processes. Rhinophores with an undulating side curtain, tipped by a clavus with 5 or 6 leaflets. Penis smooth, flagelliform. Common; found floating and crawling near the surface among kelps and *Sargassum*. *M. pellucida* Bergh, 1880, and *dalli* Heath, 1917, are synonyms. See Agersborg, 1922, *The Nautilus*, vol. 36, pp. 86–96 for habits.

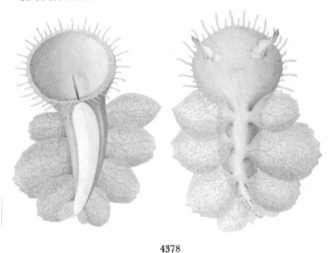

4378

Family SCYLLAEIDAE Rafinesque, 1815

Pelagic sargassum-weed dwellers, with 2 pairs of winglike expansions of the back bearing small ramified appendages. Rhinophoral sheaths very long and smooth. Brown-colored, with opaque white spots.

Genus *Scyllaea* Linné, 1758

Body elongate, compressed laterally, with 2 pairs of oarlike appendages on the sides. Posterior part of back is carinated, forming an undulating crest used in swimming. Frontal veil smooth-edged. Rhinophores with leaflets, very small, and atop very long, smooth bases. Buccal bulb with strong jaws on the sides. Radula large with finely denticulate teeth. Feeds on coelenterates. *S. pelagica* Linné, 1758 is the type.

Scyllaea pelagica Linné, 1758   4379
Sargassum Nudibranch

Massachusetts to the West Indies. Bermuda. Worldwide warm seas.

1 to 2 inches in length. Translucent cream-brown to orange-brown. With brown and white patches and numerous flecks of red-brown. Blue spots are on the flanks. Body elongate. Oral tentacles absent. 2 slender, long rhinophores. Sides of body with 2 pairs of large, clublike, foliaceous gill plumes or cerata. Common in floating sargassum weed in the Gulf Stream where it feeds on hydroids and coelenterates. For 15 synonyms, see Pruvot-Fol, 1966, p. 367: *edwardsii* Verrill, 1878; *marginata* Bergh, 1871, and others.

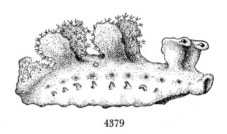

4379

Family PHYLLIROIDAE Menke, 1830

Genus *Phylliroe* Péron and Lesueur, 1810

Small pelagic nudibranchs, superficially resembling the heteropods. Body elongate, compressed laterally, and having thin dorsal and ventral edges and a thin compressed "tail." Head proboscidiform and bearing 2 long smooth tentacles or rhinophores. No foot. No gills. Anal orifice on the right side. Mandibles present. Radula with a denticulated central and needlelike marginals. These nocturnal creatures are brilliantly phosphorescent. The eyes are buried in the transparent flesh. Type: *bucephalum* Péron and Lesueur, 1810. Two species are reported from the Gulf Stream, off Florida and Bermuda: (4380) *P. bucephalum* Péron and Lesueur, 1810, and (4381) *atlantica* Bergh, 1871.

Family DOTODAE Gray, 1853

Back margin reduced; with only cerata; anus latero-dorsal (as in Dendronotidae). Rhinophore clubs smooth. Cerata with tubercles in rings and containing the disintegrated liver as diverticula. Velum smooth. Radula unseriate. Contains the genera *Gellina* Gray, 1857 (Europe); *Embletonia* Alder and Hancock, 1851; *Tenellia* A. Costa, 1866; *Doto* Oken, 1815; and *Miesea* Marcus, 1961. The family Iduliidae Iredale and O'Donoghue, 1923, is synonymous. See Opinion 697 of the I.C.Z.N.

Genus *Doto* Oken, 1815

Small body widest at front and tapering to a point; with a series of 5 to 8 papillae on each side, each bearing tubercles. Rhinophoral sheaths long and trumpet-shaped, with a thin, smoothish or circularly striate club extending from the end. Jaws very weak. Radula long, with about 60 to 150 teeth in a single row and all coarsely denticulate. Type:

coronata (Gmelin, 1791). This is *Doto* Oken, 1815, non Oken, 1807, and *Doto* of authors, such as Odhner, 1939, and Pruvot-Fol, 1954. *Melibaea* Forbes, 1838; *Dotilla* Bergh, 1879; *Dotona* Rafinesque, 1814; and *Dotona* Iredale and O'Donoghue, 1922, are synonyms, and were rejected by the I.C.Z.N., *Bull. Zool. Nomen.*, vol. 21, p. 97, opinion 697 (in which *Doto* was conserved).

Doto coronata (Gmelin, 1791)   4382
Coronate Doto

Bay of Fundy to New Jersey.
Northern Europe.

6 to 10 mm., characters of the genus. Color pale yellowish white with black dots on the tips of the tubercles and black specklings on the back. Lay their eggs on the hydroid, *Obelia commisuralis*.

Doto columbiana O'Donoghue, 1921   4383
British Columbian Doto

Vancouver, British Columbia, to central California; Gulf of California.

7 to 14 mm. (living), limaciform, whitish with brown pigment on the head, back and sides, as well as between the tubercles of the cerata. Back with 5 to 7 pairs of cerata, each of which bears 4 circles of flat tubercles. The tubercles have a black ring around the base. Rhinophore sheaths light in color, long, smooth, with the clublike, simple rhinophore retractile. Uncommon; on hydroids. Intertidal to 18 fathoms. *Doto lancei* Marcus and Marcus, 1967, is probably a synonym.

Other Atlantic species:

4384 *Doto formosa* Verrill, 1875. Nova Scotia to off Point Judith, Rhode Island, 10 to 50 fms.

4385 *Doto pita* Marcus, 1955. Southeast Florida to Brazil.

4386 *Doto divae* Marcus, 1960. Miami, Florida, to the Lesser Antilles.

4387 *Doto fragilis* (Forbes, 1838) subsp. *umia* Marcus and Marcus, 1969. Southwest of Greenland, 183 fms. *Amer. Mus. Novitates,* no. 2368, p. 27.

4388 *Doto chica* Marcus, 1960. Miami, Florida, to Curaçao, West Indies.

4389 *Doto doerga* Marcus and Marcus, 1963. West Indies and off east United States in floating *Sargassum* weed. Off Bermuda.

4390 *Doto lancei* Marcus and Marcus, 1967. Puerto Peñasco, Sonora, Mexico. 7 mm. Tan with black-brown dots. *Studies Trop. Oceanogr. Miami,* no. 6, p. 214.

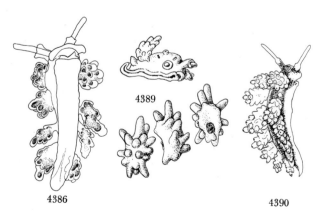

4389

4386   4390

Other Pacific species:

4391 *Doto amyra* Marcus, 1961. Monterey Bay, California. 7 mm. Common; subtidal (*ganda* and *wara* may be synonyms).

4392 *Doto ganda* Marcus, 1961. Dillon Beach to Ensenada, Baja California, on hydroids. 4 mm.

4393 *Doto kya* Marcus, 1961. Moss Beach to Shell Beach, California. 6 mm.

4394 *Doto wara* Marcus, 1961. Dillon Beach to Monterey Bay, California. 6 mm. Above 4 species in *The Veliger,* vol. 3, supplement, pp. 38–41.

4395 *Doto varians* (MacFarland, 1966). Monterey Bay, California. Tide pools. (Is *kya* Marcus, 1961, in part, see *The Veliger,* vol. 12, p. 373.)

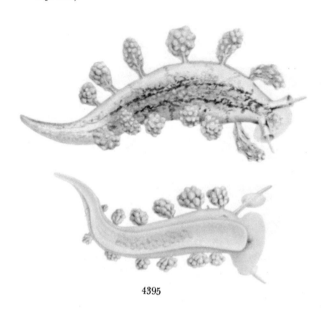

4395

Genus *Miesea* Marcus, 1961

Dotos without sheaths around the rhinophores; cerata simple. No gills. Oral veil small. Radula uniserial. Penis unarmed. Extremely delicate jaws. Type: *evelinae* (Marcus, 1961).

4396 *Miesea evelinae* (Marcus, 1957). Beaufort, North Carolina, to Brazil.

Genus *Embletonia* Alder and Hancock, 1851

Papillae in a series along each side, usually reflected backwards. No cnidosacs. Rhinophores short, conical, without sheaths. Penis ending in a short little tube. Jaws large, triangular and denticulate. Teeth extremely numerous, small, with a small cusp. Eggs in a spaghetti-shaped coil. North Atlantic. Type: *pulchra* Alder and Hancock, 1844 (*Pterochilus*). See Opinion 782, *Bull. Zool. Nomen.*, vol. 23, p. 106. Not reported from North America, as yet.

Genus *Tenellia* A. Costa, 1866

Differing from *Embletonia* in having clusters of 2 to 4 cerata on each side. With cnidosacs. Rhinophores without sheaths, long, simple. Radula and jaws present. Type: *ventilabrum* Dalyell, 1853. Two eastern American cool-water species reported:

4397 *Tenellia ventilabrum* (Dalyell, 1853). Massachusetts to Delaware Bay; western Europe. 7 mm. Found on bryozoans, *Amathia* and *Obelia*. (Synonym: *pallida* Alder and Hancock, 1854.)

4398 *Tenellia fuscata* Gould, 1870. Massachusetts and New York. (Synonyms: *lanceolata* and *remigata* Gould, 1870.)

4398

Superfamily Arminacea Rafinesque, 1814

Family ARMINIDAE Rafinesque, 1814

Genus *Armina* Rafinesque, 1814

Body elongate-oval with the broad head separated above from the mantle, below continuous with the sole. Rhinophore small and close together in a notch in the front end of the mantle which is longitudinally and deeply furrowed. Paired lamelliform branchiae on the under surface of the mantle. Mantle margin with numerous thick-set cnidosacs. Jaws present. Radula with 30 to 50 rows. Central tooth denticulate; numerous laterals pectinate. Type: *tigrina* Rafinesque, 1814. Synonyms: *Pleurophyllidia* Stammer, 1816; *Diphyllidia* Otto, 1820. For an annotated catalog of the 19 known species, see E. and E. Marcus, 1966, *Studies in Trop. Oceanogr. Miami*, no. 4, pt. 1, pp. 187–198.

Armina tigrina Rafinesque, 1814   **4399**
Tiger Armina

North Carolina to Texas and Yucatan. Europe.

2 inches, somewhat lanceolate in shape. Notum smoothish, brownish black with about 25 to 45 narrow, yellowish or whitish, longitudinal stripes. 25 to 35 lamellae on the underside of the margins of the notum on each side of the animal. Anterior to these, there is a more compact series of short, very thin, crowded gill-lamellae. Rhinophores short, with a laminated dark-brown surface. Radula with 30 to 50 rows of teeth, the central one being quadrate and denticulate, and flanked by 60 to 70 smaller, sickle-shaped laterals. Live buried in sand from the low-tide mark to 20 fathoms. Locally common. Feed on sea pansies. See Abbott, 1954, *The Nautilus*, vol. 67, p. 83. *Armina wattla* Marcus and Marcus, 1967, is possibly a synonym.

Armina californica (Cooper, 1862)   **4400**
Californian Armina

Vancouver, British Columbia, to Panama.

2 to 3 inches in length, and resembling *tigrina* from the Atlantic coast. Common; intertidal sand flats to 126 fathoms.

Feed, in part at least, on sea pansies which causes bioluminescence (*The Veliger*, vol. 10, p. 440). *A. columbiana* O'Donoghue, 1924 (Vancouver to central California) has the notal ridges running the entire length of the back, but is a synonym. *A. vancouveriensis* Bergh, 1876, is also an earlier name for it. *A. digueti* Pruvot-Fol, 1955, is also probably a synonym.

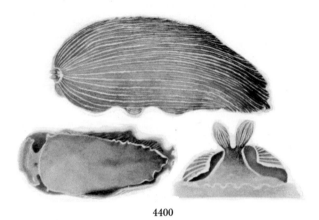

4400

Other species:

4401 *Armina convolvula* Lance, 1962. 15 miles south of San Felipe, Gulf of California. 75 mm. Brown with white spots. *The Veliger*, vol. 5, p. 51. Belongs to the subgenus *Histiomena* Mörch, 1859, of which *Camarga* Bergh, 1866, is a synonym.

Superfamily Dironacea MacFarland, 1912

This is sometimes referred to as the infraorder Pachygnatha Odhner, 1939.

Family ZEPHYRINIDAE Iredale and O'Donoghue, 1923

A multidentate radula in combination with nonretractile laminar, rhinophores and an aeolid papillation on the back is characteristic of this family. No dorsal margin. Oral tentacles present. Liver extends into dorsal cerata. Jaws thick. The family Antiopellidae Hoffmann, 1938, is a synonym.

Genus *Antiopella* Hoyle, 1902

Crest present between the rhinophores. Having jaws which are denticulated. Type: *splendida* Alder and Hancock, 1849, which could be a synonym of *cristata* (Delle Chiaje) from western Europe. *Antiopa* Alder and Hancock, 1848, non Meigen, 1800 (a fly), is a synonym. Two American species reported:

4402 *Antiopella fusca* (O'Donoghue, 1924). Vancouver, British Columbia, to Monterey Bay, California.

4403 *Antiopella mucloc* Marcus, 1958. In algae; Ubatuba, Brazil. *Amer. Mus. Novitates*, no. 1906, p. 37.

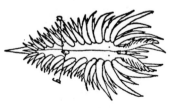

4403

Genus *Janolus* Bergh, 1884

Rhinophores have lamellae nearly vertical, irregular and indistinct in the young. The cerata have well-marked tubercles. The lobes of the digestive gland are not ramified. Tail very short. There is a crest between the rhinophores. Eggs ribbon-shaped. Type: *australis* Bergh, 1884.

4404 *Janolus barbarensis* (Cooper, 1863). Southern California to west Baja California. *J. coeruleopictus* Cockerell and Eliot, 1905 (*nomen dubium,* fide Steinberg, 1963), is probably a synonym. *Antiopella aureocincta* MacFarland, 1966, Johnson and Snook, 1927, are synonyms. 13 to 21 mm. General color gray; rhinophores yellow; dorsal crest yellow-orange; cerata tips white, remainder orange. Dorsum flecked with blue; tail with blue median band.

Family DIRONIDAE Eliot, 1910

Genus *Dirona* MacFarland, in Cockerell and Eliot, 1905

Body broad and flattish; head with a broad, thin veil with anterior undulations. Rhinophores with leaflets, but no sheaths. Dorsal papillae large along the back, and with a few around the rhinophores. Tail narrow. Cnidosacs absent. Mandibles massive. Radula narrow, with 1 central and 2 laterals on each side. Type: *picta* Cockerell and Eliot, 1905.

Dirona albolineata Cockerell and Eliot, 1905   **4405**
White-streaked Dirona

British Columbia to Laguna Beach, California. Japan.

30 to 150 mm.; color translucent-gray, with a white edge on the veil, the margins of the dorsal papillae and the median crest of the tail. Dorsal papillae lanceolate, inflated, smooth, flattened antero-posteriorly, closely set. 29 to 32 rows of radulae with formula 2·1·2. Glans penis elongate, with thickly set papillae bearing thornlike points; spermatotheca large; oviduct short and slender. Common; extreme low tide on mud flats to 15 fathoms. Sometimes feeds on small gastropods. MacFarland redescribed this species in 1912.

4405

Dirona picta Cockerell and Eliot, 1905   **4406**
Painted Dirona

Crescent City to San Diego, California; Gulf of California.

13 to 40 mm.; color yellowish to orange-brown, with darker or lighter spots. A pale-red spot on the outer side of the cerata, 1/3 distant from the base, is a constant mark. Green liver shows through body. Surfaces of dorsal papillae are tuberculate. 32 to 35 rows of radulae with formula of 2·1·2. Glans penis short, conical; spermatotheca very small; oviduct long and wide. Abundant in summer in pools on brown kelp, or floating at surface. Also down to 5 fathoms.

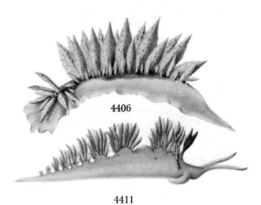

4406

4411

Dirona aurantia Hurst, 1966   **4407**
Golden Dirona

Puget Sound, Washington.

1 to 5 inches, broad in front, pointed behind; orange throughout with a few white granular spots and numerous, long, narrow, white cerata (which readily drop off). Terminal 1/2 of rhinophores with about 20 leaves. Radular typical of the genus, with 24 to 26 rows. Penis large and wide with a short, smooth papilla at the end. Moderately common; 11 to 30 fathoms. *The Veliger,* vol. 9, no. 1, p. 9.

Suborder *Aeolidoidea* Odhner, 1934

Superfamily Coryphellacea Cooke, 1899
(Pleuroprocta of authors)

Family CORYPHELLIDAE Bergh, 1889

Aeolids with simple, slender, tentacle-shaped oral palps and simple, similar-looking rhinophores. Anterior corners of foot pointed. With numerous long cerata or papillae arranged in clumps. Jaws with denticulate edge. Penis unarmed. Ptyaline glands present. Many species, usually in cold waters. *Flabellinidae* Dall and Simpson, 1901, is a synonym. See Opinion 781 of the I.C.Z.N.

Genus *Coryphella* Gray, 1850

Branchiae numerous, clustered, elongate or fusiform. Foot narrow, with the anterior angles much produced. Rhinophores simple or perfoliate. Radula with a single longitudinal series of central teeth which bear a large central denticle and several marginal denticles. There is a denticulated lateral tooth on each side of the central. *Himatella* Bergh, 1890, and *Himatina* Thiele, 1931, are synonyms. Type: *Eolis rufibranchialis* Johnston, 1832. See Opinion 781, *Bull. Zool. Nomen.,* vol. 23, p. 104.

Coryphella verrucosa rufibranchialis (Johnston, 1832)   **4408**
Red-fingered Eolis   **Color Plate 16**

Arctic Seas to New York. Europe. Japan.

1 inch in length, characters as shown in figure and in the generic description. Common in New England. Minor varieties are *mananensis* Stimpson, 1854, and *chocolata* Balch, 1909. Typical *verrucosa verrucosa* (M. Sars, 1829) is limited to Norwegian waters. For feeding habits, see M. P. Morse, 1969, *The Nautilus,* vol. 83, p. 37.

Coryphella salmonacea (Couthouy, 1839)   **4409**
Salmon Eolis

Arctic Seas to Cape Cod, Massachusetts; northern Europe. Point Barrow, Alaska and Bering Strait.

1 to 2 inches broad and depressed, yellowish white, with 100 or more, close-set, long, pointed, deep salmon-colored branchiae on each side of the dorsum. Foot broad, the anterior angles prolonged into tentacular appendages, the tail narrowing rather suddenly to an acute point. Dorsal tentacles are serrated. Moderately common in shallow water. *C. papilligera* Beck, 1847, is a synonym.

Coryphella pellucida (Alder and Hancock, 1843)   **4410**
Pellucid Eolis

Maine to Massachusetts; England.

¼ to ⅞ inch, differing from *rufibranchialis* in having a translucent body (not opaque-white), longer recurved anterior angles of the foot, in having clear-red (not orange-red) cerata with a white tip (not with a white ring). Uncommon. Found north of Cape Cod. *C. stellata* (Stimpson, 1853), and *rutila* Verrill, 1879, are probably synonyms.

4410

Subgenus *Flabellinopsis* MacFarland, 1966

Rhinophores claviform, with numerous leaflets; anterior 2 tentacles long. Anterior foot with drawn-out corners. Cerata numerous, borne on small humps, the anterior group extending laterally behind the rhinophores. Mandibles strong. Radula triseriate, the laterals wide, their inner margins denticulate. Type: *Aeolis iodinea* Cooper, 1862.

Coryphella iodinea (Cooper, 1862)   **4411**
Iodine Eolis

Vancouver, British Columbia, to Sonora, Mexico.

35 to 70 mm., body brilliant-violet. Cerata orange to scarlet with violet bases, and with the core of brown liver visible. Tips of tentacles blue. Clavus of rhinophores garnet-red with 46 half-leaflets on either side. Found in the summer; on hydroids. Common; to 10 fathoms.

Other Atlantic species:

4412 *Coryphella diversa* (Couthouy, 1839). Grand Manan Island, Maine, to Massachusetts Bay.

— *Coryphella stimpsoni* Verrill, 1880. Nova Scotia to Massachusetts Bay, 1 to 51 fms. (Is a *Cuthona*.)

4414 *Coryphella nobilis* Verrill, 1880. Off Cape Cod, Massachusetts, 75 fms. North Europe.

4415 *Coryphella gracilis* (Alder and Hancock, 1844). Greenland; Iceland; Faroes; Norway. (Synonyms: *smaragdina* Alder and Hancock, 1851; *Coryphella bostoniensis* Bergh, 1864, non *Eolis bostoniensis* Couthouy, 1839; *C. borealis* Odhner, 1922.) See H. Lemche, 1935, *Vidensk. Medd. Dansk. nat. Foren,* vol. 99 p. 131.

Other Pacific species:

4416 *Coryphella californica* Bergh, 1904. Gulf of California.

4417 *Coryphella cooperi* Cockerell, 1901. San Pedro, California, to west Baja California.

4418 *Coryphella fusca* O'Donoghue, 1921. Vancouver Island, British Columbia.

4419 *Coryphella longicaudata* O'Donoghue, 1922. Vancouver Island, British Columbia; Friday Harbor, Washington. 17 to 21 mm. *Trans. Royal Canadian Institute,* vol. 14.

4420 *Coryphella trilineata* O'Donoghue, 1921. Vancouver Island, British Columbia, to Coronado Islands, California. (Synonyms: *piunca* Marcus, 1961, *The Veliger,* vol. 3, supplement, p. 47; *fisheri* MacFarland, 1966.)

4420

4421 *Coryphella trophina* (Bergh, 1894). Alaska.

4422 ?*Coryphella subrosacea* (Eschscholtz, 1831). Sitka, Alaska.

4423 *Coryphella pricei* MacFarland, 1966. Monterey Bay, California, in rocky tidepools. 15 mm.

4423

4424 *Coryphella cynara* Marcus and Marcus, 1967. Sonora coast, west Mexico. 12 mm. Blue, with orange-brown and red cerata. *Studies Trop. Oceanogr. Miami,* no. 6, p. 220.

4425 ?*Coryphella pallida* Verrill, 1900. Bermuda. *Trans. Conn. Acad. Sci.,* vol. 10, p. 547.

Superfamily Tergipedidacea Bergh, 1889
(Acleioprocta Odhner, 1939)

Family FLABELLINIDAE Bergh, 1889

Genus *Flabellina* Voight, 1834

Similar to *Coryphella* but the cerata are on peduncled supports. Penial stylet present or absent. Type: *affinis* (Gmelin, 1791). See ICZN Opinion 781, 1966.

4426 *Flabellina annuligera* (Bergh, 1900). Puget Sound, Washington.

4427 *Flabellina telja* Marcus and Marcus, 1967. Puerto Peñasco, Sonora, Mexico. 5 to 11 mm. *Studies Trop. Oceanogr. Miami,* no. 6, p. 223.

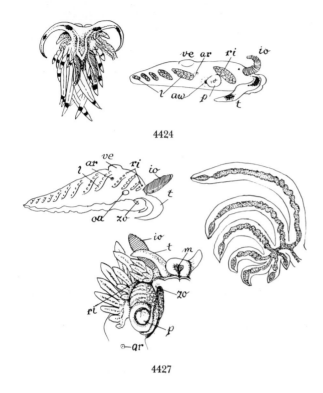

4424

4427

Family EUBRANCHIDAE Odhner, 1934

With a small tail. Cerata not numerous. Radula long and triseriate. Jaws with a denticulate edge. Penis with or without a stylet. Rhinophores simple and without sheaths.

Genus *Eubranchus* Forbes, 1838

Radula triseriate, 1 lateral which is a thin plain plate. Penis may bear a chitinous tip. Type: *tricolor* Forbes, 1838. *Amphorina* Quatrefages, 1844, and *Galvina* Alder and Hancock, 1855, are synonyms. For a review of North Atlantic species, see H. Lemche, 1935, *Vidensk. Medd. Dansk. naturh. Foren.*, vol. 99, p. 142.

Eubranchus tricolor (Forbes, 1838)   **4428**
Painted Balloon Eolis

Arctic Seas to Boston, Massachusetts. Europe.

½ inch in length, characters as shown in our figure and the generic description. 65 to 70 rows of radulae. Cerata inflated. Uncommon (?). *E. picta* Alder and Hancock, 1847, is a synonym. A number of color varieties have been named: *flava* (Trinchese, 1881); *amethystina* Alder and Hancock, 1844; *farrani* Alder and Hancock, 1844; *adelaidae* Thompson, 1860; *andraeopolis* McIntosh, 1864; *robertianae* McIntosh, 1865.

Eubranchus olivaceus (O'Donoghue, 1922)   **4429**
Greenish Balloon Eolis

Northern New England.
Puget Sound, Washington.

9 to 11 mm., aeolidiform, elongate, with 18 large, green-cored, white-dotted cerata on each side of the olive-green and brown-lined back. Rhinophores long and nonretractile, not perfoliate, and lie well in front of the first ceras. Radula triseriate with about 33 rows. Found on the hydroid, *Obelia longissima*. Common in the summer; intertidal pools.

Other species:

4430 *Eubranchus columbiana* (O'Donoghue, 1922). Gabriola Pass, British Columbia, 7 to 12 fms. *Trans. Royal Canadian Inst., Toronto*, vol. 14, p. 160.

4431 *Eubranchus pallidus* (Alder and Hancock, 1842). North Atlantic. 3 to 10 cerata in the largest rows; 35 to 40 rows of radulae. (Synonyms: *Eolis picta* and *cingulata* Alder and Hancock, 1847; *exigua* Alder and Hancock, 1848; *Eolis flavescens* Friele and Hansen, 1876; *Tergipes fustifer* Lovén, 1846.) Color Plate 16.

4432 *Eubranchus rupium* Möller, 1842. Greenland. 1 to 2 cerata in every single row; 35 to 40 rows of radulae.

Genus *Capellinia* Trinchese, 1874

Not differing from *Eubranchus* significantly, except that the penis has a cuticular stylet (usually about 50 to 100 μ). The papillae may be constricted around the middle in one or two places, and the apex of each papilla is conical or mucronate. This may only be a subgenus of *Eubranchus*. Type: *C. doriae* Trinchese, 1874.

Capellinia exigua (Alder and Hancock, 1848)   **4433**
Dwarf Balloon Eolis   **Color Plate 16**

Arctic Seas to Massachusetts; Europe.

5 to 6 mm., with 5 groups of papillae. Characters as shown in our figure. Seasonally uncommon. Synonyms: *Eubranchus exiguus* A. and H., 1848; *C. capellinii* Trinchese, 1874; *Tergipes fustifer* Lovén, 1846.

Capellinia conicla Marcus, 1958   **4434**
Conicla Eolis

Southeast Florida to Brazil.

2 to 3 mm., transparent, with opaque-white dots and some greenish to brown mottlings. Cerata large, tuberculate, pointed tips, 5 to 7 on each side. Tentacles short. Rhinophores smooth. Jaws present. Radula with 70 rows. Median tooth horseshoe-shaped and with 7 denticles. Lateral tooth quadrate and with 1 triangular tooth. Penis with a colorless stylet. Uncommon; shallow water.

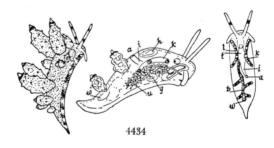

4434

Capellinia rustya Marcus, 1961   **4435**
Rustya Eolis

Monterey Bay, California, to Baja California.

5 to 10 mm., translucent white with pink, greatly inflated cerata pointed at their tips. They form 4 or 5 groups on either side and contain digestive diverticula arranged in 2 or 3 rings of tubercles. First group of 6 cerata form an arch, posterior groups with only 2 cerata. 2 long rhinophores smooth and slender, with small black eyes behind their bases. 2 tentacles, smooth. The penis is armed with a cuticular stylet. Uncommon; on the hydroid, *Obelia*. Synonym: *Eubranchus occidentalis* MacFarland, 1966.

Family TERGIPEDIDAE Bergh, 1889

Genus *Tergipes* Cuvier, 1805

Body slender; tentacles simple, the oral pair very short. Branchiae not very numerous, fusiform, inflated, set in a single series on each side of the back; foot narrow, anterior angles rounded. Eggmass kidney-shaped. Radula with a single row of plates, each with a stout central denticle and numerous delicate marginal denticles. Type: *tergipes* (Forskål, 1775).

Tergipes tergipes (Forskål, 1775)   **4436**
Johnston's Balloon Eolis   **Color Plate 16**

Arctic Seas to New Jersey. Europe. Brazil.

3 to 11 mm. in length, characters as shown in our figure and in the generic description. Gregarious on the hydroid, *Laomedia loveni*. Shore to 8 fathoms. Common. This is *T. despectus* Johnston, a widely dispersed species probably introduced by means of algae-covered ships' bottoms.

Family CUTHONIDAE Odhner, 1934

Genus *Catriona* Winckworth, 1941

Rhinophores smooth; oral tentacles similar; anterior end of foot rounded. Penis with a short, straight stylet. Denticles on the cutting edge of the jaw are in the form of fine bristles. Type: *aurantia* (Alder and Hancock, 1842).

Catriona maua Marcus and Marcus, 1960   **4437**
Maua Eolis

Southeast Florida and the Caribbean.

8 to 12 mm., smooth rhinophores; 28 to 30 sausage-shaped cerata on each side stand on cushions. End ⅓ of rhinophores and tentacles flecked with white. Red streak runs up the posterior basal ½ of each rhinophore. Cerata with 2 white bands near the halfway mark and tip. Liver in the cerata is pale-red with cream and red blotches. Oesophagus bright-red. Body silver-gray, faintly suffused with orange. Moderately common. See M. Edmunds, 1964, *Bull. Marine Sci.*, vol. 14, p. 2.

Catriona aurantia (Alder and Hancock, 1842)   **4438**
Orange-tipped Eolis   **Color Plate 16**

Arctic Seas to New Jersey. Europe.

½ inch in length. Branchiae numerous, occurring in 10 or 11 close, transverse rows, anteriorly with 5 to 6 papillae per row, posteriorly with 2 to 4. Radula of 80 plates which are horseshoe-shaped and with 6 strong, straight denticles. Common; in *Tubularia* growths on wharf pilings. *Montagua gouldii* Verrill and Smith 1873, is said to be a synonym. *Eolis aurantiaca* Alder and Hancock, 1854, is the same.

Catriona alpha (Baba and Hamatani, 1963)   **4439**
Alpha Eolis

Santa Barbara to Mission Bay, California. Japan.

10 mm. long, 2 mm. wide, with the cerata in 5 or 6 overlapping but distinct groups. The anterior 2 groups each contain 3 or 4 rows; those following, a single row each. Cerata cylindrical, tapering to rounded tips. Rhinophores simple tapering rods about ⅓ longer than the cephalic tentacles. Ground color of body translucent-white with brown cores in the cerata. Front of each ceras is crustose, opaque, iridescent-white. The third distal end of each rhinophore is bright

4439

orange. Uncommon; subtidal. *C. spadix* (MacFarland, 1966), is a synonym.

Genus *Trinchesia* von Ihering, 1879

Radula uniseriate with cusp as long as, or longer than, the lateral denticles. No small accessory denticles among the lateral denticles of the radular tooth. Denticles on cutting edge of jaw not bristled. Have an acleioproctic anal position. Opinion no. 777 of the I.C.Z.N. (1966) placed *Trinchesia* on the "Official List." For a list of species presently and formerly belonging to *Trinchesia*, see Marcus, 1958, *Amer. Mus. Novitates,* no. 1906, pp. 49–52. Further notes were made by R. A. Roller, 1967, *The Veliger*, vol. 11, p. 421. I have not always assigned the various species to their proper places, since the differences between *Catriona*, *Cuthona* and *Trinchesia* are not understood by me.

Other Atlantic species:

4440 *Trinchesia veronica* Verrill, 1880. South Harpswell, Maine, to Cohasset, Massachusetts. Off Cape Cod, Massachusetts, 23 to 31 fms., among hydroids. Also on the sponge *Halichondria*.

4441 *Trinchesia tina* (Marcus, 1957). Beaufort, North Carolina, to Miami, Florida, to Brazil. 2.5 mm. Translucent-white. (A *Catriona?*)

4442 *Trinchesia perca* (Marcus, 1958). Brazil. 12 mm. *Amer. Mus. Novitates,* no. 1906, p. 45.

Other Pacific species:

4443 *Trinchesia albocrusta* (MacFarland, 1966). Pacific Grove; Monterey, California. 12 mm.

4443

4444 *Trinchesia columbiana* (O'Donoghue, 1922). Vancouver Island, British Columbia, subtidal to 12 fathoms.

4445 *Trinchesia langunae* (O'Donoghue, 1926). Moss Beach to west Baja California. Common; intertidal; summer. (Synonyms: *ronga* Marcus, 1961; *rutila* (MacFarland, 1966).)

4446 *Trinchesia flavovulta* (MacFarland, 1966). Monterey Bay, California. 3 to 10 mm.

4447 *Trinchesia fulgens* (MacFarland, 1966). Monterey Bay, California. 6 mm.

4448 *Trinchesia virens* (MacFarland, 1966). Monterey Bay, California. 5 mm.

4449 *Trinchesia abronia* (MacFarland, 1966). Monterey Bay, California. 5 mm.

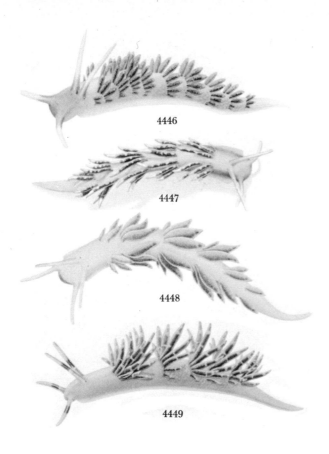

4446

4447

4448

4449

Genus *Cuthona* Alder and Hancock, 1855

Elongate; tentacles slender and simple. Rhinophores simple, no leaflets. Papillae fusiform, arranged in 11 or 12 rows. Head greatly enlarged on the sides. Foot large, with rounded anterior corners. Radula a simple plate with 1 median cusp and with denticulate laterals. No stylet present on penis. Type: *nana* (Alder and Hancock, 1842).

Cuthona concinna (Alder and Hancock, 1843) **4450**
Concise Cuthona

Nova Scotia to New Jersey; Europe.
Vancouver, British Columbia.

10 mm.; papillae purplish brown, tipped with white. Right liver with 4 rows. Total number of papillae rows 9 or 10. Head not expanded. Radula with 29 rows; 5 denticles on each side of the cusp. Intertidal. Lives on hydroid, *Thuiaria*.

Other species:

— *Cuthona rosea* MacFarland, 1966. Monterey Bay, California. On pink hydroids on brown kelps. Mem. 6, Calif. Acad. Sci., p. 326. 34 mm. (is a synonym of *Precuthona divae* Marcus).

4453

4451 *Cuthona pustulata* (Alder and Hancock, 1854). Europe; possibly New England, but only if *Eolis purpurea* Stimpson, 1853, is a synonym.

4452 *Cuthona gymnota* Couthouy, 1838. Massachusetts Bay to New York. May be *concinna* (Alder and Hancock, 1843).

4453 *Cuthona stimpsoni* Verrill, 1880. Nova Scotia to Massachusetts. (*Amer. Jour. Sci. and Arts,* vol. 17, p. 314.)

Genus *Precuthona* Odhner, 1929

Similar to *Cuthona*. Rows of cerata are double. Cerata very numerous, subclaviform. Head large; rhinophores as long as the oral palps. Most of the anterior cerata of the posterior liver may arise in front of the anus. Each liver branch divides into several irregular rows, 1 or 2 of which may lie just anterior to the anus. Type: *peachii* (Alder and Hancock, 1848). Some workers consider this a subgenus of *Cuthona*.

4454 *Precuthona divae* Marcus, 1961. Monterey to Santa Barbara, California. 10 fms. *The Veliger,* vol. 3, supplement, p. 50, pl. 10. 8 to 10 mm. (Synonym: *Cuthona rosea* MacFarland, 1966.)

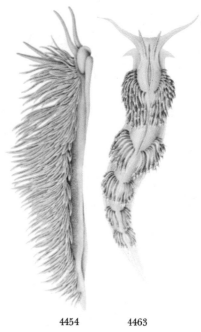

4454   4463

Family FIONIDAE Gray, 1857

Body flattish with a thin undulating edge to the foot. 2 tentacles and rhinophores simple, conical, similar in appearance. Head large. Papillae numerous over the back. Without cnidosacs. Radula uniserial, teeth wishbone shaped and denticulate. Jaws denticulate. One genus.

Genus *Fiona* Forbes and Hanley, 1851

Characters of the family. Type: *pinnata* (Eschscholtz, 1831). *Fidona* H. and A. Adams, 1854, is a synonym.

Fiona pinnata (Eschscholtz in Rathke 1831) **4455**
Atlantic Blue Fiona

Newfoundland to the Caribbean. Gulf Stream.
Alaska to Peru. Pelagic.

1 to 2 inches in length (living), color whitish, yellowish or reddish above. Penis long and filamentous. Feeds on the top surfaces of the siphonophore, *Velella*. After feeding on the stalks of the goose barnacle, *Lepas*, this nudibranch assumes a rich-pink to pastel-red color. The egg capsules are attached on small stalks to *Velella* (see F. M. Bayer, 1963, *Bull. Marine Sci. Gulf and Caribbean*, Miami, vol. 13, no. 3). Synonyms: *nobilis* (Alder and Hancock, 1848); *atlantica* Bergh, 1858; *Limax marinus* Forskål, 1775, non Gunnerus, 1770; *longicauda* Quoy and Gaimard, 1832.

Superfamily Aeolidiacea Orbigny, 1837

Anus dorsal, on right side, behind or among a group of cerata. Known by some as the infraorder Cleioprocta Odhner, 1934. Eolidacea is an unjustified emendation.

Family FACELINIDAE Bergh, 1889

This family includes the genera *Facelina* Alder and Hancock, 1855; *Acanthopsole* Trinchese, 1874; *Phidiana* Gray, 1850; *Phestilla* Bergh, 1874; *Hermissenda* Bergh, 1879; *Caloria* Trinchese, 1888; *Facelinella* Baba, 1949; *Rolandia* Pruvot-Fol, 1951; *Facelinopsis* Pruvot-Fol, 1954; *Learchis* Bergh, 1889; and *Moridilla* Bergh, 1889. The family Phidianidae is synonymous.

Genus *Facelina* Alder and Hancock, 1855

With numerous, slender papillae along each side in 2 or 3 rows. Anterior corners of the foot drawn out into tentacle-shaped processes. 2 tentacles long and slender. Rhinophores without sheaths, with lamellae. Penis armed or unarmed. Radula uniserial, denticulate. Jaws present. Type: *auriculata* (O. F. Müller, 1806) (*coronata* Forbes and Goodsir, 1839 is a synonym). A world list of species is given by Marcus, 1958, *Amer. Mus. Novitates*, no. 1906, pp. 56–59.

Facelina bostoniensis (Couthouy, 1838)   4456
Boston Facelina

Nova Scotia to Connecticut.

25 mm., rhinophores brown with 24 alternatingly large and small lamellae. Body tan with a rosy head. Tentacles with a bluish white dorsal line. Cerata lanceolate, clumped into 5 or 6 oblique rows, each with 8 to 12 cerata. Second posterior cluster of 3 to 6 clumps, each with 5 to 10 cerata. Dorsal cerata longest, internally brown and with a white ring at the tip. Radular teeth 14, tricuspid in the center and flanked by 6 or 7 smaller denticles. On hydroid, *Obelia*, on the kelp, *Laminaria*. 1 to 20 fathoms. (Information courtesy of K. H. Bailey.)

4456

Other species:

4457 *Facelina agari* Smallwood, 1910. Bermuda. Proc. Zool. Soc. London, p. 141. May be a *Palisa*, fide Edmunds, 1964, *Bull. Marine Sci.*, vol. 14, p. 12.

4458 *Facelina goslingii* Verrill, 1901. Hungry Bay, Bermuda. *Trans. Conn. Acad. Sci.*, vol. 11, p. 34.

Genus *Learchis* Bergh, 1889

With an unarmed penis. Radula with a prominent cusp. Rhinophores with a few distantly spaced rings. Type: *indica* Bergh, 1889.

4459 *Learchis poica* Marcus and Marcus, 1960. Virginia Key, Miami, Florida; Curaçao, on hydroids on wharf pilings. Jamaica 9 to 12 mm. Silvery-gray with white flecks, yellow suffusion and orange cheek patch.

Family FAVORINIDAE Bergh, 1889

Rhinophores are not perfoliate, the corners of the foot are not developed into tentacular processes and the radula is uniseriate.

Subfamily FAVORININAE Bergh, 1889

Has only 1 row of cerata to each liver branch. Includes the genera *Cratena* Bergh, 1864; *Favorinus* Gray, 1850 (*Matharena* Bergh, 1875, is a synonym); *Pteraeolidia* Bergh, 1876; *Herviella* Baba, 1949; *Amanda* Macnae, 1954; *Nanuca* Marcus, 1957; *Austraeolis* Burn, 1962; and *Dondice* Portmann and Sandmeier, 1960.

Genus *Cratena* Bergh, 1864

Anus lying between the right and left liver branches. Papillae in rows (not on vertical stalks). Liver canals simple, each with a single row of papillae. Mandibular margin with a single series of denticles. Radula uniseriate; teeth arch-shaped with cusp retracted or similar to the denticles in length. The type is *coerulea* Montagu, 1804. Synonyms: *Montagua* Fleming, 1822, non Leach, 1814; *Amphorina* of Bergh, 1885, not Quatrefages, 1844.

Cratena kaoruae Marcus, 1957   4460

North Carolina to Texas; to Brazil.

15 mm. in length; brownish gray in the dorsal side with green, brown and gray liver diverticula in the cerata. Underside white. Scattered low tubercles on the rhinophores. Cerata in 9 groups, the 4 anterior ones (14 cerata) are in horseshoes, the 5 posterior ones of 2 cerata each are in rows. The innermost cerata are longest. Jaws with minute spines on the denticles of the chewing border. Radula with 17 rows of median and lateral teeth. Penis with a dorsal appendage. Uncommon to common. Shallow water.

Cratena pilata (Gould in Binney, 1870)   4461

Nova Scotia to North Carolina.

20 to 30 mm., elongate, narrow, arched above, of a drab color, margined above with light-yellow. Along the back, beginning between the tentacles and between each tuft of tentacles, is an elongate stripe of carmine-red, margined with silvery

dots. Cerata stout, urn-shaped, with 2 white rings. Cerata arranged in distantly spaced clumps, usually 5 to 7 on each side. Posterior tentacles smooth. Moderately common; shallow water. 1 to 146 fathoms.

Genus *Favorinus* Gray, 1850

Animal with slender cephalic tentacles knobbed at the extremities. With 2 pairs of oral tentacles. Cerata arranged in several oblique rows. Jaws denticulate. Radula uniserial. Type: *albidus* Iredale and O'Donoghue, 1923. *Matharena* Bergh, 1875, is a synonym.

4462 *Favorinus auritulus* Marcus, 1955. Miami, Florida, to Jamaica, to Brazil. 8 mm.

Genus *Hermissenda* Bergh, 1879

Body elongate; rhinophores lamellate; tentacles long and simple. Dorsal cerata arranged in oblique, transverse rows. Anterior angles of the foot elongate. Margins of mandibles denticulate. Radula uniseriate. Penis unarmed. Type: *crassicornis* (Eschscholtz, 1831).

Hermissenda crassicornis (Eschscholtz in Rathke, 1831)   **4463**
Large-horned Hermissenda

Sitka, Alaska, to Baja California.

1 to 2 inches, color variable: anterior end and edges of foot a vivid blue-green; translucent yellow to gray to grass-green. Cerata reddish with blue, green, orange or white specks. Cnidosacs in the tips of the cerata are 0.5 mm. and separated by a white, orange, purple or blue ring. A brilliant white, yellow or blue line runs forward from the tip of the tail. Orange spots sometimes at the bases of the rhinophores which have 24 perfoliations. Anterior border of broad foot is bilabiate and notched. Tail is pointed. Anus is dorsal and situated behind the second group of cerata. Conical cerata arranged in about 11 groups. Spiral egg ribbon white to pink; spawning in the spring. One of California's most common nudibranchs, generally on mud flats. *Flabellina opalescens* Cooper, 1863, is a synonym. For a study of its color patterns, see U. F. Bürgin, 1965, *The Veliger*, vol. 7, p. 205.

Genus *Moridilla* Bergh, 1889

Penis unarmed; cerata in rows; 2 lateral denticles on each radula. Type: *brocki* Bergh, 1889. One Caribbean species recorded:

4464 *Moridilla kristenseni* Marcus and Marcus, 1963. Curaçao, Netherlands West Indies, in algae.

Genus *Phidiana* Gray, 1850

Tentacles long. Rhinophores with leaflets, and nonretractile. Penis with a spine. Members of this genus are cannibalistic. Type: *inca* Orbigny, 1846.

Phidiana pugnax Lance, 1962   **4465**
Pugnaceous Nudibranch

Monterey, California, to the Coronado Islands.

37 to 63 mm., eolidiform. Bright-orange streak on simple head tentacle, head and ½ of clavus. Rhinophores nonretractile, robust, with 11 complete folds. Cerata in 6 major groups,

mostly green with large white spots, but larger, central cerata are orange with white tips. Jaws serrated. Radula consists of 19 single, central, denticulate, horseshoe-shaped centrals. Seasonally common; intertidal to 10 fathoms. *Ph. nigra* MacFarland, 1966, is a synonym. *The Veliger*, vol. 4, p. 157.

4465   4477

Other species:

4466 *Phidiana lynceus* Bergh, 1867. Southeast Florida to Brazil. Pacific side of the Canal Zone. See Marcus and Marcus, 1967, p. 11. (Synonyms: *selencae* Bergh, 1879; *brevicauda* Engel, 1925.) 8 mm. Silvery-gray with suffusion of orange and with vermilion lines. Abundant in mangroves.

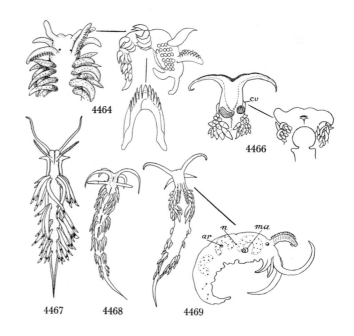

4464

4466

4467   4468   4469

Subfamily FACALANINAE Bergh, 1889

All with produced foot corners. Includes the genera *Facalana* Bergh, 1888; *Godiva* Macnae, 1954; *Echinopsole* Macnae, 1954; *Dondice* Marcus, 1958. (Do not confuse the name with Facelinidae; see above.)

Genus *Dondice* Marcus, 1958

No penial armature; no penial gland. Jaws simple. Rhinophores annulated. Type: *Caloria occidentalis* Engel, 1925.

Dondice occidentalis (Engel, 1925)   4467
Western Dondice

North Carolina to Florida to Brazil.

20 to 30 mm. in length (living), white with a median red stripe along the head and along each side. Under the cnidosacs it is brilliantly white or pink; the cerata bear an orange ring. Digestive gland reddish brown. Tentacles slender. Rhinophores with 15 to 18 rings. Anterior corners of foot with tentacular extensions. Cerata in 6 groups on each side, the first 3 or 4 horseshoe-shaped. The first arch has 25 cerata, the second has 19. Jaws present; black. Radula with denticulate median and 5 to 9 denticles on the lateral tooth. Lives and feeds on hydroids. Moderately common. F. Haas (1920) called Jamaica specimens of this species, *Facelina bostoniensis*.

Genus *Godiva* Macnae, 1954

Penis with a hooklike spine; no penial gland. The vas deferens is in part of the prostate gland. Type: *quadricolor* Barnard, 1927.

4468 *Godiva rubrolineata* Edmunds, 1964. Biscayne Bay, Florida; Port Royal, Jamaica. Mangrove roots. *Bull. Marine Sci. Gulf and Caribbean*, vol. 14, p. 23. 9 mm.; white with orange-red markings.

Genus *Austraeolis* Burn, 1962

Penis with fleshy filaments. Type: *ornata* (Angas, 1864).

4469 *Austraeolis catina* Marcus and Marcus, 1967. Biscayne Bay, Florida, algae on seawall. 5 mm. Grayish and brown.

Family AEOLIDIIDAE Orbigny, 1834

Genus *Aeolidia* Cuvier, 1797

Body depressed, rather broad; branchiae a little flattened, set in numerous, close, transverse rows; 4 tentacles simple; foot broad, anterior angles acute. Radula of a single, broad, pectinate plate. Type: *papillosa* (Linné, 1761). See Opinion 779, *Bull. Zool. Nomen.*, vol. 23, p. 100.

Aeolidia papillosa (Linné, 1761)   4470
Papillose Eolis

Arctic Seas to Maryland. Europe.
Arctic Seas to Santa Barbara, California.

1 to 3 inches in length. Color variable: brown, gray or yellowish, always more or less spotted and freckled with lilac, gray or brown and opaque-white. Number of papillae fewer in young specimens. 30 rows in radula of a single, broad, arched tooth bearing about 46 denticles. Common; offshore from 1 to 208 fathoms. *A. farinacea* (Stimpson, 1854) is a synonym. In the Northwest Pacific it is uncommon intertidally, and referred to the subspecies (**4471**) *herculea* Bergh, 1894. In New Jersey it is common in estuaries.

Genus *Aeolidiella* Bergh, 1867

Dorsal papillae packed closely, fusiform, and with cnidosacs. Genital organs opening between the third and fourth row of papillae (cerata). Rhinophores obliquely wrinkled. Type: *glauca* Alder and Hancock, 1848. *Eolidina* Quatrefages, 1847, is dubious. Two Western Atlantic species:

4472 *Aeolidiella lurana* Marcus and Marcus, 1967. Brazil.

4473 *Aeolidiella occidentalis* Bergh, 1875. St. Thomas, Virgin Islands.

4474 *Aeolidiella takanosimensis* (Baba, 1937). Palos Verdes Peninsula to San Diego, California.

Genus *Spurilla* Bergh, 1864

Aeolid-shaped, elongate, with 2 long smooth anterior tentacles; foot squarely truncated in front. Rhinophores with oblique leaflets. Papillae or cerata pointed, recurved backwards and placed in transverse rows. Jaws present and without denticles. Radula with an arched median bearing many minute, comblike denticles. Type: *neapolitana* (Delle Chiaje, 1823).

Spurilla neapolitana (Delle Chiaje, 1823)   4475
Neapolitan Spurilla

Florida to Texas; Caribbean to Brazil; Eastern Atlantic.

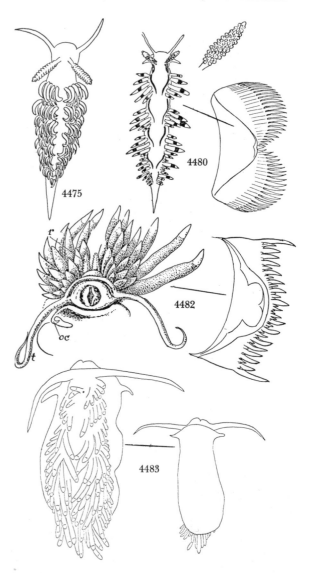

20 to 40 mm., long, ivory, yellowish rose or pinkish color with reddish brown to olive-green diverticula within the cerata which are tipped with white and keeled on the side facing the midline. Opaque white dots and spots on the cerata, head and back. Median radula tooth with 42 to 90 minute denticles on either side of the indented middle. Found on weeds attached to floating logs. Feeds on sea anemones. Synonyms: *sargassicola* Koyer, 1861; *braziliana* MacFarland, 1909; *mograbina* and *dakariensis* Pruvot-Fol, 1953.

Spurilla chromosoma (Cockerell and Eliot, 1905)   **4476**
Chromosome Spurilla

Santa Barbara, California, to Baja California.

15 to 20 mm.; body reddish with a row of white marks on the back; cerata greenish with white tips. Oral tentacles strong and large. Cerata deciduous. Short, thick, pinkish rhinophores bear about 10 oblique perfoliations. Jaws without denticles. About 19 to 21 pectiniform radulae, each bilobed, arched, with 1 or 2 central denticles lower than the rest, and on each side of them from 25 to 32 long, thin lateral denticles. This species is uncommon in California, but abundant in the Gulf of California as far north as San Felipe. Intertidal; on rocks.

Other species:

4477 *Spurilla oliviae* (MacFarland, 1966). Monterey Bay to Santa Barbara County, California. 25 mm. Rare; intertidal.

4478 *Spurilla alba* (Risbec, 1928). Nayarit, west Mexico (G. Sphon, 1971); Indo-Pacific.

Genus *Berghia* Trinchese, 1877

Rhinophores with lamellae and also tuberculate on the back side. Anterior end of foot tentaculiform. Jaws elongate. Radula teeth bilobed, with numerous comblike denticles. Cerata arranged in groups, the genital orifices being in the center of the anterior right group. Type: *coerulescens* (Laurillard, 1830).

4479 *Berghia amakusana* (Baba, 1937). Puertecitos, Gulf of California, Mexico; Sagami Bay and Amakusa, Japan.

4480 *Berghia coerulescens* (Laurillard, 1830). Beaufort, North Carolina, to Texas; West Indies to Brazil; Eastern Atlantic. 10 to 40 mm., yellowish white with scarlet and blue trimmings. For description, see Marcus, 1957, *Jour. Linn. Soc.*, vol. 43.

4481 *Berghia* (*Baeolidia* Bergh, 1891) *benteva* (Marcus, 1958). Cape Hatteras, North Carolina, to São Paulo, Brazil. 15 to 20 mm.

Genus *Cerberilla* Bergh, 1873

Rhinophores foliated, minute. Tentacles long. Cerata numerous and arranged in several transverse rows. Anterior side of the cerata bears small basal outgrowths. Cnidosacs short (0.4 mm). Gonopore lies under the third row; the anus under the sixth. Radula small, with about 15 rows of well-denticulated broad central tooth. Type: *longicirrha* Bergh, 1873, from the Indo-Pacific. *Fenrisia* Bergh, 1888, is a synonym.

4482 *Cerberilla tanna* Marcus and Marcus, 1959. Sabine Jetties, Texas. 25 mm. long. (*Publ. Inst. Marine Sci. Univ. Texas*, vol. 6, p. 259.)

4483 *Cerberilla pungoarena* Collier and Farmer, 1964. Gulf of California. Burrows in sand. 20 mm.

Family GLAUCIDAE Menke, 1828

Aeolids living in warm seas as free-swimmers near the surface; blue in color; with 3 large peduncles on each side bearing numerous thin filaments.

Genus *Glaucus* Forster, 1777

Pelagic. Animal elongated, slender. 4 tentacles conical and very short. Gills slender, cylindrical, supported on 3 pairs of lateral lobes. Penis with horny hook. Type: *atlanticus* Forster, 1777. Synonyms include *Eucharis* Péron, 1807; *Laniogerus* Blainville 1816; *Dadone* and *Nausimacha* both Gistel, 1848.

Glaucus atlanticus Forster, 1777   **4484**
Blue Glaucus

Worldwide, pelagic in warm waters.

2 inches in length, body elongate, head small and without eyes. Tentacles and rhinophores very small. 4 clumps of vivid-blue frills on each side of the body. Dorsal side smooth, striped with dark-blue, light-blue and white. Underside pale grayish blue. With strong jaws and a radula of a center row of about 10 denticulated teeth. Moderately common at certain seasons. Washed ashore with *Janthina*, the purple sea-snail. *G. radiata* Gmelin, 1791, *marginata* Bergh, 1868, and *G. forsteri* Lamarck, 1836, are this species. The name *G. marinus* du Pont, 1777, is invalid. The anatomy is given by W. Macnae, 1954, *Ann. Natal Mus.*, vol. 13, pp. 41–48. Also see *Proc. Linn. Soc. London*, vol. 178, p. 107 (1967). The animals swallow air to gain buoyancy. They store nematocysts obtained from their *Physalia* siphonophore food, and can sting bathers. 12 to 20 eggs laid in each short (20 mm.) uncoiled strings.

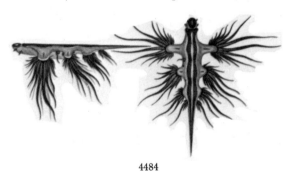

4484

Order *Onchidiata* Abbott, 1954
(Soleolifera)

Family ONCHIDIDAE Gray, 1824

For a review of the family, see R. W. Bretnall, 1919, *Rec. Australian Mus.*, vol. 12. For nomenclature, see H. B. Baker, 1938, *The Nautilus*, vol. 51, p. 85.

Genus *Onchidella* Gray, 1850

Without a shell, animal sluglike, low, oval, with 2 short tentacles or eyestalks at the end of which are the eyes. Mantle entirely covering the back; respiratory, anal and female genital pores at the posterior underside; male pore below the right tentacle and above the sensory lobe. Shallow water to intertidal. Formerly placed in the pulmonates, but now believed to be an early offshoot of the opisthobranchs.

Type: *nigricans* Quoy and Gaimard, 1832 (see Fischer and Crosse, 1878, p. 687). *Onchidiella* Fischer and Crosse, 1878, is a synonym.

Onchidella floridana (Dall, 1885)　　4485
Florida Onchidella

Both sides of Florida and Bermuda.

½ to 1 inch in length, uniform slaty-green to dark-gray; underside bluish white, with a greenish tinge to the veil. Dorsal surface velvety. Mantle margin with about 100 whitish, minute, elongate tubercles. Common along the shore at low tide. Lives in rock crevices in nests, returning home after browsing at low tide. See W. J. Clench, 1937, *The Nautilus*, vol. 50, p. 85. For biology, see Arey and Crozier, 1918, *Bermuda Biol. Station Contri. O. transatlanticum* Heilprin, 1888, is a synonym.

Onchidella carpenteri Binney, 1860　　4486
Carpenter's Onchidella

Puget Sound, Washington, to southern California.

5 mm. in length; body oblong, with its ends circularly rounded; upper surface regularly arched; uniform smoke-gray in color. Fresh specimens are needed to make a better description. Littoral to shallow water. Habits not known. The male copulatory organ with an external caecum and a retractor originating near the hind end separates *carpenteri* from the northern *borealis*. Superficially *carpenteri* is recognizable by the dark, yellow-green back, yellowish brown hyponotum, small head and short sensory lobes.

Onchidella binneyi Stearns, 1893　　4487

Gulf of California.

20 to 30 mm.; broader just behind the middle. Olive-brown to black-brown. Largest warts bear up to 30 minute warts. Other smaller warts single. Abundant; at mid-intertidal zone among protective rocks.

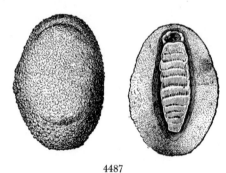

4487

Onchidella borealis Dall, 1871　　4488
Northwest Onchidella

Alaska to northern California.

8 to 12 mm. (½ inch) in length; back regularly arched but a little pointed in the middle, smooth or very finely granulose, tough and coriaceous. Color black or gray, with dots and streaks of yellowish white; foot light-colored, also the head and tentacles. The opening of the mantle cavity is covered by the posterior tip of the foot. Eyes are on the distal end of the short stubby tentacles. On rocks near high-tide mark. Gregarious. Common.

Other species:

4489 *Onchidella hildae* (Hoffmann, 1928). Sonora, Mexico, to Panama and to Ecuador. 26 mm. 20 to 24 long marginal papillae. Greenish white with brown warts on notum.

Genus *Hoffmannola* Strand, 1932

Synonym: *Watsoniella* Hoffmann, 1928. Type of the genus is *lesliei* Stearns, 1872, from the Galapagos Islands.

4490 *Hoffmannola hansi* Marcus and Marcus, 1967. Kino Bay, Sonora, Mexico. High intertidal; abundant. 48 mm. Black with brownish edges. Large, flat, whitish warts on notum. *Studies Trop. Oceanogr. Miami*, no. 6, p. 232.

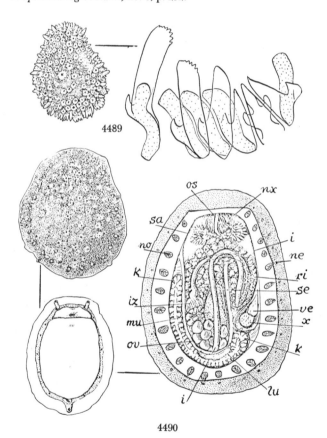

4489

4490

III

Dentaliums and other Tusk-shells

Class SCAPHOPODA Bronn, 1862

Simple mollusks producing a tubular calcareous shell, open at both ends, the larger one for the long conical foot, genital openings, mouth and clusters of feeding appendages known as "captacula." Radula present. No eyes or tentacles. All are benthonic marine species, most deep water. For Atlantic species, see J. B. Henderson, 1920, *Bull. 111, U.S. Nat. Mus.*, pp. 1–177, pls. 1–20. Also see Ruth Turner, 1955, Scaphopods of the *Atlantis* Dredgings in the Western Atlantic . . . , *Papers in Marine Biol. and Oceanogr.*, Pergamon Press, London, pp. 309–320.

Family DENTALIIDAE Gray, 1834

Shell with the greatest diameter at the aperture. Foot conical and with epipodial processes. Median tooth of the radula twice as wide as long.

Genus *Dentalium* Linné, 1758

The shell is an elongate, curved tube open at both ends, and somewhat resembles an elephant's tusk. The diagnostic characters are the type of sculpturing (ribs, riblets and circular threads or incised lines), the form of the apex, the degree of curvature, the size and thickness of shell and the position and form of the apical slit. The 10 or so subgenera are nebulous in character and definition, and one should consult the works of J. B. Henderson (1920, Bull. 111, U.S. Nat. Mus.), H. A. Pilsbry, W. H. Dall and W. K. Emerson. Type: *elephantinum* Linné, 1758. *Paradentalium* Cotton and Godfrey, 1933, is a synonym.

For details on the use of *Dentalium* by Northwest American Indians, see R. B. Clark, 1963, "The Economics of *Dentalium*," *The Veliger*, vol. 6, pp. 9–19.

Subgenus *Dentalium* Linné, 1758

Apex modified by a slit or notch.

Dentalium laqueatum Verrill, 1885   4491
Paneled Tusk

North Carolina to south Florida and the West Indies.

1 to 2½ inches in length, thick-shelled and dull-white in color. Apex sharply curved; anterior ⅔ of shell slightly arched. 9 to 12 strong, elevated, primary longitudinal ribs with equally spaced, concave intercostal (space between ribs) spaces. Ribs fade out at the anterior ⅓. There are fine reticulations over the entire shell. A supplemental tube is present in the young shells. Abundant in sandy mud from 4 to 200 fathoms. *D. regulare* Henderson, 1920, is a synonym.

Dentalium texasianum Philippi, 1848   4492
Texas Tusk

North Carolina to Texas.

¾ to 1½ inches in length, thick-shelled, well-curved, hexagonal in cross-section and dull, grayish white in color. The broad spaces between the ribs are flat. Common from 3 to 10 fathoms. The subspecies *cestum* Henderson, 1920, from Texas has numerous, cordlike riblets between the 6 main ribs, thus obscuring the hexagonal section of the shell. This latter may only be a form.

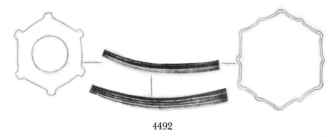

4492

Dentalium neohexagonum Pilsbry and Sharp, 1897   4493
Six-sided Tusk

Monterey, California, to the Gulf of California.

25 to 35 mm. (1 inch), moderately curved, slender (length 12 to 14 times the greatest diameter); white; sculpture of 6 strong, rounded ribs, which on the larger ½ or ⅓ of the adult shell become reduced to mere rounded angles. Aperture

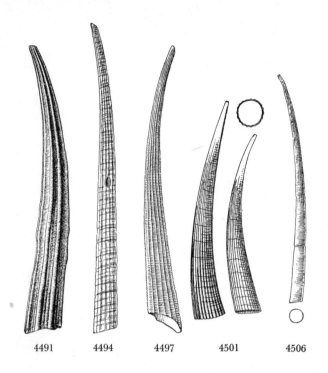

4491   4494   4497   4501   4506

6-sided, but the angles so rounded as to appear almost circular. Anal orifice roundly oval, without a notch or slit. Larger specimens occur in deeper water. Common; 1 to 100 fathoms. (Synonym: *pseudohexagonum* Arnold, 1903.)

Other *Dentalium*:

4494 *Dentalium gouldii* Dall, 1889. Off North Carolina, 12 fms. **(4495):** subsp. *portoricense* Henderson, 1920. Puerto Rico, 33 to 160 fms.

4496 *Dentalium rebeccaense* Henderson, 1920. Florida Keys, 7 to 16 fms.

4496

4497 *Dentalium (Coccodentalium* Sacco, 1896) *carduum* Dall, 1889. Off Fort Lauderdale, Florida, 155 fms., to the West Indies, 116 to 338 fms.

4498 *Dentalium (Dentalium) agassizi* Pilsbry and Sharp, 1897. Off Santa Barbara, California, to Panama, in 589 to 2,320 meters.

4499 *Dentalium (Dentalium) oerstedii* Mörch, 1860. Gulf of California to Ecuador, in 4 to 145 meters. (Synonym: *numerosum* Pilsbry and Sharp, 1897.)

Subgenus *Antalis* H. and A. Adams, 1854

Weakly ribbed; apex with a V-shaped notch on or near the convex side and a solid plug with a central pipe or orifice. Type: *entalis* Linné, 1758. Rejected names include *Dentale* Da Costa, 1778; *Antalis* Hermannsen, 1846.

Dentalium entale stimpsoni Henderson, 1920   **4500**
Stimpson's Tusk

Nova Scotia to Cape Cod, Massachusetts.

1 to 2 inches in length, round in cross-section and dull, ivory-white in color. Region of the apex always very eroded and chalky. Surface uneven and with some longitudinal wrinkles in better preserved specimens. A poor subspecies of the north European *D. entale* Linné. Common from 8 to 1,200 fathoms. *D. striolatum* Stimpson, 1851 (non Risso, 1826) is a synonym.

Dentalium occidentale Stimpson, 1851   **4501**
Western Atlantic Tusk

Newfoundland to off Cape Hatteras, North Carolina.

1 to 1½ inches in length. Primary ribs 16 to 18, fairly distinct in the young stages; sculptureless in the senile stage. Round in cross-section. Chalky-white when eroded. Common from 20 to 1,000 fathoms. *D. georgiense* Henderson, 1920, is probably a synonym.

Dentalium antillarum Orbigny, 1842   **4502**
Antillean Tusk

South half of Florida and the West Indies.

About 1 inch in length, roundish in cross-section. Primary ribs 9, but increasing to 12 near the middle and finally with 24 near the aperture. Microscopic, transverse lines between the ribs. Color opaque-white, rarely reflecting a greenish tint. Encircled with weak, zigzag bands or splotches of translucent-gray. Common; shallow water.

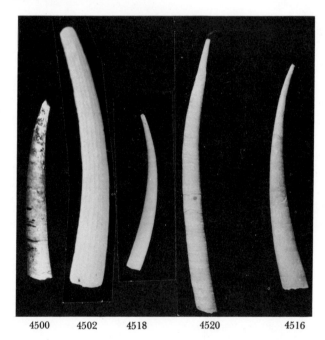

4500   4502   4518   4520   4516

Dentalium pilsbryi Rehder, 1942   **4503**
Pilsbry's Tusk

West Florida and Brazil only.

¾ to 1½ inches in length, roundish in cross-section. Primary ribs 9, with a smaller, weaker, rounded, secondary rib appearing between each. All ribs fade out toward the anterior end. Intercostal spaces flat, crossed by coarse growth lines. No transverse microscopic sculpture. Color opaque-white; without gray splotches. Formerly known as *D. pseudohexagonum* Henderson, 1920, not Arnold, 1903. Moderately common from 1 to 5 fathoms. (*The Nautilus*, vol. 56, p. 69.)

Other Atlantic *Antalis*:

4504 *Dentalium subagile* Henderson, 1920. Prince Edward Island, Canada, to Maine, 57 to 134 fms.

4505 *Dentalium disparile* Orbigny, 1842. Tampa Bay (?) to the West Indies and Brazil.

4506 *Dentalium ceratum* Dall, 1881. East and west Florida to the West Indies, 22 to 805 fms. (Synonym: *flavum* Henderson, 1920.)

4507 *Dentalium taphrium* Dall, 1889. Off Cape Hatteras, North Carolina, to Yucatan and the Florida Keys, 8 to 640 fms. Common. 17 mm.

4508 *Dentalium bartletti* Henderson, 1920. Off Cape Fear, North Carolina, to the Gulf of Mexico and West Indies, 17 to 600 fms. Common.

4509 *Dentalium tubulatum* Henderson, 1920. Off Fernandina, Florida, 294 fms.; off Cuba, 220 fms.

4510 *Dentalium sericatum* Dall, 1881. Gulf of Mexico, off Yucatan, 640 fms.

4511 *Dentalium* (*Heteroschisma* Simroth, 1885) *callithrix* Dall, 1889. Off North Carolina to the Gulf of Mexico and the West Indies, 220 to 1,591 fms.

Dentalium pretiosum Sowerby, 1860 **4512**
Indian Money Tusk

Alaska to Baja California.

About 2 inches in length, moderately curved and solid; opaque-white, ivory-like, commonly with faint dirty-buff rings of growth. Apex with a short notch on the convex side. A common offshore species which was used extensively by the northwest Indians for money. Depth of 1 to 80 fathoms. A southern subspecies (Monterey, California, to the Gulf of California, in 37 to 298 meters) **(4513)** was named *berryi* A. G. Smith and Gordon, 1948.

Dentalium semipolitum Broderip and Sowerby, 1829 **4514**
Semi-polished Tusk

Monterey, California, to Costa Rica.

1 to 1½ inches, slender, round in cross-section, curved, thin-shelled, very glossy, and milk-white. Sculpture of numerous, fine, close, subequal, longitudinal striae, extending from the apex to ⅓ to ⅔ the shell's length. Large ⅓ of shell smooth, brilliantly polished. Aperture circular, peristome thin. Anal orifice minute, round and unnotched. *D. hannai* Fred Baker, 1925, is a synonym. Common; shore to 35 fathoms.

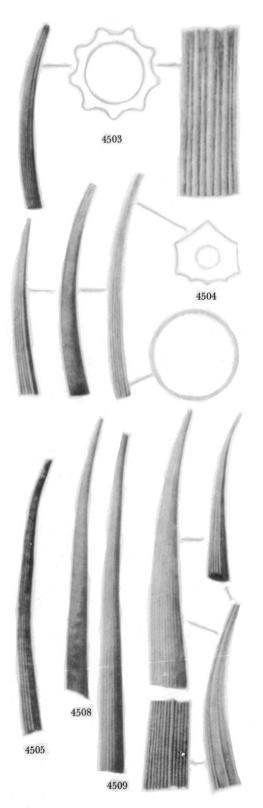

4503

4504

4508

4505

4509

4515

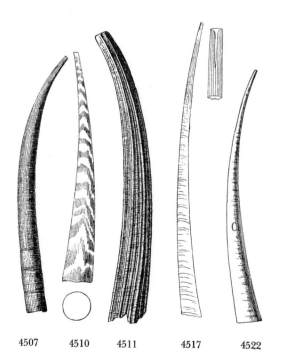

4507   4510   4511   4517   4522

Subgenus *Fissidentalium* Fischer, 1885

Numerous longitudinal riblets; apex typically with a long slit, commonly simple. Type: *ergasticum* Fischer, 1885.

Dentalium floridense Henderson, 1920　　　4515
Florida Tusk

　Southeast Florida and the West Indies.

　2 to 3 inches in length, roundish in cross-section. Shell hard and yellowish white. Apex hexagonal with concave spaces between. Ribs increase to 24 anteriorly and are rounded, equal-sized and crowded. There is a long, narrow apical slit on the convex side. Rare from 35 to 100 fathoms. Erroneously called *capillosum* Jeffreys by earlier workers.

Dentalium meridionale Pilsbry and Sharp, 1897　　　4516
Meridian Tusk

　Massachusetts to Brazil.

　3 to 4 inches, mouse-gray, oval in cross-section. Tip 16 ribbed, maximum of 90 riblets. Glossy mouse-gray periostracum. Apical notch deep, on the convex side. The northern subspecies **(4081)**, *verrilli* Henderson, 1920, extends from Massachusetts to North Carolina in 705 to 1,500 fathoms and has the riblets fading out at the anterior third.

Subgenus *Graptacme* Pilsbry and Sharp, 1897

Shell with close, fine, deeply engraved, longitudinal striae near the apex, remainder smooth. Type: *eboreum* Conrad, 1846.

Dentalium eboreum Conrad, 1846　　　4517
Ivory Tusk

　North Carolina to Florida, Texas and the West Indies.

　1 to 2½ inches in length, glossy, ivory-white to pinkish. Apical slit deep, narrow and on the convex side. Apical end with about 20 very fine longitudinal scratches. Common in sandy, shallow areas. Synonyms are: *leptum* Bush, 1885; *matara* Dall, 1889.

Dentalium semistriolatum Guilding, 1834　　　4518
Half-scratched Tusk

　South Florida and the West Indies.

　About 1 inch in length. Similar to *eboreum,* but curved more, with apical slits on the side, and its color translucent-white with milky patches. Some specimens may be reddish near the apical end. Common from 1 to 90 fathoms.

Dentalium calamus Dall, 1889　　　4519
Reed Tusk

　North Carolina to east Florida and the Greater Antilles.

　¾ to 1 inch in length, almost straight and glassy-white. Most of the shell has minute, longitudinal scratches (about 16 per mm.). The apical end is sealed over by a bulbous cap which bears a small slit. Uncommon; 1 to 25 fathoms.

Dentalium vallicolens Raymond, 1904　　　4520
Raymond's Tusk

　Washington to the Gulf of California.

　2 to 2½ inches, rather slender, moderately curved posteriorly, the latter ½ nearly straight; cream-white, often yellow-

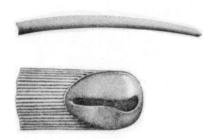

4519

ish toward the mouth. Shiny where not eroded, earlier portion usually dull and chalky. At the apex there are low, rounded, longitudinal threads, of which 7 or 8 are more prominent and 3 to 6 in each interspace are less prominent. Sculpture fades anteriorly, except for microscopic numerous striae. Common in fine gravel and sand; 50 to 265 fathoms.

Other Graptacme:

4521　*Dentalium circumcinctum* Watson, 1879. Off Bermuda, 1,075 fms.; West Indies, 470 fms.

4524　　　4514　　　4513

Subgenus *Laevidentalium* Cossmann, 1888

Shell entirely smooth, even at the apical tip, showing growth lines only. Oval to round in cross-section. Type: *incertum* Deshayes, 1825.

Dentalium callipeplum Dall, 1889　　　4522
Glory-of-the-Atlantic Tusk

　Off both sides of Florida; off Mississippi to the West Indies.

　2 to 3 inches; exceptionally brilliant surface is ivory with a salmon tint on the tip. Evenly curved like a scimitar, rapidly increasing in size, round in cross-section. Apical notch barely indicated on the concave side. A gorgeous, uncommon species; 25 to 2,075 fathoms.

Dentalium perlongum Dall, 1881　　　4523
Very Slender Tusk

　North Carolina to Uruguay.

　3½ inches, extremely long and slender (90 mm. in length, 4 mm. in diameter) and almost straight. Entirely smooth and glossy, opaque-white. Apical notch on the convex side. Uncommon; 11 to 1,330 fathoms.

Dentalium rectius Carpenter, 1864   **4524**
Western Straight Tusk

Alaska to Panama.

1 to 1½ inches, almost straight, glossy, smoothish, slender and long, attenuated toward the apex. Thin-shelled and fragile; bluish white, somewhat translucent, with some opaque white flecks or rings, often encrusted near the aperture with a reddish deposit. Apical orifice small, circular, without a notch. *D. watsoni* Pilsbry and Sharp, 1897, are extra-long *rectius* (length is 16 to 19 times the diameter).

Subgenus *Bathoxiphus* Pilsbry and Sharp, 1897

Shell thin, conspicuously compressed laterally and nearly smooth. Broad apical slit on the convex side. Type: *ensiculus* Jeffreys, 1877.

Dentalium didymum Watson, 1879   **4525**
Watson's Flattened Tusk

Lower Florida Keys to Barbados.

1 inch, extremely attenuated, slightly curved, and a little flattened laterally, especially toward the convex curve, giving a trigonal section rather than that of a flattened oval. Glossy-white and sculptureless. Uncommon; 3 to 390 fathoms.

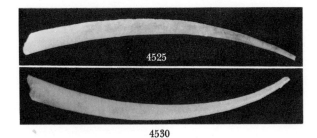

4525

4530

Dentalium ensiculus Jeffreys, 1877   **4526**
Jeffrey's Razor Tusk

Georges Banks, Massachusetts, to the West Indies. Western Europe.

1 inch, arched, flattened laterally, with a slight keel on both the convex and concave sides. Aperture oval. Grayish white. Apical notch very wide and on the convex side. Uncommon; 193 to 1,813 fathoms.

4526   4528

Subgenus *Compressidens* Pilsbry and Sharp, 1897

Conspicuously compressed between the convex and concave sides (not laterally). Shell small, tapering and weakly sculptured. Type: *pressum* Pilsbry and Sharp, 1897. Two Atlantic species:

4527 *Dentalium pressum* Pilsbry and Sharp, 1897. East Florida to the Gulf of Mexico. Puerto Rico, 280 fms. ½ inch, strongly compressed, moderately curved. (Synonym: *compressum* Watson, 1879, non Orbigny, 1850.) Uncommon.

4527

4528 *Dentalium ophiodon* Dall, 1881. East and west Florida to Barbados, 70 to 300 fms. ½ inch, less compressed, considerably curved; white. Moderately common.

Subgenus *Fustiaria* Stoliczka, 1868

Apical slit very long, about ¼ the length of the shell. Surface glossy smooth. Type: *circinatum* Sowerby, 1823. Only one American species:

4529 *Dentalium stenoschizum* Pilsbry and Sharp, 1897. Florida Keys to Barbados, 100 to 110 fms. 1 inch, strongly curved, round in section, shiny white. Uncommon.

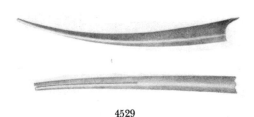

4529

Subgenus *Episiphon* Pilsbry and Sharp, 1897

Shells very small, needlelike, wholly lacking longitudinal sculpture and, as in some other subgenera, having a projecting, thin tube at the posterior end after the tip is broken or lost. Only one species in the Western Atlantic, the type of the subgenus: *sowerbyi* Guilding, 1834.

Dentalium sowerbyi Guilding, 1834   **4530**
Sowerby's Tusk

North Carolina and Texas to Florida and the Lesser Antilles.

10 to 15 mm. in length. Needlelike, not fragile, curved, glossy-white. Crowded rings of growth microscopic on tip. Apex without slit and from it projects a very thin inner tube. Erroneously known previously as *D. filum* Sowerby. Commonly dredged from 17 to 180 fathoms.

Other species:

4531 *Dentalium (Episiphon) johnsoni* Emerson, 1952. Off Puerto Rico, 180 to 470 fms. 24 mm. *Smithsonian Misc. Coll.,* vol. 117, no. 6, p. 5.

Subgenus *Tesseracme* Pilsbry and Sharp, 1898

Posterior end quadrangular in cross-section, anterior subcircular, primary ribs rounded, rarely serrate. 4 sides with longitudinal riblets increasing anteriorly by intercalation. Apical orifice unnotched, commonly with a short terminal pipe. Type: *D. quadrapicale* Sowerby, 1860, from the Indian Ocean. Three species in the Eastern Pacific:

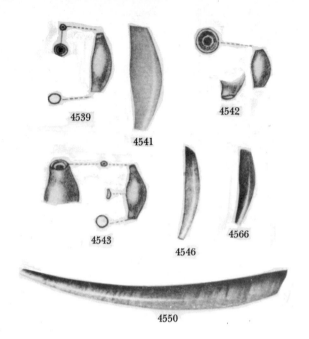

4539   4542

4541

4543   4566

4546

4550

4532 *Dentalium hancocki* Emerson, 1956. Both sides of Baja California, 5 to 20 fms. Also Pleistocene. 14 mm. Has numerous microscopic pits in the interstitial grooves separating the transverse striae. 16 to 20 riblets. *Amer. Mus. Novitates,* no. 1787.

4533 *Dentalium quadrangulare* Sowerby, 1832. Los Animas Bay, Baja California to Ecuador. 24 to 30 riblets on mid-portion. (Synonym: *fisheri* Pilsbry and Sharp, 1897.)

4534 *Dentalium tesseragonum* Sowerby, 1832. Acapulco, Mexico, to Ecuador. 4 to 8 weak longitudinal threads on each of the 4 sides at the mid-section.

Other Pacific *Dentalium*:
(subgenera not determined)

4535 *Dentalium inversum* Deshayes, 1825. Bering Sea to Panama. 15 to 30 fms., to very deep water.

4536 *Dentalium dalli* Pilsbry and Sharp, 1897. Bering Sea, to Peru. 15 to 30 fms. in Alaska.

4537 *Dentalium* (*?Fissidentalium*) *megathyris* Dall, 1890. Off California to the Gulf of California and to Chile, in 1,467 to 2,080 meters.

Family SIPHONODENTALIIDAE Simroth, 1894

Shell either like *Dentalium,* a tusk-shaped shell open at both ends, or swollen in the middle and entirely smooth. The foot is wormlike and can be expanded at the end into a round disc. The median tooth of the radula is almost as long as wide.

Genus *Cadulus* Philippi, 1844

Shell small, white, without sculpture and swollen in the middle somewhat like a cucumber. Aperture constricted and very oblique. Type: *ovulum* Philippi, 1844. The genus is divided into four subgenera as follows:

1. Apex with 2 deep slits . . . *Dischides* Jeffreys, 1867
2. Apex with 4 deep slits . . .
       . . . . . *Polyschides* Pilsbry and Sharp, 1897
3. Apex with 2 or 4 shallow slits . . . . .
       . . . . *Platyschides* Henderson, 1920
4. Apex with slits:
   a, Obese, convex on both sides . . . *Cadulus s. s.*
   b, Slender, almost flat on one side . . . . .
       . . . . *Gadila* Gray, 1847

Subgenus *Cadulus* Philippi, 1844

Shells small, very obese and bulbous in the center. Type: *ovulum* Philippi, 1844, from the Mediterranean. The 6 east American species are in very deep water:

4538 *Cadulus transitorius* Henderson, 1920. Off Fernandina, Florida, to the Gulf of Mexico; Bahamas, 90 to 660 fms.

4539 *Cadulus obesus* Watson, 1879. Off Fernandina, Florida, 294 fms.; off Cuba, 200 fms.

4540 *Cadulus cucurbitus* Dall, 1881. Off Bahia Honda, Cuba, 310 fms.

4541 *Cadulus platensis* Henderson, 1920. Off Georgia and Fernandina, Florida, 294 to 440 fms.

4542 *Cadulus ampullaceus* Watson, 1879. Off north and southwest Cuba, 190 to 1,650 fms.; Barbados.

4543 *Cadulus exiguus* Watson, 1879. Off the Isle of Pines, Cuba, 2,050 fms., to Barbados, 180 to 300 fms.

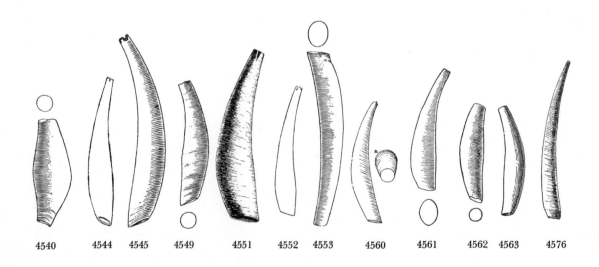

4540   4544   4545   4549   4551   4552   4553   4560   4561   4562   4563   4576

Subgenus *Polyschides* Pilsbry and Sharp, 1898

Apex with 4 prominent slits producing 4 prominent triangular lobes. Type: *tetraschistus* Watson, 1879.

Cadulus carolinensis Bush, 1885   **4544**
Carolina Cadulus

North Carolina to Florida and to Texas.

10 mm. in length. Slightly swollen. Apex with 4 shallow slits. In cross-section the shell is roundish. Commonly dredged from 3 to 100 fathoms.

Cadulus quadridentatus (Dall, 1881)   **4545**
Four-toothed Cadulus

North Carolina to both sides of Florida and the West Indies. Bermuda.

5 to 10 mm. in length, swollen behind the aperture. Apex with 4 well-defined slits. In cross-section the shell is roundish. Commonly dredged from 3 to 50 fathoms. *C. acompsus* Henderson, 1920, from Central America (Atlantic side) is a synonym.

Cadulus tetrodon Pilsbry and Sharp, 1898   **4546**
Tetrodon Cadulus

North Carolina to Florida; Bahamas.

5 to 7 mm., slender and slightly curved. Convex side evenly arched, concave side straighter, with a slight convexity at a point about ⅓ the total distance from the aperture. This amounts to but a slight swelling, from which the shell tapers

4548

gradually to the two ends. Aperture oblique, round and its peristome blunt and rounded. Apex cut by 4 slits, leaving 4 triangular lobes, the one on the convex side being slightly longer and rounded, the lobe opposite it broader and flat-topped, while the 2 lateral lobes are smaller. 1 to 100 fathoms; common.

Cadulus quadrifissatus Pilsbry and Sharp, 1898   **4547**

Monterey, California, to Baja California.

6 to 9 mm., slender, arcuate, not swollen; aperture oval, oblique, slightly constricted. Apex cut into 4 conic teeth by 4 slits. Variable in shape. Common; sandy bottoms; 6 to 30 fathoms.

Subgenus *Platyschides* Henderson, 1920

Midway between *Polyschides* and *Gadila* in that the apical teeth are greatly reduced to 2 or 4 small, shallow dips or lobes. This is probably not a good subgenus. Type: *grandis* Verrill, 1884.

Cadulus grandis Verrill, 1884   **4548**
Verrill's Giant Cadulus

Off Massachusetts to North Carolina.

12 to 15 mm., sometimes thick and heavy, moderately curved. Posterior aperture large, giving the shell a "stumpy" appear-

ance. 4 rounded, shallow apical slits, leaving broad flat lobes, the widest of which is on the convex side. Edge of this lobe is serrated. A common dweller in gray mud from 820 to 1,467 fathoms.

Cadulus agassizii Dall, 1881   **4549**
Agassiz's Cadulus

Georges Banks, Massachusetts, to both sides of Florida.

7 to 11 mm., moderately curved; maximum diameter at about the anterior ⅓. Anterior aperture constricted, oblique and round. Apical orifice slightly flattened and large. 4 apical slits, broad and shallow. Lateral 2 lobes reduced to mere points. *C. hatterasensis* Pilsbry and Sharp, 1898, is a synonym. Abundant; sand bottom; 17 to 293 fathoms. (See R. L. Wigley, 1966, *The Nautilus*, vol. 79, p. 90.)

Cadulus elongatus Henderson, 1920   **4550**
Elongate Cadulus

Mississippi to south Texas.

14 mm. long, but its diameter only 1.5 to 1.6 mm., long and slender. Apical orifice small (0.4 mm.). On the convex side the shell is much flattened between the equator and the very oblique aperture, causing the latter to be an irregular oval. Rare and a collector's item; 50 to 68 fathoms.

Cadulus californicus Pilsbry and Sharp, 1898   **4551**
Californian Cadulus

Alaska to Ecuador.

14 to 16 mm., solid, stout (4 times as long as wide), swollen anteriorly. Bluish white. Equator about at the anterior ¼. Aperture subcircular, oblique. Anal orifice large, slightly oval,

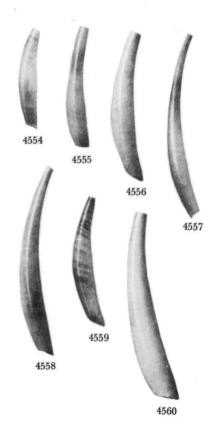
4554
4555
4556
4557
4558
4559
4560

its edge irregular from breakage, but 2 lateral nicks are normally present. Common; 60 to 634 fathoms.

Other species of *Platyschides*:

4552 *Cadulus spectabilis* Verrill, 1885. Off Georges Bank to 125 miles east of Virginia, 1,525 to 1,859 fms.

4553 *Cadulus aequalis* Dall, 1881. East coast of Florida to Dry Tortugas, 339 fms.

4554 *Cadulus parvus* (Florida–West Indies, 33 to 1,002 fms.); *foweyensis* (**4555**) (Florida Keys, 55 to 99 fms.); *providensis* (**4556**) (Florida, 390 fms.); *greenlawi* (**4557**) (Florida Keys, 130 fms.); *arctus* (**4558**) (Gulf of Mexico, 60 fms.); *miamiensis* (**4558a**) (South Carolina to Florida, 85 to 209 fms.) all Henderson, 1920.

4559 *Cadulus rushii* Pilsbry and Sharp, 1898. Gulf of Maine to North Carolina, 35 to 1,060 fms. Subspecies *arne* Henderson, 1920. Massachusetts to Florida, 197 to 984 fms.

4560 *Cadulus pandionis* Verrill and Smith, 1880. Martha's Vineyard, Massachusetts to Florida Keys, 67 to 386 fms. Common. 10 mm.

4561 *Cadulus watsoni* Dall, 1881. Cuba and Bahamas, 382 to 413 fms.

4562 *Cadulus lunula* Dall, 1881. Off Havana, Cuba, 805 fms.

4563 *Cadulus amiantus* Dall, 1889. Off Bahia Honda, Cuba, 310 fms. Off west Florida, 220 fms. 5 mm.

4564 *Cadulus poculum* Dall, 1889. Gulf of Mexico, 640 fms. to the West Indies, 424 fms.

4565 *Cadulus austinclarki* Emerson, 1951. Gulf of California to Panama; Galapagos Islands.

Subgenus *Gadila* Gray, 1847

Shells minute, the simple rim of the apical orifice has no slits or lobes. Usually rather slender. Type: *gadus* Montagu, 1803.

Cadulus mayori Henderson, 1920   **4566**
Mayor's Cadulus

Southeast Florida to Texas; Cuba.

3 to 4 mm. in length, swollen just anterior to the middle of the shell. Apical opening ⅔ the size of the aperture and usually has 1 or 2 callus rings within the opening. Fairly common from 16 to 100 fathoms.

Cadulus fusiformis Pilsbry and Sharp, 1898   **4567**
Fusiform Cadulus

Monterey, California, to Baja California.

10 mm., long and slender at the anterior ½; greatest diameter 9 times the shell's length. Glossy bluish white. Anal orifice circular, smooth, without notches. Common; 5 to 100 fathoms.

Other Atlantic *Gadila*:

4568 *Cadulus rastridens* Watson, 1879. Off Fernandina, Florida, 294 fms. Off Georgia, 440 fms.

4569 *Cadulus minusculus* Dall, 1889. Off North Carolina, 63 fms.

4570 *Cadulus iota* Henderson, 1920. Off Key West and Miami, Florida, 37 to 45 fms.; Gulf of Mexico, 30 to 100 fms., common. 2 mm. Puerto Rico, 25 fms. (**4571**) Subspecies *nanus* Clench and Aguayo, 1939. Cuba, 40 fms.

4572 *Cadulus verrilli* Henderson, 1920. Off Martha's Vineyard, Massachusetts, 115 fms., to 76 mi. south of Block Island, Rhode Island, 180 fms. (Turner, R. D. 1955). (Synonym: *propinquus* Verrill, 1880, non Sars, 1878.)

4573 *Cadulus regularis* Henderson, 1920. Off Cape Canaveral and off Fernandina, Florida, 294 and 504 fms.

4574 *Cadulus atlanticus* Henderson, 1920. Off Nantucket, Massachusetts, to Cape Hatteras, North Carolina, 428 to 1,004 fms. (Synonym: *gracilis* Dall, 1889, non Jeffreys, 1877.)

4575 *Cadulus cylindratus* Jeffreys, 1877. Off south Massachusetts, 1,525 to 1,608 fms.

4576 *Cadulus* (*Gadilopsis* Woodring, 1925) *acus* Dall, 1889. Off Fernandina, Florida, 294 fms., to Puerto Rico, 17 to 25 fms.

Other Pacific *Gadila*:

4577 *Cadulus tolmei* Dall, 1897. Vancouver Island, British Columbia, to Baja California. ((**4578**) var. *newcombei* Pilsbry and Sharp, 1898, British Columbia.)

4579 *Cadulus aberrans* Whiteaves, 1887. British Columbia to San Clemente Island, California. Offshore to 25 fms. *Trans. Royal Soc. Canada*, vol. 4, sec. 4, p. 124.

4580 *Cadulus hepburni* Dall, 1897. Knight Island, Alaska, to Catalina Island, California.

4581 *Cadulus stearnsii* (Pilsbry and Sharp, 1898). British Columbia to Baja California. (Synonym: *Dentalium simplex* Pilsbry and Stearns, 1897, non Michelotti, 1861.)

4582 *Cadulus perpusillus* Sowerby, 1832. Monterey, California, to Ecuador. Scarce. In California, 25 to 50 fms.

Subgenus *Striocadulus* Emerson, 1962

About 1 inch, moderately curved, surface sculptured with many longitudinal striae; aperture as in *Gadila;* orifice simple. Type of the subgenus: *albicomatus* Dall, 1890.

4583 *Cadulus* (*Striocadulus*) *albicomatus* Dall, 1890. Off California to Ecuador, in 733 to 3,050 meters. The date 1889 is incorrect for this species.

4584 *Cadulus* (*Striocadulus*) *striatus* Pilsbry and Sharp, 1898. Off Acapulco, Mexico, to the Gulf of Panama, in 589 to 1,210 meters.

Genus *Entalina* Monterosato, 1872

Foot of animal expands into a disc, with a digitate periphery, and has a median filament. Shell like *Dentalium*, 4- or 5-sided and strongly ribbed. Type: *tetragonum* Brocchi, 1814. One deep-water Western Atlantic species:

4585 *Entalina platamodes* Watson, 1879. Off East Florida to Cuba and Puerto Rico, 294 to 360 fms. (Synonym: *Entalina quadrata* Henderson, 1920.)

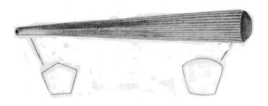

4585

Genus *Siphonodentalium* Sars, 1859

Animal with unique characters common to the family. Shell strongly arcuate, slightly tapering with the apertural end the largest. Circular in section and without sculpture. Apex large, simple or with slits or lobes. Type: *lobatum* Sowerby, 1860. The subgenus *Pulsellum* Stoliczka, 1868 (with type, *lofotense* Sars, 1864) has no apical slits.

4586 *Siphonodentalium lobatum* (Sowerby, 1860). Arctic Ocean to off North Carolina; southern Ireland; Portugal. 15 to 1,813 fms. (Synonym: *D. vitreum* M. Sars, 1851, non Gmelin, 1791.)

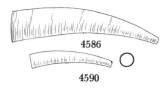

4587 *Siphonodentalium verrilli* Henderson, 1920. Off Massachusetts to Cape Hatteras, North Carolina, 384 to 1,290 fms.

4588 *Siphonodentalium striatinum* Henderson, 1920. Off Fernandina, Florida, 294 fms.

4589 *Siphonodentalium* (*Pulsellum*) *occidentale* Henderson, 1920. Martha's Vineyard, Massachusetts, to off New Jersey, 86 to 984 fms. (*lofotensis* of authors, non Sars).

4590 *Siphonodentalium* (*Pulsellum*) *bushi* Henderson, 1920. Off Georges Banks, Massachusetts, to off Cape Hatteras, North Carolina, 319 to 984 fms. (Synonym: *affinis* Verrill, 1880, non Sars.)

4591 *Siphonodentalium quadrifissatum* (Pilsbry and Sharp, 1898). Monterey Bay, California, to Baja California, in 4 to 365 meters.

IV

Chitons or Coat-of-Mail Shells

Class POLYPLACOPHORA Blainville, 1816

Bilaterally symmetrical, elongate mollusks with an external mantle bearing shelly plates or calcareous spicules. Foot expanded or aborted. Gills lateral. Mouth anterior, usually with a radula. Anus posterior and median. No cephalic eyes or tentacles. All species are marine. The outlines and descriptions of the higher categories have been largely modified from A. G. Smith, 1960, section on Amphineura, Part I (letter, not number), Mollusca, no. 1, "Treatise on Invertebrate Paleontology," Geological Society of America, New York. Also consult Pilsbry's "Manual of Conchology," series 1 on marines, vols. 14 and 15.

Chitons, usually with 8 shelly plates, encircled by a girdle. Foot flat. Gills numerous and on either side between the girdle and foot. Head present. Radula present. Adults range in size from ¼ to 14 inches. Synonyms: *Loricata* Schumacher, 1817; *Crepipoda* Goldfuss, 1820.

For a review of the Caribbean species, see P. Kaas, 1972, Caraibisch Marien-Biol. Inst. Studies Fauna Curaçao, no. 41, pp. 1–162.

Order *Neoloricata* Bergenhayn, 1955

Valves have an inner crystalline shell layer which extends into insertion plates.

Suborder *Lepidopleurina* Pilsbry, 1892

Family LEPIDOPLEURIDAE Pilsbry, 1892

Gills short and posterior. Animal usually 1 to 2 inches long. Insertion plates absent or very weak and unslit. Girdle narrow, minutely spiculed or scaly. Sculpturing usually granular or quincunx.

Genus *Lepidopleurus* Risso, 1826

Small, elongate-ovate, round-backed, sometimes with an arched jugal area. Sculpture granular. Type: *cajetanus* Poli, 1791. Synonyms: *Lepidochiton* Gray, 1847, not *Lepidochitona* Gray, 1821; *Rhombichiton* DeKoninck, 1883.

Lepidopleurus cancellatus (Sowerby, 1839)   4592
Arctic Cancellate Chiton

Greenland to the Gulf of Maine.
Bering Sea to Oregon.

½ inch in length, arched; color exterior an orange-gray to whitish gray; interior white. Anterior valve microscopically granulated in radial rows. Central areas of the intermediate valves very finely granulated with densely placed, round pimples. Posterior valve with a smooth, slightly elevated central apex. Girdle narrow, same color as the valves and densely packed with tiny, split-pea scales. Some scales are commonly club-shaped, especially at the margins of the girdle, or sometimes so irregular and crowded as to give the appearance of fine moss. Moderately common on gravel bottoms from 20 to 100 fathoms. *L. arcticus* Sars, 1878, is probably a synonym.

Lepidopleurus pergranatus (Dall, 1889)   4593
Granate Chiton

Off Florida to the Lesser Antilles.

12 mm. long, 6.5 mm. wide, pale-waxen to white. Moderately elevated. Valves wide, without a jugum and without apices. Front and back valves concave. Sculpture as in *cancellatus*, but the granules are larger; the lateral areas are less defined. Sutural plates elongated. Girdle wide, densely beset with delicate scales. 12 gills on each side, reach forward to about the middle of the sixth valve. Moderately common; 114 to 1,181 fathoms.

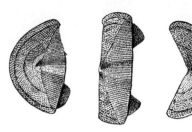

4593

Lepidopleurus alveolus (Sars, in Lovén, 1846) 4594

Arctic Seas to the Gulf of Maine.

16 mm., quite convex, back equally arched, without a trace of a keel or well-defined lateral areas. Valves rather elongate, the posterior larger than the anterior valve, half-round, truncated by a straight line in front. Median valves subequal, their posterior margins straight, anterior margins lightly emarginate in the middle. Entire surface sculptured with minute, ovate tubercles, regularly disposed. Girdle beset with needlelike and paddlelike scales. Not uncommon; 120 to 664 fathoms.

Lepidopleurus rugatus (Pilsbry, 1892) 4595

Monterey, California, to Baja California.

10 to 15 mm., white, oval, rather convex or arched; the lateral slopes nearly straight, dorsal ridge broadly arched. Front and back valves and lateral areas of the intermediate valves sculptured with excessively fine radiating striae, which are feebly granose, and having well-marked, coarse, concentric striae. Mucro subcentral and prominent. Lateral areas slightly raised. Moderately common; sublittoral on rocks buried under sand. Sometimes placed in the subgenus *Lepidochiton* Gray, 1847 (not *Lepidochitona* Gray, 1821).

Lepidopleurus oldroydi Dall, 1919 4596
Oldroyd's Black-spotted Chiton

Alaska to Catalina Island, California.

5 mm. long, 2 mm. wide, white with a blackish spot on either side of the jugal area. Strongly sculptured. Back moderately arched. Anterior valve semicircular, with numerous, irregularly placed, small, prominent, round pustules. Posterior valve with subcentral prominent mucro, the central area granulose, the periphery pustulose. Intermediate valves with axially punctostriate jugal area; lateral areas with large pustules. Girdle with crowded minute spines of equal length, giving a sandy effect. Interior of valves whitish. Jugal sinus wide with a straight edge. Sutural laminae small, subtriangular. Uncommon.

Lepidopleurus nexus (Carpenter, 1864) 4597

Monterey, California, to the Gulf of California.

7.5 mm. long, 4.5 mm. wide, whitish ashen, valves gothic-arched; lateral areas scarcely defined; entire surface ornamented with series of subquadrate granules, the series longitudinal upon the central, radiating upon the lateral areas and end valves, very close, scarcely interrupted. Jugum elevated, subacute; umbones inconspicuous. Mucro conspicuous. Girdle having narrow, close, striated scales, and needle-shaped, crystalline bristles here and there and around the margin. Not uncommon; shallow water to 15 fathoms. Placed in a subgenus (*Xiphiozona* Berry, 1919) by some workers. Synonyms: *heathi* S. S. Berry, 1919; *ambustus* Dall, 1919.

Other Atlantic species:

4598 *Lepidopleurus asellus* (Gmelin, 1791). Greenland to Norway to France. For life history, see M. E. Christiansen, 1954, *Nytt Mag. Zool.*, vol. 2, pp. 52–72.

— *Lepidopleurus carinatus* Dall, 1927, see *Hanleya*.

4599 *Lepidopleurus binghami* Boone, 1928. British Honduras. *Bull. Bingham Oceanogr. Coll.*, vol. 1, art. 3, p. 114.

Other Pacific species:

4600 *Lepidopleurus internexus* Carpenter, 1892. Alaska to San Diego, California.

4601 *Lepidopleurus farallonis* Dall, 1902. Farallon Islands, California, to Panama; deep water.

4602 *Lepidopleurus belknapi* (Dall, 1878). Off Aleutian Islands, 1,006 fms.

4603 *Lepidopleurus lycurgus* Dall, 1919. Catalina Island, California.

4604 *Lepidopleurus luridus* Dall, 1902. Puget Sound, 48 fms.; Panama, 1,270 fms.

4605 *Lepidopleurus alascensis* Thiele, 1909. Alaska.

4606 *Lepidopleurus mesogonus* Dall, 1902. Bering Sea, 688 fms.; British Columbia, 1,588 fms.

Genus *Oldroydia* Dall, 1894

Valves strongly sculptured, with parts of the girdle showing between the valves all the way up to the jugum. Jugal area prominent, sculptured differently from the pleural tracts and extending in front of them between the sutural laminae. Lateral areas not differentiated. Type and only American species: *percrassa* Dall, 1894.

Oldroydia percrassa (Dall, 1894) 4607
Oldroyd's Thick Chiton

Monterey, California, to Baja California.

14 mm., solid, pale pinkish brown with a darker brown, narrow girdle which extends between the valves. Girdle sandy with scales on the base and occasional glassy spicules everywhere. Valves thick, white underneath. Head valve with minutely nodulous concentric ridges. Sculpture of the jugum in the central valves is punctate and with elevated transverse ridges. Sides of valves with 6 to 8 wiggly ridges divaricating towards the edge, crossed by fine, elevated lamellae. Uncommon; offshore to 75 fathoms.

Family HANLEYIDAE Bergenhayn, 1955

Genus *Hanleya* Gray, 1857

Anterior valve has an insertion plate which is without slits, but is roughened. Intermediate and posterior valves without insertion plates; eaves small; girdle with fine spines. No girdle pores. Type: *hanleyi* (Bean, 1844).

Hanleya hanleyi (Bean, 1844) 4608
Hanley's Chiton

Arctic Seas to Massachusetts Bay.
Bering Strait to Monterey, California.

10 to 20 mm., oblong, convex, lateral slopes nearly straight, the dorsal ridge rather angular. Sculpture of numerous, rounded tubercles, in rows, on the central areas, but finer and closer on the jugum. Head plate and lateral areas with larger,

irregular tubercles. Lateral areas not raised. Mucro median and elevated. Sinus between sutural plates side and denticulated by the sculpture of the outside. Eaves very small. Girdle narrow, beset with short and long spicules. In the young form (*spicata* S. S. Berry, 1919) the long spicules may be clumped between the edges of the intermediate valves. Common; 25 to 300 fathoms. Synonym: *debilis* Gray, 1857.

Other species:

4609 *Hanleya mendicaria* (Mighels and Adams, 1842). Grand Manan; Eastport to Casco Bay, Maine. Georges Bank. 25 to 30 fms.

4610 *Hanleya tropicalis* Dall, 1879. Off Sand Key, Key West, Florida, 128 fms.

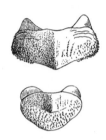

4610

4611 *Hanleya dalli* Kaas, 1957. Gulf of Maine, 12 fms.; Georges Bank, off Massachusetts. (Synonym: *carinatus* Dall, 1927, non Leach, 1852.) *Basteria,* vol. 21, p. 83.

Suborder *Ischnochitonina* Bergenhayn, 1930

Family ISCHNOCHITONIDAE Dall, 1889

Tegmentum of intermediate valves divided into lateral and central areas by a diagonal rib (commonly indistinct) extending from the apex to the anterior outer angle of the valves on each side. Articulamentum of head and tail valves multislitted, that of the intermediate valves with 1 or 2 slits in each side. Insertion plate teeth sharp-edged, not grooved, pectinated or buttressed on the outside. Eaves not porous. Sutural laminae sharp and well-developed.

Genus *Ischnochiton* Gray, 1847

Rarely over 1 or 2 inches; subovate. Surface of lateral areas usually raised, with smooth or beaded radial ribs. Central areas finely granular or with longitudinal ribs. Tail valve with large mucro. Articulamentum of head and tail valves usually with more than 8 slits in insertion plates, and that of intermediate valves with 1, or rarely 2 or more, slits. Type: *textilis* Gray, 1828. This is *Lepidochiton* Carpenter, 1857; *Ischnoradsia* Carpenter in Dall, 1879 (non Shuttleworth, 1853); *Beanella* Thiele, 1893 (not Dall, 1882); *Lophyriscus* and *Rhodoplax* Thiele, 1893; *Levicoplax* Iredale and Hull, 1925. 17 subgenera are recognized, and I have not allotted every species to its proper place.

Ischnochiton papillosus (C. B. Adams, 1845)   **4612**
Mesh-pitted Chiton

Tampa to the Lower Keys and the West Indies.

⅓ to ½ inch in length, oval. Moderately sculptured and without very distinct lateral areas. It has microscopic, even,

quincunx pittings on the upper surfaces of the valves. End valves with concentric rows of fine, low beads. Lateral areas with fine, wavy, longitudinal, incised lines. Posterior slope of posterior valve is concave and with 9 slits. Color whitish with heavy mottlings of olive-green; rarely with white spots. Girdle narrow, colored with alternating bars of white and greenish brown. Scales like microscopic split peas which are finely striated. A fairly common shallow-water species. Placed in the subgenus *Ischnoplax* Carpenter in Dall, 1879, by most workers.

Subgenus *Rhombochiton* S. S. Berry, 1919

Type of this subgenus is *regularis* Carpenter, 1855.

Ischnochiton regularis (Carpenter, 1855)   **4613**
Regular Chiton

Southern California.

1 to 1½ inches in length, oblong, appears smooth to the naked eye. Color an even slate-blue or uniform olive-blue. Valves slightly carinate. Central areas with very fine, longitudinal threads. Lateral areas slightly raised and with radial threads. Interior of valves gray-blue. Girdle with very tiny, closely packed, low, round scales, and colored blue or purple. Moderately common between tides.

Subgenus *Stenosemus* Middendorff, 1847

Type of this subgenus is *albus* (Linné, 1767). *Lepidopleuroides* Thiele, 1893, is a synonym, as is *Lophyrochiton* Yakovleva, 1952, p. 99 which also has *albus* as the type.

Ischnochiton albus (Linné, 1767)   **4614**
White Northern Chiton

Arctic Seas to Massachusetts. Europe.
Arctic Seas to off San Diego, California.

About ½ inch in length, oblong, moderately elevated. Upper surfaces smoothish except for irregular, concentric growth ridges and a microscopic, sandpapery effect. Color whitish, cream, light-orange or rarely marked with brown. Interior of valves white. Posterior valve with 12 or 13 weak slits. 17 to 19 gill lamellae on each side, beginning about halfway alongside the foot. Girdle sandpapery, with tiny, closely packed, gravelly scales. Common from shore to several fathoms in cold water. Distinguished from *ruber* by the anterior slope of the anterior valve which is straight to slightly concave in *albus*, but convex in *ruber*.

Other Atlantic species:

4615 *Ischnochiton striolatus striolatus* (Gray, 1828). Off Fernandina, Florida, 294 fms., to the Lesser Antilles. (Synonym: *squamulosa* C. B. Adams, 1845.)

4616 *Ischnochiton striolatus funiculatus* Pilsbry, 1892. Key West, Florida, and the West Indies.

4617 *Ischnochiton (Ischnoplax* Carpenter in Dall, 1879) *pectinatus* Sowerby, 1832. Florida Strait and the West Indies. Type of the subgenus.

4618 *Ischnochiton (Chondropleura* Thiele, 1906) *exaratus* Sars, 1878. Off Martha's Vineyard, Massachusetts, 101 to 194 fms.; off Fernandina, Florida, 294 fms. 20 mm. Type of the subgenus.

4618a *Ischnochiton hartmeyeri* Thiele, 1910. 5 mm. Florida Keys. See Kaas, 1972, p. 91.

Other Pacific species:

4619 *Ischnochiton (Tripoplax* S. S. Berry, 1919) *trifidus* Carpenter, 1864. Alaska to Puget Sound, Washington. Type of the subgenus. See A. G. Smith and Cowan, 1966, *Occ. Papers Calif. Acad. Sci.,* no. 56, fig. 20.

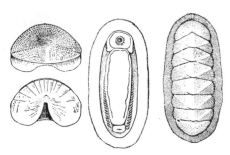

4618

4620 *Ischnochiton marmoratus* Dall, 1919. Monterey Bay, California.

4621 *Ischnochiton bryanti* Dall, 1919. "California."

4622 *Ischnochiton brunneus* Dall, 1919. San Diego, California.

4623 *Ischnochiton listrum* Dall, 1919. San Diego, California.

4624 *Ischnochiton ritteri* Dall, 1919. Juneau Channel, Alaska.

4625 *Ischnochiton retiporosus* Carpenter, 1864. Victoria, British Columbia, to San Pedro, California.

4626 *Ischnochiton retiporosus punctatus* Whiteaves, 1886. Duncan Bay, British Columbia.

4627 *Ischnochiton venezius* Dall, 1919. Venice, California.

4628 *Ischnochiton scrobiculatus* Middendorff, 1847. Bodega Bay, California.

4629 *Ischnochiton lividus* Middendorff, 1847. Sitka, Alaska.

4630 *Ischnochiton alascensis* Thiele, 1910. Sitka, Alaska.

4631 *Ischnochiton interstinctus* Gould, 1852. Aleutian Islands, to Sitka and Catalina Island, California.

4632 *Ischnochiton radians* Carpenter, 1892. Prince of Wales Island, Alaska, to San Pedro, California.

4633 *Ischnochiton scabricostatus* Carpenter, 1864. San Pedro and Catalina Island, California, to Cerros Island, Baja California.

4634 *Ischnochiton aureotinctus* Carpenter, 1892. Catalina Island, 80 fms., to Cerros Island, Baja California.

4635 *Ischnochiton veredentiens* Carpenter, 1864. Point New Year to Catalina Island, California.

4636 *Ischnochiton newcombi* Carpenter, 1892. Catalina Island, California.

4637 *Ischnochiton corrugatus* Carpenter, 1892. Catalina Island, California, to Todos Santos Bay, Baja California.

4638 *Ischnochiton abyssicola* A. G. Smith and Cowan, 1966. Semidi Islands, Alaska, to off Oregon. *Occ. Papers Calif. Acad. Sci.*, no. 56.

Genus *Lepidozona* Pilsbry, 1892

Similar to *Ischnochiton*, but the last 7 valves have a delicately denticulate lamina across the sinus. Insertion teeth sharp, somewhat rugose, fairly thick. Girdle scales strongly convex, and are smooth or striated. Type: *mertensii* (Middendorff, 1846).

Lepidozona mertensii (Middendorff, 1847)   4639
Merten's Chiton

Aleutian to Baja California.

1 to 2 inches in length, rather oval in shape. Color variable: commonly yellowish with dark reddish brown streaks and maculations. Central areas with strong, longitudinal ribs and smaller, lower cross ridges which give a netted appearance. Jugal area V-shaped and with 5 to 6 smooth longitudinal ribs. Lateral areas raised, smoothish and with a few prominent warts. Anterior valves with 30 or more radial rows of warts which are largest near the girdle. Interior whitish, or rarely tinged with pink. Girdle with alternating yellowish and reddish bands; covered with tiny, low, smooth, split-pea scales. Very abundant just offshore, especially in the northern part of its range.

Lepidozona californiensis S. S. Berry, 1931   4640
Trellised Chiton

Santa Barbara, California, to Baja California.

1 to 1½ inches in length, oval to oblong, heavily sculptured. Color a dull-greenish with yellowish splotches and with a dark-brown area on the top of each valve. Central area with longitudinal and cross ribs which give a strong netted appearance. Lateral areas raised and with 4 rows of prominent beads. Posterior edge of valves serrated with about 20 small tooth-shaped beads. Anterior valve with 20 to 27 strongly granular ribs. Girdle closely packed with convex, tiny, split-pea scales. Moderately common and formerly thought to be *L. clathrata* Reeve which, however, is only from the Panamic Province to the south. Found under stones at low-tide mark.

Lepidozona cooperi (Carpenter, in Dall, 1879)   4641
Cooper's Chiton

Southern California.

1 to 1½ inches in length, rather oval in shape. Color olive-green to olive-brown and clouded with light-blue. Central area with closely packed, sharp, longitudinal ribs which are finely striated. Jugal area with the same type of ribs and with its anterior end having about 10 notches. Lateral areas raised and with 4 to 8 irregular rows of prominent, rounded warts. Interior of valves bluish. Girdle covered with tiny, flat, striated, split-pea scales. Uncommon. *L. acutior* Dall, 1919, is a synonym.

Other species:

4642 *Lepidozona willetti* S. S. Berry, 1917. Forrester Island, Alaska.

4643 *Lepidozona decipiens* (Carpenter, in Pilsbry, 1892). Catalina Island to Monterey, California. (Synonym: *gallina* S. S. Berry, 1925.) See *The Nautilus*, vol. 49, p. 43.

4644 *Lepidozona sinudentatus* (Carpenter, in Pilsbry, 1892). Monterey, California. (?Synonym: *berryi* Dall, 1919.)

4645 *Lepidozona asthenes* S. S. Berry, 1919. Los Angeles, California, to Guadalupe Island, Mexico.

4646 *Lepidozona stearnsii* (Dall, 1902). Farallon Islands, California, 391 fms.; off Santa Catalina Island, 255 fms. (An *Ischnochiton (Tripoplax)* fide A. G. Smith and Cowan, 1966.)

4613   4639   4641

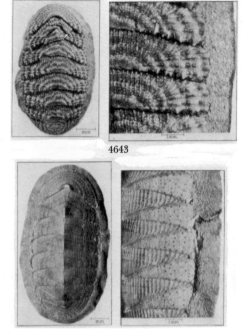

4643

4647

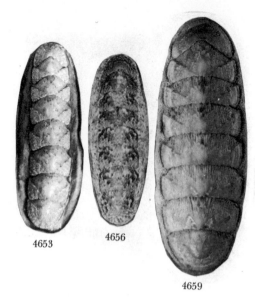

4653   4656

4659

4647 *Lepidozona golischi* S. S. Berry, 1919. Off Santa Monica, California, 100 fms.

4648 *Lepidozona subtilis* S. S. Berry, 1956. Northern end of Gulf of California. Tidepools and reefs. Uncommon. Leaflets in Malacology, vol. 1, no. 13. (Synonym: *pella* S. S. Berry, 1963.)

4649 *Lepidozona serrata* (Carpenter, 1864). Cape San Lucas, Baja California.

4650 *Lepidozona crockeri* (Willett in Hertlein and Strong, 1951). Gorda Banks, Gulf of California, 60 fms. Rare.

4651 *Lepidozona catalinae* Willett, 1937. Catalina Island, California.

4652 *Lepidozona inefficax* S. S. Berry, 1963. Sebastian Vizcaino Bay, Baja California. *Leaflets in Malacology*, no. 22, pp. 137, 138.

Genus *Stenoplax* Carpenter in Dall, 1879

About 6 or 7 times as long as wide. Tail valve large. 9 to 15 slits in end valves, 1 to 3 in intermediate valves. Girdle narrow with minute scales. Type: *limaciformis* (Sowerby, 1832).

Stenoplax floridana (Pilsbry, 1892)   4653
Florida Slender Chiton

Miami to Dry Tortugas, Florida.

1 to 1½ inches in length, about 3 times as long as wide, elevated, with the valves roundly arched, not carinate. Color whitish to whitish green with markings of olive, blackish olive or gray. Lateral areas raised and with wavy, longitudinal riblets which are commonly strongly beaded. Central areas with wavy, longitudinal ribs. Interior of valves mixed with white, blue and pink, rarely all pink or all white. End valves concentrically (or rarely axially) beaded. Intermediate valves with 1 slit, posterior valve with 9. Girdle marbled with bluish and gray, and densely covered with round, solid, finely striated scales. Moderately common.

(4654) *Stenoplax purpurascens* (C. B. Adams, 1845) (purplish slender chiton) from the Florida Keys, Bermuda and the West Indies is very similar, but the end valves and lateral areas have smooth, instead of beaded, wavy, concentric riblets. Common.

Stenoplax erythronota (C. B. Adams, 1845)   4655
Ashy Slender Chiton

Florida Keys and the West Indies.

12 to 15, elongate, valves carinate at the top. Color ash-gray or brownish, either with 2 small twin streaks of green on the jugum of each intermediate valve, or rarely sparsely mottled with green or brown. Girdle brown, with green spots, and covered with tiny, distinct, radially striated scales. Lateral areas raised and pustulose. Central area with about 15 raised longitudinal lirae on each side. Jugum smooth or weakly pitted. Head valve with about 8 concentric irregular rows of pustules. Tail valve with mucro in the middle; behind it minutely pimpled; in front of it longitudinally and rugosely ribbed. Common; rocks, intertidal.

Stenoplax limaciformis (Sowerby, 1832)   4656
Slug-shaped Chiton

Florida Keys and the West Indies.
Gulf of California to Peru.

15 to 25 mm., elongate, similar to *erythronota*, but the valves are roundly swollen, grayish green with flecks of dark-green and white. Girdle granular and coming between the side edges of the intermediate valves, and cream with marbled moss-green and white flecks. Common; under rocks, shallow water. Synonyms are *productus* and *sanguineus* Reeve, 1847.

Stenoplax fallax (Carpenter in Pilsbry, 1892)   4657
Fallax Chiton

Vancouver Island, British Columbia, to Todos Santos Bay, Baja California.

2 to 2½ inches, brightly colored in rose, lavender or mottled buff. Girdle scales are extremely small, giving the girdle a leathery appearance. Foot is reddish orange, while in *heathiana* it is light-yellow. The roundly arched outline; and single slitting of the central valves; the terraced appearance of the lateral and terminal areas due to the very strong growth lines; the fine, wrinkly, radiating sculpture of these regions; and the very even and delicate ribbing and pitting of the central areas, all serve to separate *fallax* from *heathiana* with which it lives from shore to 15 fathoms.

Stenoplax corrugata (Carpenter in Pilsbry, 1892)   4658
Corrugated Chiton

Catalina Island, California, to Baja California.

10 to 15 mm., surface granulose. Central areas have punctate wrinkles. Lateral areas strongly longitudinally corrugated. Color variable from white to gray with greens. Girdle narrow, beset with numerous short overlapping, striated scales. Uncommon. Synonym: *Ischnochiton biarcuatus* Dall, 1903.

Subgenus *Stenoradsia* Carpenter in Dall, 1879

Type of the subgenus is *magdalenensis* (Hinds, 1844). *Maugerella* Dall, 1879, is a synonym. Intermediate valves have several side slits.

Stenoplax heathiana S. S. Berry, 1946   4659
Heath's Chiton

Northern California to Santo Tomas, Baja California.

2 to 3 inches in length, elongate. Color a drab-greenish, often mottled with brown or buff, commonly whitish due to wear. Central areas with fine, irregular, longitudinal cuts and with diamond-shaped pits near the lateral areas. Lateral areas prominently raised and with 13 to 18 sharply raised radial ribs. Front slope of anterior valve straight, and with 60 to 70 low riblets. Posterior slope moderately concave. Interior of valves bluish with the posterior end of each whitish. Girdle rather narrow and with alternating, faint bars of brown and yellowish brown. Scales round, finely striate, and so small that the girdle has the texture of fine sandpaper. 33 ctenidia on each side. Foot light-yellow. Common. Also see the more southerly *magdalensis*.

4659

Stenoplax conspicua (Dall, 1879)   4660
Conspicuous Chiton

Santa Barbara, California, to the Gulf of California.

2 to 6 inches in length, very similar to *heathiana,* but the front slope of the anterior valve is very concave; the central areas are practically smooth and with flecks of green. The striated scales are elongate, hard, and so densely packed that the girdle feels velvety. Interior of valves pinkish and blue. Moderately common between tides. *I. sarcosa* Dall, 1902 is a synonym. S. S. Berry described the Gulf of California colonies as the subspecies *sonorana* Berry, 1956. 4 mollusks live commensally under the edge of the mantle of *S. conspicua: Teinostoma invallata* Carpenter, *T. supravallata* Carpenter, *Vitrinella oldroydi* Bartsch and the bivalve *Serridens oblonga* Carpenter.

Stenoplax magdalenensis (Hinds, 1844)   4661
Magdalena Chiton

Baja California and the Gulf of California.

2 to 4 inches in length, elongate-ovate, very similar to *heathiana.* Color brown or gray-green. Instead of diamond-shaped pittings on the sides of the central areas, there are wavy ribs. The front slope of the anterior valve is very concave, as in *conspicua.* The scales are similar to those in *heathiana,* but much larger. Interior of valves pinkish with a blue spot at the anterior end of each valve. Moderately common. *Ischnochiton acrior* Carpenter, in Pilsbry, 1892, is a synonym. For details, see S. S. Berry, 1946, *Proc. Malacol. Soc. London,* vol. 26, p. 161.

Other species:

4662 *Stenoplax petaloides* (Gould, 1846). Gulf of California to Guaymas, Mexico. (Synonyms: *histrio* S. S. Berry, 1945, *Amer. Midland Nat.,* vol. 34, p. 491; *mariposa* Dall, 1919.)

4663 *Stenoplax rugulata* (Sowerby, 1832). Bermuda, Florida and West Indies. Gulf of California to Peru. (Synonyms: *bermudensis* Dall and Bartsch, 1911; *roseus* Sowerby, 1832, non Blainville, 1825; *boogii* Haddon, 1886; *aethonus* Dall, 1919; *pallidulus* Reeve, 1847; *isoglypta* S. S. Berry, 1956.)

Genus *Lepidochitona* Gray, 1821

Similar to *Ischnochiton,* but generally elevated and with a subangular jugal ridge. Tail valve usually smaller than the head valve. End valves with 9 to 12 slits in the insertion plates. Intermediate valves single-slitted; eaves generally somewhat spongy. Lateral areas not prominent. Girdle with minute, scalelike processes. Type: *cinerea* (Linné, 1767). *Trachydermon* Carpenter, 1864, and *Spongioradsia* Pilsbry, 1894, are synonyms.

Lepidochitona liozonis (Dall and Simpson, 1901)   4664
Puerto Rican Red Chiton

Lower Florida Keys to Puerto Rico.

8 to 12 mm., about 7 mm. wide. Colored scarlet to dark-red with some flecks of gray or white. Surface of valves nearly smooth, minutely granulose. Girdle naked, leathery, brownish. Rare; offshore. *L. tropica* "Pilsbry, manuscript" is the same.

Lepidochitona dentiens (Gould, 1846)   4665
Gould's Baby Chiton

Alaska to Monterey County, California.

½ inch or slightly more in length, oval, slightly elevated. Color tawny, olivaceous, slaty or brownish, usually covered with specklings of a darker hue. Upper surface of valves covered with microscopic, sharp granulations which are rarely aligned in any direction. Lateral areas may be slightly raised, and may be bounded in front by a very low rib. The apex of the posterior valve is near the center and is raised; behind the apex, the valve is concave. Girdle very narrow, same color as the valves, and with very minute, gritty granules. Common intertidal form; sometimes on stems on brown algae.

Lepidochitona keepiana S. S. Berry, 1948   4666
Keep's Chiton

Cayucos, California, to Baja California.

10 to 14 mm.; girdle with flat-topped, crowded spicules. Color variable—grays, greens, white, orange and brownish mottlings. In the similar *dentiens,* the insertion teeth are prominently developed, their bounding slits in general *widely V-shaped,* and the teeth of the·posterior valve very acute on the sides; the eaves are wide and conspicuously porous or "spongy." In the latter species, there are numerous short, *narrowly slitted* teeth in the terminal valves, and extremely thin, narrow, less openly porous eaves. (See S. S. Berry, 1948, *Leaflets in Malacology,* vol. 1, no. 4.) Common; midtide level.

Other species:

4667 *Lepidochitona sharpei* Pilsbry, 1896. Unalaska, Aleutian Islands. *The Nautilus,* vol. 10, p. 50.

4668 *Lepidochitona aleutica* (Dall, 1878).

Key to the Small Red Chitons (⅓ to 1 inch)

Girdle naked:

Valves dull, color pattern maculated . . . . .

. . . . *Tonicella marmorea* Fabricius

Valves glossy, with bright, black and white lines .

. . . . *Tonicella lineata* Wood

Girdles with scales:

With overlapping, split-pea scales . . . . .

. . . . *Lepidopleurus cancellatus* Sowerby

With tiny, granular scales:

Interior of valves bright-pink . . . . . .

. . . . *Tonicella ruber* Linné

Interior of valves white . . . . . .

. . . . *Ischnochiton albus* Linné

Genus *Tonicella* Carpenter, 1873

Shells less than 1½ inches, low-arched, with rounded backs. Valves with spongy eaves and the intermediate valves single-slitted. Lateral areas poorly developed. Surface of valves smooth or microscopically granular. Girdle smoothish or with very small scales. Many species are bright-red, less than ½ inch and live in cold waters. Type: *marmorea* (Fabricius, 1780).

Tonicella rubra (Linné, 1767)　　　4669
Red Northern Chiton

Arctic Seas to Connecticut. Europe.
Alaska to Monterey, California.

½ to 1 inch in length, oblong, moderately elevated and with the valves rather rounded. Upper surfaces smooth except for growth wrinkles. Colored a light-tan over which is a heavy suffusion of orange-red marblings, or entirely suffused with red. Interior of valves bright-pink. Posterior valve with 7 to 11 slits. Girdle reddish brown with weak maculations; covered with minute, elongate scales which do not overlap each other. 15 to 18 gill lamellae, similar to those in *albus*. Common from 1 to 80 fathoms. Do not confuse with *Tonicella marmorea* Fabricius, whose girdle is naked. Balch, 1906, described a form, *index*, from Blue Hill Bay, Maine, from 12 fathoms. *T. submarmorea* (Middendorff, 1846) is a synonym.

Tonicella marmorea (Fabricius, 1780)　　　4670
Mottled Red Chiton

Greenland to Massachusetts.
Japan and the Aleutian Islands to Washington.

About 1 inch in length, oblong to oval, elevated and rather acutely angular. Colored a light-tan over which is a heavy suffusion of dark-red maculations and specks. Upper surface appears smooth, although under high magnification it is seen to be granulated. Lateral areas of intermediate valves not very distinctly outlined. Interior of valves tinted with rose. Posterior valve with 8 to 9 slits. Girdle is leathery and without scales or bristles. Superficially this species resembles *Tonicella rubra* which, however, has scales on its girdle. Common from 1 to 50 fathoms. Rare in the area of Puget Sound. *T. caerulea* Winkley, 1894, and *laevigatus* Fleming, 1828, are synonyms.

Tonicella lineata (Wood, 1815)　　　4671
Lined Red Chiton

Japan to the Aleutians to San Diego, California.

About 1 to 2 inches in length, similar to *T. marmorea*, but with its valves orange to deep-red, smooth and shiny, and it is brightly painted with black-brown lines bordered with white which run obliquely backwards on the intermediate valves. The end valves have these same color lines concentrically arranged. Common on the rocky shores of Alaska. The young live in waters off the shore from 10 to 30 fathoms, but as they mature they migrate toward shore.

Other species:

4672 *Tonicella blaneyi* Dall, 1905. Frenchman's Bay, Maine, 20 fms. *Proc. Biol. Soc. Wash.*, vol. 18, p. 204.

4673 *Tonicella insignis* (Reeve, 1847). Alaska to Washington. See Leloup, 1945, *Bull. Mus. Royal Hist. Nat. Belgique,* vol. 21, no. 4.

4674 *Tonicella sitkensis* (Middendorff, 1846). Sitka, Alaska.

4675 *Tonicella saccharina* Dall, 1878. Aleutians, shallow water; San Diego, California, 101 fms.

Genus *Cyanoplax* Pilsbry, 1892

Resembles *Tonicella*, but the valves are thicker and somewhat beaked anteriorly. Insertion-plate teeth stout, crenulate, and bi- or tri-lobed at the tops. Eaves wide, spongy or pitted. Girdle leathery, minutely papillose. Interior of valves may be blue or green. Intermediate valves single-slitted. Type: *hartwegii* (Carpenter, 1855).

Cyanoplax hartwegii (Carpenter, 1855)　　　4676
Hartweg's Baby Chiton

Washington to Baja California.

1 to 1½ inches in length, oval, rather flattened, drab grayish green, the jugal area lighter, bordered by dark longitudinal stripes. Girdle narrow and finely granulated. Sculpture of the end valves and the lateral areas of the middle valves differs from the microscopic granulations of *dentiens* in bearing easily seen, but very tiny, warts. It also differs in having the area behind the apex of the posterior valve convex instead of concave. Moderately common in intertidal areas. *C. nuttalli* (Carpenter, 1855) is a synonym.

Other species:

4677 *Cyanoplax raymondi* (Pilsbry, 1894). Seward, Alaska, to San Pedro, California.

4678 *Cyanoplax fackenthallae* (S. S. Berry, 1919). Monterey Bay, California.

4679 *Cyanoplax lowei* (Pilsbry, 1918). San Pedro, California.

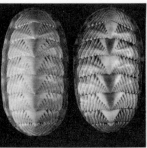

4670　　　　4671　　　　4676

Family SCHIZOPLACIDAE Bergenhayn, 1955

Genus *Schizoplax* Dall, 1878

The only genus in the family; occurring only in the North Pacific arctic waters. Shell and girdle like *Tonicella*. The central jugal slit is filled by a narrow wedge of horny cartilage. Tegmentum porous where exposed at the small eaves and jugal sinus. 11 slits in both the head and tail valves; only 1 in the intermediate valves. Type: *brandtii* (Middendorff, 1846). Two species:

4680 *Schizoplax brandtii* (Middendorff, 1846). Okhotsk Sea; Aleutian Islands to Cape Fox, Alaska.

4681 *Schizoplax multicolor* Dall, 1920. Bering Sea. *The Nautilus,* vol. 34, p. 22.

Family CALLISTOPLACIDAE Pilsbry, 1893

Surface sculpture usually with heavy radial ribs on the end valves and lateral areas of the intermediate valves. Insertion plates in all valves cut by teeth that correspond in number and position to the radial ribs. Teeth usually thickened at the edges of the slits. No shell eyes present. Girdle narrow, with scales or bristles.

Genus *Callistochiton* Carpenter in Dall, 1879

Surface strongly sculptured; central areas smooth in the middle but with a netted or pitted surface towards the apex, or sculptured throughout with parallel lirae. Insertion plates short, smooth, thickened outside at the edges of the slits. End valves multislitted, intermediate valves single-slitted. Sinus square. Girdle narrow, with minute striated or smooth scales. Type: *palmulatus* (Carpenter, 1893).

Callistochiton shuttleworthianus Pilsbry, 1893   **4682**
Shuttleworth's Chiton

Lower Florida Keys and the West Indies.

10 to 14 mm. long, oval, girdle wide, covered with orange-yellow or rust-colored, closely packed scales which are grooved at the base, but smooth elsewhere. Valves also rust-colored, dorsally carinated, the side slopes nearly straight. Lateral areas strongly raised, bicostate, the ribs nodulose. Central areas with raised network in the middle, longitudinal ribs on the sides. Posterior valve depressed, its back slope concave. Anterior valve with 10 slits, intermediate valves with 1. Sutural plates low, rounded; sinus shallow and flat. Common; under subtidal rocks.

Callistochiton palmulatus (Carpenter in Pilsbry, 1893)   **4683**
Big-end Chiton

Monterey, California, to Baja California.

10 to 15 mm., oval-elongate, light-brown to grayish green. Mucro on tail valve subcentral, depressed, the posterior area strongly swollen. Central area with about 10 lirae on each side which pectinate the sutures. Lateral areas with 2 strong ribs bearing tubercles, the interstices deeply punctate. Anterior valve with 11 distinct ribs, the outer 2 joined. Posterior valve with 7 very strong ribs bifurcating behind. Posterior valve with 26 slits. Eaves very strong; sinus small, strongly laminate, the lamina deeply slit on each side. Girdle with minute striated scales. Moderately common; low-tide area. A form (4683a) *mirabilis* Pilsbry, 1893, has the posterior area of the tail valve enormously swollen.

Callistochiton crassicostatus Pilsbry, 1893   **4684**

Monterey, California, to Baja California.

20 to 30 mm., elongate, similar to *palmulatus*, but with only 7 ribs on the anterior valve, lateral areas with only 1 massive rib, and on the tail valve the mucro is directly over the posterior slope. Posterior valve with 13 to 20 slits. Color tan to gray. Girdle with minute, striated scales. Moderately common; under rocks; subtidal.

Callistochiton decoratus Pilsbry, 1893   **4685**
Decorated Chiton

Santa Barbara, California, to Baja California.

15 to 23 mm., elongate-oval, similar to *palmulatus*, shiny-green to gray with flecks; head valve with 11 ribs; ridge or central area of the intermediate valves have a V-shaped smooth area. Lateral area with 2 ribs. Posterior valve is highest at the front margin, the mucro depressed, and with 9 or 10 rounded ribs. Posterior valve with 9 to 12 slits. Girdle narrow, ashy-brown, with minute, striated scales. Uncommon; under rocks at low-tide level. A form (4685a) with the jugal tract punctate was named *punctocostatus* Pilsbry, 1896. *C. ferminicus* S. S. Berry, 1922; *chthonius* Dall, 1919; and *cyanosus* Dall, 1919 are synonyms.

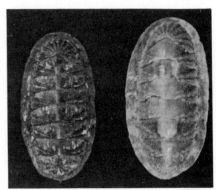

4685

Other species:

4686 *Callistochiton acinatus* Dall, 1919. San Pedro, California.

4687 *Callistochiton celetus* Dall, 1919. San Pedro, California.

4688 *Callistochiton aepynotus* Dall, 1919. Puget Sound, Washington.

4689 *Callistochiton fisheri* Dall, 1919. Tanaga Island, Aleutians. Above 4 species in *Proc. U.S. Nat. Mus.,* vol. 55, pp. 510–512. (Not 1908.)

4690 *Callistochiton diegoensis* Thiele, 1910. San Diego, California.

4691 *Callistochiton gabbi* Pilsbry, 1893. Northern part of the Gulf of California.

4692 *Callistochiton infortunatus* Pilsbry, 1893. Gulf of California to Ecuador. Uncommon.

4693 *Callistochiton connellyi* Willett, 1937. Pacific Grove, California, to Ensenada, Baja California. *The Nautilus,* vol. 51, p. 25.

Genus *Nuttallina* Dall, 1871

Elongate-oval, with a granulose surface; head valve with radial ribs; intermediate valves with 2 ribs on the lateral areas. Insertion plates sharp, cut by slits corresponding to the dorsal ribs. Teeth of tail valve directed forward. Sutural laminae well-developed, elongate, with a deep jugal sinus between. Girdle with minute striated, flattened scales and with a marginal row of bristles. Type: *fluxa* (Carpenter, 1864). *Nuttalina* is a misspelling. On rare occasions specimens with an extra intermediate valve have been found. (S. S. Berry, 1935, *The Nautilus*, p. 89.)

Nuttallina californica (Reeve, 1847)   4694
Californian Nuttall Chiton

Vancouver Island, British Columbia, to San Diego, California.

About 1 inch in length, almost 3 times as long as wide. Color dark-brown to olive-brown. Upper surface of valves finely granulated and with a shallow furrow on each side of the smooth dorsal ridge. Apex, or mucro, of posterior valve so far back that it extends beyond the posterior margin of the eaves. Interior of valves bluish. Posterior valve about as wide as long with 8 to 9 slits. Girdle with short, rigid spinelets mostly brown in color and with a few white ones intermingled. The girdle looks mossy. Moderately common.

Nuttallina fluxa (Carpenter, 1864)   4694a
Rough Nuttall Chiton

Point Conception, California, to Baja California.

25 to 35 mm., very similar to *N. californica*, but the posterior valve twice as wide as long. Color of valves brown, pink and white with an alternating dark and light pattern on the girdle. Girdle spines white and much less numerous. This is Carpenter's *Acanthopleura fluxa*. Some workers consider this a form of *californica*. Synonym: *scaber* Reeve, 1847, non Blainville, 1825. Common; tidepools and sublittoral.

4694   4694 a

Other species:

4695 *Nuttallina thomasi* Pilsbry, 1898. Monterey, California. *Proc. Acad. Nat. Sci. for 1898*, p. 289.

4696 *Nuttallina crossota* S. S. Berry, 1956. Northern Gulf of California. Tidepools. Rare. *Leaflets in Malacology*, vol. 1, no. 13.

Genus *Ceratozona* Dall, 1882

Girdle tough, bearing corneous spines, generally sparsely bunched at the sutures, with the larger ones deeply embedded in the girdle. Back broadly arched and the valves

somewhat beaked. Insertion plates of head valve long, sharp, rugose outside, thickened at the slits. Type: *squalida* (C. B. Adams, 1845).

Ceratozona squalida (C. B. Adams, 1845)   4697
Rough Girdled Chiton

East Florida to the West Indies.

1 to 2 inches in length, oblong, slightly beaked. Surface commonly eroded, whitish gray with blue-green to moss-green mottlings on the sides of the valves. Surface roughly sculptured. Anterior valve with 10 to 11 strong, rugose, radiating ribs. Lateral areas bounded in front and behind by a large, rugose rib. Central area with low, rough, longitudinal ribs. Interior of valves bluish green. Girdle leathery, yellowish brown and with numerous, yellowish brown clusters of strap-like hairs. Posterior valve rather small and with 8 to 10 slits. 35 to 36 gill lamellae. The gills extend the length of the foot, but do not go as far as the posterior end. Very common, especially in the Greater Antilles. *C. rugosa* Sowerby, 1841, is preoccupied by *Chiton rugosa* Gray, 1826. *C. guildingi* Reeve, 1847, is also a synonym. Compare with *Acanthopleura granulata* Gmelin, 1791.

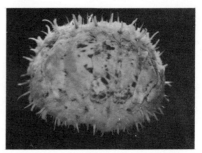

4697

Family CHAETOPLEURIDAE Plate, 1899

Valve structure with a well-developed mesostracum between the complicated tegmentum and basal articulamentum. Radial rows of pustules on the end valves. Girdle with scales or hairy processes.

Genus *Chaetopleura* Shuttleworth, 1853

Less than 1 inch, valves resembling those of *Ischnochiton;* ventral side porcelaneous, having rather long sharp teeth and a squared sinus. Sculpture of rows of fine beads. Girdle with variously sized spicules. Type: *peruviana* (Lamarck, 1819). Synonyms: *Hemphillia* Carpenter in Dall, 1879; *Pallochiton* Dall, 1879; *Arthuria* Carpenter in Dall, 1882; *Helioradsia* Thiele, 1893; *Variolepis* Plate, 1899; *Pristochiton* Clessin, 1903.

Chaetopleura apiculata (Say, 1830)   4698
Common Eastern Chiton

Cape Cod, Massachusetts, to both sides of Florida.

7 to 20 mm. in length, oblong to oval. Valves slightly carinate. Central areas with 15 to 20 longitudinal rows of raised, neat beads. Lateral areas distinctly defined, raised, and bear numerous, larger, more distantly spaced beads which may or may not be present on the more dorsal region. Interior white or grayish. Slits of anterior valve 11, central or middle valves 1, posterior valve 9 to 11. Girdle narrow, mottled cream and brown, microscopically granulose and with sparsely scattered,

transparent, short hairs. 22 to 24 gill lamellae in each gill which start just behind the juncture of the head and foot and extend all the way back to the posterior margin of the mantle where there is located a small, single-lobed lappet. Common from 1 to 15 fathoms.

In the north, the exterior color is buff to ashen, rarely reddish. On the west coast of Florida, where they are commonly found attached to *Pinna* shells, the colors vary from light-gray, mauve, yellow to white, and are commonly with a darker or lighter streak down the center or rarely with longitudinal blue stripes. For life history, see B. H. Grave, 1932, *Jour. Morphology*, vol. 54, pp. 153–160.

4698

Chaetopleura gemma Dall, 1879   4699

Vancouver Island, British Columbia, to Baja California.

12 to 18 mm., ½ as wide, oblong; red, olive or yellow. Mucro of tail valve depressed, situated behind the middle. Lateral areas decidedly raised, coarsely radiately tuberculate. Central areas with 30 longitudinal, beaded lines. Girdle narrow, leathery, sparsely clothed with short hyaline hairs which are readily rubbed off. Valves elevated. Lateral areas with 5 to 7 radiating rows of distinct, clear-cut tubercles. Interior of valves tinted with red, sometimes having a red or black spot at the jugal sinus. Sutural plates well-rounded, sinus deep and angular. Head valve with 9 to 12 slits; tail valve with 7 to 8 slits. Uncommon. *Ischnochiton marmoratus* Dall, 1919, is a synonym. This is *gemmea* Carpenter in Pilsbry, 1892.

4699

Other species:

4700   *Chaetopleura beanii* (Carpenter, 1857). Unalaska, Alaska, to Mazatlan, Mexico.

4701   *Chaetopleura parallela* (Carpenter, 1864). San Diego, California, to Colombia. Is *lurida?*

4702   *Chaetopleura prasinata* (Carpenter, 1864). San Diego, California, to Baja California. Is *lurida?*

4703   *Chaetopleura lactica* Dall, 1919. Catalina Harbor, California.

4704   *Chaetopleura* (*Pallochiton* Dall, 1879) *lanuginosa* Carpenter in Pilsbry, 1892. (?species dubium: *lanuginosa* Dall, 1879).

4705   *Chaetopleura felipponei* (Dall, 1921). Montevideo, Uruguay, or southern Brazil. *The Nautilus*, vol. 35, p. 4.

4706   *Chaetopleura lurida* (Sowerby, 1832). Gulf of California to Peru. Moderately common. (Synonym: *jaspideus* Gould, 1846.)

4707   *Chaetopleura magdalena* (Dall, 1919). Baja California. (Originally described as a *Nuttallina*.)

4708   *Chaetopleura mixta* (Dall, 1919). Gulf of California. Uncommon.

4709   *Chaetopleura euryplax* S. S. Berry, 1945. Gulf of California.

Genus *Calloplax* Thiele, 1909

Girdle with wide, ribbed scales and individual spines and minute spicules. Shell oblong, elevated; lateral areas raised, sculptured with 4 coarse granulose ribs. Sutural laminae rounded, with a shallow sinus between. 10 slits in head valve, 11 in tail valve. Insertion teeth solid; eaves wide and solid. Type: *janeirensis* (Gray, 1828).

Calloplax janeirensis (Gray, 1828)   4710
Rio Janeiro Chiton

Palm Beach, Florida, to the West Indies to Brazil.

½ to ¾ inch in length, oblong, gray to greenish brown, or speckled with red. Very strongly sculptured. Lateral areas strongly elevated by 3 or 4 very coarse large, beaded ribs; anterior valve with 12 to 18 such ribs. Central ridge (or jugal tract) with longitudinal rows of fine beads; apex elevated, smooth and rounded. Central area with about 12 very sharp, granulose, longitudinal ribs. Interior white. Anterior valve with 10, middle valves with 1 and posterior valve with 9 slits. Girdle with very fine "sugary" scales and an occasional single hair. Gills start ⅓ the way back from the head and extend posteriorly to a large, fleshy lappet on the posterior margin of the girdle. An uncommon species found in the sublittoral.

Family MOPALIIDAE Dall, 1889

Insertion-plate slits in all valves generally corresponding in position with the ends of the external ribs (8 in the head valve). Tail valve with a median sinus at the back. Girdle without scales; with simple or dendritic bristles.

Genus *Mopalia* Gray, 1847

Girdle wider at the sides than in front, leathery, more or less hairy or with bristles; with or without sutural pores. Girdle somewhat encroaching at the sutures between valves. Jugal area of intermediate valves flattish. Tail valve depressed. Type: *hindsii* (Reeve, 1847). For feeding habits of the omnivorous *Mopalia*, see Barnawell, 1960, *The Veliger*, vol. 2, pp. 85–88. For spawning, see S. R. Thorpe, Jr., 1962, *The Veliger*, vol. 4.

Mopalia ciliata (Sowerby, 1840)   4711
Hairy Mopalia

Alaska to Monterey, California.

1 to 2 inches in length, oblong, usually colored with splotches of black and emerald-green, although sometimes having cream-orange bands on the sides of the valves. Sometimes grayish green with grayish black or white mottlings. Girdle colored yellowish brown to blackish brown. Valves slightly beaked; lateral area separated from the central area by a prominent, raised row of beads. Central areas with many

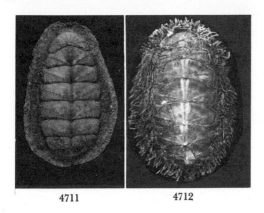

4711   4712

coarse, wavy, longitudinal riblets, which are sometimes pitted between. Lateral areas coarsely granulated or wrinkled. Posterior valve small, with a deep slit on each side and a broad, deep notch at the very posterior end. Girdle fairly wide, generally notched at the posterior end and clothed with curly, straplike brown hairs between which are much smaller, glassy-white hairs or spicules. Interior of valves greenish white. Anterior valve granulated and with 8 to 9 coarse, raised rays of beads. A common, variable intertidal species. The subspecies *wosnessenski* Middendorff, 1847 (**4711a**) (Alaska to Puget Sound) is supposed to be without the tiny white spicules in the girdle. The form *elevata* Pilsbry, 1892, is a synonym.

Mopalia muscosa (Gould, 1846)   4712
Mossy Mopalia

Alaska to Baja California.

1 to 2 inches in length, oblong to oval. Head valve with 10 beaded ribs; intermediate valves angular, flat-sided. Central areas with curving longitudinal sculpture, separated from the lateral areas by a pustulose rib. Very similar to *M. ciliata*, but differing in having a very shallow and small notch at the very posterior end. Color usually a dull-brown, blackish olive or grayish. Interior of valves blue-green, rarely stained with pinkish. Girdle with stiff hairs resembling a fringe of moss. The following species have been considered by some workers as varieties of *muscosa*, and perhaps with some justification: *porifera* Pilsbry, 1893; *lignosa* Gould, 1846; *hindsii* Reeve, 1847; *laevior* Pilsbry, 1918; *kennerlyi* Carpenter, 1864; *swanii* Carpenter, 1864 (see Berry, 1951, *Proc. Mal. Soc. London*, pl. 26); *chacei* S. S. Berry, 1919; *acuta* Carpenter, 1864; the last having also been named *plumosa* and *fissa* by Carpenter. A common intertidal species.

Mopalia lignosa (Gould, 1846)   4713
Woody Mopalia

Alaska to Lower California.

1 to 2½ inches in length, oblong. Color a grayish green or blackish green, rarely with whitish cream and brown, feathery markings. The sculpturing on the valves is very delicate and may consist only of numerous small pittings near the center. Radial ribs absent on the end valves. Girdle solid or maculated with browns and yellows. Straplike, brown hairs not numerous. Interior of valves greenish white to white. Moderately common. May be a form of *muscosa* (Gould, 1846).

Mopalia acuta (Carpenter, 1855)   4714
Acute Mopalia

Monterey, California, to Baja California.

1 to 1½ inches, dorsal ridge fairly acute, valves flat-sided, sculpture weak, the surface with fine rhomboidal pittings.

Girdle encroaching between the valves on the sides. Brown hairs on the girdle thin, curved and finely branched. Color greenish gray with black and white flecks. Interior of valves bluish. Common; on the sides of boulders in sandy intertidal areas.

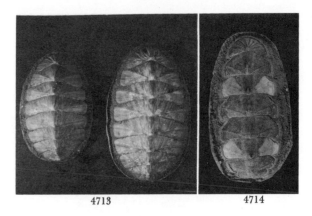

4713   4714

Mopalia hindsii (Reeve, 1847)   4715
Hinds' Mopalia

Alaska to the Gulf of California.

2 to 4 inches in length, oblong, flattened and resembling *M. ciliata* Sowerby, but generally smoother. Girdle brown, rather thin and fairly wide, and almost naked except for a few short hairs. Interior of valves white with short crimson rays under the beaks. Moderately common; intertidal, on undersides of rocks.

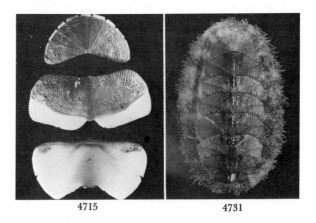

4715   4731

Mopalia recurvans Barnawell, 1960   4716
Barnawell's Chiton

Bodega Bay to San Francisco Bay, California.

Similar to *hindsii*, but the amber-yellow mantle bristles have recurving chitinous spines over the proximal ⅓, while in *hindsii* the spines are not recurved, are on the proximal ¼ of the more flattened bristles. Median valves without the basketlike pattern (of *hindsii*) between the anterior granose rib and the posterior sutural thickening. Tegmentum green with reddish flames and sometimes with white splotches. Moderately common; with *hindsii*. *The Veliger*, vol. 3, p. 39.

Subgenus *Dendrochiton* S. S. Berry, 1911

Less than 1 inch, arched, carinate, with straight-sided slopes. Surface minutely granulose, with obscure ribbing on poorly defined lateral areas. Head valve with 8 slits, intermediate valves single-slitted and tail valve with a regular crescentic insertion plate cut by 5 to 8 slits. Insertion plates continuous across sinus of intermediate valves, sinus being very spongy. Girdle microscopically spiculose, bearing rather long, branching bristles at each suture, 2 to 5 in front of the head valve and 2 behind the tail valve. Synonyms: *Lophochiton* S. S. Berry, 1925; *Ploiochiton* S. S. Berry, 1926; *Basiliochiton* S. S. Berry, 1918. We follow Leloup (1942, p. 59) in placing this group in *Mopalia*, rather than in the *Ischnochitonidae*. Type: *thamnopora* Berry, 1911.

Members of the subgenus are:

4718 *Mopalia flectens* Carpenter, 1864. Vancouver, British Columbia, to San Diego, California. (Synonyms: *heathi* Pilsbry, 1898, *semiliratus* S. S. Berry, 1927.) Surface with blue spots.

4719 *Mopalia lobium* S. S. Berry, 1925. La Jolla, California.

4720 *Mopalia gothicus* (Carpenter, 1864). Monterey to Guadalupe Island, Mexico.

4721 *Mopalia thamnopora* S. S. Berry, 1911. Alaska to Baja California.

4722 *Mopalia psaltes* (S. S. Berry, 1963). Mission Bay, San Diego County, California.

4723 *Mopalia laurae* (S. S. Berry, 1963). Bahia de los Angeles, Baja California.

4724 *Mopalia lirulatus* (S. S. Berry, 1963). Punta San Felipe, Baja California. Above 3 from *Leaflets in Malacology*, no. 22, pp. 135 and 136.

Other *Mopalia*:

4725 *Mopalia cirrata* S. S. Berry, 1919. Dundas Bay, Alaska. *Lorquiniana*, vol. 2, p. 45.

4726 *Mopalia imporcata* Carpenter, 1864. Alaska to San Pedro, California (form **(4727)** *lionotus* Pilsbry, 1918. *The Nautilus*, vol. 31, p. 126).

4728 *Mopalia egretta* S. S. Berry, 1919. Forrester Island, Alaska, 20 fms., to Puget Sound, Washington.

4729 *Mopalia sinuata* Carpenter, 1864. Alaska to Monterey, California. (Synonym: *goniura* Dall, 1919.)

4730 *Mopalia phorminx* S. S. Berry, 1919. Bowen Island, California, to off Pt. Piños, California.

4731 *Mopalia lowei* Pilsbry, 1918. San Pedro to La Jolla, California. 25 mm.

4732 *Mopalia chloris* Dall, 1918. San Diego, California.

4733 *Mopalia chacei* S. S. Berry, 1919. La Jolla, California.

4734 *Mopalia celetoides* Dall, 1919. Forrester Island, Alaska.

4735 *Mopalia pedroana* Willett, 1932. San Pedro, California. *The Nautilus*, vol. 45, p. 101.

4736 *Mopalia cithara* S. S. Berry, 1951. Vancouver Island, British Columbia. *Proc. Malacol. Soc. London*, vol. 28, p. 222.

Genus *Placiphorella* Carpenter in Dall, 1879

Oval, valves wide and short. Head valve narrowly crescent-shaped; tail valve much smaller, with a shallow posterior sinus and posterior mucro. Girdle very wide and extended in front, bearing sparse, scaled bristles or hairs. Type: *velata* Dall, 1879. Synonyms or subgenera: *Euplacophora* Verrill and Smith, 1882; *Placophora* Dall, 1889; *Langfordiella* Dall, 1925; *Placophoropsis* Pilsbry, 1893. Some members of this genus, such as *velata*, capture amphipods

and worms with their fore-girdle and eat them (see J. McLean, 1962, *Proc. Malacol. Soc. London*, vol. 35, p. 24).

Placiphorella velata Dall, 1879   4737
Veiled Pacific Chiton

Monterey, California, to Baja California.

1 to 2 inches in length, readily recognized by its flat, oval shape and wide girdle which is very broad in front. There are a few hairs on the girdle which, if viewed under a lens, will be seen to be covered by a coat of diamond-shaped scales.

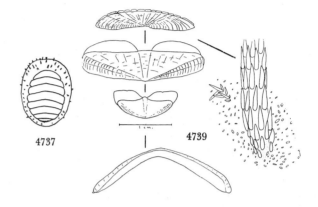

4737   4739

Girdle reddish yellow. Valves colored a dull olivaceous brown with streaks of buff, blue, pink or chestnut. Interior of valves white with a slight bluish tint. Posterior valve with 1 or 2 slits and with very large sutural plates. Fairly common intertidally. Authorship erroneously credited to Pilsbry, 1893.

Other species:

4738 *Placiphorella rufa* S. S. Berry, 1917. Forrester Island, Alaska, 15 to 25 fms.

4739 *Placiphorella borealis* Pilsbry, 1893. Bering Sea and Kuril Islands, 228 to 283 fms. 50 mm. See S. S. Berry, 1917, *Proc. U.S. Nat. Mus.*, no. 2223.

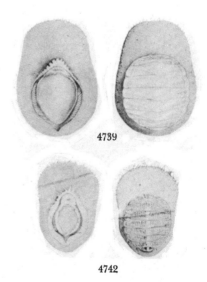

4739

4742

4740 *Placiphorella* (*Placophoropsis* Pilsbry, 1893) *pacifica* S. S. Berry, 1919. Kasaan Bay, Alaska, 95 to 98 fms.; Pioneer Seamount, off central California, 500 to 650 fms.

4741 *Placiphorella* (*Placophoropsis*) *atlantica* (Verrill and Smith, 1882). Off Georges Bank, Massachusetts, 122 fms. 30 mm.

4742 *Placiphorella stimpsoni* (Gould, 1859). Japan to Alaska; Baja California. 30 mm. See S. S. Berry, 1917, *Proc. U.S. Nat. Mus.*, no. 2223.

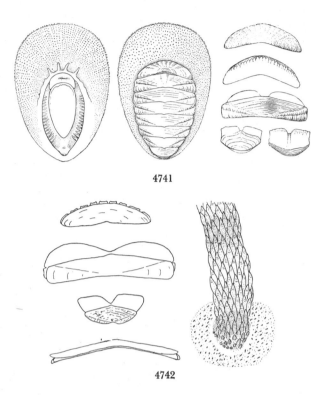

4741

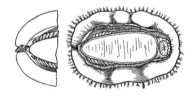

4742

Genus *Katharina* Gray, 1847

Mantle black, naked and covering ⅔ of the valves. Head valve with 7 or 8 slits. Type: *tunicata* (Wood, 1815). Synonyms: *Katherina* Carpenter, 1857; *Catharina* Dunker, 1882. One American species:

Katharina tunicata (Wood, 1815)   4743
Black Katy Chiton

Aleutian Islands to southern California.

2 to 3 inches in length, oblong and elevated. Characterized by its shiny, naked, black girdle which covers ⅔ of each gray

4743

valve. Valves usually eroded. Anterior valve densely punctate. Interior of valves white. Foot salmon to reddish. Very common between tides, especially in the north.

Genus *Amicula* Gray, 1847

The name *Amicula* has been used in different families, and the present usage seems to be the most acceptable. Valves almost entirely buried in a thick, pilose girdle, leaving very small, heart-shaped areas exposed at each apex. Tail valve with a posterior sinus and a single slit on each side. Type: *vestita* (Broderip and Sowerby, 1829). Synonyms: *Symmetrogephrus* Middendorff, 1847; *Stimpsoniella* Carpenter, 1873; *Chlamydochiton* Dall, 1879; *Chlamydoconcha* Pilsbry, 1893, non Dall, 1884, a bivalve. *Amicula* Gray, 1840, is a nomen nudum.

Amicula vestita (Broderip and Sowerby, 1829)   4744
Concealed Arctic Chiton

Arctic Seas to Massachusetts Bay.
Arctic Seas to the Aleutian Islands.

1 to 2 inches in length, oval, rather elevated. Valves covered with a thin, smooth, brown girdle except for a small, heart-shaped exposure at the center of each valve. Girdle may have widely scattered tufts of hair. Interior of valves white. Common from 5 to 30 fathoms. *S. emersonii* Winkley, 1896, is a synonym.

4744

(4745) *A. pallasii* Middendorff, 1846 (concealed Pacific chiton), is very similar to *vestita*, but the girdle is much thicker and the bunches of reddish hairs more numerous. Uncommon from 3 to 10 fathoms from Okhotsk Sea to Alaska.

Other species:

4746 *Amicula amiculata* (Pallas, 1786). Japan to Alaska. See Leloup, 1942, *Mem. Mus. Royal Hist. Nat. Belgique*, ser. 2, fasc. 25, p. 6.

Family CHITONIDAE Rafinesque, 1815

Insertion plates have well-developed pectinations which may be coarsely grooved to fine and comblike. Girdle may be scaly, nude, with spicules or bristles.

Subfamily CHITONINAE Rafinesque, 1815

Genus *Chiton* Linné, 1758

Outer edges of insertion plates with tiny, sharp teeth or pectinations; girdle with small, hard scales that look like overlapping split peas; girdle colored with alternating bars of grayish green and black; interior of valves blue-green. Type: *tuberculatus* Linné, 1758.

4747   4748   4749   4750   4751

Key to the American *Chiton*

Valves entirely smooth:

Girdle scales glassy; sinus of valves narrow . . . .
. . . . *articulatus* (Pacific)

Girdle scales dull; sinus of valves wide . . . . .
. . . . *marmoratus* (Atlantic)

Valves with sculpturing:

Central area with longitudinal ribs:

Posterior valve with round pimples . . . .
. . . . *tuberculatus* (Atlantic)

Posterior valve with radiating ribs . . . .
. . . . *virgulatus* (Pacific)

Central area smoothish and dull . . . .
. . . . *squamosus* (Atlantic)

Chiton tuberculatus Linné, 1758   4747
Common West Indian Chiton

Southeast Florida and West Indies. Bermuda.

2 to 3 inches in length. Color a dull grayish green or green-ish brown. Some or all of the valves may have a smooth, dark-brown, arrow-shaped patch on the very top. Girdle with alternating zones of whitish, green and black. The scales are placed on the girdle, so that they appear slightly higher than broad, while in *squamosus* they appear to be broader than high. Lateral areas with 5 irregular, radiating cords. Central areas smooth at the top and with 8 to 9 long, strong, wavy, longitudinal ribs on the sides. End valves with irregular, wavy radial cords. Gills beginning at the juncture of the head and foot and with 46 to 48 lamellae. A very common species on wave-dashed, rocky shores. Texas records have not been confirmed. Rare at Boca Raton, Florida. For ecology and reproduction, see P. W. Glynn, 1970, *Smithsonian Contrib. Zool.*, no. 66.

Chiton squamosus Linné, 1764   4748
Squamose Chiton

West Indies.

2 to 3 inches in length. Color a dull, ashen-gray with wide, irregular, dull-brown, longitudinal stripes. Posterior edge of first 7 valves marked with 4 or 5 squares of blackish brown. Girdle with alternating pale stripes of grayish green and grayish white. Posterior valve with minutely pimpled ribs. Lateral areas of middle valves with 6 to 8 rows of small beads between which are microscopic pinholes. Central areas smooth-ish, with fine, transverse scratches. Common. Florida records not confirmed.

(4748a) *Chiton viridis* Spengler, 1797, from the West Indies is very similar, but the margins of the central areas have 6 to 11 very short, wavy ribs, and the lateral areas have 3 or 4 strong ribs of rounded pustules. Interior of valves whitish. Moderately common; intertidal.

Chiton marmoratus Gmelin, 1791   4749
Marbled Chiton

Southeast Florida and the West Indies.

2 to 3 inches in length. Color variable: (1) entirely blackish brown; (2) olive with flecks, patches and lines of whitish merging together towards the middle; or (3) purplish brown or light-olive with zebralike stripes on the sides. Entire surface of valves smooth except for a microscopic, silky texture. Lateral areas a little raised. Underside of posterior valve with 14 to 16 slits. A common West Indian littoral species. Texas records not confirmed. Rare in Biscayne Bay, Florida.

Chiton articulatus Sowerby, 1832   4750
Smooth Panama Chiton

Gulf of California to Panama.

2 to 3 inches in length, similar to *marmoratus*. Girdle scales broader than high and not very glossy. Valves smooth, with a silky sheen and colored a grayish green over which are radiating rays of dark-brown. Similar coloration in the other valves. Underside of posterior valve with 21 narrow slits. Sutural plates on underside of middle valves with a dark blotch at the base. Common. *C. laevigatus* Sowerby, 1832, non Fleming, 1813, is the same.

(4750a) *Chiton albolineatus* Broderip and Sowerby, 1829 (western Mexico, uncommon) is also a smooth species, but differs in having only 16 or 17 slits in the posterior valve and has snow-white, radiating lines on the lateral areas and on the end valves. The girdle scales are light blue-green and edged with white. Uncommon, on intertidal rocks.

Chiton virgulatus Sowerby, 1840   4751
Virgulate Chiton

Magdalena Bay, Baja California, to Panama.

2 to 3 inches in length, similar to *tuberculatus*. Girdle scales glassy. End valves with rather even, raised, radiating threads. Middle valves with the lateral areas bearing about 8 raised threads which split into 2 at the margin of the valve. Central area with about 60 to 70 even, longitudinal threads. Posterior valve with 19 or 20 prominent slits. Common. The closely resembling species (4751a), *C. stokesii* Broderip, 1832, from Mexico to West Colombia has only 15 or 16 slits in the posterior valve. (Synonym: *patulus* Sowerby, 1840.)

Genus *Tonicia* Gray, 1847

Resembling *Chiton* in having pectinate or toothed sutural plates, but the girdle is naked and the upper surface of the valves has microscopic eyes. The second valve is usually larger than the others. Black eye dots present on valves. Type: *elegans* Frembly, 1827. *Tonichia* Gray, 1840, is a dubious name.

Tonicia schrammi (Shuttleworth, 1856)   4752
Schramm's Chiton

Southeast Florida (rare) and the West Indies. Bermuda.

About an inch in length, colored a brownish red to buff and with darker mottlings and speckles. Upper surface of valves glossy; interior white with a crimson stain in the center. Lateral areas separated from the smooth central area by a strong, rounded rib. The central area has a peppering of about 75 tiny, black eyes. Head valve smooth except for 8 to 10 broad rays of tiny, black eyes. Girdle naked, leathery and brownish to flesh-colored. Posterior valve with 14 slits. 36 lamellae in each of the 2 gills. They begin just behind the juncture of the head and the foot and extend back almost to the posterior end where there is a bilobed, small, fleshy lappet. A moderately common, intertidal species in the West Indies.

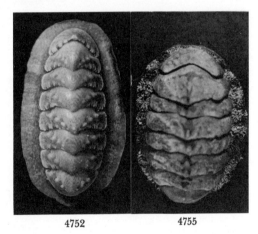

4752   4755

Other species:

4753 *Tonicia pustulifera* Dall, 1919. San Pedro, California. *Proc. U.S. Nat. Mus.*, vol. 55, p. 516.

4754 *Tonicia forbesii* Carpenter, 1856. Gulf of California to Panama. (Synonym: *crenulata* (Broderip, 1832), not Risso, 1826.)

Subfamiíy ACANTHOPLEURINAE Thiele, 1910

Genus *Acanthopleura* Guilding, 1829

Girdle thick, densely covered with calcareous spines. Tail valve with very long teeth projecting forward, instead of outward; with numerous slits in the intermediate valves. Type: *spinosa* (Bruguière, 1792). Synonyms or subgenera: *Maugeria* Gray, 1857; *Canthapleura* Swainson, 1840; *Rhopalopleura* Thiele, 1893.

Acanthopleura granulata (Gmelin, 1791)   4755
Fuzzy Chiton

South half of Florida and the West Indies.

2 to 3 inches in length, usually so worn and eroded as to eliminate the brown color and granulated sculpturing. Girdle thick, ashy white with an occasional black band, and matted

with coarse, hairlike spines. Underside of valves colored a light-green, with the middle valves having a rather large, black splotch behind the sinus. Posterior valve with about 9 slits. Compare with *Ceratozona rugosa* whose gills do not extend to the very posterior end as they do in this species. Common; on upper intertidal rocks. For ecology and reproduction, see P. W. Glynn, 1970, *Smithsonian Contrib. Zool.*, no. 66; and S. H. Hamilton, 1903, *The Nautilus*, vol. 16, p. 138.

Suborder *Acanthochitonina* Bergenhayn, 1930

Valves are partially to completely buried in the girdle. Tail valve with a small triangular jugal area and 2 adjacent subtriangular pleural areas all radiating forward from the mucro, and a relatively large posterior area behind these.

Family ACANTHOCHITONIDAE Pilsbry, 1893

Valves partially to completely buried in the girdle. Tegmental area narrow. Girdle with tufts of long bristles at the juncture of the valves.

Genus *Acanthochitona* Gray, 1821

Narrow and elongate; sculpture of rows of pustules, usually teardrop-shaped. Girdle nude or hairy, and with 18 clumps of long spicules arising from pores between the sides of the valves. 5 slits on head valve, 1 on each side of intermediate valve, and 1 on each side of the wide sinus of the tail valve. Type: *fascicularis* (Linné, 1766). Synonyms: *Acanthochites* Risso, 1826; *Acanthochetes* Gray, 1840; *Acanthochiton* Iredale, 1915.

Acanthochitona spiculosa (Reeve, 1847)   4756
Glass-haired Chiton

Southeast Florida and the West Indies.

1 to 1½ inches in length, elongate, with the girdle covering most of the valves. There are 4 clumps of long, glassy bristles near the anterior valve and 1 on each side of the other valves. The clumps are set in cuplike collars of the girdle skin. End valves and lateral areas of middle valves covered with tiny, round, sharply raised pustules. The dorsal, longitudinal ridge is raised, narrow, distinct and smoothish except for microscopic pinpoints. Lower edge of girdle with a dense fringe of brown or bluish bristles. 32 gill lamellae. The gills begin about ⅓ back along the side of the foot and do not extend quite so far back as the posterior mantle margin. A moderately common species in shallow water. *A. astriger* (Reeve, 1847) is the same species.

Acanthochitona pygmaea (Pilsbry, 1893)   4757
Dwarf Glass-haired Chiton

West coast of Florida to the West Indies.

½ to ¾ inch in length, moderately elongate and colored cream, green, brown or variegated with these colors. Similar to *A. spiculosa*, but smaller, and with its dorsal ridge triangular, less elevated and cut by longitudinal grooves. The pustules on the lateral areas and the end valves are round or oval. The clumps of bristles are the same. Not uncommonly found among rocks and dead shells at low tide.

(4758) *Acanthochitona balesae* "Pilsbry," Abbott, 1954 from the Lower Florida Keys is very elongate, only ⅓ inch in length; the pustules on the lateral areas are proportionately larger and fewer, and the dorsal ridge is rounded and covered with small, granulose pustules. Rare at Bonefish Key, Florida. *A. elongata* Kaas, 1972, is a synonym.

Acanthochitona avicula (Carpenter, 1864)   **4759**
Pacific Glass-haired Chiton

Redondo Beach, California, to Baja California.

10 to 20 mm., elongate. Jugal area has fine, longitudinal lines; lateropleural areas have teardrop-shaped, raised pustules. Color finely mottled in whites, browns and grays. Girdle with numerous fine bluish spicules and 18 tufts of long, glassy spines. *A. diegoensis* Pilsbry, 1893, is a synonym. Uncommon; intertidal among boulders.

Other species:

4760 *Acanthochitona exquisita* (Pilsbry, 1893). Gulf of California. Uncommon.

4761 *Acanthochitona angelica* Dall, 1919. Angeles Bay, Gulf of California. May be *A. avicula* Carpenter, 1864.

4762 *Acanthochitona arragonites* (Carpenter, 1857). Gulf of California.

Genus *Craspedochiton* Shuttleworth, 1853

Similar to *Acanthochitona,* but the head valve has low to prominent, radial ribbing, and the tail valve has 6 to 10 slits. Girdle variously spiculose or minutely scaled, as well as having sutural tufts. Type: *laqueatus* (Sowerby, 1841). Among the synonyms or subgenera are: *Notoplax* H. Adams, 1862; *Spongiochiton* Carpenter in Dall, 1882; *Loboplax* Pilsbry, 1893; *Phacellozona* Pilsbry, 1894; *Thaumastochiton* Thiele, 1909; and *Ikedaella* Taki and Taki, 1929.

Craspedochiton hemphilli (Pilsbry, 1893)   **4763**
Hemphill's Chiton

Lower Florida Keys and West Indies.

20 to 30 mm., girdle covering most of each valve. Color red, occasionally spotted with white. Girdle rusty-brown, lower edge with a spiculose fringe. There are 4 tufts of long, glassy bristles near the anterior valve and 1 on each side of the other valves. The tufts are set in cuplike collars. Valves heart-shaped, covered with minute, round, sharply raised pustules, except on the narrow dorsal longitudinal ridge. Moderately common; sublittoral on dead *Porites* coral.

4763

Genus *Cryptoconchus* Burrow, 1815

Elongate, narrow, valves almost completely buried in the large fleshy girdle with only the jugal area of the valve showing. Girdle with 18 sutural tufts of glassy spines. Type: *porosus* Burrow, 1815.

Cryptoconchus floridanus (Dall, 1889)   **4764**
White-barred Chiton

Southeast Florida to Puerto Rico and Aruba.

Rarely over an inch in length, long and narrow, and characterized by its thin, black, naked girdle which extends on to the valves except over the narrow, beaded dorsal area. These exposed bands in the valves make it appear as if a streak of white paint had been applied along the top of the animal. The side of the girdle at each valve-suture has a minute pore bearing short bristles, but these 2 features are commonly difficult to see. A variety is found with a brown-colored girdle. 16 gill lamellae. They begin halfway back along the side of the foot. Uncommon.

Genus *Cryptochiton* Middendorff, 1847

Very large; white "butterfly-shaped" valves completely buried in the tough, thick girdle. Type: *stelleri* (Middendorff, 1847). Synonym: *Amicula* of authors: *Cryptochiton* Gray, 1847, in part. One American species:

Cryptochiton stelleri (Middendorff, 1847)   **4765**
Giant Pacific Chiton

Japan and Alaska to California.

6 to 14 inches in length, oblong and flattened. The large, white, butterfly-shaped valves are completely covered by the large, leathery, firm girdle which is reddish brown to yellowish brown. Minute red spicules make the girdle feel gritty. Common in the northern part of its range. Found among rocks at low-tide area. Formerly called *Amicula stelleri.* For notes on biology, see MacGinitie and MacGinitie, 1968, *The Veliger,* vol. 11, p. 59.

4765

Class APLACOPHORA von Ihering, 1876
The Solenogasters

Wormlike animals, without shells, having a thickened cuticle in which spicules may be embedded. Mantle almost envelops the entire animal, except for a ventral groove. A reduced radula may be present. They live in mud from shallow to very deep water. We do not treat them in detail. Consult: Pilsbry, H. A., 1898, "Manual of Conchology," Philadelphia, vol. 17, pp. 281–310; and Heath, H., 1911, *Mus. Comp. Zool., Harvard, Memoirs,* vol. 45, pp. 1–260, 54 pls.

Order *Neomenioidea* Simroth, 1893

Family NEOMENIIDAE von Ihering, 1876

4766 *Neomenia carinata* Tullberg, 1875. North Atlantic.

4767 *Pachymenia abyssorum* Heath, 1911. Off southern California, 2,228 fms.

4768 *Alexandromenia valida* Heath, 1911. Off southern California, 603 to 1,350 fms.

4769 *Alexandromenia agassizi* Heath, 1911. 841 meters, off Revillagigedo Islands, west of Mexico.

Family PRONEOMENIIDAE Simroth, 1893

4770 *Dorymenia acuta* Heath, 1911. Off southern California, 302 to 638 fms.

4770a *Dorymenia peroneopsis* Heath, 1918. South of Martha's Vineyard, Massachusetts, 1,753 fms. *Memoirs Mus. Comp. Zool.,* vol. 45, no. 2, p. 222.

4771 *Proneomenia sluiteri* Hubrecht, 1880. Barents Sea; Arctic Seas.

4771a *Proneomenia acuminata* Wirén, 1892. Off Massachusetts to the Florida Straits.

Family LEPIDOMENIIDAE Pruvot, 1890

4772 *Nematomenia platypoda* Heath, 1911. Off Agattu Island, Aleutians, 482 fms.

4773 *Heathia porosa* (Heath, 1911). Off San Diego, California, 500 to 542 fms.

4774 *Dondersia californica* Heath, 1911. Off San Diego, California, 21 fms.

Order *Chaetodermatida* Simroth, 1893

Family CHAETODERMATIDAE von Ihering, 1876

4775 *Chaetoderma attenuatum* Heath, 1911. Alaska, 50 to 200 fms.

4776 *Chaetoderma montereyense* Heath, 1911. Monterey Bay, California, 39 to 356 fms.

4777 *Chaetoderma argenteum* Heath, 1911. Alaska, 82 to 113 fms.

4778 *Chaetoderma eruditum* Heath, 1911. Alaska, 282 to 313 fms.

4779 *Chaetoderma californicum* Heath, 1911. Coronado Islands, California, 618 to 667 fms.

4780 *Chaetoderma robustum* Heath, 1911. Alaska, 483 fms.

4781 *Chaetoderma scabrum* Heath, 1911. Monterey Bay, California, 795 to 871 fms.

4782 *Chaetoderma nanulum* Heath, 1911. Off San Diego, California, 260 to 284 fms.

4783 *Chaetoderma nitidulum* (Lovén, 1844). Greenland, 10 to 250 fms.; Nova Scotia, 10 to 100 fms.; Casco Bay, Maine, 48 to 64 fms. Common.

4783a *Chaetoderma vadorum* (Maine), **(4783b)** *lucidum* (New Jersey), **(4783c)** *bacillum* (Cape Cod, Massachusetts), **(4783d)** *squamosum* (south of Cape Cod), all Heath, 1918. *Memoirs Mus. Comp. Zool.,* vol. 45, no. 2.

4784 *Chaetoderma canadense* Nierstrasz, 1902. Newfoundland to south of Martha's Vineyard, Massachusetts, to 100 meters. Burrows in fine sandy silt.

4784a *Falcidens* (Salvini-Plawen, 1968) *caudatus* (Heath, 1918). Off Cape Cod, Massachusetts, to off New Jersey, down to 479 fms. For radula, see A. H. Scheltema, 1972, Zeit. Morph. Tiere, vol. 72, pp. 361–370.

4785 *Limifossor talpoideus* Heath, 1904. Alaska, 282 to 313 fms.

V

The Bivalves

Class BIVALVIA Linné, 1758

The bivalves or clams are dwellers of fresh, marine or brackish waters. They lack a head and are without jaws or radular teeth; they are protected by a pair of shelly valves which are connected dorsally by a horny ligament. The valves are moved by the contraction of 1 to 3 muscles attached to the inner sides of the valves. The class is also known as the Pelecypoda (Goldfuss, 1820) and Lamellibrachiata (Blainville, 1814 and 1824). We are adopting in this revision the classification of the modern paleontologists who have tempered their arrangements with knowledge of the soft parts as well. Consult: H. E. Vokes, 1967, "Genera of the Bivalvia: a systematic and bibliographic catalogue," *Bull. Amer. Paleontol.*, vol. 51, no. 232; N. D. Newell, 1965, "Classification of the Bivalvia," *Amer. Mus. Novitates*, no. 2206; R. C. Moore, editor, 1969, "Treatise on Invertebrate Paleontology," part N, vols. 1–3, Geol. Soc. Amer.

Outline of Major Divisions
Class BIVALVIA Linné, 1758

Subclass *Palaeotoxodonta* Korobkov, 1954

The nuculoids; nacreous or crossed lamellar shell structure; equivalved. Taxodont teeth.

Order *Nuculoida* Dall, 1889

Hinge with numerous, similar teeth. Usually nacreous inside. Margins not gaping. Ligament usually on both sides of the umbones. Gills protobranchiate; foot grooved, usually without a byssus. The Paleozoic superfamily Ctenodontacea Wöhrmann, 1893, is not treated here.

Superfamily Nuculacea Gray, 1824

Posterior end usually short and truncated. Pallial sinus absent. Resilifer present or absent.

Key to Families

a. Chondrophore below hinge present:
    b. External ligament absent; shell ovate . . . .
        . . . . *Nuculidae*
    bb. External ligament present; shell elongate .
    . . . . *Nuculanidae*
aa. Chondrophore absent; shell ovate . . . *Malletiidae*

Family NUCULIDAE Gray, 1824

Genus *Nucula* Lamarck, 1799

Shell ovate, usually less than ⅓ inch in size; interior pearly; ventral margins usually with fine denticulations. Surface smooth or with radially cancellate sculpture. Pallial line simple. Chondrophore narrow, inclined obliquely forward. Type: *nucleus* Linné, 1758.

Subgenus *Nucula* Lamarck, 1799

Inner layer of shell with radially ribbed structure which shows on the surface as fine radial striations. Inner ventral margin of valves crenulate. *Lembulus* Sowerby, 1842, is a synonym.

Nucula proxima Say, 1822   4786
Atlantic Nut Clam

Nova Scotia to Florida and Texas. Bermuda.

¼ inch in length, obliquely ovate, smooth. Color greenish gray with microscopic, embedded, axial, gray lines and prominent, irregular, brownish, concentric rings. Outer shell overcast with oily iridescence. Anterior end often with microscopic, axial lines. Ventral edge minutely crenulate. Common just offshore in mud and sand.

The size, shape and coloration of this species varies according to the substrate and temperature of water. Among the probable forms are: **(4787)** *ovata* Verrill and Bush, 1898 (non Deshayes, 1824); **(4788)** *truncula* Dall, 1898; **(4788a)** *annulata* Hampson, 1971 (*Proc. Mal. Soc. London,* vol. 39, p. 333). The latter ranges from Nova Scotia to Virginia, lives in mud, usually has a tear-shaped resilium, usually has fewer than 16 anterior teeth, and dark growth rings.

4786

Nucula atacellana Schenck, 1939   4789
Cancellate Nut Clam

Massachusetts to Maryland.

⅛ inch in length and oval in outline; the cancellate sculpturing is due to the crossing of numerous radial and concentric threads. Interior scarcely pearly and with a thick, transparent glaze. Color yellowish brown to light-tan. Commonly dredged offshore down to 500 fathoms. Formerly known as *N. reticulata* Jeffreys, 1876, non Hinds, 1843, and *cancellata* Jeffreys, 1881, non Meek and Hayden, 1856.

Nucula crenulata A. Adams, 1856   4790
Atlantic Crenulate Nut Clam

South Carolina to Florida; Texas and to Brazil.

¼ inch in length, ovate, internal margin finely crenulate. With numerous concentric, fine ribs which have numerous microscopic crenulations between them. Interior scarcely pearly; overlaid by a thick, transparent layer of shell matter, through which radial fractures or lines are discernible; color yellowish. Dredged in shallow water. Similar in outline to *N. tenuis* Montagu, 1808. *N. culebrensis* E. A. Smith, 1885, is a synonym. A more compressed, smoother, triangular form **(4790a)** was named *obliterata* Dall, 1881.

Nucula aegeensis Jeffreys, 1879   4791
West Indian Nut Clam

North Carolina to the West Indies. Brazil.

¼ to ⅓ inch, ovate; exterior white; interior pearly. Surface sculptured with fine concentric lines. Ventral margin minutely crenulate. Uncommon; 5 to 464 fathoms.

 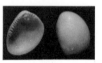

4792

4791

Nucula exigua Sowerby, 1833   4792
Pacific Crenulate Nut Clam

Southern California to Peru.

³⁄₁₆ inch in length (5 mm.), plump, with high, wide umbones, shaped like *tenuis* and the preceding species. Concentric rings strong with radial crenulations between. Strongly projecting lunular area just under the beaks. Color yellowish. Dredged commonly in shallow water. Hinge with 18 anterior and about 9 posterior teeth.

Nucula delphinodonta Mighels and Adams, 1842   4793
Delphinula Nut Clam

Labrador to Maryland.

1⁄16 inch (3 mm.) in length, ovate, fat and smooth except for coarse concentric growth lines. Anterior end slightly pushed in under beaks and bordered by slight carination. Ventral edge smooth. Color olive-brown. 9 teeth posterior to and 4 teeth anterior to chondrophore.

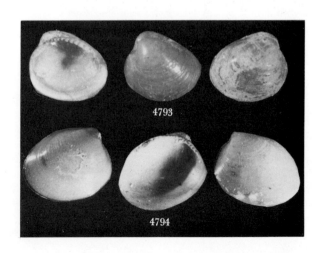

4793

4794

Subgenus *Leionucula* Quenstedt, 1930

Ventral margin of valves smooth. Type: *albensis* Orbigny, 1844. Synonyms: *Nuculopsis* Woodring, 1925, non Girty, 1911; *Ennucula* Iredale, 1931.

Nucula tenuis Montagu, 1808   **4794**
Smooth Nut Clam

Labrador to Florida. Greenland; Europe.
Alaska to Baja California.

Usually ³⁄₁₆ inch in length (up to ⅜ inch in Alaska), ovate, smooth except for irregular growth lines. Color a shiny olive-green, sometimes with darker lines of growth. No radial lines. Ventral edge smooth. Moderately common; offshore. Synonyms or forms: *inflata* Hancock, 1846; *subovata* Verrill and Bush, 1898, non Orbigny, 1850; *expansa* Reeve, 1855; *bushae* Doll-fuss, 1898.

Other species:

4795 *Nucula zophos* A. H. Clarke, 1960. North Canadian Basin, Arctic Ocean. 290 to 1,245 fms. *Breviora*, no. 119, pl. 1, figs. 15–18.

4796 *Nucula fernandinae* Dall, 1927. Off Fernandina, Florida, 294 fms.

4797 *Nucula cymella* Dall, 1886. Florida Strait and Gulf of Mexico, 205 to 1,100 fms. Brazil. 5 mm.

4798 *Nucula granulosa* Verrill, 1884. Off Massachusetts to Virginia, 384 to 1,061 fms. 2 mm.

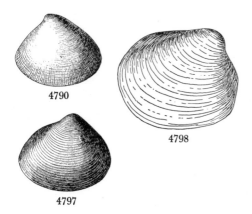

4790

4798

4797

4799 *Nucula carlottensis* Dall, 1897. Queen Charlotte Islands, British Columbia, 876 fms., to Anacapa Island, California.

4800 *Nucula darella* Dall, 1916. Off San Diego, 822 fms. See Schenck, 1939, *Jour. Paleontol.*, vol. 13, pl. 5, figs. 1, 3, 7, 8.

4801 *Nucula linki* Dall, 1916. British Columbia to Guaymas, Mexico. Illus. in Schenck, 1939, *Jour. Paleontol.*, vol. 13, pl. 5.

4802 *Nucula cardara* Dall, 1916. Monterey, California, to Baja California; deep water. Illus. in Schenck, 1939, *Jour. Paleontol.*, vol. 13, pl. 5.

4803 *Nucula petriola* Dall, 1916. Off Santa Rosa Island, California, 53 fms.

4804 *Nucula quirica* Dall, 1916. Chugachik Bay, Cooks Inlet, Alaska, 60 fms. Last 5 described in *Proc. U.S. Nat. Mus.*, vol. 52, no. 2183. Illus. in Schenck, 1939, *Jour. Paleontol.*, vol. 13, pl. 5. Probably is *bellotii* A. Adams, 1856.

4805 *Nucula groenlandica* Posselt, 1898. West Greenland, 132 fms. *Med. om Groenland*, vol. 23, p. 47.

4806 *Nucula (Nucula) marshalli* Schenck, 1939. Rio de la Plata, Uruguay. (Synonym: *uruguayensis* Marshall, 1928, non E. A. Smith, 1880.) *Jour. Paleontol.*, vol. 13, p. 29.

4807 *Nucula (Leionucula) puelcha* Orbigny, 1842. Rio de la Plata, Uruguay, 28 fms. (Synonyms: *uruguayensis* E. A. Smith, 1880; *felipponei* Marshall, 1928.)

4808 *Nucula (Leionucula) bellotii* A. Adams, 1856. Arctic Seas; Greenland; Cooks Inlet, Alaska. (Synonyms: *inflata* Hancock, 1846, non Wissmann, 1841; *expansa* Reeve, 1855, non Wissmann, 1841.)

4808

4809 *Nucula (Nucula) schencki* Hértlein and Strong, 1940. Gulf of California to Port Guatulco, Mexico, in 13 to 45 meters.

Genus *Brevinucula* Thiele, 1934

Shell small, trigonal, with both ends short. Type: *guineensis* Thiele, 1931. One deep-water East American species:

4810 *Brevinucula verrillii* (Dall, 1886). Massachusetts to the Gulf of Mexico, 274 to 1,685 fms.; off Bermuda. 1,700 fms. off West Africa, 3,196 meters. 2 mm. (Synonym: *trigona* Verrill, 1885, non Seguenza, 1877.)

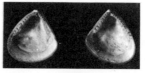

4818

Genus *Acila* H. and A. Adams, 1858

Shells up to 2 inches, similar to *Nucula,* but characterized by the presence of divaricate sculpture on the outside of the shell. One common species in North American waters. Type: *divaricata* Hinds, 1843.

Subgenus *Truncacila* Grant and Gale, 1931

Shell without the shallow sinus as seen in true *Acila,* and the posterior end of the shell nearly at right angles. Type: *castrensis* Hinds, 1843.

Acila castrensis (Hinds, 1843)   **4811**
Divaricate Nut Clam

Bering Sea to Baja California.

½ inch in length, abruptly truncate at the anterior end. Divaricate, radiating ribs plainly visible. Commonly dredged from 4 to 100 fathoms in sandy mud. For a biometric study, see D. L. Frizzell, 1930, *The Nautilus*, vol. 44, pp. 50–53.

4811

Superfamily Nuculanacea H. and A. Adams, 1858

Family MALLETIIDAE H. and A. Adams, 1858

Shell not pearly inside, oval, compressed, gaping at both ends; ligament external, elongated, resting on nymphs; numerous teeth; no resilium. A linear depression extends from the umbonal cavity to the anterior muscle scar. Worldwide, usually deep water. Includes several genera and subgenera including *Tindaria* Bellardi, 1875; *Malletia* Desmoulins, 1832; and *Saturnia* Seguenza, 1877.

Genus *Malletia* Desmoulins, 1832

Shells thin, small, glossy, with twice as many taxodont teeth in the posterior end of the hinge. Ligament external. Lacks rostrum of *Yolida*, and has no lunule or escutcheon. Pallial sinus present. Type: *chilensis* Desmoulins, 1832. Synonyms: *Solenella* Sowerby, 1832; *Ctenoconcha* Gray, 1840; *Nucularia* Conrad, 1869; *Pseudomalletia* Fischer, 1886.

Other Atlantic species:

4812 *Malletia abyssorum* Verrill and Bush, 1898. Off Chesapeake Bay, 2,620 fms.

4813 *Malletia obtusa* G. O. Sars, 1872. Off Massachusetts to North Carolina, 516 to 1,781 fms. Norway to off West Africa. (Synonym: *abyssicola* M. Sars 1858, non Thorell, 1868.)

4814 *Malletia polita* Verrill and Bush, 1898. Off Virginia, 1,569 fms.

4815 *Malletia abyssopolaris* A. H. Clarke, 1960. North Canadian Basin, Arctic Ocean; Alaska. 924 to 1,370 fms.

4816 *Malletia (Neilo* A. Adams, 1852) *dilatata* (Philippi, 1844). Off Florida, 292 to 294 fms.

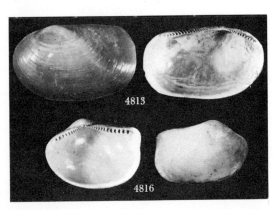

4817 *Malletia (Malletia) bermudensis* Haas, 1949. 1,700 fms., off Bermuda. *Bull. Inst. Catalana Hist. Nat.*, vol. 37, p. 72.

4818 *Malletia veneriformis* E. A. Smith, 1885. Off Bermuda, 435 fms. "Challenger" Station 33.

Other Pacific species:

4819 *Malletia faba* Dall, 1897. Queen Charlotte Islands, British Columbia, to Baja California, 581 to 627 fms.

4820 *Malletia cuneata* (Jeffreys, 1876). Off Sitka, Alaska, in 1,569 fms. Atlantic and Indian Ocean. (Synonyms: *fiora* Dall, 1916; *kolthoffi* Hägg, 1904; *pellucida* Thiele, 1912.)

4821 *Malletia pacifica* Dall, 1897. Chignak Bay, Alaska, to Monterey, California, 152 to 329 fms.

4822 *Malletia talama* Dall, 1916. Bering Sea to Oregon.

Genus *Pseudoglomus* Dall, 1898

Shell thin, circular, with weak concentric sculpture; constricted or rounded posteriorly. Few teeth. Slight pallial sinus. Type: *pompholyx* Dall, 1889.

4823 *Pseudoglomus pompholyx* Dall, 1889. Off Fernandina, 294 fms., and off Fowey Rock, Florida, 205 fms.

4823

Genus *Saturnia* Seguenza, 1877

Shell thick, strong concentric sculpture, constricted posterior end; strong to weak pallial sinus. Type: *pusio* Philippi, 1844. Synonyms: *Austrotindaria* Fleming, 1948; *Neilonella* Dall, 1881. Several deep-water American species:

4824 *Saturnia corpulenta* (Dall, 1881). Florida Straits to the West Indies. 190 to 450 fms. 9 mm.

4825 *Saturnia quadrangularis* (Dall, 1881). North of Yucatan, Gulf of Mexico. 4.6 mm.

4824   4825

4826 *Saturnia subovata* Verrill and Bush, 1897. Nova Scotia to North Carolina. 125 to 1,731 fms. 5 mm.

4826

4827 *Saturnia menziesi* (A. H. Clarke, 1959). 185 miles west of Bermuda; 2,843 fms. *Proc. Malacol. Soc. London*, vol. 33, p. 235, fig. 1.

4828 *Saturnia (Tindariopsis* Verrill and Bush, 1897) *acinula* Dall, 1889. Off Fernandina, Florida, to Brazil. Deep water.

Subgenus *Spinula* Dall, 1908

Strongly rostrate. Type: *Leda calcar* Dall, 1908. Synonym: *Bathyspinula* Filatova, 1958, *nomen nudum*. Deepsea.

4829 *Saturnia (Spinula) calcar* (Dall, 1908). Northwest Pacific, off the Kurils and Japan, between 28° and 43°N, and from the southwest Pacific (Kermadec region at about 37°S), 4,063 to 6,096 meters. (Synonym: *Leda calcarella* Dall, 1908.) See J. Knudsen, 1970, *Galathea Report*, vol. 11, p. 37.

4830 *Saturnia* (*Spinula*) *oceanica* Filatova, 1958. Northwest Pacific, Kuril-Kamchatka Trench, 5,450 to 5,582 meters; Tasman Sea, 4,670 meters (J. Knudsen, 1970, p. 39. See Filatova, 1958, *Trudy Inst. Okeanol.*, vol. 23, p. 213).

4831 *Saturnia* (*Spinula*) *vityazi* Filatova, 1964. Northwest Pacific, Kuril-Kamchatka Trench, 7,220 meters. *Zool. Zh.*, vol. 43, p. 1867.

Genus *Tindaria* Bellardi, 1875

Shell small, resembling a tiny Venus clam; fat; beaks facing slightly forward; ligament minute, external; hinge smooth, continuous just below beaks. No pallial sinus. Generally deep water and rare. Type: *arata* Bellardi, 1875.

Tindaria brunnea Dall, 1916   **4832**
Brown Tindaria

Bering Sea, Alaska, to Tillamook, Oregon.

¼ inch in length, fat, moderately pointed at posterior end. Very fine concentric scratches. Exterior dark olive-brown. Interior glossy-cream. Has been dredged abundantly in a few places in deep water. There are 8 other rare species on the West Coast of America.

Other Atlantic species:

4833 *Tindaria amabilis* Dall, 1889. Gulf of Mexico and West Indies. 169 to 940 fms.

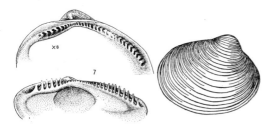

4832   **4833**

4834 *Tindaria callistiformis* Verrill and Bush, 1898. Off Virginia, 1,826 to 2,620 fms. Off Bermuda.

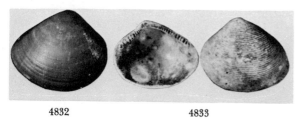

4834

4835 *Tindaria cytherea* Dall, 1889. Off Fernandina, Florida. 294 fms. Gulf of Mexico, 724 fms. (Synonym: *veneriformis* E. A. Smith, 1885.) 8.6 mm.

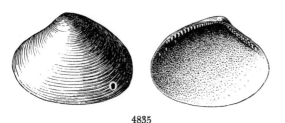

4835

4836 *Tindaria erebus* A. H. Clarke, 1959. 185 mi. west of Bermuda, 2,843 fms. *Proc. Malacol. Soc. London*, vol. 33, p. 236.

4837 *Tindaria lata* Verrill and Bush, 1897. Gulf of Mexico. 730 fms.

4838 *Tindaria smithii* Dall, 1886. Off Florida, 338 to 450 fms. (Synonym: *cuneata* E. A. Smith, 1885, not Jeffreys, 1876.)

4839 *Tindaria acinula* (Dall, 1890). West Indies; northeast Brazil.

Other Pacific species:

4840 *Tindaria californica* Dall, 1916. Santa Barbara to San Diego, California. Deep water.

4841 *Tindaria cervola* Dall, 1916. Off San Diego, California, 822 fms.

4842 *Tindaria dicofania* Dall, 1916. Off San Diego, California, 822 fms.

4843 *Tindaria ritteri* Dall, 1916. Off La Jolla, California, 293 fms.

4844 *Tindaria martiniana* Dall, 1916. Cape St. Martin to Santa Barbara Islands in deep water. Last 5 species in *Proc. U.S. Nat. Mus.*, vol. 52, no. 2183.

4845 *Tindaria mexicana* Dall, 1908. San Diego, California, to off Acapulco, Mexico. Deep water.

4846 *Tindaria gibbsii* Dall, 1897. Queen Charlotte Islands to Coronado Islands.

4847 *Tindaria kennerlyi* Dall, 1897. Off Sitka, Alaska, to Santa Barbara Islands in deep water.

Family NUCULANIDAE Meek, 1864

Shells porcelaneous, elongate, thin, posterior end longer and usually somewhat rostrate. Periostracum thin, varnishlike and dark. Ledidae is a synonym.

Genus *Nuculana* Link, 1807

Shell with the posterior end narrower and somewhat rostrate. Type: *pernula* Müller, 1771. *Leda* Schumacher, 1817, is a synonym. The various subgenera are not very distinctive.

Nuculana pernula (Müller, 1771)   **4848**
Müller's Nut Clam

Arctic Ocean to Cape Cod, Massachusetts.
Northern Alaska to Chatham Sound, British Columbia.

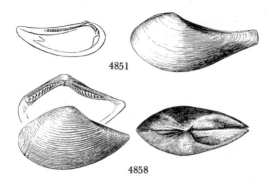

½ to 1 inch in length, elongate and truncate posteriorly, moderately fat, slightly gaping at the rounded anterior end. Numerous raised, concentric growth lines. Periostracum light-brown to dark green-brown, semiglossy. Shell dull-white, interior shiny-white. Interior of rostrum (posterior end of shell) reinforced by a strong radial roundish low rib. Lunule long, prominent, with sharp edge. Commonly dredged off shore in cold water. *N. conceptionis* Dall is much more elongate, smoother and glistening-brown.

Nuculana minuta (Fabricius, 1776)   4849
Minute Nut Clam

Arctic to off San Diego; and to Maine.

½ inch in length, ⅔ as high, rather plump and with a short rostrum whose smoothish lunule is bounded by a rather coarse rib. Concentric, raised threads are numerous and crowded. Beaks are ⅓ to almost ½ the way back from the rounded anterior end. Uncommon offshore. *N. lomaensis* (Dall, 1919) is a synonym.

Nuculana tenuisulcata (Couthouy, 1838)   4850
Thin Nut Clam

Arctic Seas to Rhode Island.

Up to ¾ inch in length, elongate, moderately compressed; rostrum moderately long with a sharp, high keel down the dorsal center (margin of the valves). Concentric ribs fairly even, well-developed, numerous. Periostracum light- to dark-brown. Commonly found in mud just below low-tide mark to 150 fathoms.

Nuculana carpenteri (Dall, 1881)   4851
Carpenter's Nut Clam

North Carolina to West Indies.

About ¼ inch in length, compressed, thin, translucent yellow-brown, with a long, slightly upturning rostrum. Anterior end round. Umbones very small, close together. Almost smooth except for minute, concentric growth lines and microscopic axial scratches which are absent in dead, white valves. Commonly dredged offshore from 10 to 287 fathoms.

Nuculana fossa (Baird, 1863)   4852
Fossa Nut Clam

Alaska to Puget Sound, Washington.

¾ to 1 inch in length, elongate, moderately fat and smoothish except for small, pronounced, concentric ribs at the anterior end and on the beaks. Dorsal area of rostrum smoothish, depressed and bounded by 2 weak radial ribs. Periostracum dark- to light-brown. Dredged offshore in shallow water to 69 fathoms. Some workers consider the following forms or variations as subspecies: *sculpta* (Dall, 1916), *vaginata* (Dall, 1916), and *curtulosa* (Dall, 1916).

Subgenus *Ledella* Verrill and Bush, 1897

Shell small, short, ovate, obese, with a small, short unicarinate rostrum. Escutcheon not sunken, but bounded by a carina. Chondrophore small. Type: *messanensis* Seguenza, 1877. Synonym: *Junonia* Seguenza, 1877, non Hübner, 1819.

Nuculana messanensis (Seguenza, 1877)   4853
Messanean Nut Clam

Cape Cod, Massachusetts, to the West Indies. Bermuda.

⅛ to ¼ inch in length, moderately elongate, with a very short, slightly pinched rostrum. Almost smooth except for a few very small concentric growth ridges near the base of the shell. When alive, glistening light-brown with a slight oily iridescence. When dead, grayish white with concentric chalky streaks. Commonly dredged in moderately deep water to 2,620 fathoms. One of our smallest species. Synonyms are *acuminata* Jeffreys, 1879; *sublevis* Verrill and Bush, 1898.

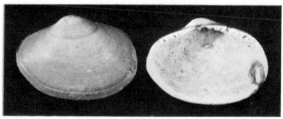

4853

Other species (*Ledella*):

4854   *Nuculana* (*Ledella*) *tamara* Gorbunov, 1946. New Siberian Islands, 2,023 fms.; Queen Elizabeth Islands, 258 fms.; North Canadian Basin, 389 to 1,210 fms.

4855   *Nuculana* (*Ledella*) *orixa* (Dall, 1927). Off Fernandina, Florida, 294 fathoms.

4856   *Nuculana* (*Ledella*) *bipennis* (Dall, 1927). Off Fernandina, Florida, 294 fathoms.

4857   *Nuculana aspecta* (Dall, 1927). Off Martha's Vineyard, 252 fms.; off Georgia and Florida, 294 and 440 fms. (Synonym: *parva* Verrill and Bush, 1897, not Sowerby.)

Subgenus *Saccella* Woodring, 1925

Shell small; a shallow groove extends from the umbo to the ventral margin at both ends of the valve. Sculpture of strong concentric rugae. Apex of pallial sinus broadly U-shaped. Type: *commutata* Philippi, 1844. Synonym: *Ledina* Sacco, 1898, non Dall, April, 1898.

Nuculana acuta (Conrad, 1831)   4858
Pointed Nut Clam

Cape Cod, Massachusetts, to Texas and the West Indies. Brazil.

¼ to ⅜ inch in length, moderately elongate, with a sharp-pointed posterior rostrum. Concentric ribs evenly sized and evenly spaced and extending over the rib which borders the dorsal surface of the rostrum. Shell usually dredged dead in a white condition. Periostracum thin, very light yellowish. Common offshore.

Nuculana concentrica (Say, 1824)   4859
Concentric Nut Clam

Northwest Florida to Texas. Brazil.

½ to ¾ inch in length, strong, rather obese and moderately rostrate. Yellow-white, semiglossy and with very fine, concentric grooves which are evident in adults on the ventral ½ of the valves. Beaks and the area just below smooth. Radial ridge on rostrum smoothish, not crossed by strong threads. Differing from *acuta* in being more obese, in having a smooth beak area, smooth rostral ridge and in having much finer, more numerous, concentric threads or cut lines. Moderately common in 1 to 3 fathoms. *N. eborea* Conrad, 1848, is a synonym.

Nuculana taphria (Dall, 1897)   4860
Taphria Nut Clam

Bodega Bay, California, to Baja California.

About ⅓ to ¾ inch in length, shiny green-brown, with prominent concentric sculpture and characterized by the nearly central umbones. Rostrum bluntly pointed, slightly upturned at the end. Concentric ribs disappear just anterior to the carinate border of the dorsal area which is strongly wrinkled. Adults over ½ inch become quite fat. Commonly dredged off southern California in shallow water. Sorensen reports that this species is found commonly in fish stomachs off Monterey. *Nucula caelata* Hinds, 1843 (not Conrad, 1833) is a synonym.

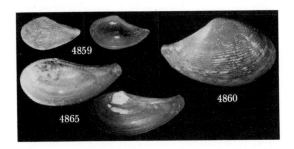

Nuculana penderi (Dall and Bartsch, 1910)   4861
Pender's Nut Clam

Forrester Island, Alaska, to Santa Barbara, California.

¼ to ⅜ inch in length, moderately elongate, very fat; rostrum short and pointed; concentric ribs prominent and evenly developed. Dorsal area of rostrum oval, finely ribbed and bounded by a sharp, smooth, large rib. Periostracum light-brown. Moderately common offshore.

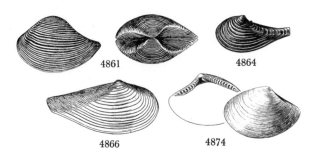

Nuculana hindsii (Hanley, 1860)   4862
Hinds' Nut Clam

Nazan Bay, Alaska, to Costa Rica.

¼ inch in length, moderately elongate (example: 7.8 mm. long; 4.4 mm. high; both valves 3.0 mm. wide); posterior end rostrate, slightly turned up at the end. Dorsal area of rostrum smoothish except for faint axial threads bounded by smooth carinate rib. Sculpture of evenly sized, closely spaced, distinct, concentric ribs which become obsolete just before the rostral rib. Exterior light yellowish brown. Interior white with faint pearly sheen. *N. acuta* Conrad, a name often given to this Pa-

cific Coast species, has its rostral rib crossed by concentric ribs. *N. penderi* Dall and Bartsch is twice as fat, with a very ovate lunule and is more rounded at the ventral margin. Hanley in 1860 first reported this species from "Gulf of Nicoya, Costa Rica." This is "*N. redondoensis*" Burch, 1944, and Woodring, 1951. It is dredged commonly off the West Coast from 15 to 600 fathoms.

Nuculana burchi Willett, 1944   4863
Burch's Nut Clam

Southern California.

12 to 14 mm. long, trigonal, brownish to olivaceous, compressed. Rounded anteriorly, bluntly pointed posteriorly. Exterior sculpture of flattened ribs with narrower interspaces. Anterior teeth 20 to 22, posterior teeth about 15. Resilium triangular. Differs from *taphria* Dall in being flatter, blunter posteriorly, with a straighter posterior dorsal margin and very much finer ribbing. Uncommon; 50 fathoms. *Bull. So. Calif. Acad. Sci.*, vol. 43, p. 71.

Subgenus Thestyleda Iredale, 1929

Surface with strong concentric ripples. 2 radial ridges along the long rostrum. Chondrophore large. Type: *ramsayi* E. A. Smith, 1885, from Australia.

Nuculana hamata (Carpenter, 1864)   4864
Hamate Nut Clam

Puget Sound to off Cedros Island.

Under ½ inch in length, moderately compressed, exterior with strong concentric ribs; characterized by the squarely truncated posterior end of the long rostrum. Fairly commonly dredged off Californian shores from 20 to 200 fathoms. *N. limata* (Dall, 1916) is a synonym.

Other Atlantic species:

4865   *Nuculana caudata* (Donovan, 1801). Gulf of Maine to Virginia. 102 to 641 fms.

4866   *Nuculana bushiana* (Verrill, 1884). Off North Carolina to Florida Strait, 120 to 516 fms. (Same as *cerata* Dall, 1881?) 15 mm.

4867   *Nuculana jacksoni* (Gould, 1841). Labrador, 10 to 80 fms.

4868   *Nuculana jamaicensis* (Orbigny, 1842). Off North Carolina to the West Indies, 54 to 640 fms.

4869   *Nuculana subaequilatera* (Jeffreys, 1879). Off North Carolina and the Gulf of Mexico, 92 to 1,731 fms.

4870   *Nuculana hebes* (E. A. Smith, 1885). Gulf of Mexico and the West Indies, 196 to 805 fms.

4871   *Nuculana pusio* (Philippi, 1844). Off Key West, Gulf of Mexico and the West Indies, 856 to 1,591 fms.

4872   *Nuculana quadangularis* (Dall, 1881). North Carolina, Florida Strait and West Indies, 683 to 1,568 fms.

4873   *Nuculana solidula* (E. A. Smith, 1885). North Carolina, Florida Keys and the Gulf of Mexico, 640 to 1,002 fms.

4874   *Nuculana solidifacta* (Dall, 1886). Florida Strait, 287 fms. 12 mm.

4875   *Nuculana vitrea cerata* (Dall, 1881). Florida Strait and West Indies, 100 to 540 fms. 6.5 mm.

4875

4876   *Nuculana verrilliana* (Dall, 1886). Florida Keys, Gulf of Mexico and West Indies.

4877   *Nuculana buccata* (Steenstrup, 1842). Greenland; Bering Straits.

4878   *Nuculana parva* (G. B. Sowerby, 1833). Melville Island, Canadian Arctic.

4879   *Nuculana cestrota* (Dall, 1889). Southwest Caribbean. 25 mm. Northern Brazil, 48 fms.

4879

Other Pacific species:

4880   *Nuculana austini* Oldroyd, 1935. Gabriola Pass, British Columbia, to Puget Sound, 25 to 100 fms. *The Nautilus,* vol. 49, p. 13.

4881   *Nuculana cellulita* (Dall, 1896). Puget Sound, Washington.

4882   *Nuculana navisa* (Dall, 1916). Farallones Islands to San Diego, California.

4883   *Nuculana amiata* (Dall, 1916). Off San Diego, California, 488 fms.

4884   *Nuculana gomphoidea* (Dall, 1916). Off Tillamook, Oregon, 786 fms.

4885   *Nuculana oxia* (Dall, 1916). Santa Rosa Island, to Gulf of California.

4886   *Nuculana phenaxia* (Dall, 1916). Off San Diego, California, 822 fms.

4887   *Nuculana fiascona* (Dall, 1916). Off San Diego, California, 822 fms.

4888   *Nuculana spargana* (Dall, 1916). Santa Barbara Islands, to Point Loma, California.

4889   *Nuculana liogona* (Dall, 1916). Bering Sea in 1,401 fms. Above 8 species in *Proc. U.S. Nat. Mus.,* vol. 52, no. 2183.

4890   *Nuculana radiata* (Krause, 1885). Arctic Seas; Bering Sea.

4891   *Nuculana pontonia* (Dall, 1890). Santa Barbara Islands, California, to Peru. Deep water.

4892   *Nuculana leonina* (Dall, 1896). British Columbia, Canada.

4893   *Nuculana amblia* (Dall, 1905). Monterey Bay, California.

4894   *Nuculana conceptionis* (Dall, 1896). Aleutian Islands to off San Diego, California.

4895   *Nuculana extenuata* (Dall, 1897). Off Sitka, Alaska, 1,569 fms., to off Dixon Entrance, British Columbia.

4896   *Nuculana dalli* Krause, 1885. Unalaska, Aleutian Islands.

4897   *Nuculana (Saccella) acrita* (Dall, 1908). Baja California to Ecuador. (Synonym: *laeviradius* Pilsbry and Lowe, 1932.)

4898   *Nuculana (Saccella) impar* (Pilsbry and Lowe, 1932). Baja California to Costa Rica, 21 fms.

4899   *Nuculana (Costelloleda* Hertlein and Strong, 1940) *costellata* (Sowerby, 1833). Baja California to Panama.

4900   *Nuculana (Costelloleda) marella* Hertlein and Strong, 1940. Gulf of California to Panama, 35 to 40 fms.

Genus *Adrana* H. and A. Adams, 1858

Shells gracefully elongate to lanceolate, compressed, the dorsal margin nearly straight, with small, subcentral beaks. No rostral area, but a narrow, flattened lunule and escutcheon is usually well-developed. Ligament internal and attached to a wide, often weakly bilobed, chondrophore. Sculpture of fine concentric, rarely oblique threads. Type: *lanceolata* (Lamarck, 1819). About 8 species known from the Tropical Eastern Pacific. Three Caribbean species:

4901   *Adrana tellinoides* (Sowerby, 1823). Canal Zone to Puerto Rico. (Warmke and Abbott, 1953, "The gross anatomy . . . ," *Jour. Wash. Acad. Sci.,* vol. 43, no. 8.) (Synonyms: *perprotracta* Dall, 1912, and *newcombei* Angas, 1878.)

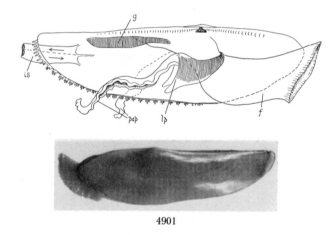

4901

4902   *Adrana notabilis* Rehder, 1939. Paraguana Peninsula, Venezuela.

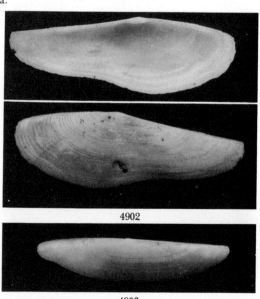

4902

4903

4903 *Adrana scaphoides* Rehder, 1939. Cartagena, Colombia. Both *The Nautilus*, vol. 53, p. 16 and 17.

4904 *Adrana electa* (A. Adams, 1846). Brazil to Argentina, 10 to 40 fms. See E. C. Rios, 1970, pl. 50.

4905 *Adrana exoptata* (Pilsbry and Lowe, 1932). West Mexico.

4906 *Adrana penascoensis* (Lowe, 1935). Northern Gulf of California, to 18 meters.

Genus *Yoldia* Möller, 1842

Somewhat similar to *Nuculana*, but the valves are much thinner and fragile, rarely with a long rostrum, usually gaping at both ends, much smoother and glistening. Type: *hyperborea* Torell, 1859. Synonyms: *Microyoldia* Verrill and Bush, 1897; *Tepidoleda* Iredale, 1939. Our key is modified from W. K. Ockelmann, 1954, *Medd. om Grönland*, vol. 107, no. 7.

Subgenus *Yoldia* Möller, 1842

Key to Northern *Yoldia*

1. Umbo somewhat behind the middle; periostracum with dense microscopic, concentric lines; length less than twice the height; muscle scars relatively large . *myalis*
   Umbo very near the middle; periostracum almost smooth; muscle scars proportionately small 2
2. No distinct antero-ventral sinuation in the shell margin 3
   With a distinct antero-ventral sinuation in the shell margin; pallial sinus with an angle nearest the umbo . 4
3. Length less than twice the height . . . . *sapotilla*
   Length more than twice the height (adults) . *limatula*
4. Posterior end not acuminate; length of adults about, or less than, twice the height . . . . . *hyperborea*
   Posterior end acuminate; tip somewhat pointed; length of adults more than twice the height. . *amygdalea*

Yoldia limatula (Say, 1831)   **4907**
File Yoldia

Nova Scotia to off North Carolina.

1 to 2½ inches in length, elongate, narrowing at the posterior end. Umbones very small, halfway between the ends of the shell. Exterior glistening greenish tan to light chestnut-brown, with only faint concentric growth lines. Interior glossy-white. A rather common species just below low-water mark. Distinguished from *Y. sapotilla* by its more elongate shape. Western records of this are *amygdalea*. For biology, see Drew, 1899, *Mem. Biol. Lab. Johns Hopkins Univ.*, vol. 4.

Yoldia amygdalea Valenciennes, 1846   **4908**
Almond Yoldia

Bering Sea to off northern California.
Iceland to the Gulf of Maine; Norway.

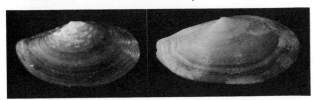

4908   4909

20 to 50 mm., very similar to *limatula*, but always having the anterior ventral margin with a small concave depression. In general shape it falls within the variations of the Atlantic specimens. Moderately common; 2 to 100 fathoms. Synonyms: *gardneri* Oldroyd, 1935; *norvegica* Dautzenberg and Fischer, 1912; *limatuloides* Ockelmann, 1954. See Cowan, 1968, *The Veliger*, vol. 11, p. 51.

Yoldia hyperborea Lovén in Torell, 1859   **4909**
Arctic Yoldia

Arctic Seas; northern Alaska and Canada.
Greenland to Norway.

16 to 35 mm., thin, fragile, compressed, gaping at both ends. Posterior end not acuminate; length of adults about, or less than, twice the height. Moderately common; deep water. Compare with *amygdalea* in the key. Synonyms: *arctica* M. Sars, 1851; *glacialis* "Gray" of authors, non Gray, 1828.

Yoldia sapotilla (Gould, 1841)   **4910**
Short Yoldia

Arctic Seas to North Carolina.

¾ to 1½ inches in length, oblong, smooth, with a moderately extended posterior end. Periostracum yellowish to greenish brown. Differing from *limatula* in being shorter and less extended and more truncate at the posterior end. Commonly dredged off New England in shallow water, 4 to 45 fathoms; often found in fish stomachs. This species can be confused with the uncommon *Y. myalis* Couthouy which is found from Labrador to Cape Cod and Alaska and which, however, is shorter and more pointed at the posterior end. *Nucula gouldii* De Kay, 1843, is probably a synonym.

4910   4911

4918

Yoldia myalis (Couthouy, 1838)   **4911**
Comb Yoldia

Labrador to Massachusetts.
Alaska to Puget Sound, Washington.

1 inch, ovate, thin-shelled, slightly gaping at both ends, moderately inflated. Surface somewhat wavy, with distant concentric ridges, and covered with an olive, dull periostracum, arranged in alternating dark and light zones. Umbones not elevated, a little behind the middle of the shell. Anterior end semi-elliptical; posterior end subtriangular. Interior yellowish white, glossy, with greenish zones and minute radiating lines. About 12 teeth on either side of the deep, triangular cartilage pit. Common; 7 to 80 fathoms. Synonyms: *oblongoides* Wood, 1840; *cascoensis* Mighels and Adams, 1844.

Other Atlantic species:

4912 *Yoldia liorhina* Dall, 1881. Florida Strait and Gulf of Mexico, 182 to 190 fms. 13 mm.

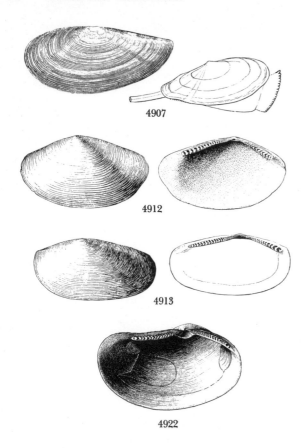

4907

4912

4913

4922

4913 *Yoldia solenoides* Dall, 1881. Gulf of Mexico, off the Mississippi Delta, 118 fms. 12.5 mm.

4914 *Yoldia regularis* Verrill, 1884. Off Martha's Vineyard, Massachusetts. 3 mm.

Other Pacific species:

4915 *Yoldia* (*Kalayoldia* Grant and Gale, 1931) *cooperi* Gabb, 1865. Southern California.

4916 *Yoldia* (*Cnesterium* Dall, 1898) *scissurata* Dall, 1897. Arctic Ocean to off San Diego, California, 8 to 75 fms. Common. (Synonym: *ensifera* Dall, 1897; *plena* Dall, 1908.)

4917 *Yoldia* (*Cnesterium*) *seminuda* Dall, 1871. Bering Sea to off Monterey, California, 21 fms.

Subgenus *Megayoldia* Verrill and Bush, 1897

Broad, compressed; chondrophore large, concave, striate within. Pallial sinus large; ligament external, strongly developed. Type: *thraciaeformis* Storer, 1838.

Yoldia thraciaeformis Storer, 1838   **4918**
Broad Yoldia

Greenland to Cape Hatteras, North Carolina.
Alaska to Puget Sound, Washington.

1½ to 2 inches in length, oblong. Characterized by its squarish, upturned posterior end; coarse, dull, flaky periostracum; large circular chondrophore; and the coarse, oblique rib running from beak to posterior ventral margin. Moderately common from shallow to deep water. Found in fish stomachs.

Other *Megayoldia*:

4920 *Yoldia secunda* Dall, 1916. Wrangell, Alaska, 50 fms.

4921 *Yoldia beringiana* Dall, 1916. Bering Sea to California, 152 to 1,041 fms.

4922 *Yoldia montereyensis* Dall, 1893. Alaska to San Diego, California. 152 to 871 fms. 25 mm.

4923 *Yoldia martyria* Dall, 1897. Alaska to the Gulf of California, 100 fms. Common.

4924 *Yoldia vancouverensis* E. A. Smith, 1880. Vancouver Island, British Columbia.

Genus *Portlandia* Mörch, 1857

Shell ovate-elongate, swollen, with a short rostration posteriorly. Type: *Nucula arctica* Gray, 1824. The generic and species names were conserved by Opinion 769 of the I.C.Z.N., 1966, names no. 1705 and 2132. Synonyms: *Pseudoportlandia* Woodring, 1925; *Portlandella* Stewart, 1930. The arctic species are:

4925 *Portlandia arctica* (Gray, 1824). Greenland; Arctic Ocean.

4926 *Portlandia glacialis* (Gray, 1828). Greenland; Hudson Strait, 15 to 25 fms.; Point Barrow, Alaska.

4927 *Portlandia collinsoni* Dall, 1919. Collinson Point, Canadian Arctic, 3 fms.

Subgenus *Yoldiella* Verrill and Bush, 1897

Like *Yoldia,* small, but not gaping posteriorly, with a shallower pallial sinus, both ends rounded, the posterior end not having a keel. Type: *lucida* Lovén, 1846. J. Knudsen, 1970, treats this as a genus.

Portlandia lucida (Lovén, 1846)   **4928**
Lucid Yoldia

Greenland to North Carolina.
Norway to the Mediterranean.

2 to 3 mm., brittle, plump, oval in outline, glossy, grayish green. Lunule small, indistinct; escutcheon poorly defined, lanceolate, with a straight, elevated ridge. Surface smoothish. 10 to 12 teeth in front of tiny umbo, 12 to 16 in back. Uncommon; 22 to 800 fathoms.

Other species (*Yoldiella*):

4929 *Portlandia frigida* (Torell, 1859). Arctic Seas to Massachusetts, 3 to 1,245 fms. 4 mm.

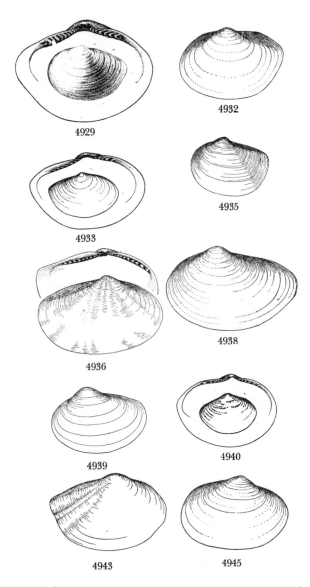

4929

4932

4933

4935

4936

4938

4939

4940

4943

4945

4930 *Portlandia intermedia* (M. Sars, 1865). Arctic Seas; Alaska; Canada, Europe. 5 to 2,078 fms.

4931 *Portlandia expansa* (Jeffreys, 1876). Off Newfoundland, 206 fms.

4932 *Portlandia fraterna* (Verrill and Bush, 1898). Gulf of St. Lawrence to off Georgia, 90 to 1,608 fms. Greenland. 6 mm.

4933 *Portlandia inconspicua* (Verrill and Bush, 1898). Nova Scotia to North Carolina, 100 to 705 fms. 3 mm.

4934 *Portlandia sericea striolata* (Jeffreys, 1876). Georges Bank, 906 to 1,790 fms; off Cape Hatteras, North Carolina, 842 fms.

4935 *Portlandia inflata* (Verrill and Bush, 1897). Off Massachusetts to North Carolina, 40 to 1,608 fms. 6 mm.

4936 *Portlandia iris* (Verrill and Bush, 1897). Gulf of St. Lawrence to North Carolina, 20 to 781 fms. Variety (**4937**) *stricta* (Verrill and Bush, 1897), off Cape Sable, Nova Scotia, 90 fms. 5 mm.

4938 *Portlandia jeffreysi* (Hidalgo, 1879). Southeast of the Georges Bank to off Virginia, 349 to 1,423 fms. Bermuda. 10 mm.

4939 *Portlandia lenticula* (Möller, 1842). North of Cape Cod, Mass., 110 to 122 fms. (Synonym: *amblia* Verrill and Bush, 1897.)

4940 *Portlandia minuscula* (Verrill and Bush, 1897). Off Massachusetts to Virginia, 505 to 1,290, fms. 2 mm.

4941 *Portlandia pachia* (Verrill and Bush, 1897). Gulf of Mexico, 730 fms.

4942 *Portlandia pygmaea* (Münster, 1842). Off Fernandina, Florida, 294 fms. (fide Dall, 1927).

4943 *Portlandia subangulata* (Verrill and Bush, 1898). Off Isles of Shoals, New Hampshire, 22 to 516 fms. 6 mm.

4944 *Portlandia subequilatera* (Jeffreys, 1879). Off Massachusetts to off Cape Hatteras, North Carolina, 499 to 1,731 fms. Greenland.

4945 *Portlandia dissimilis* (Verrill and Bush, 1898). Off North Carolina. 4 mm.

Pacific *Yoldiella*:

4946 *Portlandia cecinella* (Dall, 1916). Aleutians to the Gulf of California.

4947 *Portlandia capsa* (Dall, 1916). Off Tillamook Bay, Oregon. 786 fms.

4948 *Portlandia oleacina* (Dall, 1916). Arctic Ocean north of Bering Strait.

4949 *Portlandia orcia* (Dall, 1916). Tillamook Bay, Oregon, to San Diego, California.

4950 *Portlandia sanesia* (Dall, 1916). Alaska to California. Last 4 species in *Proc. U.S. Nat. Mus.*, vol. 52, no. 2183.

4951 *Portlandia siliqua* (Reeve, 1855). Norton Sound, Alaska. North Atlantic.

Genus *Pristigloma* Dall, 1900

Rounded, with a short hinge; teeth V-shaped and unequal in size. Type: *nitens* Jeffreys, 1876. Synonym: *Glomus* Jeffreys, 1876, not Gistel, 1848; *Pristoglomus* "Dall" C. W. Johnson, 1934, is a misspelling.

4952 *Pristigloma nitens* Jeffreys, 1876. Off Martha's Vineyard, Massachusetts, and off Fernandina, Florida, 294 fms.

Subclass *Cryptodonta* Neumayr, 1884

Primitive elongate clams with shell material of aragonite. Hinge without teeth.

Order *Solemyoida* Dall, 1889

Siphonate, burrowing protobranchs with homogeneous aragonite shells.

4953

4954

4955

4957

Superfamily Solemyacea H. and A. Adams, 1857

Family SOLEMYACIDAE H. and A. Adams, 1857

Shell equivalve, oblong, usually fragile, no lunule. Hinge without teeth. Ligament external or internal and posterior to the umbones. Periostracum thick, glossy, projecting beyond the edges of the shell. Solenomyadae Gray, 1840, is a synonym.

Genus *Solemya* Lamarck, 1818

The awning clams are very primitive in their characters and they have no near relatives. Their shells are fragile, elongate, with a weak, toothless hinge, gaping at both ends, and covered by a polished, horny, brown periostracum which extends well beyond the margins of the valves. Posterior adductor muscle scar bordered in front by a slightly thickened radial rib. Type: *togata* Poli, 1795. *Solenomya* Children, 1823; *Solenimya* Bowdich, 1822, are synonyms.

Subgenus *Petrasma* Dall, 1908

Ligament chiefly internal and found posterior to the beaks. Type: *borealis* Totten, 1834.

Solemya velum Say, 1822   4953
Common Atlantic Awning Clam

Nova Scotia to northern Florida.

½ to 1 inch in length, very fragile, and with a delicate, shiny, brown periostracum covering the entire shell and extending beyond the edges. Light radial bands of yellowish brown are present in some specimens. Chondrophore supported by 2 curved arms. Commonly dredged in shallow water in mud bottom. Compare Florida specimens with *occidentalis*.

Solemya borealis Totten, 1834   4954
Boreal Awning Clam

Nova Scotia to Connecticut.

2 to 3 inches in length, very similar to *velum,* but more compressed, heavier, and colored grayish blue or lead on the inside of the valves (instead of purplish white). The striking difference is in the siphonal opening of the animal. In *velum,* there are 2 small, median, low tubercles above the opening and 5 or 6 pairs of short tentacles at the lower end of the opening. In *borealis,* there are 3 pairs (one of which is large and long) of tentacles above the opening and about 15 smaller ones bordering the lower half. *S. borealis* is moderately common offshore.

Solemya occidentalis Deshayes, 1857   4955
West Indian Awning Clam

South Florida and the West Indies.

¼ inch in length, similar to *S. velum,* but much smaller, and has only 1 slender ridge or rib bordering the chondrophore. Uncommon just offshore. Commonly found in the stomachs of the bonefish, *Albula vulpes.* Described first by Deshayes in 1857, later by Fischer in 1858.

Solemya valvulus Carpenter, 1864   4956
Pacific Awning Clam

San Pedro, California, to the Gulf of California.

¾ inch in length, thin, translucent. Periostracum shiny, light-brown, with slender, radial lines of darker brown which

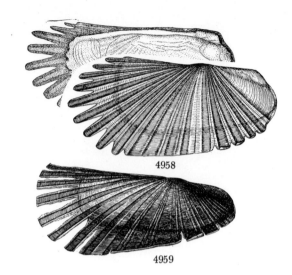

4958

4959

are finely striate posteriorly. Ligament bounded by a single, arched prop or rib. Uncommonly dredged offshore.

Solemya panamensis Dall, 1908   4957
Panama Awning Clam

Santa Barbara to Panama.

1 to 1½ inches, posterior end more pointed. Periostracum brown, glossy, recurved over the margins not produced into long processes. Anterior surface of the valves with 8 or 9 obscure radial rays, the middle zone with a few sparse rays, the posterior with 6 or 7 more closely spaced; area behind beaks smooth. Moderately common in shallow, muddy areas in the southern part of its range.

Subgenus *Acharax* Dall, 1908

Shells large, up to 6 inches; ligament posterior to the beaks, wholly internal, visible internally only where it crosses the gap between the valves. Type: *johnsoni* Dall, 1891. There are three North American species. J. Knudsen, 1970, treats this as a genus.

4958   *Solemya (Acharax) grandis* Verrill and Bush, 1898. 30 to 1,600 fms., off New Jersey and Virginia. 50 mm.

4959   *Solemya (Acharax) johnsoni* Dall, 1891. 1,005 to 1,672 fms., Forbes Island, British Columbia, to Peru. (Synonym: *agassizii* Dall, 1908.) 115 mm.

4960   *Solemya (Acharax) caribbaea* H. E. Vokes, 1970. 20 miles off Colombia (Atlantic side), 220 fms. *The Veliger,* vol. 12, p. 357.

Subclass *Pteriomorphia* Beurlen, 1944

Sedentary bivalves with free mantle margins, usually with a byssus in the adult stage or becoming cemented to the substrate; secondarily free, as in the scallops.

Order *Arcoida* Stoliczka, 1871

Ark shells and their relatives. With 2 equal-sized adductor muscles; valves equal in size. Dorsal margin bears a flat cardinal area.

Superfamily Arcacea Lamarck, 1809

Generally trapezoidal; ligament elongate and striate. Periostracum well-developed.

Key to Families

a. Shell elliptical, hinge straight . . . . . *Arcidae*
aa. Shell circular or lopsidedly circular, hinge curved:
    b. Ligament partly sunk into shell . . *Limopsidae*
    bb. Ligament external . . . . . *Glycymeridae*

Family ARCIDAE Lamarck, 1809

The ark shells have numerous similar taxodont teeth in a straight hinge. Many of the subgenera listed here are considered by some authorities as full genera.

Subfamily ARCINAE Lamarck, 1809

Genus *Arca* Linné, 1758

Characterized by the long, narrow hinge line with numerous small teeth, by the large byssal notch on the ventral side, and the wide ligamental area between the beaks. Color brown, commonly with zigzag markings. Type: *noae* Linné, 1758, from the Mediterranean. Synonym: *Navicula* Blainville, 1825.

Arca zebra (Swainson, 1833)   **4961**
Turkey Wing

North Carolina to Florida, south Texas and to Brazil. Bermuda.

2 to 3 inches in length, about twice as long as deep. Color tan with flecks and zebra-stripe markings of reddish brown. Periostracum brown, matted. Ribs of irregular sizes. No concentric riblets. Do not confuse with *A. imbricata*. A common species which attaches itself to rocks with its byssus. Used extensively in the shellcraft industry. In Bermuda, it is minced and baked in a pie crust with raisins, potatoes, carrots and curry powder. Formerly *A. occidentalis* Philippi, 1847, and *barbadensis* Orbigny, 1845.

The Baja California to Peru counterpart of this species is *Arca pacifica* (Sowerby, 1833) **(4962)**.

Arca imbricata Bruguière, 1789   **4963**
Mossy Ark

North Carolina to Texas and the West Indies to Brazil. Bermuda.

1½ to 2½ inches in length. Similar to *A. zebra,* but differing in having beaded ribs and a very large byssal opening, usually having the posterior end much larger, and in lacking the zebra stripes. Periostracum sometimes quite heavy and foliated. Commonly attached to underside of rocks in shallow water. *A. umbonata* Lamarck, 1819, is a synonym.

The Gulf of California to Ecuador counterpart of this species is *Arca mutabilis* (Sowerby, 1833) **(4964)**.

Genus *Barbatia* Gray, 1842

Subgenus *Barbatia* Gray, 1842

Shell compressed; byssal gape small. Type: *barbata* (Linné, 1758), from the Mediterranean. Synonyms: *Modioliformis* Deshayes, 1860; *Plagiarca* Conrad, 1875; *Obliquarca* Sacco, 1898; *Destacar* Iredale, 1936, and others.

Key to the Genus *Barbatia* in the Western Atlantic

a. Interior of valves reddish brown . . . *cancellaria*
aa. Interior of valves whitish . . . . b
b. Ligament very small, between umbones . . . .
    . . . . *Arcopsis adamsi*
bb. Ligament very elongate c
c. Prodissoconch, glossy, visible . . . . . *candida*
cc. Prodissoconch not visible . . . . d
d. Shell thick, coarsely beaded . . . . *domingensis*
dd. Shell moderately thin, radial ribs finely beaded . *tenera*

Barbatia candida (Helbling, 1779)   **4965**
White Bearded Ark

North Carolina to Texas to Brazil.

1½ to 2½ inches in length; fairly thin, not heavy. Irregular in shape. Byssal opening at base of shell. Numerous weak, slightly beaded ribs, those on the posterior dorsal area being very strongly beaded. Periostracum brown, longest at posterior end. Exterior and interior of shell white. Ligament moderately developed. This species was also named *candida* Gmelin, 1791; *jamaicensis* Gmelin, 1791; and *helblingi* Bruguière, 1792. Common, attached under stones.

Barbatia cancellaria (Lamarck, 1819)   **4966**
Red-brown Ark

Southern Florida and off Texas and the West Indies to Brazil.

1 to 1½ inches in length, similar to *B. candida,* but with weak, beaded or cancellate sculpture and colored a dark, pur-

4961

4963

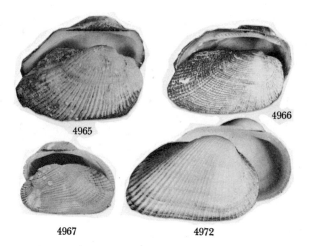

4965

4966

4967   4972

plish brown. Periostracum arranged in radial rows of fine tufts. This is a common species found among rocks from low tide to 12 feet.

Subgenus *Acar* Gray, 1857

Shell subquadrate, usually somewhat distorted; posterior end longer, crossed by an oblique angulated ridge. Periostracum thin. Cardinal area narrow, brown, with the grooves mostly in the region posterior to the beaks. Type: *gradata* Broderip and Sowerby, 1829.

Barbatia domingensis (Lamarck, 1819)   4967
White Miniature Ark

North Carolina to Texas to Brazil. Bermuda.

½ to ¾ inch in length, somewhat box-shaped, whitish in color and with no appreciable periostracum. Similar in shape and sculpture to *Arcopsis adamsi*, but instead of having a small, triangular ligament between the beaks, *domingensis* has a very narrow, long ligament posterior to the beaks. The posterior end is usually larger than the anterior end and characteristically dips slightly downward. Common at low tide under rocks. Erroneously called *Arca reticulata* Gmelin, 1791, by Dall and others (see Lamy and Woodring).

Barbatia bailyi (Bartsch, 1931)   4968
Baily's Miniature Ark

Santa Monica, California, to Gulf of California.

A little over ¼ inch in length, oblong to squarish, fat; cancellate sculpture in which the beads become foliate at the posterior end. Ligament small, narrow and placed well posterior to the fairly close beaks; about 15 teeth. Color white to brownish white.

Very common in certain localities under stones at low tide. *A. pernoides* Carpenter was thought to be this shell but is apparently some other much larger species of unknown identity.

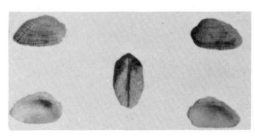

4968

Other Pacific species:

4969 *Barbatia (Acar) gradata* (Broderip and Sowerby, 1829). Baja California to Peru; Galapagos. (Synonyms: *pholadiformis* C. B. Adams, 1852; *panamensis* Bartsch, 1931; *rostae* S. S. Berry, 1954.)

4969

4970 *Barbatia (Cucullaearca* Conrad, 1865) *reeveana* (Orbigny, 1846). San Diego, California, to Peru. (Synonyms: *nova* Mabille, 1895; *lasperlensis* and *velataformis* Sheldon and Maury, 1922.)

4971 *Barbatia ectocomata* (Dall, 1886). Off Martinique and Barbados. 82 to 169 fms. 46 mm.

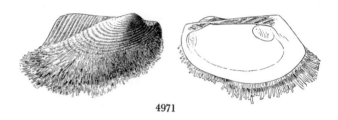

4971

Subgenus *Fugleria* Reinhart, 1937

Shells rather thin, white under a fairly thick periostracum. Type: *pseudoillota* Reinhart, 1937, from the Pliocene of Florida.

Barbatia tenera (C. B. Adams, 1845)   4972
Doc Bales' Ark

South Florida to Texas and the Caribbean.

1 to 1½ inches in length, thin-shelled, rather fat and evenly trapezoidal in shape and with numerous rather evenly and finely beaded, threadlike ribs. Ligamental area fairly wide at the beak end, becoming narrow at the other. Small byssal gap present on the ventral margin. Moderately common. *Arca balesi* Pilsbry and McLean, 1939, is this species.

The tropical Eastern Pacific counterpart is the similar, common (4973) *Barbatia (Fugleria) illota* (Sowerby, 1833) found intertidally to 40 fathoms from the Gulf of California to Peru. *B. tabogensis* (C. B. Adams, 1852) is a synonym.

Subfamily ANADARINAE Reinhart, 1935

Genus *Anadara* Gray, 1847

Shell heavy, ribbed, with the valves tightly closed along the margins, and one valve slightly larger than the other. Cardinal area usually smooth. Type: *antiquata* (Linné, 1758).

Subgenus *Larkinia* Reinhart, 1935

The type of this subgenus is *larkinii* Nelson, 1870, a Pliocene species from Peru and Ecuador.

Anadara multicostata (Sowerby, 1833)   4974
Many-ribbed Ark

Newport Bay, California, to Panama.

Shell large, 3 to 4 inches in length, very thick and squarish. 31 to 36 radial ribs. The left valve slightly overlaps the right valve. Found in sandy areas by dredging in depths over 12 feet. *A. grandis* Broderip and Sowerby in Mexican and Panamic waters is larger, heavier and has 25 to 27 ribs.

Anadara notabilis (Röding, 1798)   4975
Eared Ark

Off North Carolina to Florida to Brazil. Bermuda.

1½ to 3½ inches in length. 25 to 27 ribs per valve. Fine concentric threads cross the ribs and are prominent between the ribs. The ribs never split. Common in Florida and the

4974   4978

West Indies. Formerly called *auriculata* Lamarck which is from the Red Sea. *A. deshayesi* Hanley, 1843, is a synonym of *notabilis*.

Anadara baughmani Hertlein, 1951   4976
Baughman's Ark

Off Texas, Louisiana and Mississippi. Brazil.

1½ inches in length, similar to *A. floridana,* but much fatter, with 28 to 30 weakly noduled ribs which are not split, and with a strongly posterior-sloping anterior ventral margin. Common offshore down to 50 fathoms. *A. springeri* Rehder and Abbott, 1951, published a month later, is this species.

Anadara transversa (Say, 1822)   4977
Transverse Ark

South of Cape Cod to Florida and Texas.

½ to 1½ inches in length. Left valve overlaps right valve. Ligament fairly long, moderately narrow, rough or pustulose. Ribs on left valve usually beaded, rarely so on right valve; 30 to 35 ribs per valve. Periostracum grayish brown, usually wears off except along base of valves. Fairly common in mud below low water. The smallest of the Atlantic Anadaras. Distinguished from *ovalis* by its longer, wider, more distinct external ligament. *A. sulcosa* van Hyning, 1946, is this species.

Subgenus *Grandiarca* Olsson, 1961

Has relatively few, very heavy square ribs. Type: *grandis* Broderip and Sowerby, 1829.

Anadara grandis (Broderip and Sowerby, 1829)   4978
Grand Ark

Baja California to Peru.

4 inches, thick and heavy; 26 or 27 ribs, square in section, elevated, mostly smooth except on the anterior slope where they are crudely noded. Periostracum thick and black. Common in black mud near mangrove trees. Used for food in South America where it is called the "Pato de Buro."

Subgenus *Sectiarca* Olsson, 1961

Valves alike and not overlapping. Ribs grooved in both valves. Type: *floridana* (Conrad, 1869).

Anadara floridana (Conrad, 1869)   4979
Cut-ribbed Ark

North Carolina, Florida to Texas and the Greater Antilles.

2½ to 5 inches in length, elongate. Ribs 30 to 38 in number, square, faintly divided by a fine-cut line. Fine, raised, concentric lines seen between weakly beaded ribs. Left valve very slightly larger than right valve. Periostracum light- to dark-brown. Not very common; shallow water.

The related *Anadara lienosa* (Say) **(4980)** is fossil and very close in characters to *floridana,* but considered different. This species has often been called *A. secticostata* Reeve, 1844, which is not so elongate and whose origin is unknown.

Other species:

4981 *Anadara (Sectiarca) concinna* (Sowerby, 1833). Gulf of California to Ecuador. Common; offshore to 10 fathoms.

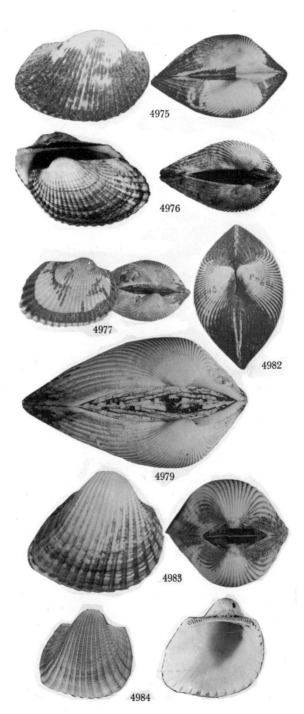

4975

4976

4977

4982

4979

4983

4984

Subgenus *Lunarca* Gray, 1857

The subgenera *Argina* Gray, 1842, and *Arginarca* Mc-Lean, 1951, are the same. The type of *Lunarca* is *ovalis* Bruguière, 1789.

Anadara ovalis (Bruguière, 1789)   4982
Blood Ark

Cape Cod to Texas and the West Indies. Brazil.

1½ to 2⅓ inches in length, not very thick, roundish to ovate; square, smooth ribs; ligament very narrow and depressed; beaks close together. Periostracum black-brown, hairy. Ribs 26 to 35 in number.

Dall considered the forms "*pexata* Say" and "*americana* Wood" too indistinct for recognition. This species was known for a long time as *campechiensis* Gmelin, 1791, and is common.

Subgenus *Cunearca* Dall, 1898

Shell trigonal, inflated, with full umbones facing slightly forward over a triangular cardinal area covered completely by the ligament. Left valve usually slightly larger. Type: *brasiliana* Lamarck, 1819.

Anadara brasiliana (Lamarck, 1819)   4983
Incongruous Ark

North Carolina to west Florida to Texas and to Brazil.

1 to 2½ inches in length; almost as high as long. Beaks facing each other at center of short, transversely striate ligamental area. Left valve overlaps right valve considerably. Ribs 26 to 28, square with strong barlike beads. Periostracum thin, light-brown. *A. incongrua* Say, 1822, is this species.

Anadara chemnitzii (Philippi, 1851)   4984
Chemnitz's Ark

Texas and the West Indies to Brazil.

1 inch, similar to *brasiliana*, but quite thick-shelled; the beaks are slightly forward of the center of the ligamental area. Uncommon. The counterpart to this species in the Pacific in *Anadara nux* (Sowerby, 1833) **(4985)**, ranging from the Gulf of California to Peru.

Genus *Bathyarca* Kobelt, 1891

Shells small, 5 mm., thin, compressed, obliquely elongate, with a proportionately large central umbo. Sculpture of fine, beaded, radial threads. Hinge line straight, with about a dozen or less fine teeth. Type: *pectunculoides* (Scacchi, 1833). Deep-water species, down to 2,000 fathoms.

4986 *Bathyarca orbiculata* (Dall, 1881). Off Delaware Bay, Delaware; off Santa Barbara to the Galapagos; West Africa; Kermadecs, 2,030 to 5,303 meters. (Synonyms: *Arca corpulenta* E. A. Smith, 1885; *imitata* E. A. Smith, 1885; *B. abyssorum* Verrill and Bush, 1898; *strebeli* Melvill and Standen, 1907; *nucleator* Dall, 1908; *pompholyx* Dall, 1908).

Atlantic species:

4987 *Bathyarca anomala* (Verrill and Bush, 1898). Off Cashes Ledge, Gulf of Maine, 27 fms. 8 mm.

4988 *Bathyarca inaequalis* (Dall, 1927). Off Fernandina, Florida, 294 fms.

4989 *Bathyarca glacialis* (Gray, 1824). Arctic Seas to the Gulf of St. Lawrence.

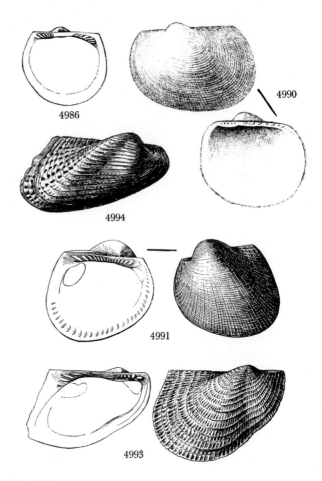

4986   4990   4994   4991   4993

4990 *Bathyarca pectunculoides* (Scacchi, 1833). Gulf of St. Lawrence to off Cape Cod, Massachusetts, 27 to 506 fms. Greenland. (Synonyms: *crenulata* Verrill, 1882; *grandis* Leche, 1878; *septentrionalis* G. O. Sars, 1878.)

4991 *Bathyarca glomerula* (Dall, 1881). North Carolina to the West Indies, 100 to 683 fms. West Florida, 60 to 220 fms., common. 6 mm.

4992 *Bathyarca frielei* (Friele, 1879). Arctic Seas; Greenland.

Genus *Bentharca* Verrill and Bush, 1898

Similar to *Bathyarca,* but extremities quadrate and anterior end of shell more reduced. Teeth not divided or serrated. Type: *Macrodon asperula* Dall, 1881. J. Knudsen, 1970, considers this a subgenus or synonym of *Acar.*

Species:

4993 *Bentharca asperula* (Dall, 1881). Virginia; off both sides of Florida and the West Indies, 180 to 1,568 fms. Bermuda. 815 mm. (Synonyms: *Bentharca profundicola* (Verrill and Smith, 1885); *Arca pteroessa* E. A. Smith, 1885; *endemica* Dall, 1908; and *Arca culebrensis* E. A. Smith, 1885.)

4994 *Bentharca sagrinata* (Dall, 1886). Off Georgia and Fernandina, Florida, 294 to 440 fms.; Florida Strait, 80 fms. 6 mm.

Subfamily NOETIINAE Stewart, 1930

Genus *Noetia* Gray, 1857

Beaks point posteriorly; valves the same size; ligament transversely striate; posterior muscle scar raised to form a weak flange. Type: *reversa* (Sowerby, 1833).

Subgenus *Eontia* MacNeil, 1938

The subgenus *Eontia* is mainly an Atlantic group. *Noetia* s. str. differs in having decidedly more regular sculpture, the ribs smoother and never divided; deeper and long crenulations on the inner margin. There is only one Recent American true *Noetia* (*reversa* Sowerby, 1833) which occurs from the Gulf of California to Peru. The type of *Eontia* is *ponderosa* Say, 1822.

Noetia ponderosa (Say, 1822)   4995
Ponderous Ark

Virginia to Florida and to Texas.

2 to 2½ inches in length, almost as high as long. Ribs raised, square and split down the center by a fine incised line; 27 to 31 ribs per valve. Posterior muscle scar raised to form a weak flange. Periostracum thick, black, but wears off at the beaks. A common shallow-water sand-dweller. Fossil specimens are rarely found on Nantucket, Massachusetts, beaches.

4995

Other species:

4996 *Noetia bisculcata* (Lamarck, 1819). Brazil to Uruguay. See E. C. Rios, 1970, pl. 51.

4997 *Noetia reversa* (Sowerby, 1833). Gulf of California to Peru. (Synonyms: *Arca hemicardium* Philippi, 1843; *N. triangularis* Gray, 1857).

Subfamily STRIARCINAE MacNeil, 1938

Genus *Arcopsis* von Koenen, 1885

Ligament limited to a very small, triangular or barlike area between the umbones. Type: *limopsis* von Koenen, 1885. Synonyms: *Fossularca* Cossmann, 1887; *Gabinarca* Iredale, 1939; *Spinearca* Iredale, 1939; and others.

Arcopsis adamsi (Dall, 1886)   4998
Adams' Miniature Ark

North Carolina to west Florida and Texas to Brazil. Bermuda.

¼ to ⅓ inch in length, oblong in shape, moderately fat, flattened sides; white to cream in color. Periostracum very thin. Sculpture cancellate. Ligament limited to a very small, triangular, black patch between the umbones. The muscle scars are usually bordered by a calcareous ridge. Inner margin of valves smooth. Common under rocks from below tide to 3 fathoms. *A. conradiana* Dall, 1886, and *adamsi* "Shuttleworth" E. A. Smith, 1888, are synonyms.

4998

The Baja California to Peru counterpart of this species is *Arcopsis solida* (Sowerby, 1833) (**4999**), a common species.

Superfamily Limopsacea Dall, 1895

Family LIMOPSIDAE Dall, 1895

Genus *Limopsis* Sassi, 1827

Rather small, obliquely oval clams with tufted, velvety-brown periostracum. Hinge line curved, with a series of oblique teeth. The hinge resembles that of the Glycymeridae. Ligament external, small, central, triangular. Mostly deep water. 4 species on the Pacific Coast, about 6 on the Atlantic side. Type: *aurita* (Brocchi, 1814). Synonyms: *Trigonocaelia* Nyst and Galeotti, 1835; *Lunopsis* Orbigny, 1850; *Cnisma* Mayer, 1868; *Limopsilla* Thiele, 1923; *Loringella* Iredale, 1929; *Phrynelima* Iredale, 1929; *Glycilima* Iredale, 1931.

Limopsis cristata Jeffreys, 1876   5000
Cristate Limopsis

Cape Cod, Massachusetts, to southeastern Florida.

¼ inch in length, similar to *sulcata* but much smaller, less tufted with periostracum, with the inner margin of the valves having a series of strong, pimplelike nobs or teeth, and the outside of the shell having its radial sculpture stronger than its faint concentric sculpture. Commonly dredged off Florida.

(**5001**) *Limopsis minuta* Philippi, 1836 (Newfoundland to both sides of Florida), is very close to this species but has cancellate or beaded sculpture and attains a length of ½ inch. The shells of *L. antillensis* Dall, 1881 (**5002**) (Florida to the Lesser Antilles), are ¼ inch in size and unique in being brightly colored with pink, orange or yellow; 80 to 683 fathoms. The latter may actually belong to the Philobryidae.

Limopsis sulcata Verrill and Bush, 1898   5003
Sulcate Limopsis

Cape Cod, Massachusetts, to Florida, the Gulf States and the West Indies.

½ inch in length, strongly oblique, with prominent, rounded ribs which are finely cut on the upper edge by short radial grooves. Inner margin of valves smooth. Shell dull-white. Periostracum thick, tufted, extending beyond the ventral edge of the shell. Commonly dredged in moderately shallow water to 349 fathoms.

Other Atlantic species:

5004 *Limopsis affinis* Verrill, 1885. South of Martha's Vineyard, Massachusetts, 197 fms.

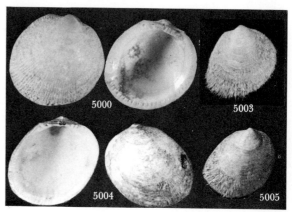

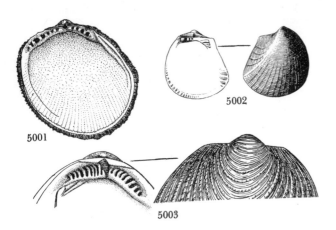

5001

5002

5003

5005 *Limopsis aurita* Brocchi, 1814. Off Mississippi to West Florida, 30 to 730 fms.; off Bermuda, 1,075 fms.

5006 *Limopsis onchodes* Dall, 1927. Off Fernandina, Florida, 294 fms.

5007 *Limopsis radialis* Dall, 1927. Off Fernandina, Florida, 294 fms.

5008 *Limopsis tenella* Jeffreys, 1876. Off New Jersey to Gulf of Mexico, 197 to 2,033 fms.

5009 *Limopsis pelagica* E. A. Smith, 1885. (Synonyms: *transversa* Locard, 1898; *guineensis* Thiele, 1931; *Limopsis profundicola* Verrill and Bush, 1898.) Off Massachusetts to North Carolina, 1,525 to 1,859 fms. 20 mm.; Indian Ocean, 3,115 to 4,020 meters.

5010 *Limopsis plana* Verrill, 1882. Off Virginia, 1,825 to 2,221 fms. (May be *pelagica* E. A. Smith, 1885.)

Limopsis diegensis **Dall, 1908**   **5011**
San Diego Limopsis

Santa Barbara Islands to Coronado Island, California.

⅓ to ½ inch in length, obliquely oval. Shell white; exterior glossy-white with concentric striae, often studded by tiny pinpoint holes. Radial scratches present. Periostracum heavy, tufted with hairs, and often with a cancellate pattern. Uncommonly dredged below 20 fathoms.

Other Pacific species:

5012 *Limopsis skenea* Dall, 1916. Bowers Bank, Bering Sea, 30 fms.

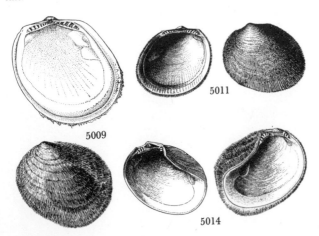

5009

5011

5014

5013 *Limopsis okutanica* Dall, 1916. Off Akutan Island, Aleutians, 72 fms.

5014 *Limopsis* (*Empleconia* Dall, 1908) *vaginatus* Dall, 1891. Bering Sea and Aleutian Islands.

Family GLYCYMERIDIDAE Newton, 1922

Genus *Glycymeris* Da Costa, 1778

Shell heavy, usually orbicular, equivalve, porcelaneous, usually with a soft, velvety periostracum. Beaks slightly curved inward. Hinge heavy, with numerous, small, similar teeth. Ligament external, its area distinct and with diverging grooves. The largest muscle scar is at the anterior end. Often misspelled *Glycimeris* or *Glicymeris*. Type: *orbicularis* Da Costa, 1778, from Europe, which is *glycymeris* (Linné, 1758). Synonyms: *Axinaea* Poli, 1791; *Pectunculus* Lamarck, 1799.

Glycymeris undata **(Linné, 1758)**   **5015**
Atlantic Bittersweet

North Carolina to east Florida and the West Indies. Brazil.

2 inches in length, heavy, smoothish, except for microscopic radial scratches and somewhat larger concentric scratches, giving a silky appearance. There are numerous very weak and hardly discernible radial ribs separated by lines of white. Beaks at about the middle of the ligamental area. Color cream to white with bold splotches of nut-brown. Interior all white or well-stained with brown. This is *G. lineata* Reeve. Common; 1 to 28 fathoms.

In the region of the Carolinas to east Florida and Texas, an inch-long species (*spectralis* Nicol, 1952) **(5016)** is found which is more oval, its beaks face slightly toward the rear and the color is almost a uniform light-brown. Common.

Subgenus *Glycymerella* Woodring, 1925

Costae numerous, low, bearing fine costellae. Type: *decussata* (Linné, 1758).

Glycymeris decussata **(Linné, 1758)**   **5017**
Decussate Bittersweet

Southeast Florida and the West Indies and Brazil.

2 inches in size, very similar to *undata,* but differs in the posteriorly pointing beaks, and in having nearly all the ligamental area in front of the beaks. The radial scratches are

5015   5017

5019   5021   5018

stronger. This is *G. pennacea* (Lamarck, 1819). Moderately common. Rare in Florida.

Glycymeris americana (DeFrance, 1829) 5018
Giant American Bittersweet

North Carolina to Florida to Texas. Brazil.

Up to 5 inches in length, rather compressed, always much flatter than *undata*. The dorsal or hinge side of large specimens is quite long. Beaks point toward each other and are located at the midpoint of the hinge. Color drab-gray or tan, rarely with weak mottlings. Found offshore; 1 to 20 fathoms. Rare.

Subgenus *Tucetona* Iredale, 1931

Radial ribs raised and simple or divided. Radial striae absent; periostracum very thin. Type: *flabellata* Tenison-Woods, 1878, from southern Australia. Two Western Atlantic species.

Glycymeris pectinata (Gmelin, 1791) 5019
Comb Bittersweet

North Carolina to both sides of Florida, Texas and to Brazil.

½ to 1 inch in size; characterized by 20 to 40 raised, radial ribs which have no fine radial striae or scratches on them. Color grayish and commonly splotched with brown. A common shallow-water species. *G. carinata* Dall, 1898, is a synonym.

Glycymeris subtilis (Nicol, 1956) 5020
Bermuda Bittersweet

Off Bermuda.

⅓ to ½ inch in size; length greater than height in larger specimens; number of ribs ranges from 44 to 58, averaging 50; ribs raised, nearly equal in size, gently rounded on top, crossed by fine concentric striae which tend to give an imbricated appearance. Crenulations on interior ventral border from 24 to 39, while in *pectinata* they are 11 to 23, averaging 18. 50 to 100 fathoms offshore. (*The Nautilus*, vol. 70, p. 51.)

5020

Pacific Coast *Glycymeris*

For a review of this group, see G. Willett, *Bull. So. Calif. Acad. Sci.*, vol. 42, p. 107.

Glycymeris subobsoleta (Carpenter, 1864) 5021
West Coast Bittersweet

Aleutian Islands to Baja California.

1 inch in size, subtrigonal, texture chalky. Periostracum velvety, but usually worn away. Ligament area short. Radial ribs flat, with narrow interspaces; usually white, but may be

with light- to medium-brown markings. A rather common shallow- to rather deep-water species (down to 40 fathoms). It is the flattest and most thin-shelled of the Pacific coast *Glycymeris*.

Other Pacific species:

5022 *Glycymeris keenae* Willett, 1943. Forrester Island, Alaska. *Bull. So. Calif. Acad. Sci.*, vol. 42.

5023 *Glycymeris septentrionalis* (Middendorff, 1849). Aleutians to Forrester Island, Alaska.

5024 *Glycymeris profunda* (Dall, 1878). Off Catalina and Los Angeles, California, 200 fms.

5025 *Glycymeris corteziana* Dall, 1916. Forrester Island, Alaska, to Cortez Bank, California.

5026 *Glycymeris migueliana* Dall, 1916. Oregon to Baja California.

5027 *Glycymeris gigantea* (Reeve, 1843). Gulf of California. Smoothish. Maculated, up to 4 inches. Common.

5028 *Glycymeris maculata* (Broderip, 1832). Gulf of California to Peru. Common.

5029 *Glycymeris* (*Tucetona*) *multicostata* (Sowerby, 1833). Gulf of California to Ecuador, 2 to 50 fms. Uncommon. (Synonyms: *bicolor* Reeve, 1843; *chemnitzii* Dall, 1909.)

5030 *Glycymeris* (*Axinactis*) *inaequalis* (Sowerby, 1833). Gulf of California to Peru, 2 to 13 fms. Uncommon. (Synonym: *assimilis* Sowerby, 1833.)

Family MANZANELLIDAE Chronic, 1952

Genus *Nucinella* Wood, 1851

Shells very small, 3 or 4 mm., porcelaneous, subtrigonal; hinge plate short and with a small resilifer, bordered on each side by a few large teeth. Left valve has a prominent posterior lateral tooth which is received in the right valve by an indentation. Surface smooth or sculptured. Posterior adductor muscle absent. Type: *ovalis* Wood, 1851. Synonyms: *Pleurodon* Wood, 1840, not Harlan, 1831; *Nuculina* Orbigny, 1844; *Cyrillona* Iredale, 1929; *Neopleurodon* Hertlein and Strong, 1940 (see Vokes, 1956, *Jour. Paleontol.*, vol. 30, no. 3). Four American species, both rare and deep-water.

5031 *Nucinella adamsi* (Dall, 1898). Florida Strait, 205 fms.

5031a *Nucinella serrei* Lamy, 1912. Off North Carolina, 450 meters, and Bahia, Brazil. 1 to 3 mm. For anatomy and close affinities to the Solemyidae, see J. A. Allen and H. L. Sanders, 1969, *Malacologia*, vol. 7, pp. 381–396.

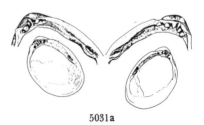

5031a

5032 *Nucinella subdola* (Strong and Hertlein, 1937). 12 fms., off Mazatlan, Mexico.

5033 *Nucinella* (*Huxleyia* A. Adams, 1860) *munita* Dall, 1898. Santa Barbara Islands to the Gulf of California. *Cyrilla* A. Adams, June 1860, is a synonym of *Huxleyia*.

Family PHILOBRYIDAE Bernardi, 1897

Genus *Philobrya* Cooper, 1897

Shell very small, mytiloid; beaks tipped with a large, flat, depressed prodissoconch with upraised margins. Periostracum heavy, spinose. Hinge line straight. Type: *setosa* Carpenter, 1864. Synonym: *Bryophila* Carpenter, 1864, not Trietschke, 1825. Produces a byssus. Broods its young inside. Authorship is attributed to Carpenter, 1872, by some workers.

Philobrya setosa (Carpenter, 1864)   **5034**
Hairy Minute Pen-shell

Alaska to the Gulf of California.

Very small, 4 or 5 mm. (⅛ inch), resembling a young *Pinna* pen-shell, but plump and covered with foliated riblets of periostracal hairs. A large, flat embryonic shell caps the beaks. Valves solid, whitish to orange-brown, iridescent inside. Locally common; attached to broken shells by a byssus.

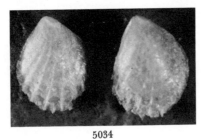

5034

Other species:

5035 *Philobrya atlantica* Dall, 1895. Off southern end of Argentina.

5036 *Philobrya inconspicua* Olsson and McGinty, 1958. Bocas del Toro Atlantic Panama. Bull. 177, *Amer. Paleontol.*, p. 46.

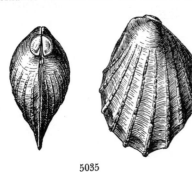

5035

Genus *Cosa* Finlay, 1926

Shell minute, only 2 mm., quadrate, moderately inflated; translucent-white. Prodissoconch raised, granulated, ¼ the length of the straight hinge which has a small, narrow, slightly oblique ligamental pit, in front of which are 8 or 9 vertical teeth, and back of which are 20 similar teeth. Exterior weakly, radially beaded. Type: *costata* Bernard, 1878. Two Western Atlantic species:

5037 *Cosa caribaea* Abbott, 1958. Grand Cayman Island, Caribbean. 1.7 mm. Monograph 11, Acad. Nat. Sci. Philadelphia, p. 112.

5038 *Cosa brasiliensis* Klappenbach, 1966. East Brazil. Rev. Bras. Biol., vol. 26, p. 23. See E. C. Rios, 1970, pl. 51.

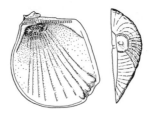

5037

Order *Mytiloida* Férussac, 1822

Contains the true mussels and *Pinna* pen shells. Shells equivalve and very inequilateral; usually producing strong byssi. With several muscle scars of various sizes. Without well-developed siphons. Shell material prismato-nacreous.

Superfamily Mytilacea Rafinesque, 1815

True mussels; thin-shelled; valves the same size; umbones curled towards the front end. Hinge smooth or with a few minute teeth. Anterior adductor muscle small or absent. Prodissoconch with microscopic teeth in the median part of the hinge. Periostracum prominent, brown or black, usually hairy; byssus usually strong in adults.

Family MYTILIDAE Rafinesque, 1815

Characters of the superfamily.

Subfamily MYTILINAE Rafinesque, 1815

Mussel-shaped, with anterior beaks; anterior margin usually twisted, with small weak teeth posterior to the ligament. Smooth or with radial sculpture.

Genus *Mytilus* Linné, 1758

Hinge with 4 to 6 small, whitish teeth below the beak. Ligamental foundation chalky-white and finely pitted. Cold-water species. Type: *edulis* Linné, 1758.

Mytilus edulis Linné, 1758   **5039**
Blue Mussel

Arctic Ocean to South Carolina.
Alaska to California.

1 to 3 inches in length, no ribs but often with coarse growth lines. Ventral margin often curved. Color blue-black with

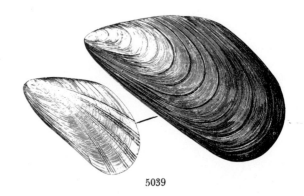

5039

eroded areas of chalky purplish. Periostracum varnishlike. Interior slightly pearly-white with deep purple-blue border. Occasionally specimens have radial rays of brown-yellow. Very common in New England. Sometimes found in more southerly waters attached to floating wood. Rarely, there occurs a brown color form, *pellucidus* Pennant, 1777, and a radially striped form. For biology, see I. A. Field, 1922, *Bull. U.S. Bur. Fish.*, vol. 39.

(5040) *Mytilus edulis diegensis* Coe, 1946 (Northern California to Baja California), is indistinguishable from specimens of *edulis* found in Alaska and New England, and probably only represents an ecological or physiological race (see W. R. Coe, 1946).

(5041) *Mytilus edulis platensis* Orbigny, 1846, is the introduced representative occurring from south Brazil to Argentina. See E. C. Rios, 1970, pl. 53.

Subgenus *Crenomytilus* Soot-Ryen, 1955

Margins of valves minutely serrated; ridge for the resilium is compact. Type: *grayanus* Dunker, 1853.

Mytilus californianus Conrad, 1837   **5042**
Californian Mussel

Aleutian Islands to Socorro Island, Mexico.

2 to 10 inches in length, thick, inflated; ventral margin nearly straight; with less than a dozen or so fairly· broad, weak radial ribs which are best seen on the middle part of the shell. Growth lines very coarse. An abundant species found between tides attached to rocks. *M. gigantea* Nordmann, 1862, is probably a synonym. For pycnogonid infestations, see Benson and Chivers, 1960, *The Veliger*, vol. 3, p. 16, pl. 3.

5042

Genus *Brachidontes* Swainson, 1840

Radially sculptured with bifurcating ribs. Margins of the shell strongly crenulated all around. Dysodont teeth strong along anterior margin. Ligament subinternal, its resilifer forming a narrow scar about ½ the length of the posterior-dorsal margin. Type: *citrinus* (Röding, 1758), which is a synonym of *modiolus* Linné. *Brachydontes* is a misspelling.

Subgenus *Brachidontes* Swainson, 1840

Brachidontes modiolus (Linné, 1767)   **5043**
Yellow Mussel   **Color Plate 20**

Southern Florida and West Indies.

1¼ inches in length, elongate, with numerous wavy, fine axial ribs, colored a light brownish yellow outside, and inside mottled a metallic purple and white. Anterior end has 4 very tiny white teeth. Bordering the ligament are about 30 very small, equal-sized teeth on the edge of the shell. Compare

with *B. exustus* Linné which is wider. The genus is commonly misspelled *Brachydontes*. *B. cubitus* (Say, 1822) and *B. citrinus* (Röding, 1798), are synonyms. Do not confuse the species name with that of the horse mussel, *Modiolus modiolus* (Linné, 1758), which was originally described as a *Mytilus*, while the yellow mussel was originally described as an *Arca*.

Subgenus *Hormomya* Mörch, 1853

Dorsal margin straight or slightly convex, the ligament and its scar narrow and lying along nearly its whole length. The posterior adductor muscle is large. Type: *exustus* (Linné, 1758).

Brachidontes exustus (Linné, 1758)   **5044**
Scorched Mussel   **Color Plate 20**

North Carolina to Texas and the West Indies.
Brazil to Uruguay.

¾ inch in length, rather elongate with numerous fine axial ribs, which form 90 to 140 fine denticles along the edge of the valves. Colored a yellowish brown to dark-brown outside, and inside mottled with a metallic purplish and white. Anterior end has 1 to 4 very tiny purplish teeth. Beyond the ligament (posterior end) there are 5 or 6 very tiny, equal-sized teeth on the edge of the shell. Prefers slightly brackish waters. Abundant on rocks and pilings. Compare with *B. modiolus*, which is more elongate, and with *domingensis*.

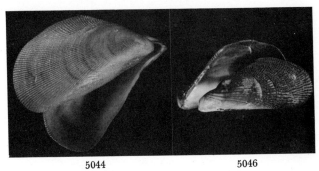

5044   5046

Brachidontes domingensis (Lamarck, 1819)   **5045**
Domingo Mussel

Southeast Florida, Bahamas, Caribbean; Bermuda.

Very similar to *exustus* (Linné), but generally thicker, darker, more arched and with much fewer radial riblets (forming about 50 to 70 small denticles on the edge of the valves). The papillae on the edge of the posterior portion of the mantle are simple, bi- or trilobed as in *exustus*. Attached to rocks and shells in oceanic, subtidal, wave-tossed areas. *M. lavalleanus* Orbigny, 1842, is a synonym.

5043   5045

Subgenus *Aeidimytilus* Olsson, 1961

Brachidontes adamsianus (Dunker, 1857)   **5046**
Stearns' Mussel

Santa Barbara, California, to Oaxaca, Mexico.

½ to 1 inch in length, obtusely carinate, with numerous coarse, beaded, radial ribs which bifurcate. Color brownish purple on the dorsal half, straw-yellow to brownish yellow on the flattened ventral half. Hinge on dorsal edge with about a dozen very tiny barlike teeth. Usually found in colonies in crevices of stones. Two small clams, *Lasaea cistula* Keen and *L. subviridis* Dall, attach themselves to the byssus of this species. Do not confuse with *Septifer bifurcatus,* with which it often lives. *B. stearnsi* Pilsbry and Raymond, 1898, is a synonym, according to Myra Keen.

Other species:

5047 *Brachidontes multiformis* (Carpenter, 1855). Northern Gulf of California to Ecuador. Common; intertidal to 17 fms.

Genus *Ischadium* Jukes-Brown, 1905

Anterior adductor muscle absent. Margins crenulate; sculpture of radial bifurcating ribs. Lunule and anterior margin bent inward, forming 1 or 2 toothlike ridges. Ligament relatively short. Type: *recurvum* (Rafinesque, 1820).

Ischadium recurvum (Rafinesque, 1820)   **5048**
Hooked Mussel

Cape Cod to the West Indies.

1 to 2½ inches in length, flattish, rather wide, with numerous wavy axial ribs. Color outside a dark grayish black, inside a purplish to rosy brown with a narrow blue-gray border. At the umbonal end there are 3 or 4 extremely small, elongate teeth on the edge of the shell. The anterior end of the shell is strongly hooked. This was known as *M. hamatus* Say, 1822. For salinity tolerances, see J. F. Allen, 1960, *The Nautilus,* vol. 74, p. 1.

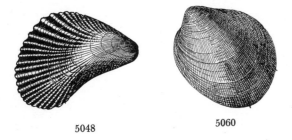

5048   5060

Genus *Septifer* Récluz, 1848

Radial sculpture strong. Having a distinct shelly plate or umbonal deck near the anterior end. Type: *bilocularis* (Linné, 1758), from the Indo-Pacific.

Septifer bifurcatus (Conrad, 1837)   **5049**
Bifurcate Mussel

Crescent City, California, to Gulf of California.

1 to 2 inches in size, subtriangular in outline, inflated. With a couple of dozen strong, wavy radial ribs. Inner margin crenulated. Periostracum black, although often worn white between the ribs. Interior pearly-white, often stained bluish brown on one half of the inner side. The subspecies *obsoletus* Dall, 1916, from San Diego is mostly black on the interior and is a quite elongate form.

5049

Other species:

5050 *Septifer zeteki* Hertlein and Strong, 1946. Gulf of California to Peru.

Subfamily CRENELLINAE Gray, 1840

Shells round to ovate, beaks at the anterior end. Anterior hinge margin thickened and vertically striated or with dysodont teeth. Dorsal hinge margin usually finely vertically striated. Median area of outer surface of valve without radial sculpture.

Genus *Crenella* Brown, 1827

Shells small, oval to oblong-oval, thin, brownish periostracum and with fine decussate radial ribs. Ligament weak, internal. Margins crenulated. Interior of shell glossy-white with a faint trace of iridescence. Mantle open in front, and folded at the posterior end into a sessile excurrent siphon. Foot worm-shaped with a disc-shaped end. Hinge finely dentate. Type: *decussata* (Montagu, 1808). Synonyms: *Stalagmium* Conrad, 1833; *Hippagus* Lea, 1833; *Nuculocardia* Orbigny, 1845.

Crenella glandula (Totten, 1834)   **5051**
Glandular Crenella

Labrador to North Carolina.

¼ to ½ inch in length, squarish, with the beaks near one corner. Radial ribs are fine, numerous, slightly beaded and often crossed by much finer, concentric threads. Color olive-brown. A very common offshore, cold-water species. The smaller *decussata* has its beaks at the center of its more symmetrical shell.

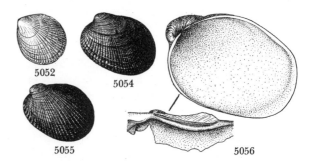

5052   5054
5055   5056

Crenella decussata (Montagu, 1808)   **5052**
Decussate Crenella

Bering Sea to San Pedro, California.
Greenland to North Carolina.

Less than ⅛ inch in size, oval, with numerous fine, decussated radial ribs. Color tan to yellowish gray. Dredged from

3 to 150 fathoms. A food of many marine fishes. Compare with *glandula*. Also in northern Europe where it is littoral.

(5053) *Crenella divaricata* (Orbigny, 1845) (North Carolina to southeast Florida and the West Indies; and Santa Barbara Islands to Panama Bay), is even smaller than *decussata*, is pure-white and very inflated. Dredged in deep water. Some workers feel that this is a southern subspecies of *decussata*.

Subgenus *Arvella* Bartsch in Scarlato, 1960

Similar to true *Crenella*, but with regular radiating ribs over the whole surface, dorsally more or less parallel to the margin. Type: *Mytilus faba* (Müller, 1776).

Crenella faba (O. F. Müller, 1776)   5054
Faba Crenella

Arctic Seas to Nova Scotia.

¼ to ½ inch in length, oval-oblong, with numerous radial ribs. Color reddish brown. Periostracum varnishlike. Byssus golden-brown. Common offshore.

5054

5051

Other species:

5055 *Crenella pectinula* (Gould, 1841). Gulf of St. Lawrence to Georges Bank, Massachusetts. 16 mm.

5056 *Crenella fragilis* Verrill, 1885. Off Chesapeake Bay, 70 fms. 20 mm.

5057 *Crenella leana* Dall, 1897. Aleutian Islands, eastward to Middleton Island, Alaska.

5058 *Crenella rotundata* Dall, 1916. Santa Cruz Island, California, 155 fms.

5059 *Crenella skomma* (Schwengel, 1944). Off Lantana, Florida, 70 fms. *The Nautilus*, vol. 58, p. 16.

Genus *Rhomboidella* Monterosato, 1884

Shell small, inflated, thin; umbones inflated and anterior; prodissoconch large, smooth; surface striated by radial diverging and bifurcating ribs, usually with a narrow, smooth area near the anterior end; hinge with thickened and vertically striated anterior margin below the umbo and with vertical striations behind the short ligament; margins finely crenulated. Type: *Modiola pridauxi* Leach, 1815, (synonym: *rhombea* Berkeley, 1827) from northwest Europe.

Subgenus *Solamen* Iredale, 1924

Ovalish, fat, translucent-white or brown, with very fine divaricate radial ribbing. No toothlike, crenulated resilifer present, and the ligament is nearly marginal. Type: *rex* Iredale, 1924, from Australia. *Megacrenella* Habe and Ito, 1965 (type: *columbiana* Dall, 1897) is a synonym.

Rhomboidella columbiana (Dall, 1897)   5060
British Columbia Crenella

Aleutian Islands to Mexico.

A little over ½ inch in length, oval-oblong, inflated, with numerous very fine, decussate radial ribs. The largest *Crenella* on the Pacific coast. Color greenish yellow-brown. 10 to 100 fathoms. Common. *Crenella rotundata* Dall, 1916, is a synonym.

Other species:

5061 *Rhomboidella grisea* (Dall, 1907). Bering Sea to Sitka, Alaska, 30 fms.

Genus *Gregariella* Monterosato, 1884

Shell fat, elongate, fairly fragile; beaks almost terminal at the front end; posterior end pointed. Umbonal slope with divaricating riblets. Fine radial riblets over the surface, except for the smooth middle part of the valve. Periostracum brown, smooth anteriorly, but long, bifurcating tufts of hairs posteriorly. A few teeth present on the hinge at each end. Live in burrows in rocks. Type: *opifex* (Say, 1825). *Tibialectus* Iredale, 1939; *Trichomusculus* Iredale, 1924; *Botulina* Dall, 1889, are synonyms.

Gregariella coralliophaga (Gmelin, 1791)   5062
Common Gregariella

North Carolina to Texas; West Indies. Bermuda. Brazil. Monterey, California, to Peru.

¾ inch, inflated, with a high posterior ridge from which radiating striae curve backward and downward. Shell white outside, iridescent bluish white inside. The beaks, placed at the extreme anterior end, curve inward and forward. Inner edges of the shell finely serrate. Periostracum brown, heavy, and hairy on the posterior dorsal slope. Bores in coral rocks. Fairly common. *G. sulcatus* Risso, 1826, and *G. opifex* (Say, 1825), are probably synonyms, as are *chenui* Récluz, 1842, and *coarctata* Carpenter, 1856. *G. denticulata* Dall, 1871, may also be a synonym. A widespread, variable species which some workers persist in splitting into many species.

Genus *Lioberus* Dall, 1898

Similar to *Modiolus*, but the animal has long siphons. Beaks near anterior end; smooth or with obsolete radial sculpture, without teeth; periostracum smooth. Type: *castaneus* (Say, 1822).

Lioberus castaneus (Say, 1822)   5063
Say's Chestnut Mussel

Both sides of Florida and the West Indies. Brazil.

5062   5063

¾ inch in length, oval-elongate, well-inflated and thin-shelled. **Exterior** chestnut- to dark-brown, the anterior half glossy, the posterior half dull and commonly with a fine grayish matting of periostracum. Interior bluish white and with an irregular surface. Hinge simple with a slight swelling or pad under the beaks. Moderately common in shallow water.

Genus *Musculus* Röding, 1798

Mussellike shells with the sculpturing divided into 3 oblique areas, the center one being smooth or almost so, and the 2 end areas having radial ribs. The ligament is much longer than that in *Crenella*. Hinge finely dentate. These are moderately deep-water clams. Mantle folded in front into a wide, incurrent siphon and behind into a conical excurrent siphon. Foot strap-shaped. They often encase themselves in a "cocoon" of byssal threads where the young clams are harbored. Type: *discors* (Linné, 1767). Synonyms: *Modiolaria* Beck, 1838; *Modiolarca* Gray, 1843; *Lanistina* Gray, 1847; *Planimodiola* Cossmann, 1887.

Musculus niger (Gray, 1824)   **5064**
Black Musculus

Arctic Seas to North Carolina.
Alaska to Puget Sound.

About 2 to 3 inches in length. Similar to *M. discors,* but much more compressed and with strongly developed axial, decussated ribs on the posterior and anterior thirds. Center section with microscopic concentric, wavy threads and pimples. Often pinkish on the inside. Common offshore; 1 to 60 fathoms. Also found in northern Europe. *M. obesus* Dall, 1916, is a synonym.

Musculus discors (Linné, 1767)   **5065**
Discord Musculus

Arctic Seas to Long Island Sound.
Arctic Seas to Puget Sound.

1 to 2 inches in length, oblong, fairly fragile. Anterior and posterior thirds of outer shell with very weak radial ribs; center section smooth except for irregular growth lines. Periostracum shiny and either dark black-brown or light-brown. Anterior bluish white with slight iridescence. Commonly dredged.

A large ecologic form **(5066)** that has no pronounced radial riblet separating the posterior third from the middle area has been called *laevigatus* Gray, 1824, and *substriata* Gray, 1824.

5064   5065

Musculus senhousei (Benson, 1842)   **5067**
Senhouse's Musculus

Washington to central California; eastern Asia.

1 inch, rather fragile, thin-shelled, green to bluish green, with delicate zigzag brownish flames. Posterior and anterior ends with fine radiating striae. Lunule with crenulated margin. Crenulations along and behind the ligament. Introduced from

Japan prior to 1944. Locally common; makes byssal nests on mud flats and attaches to pilings. Named after Mr. Senhouse. The original spelling of the species name, *senhousia,* has been legitimately modified. Formerly placed in *Modiolus* and *Arcuatula.*

5067

Musculus olivaceus Dall, 1916   **5068**
Olivaceous Musculus

Bering Sea to Catalina Island, California.

8 to 11 mm., brown with a white unsculptured prodissoconch. Moderately inflated, with 10 or 11 anterior radial ribs, and with numerous posterior ribs which are not bordered in front by a furrow, but increase in strength backward until the ribs are slightly broader than the interspaces. 4 strong, tooth-like crenulations below the umbo. Periostracum olive-colored and with a silky luster. Uncommon; offshore in shallow waters.

Subgenus *Ryenella* Fleming, 1959

Type of subgenus is the Miocene *Mytilus impactus* Hermann, 1915.

Musculus lateralis (Say, 1822)   **5069**
Lateral Musculus

North Carolina to Florida, Texas, and the West Indies.
Brazil.

⅜ inch in length, oblong, fragile, with a center area on the valve with concentric growth lines only. Remainder of shell with radial ribs. Color light-brown with a strong blush of blue-green. Interior slightly iridescent. Common offshore from 1 to 30 fathoms.

Other species:

5070 *Musculus corrugatus* (Stimpson, 1851). Arctic Seas to off North Carolina, 1 to 60 fms.; Alaska to Puget Sound; northern Europe.

5071 *Musculus protractus* Dall, 1916. Nunivak Island, Bering Sea, to Monterey, California.

5072 *Musculus impressus* (Dall, 1907). Petrel Bank, Bering Sea.

5073 *Musculus taylori* (Dall, 1897). Victoria, Vancouver Island, British Columbia.

5074 *Musculus vernicosus* Middendorff, 1849. Bering Sea and Okhotsk Sea to Sitka, Alaska.

5075 *Musculus phenax* Dall, 1915. St. George Island, Bering Sea.

5076 *Musculus marmoratus* (Forbes, 1838). Arctic Seas.

5077 *Musculus seminudus* (Dall, 1897). Bering Sea to Forrester Island, Alaska.

5078 *Musculus pygmaeus* Glynn, 1964. Monterey Bay, California. High intertidal zone. 1 to 4.5 mm. *The Veliger,* vol. 7, p. 171.

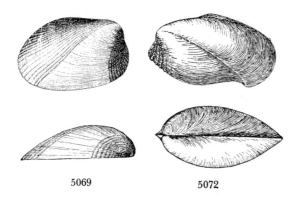

5069   5072

Subfamily LITHOPHAGINAE H. and A. Adams, 1857

Long, cylindrical, boring mussels with beaks at the anterior end. Hinge margins usually smooth. Periostracum may have calcareous incrustations.

Genus *Lithophaga* Röding, 1798

Shells long, round, resembling date seeds. Umbones nearly terminal. Periostracum heavy, black or brown. Ligament long, narrow and internal. Bore into rock by means of a chemical process, probably by a weak carbonic acid. Type: *lithophaga* (Linné, 1758). For their biology and taxonomy, see "The genus *Lithophaga* in the Western Atlantic" by R. D. Turner and K. J. Boss, 1962, Johnsonia, vol. 4, no. 41. For remarks on boring see F. Haas, 1943, *Zool. Series, Field Mus. Nat. Hist.*, vol. 29, p. 7.

Lithophaga nigra (Orbigny, 1842)   **5079**
Black Date Mussel

Southeast Florida to Brazil. Bermuda.

1 to 2 inches in length, elongate and cylindrical. Black-brown outside and an iridescent bluish white inside. Anterior lower 1/3 of each valve with strong, vertical, smooth ribs; remainder of shell smoothish with only irregular growth lines. Commonly found boring into soft coral blocks. Uncommon in Florida. Synonyms are: *antillarum* Philippi, 1847; *caribaea* Philippi, 1847; and *crenulata* Dunker, 1848.

Lithophaga antillarum (Orbigny, 1842)   **5080**
Giant Date Mussel

Southeast Florida and the West Indies. Brazil.

2 to 4 inches in length, elongate, cylindrical and colored a light yellowish brown on the outside and iridescent-cream inside. Sides of valves marked with numerous, irregular, vertical riblets. Fairly common in soft rocks in shallow water. Synonyms are: *corrugata* (Philippi, 1846); *stramineus* ("Dunker" Reeve, 1857). Also known from the Western Pacific.

Subgenus *Diberus* Dall, 1898

Shell with 1 or more radial sulci extending along the posterior umbonal slope with the calcareous encrustations covering the space between them coarse and heavy. Calcareous ends of the valves resemble short wedges. Type: *plumula* Hanley, 1843. Synonyms: *Exodiberis* Iredale, 1939; *Salebrolabis* Iredale, 1939.

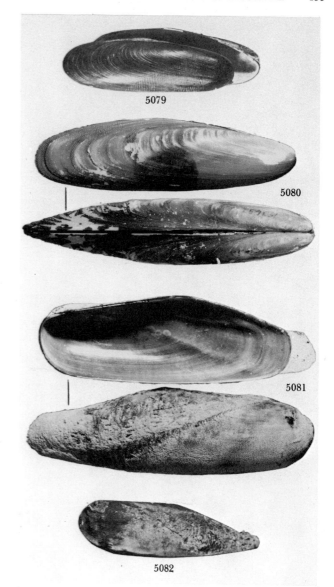

5079

5080

5081

5082

Lithophaga bisulcata (Orbigny, 1842)   **5081**
Mahogany Date Mussel

North Carolina to Texas, Florida and to Brazil. Bermuda.

1 to 1½ inches in length, elongate, cylindrical and coming to a point at the posterior end. A sharp, oblique, indented line divides each valve into two sections. Anterior ½ of valve smooth, mahogany-brown, but commonly encrusted with porous, gray calcium deposits. Posterior end more heavily encrusted with a gray, porous covering which projects beyond the edge of the shell. A fairly common rock-boring species. Scarce in Florida. Synonyms are *appendiculata* (Philippi, 1846); *biexcavata* (Reeve, 1857).

Lithophaga plumula kelseyi Hertlein and Strong, 1946   **5082**
Kelsey's Date Mussel

San Diego north to Mendocino County, California.

1 to 2 inches in length, similar to *L. bisulcata,* but the calcareous matter on the posterior end is strongly pitted and furrowed to look like a wet, ruffled feather. Typical *plumula* Hanley, 1844, ranges from Baja California to Peru. Both fairly common in rocks and *Spondylus* shells.

Subgenus *Myoforceps* P. Fischer, 1886

Posterior end with 2 crossed blades of shelly material. Type: *aristata* (Dillwyn, 1817).

Lithophaga aristata (Dillwyn, 1817)   5083
Scissor Date Mussel

North Carolina to Texas, Florida and the West Indies. La Jolla, California, to Peru.

½ to 2 inches in length. Characterized by the pointed tips at the posterior end being crossed like fingers. Color yellowish brown, but generally covered by a smooth, gray, calcareous encrustation. Moderately common in soft rock. Synonyms are: *caudigera* (Lamarck, 1819); *calyculatus* (Carpenter, 1857); *forficata* (Ravenel, 1861); *caudatus* (Gray, 1827); *ropan* Deshayes, 1836; *gracilior* (Carpenter, 1856); *tumidior* (Carpenter, 1856); *carpenteri* (Mörch, 1861); *bipenniferus* (Guppy, 1877); and *curviroster* Schröter, 1786 (a rejected name). Also found in the Eastern Atlantic and Mediterranean.

5083

Other species:

5084   *Lithophaga* (*Labis* Dall, 1916) *attenuata* (Deshayes, 1836). Baja California to Sonora, Mexico. (Synonym: *inca* Orbigny, 1846.) The form or subspecies *rogersi* S. S. Berry, 1957, occurs from southern California to south Mexico.

Genus *Adula* H. and A. Adams, 1857

Similar to the true date mussels, but lacking rough chalky encrustations and having the beaks near the center of the shell. Ventral margin broadly impressed along the midzone. Periostracum brown to black. Type: *soleniformis* (Orbigny, 1846). Rock borers.

Adula falcata (Gould, 1851)   5085
Falcate Date Mussel

Coos Bay, Oregon, to Baja California.

2 to 4 inches in length, very elongate, slightly curved. Beaks rounded and about ⅛ the length from the anterior end; a strongly marked angle occurs from the beaks to the base of the posterior extremity; numerous vertical, wavy ribs over all the shell. Color a shiny chestnut-brown. Common; bores in rocks.

Adula californiensis (Philippi, 1847)   5086
Californian Date Mussel

British Columbia to southern California.

1 to 1¼ inches in length, elongate, curved and smooth, except for a velvety hairlike covering over the posterior end. Shiny, chocolate-brown in color. It is shorter and more angular than falcata. Moderately common; lives in rocks.

5085

5086

Adula diegensis (Dall, 1911)   5087
Non-boring Date Mussel

Oregon to Sonora, Mexico.

1 to 1½ inches, mytiloid in shape, with rather large, fat, inrolled umbones near the anterior end. Chestnut-brown in color. Abundant in colonies attached to wharf pilings and breakwaters, sometimes among *Mytilus* beds.

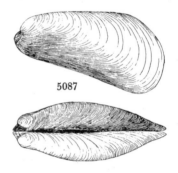

5087

Subfamily MODIOLINAE Keen, 1958

Obese, elongate with hinge margin smoother or finely striated vertically. Radial sculpture mostly absent. Periostracum hairy.

Genus *Modiolus* Lamarck, 1799

This group of mussels have shells of various forms in which the hinge is without teeth. The anterior end of the shell extends in front of the beaks, while in *Mytilus* 3 to 5 tiny teeth are present on the hinge and the beaks are at the very anterior end of the shell. *Volsella* Scopoli, 1777, has been rejected (Opinion 325, I.C.Z.N.). The genus name is pronounced "moe-*dye*-oh-lus." *Modiola* Lamarck, 1801, and *Eumodiolus* von Ihering, 1900, are synonyms.

Modiolus modiolus (Linné, 1758)   5088
Northern Horse Mussel

Arctic Seas to New Jersey; Europe. Arctic Seas to San Pedro, California.

2 to 6 inches in length, heavy, with a coarse, rather thick, black-brown periostracum with long, smooth hairs. Beaks always white and not swollen. Adult shells chalky, mauve-white. One of the largest and most common mussels found in cold waters below low-tide mark. South of New York, the subspecies *squamosus* becomes fairly common in warm water.

5088

Modiolus modiolus squamosus Beauperthuy, 1967   5089
False Tulip Shell

North Carolina to Texas to the Caribbean.

Almost identical to half-grown *M. modiolus* from New England, but rarely over 2 inches in length. Exterior brownish to purple, with an oblique whitish ray, and with coarse, flat, triangular periostracal hairs. Interior purplish to whitish. Umbones small, not swollen, always white. Is more compressed than *M. americanus*, lacks the rose coloring and never has radial rays of purple or brown. A common species in shallow water, especially in the northern ⅔ of Florida. See Bol. Inst. Oceanogr., Cumaná, Venezuela, vol. 6, no. 1.

Modiolus americanus (Leach, 1815)   5090
Tulip Mussel   **Color Plate 20**

South Carolina to Florida to Brazil; Bermuda.
Gulf of California to Peru.

Adults 2 to 4 inches, very fat, rather fragile; with swollen umbones which are either pinkish or purple (never white). Color chestnut-brown at the anterior end; with a white oblique streak in the middle and with rose, purple-rayed and brownish behind. Periostracal hairs stringlike and smooth. Interior of shell pearly-whitish, usually rose-tinted, rarely purplish. A common coral-water, tropical species occurring in 3 to 18 feet of water. *M. pseudotulipus* Olsson, 1961, is a Pacific coast synonym. *M. tulipa* Lamarck, 1819, is a synonym. In southeast United States do not confuse with *Modiolus modiolus* subspecies *squamosus* Beauperthuy, 1967.

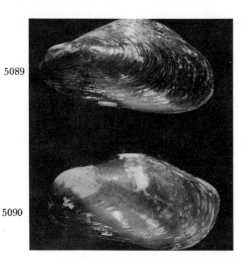

5089

5090

Modiolus carpenteri Soot-Ryen, 1963   5091
California Horse Mussel

Monterey to San Pedro, California.

About 1 inch in length, smoothish, inflated, light-brown periostracum which wears white at the beak end. Beaks curved strongly forward. Interior dull-white. Found in moderately deep water, 15 to 25 fathoms, and rarely cast ashore. Associated with *Haliotis rufescens. M. fornicata* Carpenter, 1864, non Roemer, 1836, is a synonym (see *Proc. Malacol. Soc. London*, vol. 35, p. 127).

Modiolus sacculiferus (S. S. Berry, 1953)   5092
Pouting Horse Mussel

Southern California.

2 to 3 inches, broadly almond-shaped; highest at the midpoint, thin, smooth, moderately inflated. Valves rounded and with a lobelike flare under the umbones, thus producing a "pouting" appearance. Periostracum brown, somewhat polished and shiny. Do not confuse with *carpenteri*. Uncommon; offshore down to 25 fathoms. (*Trans. San Diego Soc. Nat. Hist.*, vol. 11, p. 407.)

5092

Modiolus rectus (Conrad, 1837)   5093
Fan-shaped Horse Mussel

Vancouver, British Columbia, to the Gulf of California.

4 to 6 inches, somewhat rectangular in shape, thin-shelled, well-inflated. Shell bluish white with a yellowish brown periostracum. Interior pearly-white, tinged with pink. Lives burrowed in mud with the broad posterior margin exposed. Intertidal to 45 meters. *M. flabellatus* Gould, 1850, is a synonym.

Modiolus neglectus Soot-Ryen, 1955   5094
Neglected Horse Mussel

Monterey, California, to Baja California.

3 to 5 inches, closely resembling *M. rectus,* but *neglectus* is much more inflated. The height is 35 to 40% of the total length in *neglectus* and 25 to 30% in *rectus. M. neglectus* has a distinct posterodorsal angle in which the posterior convex margin meets the dorsal margin at an angle of about 140 degrees. This deep-water species lives at depths of 10 to 57 fathoms in sand or mud, and spins a "nest" of threads. It is common.

Modiolus capax (Conrad, 1837)   5095
Capax Horse Mussel

Santa Cruz, California, to Peru.

2 to 6 inches in size. Periostracum thick, often with coarse hairs, chestnut-brown in color. Worn shell brick-red with bluish mottlings. Interior half white, half (ventral) brownish purple. Common; shallow water. Resembles *Modiolus americanus*. Synonyms are *spatula* Menke, 1849, and *subfuscata* Clessin, 1889.

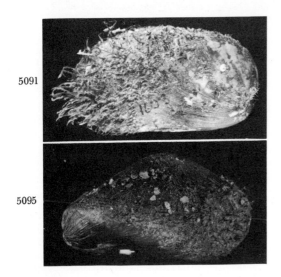

5091

5095

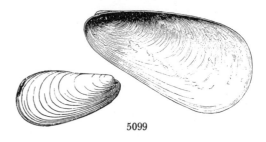

5099

Genus *Botula* Mörch, 1853

Shell 1 inch or less, oblong, fat, rather fragile, the surface with a glossy-brown periostracum. Umbones prominent, ending in small, coiled beaks. Concentric growth lines sometimes wavy. Interior bluish, nacreous. Type: *fuscus* (Gmelin, 1791). Synonym: *Botulopa* Iredale, 1939.

Other species:

5096 *Modiolus eiseni* Strong and Hertlein, 1937. Gulf of California to Ecuador, in 4 to 360 meters.

Genus *Amygdalum* Mühlfeld, 1811

Shell thin, very smooth, often with colored, cobwebby designs. These clams build nests for themselves with a copious supply of byssal threads. Type: *dendriticum* Mühlfeld, 1811.

***Amygdalum papyrium* (Conrad, 1846)   5097**
Paper Mussel

Texas and Maryland to Florida.

1 to 1¼ inches in length, elongate, smooth, glistening, fragile, and colored a delicate two-tone of bluish green and soft yellowish brown. Interior iridescent-white. The ligament is very weak and thin. Attaches to the marine tassel grass, *Ruppia maritima* (L.). For ecological and habitudinal details, see J. Frances Allen, 1955, *The Nautilus*, vol. 68, p. 83. *Modiola pulex* H. C. Lea, 1842 (not Lamarck, 1819) is a synonym.

(**5098**) *A. sagittatum* Rehder, 1934, sometimes dredged off Florida and Mississippi, is ¾ inch, very shiny, ivory-white, ½ of each valve with fine, gray, cobwebby streaks. The umbo is reinforced inside by a very small, smooth column or rib.

5097

***Botula fusca* (Gmelin, 1791)   5101**
Cinnamon Mussel

North Carolina to Florida; West Indies. Bermuda. Brazil.

¾ inch, oblong, arcuate, grayish brown, with thick concentric ridges on the outside. Inrolled beaks at the anterior end. Tiny vertical threads are on the hinge just posterior to the ligament. Found in rock burrows. Uncommon. Do not confuse with Say's chestnut mussel, *Lioberus*, which is glossy, chestnut-brown. *B. cinnamomea* Lamarck, 1819, is a synonym.

5101

Genus *Dacrydium* Torell, 1859

Shell fragile, hyaline, smooth; beaks anterior; hinge crenulated anteriorly and with vertical striations posteriorly; resilium internal. Anterior adductor on a thickened support; no grooved teeth. Type: *vitreum* (Holböll, 1842).

***Dacrydium vitreum* (Holböll in Möller, 1842)   5102**

Greenland to the Gulf of Mexico; Norway.

Shell minute, 4 to 6 mm., fragile, fat, translucent-white, with an oily, iridescent surface which may have fine particles of silt or sand attached to it. Umbones swollen, curved towards each other and touching. Posterior end of valves much broader

Other species:

5099 *Amygdalum politum* (Verrill and Smith, 1880). North Atlantic to the West Indies, 111′ to 1,000 fms. Gulf of Mexico, 220 fms. 42 mm.

5100 *Amygdalum pallidulum* (Dall, 1916). Bodega Head, California, to Baja California, 25 to 75 fms. Abundant.

5102

and more compressed than the anterior end. Common; 6 to 1,555 fathoms.

Other species:

5103 *Dacrydium pacificum* Dall, 1916. Bering Sea; 1,401 fms.

5104 *Dacrydium panamensis* Knudsen, 1970. Off Acapulco, Mexico, 3,570 meters; Gulf of Panama, 3,670 meters. *Galathea Report,* vol. 11, p. 91.

Genus *Geukensia* Poel, 1959

Radially ribbed; weaker striations on the antero-ventral area. Margins crenulated but no toothlike crenulations near the deepset ligament; 2 anterior retractor muscles in and in front of the umbonal cavity. Type: *demissa* (Dillwyn, 1817). Synonym: *Arcuatula* Soot-Ryen, 1955, not Jousseaume in Lamy, 1919.

Geukensia demissa (Dillwyn, 1817)   **5105**
Atlantic Ribbed Mussel   **Color Plate 20**

Gulf of St. Lawrence to northeast Florida.
Introduced to California (San Francisco Bay).

2 to 4 inches in length, black-brown in color, often shiny, and with strong, rough, radial, bifurcating ribs. Interior bluish white with the posterior end flushed with purple or purplish red. This is the only ribbed modiolinid mussel in our waters, but do not confuse it with *Ischadium recurvum,* which has a strongly curved beak, tiny teeth at the umbo, and is a solid rosy-brown on the inside. *Mytilus plicatulus* Lamarck, 1819, is this species. It is abundant in salt marshes intertidally.

(5106) The subspecies *demissa granosissima* (Sowerby, 1914) (both sides of Florida to Texas and Yucatan), is very similar but with almost twice as many ribs which are finely and neatly beaded. Common.

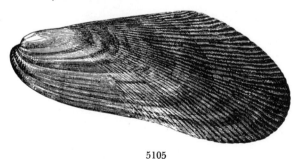

5105

Genus *Idasola* Iredale, 1915

Small, thin-shelled, silvery shells with rounded anterior and posterior margins. Beaks in front of the middle. Hinge with vertical striations on thickened anterior margin below the umbo and behind the ligament. Periostracum thin, with minute hairlets. Type: *argenteus* (Jeffreys, 1876). Synonym:

Idas Jeffreys, 1876, non Mulsant and Verreaux, 1875. One species in the North Atlantic:

5107 *Idasola argentea* (Jeffreys, 1876). 103 miles south southwest of Martha's Vineyard, Massachusetts, 335 fms. Northern Europe; Iceland.

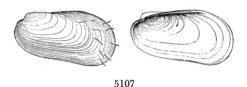

5107

Superfamily Pinnacea Leach, 1819

Fan-shaped, fragile, large shells with the narrow beak end sunken into the substrate. Byssus well-developed and serving as an anchor. Anterior adductor muscle small. Posterior muscle large. No hinge teeth. Shell material of prismatic calcite, partially nacreous. Formerly placed with the winged oyster, order Pterioida.

Family PINNIDAE Leach, 1819

Two recent monographs treat with the pen shells. "The Family Pinnidae in the Western Atlantic," by R. D. Turner and Joseph Rosewater, 1958, *Johnsonia,* vol. 3, no. 38; and "The Family Pinnidae in the Indo-Pacific," by Joseph Rosewater, 1961, *Indo-Pacific Mollusca,* vol. 1, no. 4, pp. 53–501. The muscles are edible and not unlike those of scallops.

Genus *Pinna* Linné, 1758

The pen shells are large, fragile, fan-shaped clams which live in sandy or mud-sand areas, usually in colonies. The apex or pointed end is deeply buried, and there is a mass of byssal threads attached to small stones or fragments of shells. The broad end of the shell projects above the surface of the sand an inch or so. In the genus *Pinna,* there is a weak groove running down the middle of each valve. In *Atrina,* this character is absent. The type of *Pinna* is *rudis* Linné, 1758. Synonyms include: *Pinnarius* Duméril, 1806; *Pinnula* Rafinesque, 1815; *Sulcatopinna* Hyatt, 1892; *Quantulopinna* Iredale, 1939; *Subitopinna* Iredale, 1939; *Exitopinna* Iredale, 1939.

Pinna carnea Gmelin, 1791   **5108**
Amber Pen Shell

Southeast Florida, Texas and the West Indies. Bermuda. Brazil.

4 to 11 inches in length, relatively narrow, thin-shelled and with a central, radial ridge in the middle of the valve which is more conspicuous at the pointed or hinge end. With or without 8 to 12 radial rows of moderately large, scalelike spines. Ventral lobe of nacreous layer equal to or longer than the dorsal (or hinge side) lobe. Color usually a light-orangish to translucent-amber. Rare in Florida but common in the Bahamas. *P. rudis* Linné is heavier, darker red, has the dorsal lobe of shiny, nacreous layer longer than the ventral lobe, and has 5 to 7 ribs. Synonyms are *degenera* Link, 1807, and *flabellum* Lamarck, 1819.

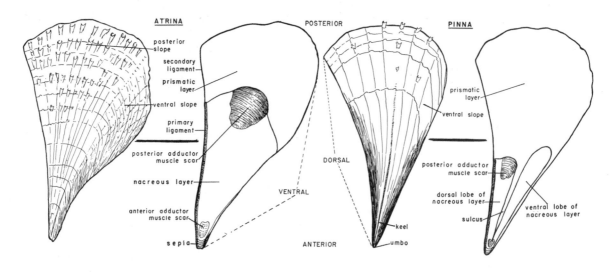

5108

5109

Pinna rudis Linné, 1758   5109

Rude Pen Shell

Puerto Rico to Trinidad.
Mediterranean to Angola.

7 to 22 inches, reddish brown, heavy-shelled, with 5 to 8 ribs bearing tubular spines. The ventral lobe of the nacreous section is shorter and more truncate than the dorsal lobe. Uncommon in the Caribbean. Synonyms include: *ferruginea* Röding, 1798; *elongata* Röding, 1798; *pernula* Röding, 1798; *varicosa* Lamarck, 1819; and *paulucciae* Rochebrune, 1883. Found at depths of 40 to 80 ft., in crevices of the corals, *Diploria* and *Montastrea annularis*.

Pinna rugosa Sowerby, 1835   5110

Rugose Pen Shell

Lower Baja California to Panama.

10 to 24 inches, with a sulcus running longitudinally down the middle of the lower ⅓ of the valve. Color amber to greenish black. 6 to 10 rows of tubular spines. Moderately common offshore, buried in sand.

Genus *Atrina* Gray, 1842

Nacreous internal layer extending about ⅔ to ¾ the length of the shell, and not divided by a longitudinal sulcus, as in *Pinna*. Type: *vexillum* (Born, 1778).

Atrina rigida (Lightfoot, 1786)   5111

Stiff Pen Shell

North Carolina to south half of Florida and the Caribbean.

5 to 11 inches in length, relatively wide, moderately thick-shelled and with 15 to 25 radial rows of tubelike spines; rarely smoothish. Color dark- to light-brown. Externally about the same as *seminuda*, and can be distinguished only by the large muscle scar at (or slightly protruding above) the border line of the nacreous section of the inside of the shell. Mantle of *rigida* is bright golden-orange, while in *seminuda* the same area is pale-yellow. Commonly washed ashore. Live buried in

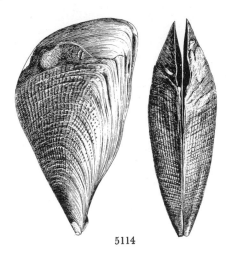

5114

5111

sandy mud, low tide to 15 fathoms. A small, commensal crab lives inside the mantle cavity. A number of unusual snails and chitons are found in or on dead or live *Pinna* shells.

Atrina tuberculosa (Sowerby, 1835)   5112
Tuberculate Pen Shell

Gulf of California to Panama.

5 to 10 inches, with 13 to 16 rows of coarse, scalelike spines. Color black-brown. Muscle scar at the edge of the nacreous area. Abundant; shallow water.

Subgenus *Servatrina* Iredale, 1939

Externally the same as *Atrina*. Inside, the muscle scar is well-enclosed within the nacreous area. Type: *pectinata* (Linné, 1767).

Atrina seminuda (Lamarck, 1819)   5113
Half-naked Pen Shell

North Carolina to Texas, to Argentina.

5 to 9 inches, externally similar to *rigida*. However, on the inner surface the relatively small, round muscle scar is well-surrounded by the nacreous area. See under *rigida*. A variable and common species just offshore. Absent from south Florida. Synonyms are: *alta* (Sowerby, 1835); *listeri* (Orbigny, 1846); *patagonica* (Orbigny, 1846); *subviridis* (Reeve, 1858); *ramulosa* (Reeve, 1858); and *dorbignyi* (Reeve, 1858).

Atrina maura (Sowerby, 1835)   5114
Maura Pen Shell

Baja California to Peru.

5 to 9 inches, blackish brown, with about 18 rows of spines. Common; on offshore mud flats. *A. oldroydi* (Dall, 1901), from off southern California is probably a synonym.

Atrina serrata (Sowerby, 1825)   5115
Saw-toothed Pen Shell

North Carolina to Florida, Texas and the West Indies.

6 to 12 inches, thin-shelled, moderately inflated, with about 30 finely scaled ribs. Posterior adductor muscle scar nearly circular and set will within the nacreous area. Lives in sandy mud in water from a few feet in depth to several fathoms. Commonly washed ashore with *rigida*.

Other species:

5116 *Atrina* (*Servatrina*) *texta* Hertlein, Hanna and Strong, 1943. Off the west side of Baja California.

5113

5112

5115

Order *Pterioida* Newell, 1965

Includes the pearly and winged oysters, scallops, the *Lima* file clams, and the true oysters. Pallial sinus absent. Ligament broken into 1 or several small parts. Shell material pearly or porcelaneous.

Suborder *Pteriina* Newell, 1965

Superfamily Pteriacea Gray, 1847

Inequivalve and inequilateral, with the left valve usually more convex than the right. Ligament external. Adults usually with a large byssus coming from a distinct notch in one valve.

Family PTERIIDAE Gray, 1847

Usually with a triangular winglike projection of the straight hinge line. Interior pearly. Strong byssus extends from an indentation in the less convex right valve below the front winglike extension. Borders of the valves usually thin and flexible.

Genus *Pteria* Scopoli, 1777

Fairly thin-shelled, moderately fat and with the hinge ends considerably drawn out. Pearly inside. The right and left valves bear 1 or 2 small denticles which fit into shallow sockets in the opposite valve. Type: *hirundo* (Linné, 1758). Synonyms: *Avicula* Bruguière, 1792; *Austropteria* Iredale, 1939; *Magnavicula* Iredale, 1939.

Pteria colymbus (Röding, 1798)   **5117**
Atlantic Wing Oyster   **Color Plate 20**

North Carolina to Florida, Texas and the West Indies and to Brazil. Bermuda.

1½ to 3 inches in length, obliquely oval with a long extension of the hinge line toward the posterior end. Left valve inflated. Right valve somewhat flatter and with a strong anterior notch for the byssus. Periostracum matted, brown and with cancellate fimbrications. Exterior color variable: brown, black or brownish purple with broken, radial lines of cream or white. Interior pearly with a wide, nonpearly margin of purplish black with irregular cream rays. Common from low water to several fathoms, usually attached to alcyonarians. *P. atlantica* Lamarck, 1819, is a synonym.

Pteria sterna (Gould, 1851)   **5118**
Western Wing Oyster

Southern California to Peru.

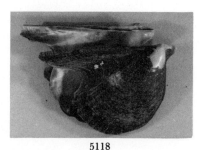

5118

3 to 4 inches, deep purplish brown with occasional paler rays. Periostracum matted, spiny. Serves as a source of pearls in tropical waters. Common. Synonyms include *Avicula fimbriata* Dunker, 1852; *peruviana* Reeve, 1857; *vivesi* Rochebrune, 1895; and *beiliana* Olsson, 1961.

Other species:

5119 *Pteria viridizona* Dall, 1916. Long Beach and San Pedro, California. From the backs of crabs. Is *longisquamosa* (Dunker, 1852) from Caribbean, fide Helen Hayes (*in thesis*).

5120 *Pteria vitrea* (Reeve, 1857). South of Martha's Vineyard, Massachusetts, on hydroids, 65 to 192 fms; off Florida Keys and Gulf of Mexico. (Synonyms: *nitida* Verrill, 1880; *verrilli* Cossmann, 1912.)

— *Pteria xanthia* Schwengel, 1942. *The Nautilus*, vol. 56, p. 64. Is *longisquamosa* Dunker, 1852, fide Helen L. Hayes (*in thesis*).

5121 *Pteria longisquamosa* (Dunker, 1852). Florida Keys, Caribbean and Bermuda.

Genus *Pinctada* Röding, 1798

This is the famous genus of pearl oysters. The byssal gape is in the right valve below the small, triangular auricle. Type: *margaritiferus* (Linné, 1758). *Margaritifera* Schumacher, 1817; *Margaritiphora* Mühlfeld, 1811, and *Meleagrina* Lamarck, 1819, are synonyms.

Pinctada imbricata Röding, 1798   **5122**
Atlantic Pearl Oyster   **Color Plate 20**

South Carolina to Florida, Texas and the West Indies to Brazil. Bermuda.

1½ to 3 inches in length, moderately inflated to flattish, thin-shelled and brittle. There is a small, thin, flat ligament at the center of the hinge. Exterior tan with mottlings or rays of purplish brown or black. Rarely tinted with dull-rose or greenish. In quiet waters, thin, scaly and very delicate, periostracal spines may be developed. Interior a beautiful mother-of-pearl. Common in shallow water attached to rocks. *P. alaperdicis* (Reeve, 1857); *squamulosa* (Lamarck, 1819); *radiata* (Leach, 1814) are synonyms.

Pinctada mazatlanica (Hanley, 1856)   **5123**
Panamanian Pearl Oyster

Baja California, the Gulf, to Peru.

4 to 5 inches, rather heavy, externally blackish purple to olive-brown with a spiny, brittle periostracum occurring in concentric layers. This is the common pearl oyster of the tropical Eastern Pacific. Synonym: *barbata* (Reeve, 1857).

Family ISOGNOMONIDAE Woodring, 1925

Genus *Isognomon* Lightfoot, 1786

Shell thin and greatly compressed; interior pearly; anterior margin with a narrow byssal gape near the dorsal margin. Hinge with numerous parallel grooves perpendicular to the dorsal margin of the valve. Type: *perna* (Linné, 1767), of the Indo-Pacific. *Pedalion* Solander, 1770, is invalid. Synonyms: *Melina* Retzius, 1788; *Perna* Bruguière, 1789. Do not confuse the name *Isognomon* with the Indo-Pacific subgeneric name *Isogonum* Röding, 1798, whose type is *isognomum* Linné, 1758.

Isognomon alatus (Gmelin, 1791)　5124
Flat Tree Oyster　**Color Plate 20**

Florida to Texas and the West Indies. Bermuda. Brazil.

2 to 3 inches in length. Hinge has 8 to 12 oblong grooves or sockets into which are set small, brown resiliums. Exterior with rough or smoothish growth lines. External color drab purplish gray to dirty-gray. Interior moderately pearly with stains of purplish brown or mottlings of blackish purple. This very flat, oval bivalve is commonly found in compact clumps on mangrove tree roots. Distinguished from *I. radiatus* by its flat, more regularly fan shape and darker color.

Isognomon radiatus (Anton, 1839)　5125
Lister's Tree Oyster　**Color Plate 20**

Southeast Florida, Texas and the West Indies. Bermuda. Brazil.

½ to 2 inches in size, very irregular in shape, commonly elongate. Sometimes twisted and irregular. Hinge short, straight and with 4 to 8 very small, squarish sockets. Exterior rough with weak, flaky lamellations. Color a solid, translucent yellowish, but commonly with a few wavy, radial stripes of light purplish brown. Common on rocks at low tide. Formerly *I. listeri* Hanley, 1843.

Isognomon bicolor (C. B. Adams, 1845)　5126
Two-toned Tree Oyster　**Color Plate 20**

Florida Keys, Texas, Bermuda and the Caribbean.

1 to 1½ inches, similar to *radiatus,* but is heavier, more oval and commonly with strong lamellations on the outside. It is usually darkly and heavily splotched with purple inside and out. Common. According to Lamy, *I. vulsella* Lamarck is a different species which is limited to the Red Sea. *I. chemnitzianus* (Orbigny, 1846) is a synonym.

(5127) The western tree oyster, *Isognomon recognitus* (Mabille, 1895), from the Coronado Islands to Chile, lives in crowded colonies under stones in shallow water. It resembles the above 2 species, is about 1 to 2 inches in size; its right valve flattish, left valve slightly swollen. It is the only Californian *Isognomon. Perna chemnitziana* of authors, not Orbigny, 1846, is a synonym.

Other species:

5128 *Isognomon janus* Carpenter, 1857. San Ignacio Lagoon, Baja California, to Oaxaca, Mexico.

Family MALLEIDAE Lamarck, 1819

Formerly called Vulsellidae H. and A. Adams, 1857.

Genus *Malleus* Lamarck, 1799

Wings of hinge usually drawn out and long or quite reduced. Ligamental pit deep, oblique, semiconical, with a projecting lower margin. Type: *malleus* (Linné, 1758) of the Indo-Pacific.

Subgenus *Malvufundus* de Gregorio, 1885

Posterior wing short and blunt; anterior wing short or absent. Byssal notch relatively broad. No internal crenulations of the dorsal margins. Early growth stages of valves with irregular concentric lamellae. Type: *regula* (Förskal, 1775). Synonyms: *Fundella* de Gregorio, 1884, non Zeller, 1848; *Parimalleus* Iredale, 1931; *Parvimalleus* Salisbury, 1932; *Brevimalleus* R. McLean, 1947.

Malleus candeanus (Orbigny, 1842)　5129
American Malleus

South Florida to off Texas and the West Indies. Bermuda. Mazatlan, Mexico, to west Colombia.

½ to 11.2 inches in length, brittle, elongate, usually distorted, but characterized by a median, longitudinal ridge on the inside of the valves. Color dark purple-black with areas of yellowish white. Early part of exterior has small, prominent, concentric laminae. Lives in old cracks and crevices of coralline rock or corals in protected areas from 7 to 300 fathoms. Moderately common, but usually overlooked by collectors. Mrs. Sue Abbott has collected it in 70 feet at Andros Island, Bahamas. See also Boss and Moore, 1967, *Bull. Marine Sci.,* vol. 17; and C. W. Johnson, 1918, *The Nautilus,* vol. 32, p. 37. Synonyms: *rufipunctatus, aquatilis* and *vesiculatus* all Reeve, 1858; *panamensis* Mörch, 1861; *obvolutus* de Folin, 1867.

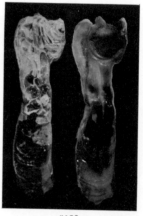

5129

Superfamily Pectinacea Rafinesque, 1815

Includes the true scallops, Pectinidae, the Propeamussidae, the Plicatulidae and the thorny oysters, Spondylidae.

Family PECTINIDAE Rafinesque, 1815

Because of the great number of fossil and living species of scallops and the almost limitless modifications exhibited by them, there have been no less than 50 genera and subgenera proposed in this family by various authors. Doubtlessly, many more will be invented. Most, if not all, of these genera are closely integrated by connecting species. Workers have a choice of using the single genus, *Pecten,* or employing a genus for nearly every species. We are arbitrarily employing only 12 genera, and we cannot justify these on biological grounds.

There are about 350 living species of Pectinidae. A detailed higher classification by Leo G. Hertlein, 1969, appears in the "Treatise on Invertebrate Paleontology," part N, vol. 1, Bivalvia, pp. 348–372. For western species, see Gilbert Grau, 1959, "Pectinidae of the Eastern Pacific," Allan Hancock Pacific Exped., vol. 23, 308 pp.

Subfamily PECTININAE Rafinesque, 1815

Genus *Pecten* Müller, 1776

Shells large, ribbed, with the lower or right valve very inflated and the upper valve usually flat or rarely concave. Type: *maximus* (Linné, 1758). Numerous synonyms include: *Vola* Mörch, 1853; *Notovola* Finlay, 1926.

Subgenus *Flabellipecten* Sacco, 1897

Type of the subgenus is *Ostrea flabelliformis* Brocchi, 1814.

Pecten diegensis Dall, 1898   5130
San Diego Scallop

Bodega Bay, California, to Baja California.

2 to 3 inches in size. Right valve convex with 22 or 23 flat-topped ribs which are generally longitudinally ridged on top. Left valve much flatter, with 21 or 22 narrow, rounded ribs. Dredged from 10 to 75 fathoms. A symbiotic species of *Capulus* is sometimes found attached to the upper valve.

5130

Pecten sericeus Hinds, 1845   5131
Silken Scallop

Gulf of California to Ecuador.

2 to 3 inches, very similar to *diegensis*. Both valves have about 24 ribs. In the lower or right valve the ribs are at first rounded, then trigonal in section, and finally becoming tricarinated at the margin of the valve. Color pale-brown to cream with a faint red blush on the umbones. Rare; down to 50 fathoms.

Subgenus *Patinopecten* Dall, 1898

Shell large, circular, weakly ribbed, valves nearly equal in slight convexity. Byssal notch deep, cardinal crura weak. Type: *caurinus* Gould, 1850.

Pecten caurinus Gould, 1850   5132
Giant Pacific Scallop

Aleutian Islands to Point Reyes, California.

5132

6 to 8 inches in size, roughly circular; upper valve almost flat, reddish gray and with about 17 low, rounded ribs; lower valve deeper, whitish and with a few more, stronger, rather flat-topped ribs. This is the common, edible deepsea scallop of Alaska.

Subgenus *Euvola* Dall, 1897

Lower or right valve very convex and with weak, rounded ribs; upper or left valve quite flat, more heavily sculptured. Type: *ziczac* (Linné, 1758).

Pecten ziczac (Linné, 1758)   5133
Zigzag Scallop   Color Plate 18

North Carolina to Florida and to Brazil. Bermuda.

2 to 4 inches in size. Upper (left) valve flat; lower valve very deep and convex. There are 18 to 20 broad, very low, rather indistinct ribs on the deep valve, which is generally colored a brownish red (rarely orange). The ribs fade out or are not present near the side margins of the valve. Flat valve with a bright mosaic of whites and browns. A fairly common species. Do not confuse with *raveneli*. Albino specimens are found in Bermuda on rare occasions.

Pecten raveneli Dall, 1898   5134
Ravenel's Scallop   Color Plate 18

North Carolina to Florida, south Texas and the West Indies.

1 to 2½ inches in size, similar to *ziczac*, but the deep valve has about 25 very distinct ribs which are commonly whitish in color. Between them are fairly wide, tan or pinkish grooves. Rarely lemon-yellow or all-orange. In the flat valve, the 25 or so ribs are rounded in cross-section whereas in *ziczac* they are flat-topped and much closer together. A rather uncommon species. Large specimens are sometimes dredged up commercially with *gibbus* off Beaufort, North Carolina.

Pecten chazaliei Dautzenberg, 1900   5135
Tereinus Scallop

Southern Florida and the Caribbean to Brazil.

5135

1 inch in size, quite fragile. Upper (left) valve flat, with about 20 small, narrow ribs; lower valve deep to moderately deep and with low, irregularly defined, roundish ribs. Color whitish tan, slightly translucent, with faint mottlings of pink near the beaks. Rarely, the flat valve may be flecked with brown, zigzag, fine lines. A rare species uncommonly dredged by private collectors in 10 to 75 fathoms. *P. tereinus* Dall, 1925, is a synonym.

Genus *Amusium* Röding, 1798

Shell generally thin, slightly swollen, usually smooth externally. Internally with radial riblets. Type: *pleuronectes* (Linné, 1758). Synonyms: *Pleuronectes* Bronn, 1831; *Campitonectes* Salisbury, 1939; *Amussium* Herrmannsen, 1846. The subfamily Amusiinae Ridewood, 1903, is a subjective synonym of Pectininae.

Amusium papyraceum (Gabb, 1873)   5136
Paper Scallop   Color Plate 18

The Gulf of Mexico, south Texas to the West Indies. Brazil.

About 2 inches in size, oily-smooth, glossy, exterior without ribs, but internally with about 22 very fine ribs which are commonly arranged in pairs. Both valves moderately convex to flattish. Upper valve light-mauve to reddish brown with darker flecks. Lower valve whitish at the center with yellow to cream margins, or all-white. Hinge line strongly arched. Uncommon in collections, but commonly brought up from 30 to 60 fathoms by shrimp fishermen off Alabama and Louisiana. Bermuda record is erroneous.

(5137; Color Plate 18) *Amusium laurenti* (Gmelin, 1791), from Honduras to the Greater Antilles is larger, with a straight hinge line and with the lighter-colored valve more convex than the darker valve. Uncommon.

Subfamily CHLAMYDINAE Korobkov, 1960

Genus *Chlamys* Röding, 1798

Valves elongate, rounded, somewhat oblique, with unequal ears and a large, deep byssal notch. Strong primary ribs and smaller interpolated ribs. Type: *islandica* (Müller, 1776). The word *Chlamys* is feminine.

Chlamys sentis (Reeve, 1853)   5138
Sentis Scallop   Color Plate 19

North Carolina to southeast Florida and to Brazil.

1 to 1½ inches in length, but not so wide (like a fan opened only 80 degrees). Valves rather flat. One hinge ear small, the

other twice as large. With about 50 ribs of varying sizes, each with tiny, closely set scales. There are 2 to 4 smaller ribs between the slightly larger ones. Color commonly brilliant: purple, red, vermilion, orange-red, brownish, white or mottled (especially near the beaks). Common under rocks below low-tide mark. Do not confuse with *ornata*, *mildredae* or *benedicti*.

Chlamys mildredae (F. M. Bayer, 1943)   5139
Mildred's Scallop

Southeast Florida and Bermuda.

1 to 1½ inches in length, similar to *sentis* and *ornata*, but the ribs of the upper valve (one without the byssal notch) 30 in number and every third or fourth one larger. Sculpture of rather large, erect scales set about 1 mm. apart. Ribs of lower valve about 30, in groups of 2 or 3. Exterior color much like *sentis*; interior yellowish with purple stains near the margins. Rare under rocks at low tide. It is possible that this form may be a hybrid between *sentis* and *ornata*. (*The Nautilus*, vol. 56, p. 110.)

Chlamys ornata (Lamarck, 1819)   5140
Ornate Scallop   Color Plate 19

Southeast Florida to the West Indies to Brazil. Bermuda?

1 to 1½ inches in length, similar to *sentis*, but with about 18 high, major, slightly scaled ribs separated by 2 small, scaly cords on the upper valve. Ribs of lower valve are in 18 groups of 3 closely spaced riblets. Exterior ivory to yellowish cream with strong maculations of maroon or purplish. Interior usually white. Compare with *mildredae*. An uncommon and favorite collector's item. Rare in Florida.

Chlamys multisquamata (Dunker, 1864)   5141
Many-ribbed Scallop   Color Plate 19

Southeast Florida, Bahamas and Cuba.

Up to 2⅓ inches, delicate, irregularly flattish, with relatively small ears, the larger one having 6 or 7 weak radial ribs in the middle. Characterized by 120 to 170 fine irregularly sized, fine radial riblets which become microscopically imbricated near the margin of larger specimens. Umbonal area smoothish, except for microscopic radial lines. Byssal area with 7 to 9 short, sharp, comblike teeth. Color varies: all lemon-yellow; whitish with brown-rose maculations; or strongly flushed with dark-red. Rare; in rock crevices from 20 to 185 feet. See *Malakozool. Blätter für 1864*, p. 100. *Pecten effluens* Dall, 1886, is a synonym.

Chlamys benedicti (Verrill and Bush, 1897)   5142
Benedict's Scallop

South half of Florida to off Texas.

Rarely over ½ inch in length. Very similar to *sentis*, but with a greater range of colors and having 2 color variations

5142

not found in *sentis* (pure lemon-yellow or mottled with chalk-white zigzag stripes). There is a lemon-yellow spot found inside the valves up under the beaks. With about 22 strong ribs alternating with weaker ribs, total about 45. Shorter ear has a sharp, 90-degree corner and bears prominent spines, while in *sentis* it is more rounded or considerably more than 90 degrees and is smoother. Hinge margin of longer ear has small projecting scales. Color pink, pinkish red, light-purple or yellow, and commonly with pronounced whitish zigzag markings. A moderately common species usually misidentified as young *sentis* or *muscosa*.

Chlamys imbricata (Gmelin, 1791)   5143
Little Knobby Scallop   Color Plate 19

Southeast Florida and the West Indies. Bermuda.

1 to 1¾ inches in length, but not quite so wide. Lower valve (the one with the byssal notch) slightly convex. Upper valve almost flat and fairly thin. 8 to 10 ribs, uncommonly with smaller cords between. They have prominent, cup-shaped, delicate, distantly spaced scales. Color dirty-white or pinkish with small, squarish, red or purplish blotches. Interior yellowish, commonly with purplish stains. Mantle blackish to brown, with 26 minute brown eyes, numerous reddish short tentacles and about 10 very long, white-speckled, retractable tentacles. Attached to the underside of rocks in 10 to 20 feet of water. Capable of swimming.

Chlamys hastata hastata (Sowerby, 1843)   5144
Pacific Spear Scallop

Monterey to Newport Bay, California.

2 to 2½ inches in length; without microscopic reticulations; right valve (with byssal notch) with about 18 to 21 primary, strongly spined ribs which have 5 to 7 much smaller, weakly spined, secondary ribs in between; left valve with 10 or 11 distantly spaced, strongly scaled primary ribs, with 12 to 16 very weak, beaded secondary ribs in between. This is not so common as the subspecies *hericia,* and is much more colorful, commonly being bright-orange, red or lemon.

Chlamys hastata hericia (Gould, 1850)   5145
Pacific Pink Scallop   Color Plate 19

Alaska to Newport Bay, California.

2 to 2¾ inches in size; without microscopic reticulations; right valve (with byssal notch) with about 18 to 21 primary, moderately scaled ribs which have 5 to 7 much smaller-spined ribs between; left valve with about 10 or 11 primary, spined ribs which have a single, rounded, almost-as-large secondary rib in between. Between these large ribs there are 15 to 18 tiny, spined ribs, 3 of which are on the large secondary rib. Color variable: solid rose, pink, white, light-yellowish and blends of all these. Commonly dredged in shallow waters down to 20 fathoms. A white form was named *albida* Dall, 1906. A form or subspecies was named *pugetensis* Oldroyd, 1920, but I do not understand its status.

Chlamys rubida (Hinds, 1845)   5146
Hinds' Scallop   Color Plate 19

Alaska to off San Diego, California.

2 to 2½ inches in length; with microscopic reticulations between the ribs near either the beaks or the margins of the valves. Left valve (without the byssal notch) with numerous primary ribs, each bearing 3 rows of spines, and with a secondary spined rib between. Right valve flattish, usually lighter-colored, and with fewer ribs which are smoothish, rounded and inclined to be grouped in pairs. The reticulate sculpturing is best seen on this side. Color variable: light-rose, mauve, lemon-yellow, pale-orange and blends of these. A rather com-

mon species dredged in shallow water down to 822 fathoms. *C. navarcha* Dall, 1898; *jordani* Arnold, 1903; and *kincaidi* Oldroyd, 1920, are probably synonyms. G. Grau (1959) has shown that *hindsii* Carpenter, 1864, is a synonym.

Chlamys islandica (Müller, 1776)   5147
Iceland Scallop   Color Plate 19

Arctic Seas to Buzzards Bay, Massachusetts.
Alaska to Puget Sound, Washington.

3 to 4 inches in length, not quite so wide. Long hinge ear is twice the length of the short one. Valves moderately convex to flattish. With about 50 coarse, irregular ribs which split in two near the margin of the valve. Rarely, the ribs are grouped more or less in groups of twos, threes or fours. Color usually a dirty-gray or cream, but some are quite attractively tinged with peach, yellow or purplish both inside and out. A very common hermaphroditic species offshore on the continental shelf in 1 to 175 fathoms.

Several subspecies or forms of this well-known species have been described: *insculptus* (Verrill, 1897); *costellatus* (Verrill and Bush, 1897); *beringianus* (Middendorff, 1849).

Other species:

5148   *Chlamys lowei* (Hertlein, 1935). Catalina Island, California, the Gulf of California to Ecuador, 30 to 55 fms. Appears to be the only *Chlamys* in the Gulf of California.

5149   *Chlamys lioicus* (Dall, 1907). Norton Sound, Alaska. *Amer. Jour. Sci.*, vol. 23, p. 457, fig. 1.

5150   *Chlamys liocymatus* (Dall, 1925). Off Cape Hatteras, North Carolina, 34 fms. *The Nautilus*, vol. 38, p. 119.

Genus *Hinnites* Defrance, 1821

Biologically speaking, this genus is really a *Chlamys* in which the adults are attached to rocks and become quite massive like *Spondylus*. For convenience, we are considering it a full genus. Type: *cortezi* Defrance, 1821.

Hinnites multirugosus (Gale, 1928)   5151
Giant Rock Scallop

British Columbia to Baja California.

Up to 8 inches in length. A heavy, massive shell characterized by the early "*Chlamys*-like" shell at the beaks. Interior white with a purplish hinge area. Attached to rocks by the

5151

right valve. The ½-inch long young are almost impossible to separate from some species of *Chlamys*, except when they show a mauve spot on the inside of the hinge line on each side of the resilium pit, or if they show signs of distortion or a mottling pattern of color on the outside of the valves. Some young are bright-orange. A common species from low-tide mark to 30 fathoms. Formerly known as *Hinnites giganteus* Gray, 1825, not Sowerby, 1814.

Other species:

5152 *Hinnites adamsi* Dall, 1886. Off St. Vincent, West Indies, 573 fms. 28 mm.

Genus *Aequipecten* P. Fischer, 1886

Upper or left valve convex, the right valve flattish. Valves widely gaping. The type of this genus is *opercularis* (Linné, 1758), from Europe. This genus is considered a subgenus of *Chlamys* by some workers.

Aequipecten glyptus (Verrill, 1882)	**5153**
Tryon's Scallop	Color Plate 18

South of Cape Cod to Florida to off Texas.

1 to 2½ inches in size. Both valves rather flat. Shell somewhat lopsided and spathate in shape. About 17 ribs which start out as fine, sharp, slightly prickled ribs, but become flattened and indistinct or absent near the margin of the valve. One valve pure-white, the other with broad, rose rays corresponding to the ribs. Internally white and with weak, fine ribs. Rare, but has been brought in by commercial trawlers from 83 to 232 fathoms. This is *P. tryoni* (Dall, 1889).

Aequipecten phrygium (Dall, 1886) **5154**
Spathate Scallop

Off Cape Cod to Florida and the West Indies.

About 1 inch in size. Characterized by its peculiar spathate or open-fan shape. With 17 sharp ribs. On closer inspection, it will be seen that each rib is composed of 3 rows of very fine, closely packed scales which are welded together to form a single rib. In cross-section, this would give the rib the shape of the letter M. Hinge-line straight with one ear slightly shorter than the other. Color dull-gray with indistinct blotches of dull-pink. Uncommonly dredged off Miami and the Lower Keys.

Aequipecten lineolaris (Lamarck, 1819) **5155**
Wavy-lined Scallop

Southeast Florida to the Caribbean.

1 to 2 inches in size, ears about equal. Valves moderately inflated. Surface highly glossy, the colored valve with about 18 very low, rounded ribs. Bottom valve white. Top valve rosy-tan with characteristic, numerous small, wavy, thin lines of pink-brown running concentrically. A few brown mottlings may be present. A very gorgeous and rare species dredged from 7 to 50 fathoms. *A. mayaguezensis* Dall and Simpson, 1901, is this species.

Aequipecten muscosus (Wood, 1828)	**5156**
Rough Scallop	Color Plate 19

North Carolina to Florida and Texas to Brazil. Bermuda.

¾ to 1¼ inches in size, both valves inflated and fairly deep. Hinge-ears equal to the width of the main part of the shell. 18 to 20 ribs, the center part of each bearing prominent, erect, concave scales, and on each side 2 rows of much smaller scales.

Color orange-brown, red, lemon-yellow, orange or commonly mottled with purple. Beachworn specimens may lose most of their scaliness. Moderately common just offshore to 90 fathoms. Formerly called *exasperatus* Sowerby, 1842, and *fuscopurpureus* Conrad. The famous "lemon-yellow Pecten," much sought after by collectors, is this species. This species may belong more closely to the genus *Chlamys*.

Aequipecten acanthodes (Dall, 1925)	**5157**
Thistle Scallop	

Bermuda, southern Florida to the lower Caribbean.

½ to 1 inch in length, slightly broader than long, rather flat, minutely spined, and very variable in color, sometimes being mottled, dark-brown, gray, orange, reds and yellows. Often confused with *muscosus*, but the latter is much fatter, has ribs that bear a central row of large scales with 2 or 3 rows of much smaller scales on each side. In *acanthodes*, the ribs bear 3 rows of smaller scales, all about the same size; between the dark ribs are sometimes 2 rows of very minute prickles. The short ear on *acanthodes* is square, while in *muscosus* it is pointed to form a U-shaped connection with the main part of the valve. Common in shallow water on turtlegrass in soft sandy mud areas. Swarms in the spring time.

5157

Other species:

5158 *Aequipecten heliacus* (Dall, 1925). Southern Florida to the West Indies.

Genus *Cyclopecten* Verrill, 1897

Shells very small, thin, circular, with equal ears. Valves having different sculpture, the thin, smaller right valve having thin, elevated concentric lirae, the thicker left valve marked with weak radials, smooth or with radial rows of arched scales or pustules. No internal lirae. Type: *pustulosus* Verrill, 1873. Mostly deep water. Synonym: *Cyclochlamys* Finlay, 1926. Some species are carnivorous.

Cyclopecten nanus Verrill and Bush, 1897 **5159**
Dwarf Round Scallop

Off Virginia to off Texas and to Puerto Rico. Brazil.

5 to 7 mm., thin, nearly orbicular, compressed. Left (upper) valve more inflated. Color grayish white with milky mottlings. Left valve with fine radiating lines and its umbonal area cancellated. Right valve nearly smooth, except for very fine, regularly spaced concentric lamellae. Internal hinge-plate broad and crossed by numerous transverse lines. Byssal notch V-shaped but lacking tiny thorns like those in *Placopecten* young. Common; 22 to 294 fathoms. For details, see Arthur Merrill, 1959, *Occas. Papers on Mollusks*, vol. 2, no. 25.

Other species:

5160 *Cyclopecten pustulosus* Verrill, 1873. Gulf of Maine to off Massachusetts, 115 to 430 fms.

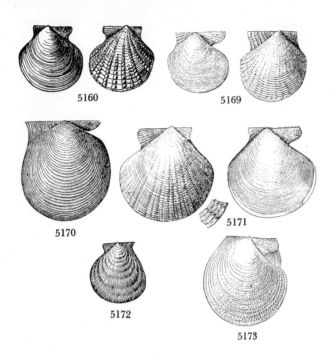

5160   5169

5170   5171

5172

5173

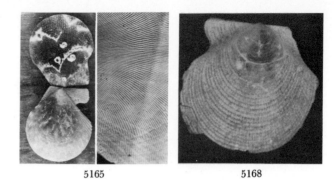

5165   5168

Subgenus *Pectinella* Verrill, 1897

Auricles unequal and oblique; surface of valves smooth except for growth lines. Type: *sigsbeei* (Dall, 1886).

Cyclopecten sigsbeei (Dall, 1886)   5161
Sigsbee's Glass Scallop

Gulf of Mexico and off Cuba.

11 mm. high, 9 mm. long; valves rather convex, the left one slightly more. Both valves polished, but with microscopic silky, concentric striae. No radiating sculpture. Anterior auricles well-marked, very small, oblique. Posterior auricles larger, with a broad shallow byssal sulcus. Color brownish with opaque-white splashes. Uncommon; 158 fathoms.

Other species:

5162 *Cyclopecten catalinensis* (Willett, 1931). Central California to the Gulf of California, in 29 to 37 meters. *The Nautilus*, vol. 45, no. 2.

5163 *Cyclopecten exquisitus* Grau, 1959. Gulf of California to Peru, in 22 to 274 meters.

5164 *Cyclopecten pernomus* (Hertlein, 1935). Baja California, the Gulf, to Ecuador, down to 355 meters.

Genus *Palliolum* Monterosato, 1884

Shells small, fragile, sculptured with very fine radial threads or with a reticulate pattern. Sculpture of opposite valves different in some species. Posterior auricle delimited or not. Hinge with one pair of small cardinal crura on each side of the ligamental pit. Type: *incomparabilis* Risso, 1826. This is *Pseudamussium* of American authors, non Mörch, 1853. Synonyms: *Pseudamusium* Verrill, 1897; *Palliorum* Oyama, 1944. Some workers, including Hertlein, consider *Hyalopecten* and *Delectopecten* as subgenera of *Palliolum*. This group is in need of a monographic species-level revision.

Palliolum striatum (Müller, 1776)   5165
Striate Glassy Scallop

Newfoundland to off Massachusetts.
Norway to the Mediterranean.

½ inch, thin, fragile, the front ear 3 or 4 times as long as the posterior one. Circular in outline. White, rose or brownish with blotches and zigzag flames. Right valve with fine concentric and radiating lines. Left valve rough with numerous, very fine, spiny riblets. Byssal notch with teeth. Moderately common; 4 to 100 fathoms.

Other species:

5167 *Palliolum subimbrifer* (Verrill and Bush, 1897). Off Martha's Vineyard, Massachusetts, 115 to 365 fms. (Synonym: *hoskynsi* Verrill, not Forbes.)

5168 *Palliolum leptaleum* (Verrill, 1884). Off Cape Hatteras, North Carolina, 142 fms. Gulf of Mexico, off Florida, 80 fms.

5169 *Palliolum reticulum* (Dall, 1886). Off Cape Hatteras, North Carolina, to the West Indies, 82 to 124 fms. Gulf of Mexico, off Florida, 130 fms. 7 mm.

5170 *Palliolum strigillatum* (Dall, 1889). Off Fernandina, Florida, to the Gulf of Mexico; Cuba. 294 to 1,181 fms. 9 mm.

5171 *Palliolum imbrifer* (Lovén, 1846). Arctic Seas to off New Jersey, 12 mm.

5172 *Palliolum undatum* (Verrill and Smith, 1885). Off New England to off New Jersey. 19 mm.

5172a *Palliolum ringnesia* (Dall, 1924). Ellef Ringnes Island, Arctic Canada. Rept. Canadian Arctic Exped., Suppl., vol. 8, part A, p. 32A, pl. 4. 12 mm.

Genus *Delectopecten* Stewart, 1930

Shell small, translucent-white; posterior ear not clearly marked off from the rest of the shell. Both valves with radial or reticulate sculpturing. Type: *vancouverensis* Whiteaves, 1893. Some consider this a subgenus of *Palliolum. Arctinula* Thiele, 1935, is a synonym.

Delectopecten vitreus (Gmelin, 1791)   5173
Vitreous Scallop

Off Newfoundland to Martha's Vineyard, Massachusetts. Northern Europe. Clipperton Island, Eastern Pacific.

¾ inch, ears about equal, the posterior one not being marked off from the rest of the valve. Byssal notch in right anterior ear is large. Ovate-circular in outline. Translucent-white. Sculpture of microscopic radiating lines and fine beaded concentric ridges. Beads may be spinose. Uncommon; 57 to 400 fathoms.

Other species:

5174 *Delectopecten greenlandicus* (Sowerby, 1842). Arctic Seas; Baffin Land, Canada. Newfoundland, 130 to 224 fms.

5175 *Delectopecten arces* (Dall, 1913). Off Santa Barbara and Santa Cruz Islands, California, 500 to 891 fms.

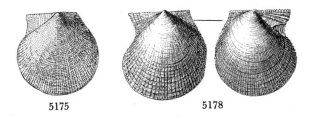

5175   5178

5176 *Delectopecten randolphi* (Dall, 1897). Bering Sea to Guaymas, Mexico, 225 to 1,064 fms. A *Leptopecten?*

5177 *Delectopecten tillamookensis* (Arnold, 1906). Pribilof Islands, to off San Diego, California. 659 fms.

5177

5178 *Delectopecten vancouverensis* (Whiteaves, 1893). Bering Sea to off San Diego, California. 10 to 200 fms. On calcareous algae.

5179 *Delectopecten bistriatus* (Dall, 1916). Off San Diego, California in 822 fms.

5180 *Delectopecten incongruus* (Dall, 1916). Off San Diego, California, 684 fms.

5181 *Delectopecten binominatus* (Hanna, 1924). Arctic Ocean. (Synonym: *andersoni* Dall, 1918, not Arnold, 1906.)

Genus *Hyalopecten* Verrill, 1897

Shell small, thin, fragile, hyaline; valves nearly equal, strongly undulated concentrically, and sculptured with fine radial striae. Posterior auricle on right valve not delimited. Type: (*dilectus* Verrill and Bush, 1897), *pudicum* E. A. Smith, 1885. Some put this as a subgenus in *Palliolum.*

Other species:

5182 *Hyalopecten undatus* (Verrill and S. Smith, 1885). Off Cape Hatteras, North Carolina; Europe; West Africa; Antarctic, 768 to 3,316 meters. (Synonyms: *fragilis* Jeffreys, 1876, non Defrance, 1825; *pudicus* E. A. Smith, 1885; *biscayensis* Locard, 1886; *dilectus* Verrill and Bush, 1897.) 19 mm.

5183 *Hyalopecten frigidus* (Jensen, 1912). Greenland; northern Europe. 1,000 to 3,000 meters.

5184 *Hyalopecten graui* (Knudsen, 1970). Gulf of Panama, 3,670 meters. *Galathea Report,* vol. 11, p. 97.

5185 *Hyalopecten neoceanicus* (Dall, 1908). Gulf of Panama, 3,670 meters.

5186 *Hyalopecten eucymatus* (Dall, 1898). Off eastern United States. (Synonym: *fragilis* of Verrill, 1884, p. 81, not Jeffreys, 1876.)

Genus *Leptopecten* Verrill, 1897

Valves thin, with a low convexity, almost equivalve; with a few, low, broad ribs. Hinge line long. Type: *monotimeris* (Conrad, 1837). Some workers place this as a subgenus under *Chlamys.*

Leptopecten latiauratus (Conrad, 1837)   **5187**
Wide-eared Scallop   **Color Plate 19**

Point Reyes, California, to the Gulf of California.

½ to ¾ inch, thin-shelled, lightweight; hinge line long; with 12 to 16 low, flat-topped ribs crossed by fine concentric lamellae. Color light-brown with diagonal streaks of opaque-white and chestnut-brown. Moderately common; attached to rocks and rubble from low tide to several fathoms. Do not confuse with the forma *monotimeris*. *L. delosi* (Arnold, 1906), is a synonym.

Leptopecten latiauratus forma **monotimeris** (Conrad, 1837) **5188**
Kelp Scallop

Santa Barbara, California, to Baja California.

¾ to 1 inch, similar to *latiauratus,* but more delicate; with rounded (not square) ribs, and the concentric lamellae are absent. Color translucent yellow or orange, with chevron-shaped lines of white and brown. Abundant; attached to giant kelp off shore or to eel-grass in bays. *L. fusicolus* (Dall, 1898) is a synonym. The observations of George R. Clark, II, in *The Veliger,* vol. 13, p. 269 (1971) seem to indicate that this ecologic form is produced by temperature and habitat differences.

Other species:

5189 *Leptopecten* (*Pacipecten* Olsson, 1961) *tumbezensis* (Orbigny, 1846). Gulf of California to Peru, 12 to 40 fms. (Synonyms: *aspersus* Sowerby, 1835, not Lamarck, 1819; *sowerbyi* Reeve, 1852; *paucicostatus* Carpenter, 1864.)

5190 *Leptopecten bavayi* (Dautzenberg, 1900). Bavay's Scallop. Uruguay and Brazil to the West Indies; common; shallow dredgings. See "Caribbean Seashells," pl. 33.

5191 *Leptopecten camerella* (S. S. Berry, 1968). La Ribera, Baja California, 35 to 40 fms. *Leaflets in Malacology,* no. 25, p. 155.

5192 *Leptopecten velero* (Hertlein, 1935). Gulf of California to Peru, in 5 to 55 meters.

Genus *Argopecten* Monterosato, 1889

Both valves convex, the right or lower one usually the more convex. Ears about equal in length. Anterior ears with a byssal notch. Valves not gaping. Radial ribs strong and squarish with fine concentric fimbriations between. Type: *circularis* Sowerby, 1835 (*solidulus* Reeve, 1853). Absolute synonym: *Plagioctenium* Dall, 1898. Members of this genus were previously placed in the genus *Aequipecten,* but the recent, extensive analysis by T. R. Waller, 1969 (*Jour. Paleontol.,* vol. 43, supplement to no. 5) strongly suggests that *Argopecten* be accepted as a genus. Other synonyms: *Plagiopecten* Hanna, 1924; *Plagiopectenium* R. Stewart, 1930; *Haumea* Dall, Bartsch and Rehder, 1938; *Corymbichlamys* Iredale, 1939.

Argopecten irradians (Lamarck, 1819)   **5193**
Atlantic Bay Scallop   **Color Plate 18**

Nova Scotia to northern half of Florida and Texas.

2 to 3 inches in size. This is the common edible scallop of our east coast. It is not a very colorful species, although its drab browns and grays are rarely enlivened with yellow. There are 4 distinct subspecies. Each has a distinct geographical range and peculiar habitat. They all live in semienclosed bays in 1 to 60 feet of water on sand bottoms usually near eel-grass. Young are attached by a byssus to eel-grass and algae.

(5194; Color Plate 18) *A. irradians irradians* (Lamarck, 1819). North shore of Cape Cod, Massachusetts, to New Jersey. 17 or 18 ribs which are low and roundish in cross-section. Each valve is about the same fatness, and the lower one is only slightly lighter in color. Drab gray-brown with indistinct, darker-brown mottlings. The most compressed of the 3 subspecies. This is *borealis* Say, 1822.

(5195) (Color Plate 18) *A. irradians concentricus* (Say, 1822). New Jersey (rare), Maryland to Georgia and Louisiana to Tampa, Florida. 19 to 21 ribs which are squarish in cross-section. Lower valve (the lightest in color and commonly all white) is much fatter than the dull bluish gray to brown upper valve. Common.

(5196) *A. irradians amplicostatus* Dall, 1898. Central Texas to Mexico and Columbia. Similar to *concentricus,* but with only 12 to 17 ribs; more gibbose; lower valve commonly white and with high, squarish to slightly rounded ribs. West of the Mississippi to south Texas.

(5197) *A. irradians sablensis* A. H. Clarke, 1965. An extinct subspecies. Valves occur on Sable Island east of Nova Scotia. Rib count: 18 to 21 (see *Malacologia,* vol. 2, no. 2, pp. 161–188, 1965). Radiocarbon analysis gives these shells an after-death age of about 1,800 years when the water was warmer at that isolated spot.

Argopecten gibbus (Linné, 1758)   5198
Calico Scallop   Color Plate 18

Off Maryland to Florida and south Texas to Brazil. Bermuda.

1 to 2½ inches. A common, colorful scallop found abundantly in southern Florida a little offshore. Both valves quite fat. Ribs usually 20 (19 to 21), quite square in cross-section. Bottom valve commonly whitish with a little color; upper valve can be of many bright hues (lavender-rose, red, whitish with purple or reddish mottlings, etc.). If collecting in southeast Florida, do not confuse with *A. nucleus.* Synonyms include: *dislocatus* Say, 1822; *liocymatus* Dall, 1925; *portusregii* Grau, 1952; *carolinensis* Grau, 1952 (*The Nautilus,* vol. 66, pp. 17 and 69). Lives in warm, open marine waters from 5 to 200 fathoms.

Argopecten nucleus (Born, 1778)   5199
Nucleus Scallop

Southeast Florida and the West Indies.

1 to 1½ inches in size. This is a difficult species to identify. It is rarely over an inch in size, has 1 or 2 more ribs than *gibbus,* is usually fatter, and is characteristically colored with small, chestnut mottlings on a cream background and commonly with snow-white specklings. Both or only one valve

5199

may be heavily colored. Never with the bright shades of orange, red, etc. The posterior ears have only 4 to 6 costae, while in *gibbus* there are about 7 to 10. Cardinal crura very strong, while in *gibbus* it is usually weak. Not uncommon in the Keys from low tide to a few fathoms on grass.

Argopecten circularis (Sowerby, 1835)   5200
Pacific Calico Scallop   Color Plate 18

Santa Barbara, California, to Peru.

2½ inches, very similar to the Atlantic *gibbus* but is slightly oblique, 20 to 22 ribs, the right valve usually white within, the left often stained brown within. Exterior colors variable, bright; but lilac and browns predominate. Common in shallow water. Synonyms are *tumidus* Sowerby, 1835, not Turton, 1822; *ventricosus* Sowerby, 1842; *filitextus* Li, 1930.

(5201) The northern subspecies, *aequisulcatus* (Carpenter, 1864), ranging from Santa Barbara to Baja California, differs from the typical more southerly *circularis* in having less convex valves, lower ears, smaller cup containing the chitinous resilium in the center of the hinge and more subdued coloration. T. R. Waller (1969) suggests this be treated as a weak subspecies. Found on subtidal sand and mud bottoms of bays and quiet bottoms of the open coast. Becoming uncommon.

Genus *Placopecten* Verrill, 1897

Without ribs; with radial striae; right valve smoother; ears unequal; valves with a slight convexity. Type: *clintonius* Say, 1824, Miocene of east United States.

Placopecten magellanicus (Gmelin, 1791)   5202
Atlantic Deepsea Scallop   Color Plate 18

Labrador to Cape Hatteras, North Carolina.

5 to 8 inches in size, almost circular. Valves almost flat to slightly convex. Interior flaky-white. Exterior rough with numerous very small, raised threads. Exterior yellowish gray to purplish gray or dirty-white. This is the common, edible, deepsea scallop fished off our New England coasts. The name *grandis* Lightfoot is nude and cannot be used. Synonyms include: *testudinarium* Röding, 1798 and 1819; *fuscus* "Linsley" Gould, 1848; *brunneus* Stimpson, 1851; *tenuicostatus* Mighels and Adams, 1841. The Labrador record is open to question, but the Gulf of St. Lawrence is verified. Young *magellanicus* (5 to 7 mm.) resemble *Cyclopecten nanus* but differ in having comb-like teeth in the byssal notch of the right valve (see Arthur Merrill, 1959. *Occ. Papers on Mollusks,* vol. 2, no. 25). For habits and anatomy, see G. A. Drew, 1906, *Univ. Maine Studies,* no. 6, 71 pp.

Genus *Lyropecten* Conrad, 1862

Type of the genus is *estrellanus* Conrad, 1856, a Pacific coast Miocene species.

Subgenus *Lyropecten* Conrad, 1862

Lyropecten antillarum (Récluz, 1853)   5203
Antillean Scallop

Southeast Florida and the West Indies. Bermuda.

½ to 1 inch in length and width. Valves fragile, both nearly flat. Only about 11 to 15 moderately rounded, low ribs. Growth lines exceedingly fine (seen with the aid of a strong lens). Ears uneven. Color either pastel-yellow, tawny-orange or light-

5203

brown, commonly with chalk-white mottlings, flecks or stripes. Found uncommonly in shallow water. *Pecten eulyratus* F. M. Bayer, 1943, is a synonym.

Other species:

5204 *Lyropecten kallinubilosus* F. M. Bayer, 1943. Off Saint Marks, west Florida, 20 fms. *The Nautilus*, vol. 56, p. 110.

Subgenus *Nodipecten* Dall, 1898

Shells large with coarse, corded, noduled ribs. Type: *nodosus* (Linné, 1758).

Lyropecten nodosus (Linné, 1758)   **5205**
Lion's Paw   **Color Plate 18**

North Carolina to Florida, Texas and to Brazil. Bermuda. Ascension Island, mid-Atlantic.

3 to 6 inches in size, rather heavy and strong-shelled. Characterized by the 7 to 9 large, coarse ribs which have large, hollow nodules. The entire shell also has numerous, much smaller but distinct, riblets. The color is commonly dark maroon-red, but may be bright-red, orange or rarely yellow. Fairly common offshore, especially on the west coast of Florida. *Pecten fragosus* Conrad, 1849, is a synonym.

Lyropecten subnodosus (Sowerby, 1835)   **5206**
Pacific Lion's Paw

Gulf of California to Peru.

Similar to the Caribbean *nodosus*, but more finely sculptured, lacking the numerous, large nodules, and usually being a deep rose-purple or wine-red. Some authors consider the form with one less rib from the northern part of the range as the subspecies *intermedius* (Conrad, 1867). Moderately common; offshore. Reported from California by Stracham, 1968, p. 49, *Calif. Fish and Game*, vol. 54, pp. 49–57.

5206

Subfamily PROPEAMUSSIINAE Abbott, 1954

Genus *Propeamussium* Gregorio, 1884

Small, thin-shelled; valves equally slightly convex. Sculptured externally with concentric lines. Left valve with radial striae. Interior with radial ribs. Type: *ceciliae* Gregorio, 1884. *Propeamusium* Dall, 1886, is a misspelling. *Paramusium* Verrill, 1897, and *Occultamussium* Korobkov, 1937, are synonyms. For shell structure in this family, see T. R. Waller, 1972, *International Geol. Congress*, 24th Session, Montreal, Sect. 7, Paleontol., pp. 48–56, 3 figs.

Propeamussium pourtalesianum (Dall, 1886)   **5207**
Pourtales' Glass Scallop

Southeast Florida and the West Indies.

½ inch in length. Valves very slightly convex. Shell extremely thin and transparent (like thin mica flakes). Each valve reinforced inside with about 9 rodlike, opaque-white ribs. Exterior of one valve is smoothish, the other valve with numerous, microscopic, concentric threads. Common offshore. 13 to 804 fathoms. *P. marmoratum* Dall, 1886, is a synonym.

Other species:

5208 *Propeamussium thalassinum* (Dall, 1886). Off Martha's Vineyard, Massachusetts, to the West Indies, 22 to 317 fms. West Florida, 25 to 40 fms. (Synonym: *fenestratum* Verrill 1881, not Forbes, 1844.)

5209 *Propeamussium holmesii* (Dall, 1886). Florida and the West Indies, 100 to 273 fms. Gulf of Mexico, 180 fms. 12 mm.

5210 *Propeamussium sayanum* (Dall, 1886). Florida Strait and the West Indies, 150 to 400 fms. Gulf of Mexico, 220 fms. 15 mm.

5211 *Propeamussium davidsoni* (Dall, 1897). Bering Sea; deep water.

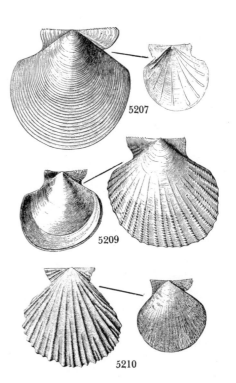

5207

5209

5210

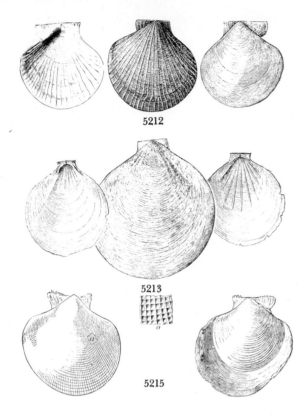

5212

5213

5215

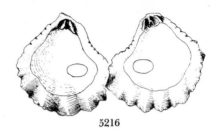

lines on the ribs. A common intertidal to offshore species. *P. ramosus* Lamarck, 1819, is a synonym, as is probably *imbricata* Reeve and "*spondyloidea* Meuschen."

5216

5212 *Propeamussium alaskense* (Dall, 1871). Pribilof Islands, Bering Sea, to the Santa Barbara Islands, California. Japan. Deep water. 22 mm.

5213 *Propeamussium dalli* (E. A. Smith, 1886). Off Texas, Gulf of Mexico, 860 fms.; West Indies, 218 to 805 fms. Bermuda. 62 mm.

5214 *Propeamussium squamigerum* (E. A. Smith, 1886). Off Bermuda, 435 fms.

5215 *Propeamussium cancellatum* (E. A. Smith, 1886). Florida and the West Indies. 26 mm.

Family PLICATULIDAE Watson, 1930

Genus *Plicatula* Lamarck, 1801

Shell trigonal or spathate, thick and attached by either valve to rocks or other shells. With only 1 adductor muscle scar. Sculpture of broad, radial ribs. Hinge with a narrow, elongate chondrophore which is flanked on each side by a fluted tooth and a socket. Type: *plicatus* Linné, 1758 (by Anton, 1839). Synonym: *Micatula* Capenter, 1859.

Plicatula gibbosa Lamarck, 1801   **5216**
Kitten's Paw   **Color Plate 20**

North Carolina to Florida, Texas and the West Indies. Bermuda. Brazil.

About 1 inch in length, somewhat cat's-paw-shaped. Shell strong, heavy, with 5 to 7 high ribs which give the valves a wavy, interlocking margin. Hinge in upper valve with two strong, equally sized teeth; lower attached valve with 2 sockets in the hinge with 2 smaller teeth set rather close together. Color dirty-white to gray with red-brown or purplish

Other species:

5217 *Plicatula spondylopsis* Rochebrune, 1895. Gulf of California to Ecuador; Galapagos. Resembles *gibbosa* from the Caribbean. (Synonym: *ostreivaga* Rochebrune, 1895.)

5218 *Plicatula penicillata* Carpenter, 1857. Gulf of California to Ecuador.

5219 *Plicatula anomioides* Keen, 1958. Guaymas, Mexico. *Bull. Amer. Paleontol.*, vol. 38, no. 172, pl. 31.

5220 *Plicatula inezana* Durham, 1950. Southern Gulf of California. Also Pleistocene.

Family SPONDYLIDAE Gray, 1826

Genus *Spondylus* Linné, 1758

Large, massive, spinose bivalves attached to a hard substrate. The ball-and-socket hinge and central ligament are characteristic. The curtainlike mantle bears minute eyes. Type: *gaederopus* Linné, 1758.

Spondylus americanus **Hermann, 1781**   **5221**
Atlantic Thorny Oyster   **Color Plate 21**

Off North Carolina to Florida, Texas and to Brazil.

3 to 4 inches in size. Spines 2 or less inches in length, usually standing fairly erect. Color variable: white with yellow umbones, red or purple; sometimes all-rose, all-cream or all-pink. The young are much less spinose, and might be confused with *Chama* which, however, do not have the ball-and-socket type of hinge. Large spinose specimens are found clinging to old wrecks and sea walls in fairly deep water from 30 to 150 feet. When cleaned, they are very beautiful. Formerly called *americanus* Lamarck, 1818; *echinatus* Martyn and *dominicensis* Röding, 1798. Sometimes called the chrysanthemum shell. Not uncommon.

Spondylus ictericus **Reeve, 1856**   **5222**
Digitate Thorny Oyster   **Color Plate 21**

Southeast and west Florida, Caribbean to Brazil. Bermuda.

1 to 2½ inches, usually brick-red or dull red-purple with white mottlings at the umbones. Rarely peach to dark-purple or with shades of yellow and orange. Spines are often digitate at the ends, even the needlelike ones in small specimens. Moderately common from low-tide mark to 6 fathoms. The larger, lighter-colored *americanus* with nonfrondose spines is more commonly found from 3 to 30 fathoms. Synonyms: *digitatus* Sowerby, 1847 (non Perry, 1811); *ramosus* Reeve, 1856; *ustulatus* Reeve, 1856; and *vexillum* Reeve, 1856.

Spondylus princeps Broderip, 1833   **5223**
Pacific Thorny Oyster   **Color Plate 21**

Gulf of California to Panama.

Up to 5 inches in size. The spines are 1½ inches or less in length and usually bent over. Color variable, and usually more brilliant than the Atlantic species. A popular, and now fairly high-priced collector's item. Often found on beaches with their spines worn off. They live in fairly deep water attached to rocks and wrecks. The name *pictorum* Schreiber, 1793, is considered invalid by most workers. *S. bicolor* Sowerby, 1847, is a synonym.

Other species:

5224 *Spondylus gussoni* O. G. Costa, 1829. Greater and Lesser Antilles, in 168 to 686 meters; Mediterranean. 1 to 3 inches, box-like, 50 to 70 radial, weakly spined ribs.

5225 *Spondylus calcifer* Carpenter, 1857. Gulf of California to Ecuador. 4 to 6 inches. (Synonyms: *radula* Reeve, 1856, not Lamarck, 1806; *smithi* Fulton, 1915; *ursipes* S. S. Berry, 1959.)

Family DIMYIDAE P. Fischer, 1887

Genus *Dimya* Rouault, 1850

Shells small, about ½ inch, irregularly orbicular, very thin, resembling a young oyster. 2 muscle scars joined by a weak, dotted pallial line. Shell subnacreous externally, sometimes laminated. Interior porcelaneous. Hinge has a row of fine denticles, as does the inner margin of the entire valve. External ligament slender. Internal ligament in a small triangular pit. Attaches by part of the right valve. Type: *deshayesiana* Rouault, 1850. 2 deep-water species on the Pacific coast; 1 Atlantic. *Deuteromya* Cossmann, 1903, is a synonym, as may also be *Dimyarina* Iredale, 1936.

5226 *Dimya californiana* S. S. Berry, 1936. 100 fms., off Santa Monica, California, on stones, to Ceralbo Channel, Gulf of California, 46 fms. *Proc. Malacol. Soc. London*, vol. 22.

5227 *Dimya coralliotis* S. S. Berry, 1944. 100 fms., 10 miles off Huntington Beach, Orange County, California. *Proc. Malacol. Soc. London*, vol. 26, p. 25.

5228 *Dimya argentea* Dall, 1886. Off North Carolina to the West Indies, 73 to 248 fms. Off west Florida, 100 to 220 fms.; common. 12 to 20 mm., attached to shells and rocks.

5228

5229 *Dimya tigrina* F. M. Bayer, 1971. Off Punta Piedras, Colombia (Atlantic), in 79 meters.

Genus *Basiliomya* F. M. Bayer, 1971

Shell about 7 mm., subcircular, translucent; attached by the right valve which is deep and has a margin of thin, tube-like frills. Hinge of right valve with a blunt, triangular tooth on each side of the internal ligament and a shallow socket at each end of the hinge line. Left valve with a shallow groove on each side of the internal ligament and a blunt tooth at each end of the hinge line. A series of interlocking small teeth and pits are around the perimeter of both valves. Anterior adductor muscle near end of hinge line; posterior adductor remote from hinge, conspicuously bilobed; pallial impression marked by a row of shallow pits. Type by monotypy: *goreaui* F. M. Bayer, 1971.

5230 *Basiliomya goreaui* F. M. Bayer, 1971. Jamaica in 170 feet, and east Andros Island, Bahamas, in 75 to 100 feet. *Studies in Trop. Amer. Mollusks*, 1971, p. 227.

Genus *Dimyella* D. R. Moore, 1969

Shell small (3 mm.), similar to *Dimya*, but the lower, cupped valve has a stout, conical tooth at each end of the hinge (somewhat like young *Spondylus*). Lower edge of upper valve has a shelly brood chamber. The small, triangular, internal resilium does not interrupt the hinge line. Posterior muscle scar twice as big as the anterior one. Prodissoconch 0.28 mm., glassy, smooth. Type: *starcki* D. R. Moore, 1969 (*Jour. de Conchyl.*, vol. 107, p. 137). One known species:

5231 *Dimyella starcki* D. R. Moore, 1969. In marine caves, shallow water, attached to coral. 5 mi. south of San Miguel, Cozumel Island, east Mexico. 3 mm., white, 10 radiating riblets crossed by concentric lamellae.

Superfamily Anomiacea Rafinesque, 1815

Family ANOMIIDAE Rafinesque, 1815

Genus *Anomia* Linné, 1758

The valve without the hole has 1 large and 2 small muscle scars. The shell is attached to a rock or wood surface by means of a calcified byssus which passes through a large notch in the right valve. Type: *ephippium* Linné, 1758. The genus *Pododesmus* differs in having only 2 muscle scars in the top or holeless valve.

Anomia simplex Orbigny, 1842   **5232**
Common Jingle Shell   **Color Plate 20**

Cape Cod, Massachusetts, to Florida, to Texas to Brazil. Bermuda.

1 to 2 inches in size, irregularly oval, smoothish, thin but strong. The upper or free valve is usually quite convex; the lower valve is flattish and with a hole near the apex. Color either translucent-yellow or dull-orange. Some with a silvery sheen. Specimens buried in mud become blackened. Very commonly attached to logs, wharfs and boats. The round, calcified base of the byssus from formerly living specimens may remain attached to stones and other shells.

Anomia peruviana Orbigny, 1846   **5233**
Peruvian Jingle Shell

Monterey, California, to Peru.

1 to 2 inches in size, variable in shape, thin, partially translucent, smooth or with irregular sculpture; colored orange or

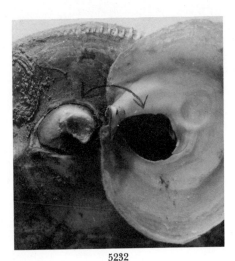

5232

yellowish green, sometimes coppery-brown. Occurs between tides attached to rocks, other shells and waterlogged wood. Common. Synonyms include: *alectus, fidenas, lampe* and *larbas* all Gray, 1850; *A. tenuis* C. B. Adams, 1852; *Calyptraea aberrans* C. B. Adams, 1852.

Some specimens may be radially ribbed with weak, irregular ridges. This may be a good species or only a form: **(5234)** *Anomia adamas* Gray, 1850 (its synonyms would be *hamillus* Gray, 1850, and *pacilus* Gray, 1850).

Anomia chinensis Philippi, 1849   5235
Chinese Jingle Shell

State of Washington (introduced from Japan).

1 to 1½ inches, thin-shelled; light-yellow to greenish to flesh-colored. Upper valve convex, coarsely sculptured, with numerous irregular radial rough lines and coarse concentric growth interruptions. Bottom valve flattish, usually bluish green with a thick white circle around the large byssal opening. Moderately common among oysters in Washington. Synonyms: *cytaeum* Gray, 1850; *lischkei* Dautzenberg and Fischer, 1907; and *argentea* Pilsbry in Stearns and Pilsbry, 1895.

5233   5235

Subgenus *Heteranomia* Winckworth, 1922

Type of this subgenus is *squamula* Linné, 1758. See *Proc. Malocol. Soc. London*, vol. 15, p. 33.

Anomia squamula Linné, 1758   5236
Prickly Jingle Shell

Labrador to North Carolina. Northern Europe.

Rarely exceeding ¾ inch in size, irregularly rounded, moderately fragile. Upper valve convex, rough, often with small prickles in older specimens. Sometimes with rather long spines. Lower valve flat and with a small hole near the hinge end.

Color drab, opaque whitish tan. A common cold-water species attaching itself to rocks and broken shells. Found from low-tide mark to 350 fathoms. For growth and spinosity, see A. S. Merrill, 1962, *The Nautilus*, vol. 75, p. 131. Formerly known as *aculeata* Müller, 1776, and Gmelin, 1791.

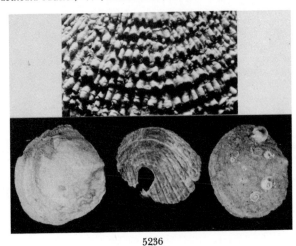

5236

Genus *Pododesmus* Philippi, 1837

The valve without the hole has 1 large and 1 small muscle scar. Type: *rudis* (Broderip, 1834).

Pododesmus macroschisma (Deshayes, 1839)   5237
False Pacific Jingle   Color Plate 20

Adak Island, Alaska, to Baja California. Japan.

1 to 4 inches in size. Radiating ribs very irregular and coarse. Color yellowish or greenish white, inner surface green and somewhat pearly. Lower valve with a large opening for the byssus. This is a very common species which is found attached to stones and wharf pilings from low-tide mark to about 35 fathoms. Often found on *Haliotis*. Larger, whitish specimens are found clinging to iron wrecks. *P. cepio* (Gray, 1850) and *alope* (Gray, 1850) are ecologic forms of this species. See E. P. Chace, 1972, *The Tabulata*, vol. 5, no. 2.

Pododesmus rudis (Broderip, 1834)   5238
False Atlantic Jingle

Off South Carolina to Florida and Texas to Brazil. Bermuda.

1 inch, brown, tan, yellowish, white or purple to lavender; exterior roughened with radial, rounded, irregular riblets. Uncommon; in crevices of rocks and on iron wrecks, from low-water mark to 25 fathoms. *P. decipiens* Philippi, 1837, and possibly *P. leloiri* Carcelles, 1941, are synonyms.

5238

Subgenus *Tedinia* Gray, 1853

With 2 muscle scars and the ligamental pedestal of *Pododesmus*. Adult with the byssal opening sealed over. Shell irregularly elongate. Type: *pernoides* (Gray, 1853).

Pododesmus pernoides (Gray, 1853)　　　　5239

Southern California to Oaxaca, Mexico.

1 to 2 inches, flattish, living in rock crevices. Irregular in shape, coarsely sculptured, dark-brown in color throughout. 2 muscle scars are rather small and set apart from each other. Ligament like a small pad in one valve, and 2 crural ridges in the other. Uncommon, or at least seldom collected.

Superfamily Limacea Rafinesque, 1815

Family LIMIDAE Rafinesque, 1815

Genus *Lima* Bruguière, 1797

Shell with a single adductor scar. Lunule present. Valves usually obliquely oblong, with small, narrow ears; usually white and radially ribbed. Hinge line straight and without teeth. Internal resilium under the beak. Type: *lima* (Linné, 1758). Synonyms: *Mantellium* Röding, 1798; *Radula* Mörch, 1853; *Austrolima* Iredale, 1929; *Meotolima* Oyama, 1943.

Nests of byssal threads and foreign matter are constructed by many species of Limidae (see details by A. S. Merrill and R. D. Turner, 1963, *The Veliger*, vol. 6, pp. 55–59).

Lima lima (Linné, 1758)　　　　5240
Spiny Lima　　　　Color Plate 20

Southeast Florida and the West Indies to Brazil. Bermuda.

1 to 1½ inches in height and pure-white in color. Sculpture of 26 to 33 even, radial ribs bearing many erect, sharp spines. The posterior ear is much smaller than the anterior one. No large posterior byssal gape as in *scabra*. Moderately common under coral stones in shallow water to 77 fathoms. This species and its various forms or subspecies (*squamosa* Lamarck, *multicostata* Sowerby, and *caribaea* Orbigny 1842), are found all over the world in tropical waters.

(5241) *Lima lima* subspecies *tetrica* Gould, 1851, occurs from Baja California to Ecuador, from 5 to 60 fathoms and is uncommon. Reaches 2 inches, coarsely and spinosely ribbed, white-shelled, and has more convex valves.

Subgenus *Ctenoides* Mörch, 1853

Shell oval-elongate, equivalve, a little oblique. Valves almost equilateral, compressed, slightly gaping behind under the posterior auricle. Type: *scabra* Born, 1778. This is given full generic status by some workers.

Lima scabra (Born, 1778)　　　　5242
Rough Lima　　　　Color Plate 20

Off South Carolina to Florida, Texas and to Brazil.

1 to 3 inches in height, ½ as long. Sculpture coarse, consisting of irregular, radial rows of short, barlike ribs, somewhat giving the appearance of shingles on a roof. Periostracum thin, dark- to light-brown. A common variation of this species (5242a; Color Plate 20) (form *tenera* Sowerby, 1843 (non Turton, 1822)) is startlingly different, in that the small radial ribs are much more numerous and much smaller. Common under rocks in shallow water at low tide to 77 fathoms.

Subgenus *Limaria* Link, 1807

Thin-shelled, inflated, oval-elongate, oblique. Widely gaping on both margins. No lunule present. Hinge line straight. Type: *tuberculata* Olivi, 1792. Synonyms: *Mantellum* Mörch, 1853; *Promantellum* Iredale, 1939.

Lima pellucida C. B. Adams, 1846　　　　5243
Antillean Lima

North Carolina to both sides of Florida; Texas and to Brazil. Bermuda.

¾ to 1 inch in height, elongate, fragile, semitranslucent, white, with a large posterior gape and with a long, narrow anterior gape. Radial ribs small, fine, uneven in size and distribution. Hinge-ears almost equal in length. Closely related to *L. hians* (Gmelin, 1791) from Europe. A fairly common species which is often misidentified in collections as *L. inflata* Lamarck, 1819 (not Gmelin, 1791). In thicker and older specimens there is a small, pinhole depression in the hinge just off to one side of the ligamental area. Gills rosy-orange; foot white; tentacles translucent-white. Makes community nests under rocks in 4 to 20 feet of water (see C. W. Johnson, 1931, *The Nautilus*, vol. 44, p. 126).

5243

Lima hemphilli Hertlein and Strong, 1946　　　　5244
Hemphill's Lima

Monterey, California, to Acapulco, Mexico.

1 inch in length, white, obliquely elliptical in shape. With fine, irregular, radial ribs which are crossed by very fine, rough threads. Gapes a little more at the anterior than at the posterior side. Anterior and posterior margins smooth. This fairly common species has been erroneously called *dehiscens* Conrad and *L. orientalis* Adams and Reeve. Occurs from 10 to 50 fathoms.

Other species:

5245　*Lima locklini* McGinty, 1955. Gulf of Mexico and the West Indies, 20 fms. *Proc. Acad. Nat. Sci. Phila.*, vol. 107, p. 84.

5244　　　　5245

5246 *Lima* (*Acesta* H. and A. Adams, 1858) *mori* Hertlein, 1952. 690 to 800 fms., Mulberry Seamount, off San Mateo County, California. *Proc. Calif. Acad. Sci.*, ser. 4, vol. 27, p. 377.

5247 *Lima* (*Acesta*) *bullisi* Vokes, 1963. Off Mobile, Alabama, 300 to 600 fms., *Tulane Studies in Geol.*, vol. 1, p. 77.

5248 *Lima* (*Acesta*) *excavata* (Fabricius, 1779). Greenland; northern Europe.

5249 *Lima albicoma* Dall, 1886. Off Florida Keys to Barbados, 115 to 121 fms. 8 mm.

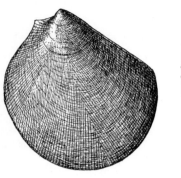

5249

5256

5250 *Lima* (*Promantellum* Iredale, 1939) *pacifica* Orbigny, 1846. Northern end of the Gulf of California to Peru, intertidal. (Synonyms: *arcuata* Sowerby, 1843, non Geinitz, 1840; *galapagensis* Pilsbry and Vanatta, 1902.)

5251 *Lima* (*Submantellum* Olsson and Harbison, 1953) *orbignyi* Lamy, 1930. Northern end of the Gulf of California to Chile, in 7 to 22 meters.

Genus *Limatula* S. Wood, 1839

Valves usually equilateral, closed. Ligamental pit small and centrally placed. Animal without eyes. Type: *subauriculata* Montagu, 1808. Synonym: *Stabilima* Iredale, 1939.

Limatula subauriculata (Montagu, 1808)   5252
Small-eared Lima

Greenland to Puerto Rico.
Alaska to Mexico.

½ inch in height, ovate-oblong, greatly inflated (having the shape of the shell of a pistachio nut) and sculptured with numerous small, longitudinal riblets. On the inside of the valves there are 2 prominent, longitudinal riblets at the center of the shell. Periostracum over the white shell is yellowish brown. Moderately common in cooler waters from just offshore to 1,000 fathoms. Also found in northwest Europe. *L. elongata* (Forbes, 1843) is a synonym.

Other species:

5253 *Limatula confusa* (E. A. Smith, 1885). North Carolina to the West Indies, 31 to 1,450 fms. (Synonym: *ovata* Jeffreys, 1876, non S. V. Wood, 1839.)

5254 *Limatula laminifera* (E. A. Smith, 1885). Off Fernandina, Florida, to the West Indies, 390 to 498 fms.

5255 *Limatula hyperborea* Jensen, 1909. Canadian Arctic; Europe; Siberia. 40 to 1,240 fms.

5256 *Limatula setifera* Dall, 1886. North Carolina to the West Indies, 52 to 450 fms. (?subspecies of *jeffreysi* Fischer, 1880). 9 mm.

5257 *Limatula regularis* Verrill and Bush, 1898. Off Delaware Bay, 70 fms.

5258 *Limatula nodulosa* Verrill and Bush, 1898. Off Mississippi Delta, 730 fms.

5259 *Limatula hyalina* Verrill and Bush, 1898. Off Apalachicola, Florida, 25 to 27 fms.

5260 *Limatula attenuata* (Dall, 1916). Southern Bering Sea to the Shumagin Islands, Alaska.

5261 *Limatula hendersoni* Olsson and McGinty, 1958. Florida; Bahamas, to Panama and Barbados. Bull. 177, *Amer. Paleontol.*, p. 47.

5262 *Limatula similaris* Dall, 1908. Baja California to Panama, in 55 to 106 meters. 4.5 mm.

Genus *Limea* Bronn, 1831

Small, suborbicular, not gaping; sculpture of radial ribs which crenulate the margin. Cardinal area narrow. Hinge with a series of short denticles on each side. Adductor impression subcentral. Type: *strigilata* Brocchi, 1814, Miocene of Europe. *Limaea* is a misspelling.

Limea bronniana Dall, 1886   5263
Bronn's Dwarf Lima

North Carolina to Florida and the West Indies.

Very small, 5 mm. in height, ovate, superficially resembeling a small *Cardium*. With about 25 to 30 strong, smooth, rounded, radial ribs. Microscopic, concentric scratches between the ribs. Inner margin of valves serrated and reinforced by small, round teeth. Shell pure-white in color. Hinge-ears with an internal set of 3 or 4 small teeth. Moderately common from 1 to 804 fathoms. Dall described a variety *lata* Dall, 1886, from off Fernandina, Florida.

5263

Other species:

5264 *Limea subovata* Jeffreys, 1876. North Atlantic to off Martha's Vineyard, Massachusetts, 100 to 500 fms.

5264

Order *Ostreina* Rafinesque, 1815

The oysters are with 1 adductor muscle (monomyarian); the foot and byssus lacking in adults; usually the lower left valve is cemented to the hard substrate; shell material chiefly calcitic and foliaceous; gills of the eulamellibranch type; pallial line without a sinus. A detailed and well-illustrated compendium of this group by H. B. Stenzel appears in "Moore's Treatise on Invertebrate Paleontology," 1971, Part N, volume 3, "Bivalvia Mollusca: Oysters," pp. 953–1217. Ostreina Ferussac, 1822, is a synonym.

Family OSTREIDAE Rafinesque, 1815

Prodissoconch hinge of planktonic larvae bearing on each valve 4 subequal tooth-precursors and their corresponding sockets split by a long smooth median gap into 2 equal groups. Adductor muscle kidney or crescent-shaped, placed almost centrally. Intestine passes by dorsum of pericardium and does not pierce the heart. Some species incubate their young.

Genus *Ostrea* Linné, 1758

This genus used to include all of the oysters, but today several valid genera are recognized, so that only 3 American species are included in true *Ostrea*. These are *O. equestris* Say and *O. permollis* Sowerby from the Atlantic coast and *O. lurida* Carpenter from the Pacific coast. The European oyster, *O. edulis* Linné is also in this group. All of these oysters are relatively small. The eggs are fertilized and developed within the mantle chamber and gills. Usually around one million eggs are produced at one spawning. The prodissoconch hinge is long, the valves symmetrical. In the adults, the muscle scar is near the center of the shell and is not colored. The name *Ostrea* was validated by the I.C.Z.N., Opinion 94 and 356. Type: *edulis* Linné, 1758.

For more details on the Californian species, see L. G. Hertlein, 1959, *The Veliger,* vol. 2, pp. 5–10, 13 figs., and P. Bonnot, 1935, *Calif. Fish and Game,* vol. 21, pp. 65–80.

Ostrea equestris Say, 1834   5265
Crested Oyster

Virginia to Texas and West Indies. Brazil.

1 to 3 inches in length, more or less oval, and with raised margins which are crenulated. The attached valve has a flat interior with a rather high, vertical margin on one side. Interior dull-grayish with a greenish or opalescent-brown stain. Margin sometimes stained a weak-violet. Muscle scar almost central. Edge of upper valve has a row of fine denticles. Not

very abundant except in some Florida bays. It lives in water that is much saltier than that in which *virginica* lives. Also named *spreta* Orbigny, 1846, and *gundata* Holmes, 1858. *O. cristata* Born, 1778, is quite different and is limited to South America. For growth and ecology, see Galtsoff and Merrill, 1962, *Bull. Marine Sci.,* vol. 12, pp. 234–244.

Ostrea permollis Sowerby, 1841   5266
Sponge Oyster

North Carolina to Florida and the West Indies.

Rarely over 3 inches in size. Lives under rock slabs or sometimes embedded in sponges (*Stellata*) with only the margins of the valve showing. The surface of the valve has a soft, golden, silky appearance. Beak weak or twisted back into a strong spiral. Exterior light-orange to tan; interior white. Inner margins smooth or with numerous small, round denticles. Common from low tide to 20 fathoms. The ecologic form *weberi* Olsson, 1951, is ovate, fragile, anomialike and has numerous fine radial, divaricate threads. It is usually attached to rocks in protected areas (*The Nautilus,* vol. 65, p. 6).

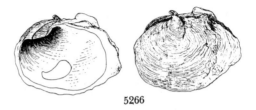

5266

Ostrea lurida Carpenter, 1864   5267
Native Pacific Oyster

Alaska to Baja California.

2 to 3 inches in length, of various shapes; generally rough with coarse concentric growth lines, but sometimes smoothish. Interior usually stained with various shades of olive-green, and sometimes with a slight metallic sheen. It occasionally has purplish brown to brown axial color bands on the exterior. This is the common intertidal native species of the Pacific coast, sometimes called the "olympic oyster." A number of ecological forms have been described: *expansa* Carpenter, 1864; *laticaudata* Carpenter, 1864; *rufoides* Carpenter, 1864.

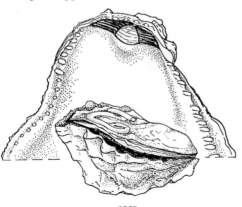

5265

5267   5276

Ostrea conchaphila Carpenter, 1857   5268
Shell-loving Oyster

Southern California to Panama.

1 to 2 inches, oval to oblong, rarely circular. Striped with broad bands of purplish brown or orange. Inside white or greenish, with minute denticles along the margin for about 1/3 of the way to the ventral edge of the valve. Usually found attached singly to dead shells, rocks and crab shells. Offshore to 20 fathoms. Uncommon. *O. mexicana* Sowerby, 1871, is a synonym.

Ostrea palmula Carpenter, 1857   5269
Palmate Oyster

Baja California to Ecuador.

2 to 3 inches, heavy, the lower valve usually deep; strongly ribbed. Interior white with a marginal band of violet-brown. Inner margin of upper valve finely denticulate all around. Common; on rocks exposed to heavy surface. Synonyms are *amara* Carpenter, 1863; *serra* Dall, 1914; *dalli* Lamy, 1930.

Ostrea megodon Hanley, 1846   5270
Megadon Oyster

Baja California to Peru.

2 to 4 inches, strongly curved to one side with 3 or 4 strong waves in the ventral margin of the shell. Lateral margins finely crenulated. Color dirty-white or greenish, tinged with purple. Moderately common offshore to 61 fathoms.

Other Pacific species:

5271 *Ostrea angelica* Rochebrune, 1895. Gulf of California to Ecuador. 2 to 4 inches.

5272 *Ostrea iridescens* Hanley, 1854. Southern Gulf of California to Peru, on rocks. (Synonyms: *lucasiana* and *turturina* Rochebrune, 1895.)

5273 *Ostrea tubulifera* Dall, 1914. Gulf of California to Panama.

Genus *Crassostrea* Sacco, 1897

This genus includes the commercially important American oyster, *C. virginica* Gmelin, which was formerly placed in the genus *Ostrea*. In *Crassostrea*, the left or attached valve is larger than the right. The inner margin is smooth. The eggs are small, produced in large numbers at one spawning (over 50 million), and are fertilized and develop in the open waters outside of the parents. The muscle scar is usually colored. The prodissoconch hinge is short, and the valves asymmetrical. The Japanese oyster (*C. gigas*), introduced to west American shores, the Portuguese oyster (*C. angulata* Lamarck) and *C. rhizophorae* Guilding from Cuba also belong to this genus. Type: *virginica* (Gmelin, 1791). *Gryphaea* Lamarck is a fossil genus which should not be associated with this genus.

Crassostrea virginica (Gmelin, 1791)   5274
Eastern Oyster

Gulf of St. Lawrence to the Gulf of Mexico and the West Indies.

2 to 6 inches in length. This is the familiar edible oyster which varies greatly in size and shape. The valve margins are

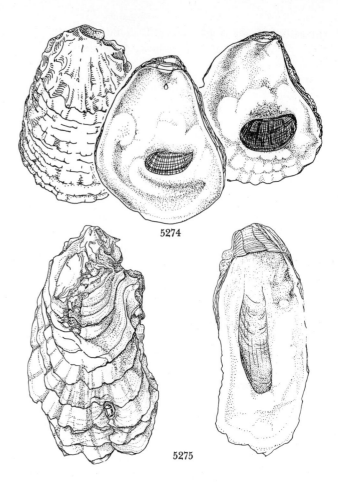

5274

5275

only slightly undulating or are straight. The muscle scar is usually colored a deep purple, the rest of the shell being white inside and dirty-gray exteriorly. Beaks usually long and strongly curved. "Blue Points," a form originally harvested at Blue Point, Long Island, are rounded in shape and with a rather deep, lower valve. "Lynnhavens" are broad, elongate forms originally harvested at Lynnhaven Bay, Virginia. These variations are due to environmental differences. *C. brasiliana* Lamarck and *C. floridensis* Sowerby are this species.

(5275) *Crassostrea rhizophorae* (Guilding, 1828) (*brasiliana* of authors) is found in the Caribbean region to Brazil, and it is a lightweight shell, deep-cupped, with a flat upper valve small and fitting well down into the lower valve. The inner margin of the lower, attached valve is splotched with bluish purple. Common.

Crassostrea gigas (Thunberg, 1793)   5276
Giant Pacific Oyster

British Columbia to California. Japan.

3 to 12 inches in length, of various shapes, but generally characterized by its large size, its coarse, widely spaced, concentric lamellae or very coarse longitudinal flutings or ridges on the outside. Interior enamel white, often with a faint purplish stain on the muscle scar or near the edges of the shell. Very rarely with a greenish stain. A common, large and marketable oyster introduced yearly into Canada and the United States from Japan. The form *laperousii* Schrenck is round. The typical *gigas* is the long, straplike form. *O. gigas* Meuschen is an invalid name and does not preoccupy that of Thunberg's. Also known as the Japanese oyster.

Crassostrea columbiensis (Hanley, 1846) 5277
Columbian Oyster

Baja California to Peru.

2 to 3 inches, irregular in shape, the attached valve deep, cup-shaped; the upper valve smaller and usually flatter. Sculpture smoothish or with coarse ribs. Color deep-purple or blackish; interior white except for a wide, purple-colored marginal band. Usually attached to the roots of mangroves. Common. Synonyms are *aequatorialis* Orbigny, 1846; and *ochracea* Sowerby in Reeve, 1871.

Other species:

5278 *Crassostrea corteziensis* (Hertlein, 1951). Head of the Gulf of California to Panama. 6 to 10 inches; common food oyster. Erroneously called *chilensis* Philippi. See Hertlein, *Bull. So. Calif. Acad. Sci.*, vol. 50, p. 68.

Subfamily LOPHINAE Vyalov, 1936

Brood young inside mantle cavity; lower valve with shelly claspers; usually with a plicated surface. Innumerable very small, slightly elongate tubercles are scattered over interior surface of valves, especially near their margins.

Genus *Lopho* Röding, 1798

Large tropical oysters, usually living in mangroves or on alcyonarian stems, with 6 to 12 deep, sharp radial plicae. Type: *cristagalli* (Linné, 1758) which has an ecomorph, *folium* (Linné, 1758), living on seawhips. Synonyms include *Dendostraea* Sowerby, 1839; *Alectronia* Logan, 1898; *Pretostrea* Iredale, 1939; and *Alectryossia* Salisbury and Edwards, 1959.

Lopha frons (Linné, 1758) 5279
Frons Oyster Color Plate 22

Florida, Louisiana and the West Indies to Brazil, Bermuda.

1 to 2 inches in size. The radial plicate sculpture and corresponding sharply folded valve margins are characteristic of this intertidal species. Inner margins of valves closely dotted with minute pimples for nearly the entire circumference of the valves. Muscle scars located well up toward the hinge. Beaks somewhat curved. Interior translucent-white, exterior usually purplish red. Rarely albinistic. Frequently elongate and attached to stems of gorgonians by a series of clasping projections of the shell, but may be also oval in shape. *Ostrea rubella* and *O. limacella* Lamarck, 1819, are this species. *Lopha folium* (Linné) is a Philippine species. Neither species lives near mangroves, nor on the roots of these trees as has been erroneously reported. Formerly called the 'coon oyster.

Family GRYPHAEIDAE Vyalov, 1936

Oysters with the prodissoconch hinge on larval shells with uninterrupted, alternating series of equal tooth-precursors and corresponding sockets. Intestine passing through the pericardium and ventricle of the heart. Adductor muscle round, usually closer to the hinge.

Subfamily PYCNODONTEINAE Stenzel, 1959

Shell structure with microscopic bubbles (vesicular).

Genus *Pycnodonte* G. Fischer, 1835

Small to large and massive. Shell material made up in part by spongy, cellularlike material. Type: *P. radiata* G. Fischer, 1835. *Pycnodonta* Sowerby, 1842, is a synonym.

Subgenus *Hyotissa* Stenzel, 1971

The type of this subgenus is *hyotis* (Linné, 1758). See "Treatise on Invertebrate Paleontology," Part N, Mollusca, vol. 3, 1971, by H. B. Stenzel. Species are associated with corals and oceanic waters.

Pycnodonte hyotis (Linné, 1758) 5280
Honeycomb Oyster

Southeast Florida and the West Indies.
Indo-Pacific.

3 to 4 inches, generally circular in outline, and sometimes with a sawtooth margin. Immediately recognized by the peculiar structure of the shell which under a lens appears to be filled with numerous bubbles or empty cells much like a bath sponge. Look along the edges of the valves or break off a piece of shell. Color cream, brownish or lavender. Uncommon; below low tide and on offshore wrecks. *Ostrea thomasi* McLean, 1941, is a synonym (*Notulae Naturae*, no. 67).

Other species:

5281 *Pycnodonte fischeri* Dall, 1914. Baja California to Ecuador; Galapagos. (Synonyms: *jacobaea* Rochebrune, 1895, not Linné, 1758.) *O. solida* Sowerby, 1871, may be an earlier name.

Subclass *Heterodonta* Neumayr, 1884

Hinge usually bearing distinct cardinal and lateral teeth. Shell material never nacreous. Ligament located posterior to the beads (opisthodetic). Without a lithodesma. Mantle lobes more or less joined along some part; siphons usually developed. Contains 2 living orders, the Veneroida and Myoida. These 2 probably contain over half of the known marine species and genera. We follow the order of superfamilies listed in the "Treatise of Invertebrate Paleontology" which is quite different from that in the first edition of "American Seashells."

Order *Veneroida* H. and A. Adams, 1856

More or less equivalve; with equal-sized muscle scars. Hinge with cardinals and laterals (rarely with only cardinals or without teeth).

Superfamily Lucinacea Fleming, 1828

Family LUCINIDAE Fleming, 1828

Genus *Linga* de Gregorio, 1884

Shell orbicular, strong and laterally compressed. Cardinal teeth small, obscure in the adults, but the laterals are well-developed. Shell margin crenulate within. Sculpture of numerous, equidistant lamellae. Type: *columbella* Lamarck,

1819. Formerly this group was called *Lucina* by American workers and in the first edition of "American Seashells."

Genus *Linga* de Gregorio, 1884

Type of this genus is *columbella* Lamarck, 1819.

Linga pensylvanica (Linné, 1758)   5282
Pennsylvania Lucina   Color Plate 22

North Carolina to south Florida and the West Indies.

1 to 2 inches in length, ovate, usually quite inflated. Concentric ridges very delicate and distinct. Color pure-white with a thin yellowish periostracum. Rarely blushed with orange (forma *aurantia* Deshayes, 1832 and *pensilvanica* Röding, 1798). Lunule heart-shaped, well-marked and raised at the center. The furrow from the beak to the posterior ventral edge of the valve is very pronounced. Beachworn specimens become smooth and shiny-white. The species name was spelled with one "n" by Linné. Moderately common in shallow water.

Subgenus *Here* Gabb, 1866

Like *Linga*, but the lunule is large, deeply excavated. Type: (*richthofeni* Gabb, 1866) *excavata* Carpenter, 1857.

Linga sombrerensis (Dall, 1886)   5283
Sombrero Lucina

Northern Gulf of Mexico to the West Indies.

¼ inch in length, oval, greatly inflated and pure-white in color. No radial sculpture. Concentric riblets numerous, sharp and irregularly crowded. Concentric growth irregularities commonly make the outer surface wavy. Lunule very small, wider than long, situated directly under the front of the prominent beaks and bounded by a fine groove. Interior of the valves' margin finely crenulate. Common; 20 to 90 fathoms.

Linga excavata (Carpenter, 1857)   5284
Excavated Lucina

San Pedro, California, to Mazatlan, Mexico.

1 inch, rounded, globose, nut-shaped. Anterior dorsal area more strongly defined than the posterior, enclosing the deep, penetrating lunule. Surface marked with strong, concentric ridges and smaller radial striae. Found in coarse gravel; 20 fms. Uncommon. Synonym: *richthofeni* (Gabb, 1866).

Subgenus *Bellucina* Dall, 1901

Shell small, suborbicular, with a coarsely cancellate sculpture, and with a small, deeply excavated lunule. Type: (*pisum* Reeve, 1850, non Orbigny, 1841) *semperiana* Issel, 1869.

Linga amiantus (Dall, 1901)   5285
Lovely Miniature Lucina

North Carolina to both sides of Florida; West Indies to Brazil.

¼ to ⅜ inch in length, not very obese, thick-shelled, pure-white in color and beautifully sculptured with 8 or 9 wide, rounded, radial ribs across which run numerous, small concentric riblets. Near the posterior upper margin of the shell there is a radial row of about 8 to 11 small, scalelike nodes. Behind the tiny, curved beaks there is an oval, heart-shaped depression. Internal margin of valves strongly crenulated with

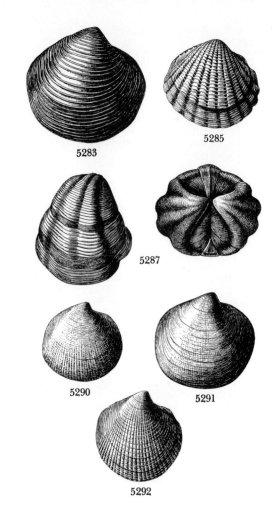

5283   5285

5287

5290   5291

5292

tiny teeth. Adults are commonly misshapen by concentric growth stops. Common from shallow water to 640 fathoms. Compare with *L. multilineata*.

Other species:

5286   *Linga (Bellucina) cancellaris* (Philippi, 1846). Gulf of California to Peru.

Subgenus *Pleurolucina* Dall, 1901

Shell small, with a few large, radial ribs crossed by concentric threads. Type: *leucocyma* Dall, 1886.

Linga leucocyma (Dall, 1886)   5287
Four-ribbed Lucina

North Carolina to Florida and the Bahamas.

¼ inch in length, roughly oval, fairly thick-shelled, inflated and white in color. With 4 conspicuous, large, rounded, radial ribs, and with numerous, small, crowded, squarish, concentric riblets. The inner margins of the valves are finely denticulate. A common, bizarrely sculptured species found from low water to several fathoms.

Other species:

5288   *Linga undatoides* (Hertlein and Strong, 1945). Gulf of California. (Synonym: *undata* Carpenter, 1865, not Lamarck, 1819.) *The Nautilus*, vol. 58, p. 105.

5289   *Linga leucocymoides* (Lowe, 1935). Gulf of California. *San Diego Soc. Nat. Hist.*, vol. 8, no. 6, p. 17.

Genus *Parvilucina* Dall, 1901

Shell small, plump; surface weakly sculptured or finely cancellate. Dorsal features obscure. Ventral margins crenulate. Hinge teeth small. Type: *tenuisculpta* Carpenter, 1865.

Parvilucina multilineata (Tuomey and Holmes, 1857)   **5290**
Many-lined Lucina

North Carolina to both sides of Florida. Brazil.

⅜ to ¼ inch in length, almost circular in shape, very obese, moderately thick-shelled, white, and very finely sculptured. Somewhat like *L. amiantus,* but without radial ribs, except for exceedingly fine threads seen best near the beaks. Concentric sculpture of numerous, rather irregular, growth threads. The shell commonly continues growth after a long rest, thus causing an irregular, concentric hump in the shell. Inner margin very finely denticulate. Differs from *Codakia pectinella* (C. B. Adams, 1852) in having the concentric sculpture stronger than the radial lines (rather than equally strong). *P. crenella* Dall, 1901, is the same species. Common from beach to 120 fathoms.

Parvilucina tenuisculpta (Carpenter, 1865)   **5291**
Fine-lined Lucina

Bering Sea to Baja California.

½ inch in length, slightly less in height, oval in outline, chalky-white and with a thin, grayish or yellowish green periostracum. Sculpture of numerous, small, weak, raised radial threads. Concentric growth lines fine and irregularly placed. Beaks fairly prominent and pressed closely together. Behind them, the narrow, depressed ligament is visible from the outside. In front is the small, heart-shaped, depressed lunule. Inner margin of valves finely toothed. Common just offshore.

Parvilucina approximata (Dall, 1901)   **5292**
Approximate Lucina

Monterey, California, to Panama.

¼ inch or less in size. Very similar to *tenuisculpta,* but smaller, almost round in outline, more inflated and with fewer and quite strong, radial riblets. Periostracum very thin, commonly worn off. Shell texture not chalky. Common in sandy mud just offshore to 48 fathoms.

Other species:

5293   *Parvilucina mazatlanica* (Carpenter, 1855). Gulf of California to Ecuador.

Subgenus *Cavilinga* Chavan, 1937

Shell about ½ inch, trigonal; dorsal areas feeble. Hinge strong, with 2 cardinals and a lateral tooth in each valve.

Surface concentrically scarred by ridges representing resting stages in the growth. Lunule deeply excavated, placed below the small, pointed beads. Ventral margins crenulated. Type: *trisulcata* (Conrad, 1841), a Miocene species.

Parvilucina blanda (Dall and Simpson, 1901)   **5294**
Three-ridged Lucina

North Carolina to Brazil.

⅓ to ½ inch, obliquely triangular in outline, strong-shelled. Color white, sometimes yellow or salmon. Surface with fine concentric lines, and 3 or 4 irregular, deep, concentric furrows. Moderately common; 10 to 20 fathoms. Formerly called *trisulcata* (Conrad, 1814), which is a fossil limited to the Miocene.

Other species:

5295   *Parvilucina (Cavilinga) lampra* (Dall, 1901). Gulf of California. 1 inch.

5296   *Parvilucina (Cavilinga) lingualis* (Carpenter, 1864). Baja California and the Mexican coast.

Genus *Codakia* Scopoli, 1777

Shell large, orbicular, moderately compressed. Hinge of right valve with a prominent anterior lateral which is typically close to the cardinals (an anterior, a posterior and a middle cardinal). Hinge of left valve with a large double anterior lateral, only 2 cardinals, and with a small, double posterior lateral. Type: *orbicularis* (Linné, 1758). Synonym: *Lentillaria* Schumacher, 1817.

Subgenus *Codakia* Scopoli, 1777

Codakia orbicularis (Linné, 1758)   **5297**
Tiger Lucina

Florida to Texas and the West Indies. Bermuda. Brazil.

2½ to 3½ inches in length, slightly less in height, well-compressed, more or less orbicular in outline, thick and strong. Beaks and ¼ inch of subsequent growth smoothish. Remainder of the shell roughly sculptured by numerous coarse radial threads which are crossed by finer concentric threads. This commonly gives the radial ribs a beaded appearance. Exterior white. Interior white to pale-lemon, commonly with a rose tinge on the ends of the hinge or along the margins of the valves. Lunule just in front of the beaks is deep, heart-shaped, small and nearly all on the right valve. A common tropical species. Do not confuse with *C. orbiculata* (Montagu, 1808).

5294   5298   5299   5300

Codakia costata (Orbigny, 1842)   5298
Costate Lucina

North Carolina to southeast Florida; West Indies to Brazil. Bermuda.

½ inch in length, variable in shape, but usually orbicular, quite obese, white to yellowish in color. With fine radial ribs, usually in pairs which are crossed by very fine concentric threads. The radial ribs become obsolete on the posterior ⅓. Beaks also with this sculpturing. Lunule small, indistinct, lanceolate, slightly more on the right valve. Compare its poorly defined lunule with those of *orbicularis* and *orbiculata*. Moderately common offshore on sandy bottoms. *C. antillarum* (Reeve, 1850) is a synonym.

Subgenus *Epilucina* Dall, 1901

Concentric sculpture only; lunule rather large, asymmetrical. Posterior laterals small. Type: *californica* (Conrad, 1837).

Codakia californica (Conrad, 1837)   5299
Californian Lucina

Crescent City, California, to Baja California.

1 to 1½ inches in length, oval to circular, moderately inflated. Exterior dull-white with numerous, crowded, rather distinct, but small, concentric threads. Lunule of right valve like a small, depressed, lanceolate shield which fits snugly into a similarly shaped recess in the left valve. A common littoral species in southern California and down to 78 fathoms. Do not confuse with large specimens of *Diplodonta*.

Subgenus *Ctena* Mörch, 1860

Shells small; hinge of *Codakia* but with the posterior lateral tooth well-developed. Lunule small, lenticular, impressed, and more or less equal in both valves. Type: *mexicana* Dall, 1901.

Codakia orbiculata (Montagu, 1808)   5300
Dwarf Tiger Lucina

North Carolina to Florida, the West Indies to Brazil; Bermuda.

1 inch or less in length, very similar to *orbicularis*, but with a large, elongate lunule in front of the beaks (instead of small and heart-shaped), and with stronger, less numerous, commonly divaricate ribs which are noticeable right up to the ends of the beaks. This species is much fatter and never has pink coloring inside. Common in sand from low water to 100 fathoms. *C. recurvata* Dall, 1901, is a synonym.

(5301) The form *filiata* Dall, 1901, has finer sculpturing much like *orbicularis*, is often yellowish in color, but can be readily distinguished from the latter by its elongate lunule. Common in the Gulf of Mexico.

Other species:

5302   *Codakia* (*Ctena*) *pectinella* C. B. Adams, 1852. Florida Keys and the West Indies. Brazil, 18 to 32 fms.

5303   *Codakia* (*Codakia*) *cubana* Dall, 1901. Gulf of Mexico and Cuba, 84 fms.

5304   *Codakia distinguenda* (Tryon, 1872). Magdalena Bay, Baja California, to Panama, shallow water. (Synonyms: *colpoica* Dall, 1901; and *pinchoti* Pilsbry and Lowe, 1932.)

Genus *Lucina* Bruguière, 1797

Shell orbicular, quite compressed. Sculpture mostly concentric. Cardinal teeth obsolete in adults, but the laterals are well-developed. *Phacoides* Blainville, 1825, is the same but is not considered valid. Type: *jamaicensis* Spengler, 1784. Synonym: *Lepilucina* Olsson, 1965.

Lucina pectinata (Gmelin, 1791)   5305
Thick Lucina   **Color Plate 22**

North Carolina to Florida, Texas and to Brazil.

1 to 2½ inches in length, ovate, compressed, white or flushed with bright-orange. Concentric ridges moderately sharp, usually unequally spaced. Ligament partially visible from the outside. Lunule strongly raised into a rather thin, rough blade. Anterior and posterior lateral tooth strong. Cardinals very weak. Moderately common in shallow water. Alias *Lucina jamaicensis* Lamarck, 1801; and *jamaicensis* "Spengler" Chemnitz, 1784. Do not confuse with *P. filosus*.

Subgenus *Lucinisca* Dall, 1901

Resembles *Codakia* in shape but has well-defined dorsal features. Cancellate sculpturing often sharply beaded. Type: *nassula* (Conrad, 1846).

Lucina nassula (Conrad, 1846)   5306
Woven Lucina

North Carolina to Florida, Texas and the Bahamas.

½ inch in length, almost circular, inflated, strong and pure-white. Sculpture of strong, closely spaced, concentric and radial ribs. These form a reticulate, rough surface. Where the ribs cross each other there is a tiny, raised scale. The ventral margin of the valve is strongly beaded by the distal ends of the axial riblets. Common in shallow water to 200 fathoms. *Lucina lintea* Conrad, 1837, is a synonym.

Lucina muricata (Spengler, 1798)   5307
Spinose Lucina

Lower Florida Keys and the West Indies. Brazil.

½ inch, almost circular, white, somewhat compressed, and characterized by about two dozen, unequal-sized, radial, very fimbriated, spiny ribs. Rare in Florida; moderately common in shallow waters over mud in the West Indies.

5306   5307   5309   5311

The Pacific counterpart is *liana* Pilsbry, 1931 (**5308**) 1 inch, Gulf of California to Peru. Common.

Lucina nuttalli (Conrad, 1837)   5309
Nuttall's Lucina

Santa Barbara, California, to Manzanillo, Mexico.

1 inch in length, circular, moderately inflated and with a fine, sharp, cancellate sculpturing. The shell is divided off at the anterior and upper portion into a slightly more compressed region which is less sculptured concentrically. Lunule very deep, short and larger in the left valve. Moderately common offshore in sand to 25 fathoms. Not uncommonly washed ashore.

(**5310**) The subspecies *centrifuga* Dall, 1901, from Baja California to Panama has stronger and distantly spaced, concentric, raised lines (*Proc. U.S. Nat. Mus.*, vol. 23, pl. 19, fig. 13). *Phacoides liana* Pilsbry, 1931, is a synonym.

Subgenus *Callucina* Dall, 1901

Shell orbicular, dosinoid, concentrically filose and sometimes with feeble radial lines. Dorsal features obsolete. Lunule small, chiefly in one valve. Hinge with one cardinal tooth in each valve, the other teeth feeble. Inner margins crenulate. Type: *radians* (Conrad, 1841).

Lucina radians (Conrad, 1841)   5311
Dosinia-like Lucina

North Carolina to Florida; West Indies. Bermuda.

¾ inch, white, nearly orbicular, superficially resembling the white clam, *Dosinia*. Surface with regular, fine concentric lines cut by faint radiating threads. Beaks high, turned forward over a deep lunule. Interior with characteristic glossy, radial rays. Ventral margin crenulated. Moderately common; 5 to 85 fathoms.

Other species:

5312 *Lucina (Callucina) bermudensis* Dall, 1901. Bermuda. New name for *Lucina lenticula* Reeve, 1850, not Gould, 1850 (*Proc. U.S. Nat. Mus.*, vol. 23, p. 810).

5313 *Lucina (Callucina) lampra* (Dall, 1901). Head of the Gulf of California to Santa Cruz Bay, Mexico, intertidal to 55 meters.

5314 *Lucina (Callucina) lingualis* Carpenter, 1864. Magdalena Bay and the Gulf of California to Acapulco, Mexico, intertidal to 24 meters.

5315 *Lucina (Lucinisca) fenestrata* Hinds, 1845. Baja California to Peru, in 13 to 73 meters.

Subfamily MYRTEINAE Chavan, 1969

Genus *Myrtea* Turton, 1822

Shells about 1 inch, usually white; thin periostracum. Dorsal edge bearing strong lamellations. Right valve with 1 cardinal tooth and 1 anterior and 1 posterior lateral tooth; left valve with 2 cardinal teeth and at least 1 anterior and 1 posterior lateral. Margins smooth. Type: *spinifera* Montagu, 1803. *Myrtaea* Dall, 1901, was an emendation.

Atlantic species:

5316 *Myrtea pristiphora* Dall and Simpson, 1901. Lamellated Lucina. Puerto Rico, 30 to 45 fms. See "Caribbean Seashells," pl. 29c.

5317 *Myrtea lens* (Verrill and Smith, 1880). Cape Cod, Massachusetts, to Brazil, 50 to 464 fms.

5318 *Myrtea (Eulopia* Dall, 1901) *sagrinata* (Dall, 1886). Florida Keys and westward to Yucatan Strait, 85 to 300 fms.

5319 *Myrtea (Eulopia) compressa* (Dall, 1881). Gulf of Mexico and the West Indies, 72 to 424 fms.

5319   5320   5325

Genus *Lucinoma* Dall, 1901

Shells up to 3 inches, circular, laterally compressed, with conspicuous periostracum and concentric lamellae. Cardinal teeth bifid, laterals obsolete. Type: *filosa* Stimpson, 1851.

Lucinoma annulata (Reeve, 1850)   5320
Western Ringed Lucina

Alaska to southern California.

2 to 2½ inches in length, oval to circular and slightly inflated. With strongly raised, concentric threads about ¹⁄₁₆ inch apart. Shell chalky-gray to white, overlaid by a thin, greenish brown periostracum. Fairly commonly dredged from 8 to 75 fathoms. The Tertiary fossil species *acutilineatus* Conrad, 1849, may be the same. *L. densilineata* Dall, 1919, is a variant.

Other species:

5321 *Lucinoma atlantis* R. A. McLean, 1936. Off Maryland, 118 to 300 fms. *The Nautilus*, vol. 49, p. 87.

5322 *Lucinoma blakeana* Bush, 1893. Massachusetts Bay to off Cape Fear, North Carolina, 18 to 464 fms.

5323 *Lucinoma aequizonata* Stearns, 1890. Santa Barbara Islands, California, 276 fms., to Chile.

5324 *Lucinoma heroica* (Dall, 1901). Southern part of the Gulf of California, in 1,836 meters.

Lucinoma filosa (Stimpson, 1851)   5325
Northeast Lucina

Newfoundland to north Florida and the Gulf States.

1 to 3 inches in length (south of North Carolina rarely over 1½ inches), almost circular, compressed, white, with a thin, yellowish periostracum. Beaks small, close together and cen-

trally located. Sculpture of sharp, raised, thin, concentric ridges each about ⅛ inch apart. The young commonly lack these ridges. No anterior lateral tooth present. Common offshore. Do not confuse with *pectinatus* which has a strong anterior lateral tooth, is tinted inside with orange and whose concentric ridges are unevenly spaced.

Subfamily MILTHINAE Chavan, 1969

Genus *Anodontia* Link, 1807

Shell large, obese, fairly thin and subcircular in outline. Hinge without distinct teeth. Anterior muscle scar short or long and parallels the pallial line. Type: *Venus edentula* Linné, 1758, according to Chavan in Moore (1969).

Subgenus *Pegophysema* Steward, 1930

Posterior end of ligament supported by a shelly ridge or nymph. Anterior muscle scar relatively long and narrow. Type: *schrammi* (Crosse, 1876) which is *philippiana* (Reeve, 1850). *Lissosphaira* Olsson, 1961, is a synonym.

Anodontia alba Link, 1807   5326
Buttercup Lucina   Color Plate 22

North Carolina to Florida, the Gulf States and West Indies. Bermuda (rare).

1½ to 2 inches in length, oval to circular, inflated and fairly strong. Hinge with very weak teeth, the posterior lateral being the most distinct. Exterior dull-white with weak, irregular concentric growth lines. Interior with a strong blush of yellowish orange. A common species used in the shellcraft business. This is *Lucina chrysostoma* Philippi, 1847.

Anodontia philippiana (Reeve, 1850)   5327
Chalky Buttercup   Color Plate 22

North Carolina to east Florida, Cuba and Bermuda.

2 to 4 inches in length, very similar to *A. alba,* but with a more chalky shell, never with orange color, interior usually pustulose, and the long, anterior muscle scar juts away from the pallial line at an angle of about 30 degrees instead of paralleling it as in *alba.* An uncommon species, commonly confused with *alba.* It lives down to 50 fathoms but at times is washed ashore. *A. schrammi* (Crosse, 1876), is this species.

Other species:

5328 *Anodontia (Pegophysema) edentuloides* (Verrill, 1870). Gulf of California to Tenacatita Bay, Mexico, in 33 to 165 meters. Similar to the Atlantic *philippiana* (Reeve, 1850).

Genus *Pseudomiltha* P. Fischer, 1885

Large, circular, compressed, with small, erect beaks. Surface concentrically ribbed or smooth. Teeth obsolete or absent. Short stout nymph. Interior punctate. Shell margin internally smooth. Type: *gigantea* Deshayes, 1825, Miocene of Europe.

Pseudomiltha floridana (Conrad, 1833)   5329
Florida Lucina   Color Plate 22

West coast of Florida to Texas.

1½ inches in length, almost circular, compressed, smoothish, except for a few weak, irregular growth lines. Pure-white with a dull-whitish, flaky periostracum. The beaks point for-

ward, and in front of them there is a deep, small pit. Hinge plate fairly wide and strong, but the teeth are weakly defined. Moderately common in shallow water to a few fathoms.

Other species:

5330 *Pseudomiltha tixierae* Klein, 1967. South Brazil, 40 to 55 fms. (*Ann. l'Inst. Oceanogr.*, vol. 45, pl. 1).

Subfamily DIVARICELLINAE Glibert, 1967

Genus *Divaricella* von Martens, 1880

Valves white, with deeply incised lines in parallel curves and sharply divaricated along the anterior-umbonal slope. Ligament posterior to the umbones. Type: *angulifera* von Martens, 1880, Indian Ocean.

Subgenus *Divalinga* Chavan, 1951

Lateral hinge teeth well-developed; inner margin of shell denticulate. Type: *quadrisulcata* (Orbigny, 1842).

Divaricella quadrisulcata (Orbigny, 1842)   5331
Cross-hatched Lucina

Massachusetts to south half of Florida and the West Indies to Brazil.

½ to ¾ inch in length, almost circular, moderately inflated, and glossy-white in color. Sculpture of fine, criss-cross or divaricate, impressed lines. Inner margin minutely impressed. A very common species washed ashore on sandy beaches. Occurs down to 52 fathoms. It is used extensively in the shellcraft business.

A similar species (**5332**), *D. eburnea* (Reeve, 1850), ranges from Baja California to Peru, and is found from shore to 55 meters. Synonyms are *lucasana* Dall and Ochsner, 1928, and *columbiensis* Lamy, 1934.

Divaricella dentata (Wood, 1815)   5333
Dentate Lucina   Color Plate 22

Southeast Florida and the West Indies. Bermuda.

½ to 1 inch, very similar to *quadrisulcata* but *dentata* has dentate edges, is flatter, has 4 or 5 growth stoppages showing prominently, has an anterior muscle scar that is elongate (cucumber-shaped), whereas *quadrisulcata* has a short stubby one (football-shaped). In *dentata* there is no tiny enlarged lunule on the right valve, as is seen in *quadrisulcata.* Moderately common; shallow water.

5333

Family THYASIRIDAE Dall, 1901

Trigonal, thin, white shells. Hinge with a protruding right lunular edge forming 1 or 2 small ill-defined tuberosities, with intermediate left and corresponding sockets. Muscle scars elongate.

Genus *Thyasira* Lamarck, 1818

Shell subglobular and of an earthy texture; umbones directed forward; posterior region of valve deeply furrowed; lunule absent; ligament in a groove and partly external; hinge without teeth and indented in front of the umbo; pallial line without a sinus. Type: *flexuosa* Montagu, 1803. *Axinus* Sowerby, 1821; *Cryptodon* Turton, 1822; and *Bequania* Brown, 1827, synonyms. *Tellimya* Brown is a synonym of the subgenus *Axinulus* Verrill and Bush, 1898.

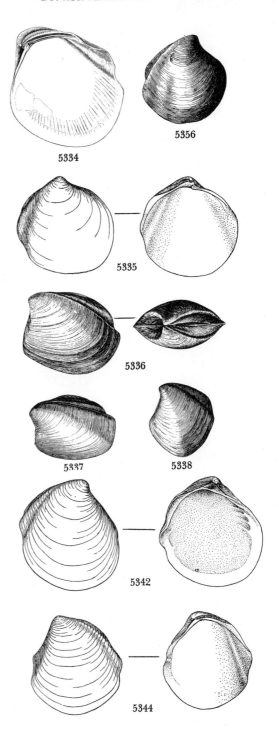

5334   **5356**

5335

5336

5337   **5338**

5342

5344

Thyasira trisinuata **Orbigny, 1842**   **5334**
Atlantic Cleft Clam

Nova Scotia to south half of Florida and the West Indies. Alaska to San Diego, California.

¼ to ½ inch in length, oblong, fragile and translucent-white. Hinge weak and with only a very long, weak posterior lateral. Posterior slope of shell with 2 strong, radial waves or rounded grooves. Moderately common in dredgings from 15 to 192 fathoms on sandy bottom. *T. polygona* Jeffreys, 1863, and *obesa* Verrill are synonyms.

Thyasira flexuosa **(Montagu, 1803)**   **5335**
Flexuose Cleft Clam

Greenland to off North Carolina. Bering Sea to off San Diego, California.

¼ inch, similar to *trisinuata*, but almost round, slightly higher, and with weak yellowish periostracum. Hinge lacks teeth. Common offshore to 60 fathoms. *T. gouldii* Philippi, 1845, and *sarsii* Philippi, 1845, are only forms of this species. *T. plana, insignis* and *inaequalis* all (Verrill and Bush, 1898) are synonyms. Also see Ockelmann, 1961, *The Nautilus*, vol. 75, p. 50.

Thyasira bisecta **Conrad, 1849**   **5336**
Pacific Cleft Clam

Alaska to Oregon. Japan.
Deep water, Caribbean Sea.

1 to 3 inches in length, almost square in side view and moderately obese. Characterized by the almost vertical, straight, anterior end which is 90 degrees to the dorsal margin. Ligament long and narrow and flush with the dorsal margin of the shell. There is a deep, prominent radial furrow on the exterior running posteriorly from the beaks. Shell chalky-white, commonly with a thin, yellowish gray periostracum. Irregular coarse growth lines present. Uncommon from 4 to 139 fathoms. We follow S. Kanno, 1971, *The Nautilus*, vol. 84, p. 96, in considering *disjuncta* Gabb, 1866, a synonym.

Other Atlantic species:

5337 *Thyasira ovoidea* Dall, 1889. Cape Fear, North Carolina, 353 fms. 25 mm.

5338 *Thyasira grandis* Verrill and Smith, 1885. Off Virginia to Yucatan, Mexico, 856 to 1,582 fms. (Synonym: *pyriformis* Dall, 1886.) 21 mm.

5339 *Thyasira granulosa* Monterosato, 1874. Gulf of Mexico and the West Indies, 60 to 116 fms.

5340 *Thyasira plicata* Verrill and Smith, 1885. Off Martha's Vineyard, Massachusetts, 1,073 to 1,122 fms.

5341 *Thyasira rotunda* Jeffreys, 1881. Off Greenland, 2 to 1,012 fms.

5342 *Thyasira croulinensis* Jeffreys, 1874. West Greenland, 199 fms. Off Bermuda, 435 fms. 3 mm.

5343 *Thyasira equalis* Verrill and Bush, 1898. Nova Scotia to Chesapeake Bay, 94 to 1,537 fms. ((5344) and var. *alta* Verrill and Bush, 1898). 6 mm.

5345 *Thyasira tortuosa* Jeffreys, 1881. Off Massachusetts to North Carolina, 500 to 1,290 fms.

5346 *Thyasira (Axinulus* Verrill and Bush, 1898) *brevis* Verrill and Bush, 1898. Georges Bank to off Cape Hatteras, North Carolina, 100 to 1,825 fms.

5347 *Thyasira (Axinulus) ferruginea* Winckworth, 1932. Arctic Seas to off North Carolina; Aleutian Islands to Alaska. (Synonym: *ferruginosa* Forbes, 1844.) *Jour. Conchol.*, vol. 19, p. 242, 251.

5348 *Thyasira cycladia* S. Wood, 1853. Baffin Bay, Canada, 1,750 fms.

5347

5351

5354

5357

5349 *Thyasira eumyaria* M. Sars, 1870. Baffin Bay, Canada, 100 fms.

5350 *Thyasira succisa* Jeffreys, 1876. North Atlantic, 92 to 1,366 fms. Off Fernandina, Florida, 294 fms.

5351 *Thyasira elliptica* Verrill and Bush, 1898. Off Martha's Vineyard, Massachusetts, 1,451 fms. 5 mm.

5352 *Thyasira* (?) *simplex* Verrill and Bush, 1898. Casco Bay, Maine, to Martha's Vineyard, Massachusetts.

5353 *Thyasira subovata* Jeffreys (?) 1881. Off Martha's Vineyard, Massachusetts, 500 fms.

5354 *Thyasira pygmaea* Verrill and Bush, 1898. Halifax, Nova Scotia, to Martha's Vineyard, Massachusetts, 206 to 499 fms. 2 mm.

Other Pacific species:

5355 *Thyasira cygnus* Dall, 1916. Cygnet Inlet, Boca de Quadra, Alaska, 160 fms.

5356 *Thyasira barbarensis* Dall, 1890. Washington to the Gulf of California, 20 to 120 fms. Locally abundant. 17 mm.

5357 *Thyasira excavata* Dall, 1901. Oregon to the Gulf of California.

5358 *Thyasira tricarinata* Dall, 1916. Off Santa Barbara Islands, California; 1,100 fms.

Genus *Axinopsida* Keen and Chavan, 1951

Suborbicular, slightly inequilateral; lunular margin concave; ventral and posterior margins rounded. Hinge with pointed projecting pseudocardinal. Muscle scars ovately elongate. Type: *orbiculata* G. O. Sars, 1878. Synonym: *Axinopsis* G. O. Sars, 1878, non Tate, 1868.

Axinopsida serricata (Carpenter, 1864)   5359
Silky Axinopsis

Aleutian Islands to Todos Santos Bay, Baja California.

Shell small, 5 mm., compressed, lenticular in shape, higher than long, yellowish to greenish, with curved beaks, and with a brownish silky periostracum. Common; 15 to 100 fathoms in mud. *A. viridis* (Dall, 1901) is probably a synonym. *A. sericata* is a misspelling.

5360 *Axinopsida orbiculata* (G. O. Sars, 1878). Greenland to Casco Bay, Maine, 10 to 30 fms. Variety *inaequalis* Verrill and Bush, 1898. Bay of Fundy and Gulf of Maine, 18 to 35 fms.

5361 *Axinopsida cordata* (Verrill and Bush, 1898). Martha's Vineyard, Massachusetts, to off Cape Hatteras, North Carolina, 43 to 202 fms.

5362 *Axinopsida viridis* (Dall, 1901). Arctic Ocean to Baja California. May be synonym of *serricata* (Carpenter, 1864).

Genus *Leptaxinus* Verrill and Bush, 1898

Small, inequilateral, rounded in front, posterior truncate; with a shallow dorsal area. Beaks almost opisthogyrate. Hinge with tuberosities. Type: *minutus* Verrill and Bush, 1898. Synonym: *Clausina* Jeffreys, 1847, non Brown, 1827.

5363 *Leptaxinus minutus* Verrill and Bush, 1898. Off Martha's Vineyard, Massachusetts, 100 fms.

5364 *Leptaxinus incrassatus* (Jeffreys, 1876). Baffin Bay and North Atlantic, 1,480 to 1,785 fms.

Family UNGULINIDAE H. and A. Adams, 1857

Hinge with 2 cardinals, medial one bifid, and with weak or absent laterals. Ligament and resilium marginal. The family was formerly called the Diplodontidae Dall, 1895.

Genus *Diplodonta* Bronn, 1831

Shell thin, white, orbicular and strongly inflated. There are 2 cardinal teeth in each valve. The left anterior and right posterior ones are split or bifid. Laterals obscure or absent. Ligament external. Adductor scars subequal in size and connected by a broad, ribbonlike pallial line. No pallial sinus. Type: *lupina* Brocchi, 1814, is a Tertiary species from Italy. *Taras* Risso, commonly used in place of the name *Diplodonta,* is a doubtful name which has been recently abandoned.

Subgenus *Diplodonta* Bronn, 1831

Diplodonta punctata (Say, 1822)   5365
Common Atlantic Diplodon

North Carolina to Florida and to Brazil. Bermuda.

1/3 to 3/4 inch in length, moderately strong, almost orbicular, well-inflated and pure-white in color. Smooth near the beaks, elsewhere very finely scratched with concentric lines and commonly with distantly spaced, coarse growth lines. Fairly common in shallow to deep water, 1 to 124 fathoms. *Mysia pellucida* Heilprin, 1889, is a synonym.

Diplodonta orbellus Gould, 1852   **5366**
Pacific Orb Diplodon

Alaska to Panama.

¾ to 1 inch in length, almost circular in outline, quite in-flated and smoothish except for moderately coarse growth lines. Beaks small, pointing slightly forward. Ligament pos-terior to beaks is long, raised and conspicuous. 2 rather large teeth in each valve below the beaks. Left anterior and right posterior teeth split. In many shallow-water localities, this clam builds a compact nest of periostracal material and detri-tus. In its more southerly range, specimens are usually more compressed, less orbicular in shape and more glossy externally (subspecies *subquadrata* Carpenter, 1856). For nest-building, see F. Haas, 1943, *Zool. Series, Field Mus. Nat. Hist.,* vol. 29, p. 10, figs. 3–6.

(5367) The Alaskan subspecies, *impolita* S. S. Berry, 1953, is subrotund, not as fat, more pointed anteriorly, with an earthy texture, heavy growth lines and with smaller, stubbier cardinal teeth. Dredged in 15 fathoms at Forrester Island. (*Trans. San Diego Soc. Nat. Hist.,* vol. 11, p. 409.)

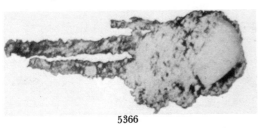

5366

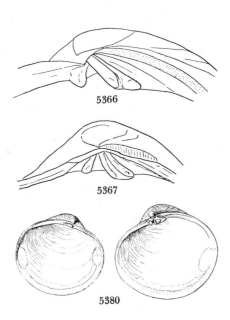

5366

5367

5380

Subgenus *Phlyctiderma* Dall, 1899

Shell subcircular, plump, rather solid, exterior granulose. Ligament deeply immersed. Anterior cardinal tooth is bifid, large and hook-shaped. Type: *semiaspera* Philippi, 1836. Considered as a genus by some workers.

Diplodonta semiaspera (Philippi, 1836)   **5368**
Pimpled Diplodon

North Carolina to Florida, Texas and the West Indies. Also Ecuador and Peru. Brazil.

Rarely over ½ inch in length, similar to *D. punctata,* but chalky-white externally and with numerous concentric rows of microscopic pimples. Moderately common in sand below low-water mark to 57 fathoms. Alias *D. granulosa* C. B. Adams; *caelata* (Reeve, 1850).

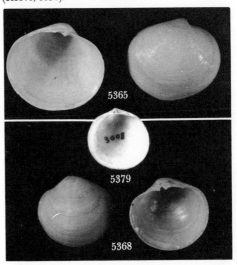

5365

5379

5368

Other species:

5369 *Diplodonta* (*Diplodonta*) *nucleiformis* Wagner, 1838. Cape Hatteras, North Carolina, to the West Indies, 15 to 52 fms. Brazil, 24 to 47 fms.

5370 *Diplodonta* (*Diplodonta*) *torrelli* Jeffreys, 1876. Southeast of Greenland, 1,450 fms.

5371 *Diplodonta* (*Diplodonta*) *aleutica* Dall, 1901. Aleutians, eastward to Sitka Bay and Drier Bay, Alaska.

5372 *Diplodonta inezensis* (Hertlein and Strong, 1947). Gulf of California to Panama.

5373 *Diplodonta* (*Felaniella* Dall, 1899) *sericata* (Reeve, 1850). Monterey, California, 15 fms., to Panama. *D. tellinoides* (Reeve, 1850) and *artemidis* Dall, 1909, are this species.

5374 *Diplodonta* (*Felaniella*) *obliqua* Philippi, 1846. Baja Cali-fornia to Ecuador.

5375 *Diplodonta* (*Felaniella*) *cornea* (Reeve, 1850). Gulf of California to northern Peru.

5376 *Diplodonta* (*Felaniella*) *candeana* (Orbigny, 1842). Florida, West Indies; Brazil, 10 to 17 fms.

5377 *Diplodonta* (*Phlyctiderma*) *soror* C. B. Adams, 1852. North Carolina to Texas; West Indies.

5378 *Diplodonta* (*Phlyctiderma*) *discrepans* (Carpenter, 1857). Gulf of California to Panama, to 18 meters. (Synonym: *semirugosa* Dall, 1899.)

5379 *Diplodonta notata* Dall and Simpson, 1901. West Florida; Puerto Rico. Shallow water.

5380 *Diplodonta* (*Timothynus* Harris and Palmer, 1946) *verrilli* Dall, 1900. (Synonym: *turgida* Verrill and Smith, 1881, not Con-rad, 1848.) Off Martha's Vineyard, Massachusetts to North Caro-lina, 15 to 69 fms. 20 mm.

5381 *Diplodonta* (*Timothynus*) *subglobosa* C. B. Adams, 1852. North Carolina to the Gulf of Mexico and West Indies, 2 to 294 fms.

5382 *Diplodonta* (*Timothynus*) *venezuelensis* Dunker, 1848. Gulf of Mexico and the Florida Strait, 19 to 80 fms.

Superfamily Cyrenoidacea Olsson, 1961

Family CYRENOIDIDAE H. and A. Adams, 1857

Genus *Cyrenoida* Joannis, 1835

A group of small, brackish-water clams, generally orbicular in shape with the small beaks nearer the anterior end. Valves rather thin-shelled, covered with a brownish periostracum. Hinge with 2 cardinals, the right valve having its anterior one bifid. No laterals. Ligament external. Large, weak lunule present. Pallial line not indented. Type: *dupontia* Joannis, 1835. Synonym: *Cyrenella* Deshayes, 1835; *Cyrenodonta* H. and A. Adams, 1857. *Cyrenoidea* Hanley, 1846, and *Cyrenoides* Sowerby, 1839, are misspellings.

Cyrenoida floridana (Dall, 1896)   5383
Florida Marsh Clam

Georgia to southern Florida.

½ to ¾ inch, rounded, thin, very delicate, whitish or translucent with a pale, silky, yellowish, dehiscent periostracum. Surface smooth, or sculptured only by fine incremental lines. Interior margin smooth, polished. Pallial line indistinct, often broken, not sinuous. Ligament short, brownish, external. Foot long, slender, filiform and with an ovate, swollen distal termination. Common; in brackish to rather fresh water.

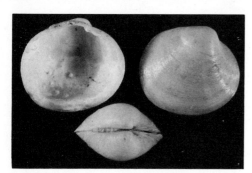

5383

Superfamily Chamacea Lamarck, 1809

Heavy, massive shells with 1 valve attached to a hard substrate. Often beautifully frondose, and resembling *Spondylus,* but the hinge has a large, single, cardinal tooth. No pallial sinus.

Family CHAMIDAE Lamarck, 1809

a. Shell equivalve, with a distinct lunule; radial rows of spines . . . . *Echinochama*
aa. Shell very inequivavle; no lunule:
    b. Umbones turning from right to left; attached by left valve . . . . *Chama*
    bb. Umbones turning from left to right; attached by right valve . . . . . . *Pseudochama*

Genus *Chama* Linné, 1758

Shell attached by its left valve, the beaks directed or coiled towards the right. Type: *lazarus* Linné, 1758, of the Mediterranean. (I.C.Z.N. Opinion 484, 1957.)

Chama macerophylla (Gmelin, 1791)   5384
Leafy Jewel Box   Color Plate 21

North Carolina to Florida; and to Brazil; Bermuda.

This is the most common and most brightly hued Atlantic species, from 2 to 3 inches in size. In quiet waters it may develop spinelike foliations to such an extent that it resembles the spiny oyster, *Spondylus.* Exterior variously colored: lemon-yellow, reddish brown, deep- to dull-purple, orange, white or a combination of these colors. Inner edges of the valves have tiny, axial ridges or crenulations. The scalelike fronds have minute radial lines. Compare with *sinuosa.*

Chama congregata Conrad, 1833   5385
Little Corrugated Jewel Box   Color Plate 21

North Carolina to Texas, to Brazil. Bermuda.

Rarely over 1 inch in size. This species closely resembles the common *macerophylla,* but in place of numerous foliations there are low axial corrugations or wavy cords. The unattached valve may have a few short, flat spines. There are fine crenulations on the inner margins of the valves. The color is usually gray with reddish specklings. In rocky areas they live in crevices and under stones. Commonly found attached to pen and ark shells.

Chama sinuosa Broderip, 1835   5386
White Smooth-edged Jewel Box   Color Plate 21

South half of Florida and the West Indies. Bermuda. Brazil.

1 to 3 inches in size. The color is always whitish, although the interior may be stained with dull-green. There are no crenulations on the inner edges of the valves. The pallial line runs directly to the anterior muscle scar and not past the end as in the other species. This is a reef species. An ecological variety of heavy shell has been named *firma* Pilsbry and McGinty 1938. *C. bermudensis* Heilprin, 1889, is also a synonym.

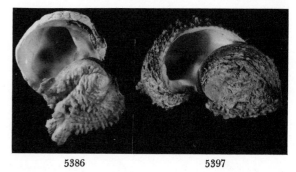

5386   5397

Chama sarda Reeve, 1847   5387
Cherry Jewel Box   Color Plate 21

Florida Keys and West Indies. East Brazil.

1 inch, with the attached valve sometimes deeply cupped. Color mostly red, inside and out. Sculpture of irregular foliations. Internal margin crenulated. Rare in Florida; common in the West Indies; attached to dead coral and shells and gorgonians.

Chama pellucida Broderip, 1835   5388
Clear Jewel Box   Color Plate 21

Oregon to Chile.

1½ to 3 inches in size, with frondlike, smoothish foliations. Color opaque to translucent-white. Interior chalk-white, the margins minutely toothed or crenulate. Commonly found attached to pilings, breakwaters and floating wood. Also dredged down to 25 fathoms.

Other species:

5389 *Chama florida* Lamarck, 1819. Southeast Florida and the West Indies. East Brazil, 5 to 44 fms.

5390 *Chama frondosa* Broderip, 1835. Gulf of California to Ecuador. Common, but perfect specimens hard to obtain.

5391 *Chama lactuca* Dall, 1886. North Carolina to the West Indies.

5391

5392 *Chama echinata* Broderip, 1835. Gulf of California to Panama. (Synonym: *coralloides* Reeve, 1846.)

5393 *Chama sordida* Broderip, 1835. Gulf of California to Colombia. (Synonym: *digueti* Rochebrune, 1895.)

5394 *Chama venosa* Reeve, 1837. Gulf of California.

Genus *Pseudochama* Odhner, 1917

These are mirror images of the chamas. According to Odhner (1919) the anatomy and prodissoconchs differ in the two genera. Type: *cristella* (Lamarck, 1819). See Odhner, 1955, *The Nautilus*, vol. 69, p. 1–6.

Pseudochama radians (Lamarck, 1819)   **5395**
Atlantic Left-handed Jewel Box   **Color Plate 21**

North Carolina to Texas and the West Indies. Bermuda. Brazil.

1 to 3 inches in size. This is the only species of *Pseudochama* in eastern America. It is not very colorful, and ranges from a dull-white to a dull purplish red. The interior is commonly stained with mahogany-brown. Crenulations are present on the inner edges of the valves. In shape, it is a mirror image of *sinuosa*. *P. ferruginea* Reeve and *variegata* Reeve are considered synonyms. Common; 1 to 42 fathoms.

Pseudochama inezae F. M. Bayer, 1943   **5396**
Inez's Jewel Box

Southeast Florida.

2 inches, alabaster-white, subcircular, thin-shelled and light-weight. Attached by right valve. Hinge teeth weakly developed. Sculpture of about 11 flared, concentric ruffles which are very thin and irregularly margined but not cut into fronds. The marginal frills are radially striate, and somewhat fluted, recurved a little downward on the upper valve. Interior margins lack any crenulations, but are finely shagreened at the edge. 10 fathoms. Uncommon.

Pseudochama exogyra (Conrad, 1837)   **5397**
Pacific Left-handed Jewel Box

Oregon to Panama.

2 to 3 inches, similar to *pellucida,* but attached by the right valve which, when viewed towards the inside, is arched counterclockwise. The opaque whitish area inside is generally not bordered by tiny crenulations. A common intertidal species.

Pseudochama granti Strong, 1934   **5398**
Grant's Chama

Central California and Catalina Island.

1 inch, with prickly spines on the underside of the attached, cup-formed valve. One end of the valve is tinted with rose inside and out. Moderately common; on gravel and shale bottoms, 20 to 75 fathoms.

Other species:

5399 *Pseudochama saavedrai* Hertlein and Strong, 1946. Gulf of California and Manganillo, Mexico.

Genus *Arcinella* Schumacher, 1817

Shell attached to the substrate only in the early stages of growth. Lunule prominent, bordered by an incised line. Sculpture of radial rows of thin spines. Type: *arcinella* Linné, 1767. *Arcinella* Oken, 1815, was rejected by I.C.Z.N. Opinion 417. Synonym: *Echinochama* P. Fischer, 1887.

Arcinella cornuta Conrad, 1866   **5400**
Florida Spiny Jewel Box   **Color Plate 21**

North Carolina to both sides of Florida to Texas.

1 to 1½ inches in length, quadrate in outline and rather obese and heavy. Lunule distinct and broadly heart-shaped. With 7 to 9 rows of moderately long, stoutish spines, between which the shell is grossly pitted. Exterior creamy-white; interior white or flushed with bright pinkish mauve. Attached to a small pebble or broken shell by the right valve. Common 3 to 40 fathoms, and commonly washed ashore. See D. Nicol, 1952, *Jour. Paleontol.*, vol. 26, p. 803.

Arcinella arcinella (Linné, 1767)   **5401**
True Spiny Jewel Box

West Indies to Brazil.

1 to 2 inches, similar to *cornuta,* but not as obese, nor as heavy, and with 16 to 35 (commonly 20) radial rows of slender spines. Uncommon; deep water to about 40 fathoms.

(5401a) The subspecies *californica* (Dall, 1903), is very similar, with slightly longer spines and with a more compressed shell. It ranges from the Gulf of California to Panama, offshore; rare.

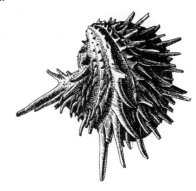

5401a

Superfamily Galeommatacea Gray, 1840

Marine; with a byssus; with 3 apertures in the mantle. Shell thin, more or less covered by the mantle. Also known as the Leptonacea Gray, 1847.

Family LASAEIDAE Gray, 1847

A group of small, fragile, inflated, translucent clams which are parasitic or commensal on other marine creatures or are active crawlers like the gastropods. Most species brood their young inside the mantle cavity. For a review of symbiotic erycinacean bivalves, see K. J. Boss, 1965, *Malacologia*, vol. 3, no. 2, pp. 183–195. The following families are synonyms: Erycinidae Deshayes, 1850; Kelliidae Forbes and Hanley, 1848; Montacutidae Clark, 1855; Mysellidae Iredale and McMichael, 1962, and Borniolidae Iredale and McMichael, 1962 (see W. F. Ponder, 1971, *Records Dominion Mus.*, New Zealand, vol. 7, no. 13).

Genus *Erycina* Lamarck, 1805

Shells 5 to 10 mm., white, ovate to oblong, thin. Resilium internal, lodged in a small, elongate resilifer behind the beak and under the shell margin. Teeth of 1 or 2 minute cardinals and 2 lateral laminae in each valve. Type: *pellucida* Lamarck, 1805. The American species attributed to this genus are questionably placed.

Other species:

5402 *Erycina linella* Dall, 1899. Off Cape Lookout, North Carolina, 31 fms. Bermuda. 4.6 mm.

5403 *Erycina emmonsi* Dall, 1899. Off North Carolina, 12 to 31 fms. 7 mm.

5404 *Erycina periscopiana* Dall, 1899. Off Cape Lookout, North Carolina, 22 fms. 5 mm.

5405 *Erycina fernandina* Dall, 1899. Off Fernandina, Florida, 294 fms. 4 mm.

None of Dall's 1916 *Erycina* seem to belong to this genus.

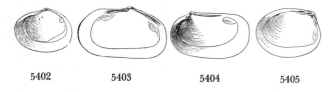

5402   5403   5404   5405

Genus *Amerycina* Chavan, 1959

Transversely elliptical, very inequilateral, with opisthogyrous beaks and with the anterior end the longest. Surface concentrically furrowed. Anterior laterals long, posterior ones short. Each valve with 1 trigonal somewhat bilobate cardinal. Well-defined resilium. Type: *colpoica* (Dall, 1913). Two species:

5406 *Amerycina colpoica* (Dall, 1913). Head of the Gulf of California to Nicaragua. *Proc. U.S. Nat. Mus.*, 1925, vol. 66, pl. 27, fig. 2.

5407 *Amerycina cultrata* Keen, 1971. Off La Paz, Baja California, in 5 to 33 meters.

Genus *Lasaea* Brown, 1827

Shell very small, beaks nearer one end. Hinge with indistinct amorphous constitution but generally with anterior and posterior laterals and a thornlike cardinal. Periostracum wrinkled (not wrinkled as in *Montacuta*). Usually colored

purplish red to yellow. A very large resilium is inserted along the under margin of the hinge-plate. Adheres by a byssus, nestling in rock cracks. Type: *adansoni* (Gmelin, 1791). Synonyms: *Cycladina* Cantraine, 1835; *Lasea* Gray, 1842; *Anapa* Gray, 1847; *Poronia* Récluz, 1843.

Lasaea adansoni (Gmelin, 1791)   5408

Southeast Florida; Bermuda; Brazil; Europe.
Monterey, California, to La Paz, Mexico.

2 to 3 mm., brittle, fat, dirty-white with tints of red. Periostracum light-yellow. Beaks prominent, in the posterior half and directed inwards and forwards. Sculpture of concentric lines and fine radial striae. Growth stages prominent. Interior white, tinted red. Right valve without cardinal tooth. Common; intertidal, where it attaches by its byssus in crevices among algae and barnacles. Do not confuse with the similar venerid clam, *Turtonia minuta*. *L. bermudensis* Bush, 1899, and *L. rubra* (Montagu, 1803), are synonyms.

The Pacific coast subspecies *subviridis* Dall, 1899 **(5409)**, is pale greenish yellow, rather flat and with low umbones. *Erycina catalinae* Dall, 1919, may be a synonym of *subviridis*.

5408

Lasaea cistula Keen, 1938   5410
Little Box Lepton

Vancouver, British Columbia, to California and to Peru.

$\frac{1}{16}$ of an inch in length (one of the smallest of our American clams), oval-oblong to quadrate, with one end slightly more rounded. Beaks slightly nearer the posterior end. Shell very obese to moderately inflated. Color light-tan with dark carmine around the dorsal margin area, and commonly blushed on the sides with light-carmine. Coarse, concentric growth lines, especially in the adults. Periostracum thin and yellowish tan. Found nestled together in great numbers attached to seaweed holdfasts and among mussels.

Subfamily KELLIINAE Forbes and Hanley, 1848

Genus **Kellia** Turton, 1822

Shell unsculptured, inflated and oval-oblong. Lateral teeth present. 2 cardinal teeth in the right valve. Pallial line wide, no sinus. Margin smooth. Type: *suborbicularis* (Montagu, 1803). Misspellings are *Kellea* and *Kellyia*. Synonyms: *Chironia* Deshayes, 1839; *Tellimya* Brown, 1827.

Kellia laperousi Deshayes, 1839   5411
La Perouse's Lepton

Alaska to Peru.

$\frac{3}{4}$ to 1 inch in length, oval-oblong, rather obese and with small beaks near the center. Shell fairly strong, chalk-white, but commonly covered with a smooth, glossy, greenish to yellowish-brown periostracum which, however, is commonly worn away in the beak area. Very common. Found attached to wharf pilings among mussels and *Chama* shells. Also down to 35 fathoms. Some workers consider this to be the same as, or a subspecies of, *suborbicularis*.

Kellia suborbicularis (Montagu, 1803)   **5412**
North Atlantic Lepton

Arctic Seas to New York.
British Columbia to Peru.
Iceland to the Mediterranean.

5 to 9 mm., brittle, fat; translucent white; beaks in front of the midline, directed inwards and forwards. Periostracum light-yellow. Internal ligament behind the cardinal teeth; a small external ligament may be visible just behind the beaks. Sculpture of fine concentric lines. Growth stage stoppages prominent. Right valve with 1 cardinal tooth and 1 posterior lateral; left valve with 2 cardinal teeth and 1 posterior lateral. Common; shallow water. *K. gouldii* (Thomson, 1867); *fabagella* Conrad, 1831; and *inflata* Philippi, 1836, are synonyms.

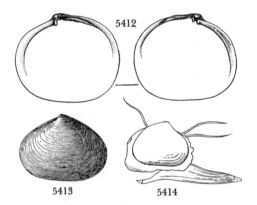

Genus *Bornia* Philippi, 1836

Shells 5 to 10 mm., translucent-tan to white, with high umbones, subtrigonal to elliptical. Surface polished, sometimes faintly ribbed along the edge. Ligament partly external, partly internal. In the left valve, the posterior lateral tooth is elongate, the anterior lateral is a small short tooth, and adjacent and behind it is a small pseudocardinal tooth. In the right valve, the posterior lateral socket is large and conspicuous, its lower rim enlarged and toothlike. Type: *sebetia* Costa, 1829.

Bornia retifera Dall, 1899   **5413**
Netted Bornia

Monterey to Santa Barbara Islands, California.

12 mm. long, 9 mm. high, 4 mm. diameter of both valves; thin-shelled, moderately obese, white; beaks distinct, not very high and set near the middle. Surface polished, with faint growth lines and minute close punctuations whose interspaces give the effect of fine netting. Uncommon; 10 to 20 fathoms.

Subgenus *Ceratobornia* Dall, 1899

Bornia longipes (Stimpson, 1855)   **5414**
North and South Carolina.

Shell 8 mm., moderately inflated, translucent-tan to whitish, beaks near the middle; valves rounded at both ends. Mantle extends beyond the borders of the shell. There are 2 long anterior-dorsal cirrhi and 1 posterior dorsal cirrhus. This is the type of Dall's *Ceratobornia*. *Lepton longipes* Kurtz, 1860, is probably a synonym.

5411   5415

Genus *Parabornia* Boss, 1965

Type of the genus is *squillina* Boss, 1965. See characters below.

Parabornia squillina Boss, 1965   **5415**
Atlantic Squilla Clam

West Florida and Mississippi to Panama.

3 to 6 mm. in length, the young oblong, adults oval-oblong; compressed and thin. Small, tan, prosogyrous umbones are slightly anterior to the middle of the thin, transparent, fragile, smooth shell. Valves slightly gaping all around, except at hinge between the umbones. Concentric lines of growth evident only in transmitted light. Internal resilium in hinge between 2 small teeth in each valve. Periostracum thin, varnish-like and extends beyond edge of shell. Foot long and with large byssal pore. Thousands of eggs are incubated between the gill curtains in the females. Inner lobe of mantle edge bears tiny papillae. As many as 15 specimens may be attached to the ventral surface of one specimen of a foot-long mantis shrimp, *Lysiosquilla scabricauda*. (illus. from K. J. Boss, 1965, *Amer. Mus. Novitates,* no. 2215).

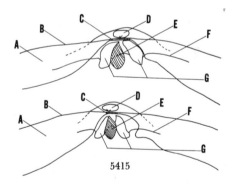

5415

Symbols: A, reflected portion of the periostracum; B, edge of the shell; C, obsolete external ligament; D, umbo; E, internal resilium; F, lateral buttress of the hinge plate; G, denticles.

Genus *Aligena* H. C. Lea, 1843

Shells 10 mm. or less, white, inflated, thin, rounded-ovate to oblong trigonal, with the anterior end longer. Umbones full, prodissoconchs small and pointing slightly forward (prosogyrate). Dorsal margins do not project across the midline. Sculpture of fine growth lines. Valves sometimes constricted in the middle. Hinge weak, with a single small anterior tooth under the beak, and behind it a wide, open notch. Ligament internal, resilifer elongated and placed in the margin of the hinge notch. Type: *aequata* (Conrad,

1843). Shallow water to 40 fathoms. Free living and commensal with polychaete worms, crustacea and echinoderms. They are hermaphrodites. Can crawl like a gastropod and produces an attachment byssus. *Kelliopsis* Verrill and Bush, 1898, is a synonym. We have taken our information from "A review of the living leptonacean bivalves of the genus *Aligena*," by Harold W. Harry, 1969, *The Veliger*, vol. 11, p. 164, and wish to thank the author for the use of his drawings.

Aligena elevata (Stimpson, 1851)   5416
Eastern Aligena

Massachusetts to North Carolina.

6 mm. long, 4.8 mm. high, 3.4 mm. wide. Umbones moderately prominent, slightly closer to the posterior end and curled forward. Growth lines give the surface a silky texture. Periostracum thin, light-tan. Prodissoconch oval, 0.4 mm. long. Hinge line narrow, with a prominent anterior cardinal tooth. Common; shore to 10 fathoms. Associated with annelid worm tubes. Do not confuse with *Neaeromya floridana* (Dall). Illustration from H. W. Harry, *The Veliger*, vol. 11, no. 3.

Aligena texasiana Harry, 1969   5417
Texas Aligena

Galveston, Texas, to Louisiana.

5 mm. long, 4 mm. high, 2.6 mm. wide. Chalky white with a thin, satiny, light-tan periostracum. Elongate-ovate, with the center ventral part of the valve pushed in. Prodissoconch oval, 0.4 mm. in length. The single tooth in the right valve is larger than, and fits in front of the left cardinal tooth. Uncommon; low tide to 7 feet in back bays, probably commensal with polychaete worms. Do not confuse with the associated *Mysella planulata* whose beaks point backward, have a much smaller prodissoconch and lack the weak sulcus on the disc.

Genus *Tomburchus* H. W. Harry, 1969

Type of the genus is *redondoensis* T. Burch, 1941. *The Veliger*, vol. 11, no. 3, p. 178.

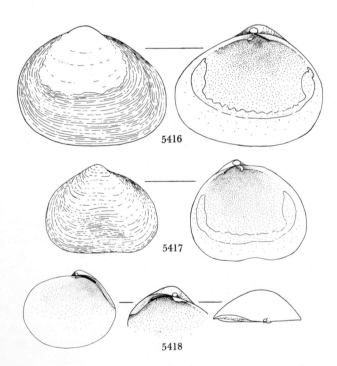

5416

5417

5418

Tomburchus redondoensis (T. Burch, 1941)   5418
Redondo Aligena

La Jolla to Santa Rosa Island, California.

2 to 3 mm., white fragile, rounded, inflated, with the beaks somewhat posterior and twisted slightly forward. Hinge with a long, narrow chondrophore; left valve without true teeth but with a laminar plate extending forward from below the umbo to the dorsal margin of the shell, leaving a depressed area below the umbo, into which fits a single large tooth from the right valve. Uncommon; in mud mixed with fine gray gravel; 48 to 129 fathoms. Described in *The Nautilus*, vol. 55, p. 50, and from which the following key is taken:

Key to Pacific Coast *Aligena* and *Tomburchus*

Shell with a median radial constriction. 7.5 mm. . . . .
. . . . *A. cokeri* Dall, 1909
Shell without a median constriction
Anterior part of shell sloping abruptly down. 8 mm.
. . . . *A. cerritensis* Arnold, 1903
Anterior part of shell not sloping abruptly down, but gently so.
Left valve with a tooth. 4 mm. *A. nucea* Dall, 1913
Left valve with a tooth. 3 mm. . . . . . .
. . . . . . *I. redondoensis* T. Burch, 1941

Other Pacific species:

5419 *Aligena cerritensis* Arnold, 1903. La Jolla, California to the Gulf of California, 1 to 5 fms. Also Pleistocene of Los Angeles County, California. 5 to 9 mm. in length.

5420 *Aligena obliqua* Harry, 1969. Guaymas to Mazatlan, Mexico, 1 to 2 fms. 6 mm.

5421 *Aligena nucea* Dall, 1913. Gulf of California to Nicaragua, 1 to 7 fms. 4 mm.

5422 *Aligena cokeri* Dall, 1909. Gulf of California to Peru, shore to 4 fms. 7.5 mm.

Subgenus *Odontogena* Cowan, 1964

Oblique right cardinal; stronger posterior laterals. Type: *borealis* Cowan, 1964.

Aligena borealis Cowan, 1964   5423
Boreal Aligena

British Columbia.

2.5 mm. in length and height, circular, compressed, beaks fairly large and centrally located. Pale yellowish. Large posterior lateral tooth in each valve. Rare; 190 fathoms. *The Veliger*, vol. 7, p. 108.

Family LEPTONIDAE Gray, 1847

Shells usually compressed; hinge plate scarcely indented by the resilium, under the very small beaks. Posterior laterals only slightly longer than the anteriors. Pallial line irregular and distant from the ventral margin.

Genus *Lepton* Turton, 1822

Shells small, polished, somewhat compressed, subquadrate; surface sometimes minutely punctate. Beaks are

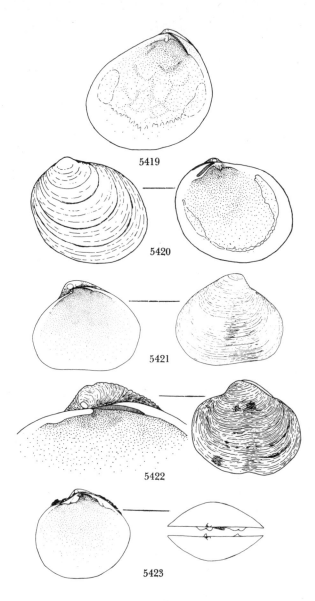

5419

5420

5421

5422

5423

nearly central and erect. Mantle reflected partially over the valves. Resilium internal, short, stout and nearly central. Anterior cardinal hook in the left valve, none in the right. Type: *squamosum* (Montagu, 1803). American species are:

5424 *Lepton lepidum* Say, 1826. South Carolina. Texas. Uncommon?

5425 *Lepton? meroeum* Carpenter, 1864. San Pedro to San Diego, California.

Genus *Solecardia* Conrad, 1849

Shell 4 to 20 mm., partly covered by the fleshy mantle. Valves subovate, moderately convex, white. Surface with a granular calcareous coating, becoming punctate when worn. Hinge plate narrow, bearing 2 diverging lamellar teeth on each side of a large, socketlike pit occupied in part by the ligament. The end margins of the hinge plate are grooved. Ligament divided, shorter part in front, longer section in back of the beaks. Resilium internal. Pallial line close to the ventral margin. Type: *eburnia* Conrad, 1849. Synonym: *Scintilla* Deshayes, 1855. Not known from the Atlantic.

5426 *Solecardia eburnia* Conrad, 1849. Baja California to Panama. (Synonym: *cumingii* Deshayes, 1855.)

Subfamily MONTACUTINAE Clark, 1855

Shells less than 10 mm., rather fragile, ovate to oblong, the anterior end the longer. Ligament mostly internal, subumbonal or lodged in an elongated, oblique resilifer in the hinge margin behind the beaks. Surface smooth or with concentric growth lines and sometimes radial undulations. Most species appear to be commensal on sand-burrowing echinoderms, such as *Echinocardium* and *Spatangus*. What Dall (1921) and some others called Pacific coast *Pseudopythina* and *Sportella* are now put in the genus *Orobitella*.

Genus *Montacuta* Turton, 1822

Mostly European clams living commensally on echinoderms, but one deep-water species is recorded from Baja California, several from the Atlantic. Shell thin, obese, transversely ovate, no pallial sinus. Anterior part of the hinge in the right valve, with a narrow lamina having minute cardinal hook at the end; left valve with a similar, reduced lamina. Resilium strong, internal, posterior, seated on nymphs of which the right one is usually the smaller.

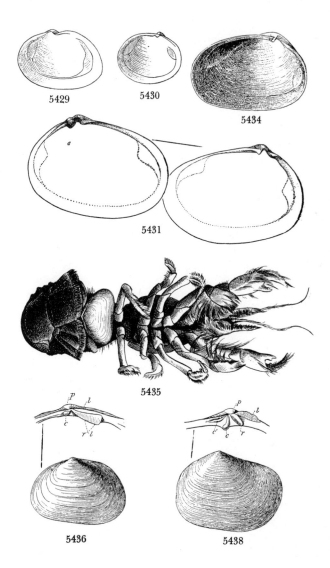

5429

5430

5434

5431

5435

5436

5438

Periostracum smooth, whereas in the similar *Lasaea* the periostracum is wrinkled. Type: *substriata* (Montagu, 1808). Synonyms: *Sphenalia* S. Wood, 1874; *Coriareus* Hedley, 1907; *Montaguia* Bronn, 1848. American species are:

5427 *Montacuta balliana* (Dall, 1916). Off South Coronado Island, Baja California.

5428 *Montacuta dawsoni* Jeffreys, 1863. Greenland to Newfoundland.

5429 *Montacuta minuscula* Dall, 1899. Cape Hatteras, North Carolina, 124 fms. 2.7 mm.

5430 *Montacuta limpida* Dall, 1899. Off both coasts of Florida, 85 to 294 fms. 3 mm.

5431 *Montacuta percompressa* Dall, 1899. Vineyard Sound, Massachusetts. 4 mm. Clings to *Leptosynapta inhaerens*.

Genus *Axinodon* Verrill and Bush, 1898

Ovately transverse, rounded, orthogyrous. Left valve hinge with 2 rounded teeth; right valve with 1 ill-defined tooth. Chondrophore oblique. Type: *symmetros* (Jeffreys, 1876). Synonym: *Kelliola* Dall, 1899. Deep-water species:

5432 *Axinodon symmetros* (Jeffreys, 1876). Baffin Bay and North Atlantic. 488 to 1,750 fms. (Synonym: *ellipticus* Verrill and Bush, 1898.)

Genus *Entovalva* Voeltzkow, 1890

Shell completely internal. Teeth distinct. Inequilateral, transversely subquadrangular. Type: *mirabilis* Voeltzkow, 1890. Mantle encloses the valves.

Subgenus *Devonia* Winckworth, 1930

Type of this subgenus is *perrieri* (Malard, 1903). The mantle is not or only partially extended over the valves. *Synapticola* Malard, 1903, non Voigt, is a synonym.

Entovalva perrieri (Malard, 1903)   5433

Cape Cod, Massachusetts; Europe.

3 to 5 mm., lives attached by a sucker to the exterior of the holothurian *Leptosynapta inhaerens*. Fragile, translucent-white to tinted-brown. Quadrate in outline. Beaks almost at the posterior end, directed inwards, touching. Sculpture of concentric lines and old growth stoppages. Hinge line without teeth. Internal ligament below and behind the beaks. Uncommon; offshore; attached to the holothurian; 2 to 3 feet of water. See Clench and Aguayo, 1931, *Occ. Papers Boston Soc. Nat. Hist.*, vol. 8, p. 5.

Genus *Neaeromya* Gabb, 1873

Transversely subquadrangular to trigonal, finely striated. With a strong right anterior laminar tooth and with an enlarged, oblique posterior margin; oblique resilium between them. Type: *quadrata* Gabb, 1873, from the Miocene.

Subgenus *Orobitella* Dall, 1900

Shells 8 to 13 mm., subovate, the posterior end short and rounded, the anterior end longer. Hinge plate relatively narrow but stout, with a single, pluglike tooth directly under the beak in each valve, and behind it, an elongated,

excavated or grooved, subumbonal resilifer which extends across the hinge-plate obliquely. Outer surface smooth, sometimes with concentric growth lines. Periostracum thin, deciduous when dry and may be weakly rayed. Differs from *Montacuta* in having the laminae obsolete, but the cardinal hooks persistent, and the sockets of the resilium elevated. *Orbitella* is a misspelling. Type: *floridana* (Dall, 1899).

Neaeromya floridana (Dall, 1899)   5434
Giant Montacuta

Both sides of Florida.

16 mm. long, 10 mm. high, 9.5 mm. both valves' diameter, white, inflated, subovate, the posterior end shorter. Beaks low, polished. Sculpture of concentric lines growing gradually stronger downward and forward until on the lower anterior third they form low, stout, evenly distributed, concentrically striated lamellae. Base nearly straight, dorsal margin arcuated. Hinge with a prominent slender cardinal in each valve, the laminae obsolete. Sockets of the resilium thickened and raised above the inner surface of the valve. Uncommon; shallow water; lives in tubes of the annelid worm, *Onuphis magnus*, deeply set in sand. Also in the Pleistocene of Florida. The small, shelled males live inside the females.

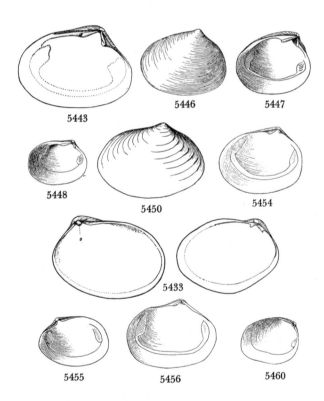

5443   5446   5447

5448   5450   5454

5433

5455   5456   5460

Neaeromya rugifera (Carpenter, 1864)   5435
Wrinkled Lepton

Alaska to Baja California.

½ to ¾ inch in length, oval-oblong, moderately obese, fairly fragile, beaks close together and located about the middle of the shell. Shell white, but in live specimens covered with a thin, light-brown, semiglossy periostracum which is feebly and concentrically wrinkled. The ventral edge of the valves is slightly indented in the middle in some specimens. May be found attached to crustaceans and the sea worm, *Aphrodita*, or be free. For anatomy and habits, see Walter Narchi, 1969, *The Veliger*, vol. 12, pp. 43–52.

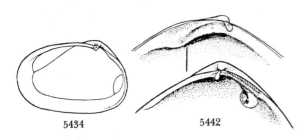

5434   5442

Neaeromya compressa (Dall, 1899)   5436
Compressed Montacuta

Alaska to Acapulco, Mexico.

18 mm. long, 13 mm. high, both valves diameter 6 mm. White, thin, compressed, subquadrate, wrinkled weakly; periostracum partly glossy and thin. Basal margin nearly straight. Beaks inconspicuous; surface with strong, irregular growth lines. The right dorsal valve margins overlap those of the left valve a little, but there are no distinct lamellae. Uncommon; on crustaceans; 4 to 28 fathoms.

Neaeromya myaciformis (Dall, 1916)   5437
Mya-shaped Montacuta

Puget Sound, Washington.

8.5 mm. long, 5.5 mm. high, diameter (both valves) 3 mm. Shaped like a *Mya* with a blackish environmentally produced coating. Anterior end considerably more attenuated than the shorter, rounded posterior end. Surface finely concentrically threaded. Shell underneath is yellowish white. Hinge weak, the ligament carrying a long, very narrow lithodesma. Uncommon; commensal with crustaceans.

Other species:

5438 *Neaeromya stearnsii* (Dall, 1899). Baja California, the Gulf, and to Nicaragua; Galapagos Islands. 13 mm.

5439 *Neaeromya bakeri* (Dall, 1916). Off south Coronado Islands, Baja California, 155 fms.

5440 *Neaeromya chacei* (Dall, 1916). Santa Rosa Island, California, to the Coronado Islands, Baja California, and the Gulf of California.

5441 *Neaeromya trigonalis* (Carpenter, 1857). Southern California to Mazatlan, Mexico.

5442 *Neaeromya* (*Isorobitella* Keen, 1962) *singularis* Keen, 1962. Baja California mud flats. 9.5 mm. *Pacific Naturalist,* vol. 3, no. 9, p. 323.

Genus *Mysella* Angas, 1877

Shells 6 to 10 mm., subquadrate, anterior end longer. Hinge plate stout and bears a short, central subumbonal resilifer; in the right valve, the resilifer is bordered by 2 small teeth, 1 on each side, the anterior tooth being larger. Above each of these there is a grooved socket in which the margin of the left valve is inserted. These clams live free or in the burrows of crustacea and polychaete worms. Type: *anomala* Angas, 1877. *Rochefortia* Vélain, 1877, is a synonym.

Mysella planulata (Stimpson, 1857)   5443
Atlantic Flat Lepton

Nova Scotia to Texas and the West Indies.

⅛ inch in length, oval-oblong in side view, well-compressed, and fairly fragile. Beaks small, ¾ the distance back from the anterior end. Dorsal margin of valves pushed in, both in front and back of the beaks. There is no thickening of the hinge line directly below the beaks. Color white, with a thin, nutbrown, smoothish periostracum. Moderately common attached to buoys, eel-grass and wharf pilings. Shore to 48 fathoms. *M. tenuis* and *fragilis* Verrill and Bush, 1898, are synonyms.

Mysella golischi (Dall, 1916)   5444
Golisch's Lepton

Southern third of California.

6 mm. in length, oval-oblong in side view, moderately compressed and rather fragile. Beaks small, ¾ the distance back from the anterior end. The dorsal margin of the valve is pushed in slightly just anterior to the beak. Shell white, semitransparent, with its glossy exterior having irregular, concentric wrinkles. In live specimens, there is a thin, yellowish brown periostracum. These clams are found attached to the gills or legs of the large sand crab, *Blepharopoda occidentalis*. Common. *M. pedroana* Dall, 1899, known from a single specimen (5445) is much more oblique in shape, but may be this species.

Mysella compressa (Dall, 1913)   5446
Compressed Lepton

Alaska to Peru.

6 mm. long, 4.6 mm. high, white, compressed somewhat, rounded quadrate. Anterior end longer and widely rounded. Hinge has a wide V-shaped notch in the middle with the beak at its tip, and directly under it in the cavity lies the scar of the resilium. In the right valve, the median notch is bordered by lamellar teeth on each side, and above each of these a grooved socket for the reception of the thickened, bevelled edge of the opposite valve. Surface with coarse lines of growth. Adductor muscles subequal. Moderately common; shallow water.

Mysella tumida (Carpenter, 1864)   5447
Fat Pacific Lepton

Alaska to Baja California.

⅛ to 3/16 of an inch in length, moderately compressed, somewhat triangular shape. The tiny beaks are almost at the very posterior end. Shell dull-white, but commonly covered with a light-brown, smoothish periostracum which is faintly marked with concentric, microscopic wrinkles. The hinge teeth are large in comparison to those in other species. Common from low water to 99 fathoms. Has been found in duck stomachs.

Mysella aleutica (Dall, 1899)   5448
Aleutian Lepton

Bering Sea to Baja California.

4.3 mm. long, 3.3 high, and both valves diameter 2 mm. White, solid, ovate, smooth, covered with a polished strawcolored periostracum with usually 3 or 4 concentric darkercolored zones. Umbones distinct, often eroded. Valves moderately convex, the ends and base rounded. Teeth strong in the right valve; anterior adductor scar narrow and irregular. 10 to 75 fathoms. Uncommon.

Other Atlantic species:

5449 *Mysella striatula* (Verrill and Bush, 1898). Off North Carolina, 5 to 50 fms.

5450 *Mysella casta* (Verrill and Bush, 1898). Off North Carolina, 14 to 17 fms.

5451 *Mysella ovata* (Jeffreys, 1881). South of Martha's Vineyard, Massachusetts, 100 to 157 fms.

5452 *Mysella triquetra* (Verrill and Bush, 1898). Off Cape Hatteras, North Carolina, in 43 fms., and off Fernandina, Florida, 294 fms.

5453 *Mysella tumidula verrilli* (Dall, 1899). Off Delaware Bay and Cape Hatteras, North Carolina. 843 to 1,091 fms.

5454 *Mysella moelleri* (Mörch, 1875). Greenland to off Halifax, Nova Scotia. (Synonym: *elevata* Mörch, 1875, not Stimpson, 1851.) 6 mm.

5455 *Mysella barbadensis* Dall, 1899. Barbados. 4 mm.

Other Pacific coast species:

5456 *Mysella planata* (Dall, 1885). Icy Cape, Arctic Ocean, south to the Shumagin Islands, Alaska.

5457 *Mysella ferruginosa* (Dall, 1916). San Francisco Bay to Santa Rosa Island, California.

5458 *Mysella beringensis* (Dall, 1916). Bering Island, Bering Sea.

5459 *Mysella grebnitzskii* (Dall, 1916). Bering Island, Bering Sea.

5460 *Mysella pedroana* (Dall, 1899). California. 9 mm.

Genus *Pythinella* Dall, 1899

Hinge of *Mysella*, shell transversely trigonal with prominent beaks and umbones, the anterior end much longer and produced. Ventral margin straight to slightly indented. Resilium restricted to the upper part of a wide arch under the beaks .Type: *cuneata* (Verrill and Bush, 1898).

Pythinella cuneata (Verrill and Bush, 1898)   **5461**
Cuneate Lepton .

South Massachusetts to Florida.

1 to 3 mm., kidney-shaped, fragile, inequilateral. Right valve with 2 prominent, rounded subtriangular cardinal teeth; anterior tooth slightly larger than the posterior. Curved ossicle situated in a triangular gap between the cardinals. 2 elongate ridges continue from the bases of the cardinals extending anteriorly and posteriorly respectively. Left valve without cardinals. Adults with tan periostracum. Sculpture of faint concentric and radial lines. Associated with sipunculid worms living in dead gastropod shells. Common. See G. R. Hampson, 1964, *The Nautilus*, vol. 77, p. 125.

Other species:

5462 *Pythinella sublaevis* (Carpenter, 1857). Mexico to Panama. Hardly separable from *cuneata*.

Genus *Thecodonta* A. Adams, 1864

Obliquely elongate, with projecting beaks. Type: *sieboldi* A. Adams, 1864.

Subgenus *Pristes* Carpenter, 1864

Shell 3 to 5 mm., quadrate, with the umbones at the anterior end. Ligament external. 2 cardinal teeth in each valve are finely serrated. Type: *oblonga* Carpenter, 1864. Synonyms: *Serridens* Dall, 1899; *Pristiphora* Carpenter, 1866, non Blanchard, 1835.

Thecodonta oblongus Carpenter, 1864   **5463**
Chiton-loving Clam

Monterey, California, to Baja California.

3 to 5 mm., quadrate, resembling a *Nucula* in shape, with the umbones at the anterior truncated end. Dorsal margin arching, below it one of the long cardinal teeth is serrated. Posterior end somewhat pointed. Lunule small and concave. Found sometimes in large numbers attached to the gills and bottom of the foot of large chitons, *Ischnochiton conspicuus* and *magdalensis*.

Family GALEOMMATIDAE Gray, 1840

Shells small, oblong-ovate, usually with an open gap along the ventral margins. Hinge without teeth or with small, weak cardinal teeth, and rarely with laterals. Ligament mostly internal, the resilium set in a small pit. Mantle extends over the external surface of the valves. Can crawl about in the manner of snails.

Genus *Aclistothyra* McGinty, 1955

Valves very thin, nearly flat, broadly gaping. Sculpture of minute granulations and pittings. Minute submedian umbones are convex and rounded. Hinge margin thick at the center, but without teeth. External ligament very thin. Type: *atlantica* McGinty, 1955.

Aclistothyra atlantica McGinty, 1955   **5464**

Off Palm Beach, Florida.

10 mm., thin, translucent and delicate; valves nearly flat. Surface microscopically pitted. Hinge straight, ventral margin evenly semicircular. Rare; under rocks, 40 fathoms. Animal white, the valves kept open at about 130 degrees.

5461   5464

"Genus" *Planktomya* Simroth, 1896

Shell minute, 0.3 to 1.5 mm. in length, smooth, equivalve, oval, translucent golden-brown. Prominent umbones directed posteriorly. Shell material chitinous, with a thin, deciduous, calcareous prodissoconch. Velum large, bilobed and half-moon-shaped. Type and only known species: *P. henseni* Simroth, 1896 (5464a).

J. A. Allen and R. S. Scheltema, 1972, *Jour. Marine Biol. Assoc. U.K.*, vol. 52, pp. 19–31, have shown that this common, holopelagic bivalve of the North Atlantic is undoubtedly a larval form of some common galeommatid bivalve living on the continental shelf. When the life history is known, the genus and species will probably fall into synonymy.

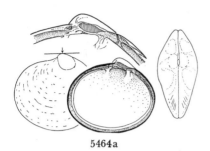

5464a

Superfamily Chlamydoconchacea Dall, 1889

Family CHLAMYDOCONCHIDAE Dall, 1889

Genus *Chlamydoconcha* Dall, 1884

Tiny oblong rudimentary, elongate shells, with minute, bubblelike prodissoconch, are entirely buried inside a sluglike, soft, brown clam. No periostracum, no pallial or muscular scars. Type: *orcutti* Dall, 1884. One Pacific coast species.

Chlamydoconcha orcutti Dall, 1884   5465
Orcutt's Naked Clam

Monterey to San Diego, California.

Animal sluglike, 10 to 13 mm., translucent-amber, the anterior end flaring, the foot, when extended, being long and pointed. Not uncommon in intertidal areas. Shell internal, rudimentary, fragile, elongate, with a tiny, bubblelike prodissoconch.

5465

Superfamily Cyamiacea Philippi, 1845

Family TURTONIIDAE Clark, 1855

Very small inequilateral clams, with prosogyrous beaks, narrow hinge plate bearing tubercular cardinals and both anterior and posterior laterals. Ligament external. 4 mantle folds and no outer demibranch gills. Placed in the Veneracea by some workers.

Genus *Turtonia* Alder, 1848

Shells about 2 to 4 mm., brown, ovate, smooth, with beaks near the anterior end. With the elongated, external resilium and ligament combined. Right valve with 2 stout and 1 weak cardinals. Left valve with 1 stout and 1 slender arched laminar cardinal tooth and an obscure lateral lamina which fits into a sulcus in the opposite valve. Pallial line distinct. Sticky eggmasses are laid on stones. Fertilization is internal. No pelagic development. Type: *minuta* (Fabricius, 1780). Formerly placed among the erycinid clams.

Turtonia minuta (Fabricius, 1780)   5466
Minute Turton Clam

Arctic Seas to Massachusetts.
Alaska to Baja California; Europe.

2 to 3 mm., plump, brown with tints of purple or rose about the umbones. Periostracum thin and glossy. Ligament a brown arched band. Lunule narrow, defined by a fine groove. Sculpture of fine growth lines. Interior of valves light-brown. Attaches to rocks, plants and animals by means of a byssus. Abundant; intertidal zone. See Kurt Ockelman, 1964, for anatomy and habits (*Ophelia*, vol. 1, pp. 121–146). Formerly placed in its own family Turtoniidae among the erycinid clams. Synonyms are: *purpurea* Montagu, 1808, and *nitida* Verrill, 1872. Do not confuse with *Lasaea adansoni* (see before).

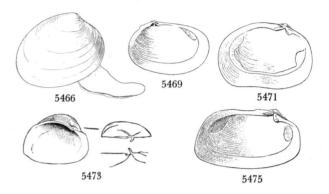

5466   5469   5471

5473   5475

Turtonia occidentalis Dall, 1871   5467
Western Turton Clam

Plover Bay, Siberia, to Rodman Bay, Alaska.

5 mm. long, subtrigonal, slightly inequilateral, purplish with a yellowish, glossy, periostracum becoming lighter towards the margin. Interior dark-purple in the middle of the valves, margins lighter, dark-brown above and behind. Dorsal and ventral margins roundly arcuate. Umbones rather prominent, usually eroded. Uncommon; intertidal.

Family SPORTELLIDAE Dall, 1899

Shells 5 to 10 mm., elongate, thin, whitish. Ligament mostly external. With a small internal resilium. Surface chalky, smooth or pustulated. Umbones prominent. The family Basterotiidae Woodring, 1925, is the same.

Genus *Ensitellops* Olsson and Harbison, 1953

Shells 5 to 10 mm., very elongate, quadrate, with the beaks near the anterior end. Valves chalky-white and with a scattering of small, sharp pustules. Left valve with 2 small, divergent anterior cardinal teeth, the right valve with only 1. The left valve has a long, slender, posterior tooth partly united with the nymph. This tooth fits into a grooved socket in the dorsal margin of the right valve. Type: *protexta* (Conrad, 1841).

Ensitellops protexta (Conrad, 1841)   5468
Pustulose Turton Clam

Off North Carolina.

10 mm. in length, elongate, quadrate, both ends gently rounded; beaks ¼ the way back from the front end. Prodisso-

conch plainly visible on the beaks. Surface pustulose, especially in adults. Rare; 22 fathoms off Cape Fear. Also in the Pleistocene to the Miocene.

Other species:

5469 *Ensitellops pilsbryi* (Dall, 1899). Off Cape Hatteras, North Carolina, 49 fms. 8 mm.

5470 *Ensitellops hertleini* Emerson and Puffer, 1957. Guaymas, Mexico. 2 fms. *Amer. Mus. Novitates*, no. 1825, p. 21.

5471 *Ensitellops californica* (Dall, 1899). Monterey, California. 6 mm.

Genus *Anisodonta* Deshayes, 1858

Subquadrangular, narrowly transverse, anterior end short, attenuated. Posterior end long and broad. Hinge with irregular anterior laterals, partly fused to the anterior cardinals. Posterior laterals long. With a very narrow resilium and a short, flat, broad nymph. The type is *complanata* Deshayes, 1858. The type of the subgenus *Fulcrella* Cossmann, 1886, is *Poromya paradoxa* Deshayes, 1857.

5472 *Anisodonta? pellucida* Dall, 1916. Monterey Bay, California.

Genus *Basterotia* Hoernes, 1859

Shell inflated, quadrate, carinate posteriorly, more or less gaping behind and ventrally. With a chalky, granular surface. Beaks curling forward. Ligament external, attached to a short, stout, platelike nymph. Hinge with a large, hook-shaped cardinal tooth in each valve, bordered behind by a deep socket. Type: *corbuloides* Hoernes, 1859. Formerly placed in the Corbulidae. May be only a subgenus of *Anisodonta*.

Basterotia quadrata (Hinds, 1843)   5473
Square Basterotia

North Carolina to Texas; West Indies. Bermuda. Brazil.

6 to 14 mm., solid, inflated, strongly carinate from the beaks to the posterior 1/3 of the shell. Color milk-white; outer surface with small granules over the anterior 2/3 of the shell. Uncommon; 5 to 640 fathoms. *Poromya granatina* Dall, 1881, is a synonym.

Subgenus *Basterotella* Olsson and Harbison, 1953

Like *Basterotia* but less carinate and without surface granulations. Type: *Pleurodesma floridana* Dall, 1903. Vokes (1967) puts this subgenus in the family Montacutidae.

Basterotia elliptica (Récluz, 1850)   5474
Elliptical Basterotia

North Carolina to the Gulf of Mexico; West Indies. Bermuda.

5 to 9 mm., resembling *quadrata*, but without the carina, more elongate, less inflated, and without granulations. Uncommon; 1 to several fathoms. *Corbula newtoniana* C. B. Adams, 1852, is a synonym.

Other species:

5475 *Basterotia* (*Basterotella*) *corbuloidea* Dall, 1899. Off Cape Fear, North Carolina, 18 to 22 fms.

5476 *Basterotia* (*Basterotella*) *hertleini* Durham, 1950. Baja California to Ecuador. (Synonyms: *californica* Durham, 1950; *ecuadoriana* Olsson, 1961.)

5477 *Basterotia* (*Basterotia*) *peninsularis* (Jordan, 1936). West Mexico and the Galapagos Islands.

Superfamily Carditacea Fleming, 1820

Shells heavy, with radial sculpture and crenulate margins. Lunule small. Hinge with 2 unequal teeth in the left valve. Byssus usually present.

Family CARDITIDAE Fleming, 1820

The genus *Calyptogena* Dall, 1891, formerly placed in this family, is now in the Vesicomyacidae (Glossacea).

Subfamily CARDITAMERINAE Chavan, 1969

Shell elongate-trapezoidal, with strong radial ribs. Cardinal tooth 3a present, 3b is V-shaped. Laterals well-developed.

Genus *Carditamera* Conrad, 1838

Transversely subrectangular, solid, somewhat compressed. Sculpture of strong ribs, posterior ones unequal. Lunule oblique. Cardinal teeth strong, laterals well-defined. Type: *arata* Conrad, 1832. Synonyms: *Lazaria* Gray, 1854; *Byssomera* Olsson, 1961.

Carditamera floridana Conrad, 1838   5478
Broad-ribbed Cardita   Color Plate 21

Southern half of Florida and Mexico.

1 to 1½ inches in length, about ½ as high, elongate, inflated, solid and heavy. Surface with about 20 strong, rounded, raised, beaded, radial ribs. In live material, the gray periostracum obscures the color of the shell. Exterior whitish to gray with small bars of chestnut-color on the ribs arranged in concentric series. Interior white with a small light-brown patch above the 2 muscle scars. Beaks close together. Lunule small, very deeply indented under the beaks. Ligament moderately large, visible from the outside. Very common on the west coast of Florida where it is washed ashore. Used extensively in the shellcraft business.

Carditamera gracilis (Shuttleworth, 1856)   5479
West Indian Cardita   Color Plate 21

Mexico to Puerto Rico.

1 inch, similar to *floridana*, but more compressed, elongate, narrow at the anterior end, with 17 larger, smoothish ribs, and the posterior lateral tooth is stained dark-brown. Common in shallow water.

Other species:

5480 *Carditamera affinis* (Sowerby, 1833). Gulf of California to Peru. Common. (Synonym: *californica* Deshayes, 1854.) Color Plate 21.

5481 *Carditamera tricolor* (Sowerby, 1833). Gulf of California to Peru; Galapagos. (Synonym: *laticostata* Sowerby, 1833.)

Genus *Cardites* Link, 1807

Heavy, ovate, with strong, broad ribs; lunule heart-shaped; posterior slope truncate. Type of the genus is *Chama antiquata* Linné, 1758.

5482 *Cardites crassicostata* (Sowerby, 1825). Gulf of California to Peru, intertidal to 55 meters. 1½ inches. (Synonyms: *cuvieri* Broderip, 1832; *michelini* Valenciennes, 1846; *sulcosa* Dall, 1908.)

5483 *Cardites grayi* (Dall, 1903). Gulf of California to Ecuador. Extreme low tide, in sand under rocks. 1 inch. Rounded ribs.

5484 *Cardites laticostata* (Sowerby, 1833). Gulf of California to Peru. Common. 1½ inches. (Synonyms: *tricolor* Sowerby, 1833; *arcella* Valenciennes, 1846.)

Genus *Glans* Mühlfeld, 1811

Shells small, elongate-quadrate, length greater than the height, beaks near the anterior end. With nodulose or squamose ribs. Lunule with a convex margin. Type: *Chama trapezia* Linné, 1767.

Glans dominguensis (Orbigny, 1845)   5485
Domingo Cardita

North Carolina to southeastern Florida.

¼ inch in length, ovate, inflated; beaks close together, pointing toward each other, located nearer the anterior end. Lunule narrow, rough, ill-defined. Numerous strong radial ribs are weakly beaded. Color whitish with a rose tint. Moderately common from 1 foot to 70 fathoms on sandy bottoms. Compare with the more common and closely resembling *Pleuromeris tridentata*.

5485

Glans subquadrata (Carpenter, 1864)   5486
Carpenter's Cardita

British Columbia to Baja California.

¼ inch, elongate, somewhat obese, with about a dozen strong rough, rounded radial ribs. Brownish and mottled; interior purplish. Common; under stones from low tide to 50 fathoms. *G. carpenteri* (Lamy, 1922) and *minuscula* Grant and Gale, 1931, are synonyms.

Genus *Miodontiscus* Dall, 1903

Small, obliquely oblong, with strong beaks pointing forward. Ill-defined lunule long. Sculpture of broad, low radial ribs. Cardinals oblique. Anterior laterals obsolete. Posterior laterals faint. Type: *prolongatus* (Carpenter, 1864). *Miodon* Carpenter, 1864, not Duméril, 1859, is a synonym.

Miodontiscus prolongatus (Carpenter, 1864)   5487
Dwarf Microcardita

Alaska to San Diego, California.

5 mm., high, oblique, anterior end longer, with broad, radiating ribs, well-marked lunule and escutcheon. Right cardinal absent, and a posterior right and anterior left lateral feebly developed. Uncommon. Dredged from 5 to 30 fathoms.

5487

Other species:

5488 *Miodontiscus meridionalis* Dall, 1916. Off Point Loma, California. 70 fathoms.

Genus *Pleuromeris* Conrad, 1867

Shell tiny, convex, triangular. Lunule small, restricted to the left valve; the escutcheon minute and poorly defined. Type: *tridentata decemcostata* (Conrad, 1867).

Pleuromeris tridentata (Say, 1826)   5489
Three-toothed Cardita

North Carolina to all of Florida.

¼ inch in length and height, trigonal in shape, inflated, with 15 to 18 heavily beaded, strong radial ribs. Beaks close together, pointing slightly forward. Lunule oval, sharply impressed, smoothish. Escutcheon small, narrow. External color grayish brown to bright-rose, sometimes with red-brown mottlings. Hinge-teeth often purplish blue. Interior of valve stained with light-brown on white background. A common, moderately shallow-water species, usually confused with *Glans dominguensis* which, however, lacks the strong tridentate hinge, is ovate in shape, whose ribs are weakly beaded and whose beaks point toward each other. Formerly placed in *Venericardia*.

5486   5489   5491

Other species:

5490 *Pleuromeris armilla* (Dall, 1902). Off northwest Florida to Mississippi, 24 to 196 fms.

Superfamily Miodomeridinae Chavan, 1969

Genus *Pteromeris* Conrad, 1862

Somewhat compressed, oblong, obliquely rounded. With concentric and radial ribbing. Lunule ill-defined. Hinge with obsolete posterior laterals. Radial ribs predominate. Ligament marginal. Type: *perplana* (Conrad, 1841).

Pteromeris perplana (Conrad, 1841)   5491
Flattened Cardita

North Carolina to southern half of Florida.

¼ inch in size, similar to *Cyclocardia borealis* but much smaller, without a periostracum, pinkish or mottled-brown, and more oblique. The ribs are wider, and close to each other. The subspecies (**5492**) *flabella* Conrad, 1842, from Tampa Bay, Florida, has fewer ribs which are squarish and separated by furrows almost equal in size to the ribs themselves. *P. perplana* is common, *flabella* only locally found at certain seasons in few numbers, 1 to 52 fathoms.

Genus *Cyclocardia* Conrad, 1867

Shell rounded-trigonal, white, with a rough blackish periostracum and with strong radial ribs which are commonly beaded; internal margins crenulate; right anterior cardinal and laterals absent. No byssus made. Type: *borealis* (Conrad, 1831). The genus *Venericardia* Lamarck, 1801, is limited to the Paleocene and the Eocene.

Cyclocardia borealis (Conrad, 1831)   5493
Northern Cardita

Labrador to Cape Hatteras, North Carolina.

1 to 1½ inches in height, rounded, obliquely heart-shaped, thick and strong; beaks elevated and turned forward. Surface with about 20 rounded, moderately rough or beaded radial ribs. Shell white, usually covered by a fairly thick, velvety, rusty-brown periostracum. Lunule small but very deeply sunk. Hinge strong; in the left valve the central tooth under the beak is large, triangular and curved. Very common on the Grand Banks where it serves as a food for fish.

(**5494**) *C. novangliae* (Morse, 1869) (Nova Scotia to New York) is similar, but is ovate, the length being slightly greater than the height of the shell. It is sometimes considered a variety of *borealis*.

Cyclocardia ventricosa (Gould, 1850)   5495
Stout Cardita

Alaska to Santa Barbara Islands.

About ¾ inch in length, rounded-trigonal, moderately fat; velvety periostracum; lunule small; with about 18 to 20 rather wide, radial ribs which are bluntly beaded. Inner margins of the valves have prominent, squarish, widely spaced crenulations which correspond to the external ribs. Beaks slightly prosogyrate. Common offshore.

A. G. Smith and M. Gordon (1948, *Proc. Calif. Acad. Sci.*, series 4, vol. 26, p. 214) recognize three subspecies:

(**5496**) *C. ventricosa ventricosa* (Gould, 1850). Alaska to northern California. Shell ovate in outline. Lower margin of hinge plate is roughly parallel to the ventral margin of the shell. Common.

(**5497**) *C. ventricosa montereyensis* (A. G. Smith and M. Gordon, 1948). Central California. Subtriangular and extended posteriorly. Shell more compressed; beaks less tumid. Common; 35 to 139 fms., mud and fine sand.

(**5498**) *C. ventricosa redondoensis* (J. Q. Burch, in Smith and Gordon, 1948). Southern California; Santa Cruz Island and Cortez Bank. Subquadrangular and quite fat. Posterior end broadly rounded. Moderately common; 40 to 250 fms.

Cyclocardia stearnsii (Dall, 1903)   5499
Stearns' Cardita

British Columbia and Washington.

¾ inch; distinguished from *ventricosa* by its greater height with respect to length; its elevated, more strongly prosogyrate beaks; its deeply impressed lunule, and its strong elevated hinge, which, in the left valve, bears a prominently developed anterior cardinal tooth behind the lunule, and a perceptibly curved posterior cardinal tooth. Periostracum radially hairy, but lacks the velvety appearance. Rare; in the inland waters, 20 to 30 fathoms.

Other species:

5500 *Cyclocardia crebricostata* (Krause, 1885). Point Barrow, Alaska, to the coast of Oregon. (Synonym: *alaskana* Dall, 1903.)

5501 *Cyclocardia crassidens* (Broderip and Sowerby, 1829). Point Barrow, Alaska, to off Eureka, California, in 100 fms. (Synonym: *paucicostata* Krause, 1885.)

5501   5503   5506

5502 *Cyclocardia barbarensis* (Stearns, 1890). Santa Barbara Channel to off San Diego, California, 1,000 fms. Rare.

5503 *Cyclocardia gouldii* (Dall, 1902). Off San Diego, California.

5504 *Cyclocardia nodulosa* (Dall, 1919). Santa Barbara to the Coronado Islands, California, 50 to 200 fms. (Synonym: *Cardita longini* Baily, 1945, *The Nautilus*, vol. 58, p. 118, an unnecessary new name.)

5505 *Cyclocardia umnaka* (Willett, 1932). Umnak Island, Alaska. Washed up on the beach. *Trans. San Diego Soc. Nat. Hist.*, vol. 7, p. 85.

5506 *Cyclocardia incisa* (Dall, 1902). Unalaska to the Semidi Islands, Alaska.

5507 *Cyclocardia spurca* Sowerby, 1833, subspecies *beebei* Hertlein, 1958. Gulf of California to Panama, in 45 to 65 meters.

5508 *Cyclocardia armilla* (Dall. 1903). Off Louisiana, 120 fms.; off Cape San Blas, Florida, 24 to 196 fms. 20 mm.

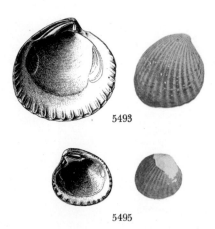

5493

5495

5508   5500

Subfamily THECALIINAE Dall, 1903

Transversely trapezoidal, ventral margin indented by an incubatory chamber in the females.

Genus *Milneria* Dall, 1881

Characterized by its obese quadrate shape and the pocket-like embayment on the ventral margin. With 2 radial carinations. Type: *minuta* (Dall, 1871). Synonym: *Ceropsis* Dall, 1871, non Solier, 1839.

Milneria kelseyi Dall, 1916   5509
Kelsey's Milner Clam

Monterey to Baja California.

⅛ to ¼ inch in length. An extraordinary clam which resembles a tiny Brazil nut. The bottom margins of the valves are pushed in to form a small cup-shaped hollow. Into this, the females put the 50 or so young whose shells are smooth and round. The hollow is covered over by a sheath of periostracum. Hinge of adult with large triangular tooth in left valve which fits snugly between 2 smaller ones in the right valve. External sculpture of scaled ribs and concentric ridges. Color light-brown. Shell thick, translucent glaze inside. Common; found in shallow water under stones. *Proc. U.S. Nat. Mus.*, vol. 52, p. 408.

5509

Other species:

5510 *Milneria minima* (Dall, 1871). Monterey to Rosario Bay, Baja California. On *Haliotis*.

Family CONDYLOCARDIIDAE Bernard, 1897

Minute, trigonal to ovate, higher than long. Usually with radial ribs. Hinge teeth somewhat ball-and-socket. Ligament internal, partly covering the cardinal. Posterior laterals long.

Subfamily CONDYLOCARDIINAE Bernard, 1897

Genus *Carditopsis* E. A. Smith, 1881

With granulose radial ribs. Beaks rounded. Internal resilium pit rounded. Laterals very long. Small cardinals. Prodissoconch saucer-shaped. Type: *flabellum* (Reeve, 1843).

Carditopsis smithii (Dall, 1896)   5511
Smith's Tiny Cardita

Bermuda and the West Indies.

1 to 2 mm., resembling a miniature *Glans dominguensis*, but solid orange-brown, with an internal resilium pit, with a raised, concentric ridge on the umbo separating the prodissoconch from the adult shell, and with 10 or 11 fimbriated or beaded radial ribs. Uncommon; 1 fathom in sand. *Condylocardia floridensis* Pilsbry and Olsson, 1946 (*The Nautilus*, vol. 60, p. 6) is a synonym.

Other species:

5512 *Carditopsis bernardi* (Dall, 1903). Lower Caribbean.

Genus *Condylocardia* Bernard, 1896

Shells minute, 2 mm., radially ribbed; prodissoconch saucer-shaped, capping the umbones. Type of the genus is *digueti* Lamy, 1916. *Hippella* Mörch, 1860, is a *nomen oblitum*.

5513 *Condylocardia digueti* Lamy, 1916. Baja California and the Gulf of California to Jalisco, Mexico.

Subfamily CUNINAE Laseron, 1953

Genus *Cuna* Hedley, 1902

Small, oblique, strong, inequilateral shells. Right valve with 3 cardinals, the anterior long and low, the central one large and triangular, the posterior one short and narrow and situated at the edge of the large ligament pit. Left valve with 3 cardinals, the posterior one a small ridge at the edge of the ligament. Pallial line without a sinus. Type: *concentrica* Hedley, 1902, from Australia.

Subgenus *Goniocuna* Klappenbach, 1962

The type of this subgenus is *dalli* Vanatta, 1904.

Cuna dalli Vanatta, 1904   5514
Dall's Cuna Clam

Northwest Florida and Mississippi.

2.5 mm. high, equally wide, 1.5 mm. width of both valves. Obliquely trigonal with the prominent umbones near the anterior end. Exterior with crowded, concentric ridges. Color tan to white with a flush of purple on the center of the disc. Adductor muscle scars rather large and prominent. Moderately common; offshore. This species is ovoviviparous (D. R. Moore, 1961, *Gulf Research Reports, Miss.*, vol. 1). The date of 1903 for this species is incorrect.

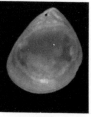

5514

Superfamily Crassatellacea Férussac, 1822

Family ASTARTIDAE Orbigny, 1844

Genus *Astarte* Sowerby, 1816

Shells compressed, heavy, subovate, with a thick, brown periostracum. 2 muscle scars joined by pallial line. No pallial sinus. Hinge tridentate in each valve. Type: *sulcatus* (da Costa, 1778).

Subgenus *Tridonta* Schumacher, 1817

Medium-sized, inequilateral; concentric ribs more or less vanishing on the disc. Lunule long and flattened. Inner margin usually smooth. Type: *borealis* (Schumacher, 1817). Considered a genus separate from *Astarte* by Chavan and other workers.

Astarte borealis (Schumacher, 1817)   5515
Boreal Astarte

Arctic seas to Massachusetts Bay.
Alaska to Japan.

1 to 2 inches in length, ovate, moderately compressed. External ligament large. Concentric ridges strong near the beaks but disappearing near the margins of the valves. Periostracum fibrous, more or less frayed, and yellowish (in young) to black (in old specimens). Differing from *subequilatera* in being more elliptical in side view, in having the beaks near the middle, with weaker concentric ribs, and with the inner surface of the valve margins smooth. A common shallow-water species. Synonyms are *semisulcata* Möller, 1842; *richardsoni* Reeve, 1855; and *saintjohnensis* Verkrüzen, 1877.

Astarte elliptica (Brown, 1827)   5516
Elliptical Astarte

Greenland to Massachusetts.
Northern Europe.

1 inch, broadly ovate in outline, inequilateral with the beaks 1/3 back from the anterior end. Characterized by a microscopic meshlike surface of the reddish brown periostracum. With 25 to 30 strong, broad concentric ridges and numerous fine concentric lines. Fine radiating lines on the prominent lunule and escutcheon. Right valve with 2 cardinal teeth, a broad anterior and thin posterior. Left valve with 3 cardinals of which the posterior one may be inconspicuous or broken off. Margin of valves smooth. Moderately common; 8 to 90 fathoms. The more elongate specimens were named *depressa* Posselt, 1895.

Subgenus *Astarte* Sowerby, 1816

Astarte crenata subequilatera Sowerby, 1854   5517
Lentil Astarte

Arctic seas to off Florida.

1 to 1½ inches in length, ovate, moderately compressed. External ligament small. Concentric ridges strong, rounded, evenly spaced. Internal margin of valves finely crenulate. Beaks turned slightly forward, often eroded. Color dull light- to dark-brown. Found in shallow water in the north and below 50 fathoms in the south. Common, 22 to 428 fathoms. Dall described a variety *whiteavesii* Dall, 1903 (5517a). *A. lens* Stimpson is a synonym. Typical *crenata* (Gray, 1824) (5518) is limited to Greenland waters according to Ockelmann, 1958, p. 90. Other forms or subspecies of Greenland waters are: *acuticostata* Jeffreys, 1881, and *inflata* Hägg, 1904. Compare with *borealis*.

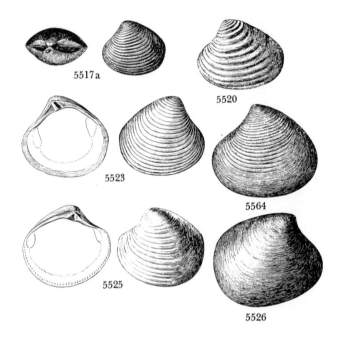

5517a

5520

5523

5564

5525

5526

Astarte undata Gould, 1841   5519
Waved Astarte

Labrador to off New Jersey.

Similar to *subequilatera,* but less elliptical, with its beaks near the center and with fewer and stronger concentric ridges. Probably the most common Astarte in New England. Dredged from 5 to 104 fathoms.

Astarte alaskensis Dall, 1903   5520
Alaskan Astarte

Bering Sea to Washington.

1 inch, with concentric, rounded ridges, resembling *elliptica,* but is more triangular and heavier. Periostracum black or dark-brown and dehiscent when dry. Moderately common; 10 to 40 fathoms.

5515   5517   5519   5521   5522

5516

Subgenus *Nicania* Leach, 1819

Type of this subgenus is *montagui* (Dillwyn, 1817).

Astarte montagui (Dillwyn, 1817)   5521
Montagu's Astarte

Arctic Seas; Greenland to Massachusetts.
Bering Sea to British Columbia.

½ to ¾ inch, equilateral, usually somewhat elongate, variable in sculpture and color but generally with fine concentric, evenly spaced riblets in the umbonal region, with less prominent, irregularly spaced ones throughout the lower ⅔ of the shell or the lines so fine that it seems smoothish. Tan to dark-brown periostracum, usually black mud-stained, and microscopically shows irregular, radiating rows of faint, dotlike depressions. A number of forms were described and were worked out by A. S. Jensen, 1912, Medd. Groenland, vol. 29. *A. compressa* Jeffreys, 1869; *banksii* Leach, 1819; *striata* Leach, 1819; *striata* Gray, 1839; *globosa* Möller, 1842; *pulchella* Jonas in Philippi, 1845; *warhami* Hancock, 1846; and *fabula* Reeve, 1855, are synonyms. Abundant; 5 to 150 fathoms.

Subgenus *Isocrassina* Chavan, 1950

Type of the subgenus is *castanea* (Say, 1822).

Astarte castanea (Say, 1822)   5522
Smooth Astarte

Nova Scotia to off New Jersey.

1 inch in length, as high, trigonal in shape, quite compressed. Beaks pointed and hooked anteriorly; external ligament small. Large, shallow lunule. Shell almost smooth, except for weak, low concentric lines. Color a glossy light-brown. Inner margin of valves finely crenulate. A commonly dredged species. Synonyms or forms are *picea* Gould, 1841; *procera* Totten, 1835.

Astarte nana Dall, 1886   5523
Southern Dwarf Astarte

North Carolina to Florida and the Gulf States.

¼ inch in length, slightly trigonal in shape, compressed. With or without about 25 well-developed, evenly spaced, rounded, concentric ridges. Ventral and inner edge of valves usually with 40 to 50 distinct small pits or crenulations. Shell cream, tan, brown or rose-brown in color, with the beaks usually whitish. A very abundant species dredged in moderately shallow water, especially off eastern Florida, from 6 to 227 fathoms.

Other species:

5524 *Astarte (Astarte) polaris* Dall, 1903. Davis Strait, 90 fms. Aleutian and Shumagin Islands, Alaska.

5525 *Astarte smithii* Dall, 1886. Gulf of Mexico to the West Indies, 54 to 450 fms. 7 mm.

5526 *Astarte globula* Dall, 1886 and 1902. Off Fernandina, Florida, Cuba, and the Gulf of Mexico, 294 to 539 fms. 8 mm.

5527 *Astarte liogona* Dall, 1903. Near the delta of the Mississippi River, off Louisiana, 118 fms.

5528 *Astarte quadrans* Gould, 1841. Gulf of St. Lawrence to Long Island Sound, New York, 6 to 40 fms.

5529 *Astarte laurentiana soror* Dall, 1903. Labrador to the Gulf of St. Lawrence, 5 to 90 fms.

5530 *Astarte (Astarte) compacta* Carpenter, 1864. Forrester Island, Alaska, to Puget Sound, Washington, 40 fms.

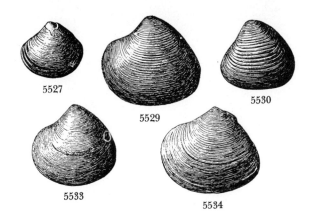

5527   5529   5530
5533   5534

5531 *Astarte (Astarte) willetti* Dall, 1917. Forrester Island, Alaska, 50 fms., to Puget Sound, Washington.

5532 *Astarte (Tridonta) rollandi* Bernardi, 1858. Pribilof and Aleutian Islands to Prince William Sound, Alaska. Variety *loxia* Dall, 1902. Semidi Islands, and Prince William Sound, Alaska.

5533 *Astarte (Tridonta) bennettii* Dall, 1903. Arctic Seas at Bennett Island, to Nunivak Island.

5534 *Astarte (Tridonta) vernicosa* Dall, 1903. Arctic and Bering Seas. Attu to Atka Islands, Aleutians. (?Subspecies of *montagui* Dillwyn, 1817.)

5535 *Astarte (Rictocyma* Dall, 1872) *esquimalti* Baird, 1863. Aleutian Islands to Puget Sound, Washington, 10 to 40 fms.

5535

Family CRASSATELLIDAE Férussac, 1822

Genus *Eucrassatella* Iredale, 1924

Shell large, thick, equivalve, posteriorly rostrate; ligament and resilium adjacent and internal in a triangular resilifer; left valve with 2 diverging cardinal teeth; right valve with 3, of which the posterior one is more or less obsolete. 3 laterals in each valve. Type: *kingicola* (Lamarck, 1805). *Crassatella* Lamarck, 1799, is fossil and not this genus. *Crassatellites* Krueger, 1823, is believed to be invalid.

Subgenus *Hybolophus* Steward, 1930

2 cardinal teeth in each valve. Umbones pointed backward. Type: *gibbosa* (Sowerby, 1832).

Eucrassatella speciosa (A. Adams, 1852)   5536
Gibb's Clam   Color Plate 22

North Carolina to both sides of Florida and the West Indies.

1½ to 2½ inches in length, ⅔ as high, heavy, beaks at the center, and the shell somewhat diamond-shaped. Concentric sculpture of neat, rather heavy, closely packed ridges (about 15 per ½ inch). Lunule and escutcheon sunken, lanceolate in shape and about the same size as each other. Exterior with a thin, persistent, nut-brown periostracum. Interior glossy-ivory with either a tan or pink blush. Moderately common just off-shore in sand. *C. floridana* Dall, 1886, is the same, being based on a young specimen. *E. gibbesi* Tuomey and Holmes, 1856, is a synonym.

Other species:

5537 *Eucrassatella gibbosa* (Sowerby, 1832). Gulf of California to Peru.

5538 *Eucrassatella diqueti* (Lamy, 1917). Gulf of California to Colombia. (Synonyms: *undulata* Sowerby, 1832, not Lamarck, 1805; *laronus* E. K. Jordan, 1932.)

5539 *Eucrassatella fluctuata* (Carpenter, 1864). Santa Barbara Islands to San Pedro, California, 25 to 30 fms. 1 to 1¾ inches, heavy, periostracum dark-brown. Posterior muscle scar brownish. Uncommon.

Subfamily SCAMBULINAE Chavan, 1952

Anterior laterals reaching beaks in front of anterior cardinals. Cardinals narrow and subparallel. Reduced resilium in narrow pit.

Genus *Crassinella* Guppy, 1874

Shell small, compressed, subtriangular, and slightly inequivalve. 2 cardinals in each valve. 1 anterior lateral in the right valve, 1 posterior lateral in the left valve. Resilium resembles two elongate, slightly spiraled antelope horns. Type: *martinicensis* (Orbigny, 1842). Synonym: *Pseuderiphyla* Fischer, 1887. For a taxonomic and biological study of the two Western Atlantic species, see H. W. Harry, 1966, *Publ. Inst. Marine Studies, Texas*, vol. 11, pp. 65–89.

Crassinella lunulata (Conrad, 1834)   5540
Lunate Crassinella

Massachusetts to Florida to Texas and to Brazil. Bermuda.

¼ to ⅓ inch in length, as high, quite compressed, solid, with the tiny, closely pressed-together beaks at the middle or slightly toward the anterior end. Dorsal margins straight and about 90 degrees, or slightly more, to each other, the anterior margin slightly longer and with a wider sunken area. The valves are peculiarly askew, so that the posterior dorsal margin of the left valve is more obvious than that of the right valve. Concentric sculpture weak or of coarse but well-developed ribs (about 15 to 17 plainly visible). Color whitish or pinkish, interior commonly waxy-brown. Sometimes faintly rayed. A common shell from 1 to 60 fathoms. The synonyms are: *C. pfeifferi* (Philippi, 1848); *parva* (C. B. Adams, 1845); *mactracea* (Linsley, 1845); *guadalupensis* Orbigny, 1842; *fastigiata* (Gould, 1862) and *galvestonensis* (Harris, 1895). It climbs over broken shell bottoms by reattaching its weak byssus.

Crassinella martinicensis (Orbigny, 1842)   5541
Martinique Crassinella

Gulf of Mexico, off Mississippi; West Indies.

Very small, 1/16 inch, triangular, somewhat inflated. Dorsal margins equal in length; the anterior and posterior slopes are at right angles to each other. Color white, shaded with brown.

Surface with 8 to 15 sharp concentric ribs. Separated from *lunulata* by its obesity, small size, having the umbonal angle usually just less than 90 degrees and having the concentric sharp ribs continuous over all parts of the valve.

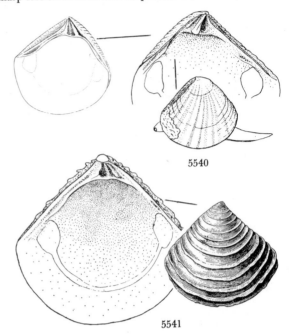

5540

5541

Crassinella oregonensis Keen, 1938   5542
Oregon Crassinella

Coos Bay, Oregon.

5.6 mm. in length, white, porcelaneous and semitranslucent. Angle of the 2 straight dorsal edges is about 95 degrees. Sculpture of 16 concentric ridges crossed by minute radial striae (25 per mm.). A rare species; dredged in 1 to 2 fathoms. *Proc. Malacol. Soc. London*, vol. 23, pp. 31 and 252.

Other species:

5543 *Crassinella mexicana* Pilsbry and Lowe, 1932. Cedros Island to the Gulf of California to Guaymas, Mexico.

5544 *Crassinella varians* (Carpenter, 1855). Gulf of California to Ecuador.

5545 *Crassinella pacifica* (C. B. Adams, 1852). Gulf of California to Peru.

Superfamily Cardiacea Oken, 1818

Family CARDIIDAE Oken, 1818

Subfamily CARDIINAE Oken, 1818

The well-known cockle family is worldwide in distribution and contains about 200 living species. The 2 siphons are short and the foot is long and muscular enough to permit the cockle to move about in short leaps. The adductor muscle scars are about the same size and joined by a weak pallial line. There is no appreciable pallial sinus. There are no true members of this subfamily in American waters today.

Subfamily TRACHYCARDIINAE Stewart, 1930

Genus *Trachycardium* Mörch, 1853

Hinge strong; shell strongly ribbed and with strong scales or nodules; higher than long. Type: *Cardium isocardia* Linné, 1758. Synonym: *Kathocardia* Tucker and Wilson, 1932.

Trachycardium egmontianum (Shuttleworth, 1856)   5546
Prickly Cockle   Color Plate 22

North Carolina to south Florida and the West Indies.

2 inches in height, with 27 to 31 strong, prickly, radial ribs. Externally whitish to tawny-gray with odd patches of weak yellow, brown or dull-purple. Interior glossy, commonly brightly hued with salmon, reddish and purple. Albino shells occasionally found on the west coast of Florida. Do not confuse with *muricatum* which is more oval, has more ribs which are not sharply scaled at the center of the shell and is commonly only yellowish inside. A common shallow-water species, especially on the Gulf side of Florida.

(5547) *Trachycardium isocardia* (Linné, 1758) from the West Indies and Bermuda has larger and slightly different scales, 32 to 37 ribs and has not been recorded from Florida. The Pacific counterpart ranging from the Gulf of California to Ecuador is (5548) *Trachycardium consors* (Sowerby, 1833). It has 30 to 34 spiny ribs.

5547

Subgenus *Dallocardia* Stewart, 1930

Hinge plate narrow; cardinal teeth slightly posterior. Thornlike spines on edge of the strong ribs. Type: *quadragenarium* (Conrad, 1837).

Trachycardium muricatum (Linné, 1758)   5549
Yellow Cockle   Color Plate 22

North Carolina to Florida, Texas and the West Indies. Brazil.

2 inches in height, subcircular, with 30 to 40 moderately scaled, radiating ribs. Externally light-cream with irregular patches of brownish red or shades of yellow. Interior commonly white, rarely yellow-tinted especially in Florida. A very common, shallow-water species. Compare with *egmontianum* and *magnum* which are both more elongate. C. *campechiense* Röding, 1798, and *gossei* Deshayes, 1854, are synonyms.

(5550) The Pacific comparable species is *Trachycardium senticosum* (Sowerby, 1833), 1½ inches, with 35 to 40 ribs. Gulf of California to Peru. Common on shallow muddy bottoms.

Trachycardium quadragenarium (Conrad, 1837)   5551
Giant Pacific Cockle

Santa Barbara, California, to Baja California.

3 to 6 inches in size, commonly slightly higher than long, inflated, and with 41 to 44 strong, closely set, squarish, radial ribs which bear small, upright, strong, triangular spines, especially at the anterior, posterior and ventral portions of the shell. Ribs on beaks smoothish. Exterior whitish tan, but commonly covered with a thin, opaque-brown periostracum. Interior dull-white. Moderately common from shore to 75 fathoms. Known locally as the spiny cockle.

5551

Subgenus *Acrosterigma* Dall, 1900

Hinge plate narrow, long and bent in the middle. Ribs in the middle of the valves have almost lost their scales or spines. Type: *Cardium dalli* Heilprin, 1887, a fossil.

Trachycardium magnum (Linné, 1758)   5552
Magnum Cockle

Lower Florida Keys and the West Indies; Bermuda. Brazil.

2 to 3½ inches in height, elongate, with 32 to 35 mostly smooth ribs. The ribs at the posterior end have small, toothlike scales. Middle ribs completely smooth and squarish. Ex-

5552

ternally light-cream with irregular patches of reddish brown. Interior china-white with the deepest part flushed with orange-buff. As a rule, the posterior margin is pale-yellow, merging into pale-purple at the extreme edge. An uncommon West Indian species which has been found on the most southerly keys of Florida. *C. marmoreum* Lamarck, 1819, is a synonym.

Other species:

5553 *Trachycardium (Mexicardia) panamense* (Sowerby, 1833). (Synonym: *Trigoniocardium eudoxia* Dall, 1916.) 3 inches. Gulf of California to Costa Rica.

5554 *Trachycardium (Mexicardia) procerum* (Sowerby, 1833). 2 inches. Mexico to Chile. (Synonyms: *laticostatum* Sowerby, 1833; *dulcinea* Dall, 1916; *parvulum* Li, 1930.)

Genus *Papyridea* Swainson, 1840

Shell longer than high, compressed, gaping at both ends; ribs spinose. Hinge short. Type: *soleniformis* (Bruguière, 1789).

Papyridea soleniformis (Bruguière, 1789)   5555
Spiny Paper Cockle   **Color Plate 22**

North Carolina to Florida and Brazil. Bermuda.

1 to 1¾ inches in length, fairly fragile, moderately compressed, and gaping posteriorly where the margin of the valve is strongly denticulated by the ends of the dozen radial, finely spinose ribs. Exterior tawny with rose flecks or mottlings. Interior glossy, mottled with violet and white, rarely a solid pastel-orange. Moderately common from low tide to several fathoms. The names *hiatus* and *spinosum* Meuschen, 1787, are not valid (ruled nonbinomial).

(5556) The Pacific comparable species, almost inseparable, is *Papyridea aspersa* (Sowerby, 1833), occurring from Baja California to Peru. *P. californica* Verrill, 1870, is a synonym.

Papyridea semisulcata (Gray, 1825)   5557
Frilled Paper Cockle

Bermuda; south Florida to Brazil.

⅓ to ½ inch, white (or rarely orange, form *petitianum* Orbigny, 1846), very obese and with 8 to 12 long denticulations at the dorsal and posterior edges. Uncommon from low-tide line to 40 fathoms.

5557

Other species:

5558 *Papyridea crockeri* Strong and Hertlein, 1937. Gulf of California, down to 175 meters.

Subfamily FRAGINAE Stewart, 1930

Genus *Americardia* Stewart, 1930

Shells quadrate, heavy, with strong flattish ribs. Hinge has the anterior laterals almost as far removed from the cardinals as the posterior laterals. Type: *C. medium* (Linné, 1758).

Americardia media (Linné, 1758)   5559
Atlantic Strawberry Cockle   **Color Plate 22**

Bermuda; North Carolina to southeast Florida and to Brazil.

1 to 2 inches in size, squarish in outline, thick, inflated, with 33 to 36 strong radial ribs which are covered with close-set, chevron-shaped plates. External color whitish with mottlings of reddish brown. Interior usually white, or may be flushed with orange, rose-brown or purple. The posterior slope is pushed in somewhat and is slightly concave. A relatively common species found in shallow water to moderately deep water.

Americardia biangulata (Broderip and Sowerby, 1829)   5560
Western Strawberry Cockle

Southern California to Ecuador.

1½ inches; thick-shelled, with 28 to 30 strong, broad, flattened ribs; exterior yellowish white; interior reddish purple. Moderately common intertidally to 85 fathoms. *Cardium magnificum* Carpenter, 1857, is a synonym.

Americardia guppyi Thiele, 1910   5561
Guppy's Cockle

Florida Keys, Bahamas to north Brazil.

6 to 15 mm., subquadrate, with 26 to 28 radial, sharp ribs bearing moon-shaped striae and swollen or fimbriated beads. Concentric threads between the riblets. Exterior cream or spotted with reddish brown or purplish. Locally common in coral sand from 1 to 55 fathoms. This is what Clench and Smith (1944) called *antillarum* Orbigny, but that species is all-white, inflated and has only 16 to 18 ribs.

5560   5561

Other species:

5562 *Americardia guanacastensis* (Hertlein and Strong, 1947). Gulf of California to Peru. (Synonym: *planicostatum* Sowerby, 1833, not Sedgwick and Murchison, 1829.)

Genus *Trigoniocardia* Dall, 1900

Shells small, obliquely ovate, alabaster-white; hinge having the anterior laterals rather close to the cardinals and the posterior laterals more remote. Type: *C. graniferum* (Broderip and Sowerby, 1829).

Trigoniocardia antillarum (Orbigny, 1842)   **5563**
Antillean White Cockle

Cuba to the Virgin Islands. Brazil.

6 to 10 mm.; inflated; subquadrate; white throughout. With 16 to 18 strong ribs, of which the 5 or 6 middle ones are extra large, with beads, and with strong, concentric, smooth bars or riblets between the ribs. Dredged from 3 to 182 fathoms. *Cardium ceramidum* Dall, 1886, is a synonym.

5563

Other species:

5564 *Trigoniocardia granifera* (Broderip and Sowerby, 1829). Baja California to Peru. 1 to 14 fms. (Synonym: *alabastrum* Carpenter, 1857.)

5565 *Trigoniocardia obovalis* (Sowerby, 1833). Baja California to Ecuador. Dredged offshore. Type of the subgenus *Apiocardia* Olsson, 1961.

5566 *Trigoniocardia ovuloides* (Reeve, 1845). Nicaragua.

Subfamily PROTOCARDIINAE Keen, 1951

Genus *Nemocardium* Meek, 1876

Shell relatively thin, posterior area merging into the central slope smoothly. Posterior margin finely crenulate. Type: *semiasperum* (Deshayes, 1858).

Subgenus *Keenaea* Habe, 1951

Smaller than *Nemocardium s. s.,* with secondary concentric lamellae on posterior ribs, the latter not sharply differentiated. Type: *samarangae* Makiyama, 1934, from east Asia.

Nemocardium centifilosum (Carpenter, 1864)   **5567**
Hundred-lined Cockle

Alaska to Baja California.

½ to ¾ inch in length, almost circular; posterior ⅓ of shell with cancellate sculpturing and separated from the finely ribbed anterior ⅔ of the shell by a single raised rib. Edge minutely serrate. Exterior with gray, greenish gray or brownish gray, thin, fuzzy periostracum. Interior dull-white. Fairly common. *N. richardsonii* (Whiteaves, 1878), is a synonym.

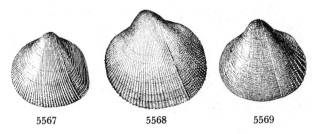

5567   5568   5569

Subgenus *Microcardium* Thiele, 1934

Shells fragile, inflated, finely sculptured. Cardinals enlarged, laterals weak. Deep pit between the 2 cardinal teeth of the left valve. Type: *peramabile* Dall, 1881. Formerly considered as a genus.

Nemocardium peramabile (Dall, 1881)   **5568**
Eastern Micro-cockle

Rhode Island to the Gulf of Mexico and to Brazil.

½ to ¾ inch in length, thin, inflated, subquadrate, white, but may be mottled-tan on the anterior slope. Sculpture prominent on the posterior ⅓ of the valve. It consists of about 90 closely packed, radial ribs (spinose posteriorly) which are crossed by minute concentric threads. The anterior ⅔ is separated from the rest of the shell by a single, crested, spinose radial rib. Very commonly dredged off eastern Florida from 18 to 350 fathoms.

(5569) *Nemocardium tinctum* (Dall, 1881) found from south Florida to Brazil is ¾ inch in length, stained with rose-red and has more than 150 minute, radial ribs. Uncommon; 7 to 225 fathoms.

Nemocardium transversum Rehder and Abbott, 1951   **5570**
Transverse Micro-cockle

Off Louisiana.

19 mm. long, 17 mm. high, 14.5 mm. wide; white with a few faint brownish maculations. Umbones rose. 80 to 90 nodulose radial riblets on the anterior ⅔ of the shell. A rib separates the posterior ⅓ which has narrower, more distantly spaced, sublamellar riblets. Hinge teeth typical, but the lateral teeth are strong. Rare; 29 fathoms, 50 miles south southwest of Marsh Island, Iberia County, Louisiana.

5570

Other species:

5571 *Nemocardium (Microcardium) pazianum* (Dall, 1916), Baja California to Panama, 25 to 95 meters.

Subfamily LAEVICARDIINAE Keen, 1936

Genus *Laevicardium* Swainson, 1840

Shell inflated, strong, smoothish, not gaping, and with prominent lateral teeth. Type: *C. oblongum* (Gmelin, 1791).

Laevicardium laevigatum (Linné, 1758)   **5572**
Common Egg Cockle   **Color Plate 22**

North Carolina to both sides of Florida and the West Indies. Bermuda. Brazil.

1 to 2 inches in size, higher than long, polished smooth, inflated, fairly thin and obscurely ribbed. Exterior generally whitish, but may be rose-tinted, mottled with brown or flushed

with purple, yellow or burnt-orange. Interior similarly colored. With about 60 very fine, subdued radial ribs. A common shallow-water species. The name *serratum* Linné has been erroneously applied to our Atlantic species by some workers. *L. vitellinum* Reeve, 1844, is a synonym.

Laevicardium mortoni (Conrad, 1830)   5573
Morton's Egg Cockle

Cape Cod, Massachusetts, to Florida and to Texas.

¾ to 1 inch in size, ovate, glossy, similar to *laevigatum*, but commonly with brown, zigzag markings and with fine, concentric ridges which are minutely pimpled. Common in southern New England from shallow water to 2 fathoms. A food of wild ducks. This species swims by sculling its extended foot (*The Nautilus*, vol. 78, p. 104).

Laevicardium pictum (Ravenel, 1861)   5574
Ravenel's Egg Cockle

North Carolina to southeast Florida and Bermuda. Brazil.

½ to 1 inch in height, obliquely triangular in shape, polished and only moderately inflated. Exterior white or cream with delicate shades of rose or brown and with a weak, iridescent sheen. A color form has strong, brown, zigzag streaks. Beaks very low and near the anterior end. Very faint radial and concentric lines present. Dredged from 75 to 85 fathoms. A common and attractive species.

Laevicardium sybariticum (Dall, 1886)   5575
Dall's Egg Cockle

North Carolina to the Lower Caribbean. Bermuda.

9 to 17 mm., inflated, subquadrate, smoothish, creamy-white with delicate shadings of pink with the umbones deep-pink. Sculpture of microscopic growth lines and fine radial riblets. Uncommon; 17 to 190 fathoms.

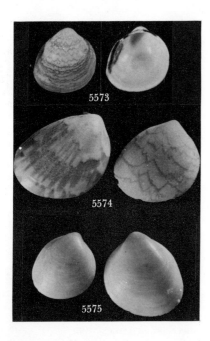

Laevicardium substriatum (Conrad, 1837)   5576
Common Pacific Egg Cockle

Ventura County, California, to the Gulf of California.

Less than 1 inch in size, obliquely ovate, smooth and slightly compressed. Color tan with closely set, narrow, radial bands

of reddish brown. These lines are commonly interrupted. Interior cream with cobwebby mottlings of purplish brown. Very common in such localities as Mission Bay and Newport.

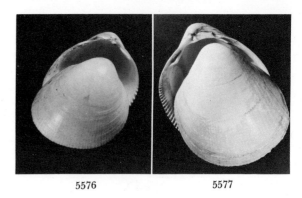

5576   5577

Laevicardium elatum (Sowerby, 1833)   5577
Giant Pacific Egg Cockle

San Pedro, California to Panama.

3 to 7 inches in height, oval, inflated, slightly oblique, with numerous, shallow, radial grooves, but the posterior and anterior regions are smooth. Exterior orange-yellow; interior china-white. This is the largest species of Recent cockles and is moderately common on the mud flats of the Gulf of California.

Other species:

5578 *Laevicardium elenense* (Sowerby, 1840). Baja California to Ecuador; shore to 50 fms. (Synonym?: *apicinum* Carpenter, 1864.) Moderately common.

5579 *Laevicardium clarionense* (Hertlein and Strong, 1947). Gulf of California. 35 to 85 fms. Rare.

Genus *Dinocardium* Dall, 1900

Shells large, ribbed, but without spines. Type: *robustum* (Lightfoot, 1786). Considered as a subgenus of *Laevicardium* by many workers.

Dinocardium robustum (Lightfoot, 1786)   5580
Giant Atlantic Cockle

Virginia to north Florida, Texas and Mexico.

3 to 4 inches in size, ovate, inflated, with 32 to 36 rounded, radial, smoothish ribs. Externally straw-yellow with its posterior slope mahogany-red shading toward purple near the edge. Interior rose, with brownish posteriorly and with a white anterior margin. This is the large, common cockle washed ashore along the Carolina and Georgia strands. It is not found in southwest Florida. Synonyms are: *obliquum* Spengler, 1799; *ventricosa* Bruguière, 1789; *maculatum* Gmelin, 1791; and *carolinensis* Conrad, 1863.

When the Florida Canal project was begun in 1935, President F. D. Roosevelt was presented with a large silver platter on which was set a specimen of *Dinocardium*, encased in gold and containing a portion of the first earth excavated as a re-

5580   5581

sult of the blast set off by the President. The canal was never completed.

Dinocardium robustum vanhyningi
Clench and L. C. Smith, 1944   **5581**
Vanhyning's Cockle

Tampa Bay to Cape Sable, Florida.

3½ to 5 inches in size, higher than long, with 32 to 36 smoothish, rounded, radial ribs. Externally straw-yellow with irregular patches and bands of mahogany-red to purplish brown. It is more elongate, glossier and more colorful than *robustum*, and is the common, large cockle on the west coast of Florida. They are popular souvenirs, being used for ash trays, melted-butter dishes, baking dishes and for holding pincushions.

Genus *Serripes* Gould, 1841

There is only one species of this peculiar genus of cockles in North American waters. The hinge is narrow, the cardinal teeth weak, and the ligament is large. The radial ribs are very weak. Type: *groenlandicus* (Bruguière, 1789).

Serripes groenlandicus (Bruguière, 1789)   **5582**
Greenland Cockle

Arctic Seas to Cape Cod, Massachusetts.
Alaska to Puget Sound, Washington.

2 to 4 inches in length, moderately thin but strong, inflated, almost round and slightly gaping at the posterior end. Exterior brownish gray and may have brown, concentric rings of growth. Interior dull-white. Beaks inflated and high. Ligament large and strong. No lunule, escutcheon or pallial sinus. Weak radial ribs seen at both ends only. Concentric growth ridges prominent near the margins. Muscle scars and pallial line deeply impressed. Foot of animal large and suffused with heavy red mottlings. Very commonly dredged in cold, northern waters, from 2 to 60 fathoms, but uncommon in New England. *S. protractus* Dall, 1900, is a synonym, as is *album* Verkrüzen, 1877.

5582

Genus *Clinocardium* Keen, 1936

Superficially resembling a venerid clam. Obliquely trigonal; obese. Umbones recurved and near the anterior end. Numerous ribs present which lack scales. Type: *C. nuttallii* (Conrad, 1837).

Clinocardium ciliatum (Fabricius, 1780)   **5583**
Iceland Cockle

Greenland to Massachusetts.
Alaska to Puget Sound, Washington.

1½ to 3 inches in size, a little longer than high, with 32 to 38 ridged radial ribs which are crossed by coarse concentric lines of growth. Externally drab grayish yellow with weak, narrow, concentric bands of darker color. Interior ivory. Periostracum gray and conspicuous. Especially abundant from Maine northward in offshore waters. The synonyms are many: *boreale* Broderip and Sowerby, 1829; *pubescens* Couthouy, 1838; *arcticum* Sowerby, 1840; *dawsoni* Stimpson, 1862 and *hayesii* Stimpson, 1868.

Clinocardium nuttallii (Conrad, 1837)   **5584**
Nuttall's Cockle

Bering Sea to San Diego, California.

2 to 6 inches in length; smaller ones being almost round, adults tending to be higher than long; moderately compressed; commonly with 33 to 37 coarse radial ribs which are creased by half-moon-shaped riblets. Older specimens worn smoothish. Exterior drab-gray, with a brownish yellow, thin periostracum. Common offshore. Once called *C. corbis* Martyn. Known locally as the basket cockle.

5583

5584

Clinocardium fucanum (Dall, 1907)   **5585**
Fucan Cockle

Sitka, Alaska, to off Monterey, California.

1 to 1½ inches in length, longer than high, moderately inflated and with 45 to 50 low, poorly developed, radial ribs which are crossed by microscopic concentric lines. No wavy, radial furrow on the upper posterior edge of the shell. Color whitish with grayish brown periostracum. Common in the Puget Sound area.

Young *C. nuttallii* are distinguished from this species by their 2 first ribs behind the ligament which are large, rounded and make a wavy edge to the shell. In small specimens of *C. ciliatum*, the top edges of the ribs are sharp; in *fucanum* they are rounded.

Genus *Cerastoderma* Poli, 1795

Shells white or gray, with a coarse periostracum; oval, obese, radially ribbed, sometimes spiny. Not gaping. Type: *Cardium edule* Linné, 1758. Usually cold-water inhabitants. Synonym: *Edulicardium* Monterosato, 1923.

5585   5586

Cerastoderma pinnulatum (Conrad, 1831)   5586
Northern Dwarf Cockle

Labrador to off North Carolina.

¼ to ½ inch in length, thin, with 22 to 28 wide, flat ribs which have delicate, arched scales on the anterior slope of the shell. Scales missing on the central portion of the valve. Externally cream; interior glossy and white, rarely tinted with orange-brown. Commonly dredged from 7 to 100 fathoms.

Other species:

5586a *Cerastoderma elegantulum* Beck, in Möller, 1842. Greenland and northern Europe. See *Johnsonia,* vol. 1, no. 13, pl. 9.

Superfamily Mactracea Lamarck, 1809

Thin-shelled, porcelaneous; hinge with inverted V-shaped cardinal tooth in left valve, 2 cardinals in right valve. Internal ligament or resilium in a socketlike resilifer.

Superfamily Mactracea Lamarck, 1809

Family MACTRIDAE Lamarck, 1809

Shells large to small, ovate to subtrigonal, strong, usually convex or obese. Posterior-dorsal area bounded by a high fringe or keel. Ligament usually internal, the resilium set in a cup-shaped chondrophore. V-shaped cardinals and several laterals present.

Subfamily MACTRINAE Lamarck, 1809

Genus Mactra Linné, 1767

External portion of the ligament is small and attached to a scar along the dorsal margin just behind the beak. Surface smoothish, covered with a dark periostracum. Type: *stultorum* Linné, 1758, of Europe.

Subgenus Mactrotoma Dall, 1894

Type of this subgenus is *fragilis* Gmelin, 1791. With a silky periostracum. Ligament long; resilifer large and shallow.

5587

5591   5597

Mactra fragilis Gmelin, 1791   5587
Fragile Atlantic Mactra

North Carolina to Florida, Texas and the West Indies.

2 to 2½ inches in length, oval, moderately thin, but strong, and smoothish. Posterior slope with 2 radial, small ridges, one of which is very close to the dorsal margin of the valve. With a fairly large posterior gape. Color cream-white, with a thin, silky, grayish periostracum. Moderately common in shallow water. Rarely reaches 4 inches in length.

Subgenus Micromactra Dall, 1894

The type of this subgenus is *californica* Conrad, 1837. Shell solid, beaks and umbones with undulating concentric ribs.

Mactra californica Conrad, 1837   5588
Californian Mactra

Puget Sound, Washington, to Costa Rica.

Up to 2 inches in length, moderately fragile, moderately elongate and with the beaks central. Characterized by peculiar, concentric undulations on the beaks. Velvety periostracum forms an angular ridge along the top posterior edge of the valve. This small species of *Mactra* is common in lagoons and bays of southern California. It lives 3 to 6 inches below the surface of the sand.

Mactra nasuta Gould, 1851   5589
Gould's Pacific Mactra

San Pedro, California, to Colombia.

Up to 3½ inches in length, similar to *californica,* but more oval at the ventral margin, without concentric undulations on the beaks, and with 2 very distinct, raised, radial ridges on the posterior dorsal margin. The whitish shell is glossy and the periostracum is shiny and yellowish tan. Not very common. Synonyms are *californica* Reeve, 1854; *deshayesi* Conrad, 1868; *hiantina* Deshayes, 1855; *revellei* Durham, 1950.

Subgenus Simomactra Dall, 1894

Type of this weak subgenus is *dolabriformis* Conrad, 1867. Sometimes placed under *Spisula.*

Mactra dolabriformis (Conrad, 1867)   5590
Hatchet Surf Clam

Lobitas, California, to Panama.

3 to 4 inches in length, rather elongate, compressed and smooth. Posterior end shorter, but more expansive than the rather drawn-out anterior end. Right valve with the posterior lateral tooth separated into 2 teeth lengthwise by a long, deep channel. Color a smooth, ivory-white, with a dull, light-tan, thin periostracum. Small gape at the posterior end. Do not confuse with *Mactra nasuta* which dips down at the ventral margin, has a shiny periostracum and a wide posterior gape, nor with *Spisula falcata* which is similar in shape, but chalky and with very convex ventral margin to the hinge just below the chondrophore. Moderately common.

Genus Spisula Gray, 1837

In the hinge, the small oval ligament which is close to the dorsal margin, does not have a shelly ridge or lamina between it and the larger chondrophore, as is the case in

Mactra. Type: *solida* (Linné, 1758), of Europe. *Spissula* is a misspelling.

Subgenus *Hemimactra* Swainson, 1840

Anterior arm of the cardinal tooth in right valve is confluent with the ventral lamina. Cardinals markedly compressed. Laterals striate. Type: *solidissima* (Dillwyn, 1817).

Spisula solidissima (Dillwyn, 1817)   **5591**
Atlantic Surf Clam

Nova Scotia to South Carolina.

Up to 7 inches in length (usually about 4 or 5 inches), strong, oval and smoothish, except for small, irregular growth lines. The lateral teeth bear very tiny, sawtooth ridges. Color yellowish white with a thin yellowish brown periostracum. Common below low-water mark on ocean beaches. After violent winter storms, these clams are cast ashore in incredible numbers, some estimates giving an approximate count of 50 million clams along a 10-mile stretch.

(5592)   The subspecies *similis* (Say, 1822) (Cape Cod to both sides of Florida and to Texas) is more elongate, its anterior slope flatter and its pallial sinus longer and not sloping slightly upward. In the left valve, the tiny double tooth, just anterior to the spoon-shaped chondrophore, is usually much larger and stronger. Moderately common, and commonly ex-

isting with the typical species in the northern part of its range. *S. raveneli* (Conrad, 1831) may be a synonym. Compare with *polynyma* which has a larger pallial sinus.

Subgenus *Spisula* Gray, 1837

Spisula planulata (Conrad, 1837)   **5593**
Fattish Surf Clam

Monterey, California, to Baja California.

1 to 2 inches in length, ⅓ as high. Beaks almost at the middle. Anterior upper margin of the shell sharp-edged and straight. Exterior smooth, yellowish, shiny with the edges commonly stained with rusty-brown. Not very common. Found from low-water line to 36 fathoms.

Spisula catilliformis (Conrad, 1867)   **5594**
Catilliform Surf Clam

Washington State to Ensenada, California.

4 to 5 inches in length, almost as high as long. An oval shell with the beaks slightly nearer the anterior end. Moderately obese. Dull-ivory, commonly stained with reddish brown. With numerous, irregularly sized and spaced growth lines. Periostracum glossy, thin and usually worn off. Pallial sinus deep, running anteriorly as far back as the middle of the shell. Rather uncommonly washed ashore. Live specimens rare.

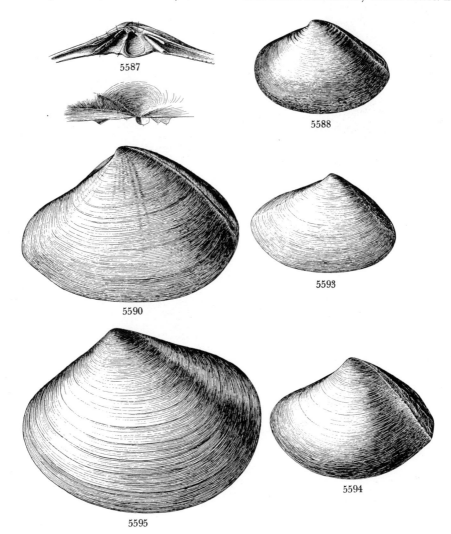

5587

5588

5590

5593

5595

5594

Spisula hemphillii (Dall, 1894)   5595
Hemphill's Surf Clam

Santa Barbara, California, to Baja California.

Up to 6 inches in length, about ¾ as high. Rather obese. Posterior end more obese and shorter than the downwardly swept, compressed anterior end. Periostracum grayish brown, dull, coarsely and concentrically wrinkled. The pallial sinus is moderately deep and inclined upward. Fairly common along the southern beaches of California.

Subgenus *Symmorphomactra* Dall, 1894

Cardinal teeth prominent, thin, with the posterior arm overhanging the resilifer. Accessory teeth present. Hinge plate flat. Type: *falcata* (Gould, 1850).

Spisula falcata (Gould, 1850)   5596
Hooked Surf Clam

Puget Sound, Washington, to California.

2 to 3 inches in length, rather elongate at the narrower anterior end. Exterior chalky with a partially worn-off, light-brown, shiny periostracum. Anterior upper margin of shell slightly concave. Moderately common in sand below low-water line.

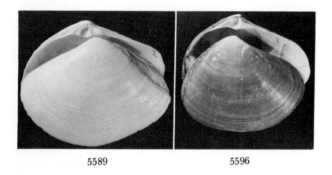

5589   5596

Subgenus *Mactromeris* Conrad, 1868

Periostracum fibrous. Lateral teeth smooth; cardinals not compressed. Type: *polynyma* Stimpson, 1860. *Mactrodesma* Conrad, 1869, is a synonym.

Spisula polynyma (Stimpson, 1860)   5597
Stimpson's Surf Clam

Arctic seas to Rhode Island.
Arctic seas to Puget Sound. Japan.

3 to 5 inches in length, beaks very near the middle of the valve. Anterior end smaller than the elliptical posterior end. Shell chalky, dirty-white and with a coarse, varnishlike, yellowish brown periostracum. Worn shells have coarse, concentric, wide growth lines. The pallial sinus is larger in this species than in *solidissima*. Moderately common from low-tide line to 60 fathoms. *Mactra ovalis* Gould, 1840 (non Sowerby, 1817) is a synonym. The form *alaskana* (Dall, 1894) is probably a synonym. The fossil, *S. voyi* (Gabb, 1868), from the Miocene or Pliocene is possibly only a subspecies.

Genus *Harvella* Gray, 1853

Shell thin; posterior slope with a strong keel. With strong concentric undulations on the disc. Anterior lateral teeth short. Type: *elegans* (Sowerby, 1825).

Harvella elegans (Sowerby, 1825)   5598
Elegant Wavy Mactra

Gulf of California to Peru.

2 inches, white, thin but strong; sides with strong, concentric undulations. Keel strong on the posterior slope. Commonly washed ashore from 14 to 40 fathoms in muddy areas. Synonyms: *pacifica* Conrad, 1867; *maxima* Li, 1930.

5598

Genus *Mactrellona* Marks, 1951

Shell large, thin, but strong, trigonal, inflated; posterior slope set off by a strong keel. Umbones prominent. Anterior laterals short. Type: *alata* (Spengler, 1802).

Mactrellona alata (Spengler, 1802)   5599
Caribbean Winged Mactra

Caribbean to Brazil.

2 to 4 inches in length, inflated, white, with a thin, flaky, yellowish periostracum. Posterior slope flattened and bounded by a distinctly elevated ridge. Common from Puerto Rico southward. 1 to 10 fathoms.

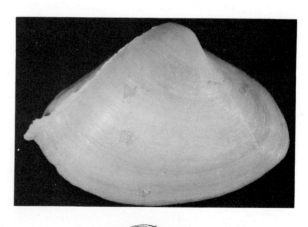

5599

Other species:

5600 *Mactrellona clisia* (Dall, 1915). Gulf of California to Ecuador. 3 inches.

5601 *Mactrellona exoleta* (Gray, 1837). Gulf of California to Peru, offshore to 13 fms. Moderately common. (Synonym: *ventricosa* Gould, 1851.)

Genus *Mulinia* Gray, 1837

Shell usually less than an inch; ligament and resilium both enclosed in a single pit and invisible from the outside. Laterals unequal in size, moderately distant. Pallial sinus short and small. Type: *edulis* (King and Broderip, 1832). *Mulinea* is a misspelling.

Mulinia lateralis (Say, 1822)   **5602**
Dwarf Surf Clam

   Maine to north Florida and to Texas.

⅓ to ½ inch in length, resembling a young *Spisula* or *Mactra*, moderately obese, beaks quite prominent and near the center of the shell and pointing toward each other. Exterior whitish to cream and smoothish, except for a fairly distinct, radial ridge near the posterior end. Concentric lines plainly seen in the thin, yellowish periostracum. Distinguished from young *Spisula solidissima* which have a proportionately much larger chondrophore in the hinge and which have tiny, sawtooth denticles on the lower anterior and lateral hinge-teeth. A very abundant species in warm, shallow water in sand. *M. corbuloides* Deshayes, 1854, is a synonym.

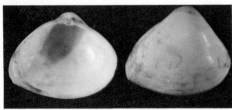

5602

Other species:

5603 *Mulinia coloradoensis* Dall, 1894. Gulf of California. 2 inches. Common. (Synonym: *modesta* Dall, 1894.)

5604 *Mulinia cleryana* (Orbigny, 1846). West Indies to Brazil, shallow water. (Synonyms: *guadelupensis* (Récluz, 1852) and *portoricensis* Shuttleworth, 1856.)

Genus *Rangia* Desmoulins, 1832

Shells 1 to 2 inches, very heavy, thick. Lateral teeth cross-striated. Pallial sinus small. Periostracum thin, grayish. Type: *cuneata* (Sowerby, 1831). Synonym: *Gnathodon* Sowerby, 1831, non Goldfuss, 1820.

Rangia cuneata (Sowerby, 1831)   **5605**
Common Rangia

   North Chesapeake Bay to Texas.

1 to 2½ inches in length, obliquely ovate, very thick and heavy. The beaks which are near the oval, anterior end are high, inrolled and pointing downward and anteriorly. Exterior whitish, but covered with a strong, smoothish, gray-brown periostracum. Interior glossy, white and with a blue-gray tinge. Pallial sinus small, but moderately deep and distinct. A com-

mon fresh-water to brackish-water species found in coastal areas. *R. nasuta* Dall, 1884, is probably only a rostrate form of this species. Compare with *R. flexuosa*. For northern range, see *The Nautilus*, vol. 83, p. 22.

Subgenus *Rangianella* Conrad, 1863

Shell rostrate, with the laterals short, straight and nearly smooth; pallial sinus obsolete. Type: *mendica* Gould, 1851.

Rangia flexuosa (Conrad, 1839)   **5606**
Brown Rangia

   Louisiana to Texas and Vera Cruz, Mexico.

1 to 1½ inches in length, resembling an elongate *cuneata*, but with no distinct pallial sinus, with much shorter laterals, with a faintly impressed, large lunule and colored light-brown inside. A rare and elusive species from marsh areas. *R. rostrata* Petit, 1853, is a synonym.

5605

5606

Other species:

5607 *Rangia (Rangianella) mendica* (Gould, 1851). Gulf of California to Mazatlan, Mexico. 1 inch. Common in mangrove brackish waters.

Subfamily LUTRARIINAE H. and A. Adams, 1856

Genus *Tresus* Gray, 1853

Shell large with a roundish posterior gape. Hinge with small cardinal teeth; lateral teeth very small and close to the cardinals. Ligament external and separated from the cartilage pit by a shelly plate. *Schizothaerus* Conrad, 1853, and *Cryptodon* Conrad, 1837, non Turton, 1822, are synonyms. Type: *nuttalli* (Conrad, 1837).

Tresus nuttalli (Conrad, 1837)   **5608**
Pacific Gaper

   Washington to Baja California.

Up to 8 inches in length. An oblongish to oval, strong, smoothish shell with a prominent gape at the posterior end. The neat, well-formed beaks are located ¼ to ⅓ from the anterior end. The pallial sinus is very large and deep. Periostracum grayish. Common. Compare with the northern species *capax* Conrad.

Tresus capax (Gould, 1850)   **5609**
Alaskan Gaper or Horse Clam

   Kodiak Island, Alaska, to Monterey, California.

5608

5609   5612

Up to 10 inches in length, differing from *nuttalli* in being much more oval, more obese and dipping downward into a well-rounded, ventral margin. This species is very common on most sand and mud beaches in Puget Sound. The pinnixid crabs, *Pinnixa faba* and *littoralis*, coexist with this species. For a lengthy account of the differences between these species, see E. F. Swan and J. H. Finucane, 1952, *The Nautilus*, vol. 66, p. 19; and J. B. Pearce, 1965, *The Veliger*, vol. 7, p. 166.

Subfamily PTEROPSELLINAE Keen, 1969

Genus *Anatina* Schumacher, 1817

Posterior slightly gaping. Shell fragile. Hinge with a prominent chondrophore. Cardinal teeth small and close to the chondrophore. Ligament submerged, except at the anterior end, and separated from the chondrophore by a shelly plate. Type: *anatina* (Spengler, 1802). *Labiosa* Möller, 1832; *Cypricia* Gray, 1847; and *Leucoparia* Mayer, 1867, are synonyms.

Anatina anatina (Spengler, 1802)   5610
Smooth Duck Clam

North Carolina to Florida and to Texas; Brazil.

2 to 3 inches in length, ¾ as high, fairly thin but strong. White to tan in color. Moderately smooth, except for irregular growth lines and tiny, but distinct, concentric ribs near the beaks. Posterior end with a distinct radial rib behind which the shell gapes with flaring edges. Uncommon in most areas of its range. Synonyms are: *pellucida* Schumacher, 1817; *lineata* Say, 1822; *recurva* Wood, 1828. See A. M. Keen, 1961, *The Veliger*, vol. 4, p. 9.

Other species:

5611 *Anatina cyprinus* (Wood, 1828). Gulf of California to Ecuador.

Genus *Raeta* Gray, 1853

Surface of valves concentrically undulated or plicated. Type: *plicatella* (Lamarck, 1818). *Lovellia* Mayer, 1867, is a synonym.

Raeta plicatella (Lamarck, 1818)   5612
Channeled Duck Clam

North Carolina to Florida, Texas and the West Indies to Argentina.

2 to 3 inches in length, ⅘ as high, eggshell thin, but moderately strong. Concentric sculpture of smoothish, distinct ribs which on the inside of the valves show as grooves. Radial sculpture of very fine, crinkly threads. Color pure-white. Formerly known as *Raeta canaliculata* Say, 1822, *R. campechensis* Gray, 1825, and *perspicua* Hutton, 1863. Commonly washed ashore, especially along the strands of the Carolinas, but rarely seen alive. Lives in 2 to 3 fathoms just offshore. For anatomy, see H. W. Harry, 1969, *The Veliger*, vol. 12, p. 1.

Raeta undulata (Gould, 1851)   5613
Pacific Duck Clam

San Pedro, California, to Peru.

3 to 4 inches, thin-shelled, white, oval. Rounded anteriorly, somewhat pointed posteriorly where the dorsal border is smoothish. Remainder of valve with numerous rounded, concentric ridges. Pallial sinus deep and narrow. Uncommon; offshore. Valves are sometimes washed ashore.

5610   5613

Family MESODESMATIDAE Gray, 1839

Genus *Mesodesma* Deshayes, 1832

Like a large *Donax* with a prominent chondrophore. Laterals with fine denticles. Periostracum fairly thick, yellowish. Ligament short, mostly external. Type: *donacium* (Lamarck, 1818), from Chile.

Mesodesma arctatum (Conrad, 1830) 5614
Arctic Wedge Clam

Greenland to Chesapeake Bay.

About 1½ inches in length, somewhat shaped like a *Donax*, fairly thick and compressed. Chondrophore fairly large and spoon-shaped. Left valve with a long anterior and posterior lateral tooth, both of which have fine, comblike teeth on each side. Pallial sinus small and U-shaped. Interior tan to cream. Exterior with a thin, yellowish, smooth periostracum. Common from low water to 50 fathoms.

Mesodesma deauratum (Turton, 1822) 5615
Turton's Wedge Clam

Gulf of St. Lawrence, Canada.

About 1½ inches in length, differing from *arctatum* in having a less truncate posterior end, and does not plunge sharply to the ventral margin. Synonyms: *Mactra denticulata* Wood, 1828; *jauresii* De Joannis, 1834. Erroneously reported from England. See J. D. Davis, 1965, *The Nautilus*, vol. 78, p. 96.

Subfamily ERVILIINAE Dall, 1895

Shells small, compressed; ligament marginal, subobsolete; resilium small; lateral teeth small, 1 cardinal in either valve, that in the left valve being bifid. Pallial sinus distinct.

Genus *Ervilia* Turton, 1822

Shell small, less than 10 mm., concentrically striate, and sometimes brightly hued. Ligament absent; resilium small and internal. Laterals small or absent. Left cardinal large and bifid or split. Right valve has a prominent cardinal just anterior to a large chondrophoral pit. Pallial sinus large. Type: *nitens* (Montagu, 1806).

Ervilia concentrica (Holmes, 1860) 5616
Concentric Ervilia

North Carolina to both sides of Florida to Brazil. Bermuda.

7 to 10 mm. in length, ⅔ as high, elliptical in outline, moderately compressed, although some are somewhat inflated. Each end is rounded to about the same degree, and the beaks are almost central. There is a pinpoint depression just behind the glossy, inrolled beaks. Sculpture of fine, numerous concentric ridges. Radial threads may be present to form tiny beads. Color white, yellow or commonly with a pink blush. Common just offshore to 50 fathoms. *E. rostratula* Rehder, 1943 (a forma), from Lake Worth, Florida, is similar, but the posterior end is slightly more pointed. As pointed out by J. D. Davis, 1967, *Malacologia*, vol. 6, p. 231, the author of *concentrica* should not be Gould, 1862.

Ervilia nitens (Montagu, 1806) 5617
Montagu's Ervilia

Florida and the West Indies. Bermuda.

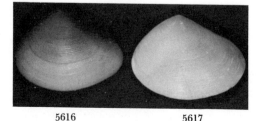

5616   5617

7 to 10 mm. in length, closely resembling *concentrica*, but differs in having the umbo about ⅓ the way back from the anterior end, whereas in *concentrica* the umbo is nearly at the midpoint. *E. nitens* is much thinner or flatter and has slightly stronger concentric threads in the region of the umbones. Color ranges from white to pink to yellow. Common; shallow water down to 22 fathoms.

(5618) The subspecies or forma *venezuelana* Weisbord, 1964, from Venezuela (fossil) to Brazil (living) is more compressed and a little more elongate. It lacks the thickening on the inner side running from the umbo to the anterior muscle scar that is found in *subcancellata*.

Ervilia californica Dall, 1916 5619
Californian Ervilia

San Pedro, California, to Baja California.

7 mm. long, 4.5 mm. high, ovate, white with a rosy flush, posterior end slightly shorter. Ends rounded, the ventral margin arcuate. Beaks inconspicuous. Sculpture of fine, closely set, regular, concentric threads over the whole surface. Hinge strong. Pallial sinus small. Moderately common; just offshore. This is the only West Coast *Ervilia*.

Other species:

5620 *Ervilia subcancellata* E. A. Smith, 1885. Bermuda. Northeast Brazil. "Challenger" Expedition. J. D. Davis, 1973, considers *E. rostratula* Rehder, 1943, as large specimens of this species.

Superfamily Solenacea Lamarck, 1809

Family SOLENIDAE Lamarck, 1809

Rapidly burrowing, elongate, thin-shelled clams, with a glossy periostracum, beaks near the anterior end, and gaping at both ends.

Subfamily PHARELLINAE Tryon, 1884

Valves wider and more compressed than in *Soleninae*; beaks mostly not terminal; hinge with 1 to 3 cardinal teeth. Cultellinae Davies, 1935, is a synonym.

Genus *Siliqua* Mühlfeld, 1811

Shell commonly 6 inches in length, oval, compressed laterally; with a rather straight, raised, internal rib ventrally directed. Hinge like *Ensis*. Type: *radiata* (Linné, 1758). Synonyms: *Leguminaria* Schumacher, 1817; *Solecurtoides* Des Moulins, 1832; *Machaera* Gould, 1841.

Siliqua costata Say, 1822 5621
Atlantic Razor Clam

Gulf of St. Lawrence to North Carolina.

2 to 2½ inches in length, ovate-elongate, compressed, fragile, smooth and with a shiny, green periostracum. Interior glossy, purplish white, with a strong, white raised rib running down from the hinge to the middle of the anterior end. Very common on shallow-water and sand flats along the New England coast.

(5622) *Siliqua squama* Blainville, 1824, found offshore from Newfoundland to Cape Cod is larger, thicker, white internally, and its internal, supporting rib slanting posteriorly instead of anteriorly as in *costata*. Uncommon.

5621

5624

5625

Siliqua lucida (Conrad, 1837)   5623
Transparent Razor Clam

Bolinas Bay, California, to Baja California.

1 to 1½ inches in length. Very thin, fragile and translucent. Moderately elongate. Shell whitish tan, with broad, indistinct, radial rays of darker tan or rosy-purplish. Periostracum thin, olive-green and varnishlike. Moderately common in sand at low tide to 25 fathoms. This species can be distinguished from the young of *S. patula* by its narrower and higher internal rib which crosses the shell at right angles, and in being more arcuate on its ventral margin. Also compare with *sloati* Hertlein.

Siliqua patula (Dixon, 1788)   5624
Pacific Razor Clam

Alaska to Monterey, California.

5 to 6 inches in length, oval-oblong in shape, laterally compressed and moderately thin. Periostracum varnishlike and olive-green. Interior glossy and whitish with a purplish flush. Internal rib under teeth descending obliquely toward the anterior end. Animal without dark coloration. The variety *nuttallii* Conrad is a synonym. Do not confuse with *S. alta*. An abundant, edible species found in mud and sand on ocean beaches. For anatomy and habits, see R. H. Pohlo, 1963, *The Veliger*, vol. 6, p. 98.

Siliqua alta (Broderip and Sowerby, 1829)   5625
Dall's Razor Clam

Arctic Ocean to Cook's Inlet, Alaska. Russia.

4 to 5 inches in length, similar to *patula*, but chalky-white inside, more truncate at both ends, a heavier shell, and with a stronger, narrower and vertical (not oblique) rib on the inside. The soft parts are chocolate-brown. *S. media* (Sowerby, 1839) from the same region may possibly be the young of this species, although it is blushed with purple inside. *S. alta* is common and edible.

Siliqua sloati Hertlein, 1961   5626
Sloat's Razor Clam

Portage Bay, Alaska, to Marin County, California.

1 to 1½ inches, similar to *lucida*, but the posterior end is more pointed, the posterior dorsal area is less expanded, more curved and not bordered by a distinct groove and the exterior of the shell is highly polished and ornamented by brightly colored bands of brown and cream in comparison to the subdued brownish and purplish color of *lucida*. Moderately common; from 10 to 86 fathoms. *Bull. So. Calif. Acad. Sci.*, vol. 60, p. 14.

Genus *Ensis* Schumacher, 1817

The jackknife clams closely resemble *Solen*, but the left valve has 2 vertical, cardinal teeth, and each valve has a long, low posterior tooth. The valves are apt to be slightly curved. Type: *ensis* Linné, 1758.

Ensis directus Conrad, 1843   5627
Atlantic Jackknife Clam

Labrador to South Carolina. Florida?

Up to 10 inches in length, 6 times as long as high, moderately curved and with sharp edges. Shell white, covered with a thin, varnishlike, brownish green periostracum. Common on sand flats in New England. Edible.

5627

Ensis minor Dall, 1900   5628
Minor Jackknife Clam

New Jersey to Florida to Texas.

Rarely exceeds 3 inches in length, is more fragile, relatively longer and more pointed at the free end (not the end with teeth). Internally it has purplish stains. Moderately common between tide marks. Some workers consider this a subspecies of *directus*. *E. megistus* Pilsbry and McGinty, 1943 (*The Nautilus*, vol. 57, p. 33) are probably 5-inch long specimens of *minor*.

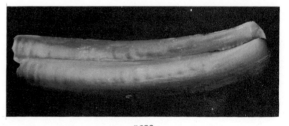

5628

Ensis myrae S. S. Berry, 1953   5629
Californian Jackknife Clam

Southern half of California.

2 to 3 inches in length, slender, polished, with much the same characters as in *directus*. Anterior end squared off. This is the only *Ensis* in California and is uncommonly dredged offshore in 10 to 25 fathoms. It has been erroneously called *californicus* Dall, 1899, which, however, is a more southerly,

smaller species with a rounded anterior end. See *Trans. San Diego Soc. Nat. Hist.*, vol. 11, pp. 395–402.

Other species:

5630 *Ensis californicus* Dall, 1899. Baja and Gulf of California to Ecuador. 1 to 2 inches.

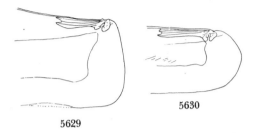

5629   5630

5631 *Ensis tropicalis* Hertlein and Strong, 1955. Gulf of California to Panama, 11 to 25 meters.

Subfamily SOLENINAE Lamarck, 1809

Beaks terminal; hinge with only 1 tooth in either valve.

Genus *Solen* Linné, 1758

Similar to *Ensis*, but with only a single tooth at the end of each valve. Beaks at the extreme anterior end. Type: *vagina* Linné, 1758.

Solen viridis Say, 1821   5632
Green Jackknife Clam

Rhode Island to northern Florida and the Gulf States.

About 2 inches in length, about 4 times as long as wide; dorsal edge straight, ventral edge curved. Hinge with a single projecting tooth at the very end of the valve. Color white; periostracum thin, varnishlike, light greenish or brownish. Moderately common in shallow-water sand flats.

Solen sicarius Gould, 1850   5633
Blunt Jackknife Clam

British Columbia to Baja California.

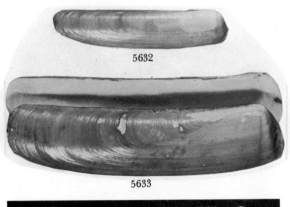

5632

5633

5634

2 to 4 inches in length, 4 times as long as wide. Exterior with a varnishlike, olive to greenish periostracum. Moderately common in certain localities in the north, especially sandy mud flats. Also dredged in 25 fathoms. Rare in the south.

Solen rosaceus Carpenter, 1864   5634
Rosy Jackknife Clam

Santa Barbara, California, to Mazatlan, Mexico.

1 to 3 inches in length, almost 5 times as long as wide. Shell fragile, with a thin, glossy, olive periostracum. Beachworn specimens are whitish with rosy stains inside and out. It is more cylindrical, the anterior extremity is more rounded and narrower than in *sicarius*. An abundant species along the sandy shores of bays. Also to 25 fathoms. For anatomy and habits, see R. H. Pohlo, 1963, *The Veliger*, vol. 6, p. 98.

Other species:

5635 *Solen obliquus* Spengler, 1794. West Indies to Brazil.

Superfamily Tellinacea Blainville, 1814

Moderately sized, usually somewhat thin-shelled clams, with 2 cardinal teeth in each hinge, tending to be bifid. Pallial sinus and unfused siphons prominent. Includes the living families Tellinidae, Donacidae, Psammobiidae, Scrobiculariidae, Semelidae and Solecurtidae.

Family TELLINIDAE Blainville, 1814

Subfamily TELLININAE Blainville, 1814

A well-known, worldwide family of laterally compressed clams, usually having a slight twist at the posterior end, and having a strong, brown ligament on the upper edge just behind the tiny umbones. Foot relatively short and not grooved. Siphons long, extensile, separated to their bases and retractible into the large pallial sinus. 2 cardinal teeth (one of which is bifid) in each valve; lateral teeth present in the subfamily Tellininae, but absent in the Macominae. All are mud or sand dwellers. For details, see the excellent series of monographs in *Johnsonia* of the Western Atlantic forms by Kenneth J. Boss (vol. 4, nos. 45, 46, 47, 1966–69) and Eugene V. Coan's "Northwest American Tellinidae," *The Veliger,* vol. 14 supplement, July 1971.

Genus *Tellina* Linné, 1758

Subgenus *Tellina* Linné, 1758

Smooth, polished surface. Hinge with 2 cardinals and 2 laterals in either valve. Type: *radiata* Linné, 1758.

Tellina radiata Linné, 1758   5636
Sunrise Tellin   **Color Plate 23**

South Carolina to south half of Florida; Bermuda to the Guianas.

2 to 4 inches in length, elongate, moderately inflated. Characterized by its oily-smooth, glistening surface and rich display of colors—either creamy-white or rayed with pale-red or yellow. Interior flushed with yellow. The beaks are usually tipped with bright-red. Pallial sinus almost touches the an-

terior muscle scar. The white to yellow, rayless form was called *unimaculata* Lamarck, 1818 (Color Plate 23). Lives off-shore down to 48 feet in sand.

Subgenus *Laciolina* Iredale, 1937

Shells 2 to 4 inches, smooth, and with a thick, calcareous, sunken portion underlying the ligament. Two species in the Caribbean, others in the Indo-Pacific. Type: *quoyi* Sowerby, in Reeve, 1868.

Tellina laevigata Linné, 1758   5637
Smooth Tellin   Color Plate 23

Bermuda; North Carolina to the Lower Caribbean.

2 to 3½ inches in length, oval to slightly elongate, moderately compressed, strong, with a smooth, glossy surface except for microscopic, radial scratches. Exterior color either whitish or usually faintly rayed, or banded at the ventral margins with soft, creamy-orange. Inside polished white to yellowish. Both anterior and posterior lateral teeth present. Rare in Florida, fairly common in the West Indies and Bermuda. In 1 to 8 fathoms; in sand. *T. concinna* Philippi, 1844; *bayleana* Bertin, 1876; and *stella* Davis, 1904, are synonyms. For information on *laevigata* × *magna* hybrids, see K. J. Boss, *The Nautilus*, vol. 78, p. 18.

Tellina ochracea Carpenter, 1864   5638
Ochraceous Tellin

Gulf of California.

2 inches, very similar to *laevigata* but is light-yellow with darker yellow at the beaks. Rare from low-tide mark to 45 fathoms; in sand.

Tellina magna Spengler, 1798   5639
Great Tellin   Color Plate 23

Bermuda; North Carolina to the south half of Florida and the West Indies.

3 to 4½ inches in length, ½ as high, quite compressed and glossy-smooth. Posterior dorsal region dull, bordered by a weak, radial ridge. Left valve glossy-white, rarely faintly yellowish; right valve glossy-orange to pinkish and microscopically cut with concentric scratches. Only the right anterior lateral tooth is developed. An uncommon and very lovely species much sought after by collectors. Found just below low tide in sand. It sometimes hybridizes with *laevigata* in Bermuda. The synonyms are: *acuta* Wood, 1815; *elliptica* Lamarck, 1818; *vitrea* Orbigny, 1842; and *sol* Hanley, 1844.

Subgenus *Tellinella* Mörch, 1853

Shells with strong concentric sculpture; right anterior lateral tooth far from the cardinals; pallial sinus connected to the front muscle scar by a linear scar; posterior end of shell twisted to right. Type: *virgata* Linné, 1758. Synonym: *Eutellina* Fischer, 1887.

Tellina listeri Röding, 1798   5640
Speckled Tellin   Color Plate 23

Bermuda; North Carolina to south half of Florida and to Brazil.

2½ to 3½ inches in length, well-elongated, moderately inflated, twisted at the posterior end where at the dorsal margin

on the right valve there are 2 rough ridges. Concentric threads numerous, evenly spaced. Color whitish with numerous small, prominent, zigzag specklings of purplish brown. Interior yellowish. Young specimens are more elongate and more rugose. Not uncommon in southeast Florida, but abundant in coarse sand in some shallow West Indian areas from 1 to 50 fathoms. The synonyms are: *interrupta* Wood, 1815; *maculosa* Lamarck, 1818; and *mexicana* Petit, 1841.

Tellina cumingii Hanley, 1844   5641
Cuming's Tellin

Baja California to Colombia.

1 to 2 inches; rayed with broken purple lines; this is the Pacific analog of *listeri* and may be considered a subspecies of it. Moderately common offshore from 5 to 40 fathoms.

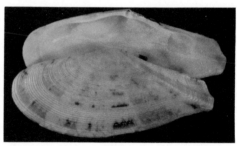

5641

Tellina idae Dall, 1891   5642
Ida's Tellin

Santa Monica to Newport Bay, California.

2 to 2½ inches in length, elongate, compressed. With strong, rather evenly spaced, concentric, lamellate threads. Posterior end narrow, slightly twisted and with a rounded, radial ridge near the dorsal margin (in right valve) or a ridge at the dorsal margin and a furrow below it (left valve). Ligament elongate and sunk deeply into the long, deep dorsal-margin furrow. Color grayish white. Moderately common.

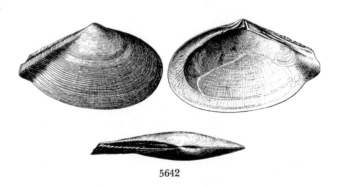

5642

Subgenus *Arcopagia* Brown, 1827

Shells large, orbicular and very thick; lateral teeth strong in the right valve. Pallial sinus deep, rounded. Type: *crassa* Pennant, 1777.

Tellina fausta Pulteney, 1799   5643
Faust Tellin   Color Plate 23

North Carolina to southeast Florida and the lower Caribbean.

2 to 4 inches in length, oval, moderately inflated, very heavy, and smoothish, except for small, rough, concentric lines of growth. Hinge strong, the posterior lateral in the right valve being long and strong. Color outside a semiglossy-white; inside highly glossed and enamel-white with a yellowish flush. Do not confuse with the thinner *T. laevigata* which is glossy outside and has orange-tinted margins. Moderately common in the West Indies where it lives in sand near eel-grass beds at 1 to 15 fathoms. Donovan gave this species the same name in 1801. Other synonyms are *laevis* Wood, 1815; *elliptica* Sowerby, 1868.

Subgenus *Phyllodina* Dall, 1900

Shells similar to those of the subgenus *Tellinella*, but without a posterior twist. The pallial sinus slants upwards and is connected to the front muscle scar by a long line. Two Atlantic species with two Pacific analogs. Type: *squamifera* Deshayes, 1855.

Tellina squamifera Deshayes, 1855   5644
Eastern Crenulate Tellin

North Carolina to Florida and Texas.

½ to 1 inch in length, elongate, concentrically and finely ridged. Characterized by the strong crenulations on the posterior dorsal margin and by the lightly hooked-down posterior ventral margin. Color whitish with a yellow or orangish tint. Moderately common from low water to 125 fathoms. Compare with *lintea* which lacks the dorsal crenulations. Formerly placed in the genus *Phylloda* Schumacher, 1817.

5644

Tellina pristiphora Dall, 1900   5645
Western Crenulate Tellin

Gulf of California to Costa Rica.

½ to 1½ inches. This is the Pacific analog of *squamifera*; yellowish white with pinkish in the upper front edge. Uncommon offshore in sand from 12 to 85 fathoms.

Tellina persica Dall and Simpson, 1901   5646
Apricot Tellin

Off northern Cuba to the Lesser Antilles.

5645   5646

¾ to 1 inch; white with a suffusion of apricot; with small pointed umbones. Similar to *squamifera* but is more oval in shape, lacks the foliations on the back top edge, and has finer, more regular concentric sculpture. Moderately rare in sand from 14 to 230 fathoms.

(5647) The Pacific analog, *Tellina flucigera* Dall, 1908, is rare (182 fathoms in Panama Bay), is 1½ inches; has flat umbones and has slight serrations on the dorsal margins.

Subgenus *Elliptotellina* Cossmann, 1886

Shell up to ⅓ inch, elliptical; posterior half with strong radial, beaded riblets; both ends rounded; 2 laterals in right valve are strong, the 2 in the left valve are very weak. One Caribbean, one Eastern Pacific, and two Indo-Pacific species only. Type: *tellinella* Lamarck, 1805.

Tellina americana Dall, 1900   5648
Dwarf American Tellin

North Carolina to Mississippi to Barbados.

4 to 8.5 mm., rounded at both ends, swollen valves of equal convexity and with a slight posterior twist. Concentric ribs are evenly spaced, raised and rounded. Radial threads on posterior ⅓. Color white or straw with a crimson or brown spot on the dorsal margin near each end. A distinct little tellin dredged from 25 to 100 fathoms.

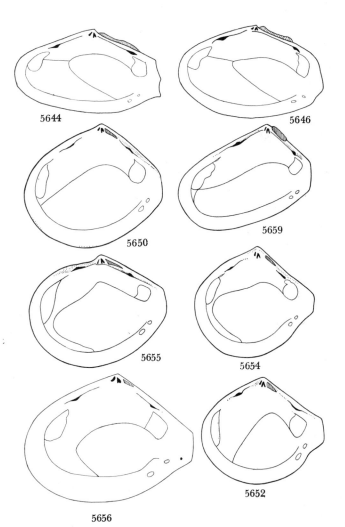

5644   5646
5650   5659
5655   5654
5656   5652

5648

5652

5650

5654

5660

Tellina pacifica Dall, 1900   5649
Dwarf Pacific Tellin

Gulf of California to Panama.

6 to 8 mm., this is the Pacific analog of *americana*. Coloration the same, but the radial riblets are found on the central, as well as the posterior, part of the valves. Uncommon from 4 to 18 fathoms.

Subgenus *Merisca* Dall, 1900

Shells dull-white, less than 1 inch, with strong, widely spaced concentric lamellations. Lateral teeth in right valve strong, weak or absent in left valve. Mostly tropical species, 5 in the Western Atlantic, 6 in the Eastern Pacific and several in the Indo-Pacific. Type: *cristallina* Spengler, 1798. *Lyratellina* Olsson, 1961, is a synonym.

Tellina aequistriata Say, 1824   5650
Lintea Tellin

North Carolina to Texas and to Brazil.

¾ to 1 inch in length, moderately oval, slightly inflated, quite strong and all-white in color. Posterior dorsal slope with 2 radial ridges in the right valve, 1 in the left. Concentric lamellae numerous, sharp and minutely raised. Left valve with 2 extremely weak, long laterals, but these are well-developed in the right valve. Dorsal line of the pallial sinus meets the pallial line not far from the anterior muscle scar. Posterior twist to the right is fairly pronounced. Commonly dredged off the Carolinas (9 to 22 fathoms), uncommonly found in a few

feet of water on the west coast of Florida. Formerly placed in *Quadrans* Bertin, 1878. Synonyms are *lintea* Conrad, 1837; *guadaloupensis* Orbigny, 1842; *quadalupensis* [sic] Orbigny, 1845; and *tumida* Sowerby, 1867.

Tellina reclusa Dall, 1900   5651
Reclusa Tellin

Baja California to Panama.

20 mm.; this is the Eastern Pacific analog of *aequistriata* Say. It is not as elongate and has a rougher, rasplike sculpture. Moderately common from 3 to 40 fathoms.

Tellina cristallina Spengler, 1798   5652
Crystal Tellin

South Carolina to the lower Caribbean.

1 inch, thin-shelled, with a short snout-shaped posterior extension which is bounded by 2 ridges. Umbones central and conspicuous. Left valve somewhat convex, right valve flat. Sculpture of strong, very widely spaced, sharp, concentric ridges. Ligament somewhat sunken. Rare in the United States, uncommon in the West Indies. *T. cristallina* Wood, 1815, and *schrammi* Récluz, 1853 are synonyms.

(5653) The Eastern Pacific analog is *Tellina* (*Merisca*) *rhynchoscuta* Olsson, 1961, which is stubbier, heavier and with coarser concentric sculpture. Uncommon, from the Gulf of California to Ecuador from low tide to 13 fathoms.

Tellina martinicensis Orbigny, 1842   5654
Martinique Tellin

South half of Florida to the Lesser Antilles. Brazil.

½ inch, slightly inflated, valves of equal convexity; with a strong twist to the right. Sculpture of crowded, strong, concentric ridges (about 4 per mm.). White or rarely iridescent. Umbones subcentral, inflated, blunt and may point forward or towards each other. Do not confuse with *aequistriata* Say which is more compressed, has weaker ridges, and has its sinus confluent at or near the base of the front muscle scar. A moderately common white tellin found in 1 to 20 fathoms.

Other *Merisca*:

5655 *Tellina juttingae* Altena, 1965. 15 to 20 fms., lower Caribbean. Brazil. *Basteria*, vol. 29; *Johnsonia*, vol. 4, no. 45, pl. 140.

5656 *Tellina alerta* Boss, 1964. 48 fms., Brazil and Uruguay. See *Johnsonia*, vol. 4, no. 45, pl. 139.

5657 *Tellina lamellata* Carpenter, 1855. Mazatlan, Mexico. See Keen, fig. 388.

5658 *Tellina proclivis* Hertlein and Strong, 1949. (Synonym: *declivis* Sowerby, 1868.) 12 to 26 fms., Baja California to Panama.

Subgenus *Acorylus* Olsson and Harbison, 1953

Shell ¾ inch, solid, obliquely subovate. Hinge line strong and heavy. 2 strong lateral teeth in the right valve, equidistant from the cardinal teeth. No true lateral teeth in the left valve. Palliate sinus large, reaching the front muscle scar and confluent with the pallial line. One living species in the Western Atlantic. Type: *suberis* Dall, 1900.

Tellina gouldii Hanley, 1846   5659
Dwarf Cuneate Tellin

Bermuda; southeast Florida to the lower Caribbean.

6 to 11 mm., milk-white, glossy, solid, strong with very weak concentric threads which are generally evident only on

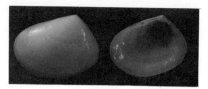

5659

the very blunt, somewhat truncated posterior end. Fairly common in sandy, weedy areas from 1 to 280 fathoms. *Tellina cuneata* Orbigny, 1842 (not Spengler, 1798) is a synonym.

Subgenus *Eurytellina* P. Fischer, 1887

Shells 1 to 3 inches, laterally compressed, elongate, without significant posterior twist, with fine concentric groovings, with a weak, interior rib radiating from the umbonal area just past the anterior muscle scar. Lateral teeth of right valve strong. Type: *punicea* Born, 1778. The resembling subgenus *Angulus* has no supportive rib internally and the posterior lateral tooth is weak, not strong.

Tellina lineata Turton, 1819   **5660**
Rose Petal Tellin   **Color Plate 23**

All of Florida to Texas and to Brazil.

1½ inches in length, moderately elongate, left valve slightly inflated, solid and with a fairly strong twist to the right at the posterior end. Smoothish and glossy, but under a lens fine, concentric, crowded grooves may be seen. The outer surface has a slight opalescent sheen. Color pure-white or strongly flushed with watermelon-red. Pallial sinus just touches the anterior muscle scar, while in the similar but more elongate *T. alternata* it does not. It differs from *T. tampaensis* which has a very weak right posterior lateral tooth and a pallial sinus that does not come near the front muscle scar. An abundant species occurring in sand from low-tide mark to several fathoms. *T. brasiliana* Lamarck, 1818, and *decussatula* C. B. Adams, 1845 are synonyms.

Tellina alternata Say, 1822   **5661**
Alternate Tellin   **Color Plate 23**

North Carolina, Florida to Texas.

2 to 3 inches in length, elongate and compressed, solid, and with a moderately pointed and slightly twisted posterior end. Sculpture of numerous evenly spaced, fine, concentric grooves. Area near umbones smooth. Color glossy and variable: whitish, yellowish or flushed with pink. Interior glossy-yellow or pinkish. The right valve has from 4 to 9 grooves more per centimeter than the left. *T. planulata* Sowerby, 1867, is a synonym. A common shallow-water species which should always be compared with *lineata*. It occurs offshore down to 70 fathoms but is frequently washed ashore.

Tellina alternata subspecies ***tayloriana*** Sowerby, 1867   **5662**
Taylor's Tellin

Texas to Tampico, Mexico.

Very similar to the typical northern subspecies, *alternata,* but is light- to bright-pink throughout, with a broad, fairly strong anterior radial rib inside, with somewhat flatter valves and with a more distinct concentric sculpture on the anterior dorsal surface of the left valve. Moderately common in sand from just below the low-tide line to 6 fathoms.

Tellina angulosa Gmelin, 1791   **5663**
Angulate Tellin   **Color Plate 23**

South half of Florida; Yucatan to Uruguay.

2 inches, similar to *alternata* Say but not as elongate, has finer grooves and a more glossy surface which is usually covered with a greenish yellow periostracum. 3 orange-red or yellow rays color the umbonal region. The pallial sinus scar does not reach the front muscle scar. It may be hybridizing with *alternata* in some areas of southern Florida. Fairly common in 1 to 4 fathoms. Synonyms are *striata* Spengler, 1798; *rosacea* King and Broderip, 1832; and *martinicensis* Lamarck, 1818.

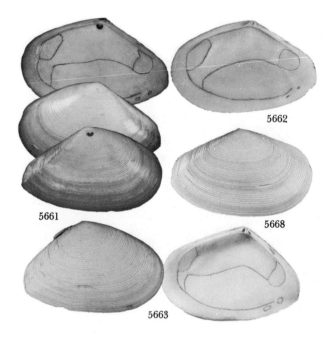

5662

5661

5668

5663

Tellina punicea Born, 1778   **5664**
Watermelon Tellin

Caribbean from British Honduras to Brazil.

1 to 2¼ inches; similar to *angulosa* but bright watermelon-red or purplish red inside and outside. Pallial sinus just touches the anterior muscle scar. A common shallow-water West Indian species.

Tellina simulans C. B. Adams, 1852   **5665**
Similar Red Tellin

Gulf of California to Peru.

1 to 2 inches; watermelon-red; this is the Eastern Pacific analog to *punicea* Born, but the pallial sinus does not touch the anterior muscle scar. Moderately common from low-tide mark to 13 fathoms in sandy mud.

5665

Tellina nitens C. B. Adams, 1845   5666
Dwarf Shiny Tellin

North Carolina to Texas; Florida to Brazil.

1 to 1½ inches; elongate, thin-shelled; left valve fatter; sculpture of finely cut, concentric grooves separated by narrow bands. Internal anterior radial rib is stronger in left valve. Pallial sinus almost touching the front muscle scar. Color externally is a vitreous-pink or apricot with concentric bands of white. The posterior ⅓ of the right valve has sharp, distinct cut lines (about a dozen in adults). Fairly common offshore from 1 to 66 fathoms. Synonyms are *georgiana* Dall, 1900, and *carolinensis* Dall, 1889.

Tellina guildingii Hanley, 1844   5667
Guilding's Tellin

Bahamas to Barbados.

2 inches, similar to *nitens*, but lacks the oily orange sheen and has pinkish rays. The shell is thicker and heavier, but lacks the strong, incised lines on the posterior slope of the right valve. A rare species.

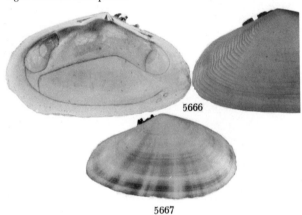

5666

5667

Other species of *Eurytellina*:

5668 *Tellina vespuciana* Orbigny, 1842. Texas (?); Central America to Trinidad. See *Johnsonia*, vol. 4, no. 46.

5669 *Tellina rubescens* Hanley, 1844. Mexico to Peru.

5670 *Tellina prora* Hanley, 1844. Gulf of California to Ecuador.

5671 *Tellina inaequistriata* Donovan, 1802. (Synonym: *leucogonia* Dall, 1900.) Gulf of California to Ecuador.

Subgenus *Angulus* Mühlfeld, 1811

Shells fragile, less than an inch, compressed, left valve slightly fatter. Sculpture of very fine, concentric cut lines. Hinge without lateral teeth in the left valve. Right valve has the anterior lateral tooth very near the cardinal. 14 species in the Western Atlantic, about 15 in the tropical Eastern Pacific. Type: *lanceolata* Gmelin, 1791. *Oudardia* Monterosato, 1884, is a synonym. *Moerella* Fischer has been erroneously used for this American group.

Tellina agilis Stimpson, 1857   5672
Northern Dwarf Tellin

Gulf of St. Lawrence to Georgia.

⅓ to ½ inch in length, moderately elongate, compressed, fairly fragile; glossy-white externally with an opalescent sheen. Interior white. Ligament external and prominent. With a large rounded pallial sinus almost extending to the anterior

muscle scar. External sculpture of faint, microscopic, concentric, impressed lines. Commonly found washed on shore from Maryland north. Formerly known as *Tellina tenera* Say, 1822 (non Schrank, 1803) and *Angulus tener* Say (not Schrank, 1803, nor Leach, 1818). *T. elucens* Mighels, 1845, might be this species, although it is very doubtful. *Macoma phenax* Dall superficially resembles this species, but is distinguished by its macomid character of lacking lateral teeth. *T. texana* is thicker-shelled and fatter.

5672

Tellina texana Dall, 1900   5673
Say's Tellin

North Carolina to Texas; Mexico to Bahamas.

½ inch, rather solid, white with a faint opalescent sheen, inflated with both valves of equal convexity and with a sharp posterior twist to the right. Color white, sometimes suffused with yellow. Some iridescence externally and generally highly polished internally. *T. polita* Say, 1822, and *sayi* Dall, 1900, are synonyms. Very close to and with its range overlapping *T. agilis*, this species replaces *agilis* in the Gulf of Mexico. The pallial sinus comes much closer to the front muscle scar, and the umbonal sculpture has oblique creases. Moderately common from low-tide mark to 25 fathoms on sand bottom.

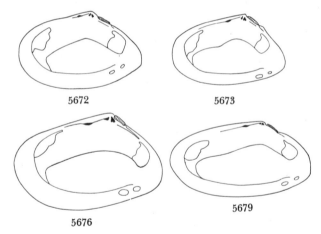

5672   5673

5676

5679

Tellina versicolor DeKay, 1843   5674
DeKay's Dwarf Tellin

Rhode Island to Key West; to Texas and the West Indies.

½ inch in length, very similar to *T. agilis*, but more elongate, colored white, red, pink or rayed, is more inflated, and has a nearly straight instead of curved ventral margin. The exterior of *versicolor* has a brighter iridescence. The pallial sinus is much closer to the anterior muscle scar. A common species dredged from 1 to 25 fathoms.

Tellina probina Boss, 1964   5675
Boss' Dwarf Tellin

Off North Carolina to Tobago, Lesser Antilles.

1 inch, oval-elongate, fragile, compressed. Sculpture of weakly cut, irregularly spaced concentric grooves; no radial sculpture present. Characterized by its subrectangular shape

and its broad, flattened and oblique posterior truncation. The exterior surface is vitreous or oily.

Tellina tenella Verrill, 1874   5676
Tenella Tellin

Southern Massachusetts to Florida and Mississippi.

½ inch, solid, subelliptical, swollen, valves of equal convexity, and without a posterior twist. Sculpture of closely spaced grooves separated by rounded bands. Left valve without lateral teeth. Right valve with only an anterior lateral. Pallial sinus distinctly separated from the front muscle scar. Interior of valves rough and thick. Exterior white, but sometimes brownish or reddish. A rather uncommon species distinguished from *agilis* by its rounded ventral margin, its long, concave anterior dorsal margin and blunter hind end. From shore to 31 fathoms. *T. modestus* Verrill, 1872 (non Carpenter, 1864) is a synonym. See C. W. Johnson, *The Nautilus*, vol. 45, p. 109.

5673

5675   5676

Tellina paramera Boss, 1964   5676a

Bermuda; southeast Florida to Barbados.

½ inch; ovate, solid, moderately inflated. Concentric sculpture of closely set, raised ridges (8 to 10 per mm.), separated by shallow grooves. Radial sculpture of weak lirations over the sides of the valves. Exterior dull-white; internal surface may have radial squiggles coinciding with external lirations. Uncommon from offshore to 50 fathoms. Very similar to *mera*, but *paramera* is thicker and heavier, bluntly truncate at the back, and the pallial sinus is very close to the front muscle scar. Common in Bermuda.

Tellina mera Say, 1834   5677
Mera Tellin

Bermuda; south half of Florida to the lower Caribbean. Brazil.

½ to ¾ inch in length, roughly elliptical, moderately inflated, pure opaque-white in color. Fairly thin but strong. Beaks fairly large for a tellin, touching and pointing toward each other and located nearer the posterior than the center of the shell. The valves show hardly any posterior bend or twist. Exterior smoothish with fine, irregular, concentric lines of growth more evident near the margins. The oval pallial sinus does not reach the front muscle scar and is not confluent with the pallial line below. Sculpture very variable in this species. Moderately common in shallow water between tides. Compare with *tampaensis*. A variable species. Boss (1968) has put *promera* Dall, 1900, and *obtusa* Sowerby, 1868, in its synonymy.

5676a

5677   5678

Tellina tampaensis Conrad, 1866   5678
Tampa Tellin

South half of Florida to Texas, and the West Indies.

½ to 1 inch in length, similar to *mera*, but more pointed posteriorly, whitish with a faint pinkish blush, rarely peach, and with very numerous, microscopic, concentric lines of growth. The pallial sinus line in this species runs forward nearly to the anterior muscle scar and then drops almost vertically toward the ventral margin of the shell before continuing posteriorly. In *mera* the pallial sinus line toward the anterior muscle scar makes a U-shaped turn, and then runs posteriorly but does not join the lower pallial line until about the middle of the ventral region of the valve. Common in shallow water.

Tellina sybaritica Dall, 1881   5679
Dall's Dwarf Tellin

Bermuda; North Carolina to Florida and Brazil.

¼ to ⅓ inch in length, very elongate, shiny, with quite strong, numerous concentric threads or cut lines. Has a slight posterior twist to the right. Color varying from translucent-white, peach, yellow, pinkish to bright watermelon-red. Our smallest and most colorful tellin, and plentiful from 1 to 60 fathoms. It somewhat resembles young *alternata*, but the latter is smoother and has a short instead of long posterior lateral tooth in the left valve, and has no lateral lamina in the left valve. *T. flagellum* Dall, 1900, and *T. rubricata* Perry, 1940, are synonyms.

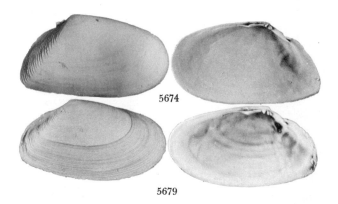

5674

5679

Other species of *Angulus:*

5680 *Tellina gibber* von Ihering, 1907. Uruguay and Brazil. 7 to 25 fms. See *Johnsonia,* vol. 4, no. 46, pl. 154.

5681 *Tellina diantha* Boss, 1964. Barbados to Brazil. See *Johnsonia,* vol. 4, no. 46, pl. 154.

5682 *Tellina euvitrea* Boss, 1964. Cuba to Puerto Rico. 1 to 12 fms. See *Johnsonia,* vol. 4, no. 46, pl. 156.

5683 *Tellina exerythra* Boss, 1964. Caribbean to Brazil. See *Johnsonia,* vol. 4, no. 46, pl. 156.

5684 *Tellina colorata* Dall, 1900. West Indies; South Carolina (?). *T. omoia* Ravenel, 1886, is a synonym. See *Johnsonia,* vol. 4, no. 46, pl. 153.

Other Pacific (*Angulus*) species:

5685 *Tellina amianta* Dall, 1900. Gulf of California to Ecuador. Shore to 15 fms. See Keen, sp. 392.

5686 *Tellina cerrosiana* Dall, 1900. Baja California to Gulf of California. 8 to 26 fms.

5687 *Tellina recurvata* Hertlein and Strong, 1949, Gulf of California to Colombia. 15 to 26 fms. Not common. (Synonym: *recurva* Dall, 1900.)

5688 *Tellina tabogensis* Salisbury, 1934. (Synonym: *T. panamensis* Dall, 1900, not Philippi, 1848.) Gulf of California to Ecuador.

5689 *Tellina coani* Keen, 1971. Gulf of California.

***Tellina modesta* Carpenter, 1864**   **5690**
Modest Tellin

Alaska to the Gulf of California.

¾ to 1 inch in length, elongate, moderately pointed at the posterior lower corner. Surface white with iridescent sheen and with fine concentric threads or grooves. These fade out at the posterior ¼ of the shell, but reappear more coarsely on the very posterior slope. There is a well-formed, radial rib inside just behind the anterior muscle scar. Common in certain sandy localities from shore to 25 fathoms. It appears that *T. buttoni* Dall, 1900, is the same species.

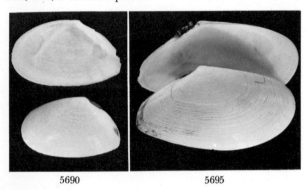

5690   5695

***Tellina nuculoides* (Reeve, 1854)**   **5691**
Salmon Tellin

Aleutian Islands to San Pedro, California.

½ inch in length, ovalish, with a short, blunt posterior end. Ligament behind the beaks prominent. Dorsal margin in front of beaks almost straight. Color chalky-white, commonly with a pinkish cast. Periostracum smooth, thin, yellowish tan. Characterized by about 4 to 7 prominent, concentric, former growth-stop lines which are usually stained dark-brown. Common from low tide to 34 fathoms in sand. Do not confuse with *meropsis. Tellina salmonea* (Carpenter, 1864) is a synonym. This species is placed in the subgenus *Cadella* Dall, Bartsch and Rehder, 1938.

5691   5693

***Tellina carpenteri* Dall, 1903**   **5692**
Carpenter's Tellin

Forrester Island, Alaska, to California.

⅛ inch in length, moderately elongate, with a rounded anterior end and rather truncate posterior end. Ligament short. Color cream, whitish and commonly blushed with watermelon-pink inside and out. It also has a faint iridescent sheen. Found very abundantly in many localities in mud and sand from shore to 369 fathoms. Synonyms are *variegata* Carpenter, 1864, not Gmelin, 1791; and *arenica* Hertlein and Strong, 1949.

***Tellina meropsis* Dall, 1900**   **5693**
Meropsis Tellin

San Diego, California, to Ecuador.

½ inch in length, ovalish, pure-white, smoothish, with exceedingly fine growth lines. Surface silky, but rarely with an iridescent sheen. Beaks slightly toward the posterior end. Ligament not prominent and light-brown. Without growth-stoppage lines. Common from shore to 15 fathoms. *T. paziana* Dall, 1900, is a synonym, as is *gouldii* Carpenter, 1864, not Hanley, 1846.

Subgenus *Peronidia* Dall, 1900

Shell without laterals, having the characters of *Angulus* and the external appearance of *Eurytellina* (Dall). Type: *albicans* Gmelin, 1791.

***Tellina lutea* Wood, 1828**   **5694**
Great Alaskan Tellin

Arctic Ocean to Cook's Inlet, Alaska. Japan.

3 to 4 inches in length, elongate, quite compressed, and with a posterior twist to the right. Worn shells chalky-white, com-

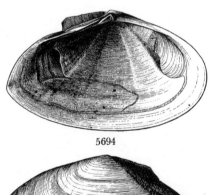

5694

monly with a pink flush. Periostracum in young is greenish yellow and glossy; in adults dark-brown. Ligament prominent. Commonly found from beach to 23 fathoms. *T. venulosa* Schrenck, 1861, is an ecologic form with brownish cracks in the shell. Other synonyms include *guildfordiae* Griffith and Pidgeon, 1833, and *alternidentata* Broderip and Sowerby, 1829.

Tellina bodegensis Hinds, 1845   5695
Bodegas Tellin

Graham Island, British Columbia, to the Gulf of California.

2 inches, white, elongated, narrowly ovate, rather thick. Similar to *lutea,* but the beaks are posterior to the center. Anterior end evenly rounded, the dorsal and ventral margins being nearly parallel. Posterior end attenuated. Pallial sinus long and narrow. In the southern part of its range the shells are flatter, thinner and less bent posteriorly, a subspecies named (5695a) *santarosae* Dall, 1900. Moderately common from shore to 15 fathoms.

Subgenus *Scissula* Dall, 1900

Shells 1 inch or less, with concentric lirations crossed by very fine cut lines or "scissulations." Only the anterior lateral tooth is present in the right valve. Type: *similis* Sowerby, 1806. 5 Western Atlantic species; about 5 in the Eastern Pacific. Do not confuse with the larger *Strigilla* tellins.

Tellina similis Sowerby, 1806   5696
Candy Stick Tellin   Color Plate 23

South half of Florida, Bermuda and the Caribbean. Brazil.

1 inch in length, moderately elongate, moderately compressed, thin but fairly strong. Color opaque-white or pink with a yellowish blush and with 6 to 12 short radial rays of red. Interior yellowish with red rays or solid pink or yellow. A red splotch commonly occurs on the hinge in front of the cardinal teeth. Pallial sinus comes near but does not touch the anterior muscle scar. Sculpture of concentric growth lines and numerous fine concentric threads which cross the middle ¾ of the shell at an oblique angle. Posterior slope of right valve with accentuated concentric threads. Common on sand flats. *T. decora* Say, 1826; *caribaea* Orbigny, 1842; and *eupareia* Ravenel, 1885, are synonyms.

5692

Tellina iris Say, 1822   5697
Iris Tellin

North Carolina to Florida, to Texas.

½ inch in length, very similar to *similis,* and often as colorful, but very thin-shelled, translucent and more elongate. The "scissulations" are microscopic. On the interior of the valves, the wavy oblique lines are evident, and there are 2 radial thickenings or weak, white, internal ribs at the posterior end. Common from intertidal flats to 20 fathoms.

Tellina candeana Orbigny, 1842   5698
Candé's Tellin

Bermuda; southeast Florida to the Lesser Antilles.

½ inch, wedge-shaped (blunter at the anterior end), shining-white, rarely suffused with yellow or pink. Minute concentric grooves limited to the posterior slope; rest of valve has close-set "scissulations" (7 or 8 per mm.). Common in sand at bases of eel-grass roots and down to 6 fathoms.

Tellina consobrina Orbigny, 1842   5699
Consobrine Tellin

Bermuda; southeast Florida to the Lesser Antilles.

½ inch; elongate, fragile, rather fat with the left valve fatter and with a posterior twist to the right. Umbones posterior to the middle. Sculpture of weak concentric lines crossed by very obscure "scissulations" (about 3 to 5 per mm.). Shell is white, variously colored with red or pink rays, rarely completely white. Offshore down to 70 fathoms; uncommon.

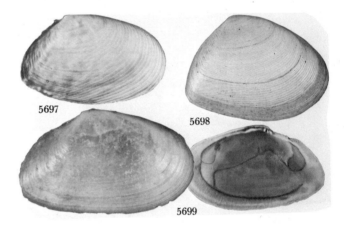

5697   5698   5699

Other species of *Scissula:*

5700 *Tellina sandix* Boss, 1968. (Synonym: *exilis* Lamarck, 1818, not Link, 1808.) Jamaica to Uruguay. See *Johnsonia,* vol. 4, no. 46, pl. 161.

5701 *Tellina virgo* Hanley, 1844. Baja California to Peru. Shore to 8 fms. Analog of *sandix* Boss. See Keen, sp. 413.

Genus *Tellidora* H. and A. Adams, 1856

Shells white, trigonal, compressed, with anterior and posterior dorsal margins bearing blunt spines. Ligament sunken and short. One species on each side of the Americas. Type: *burneti* (Broderip and Sowerby, 1829). *Tellipiura* Olsson, 1944, is a synonym.

Tellidora cristata (Récluz, 1842)   5702
White Crested Tellin

North Carolina to south Florida and Texas.

1 to 1½ inches in length, roughly ovate, compressed and all-white. The left valve is very flat, the right valve slightly inflated. Dorsal margins of valves with large, sawtooth crenulations. A bizarre clam found uncommonly in shallow water.

(5703) The similar Pacific analog is *Tellidora burneti* (Broderip and Sowerby, 1829), which reaches 2 inches, and ranges from the Gulf of California to Ecuador. Shore to 16 fathoms, common.

5702

Genus *Strigilla* Turton, 1822

Shells strong, about 1 inch in size, circular to somewhat oval; sculpture of strong, oblique, cut grooves running across the valves and taking a V- or W-shaped zigzag when they cross the upper anterior ⅓. 2 cardinal teeth in each valve, 1 being bifid or split; 2 laterals strong in the right valve, weak in the left. Do not confuse with the subgenus *Scissula* in the genus *Tellina* in which much finer lines or "scissulations" cross concentric grooves. Type: *carnaria* (Linné, 1758). The genus is worldwide and tropical with 6 species in the Western Atlantic, 8 in the Eastern Pacific, 1 in West Africa and several in the Indo-Pacific. The typical subgenus *Strigilla* has V-shaped grooves on the posterior slope. *Rombergia* Dall, 1900, is a synonym.

Subgenus *Strigilla* Turton, 1822

***Strigilla carnaria* (Linné, 1758)**   **5704**
Large Strigilla   **Color Plate 23**

South half of Florida; upper Caribbean to Argentina.

1 inch; somewhat inflated, the right valve more convex. Sculpture of strong, oblique, evenly spaced grooves which form a V-shape on the posterior slope. Pallial sinus 2 to 5 mm. from front muscle scar in adults. Color all-pink or pink with white border. A common shallow-water clam absent from the lower Caribbean. *S. rombergii* Mörch, 1853, and *areolata* Menke, 1847, are synonyms. Do not confuse with *pseudocarnaria* Boss, 1969.

***Strigilla pseudocarnaria* Boss, 1969**   **5705**
False Red Strigilla

Caribbean; exclusive of Bahamas and Cuba.

¾ inch, very similar to *carnaria* Linné but in *pseudocarnaria* the pallial sinus reaches the front muscle scar and has finer external sculpture. From *gabbi* it differs in generally being smaller, fatter, having a prominent external ligament and broader sculpture on the posterior slope. This is a common species erroneously referred to as *rombergi* Mörch by Warmke and Abbott in Caribbean Seashells (1961) and by Abbott in the first edition of American Seashells.

***Strigilla producta* Tryon, 1870**   **5706**
Ovate Strigilla

Jamaica and Panama to Brazil.

⅖ inch, elongate-oval, valves of about equal convexity, white in color with reddish around the umbone area, and with V-shaped chevrons on the posterior slope. (*pisiformis* has W-shaped chevrons). The bifid cardinal tooth in the right valve points straight down, not obliquely as in *pseudocarnaria*. A rare species.

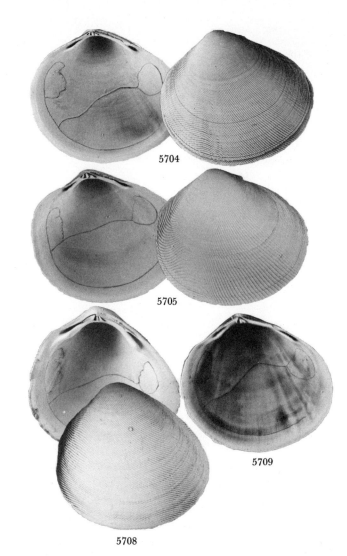

5704

5705

5709

5708

***Strigilla gabbi* Olsson and McGinty, 1958**   **5707**
Gabb's Strigilla

Key West; Bahamas to Brazil. Fossil in Texas.

1 to 1⅓ inches; pink, red or crimson with 2 posterior rays; umbones point forward; ligament sunken and almost internal; on the dorsal posterior slope of the right valve there is a radial ridge setting apart a flattened dorsal area. Locally moderately common in shallow water.

5706   5707

Subgenus *Pisostrigilla* Olsson, 1961

Posterior slope with 2 or 3 rows of chevrons. Type: *pisiformis* (Linné, 1758).

5714   5716   5718   5720   5722

Strigilla pisiformis (Linné, 1758)   **5708**
Pea Strigilla

Off southeast Florida; Bahamas to Brazil.

⅓ to ½ inch, inflated, white with pink color in the deepest part of the valve; rarely all-white. 2 sets of chevrons on posterior slope, one pointing down, the other up. An abundant offshore species, cast ashore in great numbers on Bahama beaches and brought to Florida for use in the shellcraft business. Synonyms are *discors* (Montagu, 1803), *pulchella* (C. B. Adams, 1845); *pilsbryi* Olsson and McGinty, 1958.

Strigilla mirabilis (Philippi, 1841)   **5709**
White Strigilla

Bermuda; North Carolina to Texas; Caribbean. Brazil.

⅓ to ½ inch, inflated, subovate, shiny, all-white. There may be 3 or 4 rows of chevrons on the posterior slope. The shell is thinner and has more widely spaced grooves than in red-flushed *pisiformis*. A common, widely spread species occurring just offshore to about 31 fathoms. *Tellina flexuosa* Say, 1822 (not Montagu, 1803); *carolinensis* Conrad, 1862, are synonyms.

Subgenus *Simplistrigilla* Olsson, 1961

Surface marked with strongly oblique sulci crossing from one margin to the other but without any major flexure or change in direction. Type: *Strigilla strata* Olsson, 1961, *Moll. Trop. East. Pac.*, Ithaca, N.Y.

Strigilla surinamensis Boss, 1972   **5710**
Surinam Strigilla

Alabama, Florida and Surinam.

9 mm. in length, white to yellowish, very similar to *S. mirabilis,* but the sulci extend across the posterior slope of the valve, while in *mirabilis* there are numerous, finer zig-zag lines. Known from only a few specimens. See Boss, 1972, *Zoolog. Meded., Leiden*, vol. 46, p. 25. *S. carolinensis* Conrad, 1862, may be an earlier name.

Subfamily MACOMINAE Olsson, 1961

Hinge without lateral teeth. Shell usually white, rarely blushed with rose or light-orange. Sculpture smoothish.

Genus *Macoma* Leach, 1819

The macomas are modified tellins which may be distinguished by (1) no lateral teeth; (2) usually dingy-white in color and of a chalky consistency; (3) there is a strong posterior twist; (4) the pallial sinus is larger in one valve than the other. Type of the genus: *calcarea* (Gmelin, 1791).

Macoma calcarea (Gmelin, 1791)   **5711**
Chalky Macoma

Greenland to Long Island, New York.
Bering Sea to Washington.

1½ to 2 inches in length. Oval-elongate, moderately compressed, but somewhat inflated at the larger, anterior ½. Beaks ⅗ the way toward the narrowed, slightly twisted posterior end. Shell dull, chalky-white. Concentric sculpture of fine, irregular threads. Periostracum remaining on the margins is gray. Posterior dorsal margin sharp, not indented. Pallial sinus in left valve runs from the posterior muscle scar anteriorly toward the anterior muscle scar, but does not meet the latter, and then descends posteriorly to meet the pallial line about the middle of the lower margin of the shell. A common cold-water species, distinguished from *balthica* by its larger size,

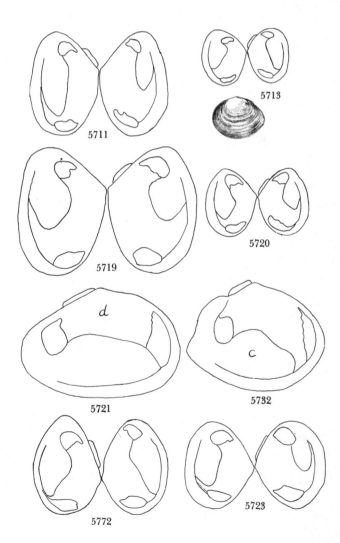

5711   5713

5719   5720

5721 *d*   5732 *c*

5772   5723

more elongate shape and pattern of the pallial sinus scar. Synonyms include *lata* Gmelin, 1791; *sitkana* Dall, 1900; *longisinuata* Soot-Ryen, 1932; *proxima* Sowerby in Gray, 1839. Do not confuse in the Pacific northwest with *Macoma elimata* (see species, 5712).

Macoma elimata Dunnill and Coan, 1968   5712
Filed Macoma

Southeast Alaska to off Los Angeles Co., California.

½ to 1 inch in length very similar to *calcarea*, but with a very narrow, flat-sided, indented escutcheon on the posterior dorsal margin. The dorsal margin is straight and steeply sloping. Shell chalky white, but covered with a greenish gray, flaky periostracum. Occurs at depths of 15 to 476 meters on silty and clayey sands. It occurs in the Pleistocene of San Pedro, California. For anatomy, sperm and other details, see Natural History Papers, no. 43, Nat. Mus. Canada, 1968, by Dunnill and Coan.

Macoma balthica (Linné, 1758)   5713
Balthica Macoma

Arctic seas to off Georgia.

½ to 1½ inches in length, oval, moderately compressed. Color dull whitish, in some with a flush of pink, and with a thin, grayish periostracum which readily flakes off. The shape is somewhat variable. A common intertidal and deep-water species. Compare muscle scars with those of *calcarea*. The type of *T. rotundata* Sowerby, 1867, proves to be this species.

Macoma inconspicua (Broderip and Sowerby, 1829)   5714
Inconspicuous Macoma

Bering Sea to off Monterey, California.

¾ inch, ovate-lenticular, anterior end shorter. Chalky-white, sometimes flushed with pink. Rather inflated. Posterior end bluntly pointed. Very close to *balthica*, but smaller and fatter. Common; offshore in mud. For ecology, see M. T. Vassallo, 1971, *The Veliger*, vol. 13, p. 279, and 1969, *ibid.*, vol. 11, p. 223. This may well be a synonym of *balthica* Linné, 1758.

Macoma moesta (Deshayes, 1855)   5715
Doleful Macoma

Circumpolar Arctic seas. Greenland.
Alaska.

1 to 2 inches, ovate, inflated, inequilateral with the beaks ⅔ back from the anterior end. Periostracum yellowish olive to greenish and flaking off when dry. Posterior end short, slightly angular, finely and concentrically wrinkled at the angle. Interior chalky-white; hinge with 2 small cardinal teeth in each valve, the anterior left one being bifid. Pallial sinus rounded behind, a little smaller in the right valve. Height-to-length ratios vary from 73 (Greenland) to 62 (Point Barrow, Alaska). Common offshore. Synonyms: *krausei* Dall, 1900; *oneilli* Dall, 1919. For biology, see Ockelmann, 1958, *Meddelelser om Grönland*, vol. 122, no. 4.

Macoma tenta (Say, 1834)   5716
Tenta Macoma

Cape Cod to south half of Florida to Brazil. Bermuda.

½ to ¾ inch in length, fragile, elongate, white in color with a delicate iridescence on the smooth exterior. Posterior and narrower end slightly twisted to the left. This small, tellin-like species is very common in shallow water in sand. *M. souleyetiana* Récluz, 1852, is the same.

(5717) *Macoma limula* Dall, 1895, is very similar in size and shape, although somewhat more elongate, and is distinguished by the finely granular external surface of the valves. Commonly dredged from North Carolina to Florida.

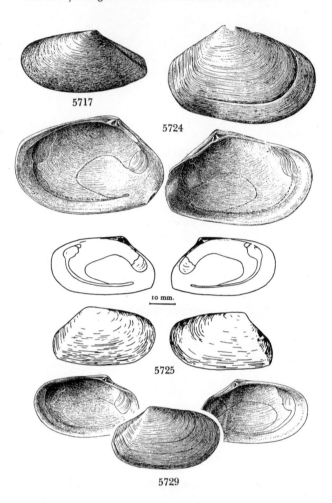

Macoma carlottensis Whiteaves, 1880   5718
Queen Charlotte Macoma

Arctic Ocean to Baja California.

About 1 inch in length, extremely fragile, inflated and with a very short, external inconspicuous ligament. Pallial sinus large, rounded at its extremity, and reaching beyond the center of the valve. Color translucent-white with a thin, greenish, glossy periostracum. *M. inflatula* Dall, 1897, and *quadrana* Dall, 1916, are synonyms.

Macoma brota Dall, 1916   5719
Brota Macoma

Arctic Ocean to Puget Sound, Washington.

3 inches in length, moderately elongate, moderately inflated and rather thick-shelled. Beaks ⅔ towards the posterior end. Resembles *calcarea* whose pallial sinus in the left valve, however, is more elongate, not as high and generally reaches nearer the anterior muscle scar. *M. brota* is larger and more truncate posteriorly than that species. Common.

Macoma lama Bartsch, 1929   5720
Grant and Gale Macoma

Arctic Ocean to Puget Sound, Washington.

About 1 inch in length. Extremely similar to *calcarea*, but porcelaneous, with a glossy, yellowish periostracum and more oval in shape. This species was thought by Dall and others to be "*carlottensis* Whiteaves," and N. MacGinitie, 1959, *Proc. U.S. Nat. Mus.*, vol. 109, no. 3412, p. 180, suggests that *praetenuis* Woodward, 1833, may be an earlier name. *Macoma planiuscula* Grant and Gale, 1931, is a synonym.

Macoma nasuta (Conrad, 1837)   5721
Bent-nose Macoma

Alaska to Baja California.

2 to 3½ inches in length, elongate, rather compressed and strongly twisted to the right at its posterior end. Beaks slightly nearer the anterior end. Can be distinguished from the other Pacific coast species by the pallial sinus in the left valve which reaches the anterior muscle scar. One of the most common species on the west coast and lives about 6 inches below the surface of the mud in quiet waters from shore to 25 fathoms. Synonyms include *tersa* Gould, 1853, and *kelseyi* Dall, 1900.

Macoma inquinata (Deshayes, 1855)   5722

Bering Strait to Los Angeles, California.

Commonly 1½ inches in length (rarely 3); oval-elongate, moderately inflated, very slightly twisted, if at all, at the posterior end. Pallial sinus in left valve almost reaches the bottom of the anterior muscle scar. Beaks slightly anterior. Erroneously identified by me in the first edition as *Gastrana irus*. See A. M. Keen, 1962, *The Veliger*, vol. 4, p. 161. For ecology, see M. T. Vassallo, 1969, *The Veliger*, vol. 11, p. 223. *M. arnheimi* Dall, 1916, is a shorter, fatter form of this species.

Macoma obliqua (Sowerby, 1817)   5723
Incongruous Macoma

Point Barrow, Alaska, to Washington.

Is similar to *inquinata*, but more oval, with a more pointed slope at the beaks, a straighter posterior dorsal edge. Its upper pallial sinus line, after nearly reaching the anterior muscle scar, turns downward and then runs anteriorly before connecting with the pallial sinus. Common in Alaska. Formerly *incongrua* of authors, not von Martens, 1865.

Subgenus *Psammacoma* Dall, 1900

Shell elongate, tagelloid in shape, valves subequal; posterior end shorter, weakly flexed. Ligament and resilium attached to a long, narrow scar which is wholly external; no nymphal ridge. Pallial sinus not extending to the anterior adductor scar. Type: *candida* (Lamarck, 1818). Synonym: *Macoploma* Pilsbry and Olsson, 1941.

Macoma tageliformis Dall, 1900   5724
Tagelus-like Macoma

Louisiana to Texas; Greater Antilles. Brazil.

1 to 2 inches, oblong, left valve fatter than the right; ligamental area long and depressed; posterior end gently rounded. Pallial sinus reaches halfway to anterior muscle scar. Moderately common offshore in the Gulf of Mexico; common in the West Indies, down to 50 fathoms.

Macoma pulleyi Boyer, 1969   5725
Pulley's Macoma

West of Mississippi Delta, Louisiana.

Very similar to *tageliformis*, but posterior comes more to a point, pallial sinus reaches ¾ the way to the anterior muscle scar, left valve not proportionately as fat and there is a weak radial exterior ridge marking off the posterior slope from central portion of the valve. Uncommon; from 6 to 25 fathoms in sandy mud. Illustration from Boyer, 1969, *The Veliger*, vol. 12, p. 41.

Other species:

5726   *Macoma (Psammacoma) lamproleuca* (Pilsbry and Lowe, 1932). Gulf of California to Peru.

5727   *Macoma (Psammacoma) acolasta* Dall, 1921. Bodega Bay to San Diego, California. (Synonym: *morroensis* T. Burch, 1945.)

5728   *Macoma (Psammacoma) elytrum* Keen, 1958. Baja California to Ecuador. (Synonym: *elongata* Hanley, 1844, not Dillwyn, 1823.)

Macoma brevifrons (Say, 1834)   5729
Short Macoma

South Carolina to Brazil.

1½ inches, oblong, polished, white, usually with a blush of iridescent-orange in the central and umbonal regions. Below the smooth umbones the shell has numerous close-set growth lines, covered by a light-brown periostracum. Anterior end is slightly shorter and rounded. Uncommon; just offshore.

Macoma yoldiformis Carpenter, 1864   5730
Yoldia-shaped Macoma

Alaska to San Diego, California.

½ to ¾ inch in length, elongate, moderately rounded at each end and with a small, but distinct, twist to the right at the posterior end. Color a uniform, glossy, porcelaneous white. Rarely translucent with an opalescent sheen. Common from shore to 25 fathoms.

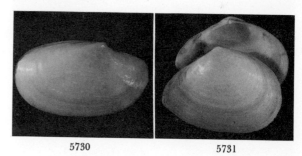

5730   5731

Subgenus *Austromacoma* Olsson, 1961

Has a large pallial sinus, high and pointed under the beak, and connected with the anterior adductor scar at its lower end. Type: *constricta* (Bruguière, 1792).

Macoma constricta (Bruguière, 1792)   5731
Constricted Macoma

Florida to Texas and the West Indies. Brazil.

1 to 2½ inches in length, moderately elongate. The posterior end is twisted to the right and is narrowed to a blunt point. Color all-white with concentric growth lines stained by the gray periostracum. Common just offshore.

Subgenus *Rexithaerus* Tryon, 1869

Dorsal margin of shell produced upward behind the large ligament. Type: *secta* (Conrad, 1837).

Macoma secta (Conrad, 1837)   **5732**
White Sand Macoma

Vancouver Island, British Columbia, to the Gulf of California.

2 to 4 inches in length. This is the largest macoma in America and is characterized by the almost flat left valve, rather well-inflated right valve and the wide and relatively short ligament which is sunk partially into the shell. There is a large, oblique, riblike extension just behind the hinge inside each valve. Color cream to white. Common in bays and beaches from shore to 25 fathoms. A small form occurs in protected waters in bays in its more southerly range. *Tellina ligamentina* Deshayes, 1843 is a synonym.

Students of the Pacific coast fauna consider *M. indentata* Carpenter, 1864 (**5733**) (same range) as a distinct species in which the shell is 1½ inches in length, a little more elongate, with a more pointed posterior end, and with a slight indentation on the posterior ventral margin. It may be a form of young *secta. M. tenuirostris* Dall, 1900, is even more elongate and may also be a form.

Other Atlantic species:

5734 *Macoma phenax* Dall, 1900. Virginia to Tampa Bay, Florida.

5735 *Macoma mitchelli* Dall, 1895. South Carolina to Texas. 15 mm.

5736 *Macoma leptonoidea* Dall, 1895. "Matagorda Bay, Texas." Pacific coast?

5737 *Macoma cerina* C. B. Adams, 1845. Florida and the West Indies.

5738 *Macoma loveni* Jensen, 1905. Greenland to Nova Scotia, 1 to 180 fms. Bering Sea.

5739 *Macoma inflata* "Stimpson" Dawson, 1872. Greenland to off Nova Scotia, 57 to 206 fms. (is *loveni?*)

5740 *Macoma (Cydippina) extenuata* Dall, 1900. Off Cedar Keys, Florida, 32 fms. 14 mm.

5741 *Macoma pseudomera* Dall and Simpson, 1901. Bermuda to Puerto Rico and Jamaica. 16 mm.

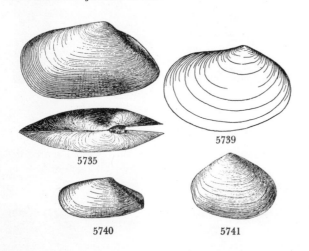

5735   5739

5740   5741

Other Pacific species:

5742 *Macoma middendorffii* Dall. 1884. Bering Sea to Chirikoff Island, Alaska.

5744 *Macoma quadrana* Dall, 1916. Alaska to Baja California.

5745 *Macoma expansa* Carpenter, 1864. Puget Sound to La Jolla, California. (Synonym: *liotricha* Dall, 1897.)

— *Macoma truncaria* Dall, 1916. Arctic Seas, Alaska. See *The Veliger,* vol. 11, p. 281. Is a *Thracia* young.

5746 *Macoma alaskana* Dall, 1900. Forrester Island, 50 fms.; Craig, 20 to 30 fms.; Izhut Bay, Afognak, Alaska. Subspecies of *moesta* Deshayes, 1845.

Genus *Cymatoica* Dall, 1889

Shells small, elongate, surface with low, wavelike undulations, transmitted through the thin shell, parallel to the growth lines over most of the valve but cutting across them on the anterior slope. Type: *undulata* Hanley, 1844.

Two American species:

5747 *Cymatoica orientalis* (Dall, 1890). Florida Strait to Santo Domingo (subspecies or forma (**5747a**) *hendersoni* Rehder, 1939, from Florida and Brazil). 9.5 mm.

5747a

5748 *Cymatoica undulata* (Hanley, 1844). Gulf of California to Ecuador. 16 mm. (Synonym: *occidentalis* Dall, 1890.)

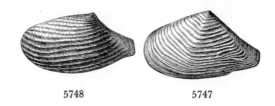

5748   5747

Genus *Psammotreta* Dall, 1900

Adductor muscle scars unequal in size, the posterior one long and narrow. Ligament external but inset in a pit in the hinge, which crowds against the cardinal tooth. Type: *aurora* (Hanley, 1844). *Cydippina* Dall, 1900; *Apolymetis* Salisbury, 1929; *Scrobiculina* Dall, 1900; and *Schumacheria* Cossmann, 1902, are synonyms.

Subgenus *Florimetis* Olsson and Harbison, 1953

Type of the subgenus is *intastriata* (Say, 1827).

Psammotreta intastriata (Say, 1827)   **5749**
Atlantic Grooved Macoma   **Color Plate 23**

South Carolina to Florida and the Caribbean. Bermuda.

2 to 3 inches in length, elliptical, fairly thin but strong and all-white in color. The shell is strongly twisted. At the posterior end, the right valve bears a strong, radial rib, while on the left valve there is a fairly strong, radial groove. Pallial sinus very large. Living specimens are not commonly collected, although shells are commonly washed ashore. Lives on its side, flatter valve up, at a depth of about 7 inches in silty black mud.

Psammotreta obesa (Deshayes, 1855)   **5750**
Pacific Grooved Macoma

Santa Barbara, California, to Ensenada, Mexico.

2 to 3½ inches in length, oval, moderately compressed, strong, and dull grayish white in color. Interior glossy-white with the central portion blushed with pastel-peach. Left valve with a shallow, radial groove near the posterior end. Right valve with a corresponding ridge at the end of which the margin of the shell is shallowly notched. Alias *A. alta* Conrad. Common; found 6 inches in fine sand from low-tide mark to 25 fathoms. *P. biangulata* (Carpenter, 1856) is a synonym, *fide* Keen, 1966, *The Veliger*, vol. 8, p. 170.

5750

Other species:

5751 *Psammotreta aurora* (Hanley, 1844). Baja California to Ecuador.

5752 *Psammotreta* (*Florimetis*) *cognata* (Pilsbry and Vanatta, 1902); subspecies *clarki* (Durham, 1950). Head of the Gulf of Mexico to Acapulco, Mexico.

Family DONACIDAE Fleming, 1828

Genus *Donax* Linné, 1758

The posterior end is the shorter and the fatter. 2 cardinal and an anterior and a posterior lateral in each valve. Pallial sinus deep. Type: *rugosus* Linné, 1758. The wedge clams, known in Florida as coquina clams, are worldwide in temperate and tropical seas where they live on sandy beaches at about the midtide line. Some live in sand offshore. The name *Donax* is masculine, not feminine as thought by Linnaeus. Most authors now use it as a masculine name.

Donax variabilis Say, 1822   5753
Coquina Shell

New York to south Florida and Texas.

½ to ¾ inch in length. Ventral margin of the valves straight and almost parallel with the dorsal margin. The thinner, anterior end is commonly smooth, but may be microscopically scratched with radial lines. From the middle of the valve to the blunt posterior end, small radial threads appear, which become increasingly larger posteriorly. Internal margin of valves minutely denticulate. Color variable and commonly very bright, especially inside: white, yellow, pink, purple, bluish,

mauve and commonly with rays of darker shades. A synonym is *protractus* Conrad, 1849 (July). *Donax variabilis* Say, 1822, is preoccupied by *Donax* (*Latona*) *variabilis* Schumacher, 1817 (a synonym of *cuneatus* Linné, 1758, of the Indo-Pacific), but the I.C.Z.N. will probably conserve Say's name.

(5754) The subspecies *roemeri* Philippi, 1849, is very common along the Texan and Mexican shores. The posterior end is usually blunter, the whole shell is not so elongate and the ventral margin sags down. The shape of this species is influenced by the density of the population. Intergrades are common.

(5755) *Donax fossor* Say, 1822, is merely a northern, seasonal form of *variabilis*, which rarely exceeds ½ inch and is less colorful and smoother. See P. Chanley, 1969, *The Nautilus*, vol. 83, p. 1. *Donax parvulus* Philippi, 1849, may be a smoothish, small, slightly fatter ecologic form found just offshore.

Donax texasianus Philippi, 1847   5756
Fat Gulf Donax

Northern Texas to Vera Cruz, Mexico.

⅓ to ½ inch in length, very obese, somewhat trigonal in shape, with its beaks swollen and posterior end strongly truncate. Threads on the blunt posterior end are slightly beaded, sometimes giving a cancellate appearance. Narrow anterior end commonly with distinct, microscopic, distantly spaced, incised, concentric lines. Color whitish with bluish, yellowish or pinkish undertones. Rarely, if ever, rayed. Uncommon. Found in 2 to 3 feet of water. Synonym: *tumidus* Philippi, 1848.

(5757) *Donax denticulatus* Linné, 1758, from the southwest Caribbean to Brazil, has been erroneously recorded from our shores. It is 1 inch in length and characterized by 2 curved, low ridges on the posterior slope of each valve and by microscopic pinpoints on the sides of the valves. Locally abundant. It is in the subgenus *Chion* Scopoli, 1777.

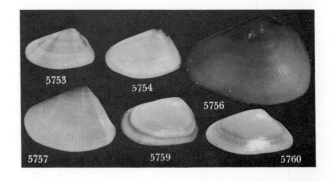

(5758) *Donax striatus* Linné, 1767, also a lower West Indian to Brazilian species, is as large, but characterized by a flat to slightly concave posterior slope which bears numerous fine radial threads. Locally abundant.

Donax gouldii Dall, 1921   5759
Gould's Donax

San Luis Obispo, California, to Baja California.

Common form: ¾ inch in length, fairly obese, truncate at the posterior end where the beaks are located. Shell glossy, smooth, except for numerous, microscopic, axial threads at the anterior end. Exterior cream to white with variable color rays of light-tan. Margins of valves commonly flushed with purple. Interior stained with purple or bluish brown. Common on beaches, especially in the north. Known locally as the bean clam.

Small form: ½ inch in length, slightly more obese, without the color rays in most cases. Common, especially in the south. Ideal habitat for this species is a very gently sloping beach with a certain amount of shallow sand-barring just beyond the median low-tide level (see P. T. Johnson, 1966, *The Veliger,* vol. 9, p. 29; and R. H. Pohlo, 1967, *The Veliger,* vol. 9, p. 330).

Donax californicus Conrad, 1837   5760
California Donax

Santa Barbara, California, to Panama.

Up to 1 inch in length, narrowly pointed at both ends. Shell glossy or oily-smooth with a tan to greenish tan periostracum. Interior white or with a purple flush, and with a strong splotch of purple at each end of the dorsal margin. Common in shallow waters of bays along the shore.

Other species:

5761 *Donax culter* Hanley, 1845. Gulf of California to Nicaragua.

5762 *Donax gracilis* Hanley, 1845. Baja California to Peru.

5763 *Donax punctatostriatus* Hanley, 1843. Gulf of California to Peru. Abundant.

Genus *Iphigenia* Schumacher, 1817

Shells 2 to 3 inches, elliptical, solid, without lunule or escutcheon. Beaks slightly towards the posterior. Hinge with 2 cardinal teeth in each valve, one of them being bifid. Lateral teeth reduced or obsolete. Ligament external. Pallial sinus large, rounded. Periostracum thin. Shell white with purplish umbones. Type: *laevigata* (Gmelin, 1791). Synonyms: *Fischeria* Bernardi, 1860; *Profischeria* Dall, 1903. *Iphegenia* and *Ephigenia* are misspellings.

Iphigenia brasiliana (Lamarck, 1818)   5764
Giant False Donax

South half of Florida to Brazil.

2 to 2½ inches in length, rather heavy, elongate, roughly diamond-shaped in side view and moderately inflated. Posterior dorsal slope flattish. Exterior smoothish, cream with a purple-stained beak area. Commonly entirely covered with a thin, glossy, brown periostracum. Moderately common in shallow water in sand. *I. brasiliensis* is a misspelling.

(5765) The Pacific counterpart (only slightly higher and the beaks more central), *Iphigenia altior* (Sowerby, 1833), ranges from the Gulf of California to northern Peru. *I. ambigua* Bertin, 1881, is a synonym.

5764

Family PSAMMOBIIDAE Fleming, 1828

Also known as Garidae (Stoliczka, 1871), Asaphidae (Winckworth, 1932) and Sanguinolariidae. Shells resembling the tellins, but red and purples predominate. 1 to 3 cardinal teeth, no laterals. Ligament large and external. Mud dwellers.

Subfamily SANGUINOLARIINAE Grant and Gale, 1931

Genus *Sanguinolaria* Lamarck, 1799

Shells large, of medium convexity, left valve slightly flatter. Hinge with 2 cardinal teeth in each valve. No laterals. Ligament external, posterior to the beaks, the resilium attached to the upper surface of a long, narrow nymph. Pallial sinus deep and ample, angular above. Adductor scars placed high in the valve. Type: *sanguinolenta* (Gmelin, 1791). Synonyms: *Lobaria* Schumacher, 1817; *Isarcha* Gistel, 1848.

Sanguinolaria sanguinolenta (Gmelin, 1791)   5766
Atlantic Sanguin   Color Plate 23

South Florida to Texas and the West Indies. Brazil.

1½ to 2 inches in length, moderately compressed, the left valve slightly flatter than the right. With a slight posterior gape. Exterior glossy, smooth, except for minute concentric scratches. Pallial sinus with a U-shaped hump at the top. Color white with the beaks and area below a bright-red which fades ventrally into white. Uncommon in the West Indies, rare in Florida. Erroneously called *cruenta* Lightfoot.

Subgenus *Psammotella* Herrmannsen, 1852

Shell elongate, left valve very flat, posterior end narrow and weakly flexed. Type: *Tellina cruenta* Lightfoot, 1786.

Sanguinolaria cruenta (Lightfoot, 1786)   5767
Operculate Sanguin   Color Plate 23

Caribbean to Brazil.

2 to 3½ inches, elongate, colored a dull purplish rose. Right valve strongly convex, the left flattened. Uncommon. Formerly *operculata* (Gmelin, 1791).

(5768) The Pacific counterpart, hardly distinguishable from the Caribbean form, is *Sanguinolaria bertini* Pilsbry and

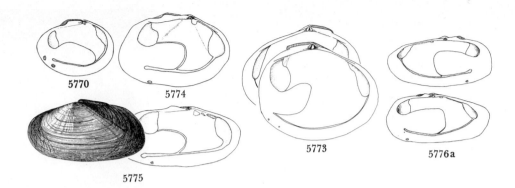

5770 5774 5773 5776a

5775

<hr />

Lowe, 1932, which ranges from the Gulf of California to northern Peru. Synonyms: *rufescens* Hanley, 1846 (not Gmelin, 1791); *hanleyi* Bertin, 1878 (not Dunker, 1853).

Other species:

5769 *Sanguinolaria* (*Sanguinolaria*) *tellinoides* A. Adams, 1850. Gulf of California to Ecuador. (Synonym: *miniata* Gould, 1851.)

Subgenus *Nuttallia* Dall, 1898

Shell large, suborbicular, right valve slightly flatter. Posterior cardinal in the left valve obsolete. Periostracum shiny-brown. Type: *nuttallii* (Conrad, 1837). Considered as a full genus by E. V. Coan (1973) and other workers.

Sanguinolaria nuttallii (Conrad, 1837) **5770**
Nuttall's Mahogany Clam **Color Plate 23**

Bodega Bay, California, to Baja California.

2½ to 5 inches in length. A handsome species characterized by its smooth, oval form, glossy nut-brown color, with its right valve almost flat and its left valve inflated. External ligament like a brown leather button. Interior whitish, commonly with rosy or purplish blush. Common near estuaries in 6 to 12 inches of mud. Locally called the purple clam.

Subfamily PSAMMOBIINAE Fleming, 1828

Genus *Asaphis* Modeer, 1793

Shell large, oblong, gaping behind, with numerous coarse, radial threads. 2 prominent cardinal teeth in each valve, the larger bifid. Pallial sinus large, rounded in front. Type: *deflorata* (Linné, 1758). Synonyms: *Capsa* Bruguière, 1797; *Corbula* Röding, 1798; *Psammocola* Blainville, 1824; *Capsula* Schumacher, 1817.

Asaphis deflorata (Linné, 1758) **5771**
Gaudy Asaphis **Color Plate 23**

Southeast Florida and the West Indies. Bermuda. Brazil.

2 inches in length, moderately inflated. Sculpture of numerous coarse, irregularly sized, radial threads. Color variable and brighter on the inside: whitish, yellow, or stained with red, rose or purple. Beaks inflated and rolled in under themselves a little. A moderately common, intertidal species, also known from the Indo-Pacific. Uncommon to rare in Florida.

Genus *Heterodonax* Mörch, 1853

Shell less than 1 inch in length, resembling a strong, oval *Tellina*. 2 cardinals and 2 lateral teeth in each valve, the laterals usually not very distinct. Ligament external. Pallial sinus extends ⅗ the length of the shell. Type: *bimaculata* (Linné, 1758).

Heterodonax bimaculatus (Linné, 1758) **5772**
Small False Donax

South half of Florida and the West Indies. Bermuda. Southern California to Panama.

½ to 1 inch in length, oval, with a truncate anterior end and moderately inflated. Exterior smoothish, with numerous fine growth lines. 2 cardinals in each valve. Anterior to the beaks (which point forward), the hinge is thick for a short distance, then followed by a thinner, concave portion. Color variable: white with 2 oblong crimson spots inside; violet with radial streaks; pink, yellow or mauve; some are speckled with black or brown. This is a common species found with *Donax* on the slopes of sandy beaches.

(5773) The western subspecies, *H. bimaculatus pacificus* (Conrad, 1837), is scarcely different from the West Indian representative. Its synonyms include *vicina* C. B. Adams, 1852; *purpureus* and *salmoneus* Williamson, 1892; and *Donax ovolina* Reeve, 1854. Drawing from E. V. Coan, 1973, *The Veliger*, vol. 16, p. 47.

5772

Genus *Gari* Schumacher, 1817

Shell fairly large, elongate-oval, beaks near the center; hinge thick and with 2 small cardinal teeth just under the beak. Posterior end with a narrow gape. Ligament large, external and attached to a prominent nymphal platform. Type: *amethystus* Wood, 1815. Synonyms: *Psammotaea* Lamarck, 1818; *Capsella* Deshayes, 1855; *Milligaretta* Iredale, 1936.

Subgenus *Gobraeus* Brown, 1844

Beaks slightly nearer the front. Surface with a rayed pattern. Posterior end with an open gap. Type: *vespertinus* (Gmelin, 1791). *Psammocola* of authors, not Blainville, 1824, is a synonym.

Gari californica (Conrad, 1849)   **5774**
Californian Sunset Clam

Aleutians to Magdalena Bay, Baja California.

2 to 4 inches in length, elongate-oval, fairly strong; the low beaks are nearer the anterior end. Sculpture of strong, irregular, concentric growth lines. Periostracum brownish gray, fairly thin and irregularly wrinkled. Exterior shell dirty-white or cream, and may have faint, narrow, radial rays of purple. Common offshore to 25 fathoms. Commonly washed ashore after storms, especially between Sea Beach and Huntington Beach, California.

5774

Gari edentula (Gabb, 1869)   **5775**
Toothed Sunset Clam

Santa Barbara to San Pedro, California.

4 to 5 inches long; relatively thin and flat; sculpture of weak concentric lines; periostracum olive-tan; surface rayed with dark lines. Uncommon; 5 to 137 meters; in fine sand. See E. V. Coan, 1973, *The Veliger*, vol. 16, p. 43.

Other species:

5775a *Gari (Gobraeus) helenae* Olsson, 1961. Gulf of California to Colombia (*regularis* of authors, not Carpenter, 1864).

5776 *Gari (Gobraeus) maxima* (Deshayes, 1855). Gulf of California to Colombia. 2 inches.

5776a *Gari (Gobraeus) regularis* (Carpenter, 1864). Both sides of Baja California. See E. V. Coan, 1973, *The Veliger*, vol. 16, p. 45.

Family SEMELIDAE Stoliczka, 1870

Resemble the tellins but in addition to the external ligament, there is an internal ligament attached to a long, oblique resilifer.

Genus *Semele* Schumacher, 1817

Resilium supported in a horizontal, chondrophore-like depression which is internal and parallel with the hinge line. 2 cardinal teeth in each valve. Right valve with 2 distinct lateral teeth, but practically absent in the left valve. The ligament is external. Pallial sinus large, entire, rounded at the end, free, its lower limb not united with the pallial line below. Type: *proficua* (Pulteney, 1799). Synonym: *Amphidesma* Lamarck, 1818. Drawings are from E. V. Coan's 1973 review of the genus in *The Veliger*, vol. 15, no. 4, pp. 314–329.

Semele proficua (Pulteney, 1799)   **5777**
White Atlantic Semele

North Carolina to Florida and Texas to Brazil. Bermuda.

5777

½ to 1½ inches in length, almost round, beaks almost central. Lunule small and pushed in. Fine concentric lines and microscopic radial striations. Externally whitish to yellowish white. Interior glossy, commonly yellowish, rarely speckled a little with purple or pink. Moderately common in shallow water. The color form (**5778**) *radiata* (Say, 1826), has a few indistinct radial rays of pink. *S. mediamericana* Pilsbry and Lowe, 1932, came from the Caribbean and is a synonym.

Semele purpurascens (Gmelin, 1791)   **5779**
Purplish Semele

North Carolina to south half of Florida and the West Indies. Brazil.

1 to 1¼ inches in length, oblong, thin-shelled, smooth except for very fine concentric growth threads over which run another set of fine, microscopic concentric lines at an oblique angle. External color variable: commonly gray or cream with purple or orangish mottlings. Interior glossy and suffused with purple, brownish or orange. A fairly common, shallow-water species.

5779

(**5780**) The Pacific counterpart, *Semele sparsilineata* Dall, 1915, is heavier and less convex. Its range is from Nicaragua to Ecuador.

Semele bellastriata (Conrad, 1837)   **5781**
Cancellate Semele

North Carolina to Florida, Texas and the West Indies. Bermuda. Brazil.

½ to ¾ inch in length, similar in shape to *purpurascens*, but a much smaller species with numerous radial and concentric riblets which cross to give a cancellate appearance.

5781

5782

5784

5785

5786

Semele rupicola Dall, 1915
5785

Rock-dwelling Semele

Santa Cruz, California, to the Gulf of California.

1 to 1½ inches in length. Irregular in shape: ovalish, oval-elongate or obliquely oval. Exterior yellowish cream with numerous, concentric crinkles and a few weak radial threads. Interior glossy, white at the center, bright purplish red at the margins and hinge. John Q. Burch finds this species common in *Chama* and *Mytilus* beds and in rocks and crevices.

Semele decisa (Conrad, 1837)
5786

Bark Semele

San Pedro, California, to Baja California.

2 to 3 inches in length, equally high. Characterized by its heavy shell, by the coarse, wide, irregular, concentric folds on the outside (resembling rotting bark or wood). Exterior yellowish gray; interior glossy-white, with a purple tinge, especially prominent on the hinge and margins. Commonly found in rocky rubble in shallow water.

Semele flavescens (Gould, 1851)
5787

Yellowish Semele

Southern California to Peru.

2 to 2½ inches, almost circular, whitish to yellowish gray, compressed, with a greenish periostracum. Surface appears to be microscopically granulose. Lunule small, deeply sunken. Interior white with yellow tints. There is a small escutcheon in both valves. Uncommon in California, common southward; intertidal sand flats. Synonyms: *proxima* (C. B. Adams, 1853), and *californica* Reeve, 1853.

Some specimens well-beaded, others have the radial ribs more prominent. External color yellowish white with reddish flecks or a solid, purplish gray. Interior white, cream or suffused with mauve or violet. Synonyms are: *cancellata* (Orbigny, 1842); *nexilis* Gould, 1862; and *lata* of Bush, 1885, not A. Adams, 1855. A form (5782) which is smooth in the center of the valve comes from Lake Worth, Florida (*donovani* McGinty, 1955) and is found fossil. *S. bellestriata* is a misspelling.

A similar species (5783) *Semele pacifica* Dall, 1915, lives from Catalina Island in the Gulf of California to Panama, is 1 inch, colored buff or cream and lightly variegated with zig-zag lines of purple. Uncommon. *S. jaramija* Pilsbry and Olsson, 1941, is a synonym.

Semele rubropicta Dall, 1871
5784

Rose Petal Semele

Alaska to Mexico.

1 to 1¾ inches in length; beaks ⅔ toward the posterior end; rather thick-shelled, especially in the south. Concentric sculpture of small, irregular growth lines. Radial incised lines numerous. Periostracum thin, smooth and yellowish brown. Exterior of shell dull grayish or tannish white with faint, radial rays of light-mauve. Interior glossy-white, with a small splotch of mauve at both ends of the hinge line. Uncommon; from 20 to 50 fathoms.

5787

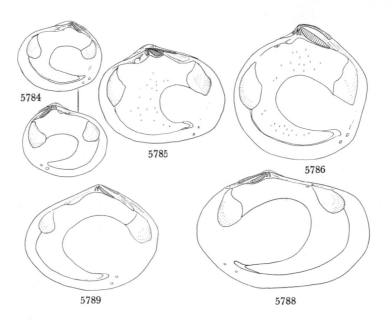

5784

5785

5786

5789

5788

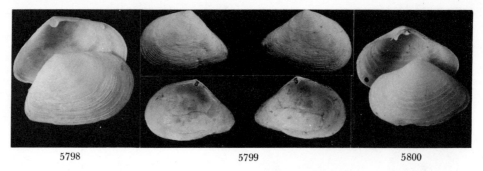

5798   5799   5800

Semele incongrua (Carpenter, 1864)   5788
Incongruous Semele

Monterey to the Coronado Islands, Baja California.

½ to ¾ inch, oblong-oval, radially and concentrically striate, thin-shelled; beaks quite posterior to the middle of the shell. Anterior dorsal margin long and only slightly curved; shorter anterior margin evenly and strongly convex. Resilium groove deep and oblique. One prominent cardinal tooth in each valve; lateral teeth indistinct. Pallial sinus large and rounded. Concentric sculpture differs in right and left valve. Not uncommon; dredged from 7 to 25 fathoms on sand bottoms. *S. montereyi* Arnold, 1903, is a form or northern subspecies (5788a).

Other Pacific species:

5789   *Semele pulchra* (Sowerby, 1832). Ventura County, California, to Costa Rica. Shore to 16 fms. 1 inch. (*S. pulchra* of authors, not Sowerby, 1832.) *S. quentinensis* Dall, 1921, is a synonym. See E. V. Coan, 1973, *The Veliger*, vol. 15, p. 323.

5790   *Semele pallida* (Sowerby, 1833). Baja California to Ecuador. (Synonyms: *regularis* Dall, 1915; *simplicissima* Pilsbry and Lowe, 1932; and *paziana* Hertlein and Strong, 1949.)

5791   *Semele verrucosa* Mörch, 1860. Gulf of California to Panama.

5792   *Semele jovis* (Reeve, 1853). Kino Bay, Sonora, Mexico, to Panama; mud flats from low tide to 15 fms. Rare. (Synonym?: *Tellina barbarae* Boone, 1928.)

5793   *Semele guaymasensis* Pilsbry and Lowe, 1932. Gulf of California to Panama.

5794   *Semele rosea* (Sowerby, 1833). Gulf of California to Panama. (Synonyms: *junonia* Verrill, 1870; *tabogensis* Pilsbry and Lowe, 1932.) Uncommon.

5795   *Semele bicolor* (C. B. Adams, 1852). Gulf of California to Panama. (Synonyms: *fucata* Mörch, 1860; *Amphidesma striosum* and *ventricosum* C. B. Adams, 1852.)

Subgenus *Semelina* Dall, 1900

Type of the subgenus is *nuculoides* (Conrad, 1841) which has a long anterior end.

5796

Semele nuculoides (Conrad, 1841)   5796
Nucula-like Semele

North Carolina to the Gulf of Mexico; West Indies.

4 to 6 mm., elongate, with the small beaks near the posterior end. Dorsal anterior margin long and only slightly curved. Surface with fine, crowded concentric ridges somewhat sharper over the posterior slope where there may also be fine radial striations. Color whitish or yellow, red or rayed. Resilium pit not conspicuous. *S. lirulata* Dall, 1900, is a synonym.

(5797)   The Pacific counterpart is *Semele* (*Semelina*) *subquadrata* (Carpenter, 1857). 5 mm. Mexico to Colombia.

Genus *Cumingia* Sowerby, 1833

Shell delicate, with concentric lamellae; slightly gaping behind. Resilium internal and supported by a spoon-shaped chondrophore. One cardinal and 2 elongate lateral teeth in each valve. Lunule small and smooth. Escutcheon largest in left valve. Pallial sinus large, confluent with the pallial line below. These clams are nestlers and become distorted as they grow. Type: *lamellosa* Sowerby, 1833. Synonyms: *Harpax* Gistel, 1848; *Mikrola* Meyer, 1887.

Cumingia tellinoides (Conrad, 1831)   5798
Tellin-like Cumingia

Nova Scotia to St. Augustine, Florida.

10 to 20 mm. in length, oblong and fairly thin. Slightly pointed at the posterior end. Exterior chalky-white, with tiny, sharp, concentric lines. This is a moderately common mud-digger which externally resembles a *Tellina*.

(5799)   The subspecies *vanhyningi* Rehder, 1939, replaces the typical species in southern Florida to Texas. It is not so high, is more elongate and more drawn out posteriorly. Common in shallow water.

Cumingia coarctata Sowerby, 1833   5800
Southern Cumingia

Southern Florida and the Caribbean. Bermuda. Brazil.

5 to 10 mm., fragile, milk-white; tellinlike in shape, but usually irregular, somewhat inflated, gaping a little behind; outer surface with distantly spaced, irregular, raised lamellae and with numerous, microscopic radial scratches. Under the beaks there is a spoon-shaped resilium pit. Left valve has 2 faint laterals, the right valve has 2 strong ones, in addition to the double cardinal behind the pit. Common; shallow water to 32 fathoms. Synonyms: *antillarum* and *petitiana* both Orbigny, 1842.

(5801)   The Pacific counterpart is the similar *Cumingia lamellosa* Sowerby, 1833, from the Gulf of California to Peru. 10 to 20 mm. and often distorted. Common; in sand, clay, and rock fissures, intertidal to 24 meters. *C. similis* A. Adams, 1850, and *C. moulinsii* De Folin, 1867, are synonyms.

Cumingia californica Conrad, 1837   5802
Californian Cumingia

Crescent City, California, to Baja California.

1 to 1⅓ inches in length, elongate-oval, moderately compressed. In front of the beaks there is a small, elongate depression. Just posterior to and partially covered by the inrolled beaks is a small, short ligament, posterior to which is a wide, flaring furrow. Concentric sculpture of numerous wavy, rather sharp, fairly large threads. Color grayish white. Pallial sinus very long. Abundant in rock crevices and wharf pilings. Also dredged down to 25 fathoms. *Cumingia densilineata* Dall, 1921, is a synonym.

5802

Subfamily SCROBICULARIINAE H. and A. Adams, 1856

Genus *Abra* Lamarck, 1818

Shell small (¼ inch), fragile, ovalish, smooth, moderately compressed. Translucent-white in color. Resilium internal and supported by a linear chondrophore. Right valve with 2 cardinals and generally with 2 lamellar laterals. Pallial sinus large and widely confluent with the pallial line below. Type: *tenuis* (Montagu, 1818), from Europe. *Habra* Agassiz, 1846, is the same. This genus was formerly placed in the subfamily Semelinae.

Abra aequalis (Say, 1822)   5803
Common Atlantic Abra

North Carolina to Texas and the West Indies. Brazil.

¼ inch in size, white, orbicular, smooth, glossy and rather inflated. Surface may show a slight iridescence. Periostracum very thin and clear-yellowish. Anterior margin of right valve grooved. A very abundant and very simple-looking bivalve. 1 to 20 fathoms. Compare with *lioica*. *A. nuculiformis* Conrad, 1867, is a synonym.

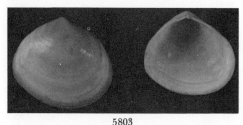

5803

Abra lioica (Dall, 1881)   5804
Dall's Little Abra

Cape Cod to south Florida and the West Indies.

¼ inch in size, white, similar to *aequalis,* but the beaks are nearer the anterior end, the shell is thinner and more elongate. Anterior margin of right valve not grooved. The prodissoconch at the beaks is large, tan and more trigonal in shape than the adult. Common from 6 to 200 fathoms. *A. tepocana* (Dall, 1915), is the Gulf of California to Ecuador counterpart.

Other species:

5805 *Abra longicallis americana* Verrill and Bush, 1898. Arctic Ocean to the West Indies, 50 to 1,467 fms. 20 mm.

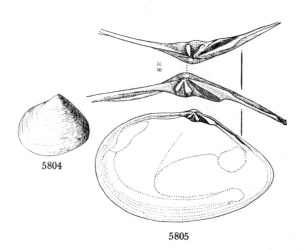

5805

5806 *Abra tepocana* Dall, 1915. Gulf of California to Ecuador. (Synonym: *A. palmeri* Dall, 1915.)

5807 *Abra pacifica* Dall, 1915. Guaymas, Mexico.

5808 *Abra californica* Knudsen, 1970. Off Baja California (24°N, 113°W), 3,481 meters. *Galathea Report*, vol. 11.

Family SOLECURTIDAE Orbigny, 1846

Elongate-quadrate clams, widely gaping at both ends; hinge plate weak and narrow. Pallial sinus present. Beaks subcentral. Includes such genera as *Azorinus* Récluz, 1869; *Solecurtus* Blainville, 1824; *Tagelus* Gray, 1847. Some workers place this group as a subfamily under the family Psammobiidae.

Genus *Solecurtus* Blainville, 1824

Quadrate to rectangular in shape; gaping at both ends. With weak, "clapboard" sculpturing. Ligament prominent, external and posterior to the small beaks. Right valve with 2 strong, horizontally jutting cardinal teeth just under the centrally located beaks. Left valve with 1 cardinal. *Psammosolen* Risso, 1826; *Adasius* Gray, 1852; and *Macha* Oken, 1835, are synonyms. *Solenocurtus* Blainville, 1826, was an emendation. Type: *strigilatus* (Linné, 1758).

Solecurtus cumingianus Dunker, 1861   5809
Corrugated Razor Clam

North Carolina to south half of Florida to Texas. Brazil.

1 to 2 inches in length with characters of the genus. Color all-white, with a dull, yellowish gray periostracum. Outer surface sculptured with coarse, concentric, irregular lines and with sharp, small, oblique, wavy threads. Uncommon offshore from 14 to 111 fathoms.

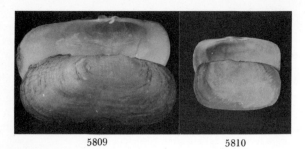

5809   5810

Solecurtus sanctaemarthae Orbigny, 1842   **5810**
St. Martha's Razor Clam

North Carolina to Florida, Texas and the West Indies. Bermuda. Brazil.

½ inch in length, differing from *cumingianus* in being twice as long as high (instead of 2½ times) and in having stronger sculpturing. Uncommon from shallow water to 25 fathoms.

Other species:

5811 *Solecurtus guaymasensis* Lowe, 1935. Cedros Island, Baja California, to Chiriqui, Panama, in 37 to 110 meters.

Genus *Tagelus* Gray, 1847

Shells cylindrical, slightly compressed, oblong, with the beaks near the middle. Periostracum brownish, wrinkled. Shell whitish with violet or brown coloring. Pallial sinus large. Hinge weak, with 2 small cardinal teeth in the right valve, only 1 in the left. No lateral teeth. Ligament external. Type: *adansonii* (Bosc, 1801).

Tagelus plebeius (Lightfoot, 1786)   **5812**
Stout Tagelus

Cape Cod to Florida, to Texas and the West Indies. Brazil.

2 to 3½ inches in length, oblong, subcylindrical, rather inflated, rounded posteriorly, obliquely truncate anteriorly. Beaks indistinct, close together and nearer the posterior end of the shell. Hinge with 2 small, projecting cardinal teeth, with a large bulbous callus just behind them. Exterior smoothish, with tiny, irregular, concentric scratches. Periostracum moderately thick, shiny, olive-green to brownish yellow. Moderately common in shallow water in mud-sand intertidal areas to a water depth of about 20 feet. Edible. *T. gibbus* Spengler, 1794, and *caribaeus* Lamarck, 1818, are later names for this species.

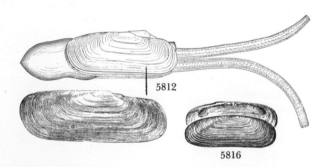

5812

5816

Tagelus californianus (Conrad, 1837)   **5813**
Californian Tagelus

Monterey, California, to northern Pacific Mexico.

2 to 4 inches in length. Pallial sinus does not extend past a line vertical to the beaks. External color yellowish white under a dark-brown periostracum which is radially striated. Interior white. Common on muddy sand flats near marshes. Lives 8 to 20 inches below the surface of the sand. Locally called a jackknife clam. A suspension feeder (see R. Pohlo, 1966, *The Veliger*, vol. 8, p. 225).

It is replaced in the south by *Tagelus* (*Solecurtellus* Ghosh, 1920) *dombeii* (Lamarck, 1818) (**5814**) from Panama to Chile.

Tagelus affinis (C. B. Adams, 1852)   **5815**
Affinis Tagelus

Southern California to Ecuador.

1½ to 2½ inches in length. Shell thin. Pallial sinus extends to a line slightly beyond the beaks, that is, about 53 to 60 percent of the total length of the shell. Dredged from 3 to 40 fathoms in mud-sand, but commonly washed ashore. Very similar to *californianus*, but stubbier and its beaks are slightly behind the center of the shell, whereas in *californianus* the beaks are central. Rare in California, more common in the Panama region. *Solecurtus cylindricus* Sowerby, 1874, is a synonym.

5812

5813

5817

Subgenus *Mesopleura* Conrad, 1868

Shell with an internal radial rib running from the beak area towards the ventral edge, strong in *divisus* (Spengler), the type, but weak in *subteres* (Conrad). *Subtagelus* Ghosh, 1920, is a synonym.

Tagelus divisus (Spengler, 1794)   **5816**
Purplish Tagelus

Cape Cod to south Florida to Texas and to Brazil. Bermuda.

1 to 1½ inches in length, elongate, subcylindrical, fragile and smooth. The valves are reinforced internally by a very weak, radial rib (commonly obscure) running across the center of the valve just anterior to the 2 small, projecting cardinal teeth. Color of shell whitish purple, covered externally with a very thin, chestnut-brown, glossy periostracum. A common shallow-water species. *T. bidentatus* (Spengler, 1794) and *centralis* (Say, 1822) are synonyms.

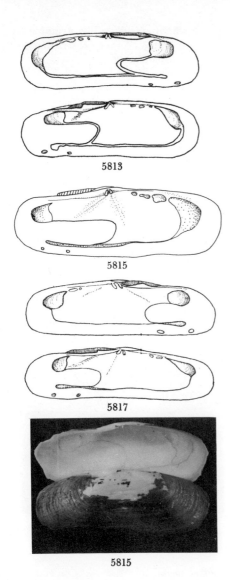

5813

5815

5817

5815

Family DREISSENIDAE Gray, 1840

Genus *Mytilopsis* Conrad, 1858

Shell mytiloid in shape; anterior adductor muscle attached to a shelflike platform in the apical section. Sometimes with a small, toothlike projection below the septum for the attachment of the byssal muscle. Fresh to brackish water. Type: *leucophaeata* (Conrad, 1831). *Praxis* H. and A. Adams, Dec. 1857, is a synonym (non Guenèe, 1852).

Mytilopsis leucophaeata (Conrad, 1831)   **5820**
Conrad's False Mussel

New York to Florida to Texas and Mexico.

½ to ¾ inch in length, superficially resembling a *Mytilus* or *Septifer* because of its mussellike shape. The *Septifer*-like shelf at the beak end has a tiny, downwardly projecting, triangular tooth on the side facing the long, internal ligament. The hinge has a long thin bar under the ligament. Exterior bluish brown to tan with a thin, somewhat glossy periostracum. Interior dirty bluish tan. This common bivalve attaches itself by its short byssus to rocks and twigs in clumps which resemble colonies of *Mytilus*. Found in brackish to fresh water near rivers.

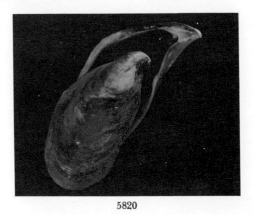

5820

Other species:

5821 *Mytilopsis dominguensis* Récluz, 1852. West Indies. See "Caribbean Seashells," pl. 35a.

5822 *Mytilopsis sallei* Récluz, 1849. Lower Caribbean.

Superfamily Arcticacea Newton, 1891

Family ARCTICIDAE Newton, 1891

Large clams with equivalves closed. Ligament on nymphs, insertion grooves incised. Valve margins smooth or weakly crenulate; adductor muscle scars subequal; pallial line entire. Cyprinidae Orbigny, 1844, and Venilidae Dall, 1889, are invalid family names (Code, article 11 e).

Genus *Arctica* Schumacher, 1817

Shell ovate, periostracum thick and black-brown, **no** pallial sinus, lunule or escutcheon. Type: *islandica* (Linné, 1767). Synonyms: *Cyprina* Lamarck, 1818; *Armida* Gistel, 1848; and *Nympha* Mörch, 1853.

Tagelus subteres (Conrad, 1837)   **5817**
Purplish Pacific Tagelus

Morro Bay, California, to Baja California.

1 to 2 inches in length, subcylindrical, slightly arcuate with the dorsal margins sloping down from the beaks. Color pale-purple inside and out. Periostracum yellowish brown and finely wrinkled. Moderately common in shallow water in sandy mud.

(5818) *Tagelus politus* (Carpenter, 1855), from Central America to Peru, does not slope down from the beaks so strongly, is a thinner shell and much more darkly colored with violet. 1½ inches long. Common. Synonyms are *carpenteri* Dunker, 1862, and *nitidissima* Dunker, 1868.

Other species:

5819 *Tagelus* (*Mesopleura*) *peruvianus* Pilsbry and Olsson, 1941. Gulf of California to Peru. 3 inches. With a flangelike wing on the posterior dorsal margin.

Superfamily Dreissenacea Gray, 1840

Mytiliform, beaks anterior. Interior of shell porcelaneous. Ligament sunken; hinge without teeth. Beak cavity bridged by a narrow shelf. Periostracum fuzzy. Byssus produced.

Arctica islandica (Linné, 1767)   5823
Ocean Quahog

Newfoundland to off North Carolina.

3 to 5 inches in length, almost circular in outline, rather strong, porcelaneous, but commonly chalky. Exterior covered with a brown to black, rather thick periostracum. The posterior laterals and the absence of a pallial sinus distinguish this clam from the true quahogs (see *Mercenaria mercenaria*). A common, commercially dredged species found in sandy mud from 5 to 80 fathoms. This is the only living species in this family. There are numerous fossil species. Also called the black clam and mahogany clam.

5823

Family BERNARDINIDAE Keen, 1963

Shells minute, ligament internal, hinge with 2 or 3 cardinals and 2 or more laterals. Pallial line entire.

Genus *Bernardina* Dall, 1910

Shell small, 3 mm., ovate in outline, beaks near the front; prodissoconch conspicuous and bounded by a raised ring. No pallial sinus. 2 right and 3 left cardinals with the resilium posterior to them. Concentric sculpture. Type: *bakeri* Dall, 1910. The young are brooded within the mantle cavity of the mother.

Bernardina bakeri Dall, 1910   5824
Fred Baker Clam

San Diego, California, to Baja California.

2 to 3 mm., tan, ovate, with the beaks ⅓ back from the anterior end. Prodissoconch conspicuous, bounded by a raised ring. Sculpture of fine concentric lines. Pallial line without a sinus. Hinge with the posterior dorsal margin of the right valve fitting into a shallow groove in the margin of the opposite valve. Anteriorly with a strong left lateral fitting between 2 prominent flexuous right anterior laterals. 2 right and 3 left cardinals with the resilium posterior to all of them. Uncommon; among rocks from low tide to 10 fathoms.

Genus *Halodakra* Ollson, 1961

Shell very small, 4 mm., externally looks like a small polished venerid clam. No lunule or escutcheon. Ligament largely internal. Resilifer represented by a small, narrow scar along the inner side of a short nymphal ridge. Cardinal teeth 2 in right valve, 3 in left. The middle left cardinal and

right anterior cardinal are bifid. The nymphal ridge in the right valve is a toothlike lamina. No pallial sinus. Type: *Circe subtrigona* Carpenter, 1857.

5825 *Halodakra subtrigona* (Carpenter, 1857). Baja California to Peru. 3.8 mm. Featherlike brown patch on posterior slope. Plentiful in beach drift. (Olsson, 1961, Paleont. Research Inst., "Panamic-Pacific Pelecypoda," pl. 27, figs. 1 to 1c).

5826 *Halodakra brunnea* (Dall, 1916). Monterey, California, to San Ipolito Point, Baja California. 20 to 50 fms. Uncommon.

Family KELLIELLIDAE Fischer, 1887

Small equivalve clams, not gaping. Ligament external, some species with the resilium under the beak. Hinge teeth weak. Inner margins smooth. Pallial sinus absent.

Genus *Kelliella* Sars, 1870

Shell minute, rounded-ovate. Lunule distinct. Hinge with 2 teeth in the left valve; 1 cardinal and 1 anterior lateral in the right valve. Deepwater species. Type: *miliaris* Philippi, 1844. *Kellyella* is an invalid emendation.

5827 *Kelliella nitida* Verrill, 1885. Off Martha's Vineyard, Massachusetts, to off Delaware, 1,525 to 2,033 fms. Off Fernandina, Florida, 294 fms. Type illustrated.

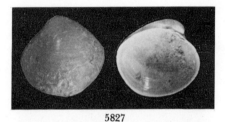

5827

Family TRAPEZIIDAE Lamy, 1920

Shell elongate, beaks near the anterior end; hinge plate narrowed, normally with 2 cardinals in either valve, 1 posterior and 1 small anterior lateral; pallial line mostly entire (Myra Keen). Libitinidae Thiele, 1924, and Lithophagellidae Cossmann, 1910, are synonyms.

Genus *Trapezium* Mühlfeld, 1811

Shell solid, oblong, laterally compressed, porcelaneous; ligament external. Exterior with radiating striae; inner margin smooth. Type: *chama oblonga* Linné, 1758. Synonyms: *Libitina* Schumacher, 1817, and *Cypricardia* Lamarck, 1819. For an excellent review, see Alan Solem, 1954, *Proc. Mal. Soc. London*, vol. 31, pp. 64–84.

Subgenus *Neotrapezium* Habe, 1951

Shell quadrangular, compressed, purple-stained within, and small, weak posterior laterals in the right valve. Type: *sublaevigatum* Lamarck, 1819.

Trapezium liratum (Reeve, 1843)   5828
Lirate Trapezium

British Columbia to California.
Introduced from Japan.

1 to 1½ inches in length, elongate-quadrate in shape, with a small umbo located anteriorly. Exterior with irregular, coarse growth lines. A strong, sharp dorsal keel extends backward from the umbones. Color pattern is in general dark-brown, occasionally with a few rays near the posterior end. Synonyms of the intertidal, crevice-dwelling species are *japonicum* Pilsbry, 1905, and *delicatum* Pilsbry, 1905. Appeared in California prior to 1935.

Genus *Coralliophaga* Blainville, 1824

Shells which resemble in shape the date mussels, *Lithophaga,* and which also bore into coral. Teeth compressed, slender, only 2 cardinals and 1 lateral in each valve. Pallial line with a faint sinuation. Exterior radially striated. Type: *coralliophaga* (Gmelin, 1791). Synonym: *Lithophagella* Gray, 1854.

Coralliophaga coralliophaga (Gmelin, 1791)   **5829**
Coral Clam

> North Carolina to Texas and the West Indies. Bermuda. Brazil.

½ to 2 inches, elongate, with the umbones at the anterior end. Color yellowish white. Very finely sculptured with radial threads. Concentric lamellations present at the posterior end. Resembles *Lithophaga antillarum,* but may be told from it by the presence of distinct teeth in the hinge. An uncommon species which lives in the burrows of other rock-boring mollusks.

5829

Superfamily Glossacea Gray, 1847

Family VESICOMYIDAE Dall and Simpson, 1901

Shell ovate to elongate; lunule incised in most; hinge with up to 3 teeth, not clearly differentiated into cardinals and laterals; with or without a pallial sinus. Vesicomyacidae is an incorrect version.

Genus *Calyptogena* Dall, 1891

Elongate, smooth, of earthy texture, with periostracum. 1 to 4 inches long. Escutcheon present; no lunule. Ligament external, deep-seated. Pallial line entire and sinuous. Deep water. Type: *pacifica* Dall, 1891. Formerly placed in the family Carditidae.

5830 *Calyptogena pacifica* Dall, 1891. Clarence Strait, Alaska, to Santa Barbara Channel, California, 322 fms. 48 mm.

5831 *Calyptogena ponderosa* Boss, 1968. 77 miles south of Mobile Bay, Gulf of Mexico, 600 fms. 93 mm. *Bull. Marine Sci.,* vol. 18, p. 737.

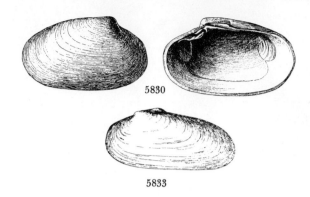

5830

5833

5832 *Calyptogena (Ectenagena) modioliforma* (Boss, 1968). 66 miles off Punta Caribana, Colombia, 641 meters. 97 mm. *Bull. Marine Sci.,* vol. 18, p. 742.

5833 *Calyptogena elongata* Dall, 1916. Santa Barbara Islands to San Diego, California, 275 fms. Type of the subgenus *Ectenagena* Woodring, 1938. 44 mm.

Genus *Vesicomya* Dall, 1886

Ovate, inequilateral, smooth. Periostracum polished; lunule bounded by a groove. Without laterals. Type: *Callocardia atlantica* E. A. Smith, 1885.

5834 *Vesicomya smithii* Dall, 1889. Lesser Antilles, 880 fms.

5835 *Vesicomya pilula* (Dall, 1881 and 1886). Off Fernandina, Florida, 294 fms., Florida Strait and West Indies. 2.6 mm.

5836 *Vesicomya venusta* (Dall, 1886). Off Cape Fear, North Carolina, 731 fms., Florida Strait, 801 fms.

5837 *Vesicomya ovalis* (Dall, 1896). Clarence Strait, Alaska, 322 fms., to Panama, 1,672 fms.

5838 *Vesicomya lepta* (Dall, 1896). Off Oregon to the Gulf of California. Deep water.

5839 *Vesicomya stearnsii* (Dall, 1895). Off La Jolla, California, to the Gulf of California, in deep water.

5840 *Vesicomya (Archivesica* Dall, 1908) *gigas* (Dall, 1896). Off Point Sur, California, to the Gulf of California. Deep water, 1,565 meters.

5841 *Vesicomya bruuni* Filatova, 1969. Kermadec Trench, 9,000 meters.

5842 *Vesicomya (Archivesica) suavis* Dall, 1913. Off Animas, Baja California, 735 fms. 35 mm.

5843 *Vesicomya (Vesicomya) cordata* Boss, 1968. 66 mi. off Punta Caribana, Colombia, 641 meters. *Bull. Marine Sci.,* vol. 18, p. 733. 63 mm. in length.

Subgenus *Callogonia* Dall, 1889

Lunule without a border; pallial sinus shallow but acute. Exterior smoothish, with only growth lines; yellow. Type: *Callocardia leeana* Dall, 1889.

5844 *Vesicomya (Callogonia) leeana* (Dall, 1889, p. 439). Off Tobago, West Indies, 880 fms. 2 inches.

5845 *Vesicomya (Callogonia) angulata* (Dall, 1896). Gulf of Panama, 2,320 to 3,050 meters.

5846 *Vesicomya (Callogonia) caribbea* Boss, 1967. Off Cabo la Vela, Caribbean Sea, Colombia, 205 fms. 4 inches. *Breviora, Mus. Comp. Zool.,* no. 266.

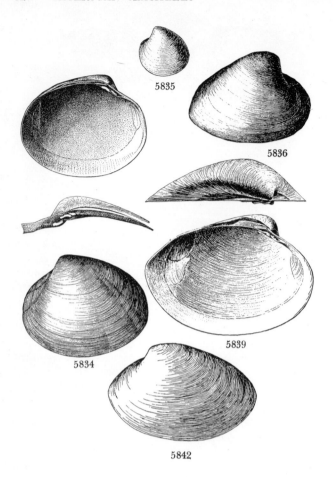

5835

5836

5839

5834

5842

Subgenus *Veneriglossa* Dall, 1886

Shell about 1 inch, resembling a venerid clam, inflated, roundly ovate. Lunule wide, short, marked by a fine inscribed line. Umbones fat, twisted away from the hinge line so that their tips are widely separated. Ligament long, in a deep groove, passing away from the hinge line under the beaks. Ventral margin of the hinge shelf upturned. Teeth somewhat like those of *Pitar*. Pallial line with a shallow wide wave just before the posterior adductor scar. Type and only known species:

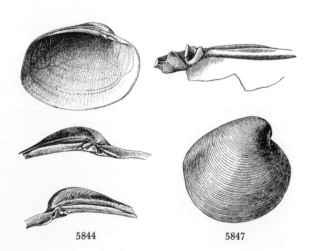

5844

5847

Vesicomya vesica (Dall, 1886)   5847
Dall's False Mya

Gulf of Mexico and the Lesser Antilles.

22 mm. long, 21 mm. high and 17 mm. in diameter, thin-shelled, ovate, fat, white, uniformly concentrically grooved; polished. Umbones curled and spaced apart. Rare; 84 to 175 fathoms.

Family GLOSSIDAE Gray, 1847

Medium-sized, subquadrate to rotund bivalves with prominent, inrolled beaks and with well-developed cardinals and laterals. Pallial sinus absent.

Genus *Meiocardia* H. and A. Adams, 1857

Shell obese, subtrapezoidal, anterior side arcuate, ventral margin gently rounded and sloping posteriorly. White to ivory, with a very thin periostracum. Beaks swollen and inrolled. Hinge thin, arched; cardinal teeth rather anterior, and closely placed to the anterior laterals, which are visible only as small protuberances; anterior cardinals short and diagonal; posterior cardinals long and rather curved; posterior laterals long and rather straight. Type: *moltkiana* Spengler, 1783.

Meiocardia agassizii Dall, 1889   5848
Agassiz's Ox-heart Clam

Off west Florida to the Caribbean. Bermuda.

1 inch in length, obese, subquadrate, polished, yellowish. Beaks small and inrolled. Sharp keel runs from the beaks to the lower posterior end of the shell. Surface with fine, even, concentric rugae. Pallial line faint, without a sinus. Rare; 80 to 100 fathoms.

5848

Superfamily Corbiculacea Gray, 1847

Ovate, porcelaneous; concentrically striate. Ligament external. Hinge with up to 3 cardinal teeth, with the largest cardinal in the right valve. Pallial line entire or with a small sinus.

Family CORBICULIDAE Gray, 1847

Brackish or marine clams, resembling venerid clams, but little or no pallial sinus, 3 cardinal teeth in each hinge and with both anterior and posterior lateral teeth. No lunule or escutcheon. Common food clam in tropical Asia and South American areas. The fresh-water *Corbicula manilensis* Philippi was introduced into the rivers of the United States in 1938, and is now a widespread nuisance. It clogs ditches, canals, pumps and water filters. For a useful bibliography see R. M. Sinclair, 1971, *Sterkiana,* no. 43, p. 11.

Genus *Polymesoda* Rafinesque, 1820

Shells roundly ovate, with a short, narrow pallial sinus. Beaks inturned. Type of the genus is *caroliniana* (Bosc, 1801). Synonyms include: *Egetaria* Mörch, 1860; *Cyprinella* Gabb, 1864; *Diodus* Gabb, 1868; *Leptosiphon* Fischer, 1872; *Americana* Clessin, 1879.

Polymesoda caroliniana (Bosc, 1801)   **5849**
Carolina Marsh Clam

Virginia to north half of Florida and Texas.

1 to 1½ inches in length, about as high, subtriangular in outline, rather obese and with a strong shell. Exterior of smoothish shell is covered with a very fuzzy or minutely scaled periostracum which is mostly glossy-brown and rather thin. Interior white, rarely stained with purple. Each hinge with 3 small, almost vertical, equally sized teeth below the beaks and each hinge with 1 anterior and posterior lateral. Ligament external, long, narrow and dark-brown. Common at the mouths of rivers where the influence of the tides is felt. Synonym: *alabamensis* Clessin, 1869. For anatomy and biology, see van der Schalie, 1933, *Occ. Papers Mus. Zool., Univ. Michigan,* no. 258.

5851

5849

(5850) A comparable species, *Polymesoda mexicana* (Broderip and Sowerby, 1829) ranges from the southern part of the Gulf of California to Puerto Vallarta, Mexico. Synonyms are *nitidula* (Deshayes, 1855) and *fragilis* (Sowerby, 1878).

Subgenus *Pseudocyrena* Bourguignat, 1854

Shells elongate; pallial sinus ill-defined. Type: *maritima* (Orbigny, 1842). *Cyrenocapsa* Fischer, 1872, is a synonym.

Polymesoda maritima (Orbigny, 1842)   **5851**
Florida Marsh Clam

Key West to northern Florida and to Texas.

1 inch in length, quite similar to *Polymesoda caroliniana,* but more variable in shape (ovalish to elongate), without the

fuzzy periostracum, with its beaks never eroded away and with 2 long, slender anterior and posterior laterals. Exterior with irregular growth lines, dull dirty-white, commonly flushed with purple or pink. Interior white with a wide margin of deep-purple or entirely purple. Brackish warm water in mud. Common. Synonyms: *protexta* Conrad, 1869; *floridana* Conrad, 1846.

Superfamily Veneracea Rafinesque, 1815

Strong, ovate shells with anteriorly placed beaks curling forward. Ligament external and posterior to the beaks. Cardinal teeth usually 3 in either valve. Pallial sinus usually large. Includes the living families Veneridae, Petricolidae, Cooperellidae and the Indo-Pacific Glauconomidae.

Family VENERIDAE Rafinesque, 1815

The classification of the family of Venus clams has been one of continual debate and rearranging for some years. Our presentation here is no better than has been suggested before, but at least it is in a form which is conservative and most likely to be accepted by the majority.

Subfamily VENERINAE Rafinesque, 1815

Sculpture usually both radial and concentric; 3 cardinal teeth in each valve; anterior lateral present, especially in the left valve, but often extraordinarily vestigial. Usually heavily sculptured. Ventral margins crenulated.

Genus *Periglypta* Jukes-Browne, 1914

Quadrate and heavy; pallial sinus rounded. Sculpture formed by strong, reflected, concentric lamellae or ridges, crenulated and with axial threads or cords in their interspaces. Type: *puerpera* (Linné, 1758) from the Indo-Pacific. Our American forms were previously put in the genus *Antigona* Schumacher, 1817 (type, *lamellaris* Schumacher, 1817) limited to the Indo-Pacific.

Periglypta listeri (Gray, 1838)   **5852**
Princess Venus   **Color Plate 24**

Southeast Florida, south Texas and the West Indies.

2 to 4 inches in length, oblong-oval, obese. Interior cream, maculated with brown. Resembling *Mercenaria campechiensis,* but characterized by numerous, fine, radial riblets which cause the sharp, concentric ribs to be serrated or beaded. Each side of the lunule is bounded by a long, deep, narrow furrow. Posterior muscle scar usually stained brown. Moderately common in shallow water in sand.

Periglypta multicostata (Sowerby, 1835)   **5853**
Many-ridged Venus

Gulf of California to Peru.

3 to 4½ inches, ovate-elongate, with numerous strong flat-topped, reflected, crenulated or beaded concentric ridges. Color white to cream with the umbones finely variegated with brown lines. Common; in sand; at low-tide mark. *Venus thouarsi* Valenciennes, 1846, is a synonym.

5853

Genus *Ventricolaria* Keen, 1954

Lunule depressed; escutcheon smooth, beveled in left valve. Interspaces between the concentric lamellae marked only with fine concentric threads. Type: *rigida* (Dillwyn, 1817). This is *Ventricola* of authors, not Römer, 1867.

Ventricolaria rugatina (Heilprin, 1887)   5854
Queen Venus

North Carolina to southeast Florida and the West Indies.

1 to 1½ inches in length, rather circular, inflated and characterized by strong, raised, lamellate, concentric ribs between which are 5 or 6 smaller, raised concentric ridges. Lunule heart-shaped, well-impressed, bordered by a fine, deep line and crossed by numerous raised threads. Escutcheon well-formed, smoothish. Color cream to whitish with light-mauve mottlings. Very uncommon; down to 70 feet.

5854

Ventricolaria rigida (Dillwyn, 1817)   5855
Rigid Venus   Color Plate 24

Florida Keys (rare); West Indies to Brazil.

1½ to 3 inches, almost circular in outline, inflated. Color cream with brown mottlings; interior white or cream. Sculptured with numerous prominent concentric ribs between which are 1 to 3 fine concentric threads. Escutcheon on left valve

marked with purplish red. Fairly common in the West Indies. Rarely dredged in Florida, 30 fathoms.

Ventricolaria rigida subspecies *isocardia* (Verrill, 1870) from the Gulf of California to Ecuador, is obese, has faint zigzag lines and large blotches of light-brown; the beak and interior sometimes flushed with pink. Moderately common; shore to 61 fathoms. *V. magdalenae* (Dall, 1902) may be a synonym.

Other species:

5856 *Ventricolaria foresti* Fischer-Piette and Testud, 1967. Northeast Brazil, 10 to 58 fms. *Annales l'Inst. Oceanogr.*, vol. 45, p. 205.

5857 *Ventricolaria listeroides* Fischer-Piette and Testud, 1967. Ile Fernando Noronha.

Genus *Circomphalus* Mörch, 1853

Lunule impressed; escutcheon larger in left valve. Sculpture of raised lamellae. A minute pustular anterior lateral present in the left valve. Type: *plicata* (Gmelin, 1791), non Barbut, 1788. Now *foliaceolamellosa* (Dillwyn, 1817).

Circomphalus strigillinus (Dall, 1902)   5858
Empress Venus   Color Plate 24

Off South Carolina to Brazil.

1½ inches in length, externally very much like a small *Mercenaria campechiensis*, but not as elongate and with more distinct, concentric riblets. Internally, it is distinguished easily by the extremely small, if not absent, pallial sinus, by the very thick margin of the shell, and in the left valve by the presence of a buttonlike anterior lateral "tooth." Exterior whitish. Dredged occasionally from 20 to 100 fathoms. Considered a collector's item.

Other species:

5859 *Circomphalus callimorphus* (Dall, 1902). Florida Strait to the Lesser Antilles, 76 to 300 fms. Bermuda, 80 fms.

5859

5860 *Circomphalus fordi* (Yates, 1890). Southern California.

Subfamily CHIONINAE Frizzell, 1936

Ovate-trigonal, inequilateral, sculpture usually cancellate; lunule impressed. Inner margins usually crenulate. Teeth strong, without the tiny, pimplelike anterior lateral. Pallial sinus short.

Genus *Mercenaria* Schumacher, 1817

The hard-shell clams or quahogs belong to this genus. The shell is large and thick; lunule large, heart-shaped and bounded by an incised line. Inner margin crenulate. 3

cardinals in each valve. Left middle cardinal split. Formerly placed in the genus *Venus* many years ago, but almost universally placed in a genus by itself by modern workers. Type: *mercenaria* (Linné, 1758). Synonym: *Crassivenus* Perkins, 1869.

Mercenaria mercenaria (Linné, 1758)   5861
Northern Quahog

Gulf of St. Lawrence to Florida and the Gulf of Mexico. Introduced to Humboldt Bay, California, and to England.

3 to 5 inches in length, ovate-trigonal, about ⅝ as high, heavy and quite thick. Moderately inflated. Sculpture of numerous, concentric lines of growth or small riblets. Near the beaks these lines are prominent and distantly spaced. The exterior center of the valves has a characteristic smoothish or glossy area. Exterior dirty-gray to whitish; interior white, commonly with purple stainings. The entire lunule is ¾ as wide as long. The form *notata* Say (5862) from the same region is externally marked with brown, zigzag mottlings. This species is very common and is used commercially for chowders and as clams-on-the-half-shell or "cherrystones." Also known as the hard-shelled clam. Do not confuse with *M. campechiensis*. *Meretrix rutila* Sternheimer, 1957, is a synonym. For biology, see M. R. Carricker, 1961, Jour. Elisha Mitchell Sci. Soc., vol. 77, no. 2.

5864

5861

(5863) *Mercenaria mercenaria texana* (Dall, 1902), is a subspecies from the northern Gulf of Mexico region. It is characterized by a glossy central area on the outside of the shell, but has large, irregular, coalescing, flat-topped, concentric ribs. It lives in the inner bays.

Mercenaria campechiensis (Gmelin, 1791)   5864
Southern Quahog

Off southern New Jersey to Florida, Texas and Cuba.

3 to 6 inches in length, very similar to *mercenaria*, but much more obese, a heavier shell, lacks the smooth central area on the outside of the valves, and the entire lunule is usually as wide as long. Nearly always white internally. Rarely it has a purplish stain on the escutcheon and brown mottlings on the side. There have been a number of forms described. In the vicinity of St. Petersburg, Florida, there is a malformed race in which there is a sharp, elevated ridge passing from the umbo obliquely backward toward the pallial sinus on the inside of each valve. The southern quahog is common but has not been exploited commercially to any great extent. It hybridizes with *mercenaria* in some areas of the Carolinas and Florida and could well be considered a subspecies. Off Delaware and Virginia it lives on sandy bottoms in 10 to 36 meters of water. Largest known specimen, from Boca Ciega Bay, Flor-

ida, weighed 6½ pounds and 6.7 inches long (H. W. Sims, Jr., 1965, *Quart. Jour. Fla. Acad. Sci.*, vol. 27, p. 348).

Genus *Chione* Mühlfeld, 1811

Shells trigonal or ovate; thick; hinge plate short and triangular; 3 cardinal teeth in each valve; left posterior cardinal long; no anterior laterals; pallial sinus small and triangular; inner margins crenulated; lunule bounded by an indented line; escutcheon smooth and bounded by a small ridge. Type: *cancellata* (Linné, 1767).

Subgenus *Chione* Mühlfeld, 1811

Surface cancellate, the radial and concentric riblets about equal in strength. Pallial sinus small; cardinal teeth smoothly or only faintly grooved.

Chione cancellata (Linné, 1767)   5865
Cross-barred Venus   Color Plate 23

North Carolina to Florida, Texas and the West Indies. Brazil.

1 to 1¾ inches in length, varying from ovate to subtriangular in shape, thick; with strong, raised, curved, leaflike, concentric ribs and numerous coarse radial ribs. Escutcheon long, smooth and V-shaped, commonly with 6 or 7 brown zebra-stripes. Lunule heart-shaped, with minute vertical threads. Color externally is white to gray; internally glossy-white with a suffusion of purplish blue. A very common, shallow-water species in Florida. Beachworn specimens have a cancellate sculpturing. The form *mazycki* Dall, 1902 (5866) has a beautiful rosy interior. Synonyms are *subrostrata* Lamarck, 1818, and *beaui* Récluz, 1852.

Chione intapurpurea (Conrad, 1849)   5867
Lady-in-waiting Venus   Color Plate 24

North Carolina to Texas and the West Indies. Brazil.

1 to 1½ inches both ways, thick, glossy-white to cream; interior white, commonly with a violet, radial band or splotch at the posterior ⅓. Exterior with crowded, smooth, low, rounded, concentric ribs and numerous radial threads. The lower edge of these ribs bears many small bars which are lined up one below the other to give the shell the impression that it has axial ribs. The concentric ribs become sharp and higher at the shell's extreme ends. Lunule brownish with raised lamellations; escutcheon with very fine, transverse lines. Uncommon. 1 to 47 fathoms. Incorrectly spelled *interpurpurea*.

Chione californiensis (Broderip, 1835)   5868
Common Californian Venus

San Pedro, California, to Panama.

2 to 2½ inches high, a little longer, subtrigonal, moderately compressed, with sharp, raised, concentric ribs whose edges turn upwards, and with low, rather wide, rounded, radial riblets. Lunule heart-shaped and striated; escutcheon V-shaped in cross-section, long and smooth. The dorsal posterior end of the right valve is not as smooth and overlaps the left valve. Exterior creamy-white with faint mauve stripes on the escutcheon. Interior white, commonly with a purple splotch at the posterior end. This is a common shore species, formerly called *C. succincta* Valenciennes, 1827 (not Linné, 1767); *leucodon* Sowerby, 1835; *nuttalli* Conrad, 1837; *gealeyi* and *durhami* Pierre Parker, 1949.

5868   5869   5870

Chione undatella (Sowerby, 1835)   5869
Frilled Californian Venus

San Pedro, California, to Peru.

Differing from *californiensis* in being more inflated, usually with more numerous and more closely spaced, thinner concentric ribs (4 to 6 ribs per centimeter on the middle of the shell), and retaining mauve-brown color splotches in the adults. Very common. Some workers consider this a subspecies of *californiensis*, and apparently additional field study is necessary. Synonyms include *Venus neglecta* Sowerby, 1835; *V. entobapta* Jonas, 1845; *Cytherea sugillata* Jonas, 1846; *V. perdix* Valenciennes, 1846; *V. simillima* Sowerby, 1853; *excavata* Carpenter, 1857; *bilineata* Reeve, 1863; and *Chione taberi* Pierre Parker, 1949.

Subgenus *Chionista* Keen, 1958

Lunule and escutcheon absent; concentric sculpture irregular and beaded. Type: *fluctifraga* (Sowerby, 1853).

Chione fluctifraga (Sowerby, 1853)   5870
Smooth Pacific Venus

San Pedro, California, to the Gulf of California.

2½ inches in height, slightly longer, moderately compressed, subtrigonal; radial grooves or ribs strong at the posterior ⅓ and at the anterior ¼ of the shell; central area with stronger, low, rather wide, concentric ribs which may have coarse, half-moon-shaped beads. Lunule not well defined; escutcheon not well seen and not sunken nor smooth as in *californiensis*. Exterior creamy-white, semi-glossy, rarely stained with blue-gray. Interior white with purple splotches near the muscle scars or on the teeth. Not uncommon along the sandy shores in southern localities. *Venus gibbosula* Reeve, 1863, is a synonym.

Other species:

5871 *Chione (Chione) subimbricata* (Sowerby, 1835). Baja California to Peru. Not an *Anomalocardia*: See Olsson, 1961, *Moll. Trop. Eastern Pacific, Pelecypoda*, p. 295.

5872 *Chione (Chione) pubera* (Bory Saint-Vincent, 1827). Florida Keys, Texas and the West Indies.

Subgenus *Lirophora* Conrad, 1863

Shell trigonal-ovate, solid, heavy, glossy, sculptured with thick, flat or round concentric folds continuing across the disk or lamellose at the ends. Type: *latilirata* (Conrad, 1841).

Chione paphia (Linné, 1767)   5873
King Venus   Color Plate 24

West Indies to Brazil.

1½ inches in length, similar to *latilirata*, but not so heavy, with 10 to 12 smaller, concentric ribs which are thin at their ends. From a side view, the dorsal margin of the lunule is very concave. Rare, if present, in the United States. Moderately common in the West Indies; 1 to 55 fathoms.

Chione latilirata (Conrad, 1841)   5874
Imperial Venus   Color Plate 24

North Carolina to Florida and to Texas. Brazil.

1 inch in length, very thick and solid, with 5 to 9 large, bulbous, concentric ribs, usually rounded, but may also be sharply shelved on top. The ribs are not thin and flattened at their ends. Lunule heart-shaped, and from a side view, its dorsal margin is almost straight. Surface of shell glossy, cream with rose and brown mottlings. Rather uncommon offshore in about 20 fathoms. *Venus varicosa* Sowerby, 1853, is a synonym.

Chione clenchi Pulley, 1952   5875
Clench's Venus

North Texas to the Gulf of Campeche, Mexico.

1 to 1⅓ inches in length, solid, similar to *latilirata*, but the ribs are narrower and number 12 to 15 in an inch-long specimen. The ribs are not reflected dorsally nor flattened into elevated plates on the posterior slope of the shell, as they are in *paphia*. Not uncommon; 10 to 30 fathoms. Beach specimens from Galveston, Texas, southward. See *Texas Jour. Sci.*, vol. 1, p. 62, and vol. 4, p. 167. This may be only a subspecies of *latilirata*.

5875

Other species:

5876 *Chione (Lirophora) kellettii* (Hinds, 1845). Gulf of California to Peru. Common; offshore.

5877 *Chione (Lirophora) mariae* (Orbigny, 1846). Baja California to Peru. (Synonym: *cypria* Sowerby, 1835, not Brocchi, 1814.) Abundant; shore to a few fathoms. May be a synonym of *paphia* (Linne, 1767).

Subgenus *Iliochione* Olsson, 1961

The type of this subgenus is *Venus subrugosa* Wood, 1828.

Chione subrugosa (Wood, 1828)   5878
Semi-rough Chione

Gulf of California to Peru.

1 to 2 inches, heavy, elongate-subtrigonal, the posterior end somewhat rostrate. Sculpture heaviest on the umbones and

consists of low, rounded, concentric folds, finely radially striate. Middle of disc and area near ventral margins smoothish. Color cream with 3 or more brown, narrow, radial rays; sometimes mottled. Periostracum thin, yellowish. Abundant; lagoonal, mudflat intertidal areas. Formerly in *Anomalocardia*.

Subgenus *Chionopsis* Olsson, 1932

Shells somewhat inflated, often with concentric lamellae frilled. Hinge with at least the right posterior and middle left cardinal teeth grooved or bifid. Type: *amathusia* Philippi, 1844. *Gnidiella* P. Parker, 1949, is a synonym. This subgenus occurs only from the Gulf of California to Ecuador.

5879 *Chione gnidea* (Broderip and Sowerby, 1829). Gulf of California to Peru. (Synonym: *ornatissima* Broderip, 1935, is considered to be distinct by some recent workers.) Common.

5879

5880 *Chione amathusia* (Philippi, 1844). Gulf of California to Peru. Common.

5881 *Chione purpurissata* Dall, 1902. Gulf of California to Ecuador.

5882 *Chione pulicaria* (Broderip, 1835). Gulf of California to Colombia.

Subgenus *Timoclea* Brown, 1827

Sculpture predominantly radial, the concentric element feeble; the middle left and 2 posterior right cardinals grooved; the escutcheon smooth. Siphons united to their orifices. Type: *ovata* (Pennant, 1777), from the Eastern Atlantic.

Chione grus (Holmes, 1858) **5883**
Gray Pygmy Venus

North Carolina to Florida to Texas.

¼ to ⅜ inch in length, oblong, with 30 to 40 fine, radial ribs which are crossed by very fine, concentric threads. The posterior dozen ribs are cut along their length by a very fine groove. Dorsal margin of right valve fimbriated and overlapping the left valve. Lunule narrow, heart-shaped, colored brown. Escutcheon very narrow and sunken. Exterior colored a dull-gray, but some Florida specimens tend to be whitish, pinkish or even orange. Interior glossy-white with purplish brown

area at the posterior end. *Purple color on hinge at both ends.* Commonly dredged in shallow water.

Chione pygmaea (Lamarck, 1818) **5884**
White Pygmy Venus

Southeast Florida and the West Indies.

¼ to ½ inch in length, similar to *grus*, but with prominent scales, 4 or 5 brown, zebra-stripes on the escutcheon, with a white lunule, and the teeth purple only on the posterior ½ of the hinge. Beaks commonly pink. Interior all-white. Fairly common in shallow water, especially in the West Indies.

5883

5884

Other species:

5885 *Chione* (*Timoclea*) *picta* Willett, 1944. Southern California to Baja California.

5886 *Chione* (*Timoclea*) *squamosa* (Carpenter, 1857). Gulf of California to Peru. (Synonym: *troglodytes* (Mörch, 1861)).

Genus *Anomalocardia* Schumacher, 1817

Lunule large and impressed. Surface with predominantly concentric sculpture; periostracum varnishlike. Type: *flexuosa* (Linné, 1758). No Eastern Pacific species.

Anomalocardia auberiana (Orbigny, 1842) **5887**
Pointed Venus

South half of Florida to Texas.

½ to ¾ inch in length, about ¾ to ½ as high, pointed into a sharp, wedgelike rostrum at the posterior end. Lunule oval to slightly heart-shaped and faintly impressed. Wide, shallow escutcheon bordered by a weak ridge. Beaks tiny and inrolled. Sculpture of small, but distinct, rounded, concentric ribs which are more prominent near the beaks. Color variable: glossy-cream, white or tan with brown or purple rays of fine specklings. Interior white, purple or brown. Brackish-water specimens are dwarfed. A common sandy-shore species. *Venus rostrata* Sowerby, 1853, and *cuneimeris* Conrad, 1846, are synonyms.

5887

Anomalocardia brasiliana (Gmelin, 1791)   **5888**
West Indian Pointed Venus   **Color Plate 24**

West Indies to Brazil.

¾ to 1½ inches, heavy, moderately elongate, with concentric, corrugated ribs extending over into the escutcheon area. Yellowish white, variously shaded and spotted with purple or brown. Common; shallow water.

Other species:

5889 *Anomalocardia leptalea* (Dall, 1894). Watling Island, Bahamas.

Genus *Protothaca* Dall, 1902

Escutcheon in left valve, narrow and flattened. Anterior cardinal tooth of right valve small. External sculpture of fine cancellations or threads. Color drab. Cold-water species. Type: *Chama thaca* Molina, 1782.

Subgenus *Callithaca* Dall, 1902

Sculpture uniformly reticulate; lunule feeble, escutcheon absent. Inner margins smooth. Type: *tenerrima* (Carpenter, 1856).

Protothaca tenerrima (Carpenter, 1856)   **5890**
Thin-shelled Littleneck

Vancouver, British Columbia, to Baja California.

About 4 inches in length and 2¾ inches high, very compressed, relatively thin, with a chalky texture, with a few raised concentric lines and numerous very small radial threads. Lunule fairly defined. Exterior light gray-brown. Interior chalky-white. A fairly common species, commonly washed ashore on Californian beaches. Specimens which may be hybrids between *tenerrima* and *staminea* were named *restorationensis* Frizzell, 1930 (*The Nautilus*, vol. 43, p. 120).

5890

Protothaca staminea (Conrad, 1837)   **5891**
Common Pacific Littleneck

Aleutian Islands to Baja California.

1½ to 2 inches in length, subovate, beaks nearer the anterior end; sculpture of concentric and radial ribs which form beads as they cross each other at the anterior end of the shell. Radial ribs stronger on the middle of the valves. Beaks almost smooth. Exterior rusty-brown with a purplish cast. A very abundant, widespread species with a number of varieties. Sometimes with a mottled color pattern. *P. orbella* (Carpenter, 1864), is an ecologic form and synonym. For a population study, see Schmidt and Warme, 1969, *The Veliger*, vol. 12, p. 193.

(5892) Variety or form: *laciniata* (Carpenter, 1864) reaches 3 inches in length, is coarsely cancellate and beaded, its color rusty-brown to grayish. *P. spatiosa* Dall, 1916, is a synonym.

(5893) Variety or form: *ruderata* (Deshayes, 1853) (typically a northern form) is chalky-white to gray, with concentric ribs large and coarse, commonly lamellate.

Compare with *Tapes philippinarum*, the Japanese littleneck (5900).

5891

Subgenus *Leukoma* Römer, 1857

Left valve with beveled escutcheon; lunule radially ribbed, incised. Type: *Venus granulata* Gmelin, 1791. Synonym: *Nioche* Hertlein and Strong, 1948.

Protothaca granulata (Gmelin, 1791)   **5894**
Beaded Venus

West Indies.

¾ to 1¼ inches, somewhat resembling *cancellata* in shape, but the surface has numerous radiating ribs crossed by concentric threads, giving the shell a granular or beaded appearance. Color variable; yellow, cream or white with mottlings of purple or brown. Absent in Florida; common on the large islands of the West Indies.

5894

Other species:

5895 *Protothaca* (*Tropithaca* Olsson, 1961) *grata* (Say, 1830). Gulf of California to Peru. Abundant. (Synonyms: *fuscolineata* Sowerby, 1835; *tricolor* Sowerby, 1835; *discors* Sowerby, 1835.)

5896 *Protothaca* (*Leukoma*) *asperrima* (Sowerby, 1835). Gulf of California to Peru. (Synonyms: *Venus histrionica* and *intersecta* Sowerby, 1835; *Tapes tumida* Sowerby, 1853.)

5897 *Protothaca* (*Leukoma*) *metodon* (Pilsbry and Lowe, 1932). Gulf of California to Panama.

5898 *Protothaca* (*Leukoma*) *pectorina* (Lamarck, 1818). Lower Caribbean.

Genus *Humilaria* Grant and Gale, 1931

Ovate-quadrate; compressed laterally, beaks anterior. Sculpture of distantly spaced, sharp, concentric lamellae. Pallial sinus short, angular, but rounded at the apex. Type: *kennerleyi* ("Carpenter" Reeve, 1863).

Humilaria kennerleyi (Reeve, 1863)   5899
Kennerley's Venus

Alaska to Carmel Bay, California.

2½ to 4 inches in length, ovate-oblong, with the beaks near the anterior end. With sharp, concentric ribs whose edges are bent upwards. Spaces between ribs. Color and texture like gray Portland cement. Interior white. Margin of shell finely crenulate, a feature that distinguishes it from worn specimens of *Saxidomus*. Dredged on mud bottoms from 3 to 20 fathoms. A collector's item, although reasonably common.

5899

Subfamily TAPETINAE H. and A. Adams, 1857

The family Paphiidae Nordsieck, 1969, is the same.

Genus *Tapes* Mühlfeld, 1811

Oblong, compressed, sculpture weak. Lunule incised. Escutcheon bordered by a low carina. Type: *literata* (Linné, 1758).

Subgenus *Ruditapes* Chiamenti, 1900

Sculpture weakly decussate posteriorly, obscurely so anteriorly. Type: *decussata* (Linné, 1758). The subgenus *Amygdala* Römer, 1857, may be used.

Tapes philippinarum (Adams and Reeve, 1850)   5900
Japanese Littleneck

Puget Sound southward.

1½ to 2 inches in length, extremely close to *Protothaca staminea* and *grata* Sowerby (the latter's range is from the Gulf of California to Panama), but differing from both in being much more elongate and more compressed. Its lunule and small escutcheon are more distinct and quite smooth as compared to those of *staminea*. The hinges are extremely simi-

lar. *P. grata* differs in having tiny, distinct crenulations on the inside of the anterior dorsal margin. *T. semidecussata* Reeve appears to be the same as this introduced species. Its colors are variable and commonly variegated or with concentric brown lines. Alias *T. bifurcata* Quayle, 1938. For a study of color variability, see R. F. Shaw, 1956, *The Nautilus*, vol. 70, p. 53.

5900

Genus *Psephidia* Dall, 1902

Shells less than 10 mm., polished, smooth or fine concentric threads. Lunule feeble. No escutcheon. Inner margins smoothish. Pallial sinus angular. 3 cardinals in each valve; no laterals. Dorsal margins, outside the hinge plate, faintly grooved. Type: *lordi* (Baird, 1863). Not found in the Atlantic.

Psephidia lordi (Baird, 1863)   5901
Lord's Dwarf Venus

Alaska to San Diego, California.

¼ inch in length, ovate, compressed or slightly fattened; beaks small; sculpture of microscopic, concentric growth lines. 3 cardinals in hinge of each valve. No laterals. On the dorsal margin there is a microscopic groove parallel to the edge. Color whitish to greenish white, commonly with darker, concentric color bands. Tiny young shells may be found inside the adult clams in the summer and spring months. Common.

5901   5902   5903

Psephidia ovalis Dall, 1902   5902
Oval Dwarf Venus

California.

⅓ inch, white, polished, oval, somewhat compressed laterally. Surface with very weak concentric threads near the anterior base, but most of the disc is smooth. Beaks small, very low, at about the anterior ⅓ of the length. Lunule elongate, narrow, nearly as long as the anterior dorsal slope. Internal margins delicately striated. Abundant offshore.

Other species:

5903   *Psephidia cymata* Dall, 1913. Santa Barbara Islands to the Gulf of California. Rare. Also Pleistocene. 6 mm.

—   *Psephidia brunnea* Dall, 1916, is a *Halodakra*.

5904 *Psephidia salmonea* (Carpenter, 1864). Farallons Islands to Baja California. 25 fms.

Genus *Irus* Schmidt, 1818

Hinge similar to *Protothaca,* but the teeth are degenerate. Internal marginal crenulations absent. Type: *irus* (Linné, 1758) from Europe. *Irus* Oken, 1815, is nonbinomial (I.C.Z.N. Opinion 417). *Irus* Oken, 1821, is *Pandora* (Keen, 1962, *The Veliger,* vol. 4, p. 180). *Notirus* Finlay, 1928, may be a synonym.

Irus lamellifer (Conrad, 1837)   **5905**
Californian Irus Venus

Monterey to San Diego, California.

1 to 1½ inches in length, usually oblong, although some specimens may be almost round. Characterized by about a dozen, strongly raised, concentric lamellae or thin ridges. Shell whitish and with a chalky texture. Moderately common. Found burrowing in gray shale from low water to several fathoms.

Genus *Liocyma* Dall, 1870

Shells less than 1 inch, thin, smooth, elongate-ovate. With concentric undulations. Periostracum polished. Pallial sinus small, rounded triangular. 3 cardinal teeth in each valve, and divaricate. Lunule faint. Type: *Venus fluctuosa* Gould, 1841. Cold-water species. *Lyocyma* is a misspelling.

Liocyma fluctuosa (Gould, 1841)   **5906**
Fluctuating Liocyma

Greenland to Nova Scotia.
Alaska to British Columbia.

½ to 1 inch, oblong-ovate, thin-shelled, laterally compressed, beaks near the center; white; periostracum thin, glossy, straw-colored. Sculpture of 20 to 25 concentric waves, not quite extending to the margin, especially anteriorly. Moderately common; offshore waters. *L. brunnea* Dall, 1902, from the Gulf of St. Lawrence, is probably a synonym.

5906

Other species:

5907 *Liocyma beckii* Dall, 1870. Plover Bay, Siberia, to Japan, and to Port Althorp, Alaska.

5908 *Liocyma scammoni* Dall, 1871. Port Simpson, British Columbia.

5909 *Liocyma viridis* Dall, 1871. Arctic Ocean to Japan and to the Kodiak Islands, Alaska, 10 to 18 fms.

5910 *Liocyma schefferi* Bartsch and Rehder, 1939. Chuginadak Island to Atka Island, Aleutians, 10 fms. *The Nautilus,* vol. 52, p. 111, pl. 8. The above four may be variants of *fluctuosa* Gould.

Subfamily MERETRICINAE Gray, 1847

Usually smooth-surfaced; hinge with strong, radiating cardinal teeth. Anterior lateral teeth in right valve usually flanked with denticles above and below.

Genus *Tivela* Link, 1807

Shells trigonal, solid, some species massive; umbones near the middle with the beaks close together. 3 cardinal teeth in each valve, supplemented by smaller auxiliary teeth. Large anterior lateral tooth in left valve fitting into socket in right valve. Ligament external. Lunule large. No escutcheon. Periostracum varnishlike. Type: *tripla* (Linné, 1771).

Tivela mactroides (Born, 1778)   **5911**
Trigonal Tivela   **Color Plate 24**

West Indies to Brazil.

1 to 1½ inches, solid, inflated, trigonal. Variously rayed and clouded with brown. Surface smooth; beaks central and prominent. Common; shallow water.

(5912) *Tivela mactroides* subspecies *bryonensis* (Gray, 1838) is almost indistinguishable from the Caribbean *mactroides.* It is common and ranges from Baja California to Peru. Synonyms include *solangensis* Orbigny, 1845; *radiata* Sowerby, 1835 (not Mühlfeld, 1811); *semifulva* Menke, 1847; *gracilior* and *intermedia* Sowerby, 1851; *pulla* Philippi, 1851.

Tivela floridana Rehder, 1939   **5913**
Florida Tivela

Palm Beach County, Florida.

⅜ inch in length, subquadrate, beaks in the center, highly polished and with microscopic growth lines near the margins. Exterior glossy, tan or purplish. Interior mottled with purplish brown. This is the only *Tivela* recorded from eastern United States. Uncommon offshore.
   Tivela abaconis Dall, 1902, from the Bahamas, is extremely close, if not the same, but "differs in having the posterior end more bluntly rounded, and in being more ovate in shape and less triangular . . . and generally roseate in color, though some others are white." (*The Nautilus,* 1939, vol. 53, p. 18.)

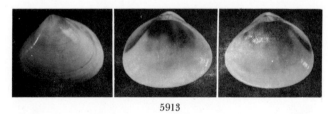

5913

Tivela abaconis Dall, 1902   **5914**
Abaco Tivela

Bahamas; Caribbean; Vera Cruz, Mexico.

5905   5914

½ inch, subquadrate, moderately inflated, subtranslucent; deep-rose at the beak and in the interior, becoming paler toward the margins. Exterior polished. Beaks high, pointed and subcentral. Uncommon.

Subgenus *Pachydesma* Conrad, 1854

Shells very large and ponderous, with smooth interior margins and a thick, shellaclike periostracum. 4 cardinal

5915

teeth in each valve. Type: *stultorum* (Mawe, 1823). Synonym: *Trigonella* Conrad, 1837, non da Costa, 1778.

Tivela stultorum (Mawe, 1823)	**5915**
Pismo Clam	**Color Plate 24**

San Mateo County, California, to Lower California.

3 to 6 inches in length, ovate, heavy, moderately inflated, glossy-smooth, except for weak lines of growth. Ligament large and strong. Color brownish cream with wide, mauve, radial bands. Bands may be absent. Posterior end marked off by a single, sharp thread. Lunule lanceolate and with vertical scratches. Periostracum thin and varnishlike. A common and edible species. This is the only West Coast *Tivela*, but it has received a number of unnecessary names, *T. crassatelloides* Stearns, 1899, being one of many.

Other species:

5916 *Tivela (Tivela) delessertii* (Sowerby, 1854). Gulf of California to Panama. Locally common. (Synonym: *arguta* Römer, 1861.)

5917 *Tivela (Planitivela* Olsson, 1961) *planulata* (Broderip and Sowerby, 1829). Gulf of California to Peru. (Synonyms: *suffusa* Sowerby, 1835; *undulata* Sowerby, 1851.)

Genus *Transennella* Dall, 1883

Left anterior lateral fitting into a socket in the right valve. Internal margins are obliquely grooved with numerous, microscopic lines. These are parallel to the growth lines at the ventral margin of the valves. Pallial sinus angular, obliquely ascending. Type: *conradina* Dall, 1883.

Transennella stimpsoni Dall, 1902	**5918**
Stimpson's Transennella	

North Carolina to southeast Florida and the Bahamas. Brazil.

¼ to ½ inch in length, glossy, rounded trigonal in shape, smooth except for fine growth lines. Inner margins of valves creased with microscopic oblique threads. Exterior cream with 2 or 3 wide, radial bands of weak brown. Interior commonly flushed with purple. Pallial sinus long. Fairly common in shallow water.

Transennella conradina Dall, 1883	**5919**
Conrad's Transennella	

South half of Florida and the Bahamas.

Shell very similar to that of *stimpsoni*, but differing in being pointed posteriorly, hence more elongate, and more variable in color. Zigzag brown lines present in some, others are solid-cream or solid-brownish. Exterior with fine, raised, concentric lines. Pallial sinus short. Common in shallow water.

Transennella cubaniana (Orbigny, 1842)	**5920**
Cuban Transennella	

Florida Keys and West Indies.

⅜ inch, trigonal in shape, pure-white, rarely flecked with brown. Sculptured with fine impressed, concentric lines. Inner margin obliquely grooved. Moderately common in the West Indies.

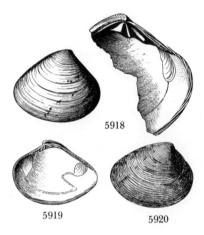

5918

5919   5920

Transennella tantilla (Gould, 1853)	**5921**
Tantilla Transennella	

Alaska to Baja California.

¼ inch in length, ovate, angle at beaks about 90 degrees, smooth except for weak, concentric lines of growth. Immediately recognized under the hand lens by the tiny grooves running on the inside of the shell margins. Exterior cream with the posterior end stained bluish. Interior white with a wide, radial band of purple-brown at the posterior end. Dredged in large numbers off California and at times found washed ashore.

It is replaced in the Gulf of California by the subspecies *humilis* (Carpenter, 1857) (**5921a**) in being only 4 mm., longer, smoother and more brightly colored.

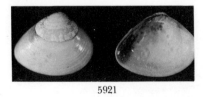

5921

Other species:

5922 *Transennella culebrana* Dall and Simpson, 1901. Puerto Rico Transennella. Puerto Rico. ⅜ inch. See "Caribbean Seashells," pl. 29b.

5923 *Transennella gerrardi* Abbott, 1958. Grand Cayman Island, West Indies. Monograph 8, Acad. Nat. Sci. Phila., p. 130.

5924 *Transennella modesta* (Sowerby, 1835). Gulf of California to Ecuador. (Synonyms: *cumingii* Orbigny, 1845, and *sororcula* Pilsbry and Lowe, 1932.)

5925 *Transennella puella* (Carpenter, 1864). Both sides of Baja California to Nicaragua, in 1 to 80 meters.

5926 *Transennella caryonautes* S. S. Berry, 1963. Southern Gulf of California to Mazatlan, Mexico. 40 mm.

Subfamily CIRCINAE Dall, 1896

Pallial line nearly entire. Anterior laterals present. Surface sculptured with radial ribs, sometimes dichotomizing. Contains the Indo-Pacific genera *Circe* Schumacher, 1817, and *Gafrarium* Röding, 1798. The subfamily name Gafrariinae Nordsieck, 1969, is a synonym.

Genus *Gouldia* C. B. Adams, 1847

Shell less than ½ inch in length; beaks minute; lunule long, bounded by an impressed line; no escutcheon. With concentric or reticulate sculpture. Anterior lateral teeth present. This genus is put in the separate subfamily Circinae by some workers. *Thetis* C. B. Adams, 1845, is a synonym (non Sowerby, 1826).

Gouldia cerina (C. B. Adams, 1845)   **5927**
Serene Gould Clam

North Carolina to Florida; the West Indies; Bermuda. Brazil.

5 to 7 mm. in length, solid, trigonal in shape, beaks in the center, high and very small; lunule long, bounded by an impressed line; no escutcheon. Sculpture reticulate in which the fine, concentric ribs predominate. The radial ribs are stronger anteriorly. Color white, uncommonly with purplish or brownish flecks. A common species from shallow water to 95 fathoms. There is a variety or similar species, *bermudensis* E. A. Smith, 1886, from Bermuda (**5928**).

5927

Gouldia californica Dall, 1917   **5929**
California Gould Clam

Gulf of California to Ecuador.

6 mm., ovate trigonal, white with touches of brown along the dorsal border. Anterior lateral tooth is large and prominent, the pallial sinus small. Sculpture reticulate, the concentric elements more prominent in the middle of the disc, the radial towards the end of the valves. Inner margins smooth. Not uncommon; offshore to 80 fathoms. *Gafrarium stephensae* Jordan, 1936, is a synonym.

Subfamily PITARINAE Stewart, 1930

Beaks generally nearer the anterior end; 3 cardinal teeth not tending to radiate; anterior laterals well-developed. Ventral margins smooth. Surface usually smoothish. Lunule cordate; escutcheon absent. Pallial sinus deep.

Genus *Pitar* Römer, 1857

Anterior left lateral fitting into a well-developed socket in the right valve. Middle left cardinal large; posterior right cardinal split. Shell cordate, convex, with large, full umbones. Type: *tumens* (Gmelin, 1791). *Pitar* is a masculine word.

Subgenus *Pitar* Römer, 1857

Pitar fulminatus (Menke, 1828)   **5930**
Lightning Venus   **Color Plate 24**

North Carolina to Florida and the West Indies. Bermuda. Brazil.

1 to 1½ inches in length, plump, umbones large and full; lunule very large and outlined by an impressed line. Anterior end broader than the posterior end. Sculpture of crowded, rather heavy lines of growth. Exterior whitish with spots and/or zigzag markings of yellowish brown. Moderately common in shallow water. ⅓-inch young are commonly dredged off Miami. Synonym: *penistoni* Heilprin, 1889.

5930

(**5931**) *Pitar albidus* (Gmelin, 1791) of the West Indies is very similar, but all-white in color, more quadrate in shape, has a narrower and more elongate lunule and is usually more compressed. Common.

Pitar morrhuanus Linsley, 1848   **5932**
Morrhua Venus

Gulf of St. Lawrence to North Carolina.

1 to 1½ inches in length, oval-elongate, moderately plump, with the lunule large and elongate. With numerous, heavy lines of growth. Color dull-grayish to brownish red. *P. ful-*

5932   5933

5946   5947

minata is similar, but is found only to the south of Cape Hatteras, is not so elongate (compare figures), and is marked with brown. Fairly commonly dredged off New England.

Pitar simpsoni (Dall, 1889)   5933
Simpson's Venus

South half of Florida and the West Indies.

¾ inch in length, plump, with fine, irregular, concentric threads; the large, ovate lunule is polished smooth. Color white to purplish white, commonly with zigzag, yellow-brown markings. Escutcheon absent. Nearest in shape to *morrhuanus*. Uncommon at low tide to 26 fathoms.

Subgenus *Pitarenus* Rehder and Abbott, 1951

Inner margins of valves crenulated. Shell inflated, with concentric riblets. Right valve with a strong, rather long, bifid posterior cardinal, and a thin, flexuous anterior cardinal. Type: *cordatus* (Schwengel, 1951).

Pitar cordatus (Schwengel, 1951)   5934
Schwengel's Venus   Color Plate 24

Off the Florida Keys and in the Gulf of Mexico to Texas. Brazil.

1½ inches in length, very similar to *morrhuanus*, but much fatter, with more distinct concentric threads on the outside, and with fine crenulations along the inside of the ventral margins of the valves. Interior white, commonly with a pinkish blush. Dredged from 30 to 50 fathoms and brought in by shrimp fishermen. Uncommon. (*The Nautilus*, vol. 64, p. 118.)

Subgenus *Nanopitar* Rehder, 1943

Shells small, round, smooth externally and with a smooth internal ventral margin. Left posterior cardinal tooth thin, high, joined in part to the ligamental nymph, the lower part free, curving away and almost reaching the end of the hinge plate; left middle and anterior cardinals united at the top. Type: *pilula* Rehder, 1943.

Pitar pilula Rehder, 1943   5935
Little Ball Pitar

Southeast Florida.

6 mm. in length and height; 4.2 mm. in breadth of both valves; shell white, covered with a thin, yellow-brown, deciduous periostracum. Broad lunule bordered by an incised line. Umbones subcentral and prominent. Pallial sinus moderately deep, rounded. Rare just offshore.

5935

Subgenus *Hysteroconcha* P. Fischer, 1887

Posterior-dorsal area is set apart sharply by the umbonal angle armed with long spines or scales. Type: *dione* (Linné, 1758). Synonyms: *Dione* Gray, 1847, non Hübner, 1819; *Hysteroconcha* Dall, 1902.

Pitar dione (Linné, 1758)   5936
Royal Comb Venus   Color Plate 24

Mexico to Panama and the West Indies.

1 to 2 inches in length, characterized by its violet and purple-white colors and 2 radial rows of long spines at the posterior end of the valve. A common species washed ashore in the West Indies. Erroneously reported from Texas beaches.

The closely resembling species, *Pitar lupanarius* (Lesson, 1830), **(5936a)**, occurs in the Pacific from Baja California to Peru.

Other species:

5937 *Pitar (Hysteroconcha) brevispinosus* (Sowerby, 1851). Gulf of California to Ecuador.

5938 *Pitar (Hysteroconcha) multispinosus* (Sowerby, 1851). Gulf of California to Peru. (Synonym: *Callista longispina* Mörch, 1861.)

5939 *Pitar (Hysteroconcha) rosea* (Broderip and Sowerby, 1829). Gulf of California to Peru. (Synonym: *lepida* Chenu, 1847.)

5940 *Pitar (Lamelliconcha* Dall, 1902) *circinatus circinatus* (Born, 1778). West Indies. Brazil.

5941 *Pitar (Lamelliconcha) circinatus alternatus* (Broderip, 1835). Gulf of California to Peru. (Synonym: *vinacea* Olsson, 1961.)

5942 *Pitar (Lamelliconcha) tortuosus* (Broderip, 1835). Gulf of California to Peru.

5943 *Pitar (Lamelliconcha) unicolor* (Sowerby, 1835). Gulf of California to Ecuador. (Synonyms: *badia* Gray, 1838; *ligula* Anton, 1839.)

5944 *Pitar (Lamelliconcha) frizzelli* Hertlein and Strong, 1948. Off the southern end of the Gulf of California, in 82 to 110 meters.

5945 *Pitar (Lamelliconcha) paytensis* Orbigny, 1845. Gulf of California to Peru. (Synonym: *affinis* Broderip, 1835, not Gmelin, 1791.)

5946 *Pitar (Pitar) zonatus* (Dall, 1902). Off North Carolina, 22 fms.

5947 *Pitar (Pitar) arestus* (Dall and Simpson, 1901). West Indies. Grayish white. Uncommon; shallow water. 50 mm.

Genus *Callista* Poli, 1791

Oblong, glossy; pallial sinus wide, horizontal, pointed. Type: *chione* (Linné 1758).

Subgenus *Costacallista* Palmer, 1927

Sculpture strong, of flat concentric ridges. Hinge plate excavated. Type: *erycina* (Linné, 1758).

5948

Callista eucymata (Dall, 1890)   **5948**
Glory-of-the-Seas Venus

North Carolina to south half of Florida, Texas, to Brazil.

1 to 1½ inches in length, fairly thin, oval, with about 50 slightly flattened, concentric ribs which have a short dorsal and long ventral slope, and separated by a narrow, sharp groove. Color glossy-white to waxy pale-brown, with clouds and zigzag markings of reddish brown. No escutcheon. Margins rounded. A beautiful and uncommon species dredged from 25 to 117 fathoms.

Genus *Macrocallista* Meek, 1876

Scar of the pedal retractor muscle on the undersurface of the hinge plate and below the cardinal tooth is distinct and deep. Shell elliptical; surface smooth; periostracum shiny; lunule narrowly cordate, defined by an incised line. No escutcheon. 3 cardinal teeth in each valve. Anterior lateral in left valve. Ventral margins smooth. Type: *nimbosa* (Lightfoot, 1786).

Macrocallista nimbosa (Lightfoot, 1786)   **5949**
Sunray Venus   **Color Plate 24**

North Carolina to Florida and to Texas.

4 to 5 inches in length, elongate, compressed, glossy-smooth with a thin varnishlike periostracum. Exterior dull-salmon to dull-mauve with broken, radial bands of darker color. Interior dull-white with a blush of reddish over the central area. Moderately common in shallow, sandy areas and not uncommonly washed ashore after storms. Concerning commercial fishing of this clam, see J. W. Jolley, Jr., 1972, Fla. Dept. Nat. Resources, *Technical Series,* no. 67.

Subgenus *Megapitaria* Grant and Gale, 1931

Type of this subgenus of ovate *Macrocallista* is *aurantiaca* (Sowerby, 1831).

Macrocallista maculata (Linné, 1758)   **5950**
Calico Clam   **Color Plate 24**

North Carolina to Florida, Texas and to Brazil. Bermuda.

1½ to 2½ inches in length, ovate, glossy-smooth with a thin varnishlike periostracum. Exterior cream with checkerboard markings of brownish red. Rarely albino or all dark-brown. Moderately common in shallow, sandy areas in certain localities. A popular food and collector's item. Also known as the checkerboard or spotted clam. Suddenly became abundant in Bermuda in the late 1950s.

Other species:

5951 *Macrocallista (Megapitaria) squalida* (Sowerby, 1835). Baja California to Peru. (Synonyms: *biradiata* Gray, 1838; *chionacea* Menke, 1847.)

5952 *Macrocallista (Megapitaria) aurantiaca* (Sowerby, 1831). Baja California to Peru.

Genus *Agriopoma* Dall, 1902

Like *Pitar,* but the upper ends of the cardinal teeth do not touch the margin above but stand free and separated from the margin by a deep, narrow slit. The much smaller anterior lateral is bordered behind by a false socket. Type: *texasiana* (Dall, 1892). *Agriodesma* Dall, 1916, is a lapsus for *Agriopoma.*

Agriopoma texasiana (Dall, 1892)   **5953**
Texas Venus

Northwest Florida to Texas and Mexico.

1½ to 3 inches in length, ¾ as high. Externally resembling *Pitar morrhuana,* but much more elongate, having the beaks rolled in under themselves, and with a more elongate, faint lunule. The posterior cardinal is S-shaped in the right valve. Uncommon, if not rare. Formerly placed in *Callocardia* A. Adams, 1864. Lives offshore in silty mud 4 to 8 cm. below the surface at water depths of 4 to 13 fathoms. (See P. S. Boyer, 1967, *The Nautilus,* vol. 80, p. 79.)

5953

Other species:

5954 *Agriopoma (Pitarella* Palmer, 1927) *catharia* (Dall, 1902). Baja California to Ecuador.

5955 *Agriopoma (Pitarella) mexicana* (Hertlein and Strong, 1948). Gulf of California to Ecuador. (Synonym: *lenis* Pilsbry and Lowe, 1932, not Conrad, 1848.)

Genus *Amiantis* Carpenter, 1863

Like *Pitar,* but shell heavy, polished, with concentric ribs. Nymphs rugose. Type: *callosa* (Conrad, 1837).

Amiantis callosa (Conrad, 1837)   **5956**
Pacific White Venus

Santa Monica, California, to south Mexico.

5956

3 to 4½ inches in length, longer than high, beaks pointing anteriorly, shell hard, heavy, glossy and with neat concentric ribs. Lunule small, heart-shaped and pressed in slightly under the beaks. Anterior end round. Color solid-ivory. A very attractive, fairly common species living just below tide line on sandy bottoms in the open surf. Commonly washed ashore alive after storms between Seal Beach and Huntington Beach. Once known as *C. nobilis* Reeve, 1850.

Genus *Saxidomus* Conrad, 1837

Shell large, slightly gaping posteriorly, hinge with 4 or 5 cardinal teeth in the right valve, 4 in the left. Pallial sinus long and fairly narrow. Type: *nuttalli* (Conrad, 1837). Synonym: *Ezocallista* Kamada, 1962. The genus name is feminine.

Saxidomus nuttalli (Conrad, 1837)   5957
Common Washington Clam

Humboldt Bay, California, to Baja California.

3 to 4 inches in length, oblong, with the beaks nearer the anterior end; heavy, with coarse, crowded, concentric ribs. Color a dull, dirty, reddish brown to gray with rust stains. Interior glossy-white, commonly with a flush of purple at the posterior margins. No lunule. Ligament large. Valves slightly gaping posteriorly. Young specimens less than 2 inches are thin-shelled, somewhat glossy and with pretty, mauve, radial streaks on the dorsal edge, both in front and behind the beaks. A very common species which is edible. Also called the butter clam.

Saxidomus gigantea (Deshayes, 1839)   5958
Smooth Washington Clam

Aleutian Islands to Monterey, California.

Possibly this is only an ecologic variation or an example of a geographical gradient within a species. It is similar to typical *nuttalli,* but generally lacks the rust-stain color and rarely, if ever, develops the prominent concentric ridges. This is the commonest and best food clam in Alaska. *S. brevis* Dall, 1916, is a synonym.

5957   5958

Subfamily DOSINIINAE H. and A. Adams, 1858

Shells round in outline, compressed, equivalve, concentrically sculptured. Hinge strong.

Genus *Dosinia* Scopoli, 1777

Nearly orbicular in outline, compressed. Lunule well-defined. Periostracum thin and usually varnishlike. Escutcheon absent. Sculpture of concentric grooves. Type: *concentrica* (Born, 1778). *Dosinidia* Dall, 1902, is a synonym.

Dosinia elegans Conrad, 1846   5959
Elegant Dosinia   **Color Plate 24**

North Carolina to Texas; Caribbean.

2 to 3 inches in length, circular, compressed, glossy, straw-yellow with numerous even, concentric ridges (20 to 25 per inch in adults). Moderately common. Do not confuse with *D. discus.*

Dosinia discus (Reeve, 1850)   5960
Disk Dosinia

Virginia to Florida, Texas and the Bahamas.

2 to 3 inches in length, similar to *elegans,* but having more and finer concentric ridges (about 50 per inch in adults), and not so circular. Commonly washed ashore in perfect condition after storms along the Carolina coasts and middle western Florida.

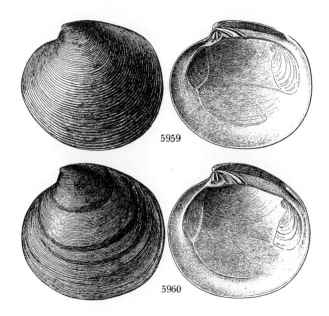

5959

5960

Other species:

5961 *Dosinia concentrica* (Born, 1778). Cuba and Mexico to Brazil. (Synonym: *floridana* Conrad, 1866.) Not found in Florida. See "Caribbean Seashells," pl. 39.

5962 *Dosinia dunkeri* (Philippi, 1844). Baja California, the Gulf, to Peru.

5963 *Dosinia ponderosa* (Gray, 1838). Baja California, the Gulf, to Peru.

5964 *Dosinia semiobliterata* Deshayes, 1853. Southern end of the Gulf of California to Panama. Uncommon. (Synonym: *annae* Carpenter, 1857.)

Subfamily CYCLININAE Frizzell, 1936

Without anterior lateral teeth or incised lunule.

Genus *Cyclinella* Dall, 1902

Margins smooth; faint lunular area present. Type: *tenuis* (Récluz, 1852).

Cyclinella tenuis (Récluz, 1852)   5965
Atlantic Cyclinella

Virginia to Texas and to Brazil.

¾ to 1 inch, circular, moderately compressed, dull-white, sculpture of numerous, irregular growth lines or waves. Often mistaken for a small *Dosinia,* but the latter has neat concentric ridges. Moderately common; 1 to 36 fathoms.

(5966) The Pacific counterpart, is *Cyclinella singleyi* Dall, 1902, 1 to 1½ inches; occurs below low-tide mark from Scammon's Lagoon, Baja California, the Gulf, to Panama.

5965

Subfamily GEMMINAE Dall, 1902

Small or minute venerids, usually with a plain or concentrically striated surface. With marginal grooves. Ovoviviparous.

Genus *Gemma* Deshayes, 1853

Shell the size and shape of a split-pea; lunule large, faintly impressed; no escutcheon; 2 large teeth in the left valve with a large, median socket between the two. A very thin ridge which might be termed a tooth occurs posteriorly beneath the ligament. 3 teeth in right valve. Pallial sinus small and triangular. The shells of the brooded young may be found inside some females. Type: *gemma* (Totten, 1834).

Gemma gemma (Totten, 1834)   5967
Amethyst Gem Clam

Nova Scotia to Florida, Texas and the Bahamas.
Puget Sound, Washington (introduced).

⅛ inch in length, subtrigonal, moderately inflated and rather thin-shelled. Exterior polished and with numerous, fine, concentric furrows or riblets. Color whitish to tan with purplish over the beak and posterior areas. Pallial sinus commonly, but not always, about the length of the posterior muscle scar. It points upward. This is a very common shallow-water species. A number of subspecies or forms have been described, but their validity needs clarification (5968): *purpurea* Lea, 1842; *manhattensis* Prime, 1862; and *fretensis* Rehder, 1939. Migrating shore birds introduce shells to various West Indian localities.

Genus *Parastarte* Conrad, 1862

Shell the size of a split-pea, very similar to *Gemma.* In *Parastarte,* the ligament is high and situated beneath the beak, occupying a very high and broad area. In *Gemma,* the ligament is very narrow and elongated, and extending posterior to the beaks. Pallial sinus much smaller in *Parastarte.* Type: *triquetra* (Conrad, 1846).

Parastarte triquetra (Conrad, 1846)   5969
Brown Gem Clam

Both sides of Florida (to Texas?).

⅛ inch in size, very similar to *Gemma gemma,* but much higher than long, with the beaks larger and elevated. Exterior highly polished and smoothish. Color usually tan to brown, but may be flushed with pink in beachworn specimens. The pallial sinus is almost absent. Moderately common on sand bars and obtained by screening the sand. The blue wing teal feeds upon this species, and during its migrations to the West Indies, sometimes drops dead valves in Caribbean lagoons.

5969

Subfamily CLEMENTIINAE Frizzell, 1936

Shell thin; without an escutcheon; sculpture reduced. Hinge without lateral teeth.

Genus *Compsomyax* R. Stewart, 1930

The right posterior cardinal is bifid. Type: *subdiaphana* (Carpenter, 1864).

Compsomyax subdiaphana (Carpenter, 1864)   5970
Milky Pacific Venus

Alaska to the Gulf of California.

1½ to 2½ inches in length, elongate-ovate, moderately inflated; beaks anterior and pointing forward. Sculpture of fine, irregular, concentric lines of growth, otherwise rather smoothish. Lunule poorly defined. 3 cardinal teeth in each valve, the most posterior one in the right valve deeply split. Color usually chalky-white, but younger specimens are yellowish white and semiglossy. Interior white. Dredged in soft mud from 5 to 25 fathoms. Abundant in some Californian localities. Formerly placed in the genus *Clementia,* Gray, 1842.

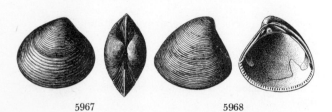

5967   5968

5970

Family PETRICOLIDAE Deshayes, 1831

Lack a lunule and escutcheon. Lack lateral teeth. Have only 2 cardinal teeth in the right valve. Pallial sinus well-developed.

Subfamily PETRICOLINAE Deshayes, 1831

Genus *Petricola* Lamarck, 1801

Sculpture pattern more or less divaricate, the radial elements forming disconnected hook-shaped bends or zigzag lines. Type: *lapicida* (Gmelin, 1791).

Petricola lapicida (Gmelin, 1791)   5971
Boring Petricola

South half of Florida, south Texas and the West Indies; Bermuda. Brazil.

½ inch in length (up to 1¾ inches in the Lesser Antilles), ovate, inflated, chalk-white, with numerous fine, beaded threads. Cardinal tooth bulbous and bifid. Purplish stains on the hinge and interior anterior edge. Posterior end gnarled or with wavy ribs consisting of fine mud particles laid down over the shell by the animal. There is an enclosed, elongate furrow between the beaks and the hinge. Color yellowish white. Found in burrow holes in coral rocks. Not uncommon. *P. divaricata* (Orbigny, 1842) is a synonym.

5971

Subgenus *Petricolaria* Stoliczka, 1870

Shell delicate, elongate, white, radially ribbed. Ligament not immersed. Type: *pholadiformis* (Lamarck, 1818).

Petricola pholadiformis (Lamarck, 1818)   5972
False Angel Wing

Gulf of St. Lawrence to Texas and south to Uruguay.

2 inches in length, elongate, rather fragile and chalky-white. With numerous radial ribs. The anterior 10 or so are larger and bear prominent scales. Ligament external, located just posterior to the beaks. Cardinal teeth quite long and pointed. The siphons are translucent-gray, large, tubular and separated from each other almost to their bases. A very common clay and peat-moss borer. Occasionally, rose-tinted ecological forms are found. *P. lata* Dall, 1925, and *Gastronella tumida* Verrill, 1872, are synonyms. Also present in the Mediterranean.

(5973) The Pacific counterpart is *Petricola parallela* Pilsbry and Lowe, 1932, from the Gulf of California to Panama.

Other species:

5974 *Petricola (Petricola) lucasana* Hertlein and Strong, 1948. Gulf of California to Oaxaca, Mexico.

Genus *Rupellaria* Fleuriau, 1802

Shell inflated, strong, rounded in front, attenuated and more compressed behind; sculpture chiefly radial, stronger anteriorly. Type: *lithophaga* (Retzius, 1786). Considered a subgenus of *Petricola* by many workers.

Rupellaria typica (Jonas, 1844)   5975
Atlantic Rupellaria

North Carolina to Florida and to Brazil.

About 1 inch in length, oblong, flattened anteriorly; compressed, usually attenuated and gaping posteriorly. Beaks point anteriorly. Exterior gray or whitish and with numerous, irregularly spaced, coarse radial ribs. Interior uneven and brownish gray. This coral borer is variable in shape and uneven in texture. It may also be truncate at the posterior end. Moderately common.

Rupellaria tellimyalis (Carpenter, 1864)   5976
West Coast Rupellaria

Santa Moncia, California, to Mazatlan, Mexico.

1 to 1⅓ inches in length. Oblong-elongate, variable in shape and outline due to crowding in the rock burrow. Shell fairly thick, white, except for purplish blotches commonly behind the hinge and at the posterior end. Radial threads are coarser to the anterior end. Growth lines are irregular and coarse. Pallial sinus broadly rounded at its anterior end. Early or nepionic shell is shaped somewhat like a *Donax*, smooth, translucent purplish brown and rarely found attached at this early stage to rocks and kelp stalks. *R. californiensis* Pilsbry and Lowe, 1932, is identical.

(5977) *Rupellaria denticulata* (Sowerby, 1834), known from the Gulf of California to Peru, has a similar nepionic shell (contrary to other reports), has a narrower, triangular pallial sinus and (contrary to reports) is a more fragile shell. Its anterior end is pointed and slightly uplifted. Interior blushed with mottlings of chestnut to purplish brown.

Rupellaria carditoides (Conrad, 1837)   5978
Hearty Rupellaria

Vancouver, British Columbia to Baja California.

1 to 2 inches in length. Very variable in shape, usually oblong; in some, squat and almost orbicular. Shell white to grayish white and very chalky in texture. Concentric growth lines quite coarse and irregular. Radial sculpture of peculiar, fine, scratched lines crowded together, but worn away in some specimens. Fairly common. Found boring into hard rock. Nepionic shell usually oblong. *R. californica* (Conrad, 1937) is the same.

5978

Other species:

5979 *Rupellaria cancellatum* Verrill, 1885. Off Chesapeake Bay, 70 fms. 10 mm.

5979

5980 *Rupellaria robusta* (Sowerby, 1834). Gulf of California to Ecuador, boring in clay. (Synonyms: *sinuosa* Conrad, 1849; *bulbosa* Gould, 1851.)

Family COOPERELLIDAE Dall, 1900

Genus *Cooperella* Carpenter, 1864

Shells small and thin. Hinge plate narrow, with 2 right and 3 left short, divaricating cardinals under the beaks. The left central cardinal is always, and the others commonly, split or bifid. No laterals. Muscle scars small and oval. Pallial line narrow, the sinus long. Type: *subdiaphana* (Carpenter, 1864). Builds a nest of agglutinated sand about itself. See F. Haas, 1943, *Zool. Series, Field Mus. Nat. Hist.*, vol. 29, p. 12, fig. 7. *Oedalia* Carpenter, 1864, and *Oedalina* Carpenter, 1865, are synonyms.

Cooperella atlantica Rehder, 1943   5981
Atlantic Cooper's Clam

Southeast Florida to the Greater Antilles. Brazil.

6 to 7 mm., broadly oval, compressed laterally, fragile, translucent-white, smooth, except for microscopic growth ridges.

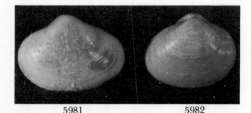

5981   5982

In the left valve there are 3 thin divergent cardinal teeth, the central one bifid; in the right valve there are 2 thin divergent cardinals. Ligament external, rather short and broad, and just posterior to the umbones. Pallial sinus broad and reaching just beyond the center of the shell. Uncommon; dredged just offshore.

Cooperella subdiaphana (Carpenter, 1864)   5982
Shiny Cooper's Clam

British Columbia to Gulf of California.

About ½ inch in length, oval-oblong, opaque-white with a brilliant gloss and slight opalescence. Fragile. Outer surface with slightly wavy concentric growth lines. Ligament tiny, short, set just behind the beaks and visible externally. Moderately common offshore to 40 fathoms. Enwraps itself with agglutinated sand grains. *Oedalia scintillaeformis* Carpenter, 1864, is a synonym.

Order *Myoida* Stoliczka, 1870

Superfamily Myacea Lamarck, 1809

Family MYIDAE Lamarck, 1809

Chalky, white shells with a wide posterior gape, coarse periostracum, long fused siphons. Chondrophore present.

Genus *Mya* Linné, 1758

These are the soft-shell or "steamer" clams which are so popular in New England. The valves are slightly unequal in size and have a large posterior gape. Resilium internal, placed posterior to the beaks and attached in the left valve to a horizontally projecting chondrophore. Type: *truncata* Linné, 1758. The genus name was placed on the official list of generic names by opinion 94, I.C.Z.N.

Subgenus *Arenomya* Winckworth, 1930

Type of this subgenus is *arenaria* Linné, 1758.

Mya arenaria Linné, 1758   5983
Soft-shell Clam

Labrador to off North Carolina. Western Europe.
Introduced to western United States to Alaska.

1 to 6 inches in length. Pallial sinus somewhat V-shaped in contrast to U-shaped in *truncata*. Shell elliptical. Periostracum very thin and light-gray to straw. Chondrophore in left valve long, spoon-shaped and shallow. This common, delectable clam, also known as the long-necked clam, steamer and nanny nose, is harvested from the readily accessible mud flats of New England in great numbers. In 1935, nearly 12 million pounds, valued at $704,000, were taken from our eastern shores. A hundred pounds of clams furnished 35 pounds of meat, while the equivalent weight of oysters would only give 13 pounds. Pollution and human overpopulation are endangering this species. Largest known specimen is 6½ inches (*The Nautilus*, vol. 74, p. 122). For burrowing habits, see G. C. Matthiesson, 1960, *Limnol. and Oceanogr.*, vol. 5, p. 381.

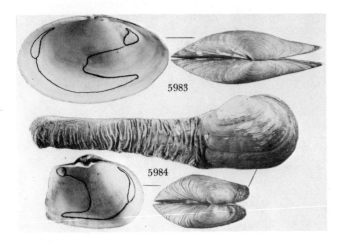

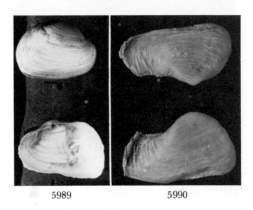

the water-filled burrow (Jenner and McCrary, 1970, *Ann. Rep. Amer. Mal. Union for 1969,* p. 42).

Subgenus *Mya* Linné, 1758

Mya truncata Linné, 1758   **5984**
Truncate Soft-shell Clam

Arctic Seas to Nahant, Massachusetts. Europe.
Arctic Seas to Washington. Japan.

1 to 3 inches in length, similar to *arenaria,* but widely gaping at its abruptly truncate, posterior end. The pallial sinus is U-shaped. In Greenland and Iceland this species is fairly common and is considered a delicacy. It is also a food for the walrus, king eider duck, arctic fox and codfish. It is uncommon in American collections. Synonyms: *ovalis* Turton, 1822; *swainsoni* (Turton, 1822); *pullus* Sowerby, 1826; *praecisa* Gould, 1850; *abbreviata* Jeffreys, 1865; *uddevallensis* Forbes, 1846.

Other species:

5985 *Mya pseudoarenaria* Schlesch, 1931. Arctic Seas; Alaska. (Synonym: *ovata* Jensen, 1900, non Donovan, 1802.)

5986 *Mya japonica* Jay, 1856. Arctic Seas. Japan to Nome, Alaska. China. (Synonym: *oonogai* Makiyama, 1935.)

5987 *Mya elegans* (Eichwald, 1871). Pribilof Islands to Kukak Bay, Alaska. (Synonyms: *crassa* Grewingk, 1850; *intermedia* Dall, 1898; *profundior* Grant and Gale, 1931.)

5988 *Mya priapus* (Tilesius, 1822). Northern Bering Sea to the Aleutians and to Homer, Alaska. Common. See MacNeil, 1965, U.S. Geol. Survey, Prof. Paper, 483-G, and Corgan, 1966, *The Nautilus,* vol. 80, p. 14.

Genus *Paramya* Conrad, 1861

Shell small, subquadrate, concentrically striate, breaks anterior to the middle line. No hinge teeth or external ligament. Hinge with a triangular, vertically directed pit for the resilium, the later borders of the pit sometimes carinated. Pallial line more or less broken up, and no sinus present. Type: *subovata* Conrad, 1845. *Myalina* Conrad, 1845, non Koninck, 1842, is a synonym. Only one species living:

5989 *Paramya subovata* (Conrad, 1845). Delaware (W. Leathem and P. Kinner, *in litt.*) to Florida and Texas, 12 to 31 fathoms. 5 to 10 mm., quadrate, gray exterior with concentric growth lines and microscopic granules near edge of valves. Also Miocene of Virginia. Lives adjacent to the burrows of the marine echiuroid worm, *Thalassema hartmani,* where it protrudes its siphons into

Genus *Sphenia* Turton, 1822

Shell less than 1 inch; fragile; surface with concentric ridges; hinge teeth absent. The elongate, flattened chondrophore in the left valve juts obliquely under the hinge margin of the right valve. These clams are nestlers, and are consequently irregular in shape in many instances. Type: *binghami* Turton, 1822. *Tyleria* H. and A. Adams, 1854, is a synonym. *Sphaena* and *Sphaenia* are misspellings.

Sphenia fragilis (H. and A. Adams, 1854)   **5990**
Fragile Sphenia

Oregon to Peru.

¾ inch in length, quite elongate, with a long, narrow, compressed, posterior snout. Anterior ½ obese and rotund. Beaks fat and close together. Shell fragile, chalky, white and with fine, concentric threads. Periostracum yellowish gray, dull and usually worn off the beak area. Chondrophore in left valve large and with 2 lobes. Socket in right valve large and round. The posterior snout is commonly twisted. Low tide to 46 fathoms in mud. Common. Synonyms are *fragilis* Carpenter, 1857, and *pacificensis* De Folin, 1867.

(**5991**) *Sphenia ovoidea* Carpenter, 1864, (Alaska to Panama) is half as large, smoother, and more ovoid in outline without the prominent snout. Uncommon. It may be an ecologic form of *fragilis.*

Other species:

5992 *Sphenia antillensis* Dall and Simpson, 1901. South Padre Island, Texas; Puerto Rico. Brazil. "Caribbean Seashells," pl. 43, fig. E.

5993 *Sphenia tumida* Lewis, 1968. Pleistocene of Flagler County, Florida; Freeport, Texas, Recent? (*Texas Conchologist,* vol. 7, no. 7, p. 71).

5993a *Sphenia pholadidea* Dall, 1916. Santa Barbara to Monterey, California. Boring in shale; fairly common. (Synonyms: *globula* Dall, 1919, and *nana* (Oldroyd, 1918).)

5993b *Sphenia trunculus* Dall, 1916. San Diego, California, to Panama.

Genus *Cryptomya* Conrad, 1848

Somewhat like a small, fragile *Mya,* but more fragile, and the right valve is larger and more obese than the left. Large chondrophore in the left valve is thin, flat-topped with an anterior ridge. Posterior gape and pallial sinus almost absent. Siphons very short. Type: *californica* (Conrad, 1837).

Cryptomya californica (Conrad, 1837)   **5994**
Californian Glass Mya

Alaska to northern Peru.

1 to 1¼ inches in length, oval, fragile, moderately obese. Right valve fatter. Right beak crowd slightly over the left beak. Posterior gape very small. Chondrophore in left valve large, tucks against a small, concave shelf under the right beak. Exterior chalky and with small growth lines. Periostracum dull-gray, faintly and radially striped at the posterior end. Interior slightly nacreous in fresh specimens. Common; in sand and gravel where it may live as deep as 20 inches. The short siphons protrude into the water-filled burrows of other marine animals. *C. magna* Dall, 1921, is a synonym.

5994

Genus *Platyodon* Conrad, 1837

Shell somewhat resembling a very fat *Mya,* with a fairly thick shell, rugose sculpturing and fairly small chondrophore. Type: *cancellatus* (Conrad, 1837). *Cryptodonta* Carpenter, 1864, is a synonym.

Platyodon cancellatus (Conrad, 1837)   **5995**
Chubby Mya

Queen Charlotte Island, British Columbia, to San Diego, California.

2 to 3 inches in length, rounded-rectangular and obese. Gaping widely posteriorly. Shell strong, rather thick and with fine, clapboardlike, concentric growth lines. Rarely with very weak radial grooves. Chondrophore in left valve quite thick and arched. Beak of right valve crowds under beak of left valve. Shell chalky and white; periostracum thin, yellowish brown to rusty, and rugose posteriorly. Moderately common near beds of pholads. Lives in hard-packed clay or soft sandstone.

5995

Family CORBULIDAE Lamarck, 1818

Genus *Varicorbula* Grant and Gale, 1931

Shell small, subtrigonal; right valve larger, concentrically plicated, convex, with a high umbo and curved forward beak. Left valve much smaller, flat, smoothish, with radial lines. Type: *gibba* (Olivi, 1792) from the Mediterranean.

Varicorbula operculata (Philippi, 1848)   **5996**
Oval Corbula

North Carolina to Florida to Texas and the West Indies. Brazil.

⅜ inch in length, ¾ as high, moderately thin-shelled and glossy. Beaks high, curled under and pointing anteriorly. Right valve subtrigonal in shape, very obese and with strong, concentric ridges. Left valve more elongate, smaller, less obese and with numerous but weaker ridges. Color white, but some may be tinted with rose near the margins. Uncommonly dredged from 12 to 250 fathoms. Live specimens very rare. This is *C. disparilis* of authors.

Other species:

5997   *Varicorbula speciosa* (Reeve, 1843). Gulf of California to Costa Rica. (Synonym: *radiata* Sowerby, 1833, non Deshayes, 1824.)

Genus *Corbula* Bruguière, 1792

Small, thick shells characterized by one valve (commonly the right) being larger than the other. Posterior end commonly rostrate. Resilium and ligament internal. The genus *Aloidis* Mühlfeld, 1811, was in current use until recently. Type: *sulcata* Lamarck, 1801.

Subgenus *Caryocorbula* Gardner, 1926

Valves almost equal in convexity, and posterior end rostrate. Type: *alabamiensis* Lea, 1833, an Eocene Alabama species.

Corbula contracta Say, 1822   **5998**
Contracted Corbula

Cape Cod to Florida and the West Indies. Brazil.

¼ inch in length, oblong, moderately to strongly obese. Both valves about the same size, except that the posterior, ventral margin of the right valve overlaps that of the left. The numerous, poorly defined, concentric ridges on the outside of the valves extend over the posterior, radial ridge on to the posterior slope. The left valve has a V-shaped notch in the hinge just anterior to the beak. Color dirty-gray. A common shallow-water species. *C. kjoeriana* C. B. Adams, 1852, is a synonym.

Corbula dietziana C. B. Adams, 1852   **5999**
Dietz's Corbula

North Carolina to southeast Florida and the West Indies. Brazil.

⅓ to ½ inch in length, like *contracta,* but larger, thicker-shelled and pinkish inside. The ventral margins are blushed or rayed with carmine-rose. Microscopic threads numerous

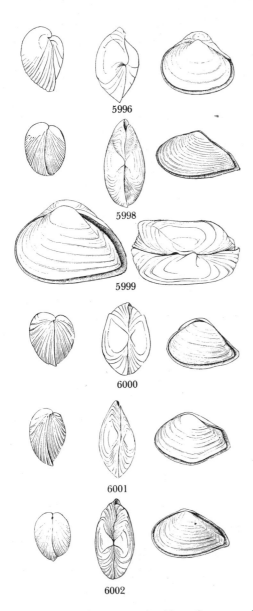

5996

5998

5999

6000

6001

6002

between the few coarse, concentric ridges. Compare with the smaller and more compressed *barrattiana*. Commonly dredged offshore in the Miami region.

Corbula chittyana C. B. Adams, 1852　　　6000
Snub-nose Corbula

North Carolina to both sides of Florida and the West Indies.

³⁄₁₆ to ¼ inch in length, oblong, obese and strongly rostrate at the posterior end. The posterior end looks as if it had been severely pinched. Right valve considerably larger than the left. Margins of valves with a thick border of dark-brown periostracum. Concentric sculpture of distinct ridges. Color yellowish to brownish white. Uncommon in shallow water. Formerly called *Corbula nasuta* Sowerby, 1833, which, however, is an Eastern Tropical Pacific species.

Corbula barrattiana C. B. Adams, 1852　　　6001
Barratt's Corbula

North Carolina to both sides of Florida to Brazil.

¼ inch in length, moderately compressed, rostrate at the posterior end, with poorly developed or without concentric

ridges in the beak area. Right valve slightly larger than the left. The posterior end of the rostrum in the right valve projects far beyond that of the left valve. Color variable: white, pink, mauve, yellow, orange or reddish. Uncommon in shallow water to 287 fathoms.

Corbula swiftiana C. B. Adams, 1852　　　6002
Swift's Corbula

Massachusetts to Florida, Texas and the West Indies.

¼ inch in length, oblong, moderately obese; right valve larger, more obese and overlapping the left valve at the ventral posterior region. Posterior slope in the right valve bounded by 2 radial ridges, one of which is close to the margin of the valve. The left valve has only 1 ridge. Shell thick, with the posterior muscle scar on a slightly raised plaftorm. External sculpture the same on each valve, consisting of irregular, concentric ridges. Color dull-white with a thin yellowish periostracum. Moderately common from 6 to 450 fathoms.

Corbula luteola Carpenter, 1864　　　6003
Common Western Corbula

Monterery, California, to Panama.

⅓ inch in length, slightly obese, with the right valve more obese and overlapping the left valve on the ventral margin. Anterior end elliptical in outline; posterior end coming to a blunt point. Beaks strong, close together and slightly nearer the posterior end. Shell porcelaneous, whitish gray and may be flushed with pinkish or purplish. Interior whitish, but commonly yellowish with purple-red staining. Sculpture of weak, concentric growth lines which are less noticeable toward the smoothish beaks. Common in some localities in sand and rocky, rubbly beaches to 25 fathoms. *C. rosea* Williamson, 1905, is merely a pink color phase of this species.

Corbula porcella Dall, 1916　　　6004
Ribbed Western Corbula

Santa Rosa Island, California, to Panama.

¼ to ⅓ inch in length, similar to *luteola,* but much fatter, chalky, gray, and with much stronger concentric riblets. The right valve overlaps the left valve very prominently on the ventral margin. Moderately common from shallow water to 53 fathoms.

Other species:

6005 *Corbula (Juliacorbula* Olsson and Harbison, 1953) *cubaniana* Orbigny, 1842. Florida Straits and the Caribbean to Panama. 2 to 100 fms. (Synonym: *knoxiana* C. B. Adams, 1852.) Type of the subgenus.

6006 *Corbula (Juliacorbula) bicarinata* (Sowerby, 1833). Gulf of California to northern Peru. (Synonym: *alba* Philippi, 1846.)

6007 *Corbula (Juliacorbula) biradiata* (Sowerby, 1833). Gulf of California to nothern Peru. (Synonyms: *rubra* C. B. Adams, 1852; *polychroma* Carpenter, 1856.)

6008 *Corbula (Caryocorbula) nuciformis* Sowerby, 1833. Catalina Island, California, to northern Peru. (Synonym: *obesa* Hinds, 1843.)

6009 *Corbula (Caryocorbula) nasuta* Sowerby, 1833. Monterey, California (?) to Baja California to Peru. (Synonyms: *pustulosa* Carpenter, 1855; *fragilis* Hinds, 1843.)

6010 *Corbula kelseyi* Dall, 1916. Esteros Bay to Catalina Island, California.

6011 *Corbula (Caryocorbula) krebsiana* C. B. Adams, 1852. Florida and the West Indies. 6 mm.

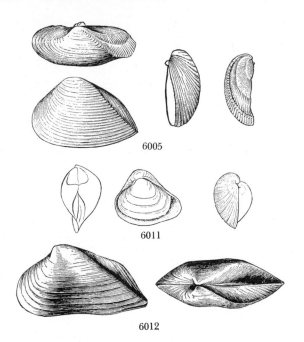

6005

6011

6012

6012 *Corbula (Caryocorbula) cymella* Dall, 1881. Off the Florida Keys, 68 fms. 13.5 mm.

Family SPHENIOPSIDAE Gardner, 1928

Shells very small, ovate-triangular; right valve with 2 hinge teeth, left valve without teeth but with a resilial pocket. Pallial sinus rounded. *Spheniopsis* Sandberger, 1863, is a European Tertiary genus.

Genus *Grippina* Dall, 1912

Shell shaped like a *Donax*, slightly inequivalve; pallial sinus rounded, ascending; right valve receiving the dorsal edge of the left in grooves beneath its own dorsal margins. Cardinal teeth in right valve only, 2, large, subequal, horizontally produced and fitting under the beak of the left valve. Resilium strong and situated between the 2 cardinal teeth. Type: *californica* Dall, 1912.

Grippina californica Dall, 1912   **6013**
Gripp's Clam

San Diego to Guadeloupe Island, Baja California.

2.5 mm. long, 1.5 mm. high, subtrigonal, whitish, solid, finely concentrically sculptured. Beaks elevated and smooth. No hinge plate in left valve. A narrow lanceolate lunule and narrow escutcheon present, each bounded by a ridge. Uncommon; offshore.

Other species:

6014 *Grippina berryana* Keen, 1971. Gulf of California, to 90 meters. "Sea Shells of Tropical West America," p. 270.

Superfamily Gastrochaenacea Gray, 1840

Burrowing shells, lying within a wood or rock cavity. Hinge without teeth. Valves broadly gaping. Shell material porcelaneous.

Family GASTROCHAENIDAE Gray, 1840

Genus *Gastrochaena* Spengler, 1783

Both valves fairly thin, somewhat chalky in substance, and equal in size and shape. The anterior gape is very large. Beaks very near the anterior end. Hinge teeth obscure. These clams form flask-shaped excavations in the rocks which they line with calcareous material. When not protected by a burrow, they form a shelly tube to which debris is attached. Type: *cuneiformis* Spengler, 1783, an Indo-Pacific species.

Subgenus *Rocellaria* Blainville, 1829

Dorsal margin of the shell not thickened and not bearing an internal lamina or myophore. Type: *dubia* (Pennant, 1777), of Europe.

Gastrochaena hians (Gmelin, 1791)   **6015**
Atlantic Rocellaria

North Carolina to Texas and the West Indies. Bermuda. Brazil.

½ to ¾ inch in length; valves rather spathate, with low, indistinct, fine, concentric ridges. Posterior end large and rounded. The entire anterior-ventral end is widely open to accommodate the foot. Color white. Common in soft coral rocks. Erroneously called *cuneiformis* Spengler, which is an Indo-Pacific species (see E. Lamy, 1924).

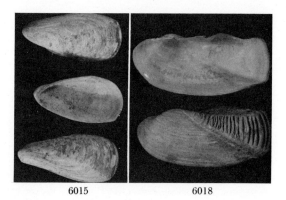

6015   6018

Gastrochaena ovata Sowerby, 1834   **6016**
Ovate Rocellaria

South Carolina to the West Indies; Bermuda.
San Diego, California, to Ecuador.

Similar to *hians*, but the beaks are at the very end of the valve, and the shell is more elongate. In *hians*, there is a very small, winglike projection of the valve in front of the beak. Uncommon; burrows in coral. *G. mowbrayi* Davis, 1904, *The Nautilus*, vol. 17, may be this species or a stenomorph of *hians*.

Other species:

6017 *Gastrochaena stimpsoni* Tryon, 1861. Beaufort, North Carolina.

Genus *Spengleria* Tryon, 1862

Similar to *Rocellaria*, but truncate at the posterior end where there are strong, concentric ridges. A deep sulcus runs from the beak to the posterior-ventral margin. Beaks at the anterior ⅓ of the shell. Type: *mytiloides* (Lamarck, 1818).

Spengleria rostrata (Spengler, 1783)   **6018**
Atlantic Spengler Clam

Southeast Florida and the West Indies. Bermuda. Brazil.

1 inch in length; valves truncate at the posterior end. There is a very characteristic, elevated, triangular area which radiates from the beaks to the large, posterior end. This area is crossed by strong, transverse lamellations resembling a washboard. Commonly found boring in soft coral rock. Uncommon in Florida.

Superfamily Hiatellacea Gray, 1824

Nestling, quadrate-elongate clams, gaping. Hinge teeth weak. Ligament on nymph. Pallial sinus present.

Family HIATELLIDAE Gray, 1824

Genus *Hiatella* Bosc, 1801

Shell irregular due to nestling and burrowing habits. Texture chalky. No definite teeth in the thickened hinge of the adults. Pallial line discontinuous; siphons naked and slightly separated at the tips. *Saxicava* Fleuriau, 1802, is the same. Type: *Mya arctica* Linné, 1767.

Hiatella arctica (Linné, 1767)   **6019**
Arctic Saxicave

Arctic seas to deep water in the West Indies.
Arctic seas to deep water off Panama.

Generally 1 inch in length, but rarely 2 to 3 inches. A very variable species in its shape. The young are rather evenly oblong, but the adults become oblong, oval or twisted and misshapen. The shell is elongate with the dorsal and ventral margins usually parallel to each other. A posterior gape may be present. Beaks close together, about ⅓ back from the anterior end. Just behind them is a conspicuous, bean-shaped, external ligament. Shell chalky, white and with coarse, irregular growth lines. Periostracum gray, thin and usually flakes off in dried specimens. With a weak or fairly strong radial rib at the posterior end. The rib may be scaled. Common in cold water. Common in California; nests in kelp holdfasts and rock crevices, from low tide to deep water. This is *H. rugosa* (Linné, 1767), and *pholadis* (Linné, 1767).

(6020) *Hiatella striata* Fleuriau, 1802, (*H. gallicana* Lamarck and *rugosa* of some authors) is almost indistinguishable in the adult form from *arctica*. However, the young, in *striata* do not have the 2 radial spinose ribs. This species breeds in winter, while *arctica* breeds in summer. The eggs are pinkish cream, while those of *arctica* are red. It nearly always bores into stone.

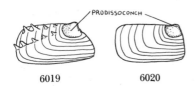

PRODISSOCONCH

6019   6020

Other species:

6021 *Hiatella azaria* (Dall, 1881). Off Florida; Gulf of Mexico. 25 mm.

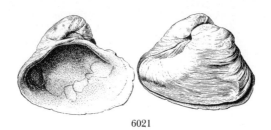

6021

Genus *Saxicavella* Fischer, 1878

Shell equivalve, inequilateral flaring, trapezoidal, obliquely angular in the rear. Hinge with a cardinal in the right valve fitting into a corresponding cavity in the left valve. Ligament short, prominent. Pallial line gently sinuous. Sinus wide but not deep. Type: *jeffreysi* Winckworth, 1930. One Pacific coast species:

6022 *Saxicavella pacifica* Dall, 1916. Los Angeles County, California, to Todos Santos Islands, Baja California, 25 to 150 fathoms. 5 mm. in length.

Genus *Cyrtodaria* Reuss, 1801

Shells fairly large, chalky-white, elongate, covered with a thick, flaky, black periostracum. Gaping at both ends. Hinge is a simple bar. Ligament external. Type: *siliqua* (Spengler, 1793). Cold-water clams. Daudin had the generic name in manuscript only.

Cyrtodaria siliqua (Spengler, 1793)   **6023**
Northern Propeller Clam

Labrador to Rhode Island.

2 to 3 inches in length, about ½ as high. Gapes at both ends, but more so posteriorly where the large siphonal snout projects out about an inch. Beaks hardly noticeable, placed slightly toward the anterior end. The strong, wide ligament is external at the very anterior end of the dorsal margin. Shell chalky, white with a bluish tint. The valves are thick-shelled, with a coarse callus inside. The valves are slightly twisted in propeller fashion. Hinge a simple bar with a fairly large, bulbous swelling under the ligament. In life, the periostracum is light-brown, glossy, smooth and covers the entire exterior. Dried valves soon lose the flaky, blackened periostracum. Moderately common offshore down to 90 fathoms. On occasion, found in fish stomachs.

6023

(6024) A similar species, *Cyrtodaria kurriana* Dunker, 1862, is found in arctic waters from Alaska to Labrador along the shores at low tide to 1,292 fathoms (dead valves). It is 3 times as long as high, hardly twisted and rarely exceeds 1½ inches in length. Uncommon in collections.

Genus *Panomya* Gray, 1857

Shell solid, large, irregular, with a single cardinal tooth under the beak in each valve; pallial line of unconnected, rounded impressions. With long, large united siphons. Type: *arctica* (Lamarck, 1818).

Panomya arctica (Lamarck, 1818)   6025
Arctic Rough Mya

> Arctic Seas to Chesapeake Bay.
> Unalaska to Point Barrow, Alaska.

2 to 3 inches in length, about ½ as high, squarish in outline. Looks somewhat like a misshapen *Mya,* but lacks teeth and a chondrophore in the hinge, and has a coarse, flaky, light-brown periostracum. Characterized by oblong or oval, sunk-in muscle and pallial-line scars. There are 2 poorly defined radial ridges near the center of the valves. Common in mud in cold waters offshore to 328 fathoms. Synonyms: *Mya norvegica* Spengler, 1793, not *norwegica* Gmelin, 1791; *turgida* Dall, 1916. Known also from northern Europe.

6025

Panomya ampla Dall, 1898   6026
Ample Rough Mya

> Aleutian Islands to Point Barrow, Alaska.

2 to 3 inches in length. A peculiarly distorted, heavy shell which is much gaping at both ends. Anterior end crudely pointed; posterior broadly truncate. With 3 to 6 depressed scars on the white interior. Exterior concentrically roughened, ash-white in color, with a border of thick, irregular, black periostracum. Hinge without definite teeth. Uncommon offshore in cold water. Puget Sound records have never been verified.

6026

Other species:

6027 *Panomya beringiana* Dall, 1916. Eastern Bering Sea.

Genus *Panopea* Ménard, 1807

Shells large, solid, rectangular, gaping on all sides except at the hinge; 1 projecting cardinal tooth in each valve. Ligament external. Pallial sinus large and broad. Siphons united and long. Type: *glycymeris* (Born, 1778). *Panope* is a later misspelling.

Panopea generosa (Gould, 1850)   6028

Geoduck (Goo-ee-duck)

> Alaska to the Gulf of California.

7 to 9 inches in length. Inflated, slightly elongate and rather thick-shelled. Gaping at both ends. Coarse, concentric, wavy sculpture present especially noticeable near the small, central, depressed beaks. Periostracum thin and yellowish. Exterior of shell dirty-white to cream; interior semiglossy and white. Hinge with a single, large, horizontal thickening. The 2 long, united siphons of the animal are ½ the weight of the entire clam. Common in mud 2 or 3 feet deep in the northwestern states. Edible but tough. Freaks have been named *solida* Dall, *globosa* Dall and *taeniata* Dall, 1898. For an interesting and well-illustrated account of this species, see *Natural History* magazine (N.Y.), April, 1948, on "We Go Gooeyducking," by the Milnes.

Panopea bitruncata Conrad, 1872   6029
Atlantic Geoduck

> North Carolina to Florida and to Texas.

5 to 6 inches, shell white with a grayish periostracum. Siphon dark. Very similar in shape and sculpture to the Pacific

6028

6029

generosa. Uncommon; low-tide mud flats to 26 fathoms. It lives as deep as 4 feet below the surface. See R. Robertson, 1963, *The Nautilus,* vol. 76, p. 75.

Other species:

6030 *Panopea globosa* Dall, 1898. Gulf of California. 6 inches. In 2 to 3 feet of mud.

Superfamily Pholadacea Lamarck, 1809

Family PHOLADIDAE Lamarck, 1809

An extensive revision of this family has been produced by Ruth D. Turner, 1954–55, in *Johnsonia,* vol. 3, nos. 33 and 34, pp. 1–160. Our treatment is taken directly from her work.

White-shelled, boring clams with an anterior pedal gape, which may or may not be closed by a calcareous covering or callum in the adult stage. Hinge teeth usually lacking. Anterior adductor muscle protected above by accessory plates or by a dorsal extension of the callum. Some forms have 2 apophyses—large, fingerlike, calcareous projections, 1 in each valve, extending from beneath the umbones to which the foot muscles are attached.

Subfamily PHOLADINAE Lamarck, 1809

Genus *Barnea* Risso, 1826

Shells up to 3 inches long, elliptical and rounded anteriorly. Accessory plate consists of a simple, calcareous and lanceolate protoplax. Space below umbonal reflection not septate. Small, short, fingerlike apophyses under the umbones. Type: *candida* Linné, 1758, of Europe. Synonyms: *Barnia* Gray, 1840; *Holopholas* Fischer, 1887.

Subgenus *Anchomasa* Leach, 1852

Shell beaked anteriorly; pedal gape almost ½ as long as the shell. Umbonal reflection simple and usually closely appressed to the surface of the umbo, free anteriorly. Type: *parva* Pennant, 1777, of Europe. Siphons united and 4 or 5 times the length of the shell.

Barnea truncata (Say, 1822)   6031
Fallen Angel Wing

Salem, Massachusetts, to Texas and Brazil.
Also Senegal to Gold Coast.

2 to 2¾ inches, somewhat resembling the true angel wing, but widely gaping at both ends. Shell white, beaked anteriorly, sculpture greatly reduced or absent in the posterior rounded end, and with a single dorsal plate, the protoplax. Internal apophyses long, narrow, curved and bladelike. Periostracum thin. Lives in strictly marine conditions. Locally

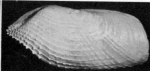

6031

abundant. Bores into mud, clay and peat. Stunted dwarfs found in wood.

Barnea subtruncata (Sowerby, 1834)   6032
Pacific Mud Piddock

Newport, Oregon, to Chile.

2 to 2¾ inches, very similar to *truncata* of the Atlantic coast, but are slightly longer at the pointed or posterior end. The protoplax is lanceolate in outline, usually truncate posteriorly and acuminate anteriorly. Sheath of siphons dark red-brown to dark-gray, with a white tip, and minutely papillose the entire length. Variable in shape. Common. Bores in mud, clay, peat and rarely in soft rock or waterlogged wood. Synonyms: *spathulata* Deshayes, 1843; *pacifica* Stearns, 1871.

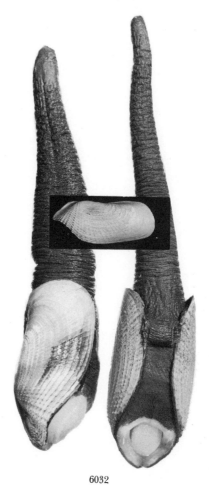

6032

Other species:

6033 *Barnea (Anchomasa) lamellosa* (Orbigny, 1846). Uruguay to Golfo Nuevo, Argentina. See *Johnsonia,* vol. 3, no. 33, pl. 10.

Genus *Cyrtopleura* Tryon, 1862

Protoplax entirely chitinous; mesoplax transverse, in 1 or 2 pieces, calcareous and solid. Type: *crucigera* (Sowerby, 1834) (*crucifera* Sowerby, 1849, is a synonym).

Subgenus *Scobinopholas* Grant and Gale, 1931

Accessory plates a T-shaped, thin, chitinous protoplax and a heavy transverse calcareous mesoplax. Internal apophyses large, broad, spoon-shaped. Type: *costata* (Linné, 1758). Synonym: *Scobina* Bayle, 1880, non Lepeletier, 1825.

Cyrtopleura costata (Linné, 1758)   6034
Angel Wing

South Massachusetts to Texas and to Brazil.

4 to 7 inches, moderately fragile; pure-white with a thin gray periostracum. With about 30 well-developed, beaded ribs which are scalelike at the anterior end of the valve. Protoplax chitinous and triangular; mesoplax "butterfly-shaped," calcareous. Siphons united, long, grayish. Rare in the north; common in deep, soft, sandy mud in west Florida. They live as deep as 2 feet and can move up and down in their burrows at will. Shells in some colonies may have pink, concentric stains due to environmental conditions.

6034

Other species:

6035 *Cyrtopleura (Scobinopholas) lanceolata* (Orbigny, 1846). Santos, Brazil, to Rio Negro, Argentina. See *Johnsonia*, vol. 3, no. 33, pl. 19.

6036 *Cyrtopleura (Cyrtopleura) crucigera* (Sowerby, 1834). Guaymas, Mexico, to Panama. (Synonyms: *crucifera* Sowerby, 1849, and *exilis* Tryon, 1870).

Genus *Pholas* Linné, 1758

Umbonal reflections with calcareous, reinforcing septa. 3 dorsal accessory plates: calcareous protoplax divided longitudinally; mesoplax calcareous, butterfly-shaped; metaplax thin, long and narrow. Apophyses short, solid, strong, ridged on the free ends. Type: *dactylus* Linné, 1758, of Europe. Synonyms: *Hypogaea* Poli, 1791; *Hypogaeoderma* Poli, 1795; *Dactylina* Gray, 1847; *Pragmopholas* Fischer, 1887.

Subgenus *Thovana* Gray, 1847

Shell rounded anteriorly; nucleus of the divided protoplax located near the anterior inner margin. Type: *campechiensis* Gmelin, 1791. *Gitocentrum* Tryon, 1862, is a synonym.

Pholas campechiensis Gmelin, 1791   6037
Campeche Angel Wing

North Carolina to Texas and to Brazil.

3 to 4½ inches, elongate, rounded at both ends. Resembling the angel wing, but the rolled-over umbonal region is supported by a dozen or so vertical, shelly plates. A double, nearly rectangular protoplax, a transverse shelly mesoplax and an elongate, narrow metaplax are present in live specimens. Lives offshore in mud; rarely in wood. Dead valves found on beaches. Absent in southern Florida.

6037

Other species:

6037a *Pholas (Thovana) chiloensis* Molina, 1782. Sonora coast, Mexico, to Chile. (Synonyms: *parva* Sowerby, 1834; *laqueata* Sowerby, 1849; *dilecta* Pilsbry and Lowe, 1932; *retifer* Mörch, 1860.)

Genus *Zirfaea* Gray, 1842

Shell oval in outline, beaked anteriorly, rounded posteriorly, widely gaping at both ends and having a sulcus extending from the umbo to the ventral margin. Only 1 dorsal accessory plate: a small, triangular, fragile, calcareous mesoplax. Apophyses solid, strongly curved, narrowly spoon-shaped. Type: *crispata* Linné, 1758. Synonyms: *Thurlosia* Catlow and Reeve, 1845; *Zirphaea* Leach, 1852.

Zirfaea crispata (Linné, 1758)   6038
Great Piddock

Labrador to Delaware; Europe.
Introduced to California.

2 to 3¾ inches, about ½ as high. Gaping at both ends, and with a radial, indented line obliquely dividing the exterior of the valves. Posterior section with irregular growth lines, the anterior section with fimbriated or scaled growth lines and a serrated edge. Siphon without small chitinous spots. This is the only *Zirfaea* on the Atlantic coast. Usually found in salt marsh peat, rarely in mud, clay, wood and red sandstone at the lower intertidal region to 40 fathoms. Burrows down 6 inches. Also introduced to Humboldt Bay, California.

6038

Zirfaea pilsbryi Lowe, 1931   6039
Pacific Rough Piddock

Bering Sea to Baja California.

6039

2 to 4¾ inches, very similar to *crispata* from the Atlantic. Color chalky-white to light-salmon. Roundly truncate at the anterior end. The siphons are usually marked with small chitinous spots. Bores into mud and clay banks, sometimes as deep as 14 inches. Rarely found in wood and sandstone.

Subfamily MARTESIINAE Grant and Gale, 1931

The large anterior gape is usually closed by a calcareous callum in the adult. Several accessory plates present, but no protoplax. Apophyses present.

Genus *Chaceia* Turner, 1955

Differs from some other piddocks in having the shell gaping at both ends and having only a partial callum in the adult stage. Lacks the ventral accessory plate, the siphonoplax, which is usually present in *Penitella*. Also lacks the metaplax and hypoplax, as well as the protoplax. Type and only known species: *ovoidea* Gould, 1851.

Chaceia ovoidea (Gould, 1851)　　　　**6040**
Wart-necked Piddock

Santa Cruz, California, to Baja California.

2 to 4½ inches, broadly oval, gaping at both ends and only partially closed at the pedal gape with a callum in the adult. Growing edge of the callum infolded over the beaks. Mesoplax small, transverse, broadly V-shaped. Protoplax absent, being replaced by a dorsal extension of the callum. Apophyses small, strong. Periostracum thin, yellowish. Siphons very large, warty and with elongate brownish orange chitinous patches. Lives in soft shale rock. May bore to a depth of 21 inches. Do not confuse with *Penitella fitchi* Turner which also has an incomplete callum but has a siphonoplax.

6040

Genus *Penitella* Valenciennes, 1846

Shells 2 or 3 inches, oval in outline, with an oblique sulcus, and producing a callum, in the adult, which covers the wide gaping anterior. Protoplax lacking. Mesoplax large. Siphoplax usually present. Apophyses short, heavy and bladelike at the free end. Type: *conradi* Valenciennes, 1846. *Navea* Gray, 1851, is a synonym.

Penitella fitchi Turner, 1955　　　　**6041**
Fitch's Piddock

Baja California.

Shell 2 inches, white, having thin, very closely appressed umbonal reflections, producing only a partial callum in the adult stage and having the siphonoplax composed of numerous chitinous leaflike layers. Mesoplax broadly rounded to truncate posteriorly, rounded anteriorly and lacking lateral wings. Siphons united, whitish, smooth and capable of complete retraction within the shell. Known only from Bahia San Bartolomé.

Penitella conradi Valenciennes, 1846　　　　**6042**
Conrad's Piddock

Vancouver, British Columbia, to Baja California.

Shell 1 inch, usually found boring into *Haliotis*, abalone shells, rarely in clay and soft stone. Umbonal reflections broad and closely appressed for their entire length. Mesoplax truncate posteriorly, pointed anteriorly and lacking lateral wings. Siphonoplax heavy, composed of a chitinous outer layer which is lined with a white, granular, calcareous deposit. Protoplax lacking. Siphons united, small, white and devoid of periostracum. Common. Synonyms: *subglobosa* Gray, 1851; *intercalata* (Carpenter, 1855); *Navea newcombii* Tryon, 1865; *Penitella parva* Tryon, 1865.

6042

Penitella penita (Conrad, 1837)　　　　**6043**
Flap-tipped Piddock

Bering Sea to Baja California.

2 to 4 inches, oval-elongate; adults with a callum which protrudes beyond the beaks. Mesoplax sharply pointed posteriorly, truncate anteriorly and having short, pointed, lateral wings. Siphonoplax composed of 2 heavy, flexible, chitinous flaps which are often diverging. Siphons smooth. Apophyses short,

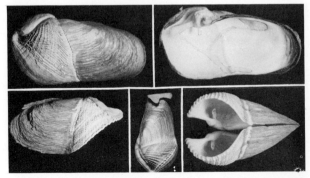

6043

solid, generally flattened at the free end and projecting from beneath the umbones anteriorly at a sharp angle. This is the most common pholad of our Pacific coast. It bores in stiff blue clay, sandstone and cement. Synonyms: *concamerata* Deshayes, 1839; *spelaea* Conrad, 1855; *curvata* Tryon, 1865; *sagitta* "Stearns" (Dall, 1916).

Penitella gabbi (Tryon, 1863)   6044
Gabb's Piddock

Drier Bay, Alaska, to San Pedro, California.

2 inches, solid; callum not protruding beyond the beaks. Umbonal reflections narrow, lightly appressed over the umbones and free anteriorly. Mesoplax in the adult with a rounded point posteriorly, pointed anteriorly and with broad lateral wings. Siphonoplax lacking. Siphons 1 to 2 times the length of the shell, devoid of periostracum, white in color and strongly pustulose. Uncommon.

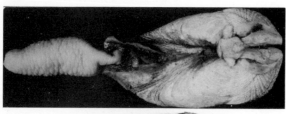

6044

Genus *Pholadidea* Turton, 1819

This genus was formerly recorded from the Pacific coast of North America, but is now known to be limited in the Americas to Central and South America on the Pacific side. The species included in this genus are: *loscombiana* (Turton, 1819) (the type from western Europe); *melanura* (Sowerby, 1834) (Baja California to Ecuador); *quadra* (Sowerby, 1834), from Ecuador, and *tubifera* (Sowerby, 1834) from Panama to Peru. For details, see Turner, 1955, *Johnsonia,* vol. 3, no. 34.

Genus *Martesia* Sowerby, 1824

Shells usually about 1 to 2 inches, pear-shaped, divided obliquely by a sulcus and producing a callum in the adult stage. Anterior end widely gaping in the young. Umbonal reflections closely appressed over the umbones, broadly recurved, forming a funnel-shaped pit below. Protoplax lacking. Mesoplax circular to arrow-shaped. Metaplax and hypoplax long, narrow and pointed anteriorly. Interior of shell with an oblique, raised ridge which becomes a condyle at the ventral margin of the valve. Apophyses long, thin and fragile. Normally bore only into wood and seeds. Type: *striata* Linné, 1758. Synonyms: *Martesiella* Verrill and Bush, 1898; *Hiata* Zetek and McLean, 1936; *Mesopholas* Taki and Habe, 1945; *Diploplax* Bartsch and Rehder, 1945.

Martesia striata (Linné, 1758)   6045
Striate Martesia

North Carolina to Texas; to Brazil. Bermuda. Mexico to Peru; Indo-Pacific.

¾ to 1¾ inches, variable in size and shape, pear-shaped, producing a round callum over the foot gape in the adult. Anterior to the oblique external sulcus there are numerous, finely denticulate riblets. The almost circular mesoplax is sculptured only by irregular wrinkles. Bores into wood; common. Turner (1955) lists over 30 synonyms for this variable species, including *funisicola* Bartsch and Rehder, 1945; *americana* Bartsch and Rehder, 1945; *infelix* Zetek and McLean, 1936; *beauiana* Récluz, 1853, and *rosea* C. B. Adams, 1850.

6045

Martesia fragilis Verrill and Bush, 1890   6046
Fragile Martesia

Pelagic, off Virginia to Texas; Brazil. Bermuda. Sonora, Mexico, to Panama. Indo-Pacific.

½ to ¾ inch, posterior portion with smooth, rounded ridges. In the adult, the pedal gape is closed by a thin callum which does not extend dorsally between the beaks. Mesoplax oval, dorsal portion depressed, with a peripheral keellike edge and with definite concentric sculpture. Found in floating nuts. Synonyms are: *minuscula* Dall, 1908; *exquisita* Bartsch and Rehder, 1945; *bahamensis* Bartsch and Rehder, 1945.

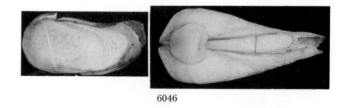

6046

Subgenus *Particoma* Bartsch and Rehder, 1945

Martesia with a thick, oval-shaped, callus over the umbones and with obliquely truncated anterior beaks. Mesoplax heart-shaped, with a central groove from which growth lines radiate. Metaplax and hypoplax long, narrow and divided posteriorly. Type: *cuneiformis* Say, 1822.

Martesia cuneiformis (Say, 1822)   6047
Wedge-shaped Martesia

North Carolina to Texas and to Brazil.

½ to ¾ inch, pear-shaped, producing a callum when adult. Beaks sinuously truncate. Mesoplax heart-shaped, and sculptured with strong growth lines radiating from a median longitudinal groove. Metaplax and hypoplax divided posteriorly. Bores into wood, and sometimes floats north of its normal range. Common.

6047

Genus *Diplothyra* Tryon, 1862

Martesialike, rock-borers or live in shells of oysters; about ½ inch in length. Mesoplax oblong and extending anteriorly in adults to cover completely the anterior adductor muscle. Metaplax and hypoplax are pointed anteriorly and forked posteriorly. Callum produced dorsally between the beaks and extending back on either side of the mesoplax. Type: *smithii* Tryon, 1862. Two American species.

Diplothyra smithii Tryon, 1862   6048
Smith's Martesia

Massachusetts to Texas.

½ inch, found boring in oyster shells and coquina rock. Anterior portion triangular in outline. Callum imbeds the beaks and extends back on either side of the mesoplax. Common.

Other species:

6049 *Diplothyra curta* (Sowerby, 1834). Gulf of California to off Ecuador. In soft stone. Close to *smithii* in appearance. See *Johnsonia*, vol. 3, no. 34, pl. 72.

Genus *Parapholas* Conrad, 1848

Shells large, 5 inches, producing a callum when adult. Valves divided into 3 distinct regions. The posterior slope with overlapping chitinous plates. Siphonoplax lacking, being replaced by a tube or "chimney" composed of fine

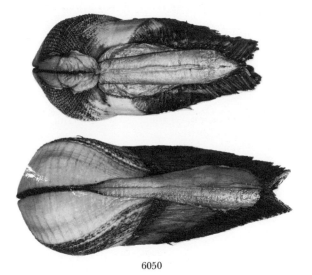

6050

particles ejected from the siphons and then cemented together. Type, and only species in California: *californica* Conrad, 1837.

Parapholas californica (Conrad, 1837)   6050
Scale-sided Piddock

Bodega Lagoon, California to Baja California.

3 to 6 inches, oval-oblong and with a callum in the adult. Hypoplax on ventral side long and arrow-shaped. Dorsal line with metaplex and mesoplax, the latter being medially divided. Posterior slope with strong, raised, concentric, chitinous, overlapping, brown plates. Common; in clay, shale or soft, friable stone from the intertidal zone to 30 feet. Bores down about twice its length.

Other species:

6051 *Parapholas calva* (Sowerby, 1834). Guaymas, Mexico, to Manta Bay, Ecuador.

Subfamily JOUANNETIINAE Tryon, 1862

Genus *Jouannetia* Moulins, 1828

Shell small, globose, with a large callum, that of left valve overlapping the right; no internal apophyses. Siphonoplax produced in the right valve only. Type: *semicaudata* Moulins, 1828. The genus *Scyphomya* Dall, 1898, and its type *semicostata* Lea, 1844, are both very dubious and so far unrecognizable.

Subgenus *Pholadopsis* Conrad, 1849

Characterized by having the siphonoplax pectinate. Type: *pectinata* Conrad, 1849. Synonym: *Triomphalia* Sowerby, 1840.

Jouannetia quillingi Turner, 1955   6052
Quilling's Borer

North Carolina to Florida and Texas.

¾ inch, white, ovate-globose, thin, fragile. Callum large, smoothish, very thin. Shell divided by an oblique keel. Sculpture of concentric ridges bearing tiny spines. Siphonoplax small, welded to posterior end, bearing a few spinelike pectinations. Found in submerged, waterlogged wood or soft rock. Common. For anatomy, see D. A. Wolfe, 1968, *The Veliger*, vol. 11, p. 126.

6052

Genus *Nettastomella* Carpenter, 1865

Shells small, about 1 inch, without a large calcareous callum, but with a chitinous membrane covering the pedal gape. Dorsal plates lacking. Calcareous siphonoplax short, welded to the back end of the valves, and widely diverging. Apophyses lacking. Type: *rostrata* Valenciennes, 1846. Synonym: *Netastoma* Carpenter, 1864, non Rafinesque, 1810.

We follow R. D. Turner, 1962, in conserving the name *Nettastomella.*

Nettastomella rostrata (Valenciennes, 1846)   6053
Beaked Piddock

Bolinas, California, to Baja California.

¾ inch, gaping anteriorly, and with a long, calcareous, white, tapering siphonoplax which, in following the contours of the clam's burrow, may be curved and misshapen. Callum exists as a band which is sculptured with high, thin flutes. Shell proper with widely spaced, delicate, high, concentric lamellae. Bores in soft shale rock from low tide to 55 fathoms. Common.

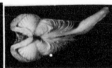

6053

Other species:

6054 *Nettastomella darwinii* (Sowerby, 1849). Uruguay to Argentina and up to Chiloe Island, Chile.

6055 *Nettastomella japonica* (Yokoyama, 1920). South British Columbia to off Washington; in blue clay and sandstone; low tide to 150 fms. Northern Japan. See R. D. Turner, 1962, *Occasional Papers on Mollusks,* vol. 2, no. 28.

Subfamily XYLOPHAGAINAE Purchon, 1941

Genus *Xylophaga* Turton, 1822

Shells small, ½ inch, globose, resembling *Teredo* ship-worm valves. Apophyses lacking. A rather large chondrophore is present in the left valve. Dorsal plates consist of a divided mesoplax. No callum developed. May produce a "chimney" of agglutinated fecal pellets. Type: *dorsalis* Turton, 1819, of Europe only. Pelagic, living in floating wood. *Xylotomea* Dall, 1898, *Protoxylophaga* Taki and Habe, 1945, are synonyms.

Xylophaga atlantica H. G. Richards, 1942   6056
Atlantic Wood-eater

Gulf of St. Lawrence to Virginia.

½ inch, fragile, globose, white to light-brown. Gaping anteriorly, closed posteriorly. Beak of valve extending ⅓ the distance to the ventral margin. Mesoplax triangular, divided, and located anterior to the umbones. Excurrent siphon nearly as long as the incurrent siphon. Found in waterlogged wood dredged from 10 to 1,000 fathoms. Moderately common. Erroneously called *dorsalis* Turton, a European species.

Xylophaga washingtona Bartsch, 1921   6057
Washington Wood-eater

Vancouver, British Columbia, to California.

¼ to ⅓ inch, very similar to *atlantica,* but smaller, with the mesoplax proportionately longer, and the posterior adductor muscle scar regularly marked with striae (rather than irregularly stippled). Common; in wood dredged from 10 to 108 fathoms. *X. californica* Bartsch, 1921, is a synonym.

Xylophaga abyssorum Dall, 1886   6058
Deep Sea Wood-eater

Off New Jersey to the West Indies.

4 to 11 mm., umbonal-ventral sulcus is shallow and bounded by a pronounced keel. Posterior adductor muscle scar broadly oval and irregularly roughened. Rare; dredged in deep water.

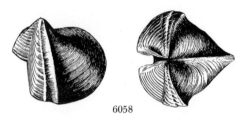

6058

Genus *Xylopholas* Turner, 1972

Very small (shell 2.5 mm.), similar to *Xylophaga,* but the fused siphonal tube is very long and is terminated by a sole pair of small, calcareous, paddlelike lateral plates and two dorsoventral, collar-like plates. The siphons are short. Young clams attached to the ventral side. Bores in waterlogged wood and coconuts found at great depths. Type: *Xylopholas altenai* Turner, 1972. One species known:

6058a *Xylopholas altenai* Turner, 1972. Shell 2.5 mm. In wood, 311 to 366 meters, off Miami and the Florida Keys, Florida; off Port Gentil, West Africa, in 239 meters; off Sao Tome Id., West Africa, 2550 meters. *Basteria,* vol. 36, p. 97, figs. 1–12.

Genus *Xyloredo* Turner, 1972

6058b *Xyloredo nooi* Turner, 1972. 1737 meters deep, off north Andros Id., Bahamas. *Breviora,* no. 397, p. 5. Type of the genus. Valves 10 mm.

6058c *Xyloredo naceli* Turner, 1972. 2072 meters deep, off Port Hueneme, California. *Breviora,* no. 397, p. 9. Valves 1.5 mm.

Family TEREDINIDAE Rafinesque, 1815

The shipworms are distributed around the world from tropical to boreal seas, although some are found in brackish conditions and at least one species lives and breeds in lakes and streams of Central and South America. Living specimens are common in floating logs and wharf pilings, and live specimens have been dredged from depths of about 4 miles where the wood became waterlogged and sank. An extensive account of the classification, anatomy and life histories is given by Ruth D. Turner, 1966, "A Survey and Illustrated Catalogue of the Teredinidae," Museum Comp. Zoology, Cambridge, Mass. 02138.

The shells are of little use in the identification of teredinids, the main useful characters being in the pair of plumelike or club-shaped pallets found at the siphonal end of the wormlike body.

Pallets and shells are generally preserved in 1 part glycerin to 4 parts alcohol (70% grain) to permit later study of the delicate cones in the pallets. Permanent slides can be made by soaking the pallets in 90% alcohol for 12 hours, then placing on a slide, covering with a few drops of diaphane or euparol, and adding a long cover slip.

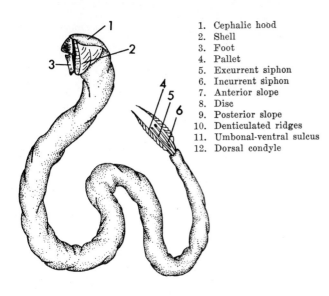

1. Cephalic hood
2. Shell
3. Foot
4. Pallet
5. Excurrent siphon
6. Incurrent siphon
7. Anterior slope
8. Disc
9. Posterior slope
10. Denticulated ridges
11. Umbonal-ventral sulcus
12. Dorsal condyle

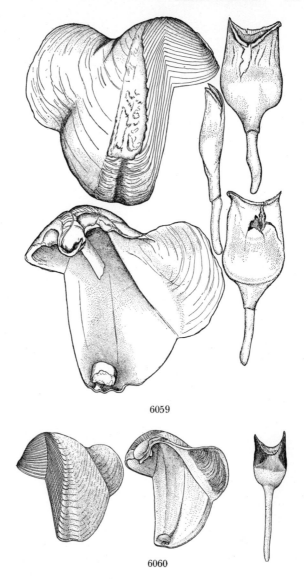

6059

6060

Subfamily TEREDINIDAE Rafinesque, 1815

Pallets not segmented. In some, the young veligers are brooded inside the mother.

Genus *Teredo* Linné, 1758

Animal long, wormlike in shape, burrowing in wood, with 2 striated shells at the boring or large end, and with 2 long separate siphons and 2 simple, paddle-shaped pallets at the small end opening into the water. Type: *navalis* Linné, 1758. The young are retained within the female until the veliger stage. Among the synonyms are *Austroteredo* Habe, 1952; *Coeloteredo* and *Zopoteredo* Bartsch, 1923; and *Pingoteredo* Iredale, 1932.

Teredo navalis Linné, 1758   6059
Common Shipworm

> Both coasts of the United States.
> Worldwide; temperate seas.

Shell like that in *Bankia* and subject to many minute variations. Each of the 2 calcareous pallets is spathate and compressed, but typically symmetrical. The leathery blade is urn-shaped, widening regularly from a stalk of medium length, then tapering somewhat toward the tip, which is decidedly excavated. The base of the blade is calcareous, but approximately the distal 1/3 is normally covered by a yellowish or brownish chitinous epidermis. A very common and destructive species found boring in wood. Synonyms include *beachi* Bartsch, 1921; *beaufortana* Bartsch, 1922; *borealis* Roch, 1931; *morsei* Bartsch, 1922; *novangliae* Bartsch, 1922; and *teredo* Müller, 1776.

Teredo bartschi Clapp, 1923   6060
Bartsch's Shipworm

> South Carolina to Texas. Bermuda.
> Introduced to California. Worldwide.

Shell close to *T. navalis*, but with the auricle of the shell typically semicircular rather than subtriangular in outline. Pallets: stalk long; blade short and deeply excavated at the top. Only the distal 1/2 of the blade is invested with periostracum, which is light horn-colored and semitransparent, permitting the calcareous portion to be seen within as an irregular, hourglass-shaped structure with a deep sinus on either side. A common species. Synonyms: *aegyptia* Roch, 1935; *balatro* Iredale, 1932; *batilliformis* Clapp, 1924; *grobbai* Moll, 1937; *hiloensis* Edmondson, 1942; *shawi* Iredale, 1932.

Genus *Teredothyra* Bartsch, 1921

Shells small (4 to 5 mm.). Pallets composed of a broad to elongate basal cup with a secondary inner cup which is divided medially. The stalk, which is sheathed by the basal cone, extends into the blade only as far as the base of the inner cup. The structure of the pallets may best be seen on young or cleared specimens using transmitted light. Siphons relatively long and separated. Synonyms: *Ungoteredo* Bartsch, 1927; *Idioteredo* Taki and Habe, 1945. One species, the type of the genus, reported from America:

6061 *Teredothyra dominicensis* (Bartsch, 1921). Cuba and Dominica, Lesser Antilles.

Genus *Teredora* Bartsch, 1921

Pallets solid; blade with prominent thumbnaillike depression. Posterior slopes of valves small, high and set at

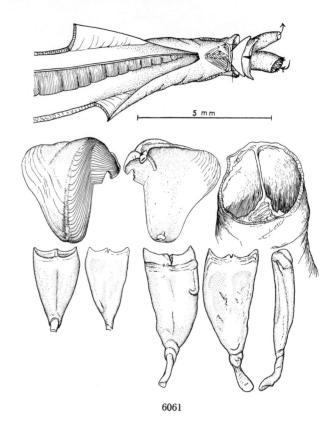

6061

nearly right angles to the dorsoventral axis; gills extending from the united siphons to the mouth without reduction. Type: *malleolus* (Turton, 1822).

Teredora malleolus (Turton, 1822)   6062
Malleated Shipworm

Temperate Atlantic Ocean; Mediterranean. Bermuda.

Blade paddle-shaped, with the stalk ⅓ the length. The blade is broad, smoothish and concave on one side with a central swelling; the other, convex side has a thumbnail depression bearing concentric lamellations. Moderately common. *T. thomsonii* Tryon, 1863, is a synonym.

Genus *Lyrodus* Gould, 1870

Pallets with a calcareous base and a pronounced brown to black periostracal cap which can readily be separated from the base, or with a periostracal cup set in the calcareous base. Siphons rather short and separate. The young are carried by the female until the late veliger stage. Type: *pedicellatus* Quatrefages, 1849. Synonyms: *Teredops* Bartsch, 1921; *Cornuteredo* Dall, Bartsch and Rehder, 1938.

Lyrodus pedicellatus (Quatrefages, 1849)   6063
Black-tipped Shipworm

Worldwide; tropical and temperate seas.

Shell similar to that of *T. navalis*, but smaller, more finely sculptured and translucent, and with numerous, closely set ridges. Pallets: blade with an oval, calcareous base, surmounted by a horny cap, amber to black in color. The horny portion commonly deeply excavated at the tip, but may be cut off bluntly. The two elements of the blade come apart very easily. Many synonyms, including *Teredo diegensis* Bartsch,

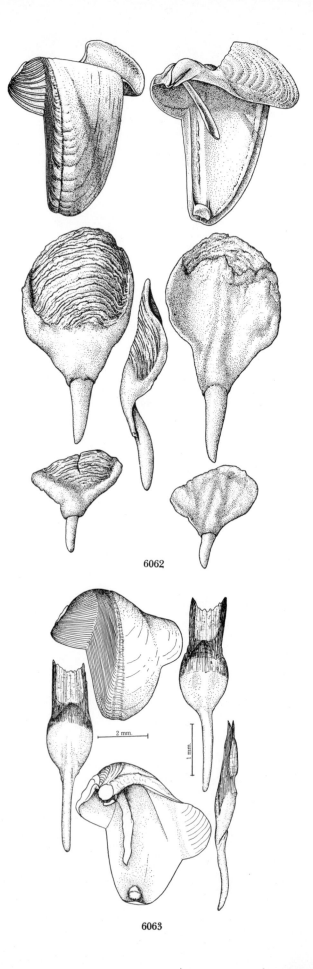

6062

6063

1916; *townsendi* (Bartsch, 1922); *chlorotica* Gould, 1870 (type of the genus); *floridana* Bartsch, 1922; *indica* Nair, 1955; *truncata* Jeffreys, 1865.

Genus *Psiloteredo* Bartsch, 1922

Blades of the pallets without periostracal sheath, similar to those of *Teredora* but the thumbnail depression is quite small and confined to about ¼ the area of the paddle. Sometimes 2 fingerlike projections extend from the blunt, distal end of the paddle. Anterior slope of the valves very large, ear-shaped and flaring. Gills reduced. Type: *megotara* (Hanley, 1848, in Forbes and Hanley). One species, *healdi* Bartsch, 1931, lives in fresh water in Central and South America. *Dactyloteredo* Moll, 1941, is a synonym.

Psiloteredo megotara (Hanley, 1848)   6064
Big-eared Shipworm

North Atlantic; New England.

Blades of the pallets as in the genus. The stalk is short, less than ¼ the length of the entire blade. In the shell the auricle is recurved and projects above the dorsal line; the dorsal condyle is very large. Common in floating wood and lobster pots. Synonyms: *dilatata* (Stimpson, 1851); *denticulata* (P. Fischer, 1856); *striator* (Jeffreys, 1865); *mionota* Jeffreys, 1865. *T. nana* Turton, 1822, is a *nomen dubium* and a name used by some for this species.

Subfamily BANKIINAE Turner, 1966

The blades of the pallets are segmented. The larval stages are planktonic.

Genus *Bankia* Gray, 1842

This is a genus of shipworms which is very common in warm American waters and which is of great economic importance. The 2 highly specialized valves of the clam are very similar to those in *Teredo,* and cannot be used as reliable identification characters. The 2 plumelike pallets at the posterior end of the wormlike animal which can close off the end of the burrow are used to distinguish species. In adult specimens, the shell at the anterior end which scrapes away the moist wood in the tunnel is about 1/40 the total length of the animal. The mantle secretes a thin, smooth calcareous lining for the burrow as an added protection to the soft body. A much fuller and technical account of this group is given by Clench and Turner in *Johnsonia* (1946). The life history is explained in detail by Sigerfoos (1908). Type: *bipalmulata* (Lamarck, 1801).

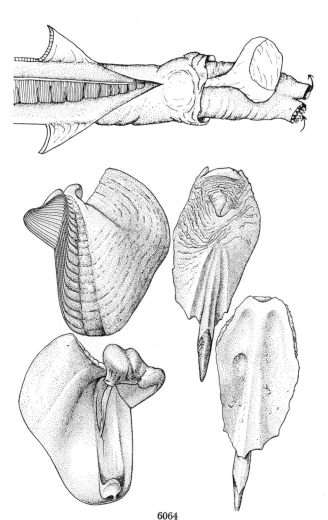

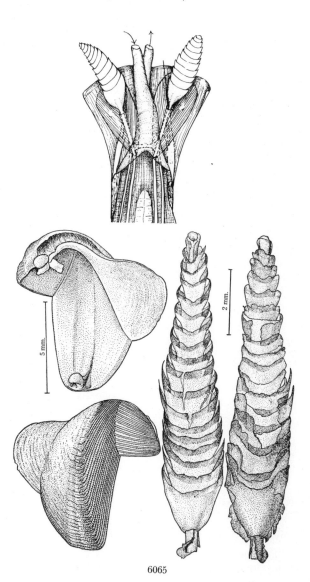

6064

6065

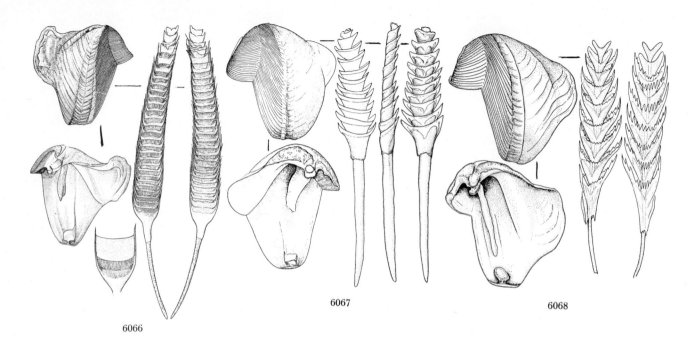

6066

6067

6068

Subgenus *Bankiella* Bartsch, 1921

Pallets have cones with smooth periostracal edges; the lateral portions of the periostracum extend to form rather short and inconspicuous awns. Type: *gouldi* (Bartsch, 1908).

Bankia gouldi Bartsch, 1908   6065
Gould's Shipworm

New Jersey to Florida, Texas and to Brazil.

Pallets about ½ inch in length. Cones deep-cupped, with smooth, drawn-out edges. Cones not very crowded at the distal end. Do not confuse with *B. carinata* Gray. Gould's shipworm is the most widespread and abundant species in this genus on the Atlantic Coast, and hence is the most destructive. It has been found on the Pacific side of the Panama Canal. *B. mexicana* Bartsch, 1921, and *schrencki* Moll, 1935, are synonyms.

Bankia setacea (Tryon, 1863)   6066
Feathery Shipworm

Alaska to San Diego, California.

Pallets feathery in appearance. Cones with long lateral awns connected by a thin membrane on inner face. Similar to *gouldi*, but with longer awns. *Bankia sibirica* Roch, 1934, is probably this species. Common.

Subgenus *Lyrodobankia* Moll, 1941

Pallets with the smooth periostracal awns produced laterally, and the distal embryonic awns are crowded and covered with a cap of periostracum. Type: *carinata* (Gray, 1827). Synonym: *Bankiopsis* Clench and Turner, 1946.

Bankia carinata (Gray, 1827)   6067
Carinate Shipworm

Indo-Pacific and Europe. West Indies. Brazil.

Pallets about ⅓ inch in length. Cones shallow-cupped, with blunt, smooth, not drawn-out, edges. Cones very crowded at

the distal end. Do not confuse with *B. gouldi* which is more abundant and a larger species. Common in the West Indies. Synonyms include: *bipalmata* (Delle Chiaje, 1829); *caribbea* Clench and Turner, 1946; *stutchburyi* (Blainville, 1828); *philippi* (Gray, 1851); *edmondsoni* Nair, 1956; *indica* Nair, 1954; *kamiyai* (Roch, 1929); *orientalis* (Roch, 1929).

Subgenus *Plumulella* Clench and Turner, 1946

Pallets with long, serrated awns. Inner margins of the cones with deep, comblike serrations. Type: *fimbriatula* Moll and Roch, 1931.

Bankia fimbriatula Moll and Roch, 1931   6068
Fimbriate Shipworm

South half of Florida and the West Indies. Brazil.

Pallets ½ to 1 inch in length. Cones deeply cupped, with beautiful, comblike serrations on the edges. It has been found on the Pacific side of the Panama Canal, and those specimens were named *Bankia canalis* Bartsch, 1944. Sometimes in wood that drifts to Europe. *Teredo fimbriata* Jeffreys, 1860, is a synonym.

Other species:

6069 *Bankia (Plumulella) fosteri* Clench and Turner, 1946. Lower Caribbean. *Johnsonia*, vol. 2, p. 24 (possibly *bipennata* (Turton, 1819)).

6070 *Bankia (Plumulella) cieba* Clench and Turner, 1946. Greater Antilles and south to Colombia. Pacific side of Panama.

6071 *Bankia (Plumulella) martensi* (Stempell, 1899). Chile and Argentina. (Synonyms: *argentinica* Moll, 1935; *chiloensis* Bartsch, 1923.)

6072 *Bankia (Liliobankia* Clench and Turner, 1946) *campanellata* Moll and Roch, 1931. Type of the subgenus. Caribbean to Brazil. India. (Synonyms: *katherinae* Clench and Turner, 1946 (subgenotype); *bengalensis* Nair, 1954.)

6073 *Bankia (Neobankia* Bartsch, 1921) *zeteki* Bartsch, 1921. Both sides of Panama. Type of the subgenus.

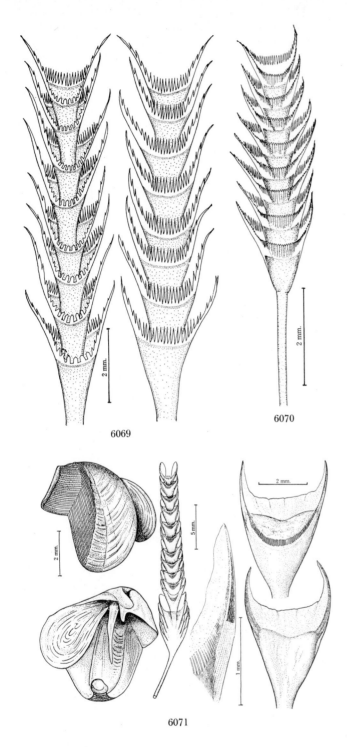

6069

6070

6071

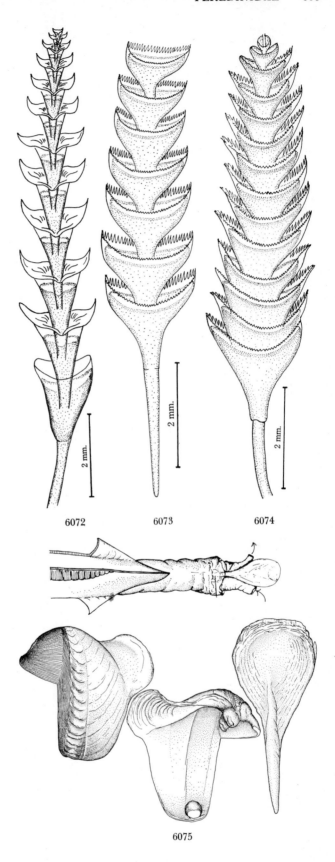

6072   6073   6074

6075

6074 *Bankia (Neobankia) destructa* Clench and Turner, 1946. Atlantic and Pacific sides of Central America. *Johnsonia*, vol. 2, p. 21.

Genus *Nototeredo* Bartsch, 1923

Pallets oval in outline, flat, with a short stalk. Blade composed of a soft, friable calcareous material laid down in closely packed segments, separated by thin layers of periostracum. Entire surface of blade covered by a pale periostracum which extends as a border distally. Two Atlantic species. Type: *edax* (Hedley, 1895) from Indo-Pacific. *Phylloteredo* Roch, 1937, is a synonym.

Nototeredo norvagicus (Spengler, 1792)   **6075**
Norwegian Shipworm

North Atlantic waters; West Africa.

Pallets irregularly paddle-shaped and having a broken or friable irregular, foliaceous outer side, and a smooth inner side which has a strong central ridge running down the middle and continuing as a short stout stem. A common cold-water borer of wooden structures. *N. norvegicus* is a misspelling. Synonyms include: *adami* Moll, 1941; *senegalensis* Laurent, 1849, non Blainville, 1828; *utriculus* "Gmelin" Hanley, 1882.

Nototeredo knoxi Bartsch, 1917   **6076**
Knox's Shipworm

North Carolina to Florida and to Brazil.

The shell and pallets are very similar to those of *norvagicus*. The blades of the pallets are much more foliaceous and the stem is shorter. The posterior dorsal "auricle" of the shelly valve is only moderately developed, not large, high and flaring. A common warm-water borer. Synonyms: *rosifolia* Moll, 1941; *bisiphites* Roch, 1931; *jamaicensis* Bartsch, 1922; *sigerfoosi* Bartsch, 1922; *stimpsoni* Bartsch, 1922; *tryoni* Bartsch, 1922.

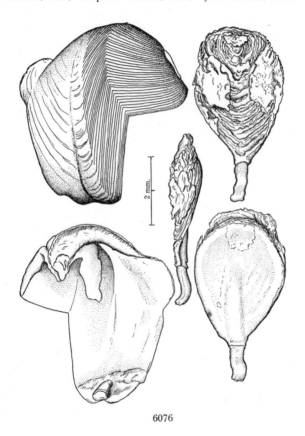

6076

Suborder *Anomalodesmacea* Dall, 1889

Superfamily Pholadomyacea Gray, 1847

Family PHOLADOMYIDAE Gray, 1847

Genus *Panacca* Dall, 1905

Shell trigonal, cuneate, nacreous, with radial sculpture, with a large lunular area and narrow dorsal area. Surface granular under a thin periostracum. Hinge with an external strong ligament seated on strong nymphs. No teeth. Indention under beaks. No pallial sinus. Type: *arata* Verrill and Smith, 1881. No lithodesma present. *Panacea*

and *Panocca* are misspellings. *Aporema* Dall, 1903 (not Scudder, 1890) is a synonym. One Atlantic species:

6077 Panacca arata (Verrill and Smith, 1881). Off Martha's Vineyard, Massachusetts, 71 to 134 fms.

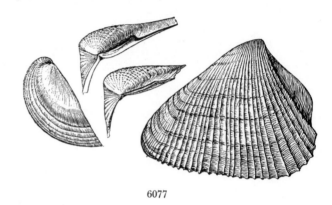

6077

Superfamily Pandoracea Rafinesque, 1815

Family LYONSIIDAE Fischer, 1887

Shells fairly fragile, oblong, nacreous, usually distorted because they are nestlers; hinge without teeth. Ligament internal, the resilium in a long, narrow slot under the dorsal margin behind the beak and further reinforced by a shelly plate or lithodesma.

Genus *Lyonsia* Turton, 1822

Shells elongate, posterior side longer, gaping at the end; thin, nacreous and with a thin yellowish periostracum. Middle section of the internal resilium is covered by a large, calcareous lithodesma. Type: *norwegica* (Gmelin, 1791). Some species attach sand grains to the outside of the valves. Synonyms: *Magdala* Brown, 1827; *Osteodesma* Deshayes, 1830 (non Blainville, 1825); *Myatella* Brown, 1833; *Arenolyonsia* Nordsieck, 1969.

Lyonsia hyalina Conrad, 1831   **6078**
Glassy Lyonsia

Nova Scotia to South Carolina.

½ to ¾ inch in length; very thin and fragile. Semitranslucent and whitish to tan. Shell elongate, with the anterior end

6078

somewhat obese and the posterior end tapering and laterally compressed. Without teeth in the weak hinge, but with a small, free, elongate, calcareous ossicle inside just under the small, inflated, anteriorly pointing beaks. Periostracum very thin, with numerous raised radial lines. Commonly has tiny sand grains attached. Common from low water to 34 fathoms. For larval development see Chanley and Castagna, 1966, *The Nautilus*, vol. 79, p. 123.

(6079) *Lyonsia hyalina floridana* Conrad, 1849, known from the west coast of Florida to Texas, is very similar, differing only in being ⅓ as high as long (instead of ½) and in having a narrower, more rostrate posterior end. Common.

Lyonsia arenosa Möller, 1842   6080
Sanded Lyonsia

Greenland to Maine.
Alaska to Vancouver, British Columbia.

½ to ¾ inch in length, resembling *hyalina,* but much less obese, with a heavier, greenish yellow periostracum, and with its posterior end more oval and higher than the anterior end. The dorsal margin of the right valve behind the beak overlaps that of the left valve considerably. There is no posterior gape as in *hyalina.* Like other species in the genus, it glues sand grains to itself. Moderately common from low water to 60 fathoms. Synonyms: *gibbosa* Hancock, 1846; *sibirica* Leche, 1883.

Lyonsia californica Conrad, 1837   6081
California Lyonsia

Sitka, Alaska, to Baja California.

1 inch in length, very thin, fragile and almost transparent. Quite elongate and moderately obese. Beak area swollen. Posterior end tapering and laterally compressed. Outer surface whitish (opalescent when worn), commonly with numerous, weak, radial, dark lines of periostracum. Interior glossy and with an opalescent sheen. Ossicle inside, under hinge, is opaque-white. Very similar to our figure of the Atlantic *hyalina.* Common in sandy mud bottoms of many California sloughs and bays down to 40 fathoms. Synonyms are *haroldi* Dall, 1915; *nesiotes* Dall, 1915; *pugetensis* Dall, 1913. For functional morphology, see Walter Narchi, 1968, *The Veliger*, vol. 10, p. 305.

6081

Subgenus *Philippina* Dall and Simpson, 1901

Shell about 1 inch, thin, polished, with the beaks at the anterior end, gaping below, compressed behind, and with a thin periostracum. Type: *beana* (Orbigny, 1842).

6082

Lyonsia beana (Orbigny, 1842)   6082
Pearly Lyonsia

North Carolina to Florida; West Indies. Bermuda. Brazil.

½ to 1 inch, irregular in shape, oblong, with a squarish anterior end where the beaks are situated; larger and swollen at the convex end. Pearly within and outside; periostracum thin. Gaping slightly at both ends. Found living in sponges; shallow water; moderately common. Synonyms: *braziliensis* Gould, 1850; *orbignyi* Fischer, 1857. May be an *Entodesma.*

Other species:

6083 *Lyonsia granulifera* Verrill and Bush, 1898. Off Labrador, 75 fms. 25 mm.

6084 *Lyonsia bulla* Dall, 1881. Off west coast of Florida, 1,920 fms.

6085 *Lyonsia* (*Allogramma* Dall, 1903) *formosa* Jeffreys, 1881. Gulf of Campeche, Mexico. (Synonym: *major* Locard.)

6086 *Lyonsia striata* Montagu, 1815. Arctic seas; to Washington. Mediterranean. (Synonyms: *coruscans* Scacchi, 1833; *montagui* Brown, 1845.)

6087 *Lyonsia gouldii* Dall, 1915. San Francisco Bay, California, to Tres Maria Islands, Baja California.

6088 *Lyonsia magnifica* Dall, 1913. Off Mazatlan, Mexico. 25 mm.

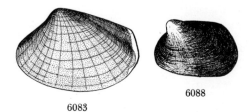
6083   6088

Subgenus *Phlycticoncha* Bartsch and Rehder, 1940

Similar to *Entodesma* but anteriorly more narrow, with the anterior ½ of the ventral margin more sinuous, and with the surface sculptured with very fine, radially arranged, white pustules which become obsolete towards the margins. Type: *lucasana* Bartsch and Rehder, 1939, *Smithsonian Misc. Coll.*, vol. 98, p. 12. Synonym: *Phlyctiderma* Bartsch and Rehder, 1939, non Dall, 1899. *The Nautilus*, vol. 53, p. 137.

6089 *Lyonsia* (*Phlycticoncha*) *lucasana* Bartsch and Rehder, 1939. 6 to 10 fms. off Cape San Lucas, Baja California.

Genus *Entodesma* Philippi, 1845

Shell large and coarse. Periostracum coarse. Lithodesma large. Type: *chilense* (Philippi, 1845).

6090

6092

Subgenus *Agriodesma* Dall, 1909

Entodesma saxicolum Baird, 1863   **6090**
Northwest Ugly Clam

Alaska to San Pedro, California.

2 to 5 inches in length. A very peculiar, ugly and misshapen clam found along the shore burrowing into rocks. Generally oblong in shape, with the posterior end flaring and gaping. Covered with a thick, rough, brown periostracum which partially flakes off when dry. Interior brownish tan to whitish with a slight opalescence. Hinge without teeth, but with a rather large, oblong, whitish ossicle lying under the internally placed ligament. Moderately common from Washington to southwest Alaska. Type of the subgenus.

Entodesma pictum (Sowerby, 1834)   **6091**
Picta Ugly Clam

Alaska to Ecuador.

1 inch, thin-shelled, elongately subovate, swollen over the umbonal middle part and compressed at the long posterior part. Anterior ½ is narrowly and regularly, concentrically undulated except for a line above the anterior-ventral gape where the undulations are wrinkled or sharply bent inward. Posterior ½ smoothish. Periostracum thin, deciduous. Interior pearly white. Common; sand and gravel bottoms, 20 to 40 fathoms. *Lyonsia inflata* Conrad, 1837, is a synonym. *Entodesma scammoni* Dall, 1871, may be this species.

Genus *Mytilimeria* Conrad, 1837

A peculiar, bladder-shaped, rounded-oval, very thin shell found embedded in compound ascidians or sea squirts. Hinge without teeth. Lithodesma present. Type: *nuttalli* Conrad, 1837.

Mytilimeria nuttalli Conrad, 1837   **6092**
Nuttall's Bladder Clam

Alaska to Baja California.

1 to 2 inches in length, obliquely oval, inflated, very fragile, opaque with a thin, brownish periostracum. Beaks small and spiral. No teeth in the weak hinge, but a small, calcareous ossicle is present. Color white with underlayers of slightly pearly material. Common under rocks at low tide to 10 fathoms, always embedded in compound ascidians or sea squirts.

Family PANDORIDAE Rafinesque, 1815

A detailed anatomical, biological and systematic account of "The family Pandoridae in the Western Atlantic" by

K. J. Boss and A. S. Merrill in 1965 appeared in *Johnsonia*, vol. 4, no. 44, pp. 181–216. The clams are hermaphroditic. Eggs are relatively large and are exuded on mucus strands. Free-swimming veligers last for about 24 hours before settling.

Genus *Pandora* Bruguière, 1797

Shells 1 to 2 inches, compressed, crescent-shaped, with the umbones nearer the front end. Left valve usually more convex and larger. Internal layers nacreous. Cardinal teeth strong. Resilium internal, sometimes supported by an internal calcareous lithodesma. Type: *inaequivalvis* (Linné, 1758). Synonyms: *Calopodium* Röding, 1798; *Trutina* Brown, 1827; and *Pandorina* Scacchi, 1836. No American representatives of the typical subgenus.

Subgenus *Clidiophora* Carpenter, 1864

A lithodesma, underneath the internal resilium, is present. 3 cardinal teeth in each valve. In the left valve, the anterior cardinal tooth is at a right angle to the dorsal margin of the shell. Type: *arcuata* (Sowerby, 1830). *Clidiphora* is a misspelling.

Pandora trilineata Say, 1822   **6093**
Say's Pandora

North Carolina to Florida and Texas.

¾ to 1 inch in length, almost ½ as high; half-moon-shaped in outline, and with a strong, squarish ridge along the hinge margin which extends posteriorly into a fairly long rostrum. Valves very flat, the entire shell very compressed. Beaks tiny, quite near the rounded anterior end. Right valve smoothish and translucent cream. Left valve more prominently divided into 2 portions by a slight radial groove, anterior to which the shell is dull cream, and posterior to which the shell is more glossy but with microscopic concentric growth lines. Sometimes slightly iridescent. Interior pearly. Moderately common in sand below low water to 60 fathoms. Synonym: *nasuta* Sowerby, 1830.

Pandora inornata Verrill and Bush, 1898   **6094**
Inornate Pandora

Nova Scotia to Cape Cod, Massachusetts.

¾ inch, similar to *gouldiana*, but more elongate, thicker-shelled, not as nacreous, and the left anterior cardinal tooth is arcuate and thickened at the end. The anterior dorsal margin is slightly convex, rounded and more or less in line with the anterior margin (in *gouldiana* it is straight or slightly concave). Uncommon; 10 to 45 fathoms.

Pandora gouldiana Dall, 1886   **6095**
Gould's Pandora

Gulf of St. Lawrence to North Carolina.

¾ to 1½ inches in length, similar to *trilineata*, but less elongate with the height slightly more than ½ the length. The posterior rostrum on the hinge line is very short, stubby and turned up. The shell is opaque, chalky and commonly worn away, showing the pearly underlayers. Margin of valves bordered with blackish brown periostracum. Pallial sinus consisting of small individual scars extending between the anterior and posterior adductor muscle scars. Common from intertidal areas to 100 fathoms.

6093

6094

6095

6097

6096

Subgenus *Pandorella* Conrad, 1863

Radial lines present on the external surface of the right valve. Right anterior cardinal tooth obsolete, left anterior cardinal tooth up against the dorsal margin of the valve. Internal lithodesma attached to the base of the resilium. Type: *arenosa* Conrad, 1834. Synonyms: *Kennerlia* Carpenter, 1964; *Kennerleya* "Carpenter" Fischer, 1887; *Kennerleyia* "Carpenter" Dall, 1903. *Panderella* is a misspelling.

Pandora arenosa Conrad, 1834   **6096**
Sand Pandora

North Carolina to Texas and Mexico.

⅓ to ¾ inch, somewhat like *trilineata,* but the shell is quite thick, has a very convex left valve, is beautifully, nacreously silver inside. Radial lines occur in the right valve. Shal-

low water to 20 fathoms alive; 640 fathoms, dead. Not uncommon. *P. carolinensis* Bush, 1885, is a synonym.

Pandora bushiana Dall, 1886   **6097**
Bush's Pandora

North Carolina to Florida, Texas and the West Indies. Brazil.

⅓ to ½ inch long, fragile, narrowly elongate; posterior dorsal margin very long, weakly convex. Living specimens are covered with a light-brownish periostracum. The left anterior cardinal tooth tends to be somewhat weakened and concave. Uncommon; 3 to 25 fathoms.

Subgenus *Heteroclidus* Dall, 1903

Left valve with the posterior cardinal ridge or tooth absent; right valve with a short posterior cardinal and a produced anterior cardinal ridge; both of the anterior teeth ridges ending in front of the anterior adductor scar. Lithodesma present. Type: *punctata* Conrad, 1837.

Pandora punctata Conrad, 1837   **6098**
Punctate Pandora

Vancouver Island to Baja California.

1 to 1½ inches, crescent-shaped with the posterior dorsal margin quite convex, and with small, distinct punctations on the inner surface of the valves. Abundant; low-tide line to 20 fathoms. Not infrequently washed ashore.

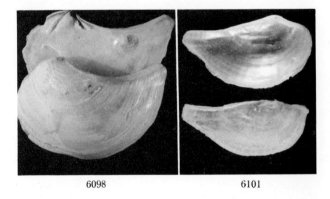

6098   6101

Pandora glacialis Leach, 1819   **6099**
Glacial Pandora

Arctic Seas to Gulf of Maine.
Alaska to British Columbia.

¾ inch, compressed, rather fragile. Left valve convex and centrally inflated, and the right valve (which overlaps the left valve along the posterior dorsal margin) is centrally flattened, sharply concave ventrally and has weak radial scratches. Left anterior cardinal tooth is coalesced with the anterior dorsal margin of the shell. Common; 3 to 130 fathoms on muddy sand bottoms. *P. eutaenia* (Dall, 1915), is a synonym.

Pandora inflata Boss and Merrill, 1965   6100
Inflated Pandora

New Jersey to both sides of Florida.

Similar to *glacialis*, ½ to ¾ inch, has 2 distinct radial carinate ridges running along the dorsal posterior region of the valve, and the middle of the ventral margin is roundly protruding downward. Common. 26 to 90 fathoms. Synonym: *brevis* Verrill and Bush, 1898, non Sowerby, 1829.

6099

6100

Pandora filosa Carpenter, 1864   6101
Western Pandora

Alaska to Ensenada, Baja California.

1 inch in length, thin but not too fragile, opalescent-white with a brownish border of periostracum. Interior opalescent. Valve semicircular in outline. Right valve almost flat and with a single, fairly large tooth which juts laterally. Left valve moderately convex. The posterior dorsal margin is almost straight, the posterior end somewhat drawn out into a rostrum. Moderately common from 10 to 75 fathoms.

(6102) *Pandora bilirata* Conrad, 1855 (Alaska to Baja California) is half as large, not rostrate posteriorly and with 2 strong radial ribs on the posterior dorsal margin of the left valve. Common; 20 to 142 fathoms on mud and clay.

(6103) *Pandora granulata* (Dall, 1915) (Guaymas to La Paz, Mexico) is much like *bilirata*, but half its size, more elongate and the 2 radial ribs are granulated. Rare.

Other Pacific species:

6104 *Pandora (Pandorella) grandis* Dall, 1877. Bering Sea to Siletz Bay, Oregon. 2 inches. 50 fms.

6105 *Pandora (Pandorella) forresterensis* Willett, 1918. Forrester Island; and Frederick Sound, Alaska, 12 to 50 fms.

6106 *Pandora (Pandorella) cornuta* C. B. Adams, 1852. Baja California to Panama. (Synonyms: *convexa* Dall, 1915; *acutedentata* Carpenter, 1864.)

6107 *Pandora (Clidiophora) arcuata* Sowerby, 1835. Baja California to Peru. (Synonyms: *cristata* Carpenter, 1864; *claviculata* Carpenter, 1855.)

6108 *Pandora (Frenamya* Iredale, 1930) *radiata* Sowerby, 1835. Gulf of California to Ecuador. (Synonym?: *Coelodon radians* Dall, 1915.)

Family THRACIIDAE E. A. Smith, 1885

Fragile shells with porcelaneous material. Surface usually granular. Hinge without teeth. Chondrophore directed obliquely toward the posterior end. Pallial sinus well-developed.

Genus *Thracia* Sowerby, 1823

Shells up to 4 inches in size, thin, chalky in texture, beaks so close that the *right one becomes punctured* by the left beak; ligament external; shell commonly moderately rostrate at the posterior end; right valve fatter than the left. Hinge without teeth. Ligament mostly internal. Resilium attached to a large slot or pit in the hinge plate. A small lithodesma may be present. Type: *pubescens* (Pulteney, 1799), from the Mediterranean. Synonyms: *Throna* Carpenter, 1859; *Osteodesma* Blainville, 1825; *Odoncineta* Costa, 1829; *Cetothrax* Iredale, 1949; *Homoeodesma* Fischer, 1887; *Crassithracia* Soot-Ryen, 1941.

Thracia conradi Couthouy, 1838   6109
Conrad's Thracia

Nova Scotia to Long Island Sound, New York.

3 to 4 inches in length, about ⅘ as high. Valves obese and chalky-white. Hinge without teeth, only thickened considerably behind the beak and below the large, wide, external ligament. Right beak always punctured by the beak in the left valve. Pallial sinus U-shaped but not very deep. Posterior end of valves slightly rostrate and with a weak radial ridge. Rarely washed ashore. Not uncommon offshore and down to 150 fathoms. Lives buried in the mud at a depth of about 6 inches on its side, with the flatter valve up. (M. L. Thomas, 1967, *The Nautilus*, vol. 80, p. 84.)

Thracia trapezoides Conrad, 1849   6110
Common Pacific Thracia

Alaska to Redondo Beach, California.

2 inches in length and not quite so high; thin, chalky and with the posterior end broadly rostrate. The beak of the right valve has a hole punctured in it by the beak of the left valve. The posterior rostrated part of the valve is set off by a broad, radial, depressed furrow which is bordered by a low, rounded, radial ridge. Color drab, grayish white. Commonly dredged off the west coast. Compare with *T. curta* Conrad.

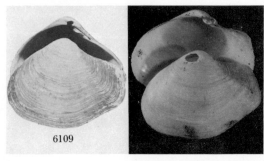

6109

6110

***Thracia curta* Conrad, 1837**   **6111**
Short Western Thracia

Alaska to Baja California, and to Ecuador.

1 to 1½ inches in length, very similar to *trapezoides*, but suboval and lacking the prominence of rostration. It is very close in shape to our illustration of the Atlantic *T. conradi*. A moderately common species. John Q. Burch reports that it is relatively abundant at San Onofre, California, in the rubbly reef at extreme low tides. It has also been taken from wharf pilings, and it is commonly dredged in over 20 fathoms on shale bottoms.

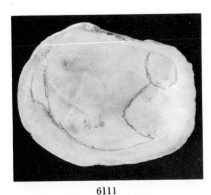

6111

***Thracia morrisoni* R. Petit, 1964**   **6112**
Morrison's Thracia

North and South Carolina.

15 to 18 mm., ovate-quadrate, white umbones slightly nearer the anterior end, inconspicuous, touching, with the right beak punctured by the left. Anterior end rounded; posterior end gaping and somewhat truncate. Right valve slightly fatter and bearing a weak radial rib on the dorsal posterior slope. Outer surface concentrically wrinkled and with sandlike granules. Pallial sinus large. Ventral margin of valves straight. Probably previously reported on our East Coast as *T. corbuloidea* Blainville, 1824. Uncommon; washed ashore. *Proc. Biol. Soc. Wash.*, vol. 77, p. 157.

Other species:

6113 *Thracia myopsis* Möller, 1842. Greenland to Massachusetts, 10 to 50 fms.

6114 *Thracia septentrionalis* Jeffreys, 1872. Greenland, 60 fms., to off Block Island, Rhode Island, 29 fms. (Synonym: *truncata* Mighels and Adams, 1842, not Brown, 1827.)

6115 *Thracia corbuloides* Blainville, 1824. North Carolina to Florida Keys, .14 to 50 fms. (is *morrisoni* Petit, 1964?); *philippiana* Nordsieck, 1969).

6116 *Thracia distorta* Montagu, 1808. Gulf of Mexico.

6117 *Thracia phaseolina* Lamarck, 1822. Florida Keys to Yucatan, Mexico. 640 fms.

6118 *Thracia stimpsoni* Dall, 1886. West coast of Florida, 640 fms. Off Freeport, Texas, 19 fms. 65 mm.

6119 *Thracia rugosa* Lamarck, 1818. West Indies.

6120 *Thracia nitida* Verrill, 1884. Off Chesapeake Bay, Delaware, 1,917 fms.

6121 *Thracia challisiana* Dall, 1915. Forrester Island, Alaska, to Monterey, California.

6122 *Thracia beringi* Dall, 1915. Bering Sea, Aleutian Islands, to Puget Sound, Washington.

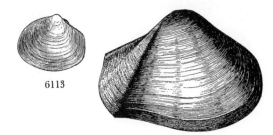
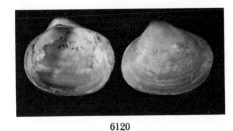

6113

6118

6120

Genus *Cyathodonta* Conrad, 1849

Similar to *Thracia*, but the right beak is without a round hole, the ligament is internal on a definite chondrophore and the valves are with oblique, concentric undulations. This is sometimes considered a subgenus of *Thracia*. There is only one species on the Pacific Coast of America, and one in the West Indies. Type: *undulata* Conrad, 1849.

***Cyathodonta undulata* Conrad, 1849**   **6123**
Wavy Pacific Thracia

Monterey, California, to Tres Marias Islands, Mexico.

1½ inches in length, subovate, very thin and fragile, white, and with obliquely concentric undulations which are largest at the anterior end, but disappear toward the posterior end of the shell. Minute, crowded, granulated, radial lines are also present. Uncommon. *C. dubiosa* Dall, 1915, and *C. pedroana* Dall, 1915, appear to be this species. Olsson, 1961, described the subspecies *peruviana*, occurring from Panama to Peru.

6123

***Cyathodonta semirugosa* (Reeve, 1859)**   **6124**
Wavy Caribbean Thracia

Greater Antilles and lower Caribbean.

¾ to 1 inch in length, subovate, very thin and fragile, white. Sculptured with concentric undulations. Posterior end gapes with flaring edges. Not uncommon; shallow water on mud bottoms.

Genus *Asthenothaerus* Carpenter, 1864

Small, elongate, earthy-white clams, with a granular surface, the ligament entirely internal and lodged just under the umbones and supported by a calcified, butterfly-shaped ossicle. Slightly gaping. Type: *villosior* Carpenter, 1864.

Asthenothaerus villosior Carpenter, 1864   6125
Little Rough Thracia

San Pedro, California, to Baja California.

10 mm. in length, 6 mm. in height, and 4 mm. in breadth of both valves; white to brownish and sometimes with orange stains on the margins. Posterior end truncated. Posterior dorsal area has a raised thread on each valve. Surface smoothish to slightly granular but irregular. Hinge feeble. Pallial sinus deep, reaching beyond the vertical of the beaks. Uncommon; in mud in shallow water. *Thracia diegensis* Dall, 1915, is a synonym.

Asthenothaerus hemphilli Dall, 1886   6126
Hemphill's Thracia

West coast of Florida.

$6 \times 6 \times 3$ mm., yellowish white, concentrically striate, with a very thin periostracum; left valve slightly smaller than the right, subovate, truncate and slightly gaping posteriorly. Surface with concentric growth irregularities and very fine granulations. Sinus deep and rounded. Ossicle bridge-shaped, wide, short and concave behind in the middle line. Uncommon; 2 to 17 fathoms. *A. balesi* Rehder, 1943 (**6126a**), from the Lower Florida Keys is a little more elongate in shape and may be a synonym.

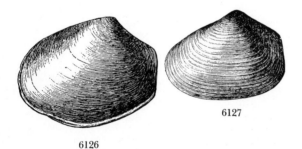

6126   6127

Genus *Bushia* Dall, 1886

Similar to *Asthenothaerus,* but porcelaneous and the toothless hinge has above it, and nestled between the beaks, a very large U-shaped shelly ossicle. Ligament also external. Shells not gaping. Right valve larger and overlapping the left. Two rare species known in American waters:

6127 *Bushia elegans* (Dall, 1886). Type of the genus. 56 to 76 fms., Florida Straits to Barbados. 12.5 mm.

6128 *Bushia panamensis* Dall, 1890. 51 fms., Panama Bay, Pacific side. Illustrated in Keen's "Sea Shells of Tropical West America," sp. 592.

Family PERIPLOMATIDAE Dall, 1895

Genus *Periploma* Schumacher, 1817

Shell small, oval, right valve fatter than the left, with a slight pearly sheen; hinge with a narrow, oblique spoon and a small, free, triangular lithodesma; ligament absent; an internal rib proceeds from under the hinge to the posterior margin. Anterior muscle scar long and narrow, the posterior one small and ovate. Type: *inaequivalvis* Schumacher, 1817, which is *margaritaceum* (Lamarck, 1801).

Periploma papyratium (Say, 1822)   6129
Paper Spoon Clam

Labrador to Rhode Island.

½ to 1 inch in length, oval, moderately compressed, thin-shelled, and dull-white with a thin, yellowish gray periostracum. Beaks slit or broken by a short, radial break. Spoonlike chondrophore faces downward and is reinforced by a sharp, curved rib which runs to the inner surface of the valve in a ventral direction. Sculpture of irregular, fine, concentric growth lines. A weak radial groove runs from the beak to the anterior part of the ventral edge. Moderately common in dredge hauls from 1 to 200 fathoms. *P. papyraceum* is an incorrect spelling for this species.

(**6130**) *Periploma fragile* (Totten, 1835), the fragile spoon clam (Labrador to New Jersey, 4 to 40 fathoms), differs in being more rostrate anteriorly, the beaks pointing more forward and placed more anteriorly. Its chondrophore is more horizontal to the hinge line.

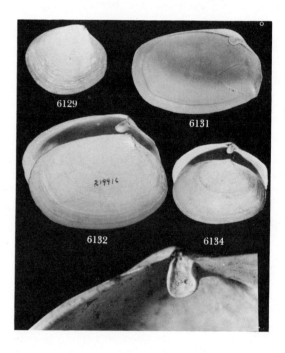

6129

6131

6132

6134

Periploma margaritaceum (Lamarck, 1801)   6131
Unequal Spoon Clam

South Carolina to Florida and to Texas.

¾ to 1 inch in length, oblong, the left valve more inflated and slightly overlapping the right valve. Fragile and pure-white. Beaks close together, each with a short, radial break or slit in the surface. An oblique, low keel runs from the beaks to the anterior ventral margin of the valve. The keel is bounded posteriorly by a groove. Sculpture consists of microscopic, concentric scratches. Hinge with a single, large, spoon-shaped tooth or chondrophore, above which is a deep slit where the small, free, triangular lithodesma fits. This species is especially abundant along certain Texas beaches. *P. inequale* C. B. Adams, 1842, is a synonym. *P. inaequivalve* Schumacher, 1817, from the West Indies has not been found in the United States, despite several erroneous records.

Periploma planiusculum (Sowerby, 1834)   **6132**
Common Western Spoon Clam

Point Conception, California, to Peru.

1 to 1¾ inches in length, ovate, thin, with weak, concentric lines of growth. Right valve fatter than the left; chondrophore ovate-trigonal, its longer diameter directed forward and reinforced posteriorly by an elongate, riblike buttress. Commonly washed ashore on southern Californian beaches. Synonyms are: *lenticulare* (Sowerby, 1834); *argentarium* (Conrad, 1837); *obtusum* (Hanley, 1842); *excurvum* (Carpenter, 1855).

Periploma discus Stearns, 1890   **6133**
Round Spoon Clam

Monterey, California, to La Paz, Baja California.

1 to 1½ inches in length, similar to *planiusculum*, but almost circular in outline, except that the posterior end is slightly lengthened into a short, broad, blunt rostrum. Uncommonly dredged in mud bottom at several fathoms; rarely washed ashore after storms.

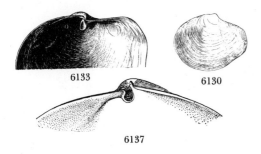

6133   6130

6137

Subgenus *Cochlodesma* Couthouy, 1839

The prominent clavicular rib supporting the chondrophore is absent, and in front of the chondrophores there is a cartilaginous dark-brown mass uniting the valves and taking the place of the lithodesma. Type: *leanum* (Conrad, 1831). Synonym: *Aperiploma* Habe, 1952.

Periploma leanum (Conrad, 1831)   **6134**
Lea's Spoon Clam

Nova Scotia to off North Carolina.

1 to 1½ inches in length, ovate, quite compressed, fairly fragile and white in color. Smoothish. The beaks located near the center of the dorsal edge of the valves have a natural, radial crease at the anterior end. Chondrophore large, points ventrally and is reinforced anteriorly by a low, sturdy ridge. The muscle scar above the pallial sinus is commonly quite silvery. Periostracum thin and yellowish. Uncommon just offshore.

Other species:

6135 *Periploma anguliferum* (Philippi, 1847). Georgia to the Florida Keys and to Texas.

6136 *Periploma tenerum* Fischer, 1882. North Carolina to the Florida Keys.

6137 *Periploma undulatum* Verrill, 1885. Off New Jersey to North Carolina, 541 to 816 fms.

6138 *Periploma affine* Verrill and Bush, 1898. Off Martha's Vineyard, Massachusetts, 100 to 115 fms.

6139 *Periploma alaskanum* Williams, 1940. Montague Island to Prince William Sound, Alaska, 25 fms. *Pomona College Journ. Entomol.* for June, 1940.

6140 *Periploma (Halistrepta* Dall, 1904) *sulcatum* Dall, 1904. San Pedro Bay, California. Type of the subgenus.

6141 *Periploma stearnsii* Dall, 1896. Gulf of California.

Order *Septibranchoidea* Pelseneer, 1889

Superfamily Poromyacea Dall, 1886

Ovate to round shells, rarely gaping. Hinge with poorly developed cardinal and lateral teeth. Resilium reinforced by a lithodesma. Mantle lobes united. Gills scantily reticulate or even absent.

Family POROMYIDAE Dall, 1886

Genus *Poromya* Forbes, 1844

Shell small, fragile; interior pearly and without a deep pallial sinus; sculpture of fine granules in radial series. Hinge of right valve with a strong cardinal tooth in front of a wide chondrophore; hinge of left valve with a small cardinal tooth behind and above the chondrophore. Ligament external and internal. No lithodesma present. Type: *granulata* Nyst and Westendorp, 1839. *Thetis* H. and A. Adams, 1856, non Sowerby, 1826, is a synonym.

Poromya granulata (Nyst and Westendorp, 1839)   **6142**
Granular Poromya

Arctic Seas to the West Indies.
Mediterranean.

¼ to ½ inch in length, ovate, inflated and fragile. Beaks inflated and turned forward. Exterior cream-white, with an irregular coating of fine granules which resemble sugar-coating. In fresh material, this granular deposit is also found on the inner margins of the valves. Slightly gaping at the posterior end. Interior of valves silvery-white. Commonly dredged in a few fathoms of water off eastern Florida. *P. rotundata* Jeffreys, 1876, is probably a synonym.

Poromya rostrata Rehder, 1943   **6143**
Rostrate Poromya

North Carolina to the West Indies.

⅓ inch, similar to *granulata*, but is distinctly rostrate posteriorly and the granules are larger, more evenly spaced and generally cover the entire outer shell. Uncommon; 60 to 100 fathoms.

6143

Subgenus *Dermatomya* Dall, 1889

Shell not granular; pallial sinus slightly developed. Hinge strong. Type: *mactroides* Dall, 1889. *Dermatomaya* is a 1908 misspelling by Dall.

Poromya tenuiconcha (Dall, 1913)   6144
Smooth-shelled Poromya

Alaska to Baja California.

½ inch, thin-shelled, obese, olivaceous, the pearly luster showing through the periostracum. Beaks nearer the rounded anterior end. Beaks prominent and curled forward, with a depression in front of them. Interior pearly, brilliant. Hinge in the left valve with a small internal resilium seated on an inconspicuous oblique chondrophore, with a notch immediately in front of it, into which fits a projecting denticle on the corresponding part of the opposite valve. Not uncommon; rocky bottom, 30 to 659 fathoms.

Other Atlantic species:

6145 *Poromya neaeroides* Seguenza, 1876. Gulf of Mexico, 114 fms. North Atlantic and off Virginia, 122 to 1,635 fms.

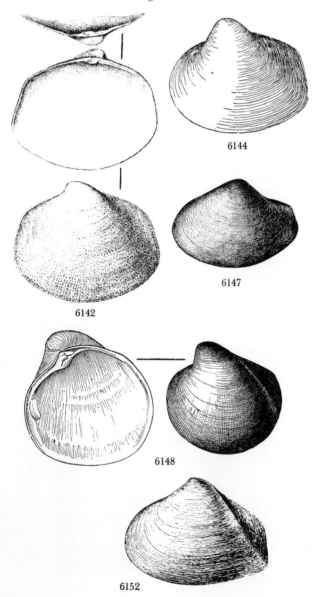

6144

6147

6142

6148

6152

— *Poromya granatina* Dall, 1881, is *Basterotia quadrata*.

6146 *Poromya* (*Cetomya* Dall, 1889) *albida* Dall, 1886. Off Cape Fear, North Carolina, and Florida Strait, 98 to 731 fms.

6147 *Poromya* (*Cetomya*) *elongata* Dall, 1886. Gulf of Mexico northwest of Cuba, 199 fms. Type of the subgenus.

6148 *Poromya* (*Cetomya*) *tornata* (Jeffreys, 1876). North Atlantic to West Indies, 1,140 to 1,785 fms. 36 miles southeast of Bermuda, 1,700 fms. Indian Ocean. (Synonyms: *microdonta* Dall, 1890; *sublevis* Verrill, 1884; *isocardioides* Dautzenberg and Fischer, 1897.) See Knudsen, 1970, p. 124. 11 mm.

Other Pacific species:

6149 *Poromya* (*Dermatomya*) *trosti* Strong and Hertlein, 1931. San Clemente Island to San Diego, California. *Proc. Cal. Acad. Sci.*, ser. 4, vol. 22, p. 163.

6150 *Poromya* (*Dermatomya*) *buttoni* (Dall, 1916). Off Monterey, California, 581 fms., in mud.

6151 *Poromya* (*Dermatomya*) *beringiana* (Dall, 1916). Aleutian Islands to Tillamook, Oregon.

6152 *Poromya* (*Dermatomya*) *leonina* (Dall, 1916). Off Washington, 877 fms.

6153 *Poromya* (*Dermatomya*) *canadensis* Bernard, 1969. Off Vancouver Island, British Columbia, 534 fms. *Jour. Fish. Research Board Canada*, vol. 26, p. 2232.

Genus *Cetoconcha* Dall, 1889

Shell differing from *Poromya* by the cartilage being almost external and the fossettes smaller in size and upturned, the external ligament consequently nearly obsolete. Teeth in hinge reduced, except for the right cardinal which is more evident in young specimens. Type: *bulla* (Dall, 1881). A genus of small, very deep-water clams, four of which are recorded from North America:

6154 *Cetoconcha bulla* (Dall, 1881). Off Virginia to the Gulf of Mexico, 1,717 to 1,920 fms. 10 mm.

6154a *Cetoconcha atypha* Verrill and Bush, 1898. Off Virginia, 1,423 fms.

6155 *Cetoconcha margarita* (Dall, 1886). Off Florida Keys and West Indies (not Bermuda), 391 to 1,019 fms. 7.3 mm.

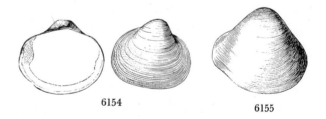

6154   6155

6156 *Cetoconcha malespinae* Ridewood, 1903. Southwest of Sitka, Alaska, 1,579 fms. Later fully described by Dall in 1916.

6157 *Cetoconcha smithii* Dall, 1908. Off Acapulco, Mexico. Deep water.

Family VERTICORDIIDAE Stoliczka, 1871

Genus *Verticordia* Sowerby, 1844

Shells about 5 mm., white, radially ribbed, suborbicular, equivalve, the beaks rolled forward above a deep, entering lunular indentation. Nacreous within. Hinge of right valve with a stout, conical cardinal tooth. No laterals. Ligament internal, supported by a lithodesma. Foot stopper-shaped. Pallial line simple. Type: *cardiiformis* Sowerby, 1844, of the Pliocene of England.

Subgenus *Trigonulina* Orbigny, 1842

Ribs widely spaced. Right valve with a long, posterior lateral pocket. Type: *ornata* (Orbigny, 1842).

Verticordia ornata (Orbigny, 1842)   6158
Ornata Verticord

Massachusetts to Florida and the West Indies. Bermuda. Brazil.
Catalina Island, California, to Panama.

¼ inch in length, oval to round, compressed and with about a dozen strong, sharp, curved radial ribs on the anterior ¾ of the valve. The ribs extend beyond the ventral margin to give a strongly crenulate margin. Exterior dull and cream-white; interior very silvery. Commonly dredged off our east coast from 5 to 200 fathoms. *V. caelata* Verrill, 1884, is a synonym.

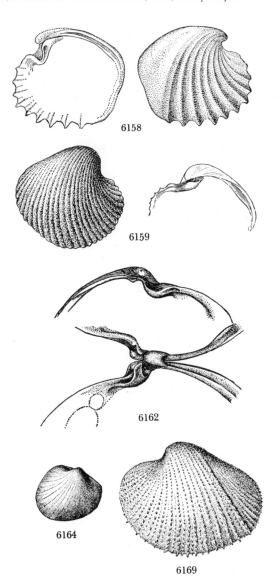

6158

6159

6162

6164

6169

Subgenus *Haliris* Dall, 1886

Shell globose, ossicle short and squarish; lunule present, but not deep; left valve with a small but distinct cardinal tooth and a short, stout lateral tooth near the umbo. Type: *fischeriana* Dall, 1881.

Verticordia fischeriana Dall, 1881   6159
Fischer's Verticord

North Carolina to Gulf of Mexico to Barbados.

⅓ to ½ inch, similar to *ornata*, but fatter, with 28 small, finely beaded or serrated, radial ribs over the entire surface of the valve. Uncommon; 10 to 229 fathoms.

Other species:

6160 *Verticordia acuticostata* Philippi, 1884. Southern Florida and the West Indies, 71 to 600 fms.

6161 *Verticordia woodii* E. A. Smith, 1885. Gulf of Mexico and the West Indies, 100 to 1,060 fms.

6162 *Verticordia granulifera* (Verrill, 1885). Off New Jersey to Virginia, 1,356 to 1,859 fms. 8 mm.

6163 *Verticordia seguenzae* Dall, 1886. Off North Carolina to the Gulf of Mexico, 124 to 640 fms.

6164 *Verticordia perversa* Dall, 1886. Off Cape Fear, North Carolina, 731 fms.

6165 *Verticordia aequacostata* Howard, 1950. Off Catalina Island, California, and Angel de la Guarda Island, Gulf of California. *The Nautilus*, vol. 63, p. 109.

6166 *Verticordia trapezoides* Seguenza, 1876. Off North Carolina to Florida, 66 to 162 fms.

6167 *Verticordia* (*Trigonulina*) *hancocki* Bernard, 1969. Off Gorgona Island, Colombia, 40 to 60 fms. *Jour. Fish. Research Board Canada*, vol. 26, p. 2233.

6168 *Verticordia* (*Haliris*) *spinosa* Bernard, 1969. Off Baja California, 150 fms. *Jour. Fish. Research Board*, vol. 26, p. 2233.

Genus *Euciroa* Dall, 1881

Similar to *Verticordia*, but more rounded, more inflated, brilliantly pearly within; exterior frosty dull and white, and with a thin, pale periostracum. Foot laterally compressed; both pairs of labial palps are present and free. For anatomy, see W. H. Dall, 1895, *Proc. U.S. Nat. Mus.*, vol. 17, p. 687. Type: *elegantissima* (Dall, 1881).

6169 *Euciroa elegantissima* (Dall, 1881). Off Cape Canaveral, Florida, to Cuba, 292 to 756 fms. 13 mm.

Genus *Halicardia* Dall, 1895

Shell wide and angular, with a granular ashy-white or pale-brown surface, showing faint traces of radiating ridges. Hinge obsolete, an obscure swelling representing the sub-lunular tooth in the right valve. Right portion of the lunule is most prominent. Lithodesma is a slender, solid, shelly arch. Type: *Mytilimeria flexuosa* Verrill and Smith, 1881. For anatomy, see W. H. Dall, 1895, *Proc. U.S. Nat. Mus.*, vol. 17, p. 697. *Halicardissa* Dall, 1913, is a synonym.

6170 *Halicardia flexuosa* (Verrill and Smith, 1881). Off Martha's Vineyard, Massachusetts, 75 to 349 fms.; off Georges Bank, Massachusetts, 677 fms. Eastern Europe, off Sahara, 2,330 meters.

6170

6171 *Halicardia perplicata* (Dall, 1890). Off British Columbia to the Galapagos Islands, 1,110 to 1,486 meters.

Genus *Lyonsiella* G. Sars, 1872

Shell small, thin; lunule faint or none; ossicle semicylindrical, forked behind; external ligament almost none; right valve without teeth, its lunular edge a little produced and thickened; left valve with an elongate obscure thickening of the hinge margin under the beak. Usually with fine radial ribs. Muscle scars unequal. Type: *abyssicola* G. Sars, 1872. Synonyms: *Laevicordia* Seguenzia, 1876; *Lysonsiela* Thiele, 1912; *Lyonciella* Friele, 1886. Most species are in very deep water:

6172 *Lyonsiella uschakovi* Gorbunov, 1946. Arctic seas; Siberia; North Canadian Basin. 807 to 1,208 fms.

6173 *Lyonsiella abyssicola* (G. Sars, 1872). South of Martha's Vineyard, Massachusetts, 192 to 500 fms. Northern Europe, 38 to 2,000 meters.

6174 *Lyonsiella cordata* Verrill and Bush, 1898. Off Massachusetts to Virginia, 1,423 to 1,825 fms. 15 mm.

6175 *Lyonsiella granulifera* Verrill, 1885. Off Chesapeake Bay, 1,423 fms.

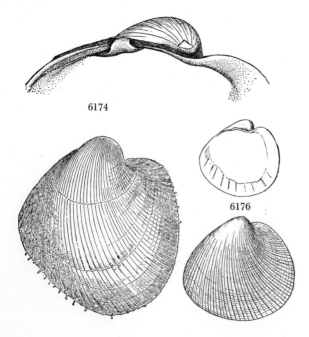

6174

6176

6176 *Lyonsiella gemma* Verrill, 1880. South of Martha's Vineyard, Massachusetts, 487 fms. 4 mm.

6175

6177 *Lyonsiella subquadrata* Jeffreys, 1881. Off Virginia, 1,825 fms.

6178 *Lyonsiella magnifica* Dall, 1913 and 1923. Off Mazatlan, Mexico, and Baja California. See *Proc. U.S. Nat. Mus.*, vol. 66, pl. 23.

6179 *Lyonsiella quaylei* Bernard, 1969. Off Vancouver Island, British Columbia, to Catalina Island, California, 363 fms. 8 mm. *Jour. Fish. Research Board Canada*, vol. 26, p. 2230.

Genus *Policordia* Dall, Bartsch and Rehder, 1939

Small, ovate, nearly smooth, with fine radial ribs; ligament in groove, with lithodesma below it; hinge edentulous; pallial sinus small, not deep. Type: *diomedea* Dall, Bartsch and Rehder, 1939.

6180 *Policordia alaskana* (Dall, 1895). Southwest of Sitka, Alaska, 1,659 fms.; off Catalina Island, California, 600 fms.

6180

6180 *Policordia insculpta* (Jeffreys, 1881). Off Martha's Vineyard, Massachusetts, 75 to 487 fms.

Family CUSPIDARIIDAE Dall, 1886

Genus *Cuspidaria* Nardo, 1840

Shell small, globose in front, rostrate behind. Hinge with a posterior lateral tooth in the right valve. External ligament elongated. Resilium in a small, spoon-shaped fossette which is posteriorly inclined and attached to the hinge margin by its posterior edge. Lithodesma distinct and semicircular. Pallial sinus absent. Type: *cuspidata* (Olivi, 1792). Synonym: *Neaera* Gray, 1834, not Robineau-Desvoidy, 1830.

***Cuspidaria glacialis* (G. O. Sars, 1878)**   **6181**
Glacial Cuspidaria

Nova Scotia and northern Europe.
Alaska and Arctic Seas.

1 to 1½ inches in length, rostrum moderately long and compressed laterally. Main part of valves fat and round. Sculpture

consists of small, irregular growth lines. Periostracum grayish white. Shell cream to white. A common species dredged from 64 to over 1,400 fathoms. *C. jeffreysi* in the south is very similar, but smaller, with a rostrum which in cross-section is much more oval and less compressed, and with hardly any periostracum. Erroneously reported by Dall from the Gulf of Mexico and off California.

Cuspidaria rostrata (Spengler, 1793)   6182
Rostrate Cuspidaria

Arctic Seas to the West Indies.

½ to 1 inch in length, with a tubelike rostrum which is ½ the length of the entire shell. The rostrum points slightly downward. Shell fairly smooth with moderately coarse, concentric growth lines. Whitish in color and sometimes with granular lumps of gray mud attached to the rostrum. Moderately common in deep water (65 to over 1,600 fathoms).

Cuspidaria jeffreysi (Dall, 1881)   6183
Jeffrey's Cuspidaria

Southern Florida and the West Indies.

⅓ inch in length, smoothish with only fine lines of growth. Rostrum moderately long; main part of shell round and fat. Similar to *glacialis* in the north which, however, is larger, more compressed, and whose rostrum points slightly downward instead of directly posteriorly as in this species. Creamy-white in color. Uncommonly dredged in water 100 to 687 fathoms off Miami.

Other species:

6184 *Cuspidaria wollastoni* E. A. Smith, 1885. 32 miles southwest of Bermuda, 1,000 fms.

6185 *Cuspidaria obesa* (Lovén, 1846). Arctic Ocean to the West Indies, 20 to 1,290 fms.

6186 *Cuspidaria pellucida* Stimpson, 1853. Gulf of St. Lawrence to Casco Bay, Maine, 40 to 95 fms.

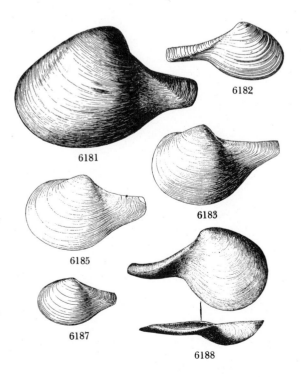

6187 *Cuspidaria microrhina* Dall, 1886. Off east and west coast of Florida, 100 to 509 fms. 40 mm.

6188 *Cuspidaria lamellosa* Sars, 1878 (*N. jugosa* Sars, 1878, not S. Wood, 1856). Off Rhode Island and New Jersey, 319 to 555 fms. 7 mm.

6189 *Cuspidaria media* Verrill and Bush, 1898. South of Martha's Vineyard, Massachusetts, 63 to 155 fms.

6189

6190 *Cuspidaria arctica* (Sars, 1878). Arctic Ocean to Nova Scotia, 190 fms., and off Fernandina, Florida, 294 fms. 20 mm.

6191 *Cuspidaria undata* Verrill, 1884. Off Chesapeake Bay, 2,221 fms.

6192 *Cuspidaria turgida* Verrill and Bush, 1898. Off Delaware Bay, 1,825 fms.

6193 *Cuspidaria arcuata* Dall, 1881. Gulf of Mexico, 640 fms.

6194 *Cuspidaria gigantea* Verrill, 1884. Off Chesapeake Bay, 1,917 fms.

6194

6195 *Cuspidaria alternata* Orbigny, 1846. Off Florida and the West Indies, 84 to 152 fms.

6196 *Cuspidaria parva* Verrill and Bush, 1898. Off Cape Cod, . Massachusetts, 515 to 1,290 fms.

6197 *Cuspidaria ventricosa* Verrill and Bush, 1898. Off Cape Cod, Massachusetts, 349 to 1,769 fms.

6198 *Cuspidaria formosa* Verrill and Bush, 1893. Off Cape Cod, Massachusetts, 1,188 fms.

6199 *Cuspidaria fraterna* Verrill and Bush, 1893. Off Cape Cod, Massachusetts, and Virginia, 302 to 984 fms. 8 mm.

6200 *Cuspidaria subtorta* (Sars, 1878). Off Nova Scotia,. 130 fms. 7 mm.

6201 *Cuspidaria subglacialis* Dall, 1913. Off California; deep water.

6202 *Cuspidaria apodema* Dall, 1916. Off Sitka, Alaska, to Panama. Common off southern California, 25 to 75 fms.

6203 *Cuspidaria chilensis* Dall, 1890. Off Oregon, 277 fms., to Chile, 1,036 fms.

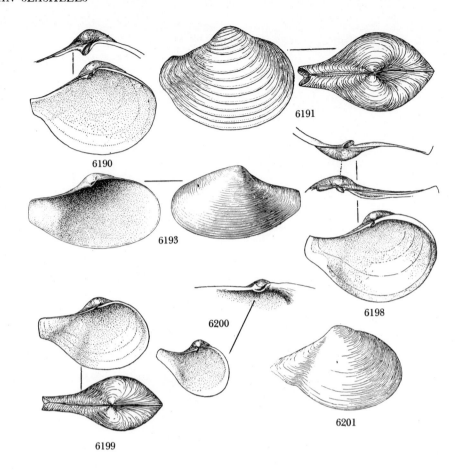

6190   6191   6193   6198   6200   6201   6199

6204 *Cuspidaria parkeri* Knudsen, 1970. Gulf of California (22°N, 109°W), 2,790 meters. *Galathea Report,* vol. 11, p. 150.

6205 *Cuspidaria parapodema* Bernard, 1969. Southern California to the Gulf of California, 53 to 275 meters.

Genus *Cardiomya* A. Adams, 1864

With strong, sharp radial ribs; fossette more vertical and prominent, otherwise like *Cuspidaria.* Type: *gouldiana* (Hinds, 1843).

Cardiomya costellata (Deshayes, 1837)   **6206**
Costate Cuspidaria

North Carolina to Florida and the West Indies.

⅓ inch in length, fragile, with a short rostrum, and with a few prominent radial ribs just in front of the rostrum which give the ventral margin of the valve in that area a scalloped edge. Anteriorly, the radial ribs are closer together, but weaker, and are rarely present at the anterior end of the shell. Additional ribs may develop in older specimens and become more even in size. (*C. multicostata* Verrill and Smith, 1898 (**6207**) may be the old form and *C. gemma* Verrill and Bush, 1898, (**6208**) the young form.) *C. corpulenta* (Dall, 1881), is a synonym. Commonly dredged off eastern Florida, 2 to 205 fathoms.

Cardiomya pectinata (Carpenter, 1864)   **6209**
Pectinate Cuspidaria

Puget Sound, Washington, to Panama.

About 6 to 10 mm. in length; anterior end globular, bearing 8 to 12 strong radial ribs (there may be smaller ribs between

the main ones). Posterior end drawn-out like a short handle and bearing 2 to 4 weak, longitudinal riblets. Color dull-gray with a glossy, grayish white interior. Commonly dredged offshore, 10 to 25 fathoms. The variety *beringensis* Leche is a synonym.

Cardiomya californica (Dall, 1886)   **6210**
Californian Cuspidaria

Puget Sound, Washington, to San Diego, California.

About 6 to 7 mm. long, differing from *C. pectinata* by its proportionately greater length; larger number of ribs (16 to 20); its straighter, longer rostrum with only 2 strong radiating lirae extending to the lower extreme (*pectinata* has none, or only several fine ones near the body of the valves); its less inflated shape, and thinner periostracum. Not uncommon; in mud bottom from 20 to 50 fathoms.

Cardiomya isolirata Bernard, 1969   **6211**
Evenly-lirate Cuspidaria

Point Loma, California, to Baja California.

8 mm., slightly inequivalve, anterior rounded; rostrum short, truncate, distally gaping; dorsal margin concave. With 12 to 20 broad, rounded ribs. Beaks inflated. Periostracum grayish. Interior shiny. Differs from *balboae* (Dall) in being a heavier shell with a denticulated margin. Dredged 55 to 183 meters. *Jour. Fish. Research Board Canada,* vol. 26, p. 2231.

Other Atlantic species:

6212 *Cardiomya ornatissima* (Orbigny, 1842). North Carolina to Yucatan and the West Indies, 2 to 124 fathoms. Brazil. (Synonym: *costata* Bush, 1885.) 9 mm.

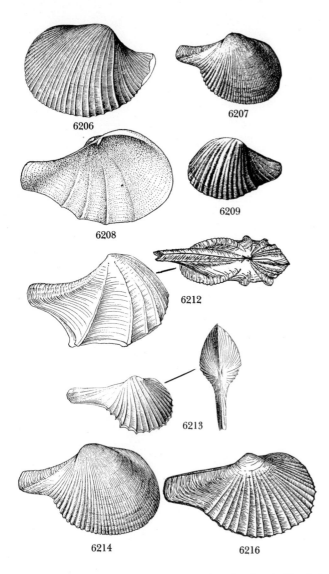

6206   6207
6208   6209
6212
6213
6214   6216

6213 *Cardiomya perrostrata* (Dall, 1881). South of Martha's Vineyard, Massachusetts, to the West Indies, 58 to 416 fms. Brazil. See "Caribbean Seashells," pl. 44, fig. d.

6214 *Cardiomya striata* (Jeffreys, 1876). Arctic Ocean to Florida Strait, 85 to 1,450 fms. Gulf of Mexico, 100 to 220 fms. 19 mm.

6215 *Cardiomya glypta* (Bush, 1885). Off Cape Hatteras, North Carolina, 48 fms.; West Indies, 2 to 124 fms.

6215

6216 *Cardiomya abyssicola* (Verrill and Bush, 1898). Off Massachusetts and North Carolina, 1,685 to 1,813 fms. 7 mm.

Other Pacific species:

— *Cardiomya beringensis* (Leche, 1883). Bering Sea to Kodiak Island, Alaska, to Panama Bay, 15 to 75 fms. is *pectinata*.

6218 *Cardiomya oldroydi* (Dall, 1924). Vancouver Island, British Columbia, to Catalina Island, California, 70 fms.

6219 *Cardiomya planetica* (Dall, 1908). Pribilof Islands, Bering Sea to Baja California, 10 to 100 fms.

6220 *Cardiomya balboae* Dall, 1916. Cortez Bank, California, 60 fms., to Catalina Island, California, 50 fms.

6221 *Cardiomya lanieri* (Strong and Hertlein, 1937). Both sides of Baja California to Ecuador, 15 to 238 meters.

Genus *Plectodon* Carpenter, 1864

Having the shape of a *Cuspidaria;* surface granular; there is a twisting of the dorsal hinge margin, just under the beaks, forming a toothlike prominence. The cartilage is inserted just behind and under the beaks. Lateral teeth laminated. Type: *scabra* Carpenter, 1864. Two American species:

6222 *Plectodon scaber* Carpenter, 1864. Puget Sound, Washington, to Panama; in 30 to 160 fms. 24 mm. long.

6222a *Plectodon granulatus* (Dall, 1881). Florida Keys to the West Indies, 54 to 118 fms. (Synonym: *velvetina* Dall, 1886.) 12 mm. long.

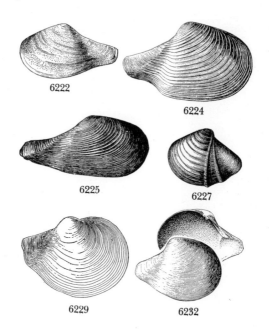

6222   6224
6225   6227
6229   6232

Genus *Myonera* Dall and Smith, in Dall, 1886

Similar to *Cuspidaria* (and considered a subgenus of it by some workers), but its hinge is without teeth. Surface of valves with radiating and concentric sculpture. Fossette vertical or posteriorly directed. Siphonal opening with numerous tentacular filaments and several ocelli. No gills or palpi known. Type: *paucistriata* Dall, 1885. Deep-water species:

6223 *Myonera tillamookensis* Dall, 1916. Off Tillamook Bay, Oregon in 786 fms.

6224 *Myonera lamellifera* (Dall, 1881). Off Cedar Keys to the West Indies, 84 to 250 fms. 12 mm.

6225 *Myonera limatula* (Dall, 1881). **Off** Nantucket, Massachusetts, 547 fms; Florida Strait, 539 fms. 11 **mm**.

6226 *Myonera gigantea* (Verrill, 1884). Off Virginia, 1,250 to 1,917 fms.

6227 *Myonera paucistriata* Dall, 1886. North Carolina to the West Indies, 193 to 880 fms. 10 mm.

6228 *Myonera undata* (Verrill, 1884). Off Chesapeake Bay, 2,221 fms. Florida Strait, 450 fms. (synonym: *lucifuga* Locard, 1898).

6229 *Myonera ruginosa* (Jeffreys, 1881). Off Nantucket, Massachusetts, 1,813 fms. 9 mm.

6230 *Myonera (?) pretiosa* Verrill and Bush, 1898. Off east coast of Florida, 338 fms.

Genus *Leiomya* A. Adams, 1864

Shell obese, ventricose, thin, semitranslucent, with a beak at the posterior end. Hinge with an external cartilage pit. Right valve with a bifid anterior cardinal; left valve with 1 cardinal. 2 strong lateral teeth. Type: *adunca* (Gould, 1861). Several deep-water species:

6231 *Leiomya* (*Rhinoclama* Dall and Smith, in Dall, 1886) *halimera* Dall, 1886. Off Cape Fear, North Carolina, 731 fms.

6232 *Leiomya* (*Halonympha* Dall and Smith, in Dall, 1886) *claviculata* (Dall, 1881). Florida Keys, Bermuda and West Indies, 100 to 339 fms. 12 mm. Type of the subgenus.

6233 *Leiomya* (*Halonympha*) *striatella* Verrill and Bush, 1898. Off east Florida, 338 fms.

The Minor Extinct Classes

Class ROSTROCONCHIA
Pojeta, Runnegar, Morris and Newell, 1972

An extinct class of bivalved mollusks having an uncoiled univalved larval shell; an untorted bivalved adult shell; no hinge teeth, ligament or adductor muscles; and a fused, almost inflexible, hinge. Occur in the Early Ordovician to the Late Permian. They were considered a class by L. R. Cox (1960), but are now regarded as a class more closely related to the Bivalvia and Scaphopoda than to other known classes of mollusks (see *Science,* vol. 177, no. 4045, 21 July, 1972, p. 264). The class contains such genera as *Eopteria* Billings, 1865, *Euchasma* Billings, 1865, from Canada, and *Conocardium* Bronn, 1835, from Europe and North America.

Class STENOTHECOIDA Yochelson, 1969

An extinct class of bivalved mollusks whose members have asymmetrical valves with pronounced beaks. The class Probivalvia Aksarina, 1968, is a synonym.

Class MATTHEVA Yochelson, 1966

A Cambrian fossil consisting of two conical shells, one anterior, the other posterior. Contains the genus *Matthevia* Wallcott, 1885, from the United States (see U.S. Geological Survey Professional Paper 523-B, 1966).

VI

Squid, Octopus and Cuttlefish

Class CEPHALOPODA Cuvier, 1797

The taxonomic literature of this group is extensive, but a very useful series of references may be found in the bibliographies in C. F. E. Roper, R. E. Young and G. L. Voss, 1969, *Smithsonian Contrib. Zool.*, no. 13; and in R. E. Young, 1972, *Smithsonian Contrib. Zool.*, no. 97. Many of our illustrations are taken from these valuable publications.

Subclass *Nautiloidea*

This subclass, which includes the chambered nautilus and about 5,000 species of fossil and extinct Ammonites, is not represented in American waters. The living species of *Nautilus* from the Indo-Pacific are characterized by a large, chambered, external shell, by 2 pairs of gills, and by their numerous suckerless arms.

Subclass *Coleoidea*

All of the living cephalopods with the exception of *Nautilus* belong to this subclass which is characterized by animals that have 1 pair of gills and 8 or 10 arms which bear rows of suckers, and whose shell is internal or entirely absent. Consult "A review of the Cephalopods of the Gulf of Mexico," by G. L. Voss, 1956, *Bull. Marine Sci.*, vol. 6, which contains keys and illustrations. See also Malcolm R. Clarke, 1966, for "A review of the systematics and ecology of oceanic squids," *Advances in Marine Biol.*, vol. 4, pp. 91–300.

With 10 arms, 2 of which are the long tentacular arms; body long and cylinder-shaped. An internal shell is present in most cases, and may be calcareous (the cuttlebone) or thin and horny (squid pen). The small suckers on the arms are usually set on small stalks or peduncles and their apertures are armed with horny rings or hooks.

Order *Sepioidea* Leach, 1817

Family SPIRULIDAE Rafinesque, 1815

Genus *Spirula* Lamarck, 1799

Type of the genus is *spirula* (Linné, 1758). Synonyms are *Lituina* Link, 1806, and *Litnus* Gray, 1849.

***Spirula spirula* (Linné, 1758)**　　　　　6235
Common Spirula

Cape Cod to the West Indies. Worldwide.
Bermuda.

The rather fragile, white shell is a chambered cone coiled in a flat spiral, usually less than 1 inch in diameter and with the coils not in contact. Each small chamber in the shell is divided from its neighbor by a nacreous-white, concave, fragile septum or wall. There is a small siphonal tube running back into the shell and piercing the septa. These shells are cast up on the beaches quite commonly. The body is short and cylindrical, and surrounds the shell completely. The 8 sessile arms and 2 pedunculated tentacular arms are very short. Lives at depths between 60 and 500 fathoms.

6235

Family SEPIOLIDAE Leach, 1817

Genus *Rossia* Owen, 1835

Short, "tubby" animals whose bodies are rounded at the end. The mantle edge is free all around. 8 arms short with 2 to 4 rows of spherical suckers which have smooth, horny rims. The 2 tentacular arms can be almost entirely withdrawn. The internal pen is slender, lanceolate and very thin and delicate. The rather large, semicircular fins are on the middle of the sides of the body. Eye with small eyelid on the lower side, none above. No sulcus or notch on front of the eye. Occasionally arms and tentacles are abnormally fused (see Voss, 1957, *Quart. Jour. Florida Acad. Sci.*, vol. 20, p. 129). Type of the genus is *palpebrosa* Owen, 1835.

Rossia pacifica Berry, 1911   6236
Pacific Bob-tailed Squid

Alaska to San Diego, California.

Total length, not including the tentacles, 3 to 4 inches. Body smooth, mantle flattened above and below, rounded behind. Fins large, semicircular or subcordate, with a free anterior lobe, their attachment more or less oblique to the general plane of the body. Color in life unknown; in alcohol, reduced to brownish buff, heavily spotted above and in less degree below with purplish chromatophore dots, which extend even over the fins, although fewer on under surfaces and margins. This is the only *Rossia* recorded on the Pacific Coast, and it is rather abundant from 9 to 300 fathoms.

Subgenus *Semirossia* Steenstrup, 1881

Type of the subgenus is *tenera* (Verrill, 1880).

Rossia tenera (Verrill, 1880)   6237
Atlantic Bob-tailed Squid

Nova Scotia to Texas, Brazil.
Northern Europe.

3 to 4 inches in length, including mantle and longest arm. A small and delicate species, very soft, translucent, and delicately rose-colored when living. Internal pen small, very thin and soft. Length of each side fin is about ⅔ of the body, and the base of attachment of the fin is about ½ the body length. Arms unequal, the dorsal ones considerably shorter. This species is characterized by the larger size of the suckers located along the middle of the lateral arms. Commonly dredged from 5 to 233 fathoms. Formerly listed as *Heteroteuthis tenera* Verrill. *R. equalis* Voss, dredged off southeast Florida, differs in having the suckers on the lateral arms about equal in size.

Rossia equalis Voss, 1956   6238
Voss' Bob-tailed Squid

Lower Florida Keys and the Gulf of Mexico.

3 to 4 inches in length, differing in the smaller relative size of the sessile suckers which decrease gradually (not abruptly) in size distally. The suckers of the tentacular club small, those of the dorsal rows not much larger than the others (in *tenera*, those of the dorsal rows are 2 to 3 times the diameter of the remaining suckers). Moderately common. (*Revista Soc. Mal. Cuba.*, vol. 7, p. 73.)

Other Atlantic species:

6239 *Rossia (Allorossia* Grimpe, 1922) *glaucopsis* Lovén, 1845. Newfoundland to Massachusetts, 7 to 640 fms. (Synonyms: *hyatti* Verrill, 1878; *sublaevis* Verrill, 1878.) Type of the subgenus.

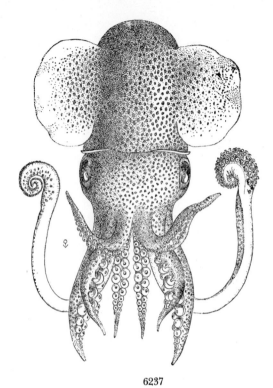

6237

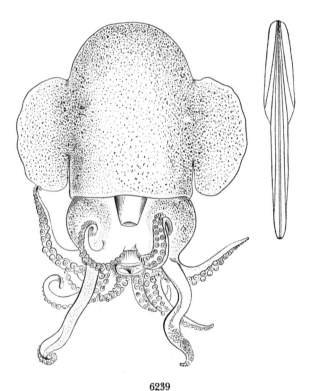

6239

6240 *Rossia (Allorossia) megaptera* Verrill, 1881. Newfoundland to Massachusetts, 150 to 640 fms.

6241 *Rossia (Allorossia) moelleri* Steenstrup, 1856. West Greenland. Arctic endemic.

6242 *Rossia palpebrosa* Owen, 1835. Elwin Bay, Prince Regent's Inlet, Arctic Canada; West Greenland.

6243 *Rossia (Allorossia) bullisi* Voss, 1956. Upper part of the Gulf of Mexico, both sides of the Mississippi. 200 to 300 fms. *Bull. Marine Sci.*, vol. 6, p. 101.

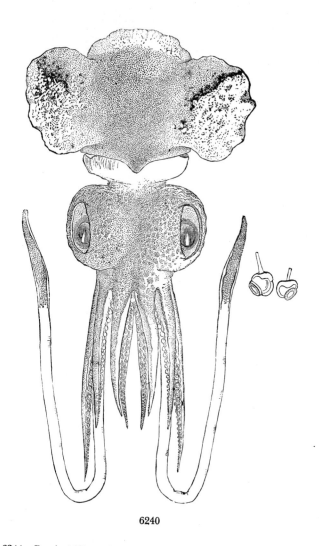

6240

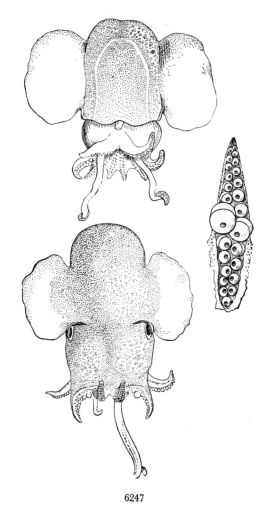

6247

6244 *Rossia (Allorossia) tortugaensis* Voss, 1956. Off Dry Tortugas, Florida, 283 to 375 fms. *Bull. Marine Sci.,* vol. 6, p. 103.

6245 *Rossia (Allorossia) antillensis* Voss, 1955. Off Tampa to off Dry Tortugas, Florida, to Cuba, 180 to 215 fms. *Bull. Marine Sci.,* vol. 5, p. 86.

6246 *Rossia brachyura* Verrill, 1883. Off Punta Alegre, Cuba, 205 fms.; St. Kitts, 208 fms.

Subfamily STOLOTEUTHINAE S. S. Berry, 1912

Genus *Stoloteuthis* Verrill, 1881

Body short, stout, rounded posteriorly. Eyes with free eyelids; pupils round. No pen. Arms united by a broad web. Fins large, narrowed at base. Tentacular club with small, pedicled suckers. Type: *leucoptera* Verrill, 1878.

6247 *Stoloteuthis leucoptera* (Verrill, 1878). Gulf of Maine, 94 to 640 fathoms. Gulf of St. Lawrence. Butterfly Squid. 1 inch.

Family SEPIIDAE Leach, 1817

Genus *Sepia* Linné, 1758

The common cuttlefish of Europe is not represented in our waters, although in rare instances the cuttlefish bone, or internal shell, has been found in Western Atlantic waters

from Florida to Texas. The cuttlefish bone is an oblong, 6-inch or so, very light slab of chalky material, rounded at one end, pointed at the other. It is used in the manufacture of toothpaste, and is tied to the bars of canary cages for the birds to peck at as a source of lime. Ink from *Sepia* was at one time a main source for durable, black writing ink. For a review of east American records, see H. W. Harry and S. F. Snider, 1969, *The Veliger,* vol. 12, p. 89, and *Texas Conchologist,* vol. 4, no. 6, Feb. 1969.

Order *Teuthoidea* Owen, 1836

An excellent illustrated key to the families of this order is given by Roper, Young and Voss, 1969, *Smithsonian Contrib. Zool.,* no. 13. Many of our illustrations are from that paper and from A. E. Verrill, 1882, "Report of the United States Commissioner of Fisheries for 1879."

Suborder *Myopsida* Orbigny, 1845

Eye covered by a transparent corneal membrane.

Family LOLIGINIDAE Lesueur, 1821

4 rows of suckers on the manus of the tentacular clubs. Back end of the fins with a concave outline. Includes the genera *Loligo, Doryteuthis, Lolliguncula, Sepioteuthis, Loliolopsis, Loliolus, Alloteuthis* and *Uroteuthis.*

Genus *Loligo* Schneider, 1784

10-armed, with elongate, tapering, cylindrical body and large, terminal, triangular fins. Arms with 2 rows of suckers provided with horny, dentated rings; fourth left arm hectocotylized in the males. Tentacular arms with 4 rows of suckers on their clubs. Internal pen horny, lanceolate with its shaft keeled on the underside. The female receives the sperm sacs of the male upon a specially developed pad below the mouth. In the genus *Lolliguncula,* the sperm sacs are received upon a callused patch within the mantle near the left gill. Type: *vulgaris* (Lamarck, 1799).

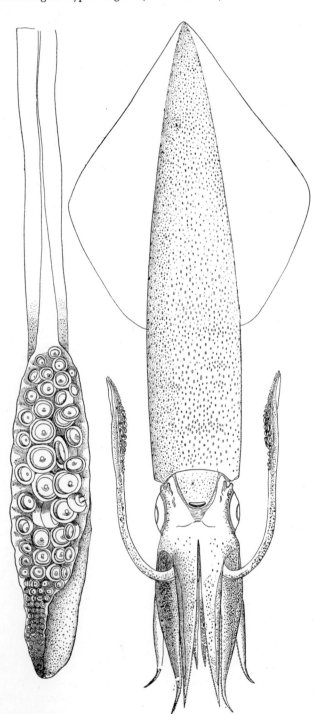

6248

Subgenus *Loligo* Schneider, 1784

Loligo pealeii Lesueur, 1821   6248
Atlantic Long-finned Squid

Nova Scotia to Florida, Texas and Venezuela. Bermuda.

Total length, including tentacular arms, 1 to 2 feet. Easily recognized by the accompanying illustration which shows the rather long, triangular fins. The proportion of fin length to mantle length varies from 1 to 1.8 and down to a ratio of 1 to 1.5. Adult males have the left ventral arm conspicuously hectocotylized. A very abundant species caught commercially for fish bait in New England. Living specimens are very beautifully speckled with red, purplish and pink. Synonyms: *borealis* Verrill, 1880; *pallida* Verrill, 1874.

Loligo opalescens Berry, 1911   6249
Common Pacific Squid

Puget Sound, Washington, to San Diego, California.

Total length, not including tentacles, 6 to 8 inches. This is the common squid of the Pacific Coast and can be readily recognized by the accompanying illustrations. At certain seasons, they occur in great schools by the thousands. See: W. G. Fields, 1965, "The structure, development, food relations, reproduction, and life history of the squid, *Loligo opalescens* Berry," *Fish Bull.* 131, *Calif. Dept. Fish and Game.*

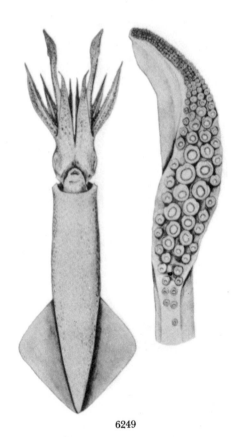

6249

Genus *Doryteuthis* Naef, 1912

Type of the genus is *pleii* (Blainville, 1823). May be considered a subgenus of *Loligo.*

Doryteuthis pleii (Blainville, 1823)   6250
Plee's Striped Squid

Georgia to Florida to Brazil. Bermuda.

Up to 8 inches in length, including the tentacular arms. Characterized by the long, narrow, slightly wavy, dark-colored, maroon bands running back along the side of the mantle. The rest of the mantle is moderately covered with small, maroon round dots. The body is long and slender, the triangular fins on the last 1/3 of the mantle and the arm suckers do not have pointed teeth on the horny circles. A common surface-living species of the Caribbean region. Eggs laid in fingerlike capsules. May be placed in the genus *Loligo*.

Genus *Lolliguncula* Steenstrup, 1881

Type of the genus is *brevis* (Blainville, 1823).

Lolliguncula brevis (Blainville, 1823)   6251
Brief Thumbstall Squid

Maryland to Texas and to Brazil. Bermuda.

Total length, including the tentacular arms, 5 to 10 inches. Characterized by its short, rounded fins, very short upper arms, and large color spots. Underside of fins white. Consult the figure. Common in warm waters. This is *brevipinna* Lesueur, 1824, and *L. hemiptera* (Howell, 1868).

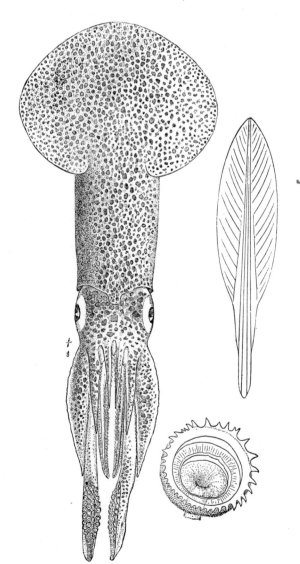

6251

Other species:

6252 *Lolliguncula panamensis* S. S. Berry, 1911. Panama to Ecuador. *Proc. Acad. Nat. Sci. Phila.* for 1911, p. 100.

Genus *Loliolopsis* S. S. Berry, 1929

Small, cylindrical squids, about 3 inches in length, with short fins joined in smoothly rounded unison past the tip of the body. Buccal lappets bearing each a few minute suckers. Both ventral arms in the male are sexually modified. Type: *chiroctes* S. S. Berry, 1929.

6253 *Loliolopsis chiroctes* S. S. Berry, 1929. Puerto Escondido, Baja California. *Trans. San Diego Soc. Nat. Hist.*, vol. 5, no. 18, p. 265. Common at surface at night.

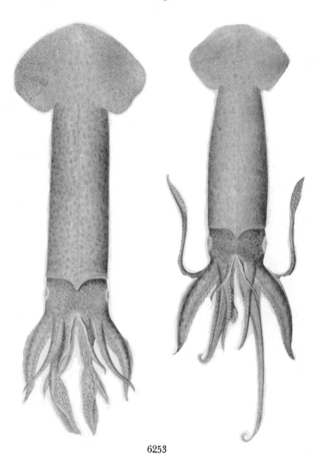

6253

Genus *Sepioteuthis* Blainville, 1824

Similar to *Loligo*, but with large, triangular fins that extend along the entire length of the mantle, thus giving the animal an oval outline. Siphonal funnel attached to the head by muscular bands. There is a strong·wrinkle behind the eye. Type: *sepioidea* (Blainville, 1823).

Sepioteuthis sepioidea (Blainville, 1823)   6254
Atlantic Oval Squid

Bermuda, Florida and the West Indies.

Total length, including tentacular arms, 4 to 5 inches. Characterized by the long fins which commence a short distance behind the mantle edge (1/4 to 1/3 inch). Internal pen thin, lanceolate and without any marginal thickenings. Skin yellow-

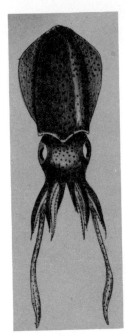

6264

ish to dark maroon-purple and regularly and closely spotted with purple dots. Two large "eye spots" may be seen at the anterior end at times. Can change body color to whitish. The eggs are large, 5 to 8 mm. in diameter, and laid in long jelly tubes. A rather common, warm-water species. Synonyms: *biangulata* Rang, 1837; *sloani* Leach, 1849; *ovata* Gabb, 1868; *ehrhardti* Pfeffer, 1884; and *occidentalis* Robson, 1926.

Family PICKFORDIATEUTHIDAE Voss, 1953

Only 2 rows of suckers on the manus of the tentacular clubs. Biserial suckers on the arms. Small, round fins with convex back edges. Buccal membranes with reduced lappets and no suckers. Photophores absent. One genus known:

Genus *Pickfordiateuthis* Voss, 1953

Animal less than 1 inch, sepiolid in shape. Fins large, round, about ½ the mantle length. Head twice as broad as long. Eyes prominent, with accessory lid and without a visi-

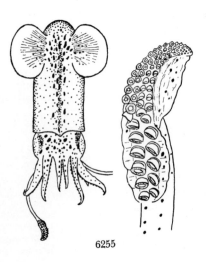

6255

ble pore. Internal pen large, thin, highly keeled, the vane being very broad. Type of the genus is *pulchella* Voss, 1953.

6255
Pickfordiateuthis pulchella Voss, 1953
Pickword's Beautiful Squid

Southeast Florida to Panama.

½ to 1½ inches. Characters of the family and genus. Males and females taken in shallow water over eel-grass beds along the shores of the Florida Keys. See *Bull. Marine Sci.*, vol. 2, no. 4, p. 602.

Suborder *Oegopsida* Orbigny, 1839

Oegopsid squids—eyes naked in front, pupils circular; eyelids present.

Family ARCHITEUTHIDAE Pfeffer, 1900

Characterized by their large size, tetraserial armature on the tentacular clubs with large suckers in the medial rows of the manus and small suckers in the marginal rows. These squid serve as food for whales.

Genus *Architeuthis* Steenstrup, 1857

For a recent listing of dubious species see M. R. Clarke, 1966, *Advances in Marine Biol.*, vol. 4, pp. 98–104. Type: *dux* Steenstrup, 1857. *Architeuthus* and *Architheuthis* are misspellings.

6256
Architeuthis harveyi Kent, 1874
Harvey's Giant Squid

Newfoundland fishing banks.

Total length 40 to 55 feet. Body stout, nearly round, swollen in the middle. Arms nearly equal in length, all bearing sharply serrated suckers. Tentacular arms 4 times as long as the 8 sessile arms. The peculiar backward-pointing tail fins separate this species from *A. princeps* Verrill, 1875 (**6257**) another giant squid found in the same area, and on one occasion from southeast Florida. A large well-preserved specimen of any giant squid is worth its weight in gold. No large specimens have been brought back from the fishing banks in many years. They may occasionally be washed ashore from Nova Scotia north. If you find one, take photographs if possible, and notify one of the leading museums. Giant squid of unknown identity have been seen in the Gulf of Mexico.

6258 *Architeuthis physeteris* (Joubin, 1899). Off Mississippi Delta, near surface. Azores and the upper Gulf of Mexico. Total length about 8 feet.

Family ENOPLOTEUTHIDAE Pfeffer, 1900

Most species possess hooks on the tentacular clubs. All possess a straight locking apparatus in the funnel; with biserial armature with at least some hooks on the arms; tetraserial armature on the clubs; photophores present; 8 buccal lappets and buccal connectives that attach dorsally to the ventral arms.

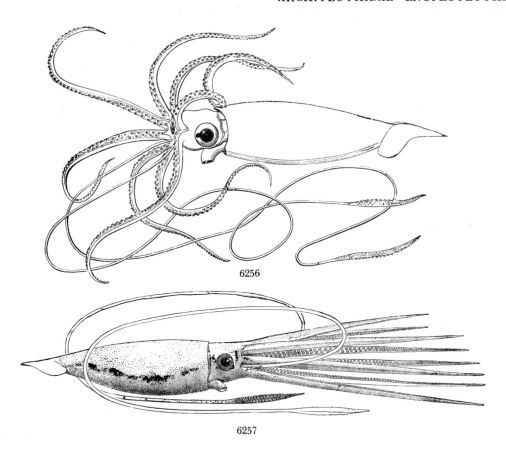

6256

6257

Subfamily ENOPLOTEUTHINAE Pfeffer, 1900

Without nidamental glands; with numerous, small photophores over the surface of the mantle head and arms. Contains the genera: *Enoploteuthis, Abralia, Abraliopsis* and *Watasenia*.

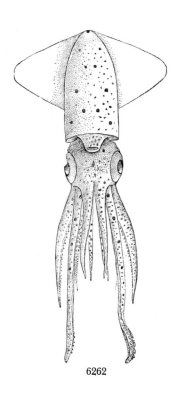

6262

Genus *Abralia* Gray, 1849

Type of the genus is *armata* (Quoy and Gaimard, 1832).

Subgenus *Asteroteuthis* Pfeffer, 1909

Type of the subgenus is *veranyi* (Rüppell, 1844).

6259 *Abralia veranyi* (Rüppell, 1844). Gulf of Mexico; Off Key West; Bahamas; Mediterranean West Africa. Type of the subgenus.

6260 *Abralia redfieldi* Voss, 1955. Bathypelagic; plankton hauls off Miami, Florida. *Bull. Marine Sci.,* vol. 5, p. 99. Mantle length 28 mm.

6261 *Abralia grimpei* Voss, 1958. Off Miami, Florida, 57 meters. *Bull. Marine Sci.,* vol. 8, no. 4, p. 375, with keys to above 3 species.

6262 *Abralia megalops* Verrill, 1882. New England to the West Indies, 137 to 173 fms. 1.5 inches.

Genus *Abraliopsis* Joubin, 1896

4 arms with 2 to 4 very large photophores at the arm tips. Tentacular clubs with 2 rows of hooks on the manus. Type of the genus: *Enoploteuthis hoylei* Pfeffer, 1884, of the tropical Eastern Pacific.

6263 *Abraliopsis morisii* (Vérany, 1837). Bathypelagic; Off Miami, Florida. Off Bermuda. *A. morissi* is a misspelling. Reddish brown; light organs on ventral surfaces, with large, black, swollen, three-parted photophores on distal part of ventral arms.

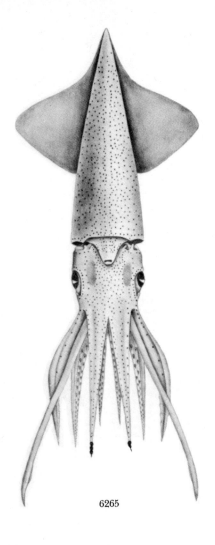

6265

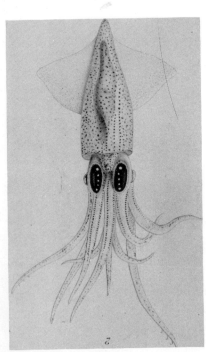

6266

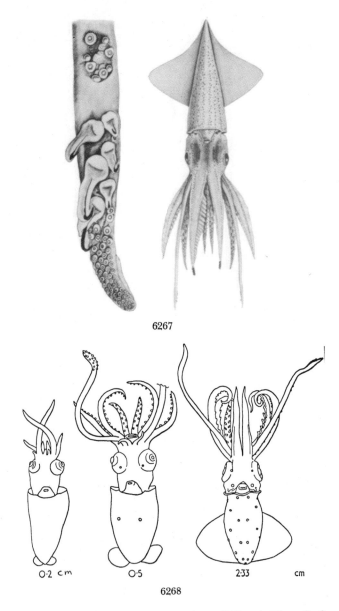

6267

6268

6264 *Abraliopsis* (*Watasenia* Ishikawa, 1913) *scintillans* S. S. Berry, 1911, Japan; Korea; Okhotsk Sea. Type of the subgenus. *The Nautilus*, vol. 25, p. 93 (Berry); *Zool. Anz.*, vol. 43, p. 162 (Ishikawa).

6265 *Abraliopsis* (*Abraliopsis*) *felis* McGowen and Okutani, 1968. Off Oregon to off central Baja California in the California Current. Mantle length, 36 to 42 mm. *The Veliger,* vol. II, p. 72.

6266 *Abraliopsis* (*Watsenia*) *affinis* (Pfeffer, 1912). Eastern Tropical Pacific 22°N to 25°S. Baja California to Ecuador. Abundant. This is *A. hoylei* of Hoyle, 1904, not Joubin, 1896, which is from the Indian Ocean.

6267 *Abraliopsis* (*Micrabralia* Pfeffer, 1900) *falco* Young, 1972. Off Guadalupe Island, west Mexico, to Hawaii. *Smithsonian Contrib. Zool.*, no. 97, p. 13.

Subfamily ANCISTROCHIRINAE Pfeffer, 1912

Nidamental glands present; with huge fins with convex posterior edges. A few large scattered photophores on the surface of the mantle, head and tentacles. Two genera: *Thelidioteuthis* and *Ancistrochirus*.

Genus *Thelidioteuthis* Pfeffer, 1900

6268 *Thelidioteuthis alessandrinii* (Vérany, 1851). Bathypelagic; Massachusetts to off Miami, Florida. Mediterranean; Indo-Pacific. Type of the genus.

Subfamily PYROTEUTHINAE Pfeffer, 1912

Nidamental glands present; large photophores embedded in the tentacles, viscera and fins which have convex posterior edges. Two genera: *Pyroteuthis* and *Pterygioteuthis*.

Genus *Pyroteuthis* Hoyle, 1904

Left oviduct present, but reduced. Hooks present on tentacular clubs.

6269 *Pyroteuthis margaritifera* (Rüppell, 1844). Bathypelagic; Europe; Off Miami, Florida. Off Bermuda. Common. See C. Chun, 1910, *Valdivia*, vol. 18, no. 1. Type of the genus.

6270 *Pyroteuthis addolux* Young, 1972. Off southern California and west Mexico to Hawaii. Mantle length 20 to 45 mm. *Smithsonian Contrib. Zool.*, no. 97, p. 22.

Genus *Pterygioteuthis* H. Fischer, 1896

Left oviduct absent. Hooks not present on tentacular clubs.

6271 *Pterygioteuthis giardi* H. Fischer, 1896. Bathypelagic to 500 meters; Off Miami, Florida. Off Bermuda. See C. Chun, 1910, *Valdivia*, vol. 18, no. 1. Type of the genus.

6272 *Pterygioteuthis gemmata* Chun, 1908. Equatorial Atlantic; South Atlantic; Indian Ocean; North Atlantic; Hawaii. (*P. microlampas* S. S. Berry, 1914, is probably a synonym.)

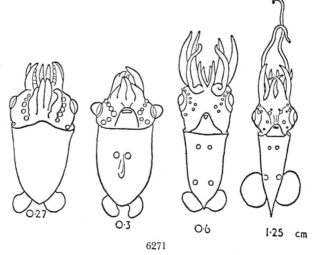

6271

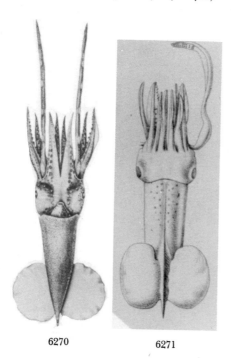

6270   6271

6272

Family GONATIDAE Hoyle, 1886

Most have no photophores, although some may have oval patches on the ventral surfaces of the eyes. Tetraserial armature on the arms. Arm tips in some may have up to 12 rows of minute suckers. Most have hooks on arms I through

III. Three genera: *Gonatus*, *Berryteuthis* and *Gonatopsis*, the last being characterized by the complete loss of tentacles, except in the early larval forms. Cold-water forms, and are among the most abundant in the higher latitudes. They are the food of whales, seals, birds and man.

Genus *Gonatus* Gray, 1849

Type of the genus is *fabricii* (Lichtenstein, 1818). Synonyms: *Lestoteuthis* and *Cheloteuthis* Verrill, 1881. *G. magister* S. S. Berry, 1913, is the type of the subgenus *Berryteuthis* Naef, 1921.

6273 *Gonatus fabricii* (Lichtenstein, 1818). Arctic seas; Nova Scotia to Rhode Island. Abundant. Main food of bottlenose whale.

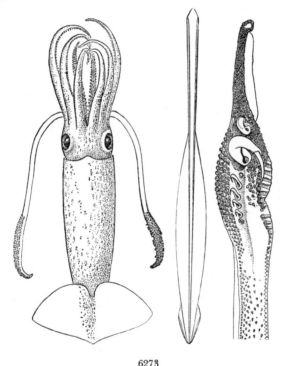

6273

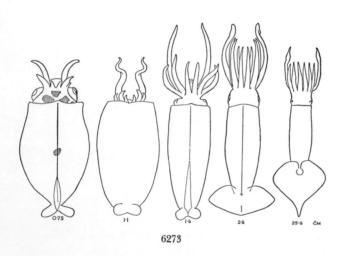

6273

6274 *Gonatus (Berryteuthis) magister* S. S. Berry, 1913. Puget Sound, Washington. *Proc. Acad. Nat. Sci. Phila.*, p. 76.

6275 *Gonatus (Berryteuthis) anonychus* (Pearcy and Voss, 1963). Off Oregon in oceanic water. *Proc. Biol. Soc. Wash.*, vol. 76, pp. 105–112.

6276 *Gonatus onyx* Young, 1972. Common off San Pedro and Santa Catalina, California.

6277 *Gonatus berryi* Naef, 1923. Off southern California and west Mexico.

6278 *Gonatus pyros* Young, 1972. Off southern California. Large photophore on the ventral surface of each eye.

6279 *Gonatus californiensis* Young, 1972. Off California and Baja California.

Genus *Gonatopsis* Sasaki, 1920

Tentacles present only in larvae. Type of the genus is *octopedata* Sasaki, 1920.

6280 *Gonatopsis borealis* Sasaki, 1923. Japan; Aleutian Islands to off Oregon and California. (Species *makko* Okutani and Nemoto, 1964, *Sci. Rep. Whales Res. Inst. Tokyo*, no. 18, p. 111, is from the Bering Sea.)

Family OCTOPOTEUTHIDAE Hoyle, 1886

No tentacles present in adults; biserial hooks on the arms (usually replaced by small biserial suckers near the arm tips). Very large fins; light organs at the tips of at least some of the arms. The family name Veranyidae is synonymous. Two genera: *Octopoteuthis* and *Taningia*.

Genus *Octopoteuthis* Rüppell, 1844

Hooks on arms; tentacles absent; pen very thin and fragile. Fins fused across the midline. *Octopodoteuthis* Krohn, 1845, and *Octopodoteuthopsis* Pfeffer,, 1912, are synonyms. Type: *O. sicula* Rüppell, 1844.

6281 *Octopoteuthis megaptera* (Verrill, 1885). North Atlantic; off Miami, Florida, and the Gulf of Mexico.

6282 *Octopoteuthis (Taningia) danae* Joubin, 1931. Bermuda. Azores; Indian Ocean.

6283 *Octopoteuthis sicula* Rüppell, 1844. Off Newfoundland; northwest Europe; western Mediterranean; Indo-Pacific. (Synonyms?: *persica* Naef, 1923; *indica* Naef, 1923.) Type of the genus.

6284 *Octopoteuthis deletron* Young, 1972. Off southern California; common; off Oregon; off Peru.

6285 *Octopoteuthis nielseni* Robson, 1948. Off west Central America.

Family ONYCHOTEUTHIDAE Gray, 1847

Tetraserial armature on the tentacular clubs, of which the 2 median rows consist of hooks and the marginal rows of suckers. 2 rows of suckers on the arms. Most are strong, active swimmers. Five genera: *Onychoteuthis*, *Onykia*, *Moroteuthis*, *Ancistroteuthis* and *Chaunoteuthis*.

Genus *Onychoteuthis* Lichtenstein, 1818

6286 *Onychoteuthis banksii* (Leach, 1817). Bermuda; Florida; Gulf of Mexico. Norway to Cape Horn. Cosmopolitan(?) in warm and temperate seas. 4 inches. Surface to 50 fms. Often "flies" onto the decks of ships. (Synonym: *bergii* Lichtenstein, 1818.) Type of the genus.

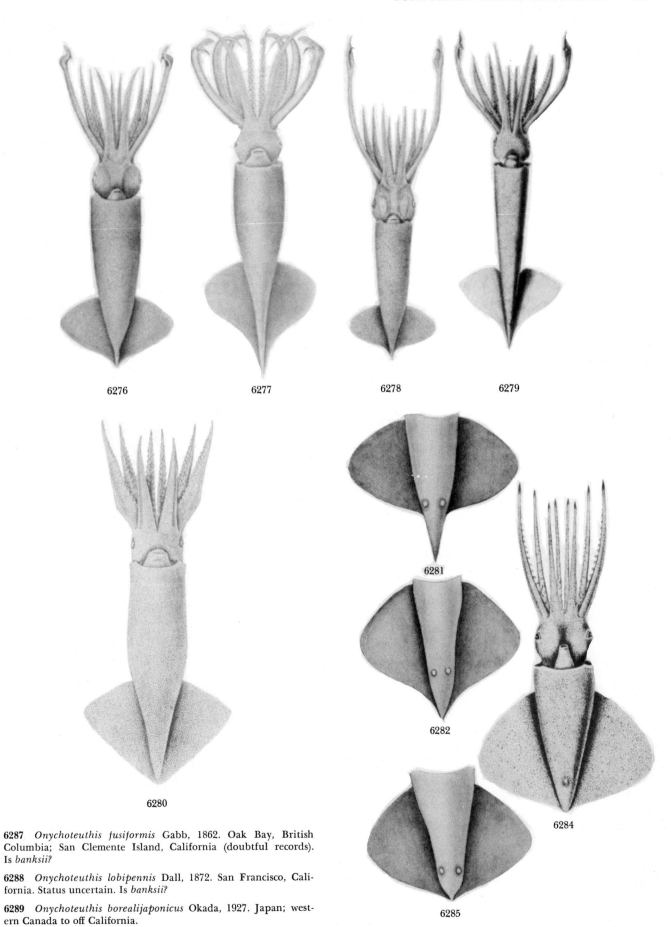

6276   6277   6278   6279

6280

6281

6282

6285

6284

6287 *Onychoteuthis fusiformis* Gabb, 1862. Oak Bay, British Columbia; San Clemente Island, California (doubtful records). Is *banksii?*

6288 *Onychoteuthis lobipennis* Dall, 1872. San Francisco, California. Status uncertain. Is *banksii?*

6289 *Onychoteuthis borealijaponicus* Okada, 1927. Japan; western Canada to off California.

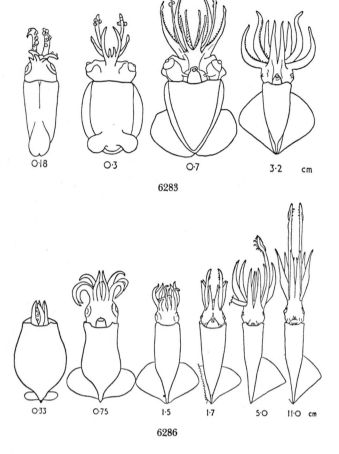

6283

6286

6289

6291

Genus *Moroteuthis* Verrill, 1881

Type: *robusta* "Dall," Verrill, 1876.

6290 *Moroteuthis robusta* (Verrill, 1876). Near Iliuliuk, Unalaska Island, Alaska, to California. Japan. Mantle length near 7 feet.

Genus *Onykia* Lesueur, 1821

Onychia is a misspelling. *O. caribbaea* Lesueur, 1821, is the type. Tentacular club with 2 central rows of hooks, rows of small suckers along each margin and a cluster of suckers and tubercles on the "wrist." Sessile arms with smooth suckers. Synonyms: *Steenstrupiola* Pfeffer, 1884; *Teleoteuthis* Verrill, 1885.

6291 *Onykia caribbaea* Lesueur, 1821. Worldwide in tropical and temperate seas; off Texas, 1,700 fms.; Bermuda. (Synonym: *agilis* Verrill, 1885, off Chesapeake Bay.) *O. carribaea* is a misspelling. Other synonyms: *laticeps* Owen, 1836; *cardioptera* Gray, 1849; *plagioptera* Eydoux and Souleyet, 1852; *binotata* Pfeffer, 1884; *atlantica* Pfeffer, 1884; *jattae* Joubin, 1900.

Genus *Ancistroteuthis* Gray, 1949

2 central rows of hooks and proximal and apical suckers on the club. Sessile arms with suckers only. Pen widest anteriorly, with a long, terminal, hollow cone.

6292 *Ancistroteuthis lichtensteinii* (Férussac and Orbigny, 1839). Mediterranean; Gulf of Mexico, at surface. Type of the genus.

Family LEPIDOTEUTHIDAE Pfeffer, 1912

Characterized by distinct "scales" on the mantle. Biserial suckers on the arms; tetraserial suckers on the tentacular clubs, except in *Lepidoteuthis* which lacks tentacles in the adults. Photophores absent. *Pholidoteuthis* and *Tetronychoteuthis* are also in this family. Pholidoteuthidae is synonymous.

Genus *Pholidoteuthis* Adam, 1950

Type of the genus is *boschmai* Adam, 1950, of the southwest Pacific.

6293 *Pholidoteuthis adami* Voss, 1956. Off Alabama to off Texas. Abundant at surface at night. *Bull. Marine Sci.*, vol. 6, p. 132. Scale-skin squid.

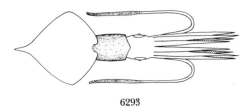

6293

Family CTENOPTERYGIDAE Grimpe, 1922

With long ribbed fins; suckers on the lappets of the buccal membrane; 8 to 14 longitudinal rows of small suckers on the tentacular clubs; 4 to 6 rows of suckers on the distal half of arms I through III. One species.

Genus *Ctenopteryx* Appellöf, 1889

6294 *Ctenopteryx siculus* (Vérany, 1851). Off Bermuda. Bathypelagic to 400 meters. Fins as long as the mantle and are fringed. North Atlantic; South Atlantic; Mediterranean. (Synonyms: *fimbriatus* Appellöf, 1889; *cyprinoides* Joubin, 1894; *nevroptera* Jatta, 1896.) Type of the genus.

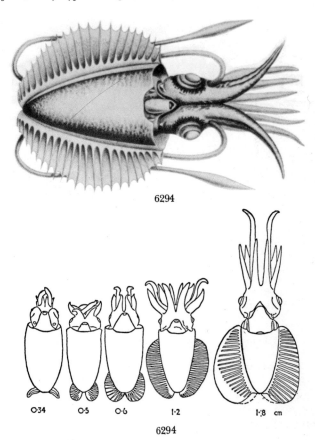

6294

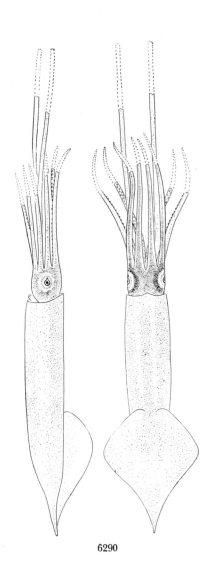

6290

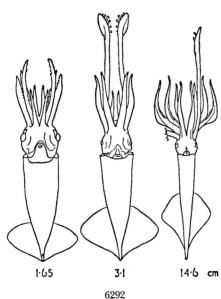

1·65   3·1   14·6 cm

6292

Family BRACHIOTEUTHIDAE Pfeffer, 1908

Characters of the only known genus, *Brachioteuthis* Verrill, 1881.

Genus *Brachioteuthis* Verrill, 1881

Caudal fin rhombic; pen with a simple, linear, anterior portion, suddenly expanding into a much broader, lanceolate, posterior portion. Arms slender; tentacular club without a spoonlike cavity at tip. Siphon with a valve and dorsal bridle. Cartilages of the mantle in simple, linear ridges. Type: *beanii* Verrill, 1881.

6295 *Brachioteuthis beanii* Verrill, 1881. Massachusetts to North Carolina, 183 to 843 fms. 5 inches.

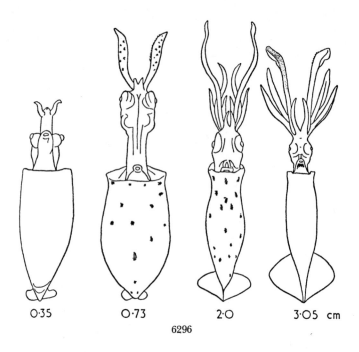

0.35   0.73   2.0   3.05 cm

6296

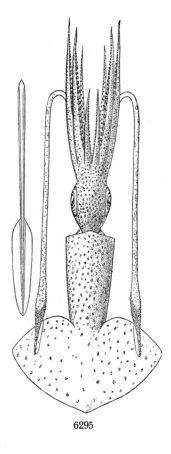

6295

6296 *Brachioteuthis riisei* (Steenstrup, 1882). Off Nova Scotia; Europe; Africa; Australia; South America. (Synonyms: *Verrilliola gracilis* Pfeffer, 1884; *nympha* Pfeffer, 1884; *velaini* and *clouei* both Rochebrune, 1884; *alicei* Joubin, 1900.)

Family LYCOTEUTHIDAE Pfeffer, 1908

Biserial suckers on the arms, tetraserial suckers on the tentacular clubs; straight simple funnel locking-cartilage; buccal membrane with 8 lappets and supports; buccal connectives that attach to the dorsal borders of arms IV and photophores at least on the viscera and the ventral surfaces of the eyes.

Subfamily LYCOTEUTHINAE Pfeffer, 1908

Genus *Lycoteuthis* Pfeffer, 1900

Thaumatolampas Chun, 1903, is a synonym.

6297 *Lycoteuthis diadema* (Chun, 1900). Off Miami, Florida; Gulf of Mexico, from the stomachs of fish or sharks; west coast of South America. Mantle length: 51 to 83 mm. (Synonyms: *jattae* Pfeffer, 1900; *planctonicum* Pfeffer, 1912; *inermis* Robson, 1924.) Type of the genus.

Genus *Oregoniateuthis* Voss, 1956

6298 *Oregoniateuthis springeri* Voss, 1956. Off Mobile, Alabama, 200 fms. *Bull. Marine Sci.*, vol. 6, p. 120. Type of the genus.

6297

Genus *Selenoteuthis* Voss, 1958

6299 *Selenoteuthis scintillans* Voss, 1958. Off Eleuthera, Bahamas. *Bull. Marine Sci.*, vol. 8, no. 4, p. 370. Type of the genus.

Family HISTIOTEUTHIDAE Verrill, 1881

Widely distributed pelagic squid from the surface to over 2,000 meters in depth. A common food of whales and sea birds. There are about 14 species and subspecies. They range in mantle length from about 2 to 12 inches. One genus now recognized. For a full account of the family, see Nancy Voss, 1970, *Bull. Marine Sci.* (Miami), vol. 19 (for 1969), no. 4, pp. 713–867.

Genus *Histioteuthis* Orbigny, 1840

Left eye usually larger than the right; surface with numerous small photophores in diagonal rows. Arms sometimes with webbing. Mantle short; 6- or 7-parted buccal membrane; 2 rows of suckers on arms; 5 to 8 rows on tentacular club. Tentacles long. Both arms no. I are hectocotylized in the male. Type: *bonnellii* Férussac, 1835. Synonyms: *Calliteuthis* Verrill, 1880; *Histiopsis* Hoyle, 1885; *Meleagroteuthis* Pfeffer, 1900; *Stigmatoteuthis* Pfeffer, 1900; *Histiothauma* Robson, 1948. The big eye is probably used at great depths, and the small eye to detect brighter light near the surface.

6300 *Histioteuthis elongata* (Voss and Voss, 1962). Mantle length 4 to 7 inches, exceptionally long (4 times width). Large black photophores on ventral surface of mantle. North Atlantic from Newfoundland to Madeira, 600 to 700 meters.

6301 *Histioteuthis reversa* (Verrill, 1880). Mantle length 1 to 2 inches. Large and small photophores on ventral mantle, 4 rows on arms IV; 18 around right margin of eyelid. Off northeast North America to west Europe and West Africa. Surface to 1,000 meters. Common.

6302 *Histioteuthis celetaria celetaria* (G. Voss, 1960). Mantle length 1½ inches. 100 miles southwest of Bermuda, 730 to 820 fms.; Madeira.

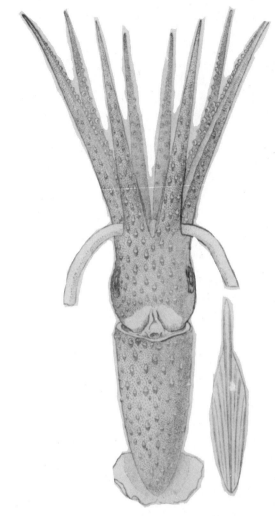

6301

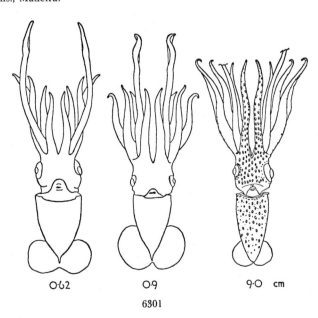

O·62   O·9   9·0 cm

6301

6303 *Histioteuthis corona corona* (Voss and Voss, 1962). Mantle length 1 to 5 inches. 17 large photophores around right eyelid, and 7 bordering the left. 3 rows on arm IV. Gulf of Mexico, Caribbean, Western Atlantic; West and East Africa.

6304 *Histioteuthis corona berryi* Nancy Voss, 1970. Mantle length 2 inches. 4 rows of photophores on arms IV. Off south California, 250 meters. Rare.

6305 *Histioteuthis dofleini* (Pfeffer, 1912). Mantle length 1½ to 4 inches. Skin with low, fleshy papillae. Worldwide; Gulf of Mexico; off East Florida; off British Columbia to southern California. Common. (Synonyms: *S. chuni* Pfeffer, 1912; *S. arcturi* Robson, 1948.)

6306 *Histioteuthis meleagroteuthis* (Chun, 1910). Mantle length 1 to 2 inches. With a median row of cartilaginous tubercles on the dorsal surface of the mantle and 3 dorsal pairs of arms. North Atlantic; South Atlantic; Indo-Pacific. (Synonym: *separata* Sasaki, 1915.)

6307 *Histioteuthis heteropsis* (S. S. Berry, 1913). Mantle length 1 to 5 inches. Densely set numerous photophores. No rows of tubercles on arms and mantle. Common; off California to Chile, 400 to 800 meters.

6308 *Histioteuthis atlantica* (Hoyle, 1885). Mantle length 1 to 3½ inches. Tips of 3 dorsal pairs of arms with 3 to 8 very large photophores. South Atlantic; New Zealand; South Africa.

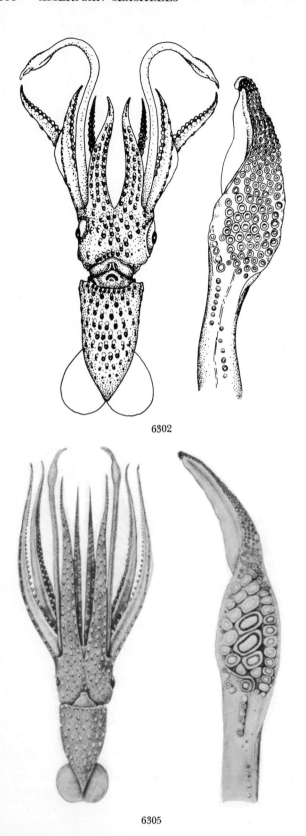

6302

6305

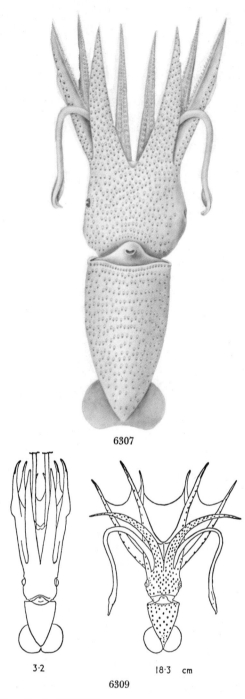

6307

3·2   18·3 cm

6309

6309 *Histioteuthis bonnellii* (Férussac, 1835). Mantle length up to 12 inches. Very large inner web connecting arms. One large photophore at tip of arms I, II and III. Abundant; 70 to 2,000 meters. Off Newfoundland and New York to Western Europe; South Africa. (Synonyms: *bonellii* and *bonelliana* Fér, misspellings of authors; *ruppelli* Vérany, 1846; *collinsii* Verrill, 1879.)

Family BATHYTEUTHIDAE Pfeffer, 1900

The Family Benthoteuthidae Johnson, 1934, is the same. Characterized by a straight funnel locking-cartilage; buccal connectives that attach to the dorsal borders of arms IV; suckers on the buccal lappets; small clubs with many rows of minute suckers; round subterminal fins; deep-maroon color. One genus.

Genus *Bathyteuthis* Hoyle, 1885

Benthoteuthis Verrill, 1885, is a synonym. For a review of the genus, see Roper, 1969, *Bull. 291, U.S. Nat. Mus.*

6310 *Bathyteuthis abyssicola* Hoyle, 1885. Cosmopolitan; Off Massachusetts, 600 to 1,073 fms.; Gulf of Mexico, 780 fms., Indian Ocean. Off Bermuda. *Benthoteuthis megalops* Verrill, 1885, is a synonym. Very abundant in the Antarctic. Type of the genus.

6311 *Bathyteuthis berryi* Roper, 1968. Off California and west Mexico.

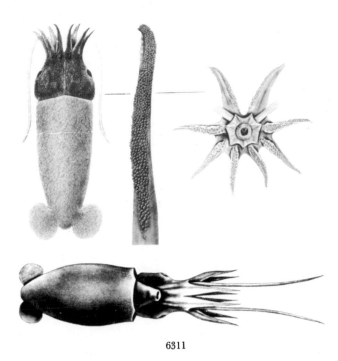

6311

Family OMMASTREPHIDAE Steenstrup, 1857

Characterized by an inverted T-shaped funnel locking-cartilage; biserial suckers on the arms; tetraserial suckers on the tentacular clubs (except *Illex* which has 8 rows of suckers on the dactylus).

Subfamily ILLICINAE Posselt, 1890

Genus *Illex* Steenstrup, 1880

Resembling *Loligo* somewhat, but with half-hidden eyes, the lids free and with a distinct notch or sinus in front. Internal pen narrow along the middle portion, and with 3 ribs. There are 8 rows of tiny suckers on the end section of the 2 long, tentacular arms. Type: *illecebrosus* (Lesueur, 1821).

Illex illecebrosus (Lesueur, 1821)   6312
Common Short-finned Squid

Newfoundland to northeast Florida.

Total length, including tentacular arms, 12 to 18 inches. A common squid characterized by the small opening to the eyes and the small, narrow sinus or notch in front of the eyes, and by the proportion of fin length to mantle length which is roughly 1 to 3. The sides of the head, back of the eyes, have a rather prominent, transverse ridge, back of which the head suddenly narrows to the neck. Under surface of head with a deep, smoothish excavation to receive the dorsal half of the siphonal tube. In males, either the left or right ventral arm is hectocotylized. A very common species used for fish bait. It

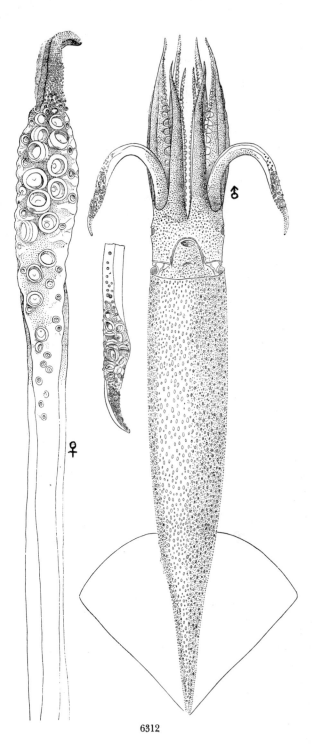

6312

may be seen in large schools near shore, especially in summer in New England. The width of the head is not greater than the width of the body, while in the similar *oxygonius* the head is wider than the body. The fin angle in *illecebrosus* is between 40 and 50 degrees. For anatomy, see A. E. Verrill, 1882, *Report of the U.S. Fish Commission for 1879*, p. 441, pl. 19.

Illex coindetii (Vérany, 1839)   6313
South Florida to Texas and the Caribbean.

Mediterranean and West Africa.

Similar to *illecebrosus*, but the mantle is proportionately shorter, widest at the anterior end; and the head is proportionately wider (especially in male *coindetii*). The dorsal triangular lobe at the mantle opening is weak (strong in *oxygonius*). Fin angle exceeds 50 degrees. The hectocotylus portion of arm IV is 30 to 35% of the total arm length and has curious, fringed and papillose, bulbous outgrowths beside the small suckers. Uncommon.

Illex oxygonius Roper, Li and Mangold, 1969   6314
Arrow-finned Squid

Off New Jersey to off Key West, Florida.

8 to 10 inches long; mantle long, slender, broadest anteriorly, drawn out posteriorly into a long attenuated tip. Fin angle about 33 degrees (25 to 40 degrees). Fin width equal to fin length. Dorsal mantle lobe conspicuous, pointed in males. Hectocotylized portion of arm IV long, with 3 papillae on the dorsal row. Uncommon; 50 to 555 meters. (*Proc. Biol. Soc. Wash.*, vol. 82, p. 295.)

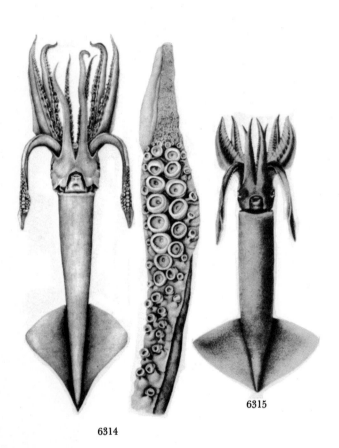

6315

6314

Subfamily OMMASTREPHINAE Steenstrup, 1857

Genus *Ommastrephes* Orbigny, 1839

Very similar to *Illex* in almost every way, but the sucker-bearing area includes less than ½ the total length of the tentacular arms. The larger suckers on the tentacular club are strongly toothed, with an additional large tooth in each of the 4 quadrants. *Sthenoteuthis* Verrill, 1880, is a synonym. Type: *bartramii* (Lesueur, 1821). *Ommatostrephes* is a modified spelling.

Ommastrephes bartramii (Lesueur, 1821)   6315
Flying Squid

Worldwide. Bermuda.

2 to 3 feet in total length, resembling the common *Illex*, but more slender, with shorter fins, and with 4, not 8, rows of tiny suckers on the end of the 2 long tentacular arms. Preserved specimens show a distinct dark, purple-brown dorsal stripe. In life, the colors are very brilliant and are continually changing. Along the middle dorsal line there is a broad violet stripe with a stripe of reddish yellow on each side of it. Body elsewhere bluish; fins rosy. Skin covered with small, red-violet chromatophore dots. On the eyes there are 2 elongated spots of brilliant blue, and below a bright spot of red. Color of ink reported to be a coffee-and-milk color. A common oceangoing species which swims with great speed, and not infrequently jumps out of water and lands on the decks of ships. Like most squid, it is attracted by artificial light.

Other species:

6316   *Ommastrephes pteropus* (Steenstrup, 1855). Bermuda. Caribbean. Mediterranean. Body and arms 30 inches.

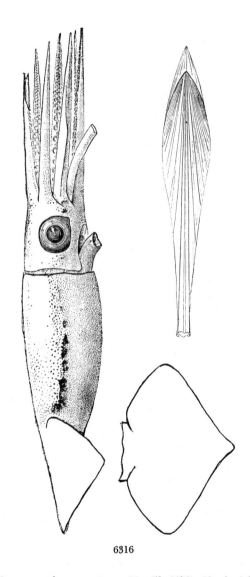

6316

6317   *Ommastrephes megaptera* (Verrill, 1878). North Atlantic. Large broad-finned squid. Body and head 19 inches.

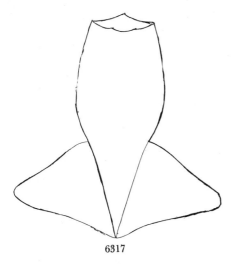

6317

Genus *Ornithoteuthis* Okada, 1927

Type of the genus is *volatilis* Sasaki, 1915.

6318 *Ornithoteuthis antillarum* Adam, 1957. Bahama Islands; Lesser Antilles. Mantle length 97 mm. See Voss, 1957, *Bull. Marine Sci.*, vol. 7, no. 4, p. 370, for details.

Genus *Symplectoteuthis* Pfeffer, 1900

Funnel locking-cartilage with typical ommastrephid inverted T-shaped, but fused in its mid-portion to the mantle locking-cartilage. Type species: *Loligo oualaniensis* Lesson, 1830.

6319 *Symplectoteuthis luminosa* Sasaki, 1915. Off Japan; Kermadec Islands; California Current; off Guadalupe Island, west Mexico. Mantle length: 136 to 163 mm.; surface with numerous small, reddish brown chromatophores; long, luminous stripes on underside of mantle.

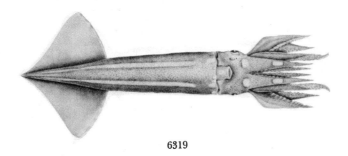

6319

Genus *Dosidicus* Steenstrup, 1857

6320 *Dosidicus gigas* (Orbigny, 1835). Monterey Bay to the Santa Barbara Islands, California. 50 inches. (Synonym: *eschrichti* Steenstrup, 1857.) Type of the genus.

Family CHIROTEUTHIDAE Gray, 1849

Genus *Chiroteuthis* Orbigny, 1839

Type of the genus is *veranyi* (Férussac, 1835). Arms IV greatly enlarged. *Bigelowia* MacDonald and Clench, 1934, is a synonym.

6320

6321 *Chiroteuthis lacertosa* Verrill, 1881. Nova Scotia to off Alabama; West Indies. 285 to 2,369 fms. 12 inches.

6322 *Chiroteuthis diaphana* Verrill, 1884. Off Massachusetts, 1,731 fms.

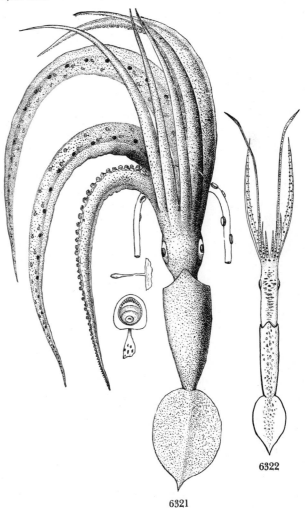

6322

6321

6323   *Chiroteuthis atlanticus* MacDonald and Clench, 1934. Off Massachusetts, deep water.

6324   *Chiroteuthis calyx* Young. Off Oregon and southern California.

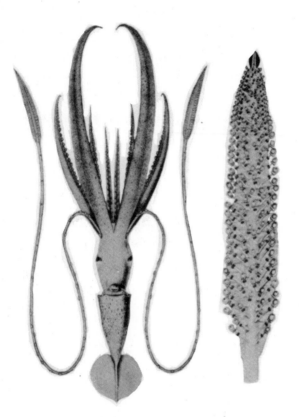

6324

6325   *?Chiroteuthoides hastula* S. S. Berry, 1920. Off East United States.

6326   *?Enoptroteuthis* (S. S. Berry, 1920) *spinicauda* S. S. Berry, 1920. Sargasso Sea, south of Bermuda.

Family JOUBINITEUTHIDAE Naef, 1922

Buccal connectives attached to ventral borders of arms IV; funnel locking-cartilage oval without tragus or antitragus; conus elongate with ventral fusion; tail elongate, filiform, longer than mantle length; arms I to III extremely long with 6 longitudinal rows of suckers; ventral arms much shorter and with 4 rows of suckers; tentacular clubs laterally compressed with 8 to 12 suckers in a transverse row; no photophores. (From R. D. Young and C. F. E. Roper, 1969, *Smithsonian Contrib. Zool.*, no. 15.)

Genus *Joubiniteuthis* S. S. Berry, 1920

Known from only one species, the type: *Joubiniteuthis portieri* (Joubin, 1912). Synonym: *Valdemaria* Joubin, 1931. Gladius extremely long and thin. Tail long and needle-shaped.

6327   *Joubiniteuthis portieri* (Joubin, 1912). Northeast Florida to Panama and the Lesser Antilles; Eastern Atlantic. 330 to 2,500 meters. Mantle length 105 mm.; tail 155 mm. (Synonym: *Valde-*

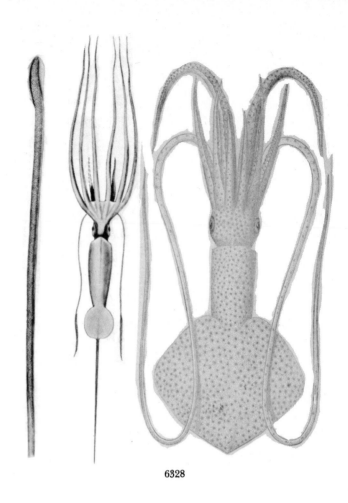

6328

maria danae Joubin, 1931.) See Young and Roper, 1969, for details.

Family MASTIGOTEUTHIDAE Verrill, 1881

Arms with biserial suckers. Tentacles whiplike, with long unexpanded clubs bearing many thousands of suckers.

Genus *Mastigoteuthis* Verrill, 1881

Tentacular arms not expanded into a club. Suckers tiny. Caudal fin ½ the length of the body.

6328   *Mastigoteuthis agassizii* Verrill, 1881. Gulf of Maine to North Carolina, 640 to 1,050 fms. Type of the genus.

6329   *Mastigoteuthis iseleni* MacDonald and Clench, 1934. Off Massachusetts, 600 fms. (a dubious species).

6330   *Mastigoteuthis pyrodes* Young, 1972. Off southern California and west Mexico.

Family CYCLOTEUTHIDAE Naef, 1923

Fins greater than 70% of the mantle length. Photophores present. Buccal connectives attach to the ventral borders of arms IV. Arms with biserial suckers; tentacular club with tetraserial suckers. With a subtriangular funnel locking-cartilage. Two genera: *Cycloteuthis* Joubin, 1919, and

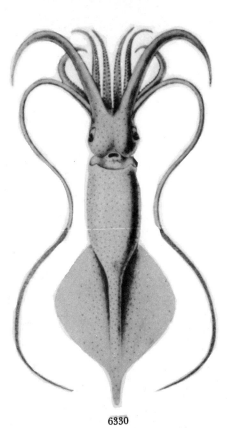

6330

Discoteuthis Young and Roper, 1969. See *Smithsonian Contrib. Zool.*, no. 5, 1969, for a review of this family.

Genus *Cycloteuthis* Joubin, 1919

Tail present; pen slightly thickened. Suckers in a transverse series on the manus are nearly equal in size. Large photophores on ink sac; a series of small photophores on eye. Fins large, but not equal to the mantle length in adults. Type: *sirventi* Joubin, 1919.

6331 *Cycloteuthis sirventi* Joubin, 1919. Bermuda; Straits of Florida to Brazil. Off northwest Africa. 100 to 500 meters. Mantle length: 27 to 42 mm.

Genus *Discoteuthis* Young and Roper, 1969

Tail lacking; pen greatly thickened. Medial suckers on manus are greatly enlarged. External photophores present; visceral photophores lacking. Fins large, round, equal to the mantle length in adults. Type: *discus* Young and Roper. 1969.

6332 *Discoteuthis discus* Young and Roper, 1969. Lesser Antilles to West Africa.

6333 *Discoteuthis laciniosa* Young and Roper, 1969. 40 miles north northeast of Bermuda, Bahamas and Northwest Africa. Mantle length up to 70 mm. Deep water. *Smithsonian Contrib. Zool.*, no. 5, p. 9.

Family GRIMALDITEUTHIDAE Pfeffer, 1900

Funnel and mantle locking-cartilages are fused, while the dorsal mantle-nuchal locking apparatus is free. Tentacles are lacking; the arms have biserial suckers. An extremely long tail bears an accessory fin. One genus.

Genus *Grimalditeuthis* Joubin, 1898

Enoptroteuthis S. S. Berry, 1920, is the young.

6334 *Grimalditeuthis bonplandii* (Vérany, 1837). Open Atlantic; Azores; Gulf of Mexico; off southern California. Type of the genus.

Family CRANCHIIDAE Gray, 1847

The mantle is fused to the head in the nuchal region and to the funnel at its two posterolateral corners. Prosch, 1849, used the same family name.

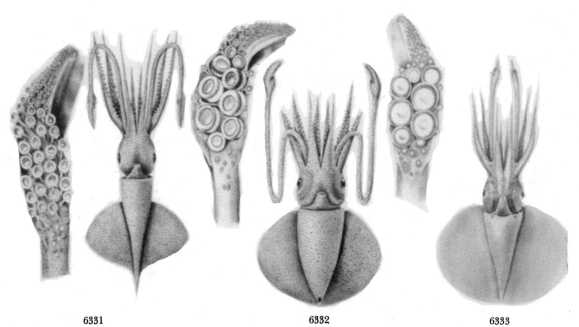

6331   6332   6333

Subfamily CRANCHIINAE Gray, 1847

Genus *Cranchia* Leach, 1817

6335 *Cranchia scabra* Leach, 1817. Cosmopolitan in warm seas; East Florida and Gulf of Mexico, in plankton tows. Type of the genus.

Genus *Leachia* Lesueur, 1821

Pyrgopsis Rochebrune, 1884, is a synonym or subgenus.

6336 *Leachia lemur* S. S. Berry, 1920. Off Cape Hatteras, North Carolina; off Puerto Rico.

6337 *Leachia cyclura* Lesueur, 1821. Off southwest Bermuda; between 700 and 1,000 meters. Type of the genus.

6338 *Leachia dislocata* Young, 1972. Off California to off Hawaii.

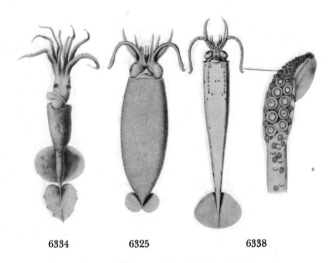

6334   6325   6338

Genus *Liocranchia* Pfeffer, 1884

6339 *Liocranchia reinhardti* (Steenstrup, 1856). Off Cuba and Puerto Rico. Type of the genus.

Subfamily TAONIINAE Chun, 1910

Desmoteuthinae Verrill, 1881, is a synonym.

Genus *Taonius* Steenstrup, 1861

Desmoteuthis Verrill, 1881, is a synonym. *T. pavo* Lesueur, 1821, is the type.

6340 *Taonius pavo* (Lesueur, 1821). Worldwide; Newfoundland and the Gulf Stream. British Columbia; Alaska; Oregon.

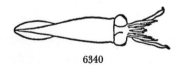

6340

Genus *Galiteuthis* Joubin, 1898

6341 *Galiteuthis armata* Joubin, 1898. Off Bermuda; upper 1,000 meters. Mediterranean. Japan to Alaska to California.

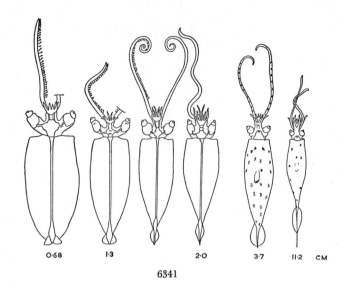

0·68   1·3   2·0   3·7   11·2   CM

6341

6342 *Galiteuthis phyllura* S. S. Berry, 1911. Type of the genus. Oregon to southern California.

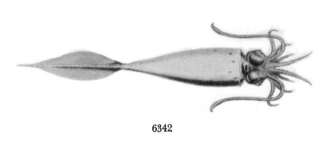

6342

6343 *Galiteuthis pacifica* (Robson, 1948). Off California (rare) and off west Central America.

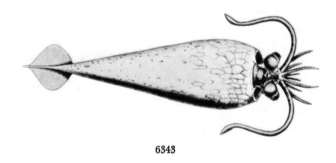

6343

Genus *Megalocranchia* Pfeffer, 1884

Type of the genus is *maxima* Pfeffer, 1884. *Verrilliteuthis* S. S. Berry, 1912, is a synonym.

6344 *Megalocranchia papillata* Voss, 1960. 100 miles southwest of Bermuda. Deep water. *Fieldiana Zool.*, vol. 39, p. 430.

6345 *Megalocranchia megalops* (Prosch, 1849). Off Bermuda; Greenland to Massachusetts, 0 to 1,346 fms. Female is illustrated. (Synonyms: *Taonius* and *Desmoteuthis tenera* Verrill, 1882; *Verrilliteuthis hyperborea* (Steenstrup, 1856).)

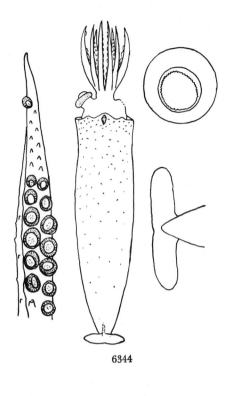

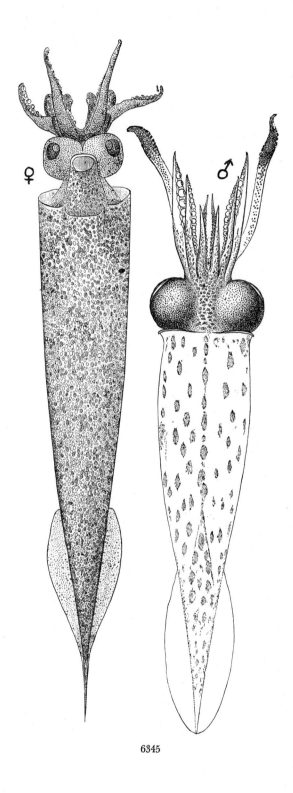

6344

6346

6347

Genus *Teuthowenia* Chun, 1910

6346 *Teuthowenia corona* (S. S. Berry, 1920). Off Massachusetts, 0 to 50 fms. May be placed in *Hensenioteuthis* Pfeffer, 1900.

Genus *Sandalops* Chun, 1906

Type of the genus is *melancholica* Chun, 1906.

6347 *Sandalops ecthambus* S. S. Berry, 1920. Little Bahama Bank.

6348 *Sandalops pathopsis* S. S. Berry, 1920. West of Bermuda.

Genus *Carynoteuthis* Voss, 1960

A taoniid squid with sessile eyes which bear 2 large, meandering light organs; funnel with a small valve, the dorsal funnel organ with 2 large, triangular flaps; arm-suckers minutely toothed; tentacular suckers toothed. Light organs present on the ink sac. Type: *oceanica* Voss, 1960.

6345

6349 *Carynoteuthis oceanica* Voss, 1960. 100 miles southwest of Bermuda. Deep water. Mantle length 76 mm. *Fieldiana Zool.,* vol. 39, p. 434.

Genus *Bathothauma* Chun, 1906

6350 *Bathothauma lyromma* Chun, 1906. Off Bermuda; deep water. North Atlantic; Indo-Pacific. Type of the genus.

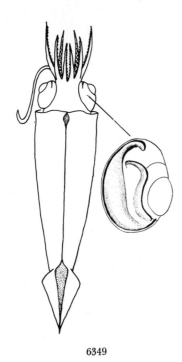

6349

Order *Vampyromorpha* Thiele, 1915

Family VAMPYROTEUTHIDAE Thiele, 1915

Genus *Vampyroteuthis* Chun, 1903

Type of the genus is *infernalis* Chun, 1903.

6351 *Vampyroteuthis infernalis* Chun, 1903. Worldwide in warm and temperate seas; Gulf of Mexico. Bermuda. For details see Pickford, 1949, Dana-Reports 29 and 32. Off Eureka, California, 380 fms.; off Santa Catalina Island, California, 2,113 fms. *Cirroteuthis macrope* S. S. Berry, 1911, is a synonym.

Order *Octopoda* Rafinesque, 1815

The octopods have only 8 arms and are without the 2 long tentacular arms that are characteristic of the squid. The suckers on the arms are without stalks and are not equipped with horny rings. No internal pen or shell. This order includes the many forms of octopus and the paper

nautilus, *Argonauta*. The female *Argonauta* secretes a shell to hold her eggs. The order was formerly called Polypoidea.

Family OCTOPODIDAE Rafinesque, 1815

Genus *Octopus* Lamarck, 1798

There are only six valid species of littoral *Octopus* so far recorded along the Atlantic coast. There are a few deep-water ones, some of which belong to closely related genera. The characters most relied upon in distinguishing species are relative length of arms, the skin surface, the nature and relative length of the small ligula (the tiny padlike extension on the end of the third right arm in the males, i.e., the hectocotylized arm). The number of gill plates and color pattern are used to a lesser extent. Type of the genus is *vulgaris* Lamarck, 1798. *Polypus* Hoyle, 1901, is a synonym.

The 8 arms have each been given a number, in order that comparisons may be made. This is done by setting the octopus down with the body up, and the arms spread-eagle out in all directions. Turn the octopus so that the 2 eyes are on the side away from you. By going from the eyes out to the mantle edge away from you, and choosing the first arm to the right, you have located the first arm. Further clockwise are the second, third and fourth right arms. Instead of counting further (fifth arm, etc.), return to the center again, and count to the left—hence, the first, second, third and fourth left arms. When giving an arm formula, only the right ones are generally given, and they are set down in order of large to smaller size. Hence, 4.3.1.2 means the fourth arm is the largest, the second one the smallest in length. It may be pointed out that on rare occasions an octopus may accidentally lose an arm.

There are two simple sets of measurements (all in millimeters) which are important in distinguishing the species of octopus. The first is the mantle-arm index which simply means the comparison of the length of the mantle (measure from the round, bulbous "head" end to a point just between the eyes) with the length of the longest arm (turn the octopus over, measure from the mouth to the tip of the longest stretched-out arm). An index is obtained by multiplying the mantle length by 100 and then dividing the result by the arm length.

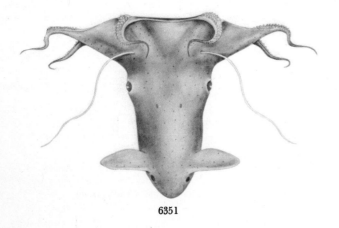

6351

6356

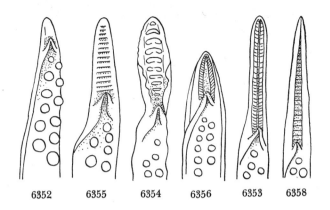

| 6352 | 6355 | 6354 | 6356 | 6353 | 6358 |

The ligula index is obtained only from males and from the third right arm which is a modified sex organ. The ligula is measured from the tip to the last sucker. The arm length is obtained as explained in the preceding paragraph. The index is: length of ligula, multiplied by 100, the result divided by the total arm length. The number of gill plates and the size of eggs are determined by cutting a deep slit in the body.

For information on octopus bites, see S. S. Berry, 1954, *Leaflets in Malacology*, vol. 1, no. 11, pp. 59–66.

Octopus vulgaris Cuvier, 1797   6352
Common Atlantic Octopus

Connecticut to Florida and the West Indies.
Europe.

Length, including the longest arm, 1 to 3 feet (the latter would give a radial spread of about 7 feet). Mantle-arm index in Florida and North Carolina is about 25 (that is, the arms are 4 times as long as the mantle). Ligula index below 2.5. Gill plates 7 to 9 (in Bermuda, usually 10 or 11). In life, skin smoothish; preserved, it is rugose with variously shaped warts. Eggs 3 mm. or less in length. A common harmless species found hidden away under large rocks and crevices near shore. This is *O. rugosus* of authors, *O. americanus* Blainville and *O. carolinensis* Verrill. In Bermuda, called the rock scuttle.

Octopus macropus Risso, 1826   6353
Grass Octopus

Off south Florida, Bermuda to Brazil.

Length, including longest arm, 3 feet. Mantle-arm index is 26 to 36. The first arm is the largest and longest. The ligula index up to 14, and with or without a wart over one side of the eye. Color in life brick-red to pinkish orange marked with numerous large white spots (½ inch across). Brownish gray when at rest. Iris of eyes dull-yellow. In Bermuda, called the grass scuttle. *O. palliata* Verrill, 1881, is believed to be this species. Preserved specimens have reddish warts. Uncommon. (See *Jour. Florida Acad. Sci.*, vol. 20, p. 223, 1957.) Other synonyms: *chromatus* Heilprin, 1888; *bermudensis* Hoyle, 1885.

Octopus briareus Robson, 1929   6354
Briar Octopus

Off North Carolina to Florida and the West Indies.

Length, including longest arm, 1 to 1½ feet. Arms fairly thick at the bases, quite long, especially the third and some-

times the second. Mantle-arm index 13 to 30, but usually about 17. Ligula index about 4. Gill plates 7, rarely 8. Skin smoothish, or finely granular in preserved material; in life, pinkish brown to red-mottled. Eggs elongate, translucent-white, 10 to 12 mm. in length and with equally long attachment stalks. Fairly common between tides under large coral blocks on the Lower Florida Keys.

Octopus burryi Voss, 1950   6355
Burry's Octopus

Both sides of Florida and the Gulf of Mexico.
Bermuda.

Length, including the longest arm, 6 to 10 inches. Characterized by a broad band of dark-purple on the top surface of the arms, and in preserved specimens, by the skin which is covered with closely set, round papillae or warts. Gill plates 8 to 10 in number. Ligula index 4 to 5. This is a recently described species named after a famous Florida collector, Leo L. Burry of Pompano Beach, Florida. It is moderately common; shallow water to 100 fathoms. See Voss, 1951, *Bull. Marine Sci.*, vol. 1, no. 3, p. 231.

Octopus joubini Robson, 1929   6356
Joubin's Octopus

South half of Florida, and the West Indies.

A small species with a length, including the longest arm, of from 4 to 6, rarely 7 inches. The arms are short, with a mantle-arm index of about 40 to 50. Ligula index about 6 to 7. Gill plates 5 or 6 usually. Skin smoothish, except for little pimples at scattered intervals. In this species, the longest arm is only 2 or 3 times the mantle length, while in *O. briareus* the longest arm is 5 or 6 times as long as the mantle. Eggs large, amber-colored, and about 7 to 10 mm. in length. Occasionally cast ashore in fair numbers on the west coast of Florida. It lives within large bivalve shells. *O. mercatoris* Adam, 1937 is the same. Formerly placed in the genus *Paroctopus* which is now considered of no value.

Octopus hummelincki Adam, 1936   6357
Seaweed Octopus

Florida Keys and the West Indies. Brazil.

4 to 6 inches in length; in life, the surface is covered with 50 to 60 wide, flat, bladelike processes which have arborescent ends. Eyes raised and surrounded by ocular cirri. Color when crawling is a rich reddish yellow-brown to whitish upon which are mottlings of light golden-yellow. When swimming, the color is changed to a uniform light-brown. Looks like a moving clump of *Sargassum* weed. Under the eye is a brown blotch, within which is a spot of pale-blue to vivid purplish blue. Moderately common in 6 to 30 feet near coral reefs and patches of *Sargassum* and *Dictyota* weeds. (See G. Voss, 1953, *The Nautilus*, vol. 66, p. 73.)

Octopus dofleini (Wülker, 1910)   6358
Common Pacific Octopus

Alaska to Baja California. Japan to south China.

Length, including longest arm, ½ to 3 feet (possibly with a radial spread of nearly 28 feet in Alaskan waters). Skin in preserved specimens covered everywhere by numerous small, pimplelike tubercles with star-shaped bases, and by many heavy, much interrupted, longitudinal wrinkles. Above each eye there is a rather small, conical wart and a very large, pinnaclelike protuberance behind it. Ligula index 4 to 7. The web between the second and third arms usually extends out to a quarter of the arms' length. Elsewhere the webs are shorter. The commonest littoral octopus on the Pacific Coast found from shore to 100 fathoms. This is *O. punctatus* Gabb, 1862, non Blainville, 1826. Formerly and erroneously called *hongkongensis* Hoyle.

A detailed study of this species was made by Grace E. Pickford (1964, *Bull. Bingham Ocean. Coll.*, vol. 19, pp. 1–70) who divided this octopus into 3 subspecies:

(6359) *O. dofleini dofleini* (Wülker, 1910). Japan and Korea. Typical order of arm length is $1 = 2 > 3 > 4$. Synonyms are *pustulosus* Sasaki, 1920, non Blainville, 1826; *O. madokai* S. S. Berry, 1921.

(6360) *O. dofleini apollyon* S. S. Berry, 1912. Bering Sea, Kamchatka, Okhotsk Sea, Kurile Islands. Typical order of arm length is $2 > 3 > 1 > 4$ or $2 > 3 > 4 > 1$. A conspicuous groove is present on the rostrum of both the upper and lower jaws in large specimens. *O. gilbertianus* S. S. Berry, 1912, is a synonym.

(6361) *O. dofleini martini* Pickford, 1964. Washington to California. Hectocotylized arm about as long as the third left arm, rarely less than 90% of the length of the latter. Lateral limbs of the funnel organ markedly shorter than the median limb, rarely exceeding 60% of the length of the median limb. Typical order of arm length $1 > 2 > 3 > 4$.

(6362) *O. californicus* Berry, 1911, an offshore species, has a large lingula with an index of 14 to 17. The skin in preserved material is covered with numerous, large stellate warts. The Californian deep-water octopus.

Octopus bimaculatus Verrill, 1883   6363
Two-spotted Octopus

Los Angeles, California, to Baja California.

Total length ½ to 2 feet. Characterized by a large, distinct, round, dark spot in front of each eye near the base of each third arm. Eggs small, 1.8 to 4.0 mm. in length with long stalks, attached in festoons. Mantle-arm index usually 22, but ranging from 14 to 29. Ligula index 2.0, not significant in separating this species from *bimaculoides*. Fairly common. Lives in the lower part of the intertidal zone down to several feet where there is rock bottom. Aside from egg size and egg clusters and mantle-arm index, there is great difficulty in separating this sibling species from *O. bimaculoides* which lives nearby in shallower water where there is mud present.

Octopus bimaculoides Pickford and McConnaughey, 1949   6364
Mud-flat Octopus

Los Angeles, California, to Baja California.

Almost identical with *O. bimaculatus* Verrill. Eggs large, 9.5 to 17.5 mm. in length with shorter stalks, attached in small clusters. Mantle-arm index 34, but ranging from 29 to 39. Fairly common in shallow water among rocks where mud is present. Adults are somewhat smaller than *bimaculatus*. See *Bull. Bingham Oceanogr. Coll.*, vol. 12, no. 4.

Other species:

6365 *Octopus leioderma* S. S. Berry, 1911. Shelikof Strait, Alaska, 106 fms., to California.

6366 *Octopus pricei* S. S. Berry, 1913. Monterey Bay, California. (*Proc. Acad. Nat. Sci. Phila.*, p. 73.)

6367 *Octopus penicillifer* S. S. Berry, 1954. 17 fms., Off Punta Arena, Baja California. (*Leaflets in Malacology*, vol. 1, no. 11, p. 66.)

6368 *Octopus rubescens* S. S. Berry, 1953. 7 to 17 fms., South Coronado Island, Baja California.

6369 *Octopus micropyrsus* S. S. Berry, 1953. Shore to 6 fms., La Jolla, California.

6370 *Octopus hubbsorum* S. S. Berry, 1953. Outer Bahia San Carlos, Sonora, Mexico.

6371 *Octopus fitchi* S. S. Berry, 1953. Punta San Felipe, Baja California.

6372 *Octopus alecto* S. S. Berry, 1953. South of Estero Soldado, Sonora, Mexico.

6373 *Octopus veligero* S. S. Berry, 1953. 50 fms., off San Juanico, Baja California. (Above 6 species described, without necessary illustrations, in *Leaflets in Malacology*, vol. 1, no. 10, pp. 51–58.)

6374 *Octopus* (*Macrotritopus* Grimpe, 1922) *equivocus* Robson, 1929. Off Nova Scotia, 1,290 fms.

6375 *Octopus bairdii* Verrill, 1873. Newfoundland to South Carolina. Norway. 7 to 524 fms.

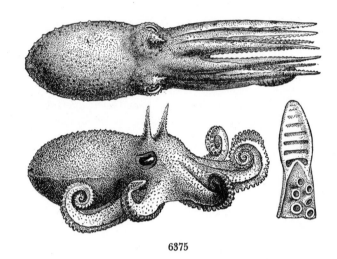

6375

Genus Benthoctopus Grimpe, 1921

6376 *Benthoctopus piscatorum* (Verrill, 1879). Newfoundland to Massachusetts, 60 to 1,362 fms. Type of the genus. See F. A. Aldrich, 1968, *The Veliger*, vol. 11, p. 70.

6377 *Benthoctopus januari* Hoyle, 1885. Gulf of Mexico, 262 fms.; Brazil, 350 fms.

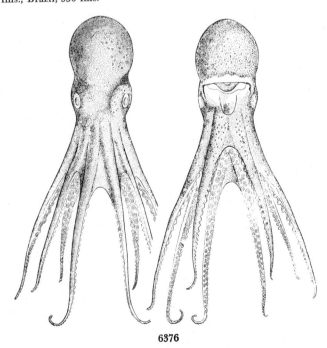

6376

Genus *Tetracheledone* Voss, 1955

With uniserial suckers; stellate skin warts, well-developed ink sac; 4 pads in the funnel organ. Type: *spinicirrus* Voss, 1955.

Tetracheledone spinicirrus Voss, 1955   **6378**
Eyebrow Octopus

Both sides of Florida to south Cuba.

6 inches (mantle length ½ that of the longest tentacle). A small deep-water octopus characterized by its surface of closely set, large, stellate tubercules and 2 large conspicuous cirri over each eye. The funnel has 4 separate, elongate parts of approximately equal size. Third right arm hectocotylized, with a ligula index of 4.3 to 9.6.

Genus *Bathypolypus* Grimpe, 1921

6379 *Bathypolypus arcticus* (Prosch, 1849). Bay of Fundy, 28 to 843 fms.; Greenland; Ipswich Bay, Massachusetts, to off Jacksonville, Florida, 105 fms. Type of the genus.

6380 *Bathypolypus lentus* Verrill, 1880. Newfoundland to South Carolina, 120 to 603 fms. 5 inches.

6380

6381 *Bathypolypus obesus* Verrill, 1880. Off Nova Scotia, 150 to 300 fms.

Genus *Danoctopus* Joubin, 1933

6382 *Danoctopus schmidti* Joubin, 1933. Off Dry Tortugas, Florida, 283 fms.; southeast of Abaco, Bahamas. Type of the genus.

Genus *Pteroctopus* P. Fischer, 1882

6383 *Pteroctopus tetracirrhus* (Delle Chiaje, 1830). Gulf of Mexico; east Florida; 40 to 60 fms. Cuba; Mediterranean. (Synonym: *Scaeurgus titanotus* Troschel, 1857.) Type of the genus.

Genus *Scaeurgus* Troschel, 1857

6384 *Scaeurgus unicirrhus* (Orbigny, 1840). Southeast Florida, 40 to 200 fms., in old beer bottle. Indo-Pacific; Azores. (Synonym: *Octopus scorpio* S. S. Berry, 1920; *S. patagiatus* S. S. Berry, 1913.) See Voss, 1951, *Bull. Marine Sci.*, vol. 1, no. 1, p. 64. Type of the genus.

Genus *Graneledone* Joubin, 1918

6385 *Graneledone verrucosa* Verrill, 1881. Nova Scotia to off Delaware Bay, 406 to 1,255 fms. Type of the genus. 8 inches.

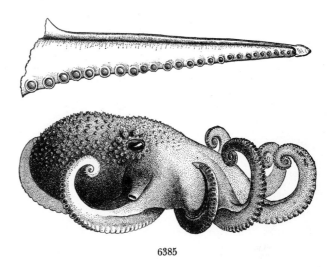

6385

Family BOLITAENIDAE Chun, 1910

Deepsea, soft, saccular squids without fins.

Genus *Eledonella* Verrill, 1884

Without fins; mantle opening very wide. A median septum present in the branchial cavity. Arms slender, the third pair much the largest. Suckers uniserial. Type: *pygmaea* Verrill, 1884.

6386 *Eledonella pygmaea* Verrill, 1884. Between New York and Bermuda, 2,949 fms.; North Atlantic; off northeast Cuba. (Synonyms: *Japetella prismatica*, Hoyle, 1885; *E. massyae* Robson, 1924; *E. ijimai* Sasaki, 1929; *E. purpurea* Robson, 1930. See S. Thore, 1949, *Dana-Report*, no. 33.)

Genus *Japetella* Hoyle, 1885

6387 *Japetella diaphana* Hoyle, 1885. Off Bermuda (young at about 200 meters; adults around 1,750 to 2,500 meters). Type of the genus.

6388 *Japetella heathi* (S. S. Berry, 1911). Off Santa Catalina Island, California. 2,228 fms.

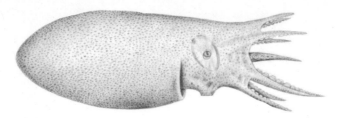

6388

Family VITRELEDONELLIDAE Robson, 1928

Body has a clear gelatinous appearance, small rectangular eyes, long slender pointed liver. Third left arm hectocotylized. Monographed in 1949 by S. Thore, *Dana-Report* no. 33.

Genus *Vitreledonella* Joubin, 1918

6389 *Vitreledonella richardi* Joubin, 1918. Off Bermuda. Young hatch below 1,000 meters, finally settle at 1,000 to 1,750 meters. Type of the genus.

Family TREMOCTOPODIDAE Tryon, 1879

The family Philonexidae Orbigny, 1845, is a synonym.

Genus *Tremoctopus* Delle Chiaje, 1830

***Tremoctopus violaceus* Delle Chiaje, 1830**   **6390**
Common Umbrella Octopus

Pelagic in warm waters. Worldwide.

Total length, including the arms, 3 to 6 feet. Deep purplish red in color. Characterized by the long skin webs between the four dorsal arms, and the 2 large holes in the body near the base of the third arm in front of the eyes. The species is gregarious, and is occasionally washed ashore on the east coast of Florida. They may use the stinging tentacles of the Portuguese man-o-war, *Physalia,* as weapons (see E. C. Jones, 1963, *Science,* vol. 139, p. 764). Type of the genus.

Family ALLOPOSIDAE Verrill, 1881

Genus *Alloposus* Verrill, 1880

Arms united by a web extending nearly to the end. In the male, the hectocotylized right arm of the third pair is developed in a sac in front of the right eye.

6391 *Alloposus mollis* Verrill, 1880. Massachusetts to the Gulf of Mexico, 252 to 1,735. North Atlantic; Azores. Off California. Type of the genus.

Genus *Heptapus* Joubin, 1929

6392 *Heptapus danai* Joubin, 1929. Off North Carolina, 100 meters. Position and status uncertain. Type of the genus.

Family CIRROTEUTHIDAE Kerferstein, 1866

Genus *Cirroteuthis* Eschricht, 1838

Type of the genus is *muelleri* (Eschricht, 1836).

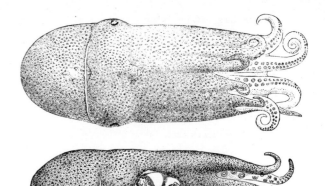

6391

6394 *Cirroteuthis muelleri* (Eschricht, 1836). West Greenland, 1,280 fms.

Family STAUROTEUTHIDAE Grimpe, 1916

Genus *Cirroctopus* Naef, 1923

Type of the genus is *mawsoni* S. S. Berry. *Grimpoteuthis* Robson, 1932, is a synonym.

6395 *Cirroctopus megaptera* Verrill, 1885. Southeast of Long Island, New York, 1,054 to 2,574 fms.

6396 *Cirroctopus plena* Verrill, 1885. Between New York and Bermuda, 1,073 fms.

Genus *Stauroteuthis* Verrill, 1879

6397 *Stauroteuthis syrtensis* Verrill, 1879. Nova Scotia to Long Island, 250 to 1,346 fms. Type of the genus.

6398 *Stauroteuthis hippocrepium* Hoyle, 1904. Off Malpelo Island, Gulf of Panama, 1,823 fms. *Bull. 43, Mus. Comp. Zool.,* p. 6.

Genus *Chunioteuthis* Grimpe, 1916

6399 *Chunioteuthis ebersbachii* Grimpe, 1916. South of Newfoundland, 400 fms. Type of the genus.

Family OPISTHOTEUTHIDAE Verrill, 1896

Genus *Opisthoteuthis* Verrill, 1883

Entire body compressed. Funnel points backward. Body and head fused. Arms with extensive webs. Body with a pair of minute muscular fins. Type: *agassizi* Verrill, 1883.

6400 *Opisthoteuthis agassizi* Verrill, 1883. North Atlantic east coast of the United States; Gulf of Mexico to Grenada, 291 fms.

6401 *Opisthoteuthis californiana* Berry, 1949. Eureka to Santa Barbara, California. *Leaflets in Malacology,* vol. 1, no. 6, p. 23.

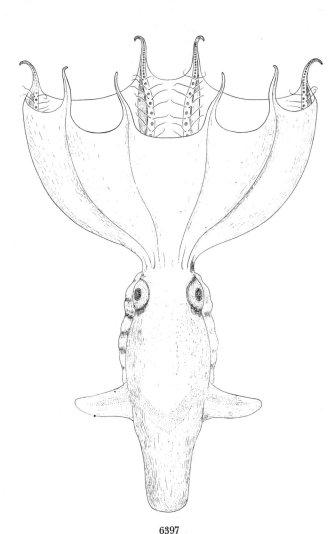

6397

6398

Family OCYTHOIDAE Gray, 1847

Genus *Ocythoe* Rafinesque, 1814

Body saclike. No shell. Hectocotylus are detachable from male. *Parasira* Steenstrup, 1861, is a synonym.

6402 *Ocythoe tuberculata* Rafinesque, 1814. Massachusetts to the West Indies. Widespread in warm seas. Type of the genus. *Parasira catenulata* Steenstrup, 1861, is a synonym.

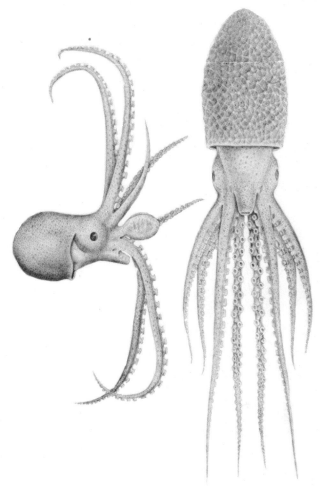

6402

Family ARGONAUTIDAE Rafinesque, 1815

Genus *Argonauta* Linné, 1758

Pelagic octopods in which the dorsal arms of the female are broadly expanded into glandular membranes that secrete and hold a delicate, calcareous shell for containing the eggs. The males are considerably smaller than the females, do not have a shell and the third right arm is modified into a detached copulatory organ which persists separately for a certain length of time in the mantle cavity of the female. Type of the genus is *argo* Linné, 1758.

Argonauta argo **Linné, 1758**   6403
Common Paper Nautilus

Worldwide in warm waters. Bermuda.

4 to 8 inches in length, quite fragile, laterally compressed with a narrow keel, numerous sharp nodules which in the

early part of the shell are stained with dark purplish brown. Rest of shell opaque, milkywhite. Occasionally washed ashore. *A. americana* Dall, 1889, is the same. Tom McGinty (*in litt.*) reports egg-laying in Florida in late March.

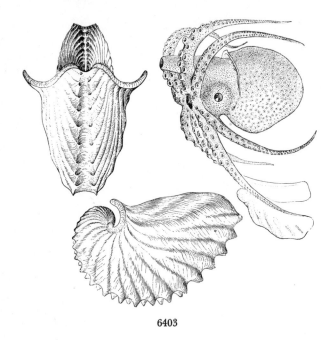

6403

Argonauta pacifica Dall, 1872   6404
West Coast Paper Nautilus

Monterey, California, to Panama.

Similar to *argo,* but the shell is more ventricose. The animal is orange, sprinkled with fine purple dots. Some workers consider this synonymous with or a subspecies of *argo.*

Argonauta hians Lightfoot, 1786   6405
Brown Paper Nautilus

Worldwide in warm waters. Bermuda.

Similar to *A. argo,* but smaller, much "fatter" with a rapidly broadening keel that bears larger and fewer nodules. Color brownish white with darker stains on the early part of the keel. Uncommonly washed ashore, and sometimes found in the stomachs of the dolphin fish. *Argonauta gondola* Dillwyn, 1817, is a synonym.

6405

Addenda

Add to page 63:

Family Neritopsidae Gray, 1847

Genus **Neritopsis** Grateloup, 1832

Shell nerite-shaped, beaded, white; columella with a squarish embayment into which fits a thick, calcareous, white operculum. Type: *Nerita radula* Linné, 1758, from the Indo-Pacific.

518a *Neritopsis atlantica* Sarasua, 1973. 13–17 mm. North coast of Cuba. Poeyana, Serie Inst. Zool. Acad. Cienc. Cuba, no. 118.

Add to page 419:

Family Pristiglomidae Sanders and Allen, 1973

4952a *Pristigloma alba* Sanders and Allen, 1973. Between Massachusetts and Bermuda, 2178 to 4892 meters; off Brazil and Argentina; off the Canary Islands. 2 mm. Bull. Mus. Comp. Zool., vol. 145, p. 244.

Genus **Microgloma** Sanders and Allen, 1973

4952b *Microgloma yongei* Sanders and Allen, 1973. Off Cape Verde Islands and off Angloa, 1964 to 2754 meters. 1 mm. Bull. Mus. Comp. Zool., vol. 145, p. 247. Type of the genus.

4952c *Microgloma turnerae* Sanders and Allen, 1973. Bay of Biscay, off France; off Canary Islands, 2351 meters. 1 mm. Bull. Mus. Comp. Zool., vol. 145, p. 250.

Index

DATE DUE

			PRINTED IN U.S.A.

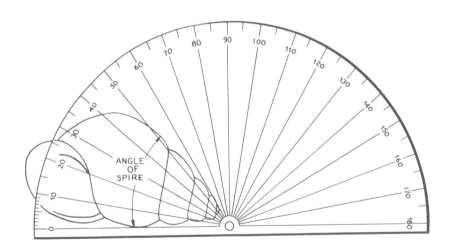